T0358908

# CONTROL APPLICATIONS IN MARINE SYSTEMS 2004
## (CAMS 2004)

*A Proceedings volume from the IFAC Conference,*
*Ancona, Italy, 7 – 9 July 2004*

Edited by


R. KATEBI
*Industrial Control Centre,*
*Department of Electronic and Electrical Engineering,*
*University of Strathclyde, Glasgow, UK*

and

S. LONGHI
*Dipartimento di Ingegneria Informatica, Gestionale e dell'Automazione,*
*Università Politecnica delle Marche, Ancona, Italy*



Published for the

INTERNATIONAL FEDERATION OF AUTOMATIC CONTROL

by

ELSEVIER LIMITED

ELSEVIER Ltd
The Boulevard, Langford Lane
Kidlington, Oxford OX5 1GB, UK

Elsevier Internet Homepage
http://www.elsevier.com

Consult the Elsevier Homepage for full catalogue information on all books, journals and electronic products and services.

IFAC Publications Internet Homepage
http://www.elsevier.com/locate/ifac

Consult the IFAC Publications Homepage for full details on the preparation of IFAC meeting papers, published/forthcoming IFAC books, and information about the IFAC Journals and affiliated journals.

---



First edition 2005

**Library of Congress Cataloging in Publication Data**

A catalogue record for this book is available from the Library of Congress

**British Library Cataloguing in Publication Data**

A catalogue record for this book is available from the British Library

ISBN 0-08-044169 6
ISSN 1474-6670



Transferred to digital print 2008
Printed and bound by CPI Antony Rowe, Eastbourne

---

**To Contact the Publisher**

Elsevier welcomes enquiries concerning publishing proposals: books, journal special issues, conference proceedings, etc. All formats and media can be considered. Should you have a publishing proposal you wish to discuss, please contact, without obligation, the publisher responsible for Elsevier's industrial and control engineering publishing programme:

Christopher Greenwell
Senior Publishing Editor
Elsevier Ltd
The Boulevard, Langford Lane
Kidlington, Oxford
OX5 1GB, UK

Phone: +44 1865 843230
Fax: +44 1865 843920
E.mail: c.greenwell@elsevier.com

General enquiries, including placing orders, should be directed to Elsevier's Regional Sales Offices – please access the Elsevier homepage for full contact details (homepage details at the top of this page).

# IFAC CONFERENCE ON CONTROL APPLICATIONS IN MARINE SYSTEMS 2004

*Sponsored by*
International Federation of Automatic Control (IFAC)
- IFAC Technical Committee on Marine Systems

*Financial support*
- Università Politecnica delle Marche, Ancona, Italy
- Dipartimento di Ingegneria Informatica, Gestionale e dell'Automazione
- Calzoni s.r.l., Milan, Italy
- Fincantieri – Cantieri Navali Italiani S.p.A., Trieste, Italy
- ISA – International Shipyards Ancona, Ancona, Italy
- National Instruments, Milan, Italy
- Rockwell Automation, Milan, Italy

*Supported by*
- Assessorato al Turismo, Regione Marche, Italy
- Fondazione della Cassa di Risparmio di Fermo, Italy

*Organized by*
Dipartimento di Ingegneria Informatica, Gestionale e dell'Automazione, Università Politecnica delle Marche, Ancona, Italy

*International Programme Committee* (IPC)
Grimble, M.J. (UK) (Chair)
Mazzoli, A. (Italy) (Vice-Chair)
Katebi, R. (UK) (Editor)

Batlle, J. (Spain)
Bertin, D. (Italy)
Blanke, M. (Denmark)
Burns, R.S. (UK)
Caiti, A. (Italy)
Casalino, G. (Italy)
Conte, G. (Italy)
Cristi, R. (USA)
Donha, D.C. (Brazil)
Dumitrescu, D. (Romania)
Fabiani, P. (France)
Fossen, T.I. (Norway)
Gerigk, M. (Poland)
Giron-Sierra, J.M. (Spain)
Haro Casado, M. (Spain)
Hasegawa, K. (Japan)
Holzhueter, T. (Germany)
Jetto, L. (Italy)
Källström, C.G. (Sweden)
Kijima, K. (Japan)
Leonard, N.E. (USA)
Longhi, S. (Italy)
Lozowicki, A. (Poland)
Mandzuka, S. (Croatia)
Ohtsu, K. (Japan)
Pascoal, A. (Portugal)
Pourzanjani, M.M.A. (UK)
Ren, G. (China)
Roberts, G.N. (UK)
Serrani, A. (USA)
Sørensen, A.J. (Norway)
Sutton, R. (UK)
Tiano, A. (Italy)
van Amerongen, J. (The Netherlands)
Veruggio, G. (Italy)
Vukić, Z. (Croatia)
Yamamoto, I. (Japan)
Yoerger, R. (USA)
Yuh, J. (USA)

# PREFACE

New technologies and methodologies for control systems play a significant role on the development of surface vessels, floating structures, sub-sea vehicles, and other devices within the marine environment. Modelling, simulation, advanced control and artificial intelligence techniques can improve performance, reliability, security, economy and ecology of marine systems. The IFAC Conference on Control Applications in Marine Systems, July 7-9 2004, Ancona, Italy, aimed at gathering the experts of the theory and applications of automatic control for the maritime systems, coming from Universities and Industries, in order to present the state of the art and the current research activities and advances in this field.

This volume contains the 71 papers presented at the Conference. The papers range from total control and modelling of vessels, underwater vehicles, off-shore structures to detailed control and modelling of ancillary and auxiliary subsystems. The conference was organised in 20 technical sessions, from which 8 sessions were focused on surface vessels, 6 sessions on underwater vehicles, 3 sessions on control and modelling of ancillary and auxiliary subsystems, 2 sessions on off-shore structure and 1 session on cooperative marine systems. The technical sessions were distributed over three days (July 7-9) together with 6 plenary sessions. The distinguished speakers for these sessions were:

- Professor Edoardo Bovio, NATO Undersea Research Centre, Italy
  "Autonomous Underwater Vehicles for Scientific and Naval Operations"

- Professor Kazuhiko Hasegawa, Osaka University, Japan
  "Some Recent Development of Next Generation's Marine Traffic Systems"

- Rear-Admiral Dino Nascetti, Italian Navy, Italy
  "Italian Navy Trends in Automation of Ship Controls"

- Professor Asgeir Johan Sørensen, Norwegian University of Science and Technology, Norway
  "Structural Issues in the Design of Marine Control Systems"

- Doctor Richard Stephens, ALSTOM Power Conversion, UK
  "Aspects of Industrial Dynamic Positioning: Reality-tolerant Control"

- Professor Edwin Zivi, U. S. Naval Academy, Annapolis, USA
  "Design of Robust Shipboard Power Automation Systems"

Finally, the technical programme was completed with a one-day tutorial (July 6):

- Professor Mogens Blanke, Technical University of Denmark, Denmark
  "Diagnosis and Fault Tolerant Control for Marine Systems"

With the participation of about 120 researchers from 20 countries, the Conference provided an excellent opportunity for control theorists and technologists coming from Universities and Industries, both senior and junior researchers, to meet and exchange ideas on the innovative topics of Control Applications in Marine Systems. The Conference was also a great opportunity to visit the Marche region for discovering its historical parts and its beautiful colours: the transparency of the sea, the green of the hills and the blue of the mountains.

The organisers would like to thank the members of the International Program Committee and the National Organising Committee for helping in reviewing the papers and supporting the organisation. Special thanks to Dr. Andrea Monteriù for his effort in organizing different technical activities, to Dr. Paola Traferro for dealing with the secretariat and to Dr. Pierluigi Antonini for managing the conference home web page.

Prof. Reza Katebi
*University of Strathclyde*
*Glasgow, UK*

Prof. Sauro Longhi
*Università Politecnica delle Marche*
*Ancona, Italy*

# CONTENTS

## *PLENARY PAPERS*

## *SHIP MODELLING AND CONTROL*



## *CONTROL DESIGN FOR MARINE SYSTEMS*



## *SHIP TRACK AND COURSE KEEPING*



## *MODELLING AND IDENTIFICATION OF AUVs – PART A*

## *SHIP INTELLIGENT CONTROL*



## *ROBUST CONTROL OF UNDERWATER VEHICLES*



## *HIGH SPEED VESSELS (Invited Session)*

## *MODELLING AND IDENTIFICATION OF AUVs – PART B*



## *POSE AND MOTION ESTIMATION OF AUVs*



## *CONTROL OF THRUSTERS AND PROPULSION SYSTEMS*



## *INTELLIGENT AND HYBRID CONTROL OF UNDERWATER VEHICLES*

## *COOPERATIVE MARINE SYSTEMS*



## *AUV MISSION PLANNING AND CONTROL*



## *DEVICES AND SIMULATION TOOLS FOR MARINE SYSTEMS*

ELSEVIER
IFAC
PUBLICATIONS
www.elsevier.com/locate/ifac

# AUTONOMOUS UNDERWATER VEHICLES FOR SCIENTIFIC AND NAVAL OPERATIONS

**E. Bovio** *,[1] **F. Baralli** * and **D. Cecchi** **

* *NATO Undersea Research Centre, Viale San Bartolomeo 400, 19138 La Spezia (SP), ITALY*
** *ISME, Interuniversity Centre of Integrated System for Marine Environment c/o DSEA University of Pisa, Via Diotisalvi 2, 56126, Pisa, Italy*

Abstract: Recognizing the potential of autonomous underwater vehicles for scientific and military applications, in 1997 MIT and the NATO Undersea Research Centre initiated a Joint Research Project (GOATS), for the development of environmentally adaptive robotic technology applicable to Mine Counter Measures (MCM) and Rapid Environmental Assessment in coastal environments. The August 2001 GOATS Conference marked the end of this 5 years project, but did not mark the end of the work. The Centre initiated in 2002 a new long term programme to explore and demonstrate the operational benefits and limitations of AUV for covert preparation of the battlespace. Recently the work addressed the evaluation of COTS (Commercial Off–The–Shelf) AUV technology for MCM operations in response to terrorist mining of port. The paper summarizes the work performed and refers to the scientific publications derived from the AUV programme at the NATO Undersea Research Centre. 

Keywords: Autonomous vehicles, Marine systems, Guidance, Navigation, Control, Accuracy

## 1. INTRODUCTION

The NATO Undersea Research Centre [2], located in La Spezia, Italy, performs basic and applied research and development to fulfill NATO's operational requirements in undersea warfare. The results of the Centre's research that can be seen at sea in many ships and submarines of the Alliance, have contributed to NATO's military capabilities over the past 41 years. Unique in its international makeup, the Centre functions as the "hub" in a virtual laboratory which brings great synergy to the research process and shortens timelines between research and development (R&D) and military applications. The Centre's own resources are therefore multiplied by collaboration and Joint Research Projects (JRP).

In response to NATO advanced planning that anticipates significant use of Autonomous Underwater Vehicles (AUVs) for Mine Counter Measures (MCM) and Rapid Environmental Assessment (REA), the Centre and the Massachusetts Institute of Technology (MIT) initiated in 1997 a 5 years joint research project, designated GOATS (Generic Oceanographic Array Technology Systems), for the development of environmentally adaptive AUV technology applicable to MCM and REA in coastal environments. The GOATS

[1] bovio@saclantc.nato.int

[2] Following the change in NATO command structure, the SACLANT Undersea Research Centre has been recently renamed NATO Undersea Research Centre

JRP grew in membership and scope and it was joined by an international host of collaborators who shared the notion that AUVs were ready to graduate from their role as research objects to a new supporting role for advanced ocean monitoring and maritime military tactics.

Between 1997 and 2001 the GOATS JRP explored and expanded the state of the art for networks of robotic ocean observers, supporting new approaches to battlespace preparation and mine hunting. The programme included a sequence of three field experiments with the participation of 14 institutions. The August 2001 GOATS Conference (Bovio *et al.* 2001) marked the end of this JRP, but not the end of the work.

Building on the success of the GOATS JRP, the Centre initiated in 2002 a new long term programme called *Battlespace Preparation (BP) with AUVs* to explore and demonstrate the operational benefits and limitations of AUVs for military battlespace preparation. Similarly to the GOATS series of experiments, the programme organizes multi–national, multi–disciplinary sea trials addressing the utilization of AUVs in coastal waters. The first experiment of the BP series took place in May–June 2002 in the Tyrrhenian and Ligurian seas. The results of this experiment have been reported during the Maritime Recognized Environmental Picture (MREP) Conference (Bovio *et al.* 2003) held in La Spezia in May 2003.

This paper summarizes the work performed and refers to the scientific publications derived from the AUV programme.

## 2. THE GOATS JRP

The GOATS JRP combined theory and modelling of the 3–D environmental acoustics with three experiments (1998, 1999, 2000) involving AUV and sensor technology.

The objective of the 1998 sea trial was to use acoustic arrays deployed on the sea floor or mounted on an AUV to characterize the spatial and temporal characteristics of the 3–D scattering from seabed targets and the associated reverberation, including the effects of multipaths. This effort was aimed at establishing the environmental acoustics foundation for future sonar concepts exploring 3–D acoustic signatures for combined detection and classification of proud and buried targets in very shallow water.

The GOATS 98 experiment provided a unique data set of three dimensional scattering and reverberation in shallow water, which has been essential for model validation and identification of features of the 3–D acoustic field. In addition, the experiment showed that small and inexpensive AUVs such as the MIT *Odyssey* can be reliably deployed, operated and recovered in shallow water from a surface vessel. It was also demonstrated that AUVs are an excellent acoustic platform for new sonar concepts for littoral MCM (Schmidt *et al.* 1998, Schmidt and Bovio 2000, Moran 1999).

The comprehensive acoustic and environmental datasets acquired during the GOATS 98 experiment have generated several scientific publications. The first papers resulting from the experiment described the physics underlying seabed penetration at sub–critical angles (Maguer *et al.* 2000*b*, Maguer *et al.* 2000*a*). Several papers deal with the processing of the bistatic synthetic aperture data acquired by the AUV, demonstrating the concept of bistatic SAS autofocusing and imaging (LePage and Schmidt 2002*a*, Edwards *et al.* 2001, Schmidt *et al.* 2000). The JRP has lead to several new developments in regard to modelling of acoustic interaction with the seabed. Specifically a unique modelling capability has been developed, providing a consistent prediction of 3–D scattering from seabed roughness and volume inhomogeneities, validated by the GOATS datasets (Veljkovic and Schmidt 2000, LePage and Schmidt 2000*c*, LePage and Schmidt 2002*b*, LePage and Schmidt 2000*a*, LePage and Schmidt 2000*b*).

Following the success of the GOATS 98 experiment, the Centre organized a workshop in January 1999 to extend the scope of the JRP to REA applications. In addition, it was also decided to assess the performance of non traditional AUV navigation algorithms based on *a priori* knowledge of the bottom topography. This required a thorough survey of Procchio bay, Island of Elba, the site of the GOATS 2000 experiment. Traditional instruments (side scan sonar, sub bottom profiler, multibeam echo sounder, underwater video camera, expandable penetrometer) were deployed from *Manning* and seafloor samples collected during the GOATS 99 experiment. The data provided a rich data set that characterizes the bathymetry and the composition of the seafloor of the area and forms the ground truth reference for comparison with data collected subsequently by AUVs.

The GOATS 2000 experiment demonstrated the capabilities of AUVs as REA platforms in shallow and very shallow water. The *Ocean Explorer (OEX)* equipped with a colour video camera and the Edgetech dual frequency DF–1000 side scan sonar and the Taipan equipped with the Applied Microsystem CTD were launched from R/V *Alliance*, to transect the bays to the east of Procchio to acquire side scan sonar data and to measure water mass properties such as current, salinity, density and temperature, for use by the nested oceanographic models. The side scan sonar data were used to generate geo–referenced acoustic images for comparison with ground truth data collected in the same area during previous experiments. The environmental information measured by the AUVs was fused in the Centre GIS

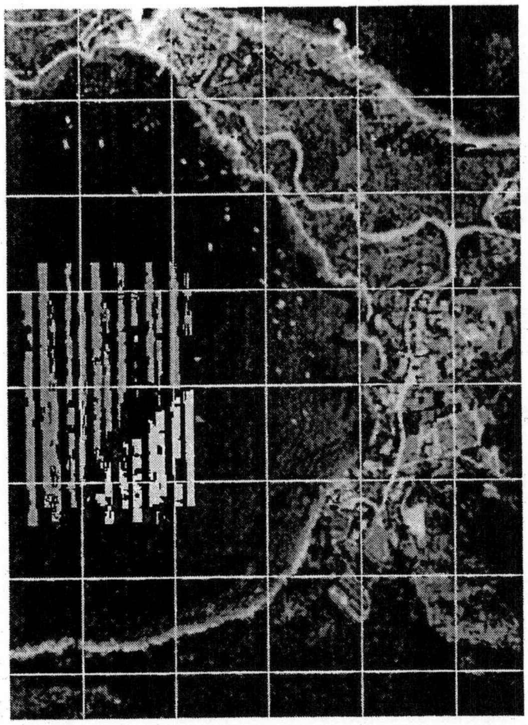

Figure 1. Boundary between sand and *Posidonia oceanica* (aerial image and unsupervised seafloor segmentation).

database. The tiled side scan sonar images were processed with unsupervised segmentation algorithms that demonstrated the capability to distinguish in a quantitative way between different types of seabeds. Figure 1 shows the *Poseidonia oceanica* to sand boundary detected by side scan sonar survey plotted over an aerial picture of the same area (Spina *et al.* 2001). The video images collected by the *OEX* were organized in a geographical database using the SeeTrack software.

A field of proud and buried targets at the main test site in Biodola Bay was insonified by the *TOPAS*[3] parametric sound source at a variety of incident and aspect angles. The *Odyssey* AUV sampled 3–D reverberation and target echoes obtaining data for validation of numerical models of mono– and bi–static seabed reverberation. A second field of proud targets including exercise mines such as the *MP80*, *Manta* and *Rockan*, was imaged at different aspects by the *OEX* instrumented with the 390 kHz Edgetech side scan sonar and video camera. The experiment demonstrated the potential of high frequency side scan sonar at multiple aspects for classification of proud targets. The results of the GOATS 2000 experiment are reported in (Bovio *et al.* 2001).

[3] TOpographic PArametric Sonar

## 3. BATTLESPACE PREPARATION WITH AUVS

Building on the success of the GOATS JRP, the Centre initiated in 2002 a new long term programme called *Battlespace Preparation (BP) with AUVs* to explore and demonstrate the operational benefits and limitations of AUVs for battlespace preparation. Similarly to the GOATS series of experiments, the programme organizes multi–national, multi–disciplinary sea trials addressing the utilization of AUVs in coastal waters for military applications. Each year the experiments are prepared at a planning meeting in January and take place in the Mediterranean sea in May–June. Initial results are discussed in the fall and every second year the Centre organizes a scientific conference to report the findings. The first experiment of the BP series took place in May–June 02 in the Tyrrhenian and Ligurian seas with three broad objectives: Oceanography, REA, MCM. The results have been reported at the MREP conference (Bovio *et al.* 2003).

### *3.1 Oceanography*

Ocean forecasting is essential for effective and efficient use of AUVs in the littoral environment. The first part of the BP02 sea trial was dedicated to carry out and quantitatively evaluate a multiscale real time forecasting experiment in support of long range AUV missions. A two–way nested HOPS[4] model was run at Harvard University to predict oceanographic parameters in local (such as around Elba) and far field (eastern Ligurian sea) regions. Adaptive sampling patterns have been determined on short notice based on forecast results for both R/V *Alliance* and *Remus*. The AUV run missions up to 8 hrs at a speed of 3–4 kt. As shown in Figure 2, acoustic communication was maintained at rendez–vous points by the WHOI[5] Utility Modem deployed from *Alliance*. The acoustic link allowed real–time data retrieval at a reduced rate and on–line programming of the vehicle mission.

*Remus* executed CTD sampling in yo–yo mode and ADCP sampling in a depth range of 0–100 m. Vehicle navigation was accomplished by dead reckoning, with GPS updates when surfaced. Heading information was obtained with a magnetic compass. Velocity information was obtained from the ADCP when in bottom lock range. When the vehicle was not in range of the bottom, velocity was based on an estimated speed derived from its propeller's rotation rate. The characteristics of the *Remus* used in BP02 are shown in Figure 3. Oceanographic data acquired by *Remus* were

[4] Harvard Ocean Prediction System

[5] Woods Hole Oceanographic Institution

| Vehicle | Sensors | Diameter/Length (cm) | Weight (kg) |
|---|---|---|---|
| REMUS #1<br>ADM #1<br>Darter | Ocean Sensors CTD 2000<br>Wet Labs LSS<br>1.2 MHz ADCP – up/down<br>900 kHz Sidescan sonar<br>Acoustic Communication<br>GPS | Vehicle - 19/158<br>Shipping Case: 38 x 38 x178<br>Auxiliary Case: 38 x 38 x178 | 36<br>67.6<br>62.6 |
| REMUS #2<br>EDM 5<br>Grudgen | Ocean Sensors CTD 2000<br>Wet Labs LSS<br>1.2 MHz ADCP- up/down<br>600 kHz Sidescan sonar | Vehicle - 19/158<br>Shipping Case: 38 x 38 x178<br>Auxiliary Case: 38 x 38 x178 | 36<br>67.6<br>62.6 |
| REMUS #3<br>EDM 6 | Ocean Sensors CT 2000<br>Wet Labs LSS<br>1.2 MHz ADCP - up/down<br>900 kHz Sidescan sonar<br>DIDSON Sonar<br>Acoustic Communication | Vehicle - 19/158<br>Shipping Case: 38 x 38 x178<br>Auxiliary Case: 38 x 38 x178 | 36<br>67.6<br>62.6 |
| PARADIGM Tracking System | Two radio buoys | Shipping Case (2)<br>53.3 x 109 x 68.6 | 63.5 each |
| Acoustic Communication | WHOI Utility Acoustic Modem<br>Acomms Radio Buoy<br>Shipboard Acoustic Receive Array<br>Tools and Electronics | Shipping Case (1)<br>53.3 x 109 x 68.6<br>Shipping Case (1)<br>25x25x150<br>Shipping Case (2)<br>30x50x90 | 65<br><br>35<br><br>35 each |
| Support Equipment | Miscellaneous electronics and mechanical equipment | Shipping Case (2)<br>53.3 x 109 x 68.6 | 65 each |

Figure 3. The *Remus* AUVs used in BP02. *Remus#1* has been used for Oceanographic, REA, MCM, and communication studies. *Remus#3* has been used for target ID, *Remus#2* has been kept as hot spare.

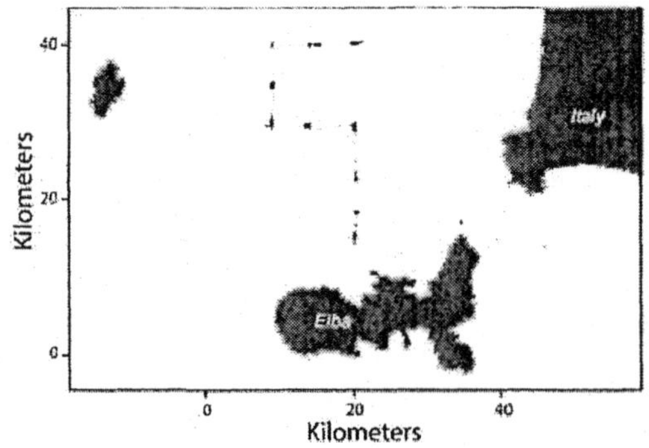


Figure 2. Oceanographic experiment in deep water. The *Remus* track is shown in light grey and the black marks indicate the locations where control information, vehicle status and oceanographic data were exchanged acoustically between vehicle and R/V *Alliance*. Typical modem range were 800–1500 m, depending on vehicle depth.

fused with with XBTs[6] and CTDs collected by R/V *Alliance* and available meteorological information. The fused data set was transmitted via Internet to the modelling team. The oceanographic team onboard *Alliance* coordinated the assimilation of oceanographic data collected by the various sources. The modelling code was run at Harvard. The model output was made available via Internet to R/V *Alliance* to perform adaptive sampling and optimize long range AUV missions.

### *3.2 REA*

The REA experiment demonstrated the capabilities of AUVs as REA platforms in shallow water. The information collected included acoustic and video images, bathymetry and water

[6] eXpandable BathyTermography

mass properties such as current, salinity, density and temperature. The AUVs were launched from R/V *Alliance*, and surveyed the bays of Levanto, Bonassola and Framura, Italy, to acquire environmental information in preparation for the MCMFORMED's Percentage Clearance (PC) trial which took place in February 2003.
The *OEX* operated the dual frequency (150/600 kHz) Marine Sonic and the video camera, the *Remus* operated the 900 kHz Marine Sonic and the DIDSON acoustic lens. All vehicles performed several missions a day, controlled from *Alliance* by acoustic and/or radio frequency communication. The sonar and video images were downloaded at the end of each mission and stored in the Centre GIS[7] database. Unsupervised segmentation software developed at the Centre divided the seafloor into areas of similar characteristic (Spina and Grasso 2003). Objects with dimensions similar to a mine were automatically extracted and marked on the GIS map. All data contributed to the production of seabed classification maps according to ATP 24 standards that were provided to the NATO mine hunters participating to the February 03 PC trial with the objective to measure the value of *a priori* environmental information in planning and conducting a PC trial.
The GESMA[8] ship R/V *Thetis* equipped with the interferometric *Klein 5400* side scan sonar conducted an independent survey of the area. The ship acquired co–registered imagery and bathymetry that has been compared with that acquired by *Manning* equipped with DF–1000 side scan and EM–3000 multi beam sonar during previous experiments. Work is in progress to assess the bathymetric capability of the sonar as an alternative to a multibeam (Zerr *et al.* 2003).

### *3.3 MCM*

The Italian Navy laid a field of exercise mines (8 *MK36* and 2 *Manta*) in two lanes selected to include a variety of different bottom types (rocks, *Poseidonia oceanica*, sand, mud) and portions of highly cluttered areas (wrecks, man made objects laid to protect cables and sewage pipes). In order to compare the performance of the experimental systems with that of the Italian Navy mine hunter *Numana*, the position of the mines was not known to all teams.
*Numana* surveyed Framura and Levanto lanes with SQQ14 sonar for detection and classification and performed visual identification with the *Pluto* ROV. *Remus* operated only in Framura due to weather. The AUV surveyed the area with "lawn mower" tracks using the 900 kHz Marine Sonic side scan sonar. Sonar images were downloaded from the vehicle upon return and analysed by the WHOI team using the Marine Sonic software. Mine like objects were successively identified with the *Remus* vehicle equipped with the DIDSON acoustic camera that provides high resolution video images and with the Centre vehicle *Ocean Explorer* equipped with sonar and video. The *OEX* that was deployed for the first time is shown in Figure 4.
*Thetis* surveyed both areas with redundant tracks providing multiple aspect of targets. The ship did not cover the north west corner of the Framura area, which was too shallow for safe towing of the *Klein*. Sonar images were received and processed in real time. Targets were detected and classified using multiple views of the objects.
Figure 5 shows the performance of the three systems. *Numana* detected classified and identified one *Manta* and two *MK36* in Framura and two *MK36* in Levanto. Two *MK36* in Framura and one *MK36* in Levanto were undetectable because they were concealed by the vegetation or masked by rocks. *Remus* detected, classified and identified two *Manta* and three *MK36* in Framura and did not operate in Levanto. *Thetis* detected and classified two *Manta* and one *MK36* in Framura and two *MK36* in Levanto. Due to water depth limitations, *Thetis* did not survey the northern corner of Framura area where three mines were present.
*Remus* and *OEX* showed great potential for MCM operations and the *Klein* demonstrated the good performance of a state–of–the–art commercial sonar.

## 4. PORTS/HARBORS SAFETY

Recently, responding to a request by SACLANT, a new project was started to evaluate the applicability of COTS (Commercial Off–The–Shelf) AUV technology to MCM operations in response to terrorist mining of ports. Four demonstrations have been successfully conducted in La Spezia and Stranraer and one more is planned in Rotterdam. Current AUV technology is sufficiently mature to complement existing MCM assets (Mine Hunters and EOD[9] divers) and improve their limitations. Of particular interest is the capability to ship overnight small AUVs anywhere a crisis might occur and to place the appropriate sensors (sonar, optical, magnetic) in close proximity of mines without risking human lives. The limited cost of COTS AUV (compared with traditional MCM assets) allows to deploy fleets of specialized vehicles to achieve large area coverage.
During exercise Northern Light 03, the *OEX*

[7] Geographic Information Systems
[8] Groupe d'Etude Sous–Marine de l'Atlantique
[9] Explosive Ordinance Disposal

| | | | |
|---|---|---|---|
| Length : | 4.5-5.5 m | Survey Speed: | 2-4 knots |
| Diameter: | .5 m | Survey Endurance: | 8 hours (at 3 knots) |
| Weight in air : | 600-800 kg | Batteries: | Nickel Metal Hidrate |
| Buoyancy: | ~+0.5 kg | Line Keeping: | +/- 2 meters |
| Maximum Depth: | 300 m | Altitude Keeping: | +/- .5 meter |
| Operation Depth: | 200 m | | |

Figure 4. The *Ocean Explorer (OEX)* is designed to accommodate various sonar, camera, and oceanographic systems in modular sections. The length and weight of the vehicle depend on the payload configuration.

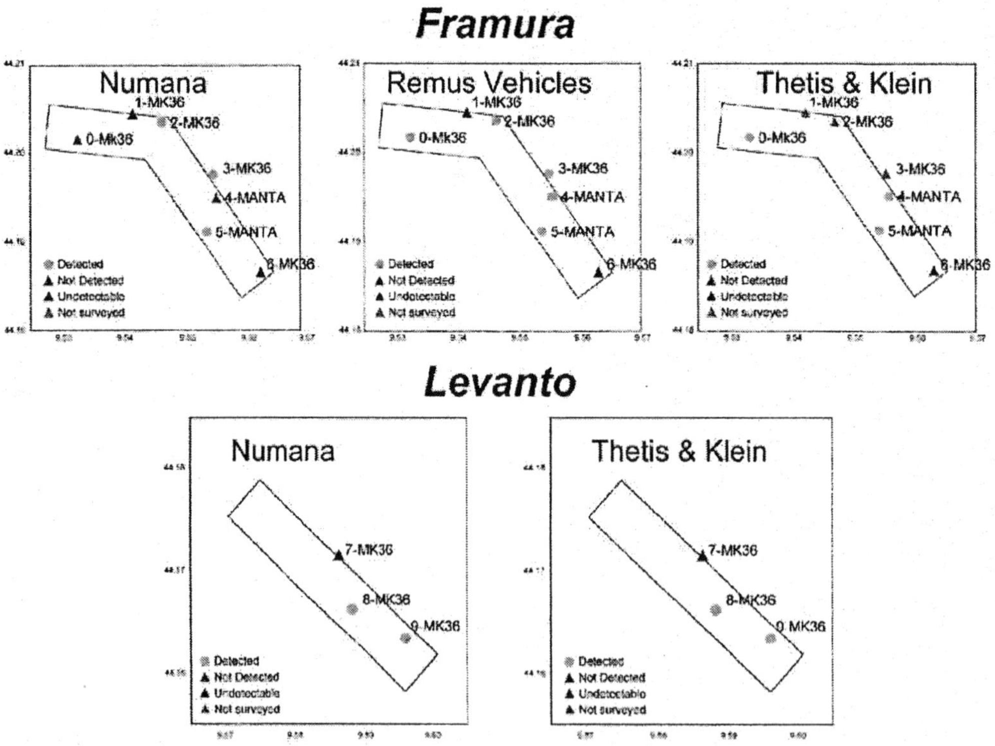


Figure 5. Results of PC trial. Due to weather *Remus* operated only in Framura. Due to water depth limitations, *Thetis* did not survey the top corner of Framura lane where three mines were laid. Two mines in Framura and one in Levanto were concealed by vegetation or masked by rocks. The experiment clearly demonstrated the potential of *Remus* and *Klein* for MCM operations.

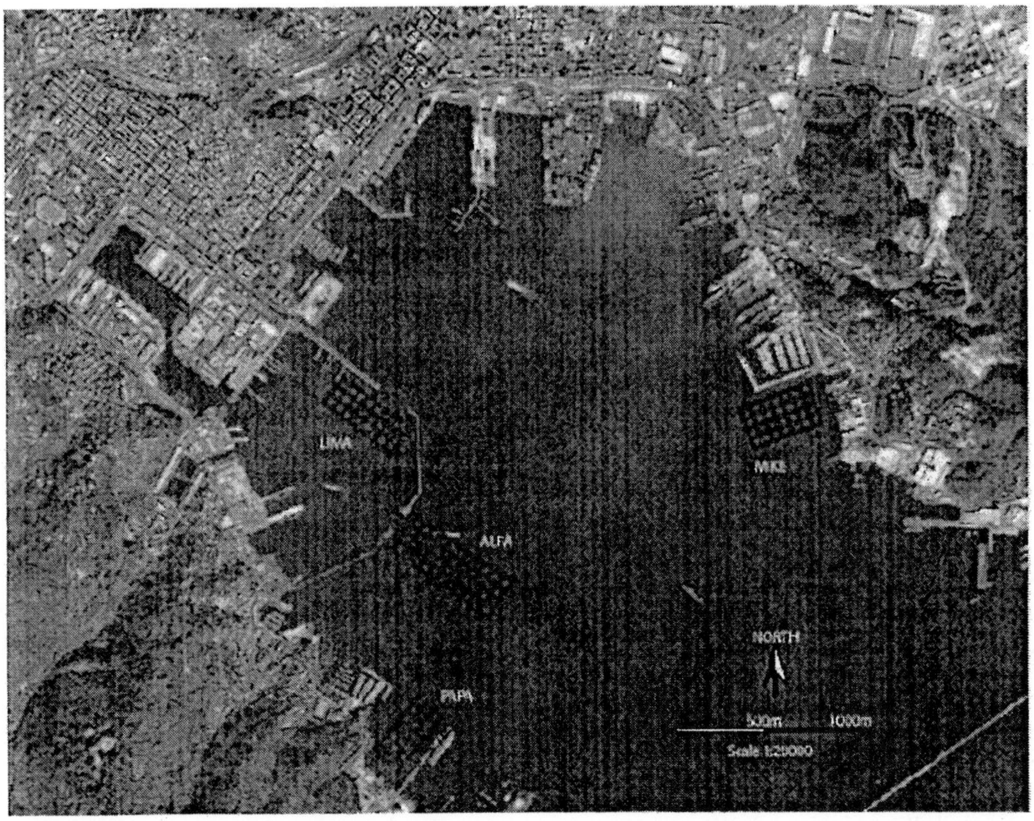


Figure 7. Working areas in La Spezia harbor.

launched from Stranraer pier, and the *Remus*, operated by Royal Navy Fleet Diving Unit 2 (FDU2), surveyed the final part of Loch Ryan, where 4 exercise Manta mines had been deployed. The AUVs covered the 3000m by 300m area with three detection missions followed by a number of classification missions. *Remus* navigated in a network of acoustic transponders and imaged the seafloor with 900 kHz Marine Sonic side scan sonar. *OEX* navigated with a GPS tow float and imaged the seafloor with 600 kHz Marine Sonic side scan sonar and with a digital video camera. The side scan sonar data acquired by the vehicles were analyzed to determine the nature of the contacts and to provide their location to FDU2 divers for identification and disposal. A team of only 5 FDU2 divers successfully completed in 3 days a task that, if performed with traditional means, would have required 20 divers working 20 days 12 hrs/day.

Figure 6. *Remus* vehicle communicating with the LBL.

The purpose of the sea trial in La Spezia harbor (March 2004) was to demonstrate the effectiveness of AUVs and ROVs in support to mine hunters and EOD divers, to counter terrorist mining of La Spezia harbor. The exercise measured the effectiveness of AUVs in detecting, classifying and correctly geo-referencing targets for further prosecution by EOD divers. *Remus*, provided by Hydroid, was configured with Marine Sonic 900 kHz side scan sonar. The vehicle launched and retrieved from a rib boat navigated within a network of acoustic long base line (LBL) transponders deployed by *Leonardo* (see figure 6). The *Remus* covered the assigned channel and the anchorage areas with orthogonal lines (see figure 7) in order to obtain multiple aspect insonification of all targets. Line spacing were designed to ensure full bottom coverage of the side scan sonar. All targets were detected and localized within 5 m of their true location. At present *Remus* communicates in real time via acoustic modem only status information. In the near future the vehicle will be able to transmit side scan sonar images of targets to a communication buoy that will relay the information to a ship or a shore installation.

## 5. CONTROL AND NAVIGATION: BASIC TOOLS FOR SUCCESSFUL MISSIONS

The success of the AUVs missions showed in previous sections depends strongly on the good performances of the control and navigation systems. Seafloor classification and mines detection are only possible when side scan sonar acquires good quality images. This requires that the vehicles are able to follow pre–progammed paths, maintain a constant heading, speed altitude or depth especially in very shallow water.
The AUV position during the missions is required with the highest precision in order to geo–reference all detected targets with minimal error, typically less than 5m.

### *5.1 Control System*

Typical requirements for the control system in AUV missions are:

- course keeping
- constant depth
- constant altitude
- noise and disturbances rejection

The first three items are satisfied when the vehicle heading, pitch, depth and altitude control loops are stable and well tuned. Moreover the controller is requested to be insensitive to sensors noise (high frequency noise) and system parameters variations and robust to external disturbances. Example of external disturbances is the presence of *Poseidonia oceanica* on the sea bottom that causes wrong altitude measurements and affects the altitude controller's behavior. The problem is more evident when the AUV passes the border between clean seafloor and *Poseidonia oceanica* navigating in constant altitude mode. It is possible that a sudden variations in altitude is measured and the controller reacts rapidly running the risk to touch the bottom depending on vehicle length and altitude.
Different strategies could be adopted for the control system design that are basically model–based or model–independent (Fossen 2002). The first approach requires the knowledge of at least reduced vehicle model and allows for the controller design, modification and first approximation tuning, by simulation. A drawback of this approach is that vehicle description could change with different payloads, so more vehicle models are needed and, perhaps, different tuning of the controller is necessary using different payloads. It is often a hard task to obtain an accurate vehicle model with tests and identification work. Model–independent controllers are, in general, not so simple to tune but have the advantage of being robust to payloads changes.

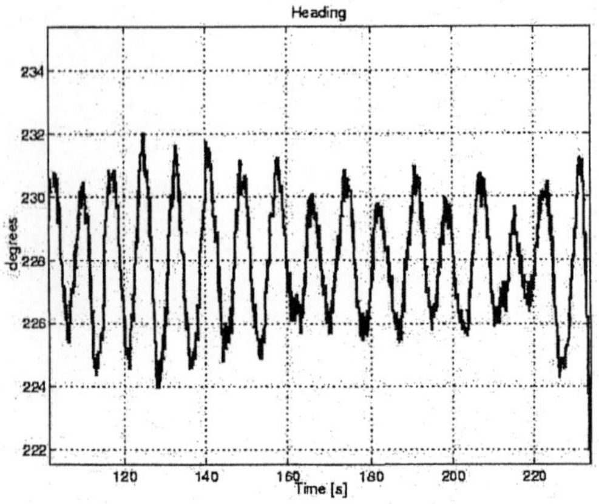


Figure 8. Vehicle oscillations: heading angle (speed = 3 knots).

The autopilots design can be approached in various methods. A standard decoupled PID controller is proposed in (Jalving 1994), multivariable sliding mode controllers are presented in (Healey and Lienard 1993), optimal controllers for AUVs are showed in (Juul *et al.* 1994, Feng and Allen 2002) (LQG/LTR and $H_\infty$ methods), a self–tuning autopilot is proposed in (Goheen and Jeffreys 1990) and fuzzy logic based controllers can be found in (Craven *et al.* 1998, Song and Smith 2000).
The control system of the *OEX–C* AUV available at the NATO Undersea Research Centre is model–independent, it is a Fuzzy Sliding Mode Controller (FSMC) (Song and Smith 2000). The vehicle dynamic was estimated through open loop at sea tests and consequently a nonlinear controller, robust to system parameters variations, have been designed. The switching curve of the sliding mode controller can be obtained by at sea measurements and then approximated by fuzzy logic. When the controllers are well tuned, the vehicle is able to track desired paths with minimum oscillations in heading (highly desirable for side scan sonar acquisitions), in pitch and in depth/altitude (depending on the mission).
The consequences of ineffective controllers can be seen in figures 8 and 9 that show oscillations in heading and pitch. The problem is more evident looking at side scan sonar images (figures 10 and 11): in figure 10 two targets are visible (highlighted with circles) and the image is good quality; in figure 11 only one of the targets is recognizable and the quality of the image is less than the previous one.

### *5.2 Navigation System*

Navigation accuracy is a key factor in the use of the AUV. As part of the global Guidance, Navigation and Control (GNC) system (Fig. 12), the choice of the navigation subsystem should be

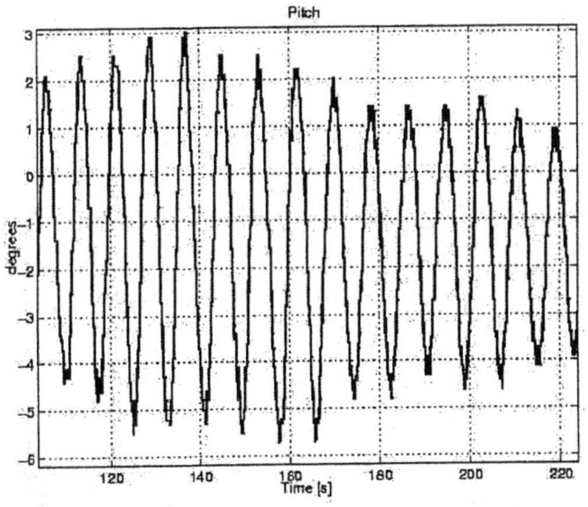


Figure 9. Vehicle oscillations: pitch angle (speed = 3 knots).

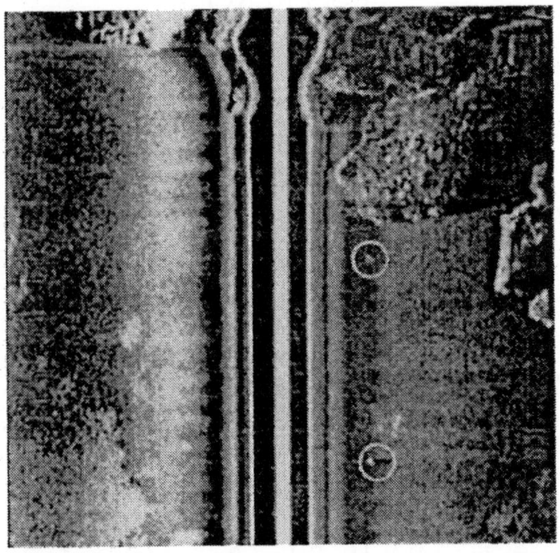

Figure 10. View of target (good control).

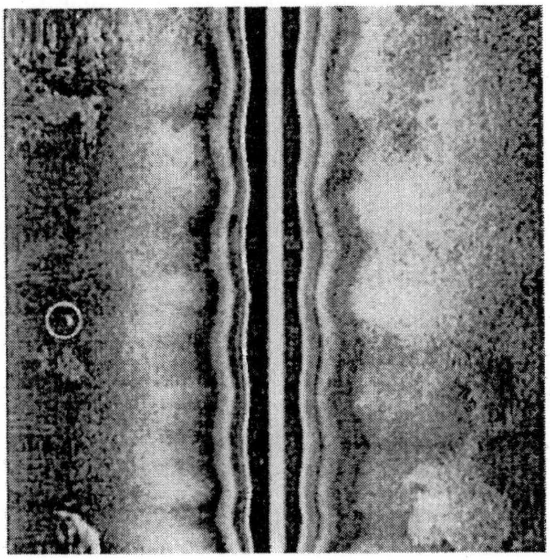

Figure 11. View of target (vehicle oscillating).

based on the global performances achieved by all the other vehicle subsystems. It is important to note that extremely high accuracy, though desirable, is not required by all the possible missions. The navigation accuracy required for an AUV collecting Oceanographic data (CTD) would be much smaller than for an AUV used for MCM or REA missions (Jalving *et al.* 2003).

Like the control system, navigation accuracy has a significant influence on the payload sensor performance, because the vehicle position is used to georefence the collected data as well as the attitude and velocity can be used to process and compensate sensor data.

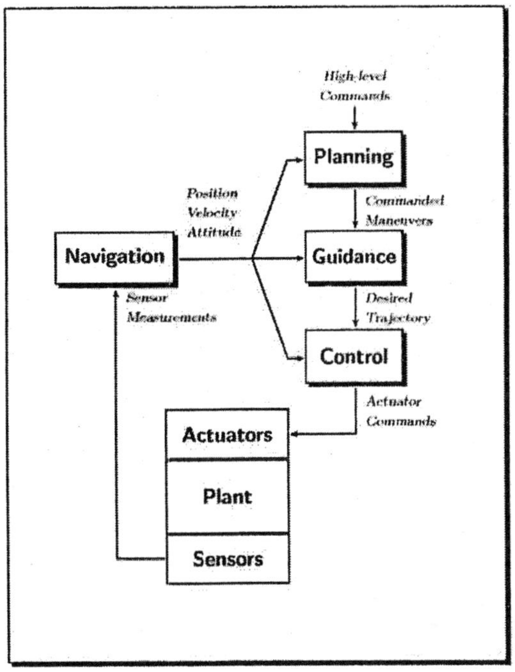


Figure 12. Block diagram of a typical GNC system

For multi–purpose AUVs the best approach to the navigation problem is the use of an Aided Inertial Navigation System (AINS), taking advantage of the reliability and high bandwidth of an Inertial Measurement Unit (IMU), using external (aiding) sensor to reduce its typical low–frequency errors. Navigation sensor data are fused using an Error State Kalman Filter rather than estimating the desired quantities (velocity, position and attitude), this filter estimates errors in measures and computed quantities. Figure 13 shows the typical scheme of an AINS, where position, velocity and attitude are calculated from IMU data (Navigation Equation) and than compensated for the errors estimated by the Kalman Filter comparing them with aiding sensor measurements.

The basic sensor set for an AINS system includes:

- Inertial Measurement Unit (IMU)
- Speed sensor, typically a Doppler Velocity Log (DVL)

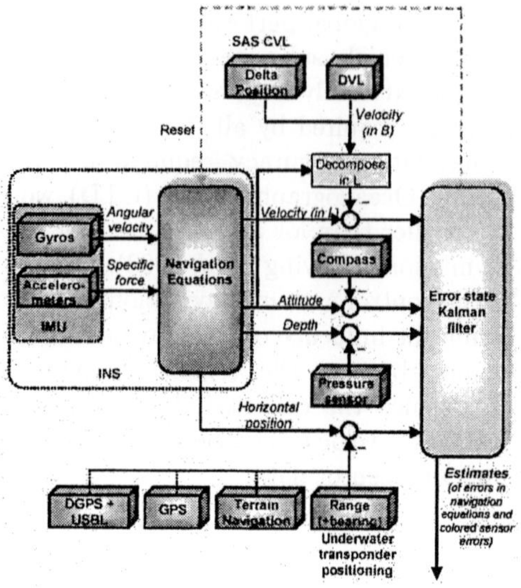


Figure 13. Aided Inertial Navigation System Structure

- Depth/Pressure Sensor
- Independent Position sensor, typically a GPS for initialization and sporadic error resets

The accuracy of the speed sensor and the availability of frequent position updates together with IMU characteristics are key factors on the overall system performances.
The availability of further aiding sensor (CVL [10], Terrain Navigation, Transponders) based on advanced techniques extends the capabilities of the AUVs to perform a wider range of missions, however the choice of the sensor set should always be a tradeoff between the desired accuracy, the system complexity and the mission requirements (covertness, environmental factors, mission duration).

## 6. CONCLUSIONS

Autonomous Underwater Vehicles (AUV) have reached sufficient maturity to be considered for military applications. After the successful completion of the GOATS Joint Research Programme (Bovio *et al.* 2001), the Centre has initiated a long term programme to explore and demonstrate the operational benefits and limitations of AUVs for battlespace preparation. This activity is based on multi–national, multi–disciplinary sea trials to evaluate the performance of commercially available AUVs in comparison with current military assets for MCM and REA applications. In addition, similarly to the GOATS series of experiments, the Centre studies with research partners the key technologies required for successful AUV deployment. The experiments carried out during the first test at sea, that took place in May–June 2002 in the Tyrrhenian and Ligurian seas, have been highly successful and demonstrated the clear advantage of using autonomous vehicles for a variety of REA and MCM missions (Bovio *et al.* 2003). In particular COTS AUV technology has been evaluated in MCM operations against terrorist mining of ports. Four experiments demonstrated that current AUV technology is sufficiently mature to complement existing MCM assets (Mine Hunters and EOD divers) and improve their limitations. Of particular interest is the capability to ship overnight small AUVs anywhere a crisis might occur and to place the appropriate sensors (sonar, optical, magnetic) in close proximity of mines without risking human lives.
The work will continue in the following years in cooperation with the research partners and NATO navies to reach the final goal of assessing the value of AUV networks for operational use.

[10] Correlation Velocity Log

## REFERENCES

Bovio, E., Coelho, E. and Tyce, R., Eds.) (2003). *Maritime Recognized Environmental Picture (MREP) Conference Proceedings.* number CP-47. NATO Undersea Research Centre. La Spezia, Italy.

Bovio, E., Tyce, R. and Schmidt, H., Eds.) (2001). *Autonomous Underwater Vehicle and Ocean Modelling Networks: GOATS 2000 Conference Proceedings.* number CP-46. NATO Undersea Research Centre. La Spezia, Italy.

Craven, P.J., R. Sutton and M. Kwiesielewicz (1998). Neurofuzzy control of a nonlinear multivariable system. In: *UKACC International Conference on Control.* Vol. 1. pp. 531–536.

Edwards, J.R., H. Schmidt and K.D. LePage (2001). Bistatic synthetic aperture target detection and imaging with an AUV. *IEEE Journal of Oceanic Engineering* **26**, 690–699.

Feng, Z. and R. Allen (2002). $H_\infty$ autopilot design for an autonomous underwater vehicle. In: *Proceedings of the 2002 International Conference on Control Applications.* Vol. 1. pp. 350–354.

Fossen, T.I. (2002). *Marine Control Systems.* Marine Cybernetics. Trondheim, Norway.

Goheen, K.R. and E.R. Jeffreys (1990). Multivariable self–tuning autopilots for autonomous and remotely operated underwater vehicles. *IEEE Journal of Oceanic Engineering* **15**(3), 144–151.

Healey, A.J. and D. Lienard (1993). Multivariable sliding mode control for autonomous diving and steering unmanned underwater vehicles. *IEEE Journal of Oceanic Engineering* **18**(3), 327–339.

Jalving, B. (1994). The NDRE–AUV flight control system. *IEEE Journal of Oceanic Engineering* **19**(4), 497–501.

Jalving, B., K. Gade and E. Bovio (2003). Integrated inertial navigation systems for AUVs for REA applications. In: *Maritime Recognized Environmental Picture (MREP) Conference Proceedings* (E. Bovio, E. Coelho and R. Tyce, Eds.). number CP-47. NATO Undersea Research Centre. La Spezia, Italy.

Juul, D.L., M. McDermott, E.L. Nelson, D.M. Barnett and G.N. Williams (1994). Submersible control using the linear quadratic gaussian with loop transfer recovery method. In: *Proceedings of the 1994 Symposium on Autonomous Underwater Vehicle Technology.* pp. 417–425.

LePage, K.D. and H. Schmidt (2000*a*). Laterally monostatic backscattering from 3-D distributions of sediment inhomogeneities. In: *Proceedings of the 5th European Conference on Underwater Acoustics* (M.E. Zakharia, P. Chevret and P. Dubail, Eds.). Lyon, Hawaii. pp. 1253–1258.

LePage, K.D. and H. Schmidt (2000*b*). Spectral integral representations of multistatic scattering from sediment volume inhomogeneities. In: *140th ASA Meeting/NOISE-CON 2000, Newport Beach, CA, USA.* Abstract published in Journal of the Acoustical Society of America, vol. 108, p. 2564.

LePage, K.D. and H. Schmidt (2000*c*). Spectral integral representations of volume scattering in sediments in layered waveguides. *Journal of the Acoustical Society of America* **108**, 1557–1567.

LePage, K.D. and H. Schmidt (2002*a*). Bistatic synthetic aperture imaging of proud and buried targets using an AUV. *IEEE Journal of Oceanic Engineering* **27**, 471–483.

LePage, K.D. and H. Schmidt (2002*b*). Spectral integral representations of monostatic backscattering from three–dimensional distributions of sediment volume inhomogeneities. *Journal of the Acoustical Society of America* **113**, 789–799.

Maguer, A., E. Bovio, W.L.J. Fox and H. Schmidt (2000*a*). In situ estimation of sediment sound speed and critical angle. *Journal of the Acoustical Society of America* **108**, 987–996.

Maguer, A., W.L.J. Fox, H. Schmidt, E. Pouliquen and E. Bovio (2000*b*). Mechanisms for subcritical penetration into a sandy bottom: Experimental and modeling results. *Journal of the Acoustical Society of America* **107**, 1215–1225.

Moran, B.A. (1999). GOATS 98 AUV network sonar concepts for shallow water mine countermeasures. In: *Proceedings of the 11th international symposium on unmanned untethered submersible technology.* Durham, NH.

Schmidt, H., A. Maguer and E. Bovio (1998). Generic oceanographic array technologies (GOATS) 98 - bistatic acoustic scattering measurements using an autonomous underwater vehicle. Technical Report SR–302. NATO Undersea Research Centre. La Spezia, Italy.

Schmidt, H. and E. Bovio (2000). Underwater vehicle networks for acoustic and oceanographic measurements in the littoral ocean. In: *MCMC2000: 5th IFAC Conference on Maneuvering and Control of Marine Crafts.* Aalborg, Denmark.

Schmidt, H., J.R. Edwards and K.D. LePage (2000). Bistatic synthetic aperture sonar concept for MCM AUV networks. In: *International Workshop on Sensors and Sensing Technology for Autonomous Ocean Systems.* Kona, Hawaii.

Song, F. and S.M. Smith (2000). Design of sliding mode fuzzy controllers for an autonomous underwater vehicle without system model. In: *OCEANS 2000 MTS/IEEE Conference and Exhibition.* Vol. 2. pp. 835–840.

Spina, F. and R. Grasso (2003). Unsupervised sea bottom classification from side-scan sonar images using multi-resolution transform features. Technical Report SR–372. NATO Undersea Research Centre. La Spezia, Italy.

Spina, F., E. Bovio and G. Canepa (2001). Seafloor classification for MCM with AUV mounted sensors. In: *Autonomous Underwater Vehicle and Ocean Modelling Networks: GOATS 2000 Conference Proceedings* (E. Bovio, R. Tyce and H. Schmidt, Eds.). pp. 237–246.

Veljkovic, I. and H. Schmidt (2000). Experimental validation of numerical models of 3-D target scattering and reverberation in very shallow water. In: *140th ASA Meeting/NOISE-CON 2000, Newport Beach, CA, USA.* Abstract published in Journal of the Acoustical Society of America, vol. 108, p. 2485.

Zerr, B., E. Bovio and F. Spina (2003). Bathymetric sidescan sonar for covert and accurate MCM REA. In: *Proceedings of the UDT 2003 Conference.* Malmoe, Sweden.

ELSEVIER
IFAC
PUBLICATIONS
www.elsevier.com/locate/ifac

# SOME RECENT DEVELOPMENTS OF NEXT GENERATION'S MARINE TRAFFIC SYSTEMS

**Kazuhiko Hasegawa**

*Osaka University, Japan*

Abstract: Some recent developments of next generation's marine traffic systems from Marine ITS project in Japan are roughly introduced. One is marine traffic simulation utilizing automatic collision avoidance system. It can simulate marine traffic simulation for any interested area based on a given OD tables etc. The safety assessment in congested area and utilization for planning of on-land AIS stations are shown as examples of its applications. The other is automatic berthing system utilizing artificial neural network. A proposal of automatic teaching data creation is introduced. A result of a model ship experiments is shown. *Copyright © 2004 IFAC*

Keywords: ship control, traffic control, automation, path planning, expert systems, artificial intelligence, neural control.

## 1. INTRODUCTION

Systemisation of automation of ship operation has started already in 1960s in Japan. It has started from labour-saving, power-saving to safing. The first national project named "Highly Reliable Intelligent Ship" (to be abbreviated as "*Intelligent Ship*") (Ohshima et al. 1989-90) (J.SNAJ ) has targeted automation for safety, including collision/aground avoidance and berthing, which were not treated before. These days are coincident with the boom of the research on artificial intelligence or soft computing such as fuzzy theory and artificial neural network. This kind of automation is a kind of neural and cognitive automation apart from the automation of muscle treated by the conventional automation. The system is realized by expert system combining rule-base or knowledge-base with several sensors.

As the second phase of this national project, national research institutes, universities or companies who participate this project continue research and development individually.

In this paper, some of these researches done by the author's group after the project will be introduced. Similar researches are continuing by other groups in Japan.

## 2. MARINE TRAFFIC SIMULATION SYSTEMS

### *2.1 Automatic Collision Avoidance System.*

The system called *SAFES* (Ship Auto-navigation Fuzzy Expert System) (Hasegawa *et al.*, 1989) is an automatic collision avoidance system with multiple-ship environment. The fundamental study was done by various researchers in the "*Intelligent Ship*" project. In the project several methods to solve the multiple-ship environment were proposed, but in *SAFES*, the combination of expert system and fuzzy theory was implemented. It was originally written in OPS83, an expert system language based on another basic system called *ACAS* (Automatic Collision Avoidance System) (Hasegawa, 1987) written in FOTRAN, which solves two-ship encounter problem.

### *2.2 Intelligent Marine Traffic Simulation System.*

*SAFES* is not only a system to be implemented for real ships, but also a system applicable for various applications such as:

- *Evaluation of automatic system.*
- *Safety assessment for harbour and waterway design.*
- *-Implementing for background traffic in ship handling simulator.*

The system originally named *SMARTS* (each-Ship-with-captain MARine Traffic System) (Hasegawa 1990) was developed as an application of *SAFES* to assess harbour and waterway design. It uses *SAFES* as decision-making engine for each ship. It contains automatic traffic flow generation according to the statistically given OD (Origin-Destination) tables for each port or gate in the gaming area. It also include data logger to be analyzed on-line and/or off-line.

The system was originally written in OPS83, and rewritten in G2, but now is completely implemented in C++ both for a PC and a workstation.

An example of the simulation result applied for Tokyo Bay, one of the most congested areas in the world, is shown in Fig. 1. The window provides system information, including the bird-eye-view of the present gaming area in any scale. For reader's convenience, the zoomed area is shown in Fig. 2, where each ship represented by a circle mark in different colour with velocity vector, which means the status of the motion such as normal and avoiding, although the colour cannot be appeared in the print. The trajectory of each ship can be traced as shown in Fig. 3, which might be useful to design the navigational lanes or regulations. The analysed result of near-miss is shown in Fig. 4 to demonstrate how it will be used to assess safety.

The system was recently utilized to plan AIS station in land. Japanese Coast Guard Agency is planning to construct AIS stations in congested area such as Tokyo Bay, Ise Bay and Osaka Bay. The system could provide useful data to predict the area distribution of AIS reports as shown in Fig. 5. We are now expanding the system for predicting the slot confliction in AIS slot reservation in realistic way.

The system can now handle various areas, just replacing a set of input setting files such as OD tables and an area map and configuration. Fig. 6 shows such an example done for Ise Bay, Japan, where Nagoya, the third biggest city in Japan exists. For the comparison an example for Osaka Bay done in the previous version of the system (Hasegawa *et al.* 2001b) is shown in Fig. 7, which was coded in G2.

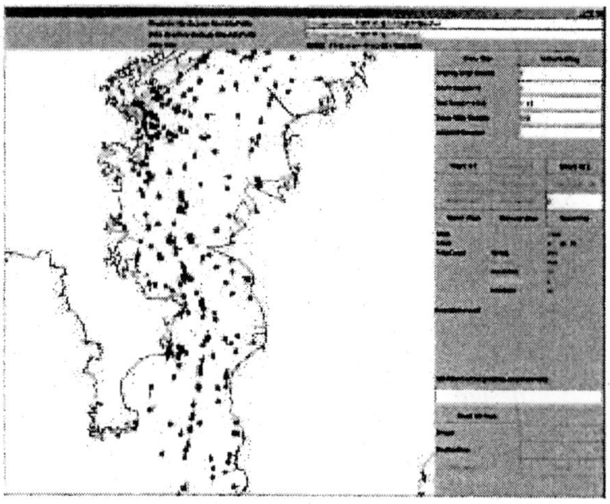

Fig. 1. Marine traffic simulation (Tokyo Bay)

The system can not only handle fully simulated traffic flow, but also any externally given ship movements. It means that the system can simulate certain modification with existing traffic flow. In the next stage the system is planning to be connected with ship path recorded by VTS (Vessel Traffic Service).

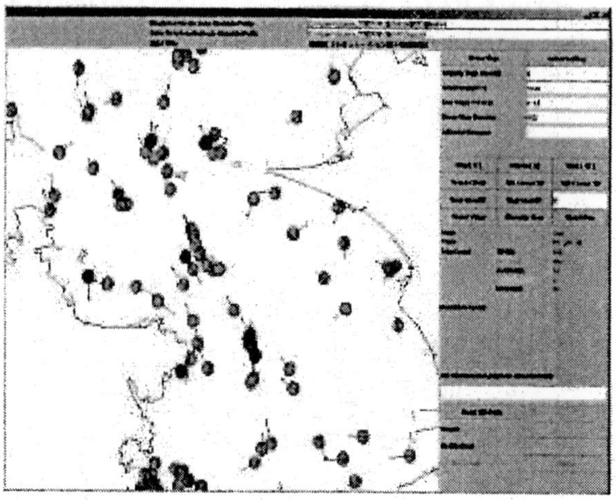

Fig. 2. Marine traffic simulation – zoomed (Tokyo Bay)

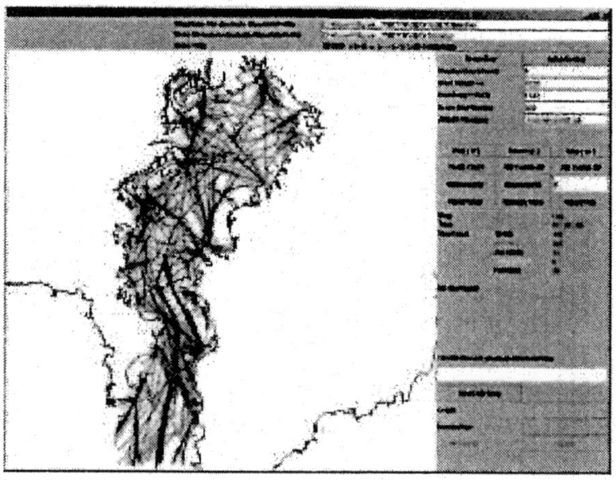

Fig. 3. Marine traffic simulation – trajectories (Tokyo Bay)

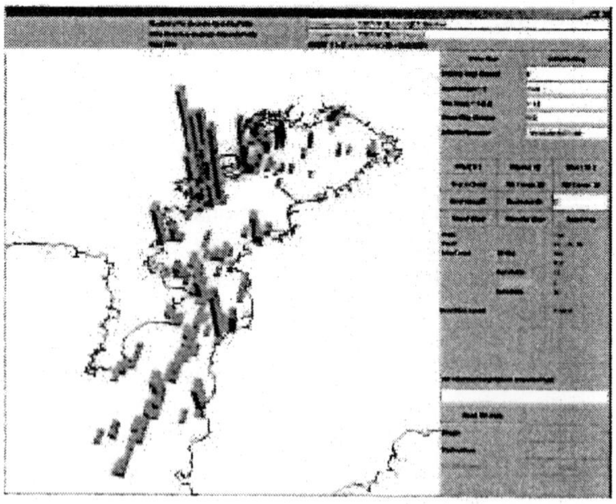

Fig. 4. Marine traffic simulation – near-miss points (Tokyo Bay)

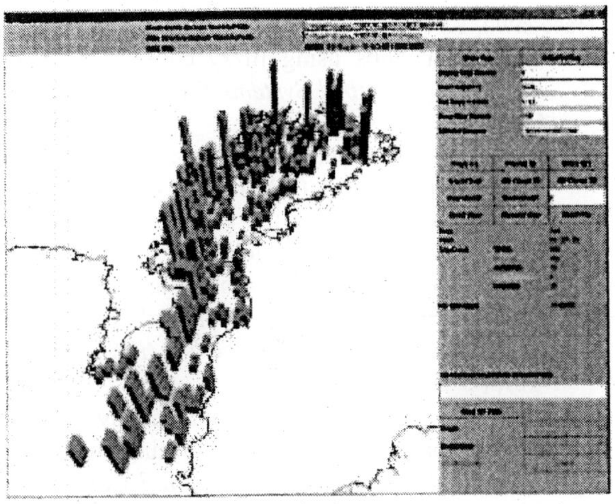

Fig. 5. Marine traffic simulation – AIS reports (Tokyo Bay)

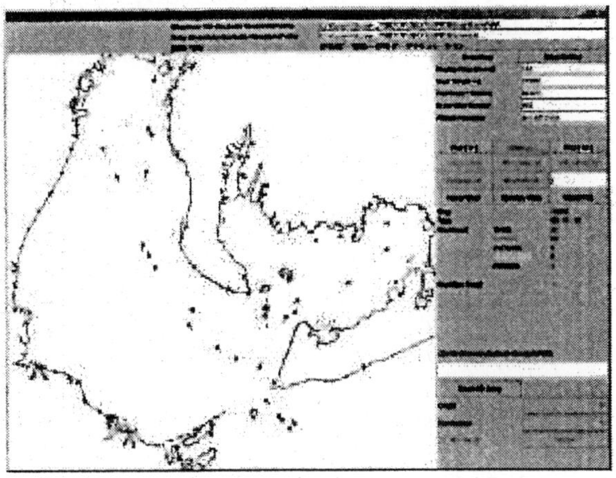

Fig. 6. Marine traffic simulation (Ise Bay)

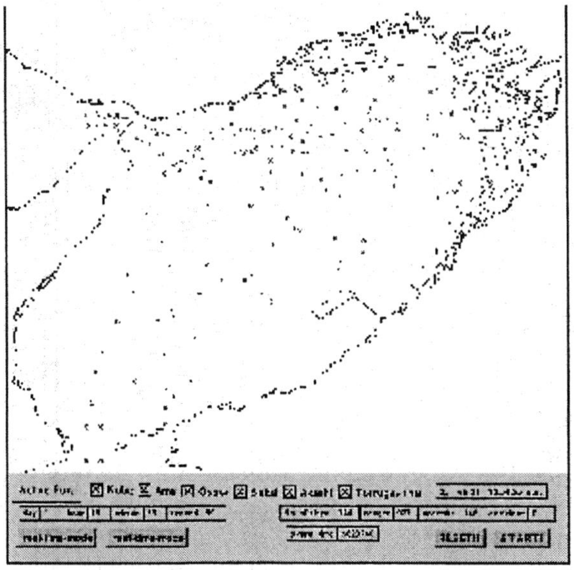

Fig. 7. Marine traffic simulation (Osaka Bay) (Hasegawa *et al.* 2001b)

## 3. AUTOMATIC BERTHING SYSTEM

Berthing is one of the most sophisticated manoeuvres in the various ship operations. A ship is normally not stable enough or sometimes unstable in low advance speed and much affected by disturbances such as wind and current. The effects of the depth of the sea bottom, bank and even the near-by ships are not ignored sometimes. Thus certain feed-forward or future-prediction is necessary. Therefore it is normally assisted by tugs, bow- and/or stern thrusters(s) as well as physical or electrical navigational aids. It will requite cautious judgement secured by long-term on-job experience. However with the decrease of number of skilled pilots and captains, it is highly requested to provide certain guidance or support system.

Several attempts of automation of berthing were done by several researchers under the "*Intelligent Ship*" project and others. The author has engaged in this problem using artificial neural network (ANN), which was first proposed by Yamato *et al.* (1990). By several researches (Hasegawa 1994, Hasegawa *et al.* 2001a, 2001c), it is verified that the artificial neural networks will work fairly for berthing control. The automatic berthing system was verified by experiments. Fig. 8 shows the teaching data obtained by manually-radio-controlled model ship and Fig. 9 shows the result of ANN controller learned from the teaching data thus obtained. In most cases it works fairly well, but several problems are also found. One is the effect of wind or current effect is not small and controller design for such disturbances is important for robustness. To improve these faults, parallel structure of the network is proposed. However, the most important factor for ANN controller design is to provide consistent teaching data. In the previous works all teaching data were provided by manual control results. Even if the operators are fully trained by themselves, human operators have various uncertainties. They will directly affect on inconsistency contained in teaching data.

Therefore teaching data is automatically created (Hasegawa, K. *et al.* 2004) based on the method proposed by Endo and Hasegawa (2003). Fig. 10 shows such results, where the origin is the berthing point with various starting points with different positions and heading angles and Fig. 11 shows the ANN controller results learned using these automatic created teaching data. It is verified that the ANN controller works consistently according to the various starting points.

The research is still undergoing for further developments such as practical tolerance against wind and current disturbances.

## 4. CONCLUDING REMARKS

The paper summarizes some recent developments of author's work aiming the next generation's marine traffic system or support system. There are still many things to be fixed or solved before going to the real-world application. However, the efforts or attempts to the collaboration between VTS or model ship experiments are continuously done. The detail of the works are mostly in the references listed below, although some of the work is still undergoing or somewhat confidential.

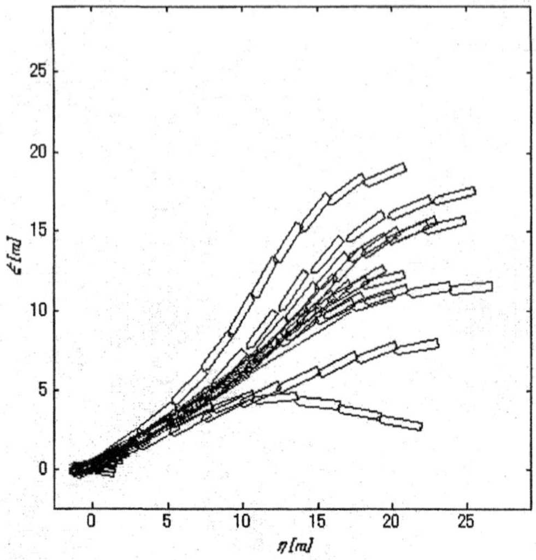


Fig. 8. Berthing manoeuvres provided for teaching data obtained by model ship experiments (Hasegawa 2003)

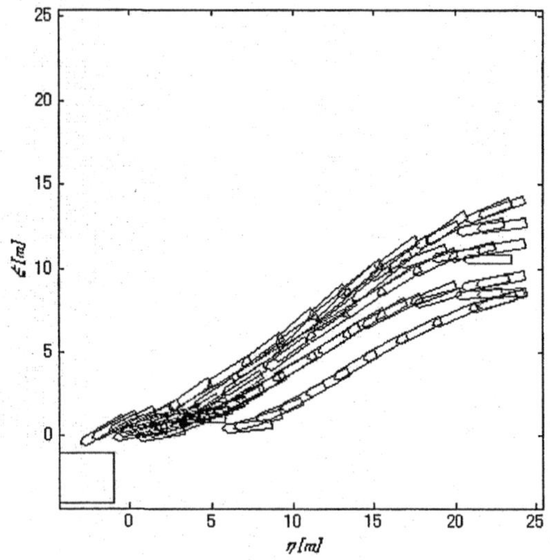


Fig. 9. Results of automatic berthing by ANN controller by model ship experiments (Hasegawa 2003)

REFERENCES


Endo, M. and Hasegawa, K. (2003). Passage Planning System for Small Inland Vessels Based on Standard Paradigms and Maneuvers of Experts. *Proc. of International Conference on Marine Simulation and Ship Maneuverability (MARSIM'03 )*, **II**, RB-19-1-RB-19-9.

Hasegawa, K. (1987). Automatic collision avoidance system for ships using fuzzy control. *Proc. of Eighth Ship Control Systems Symposium*, **2**, 34-58.

Hasegawa, K. *et al.* (1989). Ship auto-navigation fuzzy expert system (SAFES). *J. of SNAJ.* **166**, 445-452.

Hasegawa, K. (1990). An intelligent marine traffic evaluation system for harbour and waterway designs, *Proc. of 4th International Symposium on Marine Engineering Kobe '90 (ISME KOBE '90)*, (G-1-)7-14.

Hasegawa, K. (1994). On harbour manoeuvring and neural control system for berthing with tug operatio, *Proc. of 3rd International Conference Manoeuvring and Control of Marine Craft (MCMC'94)*, 197-210.

Hasegawa, K. *et al.* (2001a). Automatic ship berthing using parallel neural controller, *Proc. of CAMS'01*, (CD-ROM).

Hasegawa, K. *et al.* (2001b). Intelligent marine traffic simulator for congested waterways, *Proc. of 7th IEEE International Conference on Methods and Models in Automation and Robotics*, 631-636.

Hasegawa, K. *et al.* (2001c). An Application of ANN to Automatic Ship Berthing under Disturbances and Motion Identification, *Proc. of International Conference on Control, Automation and Systems (ICCAS 2001)*, (CD-ROM).

Hasegawa, K. *et al.* (2003). Automatic Berthing Control Using Artificial Neural Network and Its Verification by Model Ship Experiments, *Proc. of 2nd Int. EuroConference on Computer Applications and Information Technologies in the Maritime Industries, COMPIT'03*, (oral presentation.

Hasegawa, K. *et al.* (2004). Automatic Teaching Data Creation for Automatic Berthing Control System, *Proc. of 2nd Asia-Pacific Workshop on Marine Hydrodynamics (APHydro 2004)*, (to appear).

Ohshima, H. *et al.* (1989-90). Highly reliable intelligent ship (Part 1-13) (in Japanese). *Journal of SNAJ (Society of Naval Architects of Japan)*, **721,722,723,725,726,727,728,729**.

Yamato, H. *et al.* (1990). Automatic berthing by the neural controller, *Proc. of Ninth Ship Control Systems Symposium*, **3**, 183-201.


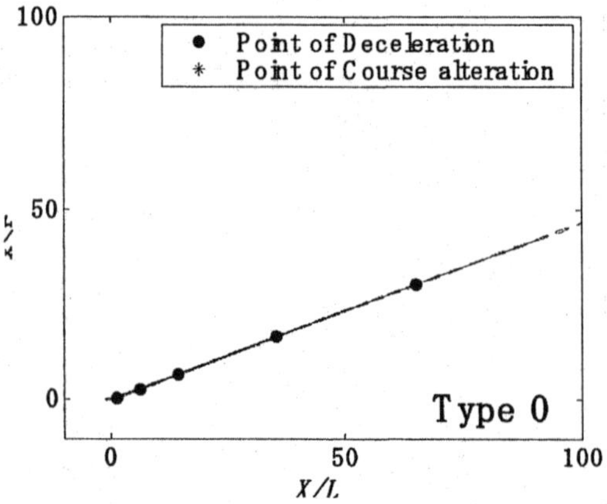

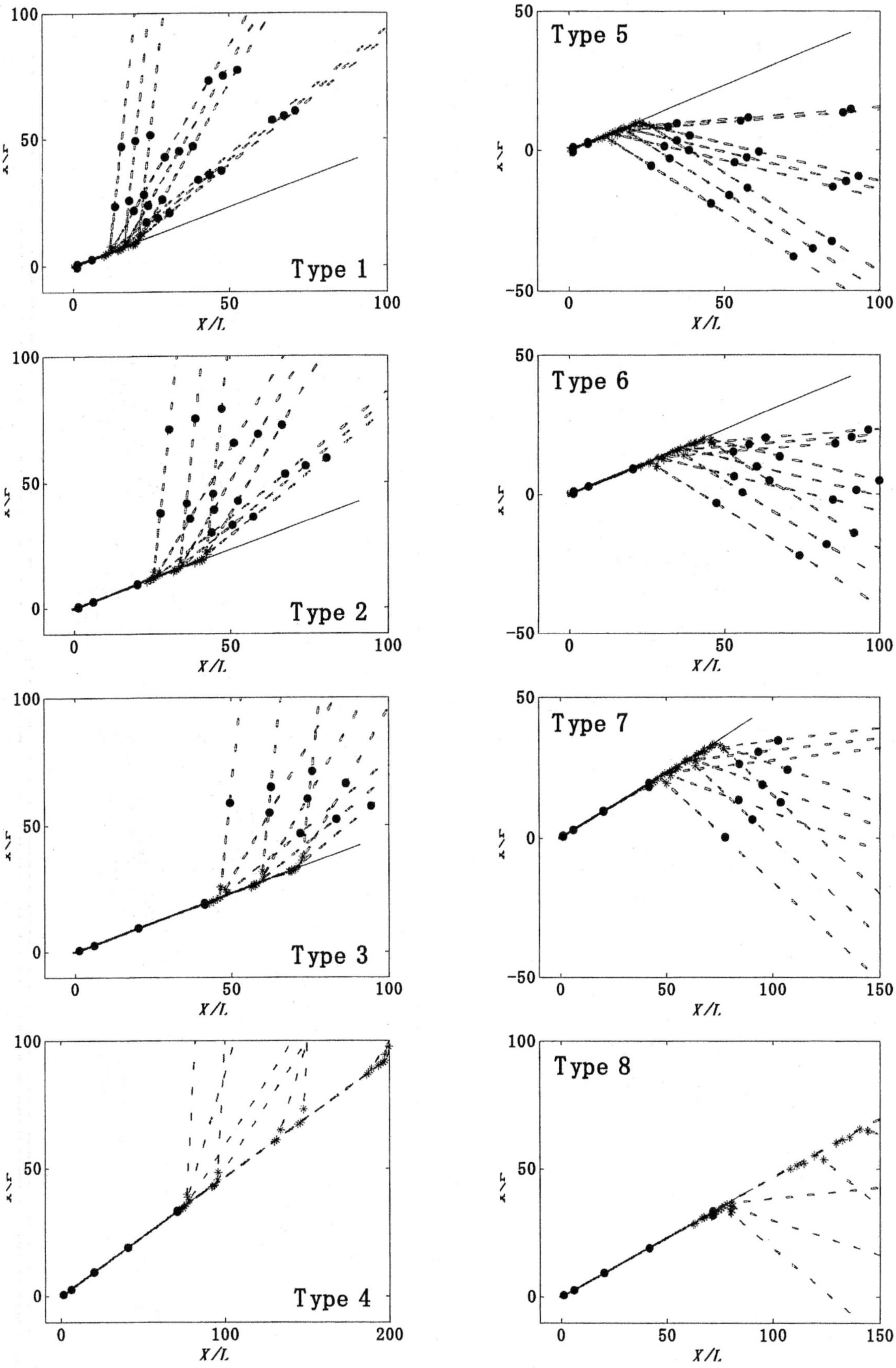


Fig. 10. Automatic created teaching data provided for ANN berthing controller (Hasegawa *et al.* 2004)

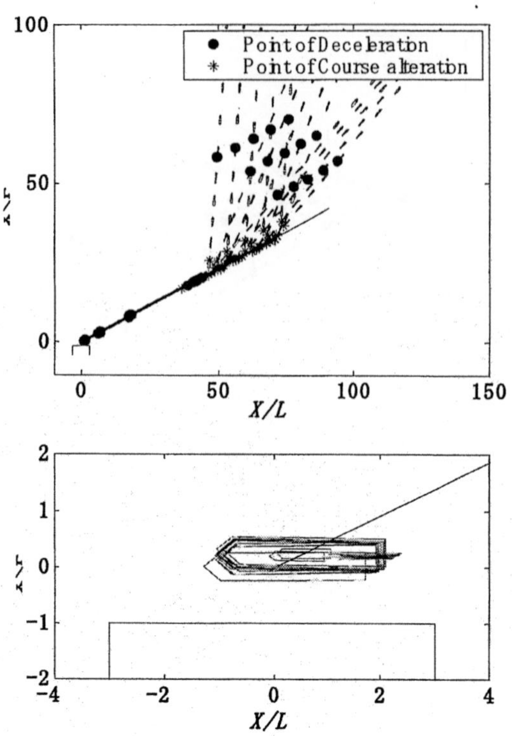


Fig. 11. Results of automatic berthing by ANN controller learned by automatic created teaching data (Hasegawa *et al.* 2004)

ELSEVIER
IFAC
PUBLICATIONS
www.elsevier.com/locate/ifac

# ITALIAN NAVY TRENDS IN AUTOMATION OF SHIP CONTROLS

**Rear Admiral Dino NASCETTI**
*Head of 4th Department- Studies, Projects and Materials- Italian Navy General Staff*

**Cdr Michele GIULIANO**
*In charge of Propulsion Systems Office- 4th Department- Italian Navy General Staff*

Abstract: This paper provides an overview of trends of the Italian Navy regarding automation of ship control systems. The first part of the paper describes the recent evolution of IPMS (Integrated Platform Management System) onboard Italian Navy Vessels and the current status of the functional integration. Following is an overview of expected capabilities of future Ship Control Systems to reduce manning on the next generation of Navy ships.


## INTRODUCTION

Italian Navy is undergoing significant changes in strategy, tactics, technology, funding, manning, and force structure. The regional conflicts, specific crisis-action scenarios, as well as humanitarian relief efforts, are all changes that shall be considered in next shipbuilding programs. Economic and social conditions will force the Navy to conduct its future missions with fewer people and lower manpower costs.

In the future, affordable automation systems and equipment will supplement or supplant many current operator functions in all shipboard routine to reduce ship's crew and minimize the number of personnel exposed to hostile actions.

## 1. THE EVOLOUTION OF IPMS SYSTEMS ONBOARD ITALIAN NAVY VESSELS.

The design and operation of naval ships have already changed radically over the last decade. Today's warships have comprehensive platform automation capabilities that allow them to achieve unprecedented levels of ship survivability and operational effectiveness. Integration of different control systems and equipments installed on board have optimized operational effectiveness and already contributed to crew reductions. As a further evolution, there has been a move from the traditional integrated automation system designed for controlling and monitoring of different functions, towards the Integrated Platform Management System concept whose capabilities are extended to information management.

### 1.1. IPMS architecture

Current IPMS architecture is based on an "Open Control System" (OCS) configuration. To achieve a seamless real time data exchange of multiple applications, the OCS is designed on well known widespread industry standards (Windows, OPC, Ethernet, TCP/IP, Profibus, Foundation Fieldbus, etc) that meet the following criteria:

- Compatibility, i.e. the control system ability to share the information with other systems in order to perform joint functions without degrading the information;
- Interoperability, i.e. the capability of a system built by one manufacturer to interact with a different one built by another manufacturer on the same network without loss of functionality;
- Interchangeability, i.e. the capability to replace a device with another built by a different manufacturer without loss of functionality or degree of integration.

A typical OCS configuration for naval applications is the so called "multi-layered" type where a number of distributed Process Controllers (PCs) manage the monitoring and control of each individual equipment or system.

The PCs connection to field equipments (transducers, sensors, actuators, etc.) can be either hardwired or via a fieldbus network. This latter solution offers considerable advantages in terms of less weight, reduced installation costs and commissioning time.

The PCs are interlinked to Windows-based operator workplaces via a dual redundant high speed fibre optic Local Area Network (LAN).

The backbone of IPMS is the Local Area Network interlinking the PCs and OWs (Operator Workplaces) for data communication and exchange. The network topology is of ring type and the physical media are fibre optics. Besides being redundant and having separate cable runs, the network is provided with switch devices that provides a separate connection for each node (computer) of the network. The use of switches enables segregation of the network into different segments. This gives the advantage of ensuring the integrity of the network in case of either mechanical or electrical damage of one or more segments of the network.

Systems (DCS) and Supervisory Control And Data Acquisition (SCADA) systems respectively. While the two systems offer more or less identical in functional character, the DCS architecture has been preferred mainly because that it is process-oriented while SCADA is data-gathering oriented.
As for any control system, the application software represents one of the most important tasks.
Until recently, standard or proprietary communication protocols have been used for a limited exchange of data between the different control systems. Now, the use of OCS features, allows the import or export of complete application programs from one system to another in real time, thus opening far and better opportunities to exploit new functionalities based on the interaction of the different systems installed onboard.

## 2. CURRENT APPLICATIONS OF THE FUNCTIONAL INTEGRATION

In the '90s, the Italian Navy launched a program for fleet renewal that included oceanographic vessels, offshore patrol vessels, frigates in co-operation with the French Navy (the "Horizon Program"), submarines, a multipurpose support vessel and an aircraft carrier. The Italian Navy focused its attention on the IPMS after performing an in-depth analysis of the ships operational requirements correlated to the capability provided by the system.

### 2.1. DCS – Damage Control System

Safety and Damage Control is an area of major concern for operator and crew due to the high potential of hazards in case of emergency situations or during real combat operations.
In accordance with Navy rules, every vessel is provided with a number of safety-related equipments like: fire detection, fire extinguishing, fire doors etc.
These safety equipments are generally based on redundant programmable controllers that operate independently from each other through their own control network. Their data (status, alarm and monitoring) are displayed to the operators in different formats such as LED panels, VDU graphics, etc. This may lead to misinterpretation or confusion for the operator thus delaying the initiation of proper actions.
The Damage Control System (DCS) is an application of functional integration for decision support based on a seamless and real time integration of all the data through a common network.

The main functions of the DCS are:

- Detecting all safety events generated by any safety related sub-system;
- Displaying to the operator where a hazard is developing;
- Supporting the operator's decisions on the possible actions to undertake (decision support function);
- Logging, printing-out and storing (black box) of all events;

The operator displays show both the longitudinal and the deck layout of the vessel's General Arrangement Plan. In case of incident or hazard (fire or flood) an alarm is automatically initiated displaying in red the zone or the compartment of the vessel where it has occurred. The operator gets more information by clicking directly on the VDU screen where a detailed view of the compartment is displayed, together with the indication of the alarm symbol (smoke, temperature, flood) and the status of doors, fans, etc. in that area.
Another feature of the DCS is the so-called Help Pages (or Kill Cards). Their purpose is to support the operator in the decision making process while control measures are being implemented. Predefined automatic control sequences to respond to specific danger conditions can be activated by the Help Page

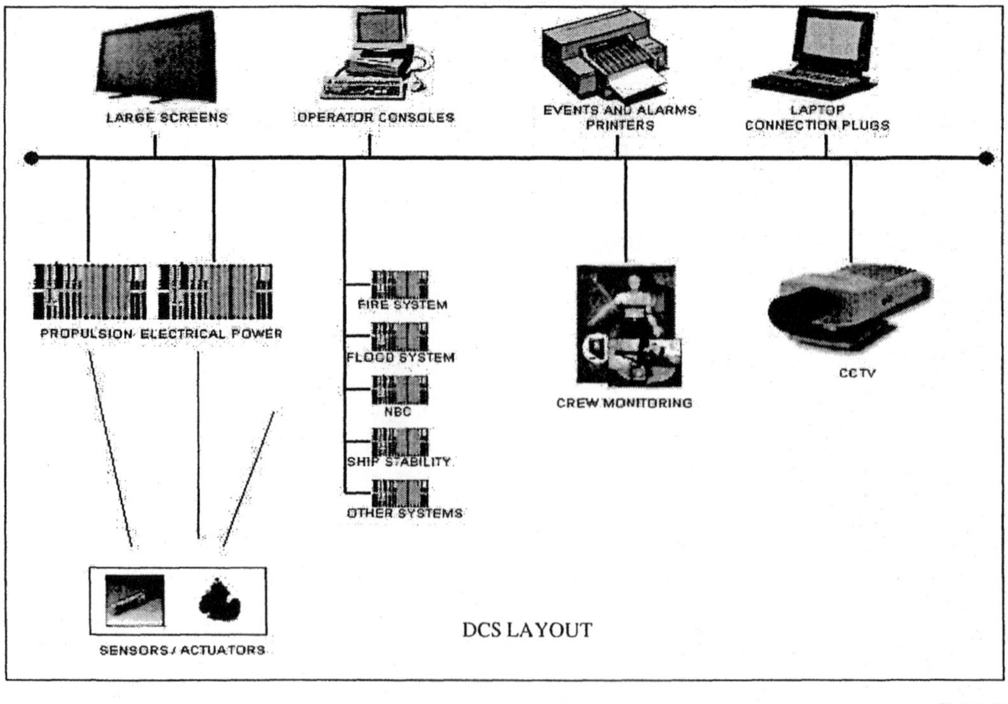

DCS LAYOUT

in addition to providing checklists for crew assignment and other damage management tasks. Each Help Page contains detailed information for each selected zone or compartment of the machinery, about the amount of and status of fire-fighting equipments, electrical switchboards, ventilation fans and dampers, dangerous materials, bilge/fire pumps and valves, etc. that are located in that zone/compartment. Moreover identification of any piping or electrical cable running through that zone/compartment are displayed.

The Help Pages have dynamic input menus to remotely initiate actions by the operator, for all the systems equipped with sensors and actuators. In other cases, the Help Pages are static and are in the form of checklists with the same information as the dynamic ones. The DCS may be linked also to a wireless crew monitoring system to track and determine the personnel position.

In addition to these features the DCS can be connected to the CCTV system to directly access on the VDU displays the CCTV image of the damaged compartment. The DCS can be configured to automatically set the CCTV camera and pop up the image as soon as an active sensor detects the hazard in the compartment.

Integrated in the DCS is the ship's stability calculation program to enable the chief engineer to evaluate in real time the impact of the damages on the stability of the vessel. The program calculates the ship's weight and stability, strength and trim, damage stability and grounding. The tank levels and ship's draughts values are made available on-line to the program through the control network.

## 2.2. CBM - Condition Base Maintenance

The main function of CBM is to optimize the life cycle costs and efficiency of the ship levering on following key aspects:

- Operating costs and their trends.
- Performance efficiency of the systems/equipments installed on board and their effect on the lifetime.
- Reduce downtimes and repair costs.

The most widely used approach for life cycle optimization is maintenance that can be either corrective (done after the fact) or planned (based on scheduled checks). A recent industry survey showed that 65% of the maintenance is corrective, 30% is preventive and only 5% is condition based.

Maintenance costs represent a considerable share of the operating costs and the adoption of a well balanced strategy can lead to significant savings.

"Intelligent" field devices, like instruments and actuators, are nowadays available off the shelf offering a number of advantages, like fieldbus communication and comprehensive diagnostic/maintenance functions. These functions can in real time be transferred to the automation system for further processing, like trending, plotting and statistical analysis.

One of the main function performed by the CBM is vibration monitoring to provide periodic check of the health of equipment and advise maintenance personnel concerning the need for maintenance. The CBM system automatically monitors online accelerometers, tachometers, and displacement probes for critical machinery such as propulsion engines, shaft bearings, and generators. For less critical machinery, an offline portable vibration measurement and recording system can be used to transfer vibration data to the system.

Another typical maintenance data measured by the fielded instrumentation is the remaining service life or the degree of wear, as a function of the operating time, temperatures and the dynamic loads.

These data and others such as present value, events, and alarms can be combined to evaluate the health of the ongoing process and to give an indication of which device or section of the process is degrading and can cause a breakdown.

This input automatically generates a fault report to alert the operator and then, through the control network, activates a work order to the Planned Maintenance System.

In addition, this information can also be transmitted, via satellite communication, to the Fleet Shore Computerized Maintenance Management System for further analysis like MTBF and MTTR comparison

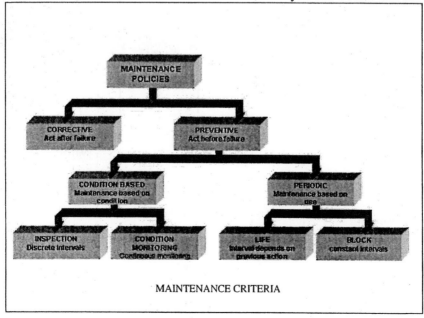

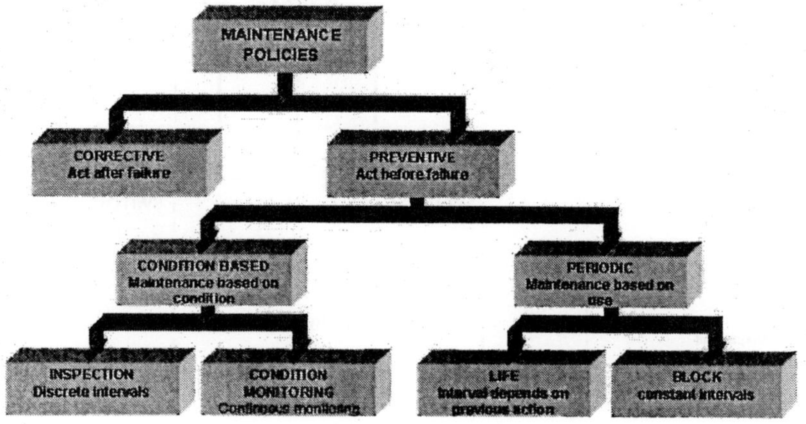


MAINTENANCE CRITERIA

between identical equipment on different ships, availability calculations, spare parts management, costs control, performance monitoring etc.

### 2.3. OBTS - On-Board Training System

Crew training is a key factor for the correct and safe operation of a vessel. But training time usually collides with the actual time and budgetary constraints of operational activities.
The usage of OBTS can be an excellent compromise to train crew and keep personnel updated and in a state of continued readiness with a significant reduction of operational costs and fixed investment (as any kind of land based training system).
The OBTS provides a similar operational training environment as full-scale simulators but using the same IPMS Human Machine Interface normally used by the operator during his daily duties.
While the functional character of the control is an exact copy of the actual platform functions, the machinery and ship systems are emulated by specific simulation functions providing a virtual replica of the real environment to the trainee personnel.
The IPMS training display pages are exactly the same as the real control display pages except for a completely different background colour, signalling that the current IPMS operator station is going in a training session, thus providing an OBTS environment as realistic as possible. While one IPMS station is used for training, other IPMS stations located in the same control room, are assigned as a backup for the real control and monitoring functions.

## 3. THE NEW AIRCRAFT CARRIER PROJECT

The "Conte di Cavour" Aircraft Carrier project is the latest IPMS that will be installed onboard an Italian Navy vessel.
The vessel has a displacement at full load of 26.700 metric tons and is powered by four gas turbines driving, through reduction gears, two cp propellers. The electrical plant consists in six diesel generators and two shaft generators.
The vessel, under construction at the Riva Trigoso and Muggiano Shipyards of Fincantieri Naval Shipbuilding Division, will be commissioned in 2007.
The high level of integration achieved by the IPMS allows us to maintain the highest level of efficiency and safety even with reduced manning.
The IPMS control & monitoring functions include:

- Remote control and monitoring of propulsion system (COGAG)
- Power Management System with automatic power restoration after a black out (controlling electric grid, diesel generators and shaft generators)
- Alarm and Monitoring
- Control of auxiliary systems
- Integration with:
  - CBM for the rotating machinery vibration monitoring system
  - OBTS for the operators' training this is achieved through any of the operator workplaces and by means of a process simulation software included in the IPMS
  - DCS including NBC subsystems
  - SDC combat system power supply interface

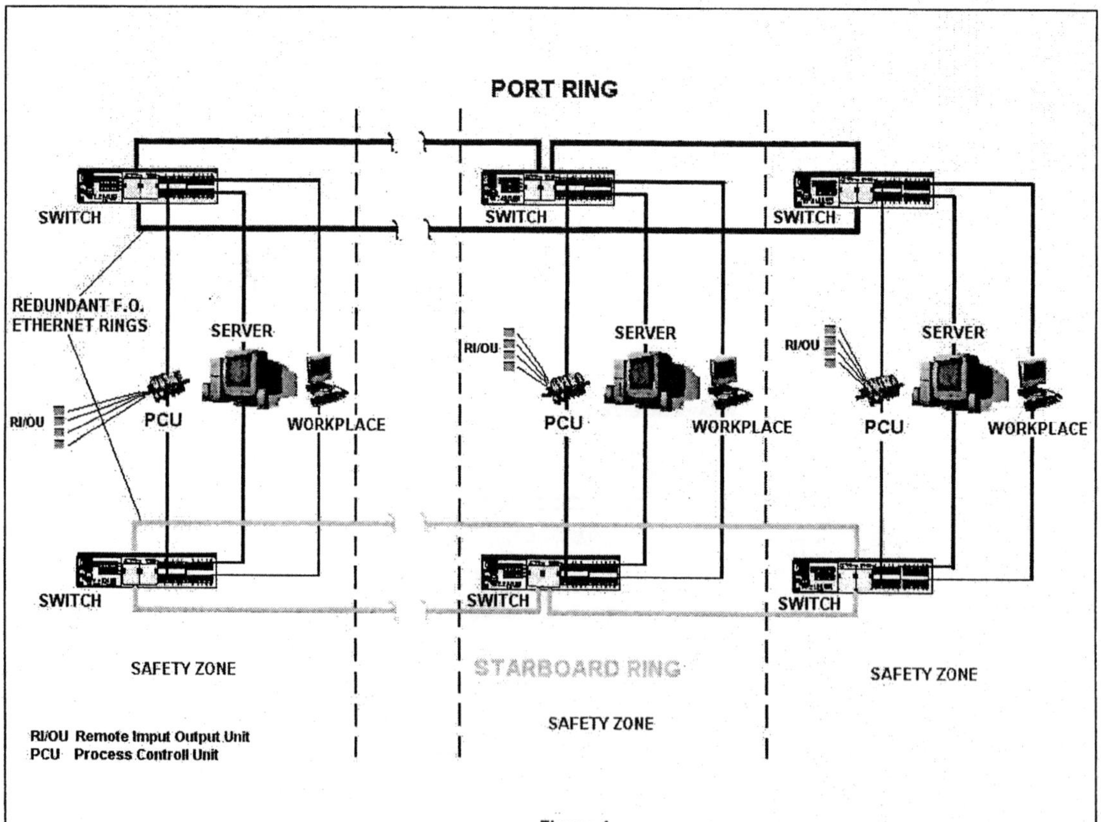


Figure 1

- TVCC subsystem
- ILS – Integrated Logistic Support

The IPMS architecture consists of distributed remote I/O units connected via fieldbus to distributed Process Controllers. Special care is given to the design of the network system and to the distribution of controllers along the circuit to minimize risks due to fire/flooding, enemy weapon effects in order to assure the performances and the survivability of the system. Other data is collected into the IPMS through Profibus and Modbus communication interfaces.

Operator Workstations for Human Machine Interface are provided in six different control rooms and enable management of the whole vessel by a single operator.

The control network is based on a Fast-Ethernet redundant fibre optic ring (100mbit/s) that, through several switches placed in the different safety zones, connects the controllers to the servers and to the operator stations (figure 1 shows an outline of a physical network). An additional network, separated from the control network, will provide the connectivity among the operator stations and the ship network for the integration of the CCTV system, the Combat system and the Ship CBM server.

All the operator stations are functionally interchangeable although their basic functionalities are assigned according to the identity/skillness of the operator logged in.

## 4. FUTURE TREND

Future Italian Navy combatants and auxiliary ships will probably be provided with different platform configuration as opposed to ships now under construction. In the field of propulsion and power generation, the trend is toward the Integrated Power System concept (IPS) with Hybrid (Mechanical-Electric) Drive or Full Electric Drive. The new IPS consists of a group of modules for power generation, to cover all the ship's needs including the electric propulsion and ship services. In fact, the electric system drive seems to offer some considerable advantages such as reduced volume, modular flexible propulsion units, lower acoustic signatures, enhanced survivability, and the enabling of new capabilities (because of huge amount of power available for the new generation of "direct energy weapons"). The fast progressing technologies, in the area of gas turbine generator units and modular permanent magnetic motors are leaning the Italian Navy towards the development of new hybrid and/or all-electric ships with associated drive, power-conditioning, and distribution systems. These type of ship will be highly automated and will be conducted with limited need of operator intervention or decision. On these new generation vessels the crew size and on board maintenance policy will be defined through a cost-effective analysis aimed at reducing operating costs while maintaining a high degree of efficiency and safety.

Current advanced combat ships have approximately a hundred of process controllers connected to the automatic control or monitoring equipment. To achieve significant manning reductions on future combatants, an higher level of shipboard automation must be achieved; therefore the number or the power of process controller will probably have to be considerably increased. Intelligent systems, system experts, advanced sensors, and the electromechanical actuators[1] will become major players in the design of future machinery systems and equipments. The systems will gradually become more modular and powerful, with "plug and play" capability, and will be logically interconnected and distributed via more advanced ship-wide networks, enabling unmanned decision-making and automated system reconfiguration. The ultimate goal will be the passage from people doing tasks to people commanding and monitoring tasks; damage control from manual to automated and intelligent control; systems with the ability to "sense", "reason", act, reconfigure themselves and share information and status. Such highly automated systems, with such a huge amount of information to process, shall be not only affordable but manageable and highly reliable.

The set of technologies required for the full automation of the new generation of ships need to improve mainly in the following fields:

- System integration.- Fully interoperable integrated ship systems, including integrated bridge system and combat management system;
- Computer and network.- High-performance capabilities, distributed open systems with more distributed intelligence and modular and scalable architectures, fault and damage tolerant, upgradable and better adaptable to next-generation technology; Extended use of Commercial Off-The Shelf (COTS) technology adapted for the marine environment;
- Power-generation and auxiliary systems.- Highly reliable and reconfigurable, fault and damage tolerant;
- Damage and decision support.- Considerable improvements in damage display and control; greater damage containment through enhanced fire fighting system automation, compartmentalization, modularity, and system interoperability.

---

[1] Many of the machinery and auxiliary systems on Navy ships heavily rely on manual intervention by operator. More extensive use of pneumatic/hydraulic actuators may be problematic because of the heavier level maintenance required. The use of electric actuators is a potential enabling technology that will reduce manning due to intensive maintenance on hydraulic/pneumatic actuators and fluid distribution systems. A wide variety of electric actuator systems are already available or under development. New types of electric actuators (such as piezo-electric, magnetostrictive, electrohydrostatic, and shape-memory alloy) are being developed and look promising.
Mature configurations such as electric motor and geared speed reducer are common in commercial and aerospace applications, but most of them have not been qualified to satisfy combatant ship service requirements (minimal acoustic signature, resistance to shock and vibration, electromagnetic interference emissions and susceptibility, corrosion resistance, power density, fail-safe operation, back-up power, reliability, and maintainability). In future a qualification program for electric actuator systems could

Among the main ways to replace human labor wherever and whenever possible to achieve shipboard manpower reduction, are:

- A higher level of automation in ship functions, with particular regard to the combat management;
- Improvements in automated conditioned-based monitoring and maintenance as well as greater integration with the logistic support shore organization. Adoption of advanced sensors capable to predict imminent failures so that maintenance is done only when really needed. Move of some maintenance ashore through the use of redundant systems or swapping out machinery or its subcomponents. Larger exploitation of information technology, so that experts can be located ashore and consulted when needed;
- adoption of systems, materials and construction techniques requiring less on board maintenance (advanced materials, coatings, and preservatives will reduce cleaning, corrosion control, and painting requirements); greater use of electromechanical actuators, advanced sensors and "self-reconfigurable" system;
- introduction of new procedures and technologies in supporting life at sea, such as material handling and hotel functions; moving as much of the workload as possible from the ship to shore support facilities/services.

With such technological advances, the future combat ship will become an integrated, interoperable platform with a crew mainly consisting of operators and decision makers, rather than maintenance technicians.

The future total ship computing architecture will benefit from the progress in information technologies to share real-time information to widely dispersed and dissimilar units.

The traditional model of information processing uses human being to receive, verify, process, correlate, and prioritize data and to determine its relevance to the situation. In the present generation of naval ships, the high degree of automation achieved has not yet solved all of the problems arising in many emergency situations. It happens in fact that Ship Control Centre operators being overloaded by an excess of information and requested to face many complex and time-critical tasks, can make mistakes and make wrong decisions. Such problems will probably increase on future ships as there will be fewer operators and therefore more tasks per operator. A possible solution to this very serious problem, is to create reliable systems expert to minimize the number of tasks for the operators and the amount of information presented to them. This system will support the operator in filtering information, resolve ambiguities, and determining their relevance. Rather than data, the operator overseeing the process, will receive basic info and suggestions to support the decision process.

All the tasks normally carried out by less experienced crewmembers are targeted to be automated. In general, the approach to manning the future ship will be to automate wherever practical, using the most advanced available and affordable technology, to minimize the ship's acquisition cost and life cycle costs.

## 5. CONCLUSION

Reduction of life cycle costs and optimization of human resources are among the most demanding requirements for future Italian Navy ships. These two objectives will lead to the optimization of the crew workload on board and to the adoption of new design to obtain a good balance between the ship's acquisition cost and life cycle costs. Among the methods for achieving this goal it may be considered:

- a higher level of automation and integration in ship functions, mainly regarding the combat management and the damage control and fire fighting systems;
- a well balanced automation technology strictly correlated with operator skills;
- the adoption of systems, materials and construction techniques that require less on board workload and maintenance.

Computer and network technology will play the most important role in the future ship control system design. As far as the platform is concerned, the development of more distributed control systems and networks with larger diffused automation will improve survivability and reduce manpower requirements for both machinery and auxiliary systems. For combat system, the introduction of netcentric warfare concept driven by advances in C4I systems will dramatically enhance the ship's fighting capabilities in an enlarged environment. Last, but not least, all the technological advances, that allow crew reduction, will most likely require a more skilled workforce and change in personnel recruiting and training policy.

## ACKNOWLEDGMENTS

The authors acknowledge the contributions of following people and organizations:
Mr. Vittorio Giuffra, Mr Giorgio Rolando of ABB
Mr. Luigi Pietro Passano of ELSAG
Mr. Stefano Michetti of FINCANTIERI.

ELSEVIER
IFAC
PUBLICATIONS
www.elsevier.com/locate/ifac

# STRUCTURAL ISSUES IN THE DESIGN OF MARINE CONTROL SYSTEMS

**Asgeir J. Sørensen**

*Department of Marine Technology*
*Norwegian University of Science and Technology*
*NO-7491 Trondheim, Norway*
*E-mail: asgeir.sorensen@ntnu.no*

Abstract: This paper addresses structural issues in the design and operation of marine control systems on ships and offshore installations. In particular, control systems for offshore vessels conducting station keeping and low speed manoeuvring operations are presented. Nevertheless, the methodologies will be valid for other applications and operational conditions. Design issues related to both real time control and monitoring systems and operational management systems will be addressed. As an example marine operations with a dynamically positioned drilling rig is presented. *Copyright©2004 IFAC.*

Keywords: Marine systems, Hydrodynamics, Structural issues, Control.

## 1. INTRODUCTION

*Marine control systems* or *Marine cybernetics* is defined to be the science about techniques and methods for analysis, monitoring and control of marine systems. It is believed that marine control systems will have its main field of applications in the three big marine industries: *Sea transportation (shipping)*, *offshore oil and gas exploration and exploitation* and *fisheries and aquaculture.* It is suggested to divide the control structure into two main areas: *real time control and monitoring* and *operational and business enterprise management,* see Fig. 1. The integration of real time systems with operational management and business transactional systems is by the automation industry denoted as *Industrial IT.* As seen in Fig. 2 the complexity level may vary both with respect to the automation and the logistics for the various shipping segments. The complexity of the management systems will increase with the number of ships in the fleet and the corresponding logistics and transport services.

The real time control structure is as shown in Fig. 1 divided into *low level actuator control, high level plant control* and *local optimization.* We will in the paper use demonstrating examples from the offshore oil and gas industry. In particular, examples with a dynamically positioned (DP) drilling rig will be used. A DP vessel maintains its position (fixed location or pre-determined track) exclusively by means of active thrusters. Position keeping means maintaining a desired position in the horizontal-plane within the normal excursions from the desired position and heading. The advantages of DP operated vessels are the ability to operate in deep water, the flexibility to quickly establish position and leave location, and to start up in higher sea states than if a mooring system should be connected. For deep water exploration and exploitation of hydrocarbons DP operated vessels may be the only feasibly solution. However, DP drilling operations at water depths less than 500 m may be more demanding as there is limitation on the riser angle offsets. Ideally, the angle should be within $\pm\ 2^{\circ}$. Riser angle larger than

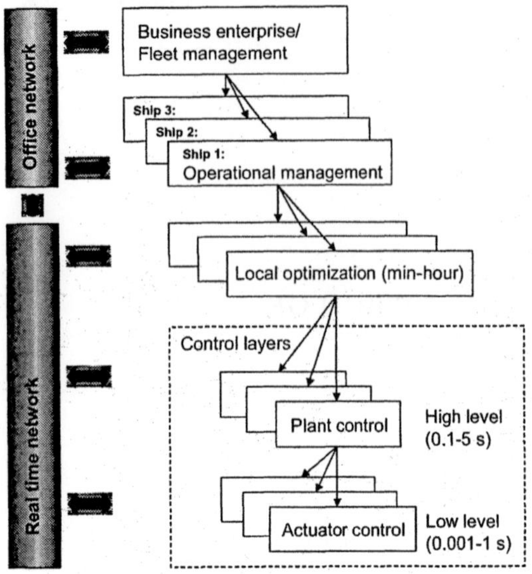


Fig. 1. Control structure.

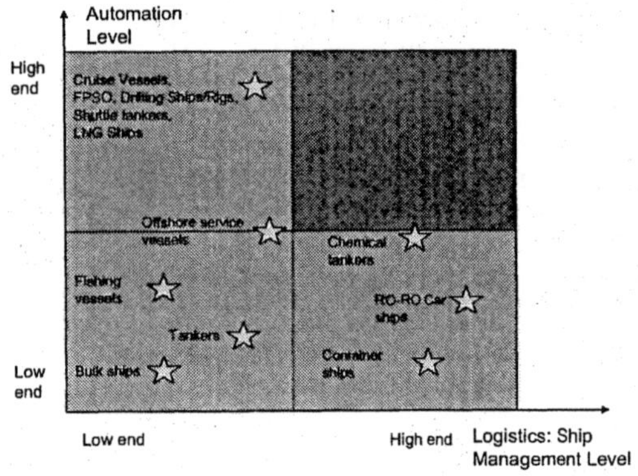


Fig. 2. Low-end and high-end market segments within automation and ship management.

5 ° to 8 ° at the sea bottom may be fatal. In case of a drive-off the available time window to decide and execute a controlled disconnect becomes more limited at shallow water compared to deep water. On the other hand, more complex riser dynamics is appearing for increasing water depth. So, by proper design of the low level thruster controllers, the plant controllers and the local optimization schemes, the control objective ensuring safe DP drilling operation subject to varying environmental loads will be met. For references on DP see Balchen *et al.* (1976, 1980), Fung and Grimble (1983), Sælid *et al.* (1983), Grimble and Johnson (1988), Fossen (1994, 2002), Sørensen *et al.* (1996), Katebi *et al.* (1997a and b), Sørensen and Strand (1999), Strand (1999), Fossen (2000), Lindegaard (2003), Sørensen *et al.* (2001), Fossen and Strand (2001), and Sørensen (2004).

## 2. OPERATIONAL MANAGEMENT

One of the main focus areas in the automation industry has been on connectivity ensuring physical integration of the various control systems. Today several vendors are in position to offer integrated automation systems. A few vendors are also in position to integrate the automation system to the electrical power generation and distribution systems. So far the integrated automation systems are proprietary. Communications with external devices, equipment and systems on control and operational management levels are provided by dedicated hard wiring, field buses and/or so-called gateways. However, some attempts to establish open communication standards between the various vendor supplied systems have been tried. The driving forces on hardware and software platform development are the commodity market for computers and the big land-based industries.

### *2.1 Maritime Industrial IT*

As reported in Rensvik *et al.* (2003) systems for operational management such as condition monitoring and diagnostics systems, supply chain management systems, enterprise management systems, etc. have increased the possibility to improve the operational performance, productivity and life cycle optimization of the assets related to the operation of the installations. Lately, the automation vendors, mainly in land-based industries, have started the next step to physically and functionally integrate the real time control systems with the operational management systems. The introduction of industrial IT into marine applications has yet started, and is still an area of research and development. Maritime industrial IT solutions for the various marine market segments will be dependent on the type of trade and charter, vessel complexity, safety and availability requirements, size of fleet, etc.

Concerning the vessel automation we will here focus on ships and vessels characterized as advanced with high number if inputs/outputs (I/O). This is denoted as the high-end market; see also Fig. 2. Examples are ships and rigs (Fig. 3) for oil and gas exploration and exploitation, passenger and cruise vessels.

As a part of enabling industrial IT solutions some vendors have installed condition monitoring and control functions locally on the power equipment and field devices as indicated in Fig. 4 with the possibility for remote monitoring and diagnostics. With reference to Fig. 1, the real time network constitutes the control network and the fieldbus network. For a drilling rig as shown in Fig. 3 the number of I/O may be up to 30.000 - 50.000

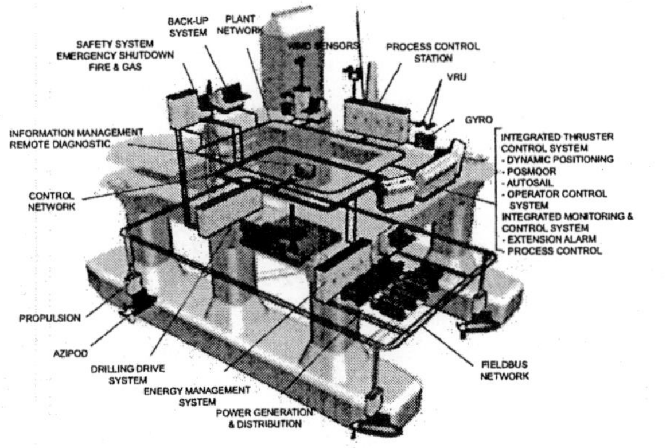


Fig. 3. Drilling rig.

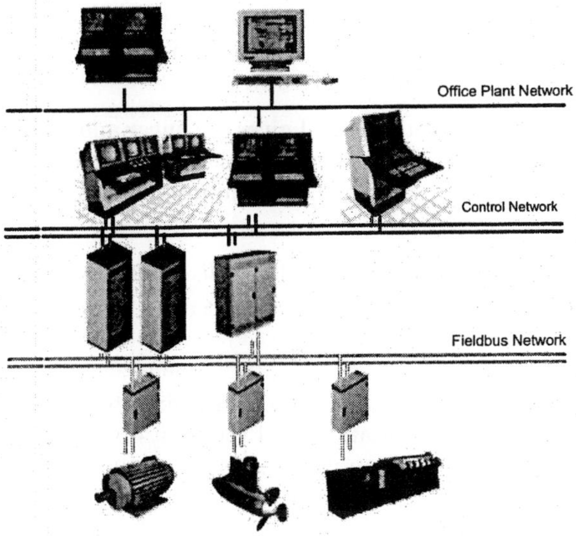


Fig. 4. Integrated automation system.

signals. For a conventional supply vessel the I/O number is in the order of 3.000-5.000 signals.

Physical integration by field bus and control networks ensures connectivity between devices, control stations and operator stations as seen in Fig. 4. The office plant network opens up for satellite communication to shore offices (Fig. 5). As cost on the vessel-to-land satellite communication is reduced and the maritime information technology architecture is improved this kind of information flow is expected to be working seamless in real time, as opposed today, where a limited amount of date is transferred at discrete events.Industrial IT is supposed to increase the integration of vessel plant data with the business management systems ensuring optimized asset management and operation of each vessel in particular and the whole fleet on corporate level, see Fig. 6.

### 2.2 Condition Monitoring

Condition monitoring and RCM (Reliability Centered Maintenance) are well known concepts in many industries. These concepts have now also been incorporated in ship maintenance management systems. Today, ships in many trades have only very short loading and discharging periods in port. This means that maintenance that only can be performed in port have to be carried out during these short periods to avoid downtime. To improve harmonization, analysis and planning methods have to be available to monitor the status of the vessel, to schedule surveys and to predict the future state of the ship based on frequent reporting and continues monitoring of technical condition of the vessel including installed systems and equipment.

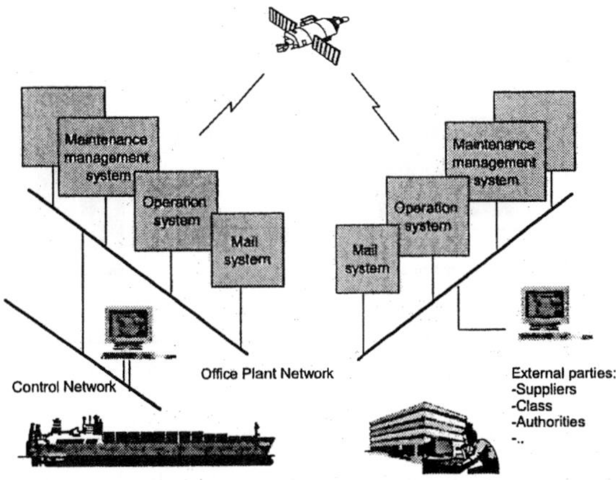


Fig. 5. Information systems for ship – shore.

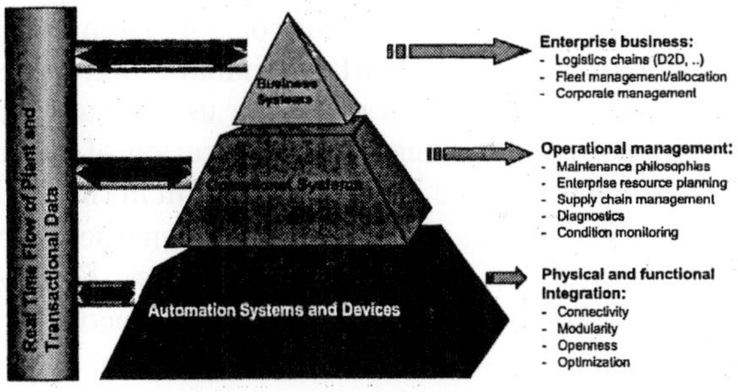


Fig. 6. Industrial IT architecture.

The last decade the use of real time observers and signal processing methods for fault diagnostics and fault-tolerant control has got an increased attention. Some important references in this field are Isermann (1997), Isermann and Balle (1997) and Blanke *et al.* (2003).

## 3. MATHEMATICAL MODELLING

Mathematical modelling of marine systems and vessels is multidisciplinary. Dependent on the operational conditions the vessel models may briefly be classified into low velocity/station keeping or high velocity/sea keeping models (Fig. 7). As we in this paper focus on DP vessels we can limit

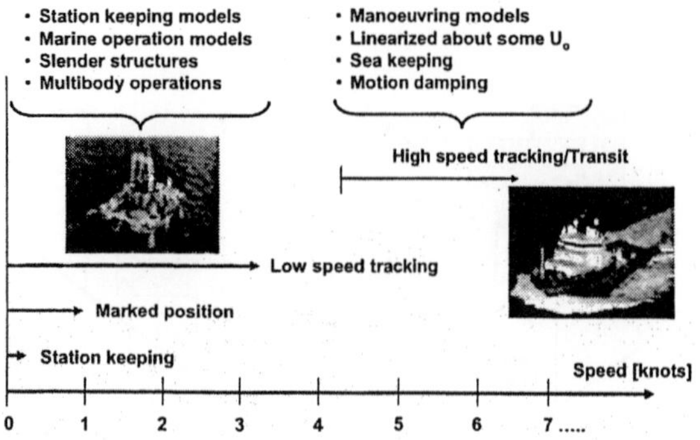


Fig. 7. Modelling properties.

the vessel modelling to consider the low velocity operation only. For high speed tracking operation the reader is referred to Fossen (2002) and the references therein. Easy access to computer capacity, and the presence of efficient control system design toolboxes, such as Matlab, Simulink and others, have motivated more extensive use of numerical simulations for design and verification of control systems. Essential in being successful in this is the ability to make sufficient detailed mathematical models of the actual plant or process. From an industrial point of view, the same tendency is driven by the fact that control system testing by hardware-in-the-loop simulations contribute to reduced time for tuning during commissioning and sea trials, and not at least, reduced risk for incidents during operation caused by software bugs and erroneous control system configurations.

Two model complexity levels are introduced:

- *Control plant model* is a simplified mathematical description containing only the main physical properties of the process or plant. This model may constitute a part of the controller. The control plant model is also used in analytical stability analysis based on e.g. Lyapunov and passivity.
- *Process plant model* is a comprehensive description of the actual process and should be as detailed as needed. The main purpose of this model is to simulate the real plant dynamics. The process plant model is used in numerical performance and robustness analysis and testing of the control systems.

Due to the lack of process knowledge and thereby proper models, control plant models are often used as process plant models.

### *3.1 Process Plant Model: Hydrodynamics*

*3.1.1. Kinematics* The reference frames used are illustrated in Fig. 8. The Earth-fixed reference frame is denoted as the $X_E Y_E Z_E$-frame. The reference-parallel $X_R Y_R Z_R$-frame is also Earth-fixed but rotated to the desired heading angle $\psi_d$.

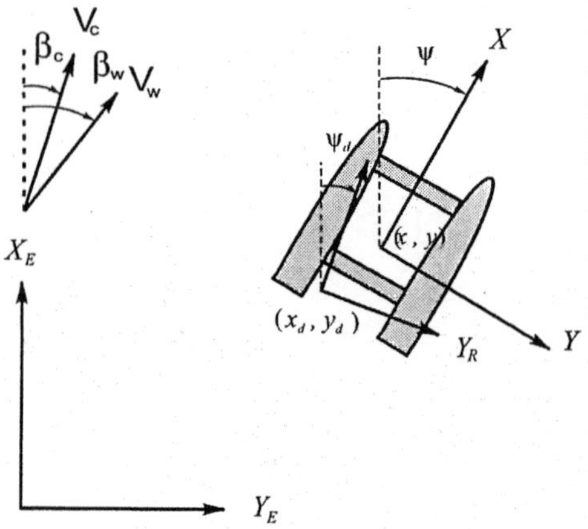


Fig. 8. Vessel reference frames.

The origin is translated to the desired $x_d$ and $y_d$ position coordinates. The body-fixed $XYZ$-frame is fixed to the vessel body.

The rotation matrix $\mathbf{J}$ gives the linear and angular velocity of the vessel in the body-fixed frame relative to the Earth-fixed frame (Fossen, 1994)

$$\begin{aligned}\dot{\boldsymbol{\eta}} &= \begin{bmatrix}\dot{\boldsymbol{\eta}}_1\\ \dot{\boldsymbol{\eta}}_2\end{bmatrix} = \begin{bmatrix}\mathbf{J}_1(\boldsymbol{\eta}_2) & \mathbf{0}_{3\times3}\\ \mathbf{0}_{3\times3} & \mathbf{J}_2(\boldsymbol{\eta}_2)\end{bmatrix}\begin{bmatrix}\boldsymbol{\nu}_1\\ \boldsymbol{\nu}_2\end{bmatrix}, \\ &= \mathbf{J}(\boldsymbol{\eta}_2)\,\boldsymbol{\nu}.\end{aligned} \tag{1}$$

The vectors defining the Earth-fixed vessel position and orientation, and the body-fixed translation and rotation velocities are given by

$$\boldsymbol{\eta}_1 = [x \quad y \quad z]^T, \boldsymbol{\eta}_2 = [\phi \quad \theta \quad \psi]^T, \tag{2a}$$
$$\boldsymbol{\nu}_1 = [u \quad v \quad w]^T, \boldsymbol{\nu}_2 = [p \quad q \quad r]^T. \tag{2b}$$

*3.1.2. Vessel model* In mathematical modeling of the marine vessel hydrodynamics it is common to separate the total model into a low-frequency (LF) model and a wave-frequency (WF) model. Hence, by superposition the total motion is a sum of the corresponding LF and the WF components. The WF motions are assumed to be caused by first-order wave loads. The LF motions are assumed to be caused by second-order mean and slowly varying wave loads, current loads, wind loads, mooring and thrust forces. The LF modelling problem can either be regarded as a three or a six DOF problem:

- **3 DOF:** For conventional ships and catamaran hulls usually only the three horizontal-plane surge, sway and yaw DOFs are of practical interest for the controller design.
- **6 DOF**: For marine structures with a small-waterplane-area and low metacentric height, e.g. rigs and SWATHS an unintentional coupling phenomenon between the vertical and the horizontal planes through the thruster action can be invoked (Sørensen and Strand,

1999). Typically natural periods in roll and pitch are 35 – 65 seconds.

Concerning the WF model it is normal to include all 6 DOF no matter how the LF model is formulated. The vertical-plane WF motions must be used to adjust the acquired position measurements provided by the vessel installed position reference systems, such as the GPS antennas and hydroacoustic position reference system transducers.

*3.1.2.1. Nonlinear low-frequency vessel model* The nonlinear six degrees of freedom (DOFs) body-fixed coupled equations of the low-frequency (LF) motions in surge, sway, heave, roll, pitch and yaw are written

$$\mathbf{M}\dot{\boldsymbol{\nu}} + \mathbf{C}_{\mathrm{RB}}(\boldsymbol{\nu})\boldsymbol{\nu} + \mathbf{C}_{\mathrm{A}}(\boldsymbol{\nu}_{\mathrm{r}})\boldsymbol{\nu}_{\mathrm{r}} + \mathbf{D}\boldsymbol{\nu}_{\mathrm{r}} + \mathbf{d}(\boldsymbol{\nu}_{\mathrm{r}}, \gamma_{\mathrm{r}}) + \mathbf{G}(\boldsymbol{\eta}) = \boldsymbol{\tau}_{\mathrm{wind}} + \boldsymbol{\tau}_{\mathrm{wave2}} + \boldsymbol{\tau}_{\mathrm{thr}}. \quad (3)$$

$\mathbf{M} \in \mathbb{R}^{6\times6}$ is the system inertia matrix including asymptotic added mass values for $\omega \to 0$, where $\omega$ is the wave frequency. $\mathbf{C}_{\mathrm{RB}}(\boldsymbol{\nu}) \in \mathbb{R}^{6\times6}$ and $\mathbf{C}_{\mathrm{A}}(\boldsymbol{\nu}_{\mathrm{r}}) \in \mathbb{R}^{6\times6}$ are the skew-symmetric Coriolis and centripetal matrices of the rigid body and the added mass. Notice that $\mathbf{C}_{\mathrm{A}}(\boldsymbol{\nu}_{\mathrm{r}})\boldsymbol{\nu}_{\mathrm{r}}$ includes the potential part of the current, whereof the Munk moments can be derived. $\mathbf{G}(\boldsymbol{\eta}) \in \mathbb{R}^6$ is the generalized restoring vector caused by the buoyancy and gravitation. $\boldsymbol{\tau}_{\mathrm{thr}} \in \mathbb{R}^6$ is the control vector consisting of forces and moments produced by the thruster system. $\boldsymbol{\tau}_{wind}, \boldsymbol{\tau}_{wave2} \in \mathbb{R}^6$ are the wind and second order wave load vectors, respectively. For details in these terms the reader is referred to Sørensen (2004).

The dissipative forces are often hard to model. We will therefore present more details here of the combined current and damping vector. The effect of current load is normally included by the definition of the relative velocity vector according to

$$\boldsymbol{\nu}_{\mathrm{r}} = \begin{bmatrix} u - u_c & v - v_c & w & p & q & r \end{bmatrix}^T. \quad (4)$$

The horizontal current components in surge and sway are defined as

$$u_c = V_c \cos(\beta_c - \psi), \quad v_c = V_c \sin(\beta_c - \psi), \quad (5)$$

where $V_c$ and $\beta_c$ are the current velocity and direction respectively, see Fig. 8. The total relative current vector is then defined as for $u_r = u - u_c$, and $v_r = v - v_c$ according to $U_{cr} = \sqrt{u_r^2 + v_r^2}$. The relative drag angle is found from the relation $\gamma_{\mathrm{r}}$ =atan2$(-v_r, -u_r)$, where the four quadrant arctangent function of the real parts of the elements of $X$ and $Y$ is such that $-\pi \leq$atan2$(Y, X) \leq \pi$. The nonlinear damping is assumed to be caused by turbulent skin friction and viscous eddy-making, also denoted as vortex shedding, Faltinsen (1990). Assuming small vertical motions, the 6-dimensional nonlinear damping vector can be formulated as

$$\mathbf{d}(\boldsymbol{\nu}_{\mathrm{r}}, \gamma_{\mathrm{r}}) = 0.5\rho_w L_{pp} \cdot \begin{bmatrix} DC_{cx}(\gamma_{\mathrm{r}})|U_{cr}|U_{cr} \\ DC_{cy}(\gamma_{\mathrm{r}})|U_{cr}|U_{cr} \\ BC_{cz}(\gamma_{\mathrm{r}})|w|w \\ B^2 C_{c\phi}(\gamma_{\mathrm{r}})|p|p + z_{py} DC_{cy}(\gamma_{\mathrm{r}})|U_{cr}|U_{cr} \\ L_{pp} BC_{c\theta}(\gamma_{\mathrm{r}})|q|q - z_{px} DC_{cx}(\gamma_{\mathrm{r}})|U_{cr}|U_{cr} \\ L_{pp} DC_{c\psi}(\gamma_{\mathrm{r}})|U_{cr}|U_{cr} \end{bmatrix}, \quad (6)$$

where $C_{cx}(\gamma_{\mathrm{r}})$, $C_{cy}(\gamma_{\mathrm{r}})$, $C_{cz}(\gamma_{\mathrm{r}})$, $C_{c\phi}(\gamma_{\mathrm{r}})$, $C_{c\theta}(\gamma_{\mathrm{r}})$ and $C_{c\psi}(\gamma_{\mathrm{r}})$ are the nondimensional drag coefficients found by model tests for the particular vessel. $B$ is the breadth, $L_{pp}$ is the length between the perpendiculars, $D$ is the draft and $\rho_w$ is the water density. The second contributions to roll and pitch are the moments caused by the nonlinear damping and current forces in surge and sway, respectively, attacking in the corresponding centers of pressure located at $z_{py}$ and $z_{px}$.

It is important to notice that for velocities close to zero, linear damping becomes more significant than the nonlinear damping. The strictly positive linear damping matrix $\mathbf{D} \in \mathbb{R}^{6\times6}$ is caused by linear wave drift damping and laminar skin friction. This kind of damping must not be confused with the frequency-dependent wave radiation damping used in the wave-frequency model. For further details about damping and current loads the reader is referred to Faltinsen (1990) and Newman (1977).

*3.1.2.2. Linear wave-frequency model* Linear theory assuming small waves and amplitudes of motion. The WF motion is in the literature calculated in the Earth-fixed frame. The hydrodynamic problem in regular waves using potential theory is solved as two sub-problems (*wave reaction* and *wave excitation*), which are added together (Faltinsen, 1990).

The coupled equations of WF motions can in the body-fixed frame be formulated as

$$\mathbf{M}(\omega)\dot{\boldsymbol{\nu}}_w + \mathbf{D}_p(\omega)\boldsymbol{\nu}_w + \mathbf{J}^T\mathbf{G}\boldsymbol{\eta}_w = \boldsymbol{\tau}_{\mathrm{wave1}}, \quad (7)$$

where $\boldsymbol{\eta}_w \in \mathbb{R}^6$ is the WF motion vector in the Earth-fixed frame. $\boldsymbol{\tau}_{\mathrm{wave1}} \in \mathbb{R}^6$ is the first order wave excitation vector, which will be modified for varying vessel headings relative to the incident wave direction. $\mathbf{M}(\omega) \in \mathbb{R}^{6\times6}$ is the system inertia matrix containing frequency dependent added mass coefficients in addition to the vessel's mass and moment of inertia. $\mathbf{D}_p(\omega) \in \mathbb{R}^{6\times6}$ is the wave radiation (potential) damping matrix. $\mathbf{G} \in \mathbb{R}^{6\times6}$ is the linearized restoring coefficient matrix.

For small angular motions, the following holds in the reference-parallel frame

$$\dot{\eta}_{Rw} = \mathbf{J}(\psi - \psi_d)\boldsymbol{\nu}_w \approx \boldsymbol{\nu}_w. \quad (8)$$

Then, (7) can be reformulated to the reference-parallel and Earth-fixed frames according to

$$\mathbf{M}(\omega)\ddot{\eta}_{Rw} + \mathbf{D}_p(\omega)\dot{\eta}_{Rw} + \mathbf{G}\eta_{Rw} = \boldsymbol{\tau}_{\text{wave1}}, \quad (9)$$

$$\dot{\eta}_w = \mathbf{J}(\psi_d)\dot{\eta}_{Rw}, \quad (10)$$

where $\eta_{Rw} \in \mathbb{R}^6$ is the WF motion vector in the reference-parallel frame. An important feature of the added mass terms and the wave radiation damping terms are the memory effects, which in particular is important to consider for non-stationary cases, e.g. rapid changes of heading angle. Memory effects can be taken into account by introducing a convolution integral or a so-called retardation function (Newman, 1977) or state space models as suggested by Kristiansen and Egeland (2003).

### *3.2 Control Plant Model: Hydrodynamics*

*3.2.1. Low-frequency control plant model* For the purpose of controller design and analysis, it is convenient to simplify (3) and to derive a nonlinear LF control plant model in surge, sway and yaw about zero vessel velocity according to

$$\dot{\eta} = \mathbf{R}\boldsymbol{\nu}, \quad (11)$$

$$\mathbf{M}\dot{\boldsymbol{\nu}} + \mathbf{D}\mathbf{v} = \boldsymbol{\tau} + \mathbf{R}^T\mathbf{b}, \quad (12)$$

where $\mathbf{v} = [u, v, r]^T$, $\eta = [x, y, \psi]^T$, $\mathbf{b} \in \mathbb{R}^3$ is the bias vector, and $\boldsymbol{\tau} = [\tau_x, \tau_y, \tau_\psi]^T$ is the control input vector. Notice the control plant model is nonlinear because of the rotation matrix $\mathbf{R}(\psi)$.

*3.2.2. Wave-frequency control plant model* In the controller design synthetic white-noise-driven processes consisting of uncoupled harmonic oscillators with damping will be used to model the WF motions. The synthetic WF model can be written in state-space form according to

$$\begin{aligned} \dot{\boldsymbol{\xi}}_w &= \mathbf{A}_w\boldsymbol{\xi}_w + \mathbf{E}_w\mathbf{w}_w, \\ \eta_w &= \mathbf{C}_w\boldsymbol{\xi}_w. \end{aligned} \quad (13)$$

$\eta_w \in \mathbb{R}^3$ is the position and orientation measurement vector, $\mathbf{w}_w \in \mathbb{R}^3$ is a zero-mean Gaussian white noise vector, and $\boldsymbol{\xi}_w \in \mathbb{R}^6$. A linear 2nd-order WF model is considered to be sufficient for representing the WF-induced motions, although higher order models may also be used, see Grimble and Johnson (1988). The system matrix $\mathbf{A}_w \in \mathbb{R}^{6\times6}$, the disturbance matrix $\mathbf{E}_w \in \mathbb{R}^{6\times3}$ and the measurement matrix $\mathbf{C}_w \in \mathbb{R}^{3\times6}$ may formulated as

$$\mathbf{A}_w = \begin{bmatrix} \mathbf{0}_{3\times3} & \mathbf{I}_{3\times3} \\ -\boldsymbol{\Omega}^2 & -2\boldsymbol{\Lambda}\boldsymbol{\Omega} \end{bmatrix},$$

$$\mathbf{C}_w = \begin{bmatrix} \mathbf{0}_{3\times3} & \mathbf{I}_{3\times3} \end{bmatrix}, \ \mathbf{E}_w = \begin{bmatrix} \mathbf{0}_{3\times3} \\ \mathbf{K}_w \end{bmatrix},$$

where $\boldsymbol{\Omega} = \text{diag}\{\omega_1, \omega_2, \omega_3\}$, $\boldsymbol{\Lambda} = \text{diag}\{\zeta_1, \zeta_2, \zeta_3\}$ and $\mathbf{K}_w = \text{diag}\{K_{w1}, K_{w2}, K_{w3}\}$. This model corresponds to

$$\frac{\eta_{w_i}}{w_{w_i}}(s) = \frac{K_{w_i}s}{s^2 + 2\zeta_i\omega_i s + \omega_i^2}. \quad (14)$$

From a practical point of view, the WF model parameters are slowly-varying quantities, depending on the prevailing sea state. Typically, the periods $T_i$, corresponding to a wave frequency $\omega_i = 2\pi/T_i$, are in the range of 5 to 20 seconds in the North Sea. The relative damping ration $\zeta_i$ will typically be in the range 0.05−0.10. As suggested by Strand and Fossen (1999) adaptive schemes may be used to update $\omega_i$ for the varying sea states. However, this should be done with care in heavy sea states with long wave lengths.

*3.2.3. Bias model* A frequently used bias model $\mathbf{b} \in \mathbb{R}^3$ for marine control applications is the first order *Markov* model

$$\dot{\mathbf{b}} = -\mathbf{T}_b^{-1}\mathbf{b} + \mathbf{E}_b\mathbf{w}_b, \quad (15)$$

where $\mathbf{w}_b \in \mathbb{R}^3$ is a zero-mean Gaussian white noise vector, $\mathbf{T}_b \in \mathbb{R}^{3\times3}$ is a diagonal matrix of bias time constants, and $\mathbf{E}_b \in \mathbb{R}^{3\times3}$ is a diagonal scaling matrix. The bias model accounts for slowly-varying forces and moment due to 2nd-order wave loads, ocean currents and wind. In addition, the bias model will account for errors in the modeling.

Alternatively, the bias model may also be modelled as random walk, i.e. *Wiener* process

$$\dot{\mathbf{b}} = \mathbf{E}_b\mathbf{w}_b. \quad (16)$$

*3.2.4. Measurements* The measurement equation is written

$$\mathbf{y} = \eta + \eta_w + \mathbf{v}, \quad (17)$$

where $\mathbf{v} \in \mathbb{R}^3$ is the zero-mean Gaussian measurement noise vector.

*3.2.5. Resulting control plant model* The resulting control plant model is written

$$\dot{\boldsymbol{\xi}} = \mathbf{A}_w\boldsymbol{\xi} + \mathbf{E}_w\mathbf{w}_w, \quad (18a)$$

$$\dot{\eta} = \mathbf{R}\boldsymbol{\nu}, \quad (18b)$$

$$\dot{\mathbf{b}} = \mathbf{E}_b\mathbf{w}_b, \quad (18c)$$

$$\mathbf{M}\dot{\boldsymbol{\nu}} = -\mathbf{D}\boldsymbol{\nu} + \mathbf{R}^T\mathbf{b} + \boldsymbol{\tau}, \quad (18d)$$

$$\mathbf{y} = \eta + \mathbf{C}_w\boldsymbol{\xi} + \mathbf{v}. \quad (18e)$$

Here, the Wiener bias model is used.

*3.2.6. Control plant model for extreme seas* As suggested in Sørensen *et al.* (2002) the wave filtering should be avoided for long waves length (low wave frequencies) appearing in extreme seas or in swell dominated seas. For such conditions the following control plant model is proposed

$$\dot{\mathbf{b}} = \mathbf{E}_b \mathbf{w}_b, \tag{19a}$$

$$\mathbf{M}\dot{\boldsymbol{\nu}} = -\mathbf{D}\boldsymbol{\nu} + \mathbf{R}^T\mathbf{b} + \boldsymbol{\tau}, \tag{19b}$$

$$\dot{\boldsymbol{\eta}} = \mathbf{R}\boldsymbol{\nu}, \tag{19c}$$

$$\mathbf{y} = \boldsymbol{\eta} + \mathbf{v}. \tag{19d}$$

### *3.3 Process Plant Model: Riser Mechanics*

A flexible system is here defined as a dynamic system, which includes discernible *elastic* or *flexible* motions in addition to the *rigid body* dynamics. When designing a control system for a flexible system it is essential to account for the flexibility in the development of the controller algorithm and in the location of sensors and actuators. As the surface vessel is one of the boundary conditions to the riser, the surface vessel motion may cause troubles on the riser, unless precautions are taken in the DP system.

As suggested in Fig. 1 local optimization may be used for calculation of the optimal setpoint for the vessel. For drilling and work-over operations the main positioning objective is to minimize the bending stresses along the riser and the riser angle magnitudes at the well head on the subsea structure, and at the top joint as well.

Flexible systems are described by *partial differential equations* (PDEs) involving two and several independent variables (often time and spatial coordinates) in contrast to *ordinary differential equations* (ODEs) that are dependent on a single independent variable only (often the time). Problems involving boundary and initial conditions are called *boundary-value problems.* Solving PDEs describing slender structures requires numerical methods using e.g. finite element method (FEM).

*3.3.1. Drilling riser mechanics* A drilling riser behaves like a tensioned beam. A key feature of a drilling riser is that the top end tension is kept close to constant under influence from floater motions by a heave compensating system at the upper end. The top tension must be sufficiently high to prevent the riser from global buckling, meaning that the effective tension must have a positive value at the lower end of the riser.

Some basic features of riser mechanics can for the 2D static case be illustrated with reference to Fig. 9. The riser is seen to have an offset position of the upper end relative to the lower end. The angles between the riser and a vertical line at the top and bottom are denoted $\alpha_t$ and $\alpha_b$, respectively. These angles are influenced by the tension at the riser ends, the vessel offset and the current forces. Compared to the 2D case, the 3D case is somewhat more complicated. The current direction may vary through the water column, meaning that riser deflections will not take place in one well-defined plane. Rotation at riser ends must hence be defined by the direction of the inclination ($\beta_b$ and $\beta_t$ ) and its magnitude ($\alpha_b$ and $\alpha_t$).

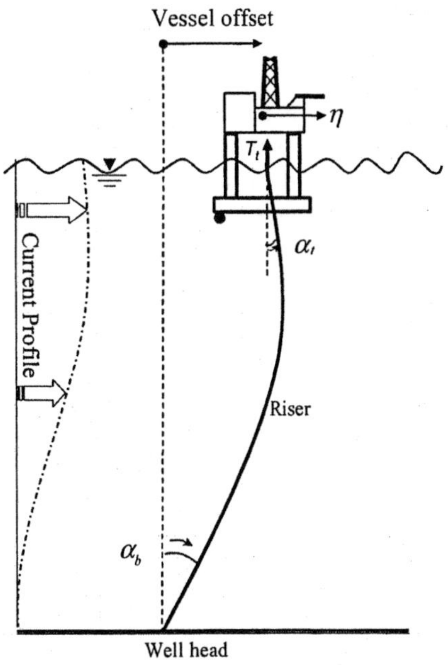


Fig. 9. Riser mechanics.

*3.3.2. Finite element method* For the purpose of computing stresses in the riser, a quite FEM is generally required. For riser motion calculations we presently apply a model with 10 elements. For water depth less than 1000 m this has shown to be sufficient. The stiffness matrix consists of two parts. The first part represents the elastic stiffness, $\mathbf{k}_{El}$, and the second represents the so-called geometric stiffness, $\mathbf{k}_G$. The latter is subsequently multiplied by the effective tension in the riser which acts at each particular element, $P_i$, where the subscript $i$ denotes element $i$. The resulting total stiffness matrix becomes for element $i$

$$\mathbf{k}_i = [\mathbf{k}_{EI} + P_i \mathbf{k}_G]_i \,. \tag{20}$$

The 6 kinematic components (i.e. 3 displacements and 3 rotations) at each end of the local elements are referred to the local coordinate system. All the 12 kinematic components for each element are collected in a vector $\mathbf{v}_i$. $E$ is the modulus of elasticity (Young's modulus), $I$ is the moment of inertia ($EI$ will then be the bending stiffness), and $l$ is the length of each element. A transformation matrix $\mathbf{T}_i$ is introduced for each element such that

$$\mathbf{r}_i = \mathbf{T}_i \mathbf{v}_i. \tag{21}$$

These sub-vectors are finally assembled into a total riser displacement vector (containing both displacements and rotations), designated by $\mathbf{r}$. The corresponding element stiffness matrix for element $i$ in the global system is then expressed by a congruence transformation as

$$\mathbf{K}_i = \mathbf{T}_i^T \mathbf{k}_i \mathbf{T}_i = \begin{bmatrix} \mathbf{K}_{11}^i & \mathbf{K}_{12}^i \\ \mathbf{K}_{21}^i & \mathbf{K}_{22}^i \end{bmatrix}. \tag{22}$$

The global system stiffness matrix will then take the form

$$\mathbf{K} = \tag{23}$$
$$\begin{bmatrix} K_{11}^1 & K_{12}^1 & & & \\ K_{21}^1 & K_{22}^1+K_{11}^2 & K_{12}^2 & & \\ & K_{21}^2 & K_{22}^2+K_{11}^3 & \cdots & \\ \vdots & & & & \\ & & & K_{22}^{N-1}+K_{11}^N & K_{12}^N \\ & & & K_{21}^N & K_{22}^N \end{bmatrix}.$$

If we assume that node 1 is located at the top of the riser, the two first DOFs correspond to the horizontal displacements of the surface floater ($x-$ and $y-$ coordinates in the global system). Furthermore, these correspond to the first two columns and rows of the system matrix $\mathbf{K}$. As observed from (23) there are at maximum 12 non-zero (here 8) elements in each of these rows or columns. The boundary conditions at the upper and lower ends of the riser are subsequently introduced. The number of restrained kinematic components eliminates the equations corresponding to fixed and prescribed DOFs, which reduces the dimension of the global displacement vector.

*3.3.3. Low-frequency riser model* Considering LF riser motion, the driving excitation mechanism is due to forced motions caused by the surface vessel at the upper end of the riser. The dynamic equilibrium equation can be written as

$$\mathbf{M}(\mathbf{r})\ddot{\mathbf{r}} + \mathbf{C}_s(\mathbf{r})\dot{\mathbf{r}} + \mathbf{K}(\mathbf{r})\mathbf{r} = \tag{24}$$
$$\boldsymbol{\tau}_{vessel_{LF}}(\mathbf{r}) + \boldsymbol{\tau}_{current}(\mathbf{u}_c, \mathbf{r}, \dot{\mathbf{r}}),$$

where the nodal displacement vector is designated by $\mathbf{r}$. The mass matrix $\mathbf{M}$, the structural damping matrix $\mathbf{C}_s$ and the stiffness matrix $\mathbf{K}$ are all considered functions of the deformed configuration found from the displacements. Note that the mass matrix must include anisotropic added mass terms. Two loading terms $\boldsymbol{\tau}_{vessel_{LF}}$ and $\boldsymbol{\tau}_{current}$ are present, one containing contributions from LF vessel motions and the other with loads from the current velocity vector $\mathbf{u}_c$ and the riser velocity vector.

*3.3.4. Wave-frequency riser model* In the WF regime, the direct wave forces on the riser must be included. This loading vector $\boldsymbol{\tau}_{hydro}$ contains a quadratic drag-term containing the relative velocity between the riser pipe and the surrounding fluid. For this case, the dynamic equilibrium equation is expressed as

$$\mathbf{M}(\mathbf{r})\ddot{\mathbf{r}} + \mathbf{C}_s(\mathbf{r})\dot{\mathbf{r}} + \mathbf{K}(\mathbf{r})\mathbf{r} = \tag{25}$$
$$\boldsymbol{\tau}_{vessel_{WF}}(\mathbf{r}) + \boldsymbol{\tau}_{hydro}(\mathbf{u}_t, \dot{\mathbf{u}}, \mathbf{r}, \dot{\mathbf{r}}).$$

Again it is seen that all matrices in principle should be functions of the displacements. The loading forces must be found from a flow velocity profile $\mathbf{u}_t$ with contributions from current and waves, and the wave induced accelerations $\dot{\mathbf{u}}$.

*3.4 Control Plant Model: Riser Mechanics*

In the control plant model we will only consider the LF riser dynamics. A generally applied simplification of (24) is to neglect the two dynamic terms related to the inertia and damping forces. The model is then said to be quasi-static, meaning that the response model is static, but external loads are calculated by taking the riser motions into account. Such models will yield acceptable accuracy for shallow water applications, but dynamic effects will become more important in deep water since the lowest eigenfrequency of the riser may approach the LF regime. This leaves only the stiffness term in (24). It is assumed that the increment in the riser displacement vector $\Delta\mathbf{r}$ is related to the in-plane increment in the surface vessel position vector $\Delta\mathbf{r}_v$ by

$$\Delta\mathbf{r} = -\mathbf{K}_I^{-1}\mathbf{K}_{Icoupl}\Delta\mathbf{r}_v. \tag{26a}$$

$\mathbf{K}_I$ is the $ndof \times ndof$ incremental riser stiffness matrix at the present instantaneous vessel position corresponding to the given water current profile. $\mathbf{K}_{Icoupl}$ is the $ndof \times 2$ incremental stiffness matrix coupling the in-plane vessel motion to the remaining DOFs. Both matrices are obtained as sub-matrices of the stiffness matrix given in (23).

The increments of top and bottom riser angles are expressed in terms of their $x$- and $y$-components. Typically, for the top angle these correspond to components 1 and 2 of the vector in (23). For the bottom angle, components number $ndof-1$ and $ndof$ in the response vector are relevant. These are accordingly expressed as

$$\Delta\alpha_{tx} = -\left(\mathbf{K}_I^{-1}\mathbf{K}_{Icoupl}\right)_1 \Delta r_{vx} = -c_{tx}\Delta r_{vx}, \quad (27a)$$

$$\Delta\alpha_{ty} = -\left(\mathbf{K}_I^{-1}\mathbf{K}_{Icoupl}\right)_2 \Delta r_{vy} = -c_{ty}\Delta r_{vy}, \quad (27b)$$

$$\Delta\alpha_{bx} = -\left(\mathbf{K}_I^{-1}\mathbf{K}_{Icoupl}\right)_{ndof-1} \Delta r_{vx} = -c_{bx}\Delta r_{vx}, \quad (27c)$$

$$\Delta\alpha_{by} = -\left(\mathbf{K}_I^{-1}\mathbf{K}_{Icoupl}\right)_{ndof} \Delta r_{vy} = -c_{by}\Delta r_{vy}, \quad (27d)$$

where the scalar constants $c_{tx}, c_{ty}, c_{bx}$ and $c_{by}$ may be computed on-line.

## 4. PLANT CONTROL

### 4.1 Observer Equations

Based on the work of Fossen and Strand (1999) a nonlinear observer copying the vessel and environmental models (18a)–(18e) is

$$\dot{\hat{\boldsymbol{\xi}}} = \mathbf{A}_w\hat{\boldsymbol{\xi}} + \mathbf{K}_1\tilde{\mathbf{y}}, \quad (28a)$$

$$\dot{\hat{\boldsymbol{\eta}}} = \mathbf{R}(\psi_y)\hat{\boldsymbol{\nu}} + \mathbf{K}_2\tilde{\mathbf{y}}, \quad (28b)$$

$$\dot{\hat{\mathbf{b}}} = \mathbf{K}_3\tilde{\mathbf{y}}, \quad (28c)$$

$$\mathbf{M}\dot{\hat{\boldsymbol{\nu}}} = -\mathbf{D}\hat{\boldsymbol{\nu}} + \mathbf{R}^T(\psi_y)\hat{\mathbf{b}} + \boldsymbol{\tau} + \mathbf{R}^T(\psi_y)\mathbf{K}_4\tilde{\mathbf{y}}, \quad (28d)$$

$$\hat{\mathbf{y}} = \hat{\boldsymbol{\eta}} + \mathbf{C}_w\hat{\boldsymbol{\xi}}, \quad (28e)$$

where $\tilde{\mathbf{y}} = \mathbf{y} - \hat{\mathbf{y}} \in \mathbb{R}^3$ is the estimation error (in the literature also denoted as the innovation or injection term), $\mathbf{K}_1 \in \mathbb{R}^{6\times3}$, and $\mathbf{K}_2, \mathbf{K}_3, \mathbf{K}_4 \in \mathbb{R}^{3\times3}$ are observer gain matrices. By proper gain settings the observer is shown to be passive. Notice that it assumed that the yaw measurement can be used directly in the rotation matrix. Full scale experience has shown that is assumption is acceptable.

### 4.2 Nonlinear Horizontal-plane PID Control Law

A nonlinear horizontal-plane positioning feedback controller of PID type is formulated as

$$\boldsymbol{\tau}_{PID} = -\mathbf{R}_e^T\mathbf{K}_p\mathbf{e} - \mathbf{R}_e^T\mathbf{K}_{p3}\mathbf{f}(\mathbf{e}) - \mathbf{K}_d\tilde{\boldsymbol{\nu}} - \mathbf{R}^T\mathbf{K}_i\mathbf{z}, \quad (29)$$

where $\mathbf{e} \in \mathbb{R}^3$ is the position and heading deviation vector, $\tilde{\boldsymbol{\nu}} \in \mathbb{R}^3$ is the velocity deviation vector, and $\mathbf{z} \in \mathbb{R}^3$ is the integrator states defined as

$$\mathbf{e} = \begin{bmatrix} e_1 & e_2 & e_3 \end{bmatrix}^T = \mathbf{R}^T(\psi_d)(\hat{\boldsymbol{\eta}} - \boldsymbol{\eta}_d), \quad (30)$$

$$\tilde{\boldsymbol{\nu}} = \hat{\boldsymbol{\nu}} - \mathbf{R}^T(\psi_d)\dot{\boldsymbol{\eta}}_d, \quad (31)$$

$$\dot{\mathbf{z}} = \hat{\boldsymbol{\eta}} - \boldsymbol{\eta}_d, \quad (32)$$

$$\mathbf{R}_e = \mathbf{R}(\psi - \psi_d) \triangleq \mathbf{R}^T(\psi_d)\mathbf{R}(\psi). \quad (33)$$

A third order stiffness term is proposed

$$\mathbf{f}(\mathbf{e}) = \begin{bmatrix} e_1^3 & e_2^3 & e_3^3 \end{bmatrix}^T. \quad (34)$$

An advantage of this is the possibility to reduce the first order proportional gain matrix, resulting in reduced dynamic thruster action for smaller position and heading deviations. Moreover, the third order term will make the thrusters to work more aggressive for larger deviations. $\boldsymbol{\eta}_d \in \mathbb{R}^3$ is the vector defining the desired Earth-fixed position and heading coordinates. $\mathbf{K}_p, \mathbf{K}_{p3}, \mathbf{K}_d$ and $\mathbf{K}_i \in \mathbb{R}^{3\times3}$ are non-negative controller gain matrices found by appropriate controller synthesis methods.

### 4.3 Roll-pitch Control Law

As proposed by Sørensen and Strand (2000) using a linear formulation the roll-pitch control law is formulated according to

$$\boldsymbol{\tau}_{\mathrm{rpd}} = -\mathbf{G}_{\mathrm{rpd}} \begin{bmatrix} \hat{p} \\ \hat{q} \end{bmatrix}, \quad (35)$$

where $\hat{p}$ and $\hat{q}$ are the estimated pitch and roll angular velocities, and the roll-pitch controller gain matrix $\mathbf{G}_{\mathrm{rpd}} \in \mathbb{R}^{3\times2}$ is defined as

$$\mathbf{G}_{\mathrm{rpd}} = \begin{bmatrix} 0 & g_{xq} \\ g_{yp} & 0 \\ g_{\psi p} & 0 \end{bmatrix}, \quad (36)$$

and $g_{xq}$, $g_{yp}$ and $g_{\psi p}$ are the corresponding non-negative roll-pitch controller gains.

### 4.4 Resulting Control Law

The resulting positioning control law is written

$$\boldsymbol{\tau} = \boldsymbol{\tau}_{wFF} + \boldsymbol{\tau}_{PID} + \boldsymbol{\tau}_{rpd}, \quad (37)$$

where $\boldsymbol{\tau}_{wFF} \in \mathbb{R}^3$ is the wind feedforward control law.

## 5. LOCAL OPTIMIZATION

### 5.1 Reference Model

A reference model is used for obtaining smooth transitions between the various setpoints. Let

$$\boldsymbol{\eta}_r = \begin{bmatrix} x_r & y_r & \psi_r \end{bmatrix}^T, \quad (38)$$

define the final Earth-fixed vector position and heading setpoint vector. This is input to a nonlinear third order reference model as presented in Fossen (1994), and is given as

$$(\mathbf{x}_{ref}, \mathbf{x}_d^{\mathrm{e}}, \mathbf{v}_d^{\mathrm{e}}, \mathbf{a}_d^{\mathrm{e}}) = \mathbf{f}(\boldsymbol{\eta}_r, \mathbf{x}_{ref}, \mathbf{x}_d^{\mathrm{e}}, \mathbf{v}_d^{\mathrm{e}}, \mathbf{a}_d^{\mathrm{e}}; t). \quad (39)$$

This model produces a smooth desired acceleration, velocity and position reference that are inputs to the positioning control law (37).

### 5.2 Optimal Setpoint Chasing

Motivated by the control plant model of the riser mechanics a quadratic loss function $L$ with the scalar weighting factors $w_t$ and $w_b$ is introduced

$$L = w_t \left[(\Delta\alpha_{tx} + \alpha_{tx})^2 + (\Delta\alpha_{ty} + \alpha_{ty})^2\right] \quad (40)$$
$$+ w_b \left[(\Delta\alpha_{bx} + \alpha_{bx})^2 + (\Delta\alpha_{by} + \alpha_{by})^2\right].$$

By solving the equation where the partial derivatives of (40) with respect to the $x$- and $y$- components of the vessel increment are set equal to zero, the optimal direction for the increment of the vessel position is derived as

$$\theta_{opt} = \tan\left(\frac{\Delta y}{\Delta x}\right),$$

where

$$\Delta y = \left(w_b c_{bx}^2 + w_t c_{tx}^2\right) * \left(w_b c_{by}\alpha_{by} + w_t c_{ty}\alpha_{ty}\right),$$

$$\Delta x = \left(w_b c_{by}^2 + w_t c_{ty}^2\right) * \left(w_b c_{bx}\alpha_{bx} + w_t c_{tx}\alpha_{tx}\right).$$

For the optimal direction, the corresponding optimal vessel incremental position is computed as

$$\Delta r_{vessel}^* = \frac{\begin{array}{c}(w_t c_{tx}\alpha_{tx}\cos(\theta_{opt}) + w_t c_{ty}\alpha_{ty}\sin(\theta_{opt}) + \\ w_b c_{bx}\alpha_{bx}\cos(\theta_{opt}) + w_b c_{by}\alpha_{bx}\sin(\theta_{opt}))\end{array}}{\begin{array}{c}(w_t c_{tx}^2\cos^2(\theta_{opt}) + w_t c_{ty}^2\sin^2(\theta_{opt}) + \\ w_b c_{bx}^2\cos^2(\theta_{opt}) + w_b c_{by}^2\sin^2(\theta_{opt}))\end{array}}.$$

Then the $x$- and $y$-components of the optimal incremental vessel position vector are expressed as

$$\Delta r_x = \Delta r_{vessel}^*\cos(\theta_{opt}),\ \Delta r_y = \Delta r_{vessel}^*\sin(\theta_{opt}).$$

Finally, in the general 3-dimensional case the updated Earth-fixed vector position and heading setpoint vector becomes

$$\boldsymbol{\eta}_r^* = \boldsymbol{\eta}_r + \Delta r_{vessel}^*\left[\cos(\theta_{opt}) \quad \sin(\theta_{opt}) \quad 0\right]^T. \quad (42)$$

A key issue in the present algorithm is the relative weighting of the top and bottom riser angles. This weighting must reflect the maximum permissible limits for these angles. The permissible bottom angle is typically much smaller than the permissible top angle. This implies that the weighting factor for the lower angle should be much larger than for the upper angle. A trade-off between the various considerations is hence required.

## 6. THRUST ALLOCATION

The high-level positioning controller produces a commanded thrust vector $\boldsymbol{\tau} \in \mathbb{R}^3$ in surge, sway and yaw. The problem of finding the corresponding force and direction of the thrusters that meets the high-level thrust commands is called *thrust allocation.* Important references in the field of thrust allocation are Sørdalen (1997), Johansen *et al.* (2003) and Fossen (2002)

### 6.1 Optimal Thrust Allocation

The relation between the control vector $\boldsymbol{\tau} \in \mathbb{R}^3$ and the produced thruster action $\mathbf{u}_c \in \mathbb{R}^r$ is defined by

$$\boldsymbol{\tau} = \mathbf{T}_{3\times r}(\boldsymbol{\alpha})\mathbf{T}_{th} = \mathbf{T}_{3\times r}(\boldsymbol{\alpha})\mathbf{K}\mathbf{u}_c, \quad (43)$$

where $\mathbf{T}_{3\times r}(\boldsymbol{\alpha}) \in \mathbb{R}^{3\times r}$ is the thrust configuration matrix, $\boldsymbol{\alpha} \in \mathbb{R}^r$ is the thruster orientation vector, and $r$ is the number of thrusters. The corresponding thrust vector $\mathbf{T}_{th} \in \mathbb{R}^r$ is given by $\mathbf{T}_{th} = \mathbf{K}\mathbf{u}_c$, where $\mathbf{K} \in \mathbb{R}^{r\times r}$ is the diagonal matrix of thrust force coefficients written $\mathbf{K} = diag\{k_i\}$. The thrust provided by the thruster unit $i$ is calculated to be

$$T_{\text{thi}} = k_i u_{ci}. \quad (44)$$

For a fixed mounted propeller or thruster the corresponding orientation angle is set to a fixed value reflecting the actual orientation of the device itself. In case of an azimuthing thruster, $\alpha_i$ is an additional control input to be determined by the thrust allocation algorithm. $\mathbf{u}_c$ is the control vector of either pitch-controlled, revolution-controlled or torque- and power-controlled propeller inputs. The commanded control action and direction provided by the thrusters becomes

$$\mathbf{u}_c = \mathbf{K}^{-1}\mathbf{T}_{3\times r}^{+}(\boldsymbol{\alpha})\boldsymbol{\tau}, \quad (45)$$

where $\mathbf{T}_{3\times r}^{+}(\boldsymbol{\alpha}) \in \mathbb{R}^{r\times 3}$ is the pseudo-inverse thrust configuration matrix.

### 6.2 Geometrical Thrust Induction

The effect of the commanded thruster action provided by the thrusters in (45) on (3) can be calculated to be

$$\boldsymbol{\tau}_{\text{thr}} = \mathbf{T}_{6\times r}(\boldsymbol{\alpha})\mathbf{K}\mathbf{u}_c, \quad (46)$$

where $\boldsymbol{\tau}_{\text{thr}} \in \mathbb{R}^6$ is the corresponding actual control vector acting the vessel as shown in (3), and $\mathbf{T}_{6\times r}(\boldsymbol{\alpha}) \in \mathbb{R}^{6\times r}$ is the thrust configuration matrix accounting for both the horizontal and the vertical contribution of the produced thruster actions. The reader should notice that (46) will introduce roll and pitch moments, that may be important to consider for rigs as they may introduce unintentional roll and pitch motions. The effect of thrust losses on (3) and the design of the local thruster controllers are presented in the next section.

## 7. ACTUATOR CONTROL

Installed power capacity on marine vessels and offshore installations is normally limited. For DP vessels the thruster system normally represents one of the main consumers of energy, and is regarded as a critical system with respect to safety. If the various consumers (thrusters, pumps, compressors etc.) of power act separately and uncoordinated from each other, the power generation system must be dimensioned and operated with larger safety margins to account for the corresponding larger mean power demands and unintentional power peaks. This section will illustrate the importance of focusing on low level actuator control exemplified on thruster control in order to achieve a thorough successful control result.

Important references in the field of low-level control of the thrusters and propulsors are Blanke (1981), Grimble and Johnson (1988), Healey *et al.* (1995), Sørensen *et al.* (1997), Whitcomb and Yoerger (1999 a and b), Bachmayer *et al.* (2000), Fossen and Blanke (2000), Blanke *et al.* (2000), Fossen (2002), Lindegaard (2003), Sørensen (2004) and Smogeli *et al.* (2004). Propulsion control in extreme conditions has been addressed by Smogeli *et al.* (2003), where methods for robust thruster control accounting for severe thrust losses due to *ventilation* and *in-and-out-of-water effects* are developed. In this work an *anti-spin thruster controller,* motivated by similar effects in wheel slip control on cars was proposed.

### 7.1 Control Problem Formulation

Traditionally, the propeller pitch ratio or the shaft speed is used to indirectly control the propeller force towards the reference setpoint $T_{\text{ref}}$ specified by the thrust allocation. In the thrust devices, a local pitch or speed controller is present. A static mapping from the reference force to the actual control input, $u_{\text{ref}}$, is used according to

$$u_{\text{ref}} = g(T_{\text{ref}}). \tag{47}$$

Since normally no sensors are available for measuring the actual force developed by the propeller, there is no guarantee for fulfilling the high-level thrust commands, and the mapping from commanded thruster force to actual propeller force can be viewed as an open-loop system. Thus, sophisticated control designs will be significantly degraded with respect to performance and stability margins if the high-level control inputs are not produced by the local controllers. Conventionally, the resulting pitch or speed set-point signals are determined from stationary propeller force-to-speed/pitch relations based on information about thruster characteristics found from model tests

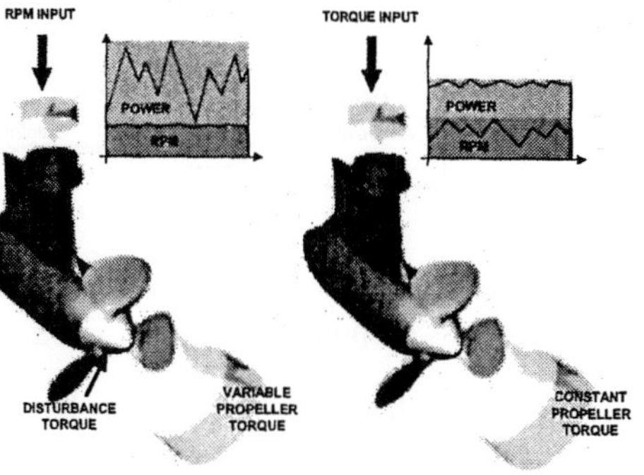


Fig. 10. Torque and power control reduce the dynamic load variations and allows for lower spinning reserves.

and bollard pull tests provided by the thruster manufacturer. These relations may later be modified during sea trials. However, they are strongly influenced by the local water flow around the propeller blades, hull design, operational philosophy, vessel motion, waves and water current. Hence, the actual developed thrust $T_a$ may differ substantially from the commanded thrust setpoint $T_{\text{ref}}$. In conventional positioning systems, variations in these relations are not accounted for in the control system, resulting in reduced positioning performance with respect to accuracy and response time. In addition, the variations may lead to deterioration of performance and stability in the electrical power plant network due to unintentional peaks or power drops caused by load fluctuations on the propeller shafts, as shown in Fig. 10. The unpredictable load variations force the operator to have more available power than necessary. This implies that the diesel generators will get more running hours at lower loads in average, which in terms gives more tear, wear and maintenance. This motivates finding improved methods for local thruster control.

### 7.2 Propeller Thrust and Torque Models

The propeller thrust $T_{\text{th}}$ and torque $Q_{\text{th}}$ are formulated according to

$$T_{\text{th}} = ku_c = \rho_w D^4 K_T \left|n\right| n, \tag{48}$$

$$Q_{\text{th}} = \rho_w D^5 K_Q \left|n\right| n, \tag{49}$$

where $\rho_w$ is the water density, $D$ is the propeller diameter, and $n$ is the propeller speed. Notice that we here have assumed a speed controlled propeller with

$$k = \rho_w D^4 K_T, \tag{50}$$

$$u_c = \left|n\right| n. \tag{51}$$

For notational simplicity the index $i$ is omitted. In Fig. 11 the relation between the output $\tau$ =

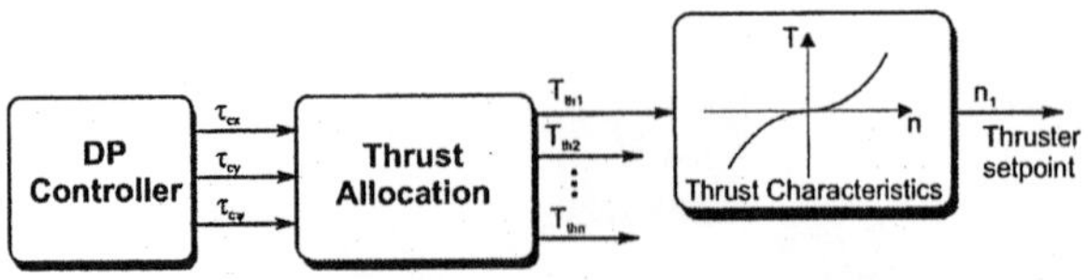


Fig. 11. Relation between the DP controller and the thrust setpoint.

$[\tau_{cx}, \tau_{cy}, \tau_{c\psi}]^T$ from the DP controller and the propeller speed setpoint $n$ is shown.

The expressions for $K_T$ and $K_Q$ are found by so-called *open water* tests.

### *7.3 Thrust Losses*

In addition to the modelled propeller losses caused by axial water inflow, several other effects (Carlton, 1994) will contribute to reduction of propeller thrust and torque:

- Water inflow perpendicular to the propeller axis caused by current and vessel speed will introduce a force in the direction of the inflow due to deflection of the propeller race. This is often referred to as *cross-coupling drag.*
- For heavily loaded propellers *ventilation* (air suction) caused by decreasing pressure on the propeller blades may occur, especially when the submergence of the propeller becomes small due to the vessel motion.
- For extreme conditions with large vessel motions the *in-and-out-of-water effects* will result in sudden drop of thrust and torque following a hysteresis pattern.
- Both thrust reduction and change of thrust direction may occur due to thruster-hull interaction caused by frictional losses and pressure effects when the thruster race sweeps along the hull. The last is referred to as the *Coanda effect.*
- Thruster-thruster interaction caused by influence from the propeller race from one thruster on neighboring thrusters may lead to significant thrust reduction, if not appropriate precautions are taken in the thruster allocation algorithm.

The sensitivity to the different types of losses depends on the type of propeller and thruster used, application of skegs and fins, hull design and operational philosophy. Main propellers are subject to large thrust losses due to air ventilation and in-and-out-of-water effects. Rotatable azimuth thrusters are subject to losses caused by hull friction and interaction with other thrusters. Tunnel thrusters are subject to losses caused by non-axial inflow due to current and vessel speed and ventilation phenomenon in heavy seas.

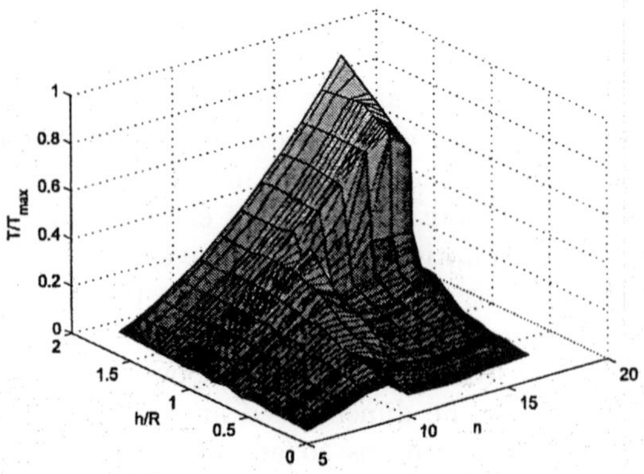


Fig. 12. Nondimensionalized thrust vs. relative submergence $h/R$ and propeller shaft speed $n$ for $J = 0.2$.

No general mathematical model is outlined for describing the thruster performance in time-varying conditions, including transient effects.

Systematic tests with a ducted propeller have been carried out at the cavitation tunnel at NTNU, Smogeli *et al.* (2003). The advance number was kept at approximately 0.2, and the thruster shaft speed and submergence varied while measuring propeller and duct thrust and propeller torque. Nondimensionalized total thrust is presented in Fig. 12 as function of relative submergence $h/R$ and propeller shaft speed. $h$ is the propeller shaft submergence, and $R$ the propeller radius. For large submergence, the thruster exhibits the expected behavior, yielding the normal thrust characteristics with $K_T$ constant and thrust proportional to $n^2$. For high propeller loads, the thrust (and load torque) drops rapidly with decreasing submergence. It is evident that for a heavily loaded propeller, proximity to the surface may lead to an abrupt loss of thrust. As can be seen in Fig. 12, the results indicate that a reduction of shaft speed in such a case will increase the thrust. Experimental results from open water tests with waves in the MCLab at NTNU confirmed that the dynamic thrust behavior followed the same tendency as in the cavitation tunnel.

### *7.4 Propeller Shaft Model*

Let $Q_m$ denote the torque generated by the thruster motor. A torque balance for the propeller shaft is written

$$I_s\dot{\omega} = Q_m - Q_a, \tag{52}$$

where $I_s$ is the moment of inertia for the shaft, and $\omega = 2\pi n$ is the angular shaft speed. However, a more sophisticated model like the two-state model presented in Blanke *et al.* (2000) will be a natural extension of the current model. A friction coefficient $K_\omega$ could also be added to the equation.

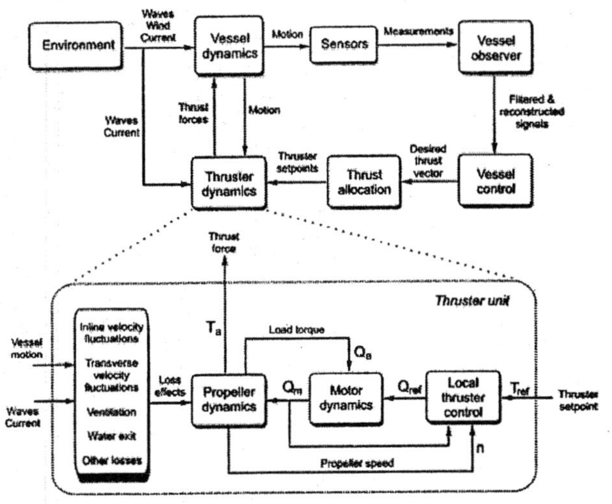


Fig. 13. Total thruster dynamics and its position in the total vessel dynamics.

The torque control is inherent in the design of most applied control schemes for variable speed drive systems. The torque is controlled by means of motor currents and motor fluxes with high accuracy and bandwidth. The closed loop of motor and torque controller may for practical reasons be assumed to be equivalent with the 1st order model, where $T_m$ is the time constant in the range of 20 – 200 *milliseconds* $(ms)$ and $Q_c$ the commanded torque

$$\dot{Q}_m = \frac{1}{T_m}(Q_c - Q_m). \tag{53}$$

The power delivered by the motor and the actual propeller shaft power accounting for the effect of thrust losses are given by

$$P_m = \omega Q_m = 2\pi n Q_m, \tag{54}$$
$$P_a = \omega Q_a = 2\pi n Q_a. \tag{55}$$

For calculation of the actual thruster load torque, $Q_a$, see Sørensen (2004).

### *7.5 Total Thruster Dynamics*

Combining the hydrodynamic thruster model, the shaft model, the motor model and the local thruster controller yields the total thruster dynamics for one thruster unit. This is shown in Fig. 13, where the position of the thruster dynamics in the total vessel dynamics also is shown.

### *7.6 Local Thruster Controller Designs*

The local thruster control scheme may be divided in two control regimes, depending on the operational conditions: 1) *Normal thruster control regime* and 2) *Extreme thruster control regime* with *anti-spin thruster control.*

Assuming normal thruster control regime we are well familiar with shaft speed feedback control or simplified, speed control as the normal way to control the thrusters. However, as shown in Sørensen *et al.* (1997) this may not be the best design choice. The following types of thruster controller designs may be considered:

- Shaft speed feedback control.
- Torque feedforward control.
- Power feedback control.
- Hybrid power and torque control.
- Anti-spin thruster control.

The purpose of the local thruster controller is to relate the desired thrust $T_{\text{ref}}$, given by the thrust allocation, to commanded motor torque $Q_c$.

*7.6.1. Shaft speed feedback control* In conventional FPP systems a speed controller is used to achieve the commanded propeller force. Given a specified force command, $T_{\text{ref}} = \rho_w D^4 K_{T0} \left| n_{ref} \right| n_{ref}$ the corresponding reference (commanded) speed in the propeller $n_{\text{ref}}$ is found by the stationary function

$$n_{\text{ref}} = g_{n0}(T_{\text{ref}}) = \text{sgn}(T_{\text{ref}}) \sqrt{\left| \frac{T_{\text{ref}}}{\rho_w D^4 K_{T0}} \right|}, \tag{56}$$

for typically $K_{T0} = K_T(J = 0)$. The speed controller calculating the torque reference $Q_{\text{ref}}$ is a PID controller operating on the shaft speed error.

*7.6.2. Torque feedforward control* In the torque control strategy for FPP drives the outer speed control loop is removed, and the thruster is controlled by its inner torque control loop with a commanded torque $Q_c$ as set-point. The torque reference is written

$$Q_{\text{ref}} = \rho_w D^5 K_{Q0} \left| n_{\text{ref}} \right| n_{\text{ref}}, \tag{57}$$

where typically $K_{Q0} = K_Q(J = 0)$, where $J$ is the advance number.

*7.6.3. Power feedback control* The torque control loop is maintained, but the commanded torque is found from a commanded power $P_{\text{ref}}$. This power reference is a signed value in order to determine the torque direction. The mapping between the reference thrust force $T_{\text{ref}}$ and the power reference becomes

$$P_{\text{ref}} = \text{sgn}(T_{\text{ref}}) \frac{2\pi K_{Q0}}{\sqrt{\rho_w} D K_{T0}^{3/2}} \left| T_{\text{ref}} \right|^{3/2}. \tag{58}$$

Based on (58) the desired torque becomes

$$Q_{\text{ref}} = \frac{P_{\text{ref}}}{2\pi |n|}, \tag{59}$$

assuming $n \neq 0$, where $n$ is the measured shaft speed.

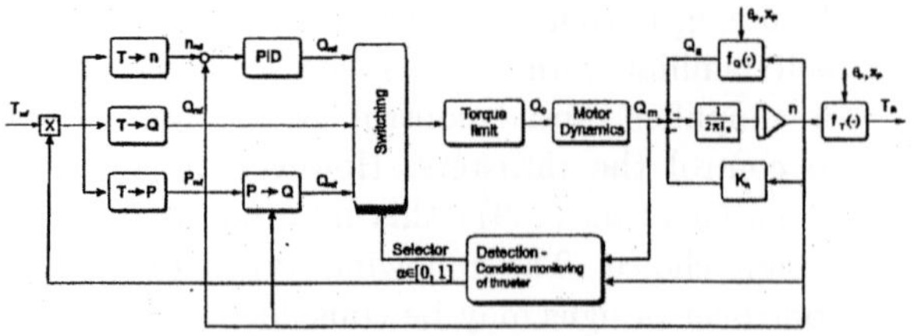


Fig. 14. Anti-spin thruster control scheme, with the mappings $T \to n$, $T \to Q$, $T \to P$ and $P \to Q$ representing the mappings defined in (56) to (58) and $f_T(\cdot)$ and $f_Q(\cdot)$ representing the thrust characteristics including losses.

*7.6.4. Hybrid power and torque control* A significant shortcoming of the power control scheme in (59) is the fact that it is singular for zero shaft speed, $n = 0$. This means that power control should not be used close to the singular point, for example when commanding low thrust or changing the thrust direction. For low thrust commands, torque control shows better performance in terms of constant thrust production than power control, since the mapping between thrust and torque is more direct than the mapping between thrust and power. For high thrust commands, it is essential to avoid large power transients, as these lead to higher fuel consumption and possible danger of power black-out and harmonic distortion of the power plant network. Power control is hence a natural choice for high thrust commands. This motivates the construction of a hybrid power/torque control scheme, utilizing the best properties of both controllers. In Smogeli *et al.* (2004) the hybrid power-torque control scheme is presented.

*7.6.5. Anti-spin thruster control* The general ideas of anti-spin control (Smogeli *et al.* 2003) are to:

- Reduce the wear and tear of the propulsion unit and transients in the power system.
- Optimize the thrust production and efficiency in transient operation regimes.

The condition of the thruster is monitored, and a switch between different control algorithms is made according to given criteria. This is motivated by the hybrid anti-spin/ABS control systems of a car. As shown in Fig. 14, the proposed anti-spin system may be divided in three: Detection, Switching and Control.

## 8. NUMERICAL SIMULATIONS

In this section a simulation study of a DP operated rig conduction drilling operations is carried out to demonstrate the effect of the setpoint chasing strategy. The rig is equipped with 4 azimuthing thrusters each able to produce 1000 kN located at the four corners at the two pontoons. The operational draught is equal to 24 m, the vessel mass at operational draught is 45000 ton, the length is 110 m, and the breadth is 75 m. Radius of gyration in roll is 30 m, in pitch equal to 33 m and in yaw 38 m. The undamped resonance periods in roll and pitch are found to be equal to 55 sec and 60 sec, respectively. The riser length used in the simulations is equal to 1000 m. The riser radius is 0.25 m, the riser wall thickness is 0.025 m and the modulus of elasticity (Young's modulus) is $E = 2.1 \times 10^8$ Pa. The top tension is 2500 Pa, and the tension of the lower part is 1200 Pa. In the simulation the riser is divided into ten elements. At the sea surface the current is varying about 1 m/s with $\beta_c = 30\,°$ . At the middle and the bottom the current is 0.75 and 0.15 of the surface current velocity, respectively. The significant wave height is set equal to 7 m with peak period equal to 14 sec and direction equal to 20 °. The mean wind velocity is set equal 15 m/ s with mean direction equal to 20 °. The chosen weighting factors are $w_t = 1$ and $w_b = 5$.

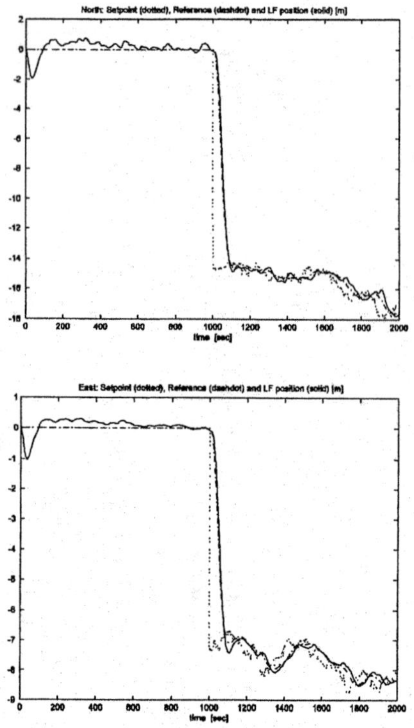


Fig. 15. North and East positions for the rig.

Initially the rig is dynamically positioned over the well head with zero (field zero point) as the constant setpoint. After 1000 sec the optimal setpoint chasing is activated.Figure 15 shows the LF position, reference and setpoint coordinates in North and East axes as function of the time. The optimal setpoint chasing forces the vessel about 16–18 m to the South and 8–9 m to the West. The

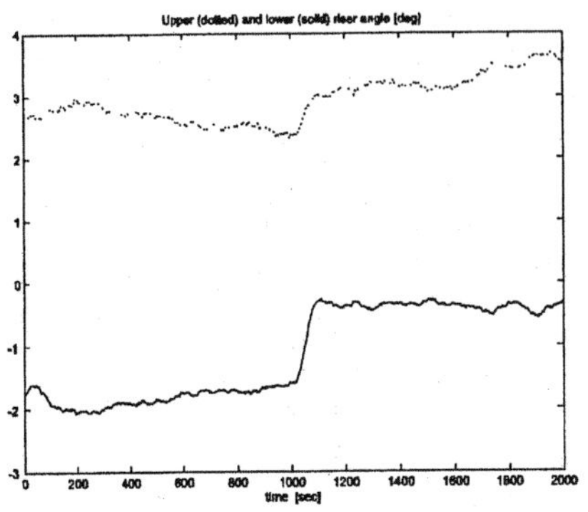


Fig. 16. Lower $\alpha_t$ (solid) and upper $\alpha_b$ (dashed) riser angles.

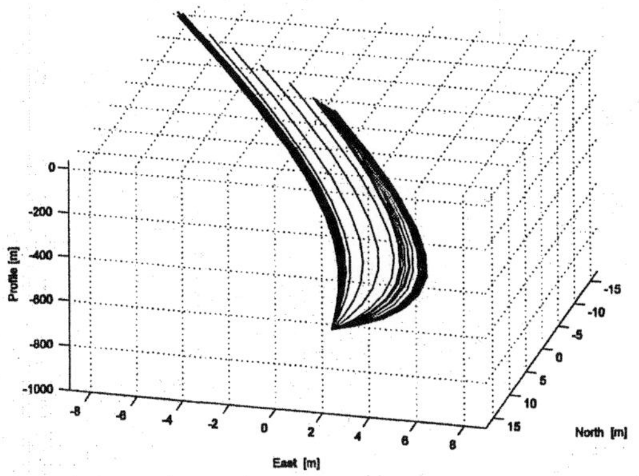


Fig. 17. Snapshots of riser profile.

heading is not affected by the optimal setpoint chasing.The effect of the optimal setpoint chasing on the riser angle offsets is clearly shown in Fig. 16. The lower riser angle decreases from 1.6 ° down to 0.4 °, while the upper riser angle increases from 2.5 ° up to 3.5 °. Since the lower riser angle offset is given higher weight compared to the upper riser angle, the opposite effect happens here.In Fig. 17 snapshots of the riser profile in 3D as function of time are shown. In the beginning the accumulation of riser profiles are found to be about the first desired setpoint defined to be over the field-zero-point equal to $(0, 0)$ in North-East coordinates. Then after 1000 sec, we observe that new accumulation areas appears corresponding to the new desired setpoint found by the optimal setpoint chasing algorithm.

## 9. CONCLUSIONS

The paper addressed structural issues in the design and operation of marine control systems. An overview of the control structure including local thruster control, plant control and local optimization was given for a DP rig conducting drilling operations. It was shown how the performance of the drilling operation was improved by using local optimization in the generation of the setpoint to the high level DP plant controller. Furthermore, the importance to consider low level thruster control was shown for operations in waves and severe sea states. The paper did also briefly discusse maritime industrial IT integrating the real time control and monitoring systems with the operational management systems.

## 10. ACKNOWLEDGMENT

This work has been carried out at the Centre for Ships and Ocean Structures (CeSOS) at NTNU. The Norwegian Research Council is acknowledged as the main sponsor of CeSOS.

## 11. REFERENCES

Bachmayer, R., L. Whitcomb and M. A. Grosenbaugh ( 2000). An Accurate Four-Quadrant Nonlinear Dynamical Model for Marine Thrusters: Theory and Experimental Validation. IEEE J. of Oceanic Engineering, 25-1, 146-159.

Balchen, J. G., N. A. Jenssen and S. Sælid (1976). Dynamic Positioning using Kalman Filtering and Optimal Control Theory. *IFAC/IFIP Symp.on Automation in Offshore Oil Field Operation,* Amsterdam, The Netherlands, 183–186.

Balchen, J. G., N. A. Jenssen, E. Mathisen and S. Sælid (1980). A Dynamic Positioning System Based on Kalman Filtering and Optimal Control. Modeling, Ident. and Control, 1-3, 135-163.

Blanke, M. (1981). Ship Propulsion Losses Related to Automated Steering and Prime Mover Control. PhD dissertation, The Technical University of Denmark, Lyngby, Denmark.

Blanke, M., M. Kinnaert, J. Lunze and M. Staroswiecki (2003). Diagnostics and Fault-Tolerant Control. Springer-Verlag, Germany, Berlin.

Blanke, M., K. P. Lindegaard and T. I. Fossen (2000). Dynamic Model for Thrust Generation of Marine Propellers. In proc. of the IFAC Conf. on Manoeuvring and Control of Marine Craft, Aalborg, Denmark.

Carlton, J. S. (1994). *Marine Propellers and Propulsion.* Oxford: Butterworth-Heinemann, UK.

Faltinsen, O. M. (1990). Sea Loads on Ships and Offshore Structures. Cambridge University Press, Cambridge, UK.

Fossen, T. I. (1994). Guidance and Control of Ocean Vehicles. John Wiley and Sons Ltd., UK

Fossen, T. I. (2000). A survey on Nonlinear Ship Control: From Theory to Practice. In proc. of the IFAC Conf. on Manoeuvring and Control of Marine Craft, Aalborg, Denmark.

Fossen, T. I. (2002). Marine Control Systems. Marine Cybernetics, Norway.

Fossen, T. I. and J. P. Strand (1999). Passive Nonlinear

Observer Design for Ships Using Lyapunov Methods: Experimental Results with a Supply Vessel. Automatica, 35-1, 3-16.

Fossen, T. I. and J. P. Strand (2001). Nonlinear Passive Weather Optimal Positioning Control (WOPC) System for Ships and Rigs: Experimental Results. Automatica, 37-5, 701-715.

Fossen, T. I. and M. Blanke. Nonlinear (2000). Output Feedback Control of Underwater Vehicle Propellers Using Feedback from Estimated Axial Flow Velocity. IEEE J. of Oceanic Engineering, 25-2, 241-255.

Fung, P. T-K. and M. Grimble (1983). Dynamic Ship Positioning Using Self-Tuning Kalman Filter. IEEE Transaction on Automatic Control, 28-3,. 339-349.

Grimble, M. J. and M. A. Johnson (1988). Optimal Control and Stochastic Estimation: Theory and Applications. Vol (1 and 2),. John Wiley and Sons Ltd. UK.

Healey, A. J., S. M. Rock, S. Cody, D. Miles and J. P. Brown (1995). Toward an Improved Understanding of Thrust Dynamics for Underwater Vehicles. IEEE J. of Oceanic Engin., 20-4, 354-360.

Isermann, R. (1997). Supervision, Fault-detection and Fault-diagnosis Methods - An Introduction. Control Engineering Practice, 5-5, 639-652.

Isermann, R. and P. Balle (1997). Trends in the Application of Model-based Fault Detection and Diagnosis of Technical Processes. Control Engineering Practice, 5-5, 709-719.

Johansen, T. A., T. P. Fugleseth, P. Tøndel, and T. I. Fossen (2003). Optimal Constrained Control Allocation in Marine Surface Vessels with Rudders. In Proc.of 6th Conference on Manoeuvring and Control of Marine Crafts, Girona, Spain.

Katebi, M. R., M. J. Grimble and Y. Zhang (1997a). $H_\infty$ Robust Control Design for Dynamic Ship Positioning. In IEE Proc. on Control Theory and Appl., **144**, 110-120.

Katebi, M. R., Y. Zhang and M. J. Grimble (1997b). Nonlinear Dynamic Ship Positioning. In Proc. of the 13th IFAC World Cingress. **Q**, 303-308.

Lindegaard, K.-P. W. (2003). Acceleration Feedback in Dynamic Positioning Systems. PhD Thesis, Report 2003:4-W, Department of Engineering Cybernetics, NTNU.

Newman, J. N. (1977) Marine Hydrodynamics. MIT Press, Cambridge, MA, US.

Kristiansen, E. and O. Egeland (2003). Frequency-dependent Added Mass in Models for Controller Design for Wave motion Damping. In Proc.of 6th Conference on Manoeuvring and Control of Marine Crafts, Girona, Spain.

Rensvik, E., A. J. Sørensen and M. Rasmussen (2003). Maritime Industrial IT. $9^{th}$ International Conference on Marine Engineering Systems (ICMES), the Helsinki University of Technology (HUT) Ship Laboratory and on board MS SILJA SERENADE, Finland, Paper B5.

Smogeli, Ø. N., L. Aarseth, E. S. Overå, A. J. Sørensen and K. J. Minsaas (2003). Anti-Spin Thruster Control In Extreme Seas. In Proc. of 6th Conference on Manoeuvring and Control of Marine Crafts, Girona, Spain.

Smogeli, Ø. N., A. J. Sørensen and T. I. Fossen (2004). Design of a Hybrid Power/Torque Thruster Controller with Loss Estimation. IFAC Conference on Control Applications in Marine Systems, Anchona, Italy.

Strand, J. P. (1999). Nonlinear Position Control Systems Design for Marine Vessels. Doctoral Dissertation, Department of Engineering Cybernetics, Norwegian University of Science and Technology, Trondheim, Norway.

Strand, J. P. and T. I. Fossen (1999). Nonlinear Passive Observer for Ships with Adaptive Wave Filtering. In "New Directions in Nonlinear Observer Design" (H. Nijmeijer and T. I. Fossen, Eds.), Springer-Verlag London Ltd., 113-134.

Sælid, S., N. A. Jenssen and J. G. Balchen (1983). Design and Analysis of a Dynamic Positioning System Based on Kalman Filtering and Optimal Control. IEEE Transactions on Automatic Control, 28-3, 331-339.

Sørdalen, O. J. (1997). Optimal Thrust Allocation for Marine Vessels. Control Engineering Practice, 5-9, 1223-1231.

Sørensen, A. J., S. I. Sagatun and T. I. Fossen (1996). Design of a Dynamic Positioning System Using Model-Based Control. Control Engineering Practice, 4-3, 359-368.

Sørensen, A. J., A. K. Ådnanes, T. I. Fossen and J. P. Strand (1997). A New Method of Thruster Control in Positioning of Ships Based on Power Control. Proc. $4^{th}$ IFAC Conf. on Manoeuv. and Control of Marine Craft, Croatia, 172-179.

Sørensen, A. J. and J. P. Strand (2000). Positioning of Small-Waterplane-Area Marine Constructions with Roll and Pitch Damping. Control Engineering Practice,.8-2, 205-213.

Sørensen, A. J., B. Leira, J. P. Strand and C. M. Larsen (2001). Optimal Setpoint Chasing in Dynamic Positioning of Deep-water Drilling and Intervention Vessels. J. of Robust and Nonlinear Control. John Wiley & Sons, UK., 11, 1187-1205.

Sørensen, A. J., J. P. Strand and H. Nyberg (2002). Dynamic Positioning of Ships and Floaters in Extreme Seas. In Proc. of MTS/IEEE Oceans 2002, Mississippi, US.

Whitcomb, L. L. and D. R. Yoerger (1999a). Development, Comparison, and Preliminary Experimental Validation of Nonlinear Dynamic Thruster Models. IEEE J. of Oceanic Eng., 24-4, 481-494.

Whitcomb, L. L. and D. R. Yoerger (1999b). Preliminary Experiments in Model-Based Thruster Control for Underwater Vehicle Positioning. IEEE J. of Ocean. Eng., 24-4, 495-506.

ELSEVIER
IFAC
PUBLICATIONS
www.elsevier.com/locate/ifac

# ASPECTS OF INDUSTRIAL DYNAMIC POSITIONING: REALITY-TOLERANT CONTROL

**Richard I. Stephens**

*Alstom Power Conversion Ltd, Rugby, UK*
*richard.stephens@powerconv.alstom.com*

Abstract: The paper discusses the concept of "reality-tolerance" within a dynamic positioning system for ships. It defines reality tolerance as an extension of fault tolerance. Two examples are highlighted. The first is the detection of excessive noise on position measurements used by the DP system. The second highlights practical problems associated with the allocation of thrust to the available thrusters.


Keywords: ship control, positioning systems, fault-tolerant systems, optimisation.

## 1. INTRODUCTION

Dynamic positioning (DP) is a method of maintaining the position and heading of a vessel without the need for anchors.

DP found its earliest applications in the early 1960s. Since then, the number of applications and number of DP systems have increased exponentially. The exploration and exploitation of oil and natural gas presented the first driving forces behind the development and deployment of DP. The expansion of oil and gas operations into increasingly inaccessible and inhospitable areas has resulted in DP being the only viable method of operation in many cases. At the same time, a wealth of new applications has arisen, including the laying of transoceanic cables and pipes, scientific research and salvage work. DP is also helping the ecology of the planet. The latest cruise ships (e.g. the Queen Mary 2) include DP systems for station keeping in ecologically sensitive areas, where the use of anchors could damage irreplaceable coral reefs.

The object of a DP system is not to hold the vessel absolutely stationary, but to maintain its station within acceptable limits (Grimble *et al*, 1979). At the same time, it is required to minimise the control effort and wear and tear on the thruster components. The magnitude of the permitted position variation is dependent not only on the application but also on the operations in progress on a particular day. For example, a supply vessel might require positioning accuracy of a few metres when alongside a rig for offloading, but will only require to be within, say, a kilometre if standing-by.

The prevailing weather has a significant effect on the vessel. Wind and current forces act as input disturbances to the system, pushing the vessel away from its station, and these must be counteracted by the control system. On the other hand, first order oscillatory wave forces are usually greater than the full thruster force available. If the control system attempts to counteract these forces, the thrusters will be completely overloaded. It is therefore important to filter out oscillatory motions due to waves.

### *1.1. DP system components*

All feedback control systems require the three fundamental elements of measurement, control calculation and actuation. These are implemented in a DP as described below.

*Measurement.* The operation of a DP system is often

critical: lives may depend upon it. It is therefore imperative to have reliable measurements of position. For safety critical operations, physical redundancy of measurement systems is a legal requirement. Specifically, three independent measurement systems must be employed for a Class III operation in which life is at risk. Where "independent" means having no common mode of failure.

The measurement of heading is usually achieved using one or more gyrocompasses. These devices have a long history, are reliable and accurate. Physical redundancy with independence is achieved by simply adding more units.

Position measurement systems utilise the global positioning systems (GPS) (Parkinson and Spilker, 1995), acoustics, physical wires, optical methods and others (Faÿ, 1990). Independence between position measuring equipment (PME) is more difficult. For example, GPS systems are susceptible to disturbances in the ionosphere, so different GPS receivers cannot be considered to be independent.

*Control calculation.* There is a wide choice of methods for control calculation. Actual and proposed systems have included PID control (Barton, 1978), Kalman filters with state variable feedback (Balchen *et al*, 1980), fuzzy control (Stephens *et al*, 1995), and $H_\infty$ robust control (Katebi *et al*, 1997).

*Actuation.* Actuation is achieved via thrust elements on the vessel. These can be tunnel thrusters, azimuth (rotatable) thrusters, main propellers, rudders or water jets. Classification of vessels requires redundancy of thrusters. It is important from a DP perspective that the thruster arrangement is capable of producing forces and turning moments suitable for the work that the vessel will perform and for the environmental conditions it will encounter. The initial stage of design, usually before the vessel has been built, includes the calculation of a "capability plot" showing the ability of the vessel to counteract environmental disturbances. Fig. 8 shows an example plot.

## 2. DP SYSTEM FEATURES

There are many challenges facing DP systems. Not least is the critical nature of the applications. Lives depend upon the consistent and safe operation of DP in the face of many unpredictable disturbances.

Equipment faults, including faults in measurement systems and actuators must be detected, isolated and corrected for.

As well as position and heading, there are other measurements available, for feedforward control. For example, suction dredger systems now utilise force measurements from the drag-head to provide feedforward control within the DP (de Keiser and van der Klugt 2000). Anchor tension and pipe tension measurements have been utilised for thruster-assisted mooring and pipe-laying respectively. Such measurements must be treated with respect. Where necessary, fault diagnosis must alert the DP system to failures. Fail-safe modes must be added to the DP control algorithms.

### *2.1. Modes of operation*

DP systems are now able to offer a multiplicity of modes designed for specific applications. A list of the names of available modes shows the range and complexity of available applications: joystick manual heading, joystick auto-heading, DP, auto-track, autopilot, auto-sail, auto-speed, ROV follow, riser follow, minimum power, weather vane, approach, loading, pick-up, hand-pilot, matrix mode.

In addition, each mode listed above has options for selection by the operator. This allows applications such as dredging, pipe-laying and barge-following and others to utilise modes auto-sail, auto-track and ROV follow respectively, with additional features specific to the application.

### *2.2. Reality-tolerance*

Much has been written on the subject of fault-tolerance (for example Blanke *et al*, 2003), which involves the reconfiguration of a control system when a fault is detected in a component. Reconfiguration of the control system may be required, however, for very different reasons, not just faults. These include changes of operating mode and operator selections. "Reality-tolerance" can be considered to be a superset of fault-tolerance, requiring a control system that are designed to reconfigure automatically depending upon the current requirements and available components.

A fundamental concept in practical DP systems it that of modularity. This has been identified as key to the ability to design and maintain a system that requires reconfiguration (Barton, 1978; van der Klugt and de Keizer, 2001). It will be shown that modularity is an enabler of reality-tolerance.

The remainder of this paper looks more closely at two aspects of reality-tolerant DP control under development. Section 3 gives an example of fault-tolerance. It is the detection and isolation of faults in PMEs and the reconfiguration of the controller. Section 4 details aspects of the allocation of force and turning moment demands to thrust devices. This includes the requirement for automatic reconfiguration under a number of conditions.

## 3. FAULT-TOLERANCE IN POSITION MEASUREMENTS

This section examines the requirements for tolerance to faults in the PMEs used by the DP system. Blanke *et al* (1997) defined fault-tolerant systems as having the following features:

- aim to prevent any simple fault developing into failure at system level
- use information redundancy to detect faults
- use reconfiguration to accommodate faults
- accept (where necessary) degraded performance due to a fault but keep plant available
- are cheap – no new hardware.

All of these are relevant to reality-tolerance, with the addition of:

- allow different modes of operation and control
- account for normal limitations of real-world equipment

### *3.1. Characteristics of position measurements*

The choice of position reference systems has important consequences for the DP controller in terms of the reference position, the noise characteristics, drift, range, reliability and suitability for particular environments. For example, a taut-wire inclinometer system has a sinker weight that determines the reference position. The levels of noise are low (< 0.5 m standard deviation) but tidal currents can cause an offset of tens of metres (Knight and Mason, 1986). The range of the device is limited by the allowable wire angle, which corresponds to about 30% of water depth. The devices are reliable under normal operating conditions and can operate in water depths up to 30 m, and with special alterations up to 60 m.

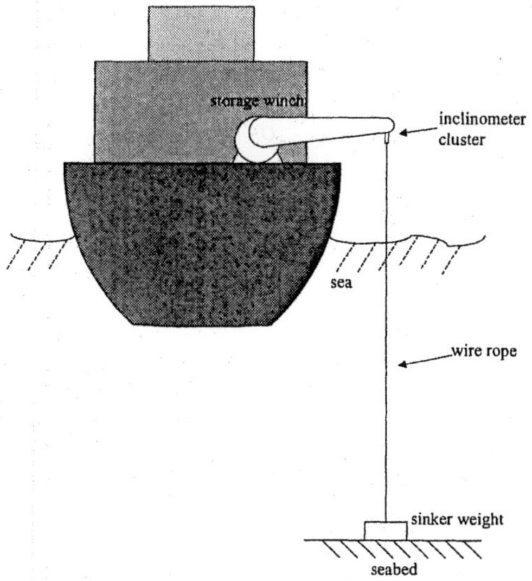


Fig. 1. Diagram of a taut-wire PME.

Other position reference systems, including acoustic systems, radio and microwave links to shore stations, direct line-of-sight laser systems and satellite navigation systems, have their own characteristics. The DP controller must be able to reconcile all the measurements available and to produce a combined position estimate.

In section 1.1 it was noted that physical redundancy of PMEs is a legal requirement. Physical redundancy alone, however, is not sufficient to provide a safe system. The redundancy must be accompanied by a strategy to detect faulty measurements and reject them. The different types of PME provide measurements of varying quality. These must be combined to form some "best estimate" of the position on the vessel. If this estimate is to be reliable, faults must be detected, isolated and either rejected or corrected (Patton *et al*, 1989).

There are a number of fault types that require detection, including excessive noise, spikes, steps and drift. Each one requires a different approach to detection and correction. The following section examines an approach to detecting excessive noise on a position measurement.

### *3.2. Detecting noise*

Noise is present on all signals. Control systems have to be able to operate in the presence of noise. For example, wave motions are treated as undesired noise on position measurements and the DP system is designed to reject them (Grimble *et al*, 1979). There are other sources of noise which can make the amount of noise on a PME unacceptable. Excessive noise on a PME can be observed on, for example, an acoustic system that has been affected by aeration of the water. Too much noise can cause the DP system to generate excessive thruster activity and ultimately it will result in loss of position. It is important, therefore, that the DP system detects a noisy signal and rejects it quickly.

Two stages of detection are required. The first is residual generation. A convenient method of residual generation is the Kalman filter, whose innovation is a zero mean estimate of the measurement noise. The adopted approach utilises a separate Kalman filter for each PME.

The second stage of fault detection is identifying the change in the residual from "normal" to "faulty". For noise detection it is clear that a simple threshold or window check will not suffice. In Fig. 3 the measured data becomes excessively noisy after 250 seconds. A threshold check set at, say, 1 metre would have rejected some of the measurements but most would have been allowed through.

In process control the CUSUM algorithm has been used to detect changes in residuals or process variables (Basseville and Nikiforov, 1993)). It considers the probability density function of the expected noise distribution. Assuming normally distributed noise on the measurement, and defining expected standard deviation before and after the change, a statistical measure of the likelihood that the signal is noisy is gained. The algorithm can be derived as follows.

A normally distributed variable, $y$, has probability density function:

$$P_\theta = \frac{1}{\sigma\sqrt{2\pi}}\exp\left(-\frac{(y-\mu)^2}{2\sigma^2}\right) \tag{1}$$

where $\mu$ is the mean and $\sigma$ is the standard deviation. If it is required to detect a change in noise, define $\sigma_0$ as the standard deviation before the change, and $\sigma_1$ as the standard deviation after the change, where $\sigma_0 < \sigma_1$, then the log-likelihood ratio for $y$ is:

$$s_i = \ln\left(\frac{P_{\theta_1}}{P_{\theta_0}}\right) = \ln\left[\frac{\frac{1}{\sigma_1\sqrt{2\pi}}\exp\left(-\frac{(y_i-\mu)^2}{2\sigma_1{}^2}\right)}{\frac{1}{\sigma_0\sqrt{2\pi}}\exp\left(-\frac{(y_i-\mu)^2}{2\sigma_0{}^2}\right)}\right] \tag{2}$$

With some simplification this becomes:

$$s_i = \ln\left(\frac{\sigma_0}{\sigma_1}\right) + \frac{(y_i-\mu)^2}{2}\left(\frac{1}{\sigma_0{}^2} - \frac{1}{\sigma_1{}^2}\right) \tag{3}$$

The basic CUSUM is formed by summing the individual $s_i$:

$$S_k = \sum_{i=1}^{k} s_i \tag{4}$$

The test for a change is based on the minimum:

$$m_k = \min_{j=1..k} S_j \tag{5}$$

from which a change is detected if $S_k$ exceeds a threshold $m_k + h$, where $h$ is a tunable parameter.

*Exponential weighting of the CUSUM.* Examination of equations (3) and (4) reveals that if there is no fault, then as $k \to \infty$ the CUSUM $S_k \to -\infty$. This will result in numerical problems if the algorithm is implemented in a digital computer. A straightforward way to avoid numerical problems is to introduce exponential weighting on the CUSUM by altering equation (4) as follows:

$$S_k = w \cdot S_{k-1} + s_k \tag{6}$$

where $w < 1$. This change ensures that $S_k$ tends towards a finite value as follows:

$$S_\infty = \frac{\ln(\gamma) + \frac{1}{2}\left(\frac{1}{\gamma^2} - 1\right)}{1-w} \tag{7}$$

where $\gamma = \sigma_0/\sigma_1$. The limiting value of the variance of $S_k$ can also be defined:

$$V_\infty = \frac{1}{2}\frac{1-\gamma^2}{1-w^2} \tag{8}$$

The results of equations (7) and (8) will be useful when considering tuning of the algorithm (section 3.3).

### *3.3. Tuning the CUSUM algorithm*

Tuning of the CUSUM algorithm is crucial to effective fault detection. Setting the parameters incorrectly may result in either a failure to detect a change or a large number of false alarms. There is a trade-off between minimising the number of false alarms and minimising the delay in detection.

There are three parameters available for tuning: the standard deviation before the change, $\sigma_0$, the standard deviation after the change, $\sigma_1$, and the threshold, $h$. The considerations for tuning each of these are described next.

*Standard deviation before the change.* If the standard deviation before the change specified in the CUSUM is less than the actual noise then the algorithm reports more false alarms. What is the effect of specifying a high $\sigma_0$? Let $y_i$ be distributed with zero mean and an actual standard deviation, $\sigma_{y0}$, less than that specified in the algorithm so that $\sigma_{y0} = r \cdot \sigma_0$ where $0 < r < 1$. Substituting for $y_i = r \cdot \sigma_0$ in equation (3) gives:

$$s_i = \ln\left(\frac{\sigma_0}{\sigma_1}\right) + \frac{r^2}{2}\left(1 - \frac{\sigma_0^2}{\sigma_1^2}\right) \tag{9}$$

Since $\sigma_0 < \sigma_1$ it follows that:

$$\ln\left(\frac{\sigma_0}{\sigma_1}\right) < 0 \tag{10}$$

and

$$\frac{r^2}{2}\left(1 - \frac{\sigma_0^2}{\sigma_1^2}\right) > 0 \tag{11}$$

and also

$$\left|\ln\left(\frac{\sigma_0}{\sigma_1}\right)\right| > \frac{r^2}{2}\left(1-\frac{\sigma_0^2}{\sigma_1^2}\right) \quad (12)$$

It follows that $s_i < 0$ as required. This shows that there should be no significant change in the efficiency of the algorithm provided the actual standard deviation before the change is less than the specified value in the algorithm.

*Standard deviation after the change.* Ideally the value for $\sigma_1$ would determine the cut-off between noise which is detected as a fault and noise which is not detected. Further manipulation of equations (1) to (3) yields the result that a fault will be detected if the standard deviation, $\sigma_{y1}$, of the noise on the actual signal satisfies the inequality:

$$\sigma_{y1} > \sigma_0\sigma_1\sqrt{\frac{2(\ln(\sigma_1)-\ln(\sigma_0))}{\sigma_1^2-\sigma_0^2}} \quad (13)$$

The expression on the right-hand side of inequality (13) is plotted in Fig. 2. This can be used to set a value for $\sigma_1$ to detect a particular noise level.

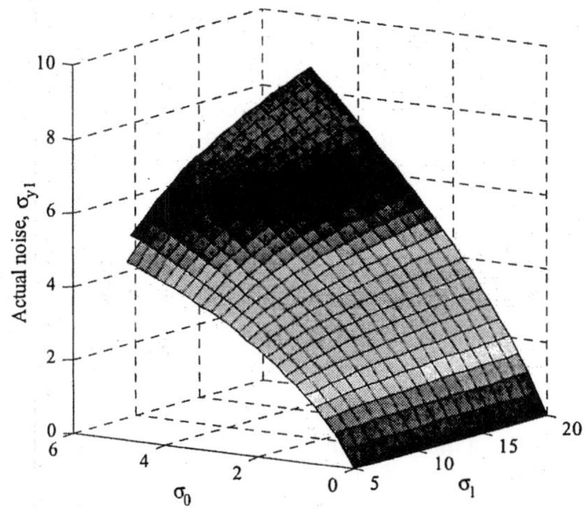


Fig. 2. Plot of fault detection. It shows the actual noise on the signal which will be flagged as a fault for given values of $\sigma_0$ and $\sigma_1$.

*Threshold.* Equations (7) and (8) define the values of the CUSUM mean and variance as time tends to infinity given known standard deviations before and after the change. These results can be used to calculate a threshold based on confidence intervals, which will ensure a given rate of false alarms. For example, a confidence interval of 99.99% is achieved with a threshold:

$$h = 5V_\infty \quad (14)$$

*3.4. Fault reset*

Academic papers on fault detection tend to go only as far as the detection. Reality often requires, however, that once a fault has been detected the controller must also determine when the fault has been cleared or has cleared itself.

The strategy adopted is to utilise the CUSUM detection algorithm again in the opposite sense to fault detection. Equation (5) becomes:

$$m_{k\_reset} = \max_{j=1..k} S_j \quad (15)$$

from which a change is detected if $S_k$ is less than the threshold $m_{k\_reset} + h_{reset}$, where $h_{reset}$ is a tunable parameter which may be different from the fault detection threshold, $h$.

*3.5. Simulation results*

The algorithm described has been extensively tested in simulation. Fig. 3 shows an example in which the standard deviation of noise on a measurement increases from 0.5 m to 1 m. The CUSUM can be seen initially to settle at a fairly constant negative value. The threshold follows above the CUSUM until it reaches its minimum as defined in equation (14).

When the noise on the signal increases, the residual noise increases and the CUSUM begins to increase. It crosses the threshold 9 seconds after the increase in noise.

Fig. 4 illustrates the reset strategy. After the fault has been detected, the algorithm determines whether the measurement remains unacceptably noisy or whether it returns to an acceptable level. In the figure, noise is reduced at 300 s and the fault flag is cleared about 80 s later.

*3.6. Other faults*

PMEs can exhibit a number of other faults. Each fault requires a separate method of residual generation. The CUSUM concept can often be reused to sensitise the change detection on the residual.

As well as increases in noise, faults that are detected include spikes, steps, freeze and drift.

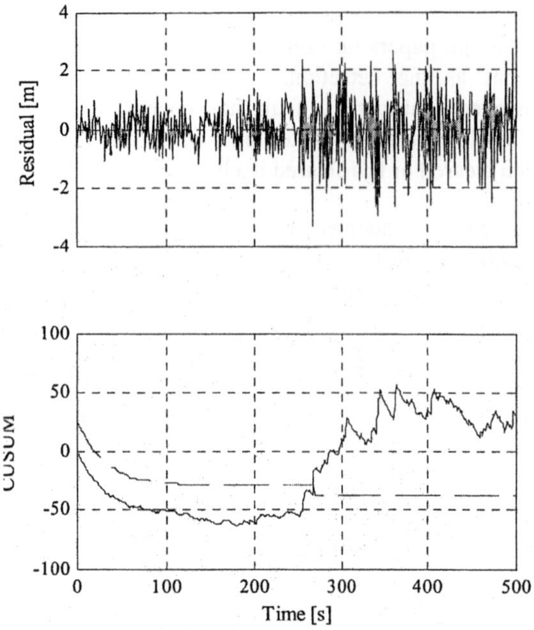


Fig. 3. Simulation results showing detection of increase in noise standard deviation from 0.5 m to 1.0 m. The top plot shows the innovation with an increase in noise at 250 s. The bottom plot shows the CUSUM (solid) exceeding the threshold (dashed) after the increase in noise, indicating a fault has been detected.

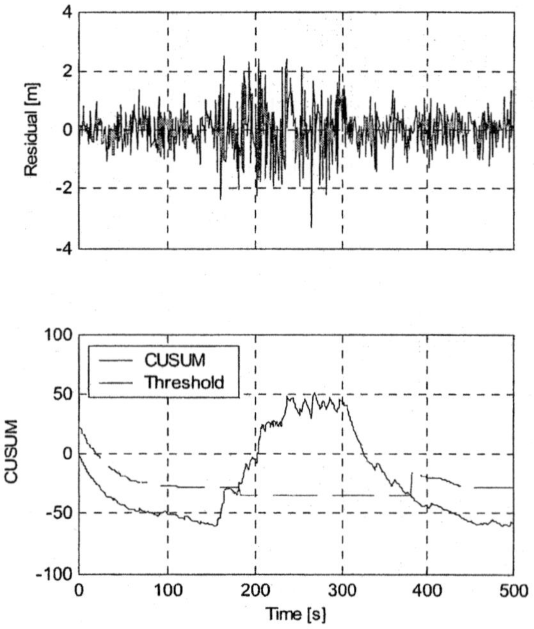


Fig. 4. Example of CUSUM algorithm including reset. Noise on the signal increases at 150 s and reduces again at 300 s.

### *3.7. Discussion*

Detection of change is an area where theoretical work is aiding the application of fault diagnosis. It has been seen in the preceding sections, however, that in the real world there are still problems to be overcome. Among these are practical considerations for continuously running applications, automatic resetting of the algorithm when the fault is removed and tuning the parameters of the algorithm.

## 4. THRUSTER TRANSFORM

The application of force to the vessel can be performed using various forms of thrust devices. The main types are: main propellers, rudders, tunnel thrusters and rotating thrusters. Each has its own characteristics and requirements, while the methods of driving them can have a major effect on the capabilities of the vessel under DP. The total available thrust places an upper limit on the performance of the vessel. Constraints on the mechanics (for example limits on the rate of change of thrust) can seriously affect the ability of the control scheme to perform to requirements.

The design of a DP system must consider all of these factors, while maintaining usability for operators.

The allocation of set points to the available thrusters and thrust elements is a transformation from the 3-dimensional space of surge-sway-yaw to an $n$-dimensional space of the force and directions for each thruster. The name used for this allocation is "thruster transform".

The DP system calculates desired X, Y, and N demands in real time. For example: if the system is in manual control, the demands are requested by the operator. If the system is in auto-position control, those demands come from the DP control algorithm. Force demands are calculated to hold position or move to a new destination. The objective of the thruster transform is to achieve the demands by allocating thrust to thrusters and calculating angles for azimuths and rudders.

### *4.1. Optimisation*

The thruster transform minimises a cost function that is a weighted sum of squares of the thrusts allocated (Sørdalen, 1996). It reduces to the solution of the equation:

$$\mathbf{A} \cdot \mathbf{z} = \mathbf{d} \qquad (16)$$

where $\mathbf{A}$ is a $3 \times n$ matrix representing the layout of the thrusters, $\mathbf{z}$ is an $n$-vector representing the thrusts in the X (surge) and/or Y (sway) axes, and $\mathbf{d}$ is a vector of length 3 of the demanded total forces in surge and sway and turning moment in yaw. The solution must minimise the weighted sum of squares of the components of $\mathbf{z}$. Using a vector of Lagrangian multipliers, $\boldsymbol{\lambda}$, the cost function is:

$$J = \mathbf{z}^T \mathbf{W}\,\mathbf{z} + \boldsymbol{\lambda}^T (\mathbf{x} - \mathbf{A}\,\mathbf{z}) \qquad (17)$$

where $\mathbf{W}$ is a triangular weighting matrix. Differentiating equation (17) with respect to $\mathbf{x}$ and $\boldsymbol{\lambda}$ and equating to zero yields the following solution for $\mathbf{z}$:

$$\mathbf{z} = \mathbf{W}^{-1}\mathbf{A}^T \left[\mathbf{A}\,\mathbf{W}^{-1}\mathbf{A}^T\right]^{-1} \mathbf{x} \qquad (18)$$

The solution, $\mathbf{z}$, consists of the thrusts in surge and sway from each thruster. These are then resolved into thrust magnitudes and directions (the latter for rotatable thrusters only).

*4.2. Thruster transform constraints*

There are a number of constraints and disturbances on the thruster allocation. Some of these are described here.

*Thruster faults.* The DP system monitors all the thrusters in order to detect faults. If a thruster is detected to be faulty, it is automatically deselected and is no longer available for the thruster transform.

*Operator interventions.* The DP operator has a number of options which affect the operation of the thruster transform. Generally, these are used for producing thruster behaviour which is non-optimal in the mathematical sense, but which achieves some other aim. For example, the operator can deselect one or more thrusters, removing them from the allocation. He might do this for planned maintenance of the thruster.

*Power system limitations.* The DP system has to take cognisance not only of thrust limitations but also of power usage on the vessel. Many new vessels use the integrated electric propulsion (IEP) concept (Benatmane, 2000). In this configuration the propulsive elements are electrical in nature and are powered from the same generation equipment as the other loads on the vessel. On a drilling ship this can lead to a conflict between the demands from the DP and the demands of the on-deck activities. It is imperative in these circumstances that some strategy is in place to prevent overloading the available generation capacity. In many cases, the DP system is required to cut back on its power requirements. This leads to additional constraints on the thruster allocation algorithm.

*Barred-zones.* With rotatable thrusters there are often constraints on the allowed direction of thrust. For example, it is undesirable for the wash from one thruster to pass over another thruster close by, since this results in a decrease in the thrust from the downstream thruster. Most rotatable thrusters are therefore given "barred zones" which define directions at which they should not be permitted to apply thrust. A typical arrangement of barred zones is shown in Fig. 5. Whilst it is possible to build barred zones into the optimisation via further constraints, and use quadratic programming to find a solution, or use gradient search techniques, the barred zones quickly make the problem "non-convex". That is, the solution space becomes split into separate, unconnected regions, and conventional minimisation techniques cannot be guaranteed to find the best solution.

One possible method is that of genetic algorithms (GAs) which introduce the possibility of jumping between regions (Goldberg, 1989). However, these are still not guaranteed to find a solution and their stochastic nature leads to unpredictable solution times, making them unsuitable for real-time operation.

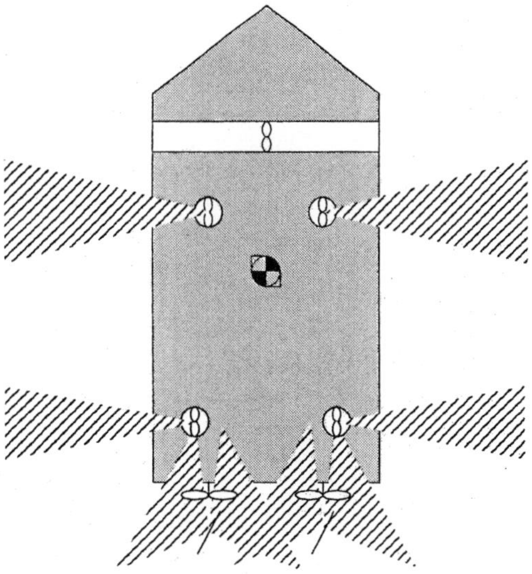

Fig. 5. Typical barred zones. The thrust vector from the rotatable thrusters must not lie within the shaded sectors.

*Thruster layout.* The popularity of DP has led to the requirement to fit DP systems to a variety of vessels. Some of these have not been designed with the requirements of DP in mind. This can lead to thruster configurations that are "awkward" for the thruster transform. One typical feature of this is the inclusion of large thrusters at the stern of a vessel, with relatively little available thrust at the bow.

The optimisation algorithm attempts to find a solution to meet the demands (equation 16). When the vessel is approaching the limits of its thrust capability (for whatever reason) the raw optimisation will use whatever thrusters it has available and this sometimes results in unwanted effects. One such effect is known as "scissoring". This is characterised by two azimuth thrusters, close together, being used in almost opposite directions in order to apply turning moment. This is illustrated in Figs. 6 and 7.

In Fig. 6, the transform has achieved a significant sway thrust demand, while maintaining small turning moment, by utilising the bow thruster. In this case none of the thruster limits has been reached. In Fig. 7 the sway force demand has increased until the bow thruster is limited. The thruster transform optimisation has used the stern thrusters at different, large angles, to achieve sway force without inducing turning moment. Note that the new angles for the stern thrusters are very different from the angles in Fig. 6. While the scissoring solution achieves a small amount of extra thrust, it is nearly always undesirable, since the rotatable thrusters are required to move through large angles and the power usage is excessive for the achievement. The thruster transform solution is therefore optionally checked to exclude scissoring.

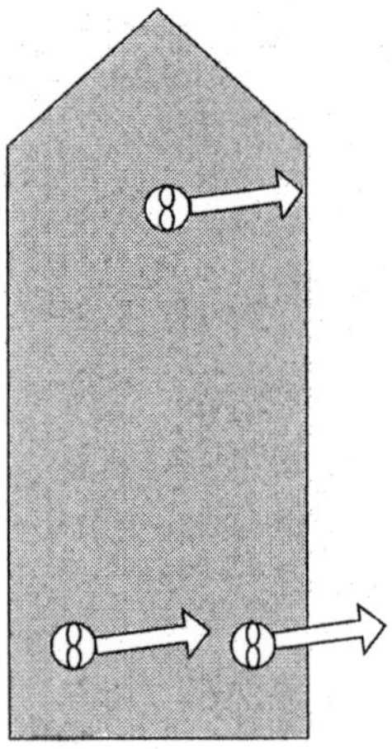

Fig. 6. Thrust allocation with no limiting

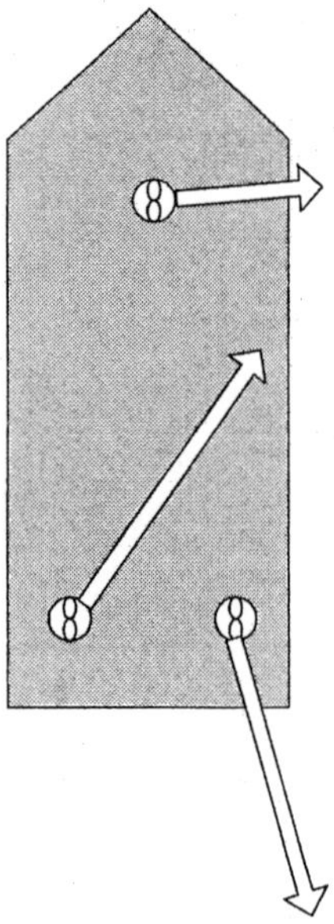

Fig. 7. Thrust allocation with limited bow thrust.

*Transients.* When the thrust demands change significantly, azimuths and rudders rotate to their new angle references and thrusters and props ramp to their new thrust references. During this transitional phase, the demands are not being met. Indeed, large transient thrusts can occur which adversely affect the control of the vessel. For example, the control system may require zero turning moment to maintain heading, and a large transient turning moment is generated which causes loss of heading.

*4.3. Solving the problems*

The previous section described various conditions and features that are detrimental to the control of the vessel. This section looks at some solutions to these problems.

*Modularity.* It has been pointed out above and by van der Klugt and de Keizer (2001) that modularity is key to success when designing and maintaining a control system. By maintaining a separation between the thruster transform and the rest of the controller, the reconfiguration of the thrusters, due to mode changes, operator interventions or faults, can be addressed without changes to other parts of the system.

*Online capability prediction.* The robustness of a DP system depends as much upon the human operator as the many of the algorithms which make it up. By better informing the operator, many of the problems can be ameliorated or avoided. One tool provided to the operator is the online capability plots. This produces a plot similar to Fig. 8 based on the modes and thruster settings at that time, or, using proposed settings. This allows the operator to make informed decisions on mode and/or thruster selections based on the prevailing conditions. Another tool available is "consequence analysis", which continuously monitors the usage of equipment and warns the operator if the failure of a single item would result in loss of position-keeping ability.

*Barred zones, thrust and power limits.* The optimisation of equations (16) to (18) is the basis of the thruster transform. It returns thrusts and angles for thrusters to achieve the axis demands. The transform utilises this optimisation in an ordered search for a solution that satisfies all the barred zone and thrust limit constraints. In essence, it performs an optimisation for each region of the divided solution space to find the global optimum.

Although it is straightforward to describe in words, this search is far from trivial even with few thrusters. Giving the algorithm the ability to recognise many of the irrelevant regions and eliminate them from the search speeds up execution.

*Axis priority.* If the demands for force and turning moment increase far enough, it will only be possible to solve the optimisation by exceeding at least one constraint (e.g. on thrust available). In this case, the thruster transform must include a strategy for graceful degradation of performance (reference the definition of fault-tolerance in section 3). The method preferred is that of axis priority, in which the three axes are given a priority order. For example the highest priority may be assigned to yaw, the next to surge and the lowest priority assigned to the sway axis. In the event that no solution can be found for all three axes, the lowest priority axis is removed from the constraints (i.e. in equation (16) the size of **d** is reduced and matrix **A** is reformed). If this fails, the next lowest priority axis is removed from the optimisation. Finally, if this produces no solution, the thrusters are set to give as near as possible to the highest priority axis.

It should be noted that the description of axis priority given above still leaves many situations where the transform effectively has a choice. It is important that such choices are recognised and a sensible decision made. For example, it may be possible to achieve a turning moment request in two ways, one which results in forward thrust in surge and the other which results in thrust astern. Clearly, one of these will be preferable, the thruster transform must recognise this.

### 4.4. *Thruster transform scheme*

The thruster transform can now be described in terms of the operations it performs and their order.

At all stages the thrusts are obtained by the solution shown in equation (18), with differing **A**, **W** and **d**.

- The first attempt is unconstrained, clearly this will meet the demands but checks are then made on the thrust limits.
- The constraints on available thrust from each thruster are then imposed and solutions found. Note that the order in which these constraints are imposed can have an effect so this may be repeated a number of times.
- The barred zones are imposed as constraints on the thrust angles by fixing rotating thrusters on the borders of barred zones that have been violated. Again, the order in which they are imposed affects the final results so more than one ordering is used.
- The requirements for biasing thrusters are imposed, and finally the anti-scissoring logic is invoked.
- If no suitable solution has been found, axis priority will be invoked by removing the lowest priority axis and the procedure repeated.

### 4.5. *Discussion*

The results presented above illustrate the problems with implementation in the real world. In fact, only a subset of the actual problems with calculating a thrust allocation has been included. The mathematical approach for optimal solution of the thrust allocation problem is elegant and straightforward. Unfortunately, it quickly becomes clear that it has limited application.

The examples above also illustrate the dangers of blindly following an "optimal" coarse. While the mathematics can achieve much, the result is not always practical – as can be seen from the scissoring example. It has been shown that the optimal solution must be carefully checked at all stages.

The algorithm must be deterministic. It must return the same answer whenever it is presented with the same problem. It must also run in the chosen hardware within a deterministic time. These requirements rule out the use of stochastic techniques such as genetic algorithms.

The solution lies in a pragmatic application of general results to a specific problem. No exciting or new mathematical results are presented, but the algorithm is guaranteed to find a solution and degrades safely and gracefully when limits are reached.

## 5. CONCLUSIONS

The paper has discussed the concept of reality-tolerance in a DP system. Reality-tolerance has been defined as the ability of a control system to perform correctly and above all safely in the presence of external constraints: whether these are normal for the system, faults or changes in available hardware. In this sense, reality-tolerance can be thought of as a superset of fault-tolerance.

Two examples have been presented. The first is the requirement to detect faults within position measurements. One such fault, excessive noise on a measurement, was examined. A scheme using Kalman filters for residual generation and a CUSUM for detection of change has been presented.

The second example is the thruster transform, or thrust allocation, which transforms the three-axis force demands of the controller into specific thrust and angle commands for individual thrusters. The example has shown that a mixture of optimisation and pragmatism, mathematics and heuristics, is the key to a practical system. Two problems have been highlighted where pure mathematical solutions fall down: the first in which barred zones split the solution space into discrete, unconnected, sectors, the second where the optimal solution was unacceptable in terms of thruster usage.

## 6. FURTHER WORK

There is much to be achieved within the areas of dynamic positioning and reality-tolerance. This paper has only been able to give a flavour of the sorts of problems facing engineers.

In the area of fault detection and isolation, work is required in the classification of fault types and causes for all PMEs. One challenge is the isolation of a PME fault (e.g. drift) when two and only two PMEs are in the pool.

Thrust allocation continues to present challenges, as new thruster types and configurations appear on DP ships. Fault detection and isolation among thrusters is also an area worthy of research.

## ACKNOWLEDGEMENTS

The author would like to thank his colleagues at Alstom Power Conversion Ltd and in particular Luke Bagnall and John Flint for invaluable help in preparing this paper.

## REFERENCES

Balchen, J. G., N. A. Jenssen, E. Mathisen, and S. Saelid (1980). "A dynamic positioning system based on Kalman filtering and optimal control", *Modelling Identification and Control*, **1** (3), pp. 135-163.

Barton, P. H. (1978). "Dynamic positioning systems", *GEC Journal for Industry*, **2** (3), October 1978, pp. 119-125.

Basseville, M. and I. V. Nikiforov, (1993). *Detection of Abrupt Changes: Theory and Application.* Prentice Hall, Englewood Cliffs, New Jersey, USA.

Benatmane, M. (2000). "The all-electric drill-ship", *3rd International Symposium on the All Electric Ship Civil or Military, AES2000, 27-29 October 2000, Paris, France*, pp. 32-49.

Blanke, M., R. Izadi-Zamanabadi, S. A. Bøgh and C. P. Lunau (1997). "Fault-tolerant control systems – a holistic view", *Doc. No, R-1997-4175*, Aalborg University, Department of Control Engineering, Aalborg, Denmark.

Blanke, M., M. Kinnaert, J. Lunze and M. Staroswiecki, (2003). *Diagnosis and Fault-tolerant Control*, Springer-Verlag, London.

Faÿ, H., *Dynamic Positioning Systems: Principles, Design and Applications*, Éditions Technip, Paris, 1990.

de Keizer, C. and P. van der Klugt (2000). "A new generation of DpDt system for dredging vessels", Marine Technology Society, Dynamic Positioning Conference, 17-18 October 2000, Houston, USA.

Goldberg, D. E. (1989). *Genetic Algorithms in Search, Optimization and Machine Learning.* Addison-Wesley, Reading MA.

Grimble, M. J., R. J. Patton and D. A. Wise (1979). "The design of dynamic ship positioning control systems using extended Kalman filtering techniques", *Proc. IEEE Oceans '79 Conf., San Diego, USA*, pp. 488-498.

Katebi, M. R., M. J. Grimble and Y. Zhang (1997). "$H_\infty$ robust control design for dynamic ship positioning", *IEE Proc.-Control Theory Appl.*, **144** (2), pp. 110-120.

Knight, P. H. and P. C. Mason (1986). "Appraisal of a taut-wire ship position reference system for offshore use", *Report ERC(W) 24.0687 January 1986*, GEC Research Ltd., Engineering Research Centre, Whetstone, UK.

Parkinson, B. W. and J. J. Spilker (Eds.) (1995). *Global Positioning System: Theory and Applications*, American Institute of Aeronautics and Astronautics, Washington DC, USA.

Patton, R., P. Frank and R. Clark (Eds.) (1989). *Fault Diagnosis in Dynamic Systems: Theory and Applications*, Prentice-Hall, New York, USA.

Sørdalen, O. J. (1996). "Thruster allocation: singularities and filtering", *Proc. 13th Triennial IFAC World Congress, July 1996, San Francisco, USA*, paper 8c-03 4, volume Q, pp. 369-374.

Stephens, R. I., K. J. Burnham and P. J. Reeve (1995). "A practical approach to the design of fuzzy controllers with application to dynamic ship positioning", *3rd IFAC Workshop on Control Applications in Marine Systems CAMS'95, 10-12 May 1995, Trondheim, Norway*, pp. 370-377.

van der Klugt, P. G. M. and C. de Keizer (2001). "A new approach towards designing ship motion control systems", *IFAC Conference on Control Applications in Marine Systems CAMS2001, 18-20 July 2001, Glasgow*, paper WA1.1.

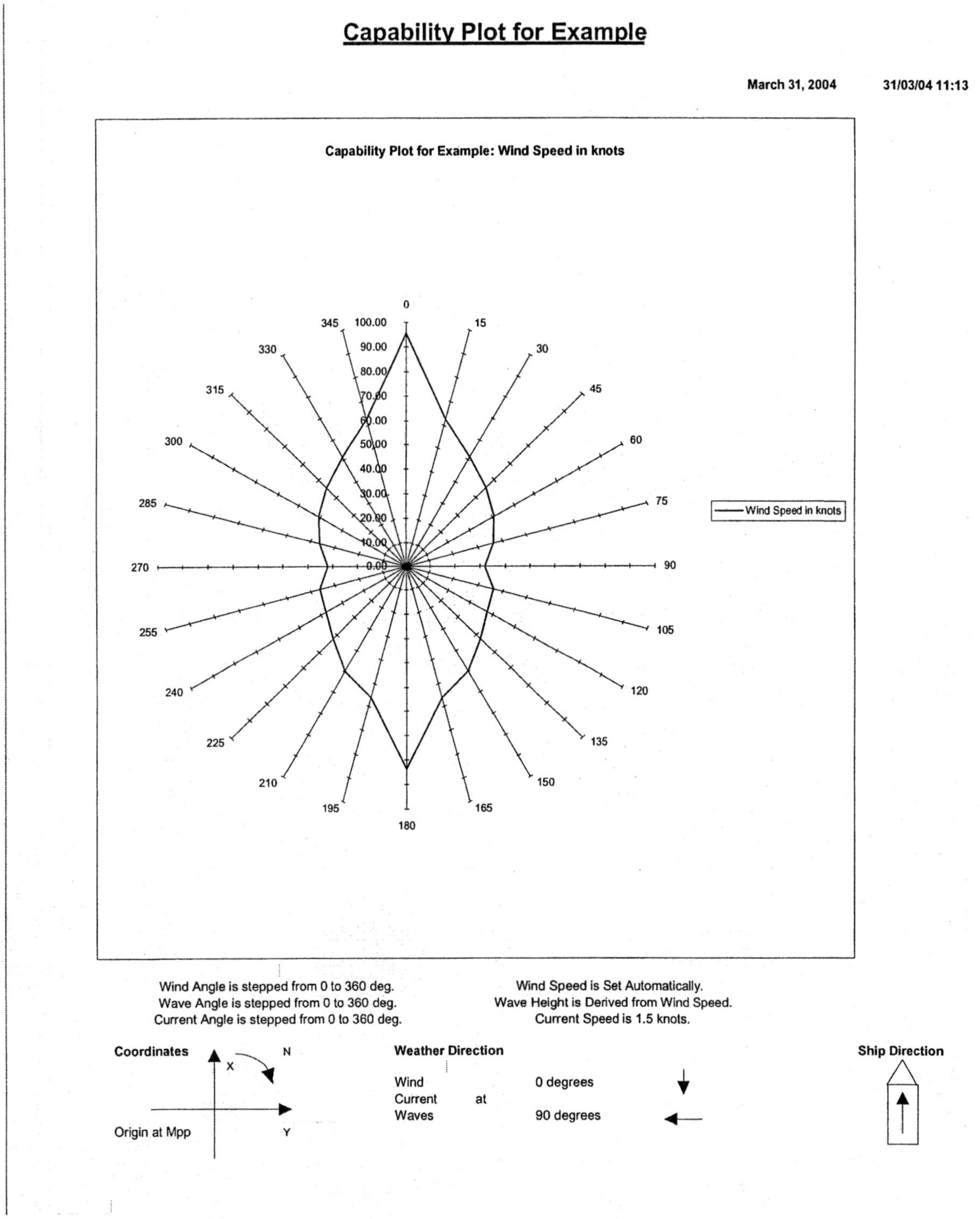


Fig. 8. Example of a capability plot showing the maximum allowable wind speed given the existing thruster ratings and selections.

ELSEVIER
IFAC
PUBLICATIONS
www.elsevier.com/locate/ifac

# DESIGN OF ROBUST SHIPBOARD POWER AUTOMATION SYSTEMS

**Edwin Zivi**

*United Statues Naval Academy*

Abstract: Emergent power and automation technologies provide new opportunities and challenges for multidisciplinary ship design. In particular, these dynamically interdependent systems require dependable, fault tolerant control to efficiently manage limited resources and to respond to casualty conditions. Design of an electric warship engineering and damage control system of systems is considered as an illustrative example. In this context, cost and survivability can be considered as either deterministic or probabilistic independent variables. In the stochastic formulation, design robustness is defined with respect to uncertainties including technology readiness, mission creep and operational environment. *Copyright © 2004 IFAC*

Keywords: Automation, Systems design, Ship control, Robust performance, Marine systems, Distributed computer control systems.

## 1. INTRODUCTION

Effective ship design is an intrinsically multidisciplinary design challenge which continues to grow in complexity as marine technologies become ever more sophisticated. Figure 1 presents the traditional ship design spiral and primary design drivers.

More specifically, the manning and automation of Hull, Mechanical and Electrical plus Damage Control (HM&E+DC) systems is considered herein. Primary design objectives include ship utility, affordability and survivability. Of particular interest is the continuity of vital services under disruptive conditions.

## 2. BACKGROUND

As noted in the United States 2001 Quadrennial Defense Review, "Transformation is not a goal for tomorrow, but an endeavor that must be embraced in earnest today". Specifically, future electric warships must provide mobility, power, and cooling management to sophisticated combat systems. These mission critical services must be continuously available despite natural and hostile disruptions. Existing U.S. Navy monitoring and control systems provide centralized, remote control with manual, crew intensive local backup systems. Intelligence resides in the human operators, not the control system. As ships become more complex, are manned more austerely, and are expected to continue to fight despite battle damage, existing control strategies will no longer be adequate.

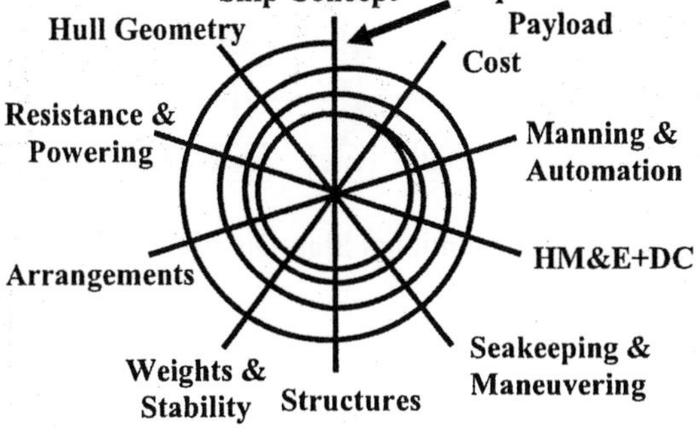


Fig. 1. Traditional Ship Design Process

The Naval Research Advisory Committee (NRAC 2000) Roadmap to an Electric Naval Force, identified four motivations for an electric warship:

1. Electric weapons & advanced sensors for superior firepower range and resolution,
2. Electric propulsion & auxiliaries for superior mobility stealth and endurance,
3. Common electric power system for real-time power allocation, reconfigurability and superior survivability,
4. Support for offboard weapons and sensors for superior reach and warfighter sustainment.

The ability to field highly available mission and life critical control systems has been demonstrated by a variety of military and commercial fly-by-wire vehicle control applications. The remaining obstacle involves engineering and fielding cost effective, survivable automation for a distributed and dynamically interdependent naval integrated power system. The Naval Research Advisory Committee

Automation of Ship Systems and Equipment study (NRAC 1989) conclusions included the following observations:

1. Automation will be essential to ensure survivable and effective surface combatants in the warfare environment of the next century.
2. High payoff automation technologies are available and reliable. They can enhance combat effectiveness, reduce manpower costs, and reduce manpower skill requirements. At the same time, they can increase systems availability, and may be introduced in a phased manner.
3. Simplification and rationalization of shipboard processes and functions are prerequisites to efficient automation. The systems engineering discipline is essential to the selection of processes and functions to be automated.
4. Models and other analytical tools to adequately assess the cost and warfighting benefits of automation do not now exist.

More recently, the Naval Research Advisory Committee Automation Optimizing Surface Ship Manning (NRAC 2000) observed that since personnel costs comprise over 50% of operating and support costs, it is imperative to reduce the number of people necessary to crew warships. A new political/military/social environment was also noted where career alternatives, quality of life issues, and family responsibilities make optimizing manning more urgent. Smart Ship was identified as a significant demonstration that technology insertion, coupled with changes in procedure, can reduce manning, maintain capability, and improve shipboard quality of life. Moreover, the obstacles encountered in fielding Smart Ship technology underscore the need for a sustained, multidisciplinary transformation of U. S. naval forces.

The Naval Research Advisory Committee Automation Life Cycle Cost Reduction (NRAC 1995) study and the Life Cycle Technology Insertion Community In Crisis (NRAC 2002) study highlight the need to predict future Life Cycle Costs (LCC), especially for systems utilizing revolutionary new technologies for which no historic cost data exist. NRAC also identified the need for a simulation based design environment to quantify LCC drivers in conceptual system designs and to project LCC implications of design alternatives. This study notes that the rapid evolution of supporting technologies, relative to Naval weapons platform acquisition cycle and service lifetimes, requires provisions for recurring technology upgrades. Concurrently, the Naval Research Advisory Committee Modelling and Simulation study (NRAC 1994) advocated advanced distributed simulations and simulation based design/manufacturing to assess new technologies and ship design options.

## 3. U.S. NAVAL PERSPECTIVE

### 3.1 U. S. Navy Control System Background

The replacement of steam propulsion plants with aeroderivative gas turbines in the 1970's marked the first time a Naval propulsion plant could not be satisfactorily controlled without automation. Period FFG-7 frigates, DD-963 destroyers, and CG-47 cruisers all use dedicated hybrid analog/digital automation for gas turbine inner loop control and sequencing. Propulsion supervisory control, along with coordination of controllable pitch propellers provides single lever bridge propulsion control. These systems employ point-to-point wiring augmented by a serial data bus for non-essential machinery monitoring.

The 1980's continued the earlier emphasis on acquisition cost over life cycle cost with few advances in propulsion automation. The representative DDG-51 Machinery Control System was marked by a strong push towards Tactical Digital Standards (TADSTANDS) including: navy standard computers, navy standard computer CMS-2 software, navy standard power supplies, navy Standard Electronic Modules (SEMs), militarized computer peripherals, and a military Data Multiplexing System (DMS) system. The primary innovation involved the replacement of mission critical point-to-point wiring with the highly redundant Data Multiplex System.

The conventional U. S. Navy electrical power distribution is presented in Figure 2. This figure indicates that a single casualty can disable both the normal and alternate power distribution paths. The flight IIA variant of the DDG-51 class ships avoids this vulnerability using a zonal distribution topology with segregated port and starboard distribution buses. Zonal distribution systems will be discussed in the following section.

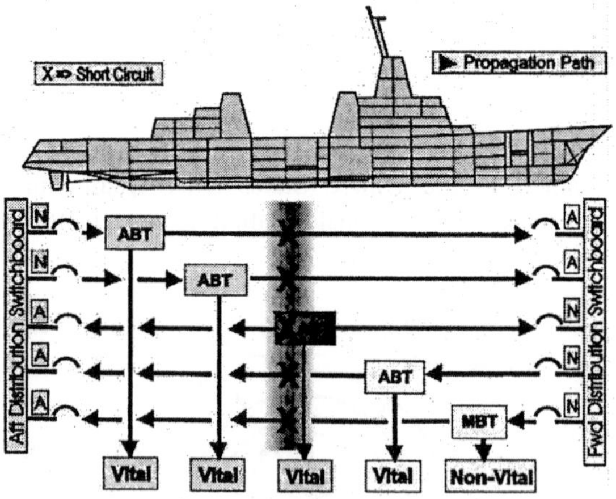


Fig. 2. Conventional Ship Distribution System

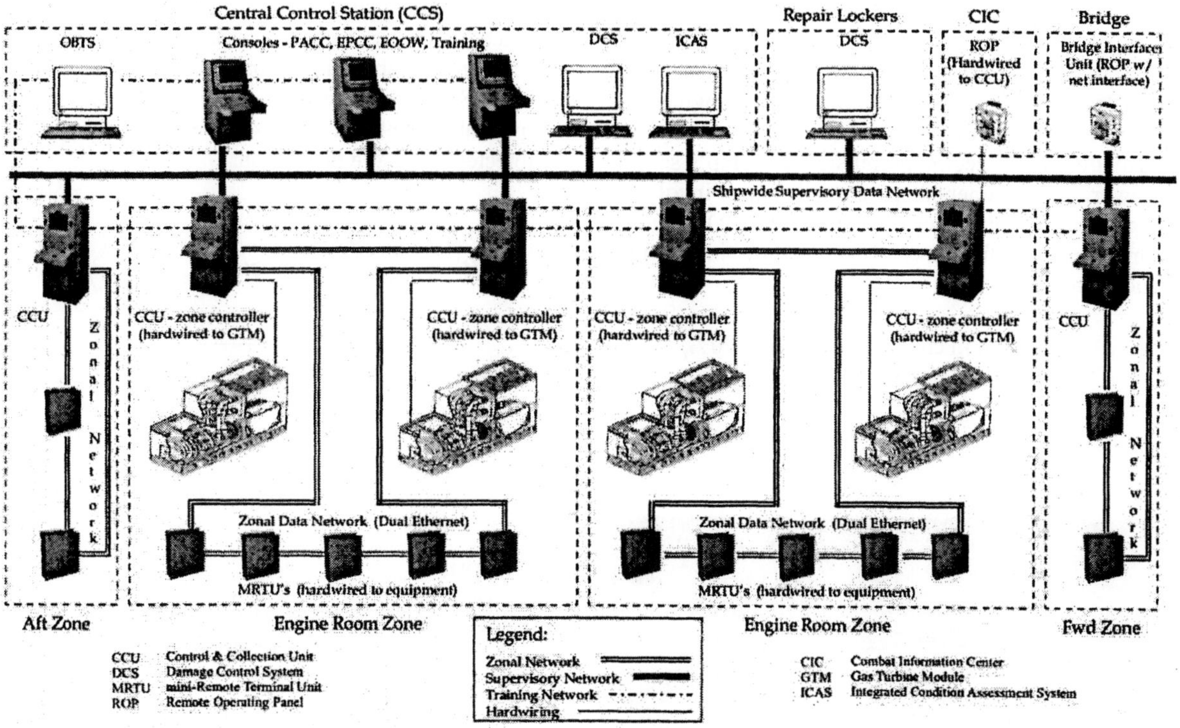


Fig. 3. Standard Monitoring and Control System

The Standard Monitoring and Control System (SMCS), in the early 1990's, marked a transition to open architecture standards based primarily on commercial technology. The primary innovations involved: standardized color, multi-watchstation consoles, a distributed digital architecture, a hierarchy of supervisory, zonal, and local controls, separation of mission/non-mission critical components, and a low Line Replaceable Unit (LRU) count. A variant of the SMCS system is being installed on the LPD 17 class amphibious assault ships. SMCS is presented in Figure 3.

3.2 Integrated Power System Introduction

As shown in Figures 4 and 5, the Integrated Power System (IPS) combines propulsion and ship's service power generation, distribution and conversion. Compared to the conventional segregated system, the Integrated Power System provides:

- The ability to redirect propulsion power to pulsed loads,
- Improved survivability,
- Improved efficiency and
- Greater operational flexibility.

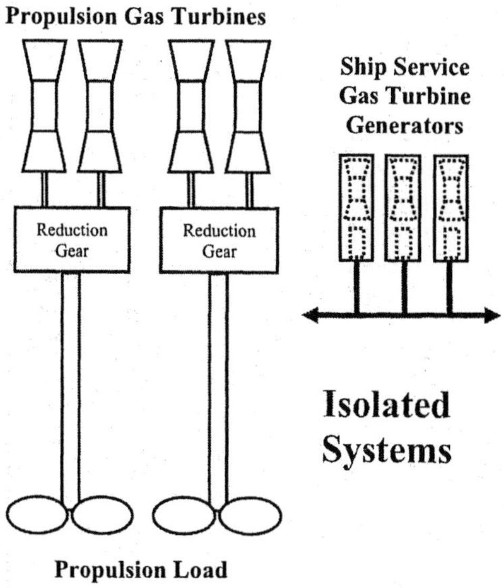


Fig. 4. Segregated Power Systems

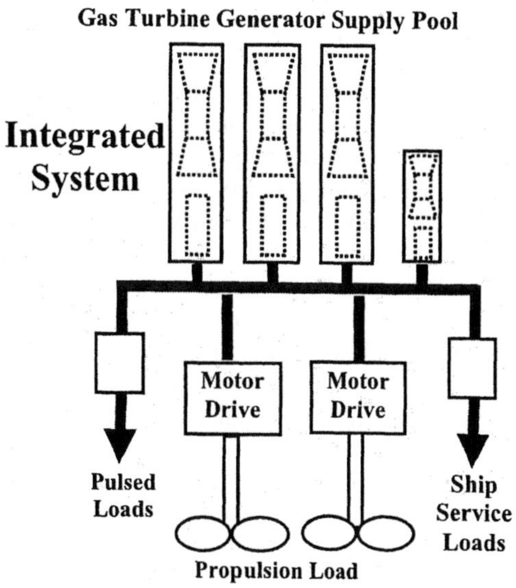


Fig. 5 Integrated Power Systems

### 3.3 U. S. Navy DDX New Ship Design

The transition to the new DDX surface combatant is illustrated in Figure 6. The fundamental engineering plant change involves an Integrated Power System providing both electric propulsion and ship's service power. A fundamental operational change involves a significant reduction in crew size. The DDX surface ships involve fundamental automation innovations including:

- A Reduction in crew size of 40% to 70%,
- Control of an advanced Integrated Power System,
- The ability to "fight through" combat damage.

Fig. 6. DDG-51 to DDX Transition

## 4. PROBLEM STATEMENT PRELIMINARIES

There is an urgent need for innovative methods and tools to formulate, design, validate, operate, and maintain dependable control systems for mission/life critical, complex, interdependent systems (Zivi 1999, 2002). This research initiative emphasizes continuity of services in lieu of traditional reliability and availability metrics. Traditional *error recovery* and *error masking* techniques are poorly suited to complex, interactive, hard real-time systems. Furthermore, the increasing dependence on complex, commercial-off-the-shelf electronic and software technology requires algorithms that minimize the susceptibility to catastrophic, common mode failures. Analogously, computer and communication systems often employ error-correcting codes to detect and correct single and multi-bit errors. Conceptually, the control system equivalent of an error correcting code is sought (Hadjicostis 2001). The desired system must maintain situational awareness and control authority despite a variety of component failures, internal errors, and exogenous disturbances.

### 4.1 Control System Background

Control theorists and practitioners have well developed methods and tools for the design and implementation of crisp, linear control systems. In the simplest context, consider the linear time invariant feedback control problem shown in Figure 7. Given a nominal plant model, *g(t)*, well known techniques can be used to design and implement the feedback controller *h(t)* to attain specified stability, performance, and robustness requirements. These techniques have been extended to accommodate certain classes of nonlinearities, uncertainties, and disruptions shown notionally in Figure 8. Insufficient theory and practice exist to dependably control complex, interdependent systems subjected to hostile disruption.

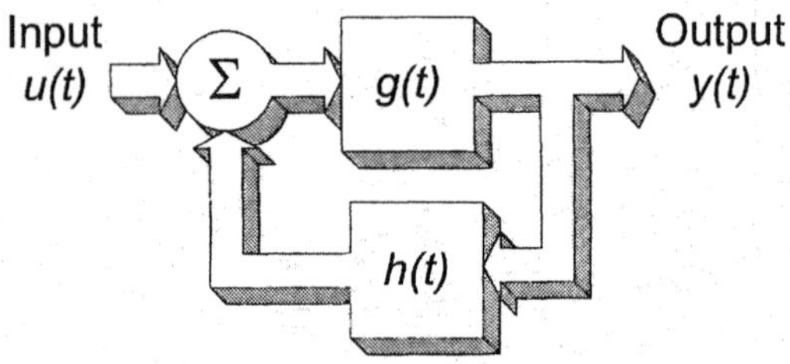


Fig. 7. Nominal System

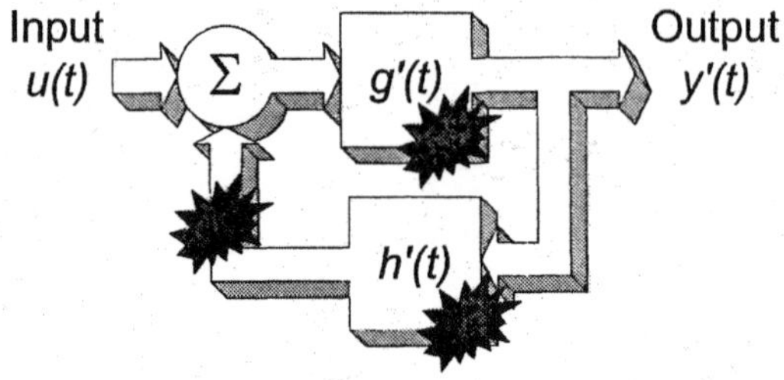


Fig. 8. Uncertain System

Consider what happens if the physical equipment, *g(t)*, is damaged through some unpredictable means. What is the appropriate change to *h(t)* such that the closed loop system remains stable? What if control system sensors and communication links are also simultaneously damaged? Can the system distinguish between physical equipment failures, sensor failures, and software errors? How can computer-controlled systems be made less susceptible to catastrophic common mode failures? Can disruptions be detected and isolated to obtain graceful degradation instead of system failure? An important related program, the DoD / AFOSR Architectures for Secure and Robust Distributed *Infrastructures* initiative explains (Lall 2002):

*"The major barrier constraining the successful management and design of large-scale distributed infrastructures is the conspicuous lack of knowledge about their dynamical features and behaviours. Up until very recently analysis of systems such as the Internet, or the national air traffic system, have primarily relied on the use of non-dynamical models, which neglect their complex, and frequently subtle, inherent dynamical properties. These traditional approaches have enjoyed considerable success while systems are run in predominantly cooperative and "friendly" environments, and provided that their performance boundaries are not approached. With the current proliferation of applications using and relying on such infrastructures, these infrastructures are becoming increasingly stressed, and as a result the incentives for malicious attacks are heightening. The stunning fact is that the fundamental assumptions under which all significant large-scale distributed infrastructures have been constructed and*

*analyzed no longer hold; the invalidity of these non-dynamical assumptions is witnessed with the greater frequency of catastrophic failures in major infrastructures such as the Internet, the power grid, the air traffic system, and national-scale telecommunication systems."*

4.2 Dependable Systems

Dependable systems must continue to operate despite component failures, internal errors, and exogenous disruptions. *System reliability* is the standard metric for measuring the effect of component failures and internal errors via component mean time to failure (MTTF) statistics and static dependency analysis. By extension, *system availability* adds consideration of the mean time to repair (MTTR) statistics. This initiative focuses on the ability to provide continuity of service despite significant disruptions due to natural (earthquake, tornado, hurricane, etc.) or hostile (terrorist or military) action. Continuity of service is critical because exogenous natural and hostile disturbances are typified by multiple simultaneous or near-simultaneous failures in the controlled plant as well as the control system. Control system failure modes include damaged hardware, inadequate control/reconfiguration algorithms, software errors, communication failures, and sensor failures. These temporal and spatially clustered "bursts" of events are much more disruptive than random failures. Moreover, the consequences of outages under hostile conditions may be disastrous.

4.3 Survivable Systems

While survivability is a well-known and often studied metric in military systems, until recently, little attention has been paid to this metric in the civilian sector. Historical methods for improving survivability include redundancy, spatial separation, and manual backup systems. Traditionally, cost and technical challenges have restricted the use of high integrity, fault tolerant systems to a limited set of high-risk systems. Examples include aero vehicle control and nuclear applications. In naval application, the primary objective is continuity of vital services during the major disruptions associated with battle and damage control operations.

The ability to fight through combat damage such as an anti-ship missile detonation requires systems which can sense, isolate, and quickly compensate for major disruptions. In this context, the system includes a spatially distributed, nonlinear, variable structure physical plant and the associated hybrid sensing, communication, control, and actuation facilities (Zivi 2001). Damage is assumed to be clustered spatially and temporally resulting in concurrent disruption of both the physical plant and the control system. An Integrated Engineering Plant must be able to tolerate simultaneous disturbances to both the machinery and the control architecture system. Representative control system disruptions include loss of sensors, actuators, communication links, algorithms or software failures. The ultimate objective is to maximize system integrity and fault tolerance while minimizing interdependencies. Time domain continuity of service subject to characteristic damage scenarios will be used to quantify success.

Traditional methods for improving shipboard survivability include redundancy, spatial separation and manual intervention. In particular, crew intensive manual intervention is no longer economically viable. Furthermore, the introduction of complex, dynamically interdependent systems exceeds human-in-the-loop control capabilities. This is because the dynamic response of the controlled equipment occurs too rapidly for the normal human reaction times and the non-linear system interdependencies are too complex for even a well-trained human operator to comprehend and react to in real-time. This conclusion is consistent with the Fitts' law and functional allocation human systems integration techniques (NATO 1999).

4.4 System Hierarchy

The three level hierarchical control system presented in Figures 9, 10, and 11 correspond to the conventional shipboard command and control structure.

Fig. 9. Supervisory Control

Fig. 10. Zonal or Module Control

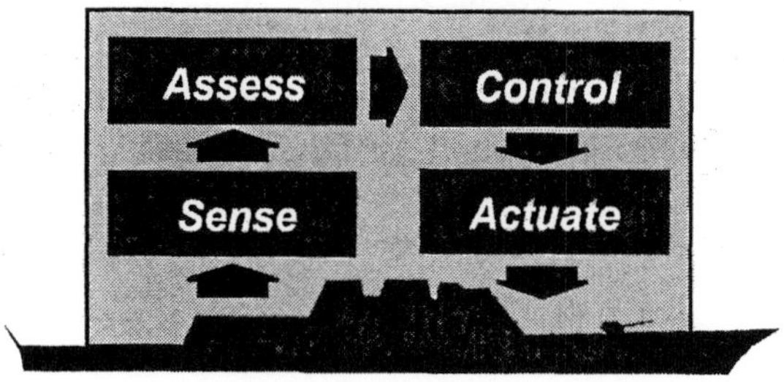


Fig. 11. Component or Device Control

4.5 Information Technology Infrastructure

One of the many shipboard control system challenges involves the wide variation in product lifecycles. In particular, typical information technology product cycles of 18 months are less than one tenth of the nominal ship lifetime of 25 years. Figure 12 presents the information infrastructure aboard the DDG-51 and LPD-17 class ships. This Local Area Network (LAN) includes:

- Industry / dual use standards and technology,
- Issues include hierarchy, protocols and connectivity,
- Monolithic programs with separate processing, information and redundancy management.

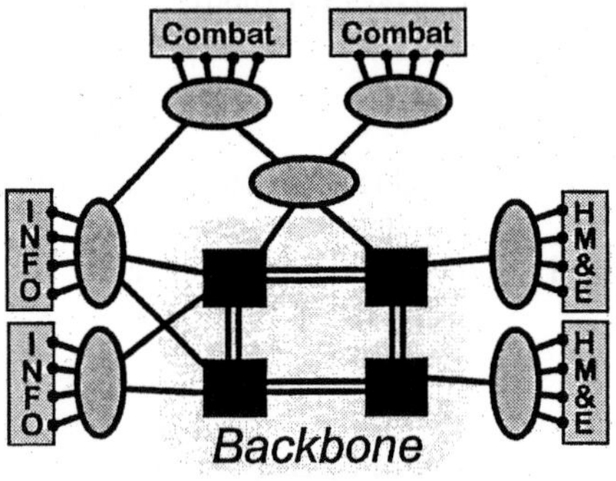


Fig. 12. DDG-51 and LPD-17 Networks.

The DDX Total Ship Computing Environment (TSCE) is expected to provide an open architecture, survivable, network centric integrated total ship command, control, and automated decision infrastructure. Figure 13 presents a schematic of a representative mesh oriented information technology infrastructure. The computing mesh includes:

- Higher redundancy and integration,
- Issues include obtaining dependable performance under disruptive conditions,
- Modular programs with a more standard interface and infrastructure.

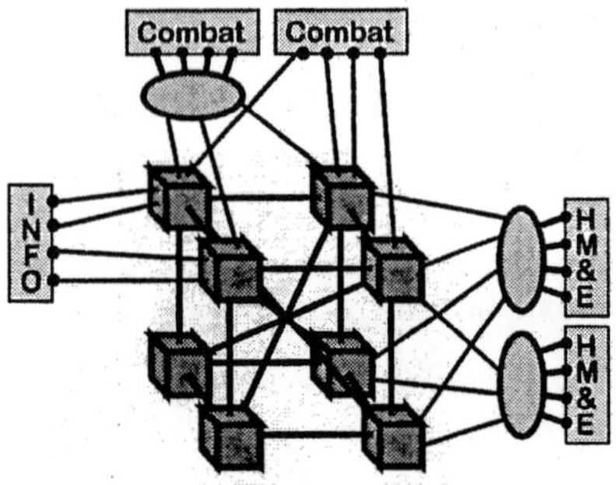


Fig. 13. DDG-51 and LPD-17 Networks.

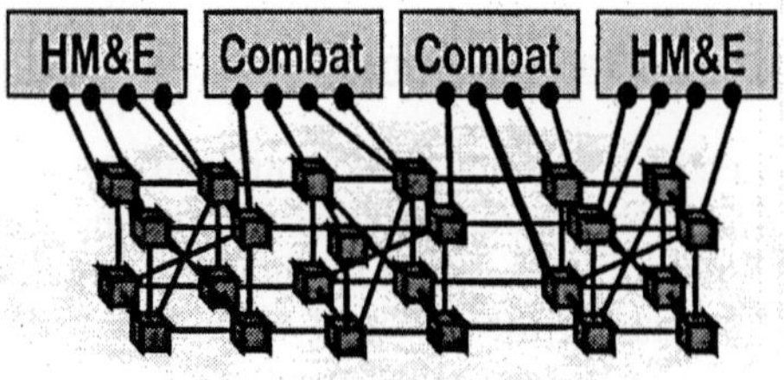


Fig. 14. Future Computing Fabric

A notional future computing network fabric is provided in Figure 14. A computing fabric provides:

- The highest redundancy and integration with global information and redundancy management,
- Ubiquitous, transparent performance,
- Smaller more modular programs with common, transparent interfaces.

## 5. POWER AND AUTOMATION REQUIREMENTS

The Office of Naval Research (ONR) Control Challenge Problem (Zivi 2001, 2003) identified the following technology shortfalls:

1. Limited ability to efficiently model and simulate complex, distributed, multidisciplinary, interactive energy, communication, and control systems
2. Incomplete theoretical basis for the robust, fault tolerant control of complex, interactive energy, sensing, and communication systems
3. Susceptibility to catastrophic common mode and common cause errors, damage, and failures
4. Limited ability to maintain situational awareness during major casualty conditions
5. Limitations in existing nonlinear stabilizing control methodologies

Automation of shipboard systems is driven by cost and performance constraints (Zivi 1999) which may be summarized as:

- The need to shift functions from humans to machines to reduce operational costs
- Control law complexity and time constraints which exceed human capabilities.

Most existing automation systems exhibit "brittle" behavior in response to non-trivial faults and failures. When subjected to significant failures or damage, brittle systems typically shut down inappropriately and/or lose stability. Conversely, dependable systems provide resilient, non-stop operation of mission/safety critical systems with provable system integrity and fault tolerance.

The control challenge objective is cost-effective, innovative control strategies which enable pervasive, industrial automation technologies to be applied to the dependable automation problem. Dependable automation strategies include:

- Extensions to stabilizing non-linear control theory,
- Generalization of analytic redundancy,
- Integrating component level intelligent distributed control into robust networks,
- Model based and non parametric estimation and fore-casting,
- Robust sensing and situational awareness,
- Analysis, design, and control algorithms for complex, interdependent systems.

Traditionally, reliable systems utilize hardware redundancy to protect against hardware-based failures. This approach is not valid for complex interactive networks and systems, is vulnerable to common mode and common cause failures, and is not affordable. A key aspect of complex dependable systems may involve the hierarchical/functional decomposition into trusted subsystems which cooperate via implicit and/or explicit communication. Modular decomposition is desired to:

- Reduce the apparent complexity,
- Build complex systems in a structured, incremental manner,
- Allow incremental replacement of obsolete technology,
- Reduce dependencies,
- Provide graceful degradation.

Highly integrated supervisory control functions are expected to operate and cooperate via a generic ubiquitous computing infrastructure. These supervisory control functions could be implemented as intelligent software agents which execute on any available computer. Component level intelligence is expected to be collocated with mission critical equipment.

Implementing supervisory control within the vehicle's information infrastructure provides a feature rich environment fully integrated with tactical decision-making, maintenance, and administration. These supervisory functions are highly available due to their ability to execute on any available generic computers. Due to complexity of the ship wide networked computer system and software-based application programs, this infrastructure may be vulnerable to hostile disruptions and common mode failures. Automated, graceful degradation, of safety and mission critical backup control functions is required. System integrity and fault tolerance is dependent on both the level of redundancy and the redundancy management strategy. Analytic redundancy may provide a fundamentally new paradigm for redundancy management.

### 5.1 ONR Control Challenge Reference Problem

The ONR control challenge reference problem is presented to motivate and focus interdisciplinary research efforts. The reference problem and associated research objectives are intended as research guidance and should not be considered definitive or exhaustive. This research thrust emphasizes continuity of vital power, fluid, and communication services in lieu of traditional reliability and availability metrics. Future electric warships must provide mobility, power, and cooling management to sophisticated combat systems. These mission critical services must be continuously available under highly lethal combat conditions. In the heat of battle, a ship that goes dark cannot fight, putting the overall mission in jeopardy, and potentially dooming the ship to complete destruction. Machinery controls for existing U.S. Naval engineering and damage control provide centralized, remote control with manual backup systems. Intelligence resides in the human operators, not the control system. As ships become more complex, are manned more austerely, and are expected to continue to fight despite battle damage, existing control strategies will no longer be adequate.

The ability to fight through combat damage such as an anti-ship missile detonation requires systems which can sense, isolate, and quickly compensate for major disruptions. Modern military and civilian systems share a fundamental challenge: to ensure continuity of service for distributed mission and life critical services despite both *natural* and *hostile* disruptions. In this challenge problem, the *system* includes a spatially distributed, non-linear, variable structure physical plant and the associated hybrid sensing, communication, control, and actuation facilities. Damage is assumed to be clustered spatially and temporally resulting in concurrent disruption of both the physical plant and the control system. In both the civilian and the military sectors, the complexity and interdependence of emerging infrastructure technologies requires new strategies, methods, and tools. The primary objective is the derivation of innovative control strategies and algorithms that maximize continuity of service despite faults and failures.

A future surface combatant with a solid-state, integrated electric power system will supply approximately 100 MW of power to propulsion, ship's service, and weapon systems. This power generation, distribution, and consumer network is also dependent upon the delivery of chill water for cooling management. The ONR reference system is based on representative, nonproprietary, reduced scale, integrated power system testbeds established at the University of Missouri-Rolla and Purdue University under a related Naval Combat Survivability initiative. The Purdue testbed provides the prototypical AC propulsion drive with a provision for a future pulse weapons system load. Ship service power supplies convert AC propulsion power to DC for zonal electrical distribution. The DC zonal electrical distribution prototypical testbed is located at the University of Missouri-Rolla.

The ONR control challenge reference system is a composite of the Purdue University and University of Missouri-Rolla testbeds. The reference system has been configured to be:

- Representative of future shipboard integrated power systems,
- Traceable to existing, non-proprietary hardware,
- Available for control system design and simulation in, mathematical and MATLAB/Simulink compatible forms.

The ONR Control Challenge reference system is presented in Figure 15.

Over the past decade, considerable effort has been made to define and refine power system architectures that are both affordable and survivable. The ONR reference system contains the minimum elements to represent an advanced integrated power system.

System characteristics include:

1. Two finite inertia AC sources and buses,
2. AC bus dynamics, stability, and regulation,
3. Redundant DC power supplies and zonal distribution buses,
4. DC bus dynamics, stability, and regulation,
5. Three zonal distribution zones feed by redundant DC power buses,
6. A variety of dynamic and nonlinear loads.

An actual ship would have a more complex configuration with additional generator capacity. The control challenge is to provide power continuity and regulation despite large variations in power demand, component failures, internal errors, and hostile system disruptions. Solid-state conversion modules introduce the potential to actively control the coupling between various system components and prevent the propagation of faults. The control challenge problem is a representative, finite inertia, tightly coupled, isolated shipboard power system.

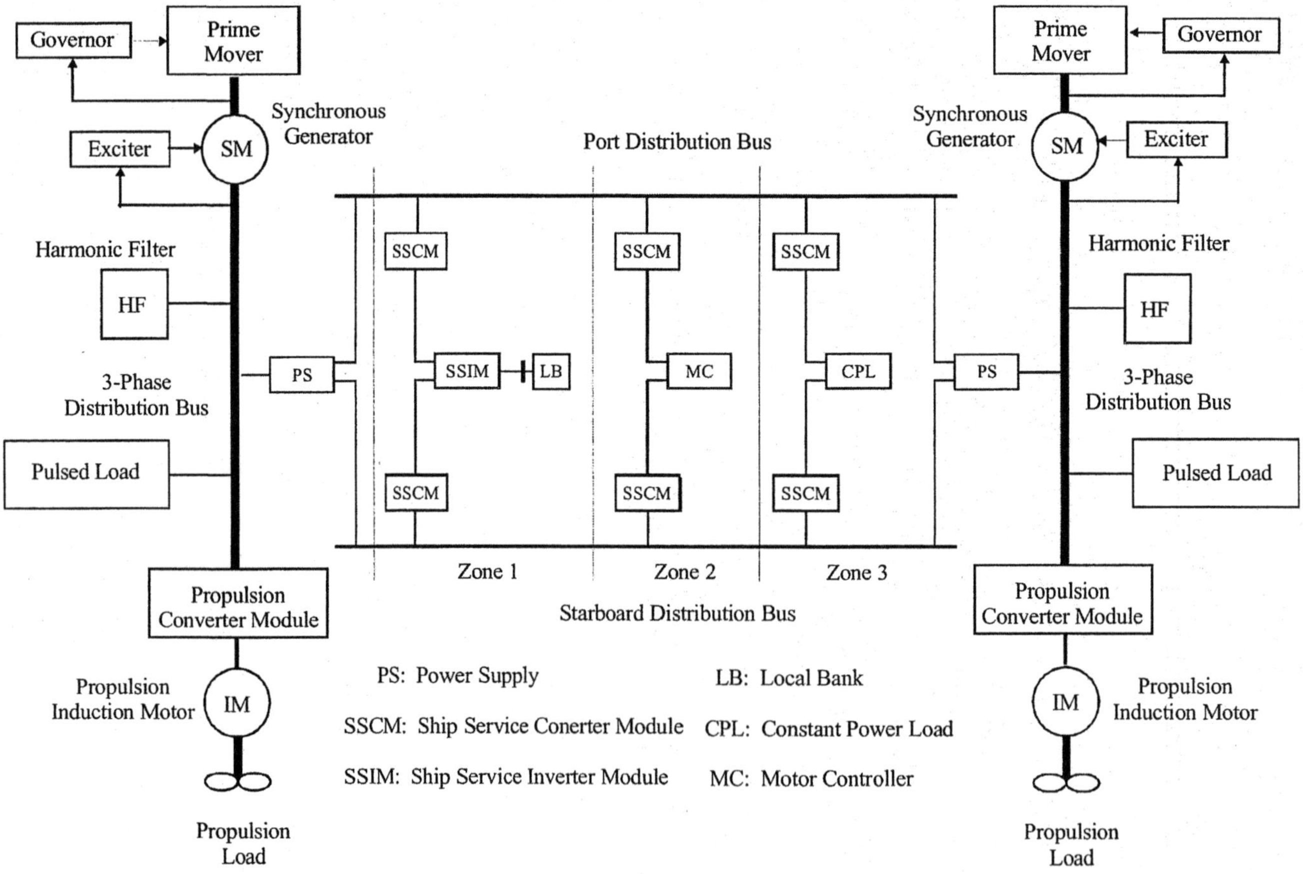


Figure 15. ONR Control Challenge Reference System

Specific research objectives include:

1. Control system integrity and fault tolerance despite component failures, internal errors, and hostile disruptions involving cascading failures,
2. Robust, real-time situational awareness with graceful degradation,
3. Continuity of power and thermal management for critical loads,
4. Self-organizing and self-healing operation which minimizes requirements for human intervention, calibration, maintenance, and repair,
5. Modular functional decomposition which leverages existing and emerging technology and standards to obtain implementation simplicity, affordability, and life cycle supportability,
6. Innovative architectures and control strategies which minimize subsystem interdependence,
7. Improved quality of service despite reductions in operating margins.

## 6. RECENT RESEARCH PROGRESS

This section presents two recent and relevant Energy Systems Analysis Consortium (ESAC) accomplishments funded under the ONR Electric Ship Research and Development Consortium program.

### 6.1 Generalized Immittance Analysis

The basic ideas of generalized immittance analysis are set forth by Sudhoff (1998, 2000, 2002, 2002b, 2003). These papers build on simple source-load systems with extensions and a MATLAB based DC Stability toolbox for large-scale systems. Single-port and two-port converters are classified and defined. Once all converters are classified, a series of mapping functions is used to reduce any given system to the single source-load equivalent presented in Figure 16.

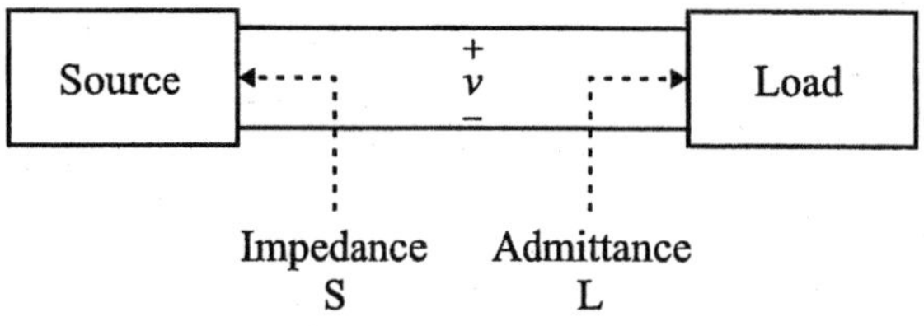


Figure 16. Single Source-Load Model

The next step is to select a frequency domain stability criteria to determine if the generalized source impedance and load admittance combination is Nyquist stable. Figures 17 and 18 present experimental validation (Sudhoff 2003) of the DC stability toolbox using the University of Missouri at Rolla Naval Combat Survivability testbed. In this example, the Ship Service Converter Module (SSCM) capacitance is modified to create stable and unstable operating conditions. As shown in Figures 17 and 18, the stability toolbox predictions match the experimental observations.

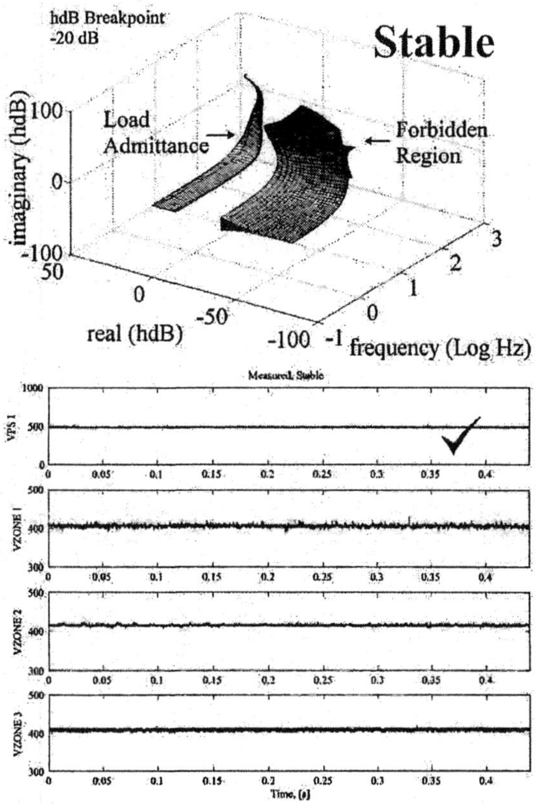


Figure 17. Single Source-Load Model

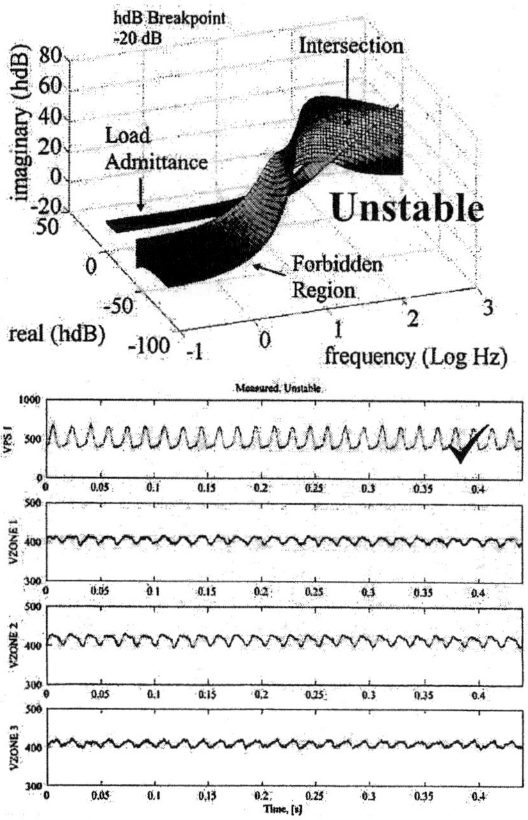


Figure 18. Single Source-Load Model

6.2 Fault Tolerant Delta-Hystersis Motor Drive

A novel Digital Delta-Hysteresis Regulation (DDHR) scheme allows motor drive systems to maintain relatively accurate control of phase currents in the event of the loss of all phase current sensors. Within the scheme, line currents are reconstructed using knowledge of the DC-link current and switching states. The generation of switching signals is based upon traditional Delta-Hysteresis Regulation (DHR). However, modifications are made to DHR to ensure the DC-link current always contains sufficient information to reconstruct phase currents. The structure of the DDHR controller is presented in figure 19. Figure 20 depicts the experimental setup where the DDHR controller has been implemented in a DSP prototype system. Figure 21 provides experimental results obtained from a vector-controlled induction machine. In particular, note the relatively small difference between the performance of the pre- and post-fault system. One advantage of this scheme is that it is independent of the load connected to the inverter. In addition, it can be readily incorporated into existing drive applications with relatively minor effort and cost.

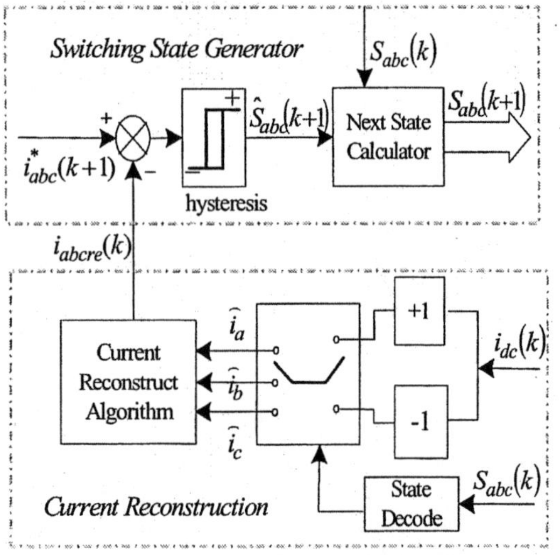


Figure 19. Digital Delta-Hysteresis Controller

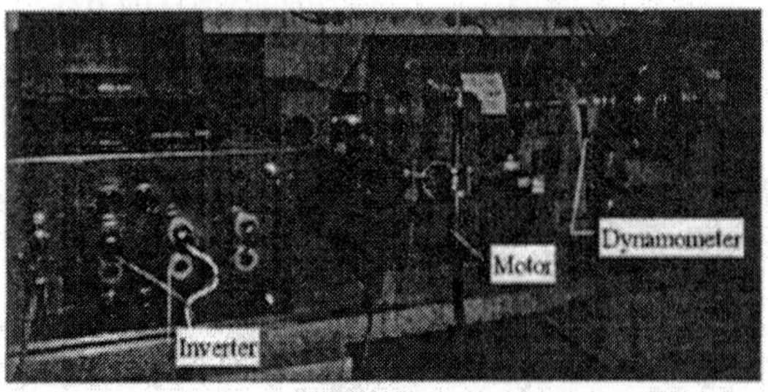


Figure 20. Experimental Setup

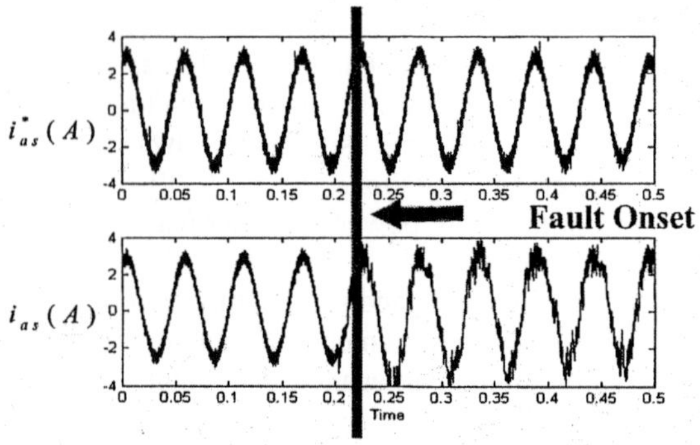


Figure 21. Fault Tolerant Operation

7. SYSTEM DESIGN

Given the complexity of modern warship design, the traditional ship design spiral presented in Figure 3 is inadequate for future systems. Note that one design spiral produces a single point design. This design cycle is repeated until an acceptable design is "discovered" through the engineering approximation, intuition, "art" and determination. The proposed design framework directly addresses the traditional design spiral limitations including:

- Labor intensive "trial and error" sequential design iterations
- Lack of insight regarding interaction of multidisciplinary design parameters
- Limited ability to analyze the dynamic interdependence of engineering plant, automation, mobility, weapon systems and casualty response.
- Inability to systematically search and optimize the ship design space.

Compared to traditional ship design disciplines including hydrodynamics, structures, and signatures, a modest HM&E+DC automation research program could achieve a disproportionably high return on R&D investment. This high research payoff potential leverages concurrent innovations in machinery systems, automation technologies and system design methodologies to produce a leaner, more agile and effective naval force. Exploration of new survivable distributed automation concepts and technologies using state-of-the-art physics-based modelling and computational design tools can support early design tradeoffs and decisions within a systematic and automated design environment. In particular, an integrated engineering plant offers new ship design flexibility which requires a more sophisticated ship design process. The integration of dynamically interdependent HM&E+DC systems presents new challenges for the ship design agent which existing design tools are not equipped to handle.

For example:

- How should integrated engineering plant components be distributed throughout the ship?
- What network topologies and spatial arrangements should be used to interconnect dynamically interdependent energy, fluid flow, information, command, and control subsystems?
- How should command, control, situational awareness, maintenance, and casualty intervention tasks be allocated between man and machine?
- Can self healing automation provide continuity of situational awareness, command, and control authority during combat induced disruptions and cascading failures?
- Can mission essential engineering and damage control continuity of service be provided despite combat induced disruptions and cascading failures?

## ACKNOWLEDGEMENTS

This paper discusses work performed by a team of researchers sponsored by the Office of Naval Research and the Naval Sea Systems Command. Support for the ONR Control Challenge and ongoing ship design efforts is provided by Ms Katherine Drew, ONR 334. Support for the Naval Combat Survivability testbeds and research was provided by Mr. Dave Clayton of NAVSEA. The recent experimental results, described in section 6, were funded by Ms Sharon Beerman-Curtin, ONR 334. The author gratefully acknowledges the contributions many fellow researchers particularly Scott Sudhoff, Steve Pekarek and Tim McCoy.

## REFERENCES

ASDL (2003) Georgia Tech Aerospace Systems Design Laboratory website, http://www.asdl.gatech.edu/.

ASDL (2004) Georgia Tech Aerospace Systems Design Laboratory ONR research website, http://www.asdl.gatech.edu/research_teams/onr.html.

Chong, E. and Żak S., (2001) *An Introduction to Optimization*, Second edition, Wiley, New York.

ESRDC (2004) Electric Ship Research and Development Consortium website, http://esrdc.caps.fsu.edu/

Hadjicostis, C., (2001) Non-Concurrent Error Detection and Correction in Discrete-Time LTI Dynamic Systems, *Proceedings of the 40th IEEE Conference on Decision and Control*, Orlando, Florida USA.

Lall, S., (2002) DoD / AFOSR Architectures for Secure and Robust Distributed Infrastructures University Research Initiative Web Site, http://element.stanford.edu/~lall/projects/architectures/.

Lee, Y. and Żak, S., (2002) Genetic fuzzy tracking controllers for autonomous ground vehicles," *Proceedings of the 2002 American Control Conference,* Anchorage, Alaska.

Massoud, A., (2001) Toward Self-Healing Energy Infrastructure Systems, *IEEE Computer Applications in Power*, pp. 20-28, Vol. 14, No. 1.

NATO (1999) Analysis Techniques for Human-Machine System Design, NATO Defense Research Group, Panel 8 Research Study 14.

NRAC (2000) *Naval Research Advisory Committee Report: Optimizing Surface Ship Manning,* http://nrac.onr.navy.mil/webspace/list/reportlistchrono.html.

NRAC (2002) *Naval Research Advisory Committee Report: Roadmap to an Electric Naval Force,* July 2002, http://nrac.onr.navy.mil/webspace/list/reportlistchrono.html.

NRAC (1989) *Naval Research Advisory Committee (NRAC) Report: Automation of Ship Systems and Equipment*, http://nrac.onr.navy.mil/webspace/list/reportlistchrono.html

NSF/ONR (2002) Partnership in Electric Power Networks Efficiency and Security (EPNES), Program solicitation NSF 02-041, http://www.nsf.gov/pubsys/ods/getpub.cfm?odskey=nsf02188.

Pekarek, S., et. al. (2003) Overview of a Naval Combat Survivability Program, *Proceedings of the 13th International Ship Control Systems Symposium (SCSS2003)*, U.S. Navy, Naval Sea System Command, April 4 - 6, 2003, Orlando, Florida.

Sudhoff, S., et. al. (1998) Stability Analysis of DC Distribution Systems Using Admittance Space Constraints, Proceedings of The Institute of Marine Engineers All Electric Ship 98, London.

Sudhoff, S., et. al. (2000) Three Dimensional Stability Analysis of DC Power Electronics Based Systems, *Proceedings of the Power Electronics Specialist Conference*, Galway, Ireland.

Sudhoff, S., et. al., (2000b) Admittance Space Stability Analysis of Power Electronic Systems, *IEEE Transactions on Aerospace and Electronics Systems*, Vol. 36. No. 3. July 2000, 965-973.

Sudhoff, S., et. al. (2002) Stability Analysis of a DC Power Electronics Based Distribution System, *Proceedings of the SAE2002 Power Systems Conference*, Coral Springs, Florida.

Sudhoff, S., et. al., (2002) Naval Combat Survivability Testbeds for Investigation of Issues in Shipboard Power Electronics Based Power and Propulsion Systems, *Proceedings of the IEEE Power Engineering Society Summer Meeting*, July 21-25, 2002,Chicago, Illinois, USA.

Sudhoff, S., et. al., (2003) Analysis Methodologies for DC Power Distribution Systems, *Proceedings of the 13th International Ship Control Systems Symposium (SCSS2003)*, U.S. Navy, Naval Sea System Command, April 4 - 6, 2003, Orlando, Florida.

Tucker, A., (2001) Opportunities & Challenges in Ship Systems & Control at ONR, presented at the 40th IEEE Conference on Decision and Control, http://www.usna.edu/EPNES/Challenge_Problem.htm.

Wang, H, et. al., (2004) "Improvement of Fault Tolerance in AC Motor Drives Using a Digital Delta-Hysteresis Modulation Scheme," *Proceedings of the 35th IEEE Power Electronics Specialists Conference,* Aaachen.

Zivi, E., and McCoy, T., (1999) Control of a Shipboard Integrated Power System, *Proceedings of the Thirty-third Annual Conference on Information Sciences and Systems*, Baltimore, MD.

Zivi, E., (2001) ONR Control Challenge Problem website, http://www.usna.edu/EPNES.

Zivi, E. and McCoy, T., (2003) ONR Ship Control Challenge Problem, *Proceedings of the 13th International Ship Control Systems Symposium (SCSS2003)*, U.S. Navy, Naval Sea System Command, April 4 - 6, 2003, Orlando, Florida.

Żak, S., (2003) *Systems and Control*, Oxford University Press, New York.

ELSEVIER
IFAC
PUBLICATIONS
www.elsevier.com/locate/ifac

# PATH FOLLOWING OF STRAIGHT LINES AND CIRCLES FOR MARINE SURFACE VESSELS

**Morten Breivik** *,[1] **Thor I. Fossen** *

* *Centre for Ships and Ocean Structures (CESOS), Norwegian University of Science and Technology (NTNU), NO-7491 Trondheim, Norway. E-mails: morten.breivik@ieee.org, tif@itk.ntnu.no*

Abstract: This paper addresses the problem of path following for marine surface vessels. A guidance-based approach which is equally applicable for land, sea and air vehicles is presented. The main idea is to explicitly control the velocity vector of the vehicles in such a way that they converge to and follow the desired geometrical paths in a natural and elegant manner. Specifically, straight lines and circles are considered. A nonlinear model-based controller is designed for a fully actuated vessel to enable it to comply with the guidance commands. The vessel is exposed to a constant environmental force, so integral action is added by means of parameter adaptation. A full nonlinear vessel model is used in the design. By introducing sideslip compensation and a dynamic controller state, the results are extended to underactuated vessels. *Copyright © 2004 IFAC*

Keywords: Marine surface vessels, Path following, Guidance, Line-of-Sight, Nonlinear model-based control, Underactuated vessels

## 1. INTRODUCTION

The ability to accurately maneuver a ship, a rig or a semi-submersible along a given path at sea is of primary importance for most applications. In this context, trajectory tracking systems for marine surface vessels are frequently proposed in the literature, see e.g. (Fossen 2002). Their task is to ensure that the vessel tracks a time-parametrized reference curve in the horizontal plane. However, there are weaknesses to the approach. Firstly, it does not take advantage of the geometric information about the desired path when it is available. It merely considers tracking an instantaneous, time-varying position signal. Hence, the way in which positional trajectory tracking is currently being performed is similar to ordinary servosystem tracking. This fact degrades the transient convergence behaviour of the position significantly, and makes it unnatural. One solution to this weakness would be to design a guidance system which utilizes the available geometric information to ensure a natural convergence behaviour.

The second weakness can be illustrated by the following: suppose that a geometrically feasible trajectory is created for a vessel to negotiate. The trajectory is based on a prespecified speed assignment which the vessel must fulfil when tracking the path to satisfy certain time constraints. To ensure the design of a dynamically feasible time parametrization, information about the weather and the initial propulsion capability of the vessel is applied. But such features can change significantly during the voyage, resulting in a dynamically infeasible trajectory. The tracking point on the original trajectory will then leave the vessel behind, saturating the actuators and making the system unstable. A reparametrization can take care of this. However, it will have to consider the new situation, which also can change rapidly, to construct a feasi-

[1] Supported by the Norwegian Research Council through the Centre for Ships and Ocean Structures, Centre of Excellence at NTNU.

ble time parametrization given the current knowledge. This is not a smart or intuitive way to act. It does not correspond to the way in which we humans adapt dynamically to changing conditions when controlling vehicles. We do not aim at tracking a conceptual point in front of us if we understand that it would be equivalent with risking lives or damaging the vehicle. Hence, so should a trajectory reparametrization be dynamically adaptable. This can be done in the framework of path following, where the task objective is first and foremost to converge to the desired path, and secondly to satisfy a given dynamical assignment along the path. If the second objective cannot be satisfied, the vessel will still be able to follow the path, though not at the initial speed profile.

The main contribution of this paper is to remove the two principal weaknesses of the traditional trajectory tracking scheme. A guidance-based path following approach which ensures a natural maneuvering behaviour is presented. The guidance and control framework in which the approach is developed easily extends from fully actuated to underactuated vessels such that path following for the latter also is achieved. The considered geometrical paths are straight lines and circles.

## 2. PROBLEM STATEMENT

In path following, the primary objective concerning a vehicle is to restrict its position to a specific manifold represented by a desired geometric path. The secondary objective is to ensure that the vehicle complies with a desired dynamical behaviour while traversing the path. Consequently, the path following problem can be expressed by the following two task objectives:

**Geometric Task:** Make the position of the vessel converge to and follow a desired geometric path.

**Dynamic Task:** Make the speed of the vessel converge to and track a desired speed.

The next section is concerned with developing a guidance law which guarantees the fulfilment of the geometric task.

## 3. GUIDANCE SYSTEM DESIGN

The theory in this section is taken from (Breivik and Fossen 2004).

### 3.1 Guidance for a General Path

Consider a point mass particle situated on a two-dimensional surface. Denote the inertial position and velocity of the particle by $\mathbf{p} \in \mathbb{R}^2$ and $\mathbf{v} \in \mathbb{R}^2$, respectively. The velocity vector has two characteristics; size and orientation. Denote the size by $U = \|\mathbf{v}\|_2 = (\mathbf{v}^\top\mathbf{v})^{\frac{1}{2}}$ (the speed) and the orientation by $\chi = \arctan(\frac{v_y}{v_x})$ (the course angle). It is assumed that

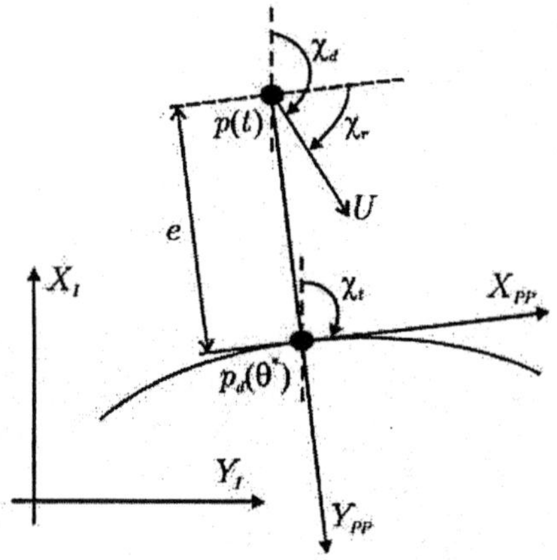


Fig. 1. A geometrical illustration of the guidance-based path following approach.

both $U$ and $\chi$ can attain any desirable value instantaneously, consequently the point mass particle will be referred to as an ideal particle hereafter. Also defined on the two-dimensional surface is a geometrical path, parametrized by a scalar variable $\theta \in \mathbb{R}$. For any given $\theta$, the inertial position of the geometrical path is denoted by $\mathbf{p}_d(\theta) \in \mathbb{R}^2$. Our main objective is to make the ideal particle converge to and live on this two-dimensional manifold.

Assume that a well-defined value of $\theta$ that minimizes the Euclidean distance between $\mathbf{p}$ and $\mathbf{p}_d(\theta)$ exists. Denote this global minimizer $\theta^*$ and define it by:

$$\|\mathbf{p} - \mathbf{p}_d(\theta^*)\|_2 \triangleq \arg\min_{\theta\in\mathbb{R}} \|\mathbf{p} - \mathbf{p}_d(\theta)\|_2 . \quad (1)$$

Define a local reference frame at $\mathbf{p}_d(\theta^*)$ and christen it the Path Parallel (PP) frame. The PP frame is rotated an angle:

$$\chi_t(\theta^*) = \arctan\left(\frac{y'_d(\theta^*)}{x'_d(\theta^*)}\right) \quad (2)$$

relative to the inertial frame, where the notation $x'_d(\theta) = \frac{dx_d}{d\theta}(\theta)$ has been used. Consequently, the x-axis of the PP frame is aligned with the tangential vector to the path at $\mathbf{p}_d(\theta^*)$.

The error vector between $\mathbf{p}$ and $\mathbf{p}_d(\theta^*)$ expressed in the PP frame is:

$$\varepsilon = \mathbf{R}_p^\top(\chi_t)(\mathbf{p} - \mathbf{p}_d(\theta^*)), \quad (3)$$

where:

$$\mathbf{R}_p(\chi_t) = \begin{bmatrix} \cos\chi_t & -\sin\chi_t \\ \sin\chi_t & \cos\chi_t \end{bmatrix} \quad (4)$$

is the rotation matrix from the inertial frame to the PP frame, $\mathbf{R}_p \in SO(2)$. By definition $\varepsilon = [0, e]^\top$, where $e$ is called the cross-track error to the path and represents the lateral distance to the path-tangent at $\mathbf{p}_d(\theta^*)$ as illustrated in Figure 1. The geometric task objective is achieved if $e \to 0$ as $t \to \infty$, which can be attained by developing a guidance law for the orientation of the velocity vector of the ideal particle.

We obtain the time-derivative of $e$ by differentiating (3) with respect to time:

$$\dot{e} = U\sin(\chi - \chi_t), \quad (5)$$

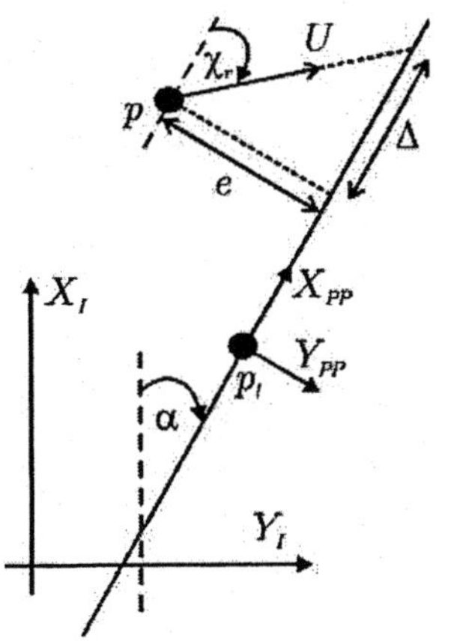

Fig. 2. LOS guidance for a straight line.

and since we are dealing with an ideal particle, we can assign the course angle of the velocity vector to a given desired course angle, i.e. $\chi = \chi_d$. Consequently, (5) can be rewritten as:

$$\dot{e} = U \sin(\chi_d - \chi_t), \tag{6}$$

where $(\chi_d - \chi_t)$ can be considered as a virtual input for stabilizing $e$. Denote this angular difference by $\chi_r = \chi_d - \chi_t$, i.e. the relative angle between the desired course and the path-tangential course. Obviously, such a variable should depend on $e$, such that $\chi_r = \chi_r(e)$. An attractive choice would be a physically motivated line-of-sight (LOS) angle like:

$$\chi_r(e) = \arctan\left(-\frac{e}{\Delta}\right), \tag{7}$$

where $\Delta > 0$ is a guidance parameter shaping the convergence to the path tangential. It is often referred to as the lookahead distance in literature treating path following of straight lines (Papoulias 1991). The physical interpretation of the LOS angle and the lookahead distance can be derived from Figure 2. Note that other shaping functions with arctan-like properties are possible candidates for $\chi_r(e)$, e.g. the tanh function.

When choosing $\chi_r(e)$ as in (7), the cross-track error dynamics becomes:

$$\begin{aligned}\dot{e} &= U \sin(\chi_r) \\ &= \frac{-Ue}{\sqrt{e^2 + \Delta^2}},\end{aligned} \tag{8}$$

from which it can be derived that the origin of $e$ is uniformly globally asymptotically and locally exponentially stable (UGAS/ULES) if the size of the velocity vector is required to be bounded from below, i.e. $U \geq U_{\min} > 0$. This serves as a theoretical justification for generally applying LOS guidance to obtain positional convergence.

To sum up, the desired course angle is given by:

$$\chi_d(\theta^*, e) = \chi_t(\theta^*) + \chi_r(e), \tag{9}$$

with $\chi_t(\theta^*)$ as in (2) and $\chi_r(e)$ as in (7).

By stabilizing the origin of $e$, the path following geometric task is achieved. The dynamic task is satisfied by making sure that $U = U_d \geq U_{d,\min} > 0$, which is a control problem and not a guidance problem. The subject of finding the global minimizer $\theta^*$ that gives $e$ for a general geometrical path is not treated in this paper. Since $e$ can easily be computed directly for straight lines and circles such geometrical topologies are considered.

### *3.2 Guidance for a Straight Line*

Denote an arbitrary point on a straight line as $\mathbf{p}_l \in \mathbb{R}^2$. The line is rotated an angle $\alpha$ and defined relative to the origin of a local reference frame in $\mathbb{R}^2$. The cross-track error can easily be calculated by extracting the second element of:

$$\boldsymbol{\varepsilon} = \mathbf{R}_p^{\mathsf{T}}(\alpha)(\mathbf{p} - \mathbf{p}_l), \tag{10}$$

from which we can calculate $\chi_d$ as:

$$\chi_d(e) = \chi_t + \chi_r(e) = \alpha + \chi_r(e), \tag{11}$$

where $\chi_r(e)$ is given by (7). See Figure 2 for an illustration of this.

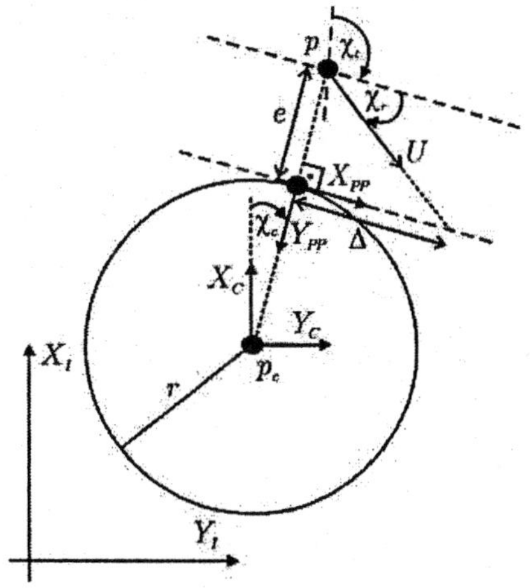

Fig. 3. LOS guidance for a circle.

### *3.3 Guidance for a Circle*

Denote the centre of a circle as $\mathbf{p}_c \in \mathbb{R}^2$. The circle is defined relative to the origin of a local reference frame in $\mathbb{R}^2$. Contrary to a straight line, a circle has non-zero curvature. Consequently, $\chi_t$ changes with time. By inspection of Figure 3, this angle can easily be deduced as either:

$$\chi_t(t) = \chi_c(t) + \frac{\pi}{2} \tag{12}$$

or:

$$\chi_t(t) = \chi_c(t) - \frac{\pi}{2}, \tag{13}$$

where:

$$\chi_c(t) = \arctan\left(\frac{y(t) - y_c}{x(t) - x_c}\right). \tag{14}$$

Equation (12) gives a clockwise circular motion, whilst (13) represents an anti-clockwise motion.

Denote the radius of the circle as $r \in \mathbb{R}_+$. The cross-track error can then be computed by:

$$e(t) = r - \|\mathbf{p}(t) - \mathbf{p}_c\|_2 , \tag{15}$$

as derived from Figure 3. The desired course angle $\chi_d$ is calculated by feeding $e$ into equation (7) and summing the result with (12) or (13).

## 4. DEFINITIONS OF COURSE, HEADING AND SIDESLIP ANGLES

The relationship between the angular variables *course*, *heading* and *sideslip* is important for maneuvering a marine surface vessel. The terms course and heading are used interchangeably in most of the literature on control of marine vessels. Consequently, definitions utilizing a consistent symbolic notation should be established and enforced. The relationship between the angular variables is illustrated in Figure 4, and defined below. In this context, the NED frame is a local geographic reference frame, while the BODY frame is the vessel-fixed reference frame, both as defined in (Fossen 2002).

*Definition 1. Course angle* $\chi$: The angle from the x-axis of the NED frame to the velocity vector of the vessel, positive rotation about the z-axis of the NED frame by the right-hand screw convention.

*Definition 2. Heading (yaw) angle* $\psi$: The angle from the x-axis of the NED frame to the x-axis of the BODY frame, positive rotation about the z-axis of the NED frame by the right-hand screw convention.

*Definition 3. Sideslip (drift) angle* $\beta$: The angle from the x-axis of the BODY frame to the velocity vector of the vessel, positive rotation about the z-axis of the BODY frame by the right-hand screw convention.

By these definitions, it is apparent that

$$\chi = \psi + \beta, \tag{16}$$

where:

$$\beta = \arcsin\left(\frac{v}{U}\right) \overset{\beta\ small}{\Rightarrow} \beta \approx \frac{v}{U}, \tag{17}$$

which is easily verified from Figure 4.

*Remark 4.* (The Society of Naval Architects and Marine Engineers 1950) defines the sideslip angle for marine vessels according to

$$\beta_{SNAME} = -\beta,$$

which can also be found in (Lewis, E.V. (Ed.) 1989). However, the sign convention chosen here follows that of the aircraft community, see e.g. (Nelson 1998) and (Stevens and Lewis 2003), which is more convenient from a guidance point-of-view.

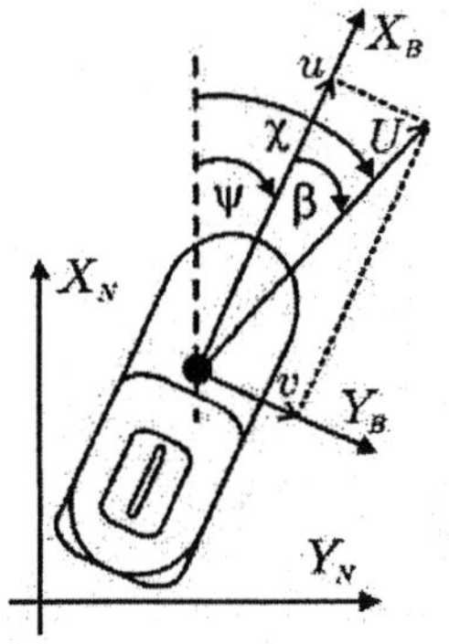


Fig. 4. The geometrical relationship $\chi = \psi + \beta$ between the course ($\chi$), heading ($\psi$) and sideslip ($\beta$) angles.

The heading angle equals the course angle ($\psi = \chi$) when the sway velocity is zero ($v = 0$), i.e. when there is no sideslip. This is generally attainable for fully actuated vessels, but not for those which are unactuated in the sway direction, which affects the design of guidance systems for such vessels. Specifically, to achieve path following for these vessels, the heading angle must be actively used to direct the velocity vector in the desired direction by using so-called sideslip compensation. This means that the desired heading angle must be computed by $\psi_d = \chi_d - \beta$.

## 5. CONTROL SYSTEM DESIGN

It is usually sufficient to consider only the 3 horizontal degrees-of-freedom (DOF) when designing control systems for marine surface vessels. The 3 DOF kinematics and dynamics can be represented as (Fossen 2002):

$$\dot{\boldsymbol{\eta}} = \mathbf{R}(\psi)\boldsymbol{\nu} \tag{18}$$

and:

$$\mathbf{M}\dot{\boldsymbol{\nu}} + \mathbf{C}(\boldsymbol{\nu})\boldsymbol{\nu} + \mathbf{D}(\boldsymbol{\nu})\boldsymbol{\nu} = \boldsymbol{\tau} + \mathbf{R}(\psi)^\top \mathbf{b}, \tag{19}$$

where $\boldsymbol{\eta} = [x, y, \psi]^\top \in \mathbb{R}^3$ represents the earth-fixed position and heading, $\boldsymbol{\nu} = [u, v, r]^\top \in \mathbb{R}^3$ represents the vessel-fixed velocities, $\mathbf{R}(\psi) \in SO(3)$ is the rotation matrix from the earth-fixed local geographic reference frame (NED) to the vessel-fixed reference frame (BODY), $\mathbf{M}$ is the vessel inertia matrix, $\mathbf{C}(\boldsymbol{\nu})$ is the centrifugal and coriolis matrix, $\mathbf{D}(\boldsymbol{\nu})$ is the hydrodynamic damping matrix, $\boldsymbol{\tau}$ represents the vessel-fixed propulsion forces and moments, and $\mathbf{b}$ describes the earth-fixed LF environmental forces acting on the vessel. The system matrices in (19) are assumed to satisfy the properties $\mathbf{M} = \mathbf{M}^\top > 0$, $\mathbf{C} = -\mathbf{C}^\top$ and $\mathbf{D} > 0$. It is further assumed that the vessel has port-starboard symmetry. This is highly valid for most cases, whereas an additional assumption of fore-aft symmetry (Pettersen and Lefeber 2001) is far more questionable because it implies uncoupled sway and yaw dynamics.

A fully actuated vehicle is able to command independent accelerations in all relevant DOFs simultaneously. Consequently, the control vector for a fully actuated marine surface vessel is given by:

$$\boldsymbol{\tau} = [\tau_1, \tau_2, \tau_3]^\top, \tag{20}$$

where $\tau_1$ represents the force input in surge, $\tau_2$ represents the force input in sway and $\tau_3$ represents the moment input in yaw.

The control law design is performed in two steps by using the backstepping technique for nonlinear systems (Krstić *et al.* 1995), and is an extension of the work in (Fossen *et al.* 2003). Specifically, environmental disturbances and a more complex vessel model is considered. Care is also taken to avoid compensating for valuable hydrodynamical damping terms. The main results of the controller design are stated in what follows.

The projection vector $\mathbf{h}$ is defined by:

$$\mathbf{h}^\top \triangleq [0, 0, 1], \tag{21}$$

while the error variables $z_1 \in \mathbb{R}$ and $\mathbf{z}_2 \in \mathbb{R}^3$ are defined by:

$$z_1 \triangleq \psi - \psi_d = \mathbf{h}^\top \boldsymbol{\eta} - \psi_d \tag{22}$$

$$\mathbf{z}_2 \triangleq [z_{2,1}, z_{2,2}, z_{2,3}]^\top = \boldsymbol{\nu} - \boldsymbol{\alpha}, \tag{23}$$

where $\boldsymbol{\alpha} = [\alpha_1, \alpha_2, \alpha_3]^\top \in \mathbb{R}^3$ is a vector of stabilizing functions to be defined later.

The following Control Lyapunov Function (CLF) is considered:

$$V = \frac{1}{2}z_1^2 + \frac{1}{2}\mathbf{z}_2^\top \mathbf{M}\mathbf{z}_2 + \frac{1}{2}\tilde{\mathbf{b}}^\top \boldsymbol{\Gamma}^{-1}\tilde{\mathbf{b}} > 0, \tag{24}$$

where $\tilde{\mathbf{b}}$ is an adaptation error defined as $\tilde{\mathbf{b}} \triangleq \mathbf{b} - \hat{\mathbf{b}}$ with $\hat{\mathbf{b}}$ being the estimate of $\mathbf{b}$. By assumption, $\dot{\mathbf{b}} = \mathbf{0}$. $\boldsymbol{\Gamma} = \boldsymbol{\Gamma}^\top > 0$ is the adaptation gain matrix.

By choosing the control input as:

$$\boldsymbol{\tau} = \mathbf{M}\dot{\boldsymbol{\alpha}} + \mathbf{C}\boldsymbol{\alpha} + \mathbf{D}\boldsymbol{\alpha} - \mathbf{R}^\top \hat{\mathbf{b}} - \mathbf{h}z_1 - \mathbf{K}_2\mathbf{z}_2 \tag{25}$$

and the parameter adaptation as:

$$\dot{\hat{\mathbf{b}}} = \boldsymbol{\Gamma}\mathbf{R}\mathbf{z}_2 \tag{26}$$

we finally obtain:

$$\dot{V} = -k_1 z_1^2 - \mathbf{z}_2^\top(\mathbf{D} + \mathbf{K}_2)\mathbf{z}_2 \leq 0, \tag{27}$$

where $k_1 > 0$, $\mathbf{K}_2 = diag(k_{2,1}, k_{2,2}, k_{2,3}) > 0$ and $\boldsymbol{\alpha} = \left[u_d, 0, -k_1 z_1 + \dot{\psi}_d\right]^\top$. Notice that the inherent damping properties of the system have been preserved.

The main result of the control design is summarized by the following proposition:

*Proposition 5.* For smooth reference trajectories $\psi_d$, $\dot{\psi}_d$ and $\ddot{\psi}_d \in \mathcal{L}_\infty$ and $u_d$, $\dot{u}_d \in \mathcal{L}_\infty$, the origin of the error system $(z_1, \mathbf{z}_2, \tilde{\mathbf{b}})$ becomes uniformly globally asymptotically and locally exponentially stable (UGAS/ULES) by choosing the control and disturbance adaptation laws as in (25) and (26), respectively.

**PROOF.** [Sketch] By collecting the error states in the vector $\mathbf{z} = \left[z_1, \mathbf{z}_2^\top\right]^\top$ and rewriting their dynamics in a manner which is motivated by Theorem A.5 in (Fossen 2002), UGAS/ULES of the origin of the error system can easily be proven.

An underactuated vehicle has fewer independent control inputs available simultaneously than there are number of DOFs to be controlled. Specifically, we consider underactuation in the sway direction:

$$\boldsymbol{\tau} = [\tau_1, 0, \tau_3]^\top, \tag{28}$$

which represents the most common actuator configuration among vessels travelling at high speeds.

Since $\tau_2$ is not available to implement the required terms for the sway direction, dynamics can be imposed on the corresponding stabilizing function such that (25) is still satisfied (Fossen *et al.* 2003). In our particular case, an analysis of the $\alpha_2$-subsystem reveals that the sway speed of the underactuated vessel remains bounded. An analysis of what happens to $v$ is seldom performed for underactuated vessels. Traditionally, this is true for literature treating autopilot design by the Nomoto model (Fossen 2002), but even nonlinear control design concepts disregard the analysis (Lapierre *et al.* 2003).

Sway-underactuated vessels is extensively treated in the literature. Most approaches only consider diagonal system matrices and no environmental disturbances. However, (Do and Pan 2003) lifts these assumptions. Unfortunately, exact path following for an arbitrary point cannot be achieved by this approach since it requires the controlled point to be located where the system matrices become diagonal. Also, $u$ is required to exceed $v$. Such restrictions are not imposed in this paper.

## 6. CASE STUDY: AN UNDERACTUATED MARINE SURFACE VESSEL

To illustrate the performance of the proposed guidance and control scheme, a simulation is performed with an underactuated marine surface vessel trying to follow a straight line path while being exposed to a constant environmental force. The vessel data is taken from the model ship Cybership 2, a 1:70 scale model of a supply vessel, which has a mass of $m = 23.8\ kg$ and a length of $L = 1.255\ m$. See (Skjetne 2004) for the exact model parameters. The vessel is considered to be unactuated in the sway direction, as in (28).

The desired path is a straight line with $\alpha = \frac{\pi}{4}$. It runs through the origin of the NED frame. The environmental disturbance acts perpendicular to the path with a size of about 2.10 $N$. Specifically, $\mathbf{b} = [-1.5\ (N), 1.5\ (N), 0\ (Nm)]^\top$. The initial states are chosen to be $\boldsymbol{\eta}_0 = [10\ (m), 0\ (m), 1.3\ (rad)]^\top$ and $\boldsymbol{\nu}_0 = [0.25\ (m/s), 0\ (m/s), 0\ (rad/s)]^\top$, where the initial surge speed is to be kept during the run. The controller gains are chosen as $k_1 = 10$, $k_{2,1} = 10$, $k_{2,2} = 1$ and $k_{2,3} = 10$ , while $\mathbf{\Gamma} = \mathbf{I}$. The LOS guidance parameter is chosen to be $\triangle = 5L$.

Figure 5 shows that the vessel converges nicely to the path, which would have been impossible without sideslip compensation. Figure 6 illustrates that the cross-track error convergences exponentially to zero.

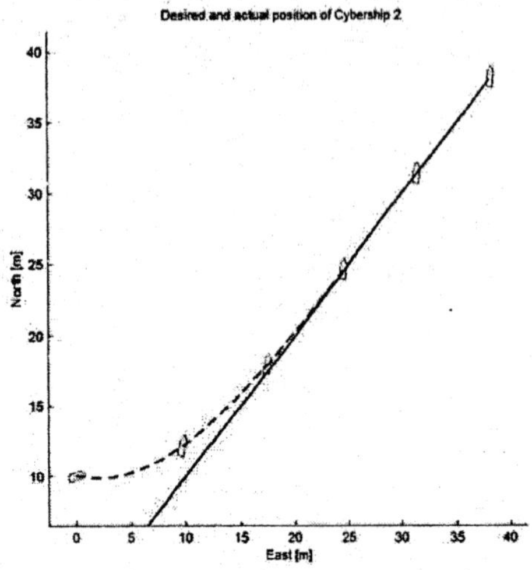


Fig. 5. Cybership 2 converges naturally to the desired path.

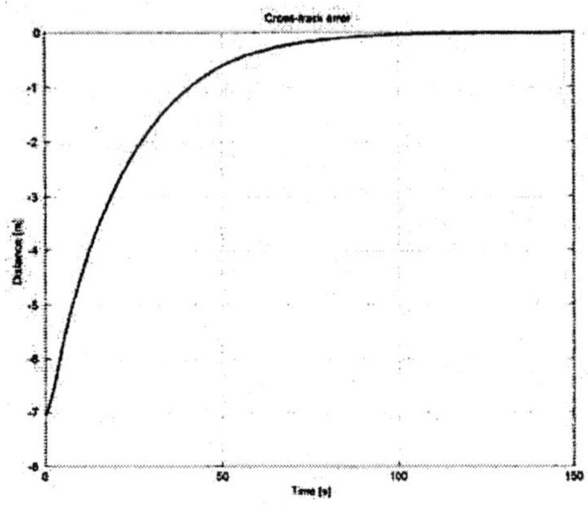


Fig. 6. The cross-track error converges exponentially to zero.

## 7. CONCLUSION

This paper has presented a guidance-based path following approach to maneuvering marine surface vessels. The approach is equally applicable for land, sea and air vehicles and eliminates the two main weaknesses associated with a traditional trajectory tracking scheme. Specifically, a line-of-sight implementation of the approach has been used to illustrate the following of straight lines and circles. The total contribution is a guidance and control scheme which fulfils the path following objective for both fully actuated and underactuated vessels. Simulation results quantitatively confirm the successful performance of the guidance and control strategy. The paper also contains unambiguous definitions of the variables course, heading and sideslip as an attempt to avoid further confusion on the subject. Finally, it should be emphasized that the guidance scheme with sideslip compensation probably could improve the path following behaviour of existing vessels equipped with standard industrial autopilots. In most cases this would only require a software fix to the existing guidance system, which is highly financially viable. This adds an immediate practical flavour to the results.

## REFERENCES

Breivik, M. and T.I. Fossen (2004). Guidance for path following and trajectory tracking. *Journal of Guidance, Control, and Dynamics*. To be submitted.

Do, K.D. and J. Pan (2003). Global tracking control of underactuated ships with off-diagonal terms. In: *Proceedings of the 42nd IEEE CDC, Maui, Hawaii, USA*.

Fossen, T.I. (2002). *Marine Control Systems: Guidance, Navigation and Control of Ships, Rigs and Underwater Vehicles*. 1st ed.. Marine Cybernetics.

Fossen, T.I., M. Breivik and R. Skjetne (2003). Line-of-sight path following of underactuated marine craft. In: *Proceedings of the 6th IFAC MCMC, Girona, Spain*.

Krstić, M., I. Kanellakopoulos and P.V. Kokotović (1995). *Nonlinear and Adaptive Control Design*. John Wiley & Sons Inc.

Lapierre, L., D. Soetanto and A. Pascoal (2003). Nonlinear path following with applications to the control of autonomous underwater vehicles. In: *Proceedings of the 42nd IEEE CDC, Maui, Hawaii, USA*.

Lewis, E.V. (Ed.) (1989). *Principles of Naval Architecture*. Vol. III. The Society of Naval Architects and Marine Engineers.

Nelson, R.C. (1998). *Flight Stability and Automatic Control*. 2nd ed.. McGraw-Hill.

Papoulias, F.A. (1991). Bifurcation analysis of line of sight vehicle guidance using sliding modes. *International Journal of Bifurcation and Chaos* **1**(4), 849–865.

Pettersen, K.Y. and E. Lefeber (2001). Way-point tracking control of ships. In: *Proceedings of the 40th IEEE CDC, Orlando, Florida, USA*.

Skjetne, R. (2004). A nonlinear ship manoeuvering model: Identification and adaptive control with experiments for a model ship. *Modeling, Identification and Control* **25**(1), 3–27.

Stevens, B.L. and F.L. Lewis (2003). *Aircraft Control and Simulation*. 2nd ed.. John Wiley & Sons Inc.

The Society of Naval Architects and Marine Engineers (1950). Nomenclature for treating the motion of a submerged body through a fluid. Technical and Research Bulletin No. 1-5.

ELSEVIER
IFAC
PUBLICATIONS
www.elsevier.com/locate/ifac

# GEOMETRIC SHIP TRACK FOLLOWING CONTROL

**G. Hearns, R. I. Stephens**

*Alstom Power Conversion Ltd, Rugby, UK*

Abstract: The development of a controller, which sails a ship along a track, and through a specified heading change is discussed. The controller uses a prediction distance obtained from ship tests to start turning in advance of the next track. A feedback controller looking ahead along the track generates a heading reference for the autopilot. The performance and robustness of the geometric ship track following controller is examined. *Copyright © 2004 IFAC*

Keywords: ship control, marine systems, tracking

## 1. INTRODUCTION

Ship control systems now include capabilities for track-following and way-point tracking in which arbitrary tracks can be followed. For operations such as cable laying at sea or channel dredging a high accuracy is required for track keeping. In these applications a mode of operation known as "auto sail" is used, where the only actuators available for track-following are the main propellers (providing forward thrust) and a rudder or rudders (providing side thrust at the stern and hence turning moment). The ability to follow a track will be determined by the turning moment available from the rudder and the rate at which the rudder can move, as well as the characteristics of the vessel. The track is usually initially defined as a series of straight lines. Clearly, it is not possible to follow such a track exactly. The aim of the auto sail controller is to construct a "feasible" track around a corner which approaches the apex as closely as possible, but which is within the capabilities of the available actuation.

Previously the authors have reported (Hearns *et al*, 2003) on a model based approach, which generates a smooth curve around a corner. The curve is feasible for the vessel to follow and physically based feedback/feed-forward controllers generate a heading reference. While this approach has proved effective, it relies on having accurate model knowledge of the vessel. In this paper, a combined approach is described which uses tests to determine the turning capabilities of the vessel, whilst geometric rules provide feedback control during the turn. Previously proposed track construction techniques have included: a circle of infinite radius, which decreases as the course changes (Husa and Fossen, 1997); cubic spline interpolation between way-points (Betin and Branca, 2000) and prediction of the distance before the entry and exit of the track to move the rudder (Holzhuter and Schultze, 1996).

## 2. WHEEL OVER DISTANCE

If the autopilot receives a step change in heading reference the vessel will sail some distance forward before settling on the new heading. If this distance can be predicted before the turn then the heading can be changed in advance such that the vessel ends up on the desired new track. The wheel over point (Fig. 1) is defined as the distance $X_{WOP}$ before the apex of the corner that the auto-pilot heading reference should be set to the heading of the second track leg such that under ideal conditions the vessel will end up sailing along the track with the correct heading. This gives a measure of the minimum distance before the corner that the vessel must be turned. It will depend on the autopilot parameters, turning moment, vessel speed and the heading change demanded. Assuming the rudder parameters are constant then carrying out vessel manoeuvres at two speeds and with two tracks will give the minimum four points to construct a lookup table. Such information is usually readily available from manoeuvrability tests and charts (Noel, 1988). When the vessel is settled on its new course then using one measured position of the vessel $(N_T, E_T)$ the wheel over distance is:

$$X_{WOP} = N_r - \frac{E_r}{\tan(\theta_{T2} - \theta_{T1})} \quad (1)$$

where $(N_r, E_r)$ is the position $(N_T, E_T)$ rotated anti-clockwise by $\theta_{T1}$:

$$N_r = (N_T - N_{WOP})\cos(\theta_{T1}) + (E_T - E_{WOP})\sin(\theta_{T1}) \quad (2)$$

$$E_r = (E_T - E_{WOP})\cos(\theta_{T1}) - (N_T - N_{WOP})\sin(\theta_{T1}) \quad (3)$$

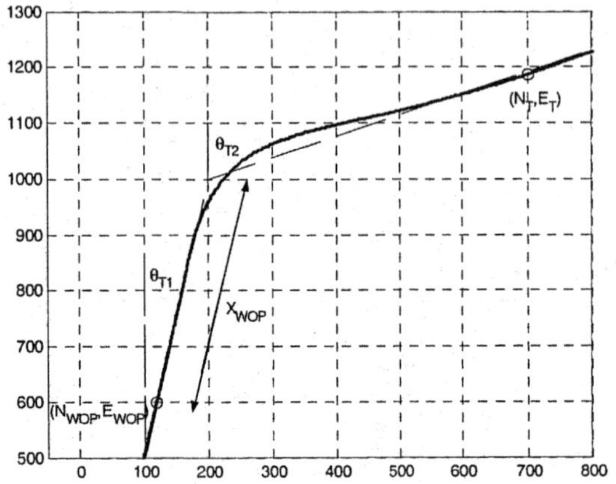


Fig. 1. Definition of wheel over point.

## 3. GEOMETRIC HEADING REFERENCE GENERATION

Using the straight-line track geometry and the position of the vessel, simple algorithms can be constructed which will calculate the heading reference necessary to guide the vessel onto the second leg of the track. The most important algorithm property is as the vessel gets closer to the track the heading reference converges with the angle of the track. Three algorithms will be defined all of which use the position of the vessel and the geometry of the tracks second leg. No information is required regarding the geometry of the first track leg and there are no assumptions about the vessel starting position. If the vessel is approaching the way-point close to the first track leg then the turning algorithm should be started at or before the wheel over point $X_{WOP}$.

### *3.1 Algorithm 1*

From the vessel, project parallel to the second leg a distance $L_p$ and then perpendicular until the track is intersected (Fig. 13). The heading reference is the angle of the line from the vessel to the track intersection. The off track distance is:

$$y_{OT} = \left(N - N_{wp}\right)\sin\left(\theta_{T2}\right) - \left(E - E_{wp}\right)\cos\left(\theta_{T2}\right) \quad (4)$$

and the heading reference:

$$\psi_{ref}(N,E) = \theta_{T2} - \tan^{-1}\left(\frac{-y_{OT}}{L_p}\right) \quad (5)$$

When $y_{OT} = 0$ then $\psi_{ref}(N,E) = \theta_{T2}$, the track heading. Figure 2 plots $\psi_{ref}(N,E)$ for all vessel positions with the contours of equal heading reference, which are parallel to the second leg of the track.

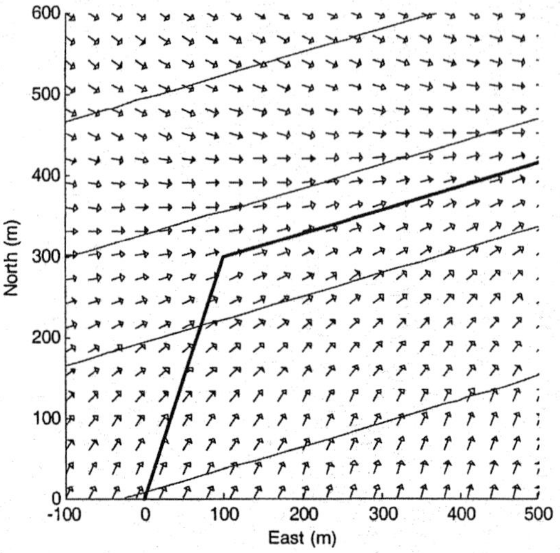


Fig. 2. Heading reference for Alg. 1 with $L_p$ =250m.

### *3.2 Algorithm 2*

The heading reference is the angle of the line of length $L_{wp} + L_p$ connecting the vessel to the second leg of the track where $L_{wp}$ is the distance from the apex of the track to the vessel and $L_p$ is a fixed predicting distance (Fig. 14).

$$L_{wp} = \sqrt{\left(N - N_{wp}\right)^2 + \left(E - E_{wp}\right)^2} \quad (6)$$

$$\alpha = \tan^{-1}\left(\left(E - E_{wp}\right), \left(N - N_{wp}\right)\right) \quad (7)$$

$$\beta = \sin^{-1}\left(\frac{L_{wp}}{L_{wp} + L_p}\sin(\alpha - \theta_{T2})\right) \quad (8)$$

$$\psi_{ref}(N,E) = \theta_{T2} - \beta \quad (9)$$

The heading reference always points at the second track leg which is defined as starting from $(N_{wp}, E_{wp})$ therefore the reference will be in the correct direction regardless of the choice of $L_p$. When the vessel is on the second leg of the track then $\beta = 0$ and $\psi_{ref}(N,E) = \theta_{T2}$, the track heading. The heading reference plotted in Fig. 3 for the vessel position shows that the contours of equal reference are focussed towards the track apex.

### *3.3 Algorithm 3*

The heading reference is the angle of the line of length $L_p$ connecting the vessel to the second leg of the track. This is equivalent to the so-called line-of-

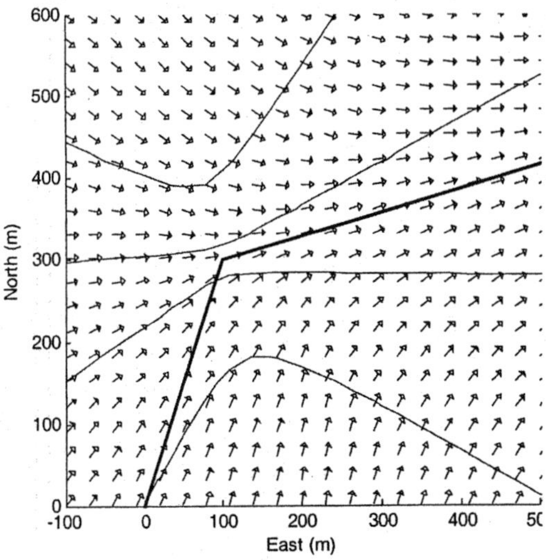


Fig. 3. Heading reference for Alg. 2 with $L_p = 50$m.

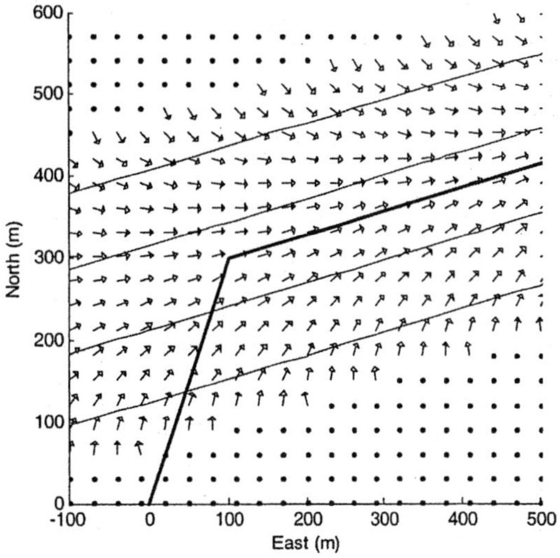


Fig. 4. Heading reference for Alg. 3 with $L_p$ =200m.

sight control (Fossen, 2002). The second leg of the track has the equation $N = m_{T2}E + c_{T2}$ where $m_{T2} = \tan(\pi/2 - \theta_{T2})$ and $c_{T2} = N_{wp} - m_{T2}E_{wp}$. $(N_i, E_i)$ can be found by finding where the track intersects the circle with centre $(N, E)$ and radius $L_p$ (Fig. 15).

$$E_i = \frac{-b + sign(\theta_{T2})\sqrt{b^2 - 4ac}}{2a} \quad (10)$$

$$N_i = m_{T2}E_i + c_{T2} \quad (11)$$

Where $a = 1 + m_{T2}^2$, $b = 2m_{T2}(c_{T2} - N) - 2E$,

$c = E^2 + (c_{T2} - N)^2 - L_p^2$

The heading reference is:

$$\psi_{ref}(N,E) = \tan^{-1}((E_i - E),(N_i - N)) \quad (12)$$

The heading reference is only defined if the off-track distance is less than $L_p$. When the vessel is on the second leg of the track then

$(N_i, E_i) = (N + L_p \cos\theta_{T2}, E + L_p \sin\theta_{T2})$

and therefore $\psi_{ref}(N,E) = \theta_{T2}$. The contours of equal reference are again parallel to the second leg of the track.

## 4. AUTOSAIL PERFORMANCE

All algorithms have one adjustable parameter, the prediction distance $L_p$. In order that $L_p$ will adjust with vessel speed it is natural to make $L_p$ a multiple of the wheel over distance $X_{WOP}$. If the turning controller is turned on when the vessel is within $X_{WOP}$ of the corner then the initial heading reference should equal the track reference if the vessel is to turn the corner in the minimum time. Figure 5 plots the simulated vessel (dredger, displacement 11000 tonnes) trajectories when it starts turning at $1.25X_{WOP}$ and with $L_p$ factors of [1.1, 1.2, 1.3, 1.5, 2, 2.5, 3] for Alg. 1, [0.4, 0.5, 0.6, 0.7, 0.9, 1, 1.25] for Alg. 2 and [0.85, 0.9, 0.95, 1, 1.25, 1.5, 2] for Alg. 3. The overall effect of decreasing $L_p$ is to decrease the time constant of the feedback control and increase any track overshoot. The vessel trajectory for Alg. 1 changes uniformly with $L_p$ whereas for Alg. 2 the initial part of the trajectory remains constant. Alg. 3 is very sensitive to $L_p$ less than the wheel over distance used. Assuming that $X_{WOP}$ is always overestimated then it is safe to use $L_p = 2X_{WOP}$ despite the conservative approach the vessel will make to the track.

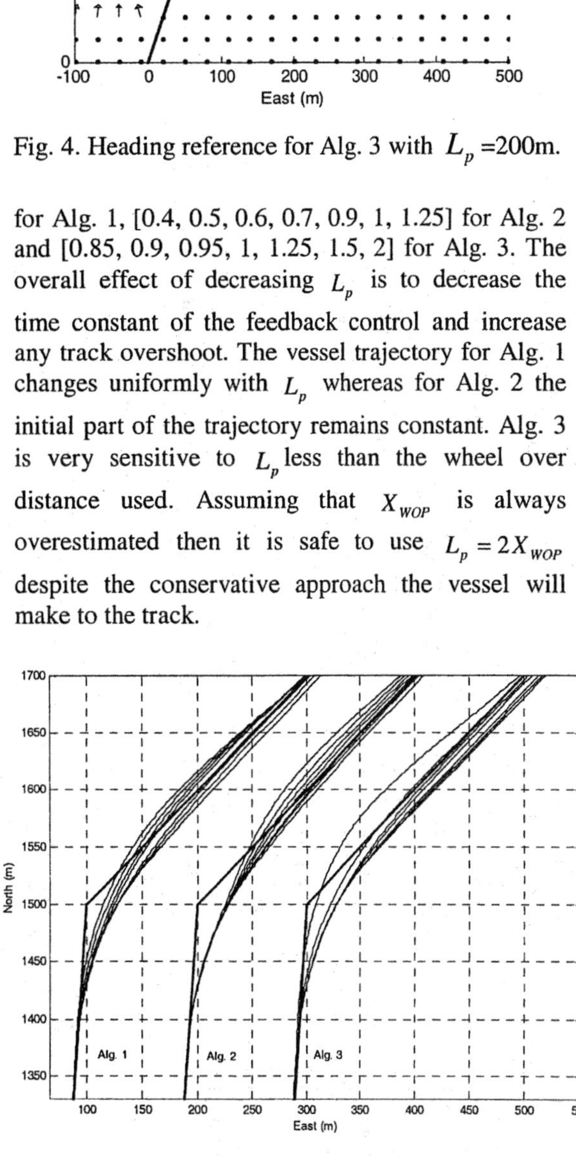


Fig. 5. Vessel tracks with varying $L_p$.

Since $X_{WOP}$ is essentially a physical measure of the turning performance of a vessel then it is useful to compare the performance with the autosail controller using the model-based track (Hearns *et al*, 2003). Only an approximate comparison can be made since the geometric controller depends on $L_p$ while the model based track depends on the maximum available turning moment that is specified. Figure 7 shows that for each controller the vessel starts turning at roughly the same point. The fact that the geometric controller overshoots while the model-based controller undershoots is simply a consequence of the controller tuning. The heading references (Fig. 6) show that the model-based reference is very close to the limitations of the vessel while the initial jump in the geometric reference will have no effect on the vessel's response but the response in the overshoot region is certainly smoother.

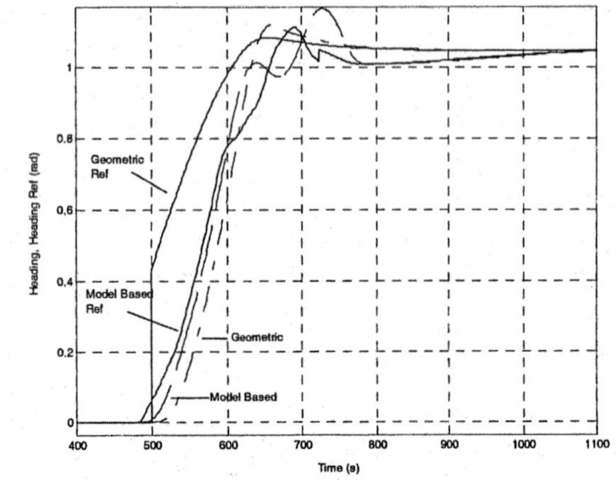


Fig. 6. Vessel heading for model based track following and geometric heading reference (Alg. 3).

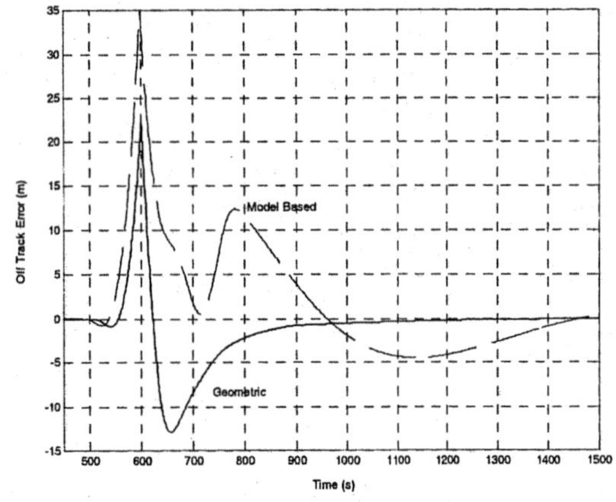


Fig. 7. Off track error for model based track following and geometric heading reference (Alg. 3).

## 5. TIDE ESTIMATION

The algorithms described are essentially proportional control leading to steady state off-track error. To compensate for steady state disturbances from wind, waves, current and tide, a "tide identifier" is used to provide a feed-forward heading reference. In one sample time $(\tau_s)$ the vessel will move forward a distance $\Delta x = \tau_s u$ (Fig. 8). The movement perpendicular to the current heading due to the vessel turning can be approximated by:

$$\Delta y_R = \Delta x \tan(\psi_{k+1} - \psi_k) \tag{13}$$

The movement due to a tide disturbance is:

$$\Delta y_T = \Delta x \tan(\alpha_{k,k+1} - \psi_k) \tag{14}$$

where $\alpha_{k,k+1} = \tan^{-1}((E_{k+1} - E_k),(N_{k+1} - N_k))$ (15)

With a suitable observer gain $L$ the across track tide velocity estimate can be updated with:

$$v_T^{k+1} = v_T^k + L(\Delta y_T - \Delta y_R - \tau_s v_T^k) \tag{16}$$

and the crab angle to counteract the tide is:

$$\beta = \tan^{-1}\left(\frac{v_T^k}{u_k}\right). \tag{17}$$

This simple tide identifier is independent of any track being followed but does not compensate for any vessel movement due to the sway thrust applied by the rudder. Holding the tide estimate and crab angle constant while changing heading does slow down the off track error correction but has the advantage that any transients in the tide estimation, especially when there is no physical tide to estimate, will not be passed to the autopilot.

Figures 9 and 10 show the performance of the held tide identifier. The step in the heading reference when the crab angle is switched to the new steady-state tide estimate can be seen. There is obviously much scope to improve the tide estimation going around a corner by using the full vessel dynamics and an optimal observer. If the magnitude and heading of the tide remains constant then knowing the steady-state tide magnitude estimate (relative to the vessel) for two or more vessel headings will enable the absolute tide magnitude and heading to be calculated. Thus the tide magnitude relative to the vessel would be known for further vessel manoeuvres.

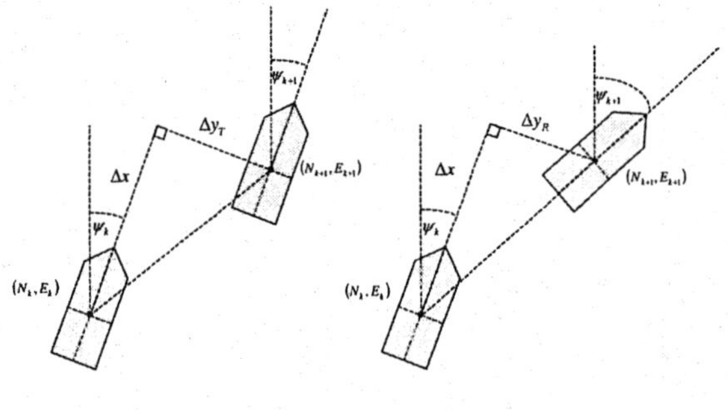

Fig. 8. Vessel translation due to tide and due to heading change

## 6. DOUBLE CORNERS

So far only a track with two legs has been considered. When there are two corners close together then the autosail controller should still work seamlessly. Figure 11 shows the heading reference generated by algorithm 3 with $X_{WOP} = 200m$ and $L_P = 2X_{WOP}$. The intersection between the two direction changes occurs $X_{WOP}$ before the second apex. There would be a discontinuity between the references if $L_p$ was not large enough. Figure 12 verifies that the vessel can turn the second corner while still finishing the first corner.

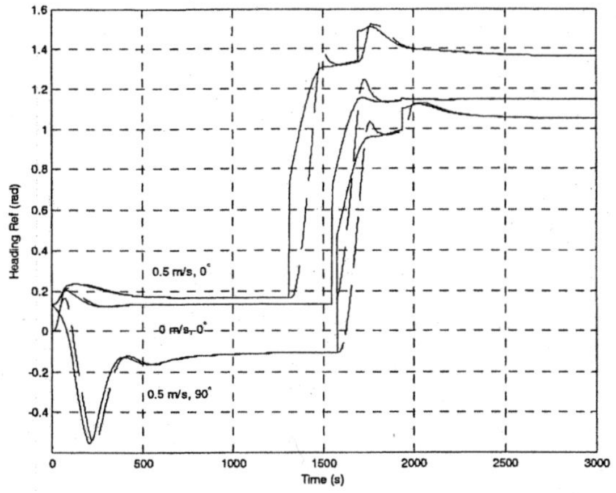


Fig. 9. Heading Reference (-) and heading (---) with tide identifier held during turn.

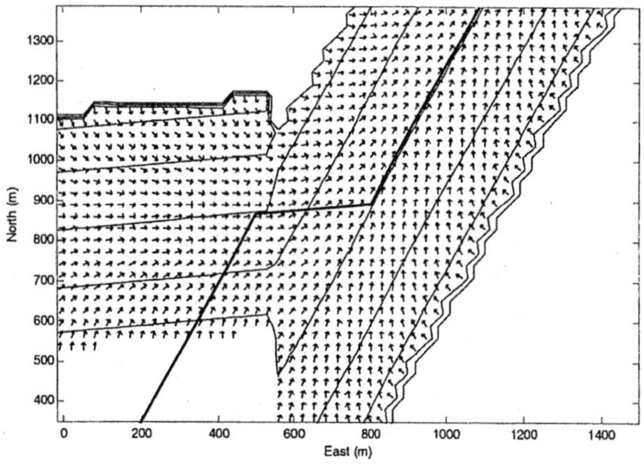


Fig. 11. Heading reference for Alg. 3 with $L_p$ =200m.

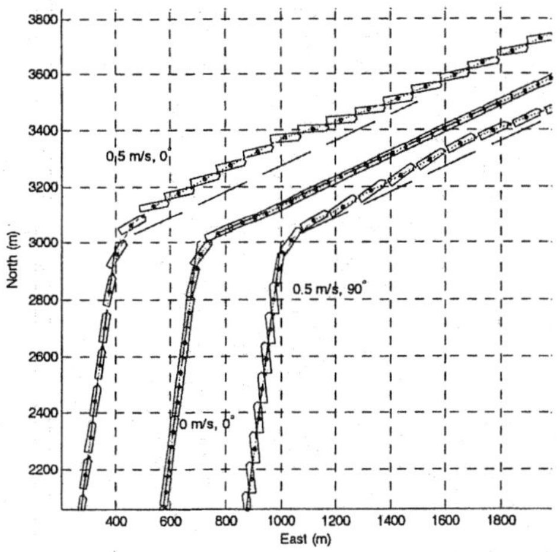


Fig. 10. Vessel Tracks with tide identifier held during turn.

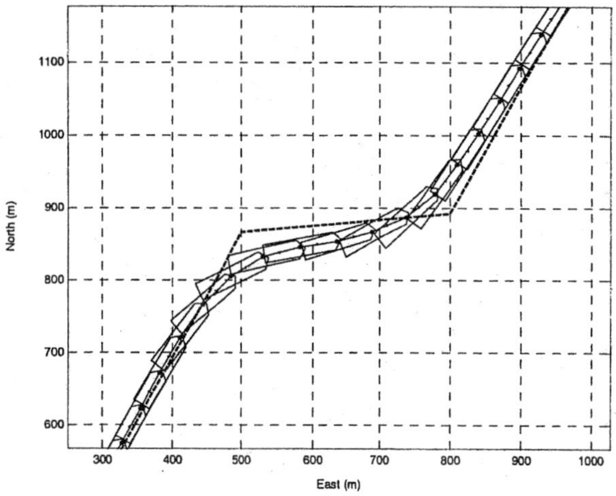


Fig. 12. Vessel positions every 25s for speed 2.5m/s.

## 7. CONCLUSIONS

While the previous work on generating a feasible track and autosail controller using vessel model parameters has provided a valuable insight it is obvious that the geometric approach of predicting ahead some distance is inherently more robust due to its simplicity. Since it is based on vessel tests then it also provides a controller that works within the physical limitations of the vessel. Of the three geometric algorithms the third one would appear to offer good performance and the flexibility to work on any track and not just one defined by straight lines. Only two parameters, $X_{WOP}$ and $L_p$ affect the performance of the geometric controllers. Future investigation should address the selection of $L_p$ and how accurate the lookup table for $X_{WOP}$ should be. Compared to standard Line Of Sight algorithms this approach incorporates data from ship tests for the wheel over distance, integrates with a tide identifier and provides tuning flexibility.

## REFERENCES

Bertin, D. and Branca, L., "Operational and design aspects of a precision minewarfare autopilot", *Warship 2000*, London, 2000.

Husa, K. E. and Fossen T. I., "Backstepping Designs for Nonlinear Way-point Tracking of Ships", *4th IFAC Conference on Manoeuvring and Control of Marine Craft*, 1997.

Hearns G., Stephens R. I. and Wilkins A. J. H., "Ship track-following control and trajectory generation", *16th Int. Conference in Systems Engineering*, Coventry, September 9-11, 2003, pp. 249-254.

Holzhuter, T. and Schultze R., "Operating experience with a high-precision track controller for commercial ships", *Control Engineering Practice* **4** (3), pp. 343-350, 1996.

Fossen, T. I., *Marine Control Systems: Guidance Navigation and Control of Ships, Rigs and Underwater Vehicles*, Marine Cybernetics, Trondheim, Norway, 2002.

Noel, J. V., *Knight's Modern Seamanship* 18th ed., Van Nostrand Reinhold, New York, 1988.

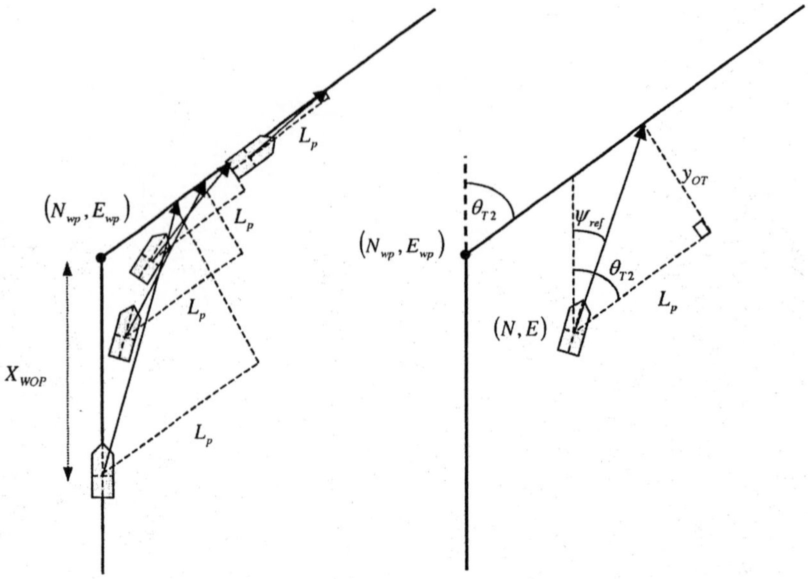


Fig. 13 Predicting ahead on second leg of track.

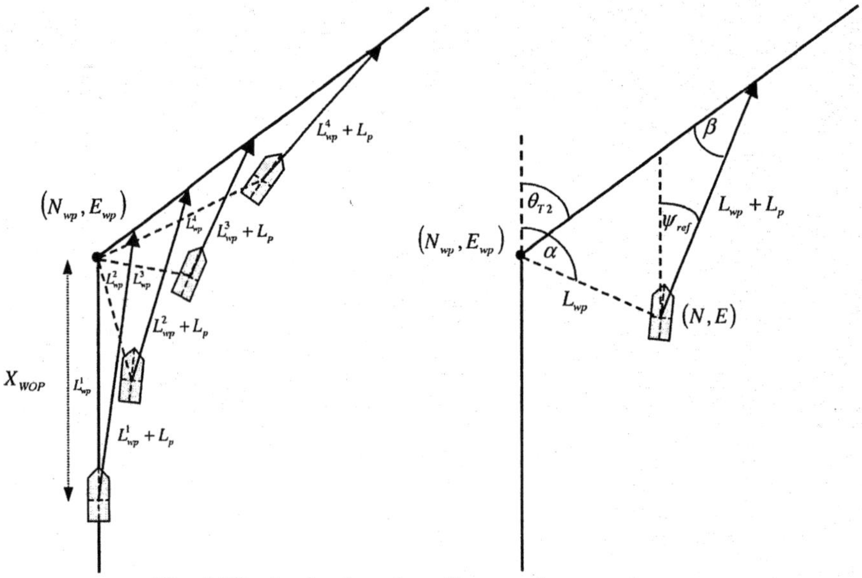


Fig. 14 Projecting based on distance from track apex.

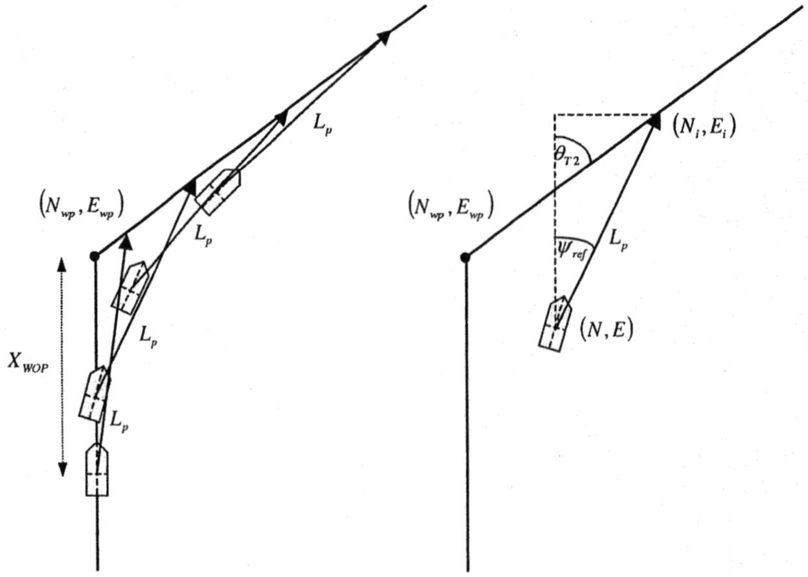


Fig. 15 Projecting shortest distance to track.

# DEVELOPMENT OF AUTOMATIC COURSE MODIFICATION SYSTEM USING FUZZY INFERENCE

**Yoshitaka Furukawa * Katsuro Kijima * Hiroshi Ibaragi ***

** Kyushu University, Fukuoka, Japan*

Abstract: Navigational safety is highly demanded in order to prevent marine accidents. The authors had already proposed collision avoidance systems using fuzzy inference to evaluate the risk of collision and to define countermeasures in order to avoid collision with other ships. In this paper, the authors introduce experimental system which was built for evaluation of automatic navigation system. As the first step of model experiment of the automatic navigation system, results of model test for course modification activated by steering control using fuzzy inference is shown. The results represent that the automatic rudder control system for course modification works well for model ships. *Copyright © 2004 IFAC*

Keywords: Automatic systems, Fuzzy inference, Manoeuvrability, Navigation systems, Ship control

## 1. INTRODUCTION

Navigational safety is highly demanded in order to prevent marine accidents. However reduction of personnel expenses is enforced recently to reduce total transportation cost and it means that securement of the sailors who have an excellent skill becomes difficult. So increase of sea disaster accident originated with degradation of skill of a sailor is concerned in the future and introduction of automatic navigation device is the one of solutions of such a problem.

In order to realize the automatic navigation device, much information concerning not only own ship but also surrounding ships are required. Recently precision of positioning using GPS becomes considerably better and guideline for the installation of the Automatic Identification System (AIS) for ships are adopted by the International Maritime Organization (IMO). Using these devices, each ship will be able to obtain her own position exactly and easily derive other ship's position, heading, speed and so on.

The collision avoidance system is the one of important functions comprised in automatic navigation system and there are many studies (Hara and Hammer, 1993) (Hasegawa and Fujita, 1993) (Lee and Rhee, 2001) (Kose *et al.*, 1998) (Zhuo and Hearn, 2003). *DCPA* (distance of closest point of approach) and *TCPA* (time to closest point of approach) are often used as parameters to evaluate the degree of collision risk.

The authors had proposed automatic collision avoidance systems for ships using *DCPA* and *TCPA* (Kijima and Furukawa, 2001) (Kijima and Furukawa, 2002) and the concept of inherent blocking area to evaluate collision risk among ships (Kijima and Furukawa, 2003). The systems were designed based on numerical simulations, then verification of effectiveness of the systems is necessary.

In this paper, the authors introduce experimental system which is developed to evaluate automatic navigation system. As the first step of model experiment of the automatic navigation system, results of model test for course modification activated by steering control using fuzzy inference is shown. The results represent that the automatic rudder control system for course modification works well for model ships.

## 2. DEVELOPMENT OF MODEL EXPERIMENT SYSTEM

The authors developed model experiment system in the seakeeping and manoeuvring basin at Kyushu University for the verification of automatic navigation system. The function of the model experiment system is to measure the position and heading of a model ship and control the motion of the model ship according to automatic control system. Figure 1 shows schematic diagram of the system. The experimental system consists of a CCD video camera, a optical fiber gyrocompass and two PCs. One PC is used for image processing to obtain model ship position and the another PC is used for control operation.

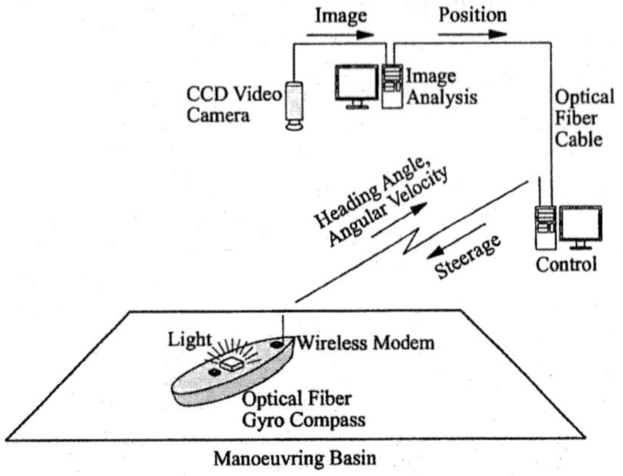


Fig. 1. Overview of the experimental system

A CCD video camera is installed at the position of 13m high from the surface of the water in the basin. The resolution of the CCD video camera is 1024 pixels × 1024 pixels and recorded bitmap picture is correspond to the area of 24m square. Then resolution of position measurement is about 2.3cm.

Light spot which has the area of 10cm square is installed on the midship of a model ship and this light spot is regarded as the position of the model ship. The center position of the light spot is obtained by image processing of the recorded bitmap picture. The heading angle and yaw rate of the model ship is provided by the optical fiber gyroscope which is installed on the model ship.

The position of the model ship which is provided by the image processing PC is transfered to the control operation PC via optical fiber cable. The heading and yaw rate of the model ship are transmitted by a radio modem from the model ship to the PC for control operation. Suitable steering is obtained by using these information and control signal for a rudder is transmitted by the radio modem.

## 3. RUDDER CONTROL USING FUZZY INFERENCE

A rudder angle which is required to lead a ship to new route smoothly and swiftly is obtained using fuzzy inference because there are many parameters which should be taken into account for rudder control.

Nondimensional lateral distance between initial and new course, $l_t/L$ ($L$ : ship length), a difference between current and new heading, $\psi_t$, and current yaw rate in nondimensional form, $r'$ ($= rL/U$, $r$ : yaw rate, $U$ : ship speed), are utilized as input parameters (Kijima and Furukawa, 2003). They are illustrated in Figure 2.

Figure 3 represents inference process of a rudder angle. The authors divided the inference process into two phases to simplify fuzzy rule. In the first phase, preferable yaw rate, $r'_d$, is reasoned using $l_t/L$ and $\psi_t$. Then a rudder angle, $\delta$, which realizes smooth and swift course modification is inferred using a difference between preferable and current yaw rate, $\Delta r' (= r'_d - r')$, in the second phase.

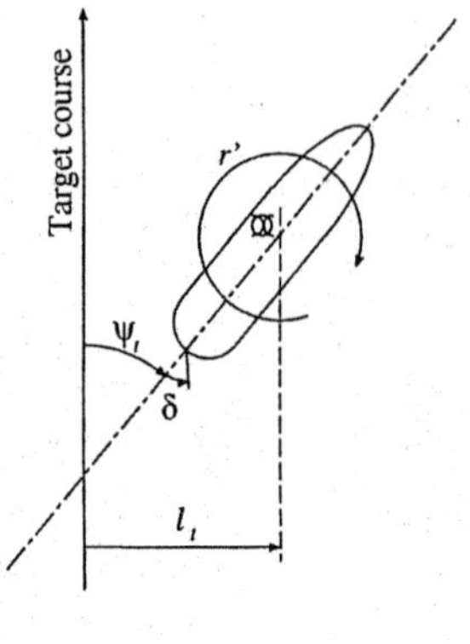


Fig. 2. Parameters used in fuzzy inference for a rudder angle

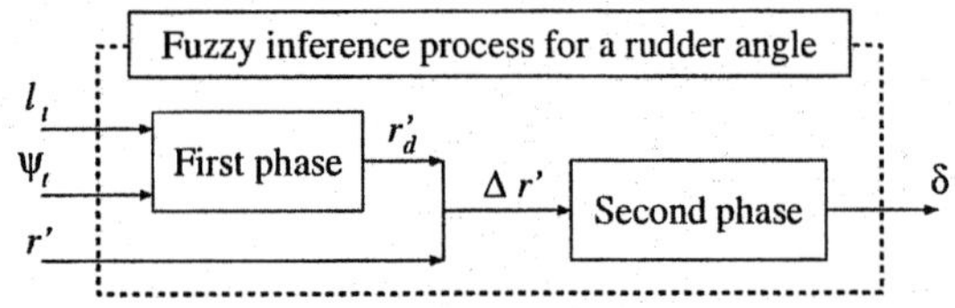


Fig. 3. Fuzzy inference process for a rudder angle

Rule table for the first phase of fuzzy inference is shown in Table 1. Three fuzzy sets such as "N : negative", "ZO : zero" and "P : positive" are used for $l_t/L$ and $r'_d$. As for $\psi_t$, five fuzzy sets indicated with "NB : negative big", "NS : negative small", "ZO", "PS : positive small" and "PB : positive big" are utilized. Membership functions for antecedent and consequent parts are shown in Figure 4. $K_y$ and $K_r$ in Figure 4 are important parameters which have much influence upon control performance, so their values are assigned after the investigation of their effect upon course modification through numerical simulations.

Table 2 and Figure 5 are rule table and membership functions for the second phase of fuzzy inference. Three fuzzy sets such as "N", "ZO" and "P" are used both for $\Delta r'$ and $\delta$. A rudder angle $\delta$ has an output range from $-35°$ to $35°$.

Table 1. Rules for the first phase of fuzzy inference

| | | $\psi_t$ | | | | |
|---|---|---|---|---|---|---|
| | | NB | NS | ZO | PS | PB |
| $l_t/L$ | N | P | P | P | ZO | N |
| | ZO | P | P | ZO | N | N |
| | P | P | ZO | N | N | N |

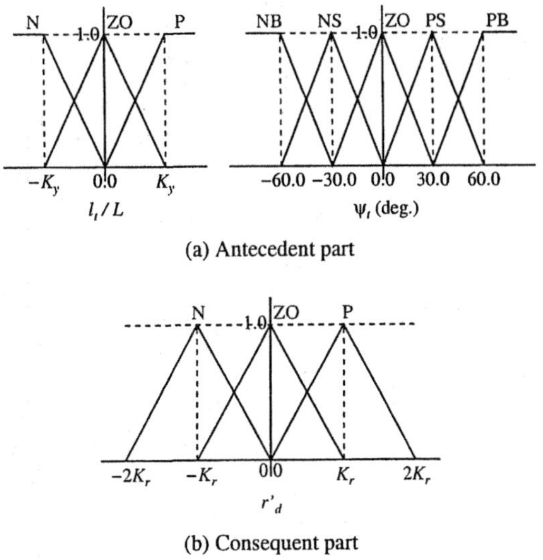


Fig. 4. Membership functions for the first phase of fuzzy inference

Table 2. Rules for the second phase of fuzzy inference

| $\Delta r'$ | | |
|---|---|---|
| N | ZO | P |
| N | ZO | P |

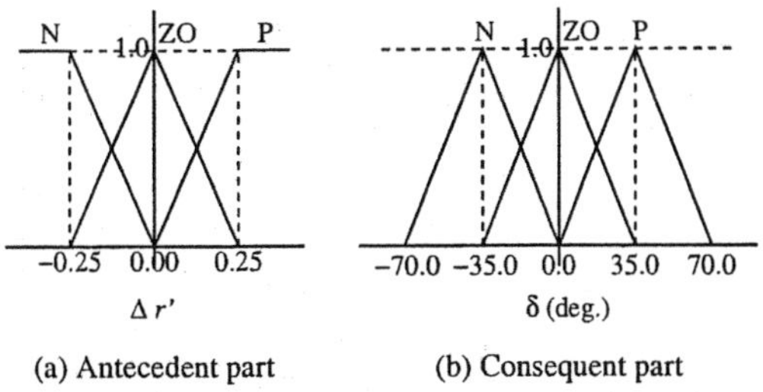


Fig. 5. Membership functions for the second phase of fuzzy inference

## 4. INFLUENCE OF FUZZY INFERENCE PARAMETER UPON COURSE MODIFICATION

The effect of the parameters $K_y$ and $K_r$ in fuzzy inference for rudder control upon course modification are investigated by numerical simulations. The equations of surging, swaying and yawing motion of a ship used in the numerical simulations can be written in the following nondimensional form.

$$\left.\begin{aligned} &(m'+m'_x)\left(\frac{L}{U}\right)\left(\frac{\dot{U}}{U}\cos\beta - \dot{\beta}\sin\beta\right) + (m'+m'_y)r'\sin\beta = X', \\ &-(m'+m'_y)\left(\frac{L}{U}\right)\left(\frac{\dot{U}}{U}\sin\beta + \dot{\beta}\cos\beta\right) + (m'+m'_x)r'\cos\beta = Y', \\ &(I'_{zz}+i'_{zz})\left(\frac{L}{U}\right)^2\left(\frac{\dot{U}}{L}r' + \frac{U}{L}\dot{r}'\right) = N', \end{aligned}\right\} \tag{1}$$

where $\beta$ is drift angle and " $'$ " indicates nondimensional quantities. $m$ is ship mass, $m_x$ and $m_y$ are longitudinal and lateral components of added mass and $I_{zz}$ and $i_{zz}$ are momoent and added moment of inertial of a ship respectively. It is assumed that nondimensional external forces $X'$, $Y'$ and moment $N'$ consist of hull, propeller and rudder components noted with subscripts "$_H$", "$_P$" and "$_R$".

$$\left.\begin{aligned} X' &= X'_H + X'_P + X'_R, \\ Y' &= Y'_H + Y'_R, \\ N' &= N'_H + N'_R. \end{aligned}\right\} \tag{2}$$

Detailed expression for each component and approximate formulae for hydrodynamic coefficients are shown in the reference (Kijima and Nakiri, 2003).

Principal particulars of the model ship which is used in the simulation study are shown in Table 3. Ship type is a container ship. Two kinds of initial angle between ship's heading and target route, $\psi_{t0}$, are specified such as 15° and 55° and initial ship speed, $U_0$, is 0.35(m/sec).

Table 3. Principal particulars of model ship

| | | |
|---|---|---|
| Length | $L$ (m) | 2.5000 |
| Breadth | $B$ (m) | 0.4294 |
| Draft | $d$ (m) | 0.1547 |
| Block Coef. | $C_b$ | 0.8252 |

Figure 6 shows change of ship trajectories depending on the value of $K_y$ and $K_r$. Initial angle between ship heading and target route is 15° in this figure. Figures (a), (b) and (c) corresponds to the condition of $K_r = 0.2$, 0.4 and 0.6 and each line type expresses trajectory for the condition of $K_y = 2.0L$, $3.0L$ and $4.0L$ respectively. Straight broken line in the figures indicates target route.

$K_y$ is a parameter which has close relation with lateral distance from target route. As the value of $K_y$ becomes smaller, larger alteration in course is made. On the other hand, remarkable difference in trajectories depending on the parameter $K_r$ which prescribe yaw rate is not seen. However, overshoot in a trajectory appears when the value of $K_r$ is small.

Ship trajectories obtained by simulations for $\psi_{t0} = 55°$ are shown in Figure 7. Tendency of the effect of $K_y$ and $K_r$ upon course changing performance found in this figure is almost same as Figure 6.

It is concluded from these investigations that $K_y$ should have small value and $K_r$ should have big value in order to achieve swift course change without overshoot. Therefore The authors decided to adopt $K_y = 2.0L$ and $K_r = 0.6$ in this paper.

## 5. VERIFICATION OF THE COURSE MODIFICATION SYSTEM BY MODEL TEST

Model tests were carried out to verify the adopted values of the parameters $K_y$ and $K_r$ in fuzzy inference for steering control. On the model test, it was assumed that target route is straight line and initial position of

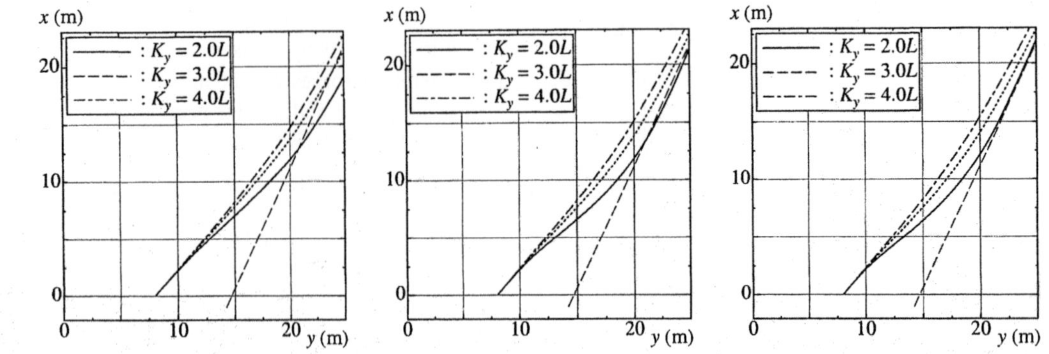


(a) $K_r = 0.2$ (b) $K_r = 0.4$ (c) $K_r = 0.6$

Fig. 6. Effects of $K_y$ on ship motion ($\psi_{t0} = 15°$)

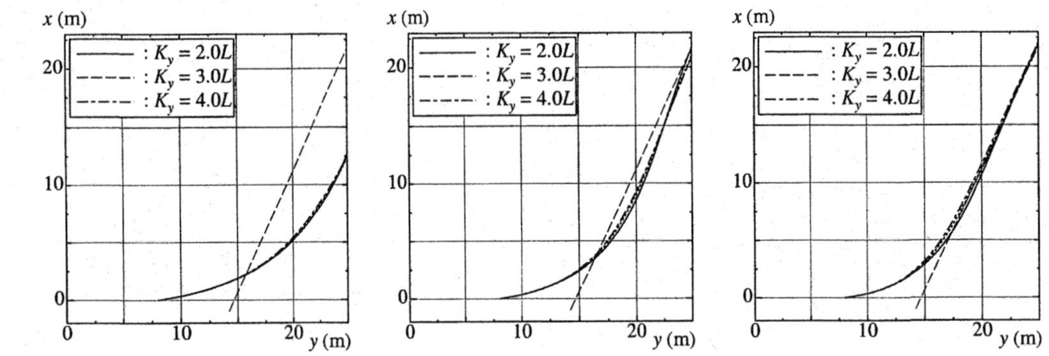


(a) $K_r = 0.2$ (b) $K_r = 0.4$ (c) $K_r = 0.6$

Fig. 7. Effects of $K_y$ on ship motion ($\psi_{t0} = 55°$)

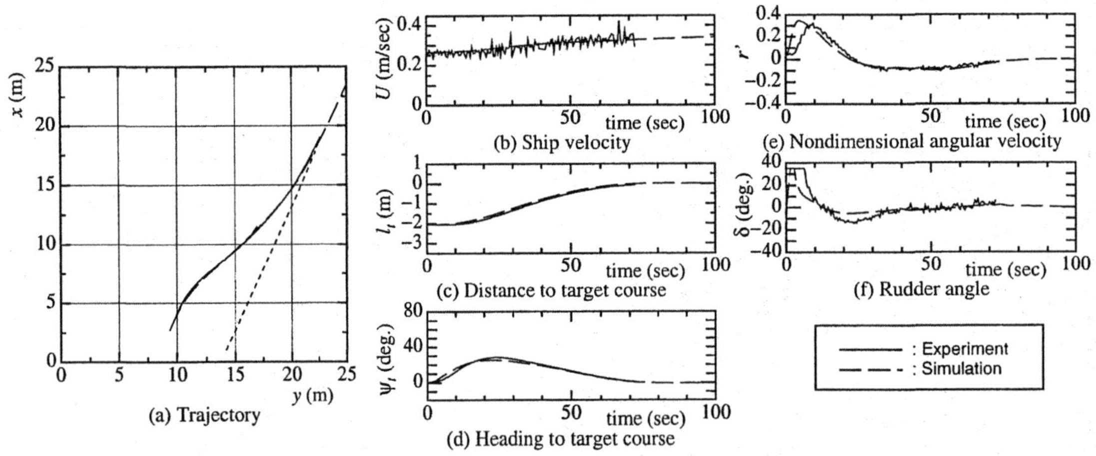


Fig. 8. Comparison between measured and simulated ship motion ($l_{t0} = 2.0L$, $\psi_{t0} = 0°$)

the model ship was specified by parameters $l_{t0}$ and $\psi_{t0}$ which indicate the initial values for lateral distance and heading to the target route.

Figures 8 to 10 indicate model test and simulation results for the conditions of $l_{t0} = 2.0L$ and $\psi_{t0} = 0°$, $45°$ and $70°$ respectively. Figure (a) in each figure shows trajectories of the midship of model ship and figures (b) to (f) display the time histories of ship speed, $U$, lateral distance to target route, $l_t$, heading to target route, $\psi_t$, nondimensional yaw rate, $r'$, and rudder angle, $\delta$. In these figures, solid line and broken line indicate model test and simulation results respectively.

It is observed that a model ship changes her course swiftly and sets her course on the target route both on model test and simulation in figure (a). It means that the parameters $K_y$ and $K_r$ with the adopted value function adequately. Though there are slight differences in yaw rate and rudder angle between model test and simulation shown in figures (e) and (f), the motion of a ship, such as trajectory or heading angle indicate good agreement between model test and simulation results.

Figures 11 to 13 show model test and simulation results for $l_{t0} = 3.0L$ and $\psi_{t0} = 0°$, $45°$ and $70°$ respectively. Model test and simulation results show

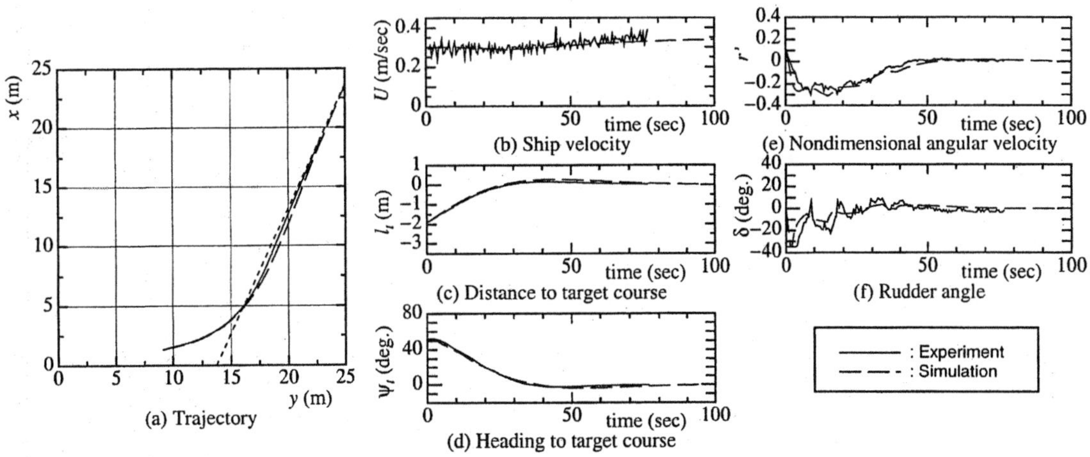


Fig. 9. Comparison between measured and simulated ship motion ($l_{t0} = 2.0L$, $\psi_{t0} = 45°$)

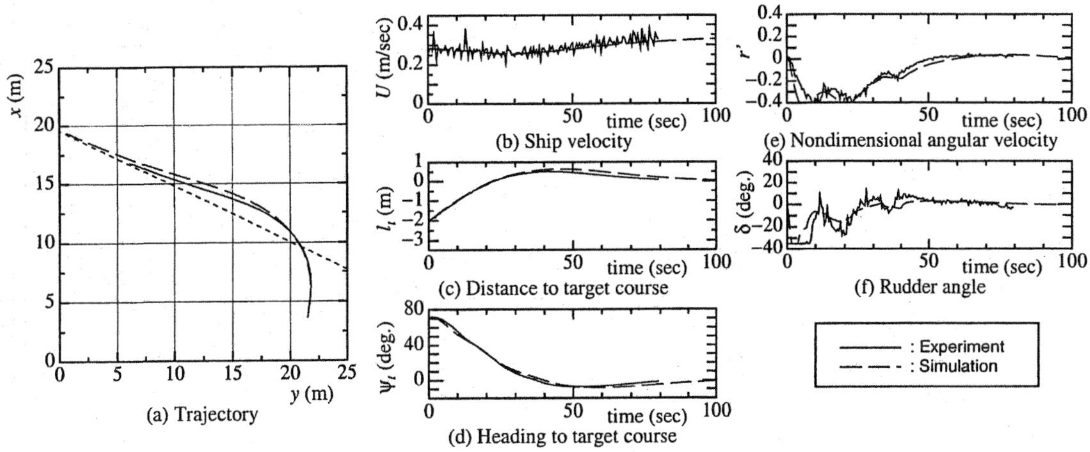


Fig. 10. Comparison between measured and simulated ship motion ($l_{t0} = 2.0L$, $\psi_{t0} = 70°$)

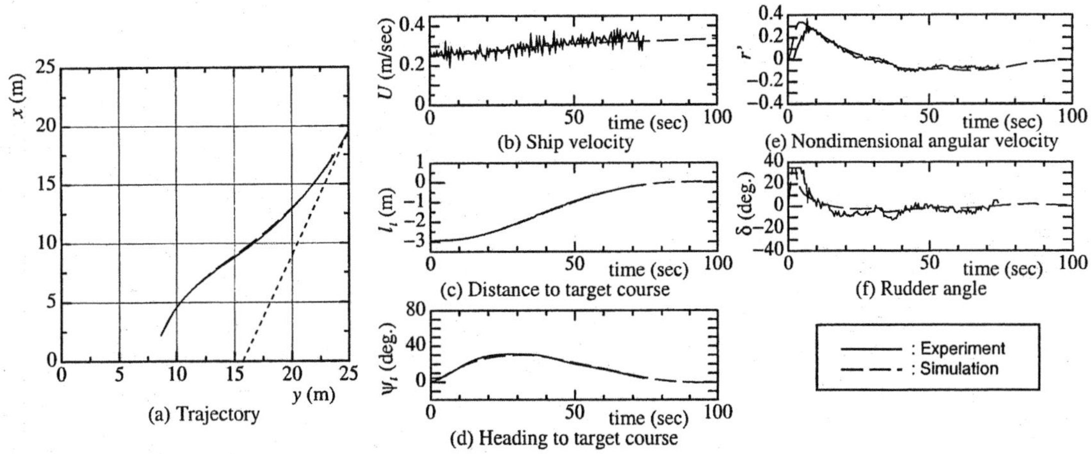


Fig. 11. Comparison between measured and simulated ship motion ($l_{t0} = 3.0L$, $\psi_{t0} = 0°$)

good agreement in these figures and it is found that $K_y$ and $K_r$ function well when initial deviation is away from target route.

It can be said that our simulation code has good accuracy for prediction of ship motion and the parameters $K_y$ and $K_r$ for fuzzy inference which were specified based on the simulation study function well for course change manoeuvre.

## 6. CONCLUSIONS

In order to evaluate automatic navigation system, experimental system using model ship were developed and it was shown that the system has enough accuracy in measurement of the position and heading of model ship. According to the comparison between model test and simulation results, our simulation code has good accuracy for prediction of ship motion and the

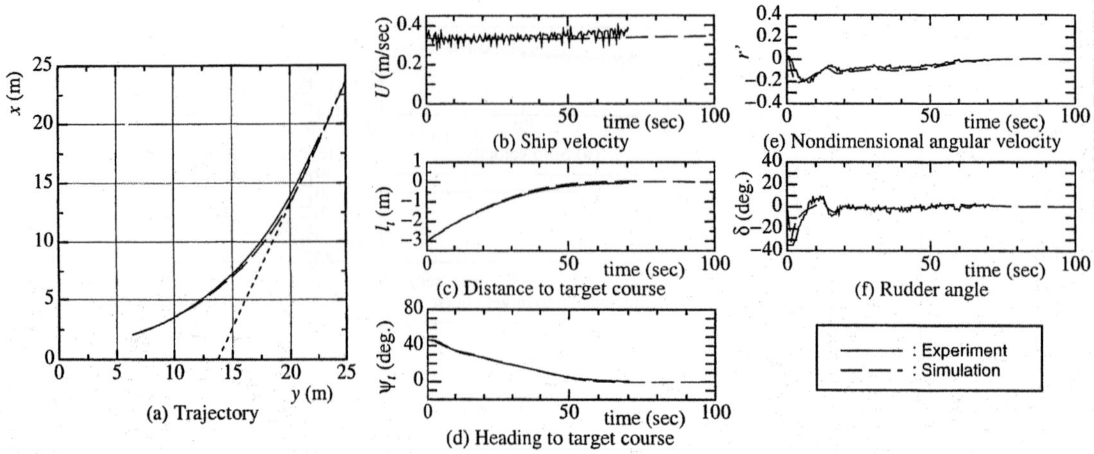


Fig. 12. Comparison between measured and simulated ship motion ($l_{t0} = 3.0L$, $\psi_{t0} = 45°$)

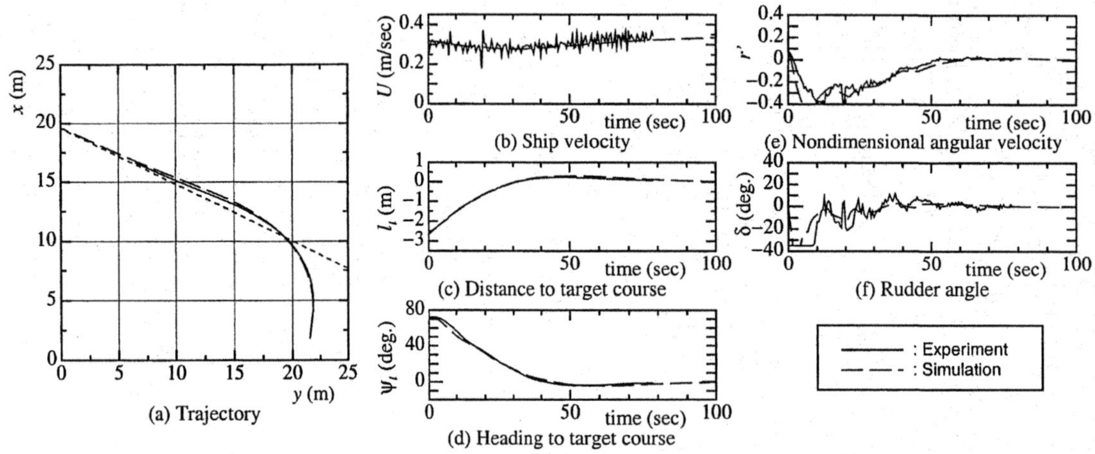


Fig. 13. Comparison between measured and simulated ship motion ($l_{t0} = 3.0L$, $\psi_{t0} = 70°$)

parameters $K_y$ and $K_r$ for fuzzy inference which were specified based on the simulation study function well for course change manoeuvre.

## 7. REFERENCES

Hara, K. and A. Hammer (1993). A safe way of collision avoidance maneuver based on maneuvering standard using fuzzy reasoning model. In: *Proc. of International Conference on Marine Simulation and Ship Manoeuvrability (MARSIM'93)*. Vol. 1. St.John's, Newfoundland, Canada. pp. 163–170.

Hasegawa, K. and Y. Fujita (1993). An extension of ship auto-navigation fuzzy expert system for safety assessment of narrow waterway navigation. *Jour. of the Kansai Society of Naval Architects, Japan* **220**, 129–133.

Kijima, K. and Y. Furukawa (2001). Design of automatic collision avoidance system using fuzzy inference. In: *Proc. of Control Applications in Marine Systems (CAMS 2001)*. Glasgow, UK.

Kijima, K. and Y. Furukawa (2002). Development of collision avoidance algorithm using fuzzy inference. In: *Proc. of ISOPE Pacific/Asia Offshore Mechanics Symposium (ISOPE PACOMS 2002)*. Daejeon, Korea. pp. 123–130.

Kijima, K. and Y. Furukawa (2003). Automatic collision avoidance system using the concept of blocking area. In: *Proc. of Manoeuvring and Control of Marine Craft (MCMC 2003)*. Girona, Spain. pp. 262–267.

Kijima, K. and Y. Nakiri (2003). On the practical prediction method for ship manoeuvring characteristics. In: *Proc. of International Conference on Marine Simulation and Ship Maneuverability (MARSIM'03)*. Kanazawa, Japan. pp. RC–6–1–RC–6–9.

Kose, K., K. Hirono, K. Sugano and I. Sato (1998). A new collision-avoid-supporting-system and its application to coasta-cargo-ship 'shoyo maru'. In: *Proc. of Control Applications in Marine Systems (CAMS'98)*. Fukuoka, Japan. pp. 169–174.

Lee, H. J. and K. P. Rhee (2001). Development of collision avoidance system by using expert system and search algorithm. *International Shipbuilding Progress* **48**(3), 197–210.

Zhuo, Y. and G. E. Hearn (2003). Ship intelligent anti-collision using slef-learning neurofuzzy. In: *Proc. of Manoeuvring and Control of Marine Craft (MCMC 2003)*. Girona, Spain. pp. 268–273.

ELSEVIER
IFAC
PUBLICATIONS
www.elsevier.com/locate/ifac

# AN INTELLIGENCE FUSION SYSTEM FOR SHIP FAULT-TOLERANT CONTROL[1]

**Tianhao Tang and Gang Yao**

*Department of Electrical & Control Engineering, Shanghai Maritime University*
*1550 Pudong Road, Shanghai, 200135, P. R. China, E-mail: thtang@ieee.org*

Abstract: This paper presents an intelligence fusion system based on hybrid neural networks for system fault detection and fault-tolerant control. The design of system scheme, intelligent fusion methods, and its applications in ocean ship fault detection and fault tolerant control will be discussed. *Copyright © 2004 IFAC*

Keywords: fault-tolerant systems, neural networks, fuzzy logic, intelligent control

## 1. INTRODUCTION

Since observer theory was applied to detect system faults in the early 1970s, fault detection and isolate (FDI) model theory went through a dynamic and rapid development. At almost the same time, integral control theory was proposed by Nideerlinski to solve the control problem in a system with faults, a lot of fault tolerant control (FTC) methods have been put forward in recent years. Now the researches in FDI and FTC are becoming an important field of automatic control theory.

With the progress of techniques in computer science and artificial intelligence (AI), there is a developing trend that the intelligent-based approaches are playing more and more significant part in whatever FDI or FTC than model-based methods to do in the early years. For example, FTC has been developed from traditional methods such as integral control, system reconfiguration and multiple model adaptive control (MMAC) to intelligent methods based on fuzzy logic or neural networks in the last decade. However, the researchers found that it is difficult if just using a single intelligent method in a complicated system.

For this reason, some new ideas such as fusion intelligence, or distributed intelligence is being researched recently (Jain and Martin, 1998). And many researchers had made some important works (Sutton, *et al*, 2001). Now there is another problem to be solved. FDI and FTC theories used to be discussed respectively from a long time. But in practice, it is required for FTC system how to detect and diagnose faults from system states and a system with FDI may need to extend the control functions at the fault conditions. So it is very important to merge FDI and FTC techniques into an integrated technology for a modern automatic control system.

This paper presents a new intelligent fusion method based on hybrid neural networks and designs a FDI and FTC integrated system for ocean ship fault tolerant control. Firstly the control system structure based on intelligence fusion will be discussed in next section. Secondly the FDI methods with fuzzy inference and fuzzy neural networks will be developed in section 3. Then the FTC controller based on fusion intelligence of fuzzy and neural networks are researched for system control in section 4. Finally, its application in ship maneuver and some simulation results will be provided in section 5.

## 2. SYSTEM SCHEME & CONTROL STRATEGY

This paper presents a new method of fault tolerant control using intelligence fusion techniques on the

[1]This paper was supported by the Project of Key Discipline of Shanghai Education Committee

basis of the research (Tang *et al.*, 2001). The scheme of the system is illustrated in Fig. 1.

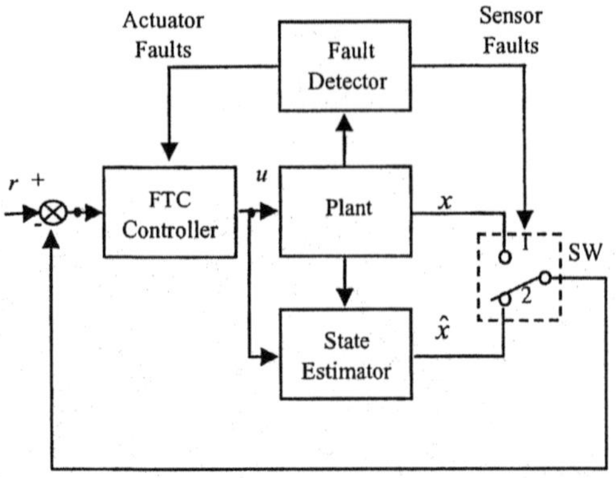


Fig.1. The scheme of fault tolerant control system

There are three parts: fault detector, state estimator and fault tolerant controller in the system.

(1) In the fault detector, two kinds of faults, actuator faults and sensor faults will be detected. If actuator faults happened, FTC controller will take an action according to the fault state. While the sensor faults have been found, the state estimator will play a part in the sensors.

(2) A state estimator based on an output recurrent neural network (ORNN) could provide the system state estimation to replace the fault sensors. Usually, the switch (SW) is on the position 1. In this case the system state feedback signals will be from sensors. If the sensor faults occur, the SW will be on the position 2. In this case, the signals from state estimator will be used as the system feedback.

(3) The fault tolerant controller will give out different control strategy according to the system state and fault cases to keep the system performance. For example, if a fault of the senor occurs, the SW in the system will be put on point 2 as above mentioned. When a fault from the actuator was be detected, the FTC controller based on adaptive fuzzy neural network could adjust the control signal to overcome the influence of the fault according to the system response on line.

## 3. FAULT DETECTION MODULES

The structure of the fault detector is shown in Fig. 2. Here a distributed data fusion method is presented for fault detection. There are two types of fault detection modules: some local data fusion modules are used to detect sensor faults itself and a global data fusion module is selected to diagnose system faults.

### *3.1 Sensor Fault Detectors*

Every local senor fault detection module is composed of fuzzy inference and multi-senor data fusion for a senor group. Each senor group adopts the same sensors to acquire one measurement of the system. If one sensor of the group has the failure, the others could provide the measurement through the local data fusion module. Another purpose of the multi-sensor local data fusion module is to remove the noises from the practical measurements.

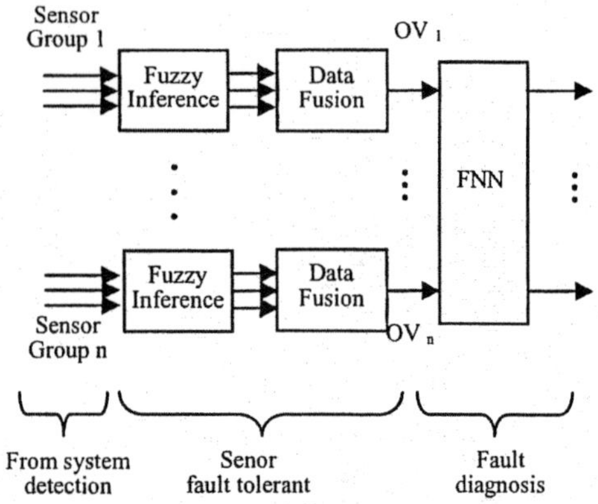


Fig.2. Fault detection modules

Here a fuzzy inference method based on distance decision-making is selected to implement the data fusion process (Luo *et al.*, 1988). The principle of the method is using measurement distance from every sensor in a group to obtain the certification of the detection value through the fuzzy inference, then to calculate the weighted sum of the measurement according to the certification. There are three steps in the process:

(1) Pre-process the measurements $x_i$ from the sensors of the group and calculate the distance $d_{ij}$ of the measurements using the formula as follows:

$$d_{ij} = \begin{cases} |x_i - x_j|, & i \neq j \\ 0, & i = j \end{cases} \quad (1)$$

(2) Input the measurement distance to fuzzy reasoning machine and obtain the certification through fuzzy inference according to some fuzzy rules as follows

$$R^k\text{: if } d_1 \text{ is } F_1^k, \ldots, d_n \text{ is } F_n^k \text{ then } w \text{ is } C^k \quad (2)$$

Where $R^k$ is a fuzzy rule in the rule base; $F_i^k$ and $C^k$ denote the fuzzy sets respectively.

(3) Calculate the observation of the sensors using the following formula

$$OV_i = \sum_{j=1}^{n} \hat{w}_j \cdot x_j \quad (3)$$

Where

$$\sum_{j=1}^{n} \hat{w}_j = 1 \quad (4)$$

And the weight values $\hat{w}_j$ had been processed through a defuzzification unit such as

$$\hat{w} = \frac{\sum_{i=1}^{n} k_i w_i}{\sum_{i=1}^{n} k_i} \quad (5)$$

Here $k_i$ is the weight value coefficients.

*3.2 System Fault Detector*

In the system fault detector, a fuzzy neural network (FNN) was used to diagnose the actuator faults or the sensor faults could not be overcome by the data fusion modules. As we known, any kind of FNN model for fuzzy reasoning is based on the calculating rules of the fuzzy union connective, the fuzzy intersection connective or a fuzzy connective between them. If a calculating rule is simple, the model based on the operator is also simple and the diagnosing process by the model will be fast. On the basis of this idea, a kind of fuzzy operators, the generalized probability sum operator and the generalized probability product operator, was presented to express the concepts of the generalized union and the generalized intersection calculating (Tang, et al., 1998a). Because of using the simple operators, the speed of fault reasoning by the model is faster than by the models based on other operators. Its structure is illustrated in figure 3.

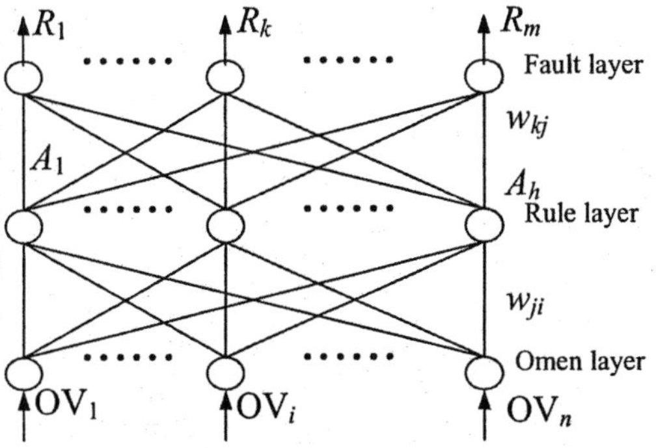


Fig.3. The structure of FNN model

The input of the model is the predictive values from RNN model. And the output of the model is

$$R_k = 1 - \left[ \prod_{j=1}^{h} (1 - A_j)^{w_{kj}} \right]^{q_k} \quad (6)$$

$$A_j = \left(\prod_{i=1}^{n} OV_i^{w_{ji}}\right)^{p_j} \quad (7)$$

Where the parameters of FNN are constrained by:

$$\sum_{j=1}^{h} w_{kj} = 1, w_{kj} \geq 0, q_k \in [0, +\infty),$$

and $w_{ji} \geq 0, p_j \in [0, +\infty)$.

The model represents the fuzzy inference rules as follows

$$R^{(l)}: \text{if } OV_1(CL_1^l), OV_2(CL_2^l), \ldots, OV_n(CL_n^l),$$
$$\text{then } R_1^l(CF_1), R_2^l(CF_2), \ldots, R_m^l(CF_m) \quad (8)$$

Where CL is a fuzzy classification defined by fuzzy membership functions and CF represents the certification of reasoning results. The details about the model were discussed in (Tang, *et al.*, 1998a).

## 4. INTELLIGENCE FUSION FTC METHODS

As mentioned above, the system FTC control strategy is divided as two parts according to sensor faults or actuator faults. Usually the senor faults will be overcome by the sensor groups with local data fusion units. If a sensor group with a serious failure that could not be tolerant by the local data fusion unit, the state estimator based on a feed forward neural network with output recurrent nodes will replace the senor to construct system feedback loop. When a fault from the actuator was detected, the fault tolerant controller based on adaptive fuzzy neural network could adjust the control signal to overcome the influence of the fault according to the system response on line. The architecture of system for fault tolerant control is simplified as shown in Fig.4.

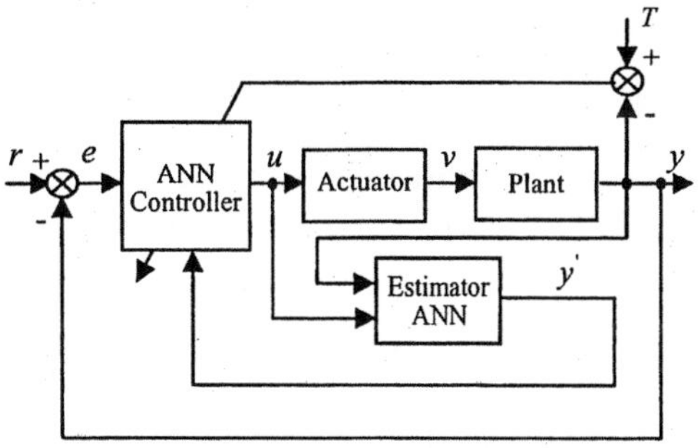


Fig.4. Hybrid ANN-based FTC model

*4.1 ORNN - based State Estimator*

Usually a class of time series methods such as linear or nonlinear ARMA models could be used for system modelling and identification. And some feed-forward neural networks or recurrent neural networks were proposed to implement the time series models for dynamic process adaptive predicting on line (Tang, *et al.*, 1998b, 2000). But considering if a sensor fault happened, it may affect the system actual output signal. In this case, a normal recurrent neural network could not be used, because it requires the actual detective value from the system as the reference value of the network feedback. For this reason, a new output recurrent neural network is designed to construct the system state estimator as shown in Fig.5. In this neural network, the network estimated output $\hat{x}_i(t)$ is directly sent to the input nodes as the recurrent feedback to replace the error feedback between the estimated value and actual detective value of the system in a normal recurrent neural network.

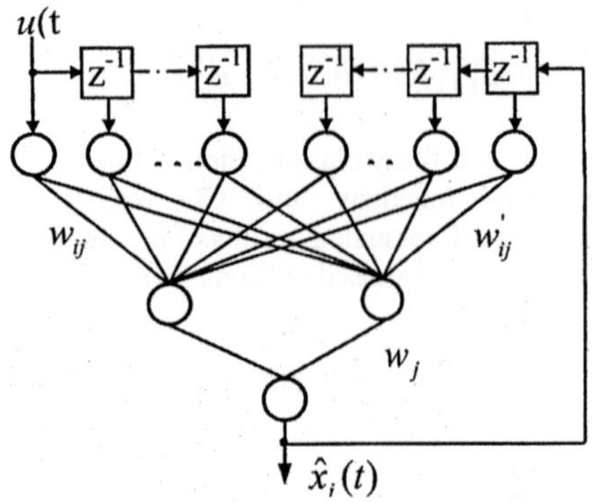


Fig.5. State estimator based on ORNN

The transfer function of this network is described as follows:

$$s_j = \sum_{i=1}^{k} w_{ij} u_{t-i+1} + \sum_{i=1}^{k-1} w'_{ij} \hat{x}_{t-i} \quad (9)$$

$$\hat{x}_i(t) = \sum_{j=1}^{h} w_j f(s_j) \quad (10)$$

Where $f(\bullet)$ is a sigmoid function.

A learning algorithm will adaptively modify the parameters of the estimator according to the errors between the estimative value and the actual detective value. Considering the following error function

$$E = \frac{1}{2}(x_t - \hat{x}_t)^2 \quad (11)$$

Where $x_t$ is the actual detective value from sensors in normal case. Then the learning algorithm could be obtained as follows:

(1) The output layer weight formula

$$\frac{\partial E}{\partial w_j} = -(x_t - \hat{x}_t) f(s_j) \quad (12)$$

(2) The hidden layer input weight formula

$$\frac{\partial E}{\partial w_{ij}} = -(x_t - \hat{x}_t) \sum_{j=1}^{h} w_j f(s_j)(1 - f(s_j)) u(t - i + 1) \quad (13)$$

(3) The hidden layer output recurrent weight formula

$$\frac{\partial E}{\partial w'_{ij}} = -(x_t - \hat{x}_t) \sum_{j=1}^{h} w_j f(s_j)(1 - f(s_j)) x(t - i) \quad (14)$$

The details on the network learning and system state estimation could be seen in another paper (Yao and Tang, 2003).

*4.2 FNN - based FTC Controller*

The structure of FTC controller based on fuzzy neural network is shown in Fig. 6. There are three parts with 5 layers. Firstly the system error signals will be converted into the fuzzy values through input layer and fuzzifying layer. Then the control strategy will be given from the fuzzy inference layer. At last the fuzzy strategy will be converted into control signals for system fault tolerant control.

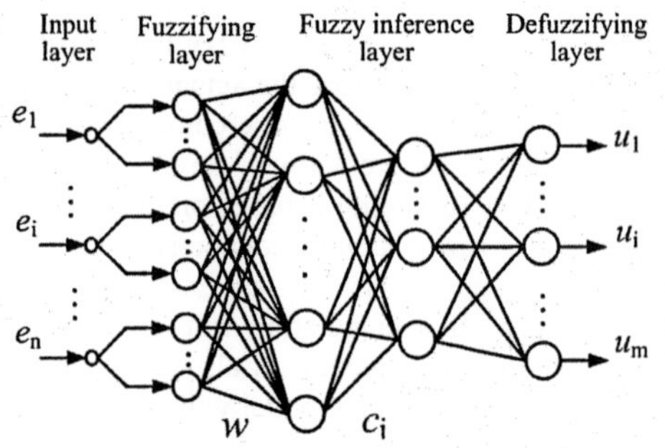


Fig.6. Diagram of fault tolerant control structure

In the FNN-based controller, Gauss function is selected as the fuzzy membership functions for input variables fuzzification. And the fuzzy inference layer is adopted the generalized probability sum operator and the generalized probability product operator as described in section 3. So the fuzzy reasoning function is the same as equation (6) and equation (7). In the defuzzifying layer, an improved centre mean defuzzification algorithm is used as follows:

$$u = \frac{\sum_{l=1}^{M} \bar{u}^l \left[ \mu_B^l(\bar{u}^l) / \delta^l \right]}{\sum_{l=1}^{M} \left[ \mu_B^l(\bar{u}^l) / \delta^l \right]} \quad (15)$$

Where $\bar{u}$ is the centre of a fuzzy output variable subsets and $\delta$ is the width of the centre.

## 5. APPLICATION AND SIMULATIONS

The intelligence fusion system above mentioned had been used in ship automatic steering system for state detection and fault tolerant control. In this paper, an application example and its simulation results will be provided. The structure of ship FTC system is shown in Fig. 7.

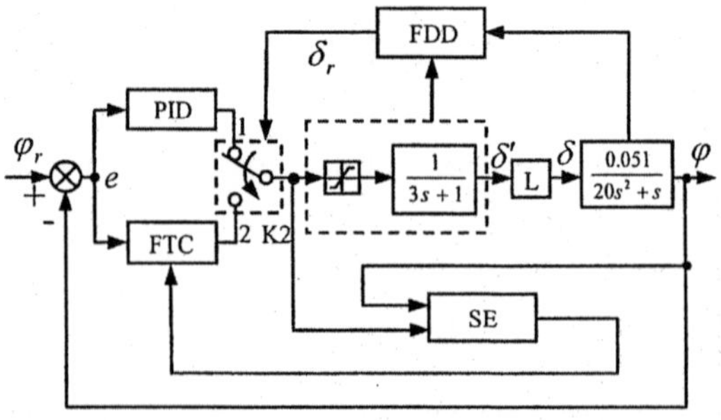


Fig.7. The structure of ship FTC system

In the system, consider a simplified ship dynamic model and the rudder servo system move within a limit by the limiters. The FDD module based on distributed data fusion and fuzzy neural network is used to detect system faults. A state estimator (SE) based on ORNN is presented to provide system feedback when the sensor fault happened. A PID controller and FTC controller construct an intelligence fusion control strategy. The PID controller will take action to keep the ship's coure at the normal condition. If the fault is detected in the rudder, the FNN-based FTC controller will control the ship. Here the FTC controller is shown in Fig. 8.

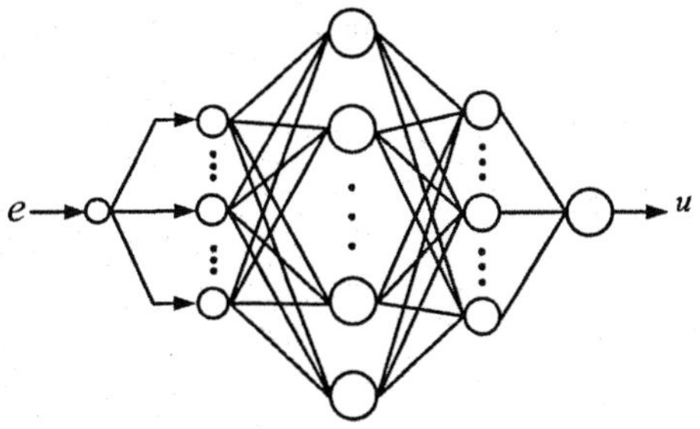


Fig. 8. FTC control network diagram

Some system simulation results are shown in Fig.9 and Fig. 10.

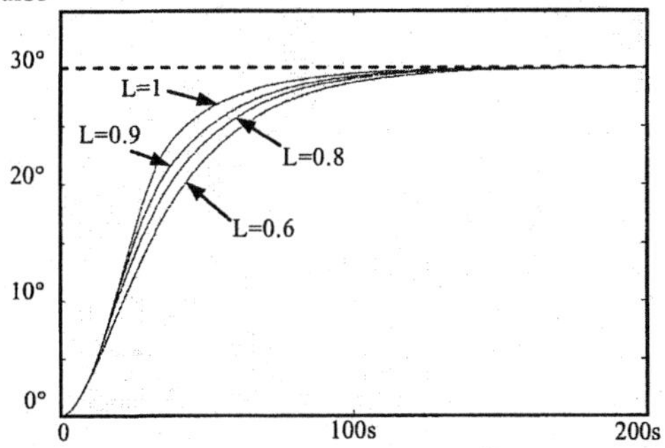


Fig.9. The system responses with PID controller

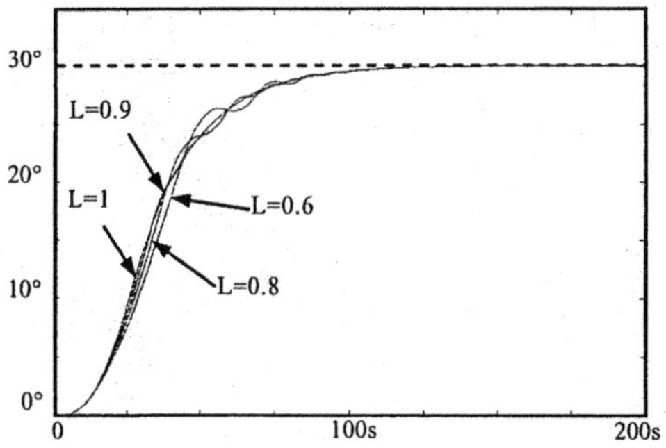


Fig.10 The system responses with FTC controller

Here the rudder servo system is selected as an example to show the system performances in the normal or fault conditions. In order to represent the rudder faults, a coefficient called as loss of efficient (LOE) is introduced to simulate the faults. When the rudder is normal, define LOE=1. And define LOE=0 if the rudder is whole failure. So the coefficient LOE expresses the fault degree of the rudder. Figure 9 shows the system responses with PID controller at normal and fault condition. Figure 10 shows the system responses with FTC controller in the same cases. Comparing the two controllers, it has the same system response at the normal condition. But in the fault cases, FTC controller can obtain the better system responses than PID controller.

## CONCLUSION REMARK

This paper presents a new intelligent FTC system based on fusion of neural networks, fuzzy sets and some integrated learning algorithms for ocean ship fault detection and fault tolerant control. The system simulation results indicate that intelligence fusion system has more functions and advantages, especially applied in complex systems such as the ocean ships and AUVs.

## REFERENCES

Jain, L.C. and N.M. Martin (1998). Fusion of neural networks, fuzzy sets, and genetic algorithms. CRC Press

Luo, R C, M. Lin and R. S. Scherp (1988). Dynamic multi-sensor data fusion system for intelligent robots. IEEE Journal of Robotics and Automation, vol.4, no.4, pp.386-396

Sutton, R., A. Pearson and A. Tiano (2001). A fuzzy fault tolerant control scheme for an autonomous underwater vehicle. Proceedings of the 7th IEEE international conference on methods and models in automation and robotics, vol. 2, pp. 595-600

Tang, T. et al (1998a). A fuzzy and neural network integrated intelligence approach for fault diagnosing and monitoring. Proceedings of the 1998 UKACC International Conference on Control, vol.2, pp.975-980. Swansea, UK

Tang, T. et al (1998b). ANN-based nonlinear time series models in fault detection and prediction, Proceedings of IFAC CAMS'98, Fukuoka, Japan, 27-30 October 1998, 335-340

Tang, T. et al (2000). A RNN-based adaptive predictor for fault prediction and incipient diagnosis, Proceedings of the 2000 UKACC International Conference on Control, Cambridge, UK

Tang, T. et al (2001). A new fault-tolerant control scheme based on hybrid neural networks. Proceedings of the 7th IEEE International Conference on Methods and Models in Automation and Robotics. vol.2, p. 625-630, Miedzyzdroje, Poland, 28-31 August

Yao, G. and T. Tang (2003). A state estimator based on output recurrent neural networks. Proceedings of the 13th China Neural Networks Academic Conference, Qingdao, China. Nov 2003, 611-616

Yao, G. and T. Tang (2003). A Fault Detection Approach Based On Hierarchy Information Fusion. Journal of Central South University of Technology, vol. 34, Suppl 2, 157-160

ELSEVIER
IFAC
PUBLICATIONS
www.elsevier.com/locate/ifac

# DIAGNOSIS OF ACTUATOR FAULTS IN AUVs BASED ON NEURAL NETWORKS

**Gianluca Antonelli** *, **Fabrizio Caccavale** **,
**Carlo Sansone** ***, **Luigi Villani** ***

* *DAEIMI - Università degli Studi di Cassino*
*Via G. Di Biasio 43, 03043 Cassino, Italy*
antonelli@unicas.it

** *DIFA - Università degli Studi della Basilicata*
*Viale dell'Ateneo Lucano 10, 85100 Potenza, Italy*
caccavale@unibas.it

*** *DIS - Università degli Studi di Napoli Federico II*
*Via Claudio 21, 80125 Napoli, Italy*
{carlo.sansone,luigi.villani}@unina.it

Abstract: A neural network approach for the diagnosis of faults affecting the propulsion system and/or the control surfaces of Autonomous Underwater Vehicles (AUVs) is presented in this paper. The approach is based on a dynamic observer which exploits the available dynamic model of the AUV as well as on neural interpolators. Namely, a Support Vector Machine (SVM) is trained off-line to achieve compensation of unknown dynamics, uncertainties and disturbances, and a Radial Basis Function (RBF) network is trained on-line at the occurrence of an actuator fault to estimate the time evolution of the failure. Simulation results are presented to show the effectiveness of the proposed approach in a case study developed for the NPS AUV II vehicle. *Copyright ©2004 IFAC*

Keywords: Underwater vehicles; fault diagnosis; neural networks; support vector machines.

## 1. INTRODUCTION

The capability to detect and tolerate faults is a key requirement for Autonomous Underwater Vehicles (AUVs) to successfully perform missions in unstructured and/or hazardous environments (Antonelli, 2003). In fact, the adoption of suitable Fault Diagnosis (FD) techniques, combined to fault tolerant strategies, allows to safely terminate the assigned task as in the case of the arctic mission of Theseus (Ferguson *et al.*, 1999).

Most of the FD techniques developed in the recent years are model-based (see, e.g, the book of Patton *et al.* (2000) and references therein) and, among them, observer-based approaches are widely adopted. Usually, the observer-based methods require a model of the system to be operated in parallel to the process (i.e., the so-called *diagnostic observer*). Then, a set of variables sensitive to the occurrence of a failure (*residuals*) can be computed by comparing the measured output variables and those predicted via the diagnostic observer. Assuming that the system dynamics is exactly known, the residuals should become nonzero only when a fault occurs. However, perfect knowledge of the model is rarely a reasonable assumption, especially for AUVs. Further, the observer has to be implemented in discrete-time, eventually working at a low sampling rate; this circumstance may introduce additional modelling errors due to the adoption of approximate discrete-time equivalents of the system.

When the model is nonlinear in the parameters and/or its structure is not exactly known, a good approximation can be obtained by resorting to neural networks (Healey *et al.*, 1992; Demetriou and Polycarpou, 1998; Zhang *et al.*, 2002). Among the different neural network models, Support Vector Machines (SVMs) allow achieving a superior ability to generalize, which is the typical goal in statistical learning (Vapnik, 1995). SVMs were initially developed to solve classification problems, but recently they have been extended to the domain of regression and function estimation problems (Vapnik *et al.*, 1997).

In this paper, a discrete-time diagnostic observer has been designed to detect, isolate and identify actuator faults (i.e., faults affecting the propulsion system and/or the control surfaces) for an AUV, inspired to the approach of Caccavale and Villani (1993) for an industrial robot. The observer has been modified by adding a term compensating for un-modelled dynamics and disturbances which is estimated by using a Support Vector Machine. The SVM is trained off-line using the data collected along a set of fault-free trajectories. Hence, the fault detection and isolation is performed by defining suitable residual variables.

When a failure occurs, the estimation of the time evolution of unknown actuator faults is computed on-line by using a Radial Basis Function (RBF) neural network. The weights of the RBF are adaptively updated during system operation, by using a discrete-time update law.

The proposed FD algorithm has been extensively tested in simulation to prove the effectiveness of the approach, for different types of actuator faults and various operating conditions. A simulation case study is presented here on the model of the NPS AUV II vehicle (Healey and Lienard, 1993), experiencing the failure of the propeller blade.

## 2. MODELLING

To describe the mathematical model of an AUV, it is convenient to introduce the vector $\boldsymbol{\eta} = [\boldsymbol{\eta}_1^{\mathrm{T}} \;\; \boldsymbol{\eta}_2^{\mathrm{T}}]^{\mathrm{T}}$, where $\boldsymbol{\eta}_1 = [x \;\; y \;\; z]^{\mathrm{T}} \in \Re^3$ is the vector of vehicle position and $\boldsymbol{\eta}_2 = [\phi \;\; \theta \;\; \psi]^{\mathrm{T}} \in \Re^3$ is the vector of vehicle Euler angles in an earth-fixed reference frame. The velocity vector expressed in a vehicle-fixed reference frame can be computed as (Fossen, 1994):

$$\boldsymbol{\nu} = \boldsymbol{J}(\boldsymbol{\eta}_2)\dot{\boldsymbol{\eta}}, \tag{1}$$

where $\boldsymbol{J}$ is a Jacobian matrix and $\boldsymbol{\nu} = [\boldsymbol{\nu}_1^{\mathrm{T}} \;\; \boldsymbol{\nu}_2^{\mathrm{T}}]^{\mathrm{T}}$, where $\boldsymbol{\nu}_1 = [u \;\; v \;\; w]^{\mathrm{T}} \in \Re^3$ is the linear velocity and $\boldsymbol{\nu}_2 = [p \;\; q \;\; r]^{\mathrm{T}} \in \Re^3$ is the angular velocity.

Using the above defined variables, the dynamic model can be written in the vehicle-fixed reference frame in the form (Fossen, 1994):

$$\boldsymbol{M}\dot{\boldsymbol{\nu}} + \boldsymbol{C}(\boldsymbol{\nu})\boldsymbol{\nu} + \boldsymbol{D}(\boldsymbol{\nu})\boldsymbol{\nu} + \boldsymbol{g}(\boldsymbol{\eta}_2) = \boldsymbol{B}_a(\boldsymbol{\nu})\boldsymbol{u}, \tag{2}$$

where $\boldsymbol{M} \in \Re^{6\times 6}$ is the mass matrix that includes both rigid body mass and added mass, $\boldsymbol{C}(\boldsymbol{\nu})\boldsymbol{\nu} \in \Re^6$ is the vector of Coriolis and Centrifugal terms including the added mass, $\boldsymbol{D}(\boldsymbol{\nu})\boldsymbol{\nu} \in \Re^6$ is the vector of friction and hydrodynamic damping terms, $\boldsymbol{g}(\boldsymbol{\eta}_2) \in \Re^6$ is the vector of gravitational and buoyant generalized forces. The quantity $\boldsymbol{B}_a(\boldsymbol{\nu})\boldsymbol{u} \in \Re^6$ is the vector of forces and moments acting on the vehicle, where $\boldsymbol{u}$ is the vector of control inputs from control surfaces, propeller speeds, thruster forces and buoyancy adjustment (Healey and Lienard, 1993).

By introducing the state vector $\boldsymbol{x} = [\boldsymbol{\eta}^{\mathrm{T}} \;\; \boldsymbol{\nu}^{\mathrm{T}}]^{\mathrm{T}}$, the dynamic model can be rewritten in the form:

$$\begin{aligned}\dot{\boldsymbol{x}}(t) = \boldsymbol{A}_c\boldsymbol{x}(t) + \boldsymbol{h}_c(\boldsymbol{x}(t)) + \boldsymbol{B}_c(\boldsymbol{x}(t))\boldsymbol{u}(t)\\ +\boldsymbol{\xi}_c(t, \boldsymbol{x}(t), \boldsymbol{u}(t))\end{aligned} \tag{3}$$

where $\boldsymbol{\xi}_c$ collects all the disturbances, un-modelled and/or uncertain dynamics. A discrete-time formulation of the dynamic model can be achieved using the first-order Euler method in the form

$$\begin{aligned}\boldsymbol{x}(k+1) = \boldsymbol{A}\boldsymbol{x}(k) + \boldsymbol{h}(\boldsymbol{x}(k)) + \boldsymbol{B}(\boldsymbol{x}(k))\boldsymbol{u}(k)\\ +\boldsymbol{\xi}(k, \boldsymbol{x}(k), \boldsymbol{u}(k))\end{aligned} \tag{4}$$

where $\boldsymbol{A} = \boldsymbol{I}_n + T\boldsymbol{A}_c$, $\boldsymbol{h} = T\boldsymbol{h}_c$, $\boldsymbol{B} = T\boldsymbol{B}_c$, $\boldsymbol{\xi} = T\boldsymbol{\xi}_c + \boldsymbol{\rho}$, and $\boldsymbol{\rho}$ represents the discretization error. Equation (4) assumes that the measurements of the state variables are sampled at a fixed time step $T$, and the inputs are constant over each time interval $\mathcal{I}_k \equiv [kT, (k+1)T[$, where the integer $k \geq 0$ denotes the discrete time variable $t_k = kT$.

The class of failures considered in this work is that of *actuator* faults, which can be represented as an unknown additive disturbance on the *nominal* input to the system $\bar{\boldsymbol{u}}$. Hence, a fault occurring at $t_k = kT$ results in a faulty input given by

$$\boldsymbol{u}(k) = \bar{\boldsymbol{u}}(k) + \boldsymbol{\delta u}(k), \tag{5}$$

where $\boldsymbol{\delta u}$ represents the unknown fault.

In the presence of faults, the dynamics (4) is

$$\begin{aligned}\boldsymbol{x}(k+1) = \boldsymbol{A}\boldsymbol{x}(k) + \boldsymbol{h}(\boldsymbol{x}(k)) + \boldsymbol{B}(\boldsymbol{x}(k))\bar{\boldsymbol{u}}(k)\\ +\boldsymbol{\xi}(k, \boldsymbol{x}(k), \bar{\boldsymbol{u}}(k)) + \boldsymbol{f}(k, \boldsymbol{x}(k)),\end{aligned} \tag{6}$$

where the fault vector $\boldsymbol{f}$ is given by:

$$\boldsymbol{f}(k, \boldsymbol{x}(k)) = \boldsymbol{B}(\boldsymbol{x}(k))\,\boldsymbol{\delta u}(k), \tag{7}$$

and the uncertain term $\boldsymbol{\xi}(k, \boldsymbol{x}(k), \bar{\boldsymbol{u}}(k))$ is assumed to depend upon the nominal input.

## 3. FAULT DIAGNOSIS

Assuming the whole state measurable, the following diagnostic observer is proposed (Caccavale and Villani, 1993):

$$\widehat{x}(k+1) = A\widehat{x}(k) + h(x(k)) + B(x(k))\bar{u}(k) + K_o e(k) + \widehat{\xi}(k, x(k), \bar{u}(k)) + \widehat{f}(k, x(k)) \quad (8)$$

where $e = x - \widehat{x}$ is the state estimation error, $\widehat{\xi}$ represents an estimate of the uncertainties $\xi$, $\widehat{f}$ represents an estimate of the fault vector $f$ and the matrix gain $K_o$ is chosen such that $F = A - K_o$ has all its eigenvalues in the unit circle.

By virtue of (6) and (8), the estimation error dynamics is given by (hereafter, dependence of $f$ and $\xi$ upon $x$ and $\bar{u}$ will be skipped)

$$e(k+1) = Fe(k) + \widetilde{\xi}(k) + \widetilde{f}(k), \quad (9)$$

where $\widetilde{\xi} = \xi - \widehat{\xi}$ represents the error between true and estimated uncertainties, while $\widetilde{f} = f - \widehat{f}$ represents the estimation error of the fault vector.

The quantities $\widehat{\xi}$ and $\widehat{f}$ are computed by using neural network interpolators. In particular, as explained in the following sections, the estimation of the uncertain vector $\xi$ is computed using a SVM trained off-line along a set of fault free trajectories, while the estimation of the fault vector $f$ is computed on-line using a RBF network.

In view of (9), the quantity sensitive to the occurrence of faults, i.e., the *residual* vector, can be chosen as:

$$r(k) = e(k) - Fe(k-1) = \widetilde{\xi}(k-1) + \widetilde{f}(k-1), \quad (10)$$

which depends both on the vector $\widetilde{\xi}$ of the uncompensated uncertainties in the absence of faults and on the vector $\widetilde{f}$ which is nominally different from zero only in the presence of faults.

A fault is declared if at least one component of this vector exceeds a given threshold, that can be set on the basis of the maximum value of $\widetilde{\xi}$ during normal operation. When a fault is declared, only the components of $\widehat{f}$ corresponding to the components of $r$ exceeding the thresholds are updated on-line to provide fault estimation.

## 4. OFF-LINE INTERPOLATION

The estimation $\widehat{\xi}(k, x(k), u(k))$ of the vector of the uncertainties $\xi(k, x(k), u(k))$ in the absence of faults can be realized by resorting to suitable interpolation functions based, e.g., on neural networks. For the problem at issue, the training of the neural network can be performed off-line along a set of fault free trajectories in order to achieve an accurate compensation of the uncertainties. A convenient model of neural network for off-line interpolation is represented by Support Vector Machines (SVMs) that, with respect to conventional neural networks, offer a superior capability of generalization (Vapnik *et al.*, 1997).

In the so called $\epsilon$-SV regression problem, given a set of training data, $\{(x_1, y_1), \ldots, (x_N, y_N)\}$, where $x_i \in \Re^m$ is an input and $y_i \in \Re$ is a target output, the goal is to find a function $f(x)$ that has at most a deviation equal to $\epsilon > 0$ from the actual targets for all the training data and, at the same time, is as flat as possible.

For the nonlinear regression problem considered in this work, the function $f(x)$ can be written as

$$f(x) = \sum_{i=1}^{N} (\alpha_i - \alpha_i')k(x_i, x) + b, \quad (11)$$

known as *support vector espansion*, where the *kernel* function $k(x_i, x)$ may be, e.g., a polynomial function or a radial basis function (Haykin, 1999).

Training the SVM is equivalent to solve the following optimization problem:

$$\max_{\alpha_i, \alpha_i'} -\frac{1}{2}\sum_{i=1}^{N}\sum_{j=1}^{N}(\alpha_i' - \alpha_i)(\alpha_j' - \alpha_j)k(x_i, x_j) - \epsilon\sum_{i=1}^{N}(\alpha_i + \alpha_i') + \sum_{i=1}^{N}(\alpha_i - \alpha_i')y_i \quad (12)$$

subject to $(i = 1, \ldots, N)$:

$$\sum_{i=1}^{N}(\alpha_i - \alpha_i') = 0, \quad \alpha_i, \alpha_i' \in [0, c]$$

where the constant $c > 0$ is a user-defined parameter which allows to enhance the "flatness" of $f$ at the expense of the violation of the constraint on the magnitude of the interpolation error.

It can be shown (Haykin, 1999) that only for $|f(x_i) - y_i| \leq \epsilon$ the coefficient $\alpha_i - \alpha_i'$ may be nonzero. Therefore, only some training examples $x_i$ called *support vectors* are required to compute $f$. Moreover, the scalar $b$ can be computed as the mean value of the quantities

$$b_i = y_i - \sum_{j=1}^{l}(\alpha_j - \alpha_j')k(x_j, x_i)$$

computed for all the training vectors $x_i$, where $l$ is the number of the support vectors $x_j$.

The kernel function used in this work is a Radial Basis Function (RBF), hence, the estimate of the $i$-th component of $\widehat{\xi}$ is given by

$$\widehat{\xi}_i(k, z) = \sum_{i=1}^{l}\widehat{\beta}_i \exp\left(-\frac{\|z - s_i\|}{\sigma^2}\right) + \widehat{\beta}_0, \quad (13)$$

where $z = [\,x^T \quad u^T\,]^T$ is the network input, $l$ is the number of the support vectors, the centroids $s_i$ are the support vectors and the values of the parameters $\widehat{\beta}_i$, $s_i$ and $\sigma$ are determined through the off-line training. To perform the training of the network, the true values of $\xi(k)$ are reconstructed off-line using the discrete-time model (4), as:

$$\xi(k, x(k), u(k)) = x(k+1) - Ax(k) \\ -h(x(k)) - B(x(k))u(k).$$

## 5. ON-LINE INTERPOLATION

The fault vector $f(k, x(k))$ is assumed to depend upon the state and the $(q \times 1)$ constant parameters vector $\theta$. If $f$ is linear in the parameters vector, it can be expressed as follows

$$f(k, x(k), \theta) = \Omega(k, x(k))\, \theta, \qquad (14)$$

where the matrix $\Omega$ is assumed to be known, while $\theta$ is unknown. On the other hand, if $f$ is not linear in the parameters and/or its structure is not exactly known, a good approximation can be obtained by resorting to on-line interpolators (Demetriou and Polycarpou, 1998; Zhang *et al.*, 2002), as neural networks.

Hence, by choosing a linear-in-the-parameters interpolator structure, the uncertain term can be expressed as (hereafter, dependence of $\Omega$ upon $x$ will be skipped for notation compactness)

$$f(k, \theta) = \Omega(k)\, \theta + \epsilon^*(k), \qquad (15)$$

where $\epsilon^*$ represents the interpolation error, which is assumed to be uniformly bounded in norm. Notice that this error is nonzero only when a generic on-line interpolator is adopted (e.g., a neural network or polynomials).

The fault vector can be evaluated through the estimation of $\theta$. Namely, and adaptive update law for the parameters estimate $\theta(k)$ can be chosen as

$$\widehat{\theta}(k+1) = \widehat{\theta}(k) + \Omega^T(k)\, \Gamma_\theta(k)\, r(k+1) \qquad (16)$$

where $r$ is the residual vector defined in (10). The gain matrix $\Gamma_\theta(k)$ is chosen as $\Gamma_\theta(k) = 2\,(\Omega(k)\Omega^T(k) + Q_\theta)^{-1}$, where $Q_\theta$ is a positive definite symmetric matrix; hence, $\Gamma_\theta(k)$ is symmetric and positive definite for all $k$. Then, the estimate of the fault vector is given by

$$\widehat{f}(k) = \Omega(k)\widehat{\theta}(k). \qquad (17)$$

Notice that equation (17) must be used as soon as a fault is detected; only the components of $\widehat{f}$ corresponding to the components of the residual vector exceeding the threshold must be updated.

The $i$-th component of the fault vector in (17) is computed using a RBF neural network trained on-line via the parameters update law (16), i.e.:

$$\widehat{f}_i(k, x) = \sum_{i=1}^{q-1} \widehat{\theta}_i \exp\left(-\frac{\|x - c_i\|}{\sigma^2}\right) + \widehat{\theta}_0, \qquad (18)$$

where the number and location of centroids $c_i$ are determined via a K-means clustering algorithm executed off line on fault-free data.

It is worth noticing that the FD approach described above is based on the same concepts as in (Demetriou and Polycarpou, 1998; Zhang *et al.*, 2002). However, in these works it is assumed that uncertain dynamics and disturbances are bounded and perhaps small with respect to the fault magnitude; otherwise, the fault estimate $\widehat{f}$ would be polarized, thus leading to an unacceptable rate of false and/or missed alarms. Here, uncertainties and disturbances (structured and unstructured) are estimated off-line using Support Vector Machines, so as to obtain an accurate compensation of such terms. Moreover, the problem is tackled in the discrete time domain as in (Caccavale and Villani, 2003); hence, effects due to the approximate discrete-time equivalent of the system's dynamics are properly taken into account.

## 6. CASE STUDY

The effectiveness of the proposed fault detection scheme has been tested on the mathematical model of the NPS AUV II in a numerical simulation developed in MATLAB/SIMULINK. The mathematical model and the control loop for the speed, steering and diving autopilot presented in (Healey and Lienard, 1993) have been used. The simulation is obtained by considering the whole six-dofs coupled and nonlinear model, while the fault detection algorithm uses a nominal model, where all the hydrodynamic terms have been neglected and the value of the parameters has been modified by an error in the range of 5%-20% of the real value.

A sampling time of $T = 0.2$ s has been considered for the training phase as well as for the on-line fault detection algorithm. All the measurements have been simulated by adding zero-mean white noise and quantization errors to the model output.

The input of the observer are the angular rate of the propeller blade, and the angles of the control surfaces for diving and steering manoeuvres. The state variables are the 6 velocities and the 3 orientation variables, since the absolute position does not give any contribution to the dynamic model. The observer gain matrix $K_o$ in (8) has been selected as $K_o = 0.1I_9$, while $Q_\theta = \text{diag}\{70I_6, 500I_3\}$.

For off-line estimation of the function $\xi$, the training set of the SVM has been chosen by considering typical maneuvers such as steering on the horizontal plane, diving, simultaneous steering and diving, variation of the cruise velocity. In total about 12000 samples have been provided as inputs for the off-line learning phase. The SVMs' training is obtained resorting to the library `libsvm-2.36` that allows to scale the inputs, training the neural network and evaluate the mean square error. The inputs' scaling has been set to the full range value, e.g., for the vehicle linear velocity and for the rotation velocity of the blade, the values of 1.3 m/s and 1500 rpm has been chosen, respectively. The initial conditions have been taken as $\boldsymbol{\nu}_1 = [.916 \quad 0 \quad 0]^{\mathrm{T}}$ m/s and $z = 2\,\mathrm{m}$, and the ocean current was taken as about $25\,\mathrm{cm/s}$ for all the training trajectories.

The on-line estimation of the faults is achieved by using the RBF network (18); 25 centers for each component have been considered, leading to 225 centers in total. The RBFs' training has been achieved resorting to the package `rbf_nets_v1` under MATLAB with 25 centers and 10 training cycles for each dof. The inputs' scaling has been set as for the SVMs.

In order to perform a proper fault detection, suitably defined thresholds on the residuals have been selected. Thresholds setting has been achieved by measuring the residuals obtained in a set of fault-free trajectories under various operating conditions.

The results of the training have been validated on a test set different from the training set and the interpolation errors, not reported here for brevity, result to be small.

The simulated failure concerns the thruster blade. In the first simulation, a couple of successive steering manoeuvres have been considered, whose planar projection is reported in Fig. 1; the corresponding angular velocity of the propeller blade ($n$) in the absence of failures is reported in Fig. 2, while the rudders' angles are reported in Fig. 3. At time $t = 60\,\mathrm{s}$ the propeller blade gets blocked, and the corresponding angular velocity is taken to zero.

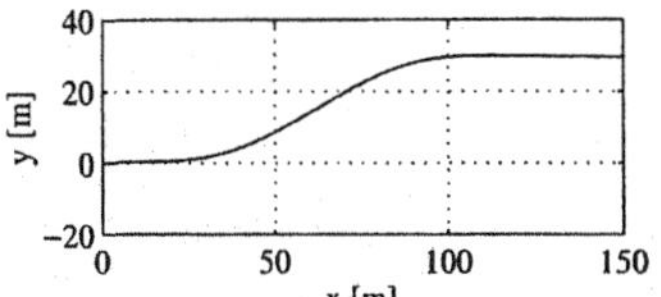


Fig. 1. Horizontal view of the path of the vehicle in the absence of failures.

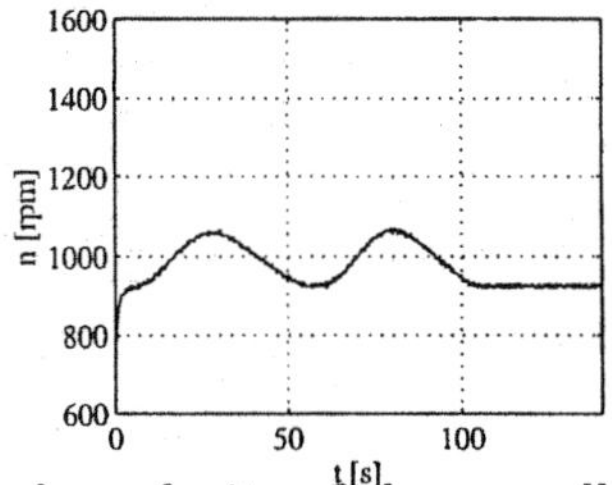


Fig. 2. Angular velocity of the propeller blade in the absence of failures.

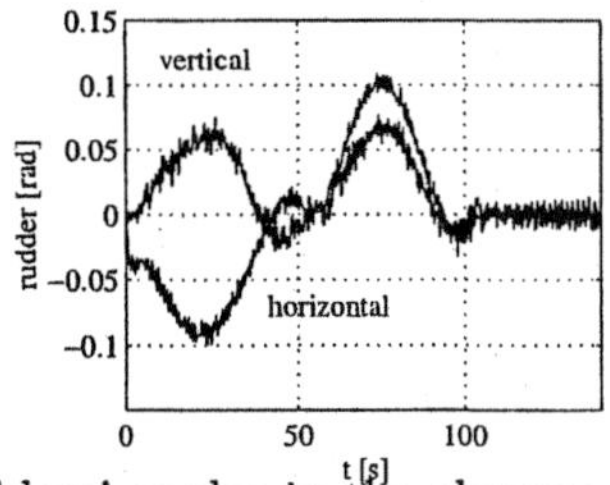


Fig. 3. Rudders' angles in the absence of failures.

Depending on the selected threshold, the detection of the fault might take different time lengths; in the simulated case, a few sample times are enough to detect and isolate the failure ($t = 60.4\,\mathrm{s}$). As expected, the fault signature is present only on the residual corresponding to the linear velocity along $x$, which is reported in Fig. 4. The time history of the estimate of the fault effect in terms of the corresponding component of the generalized force vector ($\boldsymbol{\tau} = \boldsymbol{B}_a(\boldsymbol{\nu})\boldsymbol{u}$) is reported in Fig. 5.

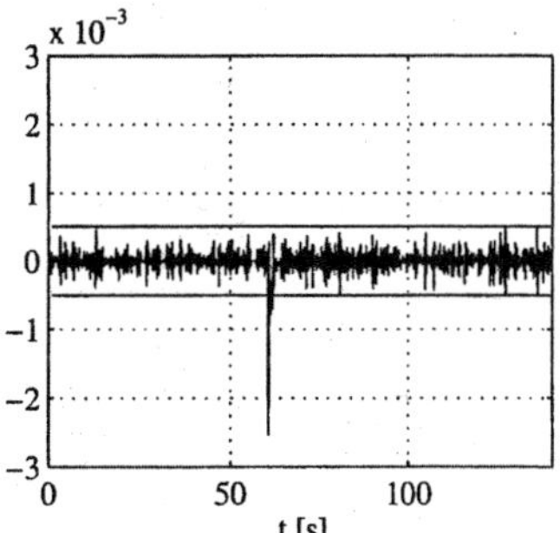


Fig. 4. Residual corresponding to the linear velocity along $x$.

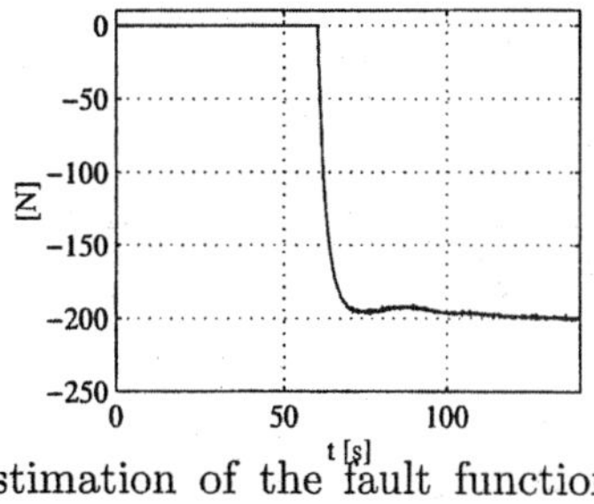


Fig. 5. Estimation of the fault function for the $x$-component of the force in vector $\boldsymbol{u}$.

As a consequence of the fault, the controller might ask more control input to the thruster; for this specific fault, in fact, the controller might increase the required propeller velocity as reported in Fig. 6. However, since the blade is blocked, the linear velocity of the vehicle is decreasing (Fig. 7).

Notice that the instant of detection depends on the dynamic influence of the fault. In other words, if a particular fault does not excite any significant dynamics, the time required for detection may be longer or the fault may be not detected at all. For example, if the rudder fails when the vehicle is moving along a straight line in the absence of current, the observer might be unable to detect the failure. In the case shown here, in

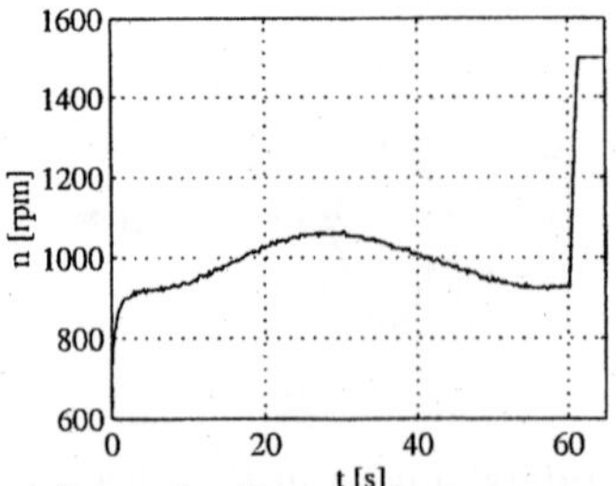


Fig. 6. Angular velocity of the commanded propeller blade in the presence of the fault (first 65 s).

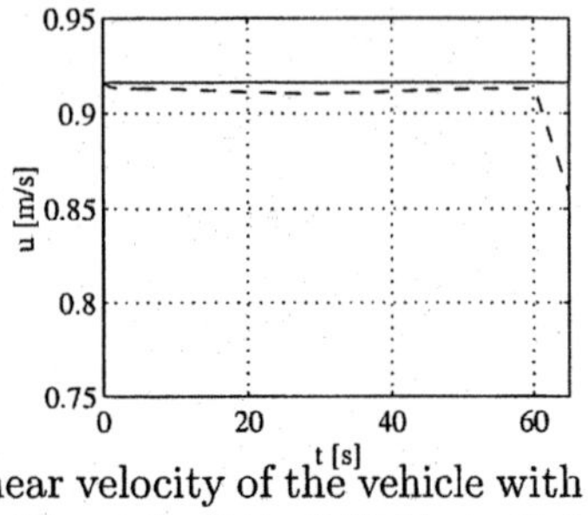


Fig. 7. Linear velocity of the vehicle with (dashed) and without fault (solid) (first 65 s).

fact, the linear force acting on the vehicle, with and without fault, is quite different as shown in Fig. 8.

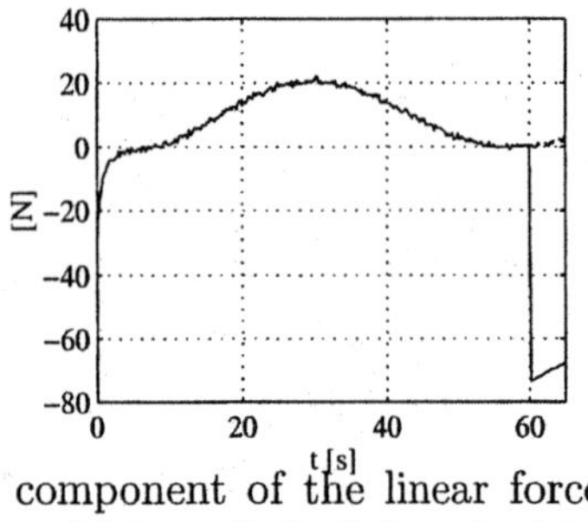


Fig. 8. $X$ component of the linear forces acting on the vehicle with (solid) and without fault (dashed) (first 65 s).

## 7. CONCLUSION

In this paper a neural network approach has ben proposed for the diagnosis of actuator faults of underwater vehicles. The diagnosis is based on a diagnostic observer, designed on the basis of the available dynamic model of the AUV, which also contains a compensation term for unknown dynamics, uncertainties and disturbances. This term is computed using a SVM neural network which is trained off-line on a set of fault free trajectories. When an actuator fault occurs, the estimation of the fault vector is performed using an on-line interpolator based on a RBF neural network. A simulation case study developed for the NPS AUV II vehicle subject to a thruster blade failure has been presented. The results confirm the effectiveness of the proposed approach on detecting, isolating and estimating the time evolution of faults.

## ACKNOWLEDGMENTS

This work was supported by MIUR and ASI.

## References

G. Antonelli, "A Survey of Fault Detection/Tolerance Strategies for AUVs and ROVs", in *Fault Diagnosis and Fault Tolerance for Mechatronic Systems: Recent Advances*, F. Caccavale and L. Villani (Eds.), Springer-Verlag, London, 2003.

F. Caccavale, L. Villani, "Fault Diagnosis for Industrial Robots," in *Fault Diagnosis and Fault Tolerance for Mechatronic Systems: Recent Advances*, F. Caccavale and L. Villani (Eds.), Springer-Verlag, London, 2003.

M.A. Demetriou, M.M. Polycarpou, "Incipient Fault Diagnosis of Dynamical Systems Using Online Approximators," *IEEE Transactions on Automatic Control*, Vol. 43, pp. 1612–1617, 1998.

J.S. Ferguson, A. Pope , B. Butler, R. Verrall, "Theseus AUV - Two Record Breaking Missions," *Sea Technology Magazine*, pp. 65–70, 1999.

T.I. Fossen, *Guidance and Control of Ocean Vehicles*, John Wiley & Sons, Chichester, UK, 1994.

S. Haykin, *Neural networks: A comprehensive foundation*, Prentice Hall, Upper Saddle River, NJ, 1999.

A.J. Healey, F. Bahrke, J. Navarrete, "Failure Diagnostics for Underwater Vehicles: A Neural Network Approach," *IFAC Conference on Maneuvering and Control of Marine Craft*, pp. 293–306, 1992.

A.J. Healey and D. Lienard, "Multivariable sliding mode control for autonomous diving and steering of unmanned underwater vehicles," *IEEE Journal of Oceanic Engineering*, vol. 18, pp. 327–339, 1993.

R.J. Patton, P.M. Frank, R.N. Clark, *Issues in Fault Diagnosis for Dynamic Systems*, Springer-Verlag, London, 2000.

V. Vapnik, *The Nature of Statistical Learning Theory*, New York, Springer-Verlag, 1995.

V. Vapnik, S. Golowich and A. Smola, "Support Vector Method for Function Approximation, Regression Estimation, and Signal Processing," in *Neural Information Processing Systems*, M. Mozer, M. Jordan, and T. Petsche (Eds.), Vol. 9, MIT Press, Cambridge, MA, 1997.

X. Zhang, M.M. Polycarpou, T. Parisini, "A Robust Detection and Isolation Scheme for Abrupt and Incipient Faults in Nonlinear Systems," *IEEE Transactions on Automatic Control*, vol. 47, pp. 576–593, 2002.

ELSEVIER
IFAC
PUBLICATIONS
www.elsevier.com/locate/ifac

# NEURO-FUZZY MODELLING OF MARINE DIESEL ENGINE CYLINDER DYNAMICS

**Radovan Antonić[*] , Zoran Vukić[**] , Ognjen Kuljača[***]**

[*] *Faculty of Maritime Studies, Univ. Split*
*Zrinsko-Frankopanska 38, 21000 Split, Croatia*
*e-mail:* antonic@pfst.hr *; tel: +385 21 380-762 ; fax: +385 21 380-779*
[**] *Faculty of Electrical Engineering and Computing, Univ. Zagreb*
*Unska 8, 10000 Zagreb, Croatia*
[***] *Dep. of Advanced Technology School of Agriculture and Applied Sciences*
*Alcorn State Univ. 1000 ASU Drive 360, USA*

Abstract: In this paper, the practical application of some well recognised fuzzy methods and neural networks techniques to modelling marine diesel engine cylinder dynamics using real-time data and expert knowledge has been considered. The simulation was done in Matlab environment with real-time data originated from 2-stroke marine diesel propulsion engine on test bed during final testing, combined with knowledge elicited from engine experts and experienced test bed operators. Takagi-Sugeno fuzzy model has been designed based on cylinder pressure data after their clustering using fuzzy subtractive method. Model parameter tuning was investigated using ANFIS with combined learning algorithms: least-squares and back-propagation gradient descent method. The model obtained can be of practical importance in engine working regime adjustment, predicting cylinder data in faulty sensor case or adaptive threshold tuning within faults detection and identification. *Copyright © 2004 IFAC.*

Keywords: marine diesel engine, test bed, expert knowledge, data pre-processing, fuzzy model, fuzzy clustering, neuro-fuzzy inference

## 1. INTRODUCTION

Identification and modelling of dynamic systems, especially of nonlinear ones, from input-output measurements becomes very important topic of scientific research with a wide range of practical applications. Artificial neural networks (ANN) and fuzzy logic (FL) combined with each other offer very promising possibilities. So in recent years, neuro-fuzzy methods take the growing interest of researchers for their application to identification, modelling and adaptive control of nonlinear systems (Takagi and Sugeno, 1985; Chiu, 1994; Nakamori and Ryoke, 1994; Jin, 2000). They provide models suitable for human expert's or plant operator's decision making and easy integration of heuristic knowledge and experience with numerical data from supervisory system in the form of familiar *If-Then* rules. Neuro-fuzzy approach offer a powerful method for converting information contained in real-time numerical data into fuzzy-logic rules which can be interpreted and optimized by human expert / operator or directly used in fuzzy system (see fig. 1).

In many practical cases the numerical measurements or experimental data have to be pre-processed, before neuro-fuzzy modelling, at least for the following reasons:

- raw data processing (filtering, normalisation,...),
- removing redundant and inconsistent data,
- resolving conflicts in the data.

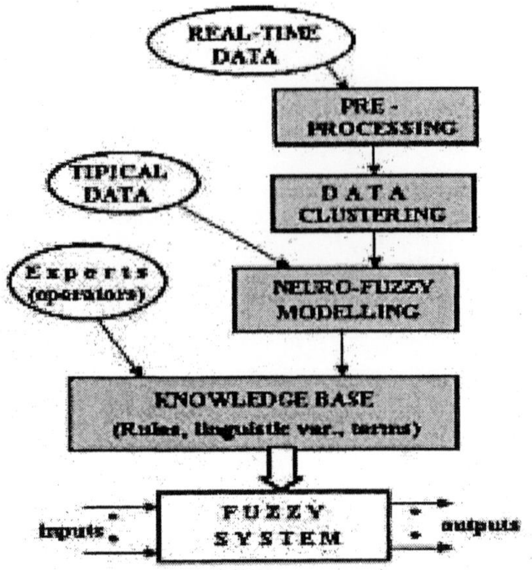


Fig. 1: Neuro-fuzzy approach to process modelling

## 2. DATA PRE-PROCESSING AND FEATURE EXTRACTION

The acquisition of relevant data to build a model of some dynamic process is sometimes rather difficult task. To build a good model, a representative data set should be obtained for all operating regimes under different conditions, both in normal and faulty engine operations. In some cases we have to use past process data, long period data trends, elicit data from experts or process operators. High risk processes, like marine diesel propulsion engine, can't be disturbed during operation, so effective means of gathering specific data are test bed experiments and simulations. We used data obtained from engine test bed for engine cylinder dynamics modelling.

Before using such data, especially those from sensors, some data conditioning i.e. pre-processing and feature extraction steps must be taken. The most common are:

- *Data filtering*: sensor data are usually noisy, so filtering can provide valuable data. Some Kalman filtering methods could be effective (Antonić, 2002).
- *Pre-processing of the data* : in this phase, the checking of data ranges, expected mean and standard deviations should be done, removing those data that originate from sensor drifts, faulty sensors, values outside the defined domains, etc. If needed, data normalisation could be done. Normalisation of data value $u$ to its normalised value $\bar{u}$ on a standard universe, say [0, 1] is done according to:

$$\bar{u} = \frac{u - u_{\min}}{u_{\max} - u_{\min}} \qquad (1)$$

where: $u_{\min}$ is the smallest value of measurements and $u_{\max}$ is the largest one. Example: normalisation of cylinder pressure signal 4 - 20 mA to the range [0, 1].

Also, scaling the values onto a particular range may be necessary. Scaling value $u$ from the range $[u_1, u_2]$ to the value $u'$ in the range $[u_1', u_2']$ can be done with the expression:

$$u' = \frac{u_2' - u_1'}{u_2 - u_1}(u - u_1) \qquad (2)$$

Example: cylinder pressure signal 4-20 mA scaled to real physical values 0-130 bar.

After pre-processing, relevant features can be selected by an expert or process operator. For a set of data $\mathbf{u} = (u_1, u_2, ..., u_K)$, some characteristic quantities, which could be used in feature extraction are:

- mean value: $$m = \frac{1}{K}\sum_{i=1}^{K} u_i \qquad (3)$$

- variance: $$v = \frac{1}{K-1}\sum_{i=1}^{K}(u_i - m)^2 \qquad (4)$$

- standard deviation: $$\sigma = \sqrt{v} \qquad (5)$$

- range: $$r = u_{\max} - u_{\min} \qquad (6)$$

- correlation: features may be correlated, so a change in one feature $X$ implies a similar change in another feature $Y$. A correlation coefficient can be calculated in this way:

$$c_r = \frac{\sum_{i=1}^{K}(x_i - m_1)(y_i - m_2)}{\sqrt{\sum_{i=1}^{K}(x_i - m_1)^2 \sum_{i=1}^{K}(y_i - m_2)^2}} \qquad (7)$$

where: $m_1$ is the mean value of all the values $x_i$ of feature *X*, and $m_2$ is the mean value of all the values $y_i$ of feature *Y*. If two features are strongly correlated, one of them is redundant i.e. unnecessary.

- *Reducing the data*: in some situations (long period of data gathering, using data from different sensors, etc.) we could have a huge amount of data with much redundancy. It may cause many problems in future data handling. One well-known method for reducing data is Principal Component Analysis (PCA). Correlation techniques can also be used to determine the influence of inputs to outputs, ranking tests using radial basis function network, mean instead of actual values. In the simulation model we used mean values during 50 engine working cycles, so reducing the data set needed for modelling cylinder dynamics.
- *Data modelling (feature extraction)*: depends on the aim the model is used for: control, identification, advice, decision making, etc. Radial basis function

networks, Kohonen or similar mappings, (Nguyen and Prasad, 1999, Antonić, et. al.,2003), some clustering methods can be used ( Chiu, 1994; Nakamori and Ryoke, 1994). In what follows, we concentrate on building neuro-fuzzy model using fuzzy clustering techniques, fuzzy modelling and adaptive neuro-fuzzy inference system (ANFIS).

## 3. BUILDING FUZZY MODEL FROM DATA

Building fuzzy models from data involves methods based on fuzzy logic and approximate reasoning (Mamdani, 1977), but also ANN techniques with their adaptive and learning possibilities. Integration of expert knowledge and numerical data can be done in two main approaches:

- The expert knowledge can be transformed into a set of *If-Then* rules creating a certain model structure. Parameters in that structure (membership functions, rule weights,.) can be fine tuned using measured input-output data. At the computational level , a fuzzy model can be seen as a layered structure to which standard learning algorithms can be applied. This approach is usually termed as *neuro-fuzzy modelling.*
- fuzzy model can be constructed from data giving the extracted rules and MF which can provide a posteriori information of the system's behaviour. An expert, with his knowledge and experience, can modify the rules or add new ones, design additional experiments etc. This approach can be termed as *rule extraction*. Fuzzy clustering is one of the techniques often applied in rule extraction from data (Nakamori and Ryoke, 1994).

### *3.1 Fuzzy clustering*

Identification methods based on fuzzy clustering originate from data analysis and pattern recognition, where the concept of graded membership is used to represent the degree to which a given object, represented as a vector of features, is similar to some prototypical object. The degree of similarity can be calculated using a suitable distance measure.

The idea is depicted in fig. 2, where the data are clustered into two groups with prototypes $c_1$ and $c_2$, using the Euclidean distance measure. The partitioning of the data is expressed in the *fuzzy partition matrix* whose elements $\mu_{ij}$ are degres of membership of the data points $[x_i\ y_i]$ in a fuzzy cluster with prototypes $v_j$.

Fuzzy *If - Then* rules can be extracted by projecting the clusters onto the axes. Each obtained cluster is represented by one rule in the Takagi-Sugeno model. The membership functions for fuzzy sets $A_1$ and $A_2$ are generated by point-wise projection of the partition matrix onto the antecedent variables. The consequent parameters for each rule are obtained as least-square estimates. Two most common used fuzzy clustering algorithms are (Bezdek, 1981):

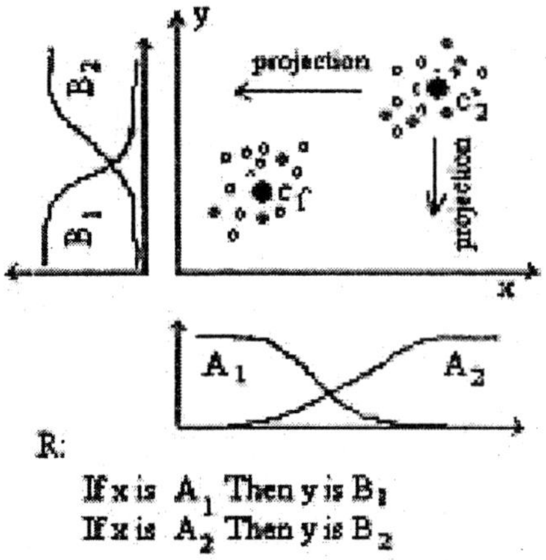


Fig. 2: Rule based interpretation of fuzzy clusters

<u>FCM (Fuzzy c-means) algorithm</u> is a supervised one, which employs fuzzy partitionig, allowing each data point to belong to a cluster according to degree specified by a membership grade, thus each data point can belong to several clusters. FCM partitions a set of K data points specified by n-dimensional vectors $\mathbf{u}_k$, (k = 1,2,...,K) into *C* fuzzy clusters and finds a center in each minimizing an objective function:

$$J(\mathbf{M},c_1,c_2,...,c_c)=\sum_{i=1}^{C}J_i=\sum_{i=1}^{C}\sum_{k=1}^{K}m_{ik}^q d_{ik}^2 \quad (8)$$

where: $m_{ik}$ is a membership grade between 0 and 1, $c_i$ is the centre of fuzzy cluster *i*, $d_{ik}=\|\mathbf{u}_k-\mathbf{c}_i\|$ is the Euclidean distance between *i-th* cluster centre and *k-th* data point, $q\in[1,\infty)$ is a weighting exponent. Membership matrix **M** is allowed to have elements in the range [0 1]. Two necessary conditions for J to reach a minimum are:

$$c_i=\frac{\sum_{k=1}^{K}m_{ik}^q\mathbf{u}_k}{\sum_{i=1}^{K}m_{ik}^q}\ ;\ m_{ik}=\frac{1}{\sum_{j=1}^{C}\left(\frac{d_{ik}}{d_{jk}}\right)^{2/(q-1)}} \quad (9)$$

The algorithm is an iteration through these two conditions. The algorithm may not converge to an optimum solution and the performance depends on the initial cluster centres.

<u>Subtractive clustering</u> is an unsupervised algorithm and it is more suited if number of cluster centres is not known, what's the case with engine cylinder pressure data. This algorithm is based on a measure of the density of data points in the feature space (Chiu, 1994). The idea is to find regions in the feature space with high densities of data points. The point with the highest number of neighbours is selected as cluster centre. The data points within a

predefined fuzzy radius are then removed i.e. subtracted and the algorithm looks for a new point with the highest number of neighbours. This continues untill all data are examined. For algorithm illustration, consider a set of K data points specified by n-dimensional vectors $\mathbf{u_k}$ (k = 1,2,...,K). Since each data point is a candidate for a cluster centre, a density measure at data point $u_k$ is defined as:

$$D_k = \sum_{j=1}^{K} \exp\left(-\frac{\|\mathbf{u}_k - \mathbf{u}_j\|}{(r_a/2)^2}\right) \qquad (10)$$

where $r_a$ is a positive constant. Only the fuzzy neghbours within the radius $r_a$ contributes to the density measure. After calculating the density measure for each data point, the point with the highest density is selected as the first cluster centre. Let $u_{c1}$ be the point selected and $D_{c1}$ its density measure. Next, the density measure for each data point $u_k$ is revised by formula:

$$D_k^* = D_k - D_{c1} \exp\left(-\frac{\|\mathbf{u}_k - \mathbf{u}_{c1}\|}{(r_b/2)^2}\right) \qquad (11)$$

where $r_b$ is a positive constant (tipically $r_b = 1.5 * r_a$) which defines a neighbourhood with reduced density measure. The process is repeated until a sufficient number of cluster centres are generated. When applying this method to a set of input-output data, each of the cluster centres represents a fuzzy rule.

*3.2 Neuro - fuzzy modelling*

Combining knowledge based fuzzy model with a data-driven tuning of the model parameters using ANN techniques is very effective way of non-linear process modelling. Fuzzy models can be seen as logical models which use "*If-Then*" rules and fuzzy logical operators to establish qualitative relationships among the variables in the model. Two *important* steps must be taken with regard to the design of fuzzy model (fig. 3):

- *structure selection*
  - determine input and output model variables,
  - structure of the rules (choice of the fuzzy model type),
  - number and type of membership functions and terms for each variable,
  - choice of the inference mechanism, connective fuzzy operators, defuzzification method.

The most common types of fuzzy models used are of Mamdani and Takagi-Sugeno's type.

Fuzzy model of Mamdani type: fuzzy linguistic rules in such ordinary models are of the following form:

$$R^{(j)}: If\ x_1\ is\ A_1^j ...\ And\ x_n\ is\ A_n^j ...\ Then\ y\ is\ B^j \qquad (12)$$

where $A_i^j, and\ B^j$ are linguistic variables - terms, $x = (x_1, x_2, \dots x_n)^T \in U \subset R^n$ and $y \in V \subset R$ are fuzzy input and output variables of the *j*-th rule, respectively, and *j* = *1,2,...J.* Both, the rule antecedents and rule consequent are defined by means of fuzzy sets $A_i^j, and\ B^j$ that are characterised by membership functions $\mu_{a\,i}^{\ j}$ and $\mu_b^{\ j}$.

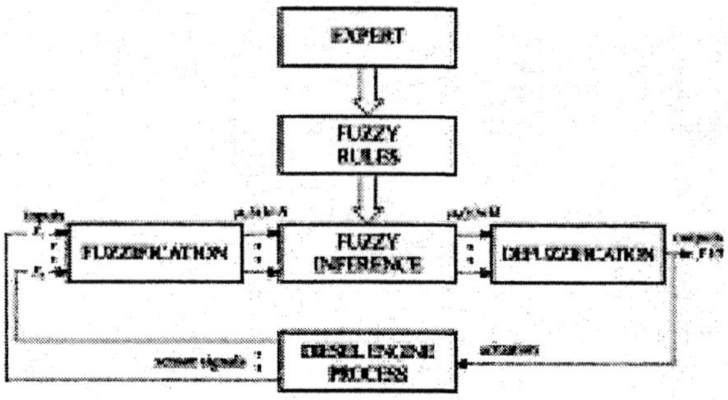

Fig. 3: Fuzzy model structure of diesel engine process using sensor data and expert knowledge

Takagi and Sugeno's (T-S) fuzzy model extends the linguistic rules to rules with consequent in the form of linear functions of antecedent or premise variables:

$R^{(j)}$: *If* $x_1$ is $A_1^j$..... *And* $x_n$ is $A_n^j$..... *Then*

$$y^j = c_0^j + c_1^j x_1 + ..... + c_n^j x_n \qquad 13)$$

$c_i^j$ is a consequent parameter, $y^j$ is the system output due to rule $R^{(j)}$, and *i* = *1,2,....,J*.
Each rule represents a locally linear model. The final output of the fuzzy model is inferred by taking the weighted average of the $y^j$ :

$$y = \sum_{j=1}^{J} w^j y^j / \sum_{j=1}^{J} w^j \qquad (14)$$

where weight $w_j$ implies overall truth value of the j-th rule premise part and is calculated as:

$$w_j = \prod_{k=1}^{n} A_k^i(x_k) \qquad (15)$$

The advantage of this fuzzy linear model is that the parameters $c_i^j$ of the model can be easily identified from numerical data (using ANN with adequate learning methods).

Singleton fuzzy model is a special case of T-S model with consequent part in the form of fuzzy singleton.

$$R^{(j)}: If\ x_1\ is\ A_1^j ...\ And\ x_n\ is\ A_n^j ...\ Then\ y^j = c_0^j \qquad (16)$$

Taking rule strength i.e. certainty grade CF ( $0 \le CF_j \le 1$ )of each rule, the model takes the form:

$$R^{(j)}: \; If \; x_1 \, is \, A_1^j ... \; And \; x_n \, is \, A_n^j ... \; Then$$

$$y^j = c_0^j \quad \text{with } CF_j \tag{17}$$

This model, to a certain degree, inherit good identification properties from T-S model and the advantages of better interpretation of the Mamdani model. It is well suited for non-linear function approximation and was chosen to be used in our simulation case.

- *parameter adjusting*

  After the structure is determined, the performance of a fuzzy model can be fine tuned by adjusting its parameters i.e. membership functions (shape and position) and rule's strength (weights). This can be done very effectively using ANN (fig. 4.).

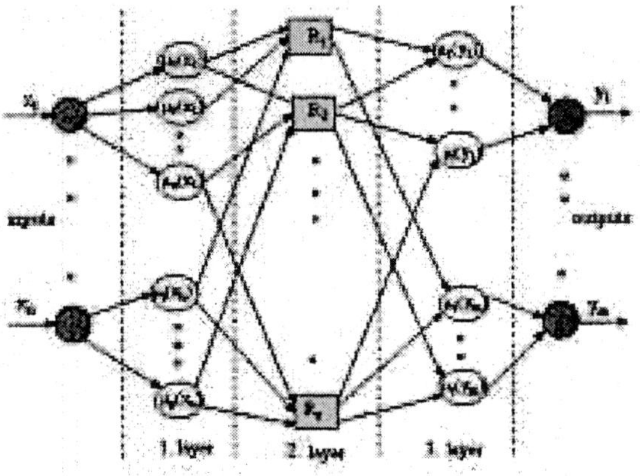

Fig. 4: Three-layer neuro-fuzzy model for parameter tuning

## 4. SIMULATION CASE: TEST RESULTS

For the modelling of marine diesel engine cylinder dynamics we used real-time data obtained during engine testing on test bed. The 2-stroke marine diesel propulsion engine was MAN-B&W of type 6S70MC, with large hydraulic brake as engine load. The measurement system (fig. 5) was based on PC with data acquisition multifunction card and software developed using C language (Antonić, 2002).

Cylinder pressure is the most informational engine working condition parameter. Its measurement is the most demanding task because of its dynamic nature and high temperatures influence, so special care was taken for its measurements and pre-processing.

Measuring and acquisition of cylinder pressure data during engine working cycles were strictly in correspondence with crankshaft angle (CA) relative to the top dead centre (TDC). Incremental encoder with 2048 discrete angle positions (0.18° resolution) was used to detect current crankshaft angle. Two piezoelectric water cooled pressure sensors were used for cylinder engine monitoring during testing cycles. 50 successive working cycles was measured for each cylinder with 2048x50=102400 data points per cylinder. Testing was done with engine load of 50%, 75%, 85% and full load (100%). Some pre-processing steps, mentioned above, were used to measured data. In the modelling of engine cylinder we used mean values during 50 working cycles of first engine cylinder.

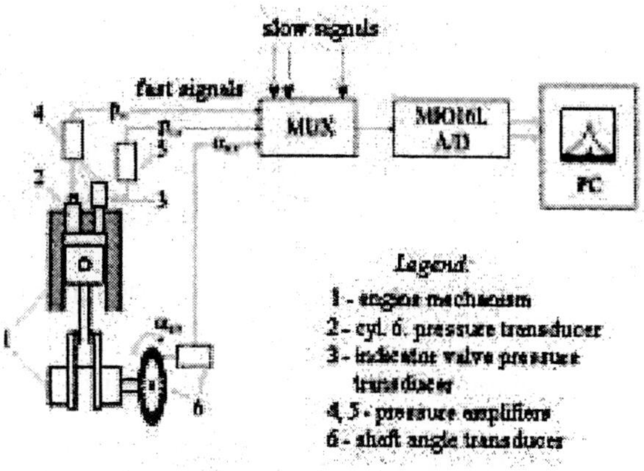


Fig. 5: Engine cylinder pressure measurement system

Simulation was used in Matlab environment with Fuzzy Logic Toolbox.

Fuzzy subtractive clustering method was used to the set of data for each of four operating points i.e. with engine load of: 50%, 75%, 85%, 100% (see fig. 6, top). Results of clustering step are shown in fig. 6, down.

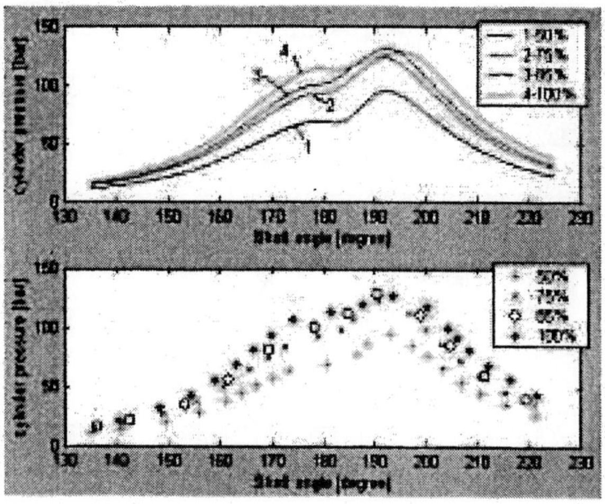

Fig. 6: Cylinder engine pressure data before and after fuzzy subtractive clustering (CA=135-225 deg.).

After data clustering, adaptive neuro-fuzzy inference method was used to improve model (T-S type) performance according to fig. 7.

A part of cylinder data within four operating points used during training procedure is given in table 1.

Combined learning algorithm was used during model training: least-squares method and back-propagation gradient descent method.

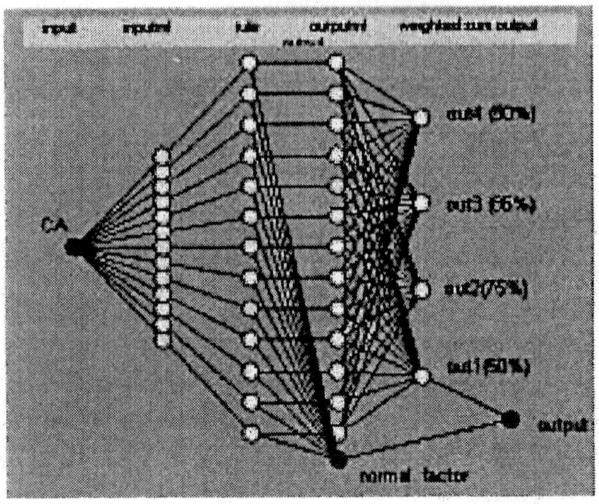


Fig. 7: Neuro-fuzzy model of engine cylinder pressure dynamics.

Table 1: A part of engine cylinder data for training

| CA [deg.] | 50% pc [bar] | 75% pc [bar] | 85% pc [bar] | 100% pc [bar] |
|---|---|---|---|---|
| 185 | 73.087 | 104.616 | 114.513 | 116.406 |
| 186 | 75.854 | 107.844 | 117.288 | 117.522 |
| 187 | 79.545 | 111.258 | 120.445 | 119.206 |
| 188 | 84.311 | 115.139 | 123.988 | 121.112 |
| 189 | 87.519 | 117.769 | 126.198 | 122.592 |
| 190 | 92.028 | 120.645 | 128.789 | 124.049 |

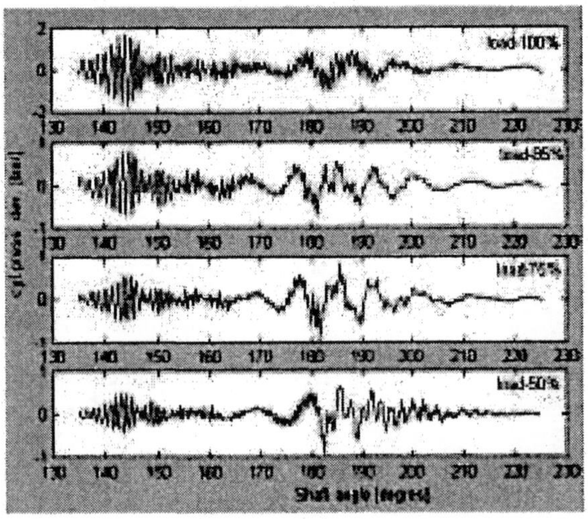

Fig. 8: Cylinder pressure deviation during training

Model performance measure was choosen to be RMSE (root-mean square error). The results obtained with training of 200 epochs were quite good. Fig. 8. shows cylinder pressure deviations between model and nominal process in four operating points. Table 2. gives RMSE without CF for each rule (first row) and with expert's given CF (second row).

The largest deviations are in the range near the TDC because of the very fast dynamic signal changes and high temperatures influence.

Table 2: RMSE with training data after 200 epochs

| RMSE [bar] | load 50% | load 75% | load 85% | load 100% |
|---|---|---|---|---|
| - | 0.231 | 0.252 | 0.261 | 0.471 |
| with CF | 0.219 | 0.223 | 0.254 | 0.452 |

## 5. Conclusion

In the paper we considered the possibilities and effectiveness of some recognised fuzzy methods and neural networks techniques to modelling marine diesel engine cylinder dynamics using real-time data and expert knowledge. The simulation was done in Matlab with real data originated from testing marine diesel propulsion engine on test bed. We used T-S model designed from data and modified by engine expert within ANFIS. The obtained model could be very effectivelly used in optimising engine working regime during testing on test bed, predicting cylinder pressure values in case of faulty sensor or in determining adaptive treshold values for better diagnosis.

## References


Antonić, R., Z. Vukić, Lj. Kuljača (2003): *Fuzzy Modelling and Control of Marine Diesel Engine Process*, Proc. of 11'th Mediterranean Conf. on Control and Automation, Greece, CD.

Antonić, R. (2002): *Fault Tolerant Control System for Large Diesel Engine in Marine Propulsion*, Ph.D. dissertation, (in Croatian).

Bezdek, J.C. (1981): *Pattern Recognition with Fuzzy Objective Function Algorithm*, Plenum Press, New Jork, 1981.

Chiu, S. (1994): *Fuzzy Model Identification Based on Cluster Estimation,* Journal of Intell. & Fuzzy Systems, Vol 2, No 3.

Jin, Y. (2000): *Fuzzy Modeling of High-Dimensional Systems: Complexity Reduction and Interpretability Improvement*, IEEE Trans. on Fuzzy Systems, Vol. 8, No 2, 212-221.

Mamdani, E.H.(1977): "*Applications of fuzzy logic to approximate reasoning using linguistic synthesis*," IEEE Trans. on Computers, Vol. 26, No. 12, 1182-1191.

Nakamori, Y., M. Ryoke (1994): *Identification of fuzzy prediction models through hyperellipsoidal clustering,* IEEE Trans. Systems, Man and Cybernetics, 24(8), 1153-1173.

Nguyen, H.T., R. N. Prasad (1999): Fuzzy Modeling and Control - Selected Works of M. Sugeno, CRC Press LLC.

Takagi, T., M. Sugeno (1985): *Fuzzy identification of systems and its applications to modeling and control,* IEEE Trans. Systems Man Cybernet. 15, 116-132.

ELSEVIER
IFAC
PUBLICATIONS
www.elsevier.com/locate/ifac

# A TWO LAYER CONTROLLER FOR INTEGRATED FIN AND RUDDER ROLL STABILISATION

**Reza Katebi**

*Industrial Control Centre, University of Strathclyde*
*Email: r.katebi@eee.strath.ac.uk*

**Abstract:** The design of integrated fin-rudder roll stabilisation (IFRS) control system has been studied in this paper. A new approach based on model based predictive control with dynamic weightings is employed to design the control system. It is assumed that the fin controller and the autopilot are designed and implemented. The controller is designed taking into account the non-minimum phase interaction from yaw to roll and the rudder angle and rate constraints. The rudder angle and rate are defined as the outputs of closed-loop ship model and outputs constraints are introduced to avoid actuator saturation. It is also shown that the weightings parameters can be used to make the controller adapt to different sea conditions. Simulation results are presented to demonstrate the effectiveness of the proposed control design technique. *Copyright © 2004 IFAC*

**Keywords:** Roll stabilisation, Autopilots, Fin, Rudder, Predictive Control

## 1. INTRODUCTION

A ship on the open sea, when subjected to the motion of waves is a highly resonant dynamic system. The ship has a dynamic behaviour similar to an inverted pendulum, where the magnitude of the swing, or roll, is dependent upon the action of the waves on the ship. The frequency and magnitude of this motion are dependent on the natural frequency $\omega_n$ and damping factor $\xi$ of the ship. It is thus necessary for ships carrying passengers or weapons platforms that this undesirable motion is controlled.

Controlling the roll motion of a ship is in essence a simple problem, the roll produced on the ship by the force of the sea wave disturbance, is counteracted either by resisting the wave or actively forcing the ship to roll in the opposite direction. Such a simple problem has produced many solutions: simple bilge keels on the hull; swinging weighted pendulums in the ship structure and the relatively complex computer controlled active use of the rudder. The ship when subjected to sea disturbances can roll over nearly two decades of frequencies. Thus, systems such as ballast tanks are good at correcting steady heel angles and counteracting low frequency wave disturbances while active fin or rudder control is used to stabilise the roll at ship natural roll frequency. Due to the single objective design and large capability to force roll the ship, fin roll control is now widely used and there are some examples of employing rudder for roll stabilisation.

The rudder roll control has also been studied in details (Cowley and Lambert, 1975, Baitis et al, 1983, der Klugt, 1987). For vessels that have fin stabilisers already fitted a logical extension is to combine RRS with control of the fins to provide an integrated control scheme. The suitability of this proposal was evaluated by Källström (1981) where a multivariable LQG controller was used. RRS has nonminimum phase behaviour and the system parameters vary extensively dependant on ship speed, this uncertainty is not explicitly handled in an LQG design. A more natural way of handling this uncertainty is by using specific robust control techniques such as H-inf (Roberts, et al, 1997) which was the method to design dual fin and rudder controllers. By combining all the aspects of ship motion control namely an autopilot controlling the rudder, a RRS and fin stabilisers a total ship control concept is established (Katebi et al 1987) which brings potential benefits to the system by co-ordinating systems so that they are operating in co-operation instead of opposition. Although the benefits of RRS have been demonstrated, the wide spread use of the technique has not occurred to the expected extent, and in the main has been limited to naval vessels. The fact that the rudder is a safety critical piece of the ship's manoeuvring system has made ship designers and operators reluctant to place any more stress on equipment that could mean the loss of the ship if it failed. Also, the naval vessels are not underwritten by the classification societies, and that optimal ship performance is essential, explains why naval vessels have made more use of RRS technology (Hickey, 2000, Hickey et al, 1999).

This paper is concerned with the design of a two layer integrated rudder-fin controller. The first layer consists of the autopilot loops and the fin controller. The second layer of the control which uses rudder to stabilise the roll is the subject of this paper. Model based predictive control with dynamic weightings is proposed to design the controller. This serves two purposes. Firstly, the MBPC can handle rudder angle and slew rate limiting and secondly by appropriate choice of the MBPC performance index and introduction of a dynamic weighting which has unit gain at low frequencies, maximum weighting at roll natural frequency and low gain at high frequencies, the effect on the autopilot loop can be minimised and the desired shape for the sensitivity function can be achieved.

The paper is organised as follows. Section two derives the models needed for control design. Section 3 introduces the state space MBPC with dynamic weighting. Simulation results and discussion on the performance of the proposed control system is presented in Section 4. Conclusion is drawn in Section 5.

## 2. SYSTEM DESCRIPTION

The basic dynamic of the ship roll motion with respects to the fin and rudder and the sea wave is shown in Figure 1. This model encapsulates the dynamic behaviour of the surface ships. The model has several structural features that can be exploited for Integrated Fin and Rudder Roll Stabilisation (IFRS) control systems.

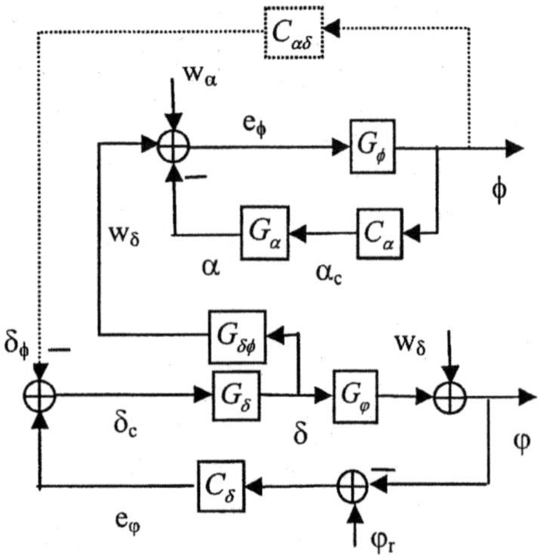


Figure 1 The integrated roll and yaw model

where the models are defined as: $G_\phi$: fin to roll, $G_\varphi$: yaw to rudder, $G_\alpha$: fin, $G_\delta$: rudder, $C_\delta$: autopilot; $C_\alpha$: fin controller, $G_{\delta\phi}$: rudder to roll interaction.

The surge motion is of low frequency nature and hence its effect is neglected in this paper. The effect of the sway motion is also neglected. The roll motion is however strongly influenced by the yaw motion and the ship speed. In addition the interaction $G_{\delta\phi}$ has a nonminimum phase dynamic which further complicates the control design.

The roll motion is of high frequency nature compared to the yaw motion and this suggests that the roll and yaw controllers, $C_\alpha$ and $C_\delta$ may be decoupled and hence designed independently. In this paper it is assumed that the fin stabilisation controller and the autopilots are designed, tuned and implemented. It is also assumed that the controllers are adapted to ship speed.

The fin and rudder actuator models are described in Figure 2 and Figure 3.

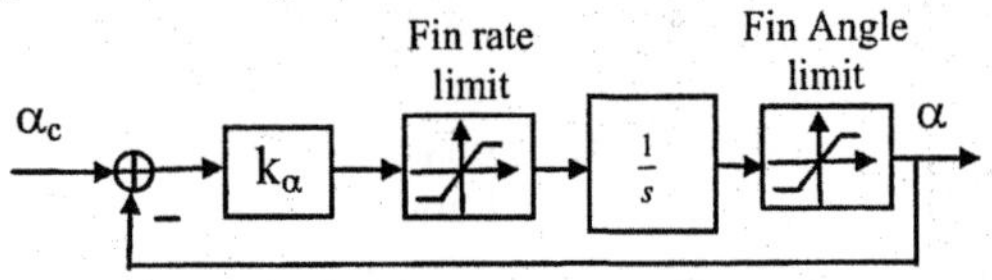


Figure 2 The fin model

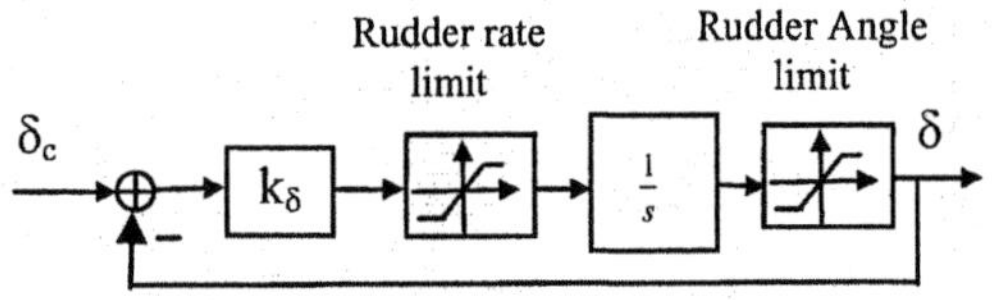


Figure 3 The rudder model

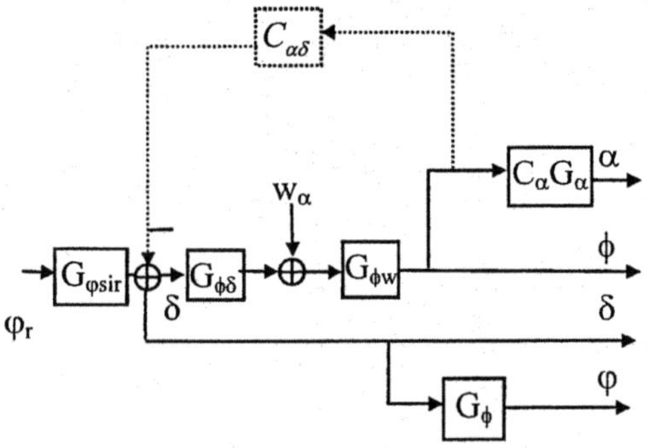


Figure 4 The closed-loop model

After some simple block diagram manipulation the system of Figure 1 can be reduced to the one in Figure 4. The different transfer functions used are given below:

*Rudder command to rudder:*

$$G_{\delta w\phi}(s) = \frac{\delta}{\delta_\phi} = \frac{G_\delta(s)}{1 + C_\delta(s)G_\delta(s)G_\varphi(s)} \qquad (1)$$

*Wave to roll:*

$$G_{\phi w}(s)=\frac{\phi}{w_\alpha}=\frac{G_\phi(s)}{1+C_\alpha(s)G_\alpha(s)G_\phi(s)} \quad (2)$$

*Yaw to roll interaction:*

$$G_{\phi\delta}(s)=G_{\delta\phi}(s)*G_{\delta w\phi}(s) \quad (3)$$

The roll stabilisation ratio is defined as the magnitude of the sensitivity function and defined as follows:

$$S(s)=\frac{1}{1+G_{\phi w}(s)G_{\delta\phi}(s)G_{\delta\delta\phi}(s)C_{\alpha\delta}(s)} \quad (4)$$

Using the closed-loop transfer function definitions (1) to (3) the model of rudder roll control system can be simplified to the one shown in Figure 5.

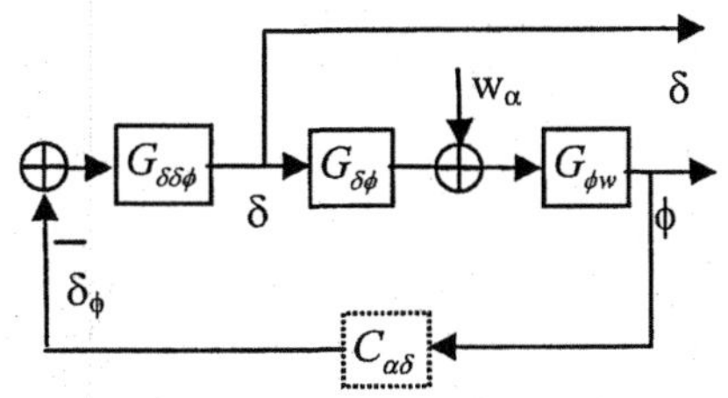


Figure 5 The closed-loop rudder roll control model

The state space models for use in model based predictive control are defined as follows:

$$G_{\delta\delta\phi}:\begin{cases}x_\delta(t+1)=A_\delta x_\delta(t)+B_\delta\varphi_r(t)\\ \delta(t)=C_\delta x_\delta(t)\end{cases} \quad (5)$$

$$G_{\delta\phi}G_{\phi w}:\begin{cases}x_\phi(t+1)=A_\phi x_\phi(t)+B_\phi C_\delta x_\delta(t)\\ \phi(t)=C_\phi x_\phi(t)\end{cases} \quad (6)$$

It is assumed the fin control loop is tuned and the rudder angle is defined as the output of the model to constraint the rudder angle and the rudder slew rate.

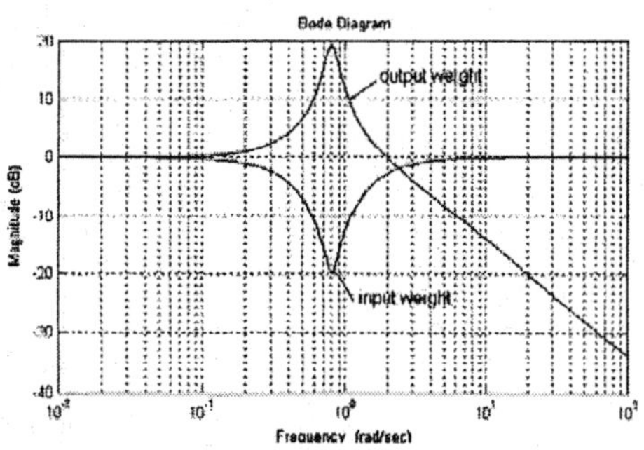


Figure 6 The input and output weightings $\omega_w$ =0.8 and $\xi_w$ = 1.0

## 3. MODEL BASED PREDICTIVE CONTROL

The MBPC formulation described here for the RRS control problem is based on the performance requirement needed to successfully use rudder for roll stabilisation. Unlike the conventional MBPC, the main control problem here is not the set point tracking but the rejection of wave disturbance in a well defined frequency range (0.3 rad/sec-1.2 rad/sec). Any attempt to reject the sea wave at the lower frequency can cause rudder angle saturation due to limited power of the rudder actuator or rejection at higher frequencies can cause rudder slew rate saturation. Furthermore, the roll sensors usually suffer from low frequency drift and high frequency noise. Using the conventional MBPC might lead to infeasible solutions or heavy computational load. In order to circumvent these problems a dynamic performance is introduced which based on the energy and noise content of the rudder and roll signals. The performance index is defined as:

$$J=\sum_{t=1}^{n}W_\phi^2(t)\phi^2(t)+\sum_{t=1}^{m}W_\delta^2(t)\delta_{\phi\delta}^2 \quad (7)$$

The rudder angle and rate constraints are defined as:

$$\begin{aligned}&|\delta(t)|\le c_1\\&|\delta(t)-\delta(t-1)|\le c_2\end{aligned} \quad (8)$$

The dynamic weightings on the output and input are shown in Figure 6. The weight at low frequencies is kept small to avoid any interaction with the autopilot or any attempt to compensate the low frequency roll motion. The high frequency output weight rolls off at -60 dB/decade to filter noise. A possible frequency dependent weighting to meet the above requirement is shown in Figure 6 with the transfer function:

$$W_\phi(s)=\frac{k_w\left(s^2+2\xi_w\omega_w s+\omega_w^2\right)}{\left(s^2+0.2\xi_w\omega_w s+\omega_w^2\right)(1+\tau_w s)} \quad (9)$$

and

$$W_\delta(s)=\frac{\rho\left(s^2+2\xi_\delta\omega_\delta s+\omega_\delta^2\right)}{\left(s^2+0.2\xi_\delta\omega_\delta s+\omega_\delta^2\right)} \quad (10)$$

Since the MBPC will be formulated in state-space domain, the above models are transformed as follows:

$$W_\phi:\begin{cases}x_w(t+1)=A_w x_w(t)+B_w C_\phi x_\phi(t)\\ y(t)=C_w x_w(t)\end{cases} \quad (11)$$

After some algebraic manipulation, the model for MBPC control design can be written as:

$$\begin{aligned}&x(t+1)=Ax(t)+Bu(t)+D\xi_i(t)\\&y(t)=Cx(t)+\xi_o(t)\end{aligned} \quad (12)$$

where

$$x=\begin{bmatrix}x_\delta & x_\phi & x_w\end{bmatrix}^T \qquad u=\delta_\phi, y=\begin{bmatrix}\delta & \phi\end{bmatrix}$$

$$A=\begin{bmatrix} A_\delta & 0 & 0 \\ B_\phi C_\delta & A_\phi & 0 \\ 0 & B_w C_\phi & A_w \end{bmatrix} B=\begin{bmatrix} B_\delta \\ 0 \\ 0 \end{bmatrix} C=\begin{bmatrix} C_\delta & 0 & 0 \\ 0 & C_\phi & 0 \end{bmatrix}$$

and $W_\delta = 1$. Using the system model in ((12)), the j-step ahead prediction of states can be calculated as:

$$\begin{aligned} x(t+j) &= A^j x(t) + \sum_{i=0}^{j-1} A^{(j-i-1)} Bu(t+i) \\ y(t+j) &= C\hat{x}_1(t+j) \end{aligned} \qquad (13)$$

For j = 1 to n the output prediction model can be written in the following matrix form:

$$Y = fx(t) + GU \qquad (14)$$

where

$$\begin{aligned} Y &= \left[y^T(t+1) \ \cdots \ y^T(t+n)\right]^T \\ U &= \left[u(t) \ \cdots \ u(t+m)\right]^T \\ f &= \left[(CA)^T \ \cdots \ (CA^n)^T\right] \end{aligned} \qquad (15)$$

$$G = [g_{ij}];\ i = 1,\dots, n_1;\ j = 0,\dots, m$$

$$\text{where } g_{ij} = \begin{bmatrix} CA^{j-i}B & j \geq i \\ 0 & j<i \end{bmatrix} \qquad (16)$$

By inserting the predictor model ((13)) into the MBPC cost index, the following optimisation problem can be formulated at time t.

$$\min J_t = \{U^T PU + q^T U + c^T c\}$$

subject to U (17)

where

$$P = (G^T G + \Lambda), q^T = -2c^T G, c = -fx(t)$$

Subject to a set of linear constraints:

$$A_i U_i \leq b_i; i = 1,\dots,r \qquad (18)$$

This optimisation problem can be solved using linear Quadratic Programming (QP). A number of fast and efficient QP algorithms are available in the literature. A method, which is widely used, is the active set method (Fletcher, 1987). This method is, however, found to be too complicated for real time application. Another class of algorithms is based on ellipsoid of decreasing volume. These algorithms are simple to code and require only the function evaluation and the direction of at least one sub-gradient. Efficient stopping criteria are also available to find the optimal solution to any degree of accuracy. An ellipsoid algorithm (Boyd and Barratt, 1991) was used in this implementation.

To implement the MBPC controller an estimator is needed to estimate the unmeasured states. In this paper, it is assumed that all the states are available

## 4. SIMULATION STUDIES

The ship is assumed to have a natural roll frequency of 0.8 rad/sec and a damping ratio of 0.2, i.e:

$$G_\phi(s) = \frac{0.64}{s^2 + 0.32s + 0.64} \qquad (19)$$

The fin dynamic is not included in this study. The fin controller is designed to give -12 dB reduction at resonance frequency. Although more fin roll reduction is possible, the controller gain is kept small to provide a more flexible system for the investigation of rudder roll control system. The fin controller is designed as:

$$C_\alpha(s) = \frac{5.1s^2 + 1265s + 40}{1000s + 1} \qquad (20)$$

The stabilisation ratio achieved is shown in Figure 7.

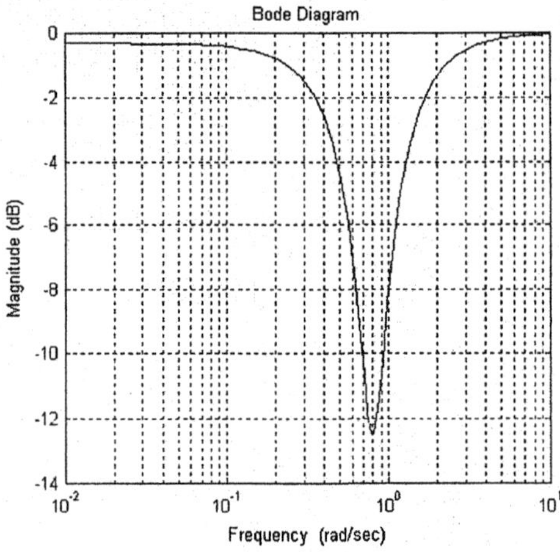


Figure 7 Fin roll stabilization ratio

The yaw angle is modelled as:

$$G_\delta(s) = \frac{0.2}{s(10s+1)} \qquad (21)$$

A PI controller is used to model the autopilot. The autopilot is designed for a bandwidth of 0.1 rad/sec. Compared with the roll natural frequency of 0.8 rad/sec, there is a factor of 8 difference in the frequency content of roll and yaw motion.

The dynamic weighting MBPC is simulated and implemented in Simulink. The roll reduction for different values of $\omega_w$ is shown in Figure 6. The roll reduction is kept to about 17 dB. It can be seen that $\omega_w$ can be used to adapt the MBPC controller to different wave encounter angles and sea states (Lauvdal and Fossen, 1998).

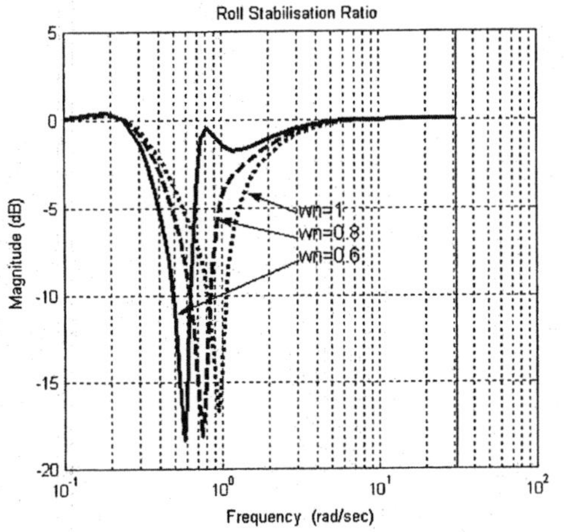


Figure 8 The roll reduction for different weights

The roll reduction for different values of $k_w$ is shown in Figure 9. A linear relationship exists between this gain and the magnitude of roll reduction for small changes in $k_w$. For large changes the sensitivity response shape will significantly change and retuning of the controller is needed.

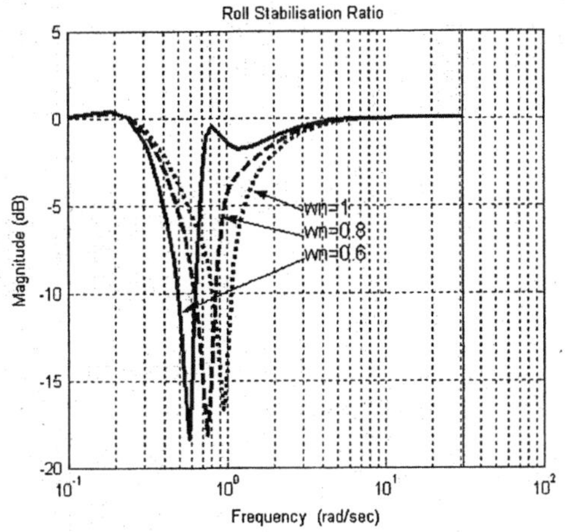


Figure 9 The roll reduction for different gains

The roll reduction calculated from the time responses is defined as:

$$R_r = 100\left(1 - \frac{\sigma_{cl}^2}{\sigma_{op}^2}\right) \quad (22)$$

This is about 69% for the ship studied in this paper. The open loop and the stabilised roll angle response is shown in Figure 10

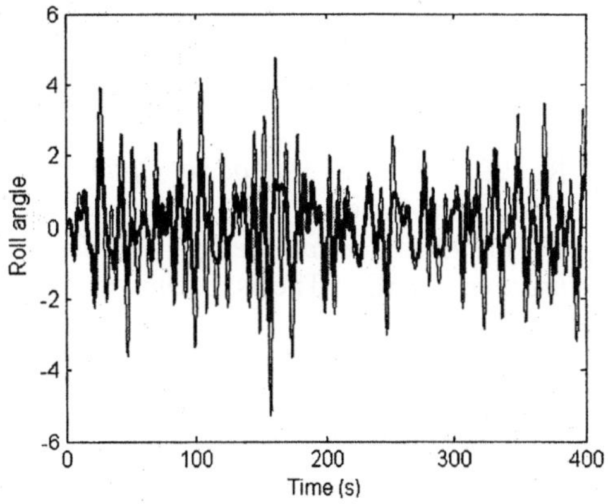


Figure 10 Open loop and closed loop roll angle

The rudder angle and rate responses are presented in Figure 11. Although there are short period of time that the limits are violated but these can be handled using automatic loop gain control. This will however leads to a suboptimal solution and the roll reduction will deteriorate due to drop in loop gain. An alternative is to use the constrained MBPC solution which is formulated in this paper.

## 5. CONCLUSION

A dynamic weighting model based predictive control was proposed for rudder roll stabilisation. The controller was designed assuming the two internal control loops of autopilot and fin roll were closed. It was shown that the weightings parameters can be used to make the rudder roll controller adaptive to different sea states and encounter angles. Simulation results were presented to show that a roll reduction of 69% can be achieved provided the rudder slew rate is sufficiently high.

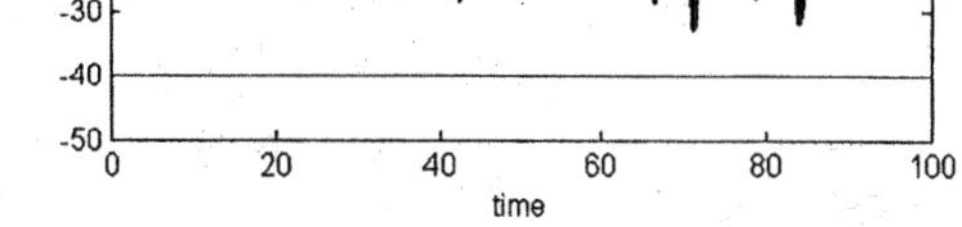

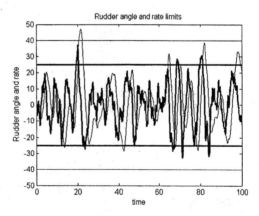

Figure 11 Rudder angle and rate

Simulation results for the constrained MBPC will be included in the final version.

## REFERENCES

Boyd S P and C Barratt, 1991, *linear controller designs: Limits of performance*, Prentice Hall Inc.

Fletcher, R, 1987, *Practical Methods of Optimisation*, 2nd edition, John Wiley.

Hickey, N (1999), *Control design for fin roll stabilisation*, PhD Thesis, University of Strathclyde.

Hickey, N.A., Grimble, M.J., Katebi M.R., Wood, D. (1995) *Hinfinity Fin Roll Control System Design. Control Applications in Marine Systems*, Trondheim, Norway.

Kallstrom, C.G.,Ottosson, P., (1983). *The Generation and Control of Ships in Close Turns*, Ship Operation Automation, IV. North-Holland Publishing Company.

Katebi M.R., Wong D.K.K., Grimble M.J., (1987), *LQG Autopilot and Rudder Roll Stabilisation Control System Design*, (SCSS'87) Proceedings of the 8th International Ship Control Systems Symposium, The Hague, The Netherlands

Roberts G.N., Sharif M.T., Sutton R., Agarwala A., (1997) "*Robust Control Methodology Applied to the Design of a Combined steering/Stabiliser System for Warships*" IEE Proceeding: Control Theory and Applications, 144(2), pp128-136.

Lauvdal T., Fossen T.I., (1998) "*Rudder Roll stabilisation of Ships Subject to Input Rate Saturation using a Gain Scheduled Control Law*" IFAC Conference on Control Applications in Marine Systems, Fukuoka, Japan, 27th-30th October.

ELSEVIER
IFAC
PUBLICATIONS
www.elsevier.com/locate/ifac

# A GAIN SCHEDULED CONTROL LAW FOR FIN/RUDDER ROLL STABILISATION OF SHIPS

**Hervé Tanguy** *,** **Guy Lebret** **

* *SIREHNA: Nantes - France - www.sirehna.com*
** *IRCCyN: Nantes - France - www.irccyn.ec-nantes.fr*

Abstract : Taking into account the variations of the environment is a means of improving performances of roll stabilisation systems. The ship behaviour is modelled as a MIMO LPV system. A methodology is presented which leads to a gain-scheduled control law. The synthesis is based on multi-objective optimisation, and on the representation of the standard system as a polytopical system, which depends on ship speed and on a stabilisation quality factor. Simulation results are given. *Copyright ©2004 IFAC*

Keywords: Ship control; roll stabilisation; $H_\infty$ control; Gain Scheduled Control; polytopic representation; LMI.

## 1. INTRODUCTION

A major improvement in the stabilisation systems' performances should be to adapt to the environmental conditions: waves (encounter angle, power, dominant frequency), ship speed, loading conditions. However, there is relatively few published documents on such control laws. Yet, the dependance on the ships speed has been described and used for many years (Lloyd, 1989; Grimble *et al.*, 1993) for PID and $H_\infty$ control laws. Manual mechanism to cope with changes in the sea state has been introduced (Blanke *et al.*, 2000).

This document details the investigation realised for roll stabilisation towards the construction of a methodology to derived gain scheduling controllers. It is presented to be as general as possible with the aim to use direct informations about the environment. However, it will be applied on a frigate to obtain, in a first step, a controller with only two varying parameters, the ship speed and a stabilizing quality factor that could be tuned by any adaptation process taking into account the environment (manually or automatically). The context aims to be as realistic as possible, and is based on industrial data. The proposed methodology is based on $H_\infty$/LMI results (Apkarian and Gahinet, 1995). This is the sequel of the study proposed in (Tanguy *et al.*, 2003), for varying conditions.

The paper is organized as follows: the process and its environment are shortly described in section 2 as a MIMO linear parameterically varying (LPV) system. The control methodology is detailed in section 3. A four steps methodology leads to the gain scheduled controller. It is applied in section 4 on a frigate type vessel. Simulation results and comparisons of performances with LTI $H_\infty$ controllers are described in section 5. Section 6 gives perspectives of improvement.

## 2. MODEL

A ship in a seaway can be modelled as a linear parameterically varying system. This section is a condensed description of the results of a chapter of (Tanguy, 2004).

Comprehensive models derived from hydrodynamics are too complex to be used in control. Thus, acceptable simplifying assumptions are made: amplitude of motions are small; the ship dynamics is independent of the swell frequency. Eventually, the roll motion is considered to be the superposition of the motions induced by the waves and the motions induced by the actuators. Perturbation motions will be taken as an additive disturbance on the outputs of the ship's dynamics, as shown on figure 1.

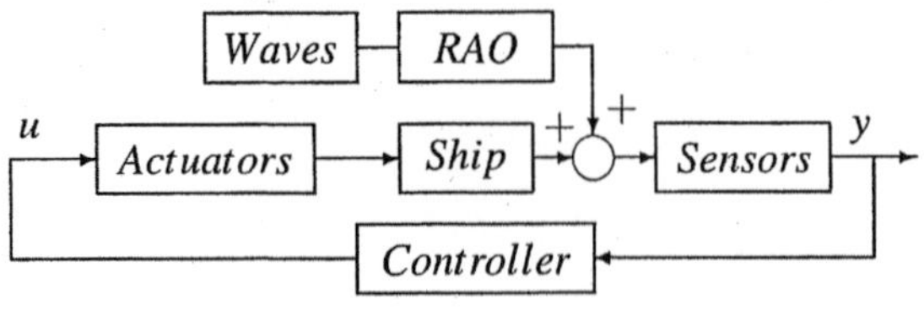


Figure 1. Control model with output disturbance.

### 2.1 *Sea disturbance*

Waves are the result of the sustained action of the wind over a wide sea surface. Complex sea states are considered to be the superposition of an infinite number of monochromatic waves. Waves will here be considered to be long crested.

Wave amplitude spectra, allow to characterise sea states all over the planet. The Bretschneider spectrum (Fossen, 1994), will be used in the simulations with parameters $H_s$ (wave height) and $\omega_P$ (the spectrum peak pulsation).

The wave spectrum encountered by a moving ship is different from the one seen by a motionless observer, due to Doppler effect. Parameters that modify it are: the ship speed $V$ and the encounter angle of waves $\psi_e$:

$$\omega_e = \omega(1 - \omega\frac{V}{g}\cos(\psi_e)) \tag{1}$$

### 2.2 *Synthesis and Simulation models*

2.2.1. *Synthesis Model* The synthesis model is linear with varying parameters. It is written as a state space model with state $x = [v,p,r,\phi,\psi,\alpha,\dot{\alpha},\delta,\dot{\delta}]^T$ where $v$, $p$, $r$ are respectively the sway, roll and yaw velocity; $\phi$ and $\psi$ are the roll and yaw angles; $\alpha$ and $\delta$ are the actual position of the actuators (fins and rudders). The control variable, $u$, is the desired position of the actuators. The measures $y$ considered for control are the roll velocity $p$ and the angle $\psi$. The model is of the following type:

$$\dot{x} = A(V)x + Bu \tag{2}$$

$$y = Cx \tag{3}$$

Only the coefficients of the matrice $A$ are dependent on $V$ as second (fins and rudders efficiency), first (damping) or zeroth (buoyancy) order polynomials. The synthesis model for the studied case of section 4 is only parameterised in speed $V$.

The coefficients are calculated from a seakeeping numerical code. Details on the construction and expressions of 2 and 3 are given in (Tanguy, 2004). Just note that the matrices are chosen for an encounter angle of 90deg, that the load is constant. Moreover the dynamics of the actuators is modelled by a second order LTI system, and is integrated in the dynamics.

2.2.2. *Simulation Model* The simulation model is obtained by adding an output disturbance (motions due to waves) to $y$. Disturbance motions are computed using the Bretschneider spectrum with $T_P = 10s$ ($\omega_P = \frac{2\pi}{T_P}$) and $H_s = 3.25m$ (Sea State 5 in the northern Atlantic) and the motion RAO of the ship.

$$y = (\ p + p_w,\ \psi + \psi_w\ )^\top \tag{4}$$

In addition, the simulations takes into account the temporal non-linear aspects of saturation (in angle and rate for both the fins and rudders) and digitalisation of the control law. A pure delay is also added in temporal simulations to make up for the information transportation effects in the ship internal network.

## 3. PROPOSED SYNTHESIS METHODOLOGY

### 3.1 *Introduction*

The motions of the ship depends on its speed ($V$), its direction ($\psi_e$) and on the environment characterized by the sea state parameters $H_s$ and $T_P$. Gain scheduled controllers that depend on these parameters should be an interesting way to tackle the roll reduction. In general, controllers implemented are PIDs, tuned at the ship roll frequency (Lloyd, 1989; Katebi *et al.*, 2000). The dependance on ship speed is rarely described in the litterature (except for (Lloyd, 1989; Grimble *et al.*, 1993; Blanke *et al.*, 2000)). The gains generally are inversely proportionnal to the square of the ship speed. No theoretical proof of the closed loop stability exists in this case; but simulation tests "show in practice" the stability and the efficiency of the method.

In this study it is proposed to use the recent $H_\infty$/LMI techniques to compute gain scheduled controllers for Linear systems with varying parameters since they guarantee the closed loop stability. This section propose a methodology for the general considered problem. In the next section, its application to particular conditions will be detailed.

### 3.2 *The general specifications for the control law synthesis*

Specifications characterizing the desired behavior of the ship are chosen from mechanics and passengers' comfort matters:

- reduce the roll motion inside the roll bandwidth and do not amplify it outside,
- keep the yaw angle as constant as possible,
- do not use too much power,
- respect a given power repartition on the actuators. The fins are used only for roll stabilisation, and should interfere very little with the heading. On the contrary, rudders have a great influence

on roll motions, but are primary used to control the yaw,

- tolerate only "acceptable" position and speed saturation of the actuators.

Others specifications are added from the control engineering point of view:

- the closed loop and the controller must be stable,
- some robustness properties are necessary against uncertainties (delay, discretisation...).

### 3.3 *A four step methodology*

In order to derive gain scheduled controllers from the now classical $H_\infty$/LMI techniques, one needs a linear parameterically varying standard model defined from the dynamics of the ship (section 2) and weight functions, which have to be a translation of the previous specifications. Note that the main difficulty here is to give a proper translation of the physical specifications into mathematics. It is proposed to procede in two stages: first, compute the weights for fixed values of the varying parameters; then, compute the varying standard model with an interpolation technique.

More generally, the following four-stage methodology is proposed to achieve the final synthesis goal:

- Stage 1: Choose the varying parameters which will be considered. All the varying parameters appearing in the model (section 3.1) should be considered; but the more parameters, the more complicated will be the computation.
- Stage 2: Choose the parameters values in a grid. At each set of the parameters, determine the weights for the standard model that result in a $H_\infty$ controller such that specifications are fulfilled. This is based on the resolution of a multi-objective optimisation problem (Tanguy *et al.*, 2003).
- Stage 3: Compute, with a classical interpolation technique, a linear standard model with varying parameters, from the fixed standard models resulting from stage 2.
- Stage 4: Compute a gain scheduled controller for the linear varying parameters model with standard numerical code [1].

Some technical aspects are now classical, and may be solved with standard existing Toolboxes: $H_\infty$ synthesis in stage 2, and $H_\infty$/LMI gain scheduled controller synthesis in stage 4. The definition and the resolution of the multi-objective optimisation problem (in stage 2) resulting in optimised weights were already introduced in (Tanguy *et al.*, 2003). The originality of the present paper lays in the whole methodology, which considers completely the variation of the parameters.

[1] Matlab LMI toolbox, for example.

## 4. APPLICATION OF THE METHODOLOGY

The studied ship is a frigate (length 120 m, displacement 3000 metric tons). The considered environmental conditions are sea state 5 ($T_P = 10s$ and $H_s = 3.25m$) for a encounter angle $\psi_e$ of 90deg.

### 4.1 *The varying parameters*

Two parameters were used: the ship speed over the water, and a tuning parameter called Stabilisation Quality Factor (SQF). Dependance in the ship speed is needed. The SQF defines roll reduction quality: it is defined by the depth of the roll sensibility transfer function between an additive roll disturbance and the roll angle. Its value is intended to be directly tuned from the bridge or by an adaption process, taking into account power consumption, actuators' saturation levels, and sea state measurements.

The values of the parameters are defined by a comprehensive gridding, with steps every 5 knots in speed from 10 to 25 knots, and every 1 unit in SQF quality from 2 to 8.

### 4.2 *Controller synthesis for fixed parameters*

A more comprehensive study (Tanguy *et al.*, 2003), premise of the present article, gives the details for the computation of controllers for constant values of the parameters. The method used was to solve a multi-objective optimisation problem with an evolution strategy (genetic algorithm), and to choose a particular solution from strict guidelines. Another solution is used here (as the optimisation process may take quite some time to reach a reasonnably good value), though very close in principles to the former: the parameters' space is comprehensively studied, and gives good results in relatively short time.

4.2.1. *Definition of the multi-objective optimisation problem* The choice of the control law (here the weights of the $H_\infty$ standard problem) is defined as the result of a multi-objective optimisation problem under contraints. The objectives and contraints are derived from the specifications introduced in subsection 3.2:

**O1 :** Reduce the roll motion. It is expressed as the minimisation of the roll RMS value on a particular sea state for the closed loop system.

**O2 :** Use the minimum power. It is necessary to ensure that the two actuators do not compensate for one another, case which may appear (for MIMO PID, for instance). The sum of the RMS values of the fins' and rudders' positions (resp. $\sigma_\alpha$ and $\sigma_\delta$) is minimised, for the same sea-state.

**O3 :** Respect as precisely as possible the repartition constraint. The objective is defined as the weighted ratio 'use of the fins' over 'total use of the actuators', the weights being the $H_\infty$ norm of the open loop transfer functions between fin and rudder position and roll.

The constraints used, in addition to these objectives, are:

**C1 :** the controller must be stable;

**C2 :** the closed loop (system + controller), given a control application delay, must be stable;

**C3 :** the delay margin must be acceptable, in order to take into account digitalisation effects, and information transfer delay... The delay margin is evaluated with sensitivity and complementary sensitivity output transfer functions analysis, as the control problem is MIMO (see (Tanguy, 2004));

**C4 :** the amplification under and over the ship roll resonance must be low (Hearns *et al.*, 2000); They are calculated from sensitivity transfer between $p_w$ and $p$;

**C5 :** the actuators may not endure too much saturation, in both position and velocity. It is not possible to determine exactly the saturations levels when working in the frequency domain, for it has only a temporal meaning. Yet, they can be evaluated from statistical considerations ((Lloyd, 1989; Price and Bishop, 1974)).

Note that this optimisation framework can be used to tune different type of controllers. It is here applied to tune $H_\infty$ controllers, but it has also been applied to MIMO PID.

4.2.2. *Definition of the $H_\infty$ problem* This part of the methodology is similar to the one detailed in (Tanguy, 2004), so it won't be precisely described here. Just note that the problem is set up as a mixed sensitivity problem (figure 2).

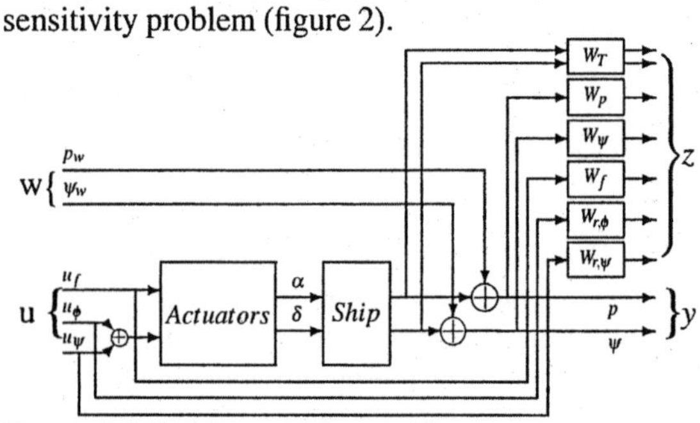


Figure 2. Mixed sensitivity problem.

Most of the parameters of the weights have a fixed value. In fact, only the gains of $W_f$ and $W_{r\phi}$ may vary[2] in the multi-objective optimisation problem. The shape of the roll derivative sensitivity weight ($W_p$) has been carefully chosen(figure 3). The depth of the well is defined by the SQF which varies from 2 to 8, its center by the roll resonance frequency of the ship; it is taken to be a constant; in a more complete study

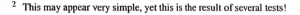

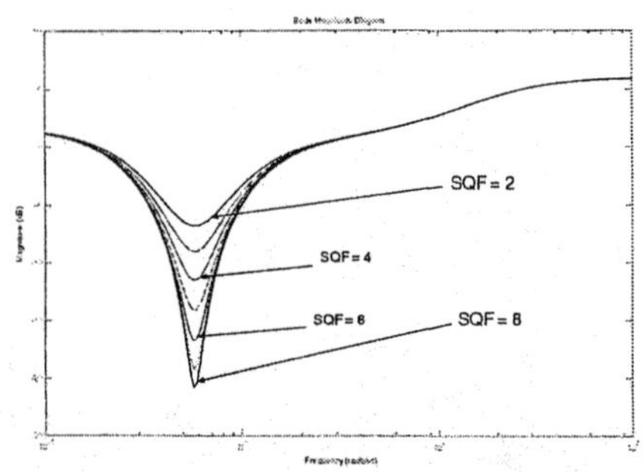


Figure 3. The roll derivative sensitivity $W_p$

it would be a fonction of the ship speed, the encounter angle of waves, the sea state...

4.2.3. *Results* The results of this step are not the $H_\infty$ controllers, even if they have to be calculated in order to assess their performances. The actual results are the weights themselves and the standard models, for every combination of parameters. They are the basis for the next step.

## 4.3 *Computation of the LPV model*

The computed standard models (dynamics augmented with frequency weights) show a dependency on $V$ and $V^2$ and $SQF$. With a classical interpolation technique, it is possible to synthesise their expression in the following LPV model:

$$P(V,V^2,SQF) = \begin{bmatrix} A(V,V^2,SQF) & B_1(V,V^2,SQF) & B_2 \\ C_1(V,V^2,SQF) & D_{11} & D_{12} \\ C_2 & D_{21} & D_{22} \end{bmatrix}$$
$$= P_c + P_V V + P_{V^2} V^2 + P_{SQF} SQF \quad (5)$$

A basic solution is to consider the model (5) as an affine model $P_a(X_1, X_2, X_3)$ obtained by replacing $V$ by $X_1$, $V^2$ by $X_2$ and $SQF$ by $X_3$. The parameters $X_1$, $X_2$ and $X_3$ are supposed to be independent of each other[3]. This leads to a very conservative model.

A polytopic model has instead been used for which, there exists dedicated control law synthesis code[4]. The chosen model is put under the following form:

$$P_p(\theta) = \left\{ \sum_{i=1}^{7} \alpha_i(\theta)\pi_i,\ \alpha_i(\theta) \geq 0, \sum_{i=1}^{7} \alpha_i(\theta) = 1 \right\}$$

where the vertices $\pi_i$ are the image of the vertices of a polytopic domain $\mathscr{P}$ which define acceptable restricted values of $\theta$: $\{\theta = (X_1, X_2, X_3) / \theta = \sum_{i=1}^{7} \alpha_i P_i,\ \alpha_i \geq 0, \sum_{i=1}^{7} \alpha_i = 1\}$. The vertices $P_i$ of $\mathscr{P}$ have been chosen such that the possible values of $(V, V^2, SQF)$ of the initial LPV model (5) are included in $\mathscr{P}$ (see figure 4). The facet $\{P_4, P_5, P_7\}$ express the constraint that for small values

[2] This may appear very simple, yet this is the result of several tests!

[3] With $X_1 \in [10\ 25]$, $X_2 \in [100\ 625]$ and $X_3 \in [2\ 8]$.

[4] LMI toolbox of matlab, for example.

of $V$ no good roll damping is possible[5]. The influence of $V^2$ remains sufficiently low to keep representation adequate (without too much conservatism).

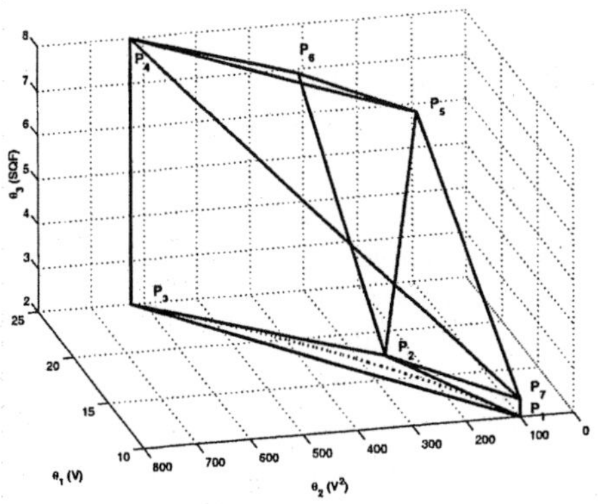

Figure 4. Polytope $\mathscr{P}$ (solid) and projection on $(\theta_1, \theta_2)$ (dashed).

### 4.4 *Computation of the gain scheduled controller*

Once a valid polytopic standard model is written, the computation of the polytopic controller is quite straightforward. The LMI Control toolbox of Matlab provides tools for such a work.

A quite similar method, and giving potentially better results, is to represent the LPV system with a LFT (Apkarian and Gahinet, 1995; Magni, 2001). The controller synthesis is clearly described in the litterature (and quite classical as for now), yet it requires a great quantity of computation and formalisation. Furthermore, controllers may exist without the theorem (and the equations) allowing to compute it.

## 5. SIMULATION RESULTS AND COMPARISONS

### 5.1 *Discrete form controller*

For the application of the controller on the simulation model, its matrices are computed at each time step. The discretisation is realised with a zero order hold approximation, which requires the computation of a matrix exponential: this method is surely not optimal, but works properly on a 800MHz computer, and it ensures stability of the controller.

### 5.2 *Simulations*

The temporal simulations are performed with $\psi_e = 90$deg and sea state 5 (figures 5, 6, 7) or 6 (figure 8). Others simulations are presented in (Tanguy, 2004).

Figure 5 and 6 illustrate the behaviour of the Gain Scheduled Controller (GS) when the conditions vary.

[5] This is the result of Stage 2.

Figure 5 presents the performances of the controller when the speed vary slowly from 10 to 20 knots and with a constant SQF (2). The activity of the actuators decrease with the speed. This physically corresponds to the fact that their efficiency is increasing with the speed (it is a function of $V^2$, see section 2.2.1), and that maximal fin deflection value decreases with speed (there exists a maximum allowed effort). This capability of the GS controller is not available with a LTI-controller, see figure 7.

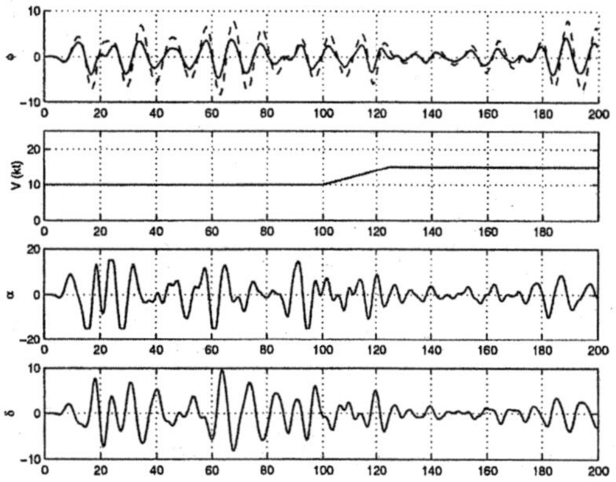

Figure 5. Variation of Speed during simulation. Closed loop (solid), open loop (dashed).

Figure 6 presents the effect of a variation of the SQF from 2 to 8, the ship having a constant speed of 15 knots. Note that the variation can be infinitely fast without destabilising the loop, thanks to the LMI derivation of the controller. The effect of SQF is clear: better roll stabilisation is obtained from t=150s, while maintaining acceptable increase of the actuators activity. In this case the increased activity was expected since the fin efficiency does not vary (constant speed).

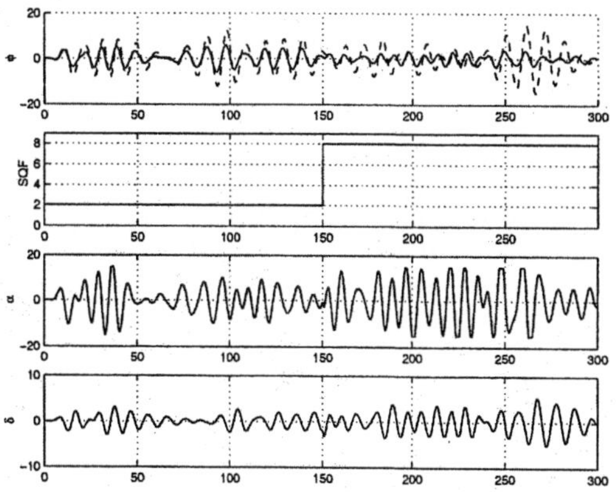

Figure 6. Variation of SQF during simulation. Closed loop (solid), open loop (dashed).

Figure 7 shows the interest of the gain scheduled controller when the ship speed vary from 15 knots to 25 knots, compared with a LTI-$H_\infty$ controller (tuned for a speed of 15 knots). The LTI-$H_\infty$ controller is optimised with the techniques of section 4.2 for a speed of $V = 15$ knots, for $\psi_e = 90$deg and for sea state 5. The SQF parameter of the GS controller is 8. The roll attenuation does not change much with this controller, and as noticed previously the activity is a little bit decreased, because of the variation of fin efficiency. On the contrary, whereas the activity of the actuators were acceptable for the LTI-$H_\infty$ controller, for 15 knots, it becomes inacceptable for 25 knots: actuators are agitated (motions have a period of 3 seconds). The LTI-$H_\infty$ shows better roll reduction than the the GS controller, but the actuators saturate too often.

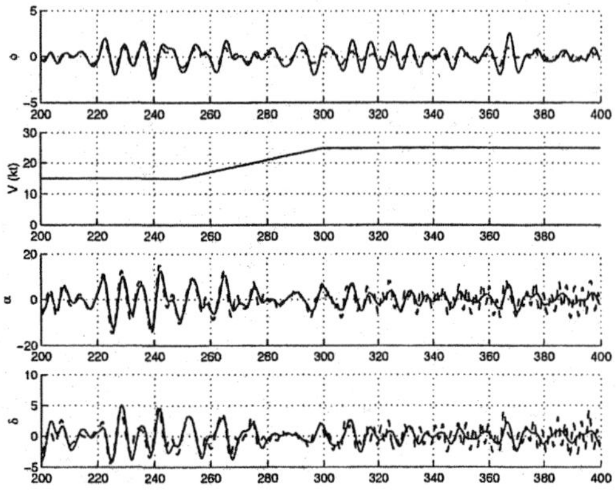

Figure 7. GS (solid) and LTI (dashed) controllers.

Moreover, figure 8 shows how an action on the SQF can be used to reduce the activity of the actuators. The two simulations shown on figure 8 correspond to the behaviour of the ship on a sea state 6. The GS controller is adapted to sea state 5. The solid curve corresponds to $SQF = 8$, and the dashed curve to $SQF = 2$. Downgrading the performances, in case of navigation on higher sea states than expected will show reduction of the saturation frequency and as a consequence will help protect of the actuators mechanics.

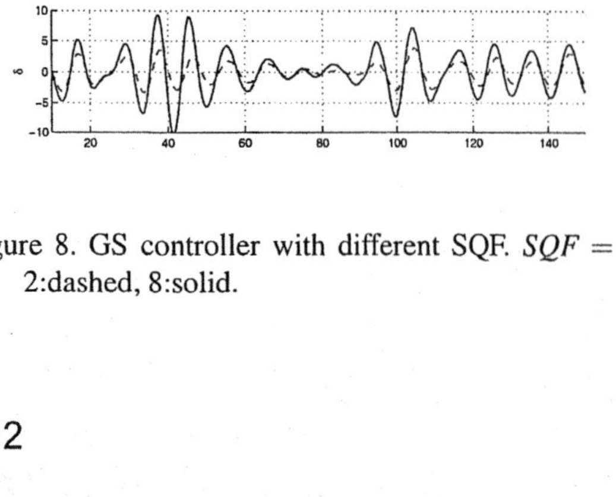

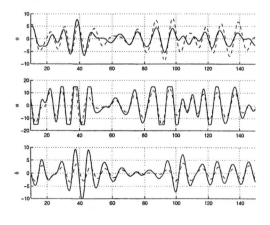

Figure 8. GS controller with different SQF. $SQF =$ 2:dashed, 8:solid.

## 6. CONCLUSION-PERSPECTIVES

A methodology is given for computing a gain scheduled controller, based on LMI and on polytopic modelisation of the process. Parameters used are the ship speed and a stabilisation quality factor. The methodology has been applied to a frigate-like ship in simulation. Evolution of the parameters during the simulation is possible, which influence the controllers matrices, and thus the behaviour of the stabilised ship, allowing for example a good adaption to ship speed and reduction of saturation frequency. Improvement are expected from introducing more parameters (encounter angle, sea state...), in order to better monitor the behaviour of the control law. Another potential way of improvement is the use of LFT representation for modelling and control law synthesis.

## 7. REFERENCES


Apkarian, P. and P. Gahinet (1995). A convex characterization of gain-scheduled $h_\infty$ controllers. *IEEE TAC* **40**(5), 853–864.

Blanke, M., J. Adrian, K.E. Larsen and J. Bentsen (2000). Rudder roll damping in coastal region sea conditions. In: *Proceedings of MCMC 2000.* Aalgorg, Danemark.

Fossen, T. I. (1994). *Navigation and Guidance of Ocean Vehicles.* John Wiley & sons. New York.

Grimble, M.J., M.R. Katebi and Y. Zhang (1993). $h_\infty$ based ship fin-rudder roll stabilisation design. In: *10th Ship Control Systems Symposium.* Vol. 5. Ottawa, Canada.

Hearns, G., R. Katebi and M.J. Grimble (2000). Robust fin roll stabiliser controller design. In: *5th IFAC Conference on Manoeuvering and Control of Marine Crafts.* Aalborg, Danmark.

Katebi, M.R., N.A. Hickey and M.J. Grimble (2000). Evaluation of fin roll stabiliser controller design. In: *5th IFAC Conference on Manoeuvering and Control of Marine Crafts.* Aalborg, Danmark.

Lloyd, A.R.J.M. (1989). *Seakeeping, Ship Behaviour in Rough Weather.* Marine Technology. Hellis Horwood.

Magni, J.F. (2001). *Linear Fractional Representations with a Toolbox for Use with MATLAB.* Toulouse, France. Technical report TR 240/2001 DCSD.

Price, W.G. and R.E.D. Bishop (1974). *Probabilistic Theory of Ship Dynamics.* Chapman and Hall. Londres.

Tanguy, H. (2004). synthèse de lois de commande à gains programmés pour la stabilisation en roulis des navires. PhD thesis. Université de Nantes, École Centrale de Nantes. Nantes, France.

Tanguy, H., G. Lebret and O. Doucy (2003). Multiobjective optimisation of pid and $h_\infty$ fin/rudder roll controller. In: *Proceedings of MCMC 2003.* Girona, Spain.

ELSEVIER
IFAC
PUBLICATIONS
www.elsevier.com/locate/ifac

# ON CONSTRAINED CONTROL OF FIN, RUDDER OR COMBINED FIN-RUDDER STABILIZERS: A QUASI-ADAPTIVE CONTROL STRATEGY

**Tristan Perez * Graham C. Goodwin ***

* *Department of Electrical And Computer Engineering, The University of Newcastle, Australia.*

Abstract: This paper considers a simple quasi-adaptive constrained control strategy that can be used for fin, rudder, or combined fin-rudder stabilizers. The strategy estimates the parameters of a linear output disturbance model for the wave induced roll motion using roll and roll rate measurements taken before closing the control loop. This model is then used to implement a constrained predictive control strategy. The strategy can thus be adaptive with respect to changes in the sea state and sailing conditions. The work also explores the benefit of penalizing roll accelerations as well as roll angle in the associated cost. *Copyright ©2004 IFAC*

Keywords: Ship Roll Stabilizers, Kalman Filtering, Constrained Predictive Control.

## 1. INTRODUCTION

In a previous work, we have proposed the use of constrained *model predictive control (MPC)* to address the control system design problem for fin and/or rudder-based stabilizers—see Perez *et al.* (2000) and Perez and Goodwin (2003). This approach offers a unified framework for minimizing the impact of roll motion on ship performance, handling input and output constraints and also provides a means for implementing adaptive strategies.

In order to implement the proposed MPC strategy, two models are necessary: a model describing the dynamic behavior of ship motion due to control action (rudder and/or fins) and a model describing the wave induced roll motion. The first model can be obtained using system identification techniques together with tests performed in calm waters—see, for example, Zhou *et al.* (1994). This model should be updated for different ship speeds. The wave induced roll motion can be modelled using a second order shaping filter, which is then used to predict the wave induced roll motion in the MPC Formulation. This model cannot be estimated before hand since it depends on the sea state and sailing conditions (speed and encounter angle.) Adaptation is necessary.

The purpose of this paper is twofold. First, to propose a simple way to estimate the parameters of the wave-induced roll model; and thus, extend our previous work. Second, to incorporate a penalty on the roll acceleration in the associated cost. The effect of roll acceleration on ship performance has long been recognized in the naval environment (Warhurst, 1969). Nonetheless, direct roll acceleration reduction has often been omitted from stabilizer control system design in literature and reported practical implementations.

## 2. CONTROL SYSTEM ARCHITECTURE AND MODELS FOR CONTROL DESIGN

The adopted architecture for the control system is shown in Figure 1. Because the control will be ultimately implemented on a computer, we will adopt a discrete-time framework to describe the models and control system design problem.

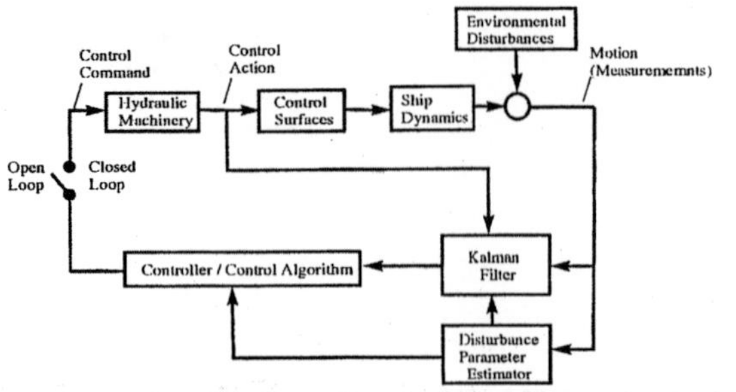


Fig. 1. Block diagram of the control system architecture.

A linear model will be used to describe the ship response due to the control command ($c$):

$$\mathbf{x}^c_{k+1} = A_c \, \mathbf{x}^c_k + B_c \, \mathbf{u}_k, \tag{1}$$

In the case of rudder (or combined fin-rudder) stabilizer design, the state vector $\mathbf{x}^c_k$ is given by

$$\mathbf{x}^c_k = \left[v^c_k \; p^c_k \; r^c_k \; \phi^c_k \; \psi^c_k\right]^{\mathrm{T}}, \tag{2}$$

where $v$ is the sway velocity (+ve to stbrd), $p$ the roll rate (+ve to stbrd), $r$ the heading rate (+ve to stbrd) and $\phi$ the roll angle. The heading deviation $\psi$ is taken relative to the desired course and filtered by a wave filter that rejects the first order wave induced yaw motion. This is standard for autopilot design (Fossen, 1994). The components of the control vector $\mathbf{u}_k$ are the rudder mechanical angle $\alpha$ and/or the fin mechanical angles. The effect of the hydraulic machinery can be incorporated in (1) as an extra state or as a time delay. The matrices $A_c$ and $B_c$ in (1) depend on the forward speed of the vessel $U$. These are updated for different ranges of operation speed; and thus, shall be considered constant in the sequel.

For the case of only fin stabilizers, the roll rate $p^c_k$ and the roll angle $\phi^c_k$ may be decoupled from the other motion components leading to a second order model. This simplification is appropriate when the fins are located close to the lateral center of gravity (LCG) of the vessel. In this case, it may also be convenient to consider two inputs; namely, the port fin angle $\alpha_{port}$ and the starboard fin angle $\alpha_{stbrd}$. In low sea states, this distinction may not be necessary, and only one angle need be considered (with the other always moving in the opposite direction). In high sea states, however, the distinction is more relevant: the weather fin (windward side) usually operates in a more disturbed flow than its counterpart fin on the leeward side—see Gaillarde (2002) This may require different constrains on the motion of each fin.

The roll motion induced by the waves ($w$) can be modelled as the output of a shaping filter:

$$\begin{aligned} \mathbf{x}^w_{k+1} &= A_w \, \mathbf{x}^w_k + \mathbf{w}_k, \\ \mathbf{y}^w_k &= \mathbf{x}^w_k + \mathbf{v}_k, \end{aligned} \tag{3}$$

where $\mathbf{x}^w_k = [\phi^w_k p^w_k]^T$, and $\mathbf{w}_k$, $\mathbf{v}_k$ are sequences of independent identically distributed Gaussian vectors with appropriate dimensions.

The complete dynamic model (for control design purposes) can then be represented as

$$\begin{aligned} \mathbf{x}_{k+1} &= A \, \mathbf{x}_k + B \, \mathbf{u}_k + W \, \mathbf{w}_k, \\ \mathbf{y}_k &= C \, \mathbf{x}_k + \mathbf{n}_k, \end{aligned} \tag{4}$$

where $\mathbf{x}^{\mathrm{T}} = [\mathbf{x}^c_k \;\; \mathbf{x}^w_k]$, and with the following measurements available to implement the control:

$$\begin{aligned} \mathbf{y}_k &= \left[p_k \; r_k \; \phi_k \; \psi_k\right]^{\mathrm{T}} + \mathbf{n}_k \\ &= \left[(p^c_k + p^w_k) \; r^c_k \; (\phi^c_k + \phi^w_k) \; \psi^c_k\right]^{\mathrm{T}} + \mathbf{n}_k, \end{aligned} \tag{5}$$

where, $\mathbf{n}$ is noise introduced by the sensors.

## 3. CONTROL OBJECTIVES AND PROBLEM STATEMENT

The basic control objectives for the particular control problem being addressed here are as follows:

(1) minimize the roll angle and acceleration;
(2) produce low interference with yaw (rudder and fin-rudder stabilizers);
(3) satisfy constraints imposed by performance and safety issues.

The above objectives can be achieved by considering the following optimal control problem:

*We seek the feedback control command* $\mathbf{u}_k = \mathcal{K}(\mathbf{y}_k)$, *that minimizes the following cost:*

$$V = \sum_{k=0}^{\infty} \|\mathbf{y}_k\|^2_Q + \|\mathbf{y}_{k+1} - \mathbf{y}_k\|^2_P + \|\mathbf{u}_k\|^2_R, \tag{6}$$

*subject to the system*

$$\begin{aligned} \mathbf{x}_{k+1} &= A \, \mathbf{x}_k + B \, \mathbf{u}_k + W \, \mathbf{w}_k, \\ \mathbf{y}_k &= C \, \mathbf{x}_k + \mathbf{n}_k, \end{aligned} \tag{7}$$

*and the following input and output constraints*

$$|\mathbf{u}_k| \leq U_M, \quad |\mathbf{u}_{k+1} - \mathbf{u}_k| \leq \Delta_U \quad \text{and} \quad |D\mathbf{y}_k| \leq Y_M \tag{8}$$

We can choose the matrices $Q$, $P$ and $R$ such that the cost function (6) becomes

$$V = \sum_{k=0}^{\infty} \lambda \, [p^2_k + \phi^2_k + \gamma_1 (p_{k+1} - p_k)^2] + (1-\lambda)[\psi^2_k + \gamma_2 \alpha^2_k \tag{9}$$

The cost (9) is a discrete-time version of the cost proposed by van Amerongen *et al.* (1987)—see also (van der Klugt, 1987). The function (9), however, incorporates an extra term that weights the roll accelerations via the difference $(p_{k+1} - p_k)$. The reason for this particular choice of the cost is that the scalar parameter $\lambda \in [0, 1)$ represents a trade-off between roll reduction and yaw interference. Hence, there is a single parameter that can be varied by the operator according to the different mission or particular scenario in which the ship is operating. The parameter $\gamma_1$ will be used to investigate the benefits of penalizing roll accelerations and $\gamma_2$ can be pre-set to penalize the control effort for the case in which only course keeping is needed (*i.e.*, , for $\lambda$=0.) In the case of fin stabilizers only, the term regarding the the yaw angle in (9) is not considered.

## 4. THE CERTAINTY EQUIVALENT MPC SOLUTION

Because of the constraints (8), the optimal control problem stated in the previous section is not easy to solve. Hence, one reverts to approximate or sub-optimal solutions. One sub-optimal solution often advocated in practical implementations is Certainty Equivalent MPC (CE-MPC)—see Perez *et al.* (2004).

The way in which the optimal solution is approximated can be described in three steps. First, the infinite horizon cost in (6) is approximated by a finite horizon cost, in which there is an additional a terminal cost that approximates the cost to go from the finite final step ($N$) to infinity. Second, the finite horizon optimal control problem is solved under the assumption that there is no uncertainty in the state, and the final solution is implemented using an estimate for the uncertain variables (Certainty Equivalence.) Third, a receding horizon control is implemented based on the solution of the finite horizon problem solves at each sampling instant.

For the problem defined in the previous section, the resulting finite horizon approximation can be formulated as follows: *Given the initial condition for the augmented state* $\mathbf{x}_0$ *we seek the vector of control moves*

$$\mathbf{U}^{\text{OPT}} = [\mathbf{u}_0^{\text{OPT}}(\mathbf{x}_0)\ldots\mathbf{u}_{N-1}^{\text{OPT}}(\mathbf{x}_0)]^T; \tag{10}$$

*that minimizes the following cost*

$$J_N(\mathbf{x}_0,\mathbf{U}) = \mathbf{x}_N^{\text{T}}\,\widetilde{S}\,\mathbf{x}_N + \sum_{j=0}^{N-1} \mathbf{x}_j^{\text{T}}\,\widetilde{Q}\,\mathbf{x}_j + \mathbf{u}_j^{\text{T}}\,\widetilde{R}\,\mathbf{u}_j + 2\,\mathbf{u}_j^{\text{T}}\,\widetilde{T}\,\mathbf{x}_j, \tag{11}$$

*subject to*

$$\begin{aligned}\mathbf{x}_{j+1} &= A\,\mathbf{x}_j + B\,\mathbf{u}_j,\\ \mathbf{y}_j &= C\,\mathbf{x}_j,\end{aligned} \tag{12}$$

*and the constraints*

$$|\,\mathbf{u}_j\,| \le U_M,\ |\,\mathbf{u}_{j+1}-\mathbf{u}_j\,| \le \Delta_U \text{ and } |D\,\mathbf{x}_k\,| \le Y_M, \tag{13}$$

with the following definitions in term of the original problem:

$$\begin{aligned}\widetilde{Q} &= (A-I)^{\text{T}}(C^{\text{T}}PC)(A-I) + C^{\text{T}}QC\\ \widetilde{R} &= B^{\text{T}}(C^{\text{T}}PC)B + R\\ \widetilde{T} &= B^{\text{T}}(C^{\text{T}}PC)(A-I)\\ \widetilde{D} &= DC,\end{aligned} \tag{14}$$

and $\widetilde{S}$ is the solution of the Riccati equation associated with the unconstrained infinite horizon problem, *e.g.*, LQR.

With some algebraic work, the above problem can be posed as a Quadratic Programme (QP) problem (this is standard in MPC (Maciejowski, 2002)):

$$\begin{aligned}\mathbf{U}^{\text{OPT}} &= \arg\min_{\mathbf{U}} \frac{1}{2}\,\mathbf{U}^{\text{T}}\,\mathcal{H}\,\mathbf{U} + \mathbf{U}^{\text{T}}\,\mathcal{F}\,\mathbf{x}_0\\ &\text{s.t.}\quad L\,\mathbf{U} \le M,\end{aligned} \tag{15}$$

where the matrices $\mathcal{H}$ and $\mathcal{F}$ have an explicit form in terms of the system matrices $A$ and $B$ and those indicated in (14). Similarly the matrix $M$ has an explicit form in terms of $U_M$, $\Delta_U$, $Y_M$, $\widetilde{\mathbf{x}}_0$, and $\mathbf{u}_{-1}$—see Perez (2003) for further details.

The initial state $\mathbf{x}_0$, is provided by a Kalman Filter (Certainty Equivalence Principle) as indicated in Figure 1, *i.e.*, $\mathbf{x}_0 = \hat{\mathbf{x}}_{k|k}$. The QP can then be solved on-line at each sampling instant and the feedback control law implicitly implemented via

$$\mathbf{u}_k = \mathcal{K}(\hat{\mathbf{x}}_{k|k}) = \mathbf{u}_0^{\text{OPT}}. \tag{16}$$

## 5. DISTURBANCE MODEL PARAMETER ESTIMATION

The solution proposed in the previous section assumes that a model is available to predict the output disturbance; namely, $A_w$ in (3) is known. As already mentioned, the entries of this matrix (parameters of the model) depend on the sea state and sailing condition (speed and heading). We present a simple approach to estimate the parameters of this model *"before"* closing the control loop. Using this approach, the proposed control strategy can be considered a *quasi-adaptive* control strategy. If the sailing conditions and/or the sea state change, it will, in general, be necessary to re-estimate the parameters of the disturbance model; *i.e.*, open the loop for a short while to re-estimate and then close it again.

If the stabilizer control loop is open (see figure 1)—*i.e.*, the rudder or fins are kept to zero angles, and only minor corrections are applied to the rudder to keep the course—we can then use the roll angle and roll rate measurements to estimate the parameters of a second order model. Indeed, the measurements can be considered the state of the following system:

$$\begin{bmatrix}\phi^w_{k+1}\\ p^w_{k+1}\end{bmatrix} = \begin{bmatrix}\theta_{11} & \theta_{12}\\ \theta_{21} & \theta_{22}\end{bmatrix} \begin{bmatrix}\phi^w_k\\ p^w_k\end{bmatrix} + \begin{bmatrix}w1_k\\ w2_k\end{bmatrix}. \quad (17)$$

By defining the vector

$$\theta_k = \left[\theta_{11}(k)\ \theta_{12}(k)\ \theta_{21}(k)\ \theta_{22}(k)\right]^{\mathrm{T}}, \quad (18)$$

we can express the available measurements as

$$\begin{bmatrix}\phi^w_k\\ p^w_k\end{bmatrix} = \begin{bmatrix}\phi^w_{k-1} & p^w_{k-1} & 0 & 0\\ 0 & 0 & \phi^w_{k-1} & p^w_{k-1}\end{bmatrix} \theta_k + \mathrm{v}_k. \quad (19)$$

By adding a parameter update equation

$$\theta_{k+1} = \theta_k + \theta_{wk}, \quad (20)$$

expressions (19) and (20) are in a form that we can apply a standard Kalman Filter to estimate $\widehat{\theta}_{k|k}$ from the measurements $\phi^w_k$ and $p^w_k$. This Kalman Filter is indicated in the block diagram of Figure 1 as *"Disturbance Parameter Estimator"*. The variable $\theta_{wk}$ represents a random disturbance that accounts for unmodelled dynamics.

## 6. SOME SIMULATION RESULTS FOR A RUDDER STABILIZER

To assess the performance of the proposed scheme we will use the benchmark model for a 360ton 50mLOA naval vessel presented by Blanke and Christensen (1993), with the addition of response amplitude operators of a vessel of similar design—see ? for the complete model. As a sea-state description we will use the ITTC spectrum parameterized by the significant wave height ($H_s$) and the mean wave period ($T$.) The rudder magnitude constraint is 22deg and the rate 18deg/sec.

### 6.1 *Parameter Estimation*

Figure 2 shows the evolution of the parameters during the estimation process for an example in bow seas and sea-state 4 (Hs=2.5m) This figure also shows the 10-step ahead predictions for roll angle and roll rate made once the parameters of the model have been identified. We can see that the simple model presents a good prediction performance. Also, the time required for the parameter estimation is relatively low: only a 3 to 5 roll periods.

### 6.2 *Control Tuning*

We performed different tests to assess the effect of the simple tuning and also incorporating a penalty for roll acceleration in the cost. Table 1 shows the average results of 10 realizations under the sailing conditions $\chi$=90 (beam seas), Hs=2.5m, T=7.5s, $\lambda$ = 0.5, and $\gamma_1$=0 and 1. As we can see, penalizing accelerations have a small effect in the roll performance but a greater

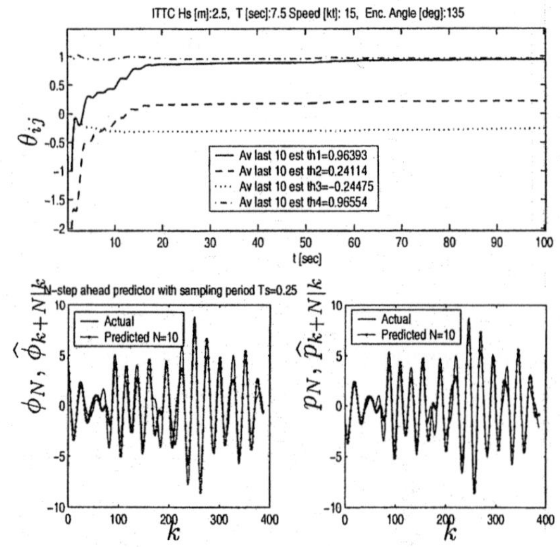


Fig. 2. Estimated Parameters and prediction in bow seas and SS4

effect on acceleration and more importantly in the reduction of Motion Induced Interruptions (which depend on lateral acceleration induced by roll-see Graham (1990)). Table 2 shows similar results for $\lambda$ = 0.9. With a high value of $\lambda$ the roll performance is increased, but in this case the improvement for penalizing accelerations is not significant.

Finally, Table 3 shows how the adaptation improves the performance of the proposed control strategy in a particular example. In Table 3, a change in course from quartering seas to beam seas was simulated. In the first column we have the performance obtained if the disturbance model is not adapted after the change in course. The second column shows the performance after opening the loop in beam seas to re-estimate the disturbance model. We can see a significant improvement in performance due to the adaptation. Table 4 shows similar results for a change in course from beam seas to bow seas.

| $\chi$=90,Hs=2.5m,T=7.5s | $\lambda = 0.5$ $\gamma_1 = 0$ | $\lambda = 0.5$ $\gamma_1 = 1$ |
|---|---|---|
| Roll red % | 29.8 | 34.0 |
| Roll Acc Red % | 27.6 | 31.7 |
| MII red % | 51.8 | 60.2 |
| Yaw rms | 0.35 | 0.41 |
| Rudder rms | 4.50 | 5.12 |
| $\chi$=90,Hs=2.5m,T=7.5s | $\lambda = 0.9$ $\gamma_1 = 0$ | $\lambda = 0.9$ $\gamma_1 = 1$ |
| Roll red % | 64.0 | 63.94 |
| Roll Acc Red % | 59.7 | 60.9 |
| MII red % | 99.0 | 99.2 |
| Yaw rms | 0.92 | 1.02 |
| Rudder rms | 10.5 | 10.7 |

Table 1. Performance at low and high gain ($\lambda$ = 0.5 and $\lambda$ = 0.9) with and without roll acceleration penalty.

Figures 3, 4 and 5 show some time series for different headings at a single speed in sea-state 4. In figure 3, we can see poor performance typical

| ITTC, Hs=2.5m,T=7.5s | 45→90NA | 45→90A |
|---|---|---|
| Roll red % | 41.5 | 65.2 |
| Roll Acc Red % | 36.0 | 60.6 |
| MII red % | 81.2 | 99.1 |
| Yaw rms | 0.70 | 0.86 |
| Rudder rms | 6.2 | 10.1 |

| ITTC, Hs=2.5m,T=7.5s | 90→135NA | 90→135A |
|---|---|---|
| Roll red % | 62.0 | 66. 4 |
| Roll Acc Red % | 46.7 | 56.6 |
| MII red % | 100 | 100 |
| Yaw rms | 0.33 | 0.4 |
| Rudder rms | 4.2 | 5.2 |

Table 2. Performance after a change in course from quatering to beam seas and from beam to bow seas with no adaptation (NA) and after adapting the disturbance predictor (A).

of rudder stabilizers when the encounter frequency is low. It has already been discussed in Perez *et al.* (2003) that there is fundamental limitation associated with reducing the variance using the rudder due the the dynamic characteristics of the ship and the frequency of the disturbance. Indeed, for a rudder stabilizer to perform well we need two things: "*the non-minimum phase zero of the rudder to roll response close to the imaginary axis and low disturbance energy in the region close to the frequency of the zero.*" Also due to the low encounter frequency, the price for roll reduction is high yaw. In the example shown in Figure 3, the dynamic characteristics of the ship dominate the achievable performance—see Perez *et al.* (2003) and Perez (2003) for further details.

Figure 4, shows the case for beam seas. In this case we have a good performance and the main limitation is the limited control action provided by the rudder. In this example we can see how MPC limits the control such that the magnitude constraint on the maximum rudder angle is satisfied. Finally, Figure 5, shows the case for bow seas. Here we have good performance and the main limitation is the rate limitation imposed on rudder command to avoid saturation of the hydraulic machinery commanding the rudder.

| ITTC, Hs=2.5m,T=7.5s | $\chi$=45 | $\chi$=90 | $\chi$=135 |
|---|---|---|---|
| Roll red % | 29.22 | 57.3 | 69.0 |
| Roll Acc Red % | 42.1 | 52.2 | 58.0 |
| MII red % | 99.2 | 99.3 | 100 |
| Yaw rms | 5.4 | 0.90 | 0.43 |
| Rudder rms | 9.3 | 9.4 | 4.27 |

Table 3. Performance for the particular realizations shown in Figures with high gain and adaptation.

## 7. RUDDER, FIN OR COMBINED RUDDER-FIN

Although we have only provided examples for rudder-based stabilizers, the case of of fins is anal-

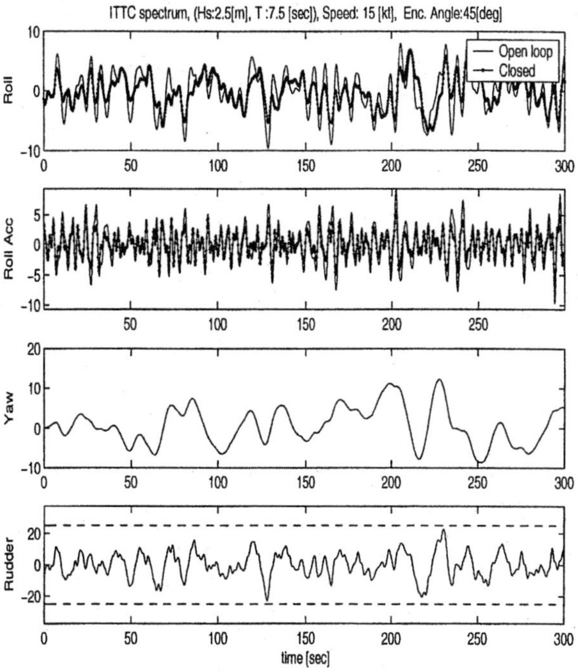


Fig. 3. Simulation SS4 in quatering seas high gain ($\lambda$=0.9, $\gamma_1$=1).

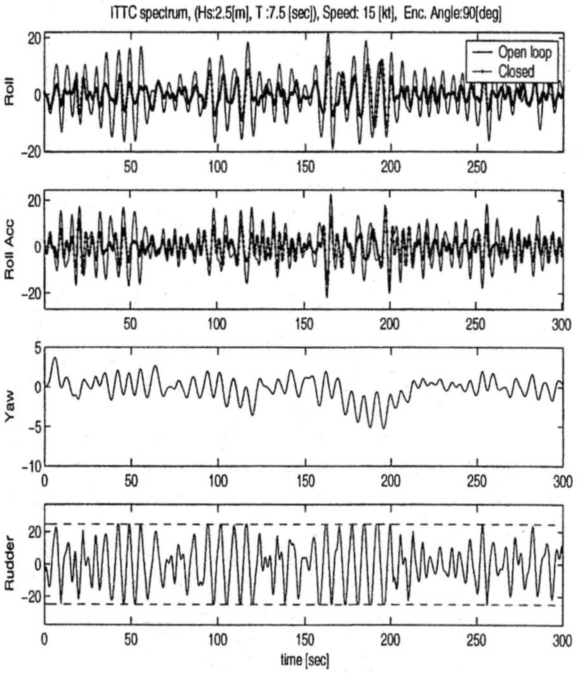


Fig. 4. Simulation SS4 in beam seas high gain ($\lambda$=0.9, $\gamma_1$=1).

ogous with the appropriate change in the model, and the cost—this has already been discusses in (Perez and Goodwin, 2003). The case of combined fin and rudder provides no additional implications to the MPC formulation presented here. Further, since the proposed MPC formulation uses rate input constraints, it already provides a bumpless transfer mechanism when switching between the different modes.

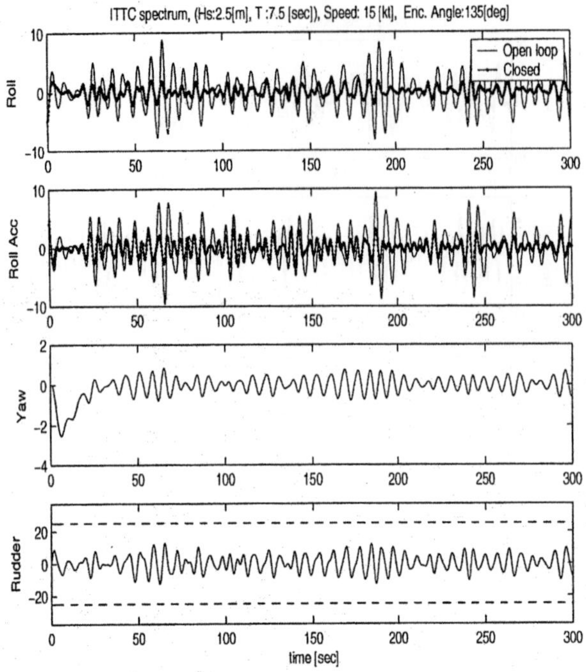


Fig. 5. Simulation SS4 in bow seas high gain ($\lambda$=0.9, $\gamma_1$=1).

## 8. CONCLUSION

We have proposed a control strategy the can be used for fin, rudder and combined fin-rudder ship roll stabilizers. The proposed scheme handles constraints and provide a simple technique for adaptation to changes in sea-state, and sailing conditions. From the presented examples it follows that the adaptation can improve the performance significantly for the proposed scheme. In regard to penalizing roll accelerations in the cost, improvement has been obtained only in some cases. However, the performance is never worse with respect to the motion induced interruption, which seems to indicate the incorporating a roll acceleration penalty could be benefici al.

## 9. REFERENCES


Blanke, M. and A. Christensen (1993). Rudder roll damping autopilot robustness to sway-yaw-roll couplings. *In: Proceedings of 10th SCSS, Ottawa, Canada* pp. 93–119.

Fossen, T.I. (1994). *Guidance and Control of Ocean Marine Vehicles.* John Wiley and Sons Ltd. New York.

Gaillarde, G. (2002). Dynamic behavior and operation limits of stabilizer fins. In: *IMAM International Maritime Association of the Mediterranean, Creta, Greece.*

Graham, R. (1990). Motion-induced interruptions as ship operability criteria. *Naval Engineers Journal.*

Maciejowski, J.M. (2002). *Predictive Control with Constraints.* Prentice Hall.

Perez, T (2003). Performance Analysis and Constrained Control of Fin and Rudder-based Roll Stabilizers for Ships. PhD thesis. School of Electrical Eng. and Computer Sc., The University of Newcastle, AUSTRALIA.

Perez, T. and G. C. Goodwin (2003). Cosntrained control to prevent dynamic stall in ship fin stabilizers. In: *6th IFAC Conference on Manouvering and Control of Marine Craft MCMC'03.* pp. 173–178.

Perez, T., G. C. Goodwin and R.E. Skelton (2003). Analysis of performance and applicability of rudder-based stabilizers. In: *6th IFAC Conference on Manouvering and Control of Marine Craft MCMC'03.* pp. 185–190.

Perez, T., G.C. Goodwin and C.Y. Tzeng (2000). Model predictive rudder roll stabilization control for ships. In: *5th IFAC Conference on Manoeuvering and Control of Marine Crafts. (MCMC2000), Aalborg, Denmark.*

Perez, T., H. Haimovich and G.C. Goodwin (2004). On optimal control of constrained linear systems with imperfect state information and stochastic disturbances. *International Journal of Robust and Nonlinear Control (IJNRC)* **14**, 379–393.

van Amerongen, J., P.G.M van der Klught and J.B.M Pieffers (1987). Rudder roll stabilization–controller design and experimental results. In: *Proceedings of 8th International Ship Control System Symposium SCSS'87, The Netherlands.*

van der Klugt, P.G.M. (1987). Rudder roll stabilization. PhD thesis. Delft University of Technology, The Netherlands.

Warhurst, F. (1969). Evaluation of the performance of human operators as a function of ship motion. Report 2828. Naval Ship Research and Development Centre.

Zhou, W-W., D.B. Cherchas and S. Calisal (1994). Identification of rudder-yaw and rudder-roll steering models using prediction error techniques. *Optimal Control Applications and Methdos* **15**, 101–114.

ELSEVIER
IFAC
PUBLICATIONS
www.elsevier.com/locate/ifac

# FIN AND RUDDER HYBRID STABILIZATION SYSTEM

**Hiroyuki Oda* ,Takashi Hyodo* , Masamitsu Kanda***
**Hiroyuki Fukushima** , Keiji Nakamura****
**Seiichi Takeda*** ,Toshifumi Hayashi*** and Hisato Fujiwara*****

* *Akishima Laboratories(Mitsui Zosen) Inc. 1-50 Tsutsujigaoka 1-chome Akishima, Tokyo, Japan*
** *Mitsui Engineering & Shipbuilding Co., Ltd. 1-1 tama 3-chome Tamano, Okayama, Japan*
*** *Tokyo University of Marine Science and Technology. 5-7 Konan 4, Minato-ku, Tokyo, Japan*

**Abstract:** Modern roll stabilization system, namely fin, anti-roll tank and rudder action are used respectively or in combination on most passengers and naval ships. This paper proposes the advanced rudder roll stabilization control system with fin control. Fin and rudder multivariate hybrid control system were designed using multivariate autoregressive model with multi-input (yaw and roll motion) and multi-output (rudder and fin angle), named MAFRCS (Multivariate Autoregressive Fin and Rudder hybrid Control System). This paper presents the results of studies which led to the development of performing mode of operation for this hybrid control system which system has full control of rudder while fin automatically reduce the roll motion and keep the yaw motion. The results of full scale experiments and simulations studies were given to illustrate the system performances.


**Keywords:** Rudder roll stabilization, Fin and rudder hybrid control, Full scale experiment

## 1. INTRODUCTION

Some successful applications of rudder roll stabilization system have shown that the adequate roll reduction is only possible with specially designed or constructed ship and rudder system. These unsatisfactory situations have led to use of fin and the rudder together for reduce the roll motions and maintaining the ship heading (Sgobbo et al., 1999).
The combination of fin and rudder is a highly attractive alternative for roll damping. However fin motions as well as high frequency rudder motions disturb the heading control system. In order to reduce this interaction to control ship's heading as well as rolling motion with fin and rudder together or alternative, this paper proposes the advanced rudder roll stabilization control system with fin control. Fin and rudder multivariate hybrid control system was designed adopting multivariate autoregressive model with multi-input (yaw and roll motion) and multi-output (rudder and fin angle), named MAFRCS (Multivariate Autoregressive Fin and Rudder hybrid Control System).
In section 2, yaw and roll control with fin and rudder was discussed. In section 3, designing fin and rudder hybrid control system based on multivariate auto-regressive model was summarized. In section 4, definition of roll reduction is explained. In section 5, results of full scale experiment were performed. In section 6, design and confirm with numerical simulation were discussed. Finally, conclusions and discussions were summarized.

## 2. YAW AND ROLL CONTROL WITH FIN AND RUDDER

The ship is steered by her rudder, the ship heels to inboard side due to the roll moment between the center of gravity and the center of the hydrodynamic force acting on a rudder control surface as shown in Fig.1 (Cowley, 1972). On the other hand, roll motion is controlled by fin as shown in Fig.1, the ships heading change due to the yaw moment between the center of gravity and the center of the hydrodynamic force acting on the fin control surface shown as white arrow at Fig.1 (Kallstrom, 1981).
The longer these distances (arm) between them and higher ship speed make the stronger yaw and roll moment. Once the roll has begun, it causes yaw moment. This coupling motion can be explained by longitudinal asymmetry of hull configuration due to roll angle. Sometimes they might induce the so called roll-yaw-rudder-fin instability.

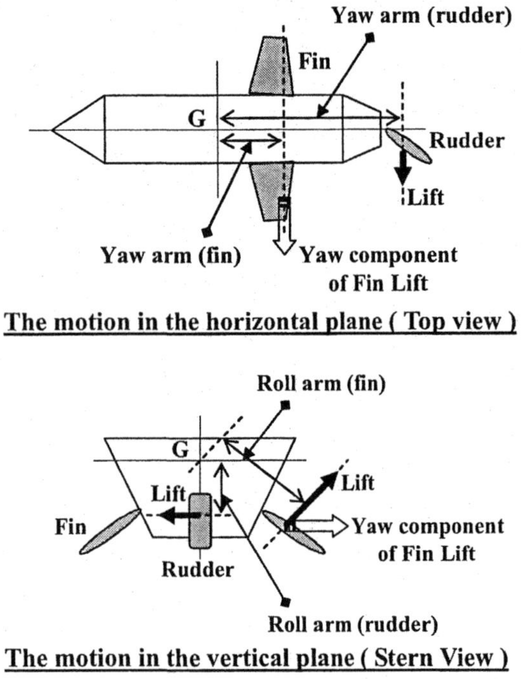


**Fig. 1 Yaw and Roll Coupling Motion with Fin and Rudder**

The basic equations of the ship motion can be written in following form. Pitch and heave can generally be neglected if the study concentrate on course keeping and roll damping. Ship motion modeling is thus considered in surge, sway, yaw and roll (Oda, et al., 1996).

$$\begin{aligned} &\text{Surge} : m(\dot{u} - vr) = X_H + X_P + X_R + X_F + X_E \\ &\text{Sway} : m(\dot{v} + ur) = Y_H + Y_P + Y_R + Y_F + Y_E \\ &\text{Yaw} : I_{ZZ}\dot{r} = N_H + N_P + N_R + N_F + N_E \\ &\text{Roll} : I_{XX}\dot{p} = K_H + K_P + K_R + K_F + K_E \end{aligned} \quad (1)$$

Where m, Izz and Ixx are mass and turning moment of inertia, u, v, r and p are ship's speed along X, Y axis and rate of turn around Z, X axis. The subscripts H, P, R, F and E denote the hydrodynamic forces from hull, propeller, rudder, fin and external forces.

## 3. HYBLID STABILIZATION SYSTEM

### *3.1 Fin and rudder hybrid controller*

As motions of heading and rolling are physically coupled during straight courses and piloting during gyrations, the rudder and fin control laws of control system have a multi- input and multi-output (MIMO) system, different from the single-input single-output (SISO) system used for classical piloting or stabilization system.
To stabilize the ship on the navigation, the image of hybrid control system has been used as shown in Fig.2. The purpose of fin and rudder hybrid control system is put on roll reduction as well as course keeping using fin and rudder. The philosophy behind the rudder roll stabilization system is that the rudder can use as the effectiveness actuator to control both steering and roll reduction. The block diagram of hybrid control system is shown in Fig.3. The purpose of multivariate hybrid control system is put on roll reduction as well as course keeping using fin and rudder.

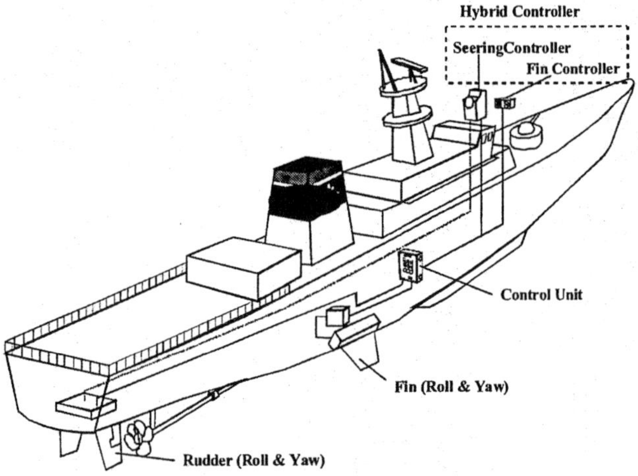


**Fig. 2 Image of Hybrid Control System**

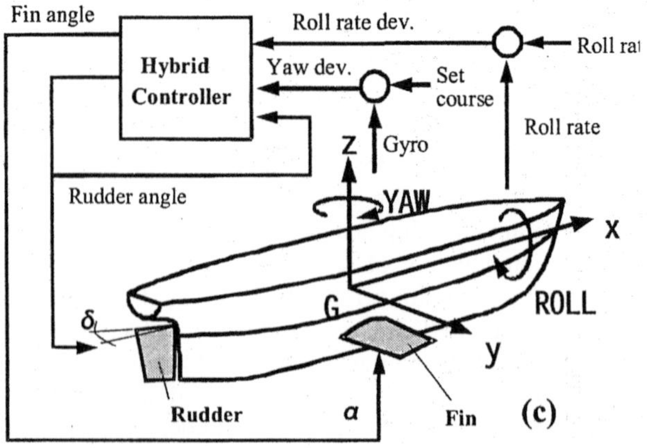


**Fig. 3 Rudder, Fin and Hybrid Control**

### *3.2 Multivariate auto-regressive model*

The basic control model adopted here for predictions of roll and yaw motion is a control type of Multivariate Auto-Regressive eXogenous model (MARX model) (Oda, et al., 1992, Oda, et al., 2003).

$$X(n) = \sum_{m=1}^{M} A(m)X(n-m) + \sum_{m=1}^{M} B(m)Y(n-m) + U(n) \quad \textbf{(2)}$$

This control law is powerful stochastic model in designing a roll reducible autopilot system (heading control), where $X(n)$ is two-dimensional vector of controlled variable (r) ; yaw and roll. $Y(n)$ is two-dimensional vector of control variable (l) ; rudder and fin angle. $U(n)$ is white noise. The order M of the model is obtained by AIC.

$$AIC(M) = N \cdot \log\{\det(dr, M)\} + 2 \cdot r(r+1)(M+1) \quad \textbf{(3)}$$

Where N is data length and optimal MARX model has the order which takes the minimum value of AIC estimation (MAICE) (Akaike, 1974). This control model is obtained by the MAICE procedure, using the actual data (roll and yaw angle, fin and rudder angle) gained from the preliminary full scale trial for the identification of the fin and/or rudder dynamics. Based on the modern control theory, this model can be transformed to a state space representation.
The optimal control law which minimizes the quadratic criterion function under constraint of the above state space equation is given by a feedback law with the stationary gain $G$. Then the optimal control law can be represented by

$$Y(n) = GZ(n) \quad \textbf{(4)}$$

The optimal problem with the criterion can be solved using the technique of dynamic programming.

## 4. DEFINITION OF ROLL REDUCTION

### *4.1 Target ship of experiment*

To confirm the effectiveness of MAFRCS, full scale experiments were carried out using prototype control system. The experiments were made on the fishery training ship "UMITAKA MARU" of Tokyo University of Fisheries. The principal particulars are shown in Table 1. The view of target ship and fin are shown in Photo 1.
In order to check the roll-yaw-rudder-fin instability and control law by means of MARX model, preliminary studies were already done using free running model experiment with radio control.

**Table 1 Principal particulars of "UMITAKA MARU"**

| | | |
|---|---|---|
| **Hull** | Length (Lpp) | 83.0 (m) |
| | Breadth | 14.9 (m) |
| | Depth | 8.9 (m) |
| | Draft | 5.95(m) |
| | Displacement | 4000 (ton) |
| | Gross Ton | 1886 (ton) |
| | GM | 1.32 (m) |
| | Roll Period | 10.3 (sec) |
| **Rudder** | Area | 9.63 (m**2) |
| | Aspect | 1.45 |
| | Slew rate | 5 (deg/s) max. |
| **Fin** | Area | 6 (m**2) |
| | Slew rate | 20 (deg/s) max |
| **ART** | Type | Variable Period |
| **Propeller** | Type | Cpp |
| | Dia. * No. | 3.8 (m) * 4 |
| **Engine** | Type | Diesel |
| | Power * rpm | 4489 (KW) * 520 (rpm) |

**Photo. 1 The view of "UMITAKA MARU"**

### *4.2 Definition of roll reduction*

To confirm the basic performance of roll reduction of MARFCS, full scale experiments were implemented that the cases from forced oscillation to free damping and fin and/or rudder stabilization control. The percentage reduction of roll motion is defined by

$$REDUCTION = \frac{Mfn(Free) - Mfs(Stabilization)}{Mfn(Free)} \cdot 100(\%) \quad \textbf{(5)}$$

The roll amplitude factors denote Mfn(free) and Mfs(stabilization) are calculated by following equation.

$$Mf\alpha = \frac{1}{\sqrt{\left\{1-\left(\frac{\omega}{\omega_n}\right)^2\right\}^2 + \left(2 \cdot C\alpha \cdot \frac{\omega}{\omega_n}\right)^2}} \quad \textbf{(6)}$$

Notations in this equation are follows;

ω : wave period , ωn : natural roll period

Cα : roll damping factor defined as follows

$$C_n = \frac{1}{2\pi} \cdot \log e \frac{\phi_n}{\phi_{n+1}} \quad , \quad Cs = \frac{1}{\pi} \cdot \log e \frac{\phi_j}{\phi_{j+1}} \tag{7}$$

$\Phi n$, $\Phi n+1$ : double amplitude of roll angle at free dumping

$\Phi j$, $\Phi j+1$ : single amplitude of stabilization control

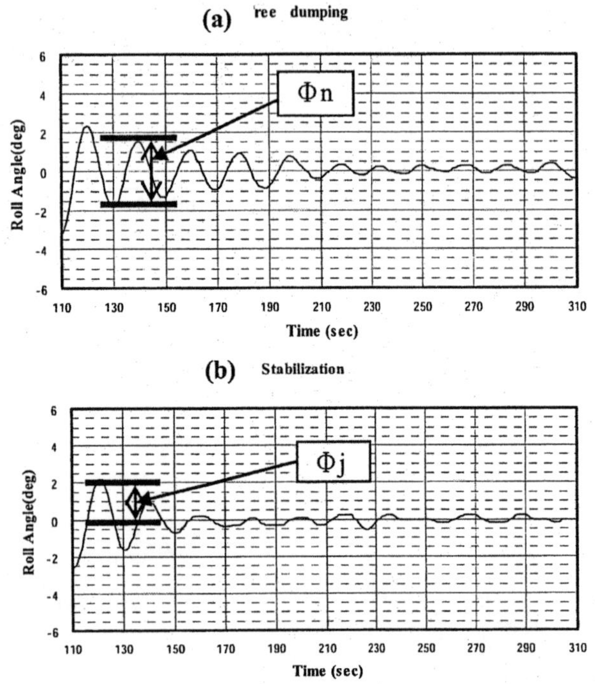


**Fig. 4 Example time series of roll angle ( Free dump and stabilization )**

Examples of definition of roll reduction are shown in Fig. 4. Fig.4 (a) shows the time series of roll angle at free dumping and Fig4 (b) shows the time series of roll angle at stabilization control (Kallstrom, 1990).

Using example data, the calculation flow of roll reduction is shown Table 2. Table 2 (a) shows double amplitude at free dumping ($\Phi n$, $\Phi n+1$). Table 2 (b) shows single amplitude at stabilization control ($\Phi j$, $\Phi j+1$ ). Table 2 (c) shows dumping factors of Cn and Cs calculated by equation (7), amplitude factors of Mfn(free) and Mfs(stabilization) calculated by equation (6) and also roll reduction defined by equation (6). In this example, the performance of roll reduction is defined as 44.83(%).

## 5. RESULTS OF FULL SCALE TRIAL

The full scale experiments on the fishery training ship "UMITAKA MARU" were made and even before updating the maneuverability modeling which is part of the hybrid control system of MAFRCS, the first trial made on the basic performances of MAFRCS were satisfactory.

**Table 2 Definition of roll reduction ( Free dump and stabilization )**

**(a)**

<table>
<tr><td>Φ 1</td><td>Φ2</td><td>Φ3</td><td>Φ4</td><td>Φ5</td><td>Φ6</td></tr>
<tr><td>-3.3deg</td><td>2.3deg</td><td>-1.8deg</td><td>1.5deg</td><td>-1.3deg</td><td>1.1deg</td></tr>
<tr><td>3.3deg</td><td>2.3deg</td><td>1.8deg</td><td>1.5deg</td><td>1.3deg</td><td>1.1deg</td></tr>
<tr><td colspan="2">Φ i</td><td colspan="2">Φ i+1</td><td colspan="2">Φ i+2</td></tr>
<tr><td colspan="2">5.6deg</td><td colspan="2">3.4deg</td><td colspan="2">2.4deg</td></tr>
<tr><td>Φ j</td><td>4.2deg</td><td>Φ j+ 1</td><td>2.9deg</td><td></td><td></td></tr>
</table>

**(b)**

| Φ 1 | Φ2 | Φ3 | Φ4 |
|---|---|---|---|
| 2.2deg | -1.6deg | 1.0deg | -0.7deg |
| 2.2deg | 1.6deg | 1.0deg | 0.7deg |

**(c)**

| Free dumping | | Stabilization | |
|---|---|---|---|
| Cn1 | 0.080 | Cs1 | 0.089 |
| Cn2 | 0.056 | Cs2 | 0.165 |
| Cn3 | 0.060 | Cn3 | 0.101 |
| Cn(mean) | 0.066 | Cs(mean) | 0.119 |
| M fn | 7.632 | M fs | 4.210 |

| REDUCTION | 44.83% |
|---|---|

The MAFRCS included measurement of roll angle, roll rate, yaw angle by vertical optical fiber gyro and measurement of rudder angle and fin angle by ship's equipment.

The performance of roll damping has been measured in different ways. From the scientific point of view, the obvious approach is to measure the roll angle with and without the rolling damping system during exactly the same condition. In order to confirm the basic performance of roll damping and course keeping of MAFRCS, full scale experiments were implemented in the cases from forced oscillation to free damp and MAFRCS. A simple approach is to move the rudder and/or fin in sinusoidal way with a frequency close to ship's natural roll period and measure rudder and/or fin angle and roll motion as shown in section 4.2.

### *5.1 Roll stabilization and course keeping by rudder and fin*

Fig. 5 and Fig 6 show the function of rolling reduction and course keeping performance in the condition of navigation under forced oscillation by fin and stabilization by rudder and fin multivariate hybrid control system ( MAFRCS ) in the calm sea. Fig. 5 is the results in the condition of ship speed 16kts and Fig. 6 is the results of ship speed 19kts.

In this figure, the first column is rudder angle, the second column is fin angle (starboard), the third column is fin angle (port), the fourth column is roll angle, the fifth column is roll rate and the last column is yaw deviation from set course.

### *5.2 Summary of measured roll reduction with MAFRCS*

The percentages of roll reduction of roll angle and roll rate were defined in section 4. Summary of measured roll reduction were performed that roll and heading control with rudder, roll and heading control with fin, roll and heading control with both fin and rudder. The roll reductions with some control mode of MAFRCS are summarized in Table 3.

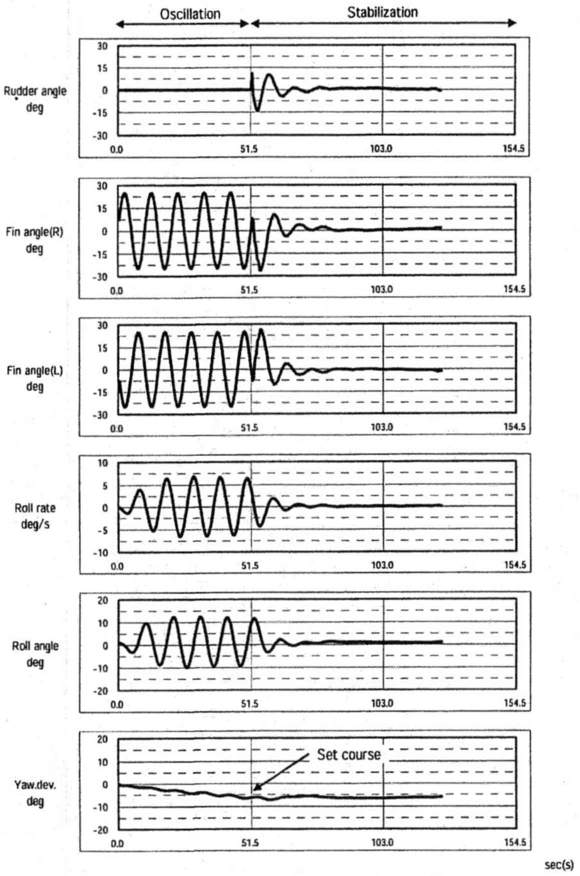


**Fig. 5 Time series of forced oscillation and stabilization ( Sea state 1with speed 16 kts: Control of rudder and fin )**

### *5.2 Summary of measured roll reduction with MAFRCS*

The percentages of roll reduction of roll angle and roll rate were defined in section 4. Summary of measured roll reduction were performed that roll and heading control with rudder, roll and heading control with fin, roll and heading control with both fin and rudder. The roll reductions with some control mode of MAFRCS are summarized in Table 3.

**Table 3 Summary of measured roll reduction**

| Control mode of MAFRCS | Reduction (%) define eq. (5) | |
|---|---|---|
| | Roll angle | Roll rate |
| Rudder only | 49 | 31~47 |
| Fin only | 71~73 | 74~76 |
| Fin & Rudder | 67~77 | 60~80 |

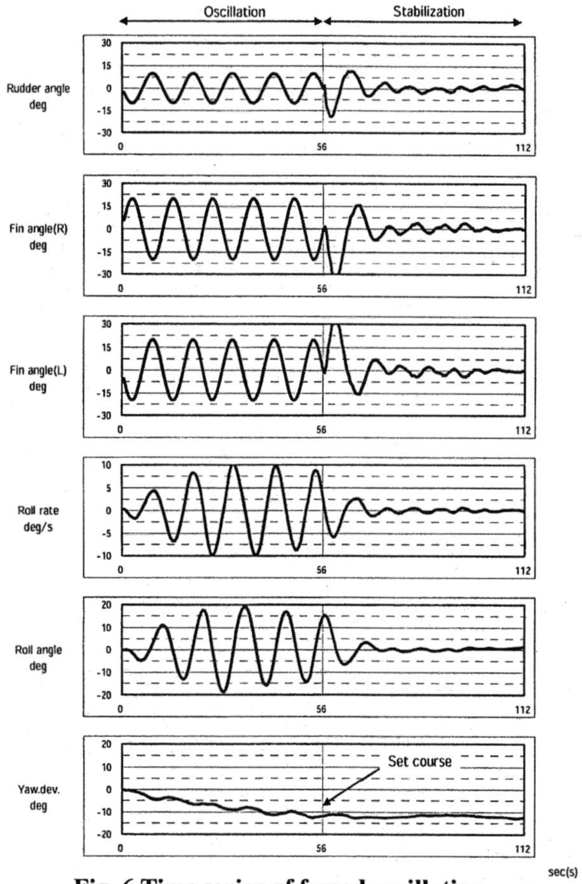


**Fig. 6 Time series of forced oscillation and stabilization ( Sea state 1with speed 19 kts: Control of rudder and fin )**

It can be concluded that in normal conditions a roll reduction in the order of 70~80(%) is achievable with MAFRCS ( fin mode and hybrid mode ), but these experiments were carried out without control gain to optimize. Also the heading control by MAFCS can maintain the desired course in allowable limit

## 6. CONFIRMATION OF HYBRID SYSTEM

Simulation techniques are presented as a possible aid in the design of fin and rudder multivariate hybrid control system (MAFRCS). The results of full scale experiment with "UMITAKA MARU" shown in section 5 were confirmed by simulation study. But these simulation studies were carried out without control gain to optimize exactly. Fig. 7 shows the simulation results of rolling reduction and course keeping performance in the condition of navigation under forced oscillation by fin and stabilization by rudder and fin multivariate hybrid control system (MAFRCS).

The results of Fig. 7 indicate that the full scale experiment and simulation were well fitted in the function of roll reduction and course keeping. In this figure, solid line shows full scale results and dotted line shows simulation results. Also the first column is rudder angle, the second column is fin angle (starboard), the third column is roll angle and the last column is yaw deviation from set course.

The results of dynamic simulation suggested that the simulation techniques were reasonable and useful tool for design and confirm the ship's dynamics and control law of MAFRCS.

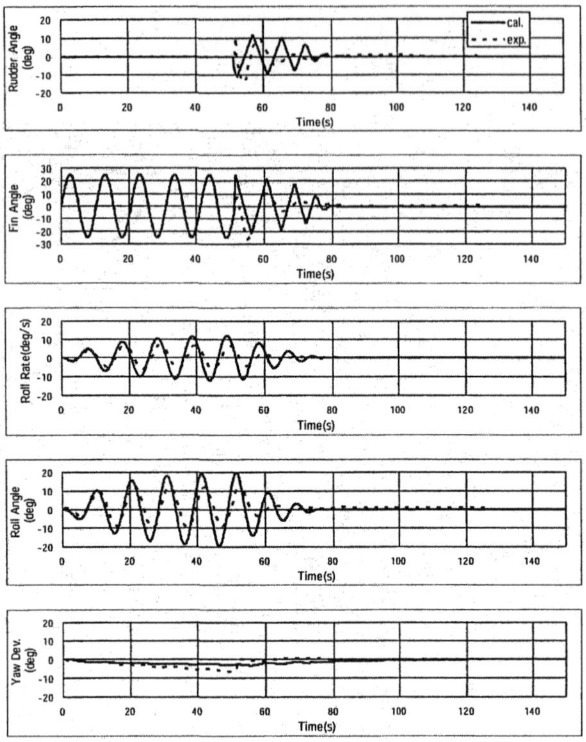


**Fig. 7 Simulation results of forced oscillation and stabilization driven by rudder and fin**

## 7. CONCLUSIONS

This paper presents the advanced rudder roll stabilization control system with fin control. Fin and rudder multivariate hybrid control system were designed using multivariate autoregressive model with multi-input (yaw and roll motion) multi-output (rudder and fin angle). This control system is called MAFRCS (Multivariate Autoregressive Fin and Rudder hybrid Control System).

The purpose of this paper is to verify the effectiveness of MAFRCS in simulation and in full scale experiments. From the results of full scale experiments, fin and rudder multivariate hybrid control system reduce the roll motion in the order of 70～80(%) with course keeping. Also the results of dynamic simulation that the simulation techniques with was useful tool for design and confirm the ship's dynamics and control law.

Through these studies, it can be expected that the MAFRCS have higher controllability than conventional rudder roll control systems or conventional fin stabilization systems. Hereafter, the authors are going to apply this hybrid control system and advanced rudder roll stabilization system (Oda, et al., 1999) to several types of ship to be of great use and to develop good effect of next generation type motion control.

## ACKNOWLEDGEMENT

In the full scale experiments described in this paper, Y.Koike captain, H.Yonemoto chief engineer and crew gave helpful support with especial thanks to "UMITAKA MARU". Helpful suggestions and encouragement were received from Prof. K.Ohtsu of Tokyo University of Mercantile Marine and Mr. S.Ueki and Mr. H.Kanehiro of Mitsui Engineering & Shipbuilding Co., Ltd.

## REFERENCES

J. N. Sgobbo, M. G. Parsons (1999).
Rudder/Fin Roll Stabilization of the USCG WMEC 901 Class Vessel, Marine Technology, Vol.36, No.3.

W. E. Cowley (1972).
The Use of the Rudder as a Roll Stabiliser, SCSS'72.

C. G. Kallstrom (1981).
Control of Yaw and Roll by a Rudder/Fin Stabilization System, SCSS'81.

H. Oda, K. Igarashi, K. Ohtsu (1996).
Simulation Study and Full Scale Experiment of Rudder Roll Stabilization System, SCSS'96.

H. Oda, K. Ohtsu, M. Sasaki, Y. Seki, T. Hotta (1992).
Rudder Roll Stabilization Control System through Multivariate Auto Regressive Model, CAMS'92.

H. Oda, T. Hyodo, M. Kanda, H. Fukushima, K. Nakamura, S. Takeda, T. Hayashi, H. Fujiwara (2003).
Designing fin and rudder multivariate hybrid control system, SCSS'2003.

H. Akaike (1974).
Statistical Analysis and Control of Dynamic System, Kluwer Academic Publishers.

C. G. Kallstrom, W. L.Schultz (1990).
An Integrated Rudder Control System for Roll Damping and Course Maintenance, SCSS'90.

H. Oda, T. Hyodo, K. Ohtsu, M. Ito, N. Hirose, J.S. Park, H. Sato (1999).
Designing Advanced Rudder Roll Stabilization System – High Power with Small Size Hydraulic System and Adaptive Control -, SCSS'99.

ELSEVIER
IFAC
PUBLICATIONS
www.elsevier.com/locate/ifac

# SWITCHED CONTROL SYSTEM FOR SHIP ROLL STABILIZATION

Anna-Zaira Engeln [*] Ali J. Koshkouei [*] Geoff N. Roberts [†] Keith J. Burnham [*]

[*] *Control Theory and Applications Centre, Coventry University, Coventry CV1 5FB, UK*

[†] *Mechatronics Research Centre, University of Wales College, Newport, Allt-yr-yn Campus, Newport, NP95XR, UK*

The stabilization of ship roll using a family of switched controllers is considered in this paper. The system includes subsystems (modes) resulting from environment changes and different orientations and ship speeds. Controllers are designed to stabilise the entire system at all situations of a ship's trajectory. The system is a parallel multi-model control (PMMC) system, which may also be regarded as a switched system. A PMMC system is a hybrid dynamical system, consisting of a family of subsystems with a set of rules (decision-making process, high-level supervisor), orchestrating between them. The activation of the switching mechanism between controllers is dependent on the dynamic behaviour of the system. *Copyright © 2004 IFAC*

Keywords: Ship roll stabilization, switched systems, parallel multi-model control system, hybrid systems, switched control design.

## 1. INTRODUCTION

The architecture of a general PMMC system involves a family of subsystems with a set of switched controllers. Various switching laws will make the behaviour of a PMMC system quite different from that of its individual components. Well-designed PMMC systems should have superior performance over single-controller systems. The stability of such switched systems and design of the decision-making algorithm have attracted a great deal of interest and resulted in many published papers (Xie *et al.*, 2001; Daafouz, *et* al. 2002; Hespanha and Morse, 1999; Liberzon and Morse, 1999).

The development of new algorithms for the evaluation of potential control procedures and the predictive cost function techniques to improve the PMMC decision making process are identified as new techniques/technologies required for a PMMC system.

Control of a ship system has been widely studied in the recent years (Belmont and Horwood, 1999; Crossland, 2000; Sharif *et al.;* Tedeschi, 1999). The behaviour of these systems is nonlinear, however to simplify the problem, a linear approximation for each system is applied. There are many modes such as roll, yaw, engine (speed), rudder as well as the sea wave behaviour, involved in the control of a ship. These modes/behaviours will all interact in some interval of time. These individual modes need to be controlled using different controllers over each interval of time. Therefore, for each mode, a control system is designed. The number of controllers is based upon the number of modes of the system.

Switching from one control to another needs rules regarding the environment and different situations of a ship. Selecting different controllers with respect to the different modes of the system needs some intelligent rules. Applying these rules may not guarantee that the controllers are simultaneously switched when the associated modes of the system are changed. The time interval that a controller is active may differ from the associated mode of the system. Therefore, in a finite time, there are sequences of the switching of system modes and a sequence of switching controllers, and it is this overall switching sequence, which is required to be optimised.

Switched systems are a class of hybrid systems consisting of a set of continuous- or discrete-time subsystems. If all the modes of the system are stable, the system may not necessarily be stable. Additional conditions are required to ensure the stability of the system (Xie *et al.*, 2001). For example, if a controller is switched (or remains active) with a non-associated mode, the behaviour of the system may be unstable. Furthermore, since the system is nonlinear, an inappropriate switching mechanism may cause the overall system to become chaotic. A controller is switched to another one depending on various modes of the system. A controller should be active for a certain period before switching to a new controller. The task of a control system supervisor is to determine which controller should be selected in a certain time depending on the environmental situations of the system modes. The control system supervisor will need to allow a sufficient time for a switched controller de-activation/activation to mitigate the transition effects.

A necessary condition for asymptotic stability of the system under arbitrary switching is that all the individual subsystems are asymptotically stable (Liberzon and Morse, 1999). However, this condition is not sufficient for stability of the switched system. The stability, analysis and control synthesis for switched model controllers have been considered in Daafouz, *et al.* (2002) in which output feedback controllers have been designed for subsystems so that asymptotic stability of the closed-loop system is achieved.

In this paper, the stabilization of ship roll as a PMMC system is considered by designing a switching mechanism for a family of PID controllers. To stabilise the system, different controllers could be activated to accommodate measured changes in the ship model dynamics and/or changes in the environmental conditions. The controller design method for ship roll stabilisation is considered utilising integrated control of rudder and stabilising fins.

The paper is organised as follows: Section 2 describes the architecture of a general PMMC system. In this section, the stabilisation of ship motion is considered as an application of PMMC. Section 3 discusses the decision maker for switching to suitable controllers based upon the system behaviour. Section 4 addresses the control design method and how the integrated system can be tuned using the controllers. In this section, two different approaches and associated algorithms are proposed for switching the controllers. Simulation results and conclusions are presented in Sections 5 and 6, respectively.

## 2. PMMC FOR SHIP MOTION STABILISATION

An indication of possible benefits from adopting a PMMC approach for warship motion stabilisation was given in (Crossland, 2003) where the potential improvements in the ability of a warship to perform an anti submarine warfare mission in the North Atlantic has been illustrated. Different control strategies for fin roll stabilisation, using a relatively *ad hoc* decision making process, were considered. An assessment of the potential benefits to operational performance of an integrated fin and rudder roll stabilisation was also presented.

In general, ship motion stabilisation primarily concerns the minimisation of roll motion (roll stabilisation), although in some instances minimisation of lateral forces (LFE stabilisation) (Sharif *et al.*, 1993) may be the goal. Such control is normally achieved using actively controlled stabilising fins. However, this motion control can also be achieved through controlled rudder movements (rudder roll stabilisation (RRS)) or by stabilising a combination of rudder and fins (integrated control). In addition, since ship motion stabilisation may be considered as a 'slow system', then low switching frequencies between controllers in a PMMC would be appropriate. The stability of the integrated system is therefore considered to be a minor concern.

As far as the application of PMMC ship motion stabilisation (including fin, rudder or rudder and fin stabilisation), is concerned, any designs for ship motion stabilisation should be considered as an integrated system. Before considering the benefits to be accrued from PMMC, it is useful to explore the reasons for considering PMMC. For simplicity, only roll stabilisation will be considered. The arguments presented are equally applicable to LFE stabilisation. The basis for using PMMC for roll stabilisation is because of the difficulty in designing a controller (the control algorithm) that achieves satisfactory roll stabilisation for the complete range of ship operating conditions.

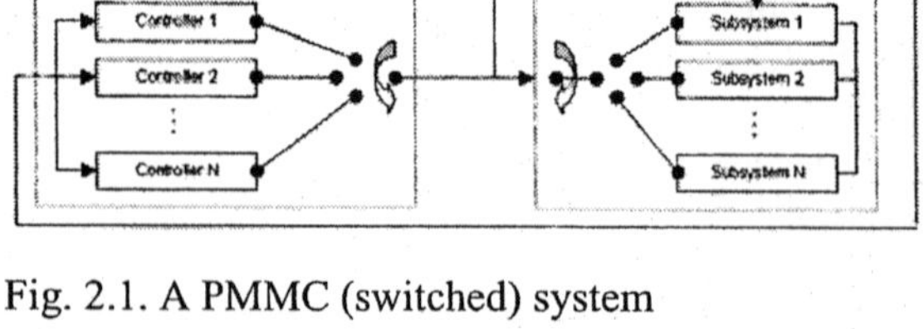

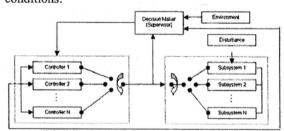

Fig. 2.1. A PMMC (switched) system

Ship roll motion is affected by the environmental conditions of wind and waves, and this is considered to be very much the heart of the problem. Sea disturbances are stochastic and their underlying energy spectrum is a function of wave height and frequency components. These factors vary quite considerably on a local basis but also have different characteristics depending where the ship is operating in mid-ocean or coastal, north or south hemisphere etc. The situation is exacerbated by the fact that even when a ship is operating in a relatively constant sea state (which in any case is rare) the encounter angle that the ship makes with the direction of the sea will change the encounter frequency. This is particularly significant for ships, which regularly and routinely undertake manoeuvres as part of their normal operations. A suitable way to demonstrate this is to consider a fin control algorithm designed to minimise ship roll motion for a sea having a typical North Atlantic wave energy.

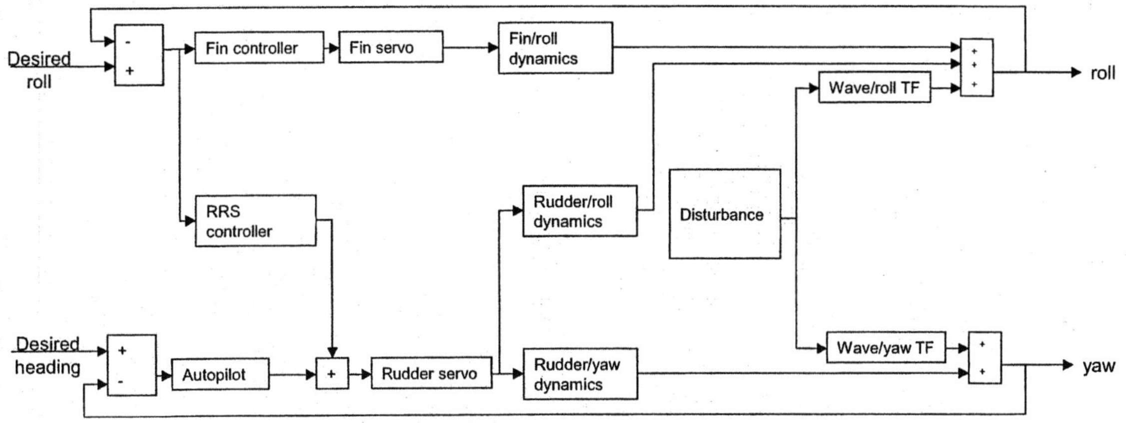


Fig. 2.2. The block diagram of the ship system

Two possible problems with controller performance are: how well the control algorithm will work with quartering and following seas and how well the control algorithm will work when the ship is operating elsewhere where wave energy spectra is very different to that of the North Atlantic. An interesting example of this problem was presented in (Blanke *et al.,* 2000), where the authors discovered that a RRS controller designed and evaluated using the universally accepted Bretschneider wave energy spectrum (based on observations of North Atlantic wave spectra) gave unsatisfactory results when implemented in a ship operating off the Danish coast. Redesign and evaluation of the RRS controller using sea spectrum obtained from local measurements resulted in successful trials.

Ship roll stabilisation (or ship motion stabilisation) may be described as designing a control algorithm for a nonlinear dynamic system operating in a changing and unpredictable environment. Ship motion stabilisation with PMMC structure, is shown in Fig. 2.1, in which different controllers are switched into operation to accommodate measured changes in the ship model dynamics (including changes of environmental conditions). The switched controllers may either be optimised to meet normal design considerations but may have specific qualities at different operating conditions or may be optimised for particular operating conditions.

The PMMC structure depicted in Fig. 2.1, is such that (subsystem) models 1 to N are connected to controllers 1 to N when the ship is subject to environmental conditions i.e. wind and waves provides the basis for initial simulation. For ship motion stabilisation the desired set point (input command) is always zero degrees. The decision making process may be accomplished using the error output between the actual system and the model or estimation of frequency of the system. In this work a control strategy is designed based on the frequency of the system, in which one appropriate controller is active depending on sea state conditions, including environmental disturbance, ship orientation and speed over a certain period.

Fig. 2.2 illustrates a block diagram of a ship control including rudder, roll, yaw and controllers.

## 3. DECISION MAKER

The decision maker block determines which controller is to be activated for the current behaviour of the system. The appropriate controller is then smoothly put into action. These transitions are achieved over a certain period of time or/and when significant changes affect the system. One can consider the following cases for making decisions for switching smoothly from one controller to another

- Sea conditions vary relatively slowly in a particular sea area. Thus, wave disturbances can be considered as being reasonably constant.
- For many operations, ships will maintain a steady course and constant speed. In this

situation, the spectrum of the wave disturbances can be assumed reasonably constant.

- Control surfaces at low speeds are ineffective. Therefore, for ship speeds less than 6 knots the PMMC system would be operating but the outputs to the actuators (stabilisers and rudder) would be inhibited
- Under steady course and constant speed conditions, initialisations of the PMMC decision maker would occur at regular time intervals, say twice per hour.
- Initialisation of the decision maker would be necessary if there were significant changes in ship speed or course, as these changes would affect ship roll motion dynamics and the effective spectrum of the waves. In this case, re-initialisation of the decision maker would occur if there were a speed change of more than a certain value (say, 3 knots) or a course change greater than a specific angle (e.g. 15°).

The decision would only be taken on current performance rather than current and past performances. In general, a drawback of predictive simulation models is the need to predict realistic input commands for the simulation mode. However, for motion stabilisation, this problem disappears as the desired set point (input command), is always zero. The decision making process again could be accomplished either by the explicit (cost function minimisation) or implicit (soft computing) approach, suitably modified to consider predicted rather that historical data.

## 4. CONTROL DESIGN AND SWITCHING ALGORITHM

Development of the PID controllers was carried out considering the fact that the roll frequency of the ship mostly depends on the following three external influences: ship speed, the angle in which the waves encounter the ship, and wave height.

PID controllers were developed considering:

- Wave encounter angle
- Ship speed
- Sea state

Due to symmetry of the vessel, the number of angles considered could be reduced.

The ship speed can be used as an input to the control system, but not the sea state or the wave encounter angle. This makes necessary the development of a system to detect the ships situation.

The roll motion of the ship can be measured using a gyroscope. For each system (or sub-model) a PID controller with the transfer function

$$G_c(s) = K_p + \frac{K_I}{s} + K_D s \qquad (1)$$

is designed. The gains $K_P$, $K_I$ and $K_D$ depend on the centre frequency of the system, $\omega_e$. The frequency $\omega_e$ is selected to damp the roll motion for all combination of external influences.

The PID controller pair with its centre frequency closest to detected roll frequency is switched into operation.

Two approaches to detect the roll frequency of the ship are considered:

(a) Detection of the extremes of the roll angle
(b) Fast Fourier Transform (FFT)

Fig. 4.1 shows the sequence deciding when the data is collected and if the switching procedure is required.

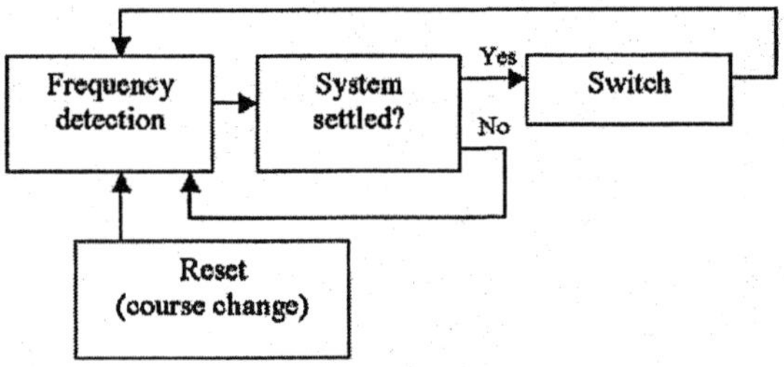


Fig. 4.1. Switching sequence

During course or speed change, the collected data is not useful as the system moves through a phase of unsteadiness. When the ship is changing course/speed, the decision maker inhibits the switching to another controller. Once a steady course/speed is achieved, the data buffer is reset and filled with new data of the settled system, and then an appropriate decision can be made.

In addition, the switching is inhibited at the beginning of a journey and only switched to another controller if the system is settled.

### *4.1. Extremes method (finding average period)*

The method finds the maximum point in the positive part of a period, stores the point in time and compares this to the last stored point of detection of a negative extreme (see Fig. 4.2).

Let $T_{roll}$ and $f_{roll}$ be ship's roll period and frequency, respectively. Then

$$\omega_{roll} = 2\pi f_{roll} = 2\pi / T_{roll}$$

After the detection and calculation of a certain number of periods, $\omega_{roll}$ is averaged for obtaining a suitable approximation of the roll frequency.

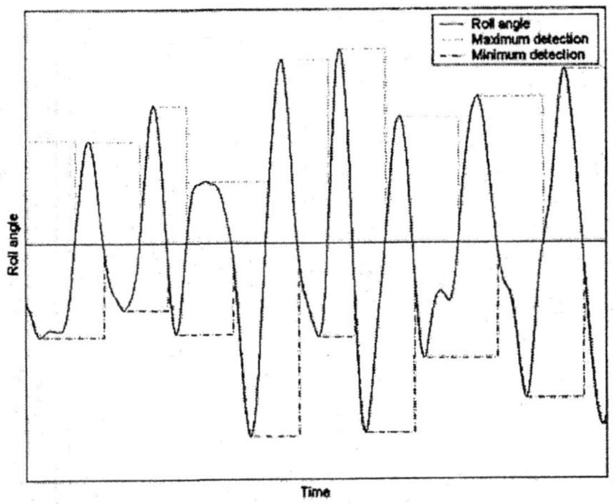


Fig. 4.2: Detecting maximums and minimums of the roll signal

### 4.2. *FFT method*

A number of samples is taken at regular time intervals from the roll signal and stored in a buffer. When the buffer is full, the Fast Fourier Transform is calculated and the power spectrum formed. The peak of this power spectrum occurs at the principal frequency of the ship's roll motion.

## 5. SIMULATION RESULTS

Simulation data are obtained for the RV Triton.

### 5.1. *Extremes method*

The following trajectory has been applied to the simulation model (see Fig.5.1).

The upper trace shows the wave encounter angle in degrees. At the start of the journey, the wave encounter angle is 90°, which stays constant for about 5 minutes. The middle trace shows the controller/angle indices (Controller No 1 for 0°, No 2 for 30°, No 3 for 60° and No 4 for 90°). The dot-dash line indicates the discrete wave encounter angle, which most closely resembles that actually experienced. The solid line illustrates the angle of the system that has been estimated and the controllers are consequently switched.

During the course change to 30°, switching is inhibited, but after the resettling of the system and re-estimation of the wave encounter angle the switching algorithm chooses controller No 3 instead of the expected No 2. After a certain period, the course changes again to 60°. A sufficient time is needed for settling the system with new control. Then, the system recalculates the roll frequency, the decision maker considers the new data, and then the current active controller is switched to another if it is required. (In this case, the controller remains i.e. No3 as expected.)

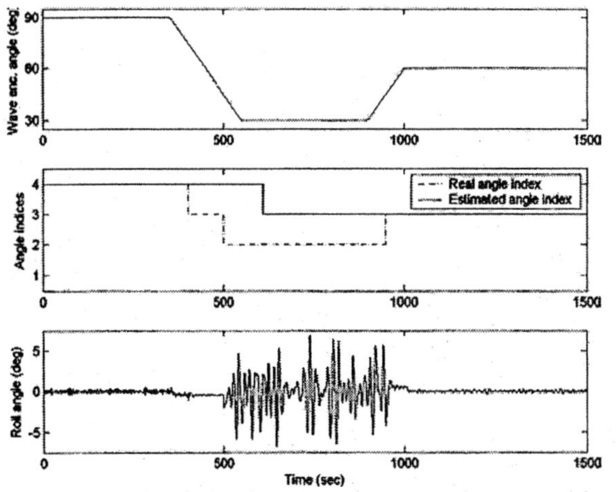


Fig.5.1: Ship behaviour on course change with constant speed using extremes switching method

The lower trace gives the roll angle of the ship in degrees. It indicates how little a ship can be stabilised when it encounters following waves (30°) in comparison with the roll angle of 60° or 90°.

Fig.5.2 shows the evolution of the estimation of roll frequencies. The dash-dotted line indicates the centre frequency of the controllers wave encounter angle (note that there are seven controllers) whilst the solid line shows the frequency, which is calculated using the algorithm for the trajectory shown in Fig.5.1.

The controller with the centre frequency closest to the calculated roll frequency of the ship is switched into the system, unless switching is inhibited due to unsteadiness of the system.

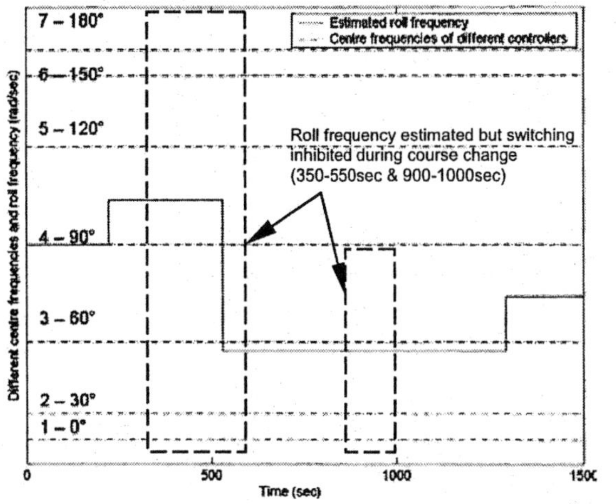


Fig.5.2: Frequency detection (Extremes)

It can be seen that the estimated frequencies of the system are quite close to the centre frequencies of the PID controllers. The active controller may not be the one which has been designed for such a situation. However, these controllers are the found to be satisfactory for implementing on the system, since their frequencies are closest to the roll frequency detected.

*5.2. FFT method*

The same scenario is obtained using the application of the FFT method. The upper trace of Fig.5.3 shows the same trajectory as in Fig. 5.1.

Unlike the extremes method, this method estimates the expected wave encounter angle after changing the first course. The frequency after the second course change (to 60°) is correctly estimated, but some uncertainties affect the system for switching between two controllers (No 2 and 3). Although if this is undesirable, the lower trace shows this does not adversely affect the roll angle.

Fig. 5.4 shows the estimated roll frequency of the ship in comparison with the centre frequencies of the PID controllers. The activity illustrated between 1000-1500 seconds explains the switching between controllers No 2 and 3 as shown in the middle trace of Fig.5.3.

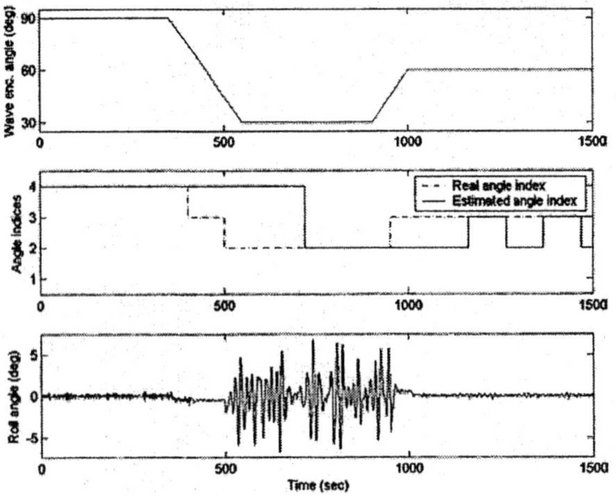


Fig.5.3: Ship behaviour on course change with constant speed using the FFT switching method

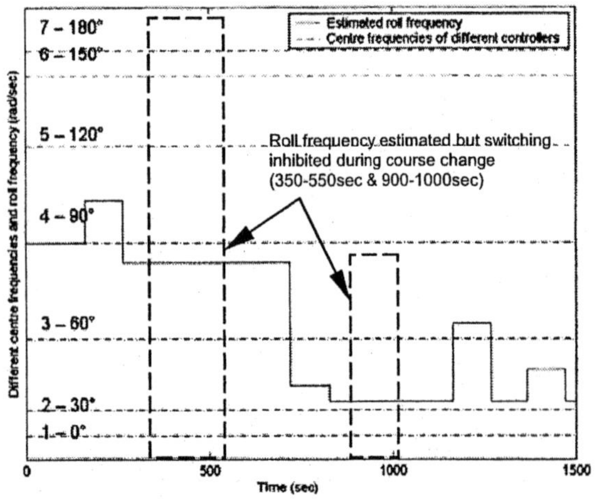


Fig. 5.4. Frequency detection (FFT)

## 6. CONCLUSIONS

Stabilisation of a ship motion using parallel multi model control (PMMC) has been studied in this paper. A family of PID controllers is used for improving the roll effectiveness and stabilising ship motion. Ship roll has been stabilised using a switching strategy based upon the estimation of system frequency using an extreme signal detection method and an FFT method.

The most appropriate controller is that with the centre frequency closest to the estimated frequency resulting from the system. Switching has been achieved by considering a set of rules, which act as a decision–maker process (supervisory control system). The switching mechanism has been demonstrated when the model of a ship is subjected to environmental changes, or changes in orientation and ship speed.

## REFERENCES


Belmont, M. R. and J. M. K. Horwood (1999). The effect of frequency distribution in sea model spectra on simulations of deterministic sea wave prediction. *International Ship Building Progress*, **46**, pp 265-276.

Blanke, M., J. Adrian, K. Larsen and J. Bentsen (2000). Rudder roll damping in coastal region sea conditions, *Proceedings of 5th IFAC Conference on Manoeuvering and Control of Marine Craft*, MCMC'2000

Crossland, P. (2003). The Effect of Roll Stabilisation Controllers on Warship Operational Performance, *Control Engineering Practice*, 423-431.

Daafouz, J., Riedinger, P. and Lung, C. (2002). Stability analysis and control synthesis for switched systems: A switched Lyapunov function Approach, *IEEE Transactions on Automatic Control*, **47**, 1883-1887.

Hespanha, J. P. and A. S. Morse. (1999). Stabilisation of nonholonomic integrators via logic-based switching, *Automatica*, **35**, 385-393.

Liberzon, D. and A. S. Morse, (1999) Basic problems in stability and design of switched systems, *IEEE Control Systems Magazine*, **19**, 59-70.

Sharif, M.T., G.N. Roberts, S.A. French, and R. Sutton, (1993). Lateral force stabilisation: a comparison of controller designs. *Eleventh Ship Control System Symposium*, Canada, **5**, 149-169.

Tedeschi, R. (1999). Sea state measurements in the Ross Sea based on ship motions, *Proceedings of the 9th International Offshore and Polar Engineering Conference*, Brest, France, **III**, 144-147.

Xie, W., C. Wen, and Z. Li, (2001). Input-to-output stabilisation of switched nonlinear systems, *IEEE Transactions on Automatic Control*, **46**, 1111-1116.

# A ROBOTIC SYSTEM FOR OFF-SHORE PLANTS DECOMMISSIONING

**E. Cavallo, R. C. Michelini, R. M. Molfino**

*PMAR Laboratory, Instrumental robot design Research group*
*Dept. Machinery Mechanics and Design, University of Genova*
*via all'Opera Pia 15A - 16145 GENOVA Italy*

The paper focuses on a robotic system conceived, designed, studied and built for cutting the legs of the off-shore platforms a few meters below the seabed. The work has been performed within the research project SBC Diamond Wire Cutting System Sub Bottom Cutter (GRD1 2000 25740) funded by the European Commission under the Fifth Research Programme and recently successfully closed.
The proposed robotic system is an innovative prototypal machine able to anchorage on the seabed soil in front of the leg to be cut, to drill the soil at the given depth by limiting the volume of removed materials and to cut the sub-sea structure.
Due to the complexity of the system and to the need of high operation reliability, the mechatronic modular approach has been adopted and the control modules have been distributed part on the sub-sea front-end and part on the support vessel remote-operated control stand. Simulation dynamic models of the robotic subsystems have been set-up and used for the knowledge base of the control system. *Copyright © 2004 IFAC*

Keywords: Robotics, Interdisciplinary design, Marine systems, Ecology

## 1. INTRODUCTION

In 1998, the OSPAR (Countries of the Convention for the Protection of the Marine Environment of the Northeast Atlantic) recognised that the 85% of the fixed oil and gas platforms in the North Sea should be removed completely. Now, decommissioning (see the Convention of Geneva, art. 5, and the "off-shore" regulations), requires: safe restoration of the marine habitat, non-interference of the removal activities with environment resources and resort to clean technologies with no risks for operators and third parties (LaBelle,1999). The European Countries involved in offshore decommissioning (Zhiguo Gao, 1997) defined that proper re-establishment imposes removal depths of the sub-sea structures (jacket legs/piles, wellheads, etc.) down to 3-5 m below the seabed soil.

The marine food industry profits of the on-duty conservativeness, and, later, of the re-establishment of natural equilibrium of the flora and fauna resources.

The new robotic system is a prototypal machine, able to fulfil the recalled offshore decommissioning regulations, limiting (Twachtman, 1997) the underwater pollution. Moreover, the people working in contact or in close vicinity with the system and the marine resources are safeguarded to the highest level of health and safety standards, as dangerous duties are accomplished under remote control.

The shearing task plays a key role in the dismantling of oil platforms and related structures, such as pipelines and loading terminals. The conventional cutting technologies, in use for sub-sea tasks, such as 'explosive' or 'high pressure water abrasive' systems, produce damages to the environment, also in consequence of the scattering at sea of dangerous materials and substances (Gerrard 1999). In this respect, the SBC robotic system is a significant improvement due to the new dig-and-saw concept, capable to avoid the effects of the suspension, by the combined actions of removing and convoying the sub-surface soil sediments, with negligible turbulence and without use of materials, resources and energy causing negative impacts on the environment. In fact a new dig approach that transfers on board the support ship all the seabed soil removed material and a saw approach based on an innovative diamond wire cutting technology have been adopted.

The qualifying objectives and key advantages of the selected approaches in comparison with other usual technologies are:

- the use of a clean technology, not interfering with the equilibrium of the marine habitat;
- the absolute guarantee of the completion of the cutting task;
- the conservativeness as for environment impact, with high overall efficiency and reliability of the technique for underwater use, with low energy consumption compared to the total power;
- the automation process of the all, by intelligent remote control/supervision station on surface;
- the integrated design of structural, power supply and underwater instrumental components (sensors, electro-hydraulic valves, data system etc...);
- the unaltered overall efficiency of the removed structures and the characteristics of the materials involved in the cutting process, thus enhancing their life-span and allowing the re-use for the same or different work-scopes (absent in other cutting techniques like explosives and flame cutting, etc...).

## 2. THE ROBOTIC SYSTEM ARCHITECTURE

A top down approach is applied to share common design skeletal guideline and system architecture, to select effective sub-systems and components and to consider the interactions among them.
To design a safe and reliable system, it is important to foresee every possible failures and to set contingencies for unplanned events that may happen during the mission, in order to set out rescue actions that avoid damages, having always under control the whole operation mission.

The main operative modules of the proposed robotic system are:

1) new concept dredging system, based on the combined use of drilling heads and new generation turbines designed to avoid pollution effects in connection with the water turbulence generated by the dredging pumps, which produce the suspension of polluted sediments (Grant, Briggs, 2002) deposited on the upper sea-bed layer (hydrocarbons, heavy metals, micro-biological pollutants);
2) new concept cutting machine different in design and characteristics from the ones available on the market, able to highly enhance all positive characteristics of the conventional diamond wires, mainly in resistance to axial loads, tolerance to the collapse of the cut materials, overall reliability, production and life-span;
3) new concept sub-sea work/deployment robotic platform, remotely operated, with its intelligent control/drive station on surface, stationed on board the vessel/barge supporting the sub-sea duty.

The robotic system has been designed, based on the fundamental concept to simplify the component structures and interfaces to improve MTBF and intrinsic operational reliability. The interfaces have been standardised from mechanic and electronic point of view, to allow a fast installation of different and dedicated equipment fit for the specific sub-water cutting missions: the equipment have been thus designed with the ability to be interfaced to different deployment configurations by simple reconfiguration (Acaccia et alii, 1998).

Figure 1 shows the robotic system architecture and its main modules.

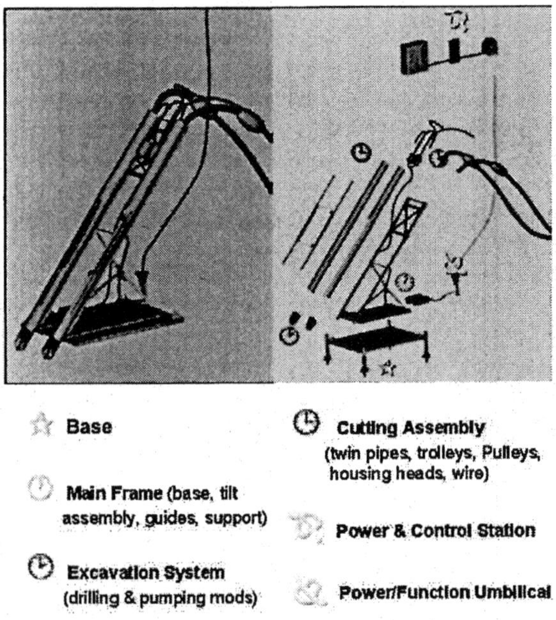


Fig.1 The robotic system CAD model.

Competing modules and components alternatives for all the core innovative mechanisms have been considered, to reach full acknowledgement of technicalities supporting the final choice (Cavallo, et al.,.2004).

## 3. MODELLING, SIMULATION AND VP ISSUES

The criticality of the operation surroundings and the deeply innovative requests of the prospected robotic system needed careful concern at the early project phases where modifications are cheaper and quicker.

Indeed, cost oriented design methodologies, machine functional and operational characteristics, hostile marine environment, rescue missions in case of derangement and easy maintenance and disassembly for recycling operations have been duly taken into account, to develop highly life-cycle effective products. In this case, deep knowledge of physical environment phenomena was needed. Knowledge-based solutions have been devised and applied for innovative automation, with balanced integration of mechanisms, actuation, sensorisation, control remote-manipulation and supervision.

Functional models for the innovative mechanisms have been implemented, taking into account the hazardous and severe operation environment, based on throughout deepened knowledge of the physical phenomena, to have *a priori* reliable idea of the realistic cinematic and dynamic behaviour. The modelling activities have been pushed to build complete mathematical descriptions and digital mock-ups of the core mechanisms and of the surroundings, in order to obtain suitable references, to be tested by simulation (Acaccia et alii,1999).

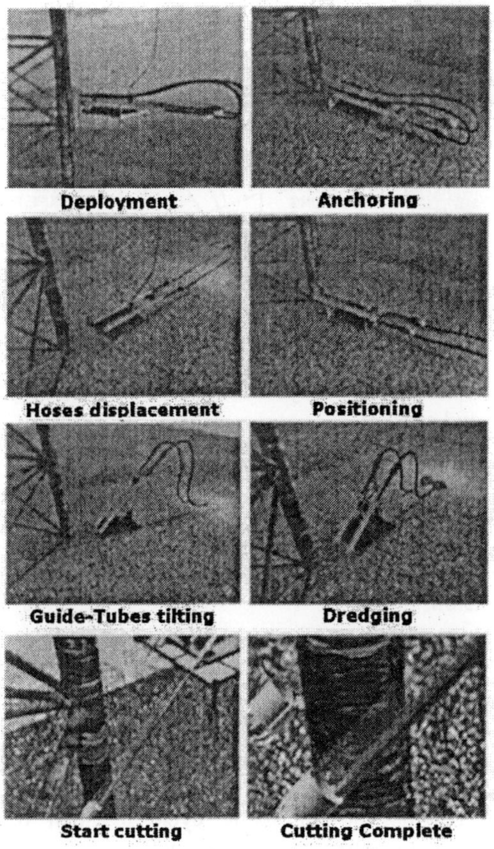


Fig.2 Snapshots from simulations on the robotic system mock-up

A special simulation environment has been implemented and used for testing and verifying the control algorithms and logics. Dextrous workspace and singular points, kinematics, operational accuracy and stiffness have been considered as main performance criteria and balanced for the competing architecture alternatives vs. cost, maintainability, MTBF, assemblability, disassemblability and other life cycle specifications. The project took advantage of different kinds of virtual prototyping and simulation techniques that allowed the validation of the analysed solutions at the subsequent design stages, as decisions support and performance verification. For this aim Simulink, Adams, Pro/Mechanica Motion and purposely developed C modules have been developed and interfaced. Figure 2 shows some snapshots of simulations performed on the robotic system digital mock-up during different tasks execution.

## 4. CONTROL SYSTEM OVERVIEW

The careful engineering of the hardware and software systems, supervising the robotic system mission, has been performed, particularly considering the equipment installed on the vehicle and hosted in the surface vessel, considering the data exchange between them, and the performance to be achieved during the different kinds of mission. Several aspects have been carefully investigated: monitoring and control concept architecture, sensory system selection, monitoring and control modules design, communication system definition, operating and contingency procedures stating.

The Figure 3 shows the standard operation layout: the machine is deployed on the sea bottom from a support vessel then, with the help of remotely operated vehicles (ROVs), is placed in front of the leg to cut and starts the dig and cut duty. An umbilical connects the submerged machine with the control unit located on the support, ship where the human operator continuously read the sensors values.

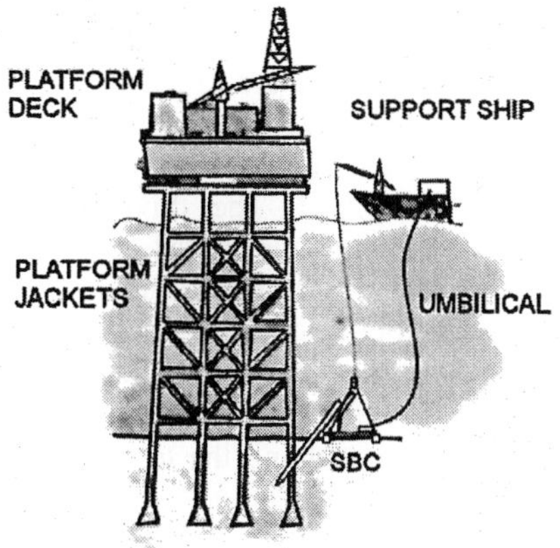


Fig. 3 – SBC machine operation scheme

### *4.1 Control Design Basic Concepts*

To enhance the system reliability and maintainability, a modular approach has been adopted in the control design. Operative missions are performed as sequence of different tasks, accomplished by pertinent control modules. Every module is designed to achieve a particular goal and a suitable control system allows the human supervisor to check, in real time, the completion of the missions.

Some modules are designed to be automatically controlled by the system, i.e. the anchoring and the cradle tilting, giving a graphic and numeric feedback to the human operator acting as remote supervisor. The overall cutting sequence, as well, is performed under local control, monitored by the operator located on the support ship.

The control architecture has been defined and specific functions were distributed to each sub-system at the lower level. The intermediate control level has been structured in few blocks, with defined goals and functions. The human/machine interface functions have been established. The previously defined models have been translated into purposely written macros, interfaced with Matlab Simulink and Pro/Mechanica tools, to accomplish algorithms checks and to allow rapid prototyping actions. Using parametric values, the cutting machine and wire string behaviours are properly set.

Full design of the control subsystems is devised, including the selection of hardware and software structure, with appropriate versatility and redundancy, to fit the software instructions according the mission targets. As far as possible and acceptable, standard equipment, software, and transmission protocols have been chosen from off-the-shelf components, in order to slightly impact the development price. The most innovative characteristics are: force control logic, position and force reflection based remote-handling, monitoring and data processing to create an expert archive modifiable and expansible, from which the user selects the mission parameters. An advanced man-machine interface is developed; it includes programming "by showing" functions (based on the system simulator issues) and task-oriented off-line programming modules.

The project main criteria are to simplify as much as possible the underwater apparatus for reliability reasons, setting on-board the surface vessel all the sophisticated equipment. The driving and control variables are timely transmitted by very quick communication channels. Man Machine Interface is very important for the mission accomplishment. A simple and robust MMI has been realised, from which the tasks are defined and overseen, the machine actions remote-operated and the task monitoring are easily performed.

The behaviour of the whole machine subjected to the defined control policy has been tested by virtual-reality simulation, in order to evaluate the system dynamical behaviour and to prevent errors.

### *4.2 Control System Implementation*

The whole machine is controlled by a supervisory unit, which coordinates the included subsystems.
The control system has been designed in order to be suitable for supervising, monitoring and controlling the many operations of modules, driving the main robotic subsystems: supporting base, mobile cradle arm, twin guide tubes and excavating heads forearm, cutting system end effector and the auxiliary systems: sea water pumping system and slurry management during five operational modes:

- stand-by mode: this is the reference (idle) state, after deployment on the sea-bed;
- emergency mode: if a failure arises, the alarm state is enabled, specifying the originating site;
- positioning (anchorage) mode: the robotic platform is located and its attitude set to start a cutting operation, the platform cradle is bent up to the selected engagement slope;
- excavating mode: the twin pipe drill-and-dig heads perform the required digging beneath the sea-bed soil;
- cutting mode: the diamond wire equipment accomplishes the planned task.

as schematically represented in figure 4.

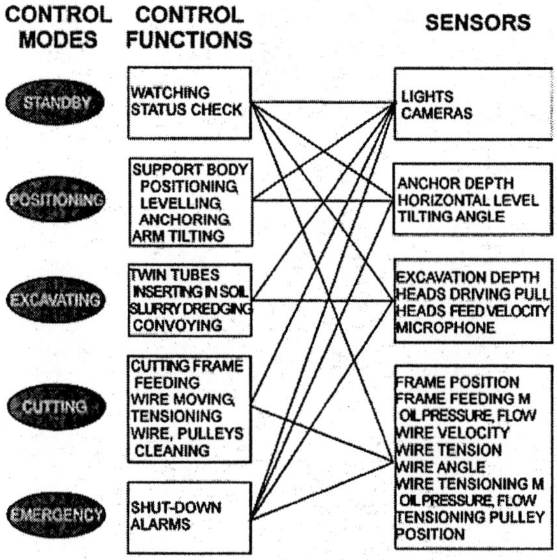


Fig.4 The robotic system control modes, functions and sensors

The different states are, subsequently, reached by referring to the stand-by state. If any trouble or wrong value arises, the control system moves the machine in the stand-by (idle) status, waiting for the supervisor decision.

The operator has access to the several monitored quantities, and disposes of a troubleshoot data-base and related hints to find out the most appropriate recovery actions. Only after the operator has checked the emergency and solved the problem, the machine can resume the automatic cycle or shall request further diagnostics, either, local actions.

The sensors were selected, and suitable specifically designed signal processing algorithms have been studied, coded and checked by simulation to satisfy the stated performance requirements in noisy environment.

The control system comprises the following units: surface control unit, including a touch screen computer, a surface-to-underwater interface and analogical/digital control unit; the sub-sea equipment, including actuators, sensors, alarms and various other equipment. The surface control unit reads all sensor signals from the sub-sea unit and presents the scaled values on the monitor. All outputs in the sub-sea unit can be activated from the surface control unit. This includes: camera control, light setting, and all hydraulic functions updating.

A few control functions are reactively performed by the under sea control unit, while the most of them are remote-operated. Focussing, e.g., on the cutting task, the control system can be subdivided into three main dynamically coupled subsystems:

- the motion of the frame supporting the diamond wire pulleys (cut feeding) ,
- the diamond wire stretching system (cut pressure),
- the diamond wire speed (cut velocity),

and a fourth one, the wire cleaning system, which is separately set.

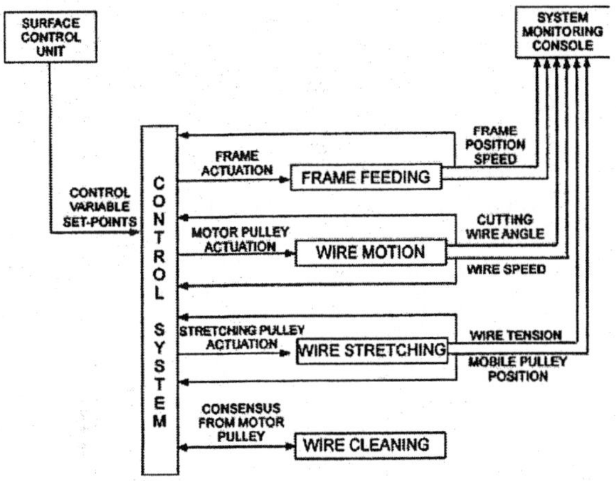


Fig.5 The reactive cutting system control

In figure 5, a block diagram of the local MIMO control system (running on board the SBC machine) is given. The observed variables are further monitored by the central system console for remote-control. The operator is helped in his choices by model-based decision supports as shown in figure 6.

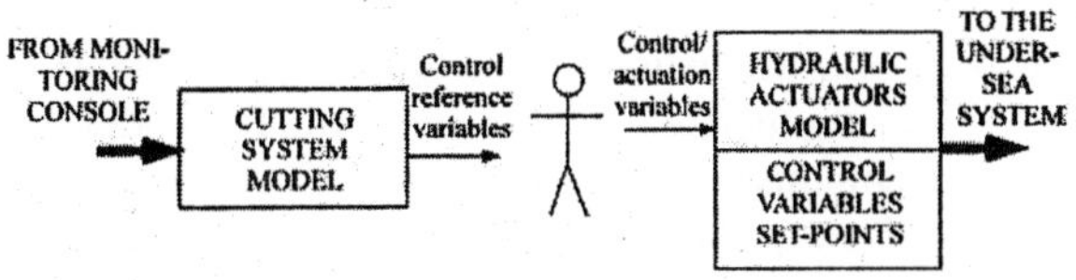


Fig.6 The cutting system remote-control scheme

The reference cutting system model considers the steady state equilibrium of the wire, taking into account the friction on the idle and driving pulleys for the chosen cut feeding, pressure and velocity figures, the hydro-dynamic effects on the free wire segments, the wire-legs engagement cutting forces and friction, together with geometrical constraints, as can be seen in figure 7 that shows the cutting system layout.

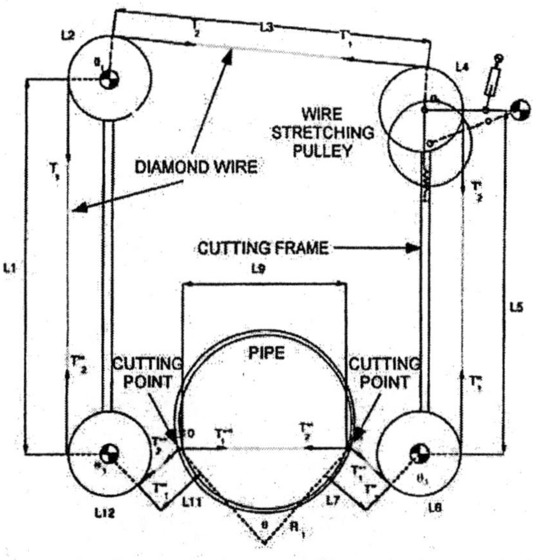


Fig.7 The diamond wire cutting system layout during a pipe cutting

The amount of information displayed and the control functions available to the operator are set individually for each control mode. In emergency mode, the value of all sensors are displayed, and all controls are available for the operator. Each mode will have direct access to the equipment necessary for its limited operation. If other equipment shall be operated, the manual override is necessary. The emergency mode has access to all functions and can automatically override current tasks by shut down procedures. The alarms are arranged for a number of parameters; the shut down is enabled only for a limited number of catastrophical accidents. This will also comprise the automatic control of the surface based high voltage starting equipment and power supply. The communication system is based upon the FIELD bus technology (PROFI bus or Can bus) and is connected through single or double twisted pairs. The video signals are directed through coaxial cables.

## 5. SBC PROTOTYPE AND SEA TESTS

Figure 8 shows the SBC robotic system physical prototype. The system is quite big: it has been sized taking into account the transportation constraints. It is over than 10 meters long and 3,5 meters large; the twin pipes diameter is around 600 mm..

Some preliminary tests on the diamond cutting system stand-alone, performed in a suitably instrumented testing bench, allowed to complete the a priori knowledge of the new wires obtained by theoretical model simulation with the a posteriori knowledge gained through experimental checks. On this basis, new pearls and cutting system were designed and the control system algorithms have been set-up. Figure 9 shows the control stand, with the touch panel and the video console by which the operator, on board the ship, drives the SBC missions.

Fig.8 The SBC physical prototype

The first sea trials were performed in July 2003 on the sea bottom along the North - West reef of the Ulstein located in Ulsteinvik. and a steel pile of 910 mm diameter, 11 mm thick, has been successfully cut. Nowadays further open sea trials are programmed.

Fig.9 A view of the SBC control room

## 6. CONCLUSIONS

The European environmental legislation, in particular, the Oil Operator and the National Authority act, often requires to convene on avoiding the resort to unsafe technologies, producing damages (explosives) or risks of contamination (water-abrasives) to the marine environment. Through the adoption of operation safe and environmentally acceptable standards, the project outcomes will benefit the oil and gas, as well as food industry by eliminating the risks of contamination of sea resources, providing an effective integration of the offshore regulations with the environment enacted rules.

As consequence of removing adverse environmental effects for the use of materials, resources and energy and of avoiding the waste of dangerous or polluting substances generated by the cutting process, the SBC guarantees that any human operators, working in contact with the system or profiting of the marine resources in close vicinity to it, are fully safeguarded, in conformity to the highest level of health and safety standards.

## 7. ACKNOWLEDGEMENT

The European Commission that funded the project SBC Diamond Wire Cutting System Sub Bottom Cutter (GRD1 2000 25740), the ITF consortium (Totalfina ELF, Shell, BP, HESS) and all the SBC partnership are gratefully acknowledged.

## REFERENCES

Acaccia G.M., E. Cavallo, E. Garofalo, R.C. Michelini, R.M. Molfino, M. Callegari (1999), Remote manipulator for deep-sea operations: animation and virtual reality assessment, in *Proc. 2nd Workshop on Harbour, Maritime & Logistics Modelling and Simulation* (HMS99), 16-18 September, Genova, Italy, pp.57-62, ISBN 1-56555-175-3.

Acaccia G.M., M. Callegari, R.C. Michelini, R.M. Molfino, R.P. Razzoli (1998), Underwater robotics: example survey and suggestions for effective devices, *4th. ECPD Intl. Conf. Advanced Robotics, Intelligent Automation & Active Systems*, Moscow, Aug. 24-26, pp. 409-416, ISBN 86 7236 013 3.

Gerrard S., Grant A., Marsh R., London C, (1999), Drill cuttings piles in the North Sea: management options during platform decommissioning *Centre for Environmental Risk, Res. Rpt.* No 31, Norwich, October, ISBN 1 873933 11 8.

Grant, A., A.D. Briggs (2002), Toxicity of sediments from around a North Sea oil platform: Are metals or hydrocarbons responsible for ecological impacts? *Marine Environmental Research*, 53, 95-116.

Cavallo E., Michelini R.C., Molfino R.M. (2004), A remote-operated robotic platform for undewater decommissioning tasks, *35th Intl. Symposium on Robotics, ISR 2004*, Paris, March 23-26, 2004, 1-6

LaBelle B. (1999), OCS Resources Management & Sustainable Development Report *US Department of Interior*, September 24.

Twachtman R. (1997), Offshore platform decommissioning perceptions change. *Houston, Oil & Gas Journal* Twachtman Snyder & Byrd Inc December 8.

Zhiguo Gao (1997), Current issues of international law on offshore abandonment, with special reference to the U K *Ocean Development and International Law* 28, pp. 59-78.

ELSEVIER
IFAC
PUBLICATIONS
www.elsevier.com/locate/ifac

# OBSERVER DESIGN FOR A TOWED SEISMIC STREAMER CABLE

**Tu Duc Nguyen and Olav Egeland**

*Department of Engineering Cybernetics,
Norwegian University of Science and Technology,
E-mail: Tu.Duc.Nguyen@itk.ntnu.no,
Olav.Egeland@itk.ntnu.no*

Abstract: In this paper, observer design for a towed seismic streamer cable is considered. Based on measurements at the boundaries, an uniformly exponentially stable observer is proposed. The stability analysis of the observer is based on Lyapunov theory. The existence and uniqueness of the solutions of the observer are based on semigroup theory. To illustrate the performance of the observer, numerical simulation results are presented. *Copyright © 2004 IFAC*

Keywords: Distributed parameter system, Cable, Observer.

## 1. INTRODUCTION

In surveying of hydrocarbon reservoirs under seabed, offshore towing of seismic sensor arrays is extensively used. These operations are accomplished by a towed cable configuration which consists of a negatively buoyant lead-in cable attached to a towing vessel at one end, and to a neutrally buoyant cable called streamer at the other end. To detect the reflected acoustic pulses from a towed acoustic source, hydrophones are embedded in the streamers. To obtain better stability and controllability of the motion of the streamers, a surface tail buoy is attached to the downstream end of the streamers. The length of the lead-in cable varies typically between 200 m to 400 m, and the length of the streamers is normally between 3000 m and 6000 m. In special cases, the length of the streamers can be as long as 10000 m. During towing, the depth of the streamers are controlled by depth controllers. The depth controllers are spaced along each streamer, typically 300 m apart. A typical towed cable configuration is shown in figure 1.

The dynamics of towed cables have been studied by several authors (Dowling, 1988),(Ersdal, 2003),(Ortloff and Ives, 1969),(Paidoussis, 1966, 1973), (Pedersen and Sørensen, 2001),(Triantafyllou and Chryssostomidis, 1988) and references therein. Most of these consider towing of neutrally buoyant cable with a free downstream end. In (Paidoussis, 1966, 1973), the equation of motion for the transverse displacement of a towed neutrally buoyant element is derived. In (Dowling, 1988), the form of the linear displacements of a neutrally buoyant cylinder is determined. In (Triantafyllou and Chryssostomidis, 1988), the authors developed a procedure for calculating the response of a towed array of seismic hydrophones when a harmonic excitation was applied at the upstream end. In (Pedersen and Sørensen, 2001), the modelling and controlling issues of towed marine seismic streamer cables are studied.

Knowing accurate position of the whole cable is of great importance, not only for precise maneuvering and control-related concerns, but also because accurate knowledge of the configuration of the cable is the first step of other tasks, e.g. in preventing that the streamers get tangled during the surveying operations. More importantly, this knowledge can be used to depress the influence of the signal noise of the recorded seismic data; which leads to a better interpretation of recorded data and thus to a more accurate depiction of the sea floor (Schlumberger, summer 2001). However, the configuration of the cable can not be measured directly. Since this requires continuous distribution of sensors along the whole cable, which is not possible in practice. Typically, there is a finite number of sensors collocated with the depth controllers. So to get informations on the position of unmeasured points on the cable, an observer is needed. The main purpose of an observer is to estimate unmeasured physical quantities e.g. position, velocity etc., based on available measurements. In this paper, one such observer is proposed.

Observer design based on Lyapunov theory is well known and widely used for both linear systems and nonlinear systems. Balas (Balas, 1998) considered observer design for linear flexible structures described by FEM. Demetriou (Demetriou, 2001) presented a method for construction of observer for linear second order lumped and distributed parameter systems using parameter-dependent Lyapunov functions. Kristiansen (Kristiansen, 2000) applied *contraction theory* (Lohmiller and Slotine, 1996) in observer design for a class of linear distributed parameter systems. Structural damping forces were included in the last two cases. Hence, exponentially stable observers can easily be designed.

In this paper, observer design for a part of a towed cable configuration shown in figure 1 is considered. We consider a streamer cable, attached to depth controllers at both ends (figure 2). The equation of motion of the seismic streamer cable is adopted from ((Ersdal, 2003),(Paidoussis, 1973)), which is a distributed parameter formulation (PDE). The dynamics of the depth controllers are described by an ordinary differential equation (ODE). Based on three measurements at the boundaries, an uniformly exponentially stable observer is designed. The stability analysis of the observer is based on Lyapunov theory. The existence and uniqueness of the solutions of the observer are based on *semigroup theory*.

The paper is organized as follows. First, a model for the towed seismic streamer cable is presented. Then, an uniformly exponentially stable observer is designed. After that, the existence and uniqueness of the solutions of the observer are studied. Finally, simulation results are presented to demonstrate the performance of the observer.

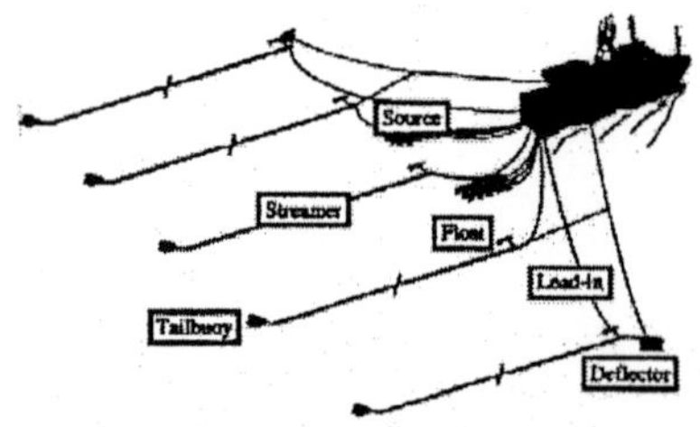


Fig. 1. A typical towed cable configuration.

## 2. MODELLING

Neglecting the bending stiffness and the material damping, the equation of motion for a neutral, flexible cylinder with small transverse excitations in the axial flow is given be the nonlinear model ((Ersdal, 2003),(Paidoussis, 1973)),

$$m w_{tt} = (T w_x)_x - (2aU w_t)_x + F_t w_x - F_n(\alpha), \quad 0 < x < L \quad (1)$$

$$M_0 w_{tt} = T w_x - 2aU w_t + \tau_0, \quad x = 0 \quad (2)$$

$$M_1 w_{tt} = -T w_x + 2aU w_t + \tau_1, \quad x = L \quad (3)$$

where

$$m = \rho_c + a = \frac{\rho_w \pi d^2}{4}(C_m + C_a) \quad (4)$$

$$T(x) = T_0 + F_t(L - x) - Bh - aU^2 \quad (5)$$

$$B = \rho_w g \pi \frac{d^2}{4} \quad (6)$$

$$F_t = \frac{1}{2}\rho_w \pi d C_f U^2 \quad (7)$$

$$\alpha \approx \frac{1}{U} w_t + w_x \quad (8)$$

$$F_n(\alpha) = \frac{1}{2}\rho_w d U^2 (C_{n1}\alpha + C_{n2}\alpha|\alpha|) \quad (9)$$

$m$ is the sum of the structural and added mass per unit length, $\rho_c$ is the density of the cable per unit length, $a$ is added mass per unit length, $C_m$ is the structural mass coefficient, $C_a$ is the added mass coefficient, $\rho_w$ is the density of the ambient water, $T(x)$ is the effective tension of the cable at $x$, $T_0$ is the aft tension, $B$ is the buoyancy force per unit length, $h$ is the distance to the free surface, $g$ is the gravitation constant, $F_t$ is the tangential hydrodynamic force per unit length of the cable, $C_f$ is the friction coefficient (see Remark 1), $\alpha$ is the angle of attach, $F_n$ is the normal hydrodynamic force per unit length of the cable, $C_{n1}$ and $C_{n2}$ are the coefficients of normal hydrodynamic force (see Remark 2), $U > 0$ is the tow speed of the towing vessel, $d$ is the diameter of the cable, $L$ is the length of the cable, $M_0$ and $M_1$ are the mass of the depth controllers at the upstream end ($x = 0$) and the downstream end ($x = L$), respectively, $w(x,t)$ is the vertical displacement of the cable at point $x$ and time $t$, $\tau_0, \tau_1 : \mathbb{R}^+ \to \mathbb{R}$ are the control forces generated by the depth controllers at the boundaries (see figure 2). Moreover, the subscripts $(\cdot)_t$ and $(\cdot)_x$ denote the partial derivative respect to $t$ and $x$, respectively.

For further discussion on this topic and the model (1)-(3), see e.g. (Dowling, 1988),(Ersdal, 2003),(Ortloff and Ives, 1969), (Paidoussis, 1973), (Triantafyllou and Chryssostomidis, 1988) and the references therein.

*Remark 1.* The friction coefficient of an axi-symmetric boundary layer is given by White's formula (White, 1972),

$$C_f = 0.0015 + \left(0.3 + 0.015\,(2L/d)^{0.4}\right) \mathrm{Re}_l^{-\frac{1}{3}}$$

where $\mathrm{Re}_l$ is Reynolds number.

*Remark 2.* In the work by Paidoussis (Paidoussis, 1966) and Dowling (Dowling, 1988), the coefficients are given as: $C_{n2} = 0$ and $C_{n1} = \gamma_{n1}\pi C_f$, where $\gamma_{n1} \in (0, 1)$. The factor $\pi$ is included because $C_f$ is scaled by the cylindrical surface, while $C_{n1}$ is scaled by cylinder diameter. Taylor (Taylor, 1952) and Triantafyllou (Triantafyllou and Chryssostomidis, 1988) consider the linear term as a viscous friction effect with $C_{n1} \leq \pi C_f$ and add a quadratic term with coefficient $C_{n2} = 1$. In (Kennedy, 1987), the quadratic term is excluded, but the linear coefficient $C_{n1}$ is almost an order of magnitude larger than $\pi C_f$. Based on the experiments carried out at Marine Technology Center, NTNU, Ersdal (Ersdal, 2003) shows that $C_{n1}$ is actually $5 - 8$ times larger than the value suggested by Taylor (Taylor, 1952) and Triantafyllou et al. (Triantafyllou and Chryssostomidis, 1988), i.e. $C_{n1} = \gamma_{n1}\pi C_f$, for some constant $\gamma_{n1} \geq 5$. The experiments also confirm that $C_{n2} > 1$.

*Remark 3.* The depth of the streamer cables are controlled by a finite number of depth control devices. Traditional concept is a wing where the angle of attack is controlled by a local controller, known as a bird. In this paper, the depth controllers are considered as a point load, which is reasonable; since the length of the bird is small compared to the length of the streamer cable.

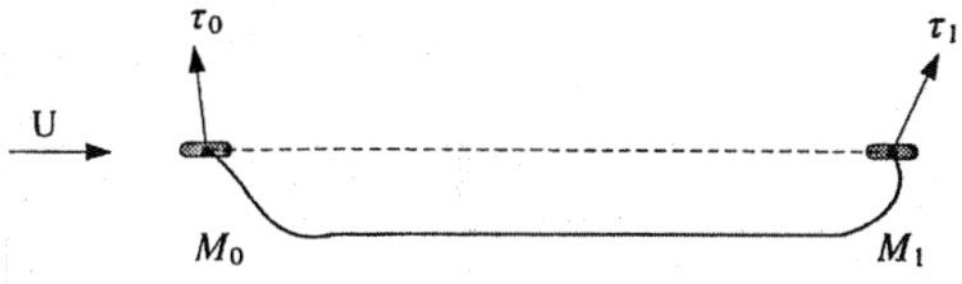


Fig. 2. A part of a towed seismic cable configuration.

### 2.1 Assumption

Without loss of generality, we assume that

*Assumption 1.* The tension $T(x)$ given by (5) satisfies the following inequality

$$0 < T_{\min} \leq T(x) \leq T_{\max} < \infty, \quad x \in [0, L]$$

and $T_{\min} \gg \{a, m, U\}$.

Since the tension along the streamer is normally much larger than the rest of the system parameters, Assumption 1 does not cause any restrictions in practice. Typically, the tow speed of towing vessel $U$ is between 0 m/s and 2.5 m/s, the added mass $a \leq 1$ and the structural mass $m \leq 1$, while the value of the tension $T(x)$ is of order $10^3$. Assumption 1 is thus reasonable.

### 2.2 Problem Statement

In this paper, we consider the following problem:

*Problem 1.* Assume that the model (1)-(3) is perfectly known. Given the measurements: $y_{01}(t) = w(0,t)$, $y_{11}(t) = w(L,t)$ and $y_{12}(t) = w_x(L,t)$, $\forall t \geq 0$. Design an observer for the system (1)-(3).

## 3. OBSERVER DESIGN

Utilizing the *coordinate error feedback* (Lohmiller and Slotine, 1996), we propose the observer

$$\bar{w}_t = \hat{w}_t - \frac{h_{01}\cdot y_{01}}{M_0}\cdot\delta(x) - \left[\frac{h_{12}Ty_{12}}{M_1} + \frac{(h_{11}+2aU)\,y_{11}}{M_1}\right]\cdot\delta(x-L), x\in[0,L] \quad (10)$$

and

$$\begin{aligned} m\bar{w}_{tt} &= (T\hat{w}_x)_x - (2aU\hat{w}_t)_x \\ &\quad + F_t\hat{w}_x - F_n(\hat{\alpha})\,, \quad x\in\,]0,L[ \quad (11)\\ M_0\bar{w}_{tt} &= T\hat{w}_x - (2aU + h_{01})\,\hat{w}_t \\ &\quad - h_{02}\,(\hat{w} - y_{01}) + \tau_0\,, \quad x = 0 \quad (12)\\ M_1\bar{w}_{tt} &= -T\hat{w}_x - h_{11}\hat{w}_t \\ &\quad - h_{12}T\hat{w}_{xt} + \tau_1\,, \quad x = L \quad (13) \end{aligned}$$

where $\hat{w}$ is the estimated of $w$, $\delta(\cdot)$ denotes the *Dirac delta function*, $F_n(\hat{\alpha})$ is given by (9), $\hat{\alpha}$ is similarly defined as $\alpha$, and $h_{01}$, $h_{02}$, $h_{11}$ and $h_{12}$ are positive observer gains, which will be determined below. Note that the coordinate error feedback has similarities with *Luenberger*'s linear reduced-order observer design (Luenberger, 1979).

Using (10) in (11)-(13), and subtracting the resulting equations by (1)-(3), we get the observer error dynamics

$$m\tilde{w}_{tt} = (T\tilde{w}_x)_x - 2aU\tilde{w}_{xt} + F_t\tilde{w}_x - \tilde{F}_n\,, \quad x \in\, ]0, L[ \qquad (14)$$

$$M_0\tilde{w}_{tt} = T\tilde{w}_x - (2aU + h_{01})\,\tilde{w}_t - h_{02}\tilde{w}\,, \quad x = 0 \qquad (15)$$

$$M_1\tilde{w}_{tt} = -T\tilde{w}_x - h_{11}\tilde{w}_t - h_{12}T\tilde{w}_{xt}\,, \quad x = L \qquad (16)$$

where $\tilde{F}_n = F_n(\hat{\alpha}) - F_n(\alpha)$ and $\tilde{w} = \hat{w} - w$ denotes the observer error.

To analyze the observer error dynamics (14)-(16), we define the Lyapunov function

$$E(t) = E_1(t) + E_2(t) + E_3(t) \qquad (17)$$

where

$$E_1 = \frac{1}{2}M_0\left(\tilde{w}_t|_0\right)^2 + \gamma_0 M_0\, \tilde{w}_t\tilde{w}|_0 + \frac{1}{2}h_{02}\left(\tilde{w}|_0\right)^2$$

$$E_2 = \frac{1}{2}mU^2\int_0^L \left(\frac{1}{U}\tilde{w}_t + \tilde{w}_x\right)^2 dx + \frac{1}{2}\int_0^L \left[T(x) + 2aU^2 - mU^2\right]\tilde{w}_x^2\, dx$$

$$E_3 = \frac{1}{2}\frac{\left(M_1\, \tilde{w}_t|_L + h_{12}\, T\tilde{w}_x|_L\right)^2}{M_1 + h_{11}h_{12}}$$

and $0 < \gamma_0 \ll \sqrt{h_{02}/M_0}$ is some suitable chosen constant which will be determined later. Differentiating $E_1$, $E_2$ and $E_3$ with respect to time along the solution trajectories of (14)-(16) gives

$$\dot{E}_1 = \tilde{w}_tT\tilde{w}_x|_{x=0} + \gamma_0\, \tilde{w}T\tilde{w}_x|_{x=0} - \left[2aU + h_{01} - \gamma_0 M_0\right]\left(\tilde{w}_t|_0\right)^2 - \gamma_0\left[2aU + h_{01}\right]\tilde{w}\tilde{w}_t|_{x=0} - \gamma_0 h_{02}\left(\tilde{w}|_0\right)^2$$

$$\dot{E}_2 = \left[\tilde{w}_tT\tilde{w}_x\right]_{x=0}^L + \frac{U}{2}\left[\tilde{w}_xT\tilde{w}_x\right]_{x=0}^L - aU\left[\tilde{w}_t^2\right]_{x=0}^L + \frac{mU}{2}\left[\tilde{w}_t^2\right]_{x=0}^L - I_\alpha + F_tU\int_0^L \left(\frac{\tilde{w}_t}{U}\tilde{w}_x + \frac{1}{2}\tilde{w}_x^2\right)dx - U\gamma_{n1}F_t\int_0^L \left(\frac{1}{U}\tilde{w}_t + \tilde{w}_t\right)^2 dx$$

$$\dot{E}_3 = -\,\tilde{w}_xT\tilde{w}_t|_L - \frac{h_{11}M_1}{M_1 + h_{11}h_{12}}\left(\tilde{w}_t|_L\right)^2 - \frac{h_{12}}{M_1 + h_{11}h_{12}}\left(T\tilde{w}_x|_L\right)^2$$

where

$$I_\alpha = \frac{U\rho_w dU^2C_{n2}}{2}\int_0^L (\hat{\alpha} - \alpha)\cdot(\hat{\alpha}\,|\hat{\alpha}| - \alpha\,|\alpha|)\, dx$$

Let $\tilde{\mathbf{q}}_0 = (\tilde{w}_t|_0\,, \tilde{w}|_0\,, \tilde{w}_x|_0)$ and $\tilde{\mathbf{q}} = (\tilde{w}_t(x,t)/U, \tilde{w}_x(x,t))$. Hence, we get

$$\dot{E} = -\left(aU + \frac{mU}{2}\right)\left(\tilde{w}_t|_0\right)^2 - aU\left(\tilde{w}_t|_L\right)^2 - I_\alpha - \tilde{\mathbf{q}}_0^T\mathbf{Q}_0\tilde{\mathbf{q}}_0 - F_t\int_0^L \tilde{\mathbf{q}}^T\mathbf{Q}_1\tilde{\mathbf{q}}\, dx - \left[\frac{h_{11}M_1}{M_1 + h_{11}h_{12}} - \frac{mU}{2}\right]\left(\tilde{w}_t|_L\right)^2 - \left[\frac{h_{12}}{M_1 + h_{11}h_{12}} - \frac{U}{2\,T|_L}\right]\left(T\tilde{w}_x|_L\right)^2$$

where

$$\mathbf{Q}_0 = \begin{bmatrix} h_{01} - \gamma_0 M_0 & \frac{\gamma_0(2aU + h_{01})}{2} & 0 \\ \frac{\gamma_0(2aU + h_{01})}{2} & \gamma_0 h_{02} & -\frac{\gamma_0}{2}\,T|_0 \\ 0 & -\frac{\gamma_0}{2}\,T|_0 & \frac{U}{2}\,T|_0 \end{bmatrix}$$

$$\mathbf{Q}_1 = \begin{bmatrix} \gamma_{n1} & \gamma_{n1} - \frac{1}{2} \\ \gamma_{n1} - \frac{1}{2} & \gamma_{n1} - \frac{1}{2} \end{bmatrix}$$

and $\gamma_{n1} > 1$ is a given constant (see Remark 2). It can easily be verified that $\mathbf{Q}_1$ is positive definite and $I_\alpha \geq 0$ for $\forall\hat{\alpha}, \alpha \in \mathbb{R}$. Selecting appropriate observer gains $h_{i1}$, $h_{i2}$, $i = 0, 1$, and $0 < \gamma_0 \ll \sqrt{h_{02}/M_0}$, the matrix $\mathbf{Q}_0$ can be made positive definite and the last two terms of $\dot{E}$ become negative. Hence, we have

$$\dot{E} \leq -\lambda_{\min}\left(\mathbf{Q}_0\right)\|\tilde{\mathbf{q}}_0\|_2^2 - \lambda_{\min}(\mathbf{Q}_1)\int_0^L \|\tilde{\mathbf{q}}\|_2^2\, dx - \left[\frac{h_{11}M_1}{M_1 + h_{11}h_{12}} - \frac{mU}{2}\right]\left(\tilde{w}_t|_L\right)^2 - \left[\frac{h_{12}}{M_1 + h_{11}h_{12}} - \frac{U}{2\,T|_L}\right]\left(T\tilde{w}_x|_L\right)^2$$

where $\lambda_{\min}(\mathbf{Q}_i) > 0$ denotes the smallest eigenvalue of the matrix $\mathbf{Q}_i$, $i = 0, 1$. Clearly, there exists a constant $C > 0$ such that

$$\dot{E}(t) \leq -C \cdot E(t)$$

Thus, the origin $(\tilde{\mathbf{q}}_0, \tilde{w}_t, \tilde{w}_x) = \mathbf{0}$ of the system (14)-(16) is uniformly exponentially stable. Moreover, by integration by parts, *Cauchy-Schwarz* inequality and the inequality, $2ab \leq a^2 + b^2$, $\forall a, b \in \mathbb{R}$, we have

$$\tilde{w}(x,t)^2 \leq 2L\int_0^L \left(\frac{\partial\tilde{w}}{\partial x}\right)^2 dx + 2\tilde{w}\,(0,t)^2$$

for $\forall x \in [0, L]$. Thus, $\tilde{w} = 0$ is uniformly exponentially stable. Hence, the origin $(\tilde{w}, \tilde{w}_t, \tilde{w}_x) = \mathbf{0}$ of (14)-(16) is uniformly exponentially stable. Problem 1 is thus solved.

## 4. EXISTENCE AND UNIQUENESS OF SOLUTIONS

In this section, the existence and uniqueness of the solutions of (10)-(13) are considered. This is based on *semigroup theory*. First, we assume that the control laws $\tau_0$ and $\tau_1$ are chosen such that the solutions of the closed loop system of (1)-(3) are well-posed. Hence, by showing the existence and uniqueness of the solutions of (14)-(16), we get the existence and uniqueness of the solutions of (10)-(13).

Let $\tilde{w}(x,t)$ be a regular solution of (14)-(16) and define $\tilde{\mathbf{w}}(t) = (\tilde{w}|_0, \tilde{w}_t|_0, \tilde{w}(\cdot,t), \tilde{w}_t(\cdot,t), \tilde{w}_t|_L)$. Equations (14)-(16) can be compactly written as

$$\dot{\tilde{\mathbf{w}}} = \mathbf{A}\tilde{\mathbf{w}} + \mathbf{F}(\tilde{\mathbf{w}}), \qquad \tilde{\mathbf{w}}_0 \in H \qquad (18)$$

where

$$\begin{aligned}\mathbf{A}\tilde{\mathbf{w}} = \Bigg[\tilde{w}_2\,,\, &\frac{T\tilde{w}_{3,x}|_0 - (2aU + h_{01})\,\tilde{w}_2 - h_{02}\tilde{w}_1}{M_0}, \\ &\tilde{w}_4\,,\ (*)\,, \\ &-\frac{T\tilde{w}_{3,x}|_L + h_{11}\tilde{w}_5 + h_{12}\,T\tilde{w}_{4,x}|_L}{M_1}\Bigg]^T, \forall \tilde{\mathbf{w}} \in D(\mathbf{A})\end{aligned}$$

$$\begin{aligned}\mathbf{F}(\tilde{\mathbf{w}}) = [0,0,0,\\ &-\frac{\rho_w d U^2 C_{n2}}{2m}\left(\hat{\alpha}\,|\hat{\alpha}| - \alpha\,|\alpha|\right), 0\Big], \forall \tilde{\mathbf{w}} \in H\end{aligned}$$

and

$$\begin{aligned}(*) = \frac{1}{m}\Big[&(T\tilde{w}_{3,x})_x - 2aU\tilde{w}_{4,x} \\ &+ F_t\tilde{w}_{3,x} - \gamma_{n1}F_t\left(\frac{1}{U}\tilde{w}_4 + \tilde{w}_{3,x}\right)\Big]\end{aligned}$$

$\alpha$ is given by (8), $\hat{\alpha}$ is the estimated of $\alpha$, $\tilde{\mathbf{w}} = (\tilde{w}_1, \ldots, \tilde{w}_5)$ and $\tilde{w}_{j,x}$ denotes $\partial\tilde{w}_j/\partial x$.

To analyze the abstract problem (18), we define the spaces

$$\begin{aligned} H &= \mathbb{R}^2 \times H^1(0,L) \times L_2(0,L) \times \mathbb{R} \\ D(\mathbf{A}) &= \{\mathbf{w} \in \mathbb{R}^2 \times H^2(0,L) \times H^1(0,L) \times \mathbb{R} | \\ &\quad w_1 = w_3|_{x=0},\ w_2 = w_4|_{x=0},\ w_5 = w_4|_{x=L}\}\end{aligned}$$

where $D(\mathbf{A})$ denotes the domain of $\mathbf{A}$, and

$$\begin{aligned} L_2(0,L) &= \left\{ f \ |\ \left(\int_0^L |f(x)|^2\,dx\right)^{\frac{1}{2}} < \infty \right\} \\ H^k(0,L) &= \left\{ f \ |\ f, f', f'', \ldots, f^{(k)} \in L_2(0,L)\right\}\end{aligned}$$

In $H$, we define the inner-product

$$\begin{aligned}\langle \mathbf{w}, \mathbf{v}\rangle_H &= M_0 w_2 v_2 + \gamma_0 M_0 w_2 v_1 + \gamma_0 M_0 w_1 v_2 \\ &+ h_{02} w_1 v_1 + \int_0^L \left(T + 2aU^2 - mU^2\right) w_{3,x} v_{3,x}\,dx \\ &+ mU^2 \int_0^L \left(\frac{1}{U}w_4 + w_{3,x}\right)\left(\frac{1}{U}v_4 + v_{3,x}\right)dx \\ &+ \frac{(M_1 w_5 + h_{12}\,T w_{3,x}|_L)\,(M_1 v_5 + h_{12}\,T v_{3,x}|_L)}{M_1 + h_{11}h_{12}}\end{aligned}$$

where $\mathbf{v} = (v_1, \ldots, v_5) \in H$ and $\mathbf{w} = (w_1, \ldots, w_5) \in H$. It can be verified that $(H, \langle\cdot,\cdot\rangle_H)$ is a Hilbert space. Note that the function (17) can be compactly expressed as

$$E(t) = \frac{1}{2}\,\langle \tilde{\mathbf{w}}(t), \tilde{\mathbf{w}}(t)\rangle_H \qquad (19)$$

where $\tilde{\mathbf{w}}(t) = (\tilde{w}|_0, \tilde{w}_t|_0, \tilde{w}(\cdot,t), \tilde{w}_t(\cdot,t), \tilde{w}_t|_L)$.

*Theorem 1.* The abstract problem (18) has a unique mild solution defined on $\mathbb{R}^+$ for every $\tilde{\mathbf{w}}_0 \in H$. The unique mild solution is given by the integral equation

$$\begin{aligned}\tilde{\mathbf{w}}(t;\mathbf{w}_0) &= Z(t)\tilde{\mathbf{w}}_0 = e^{\mathbf{A}t}\tilde{\mathbf{w}}_0 \\ &+ \int_0^t e^{\mathbf{A}(t-\tau)}\mathbf{F}\left(Z(\tau)\tilde{\mathbf{w}}_0\right)d\tau\,,\ t \geq 0 \quad (20)\end{aligned}$$

where $\{e^{\mathbf{A}t}\}_{t\geq 0}$ is the linear $C_0$-semigroup generated by $\mathbf{A}$ and $\{Z(t)\}_{t\geq 0}$ is the nonlinear semigroup generated by $\mathbf{A}+\mathbf{F}$. The solution (20) tends asymptotically to zero as $t \to \infty$ for $\forall \tilde{\mathbf{w}}_0 \in H$.

Due to the limitation of space, the proof of Theorem 1 is omitted here. In short, it is based on *Lumer-Phillips* theorem (see e.g. (Pazy, 1983)), Theorem 1.4 (p. 185, (Pazy, 1983)), and the work by (Dafermos and Slemrod, 1973).

Note that if $\tilde{\mathbf{w}}_0 \in D(\mathbf{A})$ then the mild solution (20) is the strong solution of (18), i.e. $\tilde{\mathbf{w}}(t)$ is differentiable and lies in $D(\mathbf{A})$ for $\forall t \geq 0$. From Lyapunov analysis above it follows that the solution (20) is exponentially decaying $\forall \tilde{\mathbf{w}}_0 \in D(\mathbf{A})$.

An extension of this work (to multi streamer cables) is presented in (Nguyen, 2004).

## 5. SIMULATION

To simulate the system (1)-(3) and the observer (10)-(13), the *finite element method* with *Lagrange basis functions* (of 2nd order) has been applied. A cable of length $L = 300$ meter was divided into 8 elements. The system parameters used in the simulation are: $L = 300$ [m], $\rho_w = 1024$ [kg/m$^3$], $d = 0.05$ [m], $T_0 = 1800$ [N/m], $C_m = C_a = 1$, $C_f = 0.0035$, $\gamma_{n1} = 4$, $C_{n1} = \gamma_{n1}\pi C_f$, $C_{n2} = 1$, $M_0 = M_1 = 8$ [kg], $U = 2$ [m/s], $g = 9.81$ [m/s$^2$], $h = 8$ [m]. In simulation, the following control signals are applied: $\tau_0 = 10\sin(0.5t)$ and $\tau_1 = -K_p w(L,t) - K_p \dot{w}(L,t)$ , $t \geq 0$. The controller gains and observer gains used in simulation are: $K_p = K_d = 100$, $h_{01} = h_{02} = 200$, $h_{11} = 150$ and $h_{12} = 0.25$. We turned on the observer at time $t = 25$ seconds.

The 2-norm of the observer errors $\tilde{w}(x,t)$ and $\tilde{w}_t(x,t)$ are shown in figure 3. The observer error $\tilde{w}(x,t)$ as a function of $x$ and $t$ is shown in figure 4. The proposed observer converges as expected exponentially to the plant.

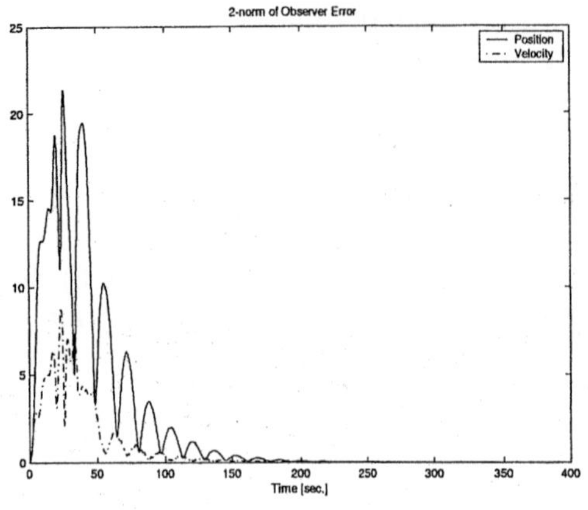


Fig. 3. 2-norm of observer errors $\tilde{w}(x,t)$ [solid line] and $\tilde{w}_t(x,t)$ [dashed line].

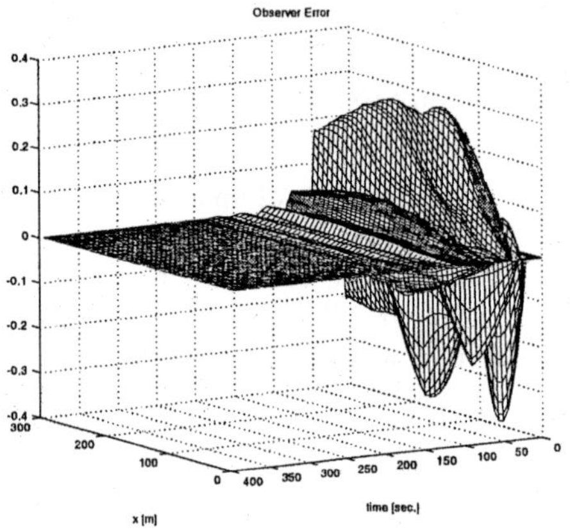


Fig. 4. Observer error $\tilde{w}(x,t)$.

## 6. CONCLUSION

In this paper, observer design for a towed seismic streamer cable is considered. Based on measurements at the boundaries, an uniformly exponentially stable observer is designed. The stability analysis of the observer is based on Lyapunov theory. The existence and uniqueness of the solutions of the observer are based on semigroup theory. The implementation of the observer requires three measurements: The position of the streamer cable at the upstream end, the position and the slope of the streamer cable at the downstream end. Simulation results are presented, and they are in agreement with the theoretical results.

## 7. ACKNOWLEDGMENTS

This work was funded by the Research Council of Norway under the Strategic University Program in Marine Cybernetics. The authors would like to thank Ph.D. student Svein Ersdal for interesting discussions on seismic cables.

## REFERENCES


Balas, M. J. (1998). Do all linear flexible structures have convergent second-order observers?. *Proc. American Contr. Conf.*

Dafermos, C. M. and M. Slemrod (1973). Asymptotic behavior of nonlinear contraction semigroups. *J. Func. Anal.* **13**, 97–106.

Demetriou, M. A. (2001). Natural observers for second order lumped and distributed parameter systems using parameter-dependent lyapunov functions. *Proc. American Contr. Conf.*

Dowling, A. P. (1988). The dynamics of towed flexible cylinders part 1. neutrally buoyant elements. *J. Fluid Mech.* **187**, 507–532.

Ersdal, S. (2003). *Control of seismic cables under tow.* Internal Report, Dep. Marine Tech., NTNU, Norway.

Kennedy, R. M. (1987). Forced vibration of a thin flexible cylinder in viscous flow. *J. Fluid Mech.* **115**, 189–201.

Kristiansen, D. (2000). *Modeling of Cylinder Gyroscopes and Observer Design for Nonlinear Oscillations.* Ph.D. thesis, NTNU, Norway.

Lohmiller, W. and J. J. E. Slotine (1996). On metric observers for nonlinear systems. *Proc. IEEE Int. Conf. on Contr. and Appl.*

Luenberger, D. L. (1979). *Introduction to Dynamic Systems.* Wiley.

Nguyen, T. D. (2004). *Observer Design for Mechanical Systems Described by PDEs.* Internal Report, Dep. Eng. Cyb., NTNU, Norway.

Ortloff, C. R. and J. Ives (1969). On the dynamic motion of a thin flexible cylinder in a viscous stream. *J. Fluid Mech.* **38**, 713–736.

Paidoussis, M. P. (1966). Dynamics of flexible slender cylinders in axial flow. *J. Fluid Mech.* **26**, 717–736.

Paidoussis, M. P. (1973). Dynamics of cylindrical structures subjected to axial flow. *Journal of Sound and Vibration* **29**, 365–385.

Pazy, A. (1983). *Semigroups of Linear Operators and Applications to Partial Differential Equations.* Springer.

Pedersen, E. and A. Sørensen (2001). Modelling and control of towed marine seismic streamer cables. *Proc. IFAC Conference on Systems and Control.*

Schlumberger (summer 2001). Oilfield review.

Taylor, G. (1952). Analysis of the swimming of long and narrow animals. *Proceedings of the Royal Society of London* pp. 158–183.

Triantafyllou, G. and C. Chryssostomidis (1988). The dynamics of towed arrays. *Proc. Int. Offshore Mechanics and Arctic Engineering Symp. 7th.*

White, F. (1972). An analysis of axisymmetric turbulent flow past a long cylinder. *J. of Basic Engineering* **94**, 200–206.

ELSEVIER
IFAC
PUBLICATIONS
www.elsevier.com/locate/ifac

# HYDRODYNAMIC PROPERTIES IMPORTANT FOR CONTROL OF TRAWL DOORS

**Karl-Johan Reite* and Asgeir J. Sørensen****

* *E-mail: karl.j.reite@sintef.no*
*SINTEF Fisheries and Aquaculture*
*N-7465 Trondheim, Norway*
** *Department of Marine Technology*
*Norwegian University of Science and Technology*
*N-7491 Trondheim, Norway*

Abstract: The paper presents a method for mathematical modelling of three-dimensional foils applied to trawl doors. Experimental data are parametrized to give the steady-state hydrodynamic coefficients. Transient effects are investigated by the use of a numerical vortex lattice method. A method for approximating these effects is suggested, along with the appropriate coefficients. The proposed model accounts for steady-state and transient hydrodynamic forces and moments in 6 degrees of freedom. It will be used for the purpose of trawl control system design and analysis. *Copyright © 2004 IFAC*

Keywords: Foil, hydrodynamics, modelling, simulation, trawl system

## 1. INTRODUCTION

Fish trawling is of great importance for both economics and food supply. At the same time its negative environmental impacts such as pollution and damage on the sea floor have gained increasing attention. More precise control of the trawl system (Fig.1) will reduce these impacts. Modelling, simulation and control of a simplified trawl system in the horizontal plane have been described by Johansen *et al.* (2002). The system itself is characterized by the strong interactions between shape and forces, nonlinearities in the hydrodynamic loads, time-varying coefficients, and lack of proper state measurements. The trawl doors are foils used to keep the horizontal opening of the trawl net. They are connected to the vessel and the trawl net through steel wires. The forces they produce may be controlled by altering a combination of the trawl doors orientation, area and shape. Control of orientation is assumed in this paper, because of efficiency and robustness considerations. The aim of the control system will in this case be to control the three orientation angles by means of local actuators installed on each trawl door.

Since the trawl doors can not be easily monitored, and the energy supply are expected to be scarce, the use of control action must be kept to a minimum. This may be achieved by the use of proper designed feed forward controllers, which emphasizes the need for thorough insight in the modelling properties of the trawl system in general and the trawl doors in particular. An adequate description of the hydrodynamic properties of the trawl doors are therefore needed, including both steady-state and transient behaviour. Such a description is the aim of the present paper.

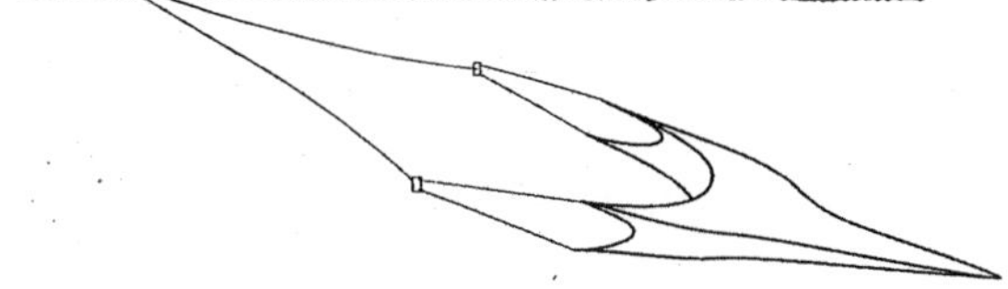

Fig. 1. Overview of the trawl system.

## 2. REPRESENTATION OF HYDRODYNAMIC MOMENTS OF LIFTING SURFACES

It is in 2D foil theory common to express the hydrodynamic steady-state forces and moments by the hydrodynamic force vector $\mathbf{F}^{2D} \in \mathbb{R}^2$ and the center of pressure, $r^{2D}$, with the coordinates $(r_1^{2D}\ r_2^{2D})$. $r^{2D}$ is confined to a straight line approximating the foil mean-camber line. The moment vector $\mathbf{M}^{2D} \in \mathbb{R}^2$ acting on the foil is then

$$\mathbf{M}^{2D} = \mathbf{p}_{r^{2D}} \times \mathbf{F}^{2D}, \qquad (1)$$

where $\mathbf{p}_{r^{2D}}$ is the vector from the origin to $r^{2D}$. This representation will be consistent as long as $\mathbf{F}^{2D}$ has a component normal to the chosen line.

In the 3D case all six force and moment components are present. It may be assumed that, as in the 2D case, the hydrodynamic moments, $\mathbf{M} = [M_1\ M_2\ M_3]^T$, are produced by the hydrodynamic force vector, $\mathbf{F} = [F_1\ F_2\ F_3]^T$, attacking a point $r$, with coordinates $(r_1\ r_2\ r_3)$ and confined to the plane approximating the mean surface of the foil. The plane is given by its normal vector $\mathbf{n} = [n_1\ n_2\ n_3]^T$ and a point $p$ with coordinates $(p_1\ p_2\ p_3)$. These assumptions yield

$$\begin{bmatrix} 0 & F_3 & -F_2 \\ -F_3 & 0 & F_1 \\ F_2 & -F_1 & 0 \\ n_1 & n_2 & n_3 \end{bmatrix} \begin{bmatrix} r_1 \\ r_2 \\ r_3 \end{bmatrix} = \begin{bmatrix} M_1 \\ M_2 \\ M_3 \\ \mathbf{n} \cdot \mathbf{p}_p \end{bmatrix}, \qquad (2)$$

where $\mathbf{p}_p = [p_1\ p_2\ p_3]^T$. This system is consistent if and only if $\mathbf{M} \cdot \mathbf{F} = 0$. In the general case, this is not fulfilled, and (2) can not be solved. Hence, it is hard to represent the moments by the point of attack of the forces. A better approach is to apply the moments directly.

## 3. KINEMATICS

### 3.1 Notation

To find the forces, positions and moments in the trawl door and the hydrodynamic force frames, transformations and notation from Fossen (2002) will be used: $\mathbf{p}_o^n$ is the position of the point $o$ decomposed in the frame $n$, $\mathbf{m}_o^b$ is the moment about the point $o$ decomposed in the frame $b$, $\mathbf{f}_o^b$ is the force in the point $o$ decomposed in the frame $b$, and $\mathbf{R}_n^b$ is the rotation matrix from the frame $n$ to the frame $b$.

Use of left-hand frames and belonging variables will be distinguished from corresponding right-hand frame variables by the use of an additional sub- or superscript $l$, like e.g. $\mathbf{f}_o^{b,l}$.

### 3.2 Trawling reference frames

*3.2.1. Trawl door frame* It is chosen to use a left-hand frame for the port trawl door and a right-hand frame for a starboard trawl door. This is done in order to use the same coefficients and definitions for the hydrodynamic orientation angles for both sides, and, thus, getting a more consistent model.

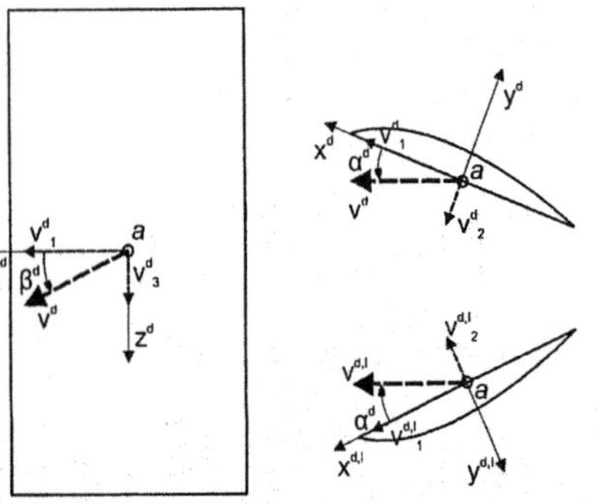

Fig. 2. Trawl door coordinate system and orientation angles

The trawl door frame and orientation angles are shown in Fig. 2. The left part shows a trawl door as seen from the port side, the upper part to the right shows the starboard trawl door as seen from above, and the lower part to the right shows the port trawl door as seen from above. The trawl door hydrodynamic orientation angles are defined by the relative speed of the door frame origin in relation to the surrounding fluid. This relative speed is given in the door frame as

$$\mathbf{v}^d = \begin{bmatrix} v_1^d\ v_2^d\ v_3^d \end{bmatrix}^T. \qquad (3)$$

The trawl door angle of attack, $\alpha^d$, and the angle of slip, $\beta^d$, are defined as

$$\alpha^d = -\ \arcsin \frac{v_2^d}{U}, \quad \beta^d = \arcsin \frac{v_3^d}{U}, \qquad (4)$$

where $U = \left\| \mathbf{v}^d \right\|$.

*3.2.2. Hydrodynamic force frame* The frame of hydrodynamic forces, $(L,\ D,\ S)$, is shown in the lower right corner of Fig. 5. It will constitute a left-hand frame for the port trawl door and a right-hand frame for the starboard trawl door. The hydrodynamic forces will be assumed to attack in the trawl door centre of area, $a$, and will be collected in the vector $\mathbf{f}_a^h = \begin{bmatrix} L\ D\ S \end{bmatrix}^T$, or $\mathbf{f}_a^d = \begin{bmatrix} f_{a,1}^d\ f_{a,2}^d\ f_{a,3}^d \end{bmatrix}^T$ in the trawl door frame. The following definitions will be used:

- Lift force, $L$, is defined as the hydrodynamic force perpendicular to both the $z^d$-axis and the relative fluid velocity. It is positive for $f_{a,2}^d > 0$.
- Drag force, $D$, is defined as the hydrodynamic force in the direction of the relative fluid velocity.
- Shear force, $S$, is defined as the hydrodynamic force perpendicular to both the relative water velocity and the $y^d$-axis. It is positive for $f_{a,3}^d > 0$.

*3.3.1. Rotation between two right-hand frames* Rotation of a vector, assuming that the frames are right-handed coordinate systems, is given by Fossen (2002). $\boldsymbol{\Theta} = \left[\phi\ \theta\ \psi\right]^T$ are the Euler angles giving rotation about the *b*-frame $x^b$-, $y^b$- and $z^b$-axes, respectively. The definitions of the principal rotations yield

$$\mathbf{R}_n^b = \mathbf{R}_x(\boldsymbol{\Theta})^T \mathbf{R}_y(\boldsymbol{\Theta})^T \mathbf{R}_z(\boldsymbol{\Theta})^T, \quad (5)$$

where $\mathbf{R}_x(\boldsymbol{\Theta})$, $\mathbf{R}_y(\boldsymbol{\Theta})$ and $\mathbf{R}_z(\boldsymbol{\Theta})$ are the principal rotation matrices about the $x^b$-, $y^b$-, and $z^b$-axes, respectively.

*3.3.2. Rotation between right-hand and left-hand frames* For rotation between right-hand and left-hand frames, the right-hand Euler angles and coordinates may be found from the left-hand Euler angles and coordinates according to

$$\boldsymbol{\Theta} = \mathbf{T}_l \boldsymbol{\Theta}^l\ ,\quad \mathbf{p} = \mathbf{R}_l \mathbf{p}^l, \quad (6)$$

where $\mathbf{R}_l$ and $\mathbf{T}_l$ are the mirroring transformation matrices. The transformation matrix from the global right-hand system into the local left-hand system is found as

$$\mathbf{R}_n^{b,l} = \mathbf{R}_l \mathbf{R}_x\left(\mathbf{T}_l\boldsymbol{\Theta}^l\right)^T \mathbf{R}_y\left(\mathbf{T}_l\boldsymbol{\Theta}^l\right)^T \mathbf{R}_z\left(\mathbf{T}_l\boldsymbol{\Theta}^l\right)^T. \quad (7)$$

## 4. WIND TUNNEL EXPERIMENTS

### *4.1 Experimental setup*

The model used for the experiments was in scale 1:5 of a $15\text{m}^2$ port trawl door, giving it an area of $0.6\text{m}^2$ and an aspect ratio of 1.45. The left part of Fig. 3 shows the suction side of a port trawl door, and the right part shows the pressure side of a starboard trawl door. The trawl door is a multifoil consisting of three foils; two small front foils followed by a larger aft foil. Its cross-section is shown in Fig. 4. The aspect ratio of the trawl door is less than what is common for pelagic and and larger than what is common for bottom trawling, and the results should therefore represent a compromise for both types of trawling.

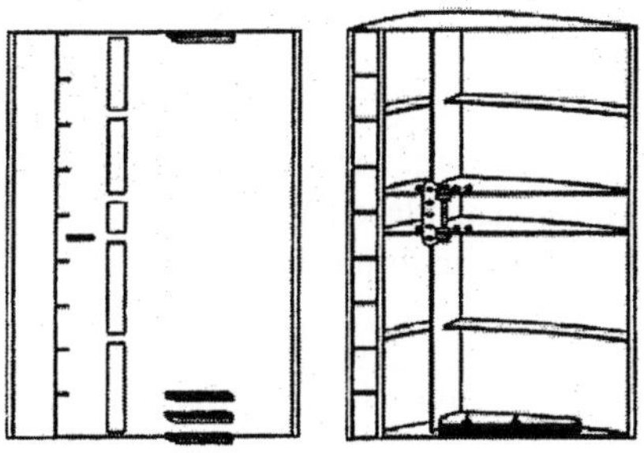

Fig. 3. Suction side of a port trawl door and pressure side of a starboard trawl door.

wind tunnel, with experimental setup as shown in Fig. 5. Left-hand frames are used for the trawl door and the hydrodynamic forces since the tested model is a port trawl door. The model was placed with the suction side upwards. It could be rotated about the $z^t$-axis to give the variations in the angle of slip $(\beta^t)$ and about the $z^{d,l}$-axis to give the variations in the angle of attack $(\alpha^t)$. The steady-state forces and moments were measured by *the scale*, a set of force transducers in the floor of the wind tunnel.

### *4.2 Experiment reference frames*

*4.2.1. Wind tunnel frame* The wind tunnel frame is shown in Fig. 5. The $z^t$-axis points strictly upwards, perpendicular to the wind. The origin is located in the centre of the scale. The $x^t$-axis points straight into the wind. The $y^t$-axis is given from the fact that this is a right-hand frame.

*4.2.2. Scale frame* The scale frame is shown in Fig. 5. The origin and the $z^s$-axis coincides with the wind tunnel frame, but the scale frame follows the model as this is rotated about the $z^t$-axis.

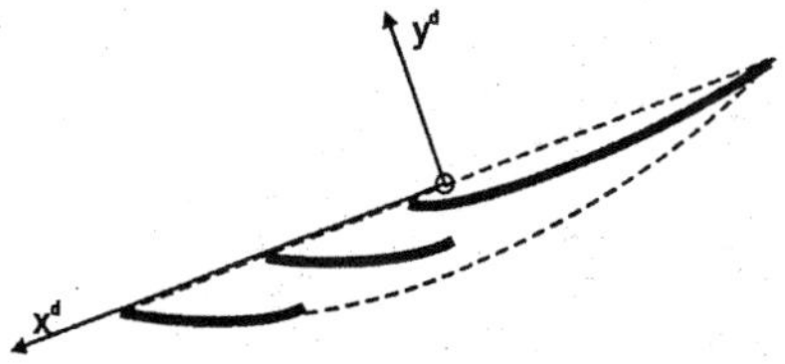


Fig. 4. Cross section of port trawl door as seen along $z_d$-axis.

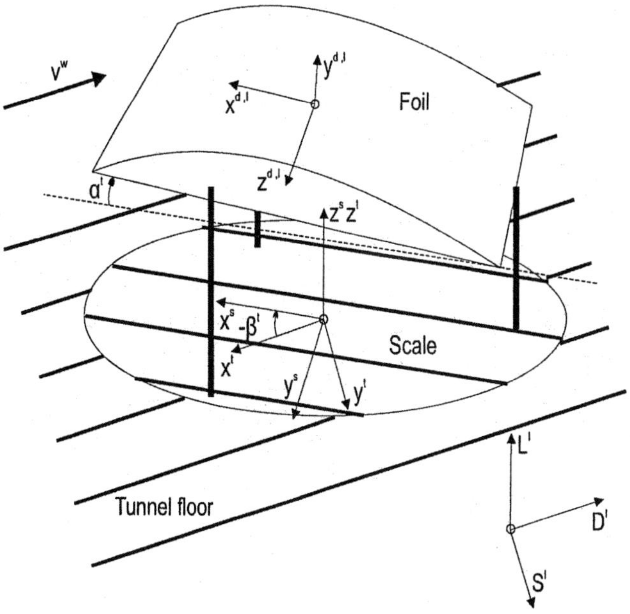


Fig. 5. Experimental setup and coordinate systems.

*4.2.3. Experiment orientation angles* The wind tunnel angle of attack, $\alpha^t$, is measured as the angle between the $x^s$- and the $x^{d,l}$-axis. $\alpha^t$ is defined to be positive in the intended range of operation. The wind tunnel slip angle, $\beta^t$, is measured as the angle between the $x^s$- and the $x^t$-axis. It is defined to be positive for counter-clockwise rotation about the $z^t$-axis.

### *4.3 Analysis*

*4.3.1. Forces in the hydrodynamic force frame* The forces in the hydrodynamic force frame can be found from the forces in the scale frame

$$\mathbf{f}_a^{h,l} = \mathbf{R}_s^{h,l}\mathbf{f}_a^s, \tag{8}$$

where $\mathbf{f}_{a,i}^{h,l} = \begin{bmatrix} f_{a,1}^{h,l} & f_{a,2}^{h,l} & f_{a,3}^{h,l} \end{bmatrix} = \begin{bmatrix} L^l & D^l & S^l \end{bmatrix}$. $\mathbf{R}_s^{h,l}$ is found from (7).

*4.3.2. Positions in scale and tunnel frame* The position of a point $a$ given in the port trawl door frame is in the scale and tunnel frame

$$\mathbf{p}_a^s = \mathbf{R}_{d,l}^s \left(\mathbf{p}_a^{d,l} - \mathbf{p}_p^{d,l}\right) + \mathbf{p}_p^s, \tag{9}$$

$$\mathbf{p}_a^t = \mathbf{R}_s^t \left(\mathbf{R}_{d,l}^s \left(\mathbf{p}_a^{d,l} - \mathbf{p}_p^{d,l}\right) + \mathbf{p}_p^s\right), \tag{10}$$

where $d$ is the door frame origin, and $p$ is a pivot point that remains fixed in both scale and door frame. $\mathbf{R}_{d,l}^s$ and $\mathbf{R}_s^t$ are found from (7) and (5).

*4.3.3. Moments in the trawl door frame* The hydrodynamic moments will be transferred to the trawl door axes. The measured force is assumed to attack in the centre of area of the trawl door, $a$. The hydrodynamic moment in the port trawl door frame about centre of area is

$$\mathbf{m}_a^{d,l} = \mathbf{R}_s^{d,l}(\mathbf{m}_s^s - \mathbf{p}_a^s \times \mathbf{f}_a^s), \tag{11}$$

where $\mathbf{m}_a^{d,l} = \begin{bmatrix} m_{a,1}^{d,l} & m_{a,2}^{d,l} & m_{a,3}^{d,l} \end{bmatrix}$. $\mathbf{m}_s^s$ is the measured moment about the scale frame origin, given in the scale frame. The hydrodynamic steady-state coefficients are found as

$$C_i^\infty\left(\alpha^d, \beta^d\right) = \frac{f_{a,i}^{h,l}\left(\alpha^d, \beta^d\right)}{\frac{1}{2}\rho A U^2}, \quad i = 1,\ 2,\ 3, \tag{12}$$

$$C_4^\infty\left(\alpha^d, \beta^d\right) = \frac{m_{a,1}^{d,l}\left(\alpha^d, \beta^d\right)}{\frac{1}{2}\rho A h U^2}, \tag{13}$$

$$C_5^\infty\left(\alpha^d, \beta^d\right) = \frac{m_{a,2}^{d,l}\left(\alpha^d, \beta^d\right)}{\frac{1}{2}\rho A \sqrt{h^2 + c^2} U^2}, \tag{14}$$

$$C_6^\infty\left(\alpha^d, \beta^d\right) = \frac{m_{a,3}^{d,l}\left(\alpha^d, \beta^d\right)}{\frac{1}{2}\rho A c U^2}, \tag{15}$$

where $A$, $h$ and $l$ are the planform area, span and cord length of the trawl door, respectively. $\rho$ is the fluid density. The superscript $\infty$ indicates that these are the steady-state coefficients.

### *4.4 Parametrization of measured coefficients*

The steady-state coefficients are parametrized to facilitate a mathematical model for simulating the trawl system. The parametrization is done by minimizing the object function

$$O_i = \sum_{n=1}^{N} w_n \left(C_{in}^\infty - \bar{C}_{in}^\infty(\mathbf{K})\right)^2, \tag{16}$$

where $C_{in}^\infty$ is the measurement $n$ of the coefficient $i$, $\bar{C}_{in}^\infty$ is the calculated coefficient $i$ for this measurement, and $\mathbf{K}$ is the parameter matrix for calculating hydrodynamic coefficients. The weighting function for this measurement is defined as

$$w_n = \frac{1}{k_1^2 + (k_2\alpha_0^d - \alpha_n^d)^2 + \left(\beta_n^d\right)^2}, \tag{17}$$

where $k_1$ and $k_2$ are factors chosen as 1 and $\frac{2}{3}$ to avoid too much weight in the central working area and for high angles of attack. $\alpha_0^d$ is the angle of attack that gives maximum hydrodynamic lift for zero angle of slip. $\alpha_n^d$ and $\beta_n^d$ are the hydrodynamic angles of attack and slip during the measurement $n$, respectively.

The resulting parametrization functions of the steady-state coefficients are

$$\bar{\mathbf{C}}^\infty\left(\alpha^d, \beta^d, \mathbf{K}\right) =$$

$$\begin{bmatrix} K_{11}c\left(K_{12}\beta^d\right)s\left(\alpha^d + K_{13}\right)\left(s\alpha^d + K_{14}c\alpha^d - K_{15}\right) \\ \dfrac{\left(\bar{C}_1^\infty\right)^2}{\pi A_R} + K_{21}c\left(\alpha^d + K_{22}\right) + K_{23}\left(K_{24} - s\alpha^d c\beta^d\right) \\ K_{31}s\beta^d s\left(K_{32}\alpha^d + K_{33}\right) \\ K_{41}s\beta^d c\alpha^d \bar{C}_1^\infty + K_{42}s\beta^d s\alpha^d \bar{C}_2^\infty \\ K_{51}s\left(4\beta^d\right)c\left(K_{52}\alpha^d\right)\bar{C}_2^\infty \\ -K_{61}s\alpha^d c\beta^d \bar{C}_1^\infty + K_{62}c\alpha^d c\beta^d \bar{C}_2^\infty \end{bmatrix} \tag{18}$$

where $cx = \cos(x)$, $sx = \sin(x)$, and $A_R$ is the trawl door aspect ratio. The coefficient matrix is

$$\mathbf{K} = \begin{bmatrix} 2.5515 & 5.0585 & 0.24602 & 0.093696 & 0.015943 & 0.56041 \\ 0.87764 & 3.5902 & 2.712 & 0.24667 & 5.8402 & 0.66617 \\ 0.14231 & 3.0528 & 4.0588 & & & \\ 4.0829 & 1.4944 & & & & \\ 2.8125 & & & & & \end{bmatrix} \tag{19}$$

The RMS values of the difference between the calculated and measured values of each coefficient are

$$\boldsymbol{\sigma}_{\bar{C}^\infty} = \begin{bmatrix} 62 & 77 & 2.7 & 10 & 0.13 & 41 \end{bmatrix}^T \cdot 10^{-4}. \tag{20}$$

As an example, Fig. 6 shows a plot of the resulting error in the calculated hydrodynamic lift. The plotted values are found as

$$p_{in} = \frac{\bar{C}_{in}^\infty(\mathbf{K}) - C_{in}^\infty}{\max\left(C_i^\infty\right) - \min\left(C_i^\infty\right)}, \tag{21}$$

where $p_{in}$ is the error in the calculated coefficient $\bar{C}_i^\infty$ for measurement $n$, in relation to the total range of measured values of this coefficient.

## 5. TRANSIENT EFFECTS

When a foil is subject to changes in speed or current, the hydrodynamic forces and moments will be different from those expected from steady-state theory and measurements. These effects will probably be of interest for trawl door controller design, since both stability and performance will be affected.

### 5.1 2D linearized theory

According to 2D linearized foil theory, the lift force will develop as the Wagner function if sudden, small speed changes are imposed (Newman 1977). The main assumptions are that the foil is thin, without camber, at small angles of attack and that the unsteady perturbations are small. These assumptions will obviously not be valid for conventional trawl doors, and a numerical 3D method is therefore utilized.

### 5.2 Numerical solution for 3D foil

An unsteady numerical vortex lattice method (VLM) resembling that described by Katz and Plotkin (1991) is used to investigate the transient properties of a 3D foil. This method will presumably predict $\frac{F_i(t)}{F_i^{\infty}}$ quite well for $i \in \{1\ 2\ 4\ 6\}$, since these forces will be dominated by the non-viscous pressure forces for reasonable angles of attack. The time dependency of the shear force and the moment about $y^d$-axis can not be found from this model, since the effects of viscosity and trawl door details, like the end plates, are not taken into account. Since these forces are small and of little importance, their transient behaviour are excluded from the model.

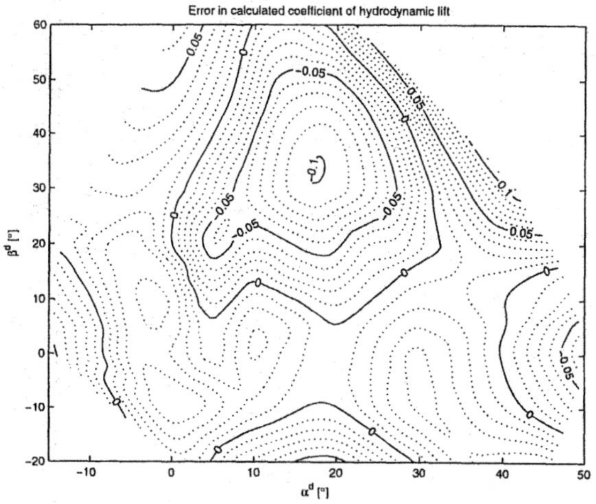


Fig. 6. Error in calculated coefficient of hydrodynamic lift.

### 5.3 Numerical results

From the formulation of the hydrodynamic coefficients, it seems natural to investigate the responses to steps in $\alpha^d$, $\beta^d$ and $U$. The effect of changes in $\beta^d$ will probably be minor and not adequately described by the VLM model. Steps in $\beta^d$ are therefore not analyzed. The initial conditions were for all cases steady state with relative speed $U^0 = 2m/s$, angle of attack $\alpha^{d0} = 20°$ and angle of slip $\beta^{d0} = 0°$. $\mathbf{F}^0 \in \mathbb{R}^6$ is the hydrodynamic forces and moments of these initial conditions. The simulated steps were both negative and positive, and the average step responses were analyzed. The following steps were simulated: $\Delta U^1 = \pm 0.2m/s$, $\Delta\alpha^{d2} = \pm 5°$ (induced by changes in velocity) and $\Delta\alpha^{d3} = \pm 5°$ (induced by changes in orientation). These steps are in the following designated by $j = 1, 2$ and 3, respectively.

The transient effects will be approximated as separate first-order dynamic systems. The input of these systems will be the steady-state hydrodynamic forces, $\mathbf{F}^{\infty}\left(\alpha^d(t), \beta^d(t), U(t)\right)$. The output will be the time dependent hydrodynamic forces $\mathbf{F}(t)$. The time constants of these systems are found according to the definition for first-order dynamic systems as the time $t$, for which

$$F_{ij}(t) = \left(1 - e^{-1}\right)\left(F_{ij}^{\infty} - F_i^0\right) + F_i^0\ , \qquad (22)$$

where $F_{ij}(t)$ is $F_i$ at a time $t$ after a step $j$ occured. $F_{ij}^{\infty}$ is the steady-state force $F_i^{\infty}$ after a step $j$. Symmetry about the $x^d$-axis dictates that $F_4(t) = 0$, and the dynamics of this force is therefore omitted in the model. The time constant matrix estimated from the VLM solutions are

$$\mathbf{T}_1 = \begin{bmatrix} 0.081 & 0.43 & 0.046 \\ 0.081 & 0.16 & 0.014 \\ 0.046 & 0.043 & 0.11 \end{bmatrix} \frac{c}{U}\ . \qquad (23)$$

The columns in (23) corresponds to the steps $j = 1$ to 3. The rows in (23) corresponds to the effect on hydrodynamic lift, drag, and moment about $z^d$-axis, respectively.

The step responses corresponding to $j = 1$ to 3, in terms of hydrodynamic forces, are presented in Fig. 7, 8 and 9, respectively. These give the value of $\frac{\mathbf{F}(t)-\mathbf{F}^0}{\mathbf{F}^{\infty}-\mathbf{F}^0}$ as a function of the reduced time $\tau = \frac{Ut}{c}$. The thin, unmarked lines are the linear approximations of the time dependency. For comparison, the 2D analytic solution is given as the Wagner function in Fig. 7. Note that the Wagner function includes the effect of added mass, while this is not included in the VLM solution.

It is seen from Fig. 9 that the responses from steps in angle of attack from orientation are not possible to approximate by a linear system. These responses seems to be proportional to the accelerations during the step, and should therefore probably be represented through the added mass formulation. The transients from such steps are therefore kept out of the model, and the following time constant matrix will be used

$$\mathbf{T} = \begin{bmatrix} 0.081 & 0.081 & 0 & 0 & 0 & 0.046 \\ 0.43 & 0.16 & 0 & 0 & 0 & 0.043 \end{bmatrix}^T \frac{c}{U}, \qquad (24)$$

where column $j$ corresponds to steps similar to the step $j$, and row $i$ adresses the hydrodynamic force $F_i$.

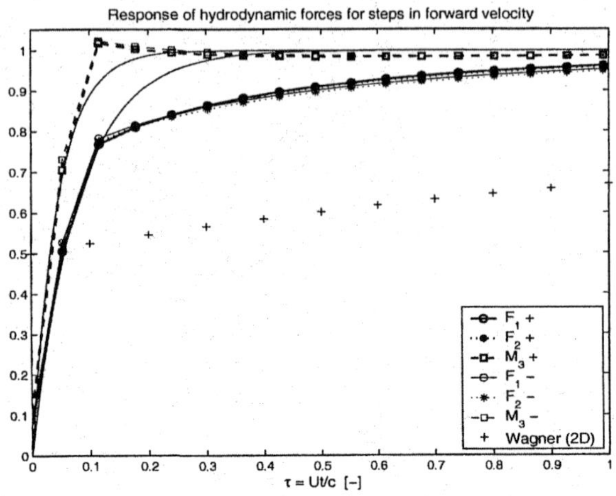


Fig. 7. Step responses for $j = 1$. $U = U^0 \pm \Delta U^1$.

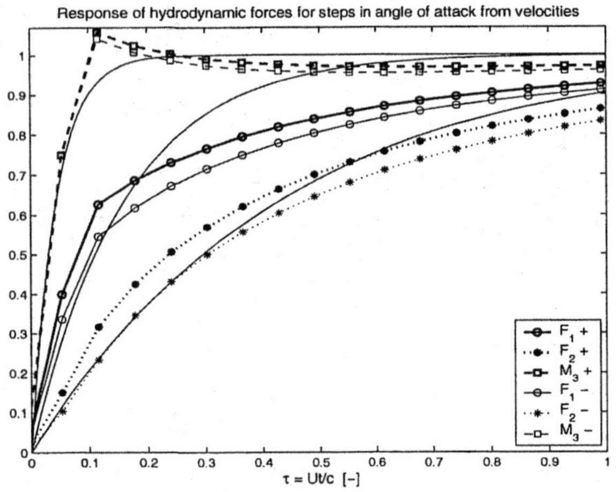


Fig. 8. Step responses for $j = 2$. $\alpha^d = \alpha^{d0} \pm \Delta\alpha^{d2}$.

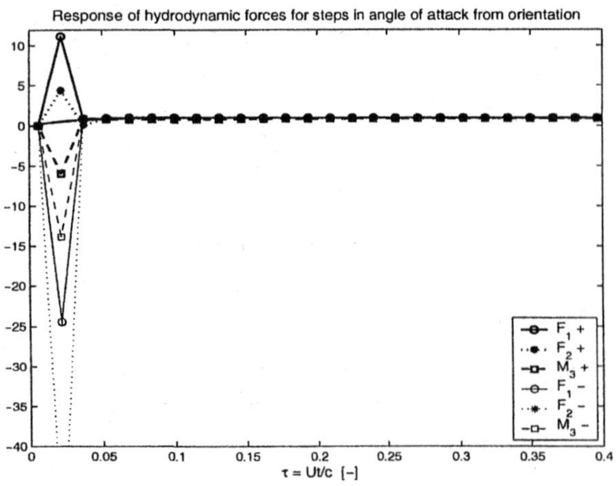


Fig. 9. Step responses for $j = 3$. $\alpha^d = \alpha^{d0} \pm \Delta\alpha^{d3}$.

## 6. RESULTING MODEL

The steady-state hydrodynamic coefficients can be found from (18). The time-varying hydrodynamic forces and moments during and after steps in relative speed and angle of attack are suggested approximated by

$$F_i(t) = F_i^{\infty}(t)\left(\frac{k_i}{T_{i1}s + 1} + \frac{1 - k_i}{T_{i2}s + 1}\right), \qquad (25)$$

for $1 \leqslant i \leqslant 6$, and

$$k_i = \left|\frac{C_i^0\left(U^2 - \left(U^0\right)^2\right)}{C_i^{\infty}U^2 - C_i^0\left(U^0\right)^2}\right|, \qquad (26)$$

where the superscript 0 describes conditions before the step occurs. By treating continuous changes in relative speed and angle of attack as small, successive steps, a continuous model is obtained.

## 7. CONCLUSIONS

Hydrodynamic coefficients for steady-state forces and moments in a 6 degrees of freedom model of a trawl door have been experimentally studied, and a parametrized model has been proposed. The memory effect and its impact on transients in the trawl door hydrodynamic forces have been considered, and it is suggested to implement this as uncoupled first-order linear models. Time constants for estimating the response in lift, drag, and moment about $z^d$-axis from changes in speed and angle of attack are presented. The effect of rapid changes in orientation and added mass are assumed to be modelled separately.

## 8. ACKNOWLEDGEMENTS


This work has been financed by the Norwegian Research Council. Birger Enerhaug at SINTEF Fisheries and Aquaculture have assisted in verification of the wind tunnel data, and is gratefully acknowledged.


## REFERENCES


Fossen, T. I. (2002). *Marine Control Systems.* Marine Cybernetics. Trondheim, Norway.

Johansen, V., O. Egeland and A. J. Sørensen (2002). Modelling and control of a trawl system in the transversal direction. *IFAC* pp. 292–297.

Katz, J. and A. Plotkin (1991). *Low-Speed Aerodynamics. From Wing Theory to Panel Methods.* McGraw-Hill. New York.

Newman, J. N. (1977). *Marine Hydrodynamics.* The MIT Press. Cambridge.

ELSEVIER
IFAC
PUBLICATIONS
www.elsevier.com/locate/ifac

# NONLINEAR OBSERVER DESIGN FOR A NONLINEAR CABLE/STRING FEM MODEL USING CONTRACTION THEORY

**Yilmaz Türkyilmaz * and Jerome Jouffroy * and Olav Egeland *,****

** Centre for Ships and Ocean Structures*
*Norwegian University of Science and Technology*
*N-7491 Trondheim, Norway*
*** Department of Engineering Cybernetics*
*Norwegian University of Science and Technology*
*N-7491 Trondheim, Norway*
*E-mail: yilmaz@itk.ntnu.no*

Abstract: Contraction theory is a recently developed nonlinear analysis tool which may be useful for solving a variety of nonlinear control problems. In this paper, using Contraction theory, a nonlinear observer is designed for a general nonlinear cable/string FEM (Finite Element Method) model. The cable model is presented in the form of partial differential equations (PDE). Galerkin's method is then applied to obtain a set of ordinary differential equations such that the cable model is approximated by a FEM model. Based on the FEM model, a nonlinear observer is designed to estimate the cable configuration. It is shown that the estimated configuration converges exponentially to the actual configuration. Numerical results and simulations are shown to be in agreement with the theoretical results. *Copyright © 2004 IFAC*

Keywords: Cables, nonlinear models, finite element method, observers, nonlinear analysis.

## 1. INTRODUCTION

Cables/strings are flexible structural elements used in a wide span of engineering applications, such as cable towing operations in marine applications. Depending on their physical properties and application areas, they exhibit vibration in the presence of disturbances. The most commonly used method for suppressing the cable vibration is to apply boundary controllers. While a variety of boundary controllers have been proposed to control the motion of different cable/string systems, observer design for nonlinear cable/string systems has received little attention, which can be of great importance in the design of boundary controllers.

Boundary control of the cable/string systems have been investigated by several authors. Among others, (Morgül, 1994) designed a boundary feedback controller for a system described by the wave equation where exponential stability of the closed loop is obtained for strictly proper transfer functions. (Baicu *et al.*, 1999) developed exponentially stabilizing controllers for the transverse vibration of a string-mass system modeled by one-dimensional wave equation. (Shahruz and Narasimha, 1997) and (Canbolat *et al.*, 1998) presented exponentially stabilizing controllers for a one-dimensional nonlinear string equation, allowing varying tension in the string. (Qu, 2000) devised a robust and adaptive controller to damp out the transverse oscillations of a stretched string, allowing nonlinear dynamics and their uncertainties in the model. The works mentioned above use a combination of the states at the boundary, namely the boundary slope, slope-rate and velocity, to design the stabilizing boundary control laws. Observer design based on Lyapunov analysis is well understood and widely used for linear systems. Recently,

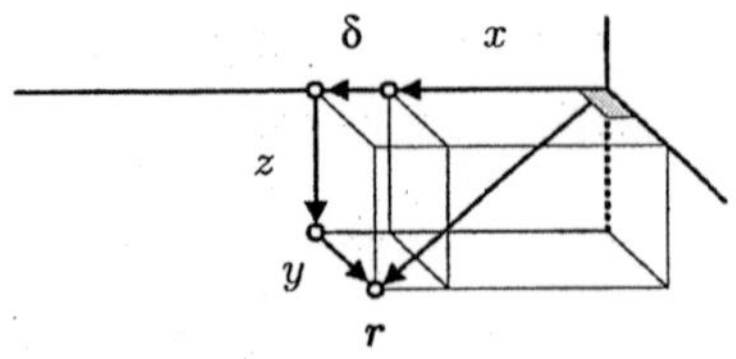


Fig. 1. Kinematic consideration of cable/string.

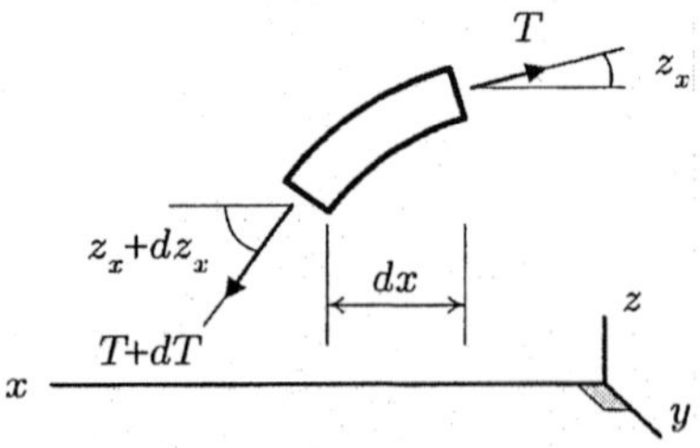


Fig. 2. The forces acting on an elemental length $dx$.

(Demetriou, 2001) presented the construction of natural observers for linear second order lumped and distributed parameter systems using parameter-dependent Lyapunov functions without resorting to a first order formulation. Unlike linear systems, construction of nonlinear observers lacked generality both from design and analysis point of view. Contraction theory is a recently developed nonlinear analysis tool which may be useful for solving a variety of nonlinear observer design problems. (Lohmiller and Slotine, 1998) have presented Contraction theory through a series of publications. (Kristiansen, 2000) has used Contraction theory for the design of nonlinear observers.

In this paper, using Contraction theory, a nonlinear observer is designed for a general nonlinear cable/string FEM (Finite Element Method) model. The cable model is presented in the form of partial differential equations (PDE). Galerkin's method (Zienkiewicz and Taylor, 2000) is then applied to obtain a set of ordinary differential equations such that the cable model is approximated by a FEM model. Based on the FEM model, a nonlinear observer is designed to estimate the cable configuration. It is shown that the estimated configuration converges exponentially to the actual configuration. Numerical results and simulations are shown to be in agreement with the theoretical results.

## 2. EQUATIONS OF MOTION

### 2.1 Kinematics

An inertial reference frame $i$ is defined with orthogonal unit vectors $\mathbf{i}$, $\mathbf{j}$ and $\mathbf{k}$ along the $x$, $y$ and $z$ axes, respectively. The spatial position of an arbitrary point on the center line of the *initially stressed* cable is given by the vector

$$\mathbf{r} = x\mathbf{i} + y\mathbf{j} + z\mathbf{k}$$

Consider an arbitrary point on the *undeformed and initially stressed* cable with the coordinates

$$\mathbf{r} = x\mathbf{i}$$

When the cable is *deformed*, the material point described by the material coordinates $\mathbf{r}$ will have the spatial coordinates

$$\mathbf{r} = (x+\delta)\,\mathbf{i} + y\mathbf{j} + z\mathbf{k}$$

as shown in Fig. 1. This gives

$$\left\|\frac{\partial \mathbf{r}}{\partial x}\right\| \approx 1 + \frac{\partial \delta}{\partial x} + \frac{1}{2}\left(\frac{\partial y}{\partial x}\right)^2 + \frac{1}{2}\left(\frac{\partial z}{\partial x}\right)^2 \quad (1)$$

where right-hand side of (1) is the binomial approximation of $\left\|\frac{\partial \mathbf{r}}{\partial x}\right\|$. A cable frame $c$ of orthonormal vectors $\mathbf{t}$, $\mathbf{b}$ and $\mathbf{n}$ is defined where $\mathbf{t}$ is the unit vector tangent to the cable such that

$$\frac{\partial \mathbf{r}}{\partial x} = (1+\epsilon)\,\mathbf{t} \quad (2)$$

where $\epsilon$ is the Lagrangian strain. Hence, from (1) and (2) the relation for the strain can be obtained as

$$\epsilon \approx \frac{\partial \delta}{\partial x} + \frac{1}{2}\left(\frac{\partial y}{\partial x}\right)^2 + \frac{1}{2}\left(\frac{\partial z}{\partial x}\right)^2 \quad (3)$$

### 2.2 Dynamics

Consider a cable of length $L$. The dynamics of the cable is assumed to be determined by the tension in the cable and the inertial forces as shown in Fig. 2. Using Hooke's law and equation (3), the tension $T$ in the cable can be expressed in the form

$$\begin{aligned} T(x) &= T_0 + E_A\epsilon \\ &= T_0 + E_A\left[\frac{\partial \delta}{\partial x} + \frac{1}{2}\left(\frac{\partial y}{\partial x}\right)^2 + \frac{1}{2}\left(\frac{\partial z}{\partial x}\right)^2\right] \end{aligned} \quad (4)$$

where $T_0$ is the constant tension in the initially stressed cable, $E_A = EA$, $E$ is the Young's modulus and $A$ is the cross section of the cable. Consider a material cable element of spatial length $dx$. Writing equilibrium of the forces in the $x$, $y$ and $z$ directions and using (4) gives the *nonlinear coupled equations* of motion

$$\begin{aligned} m\frac{\partial^2 \delta}{\partial t^2} &= \frac{\partial T}{\partial x} \\ &= E_A\left(\frac{\partial^2 \delta}{\partial x^2} + \frac{\partial y}{\partial x}\frac{\partial^2 y}{\partial x^2} + \frac{\partial z}{\partial x}\frac{\partial^2 z}{\partial x^2}\right) \end{aligned} \quad (5)$$

$$\begin{aligned} m\frac{\partial^2 y}{\partial t^2} &= \frac{\partial}{\partial x}\left(T\frac{\partial y}{\partial x}\right) \\ &= E_A\left[\frac{\partial y}{\partial x}\frac{\partial^2 \delta}{\partial x^2} + \left(\frac{\partial y}{\partial x}\right)^2\frac{\partial^2 y}{\partial x^2} + \frac{\partial y}{\partial x}\frac{\partial z}{\partial x}\frac{\partial^2 z}{\partial x^2}\right] \\ &\quad + T\frac{\partial^2 y}{\partial x^2} \end{aligned} \quad (6)$$

$$\begin{aligned} m\frac{\partial^2 z}{\partial t^2} &= \frac{\partial}{\partial x}\left(T\frac{\partial z}{\partial x}\right) \\ &= E_A\left[\frac{\partial z}{\partial x}\frac{\partial^2 \delta}{\partial x^2} + \frac{\partial y}{\partial x}\frac{\partial z}{\partial x}\frac{\partial^2 y}{\partial x^2} + \left(\frac{\partial z}{\partial x}\right)^2\frac{\partial^2 z}{\partial x^2}\right] \\ &\quad + T\frac{\partial^2 z}{\partial x^2} \end{aligned} \quad (7)$$

for $x \in (0, L)$ and $t \geq 0$, where $m > 0$ is the mass per unit length of the cable. Equations (5)–(7) can be put into matrix form

$$\mathcal{M}\frac{\partial^2 \mathbf{r}}{\partial t^2} = \mathcal{K}\frac{\partial^2 \mathbf{r}}{\partial x^2} \tag{8}$$

where

$$\frac{\partial^2 \mathbf{r}}{\partial t^2} = \left\{ \frac{\partial^2 \delta}{\partial t^2}\ \frac{\partial^2 y}{\partial t^2}\ \frac{\partial^2 z}{\partial t^2} \right\}^T$$

$$\frac{\partial^2 \mathbf{r}}{\partial x^2} = \left\{ \frac{\partial^2 \delta}{\partial x^2}\ \frac{\partial^2 y}{\partial x^2}\ \frac{\partial^2 z}{\partial x^2} \right\}^T$$

$$\mathcal{K} = E_A \begin{bmatrix} 1 & \frac{\partial y}{\partial x} & \frac{\partial z}{\partial x} \\ \frac{\partial y}{\partial x} & \frac{T}{E_A} + \left(\frac{\partial y}{\partial x}\right)^2 & \frac{\partial y}{\partial x}\frac{\partial z}{\partial x} \\ \frac{\partial z}{\partial x} & \frac{\partial y}{\partial x}\frac{\partial z}{\partial x} & \frac{T}{E_A} + \left(\frac{\partial z}{\partial x}\right)^2 \end{bmatrix}$$

and $\mathcal{M} = m\mathbf{I}_m$, $\mathbf{I}_m$ is the $3 \times 3$ identity matrix. The initial conditions for (8) are given by

$$\mathbf{r}(x, 0) = \mathbf{b}_1(x)$$
$$\partial \mathbf{r}(x, 0)/\partial t = \mathbf{b}_2(x)$$

where $\mathbf{b}_1(x)$ and $\mathbf{b}_2(x)$ are assumed to be smooth functions. The boundary conditions are given by

$$\mathbf{r}(0, t) = \mathbf{0}$$
$$T(L)\,\partial \mathbf{r}(L, t)/\partial x = \mathbf{u}(t)$$

where the function $\mathbf{u}(t)$ is the boundary control input.

### 2.3 FEM Model

In this section, the cable/string model presented in the form of PDEs in (8) will be approximated by a FEM model to obtain a set of ordinary differential equations. Without loss of generality and for the sake of simplicity in the derivations and analysis which will follow later in this paper, the motion of the cable is confined in $xy$–plane. Ignoring $\delta$ and $z$ components in (8) and choosing the parameters $m, T_0$ and $E_A$ as unity renders the following nonlinear scalar equation of motion for the cable/string (Shahruz and Narasimha, 1997), (Qu, 2000)

$$\frac{\partial^2 y}{\partial t^2} = \left[1 + \frac{3}{2}\left(\frac{\partial y}{\partial x}\right)^2\right]\frac{\partial^2 y}{\partial x^2} \tag{9}$$

The initial conditions for (9) are given by

$$y(x, 0) = b_1(x)$$
$$\partial y(x, 0)/\partial t = b_2(x)$$

where $b_1(x)$ and $b_2(x)$ are assumed to be smooth functions of $x$. The boundary conditions are given by

$$y(0, t) = 0$$
$$T(L, t)\,\partial y(L, t)/\partial x = u(t)$$

where $u(t)$ is the boundary control input. To obtain a set of ordinary differential equations, Galerkin's method is applied to (9). The cable is divided into $n$ elements, where nodal points are enumerated from 0 to $n$. Let $h = L/n$ be the length of each cable segment. For $x \in [0, L]$, Galerkin's method yields the following discretized FEM model

$$\mathbf{M}\ddot{\mathbf{y}} + \boldsymbol{\kappa}(\mathbf{y}) = \mathbf{u}(t) \tag{10}$$

where

$$\mathbf{y} = \begin{Bmatrix} y_1 \\ y_2 \\ \vdots \\ y_n \end{Bmatrix}, \quad \boldsymbol{\kappa}(\mathbf{y}) = \begin{Bmatrix} k_1(\mathbf{y}) \\ k_2(\mathbf{y}) \\ \vdots \\ k_n(\mathbf{y}) \end{Bmatrix}, \quad \mathbf{u}(t) = \begin{Bmatrix} 0 \\ 0 \\ \vdots \\ u(t) \end{Bmatrix}$$

$$\mathbf{M} = \frac{h}{6}\begin{bmatrix} 4 & 1 & 0 & & \\ 1 & 4 & \ddots & \ddots & \\ 0 & \ddots & \ddots & & 0 \\ & & \ddots & 4 & 1 \\ & & 0 & 1 & 2 \end{bmatrix}$$

and

$$\begin{aligned} k_1(\mathbf{y}) = & \frac{1}{h}(2y_1 - y_2) \\ & + \frac{1}{2h^3}y_1^3 + \frac{1}{2h^3}(y_2 - y_1)^2 y_1 \\ & - \frac{1}{2h^3}(y_2 - y_1)^2 y_2 \end{aligned} \tag{11}$$

$$\vdots$$

$$\begin{aligned} k_j(\mathbf{y}) = & \frac{1}{h}(-y_{j-1} + 2y_j - y_{j+1}) \\ & - \frac{1}{2h^3}(y_j - y_{j-1})^2 y_{j-1} \\ & + \frac{1}{2h^3}\left[(y_j - y_{j-1})^2 y_j + (y_{j+1} - y_j)^2 y_j\right] \\ & - \frac{1}{2h^3}(y_{j+1} - y_j)^2 y_{j+1} \end{aligned} \tag{12}$$

$$\vdots$$

$$\begin{aligned} k_n(\mathbf{y}) = & \frac{1}{h}(y_n - y_{n-1}) \\ & - \frac{1}{2h^3}(y_n - y_{n-1})^2 y_{n-1} \\ & + \frac{1}{2h^3}(y_n - y_{n-1})^2 y_n \end{aligned} \tag{13}$$

for $j = 2, \ldots, n-1$.

## 3. OBSERVER DESIGN

In this section, a nonlinear observer will be designed for the system given in (10) using Contraction theory (Jouffroy *et al.*, 2004), (Lohmiller and Slotine, 1998).

### 3.1 Observer dynamics

The observer structure copies the second-order nonlinear model dynamics in (10) with a correction term, and is given by

$$\mathbf{M}\ddot{\hat{\mathbf{y}}} + \boldsymbol{\kappa}(\hat{\mathbf{y}}) = \mathbf{u}(t) + \mathbf{H}\left\{\dot{\mathbf{y}} - \dot{\hat{\mathbf{y}}}\right\} \tag{14}$$

Here, $\mathbf{H}$ is the observer gain matrix, having the form

$$\mathbf{H} = \operatorname{diag}(0, \ldots, 0, K_v), \qquad K_v > 0 \tag{15}$$

where we assume that the only measurement available is the velocity $\dot{y}_n(L)$ at the boundary, $x = L$. Computing the virtual dynamics of (14) gives

$$\mathbf{M}\delta\ddot{\hat{\mathbf{y}}} + \frac{\partial \kappa}{\partial \hat{\mathbf{y}}}(\hat{\mathbf{y}})\,\delta\hat{\mathbf{y}} = -\mathbf{H}\delta\dot{\hat{\mathbf{y}}} \qquad (16)$$

where the Jacobian matrix is given by

$$\frac{\partial \kappa}{\partial \hat{\mathbf{y}}}(\hat{\mathbf{y}}) = \begin{bmatrix} \kappa_{11} & \kappa_{12} & 0 & & \\ \kappa_{21} & \kappa_{22} & \kappa_{23} & \ddots & \\ 0 & \kappa_{32} & \ddots & \ddots & 0 \\ & \ddots & \ddots & \kappa_{n-1,n-1} & \kappa_{n-1,n} \\ & & 0 & \kappa_{n,n-1} & \kappa_{n,n} \end{bmatrix}$$

with the entries

$$\kappa_{11} = \frac{2}{h} + \frac{3}{2h^3}\left[\hat{y}_1^2 + (\hat{y}_2 - \hat{y}_1)^2\right]$$
$$\kappa_{12} = -\frac{1}{h} - \frac{3}{2h^3}(\hat{y}_2 - \hat{y}_1)^2$$
$$\vdots$$
$$\kappa_{j(j-1)} = -\frac{1}{h} - \frac{3}{2h^3}(\hat{y}_j - \hat{y}_{j-1})^2$$
$$\kappa_{jj} = \frac{2}{h} + \frac{3}{2h^3}\left[(\hat{y}_j - \hat{y}_{j-1})^2 + (\hat{y}_j - \hat{y}_{j+1})^2\right]$$
$$\kappa_{j(j+1)} = -\frac{1}{h} - \frac{3}{2h^3}(\hat{y}_j - \hat{y}_{j+1})^2$$
$$\vdots$$
$$\kappa_{n(n-1)} = -\frac{1}{h} - \frac{3}{2h^3}(\hat{y}_n - \hat{y}_{n-1})^2$$
$$\kappa_{nn} = \frac{1}{h} + \frac{3}{2h^3}(\hat{y}_n - \hat{y}_{n-1})^2$$

for $j = 2, \ldots, n-1$. Note that each $\kappa$ term is a quadratic scalar function, rendering off-diagonal terms negative definite and diagonal terms positive definite for every $\hat{\mathbf{y}}$. Let

$$\mathbf{K} = \frac{\partial \kappa}{\partial \hat{\mathbf{y}}}(\hat{\mathbf{y}})$$

where $\mathbf{K}$ is the $n \times n$ matrix. Hence, the virtual dynamics in (16) can be rewritten in first order form

$$\frac{d}{dt}\begin{Bmatrix} \delta\hat{\mathbf{y}} \\ \delta\dot{\hat{\mathbf{y}}} \end{Bmatrix} = \mathbf{A}\begin{Bmatrix} \delta\hat{\mathbf{y}} \\ \delta\dot{\hat{\mathbf{y}}} \end{Bmatrix} \qquad (17)$$

where

$$\mathbf{A} = \begin{bmatrix} \mathbf{0} & \mathbf{I} \\ -\mathbf{E} & -\mathbf{F} \end{bmatrix} \qquad (18)$$

$$\mathbf{E} = \mathbf{M}^{-1}\mathbf{K}, \quad \mathbf{F} = \mathbf{M}^{-1}\mathbf{H}$$

and $\mathbf{0}$ is the $n \times n$ zero matrix and $\mathbf{I}$ is the $n \times n$ identity matrix.

### *3.2 Stability properties of the observer*

To establish contraction, it can be shown that all eigenvalues of the system matrix $\mathbf{A}$ satisfy $\operatorname{Re}\lambda_i < 0$. This is established as follows: Consider (18) and the eigenvalue problem

$$\mathbf{A}\begin{Bmatrix} \mathbf{x}_1 \\ \mathbf{x}_2 \end{Bmatrix} = \lambda\begin{Bmatrix} \mathbf{x}_1 \\ \mathbf{x}_2 \end{Bmatrix} \Leftrightarrow \begin{cases} \lambda\mathbf{x}_1 = \mathbf{x}_2 \\ \lambda\mathbf{x}_2 = -\mathbf{E}\mathbf{x}_1 - \mathbf{F}\mathbf{x}_2 \end{cases} \qquad (19)$$

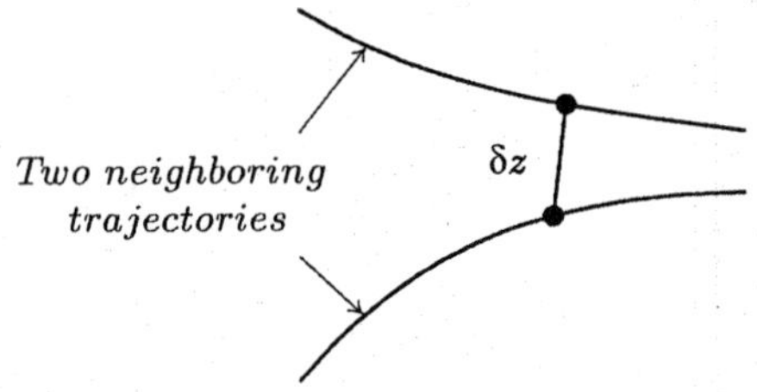


Fig. 3. The virtual displacement $\delta\mathbf{z}$.

1) Assume that $\lambda \geq 0$ and real. From (19) we have

$$\lambda\mathbf{x}_1 = \mathbf{x}_2$$
$$-\lambda^2\mathbf{x}_1 = (\mathbf{E} + \lambda\mathbf{F})\,\mathbf{x}_1$$

where $\mathbf{E}$ is positive definite and $\mathbf{F}$ is positive s definite. But since $(\mathbf{E} + \lambda\mathbf{F})$ is positive definite assumption is false by contradiction.

2) Assume that $\lambda$ is not real. From (20) we can wr

$$\lambda^2\mathbf{x}_1^T\mathbf{x}_1 + \lambda\mathbf{x}_1^T\mathbf{F}\mathbf{x}_1 + \mathbf{x}_1^T\mathbf{E}\mathbf{x}_1 = 0$$

Selecting $|\mathbf{x}_1| = 1$ gives

$$\lambda^2 + \lambda\mathbf{x}_1^T\mathbf{F}\mathbf{x}_1 + \mathbf{x}_1^T\mathbf{E}\mathbf{x}_1 = 0$$

The characteristic equation in (21) has the roots

$$\lambda = \frac{1}{2}\left\{-\mathbf{x}_1^T\mathbf{F}\mathbf{x}_1 \pm \left[(\mathbf{x}_1^T\mathbf{F}\mathbf{x}_1)^2 - 4\mathbf{x}_1^T\mathbf{E}\mathbf{x}_1\right]^{1}\right.$$

which implies that

$$\operatorname{Re}\lambda = -\frac{1}{2}\mathbf{x}_1^T\mathbf{F}\mathbf{x}_1 \leq 0$$

as $\lambda$ is not real by assumption. If $\operatorname{Re}\lambda = 0$ $\mathbf{F}\mathbf{x}_1 = 0$ and

$$\mathbf{F}\mathbf{x}_1 = 0 \quad \Leftrightarrow \quad \mathbf{x}_1 = \begin{Bmatrix} x_1 \\ \vdots \\ x_{n-1} \\ 0 \end{Bmatrix}$$

where

$$\mathbf{F} = \begin{bmatrix} 0 & \cdots & 0 & * \\ & \ddots & \vdots & \vdots \\ & & 0 & * \\ & & & K_v \end{bmatrix}$$

but from (20) we can conclude that such a vectc can not be an eigenvector of the positive definite trix $(\mathbf{E} + \lambda\mathbf{F})$. Hence, all eigenvalues of $\mathbf{A}$ sa $\operatorname{Re}\lambda_i < 0$.

Since $\mathbf{A}$ is a stability matrix with $\operatorname{Re}\lambda_i < 0$, there e a uniformly positive definite metric $\mathbf{\Gamma}$ under which observer is contracting, i.e. that the virtual displacem

$$\delta\mathbf{z} = \begin{Bmatrix} \delta\hat{\mathbf{y}} \\ \delta\dot{\hat{\mathbf{y}}} \end{Bmatrix}$$

converges exponentially to zero as shown in Fig. 3. thermore, there exist two strictly positive numbe and $\gamma$ such that for any two trajectories starting i

$\mathbf{z}(t = t_0)$ and $\hat{\mathbf{z}}(t = t_0)$, the observer error dynamics is exponentially stable in the sense that

$$\|\mathbf{z}(t) - \hat{\mathbf{z}}(t)\| \leq \alpha \|\mathbf{z}(t_0) - \hat{\mathbf{z}}(t_0)\| e^{-\gamma(t-t_0)}$$

for $t \geq t_0$ everywhere in the region in which (17) holds (Jouffroy *et al.*, 2004).

## 4. NUMERICAL RESULTS AND SIMULATION

The observer proposed in (14) is solved numerically. The results of the simulation are shown in Fig. 4 – 11. In the simulations, a cable of unit length is discretized with $n = 10$, where $n = 10$ is the node at the boundary and $n = 0$ is the stationary end point. The observer gain is chosen as $K_v = 1$. During the simulations, two different motions of the boundary node are considered. In both cases, the observer is switched on at $t = 2$ sec. First, the boundary node is assigned to follow a step reference. Fig. 4 shows the observer error at the nodes $n = 1, 4, 7$ and 10 when the boundary node follows a step reference whereas Fig. 5 – 7 are the snapshots taken during the simulation which show the current and estimated configurations at times $t = 2.2, 3.7$ and 5.2 sec. Similarly, the boundary node is then assigned to follow a time-varying reference. Fig. 8 shows the observer error at the nodes $n = 1, 4, 7$ and 10 when the boundary node follows a varying reference whereas Figures 9 – 11 show the current and estimated configurations at times $t = 2.7, 3.4$ and 6.8 sec. As seen in Fig. 4 and Fig. 8, the observer error starts decreasing rapidly and the estimated configuration converges to the actual configuration which demonstrates the contracting property of the observer dynamics. The results from the simulations are thus in agreement with the theory presented in this paper.

## 5. CONCLUSIONS

Contraction theory is a recently developed nonlinear analysis tool which may be useful for solving a variety of nonlinear control problems. In this paper, using Contraction theory, a nonlinear observer has been designed for a general nonlinear cable/string FEM model. The cable model has been presented in the form of PDEs. Galerkin's method has been then applied to obtain a set of ordinary differential equations such that the cable model has been approximated by a FEM model. Based on the FEM model, a nonlinear observer was designed to estimate the cable configuration. It is shown that the estimated configuration converges exponentially to the actual configuration. Numerical results and simulations have been shown to be in agreement with the theoretical results.

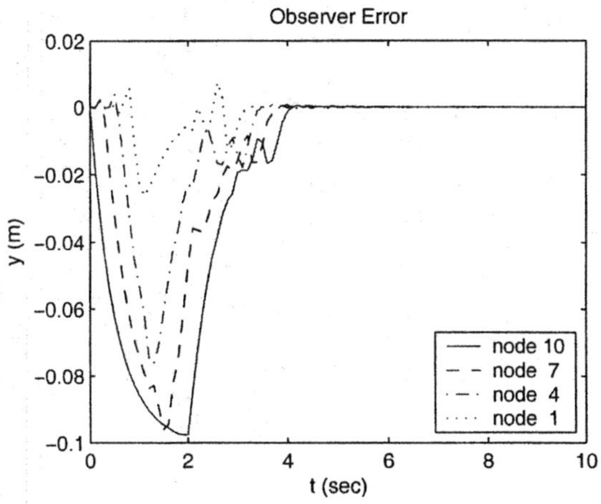


Fig. 4. The boundary node follows a step reference. The observer was switched on at time $t = 2$ sec.

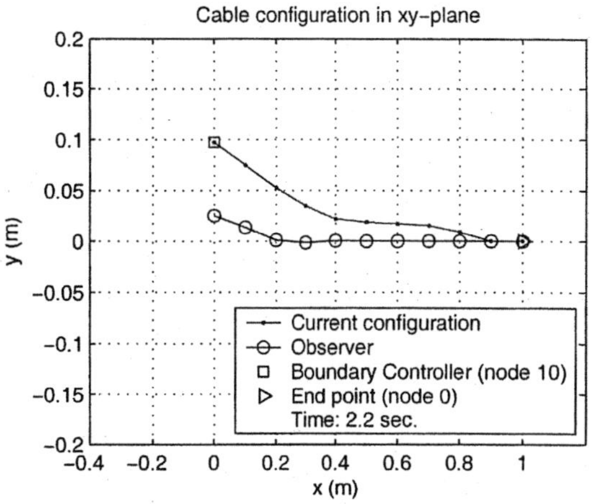


Fig. 5. Configurations at $t = 2.2$ sec.

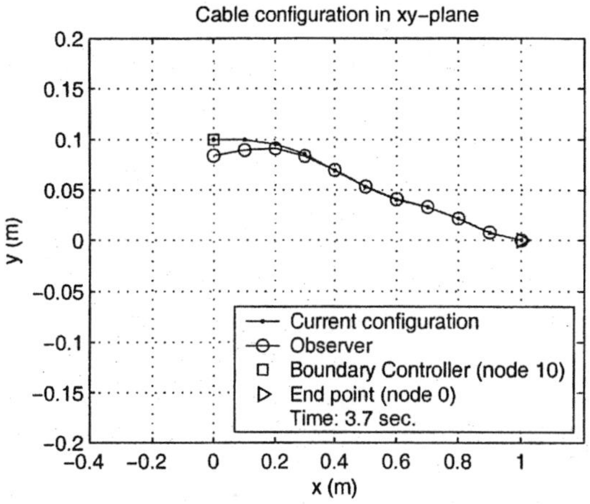


Fig. 6. Configurations at $t = 3.7$ sec.

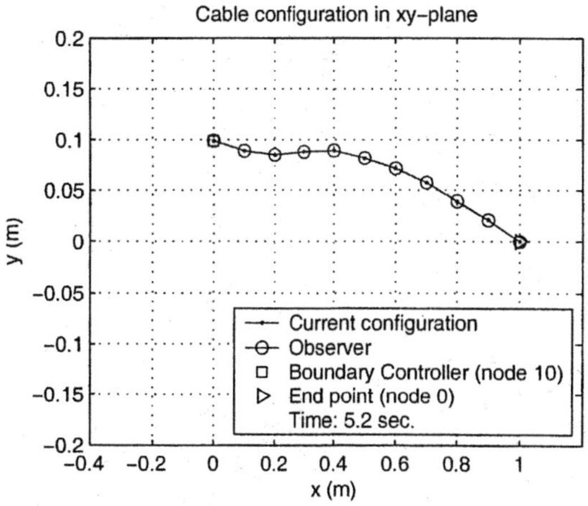


Fig. 7. Configurations at $t = 5.2$ sec.

## 6. ACKNOWLEDGEMENT

This work was funded by the Norwegian Research Council under the Strategic University Program in Marine Cybernetics.

## REFERENCES

Baicu, C. F., C. D. Rahn and D. M. Dawson (1999). Exponentially stabilizing boundary control of string-mass systems. *Journal of Vibration and Control* **5**, 491–502.

Canbolat, H., D. Dawson, S. P. Nagarkatti and B. Costic (1998). Boundary control for a general class of string models. Proceedings of the American Control Conference. Pennsylvania.

Demetriou, Michael A. (2001). Natural observers for second order lumped and distributed parameter systems using parameter-dependent lyapunov functions. Proceedings of the American Control Conference. Arlington, VA.

Jouffroy, J., T. I. Fossen and J.-J. E. Slotine (2004). Methodological remarks on contraction theory, exponentially stability and lyapunov functions. *Submitted to Automatica.*

Kristiansen, D. (2000). Modeling of Cylinder Gyroscopes and Observer Design for Nonlinear Oscillations. PhD thesis. Dept. of Engineering Cybernetics, Norwegian Univ. of Sci. and Tech.

Lohmiller, Winfried and J.-J. E. Slotine (1998). On contraction analysis for non-linear systems. *Automatica* **34**(6), 683–696.

Morgül, Ö. (1994). A dynamic control law for the wave equation. *Automatica* **30**, 1785–1792.

Qu, Z. (2000). Robust and adaptive boundary control of a strecthed string. Proceedings of the American Control Conference. Illinois.

Shahruz, S. M. and C. A. Narasimha (1997). Suppression of vibration in strecthed strings by the boundary control. *Journal of Sound and Vibration* **204**, 835–840.

Zienkiewicz, O. C. and R. L. Taylor (2000). *The Finite Element Method.* 5th ed.. Butterworth-Heinemann. Oxford, UK.

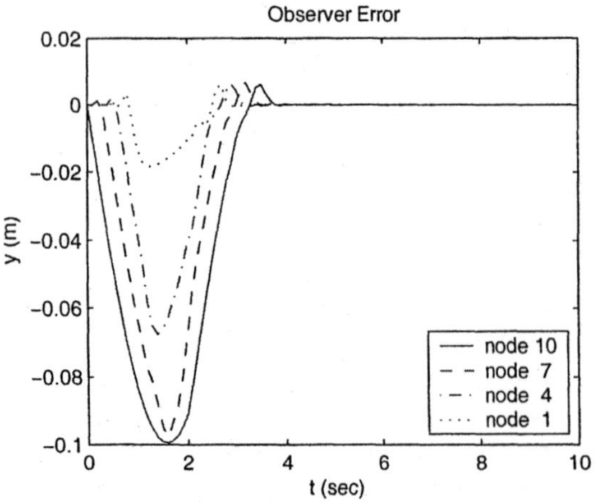


Fig. 8. The boundary node follows a varying reference. The observer was switched on at $t = 2$ sec .

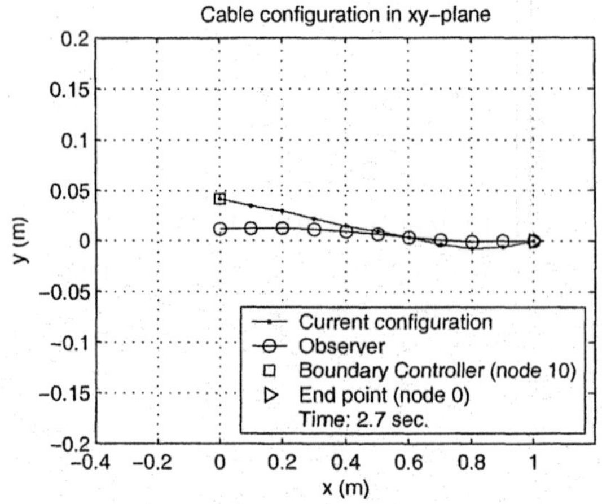


Fig. 9. Configurations at $t = 2.7$ sec.

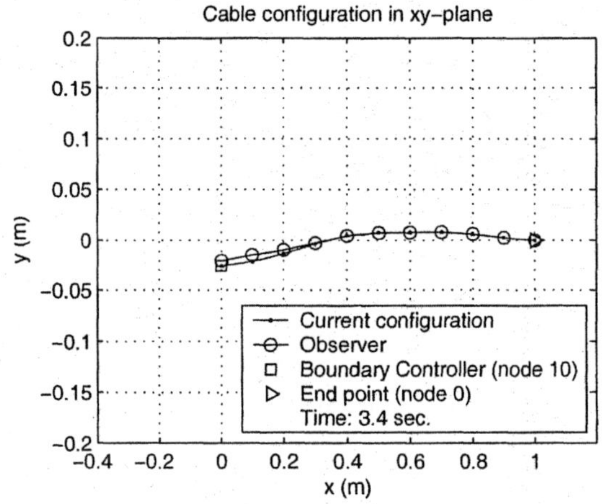


Fig. 10. Configurations at $t = 3.4$ sec.

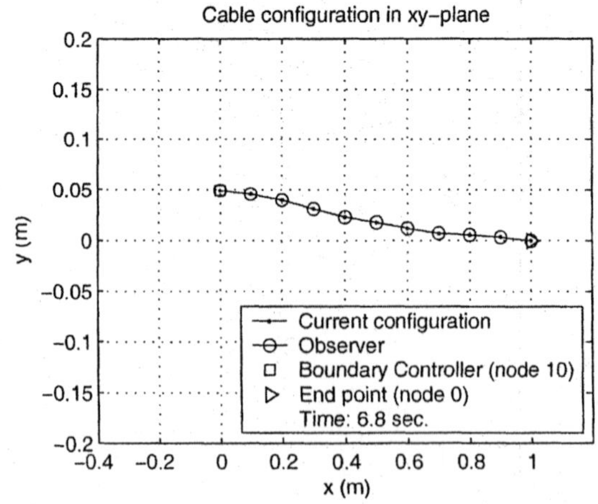


Fig. 11. Configurations at $t = 6.8$ sec.

ELSEVIER
IFAC
PUBLICATIONS
www.elsevier.com/locate/ifac

# OPTIMISATION OF A WAVE ENERGY CONVERTER

**John Ringwood** [1] **Shane Butler**

*Dept. of Electronic Eng., NUI Maynooth, Co. Kildare, Ireland*

Abstract: This paper examines a simple model of an idealised point-absorber wave energy device. The objective of the research is to examine how the device can be optimised in order to extract maximum energy from incident waves. This applies to both the broad design parameters of the buoy and also to the synthesis of the damping function, in which the wave energy is converted. The issue of the adjustment of the phase of the velocity evolution with respect to the incident force (known as phase control) is dealt with in detail, with the intention of optimising the damping force over the wave cycle. The research ultimately attempts to parameterise the optimal damping force in terms of incident wave frequency and device parameters. *Copyright © 2004 IFAC*

Keywords: Wave energy, heaving buoy, latching, optimisation

## 1. INTRODUCTION

Many researchers and practitioners have consider point absorber devices over the last two decades, including Budal and Falnes (Budal and Falnes, 1975), Wright *et al* (Wright *et al.*, 2003) and Dick (Dick, 2003). Much of the literature has focussed on the issue of device optimisation through both basic device design (shape, mass distribution and buoyancy) and also the power take-off system used to extract the wave energy. The latter is the principal focus of interest in the current study and a number of researchers have addressed this point, with a particularly extensive study by Falnes (Falnes, 2002). Falnes makes a number of important conclusions:

(1) Energy conversion is maximised if the device velocity is in phase with the excitation force, and
(2) The velocity amplitude, $|u|$ should equal $\frac{F_e}{2R_i}$, where $F_e$ is the excitation force and $R_i$ is a device resistance.

Condition 2 above has a number of difficulties:

- It requires that the wave excitation force, $F_e$, be measured,
- $R_i$ turns out to be a non-causal function, and
- The power take-off (PTO) machinery must *supply* energy during part of the wave cycle in order to achieve the optimum $|u|$.

The need to supply energy may be considered strange, but this can be likened to a person on a swing, who uses body and leg motion to increase the amplitude of the swinging, by appropriate timing of the effort. Nevertheless, the need to supply energy requires a very complex PTO system and to the best of this author's knowledge, such a system has not yet been realised. However, Condition 1, representing a *passive* requirement, has received considerable attention and several researchers have addressed the problem. In particular, a method used to delay the velocity evolution,

[1] This author would like to acknowledge the fruitful discussions with Prof. Bill Leithead of the Hamilton Institute at NUI Maynooth

called *latching*, has been employed by a variety of researchers (Falnes and Lillebekken, 2003; Babarit *et al.*, 2003; Korde, 2002; Wright *et al.*, 2003; Greenhow and White, 1997).

Note that Conditions 1 and 2 above can be alternatively formulated in terms of *complex conjugate* (Nebel, 1992) (or reactive) control, which considers the complex impedance of the device.

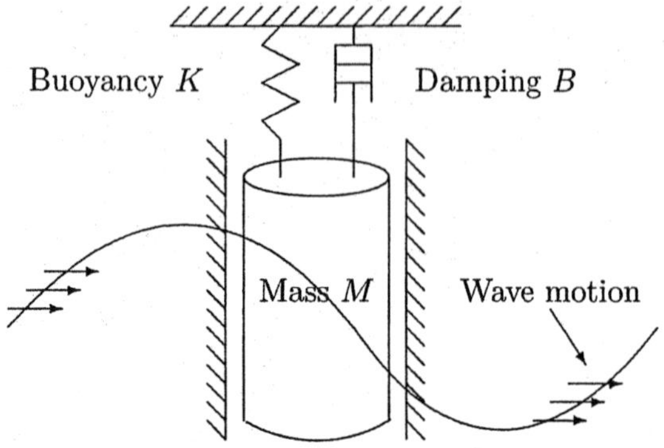


Fig. 1. Conceptual point absorber

A conceptual diagram of the device under examination is depicted in Fig.1. It consists of a cylindrical device which is constrained to move in the vertical direction (heave motion) only. The damping is notional and could be provided either internally in the device (as in (Korde, 2002)) or external to it (as in (Wright *et al.*, 2003)). One other assumption in the concept device is that the diameter of the body is very small compared with the sea wavelength, so that the sea displacement is co-incident with the wave's heave force on the device.

## 2. PRELIMINARY CALCULATIONS

### *2.1 System Model*

The dynamics of the devices described in Section 1 can be described, under some mild assumptions, by the differential equation:

$$M\ddot{x}(t) + B\dot{x}(t) + Kx(t) = F(t) \qquad (1)$$

where:

$x$ represents the displacement of the body from rest,
$M$ represents the mass (inertia) of the body,
$B$ represents the viscous friction, characteristic of hydraulic resistance used in PTO devices,
$K$ represents the buoyancy/gravity forces experienced by the body, and
$F(t)$ represents the heave force experienced by the wave-energy device.

There are a number of important assumptions inherent in the description in (1) above. In particular:

- Added mass, which affects *both* inertia and damping terms and is related to the inertia of the water surrounding the device, is largely ignored in the damping term in (1). However. it can be contained (notionally, at least) in $M$.
- Radiation damping, resulting from the energy carried away by surface waves generated by the device, is largely ignored. Some component of this can (notionally) be included as a linear term within $B$, though radiation damping is normally a nonlinear function of velocity. The linear damping component can represent skin friction, in particular (Fossen, 2002).
- The buoyancy/gravity restoring force is considered to be proportional to displacement from rest. In general, this is a small-signal approximation, since the restoring force is normally a nonlinear function of displacement.

The heave force due to incident waves will, in the first incidence, be assumed to be monochromatic, of the form:

$$F(t) = Asin(\omega_w t) \qquad (2)$$

The choice of monochromatic waves and the model in (1), is chosen for simplicity, since introduction of latching is a highly nonlinear intervention and quickly adds complexity. Equation (1) can also be easily recast in transfer function form as:

$$\frac{X(s)}{F(s)} = G(s) = \frac{1}{Ms^2 + Bs + K} \qquad (3)$$

or, in terms of transient response parameters, as:

$$G(s) = \frac{1}{K}\frac{\omega_n^2}{s^2 + 2\zeta\omega_n s + \omega_n^2} \qquad (4)$$

with

$$\omega_n = \sqrt{\frac{K}{M}} \quad , \quad \zeta = \frac{B}{2}\sqrt{\frac{1}{MK}}$$

Equation (1) can also be conveniently expressed in state-space (companion) form, as:

$$A = \begin{bmatrix} 0 & 1 \\ -\frac{K}{M} & -\frac{B}{M} \end{bmatrix} \quad B = \begin{bmatrix} 0 \\ \frac{1}{M} \end{bmatrix} \quad D = [0] \qquad (5)$$

with a state vector of:

$$X(t) = \begin{bmatrix} x(t) \\ \dot{x}(t) \end{bmatrix} \qquad (6)$$

*2.2 Power and Energy*

For a mechanical system, the power ($P$) is the product of force and velocity. In wave energy systems, the PTO device is normally represented by the damper, giving the power developed in the damper as:

$$P_d = \text{force x velocity} = B\dot{x}\,\dot{x} = B\dot{x}^2 \qquad (7)$$

The energy developed by the action on the damper over a period of time $t_1$ is:

$$E_d(t_1) = \int_0^{t_1} P_d dt = \int_0^{t_1} B\dot{x}^2 dt \qquad (8)$$

Maximum power is transferred to the damper when Equation (8) is maximised over a period of the wave force. This results in the condition:

$$\omega_n = \sqrt{\frac{K}{M}} = \omega_w \qquad (9)$$

Under this maximum condition ($\omega_n = \omega_w$), the velocity profile of the device is in phase with the wave force, consistent with Condition 1 in Section 1. Note that some adjustment of the device to achieve (9) may be possible through the use of appropriate quantity and position of water ballast.

The phase of the velocity profile (relative to the force profile) is evaluated as:

$$\angle\frac{G(j\omega)}{s} = \frac{\pi}{2} - tan^{-1}\left(\frac{\omega B}{K - M\omega^2}\right) \qquad (10)$$

Clearly, if $K = M\omega^2$, then velocity is in phase with force or, indeed, if $B = \infty$. One further consideration here is that the force *lags* the velocity when:

$$\omega_w < \omega_n \quad \text{or} \quad M\omega_w < K \qquad (11)$$

This places an upper bound on the device mass relative to the buoyancy. In addition, if a device is designed to be optimal for a given wave frequency, $\omega_w^*$, the wave force will only lag the velocity when the wave frequency, $\omega_w$ *decreases* below this value. This has important implications for the possibility of using latching to 'delay' the velocity profile in order to get it in phase with the force profile.

## 3. LATCHING BASICS

Latching can be achieved by means of a mechanical brake (applied at the appropriate latching points) or open close valves on the hydraulic lines of the PTO system. In simulation, latching can be achieved in two ways:

- The latching point is determined as the point where the device velocity goes to zero. At this point, the first and subsequent derivatives of displacement go to zero in (1) and if the wave force is replaced by a force equal and opposite to the restoring force ($Kx$), then 'latching' is achieved. This mechanism is illustrated in Fig.3.
- The closing of valve) can easily be implemented in simulation by setting the damping coefficient, $B$, to $\infty$.

Taking, for example, the case where $0.5\omega_n = \omega_w$ (with $\omega_w = 0.5$, corresponding to a wave period of just over 12 seconds), the response with latching (with $M = K = B = 1$) is shown in Fig.3.

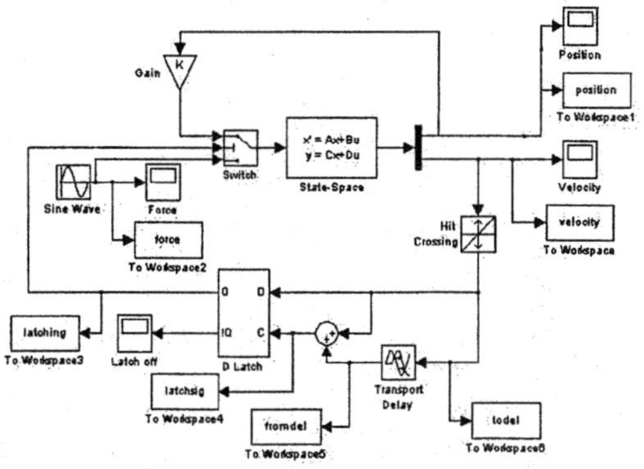


Fig. 2. Simulation configuration

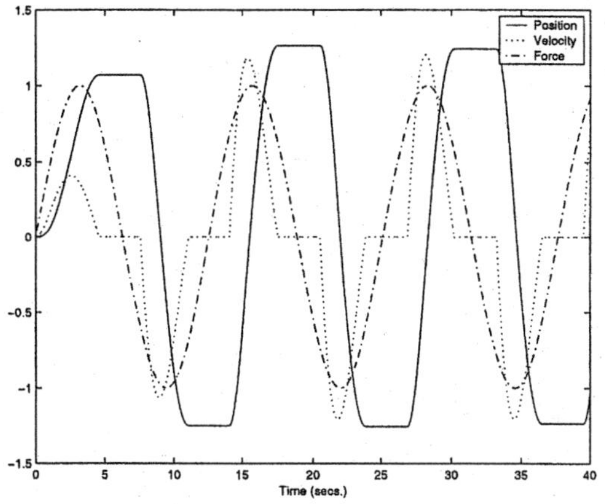


Fig. 3. Variations in $E_d$ and $T_L^{opt}$ with $B$ and $\omega_w$

A number of features can be observed from Fig.3:

- The amplitude of the position response is greater then for the unlatched case,
- The velocity response, though highly nonlinear, is now in phase with the force profile, and
- The overall energy captured from the system, via the damper, has increased from 1.94 Ws (unlatched) to 4.62 (latched) Ws per period of the incident wave.

Interestingly, the energy figure in the latching case is even greater than that achieved when $\omega_n =$

$\omega_w = 1$ (at 3.14 Ws), but this is accounted for by the fact that the wave energy is proportional to wave period (Falnes, 2002) as:

$$J = \frac{\rho g^2}{32\pi} TH^2 \quad (12)$$

where $T$ is the wave period in seconds, $H$ is the wave height (trough to crest) in meters, and $\rho$ is the water density (= 1020 kg/m$^3$ for sea water.

## 4. SOLUTION TO LATCHING SYSTEM

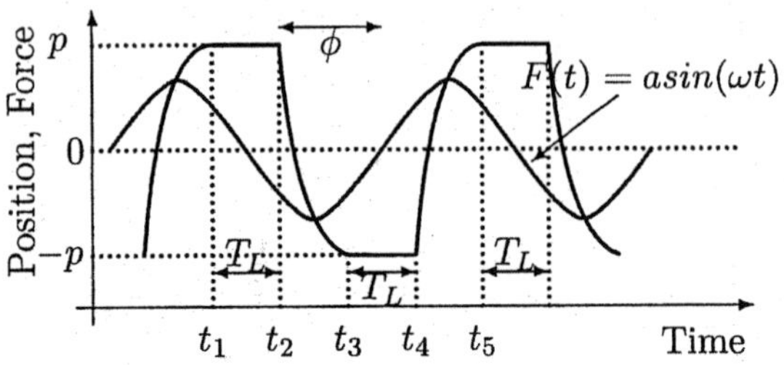


Fig. 4. Latching calculations

A solution to the latched system can be had by considering Fig.4. One period, or cycle, of the stimulus and response is given by:

$$t_5 - t_1 = \frac{2\pi}{\omega} \quad (13)$$

Given that each of the latching periods occurs consistently for $T_L$ seconds, this gives the dynamic response period as:

$$t_3 - t_2 = t_5 - t_4 = \frac{\pi}{\omega} - T_L \quad (14)$$

For the (linear) system as given, the solution over the periods $t_2 \to t_3$ and $t_4 \to t_5$ is equal and opposite (assuming the transient response has died down). Therefore, the solution need only be evaluated over a half period. The solution for $t_1 \to t_2$ is:

$$X(t) = \begin{bmatrix} p \\ 0 \end{bmatrix} \quad (15)$$

The solution for the period $t_2 \to t_3$ may be determined from the solution to (5), assuming a reference point of $t_2 = 0$, as:

$$X(t) = e^{At} \begin{bmatrix} p \\ 0 \end{bmatrix} + \int_0^t e^{A(t-\tau)} Ba \sin(\omega\tau + \phi) d\tau \quad (16)$$

Though equations (15) and (16) can be used to give an expression for the state (position and velocity) over the entire cycle, there are two unknowns:

$\phi$ , the phase offset between the force, $F(t)$, and the position response, and
$p$ , the height of the position response

However, since the response has zero mean, and the transient response has died down, we know that:

$$X(t_3) = \begin{bmatrix} -p \\ 0 \end{bmatrix} \quad (17)$$

Inserting this in (16) gives:

$$\begin{bmatrix} -p \\ 0 \end{bmatrix} = e^{A(\frac{\pi}{\omega} - T_L)} \begin{bmatrix} p \\ 0 \end{bmatrix} + a \int_0^{\frac{\pi}{\omega} - T_L} e^{A(\frac{\pi}{\omega} - T_L - \tau)} B \sin(\omega\tau + \phi) d\tau$$

Equation (18) represents 2 equations in 2 unknowns and can, in concept at least, be solved for $\phi$ and $p$. This type of solution procedure is followed in (Babarit *et al.*, 2003), using a transfer function system description.

## 5. LATCHING RESULTS

Figs.5 and 6 summarise the variations in the converted energy and optimal latching period (respectively) for variations in $B$ and $\omega_w$. Some comments are noteworthy:

- Converted energy decreases with increasing $\omega_w$ at smaller values of $B$, while it increases with $\omega_w$ at larger values of $B$.
- There is a clear optimal value for $B$, though this does seem to vary a little with $\omega_w$.
- At low $B$ values, the converted energy increases with $\omega_w$, as $\omega_w$ approaches $\omega_n$.
- The optimal latching period, $T_L^{opt}$, goes to zero as $\omega_w \to \omega_n$ (in this case $\omega_n = 1$.
- As stated above, there is little sensitivity of $T_L^{opt}$ to variation in $B$, particularly for the range of $B$ shown.
- There is a clear $\frac{1}{\omega_w}$ relation with $T_L^{opt}$ for all values of $B$. Re-plotting $T_L^{opt}$ against $\frac{1}{\omega_w}$ for (as an example) $B = 0.1$ shows a linear relationship between $T_L^{opt}$ and the wave period (slope 0.5065, intercept -3.2022). As might be expected, $T_L^{opt}$ does not appear as a consistent 'proportion' of the wave period, $T_w$, but rather is an affine function of $T_w$, with an offset of $2\pi$ in the current example ($= \omega_n$).

## 6. THE OPTIMALITY OF LATCHING

This paper focusses on latching as a solution to force the velocity profile to be in phase with

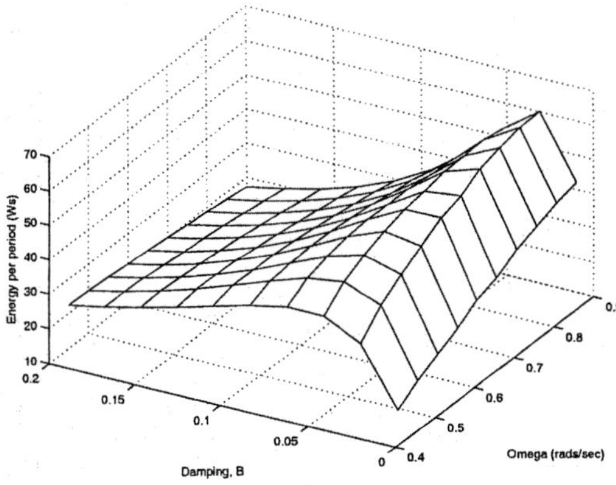


Fig. 5. Variations in $E_d$ with $B$ and $\omega_w$

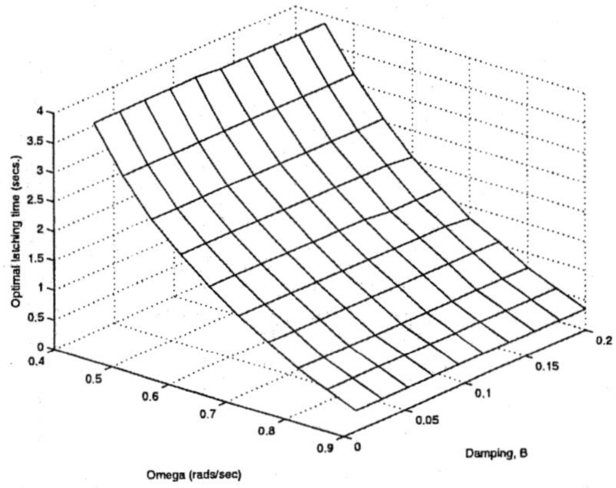


Fig. 6. Variations in $T_L^{opt}$ with $B$ and $\omega_w$

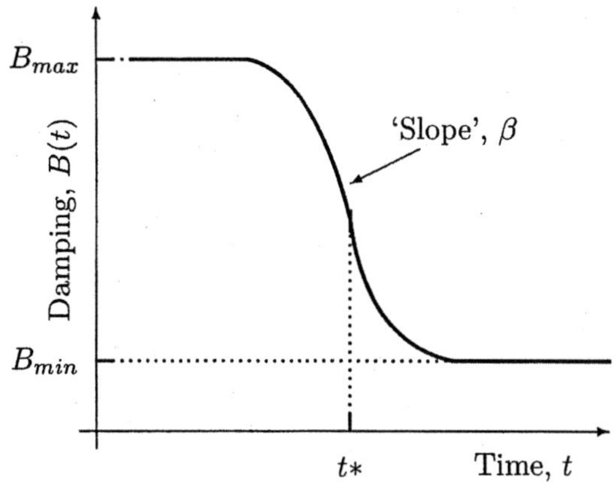


Fig. 7. Sigmoidal parameterisation of damping functions

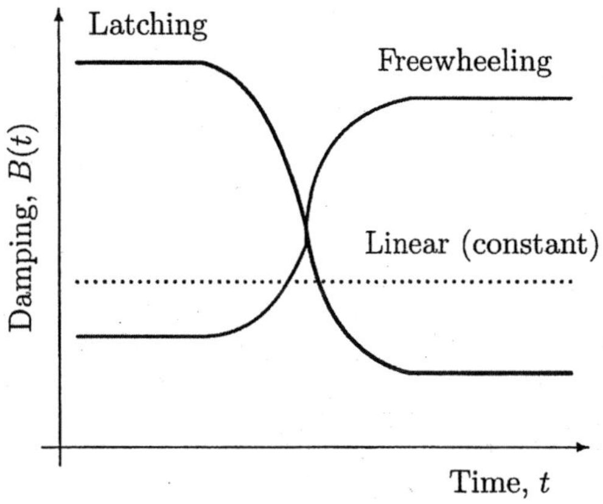


Fig. 8. Possible damping functions

the applied wave force. However, since the wave energy is converted in the damping term (see equation (8)) one can concieve of a multitude of loading possibilities, where the damping term is varied over the wave cycle, or scheduled with device velocity. Indeed, some researchers have looked at the possibility of having a very low damping 'load' at the beginning of the cycle (beginning at a point of zero velocity) and increasing the damping only after a preset velocity is reached. Such a 'freewheeling' strategy is in strong contrast to latching, where the damping is effectively infinite for the latching period i.e. a short period following the point of zero velocity.

To determine, in a limited way, the optimal loading regime, the damping term was parameterised in terms of a general sigmoid function (see Fig.7) as follows:

$$B(t) = \frac{B_{max} - B_{min}}{1 + e^{-\beta(t-t*)}} + B_{min} \qquad (18)$$

This provides for many possible damping functions, including latching, freewheeling and uniform (linear) damping, as illustrated in Fig.8.

The parameters of the sigmoid in equation(18) were adapted, using a genetic algorithm (Goldberg, 1989), in order to maximise the energy function (equation (8), Fig.??) over a wave period, where $\omega_w = 0.5$ (corresponding to a wave period of 12.56 secs.). The default values of $M = K = 1$ were used, as before. A genetic algorithm (GA), with elitism, was employed since the performance surface to be searched is non-convex with respect to the sigmoid parameters. Briefly, the parameters of the GA are given in Table 1.

The ranges allowed for the sigmoid parameters are given in Table 2.

| Parameter | Value |
|---|---|
| Chromosome coding | Binary |
| Population size | 70 |
| Number of generations | 30 |
| Generation gap | 0.7 |
| Recombination probability | 0.7 |
| Mutation probability (per bit) | 0.035 |
| Selection | Roulette wheel |

Table 1. GA parameters

| Range | $B_{min}$ | $B_{max}$ | $t*$ | $\beta$ |
|---|---|---|---|---|
| Max | 0 | 0 | -20 | 0 |
| Min | 50 | $10^{90}$ | +20 | 3000 |

Table 2. Range for sigmoid parameters

| Parameter | $B_{min}$ | $B_{max}$ | $t*$ | $\beta$ |
|---|---|---|---|---|
| Final value | 0.0546 | $10^{90}$ | 2.7 | -2530 |

Table 3. Final sigmoid parameter values

The final values attained following maximisation of the energy per wave period are as given in Table 3.

These values clearly indicate the optimality of a latching strategy. Firstly, the value for $\beta$ is negative, indicating that a 'high then low' strategy (characteristic of latching) is preferable for $B(t)$. The height of the initial damping is at the upper limit of the allowed range, indicating effectively infinite initial damping (or complete 'latching'). The magnitude of the slope parameter, $\beta$ is very large, indicating an almost instantaneous transition from latching to a subsequent finite value for $B$. The final value for $B$, $B_{min}$, is very close to the optimal damping value indicated in Fig.5, for a wave frequency of $\omega_w = 0.5$. Finally, the time for which $B(t)$ is held high (the latching time) is very close to the optimal value (for $\omega_w = 0.5$) indicated in Fig.6.

## 7. CONCLUSIONS

Optimal extraction of wave energy requires a number of device aspects to be considered. In the first instance, it is required that the resonant frequency of the device be placed in the region of dominant wave frequency. Further to this, adjustments to optimise the power absorbed by the device in the damping element can be made by attempting to get the velocity profile in phase with the incident wave force. Latching provides one mechanism of achieving this and it can be shown that latching provides the optimum adjustment of the device damping over the wave period. Given that latching is employed, some further care needs to be taken in setting the appropriate (unlatched) damping level, since the energy take depends on this damping value, with the optimal damping also dependent on the dominant wave frequency. While the analysis in this paper has assumed a simplified device model, it focusses on the salient issues in wave energy device design and should be extendible to more realistic sea/device models for heaving buoys and could also be extended to more complicated structures, such as the McCabe Wave Pump (MWP) (McCormick *et al.*, 1998).

## REFERENCES


Babarit, A., G. Duclos and A.H. Clement (2003). Comparison of latching control strategies for a heaving wave energy device in a random sea. In: *Proc. 5th European Wave Energy Conference.* Cork.

Budal, K. and J. Falnes (1975). A resonant point absorber of ocean wave power. *Nature* **256**, 478–479.

Dick, W. (2003). Report on eu eureka project e! 2278 wwec. *http://www.eureka.be/ifs/files/ifs/jsp-bin/eureka/ifs/jsps/publicShowcase.jsp?fileToInclude=ProjectProfile.jsp&docid=1483664.*

Falnes, J. (2002). *Ocean Waves and Oscillating Systems.* Cambridge University Press. Cambridge, UK.

Falnes, J. and P.M. Lillebekken (2003). Budal's latching-controlled-buoy type wave-power plant. In: *Proc. 5th European Wave Energy Conference.* Cork.

Fossen, T.I. (2002). *Marine Control Systems: Guidance, Navigation and Control of Ships, Rigs and Underwater Vehicles.* Marine Cybernetics. Norway.

Goldberg, D.E. (1989). *Genetic Algorithms in Search, Optimisation and Machine Learning.* Addison-Wesley.

Greenhow, M. and S.P. White (1997). Optimal heave motion of some axisymmetric wave energy devices in sinusoidal waves. *Applied Ocean Research* **17**, 141–159.

Korde, U.A. (2002). Latching control of deep water wave energy devices using an active reference. *Ocean Engineering* **29**(11), 1343–1355.

McCormick, M.E., J. Murtagh and P. McCabe (1998). Large-scale experimental study of a hinged-barge wave energy conversion system. In: *Proc. 3rd European Wave Energy Conference.* Patras, Greece.

Nebel, P. (1992). Maximising the efficiency of wave-energy plants using complex-conjugate control. *Journal of Systems and Control Engineering* **206**, 225–236.

Wright, A., W.C. Beattie, A. Thompson, S.A. Mavrakos, G. Lemonis, K. Nielsen, B. Holmes and A. Stasinopoulos (2003). Performance considerations in a power take-off unit based on a non-linear load. In: *Proc. 5th European Wave Energy Conference.* Cork.

ELSEVIER
IFAC
PUBLICATIONS
www.elsevier.com/locate/ifac

# GLIDER'S ROLL CONTROL BASED ON BACKSTEPPING

**L. BURLION, T. AHMED-ALI and N. SEUBE**

*Ecole Nationale Supérieure des Ingénieurs*
*des Etudes et Techniques d'Armement*
*29806 BREST Cedex 9 FRANCE*
*Email adresses : Laurent.Burlion@lss.supelec.fr ,*
*ahmedali@ensieta.fr, seube@ensieta.fr*

Abstract: This paper is devoted to controls and observers design based on backstepping and sliding modes, for non linear systems. This method is applied to the dynamics of an underwater glider that has been developed and operated by ENSIETA. The glider dynamics can be derived from flight dynamics equations. In the framework of this paper, we are interested by the lateral equation of motion which describes the lateral motion dynamics. First, we design a roll controller. Then, we consider that we don't know the derivative of the roll required by the controller and we show how to construct a sliding mode observer in presence of the controller. *Copyright © 2004 IFAC*

Keywords: Nonlinear systems, Variable structure control, Autonomous Vehicles, Observers, Robust control

## 1. INTRODUCTION

Underwater gliders use passive propulsion systems, based on the transformation of gravity (or Archimedes) forces to forward thrust by the mean of large lifting surfaces, at the price of a control of buoyancy and pitching angle. It has been shown in (Moitie *et al.*, 2001) that above a certain depth of operation gliders have lower energy requirements than propeller equipped vehicles. The limitations of such vehicle is first that they are constrained to travel along ascending or descending gliding trajectories (called cycles) and second that they operate at low speed. However, for underwater applications that do not require survey at constant depths and for which relative speed is not a limiting factor, gliders seem to be an adequate solution. In particular for oceanography applications (for instance, thermocline investigation with ctd sensors), the use of underwater glider is envisioned in short term. ENSIETA has developed a small size glider, devoted to oceanographic data collection. Its specifications are:

| | |
|---|---|
| Length: 1.4 m | Dry mass: 48 kg |
| Buoyancy control: 1 l | Max depth: 100 m |
| Width: 1.3 m | Diameter: 200 mm |
| Max. speed: 2 knts | Autonomy: 1200 cycles |

The ensieta glider has been first launched at sea on December 2002 for tests purposes. During those tests, depth, relative velocity, pitch, roll, heading, and actuator data were recorded. These data have been used for designing a simplified model for the roll. The tests also showed that it was necessary to control the roll of the glider.

## 2. MODEL DESCRIPTION

The roll angle is controlled by moving the lateral position of the center of gravity of the glider. This

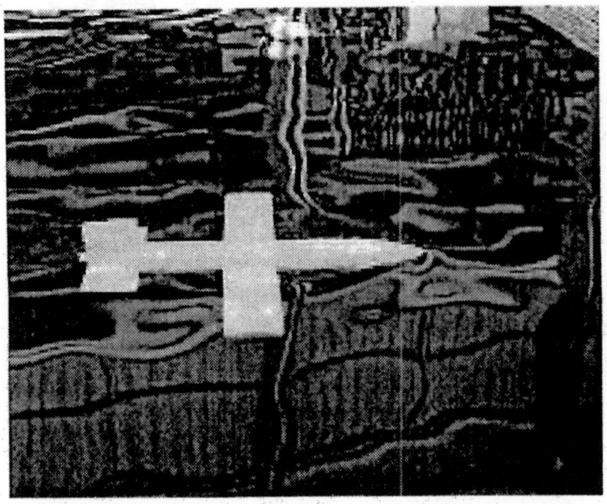

Fig. 1. The underwater glider developed by ENSIETA

is done by turning a cylindrical mass along an axis which does not coincide with the principal axis of the glider : indeed, this rotating axis is shifted from the principal axis with a small distance.

We derived the following dynamics which represents the effect of this "oscillating pendulum" :

$$\begin{cases} \dot{x}_1 = x_2 \\ \dot{x}_2 = -a\sin(x_1+x_3) - b\sin(x_1) - c\mid x_2 \mid x_2 \\ \dot{x}_3 = ku \end{cases}$$

where $a$,$b$,$c$ and $k$ are strictly positive real numbers depending of the inertial characteristics of the glider. $x_1$ denotes the glider roll, $x_2$ its time derivative and $x_3$ the angular position of the cylindrical mass.

## 3. BACKSTEPPING CONTROLLER

In this section, we present a controller which locally stabilizes the roll . Let us consider that $(x_1 + x_3)$ is very small : in this case, we have

$$\sin(x_1+x_3) = x_1 + x_3 + o((x_1+x_3)^2)$$

We note :

$$d = -c\mid x_2 \mid x_2 + sin(x_1+x_3) - (x_1+x_3)$$

We consider d as a bounded disturbance. We thus have to stabilize the following system :

$$\begin{cases} \dot{x}_1 = x_2 \\ \dot{x}_2 = -a(x_1+x_3) - b\sin(x_1) + d \\ \dot{x}_3 = ku \end{cases} \qquad (1)$$

Our initial problem boils down to the robust stabilization of a lower triangular system. Since the disturbance is no matching, we used a Backstepping strategy.

**Theorem 1:** *We assume that there exists $\gamma > 0$ such that $\forall t \geq 0$, $|d(t)| \preceq \gamma$. Then, the following controller $u$ stabilizes the roll to $]-\varepsilon, \varepsilon[$ where $\varepsilon > 0$ is as small as we like*

$$u = \frac{1}{k}[-\tilde{\alpha}(x_3 - v) + aS + \dot{v}]$$
$$= \frac{1}{k}\Bigg[-\tilde{\alpha}(x_3-v) + aS + \left(\frac{\alpha}{\lambda a} - 1 - \frac{b}{a}\cos(x_1)\right)x_2$$
$$- \frac{\lambda\alpha+1}{a}\left(a\sin(x_1+x_3) + b\sin(x_1) + c\mid x_2 \mid x_2\right)$$

where :
$\lambda > 0$
$\alpha > \frac{1}{2}$
$\tilde{\alpha} = \alpha - \frac{1}{2} > 0$
$S = x_1 + \lambda x_2$
$v = v_1 + v_2$
$v_1 = \frac{1}{a}\left(\frac{x_2}{\lambda} - ax_1 - b\sin(x_1)\right)$
$v_2 = \frac{\alpha}{\lambda a}S$

**Proof :**

**step 1 :**
Let us first consider the following sub-system :

$$\begin{cases} \dot{x}_1 = x_2 \\ \dot{x}_2 = -a(x_1+v) - b\sin(x_1) + d \end{cases}$$

where $v$ acts as a controller.

Let us now consider the following (semi-definite) Lyapunov function :

$$V_1 = \frac{1}{2}S^2$$

We note : $\tilde{\gamma} = \frac{\gamma}{\lambda} > 0$
$\forall t \geq 0$, we have :

$$\dot{V}_1 = S\left(x_2 + \lambda(-ax_1 - b\sin(x_1) - av + d)\right)$$
$$\leq -2\tilde{\alpha}V_1 + \frac{1}{2}\tilde{\gamma}^2$$

Thus, $v$ has been chosen to render the subsystem ultimate bounded according to (Khalil, 1992).

**step 2 :**
Let us now consider the full system and the following (semi-definite) Lyapunov function :

$$V = \frac{1}{2}S^2 + \frac{1}{2}(x_3 - v)^2$$

$\forall t \geq 0$, we have :

$$\dot{V} = S\,[x_2 + \lambda\cdot(-ax_1 - b\sin(x_1) - a(x_3 - v + v)$$
$$+d)] + (ku - \dot{v})(x_3 - v)$$

By using the proposed controller u, we finally get :

$$\forall t \geq 0\,, \qquad \dot{V} \leq -2\tilde{\alpha}V + \frac{1}{2}\tilde{\gamma}^2$$

By using (Khalil, 1992), we get the ultimate boundedness of V . Then, S inherits the same property and one conclude by using the fact that $S = x_1 + \lambda \dot{x}_1$ is Hurwitz.□

We used this controller in our simulations ; However, theoretically, we can prove the asymptotic stabilization of the roll :

**Theorem 2:** *The following controller u stabilizes asymptotically the roll to 0*

$$u = \frac{1}{k}[-\tilde{\alpha}(x_3 - v) + .aS + \dot{v}]$$

where :

$\lambda > 0$

$\tilde{\alpha} = \alpha - \frac{1}{2} > 0$

$K > \lambda\gamma > 0$

$S = x_1 + \lambda x_2$

$v = v_1 + v_2$

$v_1 = \frac{1}{a} \cdot \left( \frac{x_2}{\lambda} - a x_1 - b \sin(x_1) \right)$

$v_2 = \frac{1}{\lambda a} \frac{K^2 S}{|KS| + e^{-t}}$

**Proof :**

**step 1 :**

Let us first consider the following sub-system :

$$\begin{cases} \dot{x}_1 = x_2 \\ \dot{x}_2 = -a(x_1 + v) - b\sin(x_1) + d \end{cases}$$

where $v$ acts as a controller.

Let us now consider the following (semi-definite) Lyapunov function :

$$V_1 = \frac{1}{2} S^2$$

$\forall t \geq 0$, we have :

$$\begin{aligned} \dot{V}_1 &= S\Big( x_2 + \lambda(-a x_1 - b\sin(x_1) - av + d) \Big) \\ &= -\frac{K^2 S^{\,2}}{|KS| + e^{-t}} + S\lambda d \\ &\leq -(K - \lambda\gamma)|S| + e^{-t} \end{aligned}$$

By integrating the above inequality, we obtain

$$(K - \lambda\gamma) \int_0^\infty |S(t)|dt \leq V_1(0) - V_1(\infty) + \int_0^\infty e^{-t} dt$$

which means that :

$$\int_0^\infty |S(t)|dt < +\infty$$

Then, we can say that all solutions of our subsystem are bounded : this means that $S$ is uniformly continuous (because $\dot{S}$ is bounded).

Thus, $v$ has been chosen to stabilize (asymptotically) the subsystem according to Barbalat's lemma.

**step 2 :**

Let us now consider the full system and the following (semi-definite)Lyapunov function :

$$V = \frac{1}{2} S^2 + \frac{1}{2}(x_3 - v)^2$$

$\forall t \geq 0$, we have :

$$\begin{aligned} \dot{V} = S\,[x_2 + \lambda \cdot (-a x_1 - b\sin(x_1) - a(x_3 - v + v) \\ + d)] + (ku - \dot{v})(x_3 - v) \end{aligned}$$

By using the proposed controller u, we finally get :

$$\forall t \geq 0\,, \qquad \begin{aligned} \dot{V} &\leq -(K - \lambda\gamma)|S| + e^{-t} - \tilde{\alpha}(x_3 - v)^2 \\ &\leq -\Omega\Big( |S| + (x_3 - v)^2 \Big) + e^{-t} \end{aligned}$$

where $\Omega = \min(K - \lambda\gamma\ ,\ \tilde{\alpha}) \ > 0$

By integrating the above inequality, we obtain

$$\Omega \int_0^\infty |S(t)| + (x_3(t) - v(t))^2 dt \leq V(0) - V(\infty) + \int_0^\infty e^{-t} dt$$

which means that $\int_0^\infty |S(t)| + (x_3(t) - v(t))^2 dt < +\infty$. Thus,

$$\int_0^\infty |S(t)|dt < +\infty$$

Then, we can say that all solutions of our system are bounded : this means that $S$ is uniformly continuous (because $\dot{S}$ is bounded).

By Barbalat's lemma, we conclude that $S \to 0$ asymptotically.

One more time, we conclude by using the fact that $S = x_1 + \lambda \dot{x}_1$ is Hurwitz.□

## 4. VSS OBSERVER

In this section, we consider that we perfectly measure the roll and the angular position of the batteries ; however, we don't measure the time derivative of the roll : so, we need to build an observer.

*4.1 Sliding mode observer*

**Theorem:** *The following observer* $(\hat{x}_1, \hat{x}_2)$ *converges to* $(x_1, x_2)$ *in finite time*

$$\begin{cases} \dot{\hat{x}}_1 = \hat{x}_2 - K_1 sign(\hat{x}_1 - x_1) \\ \dot{\hat{x}}_2 = -a\sin(x_1 + x_3) - b\sin(x_1) - c \mid \hat{x}_2 \mid \hat{x}_2 \\ \qquad -K_2 sign(\hat{x}_2 - \bar{x}_2) \end{cases}$$

where :
$\forall t > 0, \quad K_1 > |\hat{x}_2(t) - x_2(t)|$
$\forall t > 0, \quad K_2 > c|\ \hat{x}_2(t)|\hat{x}_2(t)| - x_2(t)|x_2(t)|\ |$
$\bar{x}_2 = \hat{x}_2 - K_1 sign_{eq}(\hat{x}_1 - x_1)$

In practise, $\bar{x}_2$ is approximated by the output $x_2^\sharp$ of the following filter :

$$\tau \frac{d}{dt} x_2^\sharp + x_2^\sharp = K_1 sign(\hat{x}_1 - x_1)$$

**Proof:**

We note : $e_i = \hat{x}_i - x_i, \forall i \in \{1, 2\}$.
We first consider the sliding surface $S_1 = e_1$. We have :

$$\forall t \geq 0, \quad S_1\dot{S}_1 = S_1(\hat{x}_2 - x_2) - K_1|S_1|$$

Then, since $\forall t > 0, \quad K_1 > |\hat{x}_2(t) - x_2(t)|$,

$$\exists t_1 > 0 \ / \ \forall t \geq t_1, \quad e_1 = 0$$

Moreover it has been proved (Utkin, 1992) that :
$\forall t \geq t_1, \quad x_2(t) = \bar{x}_2(t) = \hat{x}_2(t) - K_1 sign_{eq}(e_1(t))$

We consider the sliding surface $S_2 = e_2$. We have:
$\forall t \geq t_1, \quad S_2\dot{S}_2 = -c\ S_2(\hat{x}_2|\hat{x}_2| - x_2|x_2|) - K_2|S_2|$
Since $\forall t > 0, \quad K_2 > c|\ \hat{x}_2(t)|\hat{x}_2(t)| - x_2(t)|x_2(t)|\ |$,

$$\exists t_2 > t_1 \ / \ \forall t \geq t_2, \quad e_2 = 0$$

□

*4.2 Sliding mode observer in presence of the controller*

Since the convergence of the observer is independent of the controller u, one can use the observer in presence of the controller : the observer will converge in finite time and the controller will then be able to work properly. However, the controller doesn't not properly work before the convergence of the observer and this could result in some bad effects. But, this is not so bad : since $t_2 > 0$ depends on $K_1$,$K_2$, it can be rendered as small as we like.

## 5. SIMULATION RESULTS

We used the following numerical values and we simulated our controller proposed in Theorem 1.

$a = 0.42$ ; $b = 0.74$ ; $c = 3.91$ ; $k = 15.36$ ; $\alpha = 3.5$; $\lambda = 1$

The below figure show that $x_1$ tends to 0. We remark that $S$ first tends to 0 ; since S is Hurwitz, $x_1$ and $x_2$ tend to 0 afterwards.
Meanwhile, d remains bounded and the controller u is very smooth.

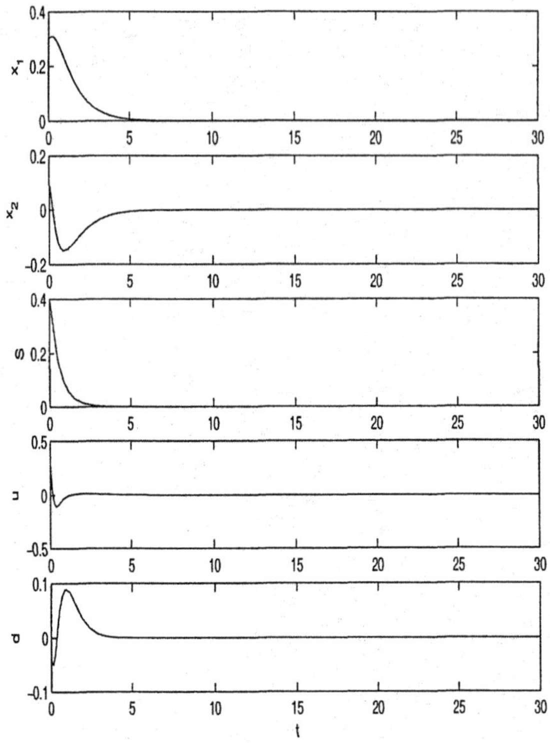


Fig. 2. $t \longrightarrow x_1(t), x_2(t), S(t), u(t)$ and $d(t)$

## 6. CONCLUSION

In this work, we have used a robust Backstepping controller in presence of a sliding mode observer. We applied this technique to a glider. Experimental results will be performed very soon. To design a more complex model in order to control the glider's yaw is under investigation.

## REFERENCES


H.K. Khalil (1992). Nonlinear Systems, Prentice Hall.

T. Ahmed-Ali and N. Seube (2002). Observation et Commande par modes glissants, F. Lamnabhi-Lagarrigue et P. Rouchon (Eds.), Commande des systèmes non linéaires, HERMES, Traité IC2.

T. Ahmed-Ali, L. Cuillerier and N. Seube (2003). Glider observer and identifier based on sliding modes control, IFAC 03

R. Moitie and N. Seube (2001) Guidance and Control of an Autonomous Underwater Glider, in UUST 12th Symposium, Durham, NH.

V.I. Utkin (1992). Sliding modes in Optimization and Control, Springer-Verlag.
J.J. Slotine and W. Li, (1991). Applied Nonlinear Control, Prentice-Hall, International Editions, Englewood Cliffs, 1991.

ELSEVIER
IFAC
PUBLICATIONS
www.elsevier.com/locate/ifac

# RUDDER–ROLL DAMPING EFFECT BY CONTROL OF THE RUDDER COMMAND TIME MOMENTS

**Viorel Nicolau, Constantin Miholca, Gheorghe Puscasu, Stelian Judele**

*Department of Automatic Control and Electronics, "Dunarea de Jos" University of Galati, 47 Domneasca Street, Galati, 800008, Galati, ROMANIA*
*Email:* Viorel.Nicolau@ugal.ro, Constantin.Miholca@ugal.ro

Abstract: The rudder commands generated to control the yaw angle affect simultaneously the roll angle. For mono-variable autopilots, there are time moments when the rudder command in one way can have positive or negative influences on roll oscillations. It is important to generate only rudder commands with damping or non-increasing effects over roll movements. The goal of this paper is to identify the time intervals when rudder commands have positive or negative influences on roll angle and to generate several expert rules which can be used to implement a more efficient yaw control law. *Copyright © 2004 IFAC*

Keywords: ship control, nonlinear systems, expert rules, rudder-roll damping effect

## 1. INTRODUCTION

The commands of the rudder act simultaneously on the yaw and roll movements of a ship. The problem of using the rudder for simultaneous heading control and roll reduction has been studied by many authors (Blanke, and Christensen, 1993; Fossen, and Laudval, 1994; Hearns, and Blanke, 1998; Son, and Nomoto, 1982; Van der Klugt, 1987). The two objectives can be separated in the frequency domain, using low frequencies for heading control and high frequencies for roll reduction.

During roll oscillations, there are time moments when the rudder command in one way can reduce the amplitude of the roll angle (which is a positive influence) or can amplify the roll movements increasing the roll's amplitude (negative influence).

Comparing the roll movements with the oscillations of a gravitational pendulum, the rudder action in one way can be viewed as a force $F$ applied to this pendulum, as shown in Figure 1.

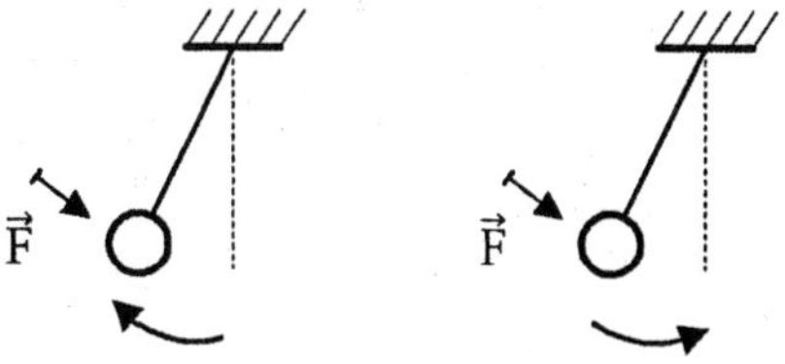


Fig. 1. The influence of a force applied to pendulum

The force influence depends on the time moments when it is applied. If the force $F$ is applied when the pendulum is moving up then a damping effect of oscillations can appear. On the contrary, applying the force when the pendulum is moving down can cause an increase in amplitude of oscillations.

A small force $F$ can produce an important damping effect, if it is applied constantly only for the right moments of time. This fact must be used not only for rudder-roll damping systems, but also for autopilots.

The heading control systems are in general of mono-variable type. Such autopilot takes into account only the yaw angle and ignores the roll movements.

Due to steering machine, even for low frequencies of the autopilot commands, the rudder movements affect the roll angle. Inevitable there are many moments of time when the command of the rudder to control the yaw angle has negative influence on roll oscillations.

It is important to generate only autopilot rudder commands with damping or non-increasing effects over roll movements. The force moment of the rudder is small, but acting continuously it can produce important roll damping effects.

The goal of this paper is to identify the time intervals when rudder commands have positive or negative influences on roll angle and to generate several expert rules which can be used to implement a more efficient yaw control law.
An example is illustrated for a course changing command with rudder-roll damping effect. Also, a conventional PID autopilot is analyzed. There are much more moments of time when the rudder command has a negative effect on roll angle than those with positive effect.

The expert rules are based on the hydrodynamic characteristics of the ship. They have fuzzy features and they are not critically dependent on the availability of accurate mathematical models of ship and disturbances.
Therefore, the expert rules can be easily applied for any conventional autopilot and for steering machines with small rudder rates, eliminating the negative influence of the autopilot on the roll angle. In addition, any roll stabilization system can be used to compensate the external disturbance influences.

The paper is organized as follows. Section2 provides mathematical models used in simulations. In section 3, the influence of steering machine model on roll movements is analyzed. In section 4, different moments of time for rudder command are studied and some expert rules are generated. Section 5 describes the simulation results. Conclusions are presented in section 6.

## 2. MATHEMATICAL MODELS

As a requisite for the simulation results, models for yaw and roll ship dynamics, steering machine and disturbances had to be generated. A linear ship model and a nonlinear steering machine model are used. Combining these models, nonlinear extended model results, as shown in Fig. 2.

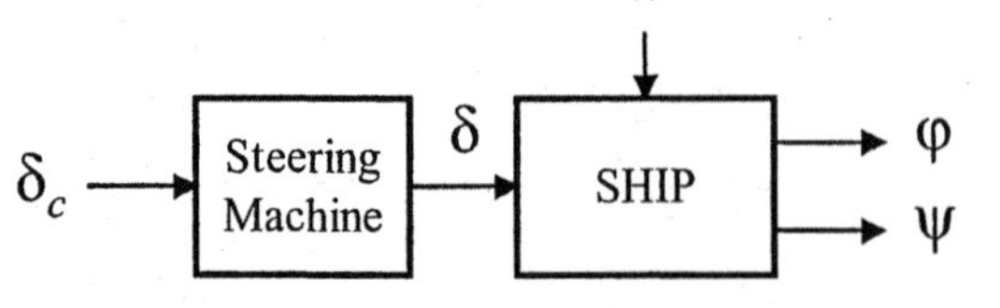


Fig. 2. Nonlinear model of a ship

### *2.1 Ship dynamics*

The equations describing the horizontal motion of a ship can be derived by using Newton's laws expressing conservation of hydrodynamic forces and moments. These equations can be approximated with Taylor series expansions about the steady-state condition (Abkowitz, 1964).

A three degree-of-freedom linear model can be obtained with coupled sway-yaw-roll equations, which can be identified (Son and Nomoto, 1982; Van der Klugt, 1987). Using the Laplace transform and eliminating the sway speed ($v$), two transfer functions result ($H_{\delta\psi}$ and $H_{\delta\varphi}$) which describe the transfer from rudder angle ($\delta$) to yaw angle ($\psi$) and to roll angle ($\varphi$) respectively. Also, wave disturbances (w) are considered (Van der Klugt, 1987).
The Laplace equations are:

$$\begin{cases} \psi(s) = H_{\delta\psi}(s)\cdot\delta(s) + H_{w\psi}(s)\cdot w(s) \\ \varphi(s) = H_{\delta\varphi}(s)\cdot\delta(s) + H_{w\varphi}(s)\cdot w(s) \end{cases} \quad (1)$$

The transfer functions have parameters depending on the speed of the ship ($u$) and the incidence angle ($\gamma$).

### *2.2 Steering machine model*

Steering machine (*SM*) is based on two-loop electro-hydraulic steering subsystem common on many ships (Van Amerongen, 1982). The model is nonlinear and is represented in Fig. 3.

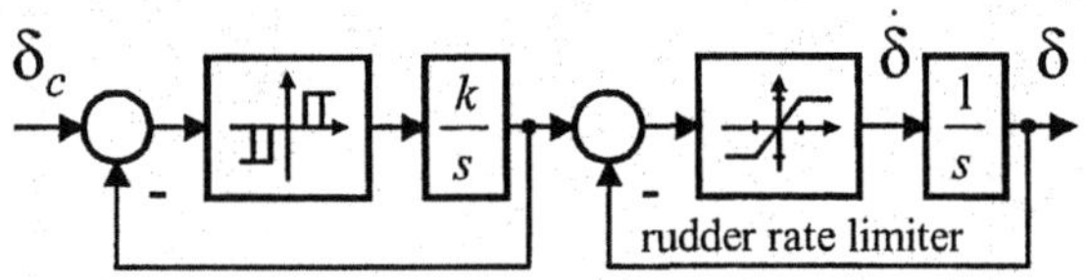


Fig. 3. Nonlinear model of steering machine

Also, the model includes the rudder limiter which is not represented in the figure to preserve the clearness of illustration. Moreover, the rudder angle is small enough and it is not limited, for all simulations.
A maximum rudder deflection of ±35 (deg) and a maximum rudder rate of ±2.5 (deg/s) are considered.

The nonlinearity of the first loop is important because it introduces a rudder positioning error and affects the efficiency of the control laws. Considering only yaw angle and mono-variable autopilots, the first loop can be disregarded. The resulting model has two nonlinearities generated by rudder limiter and rudder rate limiter (Van Amerongen, 1982).

For roll movements, the first loop increases the phase lag and decreases the rudder force moment on roll angle, as illustrated in the next section. This can affect the simulation results, when different moments of time are studied. Therefore, the first loop can not be disregarded and the nonlinear steering machine model is considered.

*2.3 Wave model*

In this paper, wave disturbances are considered. The wave can be regarded as an ergodic random process with elevation $\zeta(t)$ and zero mean. Knowing the mean square spectral density function $\phi_{\zeta\zeta}(\omega)$ of the wave elevation $\zeta(t)$, shortly called wave spectrum, the statistical parameters of the wave can be computed. Then, the wave model can be generated.

Based on wave spectrum, the wave disturbance can be modeled as the sum of a limited number of sinusoidal waves (Price, and Bishop, 1974):

$$w(t) = \sum_{i=1}^{N} A_i \cdot \sin(\omega_i \cdot t + \varphi_i) \qquad (2)$$

where $A_i$ and $\omega_i$ are the amplitude and angular frequency of the $i$-th component. $\varphi_i$ is the phase angle drawn randomly from a uniform density distribution. The relative frequency between the wave and the ship modifies the wave spectrum and this transformation must be taken into account for wave model generation (Nicolau, and Ceanga, 2001).

## 3. INFLUENCE OF THE AUTOPILOT AND STEERING MACHINE ON ROLL MOVEMENTS

The autopilot is in general of mono-variable type, taking into account only the yaw angle and ignoring the roll movements. There are many moments of time when the command of the rudder to control the yaw angle has negative influence on roll oscillations.
As a result, in the presence of external disturbances, the autopilot has a positive influence on the yaw angle in course-keeping or course-changing problem, but can increase the roll amplitude. This can be a supplemental task for the roll stabilization controller.

For example, a theoretical influence of a course-keeping autopilot is illustrated in Fig. 4, taking as reference the amplitude of the ship movements generated by external disturbances without autopilot. The reduction of the yaw amplitude is important and it is made only for low frequency components of the yaw movements.
But the autopilot increases the roll amplitude, because of the general negative influence of rudder commands on roll oscillations. This is a supplemental task for any roll stabilization system with some penalty of yaw reduction.

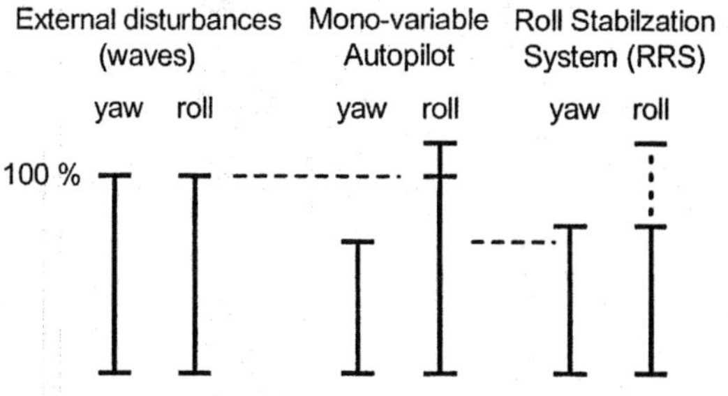


Fig. 4. Influence of the mono-variable autopilot

For every time moment, the autopilot must generate only rudder commands with damping or non-increasing effects over roll angle.
Thus, the roll amplitude can be smaller than roll movements generated by external disturbances without autopilot, as illustrated in Fig. 5. In addition, the roll reduction of any roll stabilization system can be bigger with less effort.
The roll reduction denoted (*1.*) in Fig. 5 is obtained by avoiding the rudder commands for time moments with negative effect on roll angle. The roll reduction denoted (*2.*) is generated by rudder commands with roll damping effects.

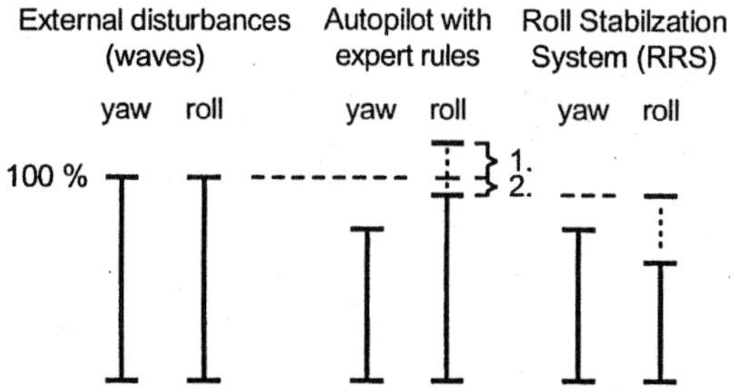


Fig. 5. Rudder-roll damping effect of autopilot

The model of the steering machine has different influences on yaw and roll angles. Two situations are considered: with and without the first loop of the model, illustrated into the next figures with solid and dotted lines respectively. Applying a rudder command ($\delta_C$) of 5 (deg), the rudder angle affects the movements of the ship, as shown in Fig. 6.

For yaw movement, only a small difference appears, caused by the positioning error of the rudder when the first loop is considered. The difference is significant after a long period and can be compensated by the control law of yaw angle. Therefore, the first loop can be disregarded when mono-variable autopilot is considered.

Roll oscillations are delayed and amplitudes are smaller if the first loop is included in the model of steering machine. An important phase lag appears from the beginning, which can affect the simulation results, when different moments of time are studied.

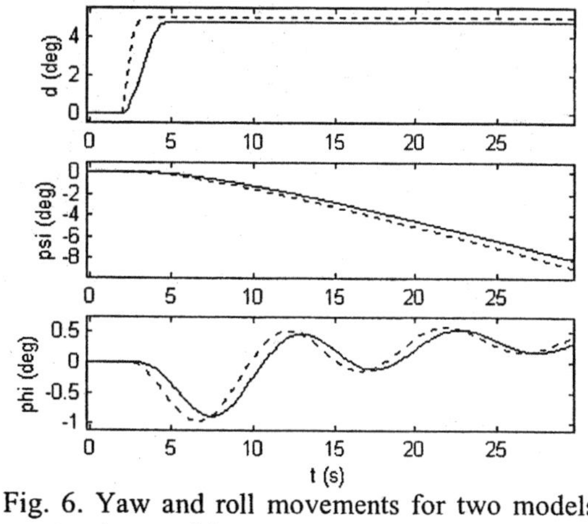


Fig. 6. Yaw and roll movements for two models of steering machine

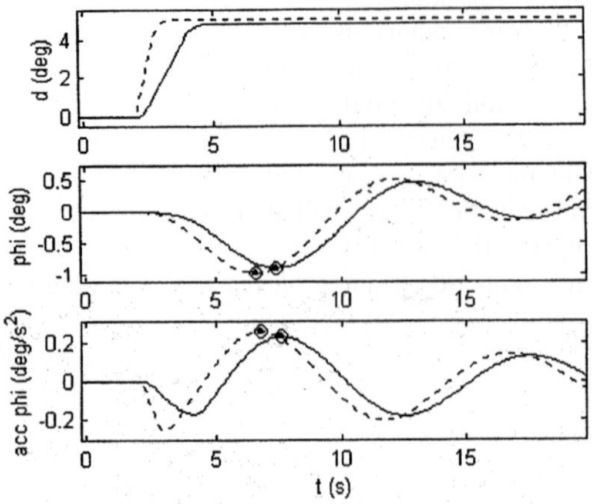

Fig. 7. Roll angle and roll acceleration

The acceleration of the roll angle is generated by the rudder force moment on roll motion. The maximum value of acceleration is obtained for the minimum value of roll angle, as shown in Fig. 7, where the points are marked distinctly.
The maximum acceleration values are different for the two models of steering machine and the difference increases as the rudder angle increases.
Starting from repose point, these values are reached after a time interval, depending on rudder angle and ship characteristics. The points correspond to the maximum value of the rudder force moment.

As a result, the maximum values of rudder moment are different, depending on rudder angle, as illustrated in Fig. 8. The linear dependency represented with dotted line corresponds to theoretical rudder moments when steering machine model is ignored. The maximum rudder moments generated by steering machine without the first loop are represented with dashed line.

If the first loop is considered, the real maximum value of the rudder moment (represented with continuous line) is smaller and decreases as rudder angle increases. Consequently, the real rudder roll damping effect is smaller than the case when steering machine is approximated without the first loop. This can affect the simulation results.
Therefore, the efficiency of a control law with rudder roll damping effect has to be tested considering the two-loop nonlinear steering machine model.

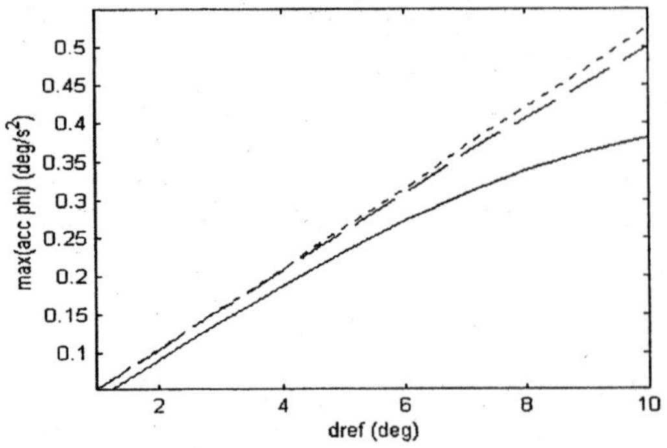

Fig. 8. Maximum rudder moments on roll angle

## 4. DIFFERENT MOMENTS OF TIME FOR RUDDER COMMAND GENERATION

The rudder command in one way acts simultaneously on the yaw and roll movements of the ship. For simulations, a pulse type rudder command of 5 (deg) is generated. The positive edge is used as a previous command, starting the roll oscillations. The negative edge represents the command that has to be analyzed.

Depending on the moment of time when the rudder command is generated (negative edge) the effect on roll movement can be of increasing or decreasing the roll angle, as illustrated in Fig. 9. The command represented with dotted line generates a bigger roll angle as the one illustrated with continuous line.

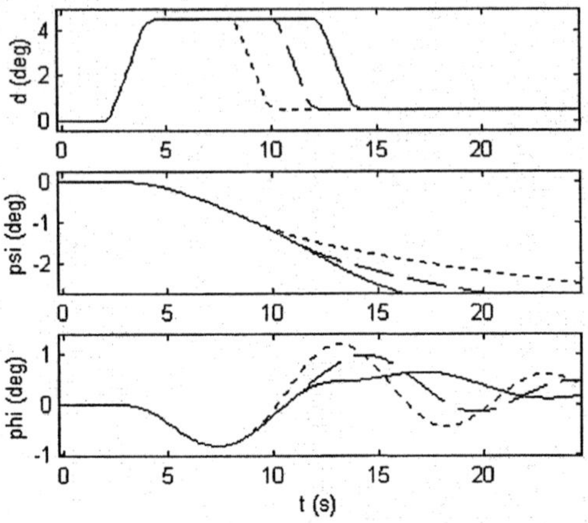

Fig. 9. Different time moments for rudder command

But a rudder command with positive effect on roll angle can generate a negative effect on yaw angle. In this case, the yaw error can be compensated by the control law, as shown in next section. Moreover, if the roll damping ratio is bigger than the yaw error ratio then a small yaw error can be accepted.

To identify specific time moments or time intervals when the rudder has positive influence on roll movements, the roll angle, rate and acceleration has to be analyzed. These roll motion characteristics are illustrated in Fig. 10, for one pulse rudder command. In simulations, the rate and acceleration are computed based on roll angle.

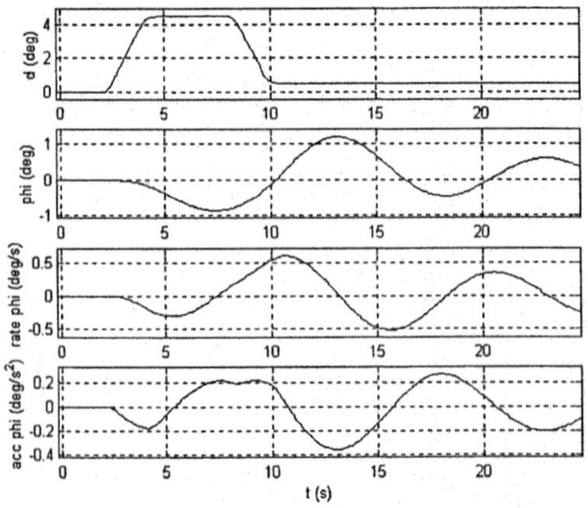

Fig. 10. The motion characteristics of roll angle

During one period of roll oscillation, the following time moments can be defined:
- the local minimum value of roll angle, which can be identified by zero roll rate and positive acceleration (maximum value);
- the zero value of roll angle, corresponding to its upward slope, identified by positive rate (maximum value) and zero acceleration;
- the local maximum value of roll angle, which can be identified by zero roll rate and negative acceleration (minimum value);
- the zero value of roll angle corresponding to its downward slope, identified by negative rate (minimum value) and zero acceleration.

The rudder command (negative edge) is generated at four different moments of time: the minimum and maximum values of roll angle and roll rate respectively. The results are compared with the original roll angle, which is obtained in the absence of rudder command (negative edge is delayed).

### *4.1 Moments of time corresponding to minimum values of roll angle and roll rate*

If the rudder angle is decreased during the minimum value of roll angle (continuous line) or roll rate (dashed line) then the roll amplitude and roll acceleration increase, as shown in Fig. 11.
It is better for roll motion if the rudder command is delayed as represented with dotted lines.

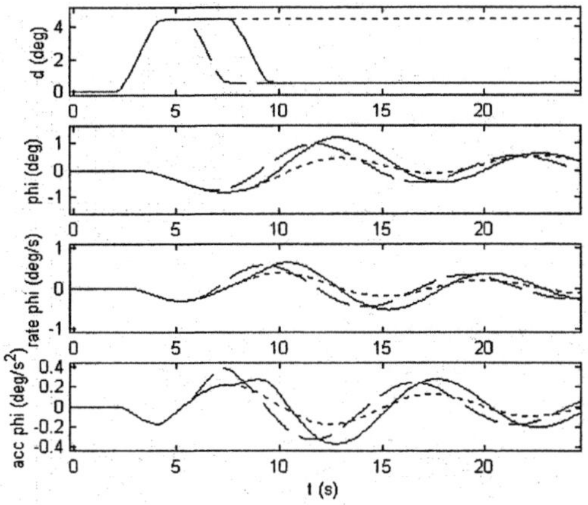


Fig. 11. Decreasing the rudder command during the minimum values of roll angle and roll rate

*Rule no. 1*: During the time moments around the minimum values of roll angle and roll rate, the rudder command must be increased or delayed.

### *4.2 Moments of time corresponding to maximum values of roll angle and roll rate*

If the rudder angle is decreased during the maximum value of roll angle (continuous line) or roll rate (dashed line) then the roll amplitude and the roll acceleration remain comparable with those of damping oscillation, as illustrated in Fig. 12.

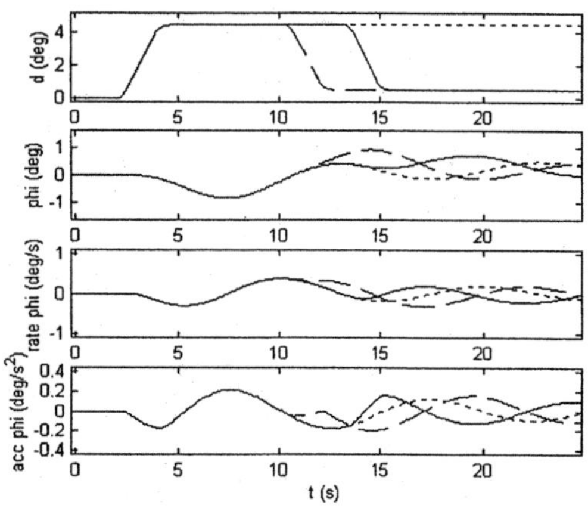


Fig. 12. Decreasing the rudder command during the maximum values of roll angle and roll rate

There is a small negative influence on roll angle, but the effect is not important. These time moments represent the bounds of time interval when the damping effect appears.
*Rule no. 2*: During the time moments around the maximum values of roll angle and roll rate, the rudder command can be decreased or delayed.

### *4.3. Optimum time moment to decrease rudder command*

The optimum time moment to decrease the rudder command depends on ship dynamics. It is somewhere between the time moments corresponding to maximum values of roll angle and roll rate, being represented with continuous line in Fig. 13.

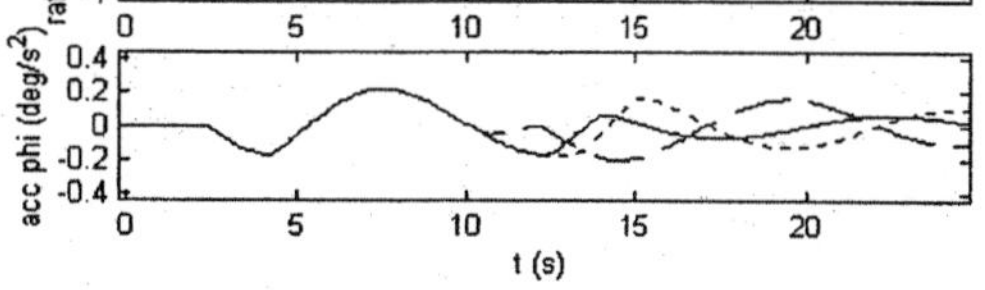

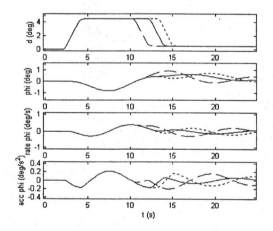

Fig. 13. Rudder command at a time moment between the maximum values of roll angle and roll rate

*Rule no. 3*: The optimum time interval to decrease rudder command is placed between maximum values of roll rate and roll angle.

Similar rules can be obtained for the time moments when the rudder command is increased.
The rules have general validity but the time intervals have fuzzy characteristics. Knowing the nonlinear extended model of the ship, more precise moments of time can be obtained.

## 5. SIMULATION RESULTS

### *5.1 Rudder command generation for course changing*

The control law of course changing system can be modified, taking into account the roll angle and using the expert rules. An example is illustrated in Fig. 14.

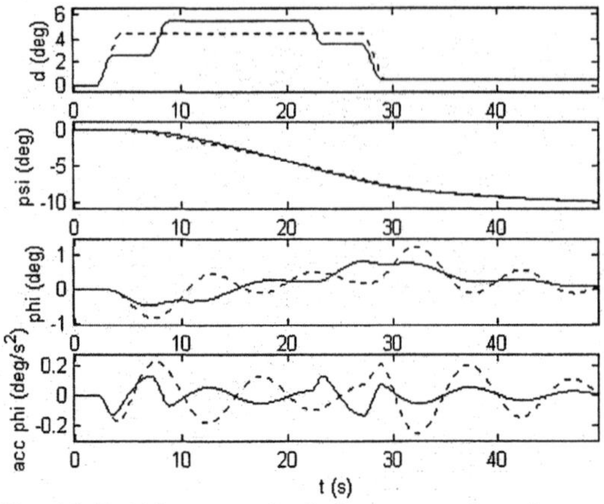

Fig. 14. Rudder command for course changing

Instead of one single pulse (represented with dotted line), the rudder command is generated in three steps. The middle pulse must be higher than the reference rudder angle. The edges of rudder commands (except the first one) are generated at certain moments of time to have positive influences on roll angle.

### *5.2 Rudder command analysis generated by a classical autopilot*

A conventional PID autopilot is considered and the rudder command sequence is analyzed. The ship's speed is 22 (knot) (Van der Klugt, 1987) and the incidence angle is $\gamma$ = 135 (deg). A wave with ITTC spectrum and $h_{1/3}$ = 4 (m) is considered.
Wave disturbances ($w$), rudder angle ($d$), yaw angle ($\psi$) and roll angle ($\varphi$) are illustrated in Fig. 15.

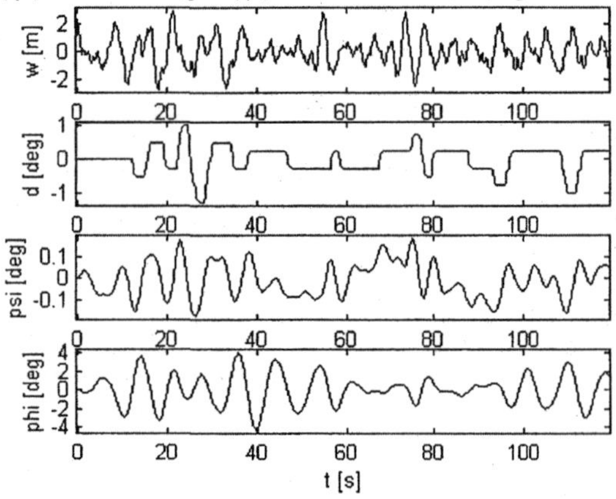

Fig. 15. Ship's motion with PID autopilot

Analyzing the rudder commands generated by the autopilot for all discrete moments of time, a new discrete function $F(\delta_C)$ is computed:

$$F(\delta_C) = \begin{cases} -1 & \text{if the effect on roll angle is negative} \\ 0 & \text{if the effect is not important} \\ 1 & \text{if the effect is positive} \end{cases}$$

For the rudder command sequence generated by autopilot in the example above, the resulting function $F$ is illustrated Fig. 16. There are many moments of time when the rudder command has a negative effect on roll angle ($F$ = -1).

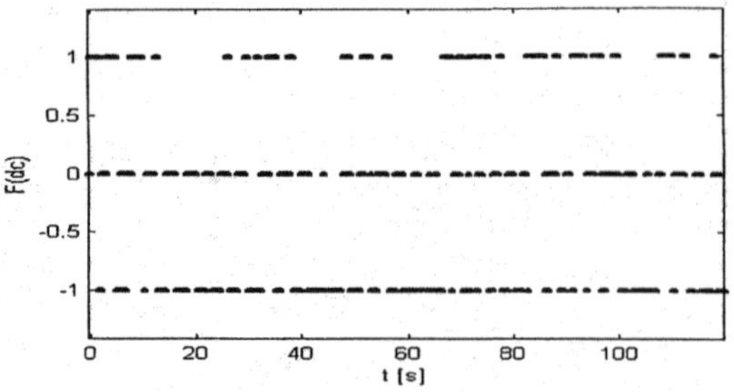

Fig. 16. The effect function of the rudder commands

## 6. CONCLUSIONS

For mono-variable autopilots, there are many moments of time when the rudder command to control the yaw angle has negative influence on roll movement. It is important to know these time intervals to generate only rudder commands with damping or non-increasing effects on roll oscillation. As a result, several expert rules are generated, which have general validity and can be used to implement a more efficient yaw control law.

## REFERENCES


Abkowitz, M. A. (1964). *Lectures on Ship Hydrodynamics – Steering and Manoeuvrability.* Technical Report, nr. Hy-5, Hydro- and Aerodynamics Laboratory, Lyngby, Denmark.

Blanke, M. and A. C. Christensen (1993). Rudder-Roll Damping Autopilot Robustness to Sway-Yaw-Roll Couplings. In: *Proc. of The 10th Int. SCSS'93*, vol. A, pp. 93-119, Ottawa, Canada.

Fossen, T. I. and T. Laudval (1994). Nonlinear Stability Analysis of Ship Autopilots in Sway, Roll, and Yaw. In: *Proc.s of The 3rd Conference MCMC'94*, Southamptom, UK.

Hearns G. and M. Blanke (1998). Quantitative Analysis and Design of a Rudder Roll Damping Controller. In: *Proceedings of IFAC Conference CAMS'98*, Fukuoka, Japan.

Nicolau V. and E. Ceangă (2001). Wave Spectrum Correction with the Ship's Speed and the Incidence Angle. In: *Proc. of IFAC Conference CAMS'2001*, pp.331-336, Glasgow, UK.

Price, W. G. and R.E.D. Bishop (1974). *Probabilistic theory of ship dynamics.* Chapmann and Hall.

Son, K. H., K. Nomoto (1982). On the Coupled Motion of Steering and Rolling of a High Speed Container Ship. *Naval Architect of Ocean Engineering*, **20**, pp. 73-83.

Van Amerongen, J. (1982). *Adaptive Steering oj Ships.* Ph.D.Thesis, Delft University of Technology, The Netherlands.

Van der Klugt, P.G.M. (1987). *Rudder Roll Stabilization.* Ph.D.Thesis, Delft University of Technology, The Netherlands.

ELSEVIER
IFAC
PUBLICATIONS
www.elsevier.com/locate/ifac

# STATE-SPACE REPRESENTATION OF MEMORY EFFECTS IN FOIL LIFT FORCE

**E. Kristiansen* and O. Egeland**

**Norwegian University of Science and Technology, email: Erlend.Kristiansen@itk.ntnu.no*

Abstract: The paper presents a method for generating a time domain formulation of the lift reduction memory effect for lifting surfaces. Previous work on this topic has relied on the use of convolution terms, whereas in this work state-space models are used. This leads to a model formulation that is well suited for controller design and simulation. *Copyright © 2004 IFAC*

Keywords: State-space realization, Model approximation, Time-Domain, Convolution Integral, Impulse Responses

## 1. INTRODUCTION

It is well known that a foil subjected to motions, or subjected to changing flow, exhibits a transient lift reduction compared to a stationary foil in steady flow. This lift reduction can be seen as a memory effect, as the lift reduction is dependent on the time history of motions. Under the assumption that the foil is subjected to sinusoidally oscillating motion with a specified frequency, this lift reduction can be calculated using the Theodorsen function (Newman 1977). The lift force will then incorporate a frequency-dependent term, which is not in agreement with models used in simulation and in automatic control. Alternatively, the lift reduction can be described in the time-domain, allowing for arbitrary motions. This formulation relies on the use of the Wagner-function as the kernel of a convolution term. Convolution terms are problematic in the sense that they are computationally expensive in real-time applications, and are not in agreement with models used in simulation and automatic control.

The main contribution of this paper is to show how the lift reduction memory effect can be reformulated into a state-space form suitable for simulation and controller design. A new method that generates a low order state-space model from a modified Wagner function will be proposed, and it will be shown that this yield an accurate and computationally efficient representation of the convolution term.

In this paper only stationary flow will be considered, and not gusting flow.

## 2. MATHEMATICAL FORMULATION

### 2.1 Introduction

In this section the mathematical model of the unsteady lift of a foil is presented. This is based on (Woods 1961), but modified into considering incompressible flow only. This model is used in Section 3.1 to derive the state-space model.

### 2.2 Physical system

Consider a thin foil with chord length $c = 2b$, where $b$ is the half-chord length, and angle of attack $\alpha$, as shown in Figure 1. A local coordinate system is used, attached to the foil with the origin situated at the mid-chord point, and following the

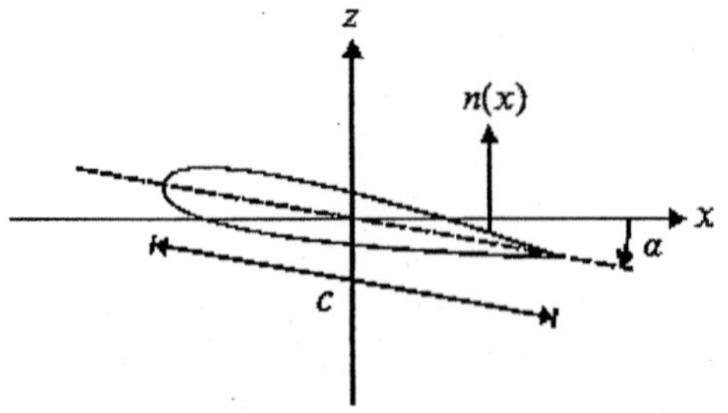


Fig. 1. Thin foil

right-hand rule with the $x$-axis pointing in the steady flow direction and the $z$-axis pointing in the upward vertical direction. The foil is moving with a steady forward speed $U$ in the negative $x$-direction, and is subject to small vertical motions. $n(x)$ is the local vertical velocity of the foil surface.

The theory in this paper is for a 2D-foil, so no 3D-effects will be considered, nor are foil camber and thickness effects considered.

### *2.3 Basic assumptions*

In this paper a closed foil is considered, or more specific, a foil moving in an incompressible, irrotational fluid, with finite wake.

Let

$$\tau(w) = \Omega + i\theta \quad (1)$$

$$w = \phi + i\psi \quad (2)$$

where $\Omega$ is a speed parameter, $\theta$ is the slope of the streamline, $\phi$ is the fluid velocity potential and $\psi$ is the stream function. Then $\tau(w)$ satisfies the Laplace equation

$$\nabla^2 \tau = 0. \quad (3)$$

in the fluid. As the foil is assumed to be a thin foil subjected to small angles of attack, it follows that the mean flow and unsteady motions can be treated with linear perturbation theory. Then $w$ can be replaced by $UZ$ where

$$Z = -\frac{1}{2}c\cos\zeta, \quad Z = x + iz \quad (4)$$

$$x = -\frac{1}{2}c\cos\gamma, \quad \eta = 0 \quad (5)$$

$$\zeta = \gamma + i\eta, \quad -\pi < \gamma < \pi, 0 < \eta < \infty \quad (6)$$

The boundary condition on the foil is

$$\theta = \theta_s + \frac{n}{U} - \alpha + \pi\lambda\delta(\gamma) \quad (7)$$

where the last term accounts for perturbations of the front stagnation point. The Joukowski condition and Kelvin's theorem is the basis for the vortex sheet formulation resulting from the unsteady motion.

The vortex sheet $\chi(x, \iota)$ is convected downstream with the speed $\bar{q} \simeq U$ and satisfies

$$\frac{1}{c}\frac{\partial \chi}{\partial \iota} + \frac{\partial \chi}{\partial x} = 0 \quad (8)$$

$$\chi(x, \iota) = f(x - c\iota) \quad (9)$$

$$\chi = \bar{\omega} U \quad (10)$$

where $\bar{\omega}$ is the jump in $\Omega$ across the sheet and $\iota$ is the reduced time given by

$$\iota = \frac{1}{c}\int_0^t U\,dt \quad (11)$$

The reduced time $\iota$ can be explained as the total length of the vortex sheet in units of the chord length for an unsteady motion that started at $t = 0$. It is thus assumed that the flow is steady for $\iota < 0$. Note that variables denoted with dot here means differentiation with respect to $\iota$, until otherwise stated.

### *2.4 Solution of boundary value problem*

The transformation in (4) to (6) maps the foil on to $\eta = 0$ and the vortex sheet on to $\gamma = \pm\pi$. The first boundary value problem is to find $\tau(\zeta)$ given $\theta$ on $\eta = 0$ and the discontinuity $\bar{\omega}$ in $\Omega$ across $\gamma = \pm\pi$. The solution to this problem is (Woods 1961)

$$\tau(\zeta, \iota) = \frac{1}{2\pi}\int_{-\pi}^{\pi}\theta(\gamma, \iota)\cot\frac{1}{2}(\gamma - \zeta)\,d\gamma + \frac{\sin\zeta}{2\pi}\int_0^{\infty}\frac{\bar{\omega}(\eta, \iota)}{\cosh\eta + \cos\zeta}\,d\eta \quad (12)$$

$$0 = \int_{-\pi}^{\pi}\theta(\gamma, \iota)\,d\gamma + \int_0^{\infty}\bar{\omega}(\eta, \iota)\,d\eta \quad (13)$$

where (12) can be expanded into

$$\tau = \frac{i}{n}\sum_{n=1}^{\infty}\frac{1}{\sigma^n}[\,(-1)^n\int_{-\pi}^{\pi}\theta e^{-ni\gamma}\,d\gamma + \int_0^{\infty}\bar{\omega}\cosh n\eta\,d\eta] \quad (14)$$

The total circulation about the foil and the wake is

$$\Gamma = \mathrm{Re}\int_C q e^{-i\theta}\,dz \quad (15)$$

where $C$ is a contour enclosing both the foil and the wake. Using linear perturbation theory it can be shown that

$$\mathrm{Re}\left(q e^{-i\theta}dz\right) = U dx - \mathrm{Re}\,\tau dZ \quad (16)$$

making it possible to use (4), (6) and $\sigma = -e^{-i\zeta}$ to write

$$\Gamma = \mathrm{Re}\frac{c}{4}\int_C \tau(\sigma)\left(1 - \frac{1}{\sigma^2}\right)d\sigma \quad (17)$$

As the flow is steady, $\Gamma$ is zero before the unsteady motion starts. Then from Kelvin's circulation theorem it follows that $\Gamma = 0$ throughout the unsteady motion. It then follows from the two previous equations and the residue theorem that the real part of $\frac{1}{\sigma}$ in the expansion for $\tau$ vanish, which implies that

$$\int_{-\pi}^{\pi} \theta \cos\gamma \, d\gamma - \int_{0}^{\infty} \bar{\omega} \cosh\eta \, d\eta = 0 \quad (18)$$

The unsteady motion of the foil will not affect the foil closure condition, $\int_C dy = 0$. In linear perturbation theory $\frac{dy}{dx} \simeq \theta$ which when using (5) yield $\int_C \theta \, dx = 0$ or

$$\int_{-\pi}^{\pi} \theta \sin\gamma \, d\gamma = 0 \quad (19)$$

Using linear perturbation theory, Kelvin's theorem, the residue theorem and ignoring second-order terms, the pressure on the foil can be written (Woods 1961)

$$p(\gamma) = G(\iota) + \rho U^2 \Omega(\gamma) + \frac{1}{2}\frac{\rho U}{c}\frac{dU}{d\iota}\cos\gamma + \frac{\rho U}{2}\int_0^{\gamma} \frac{\partial (U\Omega)}{\partial \iota} \sin\gamma \, d\gamma \quad (20)$$

*2.5 The strength of the vortex sheet*

In steady flow there is no vortex sheet, that is, $\bar{\omega} = 0$. Substituting the boundary condition (7) into the solution (13), the real part (18) of the expansion and the closure condition (19) yield the equations

$$\lambda U = 2\alpha U - \frac{1}{\pi}\int_{-\pi}^{\pi} n(\gamma,\iota)\, d\gamma - \frac{1}{\pi}\int_0^{\infty} \chi(\eta,\iota)\, d\eta \quad (21)$$

$$0 = 2\alpha U - \frac{1}{\pi}\int_{-\pi}^{\pi} n(\gamma,\iota)(1-\cos\gamma)\, d\gamma - \frac{1}{\pi}\int_0^{\infty} \chi(\eta,\iota)(1+\cosh\eta)\, d\eta \quad (22)$$

$$0 = \int_{-\pi}^{\pi} n(\gamma,\iota)\sin\gamma \, d\gamma \quad (23)$$

The first equation fixes the value of $\lambda$, the second is for the vortex strength $\chi$ and the last one is a physical restriction of the unsteady perturbations. By defining

$$\xi = \frac{1}{2} + \frac{x}{c} = \frac{1}{2}(1+\cosh\eta) \quad (24)$$

and using the functional relation

$$\chi(\xi,\iota) = \chi(1, \iota+1-\xi) \quad (25)$$

(21) and (22) can be rewritten into

$$\lambda U = 2\alpha U - 2a_0 - \frac{1}{\pi}\int_1^{\infty} \chi(1,\iota+1-\xi)\frac{1}{\sqrt{\xi(\xi-1)}}\, d\xi \quad (26)$$

$$0 = 2\alpha U - 2a_0 + 2a_1 - \frac{1}{\pi}\int_1^{\infty} \chi(1,\iota+1-\xi)\frac{2\xi}{\sqrt{\xi(\xi-1)}}\, d\xi \quad (27)$$

where

$$a_r(\iota) = \frac{1}{2\pi}\int_{-\pi}^{\pi} n(\gamma,\iota)\cos r\gamma \, d\gamma \quad (28)$$

Changing the variable by

$$\sigma = \iota + 1 - \xi \quad (29)$$

and using the fact that the end of the wake is at $x = \frac{1}{2}c + c\iota$ or equivalently at $\xi = 1 + \iota$, where $\chi = 0$ for $1 + \iota \le \xi < \infty$, makes it possible to write (26) and (27) as

$$\lambda U = 2\alpha U - 2a_0 - \frac{1}{\pi}\int_0^{\iota} \frac{\chi(1,\sigma)}{\sqrt{(\iota+1-\sigma)(\iota-\sigma)}}\, d\sigma \quad (30)$$

$$0 = \alpha U - a_0 + a_1 - \frac{1}{\pi}\int_0^{\iota} \chi(1,\sigma)\sqrt{\frac{\iota+1-\sigma}{\iota-\sigma}}\, d\sigma \quad (31)$$

Here, (31) is an equation for the vortex strength at the trailing edge of the foil. Defining

$$A(\iota) = \begin{cases} \alpha(\iota)U(\iota) - a_0(\iota) + a_1(\iota), & \iota > 0 \\ 0, & \iota < 0 \end{cases} = \mathbf{1}(\iota)(\alpha U - a_0 + a_1) \quad (32)$$

and using the Laplace-transform and hyperbolic coordinate transformations to solve the integrals, yield

$$\mathcal{L}(\lambda U) = sk(s)A(s) - 2a_1(s) \quad (33)$$

where

$$k(s) = \frac{2K_1\left(\frac{1}{2}s\right)}{s\left(K_0\left(\frac{1}{2}s\right) + K_1\left(\frac{1}{2}s\right)\right)} \quad (34)$$

is called the Wagner function, and

$$K_v(s) = \int_0^{\infty} \cosh vx \, e^{-s\cosh x}\, dx \quad (35)$$

is a cylindrical function. Further,

$$\mathcal{L}(\mathbf{1}(\iota)\chi(\iota)) = k_0(s)\, sA(s) \quad (36)$$

where

$$k_0(s) = \frac{2\pi e^{-\frac{1}{2}s}}{s\left(K_0\left(\frac{1}{2}s\right) + K_1\left(\frac{1}{2}s\right)\right)} \quad (37)$$

*2.6 The pressure distribution on the foil*

The boundary condition equations (7), (10) and the real part of the solution (12) yield

$$U\Omega(\gamma,\iota) = U\Omega_s(\gamma) - \frac{1}{2}\lambda U \cot\frac{1}{2}\gamma + \frac{1}{2\pi}\int_{-\pi}^{\pi} n(\tilde{\gamma},\iota)\cot\frac{1}{2}(\tilde{\gamma}-\gamma)\,d\tilde{\gamma} + \frac{\sin\gamma}{2\pi}\int_0^{\infty}\frac{\chi(\eta,\iota)}{\cosh\eta+\cos\gamma}\,d\eta \quad (38)$$

where

$$\Omega_s = \frac{1}{2\pi}\int_{-\pi}^{\pi}\theta_s \cot\frac{1}{2}(\tilde{\gamma}-\gamma)\,d\tilde{\gamma} \quad (39)$$

Ignoring the small term $\Omega_s\left(\frac{dU}{d\iota}\right)$, (20) can be rewritten in the form

$$p(\gamma,\iota) = p_s(\gamma) + G + \rho U^2(\Omega - \Omega_s) + \frac{1}{2}\rho U\dot{U}\cos\gamma + \frac{\rho U}{2}\int_0^{\gamma}\frac{\partial}{\partial\iota}(U\Omega - U\Omega_s)\sin\gamma\,d\gamma \quad (40)$$

By defining

$$N(\gamma,\iota) = \frac{1}{2}\int_0^{\gamma} n(\gamma,\iota)\sin\gamma\,d\gamma \quad (41)$$

and doing some integral manipulations involving results from the previous section, the pressure distribution on the foil can be written (Woods 1961)

$$p(\gamma,\iota) = p_s(\gamma) + C + \frac{\rho U}{2}\dot{U}\cos\gamma - \frac{\rho U}{2}\left(\frac{\partial}{\partial\iota}(\alpha U)\sin\gamma + \lambda U\cot\frac{1}{2}\gamma\right) + \frac{\rho U}{2\pi}\int_{-\pi}^{\pi} n(\tilde{\gamma},\iota)\cot\frac{1}{2}(\tilde{\gamma}-\gamma)\,d\tilde{\gamma} + \frac{\rho U}{2\pi}\int_{-\pi}^{\pi}\dot{N}(\tilde{\gamma},\iota)\cot\frac{1}{2}(\tilde{\gamma}-\gamma)\,d\tilde{\gamma} \quad (42)$$

### *2.7 The lift in general form*

The lift about the mid-chord point is given by

$$L = L_s - \frac{c}{2}\int_{-\pi}^{\pi}(p - p_s)\sin\gamma\,d\gamma \quad (43)$$

where $L_s$ is the lift in steady flow. Now substituting the pressure distribution and using (28) yield

$$L = L_s + \frac{\pi\rho c U}{4}\left(\frac{\partial}{\partial\iota}(\alpha U) + 2\lambda U + 4a_1\right) + \frac{\rho c U}{2}\int_{-\pi}^{\pi}\dot{N}(\tilde{\gamma},\iota)\cos\tilde{\gamma}\,d\tilde{\gamma} \quad (44)$$

where it is assumed that $n(\gamma^*,\iota)$ and $\dot{N}(\gamma^*,\iota)$ are even functions with respect to $\gamma^*$.

### *2.8 Lift for arbitrary vertical motions with steady forward speed*

The motion producing the trailing vortex sheet is

$$\dot{A} = \dot{\alpha}U + \alpha\dot{U} - \dot{a}_0 + \dot{a}_1 \quad (45)$$

Substituting for $\lambda U$, $\dot{N}(\gamma^*,\iota)$, $\dot{a}_0$ and $\dot{a}_1$, the lift can be written

$$L = L_s + \frac{\pi\rho c U}{4}\left(\dot{\alpha}U + \alpha\dot{U}\right) + \frac{\pi\rho c U}{2}\int_0^{\iota}[\dot{\alpha}(\tilde{\iota})U(\tilde{\iota}) + \alpha(\tilde{\iota})\dot{U}(\tilde{\iota}) + \frac{1}{2\pi}\int_{-\pi}^{\pi}\dot{n}(\gamma,\tilde{\iota})(\cos\gamma - 1)\,d\gamma]k(\iota-\tilde{\iota})\,d\tilde{\iota} + \frac{\rho c U}{4}\int_{-\pi}^{\pi}\int_0^{\tilde{\gamma}}\dot{n}(\gamma_1,\iota)\sin\gamma_1\,d\gamma_1\cos\tilde{\gamma}\,d\tilde{\gamma} \quad (46)$$

A general vertical motion $n$ of the point $x$ can be written

$$n(x,\iota) = v(\iota) + \frac{1}{2}c\frac{d\alpha}{dt}\cos\gamma \quad (47)$$

where $v$ is the vertical velocity of the midchord point.

Until now, the forward speed $U(\iota)$ has been a function of the reduced time. Assuming constant $U(\iota) = U$ we can establish from (11) the relation

$$\iota = \frac{Ut}{c} \quad (48)$$

which gives

$$d\iota = \frac{U}{c}dt \quad (49)$$

yielding

$$\dot{n}(\iota) = \dot{v}(\iota) + \frac{1}{2}U\ddot{\alpha}(\iota)\cos\gamma \quad (50)$$

The goal of this paper is to have a representation in agreement with models used in simulation and automatic control. Thus, after substitution of the above into the general lift equation (46), it is rewritten with proper time $t$ as the argument and substitution of $c = 2b$ as

$$L(t) = L_s + \pi\rho b^2(\dot{v}(t) + U\dot{\alpha}(t)) + \pi\rho b U\int_0^t\left(U\dot{\alpha}(\tilde{t}) - \dot{v}(\tilde{t})\right)k(t-\tilde{t})\,d\tilde{t} + \frac{1}{2}\pi\rho b^2 U\int_0^t\ddot{\alpha}(\tilde{t})\,k(t-\tilde{t})\,d\tilde{t} \quad (51)$$

and with the dot now denoting differentiation with respect to $t$.

## 3. TIME-DOMAIN LIFT IN STATE-SPACE FORM

### *3.1 Introduction*

The main procedure of the method for developing a state-space approximation of the memory effects, is to do an input-output system-identification of the impulse response, followed by a model reduction to lower the approximated system order into a suitable low-order form, as shown in (Kristiansen and Egeland 2003) for the problem of frequency-dependent added mass.

### 3.2 *Wagner function as impulse response*

The input-output dynamics of a linear time-invariant system is uniquely represented by the impulse response of the system. The unsteady lift equation for arbitrary vertical motions of a foil with steady forward speed is given by (51), and the convolution term kernel is the Wagner function $k(t)$.

Dependent of what base length variable that is used in the reduced time in the model formulation, either $b$ or $c = 2b$, the Wagner function has two different versions. Thus, the version used here based on the chord length $c$ has the range of $k(0) = 1$ to $k(\infty) = 2$, which is twice the value of the version based on the halfchord length $b$. The approximation for the Wagner function used here is the Küssner expansion

$$k(\iota) = 1 + \sum_{j=1}^{\infty} \frac{\iota^j}{(2\iota + 2)^j} \qquad (52)$$

for $j = 20$.

To be able to find a state space representation based on the impulse response, convergence to zero is needed. Therefore, (51) is rewritten into

$$\begin{aligned} L(t) &= L_s + \pi\rho b^2 (\dot{v} + U\dot{\alpha}) \\ &+ 2\pi\rho b U \left( U\alpha - v + \frac{1}{2} b\dot{\alpha} \right) \\ &- \pi\rho b U \int_0^t \left( U\dot{\alpha}(\tilde{t}) - \dot{v}(\tilde{t}) \right) \left( 2 - k(t - \tilde{t}) \right) d\tilde{t} \\ &- \frac{1}{2}\pi\rho b^2 U \int_0^t \ddot{\alpha}(\tilde{t}) \left( 2 - k(t - \tilde{t}) \right) d\tilde{t} \end{aligned} \qquad (53)$$

The convolution term is now separated into three independent terms with separate inputs $\dot{v}$, $\dot{\alpha}$ and $\ddot{\alpha}$. Letting these be a unit inpulse, yields i.e for $\dot{v}(t) = \delta(t)$

$$\int_0^t \delta(\tilde{t}) \left( 2 - k(t - \tilde{t}) \right) d\tilde{t} = 2 - k(t) \qquad (54)$$

which is positive and converges to zero as $t$ goes to infinity. It is here assumed that $v$, $\alpha$ and $\dot{\alpha}$ have zero as initial values. For further reference, let

$$\tilde{k}(t) = 2 - k(t) \qquad (55)$$

### 3.3 *State space model*

The three separate convolution terms with inputs $\dot{v}, \dot{\alpha}$ and $\ddot{\alpha}$, can be approximated as the outputs $\mu_1, \mu_2$ and $\mu_3$ of three additional state-space systems. The approximation of the lift equation can then be written as the system

$$\begin{aligned} L - L_s &= 2\pi\rho b U (U\alpha - v) + \pi\rho b^2 (\dot{v} + 2U\dot{\alpha}) \\ &- \pi\rho b U \left( U\mu_1 - \mu_2 + \frac{1}{2} b\mu_3 \right) \end{aligned} \qquad (56)$$

$$\dot{\boldsymbol{\xi}}_1 = \mathbf{A}_1\boldsymbol{\xi}_1 + \mathbf{B}_1\dot{v}, \; \mu_1 = \mathbf{C}_1\boldsymbol{\xi}_1 + \mathbf{D}_1\dot{v} \qquad (57)$$

$$\dot{\boldsymbol{\xi}}_2 = \mathbf{A}_2\boldsymbol{\xi}_2 + \mathbf{B}_2\dot{\alpha}, \; \mu_2 = \mathbf{C}_2\boldsymbol{\xi}_2 + \mathbf{D}_2\dot{\alpha} \qquad (58)$$

$$\dot{\boldsymbol{\xi}}_3 = \mathbf{A}_3\boldsymbol{\xi}_3 + \mathbf{B}_3\ddot{\alpha}, \; \mu_3 = \mathbf{C}_3\boldsymbol{\xi}_3 + \mathbf{D}_3\ddot{\alpha} \qquad (59)$$

where subsystem $\xi_i$ can be of arbitrary order $r_i$. As the convolution terms have the same kernel $\tilde{k}(t)$, the state-space matrices will be identical for the three inputs given that they are of the same order, and the suffixes of the matrices can be dropped.

The state-space matrices are found by using the system identification scheme based on the Hankel SVD method proposed in (Kung 1978), which is readily available as the function `imp2ss` in the Robust Control Toolbox of MATLAB. Here, the upper limit of $\iota$ is taken to be $\iota = 40$, ($t = 80$ when $c = 1, U = 2$). The temporal resolution were taken as $dt = 0.01$. To make it computationally efficient, only some evenly temporally spaced values of $\tilde{k}(\Delta t)$ are used as inputs, but densely enough to capture the main dynamics of the response. In this case $\Delta t = 0.4$. The output is then scaled with the time-step $\Delta t$, according to

$$\begin{aligned} \bar{\mathbf{A}} &= \tilde{\mathbf{A}}, \quad \bar{\mathbf{B}} = \tilde{\mathbf{B}} \\ \bar{\mathbf{C}} &= \tilde{\mathbf{C}}\Delta t, \quad \bar{\mathbf{D}} = \tilde{\mathbf{D}}\Delta t \end{aligned} \qquad (60)$$

`imp2ss` has a built-in model reduction option which bounds the $H^\infty$-norm of the error between the approximate realization and an exact realization. This turned out to be an unsatisfactory solution to control the accuracy and order of the state-space approximation. Thus, no specific tolerance was stated, resulting in the order of the approximate system reaching as high as 200, but with very good accuracy. This can be seen in relation with the number of measurements of $\tilde{k}(t)$ being $T/\Delta t = 200$. Therefore balanced model reduction was used by the MATLAB function `balmr` from the Robust Control Toolbox, which are based on truncation and Schur methods (Safonov and Chiang 1989). In this work, models of order 2, 3 and 4 were calculated.

## 4. RESULTS

The foil system was simulated with $c = 1\ m$, $U = 2\ m/s$, $\rho = 1025\ kg/m^3$. First, the state-space approximations are calculated. This is only

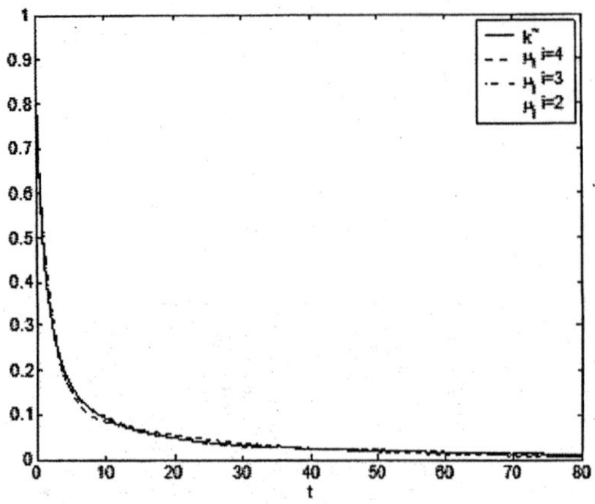


Fig. 2. Impulse response of the state-space approximations.

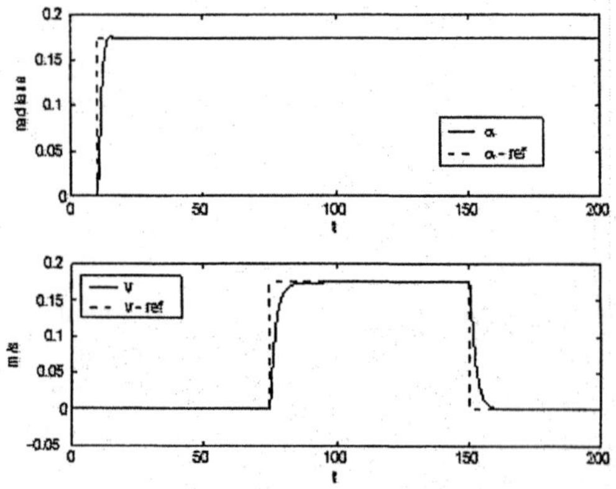


Fig. 3. Angle of attack and vertical velocity reference signals and system inputs.

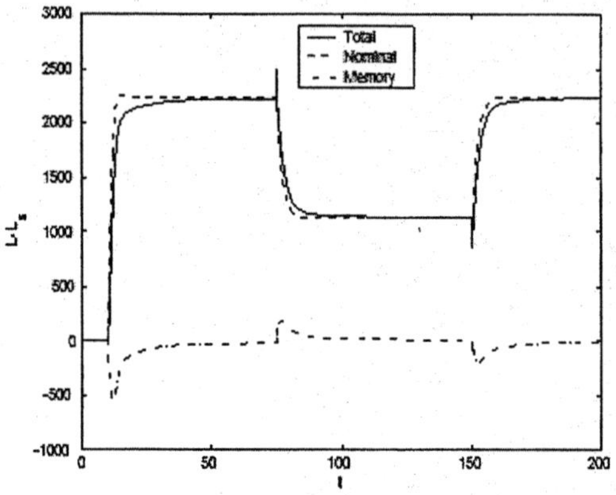


Fig. 4. Lift force including memory effects from change in vertical velocity and change in angle of attack.

needed once. The outputs $\mu^i(t)$ of the approximated memory effect systems given a unit impulse as input had very good accuracy compared to $\tilde{k}(t)$. Here, $i$ is the order of the approximated system. As can be seen in Figure 2, an approximation of order as low as 2 give good agreement with $\tilde{k}(t)$, which suggests that it is more to gain in overall accuracy by increasing the accuracy of the approximation of the Wagner-function itself than increasing the order of the state-space approximation. In further simulations, an approximation of order 4 was chosen.

A model of the total foil system with memory effects was implemented in Simulink. In order to have smooth inputs for all the derivatives in the lift equation, the vertical velocity and angle of attack was implemented as two separate PD-controlled feedback systems

$$\dot{\mathbf{x}}_i = \mathbf{A}_i\mathbf{x}_i + \mathbf{B}_i u_i \tag{61}$$

$$y_i = \mathbf{C}_i\mathbf{x}_i \tag{62}$$

where $i = v, \alpha$. Here, $\mathbf{x}_\alpha = \left(\alpha\ \dot{\alpha}\ \ddot{\alpha}\right)^T$, $\mathbf{A}_\alpha = \begin{pmatrix} 0 & 1 & 0 \\ 0 & 0 & 1 \\ 0 & 0 & 0 \end{pmatrix}$, $\mathbf{B}_\alpha = \left(0\ 0\ 1\right)^T$, $\mathbf{C}_\alpha = \left(1\ 0\ 0\right)$, and $\mathbf{x}_v = \left(v\ \dot{v}\right)$, $\mathbf{A}_v = \begin{pmatrix} 0 & 1 \\ 0 & 0 \end{pmatrix}$, $\mathbf{B}_v = \left(0\ 1\right)$, $\mathbf{C}_v = \left(1\ 0\right)$. The vertical velocity system had $P = 1, D = 1$, and the angle of attack system had $P = 1, D = 1.5$. Both input systems had zero as initial conditions. The total system was simulated for 200 seconds, with the input reference signals and system input signals as shown in Figure 3.The lift can be seen in Figure 4.The behaviour of the unsteady lift is in good agreement with the discussion in (Newman 1977), but seems to have a slightly smaller amplitude. This may result from the smoothing of the input from the feedback system implementation, opposed to the ideal, but non-physical, case of an impulse input in $v$ or $\alpha$. The spikes in the nominal lift are due to the second term in (56).

## REFERENCES


Kristiansen, E. and Egeland, O.: 2003, Frequency-dependent added mass in models for controller design for wave motion damping, *6th IFAC Conference on Manoeuvering and Control of Marine Craft*, pp. 90–95.

Kung, S.: 1978, A new identification and model reduction algorithm via singular value decompositions, *Proc. Twelfth Asimolar Conf. on Circuits, Systems ans Computers*, pp. 705–714.

Newman, J.: 1977, *Marine Hydrodynamics*, MIT Press.

Safonov, M. and Chiang, R.: 1989, A schur method for balanced model reduction, *IEEE Trans. on Automat. Contr.* **AC-34**(7), 729–733.

Woods, L.: 1961, *The Theory of Subsonic Plane Flow*, Cambridge University Press.

ELSEVIER
IFAC
PUBLICATIONS
www.elsevier.com/locate/ifac

# ASFOSS – CURRENT INFORMATION SYSTEM USING AIS[1]

**Holger Korte*, Jens Ladisch**, Matthias Wulff**, Cathleen Korte**, Jürgen Majohr****

**Institute of maritime Automation Technologies and Navigation (MATNAV)*
*Friedrich-Barnewitz-Straße 3, 18119 Warnemünde, Germany*
*e-mail: holger.korte@etechnik.uni-rostock.de*
***University of Rostock, Institute of Automation*
*Richard-Wagner-Straße 31, 18119 Warnemünde, Germany*

Abstract: The manoeuvring freedom is reduced by use of larger and larger vessels in estuarine waters. Modern steering and propulsion devices only compensate in a conditional manner this disadvantage. Especially unknown adverse conditions of winds and currents give an only hardly solvable task to ship officers. A reasonable support in the decision making can give the appropriation of externally measured surrounding values on board.
The developed system ASFOSS transmit the measured data via AIS from the estuary basis to the ship. On board of the ship the data will be presented at an integrated display device like ECDIS, conning or radar. *Copyright © 2004 IFAC*

Keywords: Ship, Automatic Identification System (AIS), Broadcast Message, Navigation System, Current, Wind

## 1. INTRODUCTION

Presently, a trend to larger and larger vessels can be observed in shipping. This is an effect of greater economical compulsions on one hand. On the other hand, the widths of estuarine waters remained overdue unchangeable. This evolution leads to a decrease of the manoeuvring freedom. At the same time the shipping companies reduced the manpower on board. A lot of ships are free of pilotage and tug assistance. More and more the responsibility over the ship motion process concentrates on ship's masters.
On board new ships e.g. ferries and car vessels we can find some modern steering gears, sometimes in quite opulent number. Such ones are for example podded drives, pump jets, Schottel propellers, bow and stern thrusters etc. Their use requires high attention of the captain during the passage of narrow waters and harbour entrances.
But such gears are excellent aids in calm waters to manoeuvre in docklands to put on and push off the pier. In this way, they contribute to the reduction of tug assistance and the ship owner achieves an economical advantage. Yet its effort proves in the state of sail to be treacherous. In dependence of ships speed they lose their steerage effect fast. Then the ship is led with traditional steering methods (equal with aft rudder propulsion device). In case of adverse weather and hydrodynamic conditions, the better steering qualities of modern large vessels result to a higher risk of misjudgements with high damage potential. Information about conditions in areas to be passed through gives a realistic help. Presently, we have a development boom of navigation devices which assist in manoeuvre planning and decision making (Korte et.al. 2001, Kruijt 2003, Zölder et.al. 2003, Zimmermann 2000).
In co-operation with the industrial partner MAR Rostock GmbH our research group works on the development of such navigation aids including external current and wind information.

## 1. PROJECT MAPSYS[2]

MAPSYS is a new module of an integrated navigation and ship handling system. It computes a motion prediction for the ship with high faith of reality. This

[1] AIS - Automatic Identification System
[2] MAPSYS - *Ma*noeuvre *P*rediction *Sy*stem for *S*hips

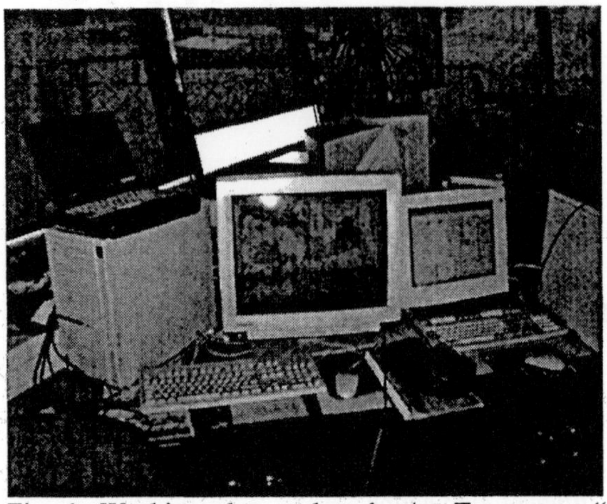

Fig. 1. Working place onboard m/v „Transeuropa" with host-system and ECDIS double.

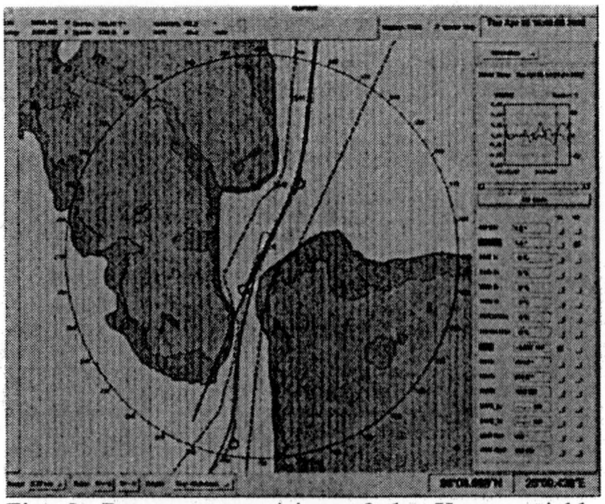

Fig. 2. Process supervision of the Kustaanmiekka passage in the simulation menu after prediction.

goal is achieved by inclusion of externally measured current data and an actually identified motion model of the ship. A vertically working Acoustic Doppler Current Profiler (ADCP) determines the current profile near a critical passage and provides it on a database system. A ship can receive these data by a call via GSM-net D2 (*G*lobal *S*ystem for *M*obile Communication).

The available current data are used for a calculation of the prediction path through a critical passage. A locally limited current model computes the current data near the planned path at first. In the second step, the motion model generates the future movement of the ship. Our navigation system is strongly connected with the surrounding technique. It needs the actual and past state values to adapt the motion model parameters on one hand, and on the other hand, the system output should be represented at an available screen of the ship handling system in order not to increase the load of mates. For the realisation of the system on board we installed the prediction computer and a D2-net modem into the ship console only.

The prediction system is controlled via specific tools in the ECDIS. So we can use the voyage planning data for our prediction module. If required, the ship mates can obtain additionally information about the approaching probable ship motion.

MAPSYS was tested onboard the m/v "Transeuropa" during a routine round trip dated 18.-22.03.2002 (Korte et.al. 2003). Fig. 1 presents the place of work of our group with host-computer and ECDIS double in the wheel house. As example, Fig. 2 shows the result of the prediction calculation during the process supervision in the passage of Kustaanmiekka (Helsinki).

After the extensive alignment of the systems on the bridge, MAPSYS worked reliably in connection with the NACOS-System of STN Marine Electronics. None system failures were found in real time in all special system states. For reasons of technical admission, the ECDIS device was duplicated. To get practical feedback, we discussed with the watching officers about the handling of our tools and the quality of system outputs. Finally, there were different opinions about representations during the process supervision. The integration of our system into the ship steering console was voted positively.

## 3. PRACTICAL PROBLEMS TO INTEGRATE OUR SYSTEM

The aims of project MAPSYS were fulfilled in the development period. Some problems yet existed for the practice introduction.

At first, it is a scientific problem to make a verification of a small scaled current model which describes situations of statistical exceptions. MAPSYS can't give a guarantee for its calculation results in such extreme hydrodynamic situations. Indeed, we get real currents resulting of the sensor location nearly the critical passage. But a risk exists for the edge field of our model caused by the extrapolation of model data. The prediction results can be different from the experiences of mates. Hence, the acceptance by ship officers would be reduced finally.

As second, the use of a GSM-based telemetry between board and estuarine components wasn't really a problem in our tests. An advantage to use such systems was the costs assignment to the end user. At every time, the connections were built punctually. All necessary data could be transmitted. But the system can't guarantee a safe connection at sea. Therefore, we search for other possibilities of telemetry.

As third, the aim of MAPSYS is the prediction of movements close to reality. This requires a continuous adjustment of the model parameters for ships with frequently changing load conditions. A lot of motion data has to be received end evaluated automatically from the periphery of the vessel. But we haven't compatible modules of ship handling devices between the different manufactures. Data busses are optimized for the requirements of individual systems. The system data, e.g. state values are not approved for the use with external devices. At least, the system MAPSYS represents a modular single solution. Only a special ship with a fixed system configuration can take up the additional technique. Nevertheless, the manufactures of several wheelhouse techniques are asked to open their systems for higher integration. So, we are able to arrange the system MAPSYS in dependence of the existing components on board.

## 4. SYSTEM ASFOSS[2]

With the introduction of horizontal ADCP technology (RD Instruments 2003) into practice and the world wide outfit obligation of AIS (Automatic Identification System) for ships in a period of the latest 3 years (MSC 74/69), suitable devices and aids are available to neglect the disadvantages of comparable MAPSYS subsystems. On one side, the horizontal measuring principle of HADCP allows calculations of local gradients directly. Extrapolation errors are prevented in peripheral areas of flow models and extreme weather conditions by the use of such technology. On the other side, a save radio data link exists onboard with the outfit obligation of AIS. The AIS radio link between shore and onboard systems can be used for safety related information transfer, because the system is certificated by the governmental authorities. The cost-benefit ratio of an AIS investment will be decreased with the additionally

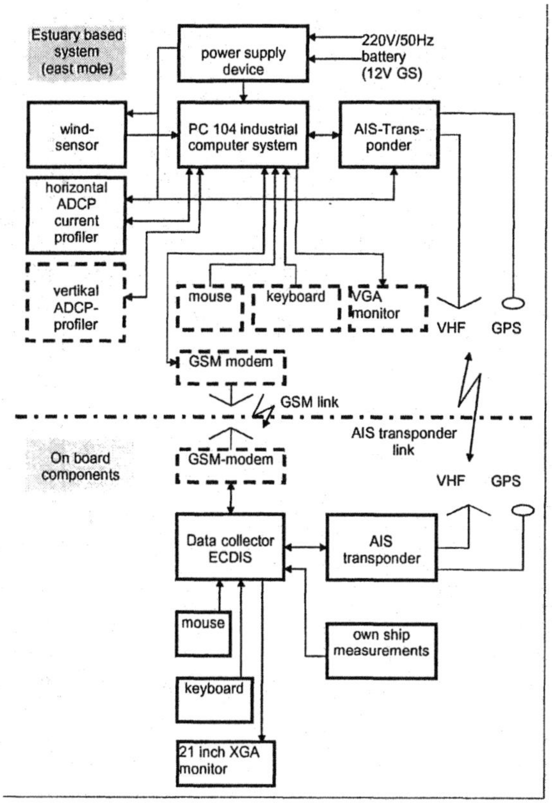


Fig. 3. Block diagram of ASFOSS components.

effort. In accordance to MAPSYS, the additional information should be represented in an available integrating navigation aid on board. In our projects we use the Electronic Sea Chart, called ECDIS, as integrating device.

Fig. 3 shows the general structure of the system, whereby the upper part consists of different sensors, a data collector, several transformer units and an AIS transmitter system. Together, these are the estuarine based station. The lower part gives an overview about the included onboard components. All components are available on a modern wheel house, so the onboard components of the system needs only additionally software tools to understand the special transmission code (section 6). All dashed components were already used in MAPSYS. If necessary, these can supplement the ASSFOSS system.

## 5. PRINCIPLES OF HORIZONTAL ADCP-MEASURING

It is a new idea to use the horizontal ADCP measuring procedure in the field of sea shipping. The main principle of measurement based on a shifting frequency of acoustic reflections on moving particles in water, the so-called Doppler Effect. By measuring of differently arranged sound impulses a multidimensional determination of the current is possible. Horizontal ADCP's works with two beams, minimally. In our investigations we used the long range HADCP of the company RD Instruments, San Diego. This profiler system works with three around 20° shifted beams. Therefore, it is possible to recognize failures of individual beams and to correct the calculation result additionally. A maximum range $R_{max}$=300 m is reached by a basic frequency of sounds $f_G$=300 kHz. The allocation into measuring cells can take place by means of a quantization of the echo signal in appropriate time windows (RD Instruments 2003).

Up to now, horizontal ADCP's were used for flow regulations in rivers and channels. Due to the expansion of the measuring section which results by

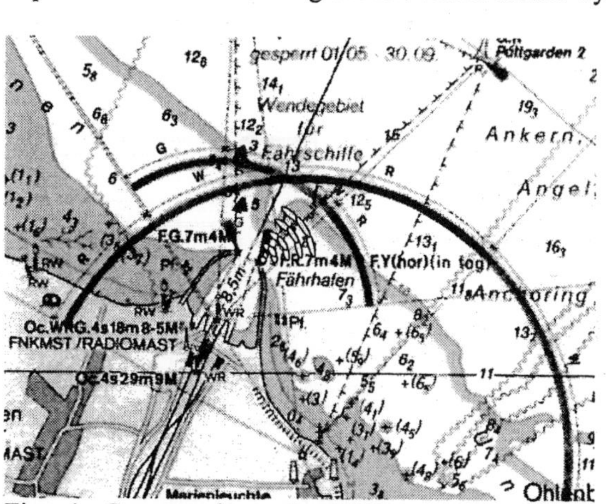


Fig. 4. Cut out from the BSH sea chart no. 31 „Waters around Fehmarn Island" with HADCP measuring sector.

---

[2] ASFOSS - *A*ssistance *S*ystem *fo*r *S*afe *S*hiphandling

the different radiation patterns homogeneity of the current is presupposed. This basic assumption is hurt in the sea-range both by swell and by change of depth. Nevertheless, the measurement principle appears useful for shipping, because the current component must be only determined transverse to the harbour approach and parallel to the shore. In this case, the HADCP procedure can be used only after different pre-investigations.

In the context of the project ASFOSS investigations should be accomplished also to this aspect.

The propagation of the sound takes place strongly bundled as beams. Caused by the beam angle $\alpha_B$, the thickness $h_B$ increases with the increasing distance $E_S$. A clear allocation of the backscatter to the flowing medium can be ensured only, if the water depth $h_{WS}$ is bigger as the local thickness of the beam. In addition, the beams may not turn out by diagonal radiations ($\Delta\vartheta$ - pitch and $\Delta\varphi$ - roll) to the water surface or the sea bottom. With a stationing of the sensor device in a diving location being the half water depth, the maximally attainable distance $E_{max}$ amounts to

$$E_{\max} = \frac{h_{WS}}{\tan[\alpha_B + 2\Delta\vartheta + 2\tan(20°)\Delta\varphi]}.$$

For instance, for the estuary of the Puttgarden harbour approach with a water depth of $h_{WS}$=7 m (see fig. 4.), a beam width of $\alpha_B$<1° and angular set accuracies of $\Delta\vartheta,\Delta\varphi$=0,5° the maximally attainable range amounts to $E_{max}$=215 m.

The range of HADCP reaches up to the measuring point of the MAPSYS flow model (Korte et.al 2001). This corresponds also with experiences, which were made during the measurements in Puttgarden, last summer.

To examine the procedure of HADCP, a comparative measurement with a vertical working ADCP was implemented. In contrast to the horizontal ones, the homogeneity of the current in the vertical flow cells represents a justified acceptance. The vertical sensor was installed onboard the small catamaran MESSIN™. MESSIN, that is a 300 kg more heavily, autonomously acting sea-measurement vehicle of approximately 3 m length (Majohr et.al. 2000). During a period from 07th to 10th May of 2003 we implemented a set of profile measurements in the sector of HADCP. Unfortunately the currents were not strongly pronounced in this period due to a stationary high pressure weather situation with light to moderate off-shore winds from southwest. Fig. 5 shows a comparison of HADCP measurements with vertical ADCP's from the 08th May against 16:25 o'clock. Caused by an alteration of the wind direction to west the current was increased. But also, the height of waves was increased due to the weather changing. So, the MESSIN works more in the sea and its roll and pitch motion were faster. The ADCP internal inclination sensor is to slowly-acting to compensate the angularly motion of the vehicle completely. Therefore, a clear change of the current direction along its path is to be seen in fig. 5. Both procedures give a nearly equal result in the waterway crossing direction. Nevertheless, we assume a permanent current measuring campaign to testing the usefulness for shipping. These were not possible in context with ASFOSS, because technical deployments were focussed.

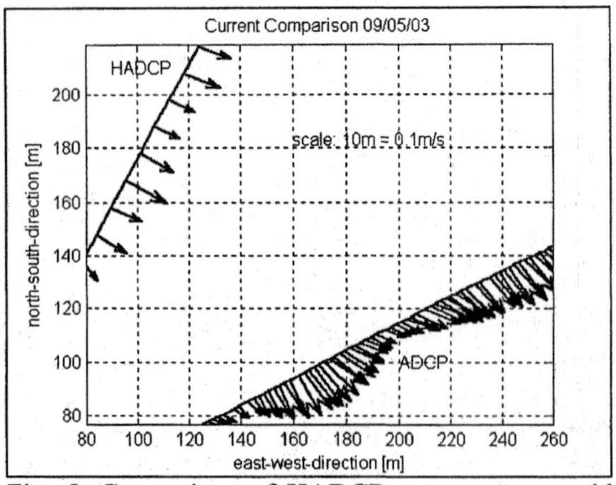


Fig. 5. Comparison of HADCP measurements with vADCP measurements outside the breakers zone. The point (0,0) is the location of the HADCP sensor device.

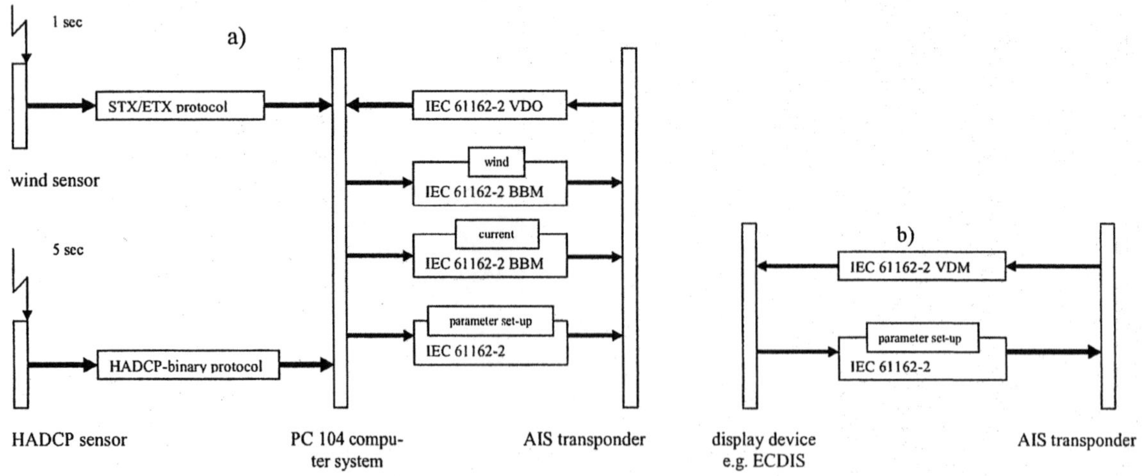


Fig. 6. Communication procedures between components of the estuarine system (a) and on board (b).

## 6. AIS AND BINARY MESSAGES

The second substantially change to the system MAPSYS exists in the realization of a secured data communication. With the introduction of the AIS system to the navigation a certified transmitting medium is available. This AIS system is a self-organizing radio link in the VHF range (canals 87B and 88B), which works in the time-slot method (ITU 2001). It serves for the change of traffic-relevant data and aims at a more objective estimate of traffic situations by ship mates.

On this field the goal of MAPSYS and ASFOSS is the same with the difference that none vehicle information are changed. But in our case, unknown environmental data are made available to get a better situation overview. Traffic-relevant environmental data are above all wind and sea current data on estuarine waters and sea areas with large ecological sensitivity. A transmission of environmental data by means of AIS appears justified, if onboard none other possibility for their determination exists. Here in particular the limited capacity of the data communication is considered, in order to block the radio link not unnecessarily. For the case of the currents in the estuary of Puttgarden the conditions for the use of AIS are given. Particularly for such cases a so-called Binary Message in the AIS was planned, which makes additional data communication possible (IEC 61993-2). However, these additional data communication deviates from the standard.

The idea of the transmission of environmental data via AIS is not new actually. In particular in the north of America realized projects are to be found in the references /12, 13/. However, the disadvantage of the Binary Message is that it becomes accessible for few users only due to missing standardization. Thus also in the project ASFOSS a data-set was developed, which is adapted to the sensor system, which consists of a wind sensor and the HADCP. The structure of this data record was implemented in accordance with the work stated above (VTSC 2002), whereby the current data one extended regarding on the possible use of profilers with several measuring cells.

Always, a Binary Message consists of maximally 968 bits, whereby the first 16 bits are reserved for the user identification. Thus the further 952 bits are available for the actual data. Especially for the current data set the structure is given by (Korte et.al. 2003).

The communication between the individual systems takes place differently in the two subsystems. These subsystems are represented schematically in the fig. 6. The principal item of the estuarine part is an industrial computer type PC 104. It receives the raw data of the sensors and prepares it for the data communication. Beside the data of the environmental sensors the process computer receives the position from the GPS-receiver of the AIS by means of a NMEA-VDO-string for the distinction from foreign positions (VDM). Thereupon the binary messages can be provided in accordance with the structures specified above. For the supply of the data in the AIS master station the Messages are totally enclosed with the NMEA format and will transferred on a serial data link. But the fig. 6a shows a further data stream of the process computer to the AIS transmitter. If necessary this data stream is used for the parameter attitude of the transmitting plant.

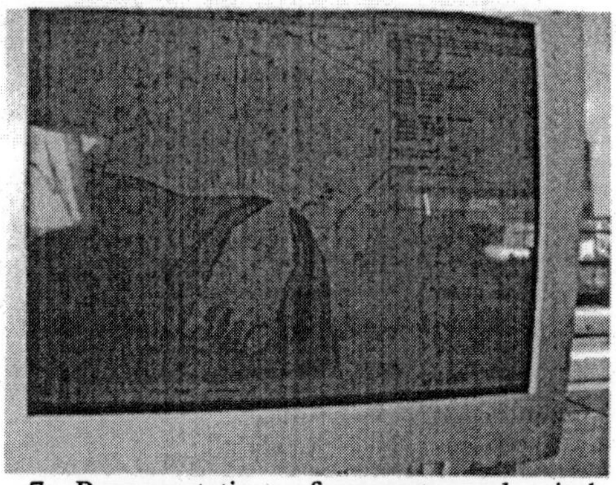

Fig. 7. Representation of currents and winds in ECDIS onboard of m/f „Schleswig-Holstein" during a test on 17th of Sept. 2003 (data transfer via AIS).

The receiver station is designed more simply (see fig. 6b). In addition in the near future all the ships will be equipped with AIS in series, so that the telemetry does not cause additional costs in the system. However, because the format of the binary messages cannot be standardized, the ships need a special software in the receiver, which decodes the message code. This software could be installed for example on the ECDIS computer, how it was realized in the project too. In addition, other visualization systems are suitable as indicator module (radar, conning display).

After a two-year development time the system could be tested on 17th of Sept. 2003 for the first time onboard the ferry liner "Schleswig-Holstein" (Scandlines Germany). Fig. 7 show the representation of the measured values of selected current cells and of a wind sensor in the ECDIS. Unfortunately only small currents prevailed forwards at this time too, so that no more noticeable improvement was to be registered in decision making. However, in the discussions with the officers under command of captain Gering it becames clearly that this current representations are very meaningful, particularly during voyages at night and with foggy conditions. In these cases the wake of the port leaving ferries cannot be constituted. However a representation of the measured values in the radar or conning display would be preferred by the mates, because these devices are particularly used during the manoeuvres.

At the same time the voyage was used to test the further developed ADCP telemeter. In the context of a student work the program for representations of the measured values of the HADCP-sensor was

Fig. 8. Representation of HADCP data onboard of m/f „Schleswig-Holstein“ after transmission via GSM in the modified ADCP Telemeter.

prepared. Additionally this is a simple current display system which use the developed estuarine components. Apart from the results of the project MAPSYS also this test shows that GSM is suitable as a transmitting medium. Finally, an information improvement exists after the repeated tests in different sea-areas. If the GSM connection cannot be made, then this error condition is indicated. However, because the current is a not prescribed safety-relevant value, this disadvantage can be accepted.

## 7. CONCLUSIONS

The phenomenon of the current makes high requirements against the ship officers. Onboard there are suitable methods for sufficiently exact regulation only for tidal waters. However, in non-tidal waters the currents can achieve relevant values of magnitude. In particular strong current gradients produce large problems for the ship's management. Because, on the one hand they produce an additional strong torque to the drift movement, and on the other hand they can not determined from the past track process. In the past this leds to strandings and collisions with buildings despite modern propulsion principles. Within the research projects MAPSYS and ASFOSS at the University of Rostock and Institute MATNAV several technical possibilities were created to transfer the information onboard of ships by means of ADCP sensor technology and a suitable telemetry devices. In a practice test the procedures could be examined for their efficiency on board. The integration of the display and control devices into existing ship guidance systems formed a substantial development focussed to obtain an ergonomically optimal information gain.

By means of the AIS Binary Messages the transfer of environmantal information represents a usual certified procedure. However it is accessible to certain users only. Therefore in particular in situations with high density of traffic the transmission of such environmental parameters should be minimized. In addition a standardization of current messages would be meaningful.

Accomplished attempts in different sea-areas led to the punctual connection and thus to an information gain for the ship managements. However, the current values determined within the measurements and system tests of the project ASFOSS don't represent a statistically relevant data set. Therefore in common efforts with the offices for navigation a continuous measuring campaign with the developed technology is recommended.

## REFERENCES


K. Berger-North (2002). AIS Environmental Monitoring System measuring winds currents and water level at Delta Port, final report, AXYS Environmental Systems, Vancouver (CAN).

IEC 61993-2. Maritime navigation and radiocommunication equipment and systems - Automatic identification systems (AIS) - Part 2: Class A shipborne equipment of the universal automatic identification system (AIS) - Operational and performance requirements, methods of test and required test results.

ITU (2001). Technical characteristics for a universal shipborne automatic identification system using time division multiple access in the VHF maritime mobile band, p. 65, International Telecommunication Union, Doc. 8/BL/5-E, 19.04.2001.

H. Korte, H.-D. Kachant, J. Majohr, Th. Buch, C. Korte, M. Wulff (2001). Concept of a modern Manoeuvre Prediction System for Ships, *Proc. IFAC Conference Control Applications in Marine Systems CAMS 2001*, No. WA4.3, Glasgow (UK). 18.-20.07.2001.

H. Korte, J. Majohr, C. Korte, M. Wulff (2003). ASFOSS – current information via AIS, *Proc. 9th Schifffahrtskolleg*, Institute of Shipping, Warnemünde, 05.-06.11.2003 (in german).

W. J. Kruijt (2003). The Integrated Bridge: more than Integration of Bridge Systems alone?, *Proc. 13th Ship Control Systems Symposium*, paper no. 229, Orlando (USA), 7-9.04.2003.

J. Majohr, Th. Buch, C. Korte (2000). Navigation and automatic Control of the Measuring Dolphin (MESSIN), *Proc. 5th IFAC Conference on Manoeuvring and Control of Marine Crafts*, pp. 405-410. Aalborg (Denmark), 23.-25.08.2000.

MSC 74/69. Recommendation on Performance Standards for a Universal Automatic Identification System (AIS), MSC 74/69.

RD Instruments Inc. (2003). http://www.rdinstruments.com/pdfs/HADCP300k.pdf ,(version from 20th Okt.).

VTSC (2002). St. Lawrence Seaway AIS Data Messaging - Formats and Specifications-, Revision 4.0A, J.A. Volpe Transportation Systems Center, Cambridge, MA (USA), 09.05.2002.

A. Zölder, K. Pankow, R. Eyrich, R. Müller (2003). Pilot's Mate – assistance system for shipping at sea. In: *Concepts for a better sea traffic safety in the North and Baltic Sea* (F. Ziemer. (Ed)), pp. 79-94, Institute of Shipping, Warnemünde, ISSN 1437-031X (in german).

R. Zimmermann (2000). Representations of ship dynamic models in a navigation system for inland shipping, dissertation, University of Stuttgart and Logos-Verlag Berlin, ISBN 3-89722-477-1 (in german).

ELSEVIER
IFAC
PUBLICATIONS
www.elsevier.com/locate/ifac

# AUTONOMOUS FAST SHIP PHYSICAL MODEL WITH ACTUATORS FOR 6DOF MOTION SMOOTHING EXPERIMENTS

**J. Recas, J. M. Giron-Sierra, S. Esteban, B. de Andres-Toro, J.M. De la Cruz, J.M. Riola**

*Dep.. A.C.Y.A., Fac. CC. Fisicas. Universidad Complutense de Madrid*
*Ciudad Universitaria, 28040 Madrid. Spain*
*e-mail: gironsi@dacya.ucm.es*

Abstract: This paper describes a new experimental system that has been recently developed for a research on vertical motion alleviation of fast ferries. The case of a fast ferry with a pair of transom flaps, lateral fins and a T-foil is considered. All these actuators should be moved in the most effective way, and an appropriate control strategy must be obtained and tested. A scaled physical model of the fast ferry has been built, with moving actuators and a distributed monitoring and control system. The research considers 6DOF motion attenuation, for any ship's heading. For this scenario, an autonomous physical model, not to be towed, is required. The physical model includes two scaled waterjets, with the jet orientation under control. The ship has no rudder. Heading and motion alleviation control includes many sensors and actuators. As a good practical control solution, a distributed architecture based on the CANbus has been devised. There is a set of microcontrollers and an embedded PC as coordinator. The microcontrollers for the actuators simulate the time behaviour of hydraulic cylinders. The microcontrollers for sensors include signal conditioning functions. There is a digital radio communication with an off-shore monitoring system. The off-shore system can display the present and the recorded experiments with animated 3D graphics. The complete experimental platform can be used as an Internet laboratory. The system can be easily tailored for use in real ships. *Copyright © 2004 IFAC.*

Keywords: Ship control. Fast ferries. Experimental systems. Monitoring and control.

## 1. INTRODUCTION

Passenger comfort is an important issue in fast ferries. In particular, seasickness is related to oscillating vertical accelerations, with a frequency around 1 rad./sec. (O'Hanlon and MacCawley, 1974). Certain combinations of sea state and ship's speed cause this kind of vertical accelerations, with negative effects on passengers. By means of submerged moving actuators, it is possible to counteract the effect of encountered waves, focusing specially in the frequency band around 1 rad./sec.

Our research deals with control design for seasickness minimization, using actuators such flaps, fins and T-foil. A particular fast ferry has been selected for case study. The research has been divided into two steps:

A. The simple case of the fast ferry with head seas. Only pitch and heave motions are considered. These two motions are smoothed by means of transom flaps and a T-foil.
B. The more general case of the fast ferry with any heading is considered. Pitch, roll and heave cause vertical accelerations. In addition to the flaps and the T-foil, lateral fins are included for roll smoothing. The actuators may also induce effects on yaw, sway and surge motions. A 6DOF study is required.

The step A has been already accomplished. A physical model of the ship, at 1/25 scale, has been built. It has been used for experiments in El Canal de Experiencias Hidrodinamicas de El Pardo

(CEHIPAR), Madrid. This towing tank facility has a long channel for quiet water experiments, and a 150m x 30m basin with wavemaker. The physical model has been towed along the channel to measure the drag and lift of actuators. It has been also towed in the basin with several sea conditions for two purposes. First to get data for pitch and heave motion modelling. Second for testing of motion alleviation control designs. This step of the research has been reported along several publications, such (Esteban, et al., 2000; Giron-Sierra, et al., 2001; Esteban, et al., 2002). A relevant reference for ship control is (Fossen, 2002).

Step B of the research is now under way. It is no longer advisable to use a towed physical model for the new experiments with any ship's heading. Consequently, a new, smaller physical model has been developed for *autonomous motion.* Stringent requirements of weight, robustness and power consumption have been considered. New technologies have been included. A distributed monitoring and control system has been developed. This system uses digital radio for distant monitoring and supervision of experiments by an off-shore unit. The off-shore unit displays the dynamic behaviour of the ship with animated 3D graphics. Both the on-board system and the off-shore unit can be used with real ships.

The paper describes the new experimental system, including the physical model with actuators, and the off-shore monitoring and supervision unit.

## 2. EXPERIMENTAL OBJECTIVES AND DESIGN

The objective of the research is to smooth vertical accelerations by using a set of controlled moving actuators. Figure 1 shows the location of the actuators: two transom flaps, two lateral fins and a T-foil near the bow. The T-foil has two independent wings.

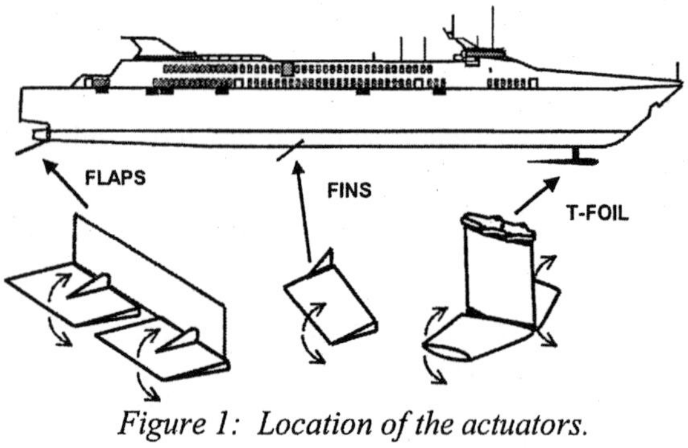


*Figure 1: Location of the actuators.*

The control of the actuators is multiobjective. Small waves do not deserve actuator action, wasting energy and system lifetime. Also, high angles of attack of the actuators induce cavitation, which destroys the actuators. Consequently, the control must try to minimize seasickness, while selecting when to react against waves and keeping cavitation as low as possible.

The experimental study for the optimal control design begins with a series of experiments to determine a 6DOF model of the motion of the ship with actuators. On the basis of the model, a simulation environment will be developed, and control design studies will be done. The second part of the experimental study will be devoted to control design validation.

The experiments are made in the basin with wavemaker, using the new physical model. In each experiment, the model runs autonomously across the basin, with a fixed heading. A set of sea states, SSN4, 5 and 6, with JONSWAP spectra is selected for the experiments. Two speeds, 30 and 40 knots, are studied. Twelve ship's headings, regularly spaced along the 180° are considered. That means 72 cases. For each case, two runs are made with fixed actuators, and two runs are made with controlled moving actuators.

The physical model is not towed. Instead, it uses *waterjets* for autonomous motion. There will be no cables connecting the replica to external instrumentation or control system. Note that from the beginning we need a first version of control, to get constant ship headings along the experiments.

The data to be obtained are records of the six motions and accelerations, and the control signals. A selection of the data will be transmitted via radio to the off-shore unit, for real-time supervision and monitoring. During an experiment, all data are saved by the on-board computer. At the end of the experiment, all on-board saved data are transmitted to the off-shore unit to be saved on hard disk.

Apart from the CEHIPAR basin, other experimental scenarios are considered, such ponds or quiet sea coast.

## 3. THE PHYSICAL MODEL

Figure 2 shows a view of the fast ferry. She has the following characteristics: 110 m. long, 1,250 passengers, aluminum-made deep-V monohull, reaching 40 knots or more. She uses powerful waterjets for propulsion and heading, having no rudder. This ferry is actually operating, between Denmark and Norway. The internal distribution of the ship, with a large market in the center, makes the passengers to be seated both sides of the ship or near the bow. These places experiment the highest vertical accelerations.

*Figure 2: A view of the fast ferry.*

The scale chosen for the new physical model was 1/40, the smallest size having results confidence. Since the real scale ship is aluminium made, the weight of the replica must be less than 29 kg. This is an implementation challenge. Sophisticated techniques have been applied to build a very light hull. There are important limitations to the weight and energy consumption of the on board system. All the energy must be obtained from on board batteries (an important contribution to total weight).

*Scaled down flaps, fins and T-foil* has been added to the model. The fins and the T-foil are made in aluminium, with a curved profile. The flaps are more simple, in plastic and flat profile. High-torque high-speed servos, to be driven by PWM signals, are in charge of moving the active surfaces. Figure 3 shows a photograph of the physical model with actuators.

*Figure 3: General view of the physical model.*

Figure 4 shows a photograph of one of the lateral fins., and figure 5 shows a photograph of the T-foil.

*Figure 4: Photograph of one of the fins.*

*Figure 5: Photograph of the T-foil.*

Figure 6 shows a photograph of the servo moving one of the lateral fins, the rest of the servos are similar.

*Figure 6: Servo moving one of the fins.*

The main sensor is an *inertial unit* located at the c.o.g. This unit weights 2 Kg. and requires 24 v., 1 amp. Since it is advisable to add some redundant sensors, three 2-axis accelerometers and a digital compass were also put in the physical model. The three accelerometers were located to measure heave and pitch acceleration, vertical acceleration due to roll, sway and surge acceleration, and yaw acceleration. The compensated digital compass is devoted to course control. It was isolated by anti-magnetic film from the rest of the system, to avoid the influence of motors (which create magnetic fields).

The physical model has two scaled down waterjets, with an additional appendage to change the jet orientation. There is no rudder. The waterjets are driven by powerful DC motors, with a maximum consumption each of 30 amps (at 6 V). Figure 7 shows the waterjets.

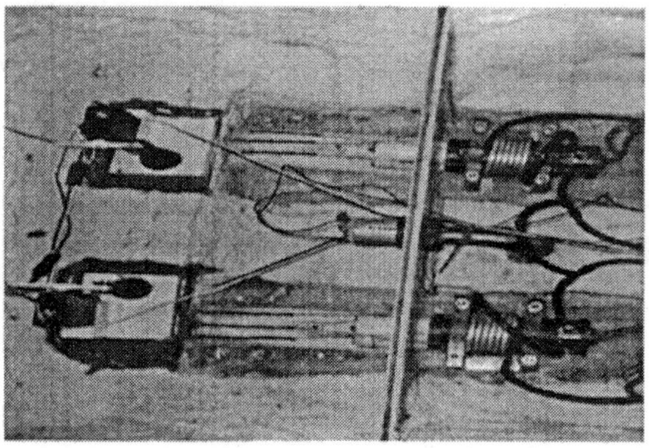

*Figure 7: The waterjets.*

Figure 8 shows a photograph of the stern with the waterjet outlets and the two flaps.

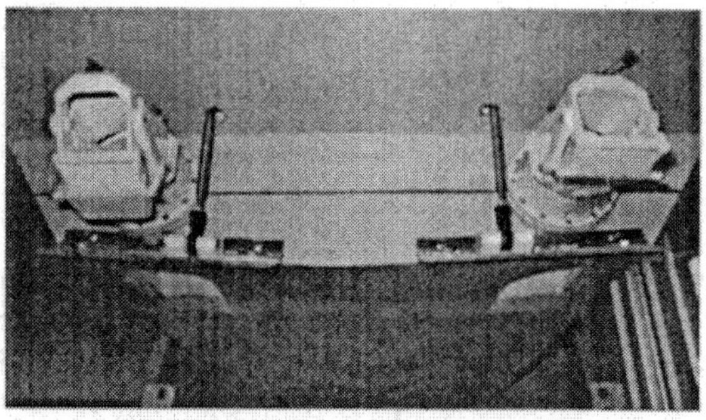

*Figure 8: A view of the model stern.*

## 4. THE ON-BOARD ELECTRONIC SYSTEM

An on-board distributed monitoring and control system was developed and put in the physical model. The system must exchange information, via digital radio, with an off-shore system which is denoted as the external support system (ESS). The missions of the on board system are the following:

- To acquire, condition and record all signals from on board sensors along experiments.
- To control the actuators of the ship.
- To transmit real-time data to the ESS.
- To transmit the complete record of data, at the end of experiments, to the ESS.
- To obey to orders given by the ESS.

Figure 9 depicts a diagram of main functions of the on board system. One of the blocks in the figure includes sensors for data acquisition and control. Another block is devoted to actuators handling. A third block is in charge of wireless communications. And finally there is a central block, with a low power embedded computer, for govern, coordination and data processing. Notice that there is a block to admit high priority remote control orders. The purpose of this last block is to be able to stop and recover the physical model when there are problems.

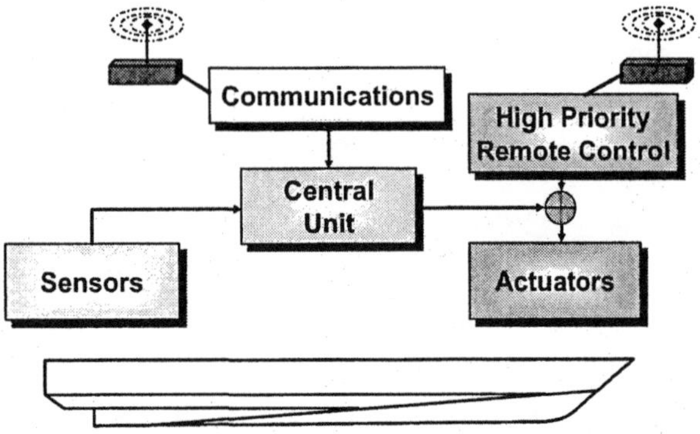


*Figure 9: Block diagram of the on-board system.*

Since there are many sensors and actuators, a distributed architecture has been devised for the on-board electronic system. Figure 10 shows a diagram of the system architecture. It consists in seven nodes connected via CANbus and a central embedded PC. The reasons for using a CANbus are related to the complexity of the system, and the long distances that can be expected for real applications on ships. This bus is well proven, and there are easy to get off-the-shelf components for CANbus based systems.

The embedded PC has been selected according with the following criteria: small size, light weight, small energy consumption, enough computing power. After considering several alternatives, the Tern 586 Engine was chosen.

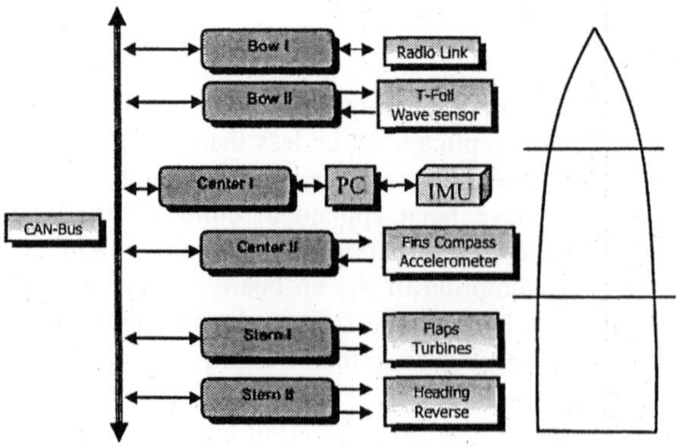


*Figure 10: Distributed architecture of the on-board system*

The nucleus of each node is a PIC 18F485 microcontroller, which includes in firmware the CANbus protocol. The embedded PC access the bus via one of these microcontrollers. The A/D channels of the embedded PC are used for a direct, fast interaction with the inertial unit.

Along experiments, the system measures all six motions and accelerations. As said before, some measurements are redundant due to a variety of reasons (safety, bias compensation, time constants, etc.).

At the stern of the ship there are two waterjets, two servos for orientation of the waterjets, two servos for reversing the water flow (for the ship to go back), and two servos to move the flaps. By means of PWM control, the speed of the motors driving the waterjets can be controlled. All these motors are controlled by two CANbus nodes. Another CAN bus node, near the bow is in charge of the T-foil. One of the CANbus nodes, located near the c.o.g. is in charge of the fins. Figure 11 shows a photograph of two CANbus nodes, mounted in the ship, inside plastic cases.

An important advantage of the distributed architecture is the easy implementation of some local functions. For instance, the microcontrollers in charge of moving the actuators include routines to *simulate the dynamic characteristics of the hydraulic cylinders*, which move the actuators in real ships. Also, the microcontrollers perform *signal conditioning* functions with respect to sensors.

*Figure 11: Two CANbus nodes in the ship.*

The heart of the wireless communication system is a digital radio unit, called MaxStream. It serves as a digital wireless transceiver. Physically is a small metallic module, shown in Figure 12 as mounted in the ship.

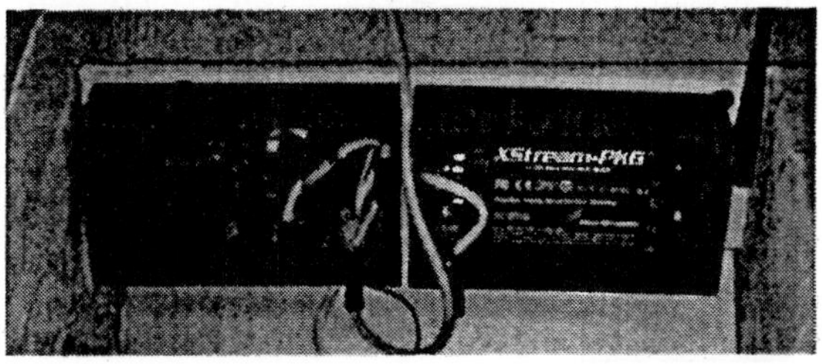


*Figure 12: The wireless communication module.*

At the border of the basin, or in another place far from the replica, the ESS is made with a conventional (portable) computer and a box containing another MaxStream and a microcontroller. Box and computer communicate via serial RS232 port.

An important feature of the onboard embedded PC is that it handles a Flash Card (like a digital camera), where several hours of experimental data, up to 1 GB, can be stored. The bandwidth of the transceiver allows for a transmission of some important real-time data along experiments, for ESS monitoring purposes. In the end of each successful experiment, all data can be transmitted from the magnetic card to the ESS (or even, data can be obtained taking directly the card). In real operational conditions it may happen that experiments take place in the sea, and it could be difficult to approach the replica to take the card: it is better a wireless transmission.

The use of a CANbus implies the design of a messaging protocol, assigning i.d. numbers to each kind of message. In this case, a simple method was devised, with the embedded PC marking sampling periods, asking to sensors, and giving orders to actuators. A code module for the CANbus interaction was developed to be handled by control and monitoring programs. These control and monitoring programs are developed on conventional PC computers. Once an application program is successfully developed, the user compiles it and gets an executable code. This code is further downloaded to the embedded PC.

The general activity of the physical model during experiments has been described as a finite automaton, which is implemented in software and executed by the embedded PC.

An advantage of the use of a fieldbus is that the several parts of the system (radio communication, servomotors, compass, etc.) can be easily distributed in different places, perhaps at relatively long distances. For instance, the compass should be placed far form magnetic fields; and radio communications should no interfere with data acquisition.

## 5. THE PHYSICAL MODEL AND THE ESS

The physical model and the ESS communicate through radio, using digital packets. A protocol has been defined for the purposes of the experiments. The ESS can be located at the border of the basin, or perhaps in a boat (in open air experiments). It was noticed that research with autonomous naval physical models could add interesting features to the experimental facility. Therefore, the architecture of the on board system must be modular and flexible. The idea is to provide a sort of *"universal" monitoring and control system*, able to be applied quickly and easily to any other marine vehicle.

From the side of functional structure and software, a modular concept is also useful for flexibility and fast application. The wireless communication paves the way for the use of internet. This is a powerful idea: *experiments from distance.*

A distribution of tasks, between replica and ESS, has been defined for normal experiments and for abnormal situations. Protocols for data monitoring and for control, along experiments and also special events, have been developed. Routines have been programmed in modular form.

## 6. THE OFF-SHORE SYSTEM (ESS)

Using the powerful tools of Builder C++, a Windows-based visualization environment has been developed for monitoring of signals coming from the physical model to the ESS. One of the reasons for the monitoring of experiments is to ensure that sensors work correctly, and the experiment is successful. Figure 13 shows a screen of the monitoring application.

A second function of the ESS is to receive the data from the physical model at the end of experiments, to record them and to analyze them.

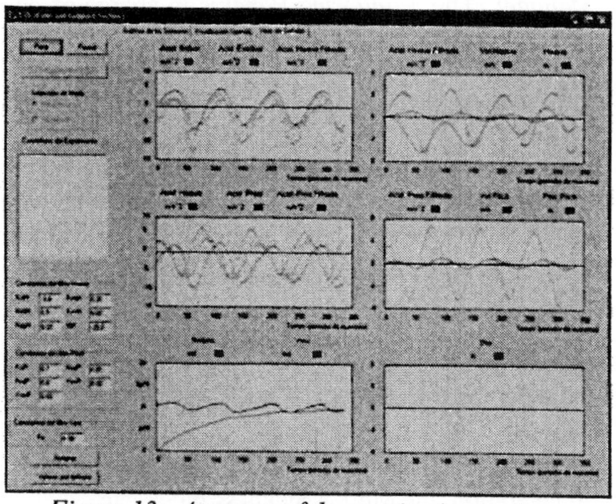

*Figure 13: A screen of the monitoring system.*

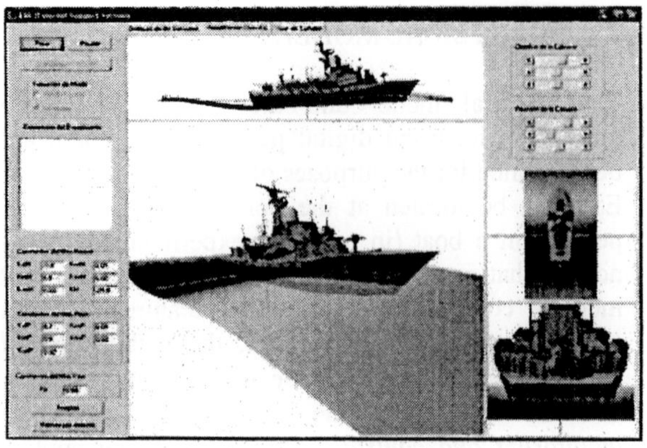

*Figure 14: The 3D animated visualization system.*

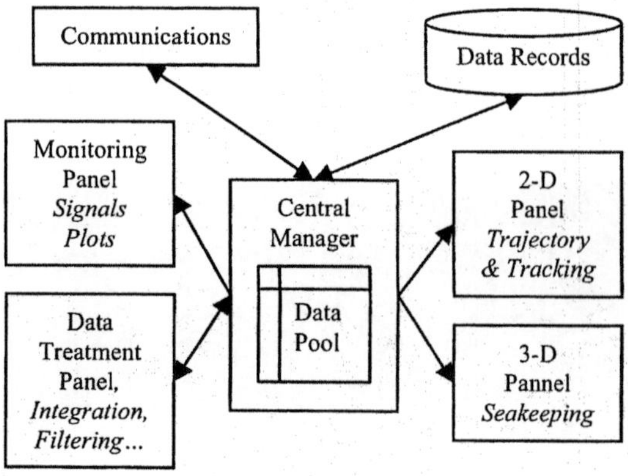


*Figure15: Software architecture of the ESS.*

Since it is difficult from the huge set of data to get a complete idea of what happened, a sort of dynamic reconstruction of events has been developed. Using OpenGL, a visualization of the ship with animated 3D graphics has been achieved. It is possible to handle a virtual camera, to see the ship from different points of view. This visualization can be linked to the recorded data (for example, to examine in detail slamming events). Moreover, the visualization can be also linked to the real time data during experiments, to offer a more complete view of what happens. Figure 14 shows a screen of the 3D visualization system.

The monitoring system is completed with a tool for programming manoeuvres to be done by the replica. In this way, experiments can be dynamically designed. Figure 15 shows the structure of blocks of the ESS software.

The monitoring software is based on a central nucleus, *Central Manager*, for coordination tasks and for data storage in a *Data Pool.* The rest of modules can access to the data stored in this pool. The *Central Manager* has two modes: Real-Time Experiment Monitoring, and Experiment Replay.

The *Data Records* module is a conventional file management system, at the service of the *Central Manager*. The *Monitoring Panel* is in charge of data visualizations tasks (plots). The *3-D Panel* visualizes in animated 3-D graphics the motions of the ship. It also includes the three orthogonal projections (views) of the ship, for seakeeping studies about the six motions of the ship. The *2-D Panel* is intended for trajectories and manoeuvring. It consist in a rectangle where the desired trajectory is programmed with the mouse, and where the real trajectory of the ship, obeying to orders and internal control, is displayed (to be compared with the reference trajectory). The *Data Treatment Panel* is in charge of several data processing functions. Experimental signals need filtering and statistical processing. Information must be extracted from signals, such the Worst Vertical Acceleration, the Motion Sickness Index, etc.

## 5. CONCLUSIONS

According with the new requirements of the research, that is starting to consider 6 DOF motions of a fast ferry and the use of several moving actuators, a new monitoring and control system has been developed. The purpose is to devise an experimental environment with an autonomous scaled model, having an on board computerized system and a wireless connection to an external computer.

Relatively strict requirements of weight, size and power, made difficult to develop the on board system. However, the results obtained are encouraging. The distributed architecture, based on the CANbus, is flexible enough to be applied to other naval physical models or real ships.

## REFERENCES

Esteban, S., J.M. De la Cruz, J.M. Giron-Sierra, B. De Andres, J.M. Diaz and J. Aranda (2000). Fast ferry vertical acceleration reduction with active flaps and T-foil, in *Proc. IFAC Intl. Symp. on Maneuvering and Control of Marine Craft*, Aalborg, Denmark.

Esteban, S., B. Andres-Toro, E. Besada-Portas, J.M. Giron-Sierra and J.M. de la Cruz (2002). Multiobjective control of flaps and T-foil in high-speed ships, *IFAC World Congress*, Barcelona.

Giron-Sierra, J.M., S. Esteban, B. De Andres, J.M. Diaz and J.M. Riola (2001). Experimental study of controlled flaps and T-foil for comfort improvement of a fast ferry, in *Proc. IFAC Intl. Conf. Control Applications in Marine Systems*, Glasgow, U.K.

Fossen, T.I. (2002). *Marine Control Systems,* Marine Cybernetics AS, Trondheim..

Lloyd, A.R.J.M. (1998), *Seakeeping: Ship Behavior in Rough Weather*, A.R.J.M. Lloyd, Gosport, Hampshire, U.K.

O'Hanlon, J.F. and M.E. MacCawley (1974), Motion sickness incidence as a funtion of frequency and acceleration of vertical sinusoidal motion, *Aerospace Medicine.*

ELSEVIER
IFAC
PUBLICATIONS
www.elsevier.com/locate/ifac

# MODELING, IDENTIFICATION, AND ADAPTIVE MANEUVERING OF CYBERSHIP II: A COMPLETE DESIGN WITH EXPERIMENTS

**Roger Skjetne**[a] **Øyvind Smogeli**[b] **Thor I. Fossen**[a]

[a] *Department of Engineering Cybernetics, Norwegian University of Science and Technology (NTNU), NO-7491 Trondheim, Norway. E-mails: skjetne@ieee.org, tif@itk.ntnu.no.*

[b] *Department of Marine Technology, Norwegian University of Science and Technology (NTNU), NO-7491 Trondheim, Norway. E-mail: oyvind.smogeli@marin.ntnu.no.*

Abstract: Good nonlinear maneuvering models, including numerical values, for control of ships are hard to find. This paper presents a complete modeling, identification, and control design for maneuvering a ship along a desired path. A variety of references have been applied to describe the ship model, its difficulties, limitations, and possible simplifications for the purpose of automatic control design. The numerical values of the parameters in the model are identified in towing tests and adaptive maneuvering experiments for a scaled model ship in a marine control laboratory. *Copyright © 2004 IFAC*

Keywords: Modeling; Maneuvering; System identification; Robust adaptive control; Experimental results

## 1. INTRODUCTION

Model-based control for steering and positioning of ships has become state-of-the-art since LQG and similar state-space techniques were applied in the 1960s. The rigid-body and hydrodynamic equations of motion for a ship are given by a set of complicated differential equations describing 6 degrees-of-freedom (6 DOF) motion. The models used in model-based control design, on the other hand, vary as much as the underlying control objectives vary. Roughly divided these control objectives are either slow speed positioning or high speed steering. The first is called dynamical positioning (DP) and includes station keeping, position mooring, and slow speed reference tracking. For DP the 6 DOF model is reduced to a simpler 3 DOF model that is linear in the kinetic part. Such applications with references are thoroughly described by Strand (1999) and Lindegaard (2003). High speed steering includes automatic course control, high speed position tracking, and path following; see for instance Holzhüter (1997) and Fossen *et al.* (2003). For these applications, Coriolis and centripetal forces together with nonlinear viscous effects are dominating and therefore make the kinetic model nonlinear. For cruising at a nearly constant surge speed and only considering first order approximations of the viscous damping, then a linear approximation of the steering dynamics is applicable. The origin of these types of models are traced back to Davidson and Schiff (1946) and Nomoto *et al.* (1957). See Clarke (2003) and Fossen (2002) for more details on these original models and their later derivations.

The contribution of this paper is a 3 DOF nonlinear maneuvering model for a ship. This model can be simplified further to either a 3 DOF model for DP, a steering model according to Davidson and Schiff or Nomoto, or it can be used as is for nonlinear control design. System identification procedures for a model ship called CyberShip II (CS2) in a towing tank facility have produced numerical values for nearly all the hydrodynamic coefficients. The other parameters are found by free-running maneuvering experiments and adaptive parameter estimation techniques.

## 2. THE 3 DOF SHIP MANEUVERING MODEL

A ship is differentially described by 6 degrees-of-freedom (6 DOF) equations of motion. The modes are $(x, y, z)$, referred to as *surge*, *sway*, and *heave*, describing the position in three-dimensional space, and $(\phi, \theta, \psi)$, called *roll, pitch*, and *yaw*, describing the orientation of the ship. By assuming the ship is longitudinally and laterally metacentrically stable with small amplitudes $\phi = \theta = \dot{\phi} = \dot{\theta} \approx 0$, one can discard the dynamics of roll and pitch. Likewise, since the ship is floating with $z \approx 0$ in mean, one can discard the heave dynamics. The resulting model for the purpose of maneuvering the ship in the horizontal plane becomes a 3 DOF model. Let an inertial frame be approximated by the earth-fixed reference frame $\{e\}$ called NED (North-East-Down), and let another coordinate frame $\{b\}$ be attached to the ship as seen in Figure 1. The states of the vessel are then taken as $\eta = [x, y, \psi]^\top$ and $\nu = [u, v, r]^\top$ where $(x, y)$ is the Cartesian position, $\psi$ is the heading (yaw) angle, $(u, v)$ are the body-fixed linear velocities (surge and sway), and $r$ is the yaw rate.

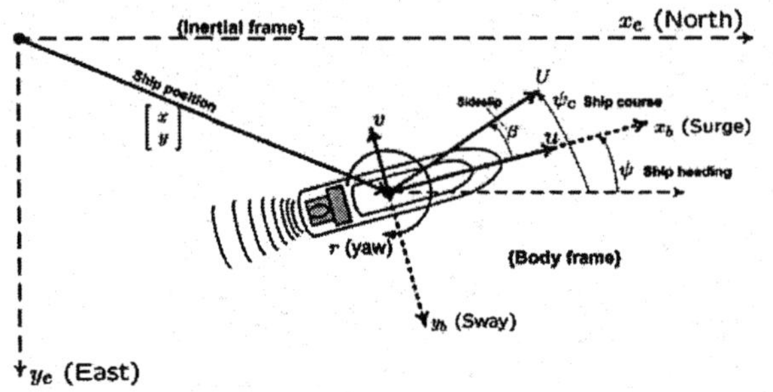


Fig. 1. The inertial earth-fixed frame $\{e\}$ and the body-fixed frame $\{b\}$ for a ship.

### 2.1 *Rigid-body dynamics*

The earth-fixed velocity vector is related to the body-fixed velocity vector through the kinematic relationship

$$\dot{\eta} = R(\psi)\nu \tag{1}$$

where $R(\psi)$ is a rotation matrix. It has the properties that $R(\psi)^\top R(\psi) = I$, $\|R(\psi)\| = 1$ for all $\psi$, and $\frac{d}{dt}\{R(\psi)\} = \dot{\psi} R(\psi) S$ with

$$R(\psi) := \begin{bmatrix} \cos\psi & -\sin\psi & 0 \\ \sin\psi & \cos\psi & 0 \\ 0 & 0 & 1 \end{bmatrix}, \quad S := \begin{bmatrix} 0 & -1 & 0 \\ 1 & 0 & 0 \\ 0 & 0 & 0 \end{bmatrix}$$

where $S$ is skew-symmetric. By Newton's second law it is shown in Fossen (2002) that the rigid body equations of motion can be written

$$M_{RB}\dot{\nu} + C_{RB}(\nu)\nu = \tau_{RB} \tag{2}$$

where $M_{RB}$ is the rigid-body system inertia matrix, $C_{RB}(\nu)$ is the corresponding matrix of Coriolis and centripetal terms, and $\tau_{RB} = [X, Y, N]^\top$ is a generalized vector of external forces $(X, Y)$ and moment $N$. Let the origin 'O' of the body frame be taken as the geometric center point (CP) in the ship structure. Under the assumption that the ship is port-starboard symmetric, the center-of-gravity (CG) will be located a distance $x_g$ along the body $x_b$-axis. In this case, $M_{RB}$ takes the form

$$M_{RB} = \begin{bmatrix} m & 0 & 0 \\ 0 & m & mx_g \\ 0 & mx_g & I_z \end{bmatrix}$$

where $m$ is the mass of the ship and $I_z$ is the moment of inertia about the $z_b$-axis (yaw rotation). Several representations for the Coriolis matrix are possible. Based on Theorem 3.2 in Fossen (2002), we choose the skew-symmetric representation

$$C_{RB}(\nu) = \begin{bmatrix} 0 & 0 & -m(x_g r + v) \\ 0 & 0 & mu \\ m(x_g r + v) & -mu & 0 \end{bmatrix}.$$

The force and moment vector $\tau_{RB}$ is given by the superposition of actuator forces and moments $\tau = [\tau_u, \tau_v, \tau_r]^\top$, hydrodynamic effects $\tau_H$, and exogenous disturbances $w(t)$ due to, for instance, waves and wind forces (Sørensen *et al.* 1996). The forces and moments in $\tau_{RB}$ are all expressed with reference to the center point (CP) such that the full set of dynamical equations is given in the body-fixed reference frame.

### 2.2 *Hydrodynamic forces and moments*

The vector $\tau_H$ is the superposition of several hydrodynamic effects. For an ideal fluid, some of these components are *added mass, radiation-induced potential damping*, and *restoring forces*. In addition there are also other damping effects such as *skin friction, wave drift damping*, and damping due to *vortex shedding* (Faltinsen 1990, Fossen 2002). For 3 DOF maneuvering, restoring forces are not important and therefore neglected.

Due to currents in the ocean fluid, the velocity $\nu$ is different than the relative velocity $\nu_r$ between the ship hull and the fluid. The hydrodynamic forces and moments depends on this relative velocity. For a nonrotational current with fixed speed $V_c$ and angle $\beta_c$ in the earth-fixed frame, the current velocity is given by

$$v_c := \begin{bmatrix} V_c \cos\beta_c \\ V_c \sin\beta_c \\ 0 \end{bmatrix}. \tag{3}$$

In the body-frame this gives the current component $\nu_c := R(\psi)^\top v_c$ and the relative velocity $\nu_r := \nu - \nu_c = [u_r, v_r, r]^\top$. With these definitions it is common (Sørensen 2002) to model the hydrodynamic effects as

$$\tau_H = -M_A\dot{\nu}_r - C_A(\nu_r)\nu_r - d(\nu_r) \tag{4}$$

where $M_A$ accounts for added mass, $C_A(\nu_r)$ accounts for the corresponding added Coriolis and centripetal terms, and $d(\nu_r)$ sums up the damping effects. By the notation of The Society of Naval Architects and Marine Engineers (1950) the matrix $M_A$ is given by

$$M_A = \begin{bmatrix} -X_{\dot{u}} & 0 & 0 \\ 0 & -Y_{\dot{v}} & -Y_{\dot{r}} \\ 0 & -N_{\dot{v}} & -N_{\dot{r}} \end{bmatrix}$$

where the assumption of port-starboard symmetry again is applied. For $\nu_r = 0$, zero frequency of motion due to water surface effects, and an ideal fluid, the added mass matrix is constant and $M_A = M_A^\top > 0$. However, under non-ideal conditions with waves and high velocity, $M_A = M_A(\omega_e) \neq M_A(\omega_e)^\top$ where $\omega_e$ is the frequency of encounter given by

$$\omega_e = \left| \omega_0 - \frac{\omega_0^2}{g} U \cos\beta \right|. \tag{5}$$

Here $\omega_0$ is the dominating wave frequency, $g$ is the acceleration of gravity, $U = \sqrt{u^2 + v^2}$ is the total ship speed, and $\beta$ is the angle of encounter defined by $\beta = 0°$

for following sea. For control design it is common to assume that $M_A = \lim_{\omega_e \to 0} M_A(\omega_e)$ is constant and strictly positive.

Since $M_A$ is not necessarily symmetric, Theorem 3.2 in Fossen (2002) is not directly applicable to find $C_A(\nu_r)$. To solve this obstacle, we observe that this theorem is deduced from the kinetic energy $T = \frac{1}{2}\nu^\top M \nu$. A modification for the added mass kinetic energy is

$$T_A = \frac{1}{2}\nu_r^\top M_A \nu_r = \frac{1}{4}\nu_r \left(M_A + M_A^\top\right) \nu_r = \frac{1}{2}\nu_r \bar{M}_A \nu_r$$

where $\bar{M}_A := \frac{1}{2}(M_A + M_A^\top) = \bar{M}_A^\top$. This means that $C_A(\nu_r)$, for a nonsymmetric $M_A$, is derived from Theorem 3.2 of Fossen (2002) using $\bar{M}_A$ instead of $M_A$, and this gives

$$C_A(\nu_r) = \begin{bmatrix} 0 & 0 & c_{13}(\nu) \\ 0 & 0 & c_{23}(\nu) \\ -c_{13}(\nu) & -c_{23}(\nu) & 0 \end{bmatrix} \quad (6)$$

where $c_{13}(\nu) = Y_{\dot{v}} v_r + \frac{1}{2}(N_{\dot{v}} + Y_{\dot{r}}) r$ and $c_{23}(\nu) = -X_{\dot{u}} u_r$.

The most uncertain component in the hydrodynamic model (4) is the damping vector $d(\nu_r)$. Let $d(\nu_r) = [X_D(\nu_r), Y_D(\nu_r), N_D(\nu_r)]^\top$. For a constant cruise speed $\nu_r = \nu_0 \approx [u_0, 0, 0]^\top$ one can fit the damping forces and moments at $\nu_0$ to the linear functions

$$\begin{aligned} X_D(\nu_r) &= -X_u(u_r - u_0) - X_v v_r - X_r r \\ Y_D(\nu_r) &= -Y_u(u_r - u_0) - Y_v v_r - Y_r r \\ N_D(\nu_r) &= -N_u(u_r - u_0) - N_v v_r - N_r r \end{aligned} \quad (7)$$

where the hydrodynamic coefficients $\{X_{(\cdot)}, Y_{(\cdot)}, N_{(\cdot)}\}$ are called hydrodynamic derivatives. Seeking in this paper a globally valid model, a nonlinear representation is considered instead of (7). Abkowitz (1964) proposed using a truncated Taylor series expansion of $d(\nu_r)$, while Fedyaevsky and Sobolev (1963) and later Norrbin (1970) gave another nonlinear representation called the *second order modulus* model (Clarke 2003). Based on experimental data and curve fitting as presented in the next section, the damping model

$$d(\nu_r) = D_L \nu_r + D_{NL}(\nu_r)\nu_r =: D(\nu_r)\nu_r \quad (8)$$

is chosen where

$$D_L := \begin{bmatrix} -X_u & 0 & 0 \\ 0 & -Y_v & -Y_r \\ 0 & -N_v & -N_r \end{bmatrix},$$

$$D_{NL}(\nu_r) := \begin{bmatrix} -d_{11}(\nu_r) & 0 & 0 \\ 0 & -d_{22}(\nu_r) & -d_{23}(\nu_r) \\ 0 & -d_{32}(\nu_r) & -d_{33}(\nu_r) \end{bmatrix},$$

and $d_{11}(\nu_r) = X_{|u|u}|u_r| + X_{uuu}u_r^2$, $d_{22}(\nu_r) = Y_{|v|v}|v_r| + Y_{|r|v}|r|$, $d_{23}(\nu_r) = Y_{|v|r}|v_r| + Y_{|r|r}|r|$, $d_{32}(\nu_r) = N_{|v|v}|v_r| + N_{|r|v}|r|$, and $d_{33}(\nu_r) = N_{|v|r}|v_r| + N_{|r|r}|r|$.

With $\tau_{RB} = \tau + \tau_H + w(t)$ the kinetic equation of motion (2) becomes

$$\begin{aligned} M_{RB}\dot{\nu} + M_A\dot{\nu}_r + C_{RB}(\nu)\nu + C_A(\nu_r)\nu_r \\ + D(\nu_r)\nu_r = \tau + w(t) \end{aligned} \quad (9)$$

where

$$\nu_r = \nu - R(\psi)^\top v_c, \qquad \dot{\nu}_r = \dot{\nu} - rS^\top R(\psi)^\top v_c.$$

For the kinetic model (9) one must decide upon using either the relative velocity $\nu_r$ or the inertial velocity $\nu$ as the velocity state. There are different practices in the literature. A simplifying technique was applied by Fossen and Strand (1999) who used $\nu$ as the velocity state and assumed that the dynamics related to the current $v_c$ (and other unmodeled dynamics) is captured by a slowly varying bias $b$ in the earth frame. This gives the simplified model

$$M\dot{\nu} + C(\nu)\nu + D(\nu)\nu = \tau + R(\psi)^\top b + w(t) \quad (10)$$

where $M := M_{RB} + M_A$ and $C(\nu) := C_{RB}(\nu) + C_A(\nu)$. The alternative, applied among others by Holzhüter (1997), is to use $\nu_r$ as the state. In this case the kinematic relationship (1) must be rewritten as

$$\dot{\eta} = R(\psi)\nu_r + v_c \quad (11)$$

which means that $v_c$ enters both the kinematic and kinetic equations of motion. Experience has shown that (1) and (10) are adequate for control design provided some type of integral action is used in the controller to compensate for the bias $b$.

### *2.3 Actuator forces*

The actuator forces and moments are generated by a set of $p_1$ thrusters with revolutions per second $n \in \mathbb{R}^{p_1}$ and a set of $p_2$ control surfaces (or propeller blade pitch) with angles $\delta \in \mathbb{R}^{p_2}$. They are related to the input vector $\tau$ through the mapping

$$\tau = Bf_c(\nu_r, n, \delta) \quad (12)$$

where $B \in \mathbb{R}^{3\times(p_1+p_2)}$ is an actuator configuration matrix, and $f_c : \mathbb{R}^3 \times \mathbb{R}^{p_1} \times [-\pi, \pi)^{p_2} \to \mathbb{R}^{p_1+p_2}$ is a function that for each velocity $\nu_r$ relates the actuator set-points $(n, \delta)$ to a vector of forces. For more details, see Skjetne *et al.* (2004*a*).

## 3. SYSTEM IDENTIFICATION

The Marine Cybernetics Laboratory (MCLab) is a Marie Curie EU training site for testing of ships, rigs, underwater vehicles, and propulsion systems. It is equipped with a towing carriage, a position measurement system, and a wave maker system, among others.

CyberShip II (CS2) is a 1:70 scale replica of a supply ship. Its mass is $m = 23.8\,\text{kg}$, its length is $L_{CS2} = 1.255\,\text{m}$, and its breadth is $B_{CS2} = 0.29\,\text{m}$. It is fully actuated with two main propellers and two rudders aft, and one bow thruster.

By running the experiments without currents or exogenous disturbances, the CS2 ship model becomes

$$M\dot{\nu} + C(\nu)\nu + D(\nu)\nu = \tau \quad (13)$$

where the parameters in $M_{RB}$, $M_A$, $D_L$, and $D_{NL}(\nu)$ must be identified.

The parameters in the rigid-body system inertia matrix $M_{RB}$ is found from straight forward measurements of the main particulars of the ship; its dimensions, weight, mass distribution, volume, area, etc. The zero frequency added mass coefficients in $M_A$ can be found from semi-empirical formulas, simple engineering "rules-of-thumb," or strip theory (Faltinsen 1990). For CS2 these parameters have all been estimated beforehand by Lindegaard (2003), and their values are given in Table 1. Seeking a nonlinear representation of the damping effects, the system identification procedure next will be concerned with the damping coefficients.

Table 1. Mass-related parameters with respect to CP for CyberShip II

| | | | | | |
|---|---|---|---|---|---|
| $m$ | 23.800 | $X_{\dot{u}}$ | - 2.0 | $N_{\dot{v}}$ | - 0.0 |
| $I_z$ | 1.760 | $Y_{\dot{v}}$ | -10.0 | $N_{\dot{r}}$ | - 1.0 |
| $x_g$ | 0.046 | $Y_{\dot{r}}$ | - 0.0 | | |

The parameters to be identified in pure surge direction are $\{X_u, X_{|u|u}, X_{uuu}\}$. Using the towing carriage, CS2 was towed both forward and backward at different constant speeds, and for each run the average pull force $X_{pull}$ was measured and recorded; see Figure 2.

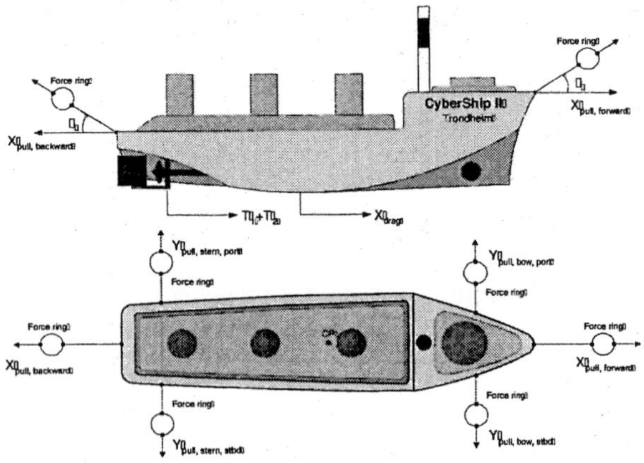


Fig. 2. Two force rings, forward and backward, were applied to measure the drag and propulsion forces when towing CyberShip II longitudinally at different speeds. Four force rings, two port and two starboard, were used to measure the drag force and moment for lateral towing.

Since $\dot{u} = v = r = 0$, we have for pure surge motion that $X_{pull} + X_u u + X_{|u|u} |u| u + X_{uuu} u^3 = 0$. Setting this up as a linear set of equations, $Ax = b$, where $x$ contains the unknown parameters, $A$ contains the towing speeds $u$, and $b$ contains the corresponding measured forces $X_{pull}$, the unknown coefficients are calculated by a least square fit. Figure 3 shows these measured forces and the corresponding interpolation. It shows in addition the linear DP curve $X_{drag} = X_u u$ fitted to those measured points that are within the slow speed region $u \in [-0.15, 0.15]$. Clearly, there is a large discrepancy for higher speeds.

By towing the ship in pure sway motion, similar techniques were used to also estimate $\{Y_v, Y_{|v|v}, N_v, N_{|v|v}\}$. See Skjetne *et al.* (2004*a*) for details. The numerical values are summarized in Table 2.

Table 2. Experimentally identified parameters for CyberShip II

| | | | |
|---|---|---|---|
| $X_u$ | -0.72253 | $Y_v$ | -0.88965 |
| $X_{\|u\|u}$ | -1.32742 | $Y_{\|v\|v}$ | -36.47287 |
| $X_{uuu}$ | -5.86643 | $N_v$ | 0.03130 |
| | | $N_{\|v\|v}$ | 3.95645 |

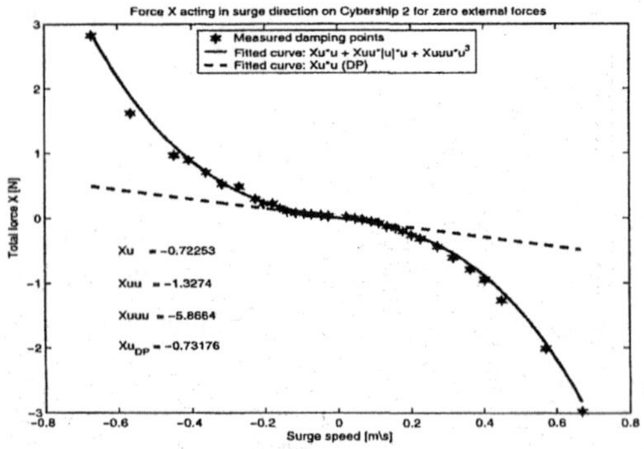


Fig. 3. Measured drag forces of CyberShip II at different speeds and the corresponding fitted nonlinear curve as well as a linear curve for DP.

Since no yaw motion was induced in these towing experiments, the parameters $\{Y_{|r|v}, Y_r, Y_{|v|r}, Y_{|r|r}, N_{|r|v}, N_r, N_{|v|r}, N_{|r|r}\}$ are yet to be identified. These are estimated by the adaptive maneuvering controller developed and experimentally tested in the next section.

## 4. ADAPTIVE SHIP MANEUVERING WITH EXPERIMENTS

We consider the dynamic ship model (1) and (13) which for $\tau = Bf_c(\nu, n, \delta)$ can be rewritten as

$$\begin{aligned} \dot{\eta} &= R(\psi)\nu \\ M\dot{\nu} &= \tau - C(\nu)\nu + g(\nu) + \Phi(\nu)\varphi. \end{aligned} \tag{14}$$

Here, $g(\nu)$ is the known part of $-D(\nu)\nu$ and

$$\varphi := \left[Y_{|r|v}, Y_r, Y_{|v|r}, Y_{|r|r}, N_{|r|v}, N_r, N_{|v|r}, N_{|r|r}\right]^\top$$

$$\Phi(\nu) := \begin{bmatrix} 0 & 0 & 0 & 0 & 0 & 0 & 0 & 0 \\ |r|\,v & r & |v|\,r & |r|\,r & 0 & 0 & 0 & 0 \\ 0 & 0 & 0 & 0 & |r|\,v & r & |v|\,r & |r|\,r \end{bmatrix}$$

are the vector of unknown parameters and the regressor matrix, respectively, so that $g(\nu) + \Phi(\nu)\varphi = -D(\nu)\nu$. The objective is to design a robust adaptive control law that ensures tracking of $\eta(t)$ to a time-varying reference $\eta_d(t)$ while adapting the parameters in $\varphi$.

The time-varying reference $\eta_d(t)$ must trace out a *desired path* on the surface as well as satisfying a desired speed specification along the path. Such problems are conveniently solved according to the methodology in Skjetne *et al.* (2004*b*) where the tracking objective is divided into two tasks, formally stated as a *maneuvering problem*:

*1.* ***Geometric Task:*** Force the ship position and heading $\eta$ to converge to and follow the desired path $\eta_d(\theta)$,

$$\lim_{t\to\infty} |\eta(t) - \eta_d(\theta(t))| = 0, \tag{15}$$

where $\theta$ is a scalar variable that continuously parametrizes the path.

*2.* ***Dynamic Task:*** Force the path speed $\dot{\theta}$ to converge to the desired speed assignment $v_s(\theta, t)$,

$$\lim_{t\to\infty} \left|\dot{\theta}(t) - v_s(\theta(t), t)\right| = 0. \tag{16}$$

The desired path will be an ellipsoid with heading along the tangent vector, that is,

$$\eta_d(\theta) = \left[ x_d(\theta),\ y_d(\theta),\ \arctan\left(\frac{y_d^\theta(\theta)}{x_d^\theta(\theta)}\right) \right]^\top \quad (17)$$

where $x_d(\theta) = 5 + 4.5\cos(\frac{\pi}{180}\theta)$ and $y_d(\theta) = -0.75 - 2.25\sin(\frac{\pi}{180}\theta)$, and the notation $x_d^{\theta^k}(\theta)$ is the $k$'th partial derivative, $\frac{\partial^k x_d(\theta)}{\partial\theta^k}$. For the speed assignment, we want the surge speed $u(t)$ to track a desired surge speed $u_d(t)$ which is adjustable online by an operator. This latter objective can be translated into a *speed assignment* for $\dot\theta(t)$ by noting the relationship

$$u_d(t) = \sqrt{x_d^\theta(\theta(t))^2 + y_d^\theta(\theta(t))^2}\dot\theta(t).$$

The corresponding speed assignment for $\dot\theta$ becomes

$$v_s(\theta, t) := \frac{u_d(t)}{\sqrt{x_d^\theta(\theta)^2 + y_d^\theta(\theta)^2}}. \quad (18)$$

The dynamic task can be solved identically by letting $\dot\theta = v_s(\theta, t)$ be a dynamic state in the control law, called a *tracking update law* (Skjetne *et al.* 2004*b*).

The maneuvering control design is based on adaptive backstepping, and the control law, the adaptive update law, and the maneuvering update law are given in Table 3 where $\hat\varphi$ is the parameter estimate, and $K_p = K_p^\top > 0$, $K_d = K_d^\top > 0$, and $\Gamma = \Gamma^\top > 0$ are controller gain matrices.

Table 3. Maneuvering control and guidance system for CyberShip II

| |
|---|
| Internal signals : |
| $z_1 := R(\psi)^\top (\eta - \eta_d(\theta))$ |
| $z_2 := \nu - \alpha_1(\eta, \theta, t)$ |
| $\alpha_1 = -K_p z_1 + R(\psi)^\top \eta_d^\theta(\theta) v_s(\theta, t)$ |
| $\sigma_1 = -K_p \left(\dot R(r)^\top R(\psi) z_1 + \nu\right) + \dot R(r)^\top \eta_d^\theta(\theta) v_s(\theta, t)$ $+R(\psi)^\top \eta_d^\theta(\theta) v_s^t(\theta, t)$ |
| $\alpha_1^\theta = -K_p R(\psi)^\top \eta_d^\theta(\theta)$ $+R(\psi)^\top [\eta_d^{\theta^2}(\theta) v_s(\theta, t) + \eta_d^\theta(\theta) v_s^\theta(\theta, t)]$ |
| Control : |
| $\dot{\hat\varphi} = \Gamma\Phi(\nu)^\top z_2$ |
| $\dot\theta = v_s(\theta, t)$ |
| $\tau = -z_1 - K_d z_2 - g(\nu) - \Phi(\nu)\hat\varphi$ $+C(\nu)\alpha_1 + M\sigma_1 + M\alpha_1^\theta v_s(\theta, t)$ |
| input $= \left\{ (\eta, \nu), \left(\eta_d(\theta), \eta_d^\theta(\theta), \eta_d^{\theta^2}(\theta)\right), \left(v_s(\theta, t), v_s^\theta(\theta, t), v_s^t(\theta, t)\right) \right\}$ |
| output $= \{\tau, \theta\}$ |
| Guidance : |
| input $= \{\theta, u_d(t), \dot u_d(t)\}$ |
| output $= \left\{ \left(\eta_d(\theta), \eta_d^\theta(\theta), \eta_d^{\theta^2}(\theta)\right), \left(v_s(\theta, t), v_s^\theta(\theta, t), v_s^t(\theta, t)\right) \right\}$ |

To find the optimal actuator set-points $(n, \delta)$ for each commanded input $\tau$ in (12) is called control allocation. The simplest approach is to use the generalized pseudo-inverse and the inverse function $f_c^{-1}$, that is,

$$(n, \delta) = f_c^{-1}\left(\nu, B^\dagger \tau\right)$$

where $B^\dagger = W^{-1}B^\top\left(BW^{-1}B^\top\right)^{-1}$ where $W$ is a weight matrix. Since CS2 must actively use the rudders together with thrust from the main propellers to generate enough lift forces at zero speed, the pseudo-inverse does not result in good control allocation. A more advanced method is to use constrained optimization techniques. For CS2 this has been developed and reported by Lindegaard (2003) and Johansen *et al.* (2003). The routine developed by the former author have been used in these experiments.

For the experiment, the controller settings were $K_p = \text{diag}(0.5, 2.0, 1.5)$, $K_d = \text{diag}(8, 25, 18)$, and $\Gamma = \text{diag}(8, 4, 8, 8, 8, 4, 8, 8)$. The initial condition for the parameter update was $\hat\varphi(0) = 0$. The ship was first put to rest in dynamic positioning (zero speed) at $\eta_d(0)$, and then the ship was commanded online to move along the path with $u_d = 0.15\,\text{m/s}$ for 22 rounds. The experiment was conducted on calm water (sea state code 0) to estimate the zero frequency hydrodynamic coefficients.

Figure 4 shows how CS2 accurately traced the path (in the time interval $t \in [808, 950]\,\text{s}$). Figure 5 shows that CS2 tracked the commanded speed quite well. The transients around $t \approx 500\,\text{s}$ is caused by problems with position measurement outages along the upper part of the path. Figure 6 shows the adaptive parameter estimates of

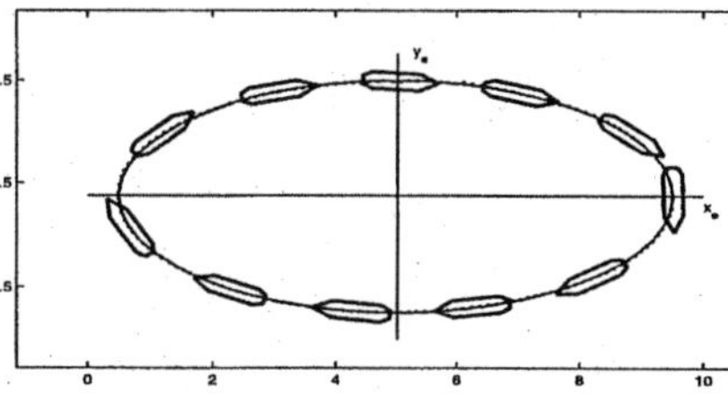


Fig. 4. CyberShip II tracing the desired path.

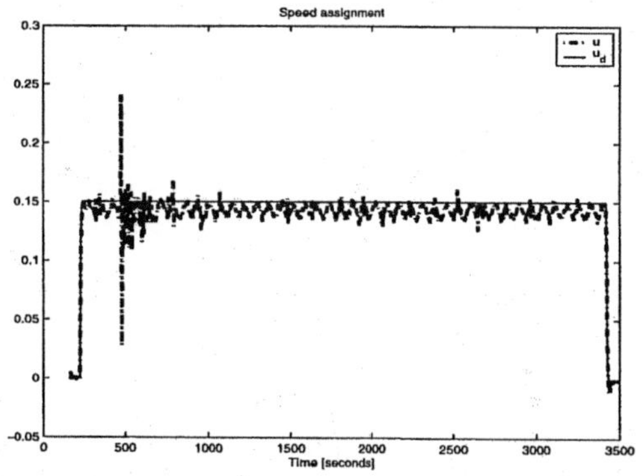


Fig. 5. The desired and actual surge speed of CyberShip II for the full experiment.

$\hat\varphi(t)$. A rapid change and a subsequent slow convergence to new values are observed. We adopt these values as approximate values for the remaining parameters; see Table 4.

Table 4. Adaptively estimated parameters for CyberShip II

| | | | |
|---|---|---|---|
| $Y_{\vert r\vert v}$ | -0.805 | $N_{\vert r\vert v}$ | 0.130 |
| $Y_r$ | -7.250 | $N_r$ | -1.900 |
| $Y_{\vert v\vert r}$ | -0.845 | $N_{\vert v\vert r}$ | 0.080 |
| $Y_{\vert r\vert r}$ | -3.450 | $N_{\vert r\vert r}$ | -0.750 |

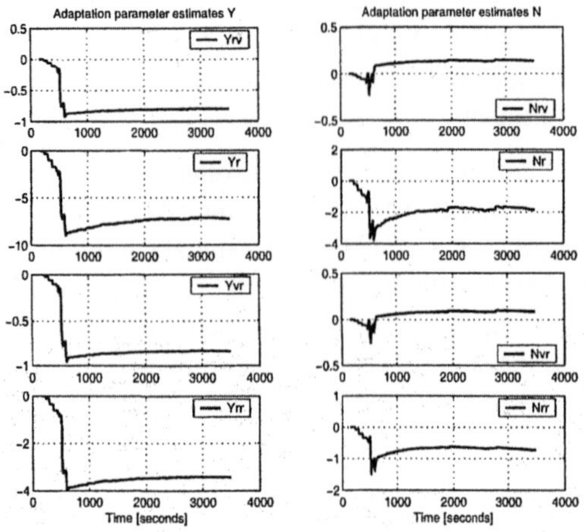


Fig. 6. Adaptive parameter estimates $\hat{\varphi}(t)$ in the free-running CyberShip II maneuvering experiment.

This robust adaptive maneuvering design with experiment also illustrates that 100% numerically correct values for the hydrodynamic parameters are not necessary to achieve accurate tracing of the path. Table 5 shows the standard deviations of the error signals in $z_1$. The most important variable for path keeping is $z_{12}$ since this is an approximate measure of the cross-track error (provided the ship is pointed along the path, $z_{13} \approx 0$). An accuracy of 2.26 cm is 7.8% of the ship breadth and acceptable. This corresponds to an accuracy of 1.58 m for the full scale ship having a breadth of 20.3 m.

Table 5. Standard deviations for CyberShip II in the free-running maneuvering experiment.

| $u_d$[m/s] | $z_{11}$[m] | $z_{12}$[m] | $z_{13}$[deg] | $u - u_d$[m/s] |
|---|---|---|---|---|
| 0.15 | 0.0350 | 0.0226 | 2.623 | 0.0080 |

## 5. CONCLUSION

A complete modeling, identification, and control design for maneuvering a ship along a desired path was presented. The identification and adaptive maneuvering procedure with experiments have provided numerical values for all parameters in the nonlinear ship model for CyberShip II (except parameters in the actuator system).

System identification procedures, using a towing carriage, were performed where the model ship CyberShip II was towed at many different velocities and the average towing forces were recorded. For zero acceleration and zero input forces these measurements are directly related to the drag of the ship hull, and they were accurately fitted to a nonlinear damping model of the ship for pure surge and sway motions. The remaining damping parameters, related to the yaw rate of the ship, were approximately found by application of an adaptive maneuvering control law.

**Acknowledgement:** We wish to thank Dag Abel Sveen for his support and patience in the towing experiments. Supported by the Centre for Ships and Ocean Structures (CESOS) at the Norwegian University of Science and Technology and partly by the Norwegian Research Council through the Strategic University Program on Marine Control.

## REFERENCES

Abkowitz, M. A. (1964). Lectures on ship hydrodynamics - steering and manoeuvrability. Technical report Hy-5. Hydro- and Aerodynamics Laboratory. Lyngby, Denmark.

Clarke, D. (2003). The foundations of steering and manoeuvring. In: *Proc. IFAC Conf. Manoeuvering and Contr. Marine Crafts*. IFAC. Girona, Spain.

Davidson, K. S. M. and L. I. Schiff (1946). Turning and course keeping qualities. *Trans. Soc. of Nav. Architects Marine Eng.* **54**, 152–200.

Faltinsen, O. M. (1990). *Sea Loads on Ships and Offshore Structures*. Cambridge University Press.

Fedyaevsky, K. K. and G. V. Sobolev (1963). *Control and Stability in Ship Design*. State Union Shipbuilding Publishing House. Leningrad.

Fossen, T. I. (2002). *Marine Control Systems: Guidance, Navigation, and Control of Ships, Rigs and Underwater Vehicles*. Marine Cybernetics. Trondheim, Norway.

Fossen, T. I. and J. P. Strand (1999). Passive nonlinear observer design for ships using lyapunov methods: full-scale experiments with a supply vessel. *Automatica* **35**(1), 3–16.

Fossen, T. I., M. Breivik and R. Skjetne (2003). Line-of-sight path following of underactuated marine craft. In: *Proc. IFAC Conf. Manoeuvering and Contr. Marine Crafts*. IFAC. Girona, Spain. pp. 244–249.

Holzhüter, T. (1997). LQG approach for the high-precision track control of ships. *IEE Proc. Contr. Theory Appl.* **144**(2), 121–127.

Johansen, T. A., T. P. Fuglseth, P. Tøndel and T. I. Fossen (2003). Optimal constrained control allocation in marine vessels with rudders. In: *Proc. IFAC Conf. Manoeuvering and Contr. Marine Crafts*. IFAC. Girona, Spain.

Lindegaard, K.-P. (2003). Acceleration Feedback in Dynamic Positioning. PhD thesis. Norwegian Univ. Science & Technology. Trondheim, Norway.

Nomoto, K., T. Taguchi, K. Honda and S. Hirano (1957). On the steering qualities of ships. Technical report 4. Int. Shipbuilding Progress.

Norrbin, N. H. (1970). Theory and observation on the use of a mathematical model for ship maneuvering in deep and confined waters. In: *8th Symp. Naval Hydrodynamics*. Pasadena, California, USA.

Skjetne, R., Ø. N. Smogeli and T. I. Fossen (2004*a*). A nonlinear ship maneuvering model: Identification and adaptive control with experiments for a model ship. *Modeling, Identification and Control* **25**(1), 3–27.

Skjetne, R., T. I. Fossen and P. V. Kokotović (2004*b*). Robust output maneuvering for a class of nonlinear systems. *Automatica* **40**(3), 373–383.

Sørensen, A. (2002). *Kompendium: Marine Cybernetics - Modelling and Control*. Marine Technology Centre. Trondheim, Norway.

Sørensen, A. J., S. I. Sagatun and T. I. Fossen (1996). Design of a dynamic positioning system using model-based control. *Contr. Eng. Practice* **4**(3), 359–368.

Strand, J. P. (1999). Nonlinear Position Control Systems Design for Marine Vessels. PhD thesis. Norwegian Univ. Science & Technology. Trondheim, Norway.

The Society of Naval Architects and Marine Engineers (1950). Nomenaclature for treating the motion of a submerged body through a fluid. Technical and Research Bulletin No. 1-5.

ELSEVIER
IFAC
PUBLICATIONS
www.elsevier.com/locate/ifac

# ADVANCES IN THE 6 DOF MOTIONS MODEL OF A FAST FERRY,

**J.M. Giron-Sierra, B. Andres-Toro, S. Esteban**
**J. Recas, J.M. De la Cruz, J.M. Riola**

*Dep.. A.C.Y.A., Fac. CC. Fisicas. Universidad Complutense de Madrid*
*Ciudad Universitaria, 28040 Madrid. Spain*
*e-mail: gironsi@dacya.ucm.es*

Abstract: In the present stage of our research on the use of moving submerged appendages for motions smoothing of a fast ferry, a 6 DOF motions mathematical model of the ship for control studies is under development. This model is needed from the very beginning of the next experimental research, because we are going to use an autonomous scaled physical model of the ship, and we need to design a basic course control. The actuators for motion smoothing of the fast ferry are two transom flaps, two lateral fins and a T-foil. These actuators must be considered by the 6DOF model. In a previous paper we discussed a first principles approach to build the mathematical control-oriented model, using data from a CFD program. On the basis of these principles, now we present new developments of the 6DOF model. In particular, we study the ship dynamic behaviour for heading angles of 180°, 150° and 90°. Some hints are obtained to predict how the model will change for other heading angles. *Copyright © 2004 IFAC*

Keywords: Seakeeping, ship modelling, fast ships.

## 1. INTRODUCTION

Our research on optimal motion smoothing in a fast ferry, by using active submerged appendages, progress along two main steps. The fist step has been already accomplished, and considers the simple case of the ship motions with head seas. The second step is now under way, considering the general case of any ship's heading.

During the first step of the research, only transom flaps and T-foil were considered, for heaving and pitching motion smoothing (Esteban, et al., 2000, 2002; Aranda et al., 2001; Giron-Sierra, et al., 2002). Models of heave and pitch motions of the ship, and models of the actuators were developed. With these models, a simulation environment for control design was developed (Esteban, et al., 2001). The mathematical models and the control designs were validated with a towed scaled down replica of the ship, using the CEHIPAR ("Canal de Experiencias de El Pardo", Madrid) facilities: in particular a 150m x 30m x5m basin with wavemaker (Giron-Sierra, et al., 2001).

For the second step of the research we add to the flaps and the T-foil a pair of lateral fins. All six motions of the ship must be considered. There will be motion coupling and asymmetries. The research procedure is modelling, simulation, control design, and experimental validation for the 6 DOF problem. A new autonomous scaled down physical model is under development, to conduct experimental aspects of the research. We need already a course control for the physical model, and for the design of this controller a first 6 DOF mathematical control-oriented model is needed.

The development of the 6DOF ship motions model started with a first principles analysis of the problem. The internal architecture of the model was defined in terms of block diagrams, and some details of the block dynamics were obtained. This initial part was the content of (Giron-Sierra, et al., 2003). The data for modelling are obtained using a CFD program handled by El Canal de Experiencias Hidrodinamicas de El Pardo (CEHIPAR), Madrid, for seakeeping prediction. The CFD program departs from a description of the geometry of the hull.

This paper describes new advances in the development of the 6DOF model. Specifically, we considered a 150° heading angle with respect to waves. Some criteria were developed to predict how the heading angle would influence in changes in the model transfer functions. The criteria were confirmed by the fitting task. Along this work, MATLAB tools for transfer functions fitting, such the routine *invfreqs*, have been very useful. We also obtained models for 180° and 90°, which are important to capture primary dynamic behaviours, being as well interesting for comparison and verification purposes.

The paper is organized as follows. First, a brief synthesis of the model architecture, as obtained in (Giron-Sierra, et al, 2003). Then a section about modelling studies for 180° and 90° heading angles. The next section describes in detail the modelling for 150° heading angle, and the criteria for the prediction of transfer function changes when the heading angle change. It is hoped that the findings of this work could be useful in general, for other modelling investigations

## 2. ARCHITECTURE OF THE 6DOF MODEL

Figure 1 shows the fast ferry with the actuators: T-foil, fins and flaps. We want a 6DOF motions model where we could study in an easy way the effect of actuators.

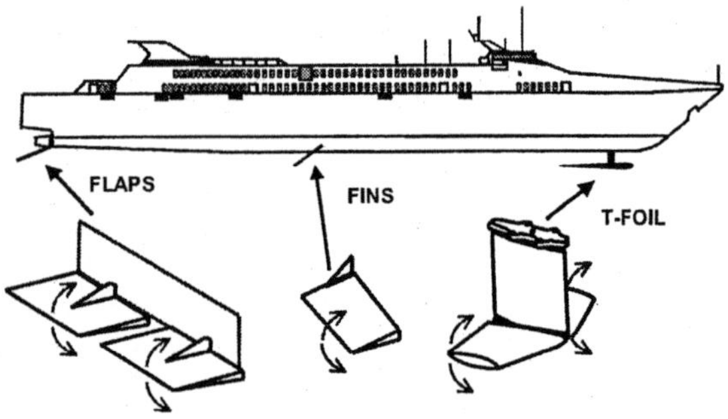


Fig.1. The fast ferry and the actuators.

Figure 2 shows the main approach for the model architecture. It may describe, for instance, the action of the T-foil to smooth pitch motion. A two-block model of the ship's motions is introduced. One of the blocks gives the pitch moment caused by waves. The other block gives the pitch motion caused by the pitch moment. The action of the T-foil is an opposite pitch moment. Notice that the decomposition of the ship's model into two blocks makes easy to include the action of the T-foil

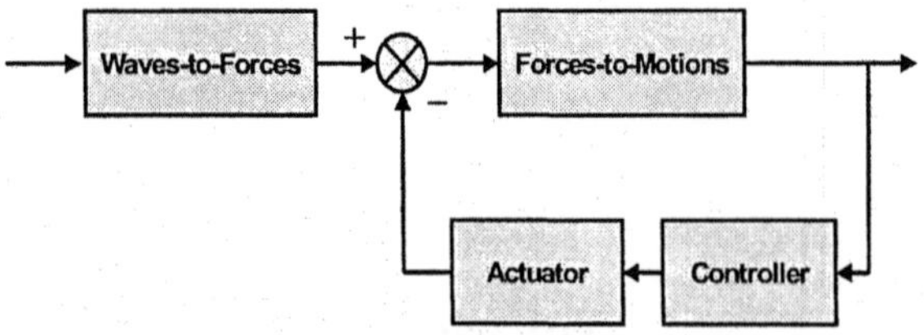


Fig.2. Main concept of model architecture.

The two-block decomposition (Waves-to-Forces, and Forces-to-Motions) will be applied to establish the complete model. The particular case of 40 knots of ship's speed will be followed along the paper (the methodology is similar for 20 and 30 knots). There is a section devoted to modelling of actuators in (Giron-Sierra, et al., 2003).

Applying the idea of two-block decomposition, the complete 6DOF model can be divided into two parts. Figure 3 shows the structure of the Waves-to-Forces part. There are blocks linking waves to torques (related to pitch, roll and yaw), and blocks linking waves to forces (related to surge, heave and sway). Notice the expected influence of actuators. Although the fast ferry has no rudder, some means of course control should be represented in the model (for instance the use of waterjets for change of heading).

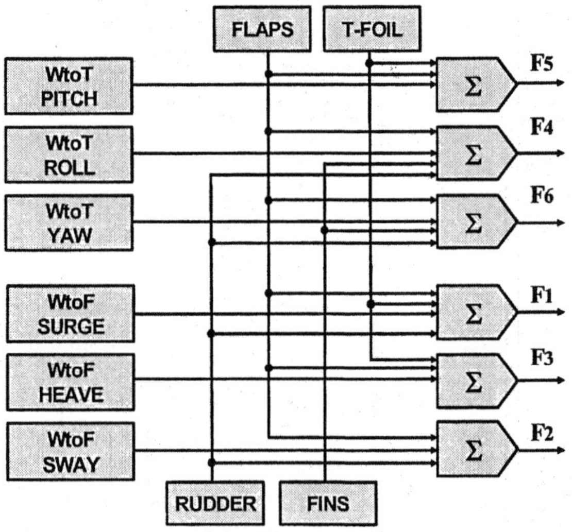


Fig. 3. Structure of the Waves-to-Forces part.

Figure 4 shows the other part of the model, Forces-to-Motions. Motions such pitch or sway are the result of a combination of forces.

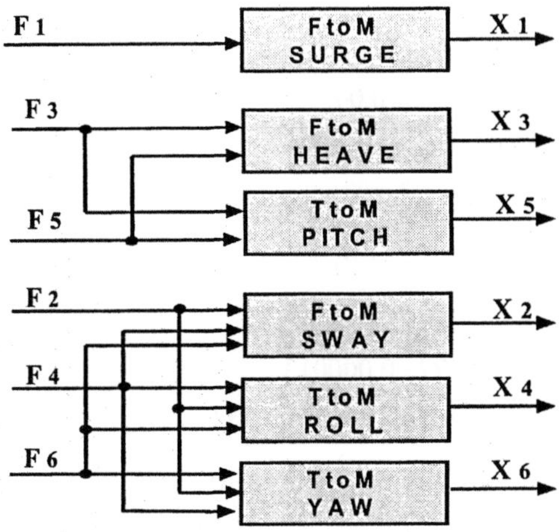


Fig. 4. Structure of the Forces-to-Motions part.

The set of equations given by Lloyd (1998) (see also Fossen, 2002), are used to describe the six ship's motions. It is a set of six differential equations. Take, for instance, one of these equations:

$$(m+a_{33})\ddot{x}_3+b_{33}\dot{x}_3+c_{33}x_3+ \\ +a_{35}\ddot{x}_5+b_{35}\dot{x}_5+c_{35}x_5=F_3 \qquad (1)$$

(this equation corresponds to heave; $a_{33}$ and $a_{35}$ are added masses, $b_{33}$ and $b_{35}$ are damping coefficients, $c_{33}$ and $c_{35}$ are restoring coefficients)

Using the Laplace transform, the equation can be expressed as

$$A(s)\,X_3+B(s)\,X_5=F_3(s) \qquad (2)$$

where

$$A(s)=(m+a_{33})\,s^2+b_{33}\,s+c_{33} \qquad (3)$$

$$B(s)=a_{35}\,s^2+b_{35}\,s+c_{35}$$

The same procedure is applied to the rest of equations. The complete set of equations, in the Laplace domain, is the following (see Giron-Sierra, et al., 2003, for more details):

$$S(s)\,X_1=F_1(s) \qquad \text{(surge) (4)}$$

$$A(s)\,X_3+B(s)\,X_5=F_3(s) \qquad \text{(heave) (5)}$$

$$C(s)\,X_3+D(s)\,X_5=F_5(s)\,. \qquad \text{(pitch) (6)}$$

$$E(s)\,X_2+F(s)\,X_4+G(s)\,X_6=F_2(s) \qquad \text{(sway) (7)}$$

$$H(s)\,X_2+I(s)\,X_4+J(s)\,X_6=F_4(s) \qquad \text{(roll) (8)}$$

$$K(s)\,X_2+L(s)\,X_4+M(s)\,X_6=F_6(s) \qquad \text{(yaw) (9)}$$

From equations (4..9) it is possible to obtain expressions linking motions to forces. For example, denoting:

$$\Gamma_{33}(s)=\frac{D(s)}{A(s)\,D(s)-B(s)\,C(s)} \qquad (10)$$

$$\Gamma_{35}(s)=\frac{B(s)}{A(s)\,D(s)-B(s)\,C(s)} \qquad (11)$$

It is possible to write:

$$X_3(s)=\Gamma_{33}(s)\,F_3(s)-\Gamma_{35}(s)\,F_5(s) \qquad (12)$$

With similar algebra, the set of equations (4..9) can be transformed to the following:

$$X_1(s)=\Gamma_1(s)\;F_1(s)$$
$$X_3(s)=\Gamma_{33}(s)\,F_3(s)-\Gamma_{35}(s)\,F_5(s)$$
$$X_5(s)=\Gamma_{55}(s)\,F_5(s)-\Gamma_{53}(s)\,F_3(s)$$
$$X_2(s)=\Gamma_{22}(s)\,F_2(s)-\Gamma_{24}(s)\,F_4(s)-\Gamma_{26}(s)\,F_6(s)$$
$$X_4(s)=\Gamma_{44}(s)\,F_4(s)-\Gamma_{42}(s)\,F_2(s)-\Gamma_{46}(s)\,F_6(s)$$
$$X_6(s)=\Gamma_{66}(s)\,F_6(s)-\Gamma_{62}(s)\,F_2(s)-\Gamma_{64}(s)\,F_4(s)$$

(13)

Equations (13) are the basis for the Forces-to-Motions model, and show clearly the cross-couplings that are represented in figure 5. A source of difficulty is that parameters (added masses, damping) vary with the frequency of encounter with waves.

The data for modelling are obtained with a CFD program, which predicts the responses of the ship for regular sinusoidal waves. A set of 25 wavelengths of incident waves has been defined, covering the frequency range of interest, given the characteristics of the ship. The ship's motions will have a resonance peak when the wavelength of incident waves is near to the ship's length. The range of interest is below and up this peak. A set of 12 headings: 0° 15°, 30°, etc., and three ship's speeds (20, 30 and 40 knots) have been studied with the CFD program. The results are expressed by a data base, and six RAOS, corresponding to the six DOF motions, for each heading and speed. Each RAO has 25 data points. The RAOS describe the motions versus wave wavelength. There are also data that could be used to plot RAOS of forces versus wave wavelength. Finally, the CFD program gives the values of added masses and damping coefficients for each frequency of encounter (these parameters change with frequency).

## 3. MODELS FOR 180° HEADING

For the Waves-to-Forces model there are not equations to help. The alternative is to use the Waves-to-Forces data offered by the CFD program and apply MATLAB tools for transfer function fitting. These tools need some preliminary information on a plausible structure of the transfer function. From basic physics some important hints can be elicited (Giron-Sierra, et al., 2003), and a structure of two complex poles and a complex zero (and a zero at the origin when applicable) is decided for all Waves-to-Forces fittings.

Table 1 shows the poles and zeros of the surge, heave and pitch Wave-to-Forces transfer functions, at 40 knots. The 180° heading angle (head seas) excites primary resonances in the surge, heave and pitch dynamics.

Table 1. Surge, heave and pitch Waves-to-Forces transfer functions at 40 knots, 180° heading.

| | SURGE | HEAVE | PITCH |
|---|---|---|---|
| KG | 0.33457 | 1.174 | 0.29584 |
| Zero at origin | yes | no | yes |
| Complex pole: wn | 1.4433 | 1.5073 | 1.4925 |
| Complex pole: delta | 0.51107 | 0.91158 | 0.94216 |
| Complex zero: wn | 2.262 | 2.2918 | 2.5807 |
| Complex zero: delta | 0.07503 | 0.00999 | 0.0724 |
| HF Complex pole: wn | 3.3617 | 4.5581 | 4.0553 |
| HF Complex pole: delta | 0.35038 | 0.26309 | 0.38292 |

Transfer functions for the $\Gamma_{ij}$ terms in equations (13) have been determined. For instance:

$$\Gamma_{11}=\frac{0.00055}{s(s+0.015)} \quad (14)$$

$$\Gamma_{33}=\frac{(KG\, w_{p1}^2\, w_{p2}^2\, / w_z^2)}{(s^2+2\delta_{p1}\, w_{p1}\, s+w_{p1}^2)}\frac{(s^2+2\delta_z\, w_z\, s+w_z^2)}{(s^2+2\delta_{p2}\, w_{p2}\, s+w_{p2}^2)} \quad (15)$$

with:

$KG=0.00008,\ w_z=1.16, \delta_z=0.47$

$w_{p1}=0.98, \delta_{p1}=0.55$

$w_{p2}=1.66, \delta_{p2}=0.195$

$$\Gamma_{55}=\frac{(KG\, w_{p1}^2\, w_{p2}^2\, / w_z^2)}{(s^2+2\delta_{p1}\, w_{p1}\, s+w_{p1}^2)}\frac{(s^2+2\delta_z\, w_z\, s+w_z^2)}{(s^2+2\delta_{p2}\, w_{p2}\, s+w_{p2}^2)} \quad (16)$$

with:

$KG=0.000000115, w_z=0.92, \delta_z=0.42$

$w_{p1}=0.77, \delta_{p1}=0.46$

$w_{p2}=1.66, \delta_{p2}=0.205$

## 4. MODELS FOR 90° HEADING

The 90° heading angle excites primary resonances in the sway, roll and yaw dynamics.

Again, transfer functions for the $\Gamma_{ij}$ terms in equations (13) have been determined. For example:

$$\Gamma_{22}=\frac{(KG w_{p1}^2\, w_{p2}^2\, / z_1 w_z^2)}{s(s^2+2\delta_{p1}\, w_{p1}\, s+w_{p1}^2)}\frac{(s+z_1)(s^2+2\delta_z\, w_z\, s+w_z^2)}{(s^2+2\delta_{p2}\, w_{p2}\, s+w_{p2}^2)} \quad (17)$$

with:

$KG=0.00035, z_1=0.027,\ w_z=0.995, \delta_z=0.39$

$w_{p1}=0.17, \delta_{p1}=0.04$

$w_{p2}=1.04, \delta_{p2}=0.4$

$$\Gamma_{44}=\frac{(KG\, w_p^2)}{(s^2+2\delta_p\, w_p\, s+w_p^2)} \quad (18)$$

with:

$KG=0.000013$

$w_p=1.04, \delta_z=0.032$

$$\Gamma_{66}=\frac{(KG w_{p1}^2\, w_{p2}^2\, / z_1 w_z^2)}{s(s^2+2\delta_{p1}\, w_{p1}\, s+w_{p1}^2)}\frac{(s+z_1)(s^2+2\delta_z\, w_z\, s+w_z^2)}{(s^2+2\delta_{p2}\, w_{p2}\, s+w_{p2}^2)} \quad (19)$$

with:

$KG=0.0000007, z_1=0.06, w_z=0.995, \delta_z=0.34$

$w_{p1}=0.17, \delta_{p1}=0.2$

$w_{p2}=1.04, \delta_{p2}=0.37$

Table 2 shows the poles and zeros of the sway, roll and yaw Wave-to-Forces transfer functions, at 40 knots.

Table 2. Sway, roll and yaw Waves-to-Forces transfer functions at 40 knots, 90° heading.

| | SWAY | ROLL | YAW |
|---|---|---|---|
| KG | 0.10188 | 0.32389 | 0.2 |
| Zero at origin | yes | yes | yes |
| Complex pole: wn | 1.103 | 0.99286 | 0.79121 |
| Complex pole: delta | 0.30205 | 0.37 | 0.6 |
| Complex zero: wn | 2.2325 | 2 | 0.91082 |
| Complex zero: delta | 0.1 | 0.4 | 0.3 |
| HF Complex pole: wn | 2.4087 | 2.7 | 2 |
| HF Complex pole: delta | 0.1 | 0.25 | 0.5 |

Indeed the other $\Gamma_{ij}$ terms exist, and have been modelled with transfer functions.

## 5. PREDICTION OF TRANSFER FUNCTION VARIATION WHEN HEADING CHANGES

Note that sections 3 and 4 focused on "pure" resonance peaks (corresponding to complex poles) of the six motions. Now let us analyze what should happen when the ship's heading change. There are resonance peaks of the Waves-to-Forces part and resonance peaks of the Forces-to-Motions part. Figure 5 shows a possible situation, with a main peak with frequency $w_F$ of Waves-to-Forces and another main peak with frequency $w_M$ of Forces to Motions.

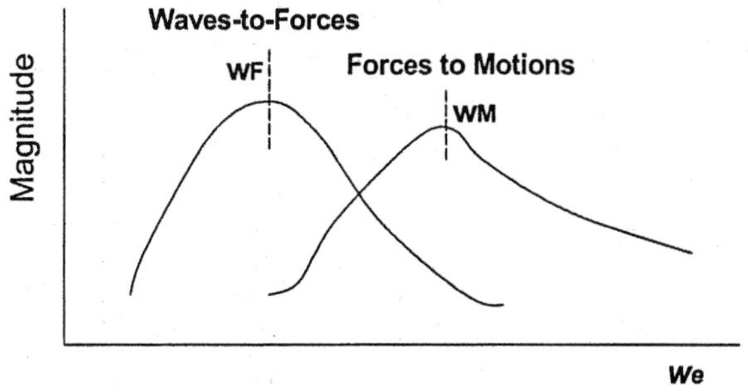


Fig.5. Resonance peaks of Waves-to-Forces and Forces-to-Motions.

The key equation to be considered relates the frequency of waves $w_o$ with the frequency of encounter with waves $w_e$,

$$w_e = w_o - \frac{w_o^2 U}{g} \cos\beta \qquad (20)$$

where $U$ is ship's speed and $\beta$ is heading angle.

An analogy for the Forces-to-Motions part could be the mass-friction-spring system. The resonance will take place for a certain frequency of encounter with waves. This frequency can change if added masses and other coefficients in equations (4..9) vary in function of heading angle. Let us denote as $w_{MH}$ the frequency of resonance of Forces-to-Motions with head seas, and $w_{MX}$ the frequency of resonance of Forces-to-Motions with another heading angle.

For heave, pitch and surge, the Waves-to-Forces part depends on the relationship between the distance of successive wave crests and the ship's length. The resonance takes place when the distance between wave crests are similar to the ship's length. Now, for example, consider a heading change from 180° to 120°. The distance between wave crests as seen by the ship increases, waves with shorter wavelength (higher frequency) are needed for resonance. Let us denote as $w_{OH}$ the frequency of waves that excites the Waves-to-Forces resonance with head seas, and $w_{OX}$ the frequency of waves that does the same with another heading angle. For non perpendicular to waves headings, we can establish the following relationship:

$$w_{OX}^2 = \frac{w_{OH}^2}{|\cos\beta|} \qquad (21)$$

Since the ship is moving, she sees a frequency of encounter with the waves that excites the Waves-to-Forces resonance. Let us denote as $w_{FH}$ the frequency of encounter with waves causing resonance of Waves-to-Forces with head seas. Like wise, let us denote as $w_{FX}$ the frequency of encounter with the waves causing resonance of Waves-to-Forces with another heading angle. Reasoning with equations (20,21) we can establish the following relationship:

$$w_{FX} = w_{FH} + \frac{w_{OH}\,(1-\sqrt{|\cos\beta|})}{\sqrt{|\cos\beta|}} \qquad (22)$$

Consequently, the peak of Waves-to-Forces in figure 5 moves to the right when heading change from 180° to another angle. Supposing the peak of Forces-to-Motions do not move, the heading angle where both peaks coincide are 131.08° for heave, and 129.19° for pitch.

Equation (22) is useful to predict where Waves-to-Forces complex poles will be at 150° heading angle, based on the values found at 180° heading angle.

## 6. MODELLING FOR 150° HEADING ANGLE

With the help of predictions made with equation (22) the Waves-to-Forces transfer functions for 150° heading angle have been obtained. Table 3 shows the results for surge, heave and pitch; compare with Table 1. The predictions have been confirmed. Table 4 shows the results for sway, roll, and yaw; compare with Table 2.

Table 3. Surge, heave and pitch Waves-to-Forces transfer functions at 40 knots, 150° heading.

| | SURGE | HEAVE | PITCH |
|---|---|---|---|
| KG | 0.38 | 1.14 | 0.31 |
| Zero at origin | yes | no | yes |
| Complex pole: wn | 1.4511 | 1.556 | 1.5038 |
| Complex pole: delta | 0.51107 | 0.91158 | 0.94216 |
| Complex zero: wn | 2.1 | 2.05 | 2.4 |
| Complex zero: delta | 0.07503 | 0.0099 | 0.0724 |
| HF Complex pole: wn | 3.4398 | 4.6535 | 4.1442 |
| HF Complex pole: delta | 0.6 | 0.4 | 0.6 |

Table 4. Sway, roll and yaw Waves-to-Forces transfer functions at 40 knots, 90° heading.

| | SWAY | ROLL | YAW |
|---|---|---|---|
| KG | 0.29 | 0.105 | 0.0565 |
| Zero at origin | yes | yes | yes |
| Complex pole: wn | 1.103 | 1.7 | 1.75 |
| Complex pole: delta | 0.43 | 0.37 | 0.235 |
| Complex zero: wn | 1.9 | 3 | 0.6 |
| Complex zero: delta | 0.012 | 0.3 | 0.125 |
| HF Complex pole: wn | 2.2 | 5 | 3.5 |
| HF Complex pole: delta | 0.2 | 0.6 | 0.25 |

We also found the transfer functions of the $\Gamma_{ij}$ terms in equations (13), for 150° heading angle. $\Gamma_{11}$ is practically the same than $\Gamma_{11}$ at 180° heading angle. The rest of $\Gamma_{ij}$ have same transfer function structure than at 180° heading angle, with modified values:

$\Gamma_{33}$ :

$$KG = 0.00008,\ w_z = 0.88, \delta_z = 0.51$$
$$w_{p1} = 0.78, \delta_{p1} = 0.56$$
$$w_{p2} = 1.7, \delta_{p2} = 0.235$$

$\Gamma_{55}$ :

$$KG = 0.000000115, w_z = 0.92, \delta_z = 0.4$$
$$w_{p1} = 0.78, \delta_{p1} = 0.44$$
$$w_{p2} = 1.7, \delta_{p2} = 0.22$$

$\Gamma_{22}$ :

$$KG = 0.00035, z_1 = 0.036,\ w_z = 1.02, \delta_z = 0.34$$
$$w_{p1} = 0.2, \delta_{p1} = 0.24$$
$$w_{p2} = 1.08, \delta_{p2} = 0.37$$

$\Gamma_{44}$ :

$$KG = 0.000015$$
$$w_p = 1.08, \delta_z = 0.024$$

$\Gamma_{66}$ :

$$KG = 0.0000007, z_1 = 0.082, w_z = 1.02, \delta_z = 0.34$$
$$w_{p1} = 0.2, \delta_{p1} = 0.25$$
$$w_{p2} = 1.08, \delta_{p2} = 0.33$$

Figure 6 shows the model and data agreement for Waves-to-Sway force. Figure 7 idem for Waves-to-Roll moment. Figure 8 idem for $\Gamma_{55}$. Figure 9 idem for $\Gamma_{53}$.

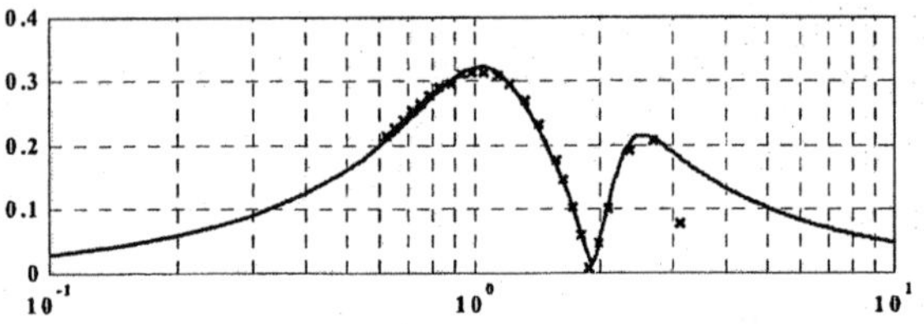

Fig.6. Model and data agreement for Waves-to-Sway force, at 40 knots and 150° heading.

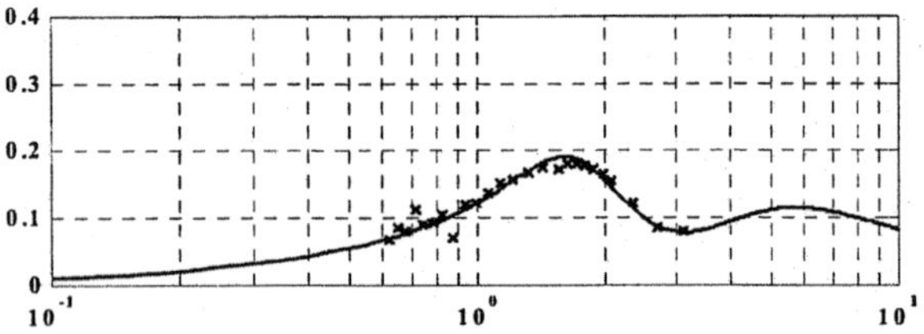

Fig.7. Model and data agreement for Waves-to-Roll moment, at 40 knots and 150° heading.

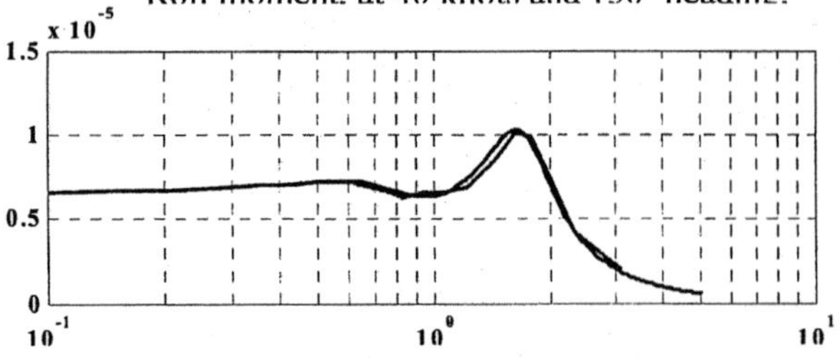

Fig.8. Model and data agreement for Pitch moment-to- Pitch motion, at 40 knots and 150° heading.

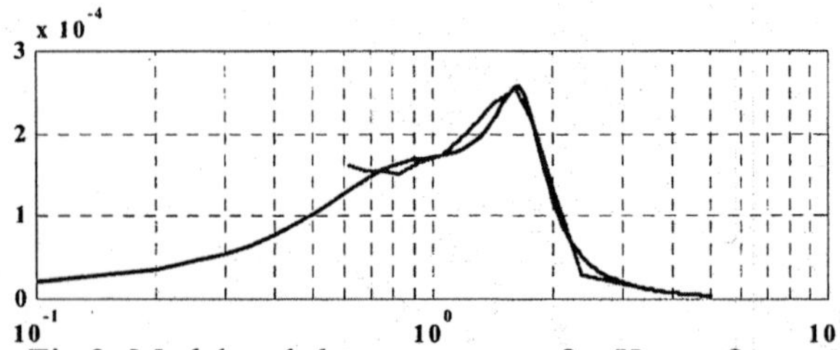

Fig.9. Model and data agreement for Heave force-to- Pitch motion, at 40 knots and 150° heading.

## 7. CONCLUSIONS

Advances in the development of a 6DOF ship's motions model have been presented. Specific emphasis has been put in 180°, 90° and 150° heading angles at 40 knots. The model is required for a good multivariable control design, taking into account possible difficulties related to motion coupling.

## REFERENCES


Aranda, J., J.M. Diaz, P. Ruiperez, T.M. Rueda and E. Lopez (2001). Decreasing of the motion sickness incidence by a multivariable classic control for a high speed ferry, In *Proceedings IFAC Intl. Conf. Control Applications in Marine Systems CAMS2001*, Glasgow.

Esteban, S., J.M. De la Cruz, J.M. Giron-Sierra, B. De Andres, J.M. Diaz and J. Aranda (2000). Fast ferry vertical acceleration reduction with active flaps and T-foil, In *Proceedings IFAC Intl. Symposium Maneuvering and Control of Marine Craft MCMC2000*, Aalborg, 233-238.

Esteban, S., B. De Andres, J.M. Giron-Sierra, O.R. Polo and E. Moyano (2001). A simulation tool for a fast ferry control design, In *Proceedings IFAC Intl. Conf. Control Applications in Marine Systems CAMS2001*, Glasgow.

Esteban, S., B. Andres-Toro, E. Besada-Portas, J.M. Giron-Sierra and J.M. De la Cruz (2002). Multiobjective control of flaps and T-foil in high speed ships, In *Proceedings IFAC 2002 World Congress*, Barcelona.

Fossen, T.I. (2002). *Marine Control Systems*, Marine Cybernetics AS, Trondheim.

Giron-Sierra, J.M., S. Esteban, B. De Andres, J.M. Diaz and J.M. Riola (2001). Experimental study of controlled flaps and T-foil for comfort improvement of a fast ferry, In *Proceedings IFAC Intl. Conf. Control Applications in Marine Systems CAMS2001*, Glasgow.

Giron-Sierra, J.M., B. Andres-Toro, S. Esteban, J. Recas, E. Besada, J.M. De la Cruz and A. Maron (2003). First principles modelling study for the development of a 6DOF motions model of a fast ferry, In *Proceedings IFAC MCMC 2003,* Gerona, Spain.

Lloyd, A.R.J.M. (1998). *Seakeeping: Ship Behaviour in Rough Weather*, A.R.M.J. Lloyd, Gosport, Hampshire, U.K.

ELSEVIER
IFAC
PUBLICATIONS
www.elsevier.com/locate/ifac

# STATE-SPACE REPRESENTATION OF FREQUENCY-DEPENDENT HYDRODYNAMIC COEFFICIENTS

**Åsmund Hjulstad * Erlend Kristiansen ** Olav Egeland *****

** Norwegian University of Science and Technology, email: Hjulstad@itk.ntnu.no*
*** Norwegian University of Science and Technology, email: Erlend.Kristiansen@itk.ntnu.no*
**** Norwegian University of Science and Technology, email: Olav.Egeland@itk.ntnu.no*

Abstract: The hydrodynamic coefficients in the equations of motion for a surface vessel are frequency-dependent. An alternative model with convolution terms includes the same effects without resorting to frequency-dependent coefficients. These convolution terms can be represented by state-space models, as proposed in a recent work. In this paper the method is applied to detailed data for an offshore support vessel generated from WAMIT, and the resulting state space model is simulated together with wave excitation from irregular sea using Simulink. A state space model of low order is sufficient to describe the dynamics without resorting to frequency varying coefficients. *Copyright ©2004 IFAC.*

Keywords: State-space realization, Frequency-dependent characteristics, Time-domain responses, Convolution integral, Marine systems

## 1. INTRODUCTION

The concept of frequency-dependent added mass and potential damping is well established in the formulation of the equations of motion for a surface ship. This formulation is used in identification experiments for ship models using motion at a single frequency, and it is used in commercially available programs like WAMIT and VERES. It was shown by (Cummins, 1962) that the frequency dependence of added mass and potential damping can be seen as a consequence of a convolution term in the radiation potential, leading to a convolution term in the equation of motion (Cummins, 1962; Ogilvie, 1964). For motion at a single frequency the convolution term in the equation of motion is equivalent to frequency-dependent added mass and potential damping parameters, making it possible to find the convolution term from either added mass or potential damping.

The formulations of the equation of motion based on the use of convolution terms, or, alternatively, on frequency-dependent parameters are not in agreement with the model formulations used in simulation and in automatic control. As a result of this it is not straightforward to apply the usual methods for simulation and for controller analysis and design. A partial solution to this problem was presented in (Kristiansen and Egeland, 2003) where a method for finding a state space representation of low order for the convolution term was described.

In this paper the results from (Kristiansen and Egeland, 2003) are further developed, and the method is applied to detailed data from WAMIT for an offshore support vessel. It is shown that the expressions are simplified in stationkeeping applications. For the offshore support vessel under investigation an accurate representation of all first-order wave effects was achieved with 8 states per convolution term for 6DOF motion. The resulting state space model was implemented together with wave excitation, thruster and restoring forces in a Simulink model, and simulation results are shown in the paper.

## 2. PRELIMINARIES

### 2.1 Equation of motion

The usual method for finding the hydrodynamic forces acting on a ship moving in waves is based on the superposition of the radiation forces due to the ship motion in an undisturbed sea, and of the wave forces on a nonmoving ship (Newman, 1977). The velocity potential $\phi$ is then described as the sum $\phi = \phi_R + \phi_W$ where the radiation potential $\phi_R$ is due to the motion of the ship on an undisturbed sea, and the potential $\phi_W$ is due to the wave excitation. The pressure on the hull is assumed to be given by

$$p = \rho g z - \rho \frac{\partial \phi_R}{\partial t} - \rho \frac{\partial \phi_W}{\partial t} \quad (1)$$

which is a sum of hydrostatic pressure, radiation pressure and diffraction pressure. The force $\boldsymbol{F}$ and the moment $\boldsymbol{M}$ on the ship hull can then be described as sums of hydrostatic terms, radiation effects and diffraction effects.

For notational simplicity the motion is described with generalized coordinates $\mathbf{q} = (q_1 \dots q_6)^T$ and associated generalized forces $\boldsymbol{\tau} = (\tau_1 \dots \tau_6)^T$. The results can be extended to the usual kinematics as given in (Fossen, 1994). The equation of motion is given by

$$\sum_{k=1}^{6} m_{jk} \ddot{q}_k = \tau_j^H + \tau_j^R + \tau_j^W + \tau_j^A + \tau_j^E \quad (2)$$

where $m_{jk}$ are the inertia parameters of the ship, $\tau_j^H$ are the hydrostatic forces, $\tau_j^R$ are the radiation forces, $\tau_j^W$ are the wave excitation forces, $\tau_j^A$ are the actuator forces, and $\tau_j^E$ are the external forces.

### 2.2 Radiation problem with memory effects

Memory effects in the radiation potential are included in the form of a convolution term (Cummins, 1962). The radiation potential is then

$$\phi_R(\boldsymbol{r}, t) = \sum_{i=1}^{6} \dot{q}_i(t) \psi_i(\boldsymbol{r}) + \sum_{i=1}^{6} \int_{-\infty}^{t} \chi_i(\boldsymbol{r}, t-\tau) \dot{q}_i(\tau) \mathrm{d}\tau \quad (3)$$

The potentials $\psi_i(\boldsymbol{r})$ represent the instantaneous response of the fluid due to the ship motion, whereas $\chi_i(\boldsymbol{r}, t)$ are impulse responses to the ship velocity, accounting for the wave propagation on the free surface. Boundary conditions and further details are found in (Ogilvie, 1964). It should be noted that the force contribution from $\psi_i(\boldsymbol{r})$ alone, will lead to a definition of added mass that can be directly related to the added mass of a double body in unbounded water.

When the radiation potential is given by (3), and neglecting forward speed and current, the radiation force is found to be (Ogilvie, 1964)

$$\tau_j^R = -\sum_{k=1}^{6} a_{jk} \ddot{q}_k - \sum_{k=1}^{6} \int_{-\infty}^{t} K_{jk}(t-\sigma) \dot{q}_k(\sigma) \mathrm{d}\sigma \quad (4)$$

where $a_{jk}$ are constant hydrodynamic coefficients, and the convolution terms in the expression for $\tau_j^R$ are due to the convolution terms in the radiation potentials (3). $K_{jk}$ is commonly referred to as a memory function or retardation function. This paper exploits its similarites to a impulse response, and refers to it using this latter term.

Assuming small motions, the hydrostatic forces are linear and the equation of motion is found to be

$$\begin{aligned} &\sum_{k=1}^{6} (m_{jk} + a_{jk}) \ddot{q}_k + \sum_{k=1}^{6} C_{jk} q_k \\ &+ \sum_{k=1}^{6} \int_{-\infty}^{t} K_{jk}(t-\sigma) \dot{q}_k(\sigma) \mathrm{d}\sigma \quad (5) \\ &= \tau_j^W + \tau_j^A \end{aligned}$$

The derivation of (5) was done without resorting to frequency analysis results. In particular, the concept of frequency-dependent added mass and potential damping has not been introduced so far.

### 2.3 Single frequency motion

The parameters of the equations of motion can be found from model experiments with single frequency forced motion $q_k(t) = q_k \cos \omega t$ with $q_k(t) = 0$ for $t < 0$. Then, according to standard arguments all terms in the equation of motion will eventually be sinusoidal and of the same frequency $\omega$. In particular, note that the convolution integral can be written (Ogilvie, 1964)

$$\int_{-\infty}^{t} K_{jk}(t-\sigma) \dot{q}_k(\sigma) \mathrm{d}\sigma \quad (6)$$

$$= -\ddot{q}_k(t) \left( \frac{1}{\omega} \int_0^{\infty} K_{jk}(\sigma) \sin \omega\sigma \mathrm{d}\sigma \right) \quad (7)$$

$$+ \dot{q}_k(t) \left( \int_0^{\infty} K_{jk}(\sigma) \cos \omega\sigma \mathrm{d}\sigma \right) \quad (8)$$

This leads to the equation of motion in the widely used form

$$\sum_{k=1}^{6} (m_{jk} + \alpha_{jk}(\omega))\, \ddot{q}_k + \sum_{k=1}^{6} \beta_{jk}(\omega) \dot{q}_k$$

$$+ \sum_{k=1}^{6} C_{jk} q_k = \tau_j^W + \tau_j^A \qquad (9)$$

where the frequency-dependent added mass $\alpha_{jk}(\omega)$ and the frequency-dependent potential damping parameters $\beta_{jk}(\omega)$ are given by

$$\alpha_{jk}(\omega) = a_{jk} - \frac{1}{\omega} \int_0^\infty K_{jk}(t) \sin \omega t dt \quad (10)$$

$$\beta_{jk}(\omega) = \int_0^\infty K_{jk}(t) \cos \omega t dt \quad (11)$$

This formulation captures the frequency dependence of the added mass and the potential damping parameters that is observed in identification experiments with single-frequency motion. However, the equation of motion in the form (9) is only valid under the assumption that $q_k(t) = q_k \cos \omega t$, and that $\tau_j^W$ and $\tau_j^A$ are sinusoidal functions at frequency $\omega$, and cannot be used to describe transient dynamics (Cummins, 1962).

The impulse response $K_{jk}(t)$ can be found from either (10) or (11) by comparison with the Fourier sine and cosine transform of $K_{jk}(t)$, resulting in two equivalent expressions for $K_{jk}(t)$ (Ogilvie, 1964).

$$K_{jk}(t) = -\frac{2}{\pi} \int_0^\infty \omega \left[\alpha_{jk}(\omega) - \alpha_{jk}(\infty)\right] \sin \omega t d\omega \quad (12)$$

$$K_{jk}(t) = \frac{2}{\pi} \int_0^\infty \beta_{jk}(\omega) \cos \omega t d\omega \quad (13)$$

When calculated numerically it appears only feasible to use (13) as (12) converges very slowly as $\omega$ increases. In addition, for $t = 0$ (12) will give $K_{jk}(t) = 0$.

## 3. LAPLACE TRANSFORMATION AND STATE-SPACE MODEL

The Laplace transformation of the convolution integral is

$$\mathcal{L}\left\{ \int_{-\infty}^{t} K_{jk}(t-\sigma) \dot{q}_k(\sigma) d\sigma \right\} = s K_{jk}(s) q_k(s) \quad (14)$$

where $K_{jk}(s) = \mathcal{L}\{K_{jk}(t)\}$, $q_k(s) = \mathcal{L}\{q_k(t)\}$, and $\mathcal{L}\{\dot{q}_k(t)\} = s q_k(s)$. Then the Laplace transformation of the equation of motion (5) is found to be

$$\sum_{k=1}^{6} \{(m_{jk} + a_{jk})\, s^2 + K_{jk}(s) s \quad (15)$$

$$+ C_{jk}\} q_k(s) = \tau_j^W(s) + \tau_j^A(s) \quad (16)$$

Convolution term $j$ may be written

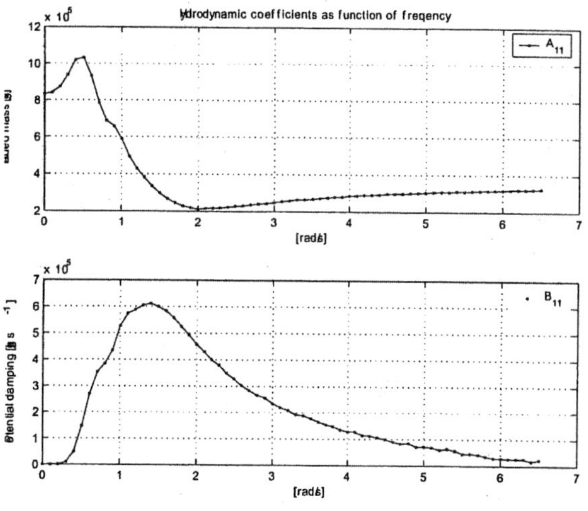


Fig. 1. Hydrodynamic coefficients in surge

$$\mu_j = \sum_{k=1}^{6} \int_{-\infty}^{t} K_{jk}(t-\sigma) \dot{q}_k(\sigma) d\sigma \quad (17)$$

Define the vector $\boldsymbol{\mu}$ and the matrix $\mathbf{K}$ by

$$\boldsymbol{\mu} = (\mu_1, \ldots, \mu_6)^T, \quad \mathbf{K}(s) = \{K_{ij}(s)\} \quad (18)$$

where $K_{ij}(s) = \mathcal{L}\{K_{ij}(t)\}$. If a state space model $(\mathbf{A}, \mathbf{B}, \mathbf{C}, \mathbf{D})$ with state vector $\boldsymbol{\xi}$, input $\dot{\mathbf{q}}$ and output $\mu$ can be found the equation of motion can then be written

$$0 = \sum_{k=1}^{6} \left[(m_{jk} + a_{jk})\, \ddot{q}_k + \mu_{jk}\right] \quad (19)$$

$$- \left(\tau_j^H + \tau_j^W + \tau_j^A\right) \quad (20)$$

$$\dot{\boldsymbol{\xi}}_{jk} = \mathbf{A}_{jk} \boldsymbol{\xi}_{jk} + \mathbf{B}_{jk} \dot{\mathbf{q}}_k \quad (21)$$

$$\mu_{jk} = \mathbf{C}_{jk} \boldsymbol{\xi}_{jk} + \mathbf{D}_{jk} \dot{\mathbf{q}}_k \quad (22)$$

From energy considerations of the radiation problem one can see that the mapping $\dot{\mathbf{q}} \mapsto \boldsymbol{\tau}^R$ is passive (Kristiansen and Egeland, 2003). This implies that the transfer function

$$\mathbf{H}(s) = \mathbf{A}s + \mathbf{K}(s) \quad (23)$$

defined by $\boldsymbol{\tau}^R(s) = \mathbf{H}(s)\dot{\mathbf{q}}(s)$ is positive real (Egeland and Gravdahl, 2002). This transfer function is seen to be positive real whenever $\mathbf{K}(s)$ is positive real. It seems reasonable that the diagonal elements, that is $\mathbf{K}_{jj}(s)$ should be passive, and therefore positive real. These properties should be conserved throughout all manipulations of the model.

## 4. GENERATION OF APPROXIMATE STATE-SPACE MODEL

This section presents a method for the calculation of a low order state-space representation of the convolution term. This paper supplements (Kristiansen and Egeland, 2003) with suggestions from experiences with processing real data.

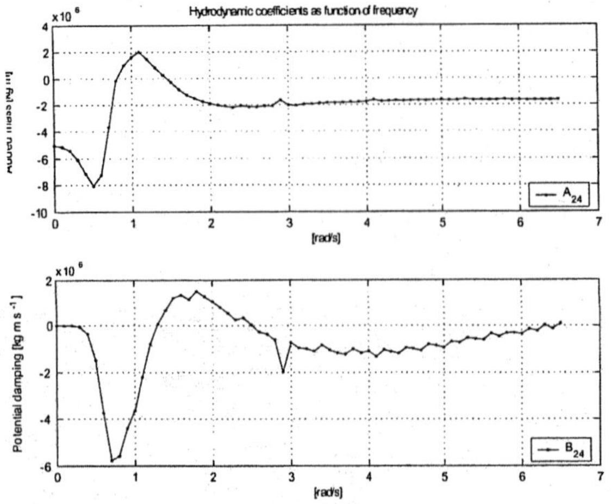


Fig. 2. Hydrodynamic coefficients in coupling mode sway–roll

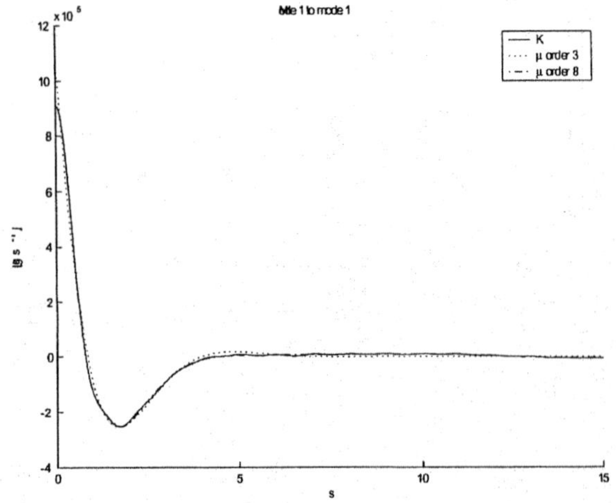


Fig. 3. Impulse response in surge

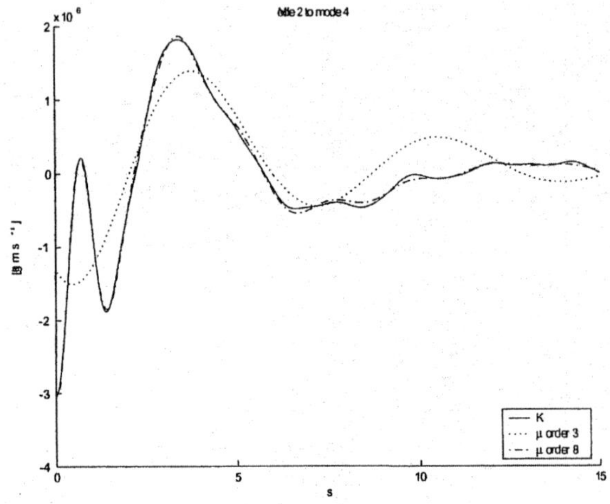


Fig. 4. Impulse response in coupling mode sway–roll

Frequency-dependent added mass $\alpha_{jk}(\omega)$ and damping $\beta_{jk}(\omega)$ for a ship can be calculated numerically using software packages such as WAMIT. In this work, a 3400 panel hull geometry of an approximately 110 m long, 6000 tonnes deadweight offshore support vessel is used with WAMIT version 6. Values for $\beta_{jk}(\omega)$ were calculated in the frequency range $0.1 < \omega < 6.5$ rad/s using 0.1 rad/s intervals. Two of the coefficients are shown in figures 1 and 2. Note the large differences between the maximum and minimum values, in some coupling modes the force even changes sign. Variations in some of the coefficients at high frequency, seen here in the coupling mode between sway and roll (figure 2), are expected to be related to the panel size in the hull geometry.

The impulse responses $K_{jk}(t)$ are then calculated from (13). $K_{jk}(t)$ is found by trapezoidal integration over $\omega$ with $\Delta\omega = 0.01$ for each $t$, spaced by the time-step $h = 0.05$ s. The upper limit for $t$ must be chosen with care. A too large value makes the subsequent system identification time consuming, whereas a too small value might cause some dynamics to be lost. Here, the upper limit was taken to be 15 s, which is just after $K_{jk}(t)$ converges to zero for all modes.

The state space model is generated from $K_{jk}(t)$ by applying the system identification scheme based on the Hankel SVD method proposed in (Kung, 1978), available as the function IMP2SS in the Robust Control Toolbox of MATLAB. After scaling with the time-step $h$, one gets system matrices $\bar{\mathbf{A}}_{jk}, \bar{\mathbf{B}}_{jk}, \bar{\mathbf{C}}_{jk}$ and $\bar{\mathbf{D}}_{jk}$ (Kristiansen and Egeland, 2003).

The state space model from IMP2SS is of high order, as no explicit bound of the $H^{\infty}$-norm of the error between the approximate realization and the exact realization was stated. The original model order varies with $T$ and $h$, and is close to 280 in several modes. The model order is reduced using truncation and Schur methods (Safonov and Chiang, 1989), implemented as Matlab function BALMR from the Robust Control Toolbox. The output is the new state-space matrices $\mathbf{A}_{jk}, \mathbf{B}_{jk}, \mathbf{C}_{jk}$ and $\mathbf{D}_{jk}$. Models of order between 3 and 12 have been calculated in relation to this work.

### *4.1 Results*

The results of the model identification and reduction can be visualised by plotting the impulse responses of the reduced model together with the original memory function. This is shown in figure 3 and 4 for two modes. Here $i$ in $\mu^i_{jk}$ denotes the order of the approximated state-space system. In figure 5 and 6 the Nyquist diagram for the diagonal element $\mu^i_{22}$ and $\mu^i_{44}$ for $i = 3, 4, 5$ are plotted. Note that $K^4_{44}(s)$ and $K^5_{44}(s)$ are positive real, while $K^3_{44}(s)$ is not positive real. In sway, however, even $K^5_{22}(s)$ is not positive real. In the simulations that follow, eighth order approximated systems are used.

## 5. CALCULATION OF WAVE EXCITING FORCES

The vessel is simulated in 6DOF excited with long-crested waves. The sea state is described using a JONSWAP spectrum with 8 s peak wave period and

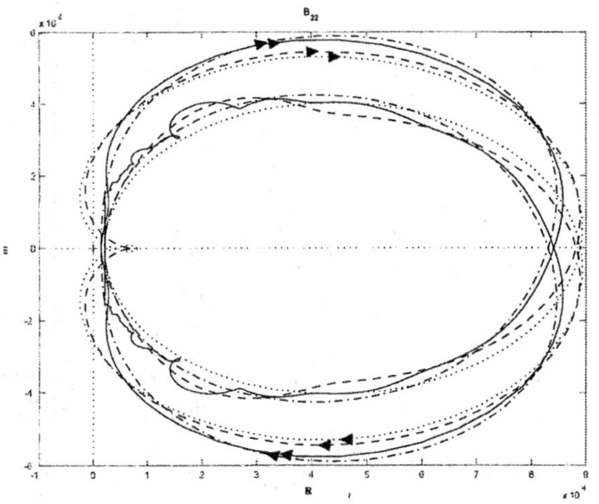


Fig. 5. Nyquist diagram of $K_{22}^{i}(s)$ where $i$ denotes the approximation order. $i = 5$ (dashed), $i = 4$ (dash-dotted) and $i = 3$ (dotted), together with impulse function before reduction

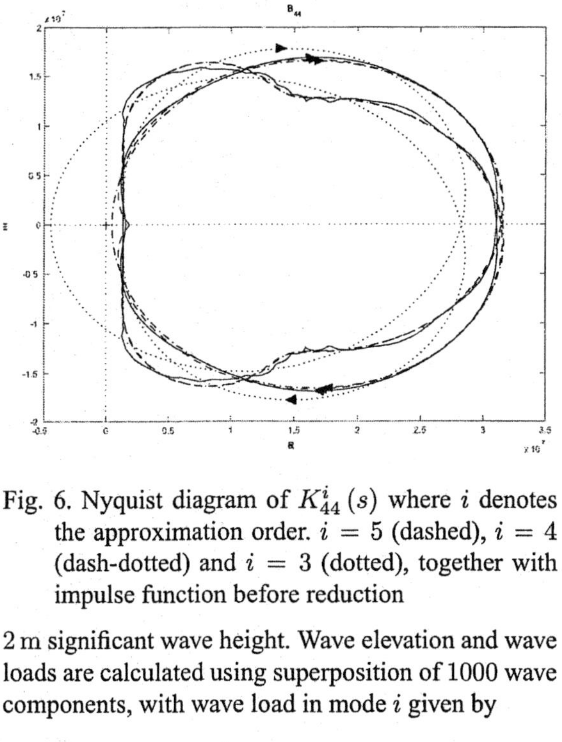


Fig. 6. Nyquist diagram of $K_{44}^{i}(s)$ where $i$ denotes the approximation order. $i = 5$ (dashed), $i = 4$ (dash-dotted) and $i = 3$ (dotted), together with impulse function before reduction

2 m significant wave height. Wave elevation and wave loads are calculated using superposition of 1000 wave components, with wave load in mode $i$ given by

$$\tau_i(t) = \sum_j K_{i,j} A_j \sin(\omega_j t - k_j x \cos\theta_j \quad (24)$$

$$-k_j y \sin\theta_j + \varphi_{i,j} + \varepsilon_j) \quad (25)$$

where $A_j$ is the amplitude of component $j$ at frequency $\omega_j$

$$A_j = \sqrt{2S(\omega_j)\Delta\omega} \quad (26)$$

$k_j$ is wave number ($k = \frac{\omega^2}{g}$), $x$ and $y$ are vessel position, $K_{i,j}$ and $\varphi_{i,j}$ represent response amplitude and phase, and $\varepsilon_j$ is a random phase for each wave component. In addition, $K_{i,j}$ and $\varphi_{i,j}$ vary with the direction of the incoming waves, relative to the vessel.

All coefficients are calculated from WAMIT, using the same frequency interval as the hydrodynamic coefficients, and for 9 evenly spaced wave headings between 0° and 180°. Values for all headings are calculating using symmetry and linear interpolation.

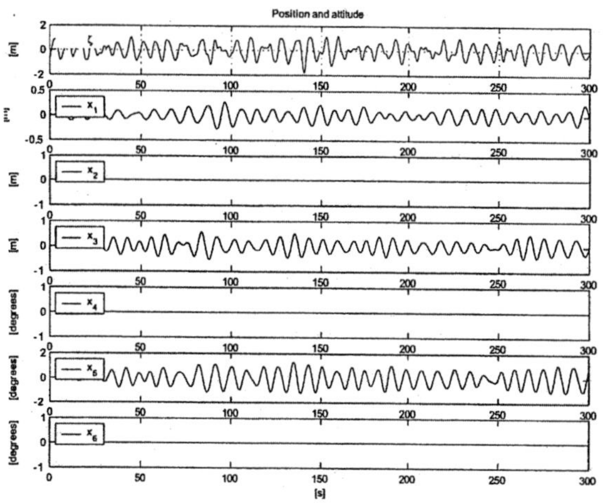


Fig. 7. Motion in head sea

## 6. SIMULATION RESULTS

Simulation is done using Matlab/Simulink on a 2.4GHz Pentium IV. Using eight order approximations of the damping force (for each coupling mode, for a total of 144 states) and recalculating the wave exciting force from 1000 wave components every 0.3 s, 300 s of simulation time took little more than 20 seconds.

The kinematics are modelled as a rigid body influenced by forces in the body frame. The kinematics block is general, using the total inertia matrix (6x6 rigid body inertia added mass) as a single parameter. The separate force contributions; forces from the radiation potential, wave excitation and restoring forces are calculated in separate blocks. The dynamics are modelled as linear, and superposition is assumed.

Simple control is added to keep the vessel near equilibrium. Position is controlled using a simple limited PD-controller and a fully actuated thruster configuration in the horizontal plane ($K_{jk} = \frac{m_{jk}}{40}$, $T_d = 15\,\mathrm{s}$ and $\alpha = 0.01$). In order to not avoid exciting the vessel with a step from suddenly activating wave loads at $t = 0$, the wave loads increase during the first 30 s of the simulation.

Figure 7 shows the responses in all modes for head sea. A simple test is to compare the responses with the response amplitude operator (RAO) from WAMIT. In heave, this is 0.3 near the peak frequency of the wave spectrum; in pitch: 1.03. The motion from the simulation compares well with this data.

Figure 8 shows the ship motion with waves coming from 45° starboard of the ships heading (quartering sea). The waves now excite all modes. The response in heave is now larger than in figure 7, as should be expected. The contributions from the various force contributions are shown in figure 9.

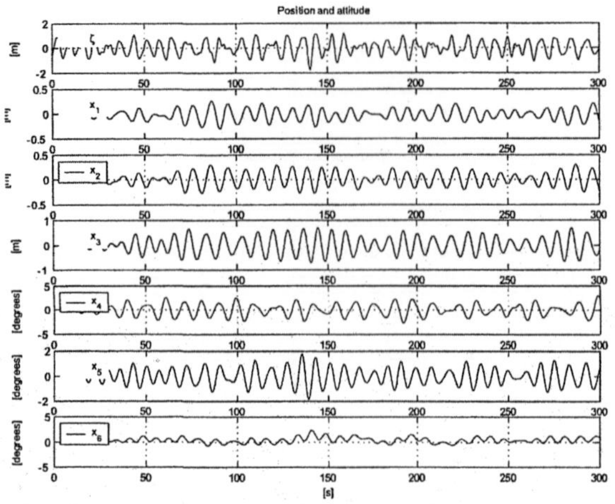


Fig. 8. Motion in quartering sea from starboard

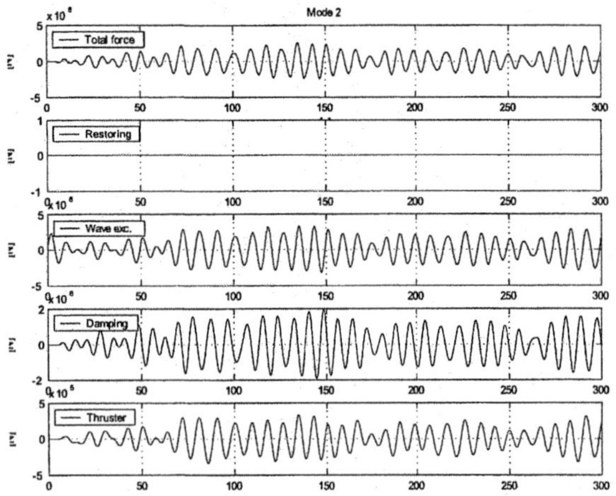


Fig. 9. Forces in sway, quartering sea

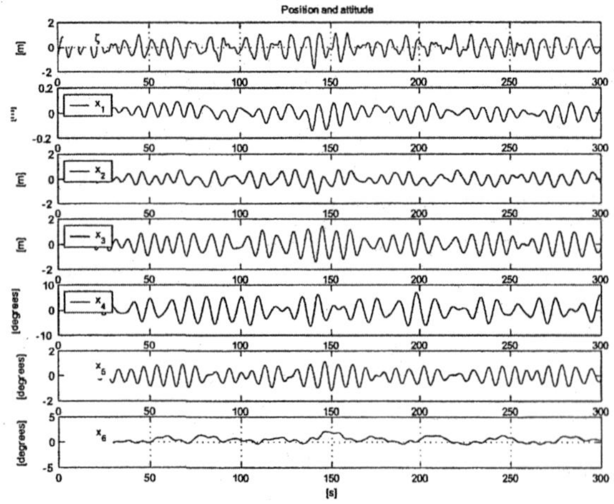


Fig. 10. Motion in sea from starboard

In figure 10 the waves approach from starboard. Comparing with quartering sea, motion in sway increases. There is still surge motion, caused by coupling effects from heave and pitch. Again, these results should not be surprising.

## REFERENCES

Cummins, W.E. (1962). The impulse response funtion and ship motions. *Schiffstechnik* pp. 101–109.

Egeland, O. and J.T. Gravdahl (2002). *Modeling and Simulation for Control*. Marine Cybernetics. Trondheim, Norway.

Fossen, T.I. (1994). *Guidance and Control of Ocean Vehicles*. Wiley.

Kristiansen, E. and O. Egeland (2003). Frequency-dependent added mass in models for controller design for wave motion damping. In: *6th IFAC Conference on Manoeuvering and Control of Marine Craft*. pp. 90–95.

Kung, S.Y. (1978). A new identification and model reduction algorithm via singular value decompositions. In: *Proc. Twelfth Asimolar Conf. on Circuits, Systems ans Computers*. pp. 705–714.

Newman, J.N. (1977). *Marine Hydrodynamics*. MIT Press.

Ogilvie, T.F. (1964). Recent progress toward the understanding and prediction of ship motions. In: *The Fifth Symposium on Naval Hydrodynamics*. pp. 3–128.

Safonov, M.G. and R.Y. Chiang (1989). A schur method for balanced model reduction. *IEEE Trans. on Automat. Contr.* **AC-34**(7), 729–733.

ELSEVIER
IFAC
PUBLICATIONS
www.elsevier.com/locate/ifac

# ELECTRONIC CONTROL OF AN UNDERWATER BREATHING APPARATUS

**Franco Garofalo** * **Sabato Manfredi** * **Stefania Santini** *

* *Dipartimento di Informatica e Sistemistica*
*Università di Napoli Federico II, Napoli, Italy.*
*E-mail: {franco.garofalo,smanfred,stsantin}@unina.it*

Abstract: In this paper we present the results of a research activity on the electronic control of the oxygen partial pressure in a underwater breathing apparatus with gas recycle. We present a mathematical model of the device useful for the design of the control system and for Hardware in the Loop real time simulations, a key tool for performance assessment of the control system under critical and dangerous conditions. A comparison among different control techniques is also shown by numerical and hardware in the loop simulations.


KeyWords: underwater systems, real time control systems, control applications, modelling, hardware-in-the-loop simulations.

## 1. INTRODUCTION AND PROBLEM MOTIVATION

Underwater breathing apparatus with the characteristics of partially or totally recycling the breathed gasses are commonly called 'rebreathers'.

Human respiration is a very complex phenomenon which involves many biochemical transformations. However the macroscopic chemical effect on breathed gasses is (partial) oxygen subtraction for metabolic use with a consequent carbon dioxide increase. All other gasses different from oxygen are inert with respect to the respiration process and flow through the lungs without being chemically transformed. They represent the greatest volumetric fraction of the breathing mixture. This means that exhaled gasses can be recycled (or better re-breathed), provided that oxygen content is restored and and carbon dioxide is removed.

To recirculate gasses, all rebreather concepts include a mouthpiece, through which the diver breathes, connected with a collapsable bag that inflates when he exhales, and deflates when he inhales. This bag is usually called counterlung. In order to prevent the diver from inhaling just exhaled gas, two one-way valves located on either side of the mouthpiece force the gas circulation in one direction. The system formed by mouthpiece, connecting hoses, valves and the counterlung is often referred to as the 'breathing loop'.

Complete collapse of the counterlung when external pressure increases is avoided by the action of a demand valve which is triggered when the counterlung is almost completely collapsed. This always guarantees the availability of a minimum breathable gas volume at any depth by the injection of a fresh gas mixture usually called 'diluent gas'. Similarly over-expansion of the counterlung when external pressure decreases is avoided by an overpressure relief valve.

To reestablish a breathable mixture, rebreathers must be equipped with a device (usually a chemical scrubber) for $CO_2$ removal and a supply valve for $O_2$ injection into the breathing loop. The schematics of a rebreather is given in figure 1.

The so called fully closed circuit rebreather (CCR) has a feed-back electronic controller that, based

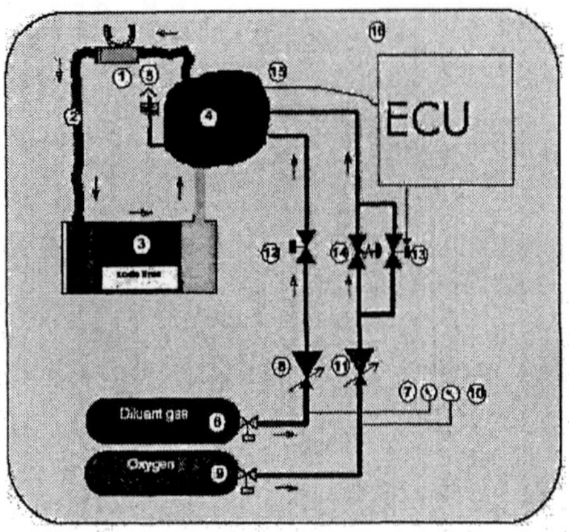


Fig. 1. Schematics of a rebreather. 1) mouthpiece; 2) connecting hose; 3) $CO_2$ scrubber; 4) inhalation breathing bag; 5) relief valve; 6) gas cylinder diluent; 7) pressure gauge diluent; 8) pressure regulator diluent; 9) gas cylinder $O_2$; 10) pressure gauge $O_2$; 11) pressure regulator $O_2$ 12) demand valve; 13) electrically actuated solenoid valve; 14) bypass valve $O_2$; 15) $O_2$ sensors; 16) electronic control unit.

on the measure of the oxygen level in the counterlung, injects oxygen by operating a solenoid supply valve, so as to regulate oxygen partial pressure to a given set-point value during the entire dive. The set point value should be selected by considering that

- the oxygen partial pressure inside the breathing loop should never fall below 0.16 [atm] to avoid hypoxia, *i.e.* the impossibility to fulfill the metabolic oxygen requirements. It can rapidly bring the diver to unconsciousness (Clark and Lambertsen, 1971).
- breathing oxygen at high partial pressure (greater than 0.5 [atm]) can be toxic (Lambertsen *et al.*, 1987). Central Nervous System (CNS) oxygen toxicity is the main concern. It limits both oxygen pressure level and duration of the exposure, (Butler and Thalmann, 1984).

In practice it is convenient to keep oxygen partial pressure to the highest values compatible with CNS toxicity exposure limits in relation to the planned duration of the dive. This decreases the partial pressure of the inert gas in the mixture and hence decompression obligations (Buehlmann, 1995).

Oxygen partial pressure in the breathing loop in subject to variations due to

- diver individual metabolism, which depends on the workload;
- internal pressure of the counterlung which changes with the dive profile.

While variations of $O_2$ partial pressure due to the depth can be easily predicted, metabolic oxygen consumption rate can vary from person to person by a factor of 6 (or more) in normal conditions, and as much as 10-fold in extreme conditions, depending on the activity level.

For these reasons controlling the oxygen partial pressure in a fully closed rebreather during all phases of a dive is a crucial task for the correct and safe use of a CCR. The use of CCR is now limited to military applications and scientific diving (Pyle, 2000). It is common opinion that the diffusion of these devices will increase as soon as these control problems will be properly and reliably solved.

## 2. REBREATHER MATHEMATICAL MODEL

A physics based analytical model is presented for predicting the oxygen level in rebreather counterlung for various dive profiles and diver activity levels. This model will be useful to devise control strategies and to asses their performances.

The model considers the counterlung as an adiabatic collapsible recipient which contains the breathable mixture with the following assumptions

- the breathable gas is composed by two species: inert gas (usually nitrogen and/or helium) and oxygen;
- the breathing gas behaves as perfect mixture;
- the recipient internal pressure instantaneously equals the external pressure.

The respiration process is simply modelled as a (partial) oxygen subtraction from the collapsible bag at a time varying rate. No carbon dioxide presence is assumed in the breathable gas, *i.e.* perfect $CO_2$ adsorption is considered.

According to these assumptions, it is possible to derive our counterlung dynamic model as a balance of volume flow rates [liter/min] (see (Garofalo *et al.*, 2003) for details)

$$\dot{V}_{O_2} = \alpha\frac{p_a}{p_e}s - \frac{p_a}{p_e}m - V_{O_2}\frac{\dot{p}_e}{p_e} + \beta s_{dv} + - \frac{V_{O_2}}{V_{O_2}+V_I}q, \tag{1a}$$

$$\dot{V}_I = (1-\alpha)\frac{p_a}{p_e}s - V_I\frac{\dot{p}_e}{p_e} + (1-\beta)s_{dv} + - \frac{V_I}{V_{O_2}+V_I}q. \tag{1b}$$

where $V_{O_2}$ and $V_I$ are the oxygen and inert partial volumes [liter], $p_e$ is the hydrostatic pressure [atm], s is the flow of the supply valve [liter/min], $\alpha$ is the oxygen fraction of the supplied gas mixture, $\beta$ is the oxygen fraction of the gas mixture supplied by the demand valve, m is the oxygen metabolic volume rate consumption [liter/min],

$p_a$ is the pressure at the sea level [atm], $s_{dv}$ is the demand valve flow [liter/min] and q is the exhaust flow through the relief valve [liter/min].

Model (1) has been validated through extensive simulations on experimental data collected during different dives. It shows a good agreement with the real scenario and some validation results can be found in (Garofalo *et al.*, 2003).

Since we are interested in controlling the oxygen partial pressure level,

$$p_{PO_2} = \frac{V_{O_2}}{V_{O_2} + V_I} p_e, \qquad (2)$$

we can rewrite model equations (1) in terms of $p_{PO_2}$ and counterlung total volume $V = V_I + V_{O_2}$ as

$$\dot{p}_{PO_2} = \left(\frac{p_a}{V} - \frac{p_{PO_2} p_a}{p_e V}\right) s + \left(\frac{p_e \beta}{V} - \frac{p_{PO_2}}{V}\right) s_{dv} + + \frac{p_{PO_2}}{p_e} \dot{p}_e + \left(\frac{p_{PO_2} p_a}{p_e V} - \frac{p_a}{V}\right) m, \qquad (3a)$$

$$\dot{V} = \frac{p_a}{p_e} s - \frac{p_a}{p_e} m - \frac{V}{p_e} \dot{p}_e + s_{dv} - q. \qquad (3b)$$

Equations (3) show that the simplified model of the rebreather is nonlinear with respect to V and $p_{PO_2}$. They also show, as expected, that counterlung volume dynamic does affect oxygen partial pressure dynamics while the converse is not true.

As already said CCR has to be controlled so as

(1) to limit the excursions of the counterlung total volume in an interval $V_m$ - $V_M$ (this both to guarantee at any time a minimal breathable volume for the diver and to avoid its damage by over-expansion);
(2) to keep oxygen partial pressure below a given value, selected trading-off between decompression obligations and oxygen toxicity risk mitigation.

The first control requirement is achieved by the 'demand valve-relief valve' pneumatic system which behaves as static nonlinear feedback of the actual volume with the characteristic given in figure 2. This system keeps the counterlang volume in the prefixed interval $V_m$ - $V_M$ in spite of the external pressure variation and oxygen subtraction/injection. The second requirement, e.g. the regulation of the $p_{PO_2}$, is achieved by controlling the oxygen flow from the injection valve, thus compensating for metabolic oxygen consumption and variations of external pressure (see equation (3a)). In next section we will address the problem of designing this control loop.

We remark that the pure oxygen supplied into the counterlung is spilled out from a pressure regulator connected to an high pressure oxygen cylinder. This means that the valve upstream pressure is $p_e + \Delta p$ with fixed $\Delta p$. Hence the mass flow through the solenoid valve orifice is a function of the hydrostatic pressure

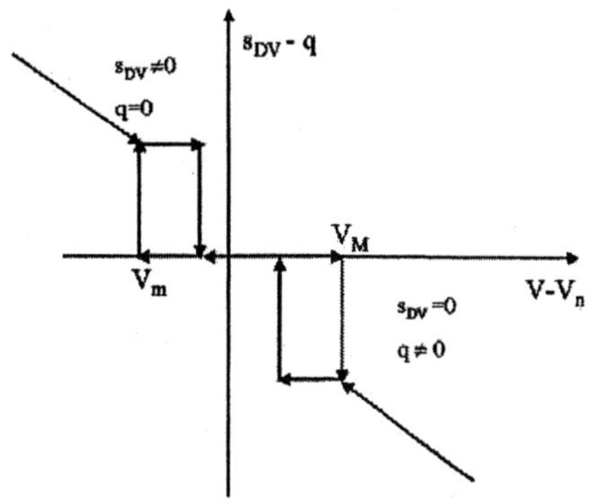


Fig. 2. Characteristic of the static nonlinear feedback of the actual volume V. $V_n$ is the counterlung volume at the working point.

$$s = s(p_e). \qquad (4)$$

Function $s$ assumes different forms depending on sonic or subsonic flow regimes (Shapiro, 1977).

## 3. FEED-BACK CONTROL OF OXYGEN PARTIAL PRESSURE

In current technology commercial CCR, the solenoidal valve is activated by a closed-loop control driven by the oxygen level in the counterlung. Measures of the oxygen partial pressure are obtained by redundant electrochemical oxygen cells located into the breathing bag. A "voting" logic is used to discriminate whether the set point level has been exceeded. The arithmetic mean among the reading of the sensors in the majority is used as an estimate of the actual value of $p_{PO_2}$.

At the beginning of each actuation period T, if the current oxygen level is below the set point ($\bar{p}_{PO_2}$) the controller switches-on the solenoid valve for a prefixed duration $\tau$. The drawback of this simple control is the injection of an oxygen mass which varies with depth (see equation (4)). Extensive simulations studies, partially reported in the following subsection 3.1, and dive experiences confirm that this control action can be ineffective in guaranteeing a suitable oxygen partial pressure regulation in all circumstances. This means that to improve control performances $\tau$ duration must be modulated over the period $T$.

A first way is that of using the above described two levels control logic with a modulation of $\tau$ to take into account oxygen flow variability at different depths.

In what follows we show that better and safer performances are obtained by modulating the $\tau$

duration using a PI feed-back of the oxygen partial pressure regulation error. Model equations (3) have been evaluated at a nominal working point (*i.e.* constant depth and metabolic consumption rate). Then, by using a step-response identification method, this nominal behavior has been approximated by a first order model

$$G_p(s) = \frac{K_p}{1 + sT} e^{-sL}. \quad (5)$$

On the basis of the previous approximation the PI parameters are tuned in order to minimize the integrated absolute error index (IAE) in the presence of a wide class of disturbances (Astrom and Hagglund, 1995)

$$\mathrm{IAE} = \int_0^{\infty} \mid p_{PO_2}(t) - \bar{p}_{PO_2} \mid dt. \quad (6)$$

Typical dives are characterized by a rapid first descend to the target depth. In this situation the $p_{PO_2}$ controller is forced to inject oxygen to reach the set point (usually far from the surface $p_{PO_2}$ value) and the demand valve is activated by counterlung volume control. The combination of all this actions can cause a relevant oxygen pressure overshoot.

To limit this over excursion expert divers carefully set the oxygen fraction in the diluent (the $\beta$ parameter value in equations (3)) according to the maximum planned depth in the dive. In order to further reduce this overshoot we simply introduce a first order filter that slows down the set-point growth thus moderating the control action during the descent phase. The filter time constant can be simply chosen on the bases of the maximum admissible descent velocity for a typical dive. Of course more sophisticated control techniques, acting in closed-loop on the reference signal, can be chosen (see for example (Bemporad, 1998), (Gilbert and Kolmanovsky, 2001) and reference there in).

### 3.1 *Simulation Results*

As already said, we designed a controller aimed to guarantee relatively constant oxygen partial pressure in breathing loop in spite of variations in oxygen consumption and in depth. Here we investigate the effectiveness of the proposed strategy on a representative dive profile. In particular, we chose a dive profile with a corresponding metabolic requirements as in figure 3. This situation is rather usual, especially in a commercial dive, where the diver has a workload during the deeper phase of the dive.

This dive profile includes three potentially dangerous situations. The first one is a rapid descent to a deep depth at low oxygen metabolic demand. The second one is the rapid increase of the oxygen consumption (from about the 500 [sec] to 1250 [sec] of the dive profile). The last one is the rapid ascent to the decompression depth at a medium level oxygen demand (at about 1400 [sec] of the mission profile). The controller performance should be evaluated with a particular attention to these situations.

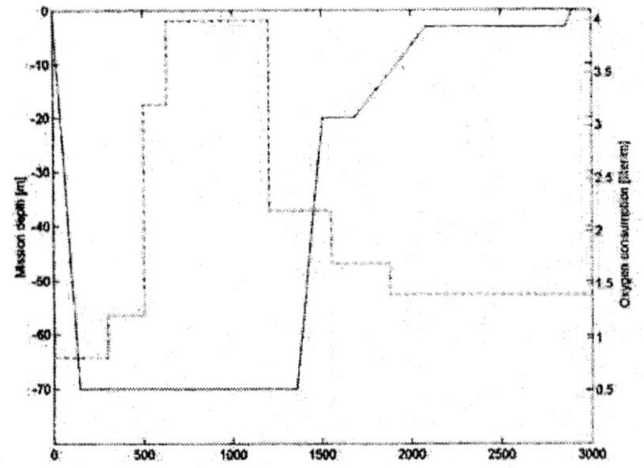


Fig. 3. Mission depth profile [meter] and oxygen consumption during the mission [liter/min].

As a representative example of fixed $\tau$ controller, we draw two different scenarios related to under and over-dimensioning $\tau$ with respect to T. As shown in figure 4, with a small $\tau$ the controller is unable to significantly reject disturbances and there is a relevant undershooting induced by the metabolism requirement growth and rapid ascent. On the other side (see figure 5), the choice of large values for $\tau$ avoids these problems but brings too large overshooting during the rapid descent phase and generates oscillations in steady state phase (from about 600 [sec] to 1200 [sec] of the mission profile).

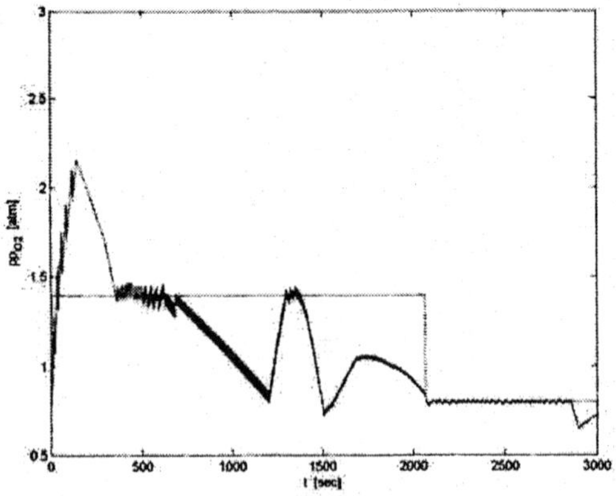


Fig. 4. Oxygen partial pressure dynamics corresponding the mission profile in figures 3 for a commercial CCR. $\tau = 0.35$ [sec], T=6 [sec].

An appropriate choice of $\tau$ is always critical. To overcome this obstacle, as previously discussed, it is possible to modulate the $\tau$ duration to take

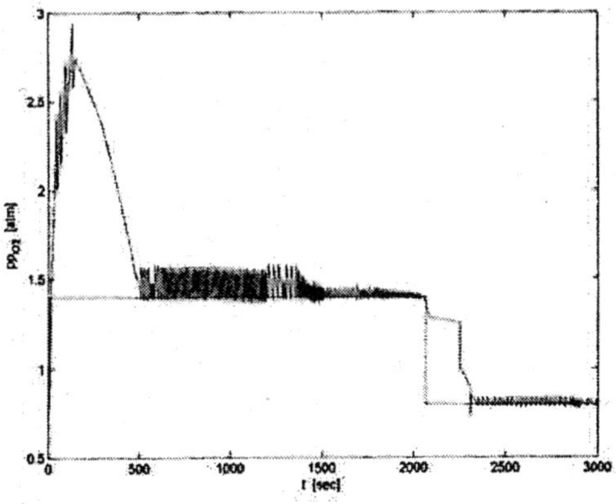


Fig. 5. Oxygen partial pressure dynamics corresponding to the mission profile in figure 3 for a commercial CCR. $\tau = 3$ [sec], T= 6 [sec].

into account oxygen flow variability at different depths. The performances obtainable by this strategy are reported in figures 6. The improvement of the achievable performances is substantial.

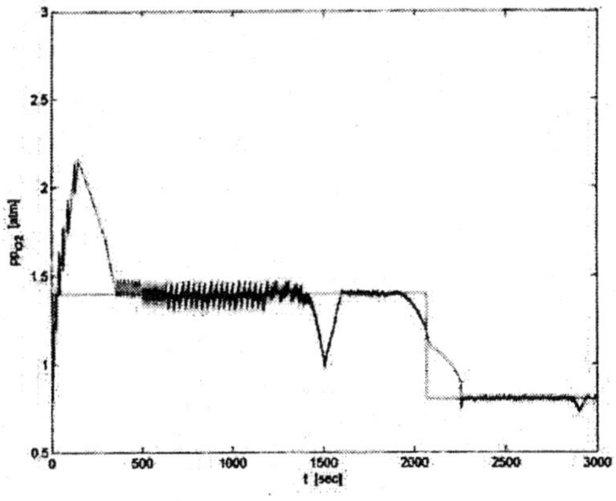


Fig. 6. Oxygen partial pressure dynamics corresponding to the mission profile in figure 3 obtainable with $\tau$ modulation guaranteeing a 4 standard liter injection in T, T=6 [sec].

Better performances can be achieved by the joint action of the PI controller to modulate $\tau$ duration and smoothing the set point signal in descent phase as described in the previous section. Simulation results in figure 7 highlight a significant overshooting reduction during the descent and the absence of significant oscillations or undershooting due to both increased oxygen demand or rapid ascents.

## 4. HARDWARE-IN-THE-LOOP VALIDATION

Rebreather systems operate in safety-critical situations and therefore need a rigorous engineering process. A useful tool in this direction is hardware-in-the-loop simulation activity (HILS). HILS are

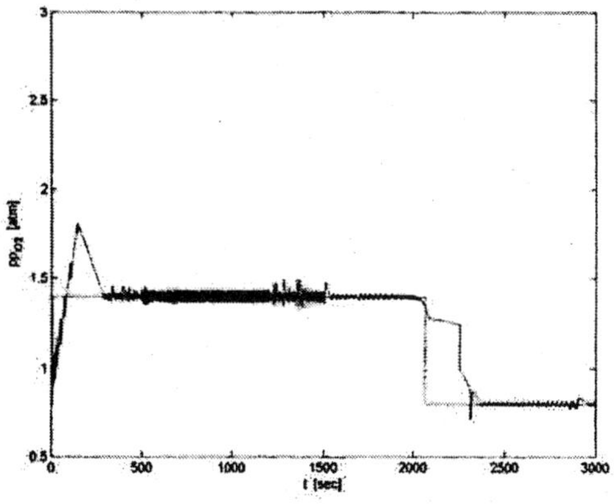


Fig. 7. Oxygen partial pressure dynamics corresponding the mission profile in figure 3 obtainable by modulating the $\tau$ duration.

real time simulations that allow to test the embedded control system under different real working loads and conditions.

In this work HILS have been used for the validation of the rebreather control unit whose design has been described in section 3. This activity is fundamental in assuring a certain level of confidence in the obtainable control performances, while reducing the risks of error in the very last stage of the on-the-field testing.

In HILS testing, the counterlung and all auxiliary components in the control loop are replaced by real-time simulations of their dynamic behavior. The hardware component in the loop is the microcontroller (PIC16F876A based board). All the real time models have been translated in a C-Code by using the MathWork's Real-Time Workshop/Windows Target. I/O management has been achieved by DAQ-Board National Instruments (200k sampling rate; inputs: 16 single-ended and 8 differential channels with 16 bit resolution; output: 2 channel with 12-bit resolution).

Some HILS results are shown in figure 8. They refer to the same mission profile described in figure 3. They confirm the good control performances obtained through numerical simulations and make us confident in the results achievable by in-the-field testing of the system.

## 5. CONCLUSIONS

In this paper we have presented the results of some research activity on the automatic control of the oxygen level in the so called CCR rebreather. With respect to the usual SCUBA device this kind of underwater breathing apparatus has the advantage of allowing variation of the oxygen composition of the breathing mixture and hence of optimizing decompression obligations. Moreover

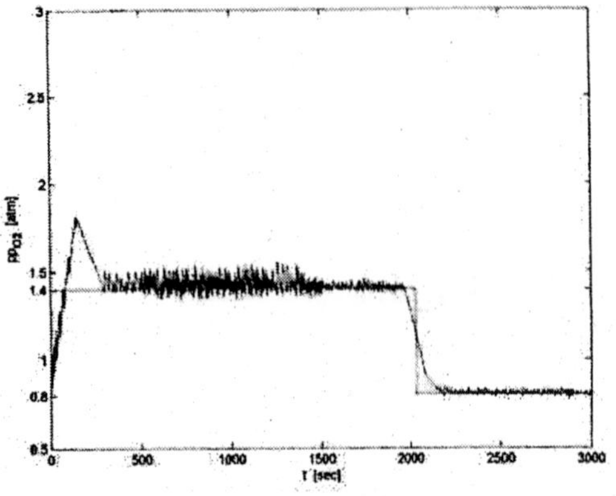


Fig. 8. Real time simulation results of oxygen partial pressure dynamics corresponding to the mission profile described in figure 3 with the proposed strategy.

it is much more efficient (and hence safer) from the point of view of gas consumption. Hence it has the potential of a great diffusion for both commercial and recreational diving as soon as the control problems will be properly solved and the electronic central unit implementing automatic control algorithms (and other vital functions like decompression calculations, dive-to-dive communication, man-machine interface etc.) will be realized in a reliable and fault-tolerant fashion.

This paper has given a contribution in this direction presenting a mathematical model of the rebreather useful for the design of the control system and for the realization of the HIL real time simulator of a rebreather, a key tool for performance assessment of the electronic central unit under critical and dangerous situations. Moreover, with the only purpose of showing the possibility of substantially improve controller performance with respect to the commercially available system, we designed a PI controller action, optimized with the aid of the model.

Future works will regard the use of extended Kalman filter and data fusion algorithms to address the problem of oxygen partial pressure control in the presence of high uncertainty on sensor signal and/or sensor failure. This, in conjunction with HIL simulations will enable the proper selection the architecture of the control system. It will be implemented and experimentally tested in undersea operation in a next future.

## 6. ACKNOWLEDGMENT

The authors would like to thank TECHNOSUB Ischia for the technical support provided during the experimental part of this work. A special thank also to Eduardo De Robbio for his competent contribution to HIL activities.

## 7. REFERENCES

Astrom, K. and T. Hagglund (1995). *PID Controllers: Theory, Design and Tuning.* Insrtument Society of America.

Bemporad, A. (1998). Reference governor for constraint nonlinear systems. *IEEE Transaction on Automatic Control, vol. 43, no. 3, pp. 415–419* **43**, 415–419.

Buehlmann, AA. (1995). *Tauchmedizin. Barotrauma. Gasembolie. Dekompression, Dekompressionskrankheit,Dekompressionscomputer.* Springer-Verlang, Berlin, 4th Edn.

Butler, F. K. and E. D. Thalmann (1984). Cns oxygen toxicity in closed-circuit scuba divers.. *In: Underwater Physiology VIII. Eds. A.J. Bachrach and M.M. Matzen. UMS, Inc., Bethesda.*

Clark, J.M. and C.J. Lambertsen (1971). Pulmonary oxygen toxicity: a review.. *Pharmacol, pp. 37–133.*

Garofalo, F., S. Manfredi and S. Santini (2003). Modelling and control of oxygen partial pressure in an underwater breathing apparatus with gas recycle. *Proceedings of IEEE European Control Conference, England, September.*

Gilbert, E.G. and I. Kolmanovsky (2001). A generalized reference governor for nonlinear systems. *Proc. of the 40th IEEE Conference on Decision and Control, pp. 4222–4227.*

Lambertsen, C.J., J.M. Clark and R. Gelfand (1987). Definition of tolerance to continuous hyperoxia in man. an abstract report of predictive studies.. *Underwater and Hyperbaric Physiology IX (A.A. Bove, A.J. Bacherach, and L.J. Greenbaum, eds), Undersea and Hyperbaric Medical Society, Bethesda. pp. 717-735.*

Pyle, R. (2000). The use of nitrox in closed circuit rebreathers for scientific purposes. *Proceedings of the American Accademy of Underwater Scientist, Nitrox Diving Workshop, Catalina Island.*

Shapiro, Ascher H. (1977). *The Dynamics and Thermodynamics of Compressible Fluid Flow, Vol. I.* John Wiley and Sons Inc.

ELSEVIER
IFAC
PUBLICATIONS
www.elsevier.com/locate/ifac

# FORCE CONTROL OF A LOAD THROUGH THE SPLASH ZONE

**Bjørn Skaare * Olav Egeland ****

* *Norwegian University of Science and Technology, Department of Engineering Cybernetics, Odd Bragstads plass 2D, 7491 Trondheim, Norway. email: bjorn.skaare@itk.ntnu.no, fax :+47 73 59 43 99, phone: +47 73 59 43 92*
** *Centre of Ships and Ocean Structures, NTNU, Marine Technology Centre, Otto Nielsens veg 10, 7491 Trondheim, Norway. email: olav.egeland@itk.ntnu.no, fax : +47 73 59 43 99, phone: +47 73 55 11 11*

Abstract: A parallel force and position controller for control of loads through the wave zone in marine operations is proposed in this article. The controller structure has similarities to parallel force/position controllers used in the robotics literature and has been developed based on linear theory and tested in simulations on a system with the highly nonlinear hydrodynamic forces that a load is exposed during water entry in waves. Wave synchronization and parallel force/position control are compared in the simulations. The parallel force/position controller gives significant reductions in the oscillations in the wire force, while hydrodynamic forces on the object are about the same with the two control strategies. *Copyright © 2004 IFAC*

Keywords: Marine systems, Force control, Synchronization, Waves, Hydraulic motors.

## 1. INTRODUCTION

A critical phase in an offshore crane operation is when the load goes through the splash zone. The performance of the crane system through the splash zone will to a large extent depend on the crane control system. Currently, crane systems are not adequate for operation at high sea states seen during winter conditions in the North Sea.

Some offshore oil and gas fields will be developed with all processing equipment on the seabed in the years to come. This means that high operability on the subsea intervention is required, which in the North Sea and other exposed areas implies underwater intervention in harsh weather conditions. Maintenance, repair and replacement of equipment are therefore also required during these weather conditions.

The compensation that is performed on marine cranes today is passive or active heave compensation, where the goal is to keep the load motion unaffected by the vessel motion. The concept of wave synchronization was introduced in (Sagatun *et al.*, 2002) where the object was lowered through the wave zone with a constant velocity relative to the waves, in order to minimize the hydrodynamic forces on the load.

Wave synchronization may give large oscillations in the wire force if the load is heavy and the waves are large.

In this paper, a parallel force/position controller that operates in parallel with a passive heave compensation system is proposed in order to reduce the oscillations in the wire force.

[1] This work is sponsored by Norsk Hydro ASA.

## 2. BACKGROUND

### 2.1 *Forces on an Object in Waves*

The forces acting on an object through the wave zone are given from Newtons second law as

$$\sum F = m\ddot{z}_L = S + f_h - mg, \quad (1)$$

where $m$ is the mass of the object, $\ddot{z}_L$ is the acceleration of the object, $g$ is the acceleration of gravity, $S$ is the wire force and $f_h$ represents the hydrodynamic forces on the object. The wire force S becomes:

$$S = m\ddot{z}_L - f_h + mg \quad (2)$$

The hydrodynamic forces on an object going through the wave zone are derived in (Faltinsen, 1990) from potential theory as

$$f_{hp} = \rho\nabla(z_r)\ddot{z}_r + \rho g\nabla(z_r) + Z_{\ddot{z}_r}(z_r)\ddot{z}_r + \frac{\partial Z_{\ddot{z}_r}(z_r)}{\partial z_r}\dot{z}_r^{\,2}, \quad (3)$$

where $f_{hp}$ has positive direction upwards, $\rho$ is the density of the water, $\nabla(z_r)$ is the instantaneously submerged volume of the object, $w$ is the velocity of the wave, $z_r$ is the relative position between the wave and the object and $Z_{\ddot{z}_r}(z_r)$ is the added mass of the object as function of submergence. The first term represents the Froude-Kriloff pressure, the second term is due to the hydrostatic pressure, while the last terms are the effect of the added mass and diffraction force. $f_{hp}$ is zero when $z_r$ is less than or equal to zero, which is the case when the object is hanging in the air. The last term is often denoted as the slamming term and is always positive. The slamming term is zero when the object is entirely submerged. Equation (3) is derived under the assumption of a large impact velocity, and the added mass terms for infinite frequency should be used. Since marine operations not are performed with that large velocities, the values of the above terms may be discussable.

The viscous hydrodynamic forces acting on an object through the wave zone may be determined from model tests, and can be written as:

$$f_{hv} = 0.5\rho C_D A_p \dot{z}_r \left|\dot{z}_r\right| + d_l \dot{z}_r, \quad (4)$$

where $C_D$ is the nonlinear drag coefficient, $A_p$ is the projected efficient drag area and $d_l$ is the linear drag coefficient.

The hydrodynamic forces on an object through the wave zone are found as the sum of (3) and (4):

$$f_h = \rho\nabla(z_r)\ddot{z}_r + \rho g\nabla(z_r) + Z_{\ddot{z}_r}(z_r)\ddot{z}_r + \frac{\partial Z_{\ddot{z}_r}(z_r)}{\partial z_r}\dot{z}_r^{\,2} + 0.5\rho C_D A_p \dot{z}_r \left|\dot{z}_r\right| + d_l \dot{z}_r \quad (5)$$

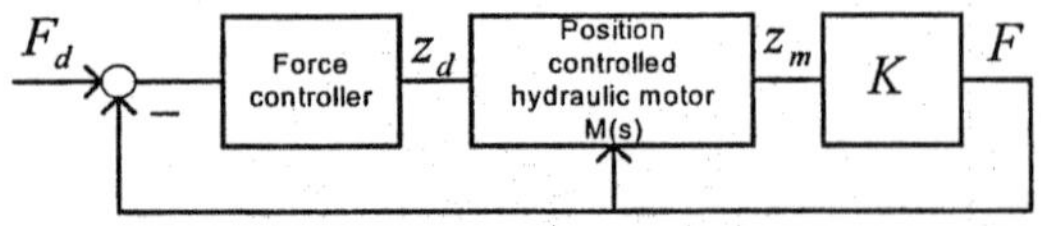


Fig. 1. Schematic overview of parallell position and force feedback control used in robotics.

### 2.2 *Present Control Strategies*

2.2.1. *Active Heave Compensation* During active heave compensation, the goal of the compensator is to keep the load motion unaffected by the vessel motion. This can be achieved with a passive hydraulic pneumatic compensator working in parallel with an active compensator. The active compensator is then compensating for the difference between the motion of the vessel and the passive compensator. Active heave compensation can be achieved with the position reference signal

$$z_{ref} = \hat{z}_{compensator} - \hat{z}_{crane\ tip} + z_d, \quad (6)$$

where $\hat{z}_{compensator}$ is the measured or estimated motion of the passive heave compensator and $\hat{z}_{crane\ tip}$ is the estimated motion of the crane tip, based on measurements of the vessel motion. $z_d$ is the desired trajectory, unaffected by the vessel motion.

2.2.2. *Wave Synchronization* During wave synchronization, the goal is to synchronize the object motion with the wave motion. This can be achieved by the position reference signal

$$z_{ref} = \hat{z}_w + \hat{z}_{compensator} - \hat{z}_{crane\ tip} + z_d, \quad (7)$$

where $\hat{z}_w$ is the estimated wave motion.

## 3. PARALLEL FORCE/POSITION CONTROL OF AN HYDRAULIC MOTOR

### 3.1 *Parallel Force/Position Control*

Parallel force and position control is discussed extensively in the robotics literature, e.g. in (De Schutter, 1988), (Chiaverini, 1993). The goal is to control both the position and the force of an end effector containing a spring with a known stiffness $K_0$. A block diagram of a one dimensional system is shown in figure 1.When the bandwidth of the force control loop is placed sufficiently below the bandwidth of the position control loop, the dynamics of the position control loop can be neglected. The force controller can be chosen as

$$z_d = \frac{K_i}{T_i s}(F_d - F) \quad (8)$$

This choice of force controller gives the force loop transfer function

$$h_{0f}(s) = \frac{K_i}{T_i s} M_p(s) K_0 \quad (9)$$

The effect of the spring can be cancelled in the force loop transfer function by choosing

$$K_i = \frac{1}{K_0} \quad (10)$$

and the integration constant $T_i$ is chosen such that the force control loop has bandwidth sufficiently below the bandwidth of the position control loop. In this case the approximation for the closed position loop transfer function is $M_p(s) \approx 1$ and the approximate force loop transfer function becomes

$$h_{0f}(s) = \frac{1}{T_i s} \quad (11)$$

In this paper a similar controller structure will be analyzed for parallel force and position control of a hydraulic motor connected with a wire to a crane load which is transferred through the wave zone. An overview of the system with linearized hydrodynamic forces is shown in figure 2. It is seen that the dynamics of the wire force is more complex in this case than for the end effector of a robotic manipulator, since the spring is connected to a mass with its own dynamics related to the wave and vessel motion. As long as the load is transferred through the wave zone, the system can allow relatively large errors in position, but not large errors in the wire force.

### 3.2 *Mathematical Model of an Hydraulic Motor*

The mathematical model of a linearized valve controlled hydraulic motor can be written as in (Egeland and Gravdahl, 2002):

$$\frac{V_t}{4\beta}\dot{p}_L = -K_{ce}p_L - D_m\omega_m + K_q x_v \quad (12)$$

$$J_t\dot{\omega}_m = -B_m\omega_m + D_m p_L - T_L \quad (13)$$

$$\dot{\theta}_m = \omega_m, \quad (14)$$

where $p_L$ is the load pressure, $\omega_m$ is the motor angular velocity, $\theta_m$ is the motor angle, $x_v$ is the spool position, $T_L$ is the load torque, $J_t$ is the motor inertia, $V_t$ is the volume of both chambers, $\beta$ is the bulk modulus, $K_{ce}$ is the leakage coefficient for motor and valve, $D_m$ is the displacement, $B_m$ is the coefficient of viscous friction, and $K_q$ is the linearized valve constant.

When it is assumed that $\frac{B_m K ce}{D_m^2} << 1$, the Laplace transformation of the model (12)-(14) can be written as

$$\theta_m(s) = \frac{\frac{K_q}{D_m}x_v(s) - \frac{K_{ce}}{D_m^2}(1 + \frac{V_t}{4\beta K_{ce}}s)T_L(s)}{s(1 + 2\zeta_h\frac{s}{\omega_h} + \frac{s^2}{\omega_h^2})}, \quad (15)$$

where

$$\omega_h = \sqrt{\frac{4\beta D_m^2}{V_t J_t}} \quad (16)$$

$$\zeta_h = \frac{K_{ce}}{D_m}\sqrt{\frac{\beta J_t}{V_t}} + \frac{B_m}{4D_m}\sqrt{\frac{V_t}{\beta J_t}} \quad (17)$$

$\omega_h$ is the hydraulic resonance frequency, while $\zeta_h$ is the relative damping ratio.

### 3.3 *Position Controller*

An electrohydraulic servo valve is assumed used, and it is assumed that the bandwidth of the electrohydraulic servo valve is high compared to the hydraulic motor (typically the bandwidth is about 50-100 Hz). For simplicity it is assumed that

$$\frac{x_v}{i_v}(s) = 1 \quad (18)$$

Further, a position controller that is recommended (Egeland, 1993) for hydraulic motors with small relative damping is proposed:

$$i_v(s) = \frac{K_v}{1 + Ts}(\theta_0(s) - \theta_m(s)), \quad (19)$$

where $\theta_0(s)$ is the position reference signal and the time constant $T$ of the low pass filter is chosen in the geometric mean value between $\zeta_h\omega_h$ and $\omega_h$. $K_v$ will be chosen in order to obtain sufficient stability margins. The position loop transfer function $h_{0p}(s)$ for the position loop becomes

$$h_{0p}(s) = \frac{\frac{K_v K_q}{D_m}}{s(1 + 2\zeta_h\frac{s}{\omega_h} + \frac{s^2}{\omega_h^2})(1 + Ts)} \quad (20)$$

The closed loop transfer function for the position loop is given by

$$M_p(s) = \frac{h_{0p}(s)}{1 + h_{0p}(s)} \quad (21)$$

The sensitivity function is given by

$$N_p(s) = \frac{1}{1 + h_{0p}(s)} \quad (22)$$

The transfer function from the load momentum $T_L$ to the motor angle $\theta_m$ is given as

$$\frac{\theta_m}{T_L}(s) = N_p(s)h_{T_L}(s), \quad (23)$$

where $h_{T_L}(s)$ is the transfer function from the load momentum to the motor angle without the position feedback loop. When it is assumed that $B_m = 0$, $h_{T_L}(s)$ can be found as

$$h_{T_L}(s) = -\frac{\frac{K_{ce}}{D_m^2}(1 + \frac{V_t}{4\beta K_{ce}}s)}{s(1 + 2\zeta_h\frac{s}{\omega_h} + \frac{s^2}{\omega_h^2})} \quad (24)$$

### 3.4 *Force Controller*

It is assumed that the hydrodynamic forces on the object in the wave zone given in (5) can be linearized as

$$f_h = a(\ddot{z}_w - \ddot{z}_L) + b(\dot{z}_w - \dot{z}_L) + c(z_w - z_L)$$
$$= F_{wave} - (a\ddot{z}_L + b\dot{z}_L + cz_L), \qquad (25)$$

where $a$ represents the added mass and Froude Kriloff force, $b$ represents the damping forces and $c$ represents the buoyancy force. $f_{wave}$ represents the forces due to the wave motion and is given as

$$F_{wave} = a\ddot{z}_w + b\dot{z}_w + cz_w \qquad (26)$$

Note that the model is valid when the object is partly submerged, since this is the most critical phase during the transfer through the wave zone. If the wave and vessel motion are treated as disturbances, the system of the hydraulic motor, the load and the linearized hydrodynamic effects can be modelled as seen in figure 2. $k$ is the stiffness in the wire and eventually a passive heave compensator. The force loop transfer

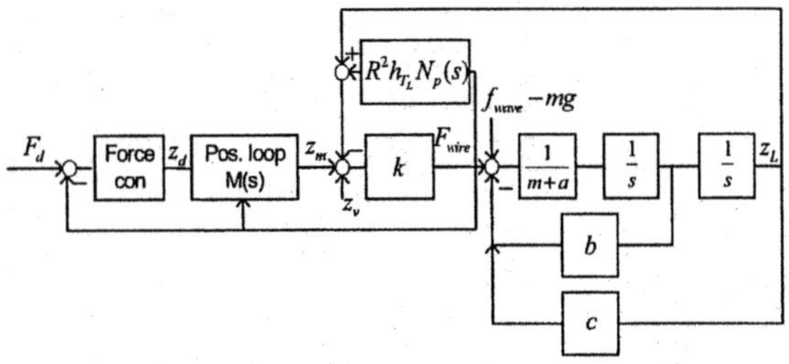


Fig. 2. Schematic overview of the parallell position and force controller and the dynamic model including the hydrodynamic forces.

function $h_{0f}(s)$ can be found after transformations of the block diagram in Figure 2 as

$$h_{0f}(s) = \qquad (27)$$
$$\frac{K_i}{T_i s} M_p(s) \frac{ck}{c+k} \frac{(\frac{s}{\omega_a}+1)(\frac{s}{\omega_1^2}^2 + 2\zeta_1\frac{s}{\omega_1}+1)}{\frac{s}{\omega_0^2}^2 + 2\zeta_0\frac{s}{\omega_0}+1-h_{N_p}(s)},$$

where

$$\omega_0 = \sqrt{\frac{c+k}{m_L+a}} \qquad (28)$$
$$\zeta_0 = \frac{b}{2\sqrt{(c+k)(m_L+a)}} \qquad (29)$$
$$\omega_1 = \sqrt{\frac{c}{m_L+a}} \qquad (30)$$
$$\zeta_1 = \frac{b}{2\sqrt{c(m_L+a)}} \qquad (31)$$
$$h_{N_p}(s) = R^2 \frac{ck}{c+k} N_p(s) h_{T_L}(s) \qquad (32)$$
$$(\frac{s}{\omega_1^2}^2 + 2\zeta_1\frac{s}{\omega_1}+1) \qquad (33)$$

Note that the transfer function $h_{N_p}(s)$ in (27) contains the the transfer functions $N_p(s)h_{T_L}(s)$, which is the transfer function from the load momentum to the motor angle in the hydraulic motor. It is desirable that these transfer functions are small within the bandwidth of the force control loop and therefore may be neglected during controller design and analysis. The Bode plot of the transfer function $h_{N_p}(s)$ is shown in Figure 3 for the simulated system in the next section. It is seen that $h_{N_p}(s)$ is small within the bandwidth of the force control loop.

Similarly as for the force controller (8), the controller gain $K_i$ is chosen according to

$$K_i = \frac{c+k}{ck} \qquad (34)$$

and the integration time constant $T_i$ is chosen in order to obtain sufficient stability margins.

There are three resonances in (27) and in the plots in Figure 3. First there is an inverse resonance at $\omega_1$, then there is a resonance at $\omega_0$, and there is a resonance at the hydraulic resonance frequency $\omega_h$. The resonances are given from expression (30), (28) and (16), respectively. Note that $\omega_0$ and $\omega_1$ is dependent on the wire and load parameters $k$, $c$, $b$, $m$ and $a$, while $\omega_h$ is dependent on the parameters of the hydraulic motor.

The effective bandwidth of the force control system will to a large extent depend on the inverse resonance $\omega_1$, since the gain becomes very small around this frequency, as seen from the Figure 3.

A lower value of the wire stiffness $k$ will give a lower resonance frequency $\omega_0$. A lower resonance frequency $\omega_0$ will result in larger gain margin, since the frequency where the phase reaches $-180°$, $\omega_{180}$, is strongly dependent of the hydraulic resonance frequency $\omega_h$. This makes it possible to choose a smaller value of the integration time constant $T_i$. A smaller $k$ will also give a smaller difference between the resonance frequency $\omega_0$ and the inverse resonance frequency $\omega_1$. As the resonance and the inverse resonance come closer together the resonances will damp each other. A smaller $k$ will also result in higher relative damping $\zeta_0$, which will give a smaller resonance at $\omega_0$. All these factors results in a higher possible bandwidth.

A larger value of the buoyancy stiffness $c$ will increase the inverse resonance frequency $\omega_1$ and to some extent also the resonance frequency $\omega_0$. It will also bring the two resonances closer together. Both these factors results in a higher possible bandwidth.

Larger value of the damping parameter $b$ gives larger damping ratios $\zeta_0$ and $\zeta_1$ resulting in smaller resonances at $\omega_0$ and $\omega_1$. Larger mass and added mass parameters results in equally scaled lower resonance frequencies $\omega_0$ and $\omega_1$ and lower damping ratios $\zeta_0$ and $\zeta_1$. The lower value of the resonance frequency $\omega_0$ does not allow us to choose a lower value of the integration time constant $T_i$ as was the case when the wire stiffness $k$ was smaller. The reason is that the resonance at $\omega_0$ are larger due to smaller damping, and the gain margin is not increased. The effective bandwidth which is determined by $\omega_1$ is therefore decreased when the mass or the added mass of the load is increased.

Many of the factors mentioned above are given for the load to be transferred through the wave zone. One factor that can be changed is the wire stiffness $k$. This can be achieved if the crane system is equipped with a passive heave compensation system. A passive heave compensation system can be considered as a spring-damper system with a relatively soft spring and damping. The spring and damper from the passive heave compensation system will then act in series with the wire stiffness.

## 4. SIMULATIONS

In this section we will consider a system with a hydraulic motor connected with a wire to a spherical load that is lowered through the splash zone. Wave synchronization and parallel force and position control will be compared.

The hydraulic motor has the following parameters: $D_m = 0.14 \frac{\text{m}^3}{\text{rad}}, V_t = 0.28 \text{ m}^3$, $J_t = 20000 \frac{\text{kg}}{\text{m}^2}$, $\beta = 7 \cdot 10^8$ Pa, $K_{ce} = 1.98 \cdot 10^{-9} \frac{\text{m}^3}{\text{sPa}}$, $K_q = 1 \frac{\text{m}^2}{\text{s}}$ and $B_m = 0 \frac{\text{kg}}{\text{ms}}$. The numerical values for hydraulic resonance frequency and relative damping given in (16) and (17) becomes $\omega_h = 99.05$ [rad/s] and $\zeta_h = 0.1$.

The position controller is the same for both control strategies, and is found according to the design rule given in the previous section, with the parameters $T = \frac{1}{\sqrt{\zeta_h \omega_h}}$ and $K_v = \frac{0.15}{T}$.

The numerical values of the parameters of the spherical load with diameter 2.7 meter and wire stiffness $k$ was: $m = 15845.5$ kg, $a = 7394.6$ kg, $b = 3523.0 \frac{\text{Ns}}{\text{m}}$, $c = 57571.9 \frac{\text{N}}{\text{m}}$, $k = 1 \cdot 10^5 \frac{\text{N}}{\text{m}}$.

The values of the parameters $a$, $b$ and $c$ are linearized for an object that is half submerged, with a relative velocity $\dot{z}_r = 1 \frac{\text{m}}{\text{s}}$. The stiffness and damping can be considered as the combined stiffness and damping in the wire, crane boom and passive heave compensation system.

The force controller was designed according to the force control design rule in the previous section. The Bode plot of the force loop transfer function $h_{0f}(s)$ as given in (27) with the controller gain (34) and integration time constant $T_i = 0.21$ s, is shown in figure 3. The effective bandwidth of the force control loop, the frequency where the force loop transfer function $h_{0f}(s)$ crosses the 0-$dB$ for the first time, was found to be at 1.41 [rad/s].

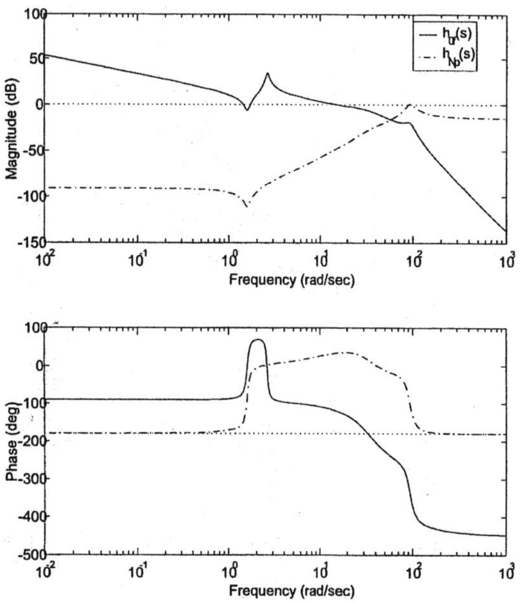


Fig. 3. Bode plot of the force loop transfer function $h_{0f}(s)$ and the transfer function $h_{N_p}(s)$.

Simulations of the system during wave synchronization and parallel force/position control are shown in nondimensional plots in Figure 4 - Figure 6. The system was exposed to irregular waves generated from the JONSWAP wave spectrum with wave period $T =$ 7 s and significant wave height $H_s = 3$ m. The corresponding vessel motions were generated JONSWAP spectrum motions with period $T = 7$ s and $H_s = 1$ m. The position control part of the force controller was equipped with heave compensation functionality. A ramp in the desired wire force was given after 100 seconds, and it is seen that the load is "dancing" on top of the wave motion before the load is transferred through the wave in Figure 4. Note that the wire force in Figure 5 is negative during the wave synchronization approach, and the operation is therefore not allowed according to DNV rules.

## 5. CONCLUSION

A parallel force/position controller can give significant reductions in oscillations in the wire force during transfer of a load through the wave zone in marine operations. A critical factor with respect to the bandwidth of the parallel force/position controller is the value of the total stiffness of the wire, crane boom and a possible passive heave compensation system.

Parallel force/position control showed best performance with respect to oscillations in wire force in the simulations, while the performance of the two strategies were approximately the same with respect to the hydrodynamic forces on the load.

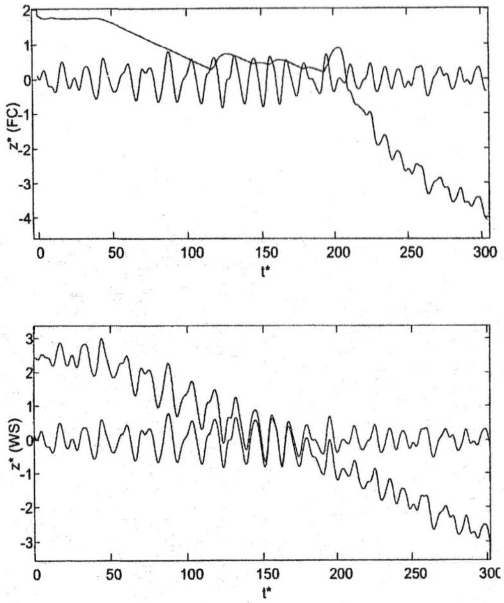


Fig. 4. Nondimensional motion of the bottom on the load and the wave during force control (on the top) and wave synchronization (on the bottom).

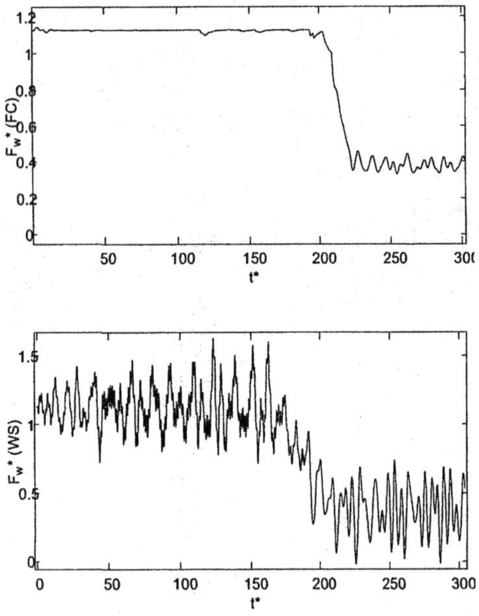


Fig. 5. Nondimensional wire force during force control (on the top) and wave synchronization (on the bottom).

## 6. REFERENCES


Chiaverini, S., Sciavicco L. (1993). The parallel approach to force/position control of robotic manipulators. *TRA* **9**, 361–373.

De Schutter, J. (1988). Improved force control laws for advanced tracking applications. In: *ROBAUT88*. Philadelphia, PA.

Egeland, O. (1993). *Servoteknikk*. Tapir. Trondheim.

Egeland, O. and J.T. Gravdahl (2002). *Modeling and Simulation for Automatic Control.*. Marine Cybernetics AS. Trondheim, Norway.


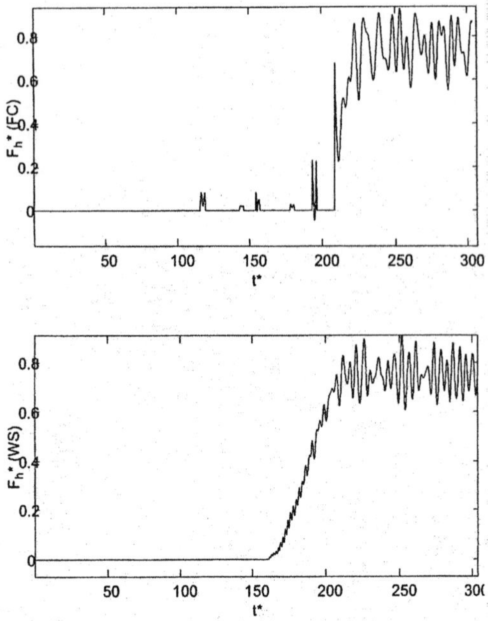


Fig. 6. Nondimensional hydrodynamic force on the load during force control (on the top) and wave synchronization (on the bottom).


Faltinsen, O. M. (1990). *Sea Loads on Ships and Offshore Structures*. 1 ed.. Cambridge, UK.

Hansen, T. R. (2002). Intelligent lasthaandtering. Kranteknisk forening.

Sagatun, S. I., T. A. Johansen, T. I. Fossen and F. G. Nielsen (2002). Wave synchronizing crane control during water entry in offshore moonpool operations. In: *Proc. IEEE Int Conf. Control Applications*. Glasgow, Scotland.

ELSEVIER
IFAC
PUBLICATIONS
www.elsevier.com/locate/ifac

# AUGMENTED REALITY SUPPORTS REAL-TIME DIAGNOSIS AND REPAIR OF SHIPBOARD SYSTEMS

**Fred M. Discenzo[†], Dukki Chung[†], Steven L. Chen[‡], Reinhold Behringer[‡], V. (Sundar) Sundareswaran[‡], Daniel Carnahan[†], Ken Wang[‡], Jose Molineros[‡], Joshua McGee[‡]**

[†] *Rockwell Automation, 1 Allen-Bradley Drive, Mayfield Hts., OH 44124, USA*
[‡] *Rockwell Scientific, 1049 Camino Dos Rios, Thousand Oaks, CA 91360, USA*
*{fmdiscenzo, dchung, dlcarnahan}@ra.rockwell.com*
*{slchen, rwbehringer, vsundareswaran, kkwang, jmolineros, jhmcgee}@rwsc.com*

Abstract: A prototype shipboard pumping system has been developed integrated with an Augmented Reality system. This system demonstrates how a service engineer can visually inspect a running system and "see" how the system is operating. In addition to stored, historical information, real-time data and analysis corresponding to the component the engineer is looking at can be viewed with the device. For example, the associated dynamic component data such as temperature, pump speed, flow rate, suction pressure, discharge pressure, cavitation state, and lubricating oil contamination data are visible. Cues are provided to the service engineer indicating potential problems and directing subsequent investigation. *Copyright © 2004 IFAC*

Keywords: diagnostic, man/machine interfaces, maintenance, man/machine interaction, machinery, augmented reality.

## 1. INTRODUCTION

There is a growing trend to deploy "smart" devices for industrial, commercial, and shipboard systems. Many devices such as actuators and sensors previously deployed as "dumb" devices are now enhanced with embedded processors, sensors, logic, and communications. Examples of smart devices include motors, drives, starters, valves, power switchgear, sensors, and pumps. The benefits include greater machinery flexibility, additional information to support system integration and distributed control, asset management, and information on the condition of system components and operating state. However, there is a commensurate increase in system complexity and the critical need to effectively manage and interpret the substantial volume of real-time data.

In recent years, Augmented Reality (AR) has matured into a technology that is suitable to provide complex and dynamic information to the user in an intuitive way, by being spatially registered to the environment [Azuma *et al.*, 2001]. Initial industrial field tests in the early 1990s like the Boeing wire bundle project [Curtis *et al.*, 1998] showed the advantages and feasibility of AR technology, but failed due to immature hardware available at that time. Due to subsequent miniaturization and progress in several hardware components (display, computing power), the use of AR technology in an industrial and shipboard environment is now feasible.

A prototype shipboard pumping system has been developed that is integrated with an AR system. This system demonstrates how a service engineer can visually inspect a running system and "see" how the system is operating. This service engineer views the operating machinery through goggles. The goggles incorporate a small camera and computer-generated graphic display within the field of view of one eye. Real-time sampled data and derived device information are displayed in the viewable image along with stored, historical information, and specifications describing the device the observer is currently looking at. For example, the associated dynamic component data such as temperature, pump speed, flow rate, suction pressure, discharge pressure, cavitation state, and lubricating oil contamination data are visible. A software image registration program that recognizes specific visible markers placed on the machinery functions to insure that the computer generated data correspond to the specific device the operator is viewing. Visual cues are provided to the service engineer indicating potential problems and directing subsequent investigation by providing spatial highlighting of the relevant problem zone of the machinery (e.g. pump impellor).

The integration of dynamic machinery information with visual inspection and maintenance & repair operations can significantly enhance the effectiveness and safety of future shipboard service activities. Integrating machinery design, database data, dynamic operating data, and context information will more fully utilize the superior capabilities provided by the next generation of smart shipboard systems.

## 2. SYSTEM COMPLEXITY

In 2000 the number of reported industrial accidents exceeded 2,100 and over 5,000 people were injured [Felton, 2001]. The predominant cause of failure, where the cause could be confidently established, was equipment failure. It is significant that 90% of these accidents occurred during normal operation and only 8% occurred during maintenance activities. Ultimate concern for improving workplace safety is addressed by standards targeted at minimizing risk of machinery hazards and protecting workers, such as ISO 14121 and OSHA 1910 respectively. However, process upsets do occur, disturbances cause unexpected stress on machinery, machines do fail, and people do get injured.. Shipboard machinery may experience an additional degree of stress and new failure modes not seen in industrial systems. This is due to the unique application characteristics of marine systems, the duty cycle and loading demand required to complete critical missions, and the consequences of potential battle damage.

The expanded use of complex machines in critical applications and coupling of interacting subsystems places a severe demand on the ability of the shipboard sailor to effectively perform machinery service activities. The service engineer must quickly understand the complex machine, how it is operating, what actions may be required, how to perform the required maintenance and quickly return the equipment back to service, as well as the implications of performing a particular activity. This requires the effective integration of machinery design information, configuration information, dynamic machinery operating data, and other control and context information. This must be done in a timely and efficient manner with 100% accuracy.

Emerging technologies such as multi-agent systems will provide unique and important capabilities for responding to machinery failures and maintaining critical functionality as much as possible [Maturana *et al.*, 2003]. This technology has previously been demonstrated to provide dynamic, autonomous re-configuration on a land-based chilled water system. However, the requirement for device-level maintenance and repair still remains.

Trends such as maintenance staff reduction and reduced shipboard manning objectives, staff turnover, the deployment of new complex machinery, continuous operating requirements, and reduced life-cycle cost requirements will continue to stretch the capabilities of maintenance and repair staff.

.

## 3. SYSTEM DESCRIPTION

### *3.1 Overview*

A laboratory system has been developed to demonstrate real time data acquisition, analysis, and presentation of machinery operating data to a service engineer. The laboratory system includes an AC motor with a variable speed drive and an instrumented centrifugal pump loop. Data from the system are presented to the service engineer in a manner that selectively shows operating data and analytical results relevant to the operating context of the machinery and within the field of view of the service engineer. The major system components are shown schematically in Figure 1.

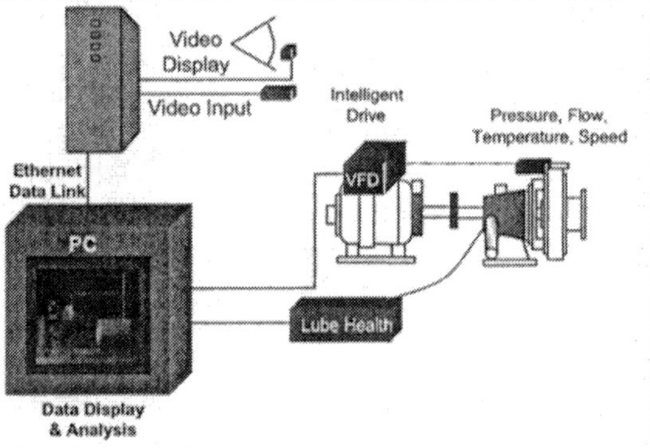


Figure 1. Demonstration System Configuration

Data captured and analyzed by the Intelligent Drive and the lubricating oil sensor (Lube Health sensor) is transmitted to an integrated industrial PC/Display System (Versaview). Through an Ethernet link captured data is provided to the AR Display Manager. The AR Display Manager functions to fuse appropriate real-time data, generated cues and icons with the actual image being currently viewed by the operator.

### *3.2 Pump System Description*

The pump system consists of a water storage tank holding roughly 80 gallons of water. An ANSI standard centrifugal pump is piped to the tank to pump water from the bottom of the tank and return water to the tank just below the water level. A 2 hp Intelligent AC motor with embedded processor and sensors is coupled to the pump. Motor operation is switch-selectable for operation directly from the 460 volt, 3-phase power line or through the variable frequency drive (VFD). The demonstration pump system is shown in Figure 2.

Figure 2. Demonstration System Hardware

During pump system operation, sensor data including suction and discharge pressure, flow rate, water and pump temperature, and motor power are input to the VFD and processed within the drive to establish the condition of the pump system. Sampled sensor data and analytical results such as cavitation state are transmitted from the drive via an Ethernet link to a local display computer [Discenzo *et al.*, 2002a]. The local display computer also receives real-time pump lubrication data such as mineral oil temperature and oxidation state from a lubrication condition sensor embedded in the pump. The display computer then sends the received data through an Ethernet link to the Augmented Reality System. Sampled data values provided to the AR Display Manager computer correspond to physical locations on the pump system and describe specific conditions or health states of the designated locations. Alternatively, the asset information can be selected for display, such as the type and serial of the equipment in the field of view. Figure 3 shows the motor-pump system with machinery locations tagged corresponding to the data values and diagnosed system components.

Figure 3. Motor-Pump System with Data Value Sources and Diagnosed Components Tagged

The tags appear as small white squares with black borders and are used by the AR Display Manager to easily register the field of view of the operator.

### *3.3 Augmented Reality System Description*

The Augmented Reality System is integrated with the operating machinery by communication over Ethernet. The objective of this integrated system is to provide machinery health (state) information and repair instructions to a mechanic by means of a head-mounted display. The head-mounted display shows the computer-generated information overlaid directly into the field of view. The Augmented Reality System consists of three components, the Head-Worn Display, the Video-based Tracking System, and the Graphics Rendition System.

*Head-Worn Display*

A MicroOptical color display (resolution: 640x480) is mounted onto a set of standard, commercially available safety goggles to create a head-worn display (HWD). This particular display is not "directly" see-through; however, the fact that only one eye is covered, provides a pseudo-see-through, with the brain fusing both images captured from the eyes into one view. In order to show the information in the HWD aligned with what the user sees, the wearer's head must be tracked, consistent with the viewing direction and the display orientation. In order to achieve this, a small pinhole video camera (Supercircuits) is mounted onto the display unit. This permits capture of live video of the scene as the user wears the display. The complete modified set of safety goggles with display and camera is shown in Figure 4.

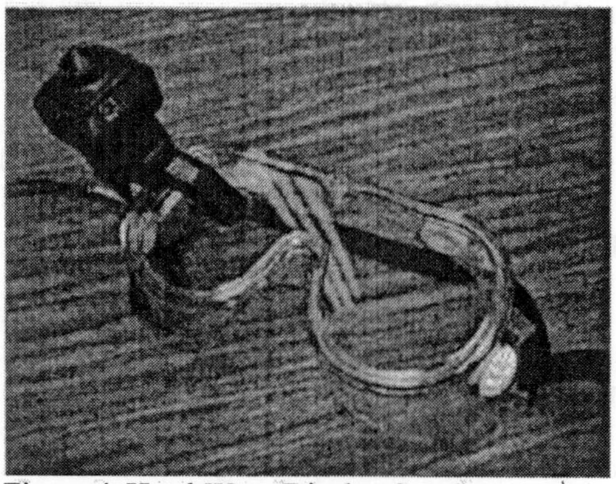

Figure 4. Head-Worn Display Goggles

Markers are placed on the motor, pump, pipes, sensors and other places of interest as shown in Figure 3. These markers serve as visual anchors with information attached in their vicinity. The video processing system is able to detect the location of these markers in the captured image sequence and is able to identify them to determine their relative orientation and position with respect to the user. This allows graphical information to be correctly prescribed and spatially located relative to the user's viewing direction.

Video processing computation runs on a Shuttle SB61G2 "Shoebox PC" with an Intel P4 processor, running at 2.8 GHz. The captured images have a size of 640x480 pixels. After capture, the image is de-interlaced and (locally) thresholded. Each marker has a square black border that is detected by a contour following algorithm. Inside this black square frame is a unique pattern which corresponds to an ID number for each marker. The vision system is able to distinguish these markers, to track their location in the image, and to determine the orientation of the camera relative to the plane of the markers. This enables the correct rendition of information that is attached to these marker anchors by a spatial offset. An example captured video image in the vicinity of such a marker is shown in Figure 5.

Figure 5. Example Captured Video Image

*Graphics Rendition*

The 6 degree-of-freedom (DOF) information provided by the visual tracking system is used by the renderer to place information into the user's view, correctly aligned with the environment. The 3D framework that was used here for the rendition is Microsoft's Direct3D. This framework controls the correct rendition of the information placement based on a sequence of coordinate transformations from the user (camera) into the real world (marker) to the display.

In order to avoid visual clutter, the actual display graphics are kept simple. The diagnostic information displayed consists of a text line, providing either maintenance instructions, live status data or asset information, an arrow pointing to the relevant location to which this information is referring to, and a ring around the pointer of this arrow. The status information text displayed on the black background is colored green, yellow, or red whether an abnormal condition is detected and the degree of abnormality or fault severity.

Figure 6 shows a view of the pump system as the operator would see the system with system health information and other diagnostic information displayed and registered in the correct location. This is achieved by using the captured video as an underlay beneath the graphical symbology.

Figure 6. Pump Image with System Diagnostic and Video Cues Displayed

In this image the operator is looking at the pump impellor casing as indicated by the red circle. The data at the top of the screen is presented in the context of the current targeted viewing location. The data displayed indicates that adequate NPSH (Net Positive Suction Head) is available to prevent cavitation. The color green for this text further indicates that the NPSH level is acceptable. The pump operating state was established by the diagnostic logic embedded in the Intelligent Drive.

A red circle is an effective targeting cue for the operator. The size of the ring is fixed in real world coordinates. This permits the size to change as the user moves closer or further from the marker. The tip of the arrow shown points to a location that is spatially fixed relative to the marker. This permits placing the marker somewhere in the vicinity of an area of interest, without obscuring the actual real scene at this location.

The AR display with the live video underlay was created on a 2$^{nd}$ computer, independent of the PC that created the HMD AR display. Since the camera captured a larger field of view (FOV) than the HMD covers, the information overlay had to be created for this larger image area. The technology of fusing live video with context-relevant graphical overlay indicates another paradigm for use of such a system: remote surveillance and tele-presence. An operator who would not be present at the actual machinery location could assist the local engineer in the diagnostic / maintenance process with remote expert knowledge. Only the video camera signal has to be transmitted from the machinery location to the remote location – no other head-tracking system is employed here. This allows such tele-presence (or tele-maintenance) system to be seamlessly integrated into any video surveillance link. Tele-maintenance can provide important benefits to enhance the timeliness, safety, and efficiency of maintenance activities for future complex shipboard systems.

## 4. DIAGNOSTICS AND MAINTENANCE

This system is intended to be another tool for the service engineer to utilize in the tasks of monitoring and servicing machinery. A user can wear the goggles and obtain real-time information on machinery operation diagnostic state. The goggles may easily be worn for initial machinery inspection. It is very useful to have stored, catalog and specification information available at the machine as well as real-time information describing the state of the machine. In many cases machinery cannot be shut down for inspection. The AR system provides a unique and valuable window into the machines dynamic operation. Figure 7 shows an operator viewing the system to see the machinery operating state. For example, when viewing the pump lubrication sensor, information regarding the oxidation state and temperature of the mineral oil in the pump is automatically displayed.

Figure 7. Operator Wearing Goggles

Human interaction with machinery will be more efficient and safer if accurate, real-time data is presented within the context of machine operation and consistent with operator duties, especially when multiple interacting subsystems are present.

Emerging Open-System Standards for machinery diagnostic information can provide a standard information interface for integrating dynamic machinery condition data with an Augmented Reality Diagnostics and Repair system [Discenzo *et al.*, 2002b].

The effects of changing machine control can be made readily presented as the service or maintenance activity is occurring. Exception conditions or abnormalities are readily highlighted to the operator and clearly shown within the operators field of view. Figure 8 shows an operator interacting with the pump loop demonstration system while monitoring system state changes.

Figure 8. Operator Changing Pump Control While Monitoring System State

In the AR system demonstrated, a tactile control interface has been implemented with application logic, so that the user -- potentially a mechanic performing machine maintenance or troubleshooting -- can proceed through a prescribed sequential procedure or check list. The focus of the augmented reality display and marker tracking system is on only a single marker at a time, corresponding to which step in the servicing/diagnostic procedure the user is currently at. Each press of the operator button changes the AR display to guide the operator to the next component or device for attention along with the appropriate data display and operator notification. The user can proceed through the different steps of the procedure by pressing keys on a simple keypad that is attached to the operator's belt.

## 5. CONCLUSION

Augmented Reality may be effective not only for integrating critical database & real-time data, but also for guiding a service engineer in understanding dynamic machinery operating condition and critical control and maintenance requirements. In addition, AR is a powerful tool for training a user in the operation of a complex machine.

When looking at various system components the engineer will see the actual machinery component in the field of view along with stored data such as the device specifications. The Augmented Reality Diagnostic and Repair System will be another tool to assist the shipboard service engineer in servicing complex operating machinery. The advantages of this system include the ability to readily integrate accurate and timely design and repair information with dynamic machinery information. Components that are too hot to touch, are pressurized or energized, or otherwise unsafe to service may be clearly identified in the augmented reality view to further improve worker safety.

"Power tools" such as the AR Diagnostic System described here will become essential for efficiently and safely servicing future complex machinery.

## REFERENCES


Azuma, R., Baillot, Y., Behringer, R., Feiner, St., Julier, S., and MacIntyre, B. (2001) Recent Advances in Augmented Reality. *IEEE Computer Graphics and Applications*, Vol. 21, No. 6, pp. 34-47. IEEE Computer Society, November 2001

Curtis, D., Mizell, D,. Gruenbaum, P., and Janin, A. (1998) Several Devils in the Details: Making an AR Application Work in the Airplane Factory. In R. Behringer, Gudrun Klinker, anmd D. Mizell, eds., *Augmented Reality – Placing Artificial Objects in Real Scenes*, A.K.Peters, 1999, pp 47-60. *Proc. of First IEEE International Workshop on Augmented Reality*.

Discenzo, F.M., Rusnak, D., Hanson, L., Chung, D., and Zevchek, J. (2002a) Next Generation Pump Systems Enable New Opportunities for Asset Management and Economic Optimization. In: *Fluid Handling Systems*, **5(3)**, pp. 35-42.

Discenzo, F.M., Chung, D., Bezdicek, J., Hejda, P., Cernohorsky D., and Flek, O. (2002b) Pump Diagnostics Using an Open Systems Architecture, In: *Proceedings of the 56th Meeting of the Society for Machinery Failure Prevention Technology*, Virginia Beach, Virginia. April 2002, pp. 157-168.

Felton, B. (2001) Equipment Failure Leading Accident Cause, *Intech*, **48(8)** , p. 77.

Maturana, F.P., Tichý, P., Šlechta, P., Staron, R.J., Discenzo, F.M., Hall, K., and Marík, V., Cost-based Dynamic Reconfiguration System for Evolving Holarchies, Holomas 2003 Conference Proceedings, Prague, Czech Republic, September 1-3, 2003.

ELSEVIER
IFAC
PUBLICATIONS
www.elsevier.com/locate/ifac

# OPTIMAL PLANNING OF CARGO OPERATIONS AT BUNKERING TANKERS WITH RESPECT TO DYNAMICAL CHARACTER OF THEIR PARAMETER RESTRICTIONS

**Y.P. Kondratenko, G. F. Romanovsky, D.M. Pidopryhora, G.V. Kondratenko**

*Ukrainian State Maritime Technical University*
*9 Geroiv Stalingrada Av., Mykolaiv, 54025, Ukraine*
*tel.: +(380) 512 400939,e-mail: kondrat@rmc.mksat.net*

Abstract: The present paper is devoted to the problem of bunkering tankers' optimal cargo planing. The brief analysis of existing approaches for solving the multiple-criterion optimisation problem is fulfilled with respect to specific features of the bunkering control processes. The complex three-component non-linear criteria taking into account restrictions at both cargo tanks capacity and tanker's loading conditions is developed. The solution of the non-linear mathematical programming problem for the model of the tanker in the form of rectangular pontoon is presented and confirms efficiency of the suggested criterion. The necessity of dynamical character of restrictions applying is proved for the problem. The fuzzy system for the restrictions issuing is also developed. *Copyright © 2004 IFAC*

Keywords: bunkering tanker, cargo planning, multiple-criterion optimisation, dynamical restrictions, fuzzy system

## INTRODUCTION

Growth of a competition in the world charters market of ships makes ship-owners look for ways for decreasing the cost of sea transportation. From this standpoint, improvement of existing ship computer-aided control system (CACS) and the development of new one are urgent problems especially for tanker fleet. Application of such systems allows to decrease a number of crew and increase safety and economical efficiency of ship's mechanisms and systems operation.

The determining factors for the choice of CACS strategy and tactics are the quantity of transported cargo and the scheme of its distribution at compartments according to the preliminary cargo plan (Kondratenko, *et al.*, 2001). Thus, the development of algorithms of the automated formation optimal cargo plan (CP) is an actual problem.

To decide the problem of optimum CP formation, the methods of linear and non-linear optimisation are traditionally used (Avrahov and Vorobjov, 1968; Avrahov, *et al.*, 1971; Volovoy, 1984; Egorov, 1989; Letnyanchic and Shemagina, 1989; Kozljako and Egorov, 1991). The analysis of the publications shows that setting of the mathematical programming problem essentially depends on the type of a vessel and cargo. The majority of the works considers optimisation of cargo plans of dry-cargo ships and is based on methods of linear programming. Such approach allows to receive allowable but not always optimum decisions which does not provide maximum effective loading of ships. In the works on optimisation of tanker CP (Avrahov and Vorobjov, 1968, Egorov, 1989) the full capacity of a vessel is the basic requirement. According to the statement, the existing algorithms can not be used for such type of tankers as "bunkering tankers" (BT), which usually operate in rather dangerous modes of partial use of cargo tanks capacity. Taking into account the growth of safety requirements to bunkering operations (Leigh-Jones, 2001), the development of models and algorithms for optimal BT cargo operations planning in a structure of multi-level hierarhical CACS (Kondratenko and Podoprigora,

2001) is a topical task.

The present paper is devoted to synthesis of complex three-component non-linear optimality criteria for solving the mentioned problem, which allows to take special modes of BT operation into account. The special emphasis is placed input data preparation (by a specially designed complex fuzzy system) for the multiple-criterion optimisation task with respect to dynamical character of restrictions for the tanker's hull surface conditions (heel, trim and maximum draft).

## 1. MATHEMATICAL MODEL OF OPTIMUM LOADING OF THE TANKER

The form of the tanker's hull can be mathematically described as a model in the form of a rectangular pontoon (Volovoy, 1984, Kondratenko and Podoprigora, 2000). Such an assumption does not bring an essential error to formation of the optimum cargo plan of the BT and allows to simplify the problem considerably. Let's consider in more detail the process of formation of the criterion and synthesis of CP optimisation algorithms.

Geometrical parameters of the simulation model of the tanker are given on Fig. 1

During the analysis, the imitating model parameters are used: displacement of an empty pontoon is $D_0 = 60\cdot 10^3$ kg, total quantity of cargo tanks is $N = 9$, co-ordinates of the gravity centre of cargo in tanks are $G_i = (X_i, Y_i, Z_i), i = 1..N$, the maximal tonnage of cargo tanks are $P_{\max} = \{P_{i,\max} \mid i = 1..N\} = \{60\cdot 10^3, 120\cdot 10^3, 60\cdot 10^3, 60\cdot 10^3, 120\cdot 10^3, 60\cdot 10^3, 60\cdot 10^3, 120\cdot 10^3, 60\cdot 10^3\}$ kg.

From the economic point of view, the basic criterion of quality of the cargo plan is distribution the given quantity of cargo in full accordance with the order (when the tanker is bunkered – mode.1 or when the tanker bunkers – mode.2). For that requirements to the tanker's hull surface conditions should be observed.

Let's write down the goal function which confirms the given criterion, as a root-mean-square deviation of the current loading of the tanker from the required value:

$$J_P(\bar{P}) = \frac{1}{2}\left\{\left(\sum_{i=1}^{N} P_i^{init} \pm \Delta P_\Sigma\right) - \sum_{i=1}^{N} P_i\right\}^2, \qquad (1)$$

where $\bar{P} = \{P_1, P_2, \ldots, P_N\}$ – a vector of quantity of cargo in tanks; $\sum_{i=1}^{N} P_i^{init}$ – initial total loading of a tanker; $\pm\Delta P_\Sigma$ – the total quantity of cargo loading (mode.1) or discharging (mode.2); $\sum_{i=1}^{N} P_i$ - total value of the current quantity of cargo in tanks. The geometrical interpretation of the goal function (1) of two independent variables under the given condition is given in Fig. 2 (a,b).

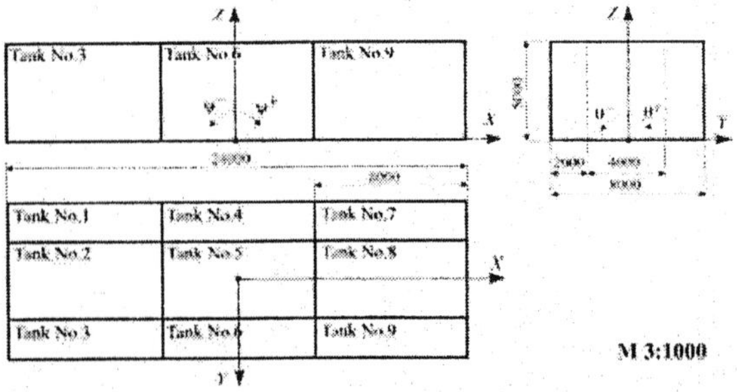


Fig. 1. Simulation model of the tanker

Fig. 2 shows that the criterion accepts the minimal value in any point, which is on straight line *MN* and is described by $P_2 + P_4 = 60000$ equation.

Further let's introduce the restriction on capacity of tanks (mode.1/mode.2):

$$P_i^{init/0} \le P_i \le P_i^{max/init}, i = 1\ldots N \qquad (2)$$

where $P_i^{\max}$ – maximum allowable quantity of cargo for loading into to the $i$-th tank.

While in service, there is often necessary to perform simultaneous loading/discharging various types of cargo (often two, e.g. black oil and diesel fuel). In this case it is necessary to take into account the restriction on the total cargo of each type:

$$\sum_{i=1}^{m_j} P_{s_i}^j = \left(\sum_{i=1}^{m_j} P_{s_i}^{j,init} \pm \Delta P_\Sigma^j\right) = const, \forall s_i \in S \qquad (3)$$

where $j \in \{1,2\}$, 1 - the first type of cargo (black oil), 2 - the second type of cargo (diesel fuel); $S^j = \{s_1^j, s_2^j, \ldots, s_{m_j}^j\}$ – set of tanks numbers with $j$-th type of cargo; $\sum_{i=1}^{m_j} P_{S_i}^j, \sum_{i=1}^{m_j} P_{S_i}^{j,init}$ - the current and initial total loading of tanks with $j$-th type of a cargo, accordingly. Thus, the search of the optimum cargo plan of tanker is the problem of non-linear programming with non-linear criterion function and linear trivial restrictions. The generalised gradient method (Himelbau, 1974, Methuse and Fink, 2001) which allows to solve problems of such class is rather effectively applied for its solution.

The results of the search of the optimum cargo plan of the tanker (in mode.1) on the basis of the criterion (1) and in view of the restrictions (2) - (3) for different initial conditions (I, II, III) is given in Table.1. The problem of the optimum distribution of black oil of the total quantity $\Delta P_\Sigma^1 = 15\cdot 10^4$ and diesel fuel of total $\Delta P_\Sigma^2 = 10\cdot 10^4$ is considered.

Table 1 shows, that the final cargo plan $\bar{P}^j = \{P_1^j, P_1^j, \ldots, P_{m_j}^j\}$, $j \in \{1,2\}$ is formed practically proportionally to initial loading of tanks with the appropriate cargo type.

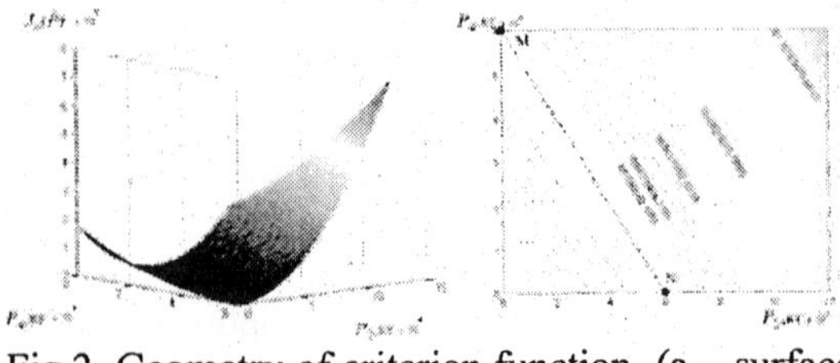

Fig.2. Geometry of criterion function (a – surface of criterion function, b – lines of the level)

Table 1 Results of search of the optimum cargo plan with the criterion $\min J_P\left(\overline{P}\right)$.

| | | Condition I | | Condition II | | Condition III | |
|---|---|---|---|---|---|---|---|
| $j$ | $i$ | $P_i^{init}$ | $P_i^j$ | $P_i^{init}$ | $P_i^j$ | $P_i^{init}$ | $P_i^j$ |
| 1 | 1 | 1000 | 31000 | 0 | 30004 | 5000 | 34097 |
| 1 | 2 | 1000 | 31000 | 5000 | 34994 | 0 | 30081 |
| 1 | 3 | 1000 | 31000 | 0 | 29994 | 0 | 30081 |
| 1 | 5 | 1000 | 31000 | 0 | 30004 | 0 | 30303 |
| 1 | 8 | 1000 | 31001 | 0 | 30004 | 0 | 30438 |
| $\sum_i P_i^1$ | | 5000 | 155000 | 5000 | 155000 | 5000 | 155000 |
| 2 | 4 | 1000 | 25999 | 2000 | 26995 | 0 | 25191 |
| 2 | 6 | 1000 | 25999 | 2000 | 27005 | 0 | 25548 |
| 2 | 7 | 1000 | 26001 | 0 | 24995 | 4000 | 28207 |
| 2 | 9 | 1000 | 26001 | 0 | 25005 | 0 | 25055 |
| $\sum_i P_i^2$ | | 4000 | 104000 | 4000 | 104000 | 4000 | 104000 |
| LC | $\psi$ | 0 | -0.49 | -0.25 | -0.74 | -0.05 | -0.52 |
| LC | $\theta$ | 0 | 0.000 | 0 | 0.002 | -1.69 | -1.14 |
| LC | $T_{max}$ | 0.55 | 1.95 | 0.59 | 2.01 | 0.676 | 2.03 |

It will cause respective alteration of parameters of the tanker's model loading conditions (LC): $\theta$ – heel; $\psi$ – trim; $T_{max}$ – maximum draft.

For Condition I, all cargo before the beginning of cargo operations is distributed uniformly among tanks. After the solution of the optimisation problem the values of $\Delta P_\Sigma^1, \Delta P_\Sigma^2$ are also distributed uniformly among the appropriate tanks in view of restrictions (3). Conditions II and III model the cases when the vessel has an initial stern trim and port side heel, accordingly. As a result of the optimisation problem solution by gradient method, the values of $\Delta P_\Sigma^1, \Delta P_\Sigma^2$ are also distributed in almost regular intervals on the appropriate tanks.

During the operation of a tanker the certain restrictions on the LC are imposed, namely: absence of heel before sailing $\theta = 0$, trim should always have negative value or to be equal to zero, that is $\psi \le 0$. In view of that the results of the previous optimisation algorithm are unsatisfactory. Table 1 shows, that the final LC of a tanker completely depends on initial loading conditions.

Taking into account that the static heel of a tanker results from displacing of the gravity centre ordinate of a tanker from equilibrium position (Clayton and Bishop, 1986) restrictions on zero value of heel can be written down as follows

$$Y_g^e - Y_g = 0, \tag{4}$$

where $Y_g^e$ – ordinate of the tanker's equilibrium gravity centre $(\theta = 0)$; $Y_g$ – ordinate of the current pontoon's gravity centre.

Thus $Y_g$ is calculated by the dependence

$$Y_g = \left(D_0 Y_0 + \sum_{i=1}^{N} P_i Y_i\right) \Big/ \left(D_0 + \sum_{i=1}^{N} P_i\right), \tag{5}$$

where $D_0, Y_0$ – weight and ordinate of the gravity centre of an empty pontoon, accordingly; $Y_i$ – ordinate of the gravity centre of cargo in the $i$-th tank.

Let's reduce the restriction (4) to the linear form relative to the vector of variables:

$$\sum_{i=1}^{N} P_i\left(Y_g^e - Y_i\right) = D_0\left(Y_0 - Y_g^e\right) = const. \tag{6}$$

Both for a real tanker and for its model (pontoon) $Y_g^e = 0, Y_0 = 0$, then the formula (6) results in

$$\sum_{i=1}^{N} P_i Y_i = 0. \tag{7}$$

Similarly, the restriction on the absence of positive trim of tanker can be written down as an inequality

$$X_g^e - X_g \ge 0, \tag{8}$$

where $X_g^e$ - abscissa of the tanker's equilibrium gravity centre $(\psi = 0)$; $X_g$ - abscissa of the tanker's gravity centre:

$$X_g = \left(D_0 X_0 + \sum_{i=1}^{N} P_i X_i\right) \Big/ \left(D_0 + \sum_{i=1}^{N} P_i\right). \tag{9}$$

With respect to (8) - (9) it is received

$$\sum_{i=1}^{N} P_i\left(X_g^e - X_i\right) \le D_0\left(X_0 - X_g^e\right). \tag{10}$$

For a pontoon $X_g^e = 0, X_0 = 0$, that allows to rewrite (9), as

$$\sum_{i=1}^{N} P_i X_i \le 0. \tag{11}$$

Restriction (11) is provided with negative value of heel within the allowable values.

However, during ship operation (including tankers) there is frequently a necessity of deliberate stern trimming $\psi \le 0$ for the best sailing performance. In this case, the value $\psi$ is defined by the captain on the basis of good sea practice (Avrahov, *et al.*, 1971) and should be taken into account at the stage of optimum cargo planning. The factor essentially influences both on the period of tanker's sailing from the current position to a place of the next bunkering implementation and profitability of operation modes of its power plant. In the mathematical form this criterion $J_X$ can be written down as a root-mean-square deviation of abscissa $X_g$ of the tanker's gravity centre from its optimum value (determined by the human-operator)

$$J_X = \frac{1}{2}\left\{X_g^{opt} - X_g\right\}^2, \tag{12}$$

where $X_g^{opt}$ – the optimum value of the gravity centre abscissa of a tanker after loading (is defined by desirable value of trim of a vessel). With regard to (9), the criterion (12) can be rewritten as

$$J_X\left(\overline{P}\right) = k_X \frac{1}{2}\left\{X_g^{opt} - \left(\frac{D_0 X_0 + \sum_{i=1}^{N} P_i X_i}{D_0 + \sum_{i=1}^{N} P_i}\right)\right\}^2, \tag{13}$$

where $k_X$ - scale factor which decreases value of the criterion $J_X$ to the order of values of the basic criterion (1).

The geometrical interpretation of the criterion $J_X(P_2, P_4)$ at $k_X = 2 \cdot 10^9$, $X_g^{opt} = -1$, $\sum_{i=1}^{N} P_i^{init} = 0$, $P_1, P_3, P_5 \ldots P_9 = const = 0$ is given in Fig. 3, where line *MN* defines the ratio between variables $P_2$ and $P_4$ in which the goal function is of its minimal value, that is $J_X = 0$. It corresponds to LC of a pontoon with a aft trim $X_g^{opt} = -1$ m.

In Table 2 the results of the solution of the pontoon's cargo plan optimisation problem for the complex criterion $J = J_X + J_P$ are given in view of the restrictions (2) - (3), (7), (11). Thus the initial variants of cargo distribution are the same as in the previous case (Table 1). It allows to estimate the evolution of optimum final cargo plans by introducing the criterion $J_X$ and the restrictions (7), (11).

From results of modelling (Table 2) it is seen: when the criterion $J_X(\bar{P})$ is used the values of trim $\psi \cong -2.2$, that corresponds to $X_g^{opt} = -1$ at the restriction (11). Besides, the total weight of the both types of cargo - $\sum_{i=1}^{N} P_i$ (black oil - $\sum_{i=1}^{m_1} P_{s_i}^1$ and diesel fuel - $\sum_{i=1}^{m_2} P_{s_i}^2$) is completely adequate to the pre-set values, according to (1) and (3). Heel $\theta$ of a pontoon after solving the optimisation problem of the three cases is equalled to zero (irrespective of initial conditions), that is provided with the restriction (7). The cargo plans for simulation model of the tanker received as a result of the optimisation (Table 2) are optimum from the point of view of LC of a pontoon (roll, trim and draft). However, it is necessary to pay attention to partial filling of all tanks in each of three variants of loading. It causes the big free surface of liquid cargo, that, in turn, negatively influences the stability of a tanker.

It is necessary to note that during BT operation one of the important criteria of the formation of cargo plans is filling a maximum quantity of tanks up to the top level (mode.1) or, on the contrary, their full unloading (mode.2) under the given LC. As mathematical interpretation of this criterion $J_S$, it is offered to apply the function:

$$J_S(\bar{P}) = k_S \sum_{i=1}^{N} \left\{ 1 / \left( 1 + e^{\pm\lambda_i (P_i - \Lambda_i P_i^{max})} \right) \right\}, \qquad (14)$$

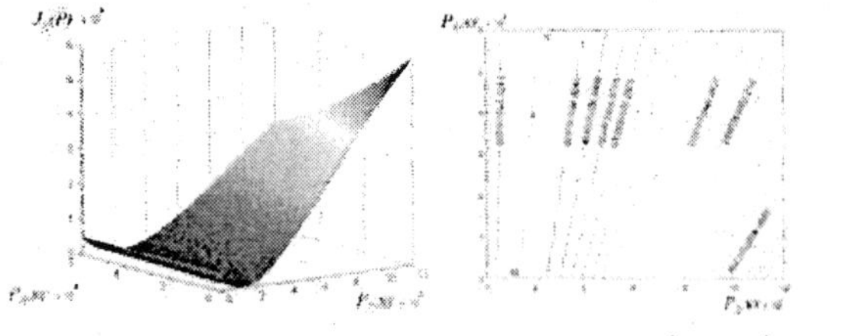

Fig.3. Geometry of criterion function $J_X(P_2, P_4)$ (a - surface of criterion function, b - lines of the level).

Table 2 Results of search of the optimum cargo plan by criterion $J(\bar{P}) = J_P(\bar{P}) + J_X(\bar{P})$

| | | Condition I | | Condition II | | Condition III | |
|---|---|---|---|---|---|---|---|
| $j$ | $i$ | $P_i^{init}$ | $P_i^j$ | $P_i^{init}$ | $P_i^j$ | $P_i^{init}$ | $P_i^j$ |
| 1 | 1 | 1000 | 40228 | 0 | 34882 | 5000 | 49018 |
| 1 | 2 | 1000 | 33378 | 5000 | 40029 | 0 | 38027 |
| 1 | 3 | 1000 | 33582 | 0 | 35947 | 0 | 28214 |
| 1 | 5 | 1000 | 16225 | 0 | 14607 | 0 | 18121 |
| 1 | 8 | 1000 | 31586 | 0 | 29535 | 0 | 21620 |
| $\sum_i P_i^1$ | | 5000 | 155000 | 5000 | 155000 | 5000 | 155000 |
| 2 | 4 | 1000 | 34477 | 2000 | 40122 | 0 | 20438 |
| 2 | 6 | 1000 | 38486 | 2000 | 27241 | 0 | 33927 |
| 2 | 7 | 1000 | 14199 | 0 | 12411 | 4000 | 21160 |
| 2 | 9 | 1000 | 16837 | 0 | 24226 | 0 | 28474 |
| $\sum_i P_i^2$ | | 4000 | 104000 | 4000 | 104000 | 4000 | 104000 |
| LC | $\psi$ | 0 | -2.23 | -0.25 | -2.23 | -0.05 | -2.20 |
| LC | $\theta$ | 0 | **0** | 0 | **-0.00** | -1.69 | **0** |
| LC | $T_{max}$ | 0.55 | **-1.00** | 0.59 | **-1.00** | 0.676 | **-0.99** |

where $P_i^{max}$ – maximum allowable cargo quantity in the $i$-th tank; $k_S$ – scale factor which decreases the value of criterion $J_S$ to the values of the basic criterion $J_P$; $\lambda_i, \Lambda_i$ - constant factors which define steepness of function $J_S(\bar{P})$ and its shift along co-ordinate $P_i$, accordingly. At Fig. 4 the geometrical interpretation of criterion $J_S$ for $i = \{2, 4\}$, $\lambda_2 = 0.00005, \Lambda_2 = 0.75$, $\lambda_4 = 0.00008, \Lambda_4 = 0.75$, $k_S = 2 \cdot 10^9$, $\sum_{i=1}^{N} P_{i,init} = 0$, $P_1, P_3, P_5 \ldots P_9 = const = 0$ is presented.

Fig.4.a shows, that every separate constituent of the criterion $J_S(P_i)$ receives the minimal value at the maximal loading of the appropriate $i$-th tank - $P_i^{max}$. The total criterion $J_S(\bar{P})$ approaches zero at the maximal filling of the appropriate tanks included in $\bar{P} = \{P_1, P_2\}$.

In view of all the above, the complex optimisation criterion of the tanker's cargo plan can be written down as the sum of the criteria (1), (13) - (14):

$$\min J(\bar{P}) = J_P(\bar{P}) + J_X(\bar{P}) + J_S(\bar{P}) \qquad (15)$$

The geometry of the complex criterion (15) is given in Fig. 5.

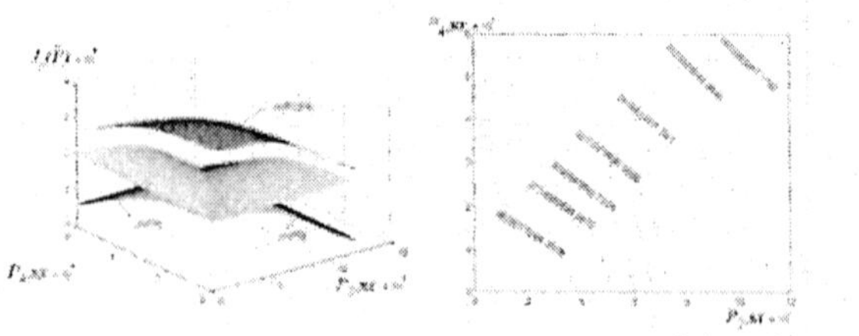

Fig.4. Geometry of criterion function $J_S(P_2, P_4)$ (a - surface of criterion function, b - lines of a level).

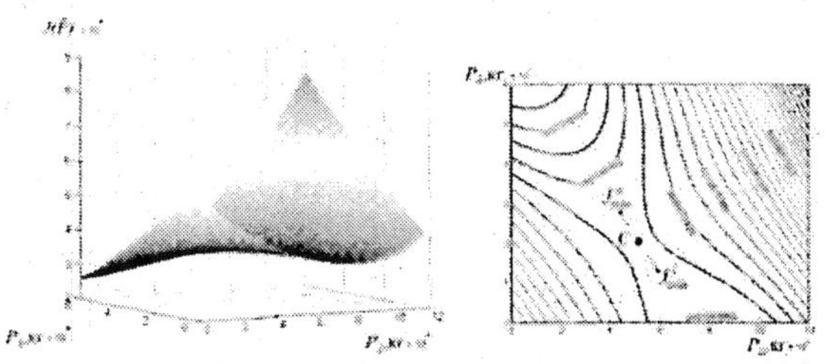

Fig.5. Geometry of complex criterion function $J(\bar{P})$ (a – surface of criterion function, b – lines of a level).

Apparently from Fig. 5 the criterion $J(\bar{P})$ is a multimodal function that causes presence of global – $J^g_{\min}$ and local $J^l_{\min}$ optimums which are on the different sides from saddle point $C$. The majority of algorithms of non-linear optimisation do not succeed in finding a global optimum, however for the problem of optimum cargo planning it is enough to find vector $\bar{P}$ which will provide the given LC of a tanker at values $J$, approached to the optimum.

The results of the calculation of the tanker's optimum cargo plan by the criterion (15) in view of the restrictions (2) - (3), (7), (11), are given in Table 3. The initial conditions and restrictions are chosen the same as in the previous cases, and the total cargo is subject to loading given as $\Delta P^1_\Sigma = 250000$ kg, $\Delta P^2_\Sigma = 150000$ kg.

The Table 3 shows the solution of the problem of optimum cargo planning for all three variants of initial loading to a pontoon total of a cargo onboard: $\sum_{i=1}^{N} P_i = \sum_{i=1}^{N} P_i^{init} + \Delta P_\Sigma = 409000$ kg; the total weight of each separate type of cargo - $\sum_{i=1}^{s} P^1_{S_i} = 255000$ kg; heel of pontoon - $\theta \cong 0$; trim – $\psi \cong -3.22$

Table 3 Results of search of the optimum cargo plan with the complex criterion of the cargo plan $J(\bar{P}) = J_P(\bar{P}) + J_X(\bar{P}) + J_S(P_2, P_4, P_6)$

| | | Condition I | | Condition II | | Condition III | |
|---|---|---|---|---|---|---|---|
| $j$ | $i$ | $P_i^{init}$ | $P_i^j$ | $P_i^{init}$ | $P_i^j$ | $P_i^{init}$ | $P_i^j$ |
| 1 | 1 | 1000 | 5330 | 0 | 4400 | 5000 | 5160 |
| 1 | 2 | 1000 | 120000 | 5000 | 120000 | 0 | 120000 |
| 1 | 3 | 1000 | 2960 | 0 | 270 | 0 | 3650 |
| 1 | 5 | 1000 | 95560 | 0 | 102820 | 0 | 94390 |
| 1 | 8 | 1000 | 31140 | 0 | 27510 | 0 | 31800 |
| $\sum_i P_i^1$ | | 5000 | **255000** | 5000 | **255000** | 5000 | **255000** |
| 2 | 4 | 1000 | 60000 | 2000 | 60000 | 0 | 60000 |
| 2 | 6 | 1000 | 60000 | 2000 | 60000 | 0 | 60000 |
| 2 | 7 | 1000 | 15810 | 0 | 14930 | 4000 | 16250 |
| 2 | 9 | 1000 | 18190 | 0 | 19070 | 0 | 17750 |
| $\sum_i P_i^2$ | | 4000 | **154000** | 4000 | **154000** | 4000 | **154000** |
| LC | $\psi$ | 0 | -3.22 | -0.25 | -3.23 | -0.05 | -3.21 |
| | $\theta$ | 0 | -0.00 | 0 | -0.00 | -1.69 | -0.00 |
| | $T_{\max}$ | 0.55 | -1.00 | 0.59 | -1.00 | 0.676 | -0.99 |

$(X_g = -1 \text{ m})$. Moreover complete loading of tanks 2, 4, 6 is provided.

Results of simulation modelling show, that the developed mathematical model of optimum loading of a tanker and application of generalised gradient method were efficient at the stage of optimum cargo planning of a rectangular pontoon. Every complex criterion achieves the optimum value within the established restrictions that allows to draw the conclusion on efficiency of the approach. Taking into account that the form of the case of a pontoon is very close to the form of the case of a tanker, it is possible to assume, with a high degree of probability, that such a model will also be effective for a bunkering tanker.

## 2. FUZZY SYSTEM FOR INPUT DATA PREPARATION

Real time terms at the preparatory stage of technological process of bunkering may be different (from several hours up to one - three days). Therefor the changes of requirements for the process are possible, in particular to the total amount of orders, the conditions of bunkering, the further disposition of tanker, the duration of process etc.

In such cases, when correcting the optimum cargo plan of a tanker before the beginning of process, the dynamic character of the component $J_X(\bar{P}) = J_X(\bar{P}, t)$ and the appropriate restrictions are taken into account. The new predicted limiting values $X_g^{opt}(t)$, $Y_g^{opt}(t)$ in expressions (6) - (7) are formed, where $t$ - current time.

For forming the dynamical restrictions the special fuzzy system for input data preparation (IDP FS) by is proposed. It is a complex fuzzy system which is based on algorithms of fuzzy logic inference (Fig. 6) for calculation of the following:

$$X_g^{opt}(t) = F_4'(Z_1, Z_2, Z_3, t),$$

$$Y_g^{opt}(t) = F_4''(Z_1, Z_2, Z_3, t),$$

where $Z_1 = F_1(x_1, x_2, x_3, x_4, x_5)$ – the generalised parameter of non-stationary external disturbing;

$Z_2 = F_2(x_6, x_7, x_8)$ – the generalised parameter of liquid cargo condition onboard;

$Z_3 = F_3(x_9, x_{10}, x_{11})$ – the generalised parameter of complexity of conditions of further tanker operation after the current bunkering operations.

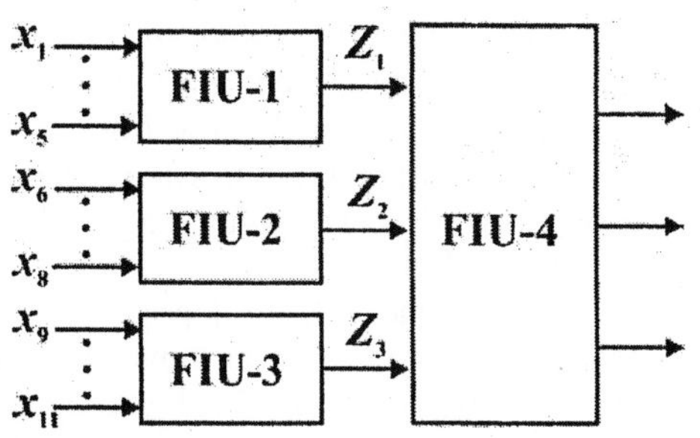


Fig.6. The structure of IDP FS

Generalised parameters $Z_1$, $Z_2$, $Z_3$ are formed by the appropriate fuzzy inference unit (FIU) which carry out processing the fuzzy information on a basis of predicting quantitative and qualitative input signals IDP FS $x_i$, where: $x_1$ – intensity of wind disturbances; $x_2$ – force of sea rolling; $x_3$ – intensity of sea current; $x_4$ – air temperature; $x_5$ – temperature of outboard waters; $x_6$ – type of the cargo ordered; $x_7, x_8$ – the current value and temperature of cargo in tanks, accordingly; $x_9$ – a quality indicator of conditions of tanker's disposition change after implementation of bunkering operations; $x_{10}$ – time restrictions on bunkering operations; $x_{10}$ – an opportunity of the new order without the change of tanker's disposition.

Fuzzy bases of rules of all FIU are generated by expert way using the experimental results obtained on the base of the virtual Tribon M1-model of bunkering tanker "Aluminy".

## CONCLUSION

Results of experiments and imitating modelling confirm efficiency of the developed algorithms and models for optimisation of bunkering operations in modes of repeated realisation of technological process of transfer of liquid cargoes without additional supply of a cargo onboard.

The following stage of researches may be the adaptation of the offered algorithms of search of optimum cargo plans with respect to dynamical restrictions to a real bunkering tanker in view of nonlinearity of the form of the hull of a tanker and its cargo tanks, ballast tanks and tanks of stocks.

## REFERENCES


Avrahov G.V., Y.L.Vorobjov and E.N.Kungurceva (1968). Calculation at PC of optimal loading of the dry-cargo vessel. *Shipbuilding and shiprepare*, **Vol. 2,** Odessa, pp. 17-26.

Avrahov G.V. and Y.L.Vorobjov (1971). Aplying the PC for determination of variont of tanker loading. *Shipbuilding and shiprepare*, **Vol. 2,** Odessa, pp. 146-152.

Egorov G.V. (1989). The questions of ship's loading and ballasting optimisation. *Shipbuilding and shiprepare*, Moscow, «Mortechinformreclama», pp. 55-59.

Himelbau D. (1974). *Applied nonlinear programming.* (Trans. from Eng. under red. of Bihovsky M.L.), 534 p., Moscow, Mir.

Kleyton B. and R. Bishop (1986). *The mechanics of seagoing vessels.* (Tr. from Eng.), 436 p., Leningrad, Sudostroenie.

Kondratenko Y. and D. Pidoprigora (2000). Modelling of the rectangular pontoon behavior for various types of the fixed gravity center cargo distribution. *Technical news,* **No.1(10), 2(11),** Lviv, pp. 111-115.

Kondratenko Y.P., D.N. Podoprigora and S.A. Sydorenko (2001). Fuzzy approach to thynthesis of online algorithms of vessel's cargo plan formation. *Proceedings of International scientific-practical conference "The theory of active systems"*, pp. 96 – 97.

Kondratenko Y.P. and D.M. Podoprigora (2002). Automatisation of the technological process of ships bunkerning. *Proceeding of Odessa Polytechnic University,* **Vol. 1 (17),** pp. 131-136.

Kozljakov V.V. and G.V. Egorov (1991). Automated calculation of cargo plan and ship's floodability at the onboard PC. *Egineering of ships and marine hardware,* Nikolaev, NKI, pp. 55-63.

Leigh-Jones C. Barging Ahead (2001). *Maritime reporter and engineering news*, **May**, pp. 38 – 41.

Letnjanchik E.Y. and L.N. Shemagina (1989). The features of calculation at PC of the optimal schemes of loading by nonlinear programming methods. *Optimisation of control by technical means at river transport,* **Vol.204**, Gorly, GIIVT, pp.37-47.

Methuse D.G. and K.D. Fink (2001). *Numerical methods. Using of MATLAB*, 3-th issue, (Trans. from Eng.), 720 p., Moscow, *Publishing house* «Vilyams».

Volovoy D.I. (1984). To the question of most advantageous distribution of cargo along the ship. *Optimisation of control by technical means at river transport,* **Vol.204**, Gorly, GIIVT, pp. 49-60.

ELSEVIER
IFAC
PUBLICATIONS
www.elsevier.com/locate/ifac

# A GEOMETRIC APPROACH TO A PURSUING PROBLEM.

**Luca Consolini * Mario Tosques ****

* *Dipartimento di Ingegneria dell'Informazione, University of Parma (Italy), e-mail: luca.consolini@polirone.mn.it*
** *Dipartimento di Ingegneria Civile, University of Parma, (Italy), e-mail: mario.tosques@unipr.it*

Abstract: This article addresses a 2-dimensional pursuing problem, namely the guidance of a ship towards another ship which will be our "target". The second order target dynamics are assumed to be known and this knowledge is used by the control system to guide the first ship and, under some hypotheses, reach the target. A geometrical analysis of the proposed control strategy is presented, together with numerical simulations. *Copyright © 2004 IFAC*

Keywords: Guidance systems, target tracking, geometric approaches.

## 1. INTRODUCTION

This article addresses a 2-dimensional pursuing problem, namely the guidance of a ship A towards a ship B which will be our "target". Such kind of problems have already been considered in the literature. For instance, studies as (Becker, 1990), (Cochran *et al.*, 1991) and (Yuan and Chern, 1992) provide complete closed-form solution for the guidance problem in some specific cases, even if the results are mostly analytical and do not give an insight on the geometry of the pursuing problem. In (Kuo *et al.*, 2001) the problem of a tactical missile hitting a target is solved in the case of a target moving on a straight line and on a circle, the idea used in that paper involves driving the missile not towards the target, but directly to the future collision point assuming that the target were to follow a straight line. This paper presents a similar approach, but addresses a more complex case in which the second order target dynamics are assumed to be known and this knowledge is used by the control system to guide a ship A to reach a ship B.

## 2. PROBLEM FORMULATION

The kinematic equations of the ship A motion are assumed to be given by the following simplified model, $\forall t \geq 0$:

$$\begin{cases} \dot{x} = v\cos\theta \, , \; x(0) = x_0 \\ \dot{y} = v\sin\theta \, , \; y(0) = y_0 \\ \dot{\theta} = \omega \, , \; \theta(0) = \theta_0 \\ \dot{\omega} = u \, , \; \omega(0) = \omega_0 \end{cases} \tag{1}$$

where $(x(t),y(t))^T$ is the position of the ship center of mass, $v$ its velocity which is supposed to be constant, $\theta(t)$ is the angle between the speed vector and the x-axis, $u \in \mathscr{C}([0,+\infty))$ is the control which is assumed to be bounded by a positive constant $K$:

$$|u(t)| \leq K,$$

where $\mathscr{C}([0,+\infty))$ denotes the set of all continuous functions defined on $[0,+\infty)$. This model takes into account the fact that the ship is controlled through its rudder, therefore the control generates a torque which is proportional to the derivative of the angular velocity $\omega$.

Analogously the kinematic model of the target B may be assumed of the same form as (1), $\forall t \geq 0$:

$$\begin{cases} \dot{x_T} = v_T \cos\theta_T \ , \ x_T(0) = x_{T0} \\ \dot{y_T} = v_T \sin\theta_T \ , \ y_T(0) = y_{T0} \\ \dot{\theta}_T = \omega_T \ , \ \theta_T(0) = \theta_{T0} \\ \dot{\omega}_T = u_T \ , \ \omega_T(0) = \omega_{T0} \end{cases} \quad (2)$$

with $u_T \in \mathscr{C}([0,+\infty))$ such that

$$|u_T| < K_T \ , K_T > 0 \ ,$$

we will always assume that

$$v > v_T \ .$$

Let

$$P = (v,\ v_T,\ K,\ K_T)^T \in \mathbb{R}^4_+ \ , \ \mathbb{R}_+ = [0,+\infty)$$

be the vector of the model parameters $X(t) = (x(t),\ y(t),\ \theta(t),\ \omega(t))^T$, $X_T(t) = (x_T(t),\ y_T(t),\ \theta_T(t),\ \omega_T(t))^T$ be the state vectors and $\Xi(t) = (X(t), X_T(t))^T$ be the state vector for the augmented system (1)+(2)) which will be denoted by

$$\begin{cases} \dot{\Xi} = F(\Xi, u, u_T),\ t \geq 0 \\ \Xi(0) = (x_0,\ y_0,\ \theta_0,\ \omega_0,\ x_{T0},\ y_{T0},\ \theta_{T0},\ \omega_{T0}) \ , \end{cases} \quad (3)$$

let $\gamma(t) = (x(t),\ y(t))^T$, $\gamma_T(t) = (x_T(t),\ y_T(t))^T$ be the trajectories followed by ship A and B.

Given a solution $\Xi$ of system (3), consider the vector map

$$\Delta_\Xi(t) = \gamma(t)^T - \gamma_T(t)^T,\ t \geq 0$$

and define

$$d_\Xi(t) = \|\Delta_\Xi(t)\| \ ,\ t \geq 0$$

which will be called the distance, at time $t$, between the ships A and B (relative to the solution $\Xi$).

*Definition 1.* Given a choice of the vector $P \in \mathbb{R}^4_+$, any map $\mathscr{S}_P \in \mathbb{R}^8 \to \mathbb{R}$ is a feedback control strategy for ship A. The strategy is said *exact* if for any choice of the control $u_T \in \mathscr{C}([0,+\infty))$ such that $|u_T(t)| \leq K_T$ there exists a reaching time $t_r$ such that $d_\Xi(t_r) = 0$, where $\Xi$ is the solution of:

$$\begin{cases} \dot{\Xi} = F(\Xi, S_P(\Xi), u_T) \\ \Xi(0) = (x_0,\ y_0,\ \theta_0,\ \omega_0,\ x_{T0},\ y_{T0},\ \theta_{T0},\ \omega_{T0}) \end{cases}$$

The general pursuing problem may be stated as follows

**Problem** Determine sufficient conditions such that, given a choice of the parameters $P$, it is possible to find a control strategy $\mathscr{S}_P$ which is exact.

Clearly this problem is very general and there have been many attempts to give an answer in specific cases, see for instance (Becker, 1990),(Cochran *et al.*, 1991),(Yuan and Chern, 1992),(Kuo *et al.*, 2001). This article proposes a new strategy which is discussed both theoretically and in simulation.

## 3. CONTROL STRATEGY

Here is presented the overall pursuing strategy.

It is supposed that the ship A controller knows the target state at any time, therefore, at any time, the target position $\gamma_T(t)$ and the osculating circle to the trajectory $\gamma_T(t)$ are known.

The ship A controller supposes that the target continues to follow its osculating circle, which is, in fact, the best approximation of the future trajectory of the target which can be made by the controller from the knowledge of the target state. In fact this second order approximation gives more information on the target behavior than the first order one that assumes that the target continues to follow the tangent to its path.

The controller computes the minimum distance between its trajectory and the target one assuming that both were to follow the osculating circles to their respective trajectories and varies its curvature radius by means of the control $u$ in such a way to reduce such minimum distance at the fastest ratio.

To this goal, set, $\forall X = (x, y, \theta, \omega) \in \mathbb{R}^4\ \forall v \geq 0, \forall \tau \geq 0$:

$$\mathscr{O}^v_X(\tau) = \begin{cases} \dfrac{v}{\omega}\left[\begin{pmatrix} \sin(\omega\tau + \theta) \\ -\cos(\omega\tau+\theta) \end{pmatrix} - \begin{pmatrix} \sin\theta \\ -\cos\theta \end{pmatrix}\right] + \\ + \begin{pmatrix} x \\ y \end{pmatrix},\ \text{if } \omega \neq 0 \\ v\begin{pmatrix} \cos\theta \\ \sin\theta \end{pmatrix}\tau + \begin{pmatrix} x \\ y \end{pmatrix},\ \text{if } \omega = 0 \end{cases}$$

remark that $\mathscr{O}^v_X(\tau)$ and $\mathscr{O}^{v_T}_{X_T}(\tau)$, as functions of $\tau$, are the trajectories followed by A and B respectively if they followed their osculating circles at the points $\gamma(t)$ and $\gamma_T(t)$, with radius $\frac{v}{\omega}$ and $\frac{v_T}{\omega_T}$, for the time $\tau$;

then the parametric equations in $\tau$ of the osculating circles to $\gamma$ and $\gamma_T$ at time $t$ are respectively:

$$\mathscr{O}^v_X(t) \text{ and } \mathscr{O}^{v_T}_{X_T}(t),\ \forall \tau \geq 0.$$

Set $\forall \Xi = (X, X_T) \in \mathbb{R}^8$

$$d(\Xi) = \inf_{\tau \geq 0}\{\|\mathscr{O}^v_X(\tau) - \mathscr{O}^{v_T}_{X_T}(\tau)\|^2\}$$

and suppose that $\Xi$ is such that there exists a local minimum $\tau_m = \tau_m(\Xi) > 0$ such that $\forall \tau \in [0, \tau_m)$

$$d(\Xi) = \|\mathscr{O}^v_X(\tau_m) - \mathscr{O}^{v_T}_{X_T}(\tau_m)\|^2 < \\ < \|\mathscr{O}^v_X(\tau) - \mathscr{O}^{v_T}_{X_T}(\tau)\|^2 \ .$$

Therefore, it is:

$$\frac{\partial d}{\partial \omega}(\Xi) = 2 < \frac{\partial \mathscr{O}^v_X}{\partial \omega}(\tau_m), \mathscr{O}^v_X(\tau_m) - \mathscr{O}^{v_T}_{X_T}(\tau_m) > + \\ +2\frac{\partial \tau_m}{\partial \omega} < \frac{\partial \mathscr{O}^v_X}{\partial \tau}(\tau_m), \mathscr{O}^v_X(\tau_m) - \mathscr{O}^{v_T}_{X_T}(\tau_m) >= \\ = 2 < \frac{\partial \mathscr{O}^v_X}{\partial \omega}(\tau_m), \mathscr{O}^v_X(\tau_m) - \mathscr{O}^{v_T}_{X_T}(\tau_m) > ,$$

being $< \frac{\partial \mathscr{O}^v_X}{\partial \tau_m}(\tau_m), \mathscr{O}^v_X(\tau_m) - \mathscr{O}^{v_T}_{X_T}(\tau_m) >= 0$, since $\tau_m$ is an interior local minimum; moreover, $\forall \omega \neq 0$, set for simplicity $\beta = \omega\tau + \theta$, then:

$$\frac{\partial \mathcal{O}_X^v}{\partial \omega}(\tau) = -\frac{v}{\omega^2}\begin{pmatrix} \sin\beta \\ -\cos\beta \end{pmatrix} + \frac{v}{\omega^2}\begin{pmatrix} \sin\theta \\ -\cos\theta \end{pmatrix} +$$
$$+\frac{v\tau}{\omega}\begin{pmatrix} \cos\beta \\ \sin\beta \end{pmatrix} = \frac{v}{\omega^2}\left\{\begin{pmatrix} -\sin\beta \\ \cos\beta \end{pmatrix}(1-\cos(\omega\tau)) + \right.$$
$$\left. +\begin{pmatrix} \cos\beta \\ \sin\beta \end{pmatrix}(\omega\tau - \sin(\omega\tau))\right\} =$$
$$= \frac{v}{\omega^2}R(\omega\tau+\theta)\begin{pmatrix} \omega\tau - \sin(\omega\tau) \\ 1-\cos(\omega\tau) \end{pmatrix}$$

where $R(\alpha) = \begin{pmatrix} \cos\alpha\,, & -\sin\alpha \\ \sin\alpha\,, & \cos\alpha \end{pmatrix}$, because $\begin{pmatrix} \sin\theta \\ -\cos\theta \end{pmatrix} = \begin{pmatrix} \sin\beta \\ -\cos\beta \end{pmatrix}\cos(\omega\tau) - \begin{pmatrix} \cos\beta \\ \sin\beta \end{pmatrix}\sin(\omega\tau)$.

Therefore

$$\frac{\partial \mathcal{O}_X^v}{\partial \omega}(\tau) = v\tau^2 R(\omega\tau+\theta)\begin{pmatrix} \dfrac{\omega\tau - \sin(\omega\tau)}{(\omega\tau)^2} \\ \dfrac{1-\cos(\omega\tau)}{(\omega\tau)^2} \end{pmatrix}$$

which implies that

$$\frac{\partial d}{\partial \omega}(\Xi) = v\tau_m^2 < R(\omega\tau_m+\theta)(\begin{pmatrix} 0 \\ 1 \end{pmatrix} + \varphi(\omega\tau_m)),$$
$$, \mathcal{O}_X^v(\tau_m) - \mathcal{O}_{X_T}^{vT}(\tau_m) > .$$

where $\varphi : \mathbb{R} \to \mathbb{R}^2$ is a $\mathscr{C}^\infty$ map such that $\lim_{s\to 0}\varphi(s) = 0$. The previous remarks suggest the following control law which reduces the minimum distance as fast as possible while satisfying the input constraints:

$$u(\Xi) = -\text{ sign } < R(\omega\tau_m(\Xi)+\theta)(\begin{pmatrix} 0 \\ 1 \end{pmatrix} +$$
$$+\varphi(\omega\tau_m(\Xi)), \mathcal{O}_X^v(\tau_m) - \mathcal{O}_{X_T}^{vT}(\tau_m) > K.$$

which can be approximated by the following one:

$$u(\Xi) = -\text{ sign } < R(\omega\tau_m(\Xi)+\theta)\begin{pmatrix} 0 \\ 1 \end{pmatrix},$$
$$, \mathcal{O}_X^v(\tau_m) - \mathcal{O}_{X_T}^{vT}(\tau_m) > K.$$

In other words, the proposed control strategy will be the following one

$$\mathscr{S}_P(\Xi) = -\text{ sign } < R(\omega\tau_m(\Xi)+\theta)\begin{pmatrix} 0 \\ 1 \end{pmatrix},$$
$$, \mathcal{O}_X^v(\tau_m) - \mathcal{O}_{X_T}^{vT}(\tau_m) > K,$$

which says that the controller has the following behavior:

-if at the time of minimum distance the target is inside the the ship A osculating circle, take the control such that $|u| = K$ and the sign is such that the radius of the osculating circle itself is reduced;

-if at the time of minimum distance the target is outside the ship A osculating circle, take the control such that $|u| = K$ and the sign is such that the radius of the osculating circle itself is increased.

-if at the time of minimum distance the target is on the ship A osculating circle, take $u = 0$.

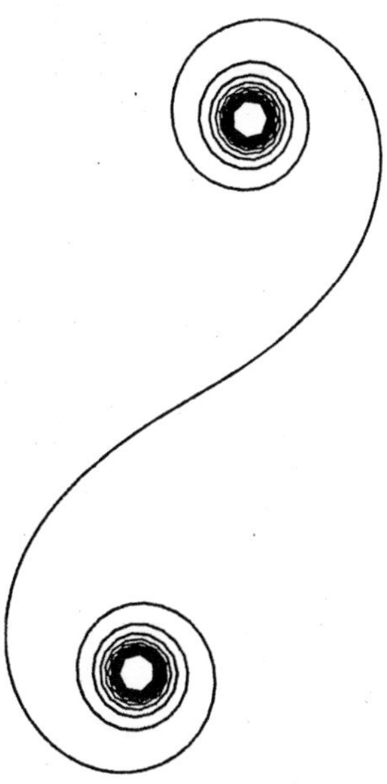

Fig. 1. A clothoid.

## 4. THEORETICAL RESULTS

Consider system (1) with a constant input $u(t) = L$, the trajectory generated is a curve whose curvature varies linearly, such a curve is called a *clothoid* and it includes an infinite number of loops of decreasing size around two limit points (see figure 1).

For every $X_0 = (x_0, y_0, \theta_0, \omega_0)^T \in \mathbb{R}^4$, $\forall u_0 \in \mathbb{R}$, call $\Phi(X_0, u_0, t) : \mathbb{R}^6 \to \mathbb{R}^4$ the solution at time $t$ of system (1) with constant control $u_0$ and initial state $X_0$; $\forall L \neq 0$, let $R_L : [0, +\infty) \to [0, +\infty]$ be the function defined by $R_L(t) = \frac{v}{|\omega_0 + Lt|}$, with the convention that $R_L(t) = +\infty$, if $\omega_0 + Lt = 0$,

$$D_L = \{(t,\tau) | t \geq 0,\ 0 \leq \tau < \frac{2\pi R_L(t)}{v}\}$$

and let $T_{X_0}^L : D_L \to \mathbb{R}^2$ be the map defined by

$$T_{X_0}^L(t,\tau) = \Pi(\Phi(\Phi(X_0, L, t), 0, \tau)) \tag{4}$$

where $\forall\, X = (x, y, \theta, \omega)^T \in \mathbb{R}^4$, $\Pi(X) = (x, y)$.

Map $T_{X_0}^L(t,\tau)$ represents the position $\gamma(t)$ of the ship A after having applied for a time $t$ the maximum control $L$ and then for a time $\tau$ the null control $u = 0$. The first movement (see figure 2) is along a clothoid, the second one is along the osculating circle to the clothoid at time $t$ if $\omega_0 + Lt \neq 0$ or along the ray tangent to the clothoid at time $t$, if $\omega_0 + Lt = 0$. In the previous notations the following theorem holds.

*Theorem 1.* Let $X_0$ be any element of $\mathbb{R}^4$, then

a) $\forall L \neq 0$, $T_{X_0}^L : D_L \to \mathbb{R}^2$ is one to one

b) $\forall K > 0$, $T_{X_0}^K(D_K) \cap T_{X_0}^{-K}(D_{-K} \backslash \{0\} \times [0, \frac{2\pi R_{-K}(0)}{v})) = \emptyset$

c) $\forall K > 0$, $T_{X_0}^K(D_K) \cup T_{X_0}^{-K}(D_{-K} \backslash \{0\} \times [0, \frac{2\pi R_{-K}(0)}{v})) = \mathbb{R}^2 \backslash (\{P_1, P_2\} \cup l_-)$

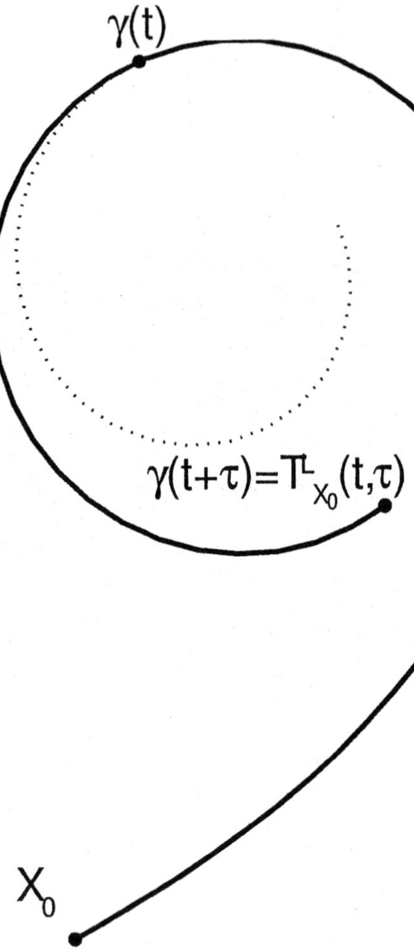


Fig. 2. Mapping $T^L_{X_0}(t,\tau)$.

when $P_1 = \lim_{t\to+\infty} T^K_{X_0}(t,0)$, $P_2 = \lim_{t\to+\infty} T^{-K}_{X_0}(t,0)$ and

$$l_- = \begin{cases} \left\{ T^{-K}_{X_0}(\frac{\omega_0}{K},0) + \tau \frac{\partial T^{-K}_{X_0}}{\partial t}(\frac{\omega_0}{K},0) | \tau < 0 \right\}, \\ \text{if } \omega_0 \geq 0 \\ \left\{ T^{K}_{X_0}(-\frac{\omega_0}{K},0) + \tau \frac{\partial T^{K}_{X_0}}{\partial t}(-\frac{\omega_0}{K},0) | \tau < 0 \right\}, \\ \text{if } \omega_0 < 0 \end{cases}$$

*Proof.* .

a) It is not restrictive to suppose that $L > 0$, we have to show that
1)$T^L_{X_0}(t_1,\tau_1) \neq T^L_{X_0}(t_2,\tau_2)$, $\forall t_1,t_2,\tau_1,\tau_2 \geq 0$, $t_1 \neq t_2$
2)$T^L_{X_0}(t,\tau_1) \neq T^L_{X_0}(t,\tau_2)$, $\forall t \geq 0$, $\forall \tau_1,\tau_2 \geq 0$, $\tau_1 \neq \tau_2$.

Point 2) is clear, to show point 1) we can suppose that $t_1 < t_2$, therefore we can have two cases:

$0 \leq \omega_0 + Lt_1 < \omega_0 + Lt_2$ and $\omega_0 + Lt_1 < 0 \leq \omega_0 + Lt_2$.

Suppose, firstly, that $0 \leq \omega_0 + Lt_1 < \omega_0 + Lt_2$ then the statement is proved if we show that the circle $T^L_{X_0}(t_2,[0,\frac{2\pi R_L(t_2)}{v}))$ is strictly contained in the circle $T^L_{X_0}(t_1,[0,\frac{2\pi R_L(t_1)}{v}))$, and this holds if we prove that the distance $\|C(t_2) - C(t_1)\|$ between their centers is strictly less than the difference $R_L(t_1) - R_L(t_2)$ between their radius. In fact since the coordinates of the center $C(t)$ are:

$$C(t) = (x(t),y(t))^T + R_L(t)\begin{pmatrix} -\sin\theta(t) \\ \cos\theta(t) \end{pmatrix}$$

and

$$\dot{C}(t) = \begin{pmatrix} \cos\theta(t) \\ \sin\theta(t) \end{pmatrix}(v - R_L(t)\omega(t)) +$$
$$+\dot{R}_L(t)\begin{pmatrix} -\sin\theta(t) \\ \cos\theta(t) \end{pmatrix} = \dot{R}_L(t)\begin{pmatrix} -\sin\theta(t) \\ \cos\theta(t) \end{pmatrix},$$

where we used the fact that $v = R_L(t)(\omega_0 + Lt) = R_L(t)\omega(t)$, then

$$\|C(t_2) - C(t_1)\| < \int_{t_1}^{t_2} \|\dot{C}(t)\| dt =$$
$$= -\int_{t_1}^{t_2} \dot{R} dt = R_L(t_1) - R_L(t_2),$$

(remark that $\dot{R}_L(t) \leq 0$, $\forall t \in [t_1,t_2]$).

If $\omega_0 + Lt_1 < 0 \leq \omega_0 + Lt_2$, let $t_0$ be such that $\omega_0 + L\bar{t} = 0$, then $\tau \rightsquigarrow T^L_{X_0}(\bar{t},\tau)$ is the parametrization of the straight line $l$ tangent to the clothoid $t \rightsquigarrow T^L_{X_0}(t,0)$ at the time $\bar{t} > 0$. If we call $F_1$ and $F_2$ respectively the two connected components of $\mathbb{R}^2 \backslash l$, it is easy to see that the two circles $T^L_{X_0}(t_1,[0,\frac{2\pi R_L(t_1)}{v}))$, $T^L_{X_0}(t_2,[0,\frac{2\pi R_L(t_2)}{v}))$ have empty intersection since they belong respectively to $F_1$ and $F_2 \cup l$. Therefore $T^L_{X_0} : D_L \to \mathbb{R}^2$ is one to one.

Now let $K > 0$, it is easy to see that b) holds, to show point c) we suppose for simplicity that $x_0 = y_0 = \theta_0 = \omega_0 = 0$, that is $X_0 = 0$, $l_- = \{(x,0)|x < 0\}$ and c) holds if we prove that
1)$T^K_0(D_K) \supset \{(x,y)|x \in \mathbb{R}, y > 0\} \cup l_+ \backslash \{P_1\}$
2)$T^{-K}_0(D_{-K} \backslash \{0\} \times [0,\frac{2\pi R_{-K}(0)}{v})) \supset \{(x,y)|x \in \mathbb{R}, y < 0\} \backslash \{P_2\}$
where $l_+ = \{(x,0)|x \geq 0\}$. To show 1), take any point $P \in \{(x,y)|x \in \mathbb{R}, y > 0\} \cup l_+ \backslash \{P_1\}$, we have to prove that there exists $(t,\tau) \in D_K$ such that $T^K_0(t,\tau) = P$. Since this is clear if $P \in l_+$, we suppose that $P \in \{(x,y)|x \in \mathbb{R}, y > 0\} \backslash (l_+ \cup \{P_1\})$ and let $S$ be the straight ray starting from $P_1$ and passing through $P$. There exists $t_1, t_2$, with $0 < t_1 < t_2$, such that $P$ belongs to the interior of the circle $T^K_0(t_1,[0,\frac{2\pi R_K(t_1)}{v}))$ and $P$ is outside the circle $T^K_0(t_2,[0,\frac{2\pi R_L(t_2)}{v}))$. Call $\forall t > 0$ $P(t)$ the unique point, intersection of the generic circle $T^K_0(t,[0,\frac{2\pi R_K(t)}{v}))$ with $S$ and $D : (0,+\infty) \to \mathbb{R}$ be the continuous function defined by

$$D(t) = < P(t) - P, P - P_1 > .$$

By construction we have that $D(t_1) < 0 < D(t_2)$, therefore there exists $t_0$ such that $D(t_0) = 0$, that is $P(t_0) = P$ which implies that $P \in T^K_0(D_K)$. In an analogous way we prove point 2).□.

Next proposition addresses the case in which the target vehicle speed is null.

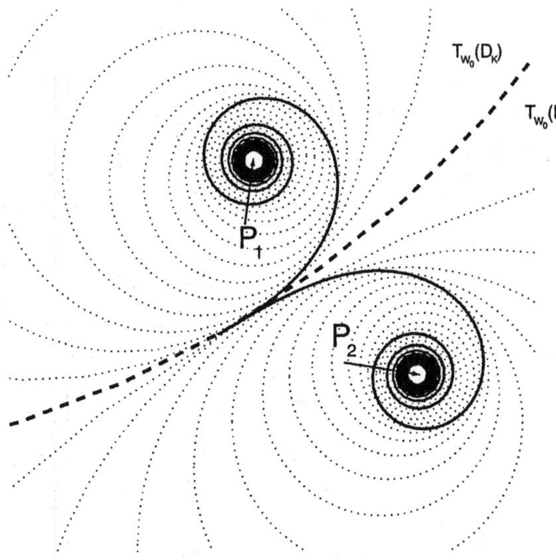


Fig. 3. Theorem 1.

*Corollary 1.* Every fixed target at position $(x_0, y_0) \in \mathcal{D}$ is reached in a finite time by system (1) with control (3) and the control action is as follows:

-if the target is inside the ship A osculating circle, the control norm is $|u| = K$ and the control sign is such that the osculating circle radius is reduced until it hits the target, then the control becomes null and the ship A goes to the target on the circle itself.

-if the target is outside the ship A osculating circle, the control norm is $|u| = K$ and the control sign is such that the osculating circle radius is increased until the target stays on it, then the control becomes null as in the previous case.

*Proof.* It is a straight consequence of Theorem 1, which proves that point $(x_{T0}, y_{T0})^T$ can be reached by a control action of the kind described by the theorem statement. □.

Consider now the case in which the target is moving but its control is null and $\omega_{T0} \neq 0$, therefore its trajectory is a circle.

*Lemma 1.* At the point $t_0$ of minimum distance between the ship A and the target it is

$$| < \hat{r}, \hat{d} > | \geq \sqrt{1 - (\frac{v_T}{v})^2} \quad (5)$$

where $\hat{r}$ is the normalized radial vector which joins the ship A osculating center to the point of minimum distance and $\hat{d}$ is the distance vector $d$ taken with unitary norm. Remember that we have supposed that $v_T < v$.

*Proof.*

At $t_0$ the distance time derivative must be zero, this means that

$$< v\tau - v_T \tau_T, d > = 0 ,$$

where $\tau$ and $\tau_T$ are respectively the ship A and the target unitary tangent vectors to their circular trajectories, then

$$< v\tau, d > = < v_T \tau_T, d >$$
$$< v\tau, d > \leq \|v_T d\| ,$$

being $< \tau, d >^2 + < \hat{r}, d >^2 = \|d\|^2$, the thesis follows. □.

*Theorem 2.* A moving target described by (2) with $u_T = 0$, $\omega_{T0} \neq 0$ (that is, the target trajectory is a circle of radius $\frac{v_T}{\omega_{T0}}$), $v_t < v$ and such that the limit points $P_1$ and $P_2$ do not belong to the trajectory, is reached in a finite time by ship A.

*Proof.* Assume without loss of generality that the target trajectory is inside $T^K_{X_0}(D_K)$. Being the target trajectory closed and bounded, there exists $t_1, t_2$ such that the trajectory is contained in the circle $C_1 = T^K_{X_0}(t_1, [0, \frac{2\pi R_K(t_1)}{v}))$ and is disjoint from the set surrounded by the circle $C_2 = T^K_{X_0}(t_2, [0, \frac{2\pi R_K(t_2)}{v}))$.

Set $\gamma(\tau) = T^K_{X_0}(t, \tau)$, $\gamma_T(t + \tau)$ the target position at time $t + \tau$,

$$d(t, \tau) = T^K_{X_0}(t, \tau) - \gamma_T(t + \tau)$$

the distance vector between ships A and B at time $t + \tau$, be

$$d_t(\tau) = < T^K_{X_0}(t, \tau) - \gamma_T(t + \tau), \begin{pmatrix} -\sin\theta(t+\tau) \\ \cos\theta(t+\tau) \end{pmatrix} > ,$$

the component of the distance vector at time $t + \tau$ normal to the ship A osculating circle and set $D(t) = \inf_\tau d_t(\tau)$.

Now $D(t_1) > 0$, $D(t_2) < 0$, so, being $D$ continuous there exists a $t_0$ such that $D(t_0) = 0$, this means that there exists a couple $t_0, \tau$ for which the normal component of $d(t_0, \tau)$ is null, then, by the lemma,

$$0 = | < \hat{r}, d > | \geq \|d\| \sqrt{1 - (\frac{v_T}{v})^2} ,$$

therefore $d(t_0, \tau) = 0$.

Now we shot that the control $u(t)$ generated by the controller is indeed

$$\begin{cases} u = k \text{ if } t < t_0 \\ u = 0 \text{ if } t \geq t_0 \end{cases} ,$$

in fact, the absolute value of the normal component of the distance vector is always reduced by the controller until it is null, so the control input $u$ must be piecewise constant with one single discontinuity point, moreover at the discontinuity point the computed minimum distance is null, therefore the discontinuity point must be $t_0$. □.

*Corollary 2.* Let be given $P = (v, v_T, k, 0)$ , with $v_t < v$, in other words the target trajectory is a circle that does not contain the limit points $P_1$ and $P_2$, then it is possible to find a control strategy which is exact.

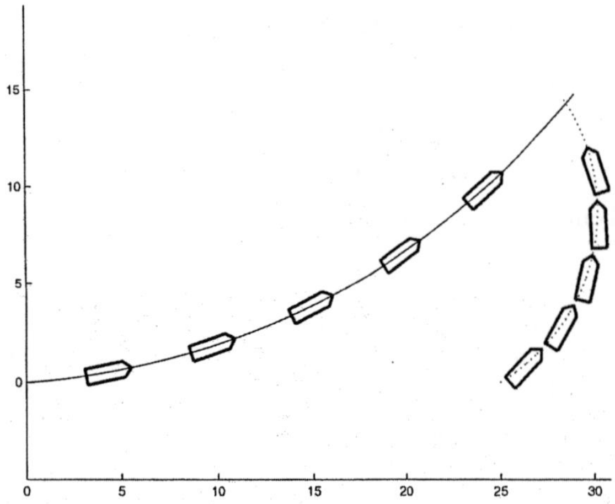


Fig. 4. Simulation 1.

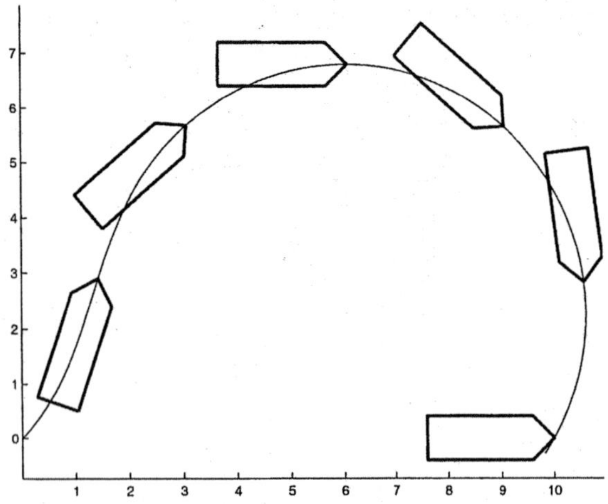


Fig. 5. Simulation 2.

## 5. SIMULATION

The controller presented in section 3, has been tested in a simulation running on MATLAB, in different cases with both fixed and moving targets. In the following $X$ and $X_T$ represent respectively the ship A and the target state vectors.

*Simulation 1.* Parameters and initial data

$$X(0) = (0,0,0,0.05);\ X_T(0) = (25,0,0.5,0.1);$$
$$v = 2;\ v_T = 1;\ K = 1.5$$

Figure 4 represents with a solid line the ship A trajectory and whit a dotted line the ship B one, at the intersection between the two lines the ship A hits the target.

*Simulation 2.* In this case the target is fixed, the parameters and the initial data are

$$X(0) = (0,0,0.8,0.3);\ X_T(0) = (10,0,0,0);$$
$$v = 1;\ v_T = 0;\ K = 0.1$$

See figure 5 for the results.

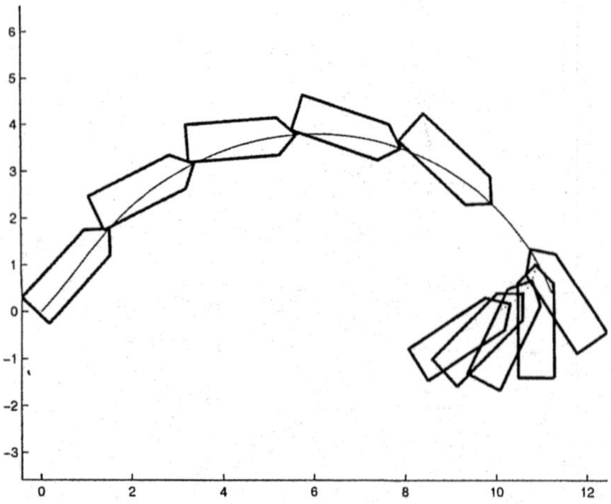


Fig. 6. Simulation 3.

*Simulation 3.* This is a more complex case in which the target is steering with a control $u(t) = 0.1cos(0.1t)$. Note that the case $u_T \neq 0$ is not address in the theoretical part of this paper. The other parameters are

$$X(0) = (0,0,0.8,0.3);\ X_T(0) = (10,0,0.5,0.001);$$
$$v = 2;\ v_T = 0.3\ K = 0.5$$

As figure 5 shows that also in this particular case the ship A is able to reach the target. In other similar cases the ship A does not seem to be able to reach its target. Finding out in which cases and under which conditions the pursuit is successful is still an open problem.

## 6. CONCLUSIONS

This article reports a work-in-progress on the geometric guidance of a ship A towards a ship B. A control law has been proposed for the simple cases of a fixed or non manouvering target. Some preliminary theoretical results have been given and simulations have been provided. Much work remains to be done to face the theoretical framework of the complex case in which the target vehicle is driven by an arbitrary control law.

## REFERENCES

Becker, K. (1990). Closed-form solution of pure proportional navigation. *IEEE Trans. Aereosp. Electrion. Syst.* **26**, 526–532.

Cochran, J.E., Jr. T.S. No and D. G. Thaxton (1991). Analytical solution to a guidance problem. *AIAA J. Guid. Contr. Dyn.* **4**, 117–122.

Kuo, C.Y., D. Soetanto and Y.C. Chiou (2001). Geometric analysis of flight control command for tactical missile guidance. *IEEE Transactions on Control Systems Technology* **9**(2), 234–243.

Yuan, P. J. and J. S. Chern (1992). Solutions of true proportional navigation for maneuvering and nonmaneuvering targets. *AIAA J. Guid. Contr. Dyn.* **15**, 268–271.

ELSEVIER
IFAC
PUBLICATIONS
www.elsevier.com/locate/ifac

# TRAJECTORY TRACKING CONTROL SYSTEM FOR SHIP

**Janusz Pomirski, Leszek Morawski, Andrzej Rak**

*Gdynia Maritime University, Department of Ship Automation,*
*Morska Str. 83, 81-225 Gdynia, Poland*
*jpomir@am.gdynia.pl*

Abstract: The conventional way-point guidance system has two main operation modes: the course-keeping mode at straight trajectory segments and the course changing mode in the way-points. This paper presents the design of a course and a turning controllers which have high steering performance and are robust due to ship dynamics' variations. The bases for the design are simple, linear ship dynamics models of the 1st and 2nd order (Nomoto models). The controllers have fixed structure and theirs coefficients are selected using the performance indices: the settling time and integral performance index ITAE (the integral of time multiplied by absolute error). *Copyright © 2004 IFAC*

Keywords: ship control, control system design, controllers, tracking systems, traction yaw control

## 1. INTRODUCTION

The route of a ship is usually specified as a sequence of way-points. Cartesian coordinates of way-points defines the desired broken line which the ship should follow as close as possible. When the ship is sailing in an open ocean a precise passing a way-point is not required. The ship is only expected to gentle change the course from one straight segment to another while passing the way-points. Therefore conventional guidance system has two main operations modes: course-keeping mode at straight trajectory segments and course-changing mode in way-points.

A course-keeping mode involves the use of a classical autopilot system, where the desired course is nearly steady and can be calculated using few simple methods (Fosen, 2002) depending on the course of the actual segment commonly with the small correction from cross-track error. The turning controller should take-over during the course-changing manoeuvre to keep the yaw rate at the desired value to switch fast and accurately from one trajectory segment to another.

The goal of the paper is to design course-keeping and turning controllers, which have high steering performance and are robust due to ship dynamics variations.

## 2. COURSE-KEEPING CONTROLLER

For the design of the course-keeping controller the simple and often-used model of a ship dynamics - Nomoto 1st order model is used (Fosen, 2002):

$$P(s)=\frac{\psi(s)}{\delta(s)}=\frac{k}{s(1+sT)} \tag{1}$$

where $\delta$ is the rudder angle, $\psi$ – the ship course, $T$, $k$ – the dynamics parameters.

The proposed control system is presented in figure 1 (Morawski and Pomirski, 2002). P(s) is the ship dynamics, and $G_P(s)$, $G_C(s)$ are two controllers to be designed. $\psi_R$ is the desired course, and $\psi$ is the ship actual course. $\delta_\psi$ is the commanded ruder angle actuator.

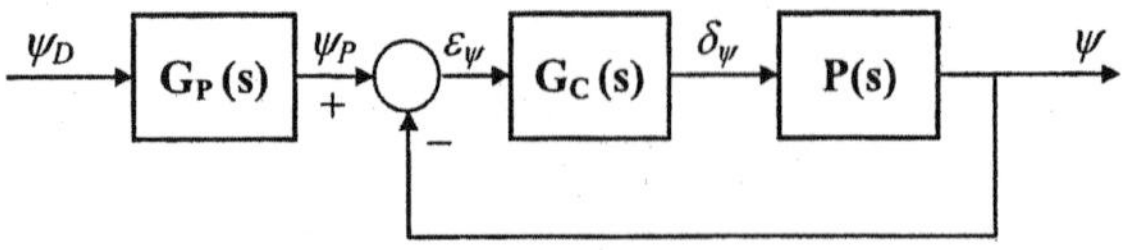


Fig. 1. Course-keeping control system for ship

A PID controller is often adopted in the ship autopilots so it is considered as $G_C(s)$. The popularity of PID controllers result from their functional simplicity and also from their robust performance in a wide range of operating conditions. Therefore the transfer function of $G_C(s)$ can be written as:

$$G_C(s) = H_1 + \frac{H_2}{s} + H_3 s = \\ = \frac{H_3 s^2 + H_1 s + H_2}{s} \quad (2)$$

where $H_1$, $H_2$, $H_3$ are coefficients for proportional, integral and differential parts of the controller.

The coefficients are selected using the performance specifications: the settling time and integral performance index ITAE. The ITAE index is the integral of time multiplied by absolute error:

$$\eta_{ITAE} = \int_0^T t \cdot |e(t)| \cdot dt \quad (3)$$

where $e_\psi(t) = \psi_D - \psi$.

Assuming that the closed-loop transfer function has the form:

$$T(s) = \frac{L(s)}{M(s)} = \\ = \frac{b_0}{s^n + b_{n-1}s^{n-1} + \ldots + b_1 s + b_0} \quad (4)$$

the ITAE optimum coefficients $b_0$, $b_1$, ..., $b_n$ which minimise the index for the unit step transient response can be calculated. For example for the 3rd order transfer function T(s) the denominator M(s) should be:

$$M(s) = s^3 + 1.75\omega_n s^2 + 2.15\omega_n^2 s + \omega_n^3 \quad (5)$$

where $\omega_n$ is a natural frequency (Dorf and Bishop, 1998).

For linear systems the natural frequency can be approximately solved using the 2% settling time $t_{s\psi}$ (time required for the system to settle within 2% of the final output signal) and the damping factor $\xi$:

$$\omega_n \approx \frac{4}{t_{s\psi}\xi} \quad (6)$$

In the designed control system the damping factor is unknown, but for ITAE optimal system is near $\xi = 0.8$. Therefore the approximation of $\omega_n$ has the form:

$$\omega_n \approx \frac{5}{t_{s\psi}} \quad (7)$$

The closed-loop transfer function of the control system (Fig.2) is:

$$T_\psi(s) = \frac{\psi(s)}{\psi_D(s)} = G_P \frac{G_C \cdot P}{1 + G_C \cdot P} = \\ = G_P \frac{\frac{k}{T}\left(H_3 s^2 + H_1 s + H_2\right)}{s^3 + s^2 \frac{1 + kH_3}{T} + s\frac{kH_1}{T} + \frac{kH_2}{T}} \quad (8)$$

The PID controller coefficients $H_1$, $H_2$, $H_3$ are selected to create the ITAE optimal transfer function denominator (5). Using the natural frequency approximation (7), the PID coefficients are selected as:

$$H_1 = \frac{53.75T}{kt_{s\psi}^2} \\ H_2 = \frac{125T}{kt_{s\psi}^3} \\ H_3 = \frac{8.75T - t_{s\psi}}{kt_{s\psi}} \quad (9)$$

The pre-filter $G_P(s)$ is formed in such manner that the closed-loop T(s) does not have any zeros, as required by equation (4):

$$G_P = \frac{H_2}{H_3 s^2 + H_1 s + H_2} = \\ = \frac{125T}{(8,75T - t_{s\psi})t_{s\psi}^2 s^2 + 25Tt_{s\psi}s + 125T} \quad (10)$$

## 3. TURNING CONTROLLER

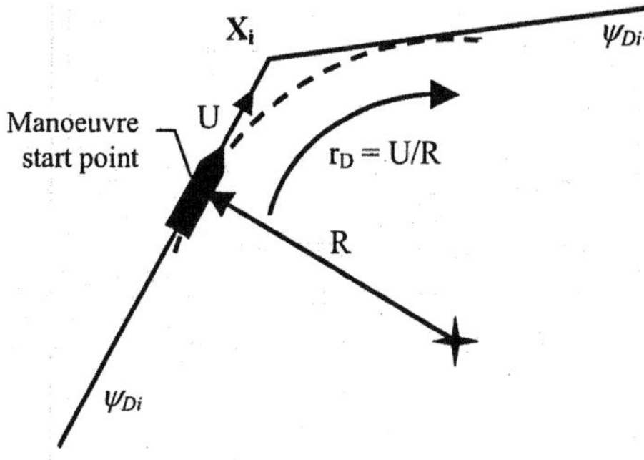


Fig.2. Passing the trajectory way-point

In trajectory turning points ship is usually required to follow the arc of desired radius R, which is tangential to both trajectory sections in the actual turning point $X_i$ (Fig. 2). Therefore during manoeuvre ship's yaw rate should be kept equal to:

$$r_D = \frac{U}{R} \tag{11}$$

where $U$ is the actual ship velocity, $R$ – the required turn radius.

The Nomoto 1st order model (1) is adequate for synthesis of the course-keeping controller, when the course error and the rudder angle are small, but when change of the course is large the ship dynamics is modelled better by Nomoto 2nd order model:

$$Q(s) = \frac{r(s)}{\delta(s)} = \frac{k_{II}(1+sT_0)}{(1+sT_1)(1+sT_2)} \tag{12}$$

where $\delta$ is the rudder angle, $r = d\psi/dt$ – the yaw rate, $k_{II}$, $T_0$, $T_1$, $T_2$ – the dynamics parameters. The parameters $T_0$, $T_1$, $T_2$ are usually related to the parameter T in the Nomoto 1st order model (1): $T \approx T_1 + T_2 - T_0$

The proposed turning control system is presented in figure 3. Q(s) is ship dynamics, and $K_P(s)$, $K_C(s)$ are two controllers to be designed. $r_D$ is the desired yaw rate and r is the ship actual yaw rate. $\delta_r$ is commanded ruder actuator.

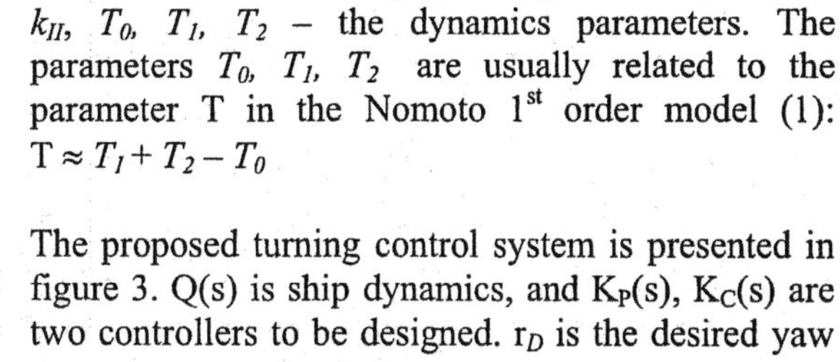

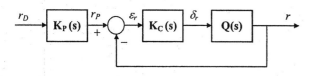

Fig. 3. Turning control system for ship

A PD controller is considered as $K_C(s)$. The integral action is omitted to simplify the resultant controller. Therefore the transfer function of $K_C(s)$ is:

$$K_C(s) = h_1 + h_3 s \tag{13}$$

where $h_1$, $h_3$ – are coefficients for proportional and differential part of the controller. The coefficients are selected using the performance indices: the settling time and integral performance index ITAE.

The closed-loop transfer function of the control system (Fig.2) is given by:

$$T_r(s) = \frac{r(s)}{r_D(s)} = K_P \frac{K_C \cdot Q}{1 + K_C \cdot Q} \tag{14a}$$

where:

$$\frac{K_C \cdot Q}{1 + K_C \cdot Q} = \frac{\dfrac{k_{II}(h_1 + h_3 s)(1 + sT_0)}{T_1 T_2 + k_{II} h_3 T_0}}{s^2 + s\dfrac{T_1 + T_2 + k_{II} h_3 + k_{II} h_1 T_0}{T_1 T_2 + k_{II} h_3 T_0} + \dfrac{1 + k_{II} h_1}{T_1 T_2 + k_{II} h_3 T_0}} \tag{14b}$$

Assuming that the closed-loop of the 2nd order transfer function $T_r(s)$ has the form (4) the denominator should be:

$$M(s) = s^2 + 1.4\omega_m s + \omega_m^2 \tag{15}$$

to minimise the ITAE index for the unit step transient response (Dorf and Bishop, 1998). The natural frequency $\omega_m$ can be approximately estimated using the 2% settling time $t_{s\psi}$ and the damping factor $\xi$ (assuming $\xi = 0.8$):

$$\omega_m \approx \frac{4}{t_{sr}\xi} \quad \Rightarrow \quad \omega_m \approx \frac{5}{t_{sr}} \tag{16}$$

The PD turning controller coefficients $h_1$, $h_3$ are selected to create the ITAE optimal transfer function denominator (15). Using the natural frequency approximation (16), the PD coefficients are selected as:

$$h_3 = \frac{(T_1 + T_2 - T_0)t_{sr}^2 - T_1 T_2 (7t_{sr} - 25T_0)}{k_{II} T_0 (7t_{sr} - 25T_0) - t_{sr}^2 k_{II}}$$

$$h_1 = \frac{25T_1 T_2 + 25k_{II} h_3 T_0 - t_{sr}^2}{k t_{sr}^2} \tag{17}$$

The pre-filter $G_P(s)$ is formed so that the closed-loop T(s) does not have any zeros, as required by (4):

$$K_P = \frac{1 + k_{II} h_1}{k_{II}(h_1 + h_3 s)(1 + sT_0)} \quad (18)$$

## 4. COMPUTER SIMULATIONS

For the computer simulations the Bech model of directionally unstable ship was used:

$$T_{1m} T_{2m} \ddot{r} + (T_{1m} + T_{2m})\dot{r} + H_B(r) = \\ = k_m(\delta + T_{0m}\dot{\delta}) \quad (19)$$

where δ is the actual rudder angle and:

$$H_B(r) = a_3 \cdot r^3 + a_2 \cdot r^2 + a_1 \cdot r + a_0 \quad (20)$$

is the nonlinear function describing the manoeuvring characteristics of the ship. For the directionally unstable vessels $a_1 = -1$ but for those stable ones $a_1 = 1$. $T_{0m}, T_{1m}, T_{1m}, k_m$ – are the model parameters.

δ is the output of the steering gear model where rudder angle and the rudder turn rate are restricted. The steering gear has also the dead zone, which cause the rudder fin insensitive to the small changes of the commanded rudder angle. Finally the model of steering gear is presented in the figure 3 (Amerongen, 1982).

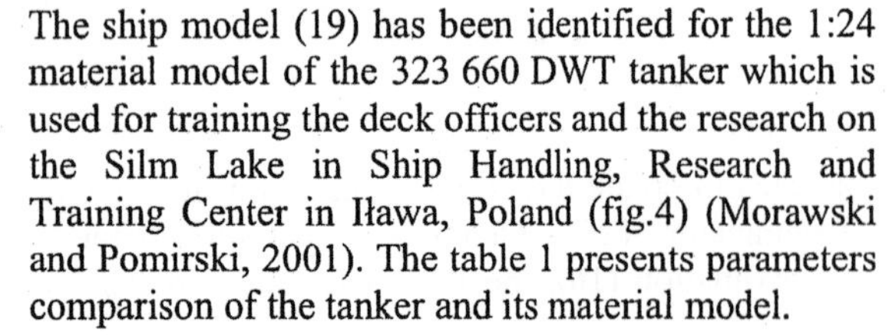

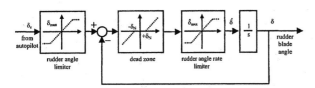

Fig. 3. Steering gear model used in simulations.

The ship model (19) has been identified for the 1:24 material model of the 323 660 DWT tanker which is used for training the deck officers and the research on the Silm Lake in Ship Handling, Research and Training Center in Iława, Poland (fig.4) (Morawski and Pomirski, 2001). The table 1 presents parameters comparison of the tanker and its material model.

Figure 5 presents the step response of the course-keeping controller. Small movements of rudder while in course-keeping mode are caused by steering gear dead zone. The step response of the turning controller is presented in figure 6. During both experiments the parameters in the model (19) are close to those used for the controllers' synthesis. Simulations show that the course-keeping and the turning controllers are robust for changes of ship dynamics. Numerous experiments with a modified ship model parameters provided similar results with almost the same values of settling time and corresponding shapes of the step response.

Table 1. Tanker and its material model

| *Main particulars* | *Tanker* | *Material model* |
|---|---|---|
| Length overall | 330.65 [m] | 13.78 [m] |
| Beam | 47.00 [m] | 2.38 [m] |
| Draft loaded condition | 20.60 [m] | 0.86 [m] |
| Displacement loaded | 323 660 [T] | 22.83 [T] |
| Speed | 15.2 [kn] | 3.1 [kn] |

Fig. 4. Blue Lady - material model of tanker

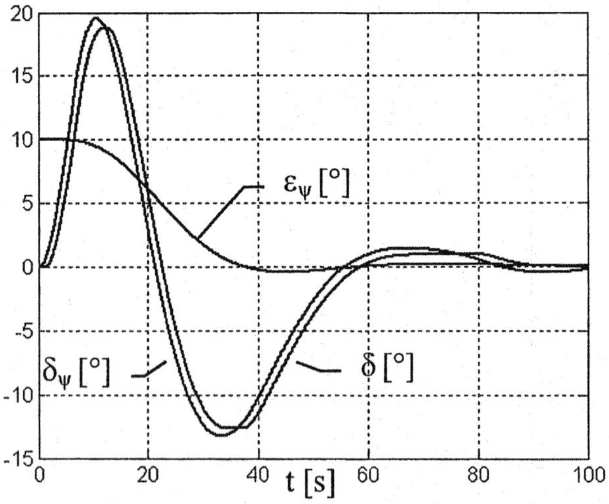


Fig.5. Step response of course-keeping autopilot $\Delta\psi_D = 10°$, $t_{s\psi} = 45s$.

Due to actuator limitation (restricted rudder angle) the proposed control system does not work properly when rudder saturates. Saving of the course controller from a big overshoot and a long settling time (fig.7) can be done by limiting the commanded turning rate to the value which does not cause the rudder saturation (fig.8 and fig.9).

In the turning controller the saturation appears when the desired yaw rate $r_D$ exceeds ±1.5°/s which is caused by the nonlinearity of the ship model (19) (fig.10), therefore the limiter for the required turn rate should be applied.

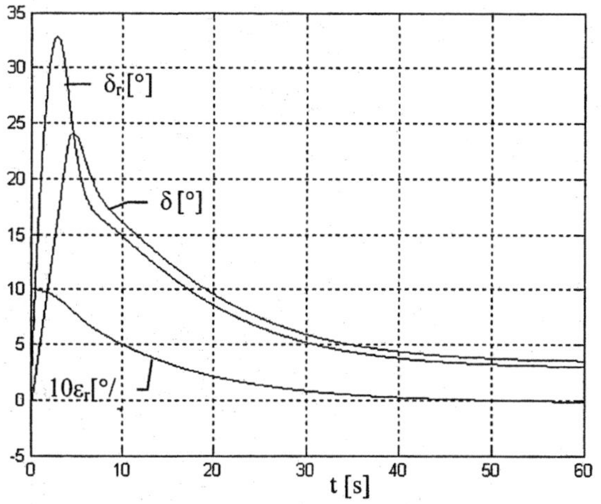


Fig.6. Step response of turning autopilot $\Delta r_D = 1$°/s, $t_{sr} = 25$s.

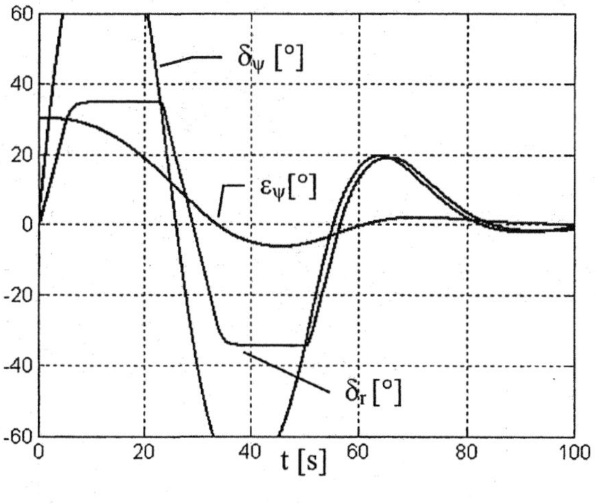


Fig.7. Overshoot and steering gear saturation during trial of course-keeping autopilot: $\Delta\psi_D = 30°$, $t_{s\psi} = 45$s.

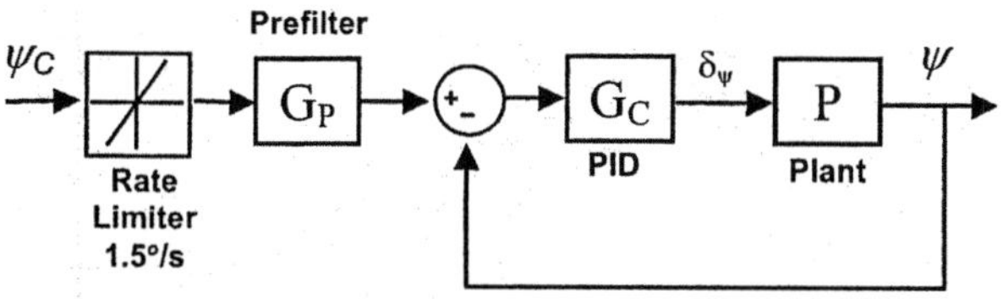


Fig.8. Final course-keeping ship autopilot.

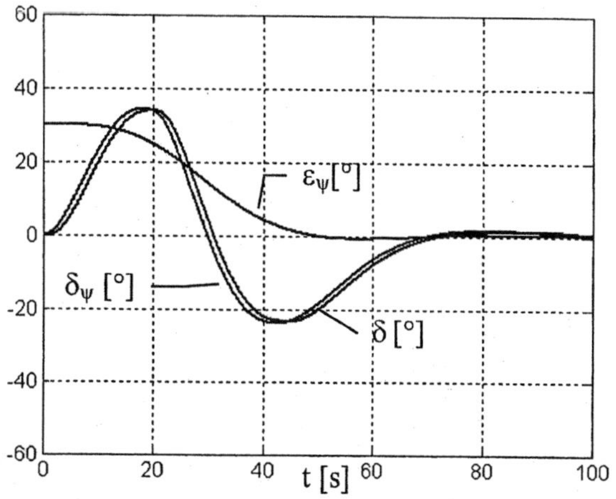


Fig.9. Step response of course-keeping autopilot with rate limiter *±1.5 °/s;* $\Delta\psi_D = 30°$, $t_{s\psi} = 45$s.

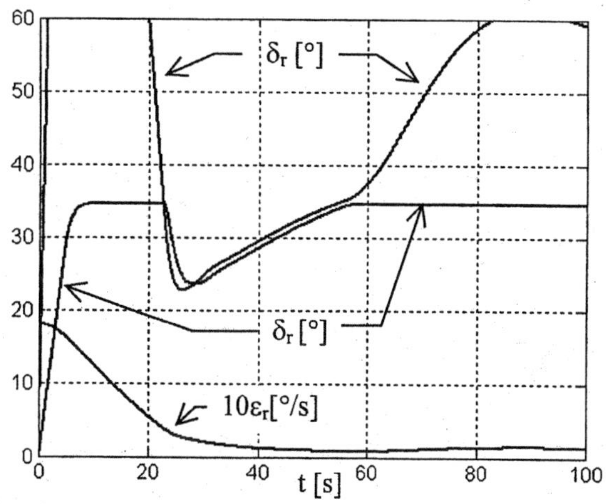


Fig.10. Staring gear saturation of the turning autopilot: $\Delta r_D = 1.8$°/s, $t_{sr} = 25$s.

## REFERENCES


Amerongen, J. (1982), *Adaptive Steering of Ships - A Model Reference Approach to Improved Maneuvering and Economical Course Keeping,* Ph.D. Thesis, Delft University of Technology.

Dorf R.C., Bishop R.K. (1998), *Modern Control Systems*, Addison-Wesley, Menlo Park.

Fosen T.I. (2002), *Marine Control Systems. Guidance, Navigation and Control of Ships, Rigs and Underwater Vehicles*, Marine Cybernetics, Trondheim.

Morawski L., Pomirski J. (2001), Identification and control of a direction unstable ship. *Problems of Nonlinear Analysis in Engineering Systems*, **1(13)**, vol.7, Kazan.

Morawski L., Pomirski J. (2002), Design of the robust PID course-keeping control system for ship, *Polish Maritime Research*, **1/2002**.

ELSEVIER
IFAC
PUBLICATIONS
www.elsevier.com/locate/ifac

# OPTIMAL DESIGN OF 2-DOF DIGITAL CONTROLLER FOR SHIP COURSE CONTROL SYSTEM

**J. Ladisch * K. Polyakov ** B.P. Lampe *,[1]**
**E.N. Rosenwasser ****

* *Department of Automation, University of Rostock, D-18051 Rostock, Germany.*
*FAX: +49 381/4983563, Phone: +49 381/4983564,*
*E-mail: bernhard.lampe@etechnik.uni-rostock.de*
** *Department of Automatic Control, State University of Ocean Technology, 190008 St. Petersburg, Russia.*

Abstract: A two-degrees-of-freedom (2-DOF) controller is designed for a ship course control systems using direct design methods based on the frequency-domain theory of linear sampled-data systems. The feedback controller was optimized to ensure minimal course fluctuation and rudder activity under stochastic wave disturbances, while the reference controller provides a quick transient response when the desired course is changed. The obtained 2-DOF controller is compared with an existing PID-controller, and its advantages over the latter are demonstrated by simulation. *Copyright © 2004 IFAC*

Keywords: Ship control, Waves, Disturbance signals, Tracking systems, Two-term-controllers, Sampled-data systems, Optimal systems

## 1. INTRODUCTION

The classical single loop control scheme with one controller has an important drawback, namely, there is a need to find a trade-off between good compensation of stochastic disturbances and small tracking errors. In particular, in the today's sea-going vessel shipping it is important to find a suitable compromise between a sufficiently fast course tracking response and an energetically favorable control signal acting on the maneuvering actuators of the ship (Majohr 1985). Two-degrees-of-freedom controllers were proposed as a way to avoid the above trade-off (Youla and Bongiorno 1976, Grimble 1994, Grimble 2001). Actually such a controller consists of two independent controllers, one of which determines the properties of the inner loop, i.e., stability, robustness and disturbance rejection, while the second one is designed with the only aim to improve the tracking performance. Most modern ship control systems are based on digital controllers. Since the plant to be controlled is a continuous-time object, the system as a whole should be investigated as a sampled-data system. Recently, direct design of digital controllers for sampled-data systems became an important topic in control theory (Chen and Francis 1995, Hagiwara and Araki 1995). One of the most promising approaches is based on linear theory of digital control in continuous time (Rosenwasser and Lampe 2000), which makes it possible to investigate sampled-data systems in the frequency domain and leads to sufficiently simple analysis and design algorithms (Rosenwasser *et al.* 1996, Rosenwasser *et al.* 1997). From numerical point of view, polynomial design algorithms developed in (Polyakov 1996, Polyakov 2001*b*) appeared to be most attractive. Based on this ideas, in (Polyakov 2001*a*, Polyakov *et al.* 2001) a polynomial method was proposed for

[1] B.P. Lampe is the corresponding author

design of optimal 2-DOF controllers for sampled-data systems. In the present report the design algorithm of (Polyakov 2001*a*, Polyakov *et al.* 2001) is applied to design a 2-DOF digital controller for ship course control. Numerical simulation shows that this controller is evidently superior to the PID-controller actually used on that ship.

## 2. DESIGN PRINCIPLE FOR 2-DOF CONTROLLER

Figure 1 shows the block diagram of the 2-DOF sampled-data system under consideration. The plant to be controlled (a ship) consists of two separate blocks with transfer functions $F_1(s)$ and $F_2(s)$. The course angle is denoted by $\psi$. The plant is acted upon by wave disturbances simulated as a stochastic signal at the output of a forming filter with transfer function $F_w(s)$ excited by zero-mean Gaussian white noise. The control loop includes also actuator (rudder machine) $H(s)$ and dynamic negative feedback (gyro-compass) $G(s)$. By $\delta$ we denote the rudder angle. Measurement noise is not taken into account. The digital controller consists of two parts: The feedback controller $C_0(\zeta)$ inside the loop, and the reference controller $C_1(\zeta)$. The controller $C_0(\zeta)$ must be designed to provide stability of the systems and minimal average variance of actuator movements under wave disturbances, while the reference controller $C_1(\zeta)$ should ensure good tracking capabilities in following a set-point signal $x$. Since the wave disturbance is a random process, it is reasonable to use a solution to the sampled-data $\mathcal{H}_2$-optimization problem (Rosenwasser and Lampe 2000, Rosenwasser *et al.* 1997, Polyakov 1996) as the feedback controller $C_0(\zeta)$. It is known (Rosenwasser and Lampe 2000), that under stationary exogenous disturbances all continuous-time signals in the system, including $\psi$ and $\delta$, are periodically nonstationary, and their variances $v_\psi(t)$ and $v_\delta(t)$, respectively, are periodic functions of time:

$$v_\psi(t) = v_\psi(t+T)\,, \quad v_\delta(t) = v_\delta(t+T)\,.$$

As measure for continuous-time performance, it was proposed in (Rosenwasser and Lampe 2000) to use the average variances:

$$\bar{v}_\psi = \frac{1}{T}\int_0^T v_\psi(t)\,dt\,, \quad \bar{v}_\delta = \frac{1}{T}\int_0^T v_\delta(t)\,dt\,.$$

In the simplest case, the quality criterion can be formed as

$$J_0 = \bar{v}_\psi + \varrho_0^2\,\bar{v}_\delta \to \min \tag{1}$$

where $\varrho_0^2$ is a nonnegative constant. A solution to this problem has been given in (Rosenwasser and Lampe 2000, Rosenwasser *et al.* 1997) on the basis of the parametric transfer functions (PTF) concept. Computational algorithms are developed in (Polyakov 1996) and realized in the *DirectSD Toolbox* for MATLAB, see (Polyakov *et al.* 1999). The reference controller should provide for good tracking capabilities, therefore, a suitable cost function is the integral quadratic error between response of actual system and that of some ideal system to a set-point signal. Let the reference signal $x(t)$ have the Laplace image $X(s)$. Then, we specify the desired signals $\psi(t)$ and $\delta(t)$ by transfer functions $Q_\psi(s)$ and $Q_\delta(s)$ of ideal operators such that the Laplace images of the ideal signals are given by $Q_\psi(s)X(s)$ and $Q_\delta(s)X(s)$, respectively. It should be noted that the matrices $Q_\psi(s)$ and $Q_\delta(s)$ must be chosen in a special way in order to guaranty the existence of a solution. This topic is analyzed more detailed in (Polyakov 2001*b*). Denote the output and control error signals by

$$\varepsilon_\psi(t) = \psi(t) - \psi(t)\,, \quad \varepsilon_\delta(t) = \delta(t) - \delta(t)\,.$$

Then, to evaluate tracking performance we employ the cost function

$$J_1 = \int_0^\infty \varepsilon_\psi^2(t) + \varrho_1^2\varepsilon_\delta^2(t)\,dt = J_{1\psi} + \varrho_1^2 J_{1\delta}\,. \tag{2}$$

The reference controller $C_1(\zeta)$ must be designed in such a way that the whole system admits a stable realization and the cost function (2) is minimal. A parameterization of all admissible controllers $C_1(\zeta)$, as well as a polynomial algorithms solving this problem, are given in (Polyakov 2001*a*, Polyakov *et al.* 2001) and realized in the *DirectSD Toolbox* for MATLAB.

## 3. CONTROLLER DESIGN FOR A MARITIME APPLICATION

Let a 1st order Nomoto model (K. Nomoto and Hiramo 1957) be used to describe a container ship

$$\dot{\psi} = r \tag{3}$$

$$\dot{r} = -\frac{1}{T_N}r + \frac{K_{IS}}{T_N}\delta\,, \qquad \delta = \delta_0 - b \tag{4}$$

where $r$ is the rate of turn, and $\psi$ is the course angle. Further, there are the main time constant $T_N$ and the rudder to yaw gain constant $K_{IS}$. The rudder bias $b$ is needed to counteract slowly-varying moments on the ship due to wave drift forces, LF-wind, and ocean currents (Fossen 2002). Consequently, the bias term $b$ ensures that $\delta_0 = b$ gives $r = 0$ and $\psi = \text{const}$ in steady-state. Based on the state space model (3), (4), we get the Nomoto rudder to course transfer function.

$$F(s) = \frac{\psi(s)}{\delta(s)} = \frac{K_{IS}}{s(T_N s + 1)} \tag{5}$$

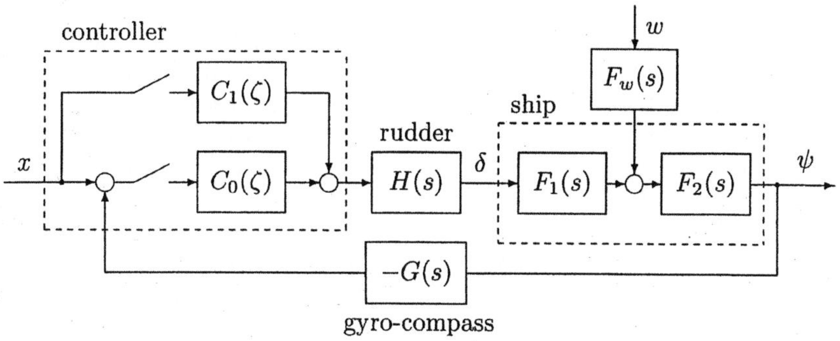


Figure 1. Sampled-data ship course control system with 2-DOF controller

For a velocity of $u = 20$ knots the considered container ship has the following characteristic data: $K_{IS} = 0.0694\,\text{sec}^{-1}$, $T_N = 18.2\,\text{sec}$. Furthermore, the linear wave-model

$$\dot{\xi}_w = r_w$$
$$\dot{r}_w = -\omega_0^2 \xi_w - 2\lambda\omega_0 r_w + K_w w$$

was used to model the wave response according to a Pierson-Moskowitz spectrum or a *JONSWAP* spectrum. The process noise term $w$ is modelled as zero mean Gaussian white noise process. Based on this model a second order wave response transfer function approximation can be calculated (Fossen 2002) and we can set the stable forming filter

$$F_w(s) = \frac{r_w(s)}{w(s)} = \frac{K_w s}{s^2 + 2\lambda\omega_0 s + \omega_0^2}.$$

The linear wave response model was taken for fully developed sea. Normally, a strong wind which has lasted for a longer time creates such a fully developed sea. After the wind has stopped, a low frequency decaying sea or swell is formed. These long waves form a wave spectrum with a low peak frequency (Fossen 2002). For this case a representative disturbance forming filter is chosen for calculating the controllers. According the modified Pierson-Moskowitz spectrum the following disturbance forming filter data are chosen: $\omega_0 = 0.3\,\text{rad}\cdot\text{sec}^{-1}$, $\lambda = 0.3$, $\sigma = 0.725$ , $K_w = 2\lambda\omega_0\sigma$. By combining the ship and wave models, the rate of turn equation can be expressed by the sum (Fossen 2002):

$$r_s = r + r_w(s).$$

Therefore, we can split the ship transfer function $F(s)$ into the following two transfer functions

$$F_1(s) = \frac{r(s)}{\delta(s)} = \frac{K_{IS}}{T_N s + 1}, \quad F_2(s) = \frac{\psi(s)}{r(s)} = \frac{1}{s}.$$

The compass measuring system can be modelled as a first order dynamic system with transfer function

$$G(s) = \frac{K_{FB}}{T_{FB} s + 1}$$

where $K_{FB} = 1$ and $T_{FB} = 6$ sec (Majohr 1979).

The complete rudder machine of the ship is approximated by a 1st-order system

$$H(s) = \frac{K_{RU}}{T_{RU} s + 1}$$

where $K_{RU} = 1$ and $T_{RU} = 2$sec (Majohr 1979). In the ship guidance system of the considered real container ship the following PID-controller was found

$$G_{PID}(s) = K \quad 1 + \frac{sT_D}{1 + sT_{VD}} + \frac{1}{sT_J} \qquad (6)$$

with the parameters $K = 0.8$, $T_J = 1000\,\text{sec}$, $T_D = T_N$, $T_{VD} = 1\,\text{sec}$. The reference signal $x(t)$ is chosen to be the unit step applied at $t = 0$, so that $X(s) = 1/s$. Then, we take yaw and rudder angle transient processes in the existing system with controller (6) as ideal output and control signals, because this controller corresponds to the nautical requirements regarding its behavior. This gives

$$Q_\psi(s) = \frac{0.03622s^2 + 0.001526s + 2 \cdot 10^{-6}}{s^5 + 1.6s^4 + 0.6s^3 + 0.06s^2 + 0.002s + 2 \cdot 10^{-6}}$$

$$Q_\delta(s) = \frac{9.51s^4 + 0.92s^3 + 0.023s^2 + 2.7 \cdot 10^{-5}s}{s^5 + 1.6s^4 + 0.6s^3 + 0.06s^2 + 0.002s + 2 \cdot 10^{-6}}.$$

The weighting coefficients $\varrho_0^2$ and $\varrho_1^2$ were set during experimental investigations under view of the nautical requirements to a course control system. The value $\varrho_0^2 = 4$ for calculation of the inner controller $C_0(\zeta)$, in order to weights the rudder control signal more strongly, to ensure an energetically favorable compensating behavior of stochastic disturbances. A weighting of the course signal would be rather senseless, because it is hardly possible to effect the stochastic ship course movement anyway. The second weighting coefficient was set to $\varrho_1^2 = 4 \cdot 10^{-6}$ for calculation of the reference controller $C_1(\zeta)$, because now in particular an accurate and dynamic tracking behavior is desired, whereas the process of the rudder moving behavior is of smaller importance only. Using the ideas described in Sec. 2, for a sampling period $T = 2$ sec we calculated, using *DirectSD* toolbox for MATLAB, the following controllers $C_0(\zeta)$ and $C_1(\zeta)$

$$C_{0,1}(\zeta) = \frac{b_p \cdot \zeta^p + b_{p-1} \cdot \zeta^{p-1} + \cdots + b_0}{a_p \cdot \zeta^p + a_{p-1} \cdot \zeta^{p-1} + \cdots + a_0} \quad (7)$$

The parameters of the controller $C_0(\zeta)$ are

| $n$ | $b_n$ | $a_n$ |
|---|---|---|
| 0 | 0.0041 | 0.0609 |
| 1 | −0.0239 | 0.3292 |
| 2 | 0.0491 | −0.0516 |
| 3 | −0.0459 | −1.3326 |
| 4 | 0.0169 | 1.0000 |

and those of $C_1(\zeta)$ are:

| $n$ | $b_n$ | $a_n$ |
|---|---|---|
| 0 | 0.0002 | −0.00002 |
| 1 | −0.0097 | −0.0001 |
| 2 | −0.0540 | −0.0004 |
| 3 | 0.0089 | 0.0130 |
| 4 | 0.6657 | 0.0531 |
| 5 | 1.5731 | −0.1423 |
| 6 | −4.9015 | −0.4741 |
| 7 | −8.1924 | 0.8706 |
| 8 | 30.120 | 0.8855 |
| 9 | −27.465 | −2.2047 |
| 10 | 8.2546 | 1.0000 |

The designed 2-DOF controller was tested using simulation under MATLAB/SIMULINK. For the given process parameters, these controllers ensure mean-square yaw angle $\overline{\sigma}_\psi = 0.98$ deg and mean-square rudder angle $\overline{\sigma}_\delta = 0.037$ deg. Under the same conditions the Tustin-transformed (Åström and Wittenmark 1997) existing PID-controller (6) gives $\overline{\sigma}_\psi = 1.08$ deg and $\overline{\sigma}_\delta = 4.73$ deg. That means that the mean-square rudder angle was reduced by the factor 127, whereas the mean-square yaw angle remains almost alike. Further, for the chosen reference signal $X(s) = 1/s$ the reference controller $C_1(\zeta)$ ensures an integral square course tracking error $J_{1\psi} = 48.29\,\text{deg}^2$ sec and a rudder angle tracking error $J_{1\delta} = 57.71\,\text{deg}^2$ sec. However, because the $\mathcal{H}_2$-optimal controller is optimized under the assumption that the exogenous signals in figure 1 are only *centered* stationary stochastic processes, it will not be able to compensate stationary disturbances, such as constant cross-winds (Korte *et al.* 2003). In order to guarantee sufficient compensation for such constant stationary disturbances, an additional discrete PI-controller is added *in parallel* to the calculated controller $C_0(\zeta)$ (Majohr 1979). In order to adapt the behavior of compensating such stationary disturbances to the appropriate behavior of the PID controller, the dimensioning of the additional PI-controller corresponds to the PID-controller. Therefore, the gain of the additional P-part is 0.8 and considering the poles of the closed loop system the time constant of the additional I-part is set to 500 sec. Consequently, thereby the reference controller $C_1(\zeta)$ must be computed again. Using the ideas described in Sec. 2, for sampling period $T = 2$ sec, and using the *DirectSD* Toolbox for MATLAB, we calculated two further controllers $C_0(\zeta)$ (extended with discrete PI-controller incorporated) and $C_1(\zeta)$ according to (7). The parameters of the extended controller $C_0(\zeta)$ are

| $n$ | $b_n$ | $a_n$ |
|---|---|---|
| 0 | −0.0524 | −0.0609 |
| 1 | −0.1839 | −0.2682 |
| 2 | −0.2311 | 0.3807 |
| 3 | 1.1092 | 1.2810 |
| 4 | −1.9209 | −2.3326 |
| 5 | 0.8169 | 1.0000 |

and those of $C_1(\zeta)$ are now:

| $n$ | $b_n$ | $a_n$ |
|---|---|---|
| 0 | 0.0002 | −0.00002 |
| 1 | −0.0096 | −0.0001 |
| 2 | −0.0533 | −0.0004 |
| 3 | 0.0011 | 0.0130 |
| 4 | 0.6318 | 0.0531 |
| 5 | 1.6948 | −0.1423 |
| 6 | −4.5454 | −0.4741 |
| 7 | −8.9256 | 0.8706 |
| 8 | 29.447 | 0.8855 |
| 9 | −25.696 | −2.2047 |
| 10 | 7.4546 | 1.0000 |

The designed 2-DOF controller was tested using simulation under MATLAB/SIMULINK and also under real time conditions on a xPC-target-platform on industrial PC-104-basis. Now, for the given process parameters, these controllers ensure mean-square yaw angle $\overline{\sigma}_\psi = 1.06$ deg and mean-square rudder angle $\overline{\sigma}_\delta = 0.78$ deg. To a repeated comparison: under the same conditions the Tustin-transformed existing PID-controller (6) gives $\overline{\sigma}_\psi = 1.08$ deg and $\overline{\sigma}_\delta = 4.73$ deg. Thus the mean-square rudder angle was reduced by the factor 6, whereas the mean-square yaw angle remains almost alike again. Further, for the chosen reference signal $X(s) = 1/s$ the reference controller $C_1(\zeta)$ ensures now an integral square course tracking error $J_{1\psi} = 5.50 \cdot 10^{-10}\,\text{deg}^2$ sec and a rudder angle tracking error $J_{1\delta} = 82.50\,\text{deg}^2$ sec. That means that the course tracking behavior could be very much improved in relation to the desired behavior of the PID controller, whereas the tracking behavior of the first 2-DOF-Design (without incorporated discrete PI-controller) is characterized by slower dynamics. In Figures 2-5 the ship course and rudder angle transient processes are represented for the same course change

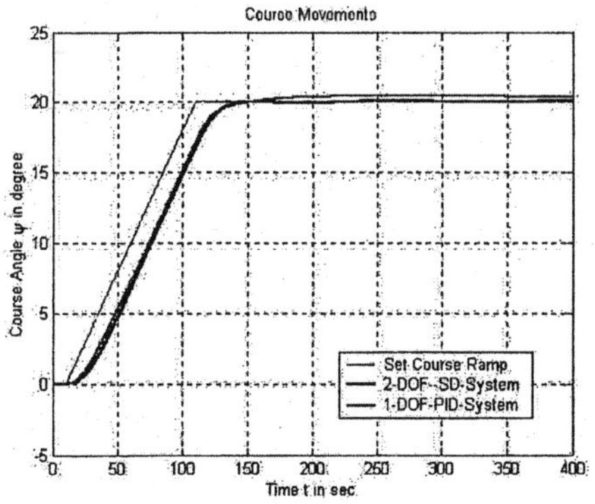


Figure 2. Course angle without disturbances

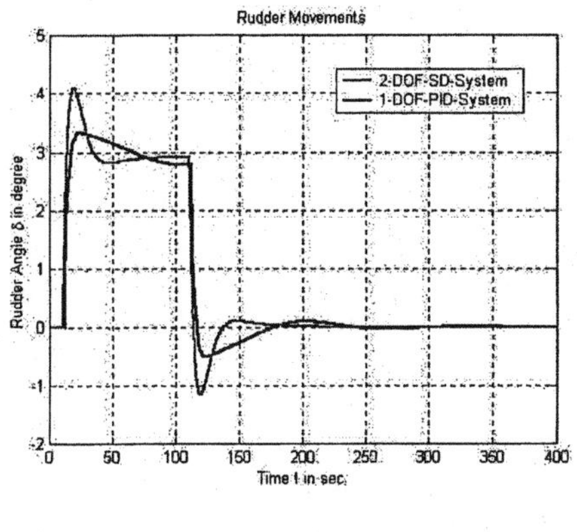


Figure 3. Rudder angle without disturbances

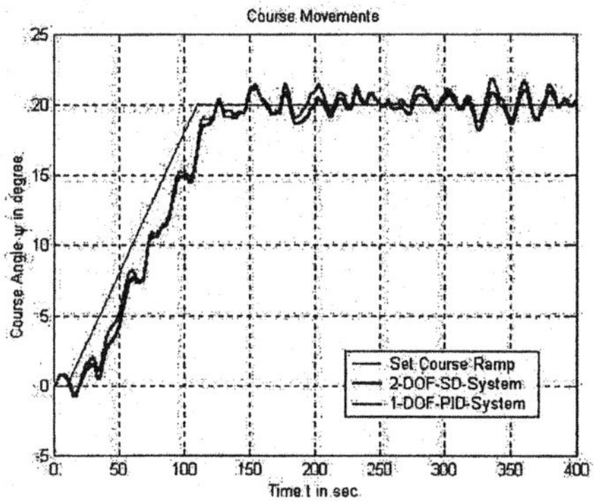


Figure 4. Course angle under disturbances

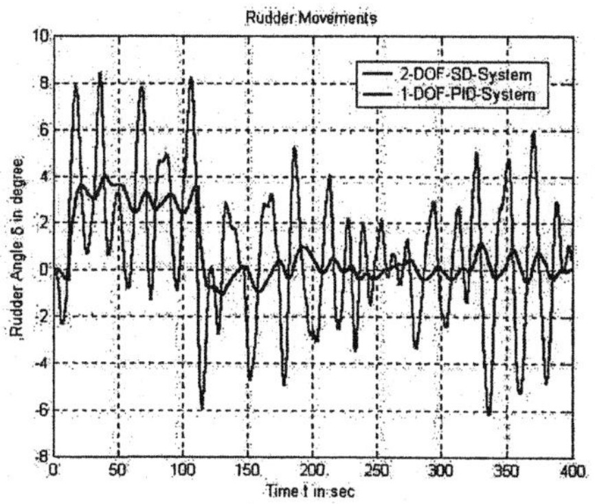


Figure 5. Rudder angle under disturbances

manoeuvre by $\Delta\psi = 20\,\text{deg}$ in 100 seconds. In order to examine compensating stationary disturbances, to the rate of turn of the ship (see

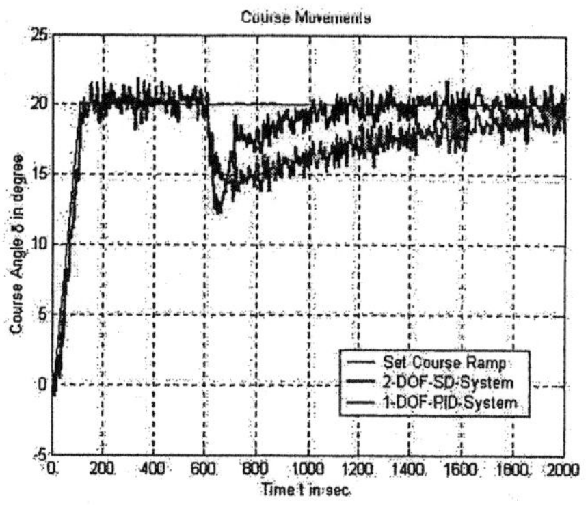


Figure 6. Course angle transients under stationary and stochastic disturbances

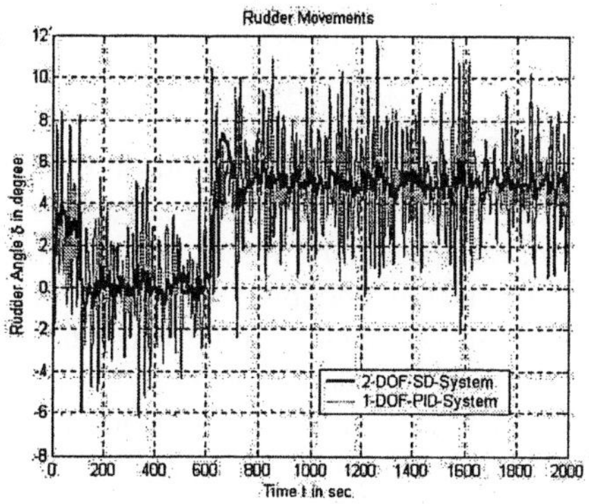


Figure 7. Rudder angle transients under stationary and stochastic disturbances

Figure 1) a constant value was added as step function, which corresponds to the effect of a constant rudder position of 5 degrees. In Figures 6 and 7 the ship course and rudder angle transient processes are represented for the already used course change manoeuvre by $\Delta\psi = 20\,\text{deg}$ in 100 seconds and for the given stationary disturbance under additional stochastic disturbing influence. The simulations under real time conditions on a xPC-target-platform was accomplished within a special test environment. This test environment consists of two separate PC's, where on the first PC the ship behavior is simulated (linear and nonlinear modelling), and on the second PC (industrial PC-104) a simplified ship guidance system with the standardized interfaces (NMEA 0183 interface standard) is realized. These simulations under real time conditions show that the new course controllers can be realized on this hardware basis very well. Moreover, the 2-DOF sampled-data system was tested with unchanged system parameters under changed disturbance parameters. For this purpose the peak frequency and the damping rate of the wave spectrum were changed from $\omega_0 = 0.3\,\text{rad} \cdot \text{sec}^{-1}$ up to $\omega_0 = 1.6\,\text{rad} \cdot \text{sec}^{-1}$ in accordance with the Pierson-Moskowitz spectral density. These additional tests resulted in similar and mostly even still better results concerning the average variances of the rudder movements. Of course the 2-DOF system was also examined regarding the robustness in relation to changes of parameters of the ship transfer function (5). For this purpose the parameters $K_{IS}$ and $T_N$ were changed within the simulation at the same time. In order to simulate a more agile ship, $K_{IS}$ was increased over up to 40 % and $T_N$ was reduced over up to 40 % at the same time and in order to simulate a slower-acting ship, $K_{IS}$ was reduced over up to 40 % and $T_N$ was increased over up to 40 % simultaneously. In all these cases the ship course processes and the rudder angle transient processes show a quite satisfying behavior. On basis of the pole positions for the closed loop, the stability for different ship parameters was proven.

## 4. CONCLUSIONS

The paper presents an application of the 2-DOF controller concept to the design of a course control system for a sea-going vessel. Using sampled-data theory makes it possible to take continuous-time performance of the system into account, while the chosen 2-DOF controller structure allows designers to meet energetically and navigationally justified control criteria at the same time. The design of the 2-DOF controller concept was performed using the *DirectSD* toolbox for MATLAB. The performance of the designed 2-DOF system was compared with that of an existing conventional PID-controller system. Using simulation, it was shown that control quality improves many times with the proposed 2-DOF controller. Simulations were performed under MATLAB/SIMULINK as well as under real time conditions on an xPC-target-platform on industrial PC-104-basis. Therefore, the developed 2-DOF controller can be used in existing ship guidance systems. Nevertheless, further investigations are necessary in order to reduce the controller order, particularly those of the reference controller.

## ACKNOWLEDGEMENT

The work has been supported by the German Federal Ministry of Education and Research. This support is very gratefully acknowledged.

## REFERENCES

Åström, K. J. and B. Wittenmark (1997). *Computer controlled systems: Theory and design.* 3 ed.. Prentice-Hall. Englewood Cliffs.

Chen, T. and B. A. Francis (1995). *Optimal Sampled-Data Control Systems.* Springer-Verlag. Berlin Heidelberg New York.

Fossen, T. (2002). *Marine Control Systems.* Marine Cybernetics. Trondheim, Norway.

Grimble, M. J. (1994). *Robust Industrial Control.* Prentice-Hall. UK.

Grimble, M. J. (2001). *Industrial Control Systems Design.* John Wiley. Chichester.

Hagiwara, T. and M. Araki (1995). FR-operator approach to the $\mathcal{H}_2$-analysis and synthesis of sampled-data systems. *IEEE Trans. Automat. Contr.* **AC-40**(8), 1411–1421.

K. Nomoto, T. Taguchi, K. Honda and S. Hiramo (1957). On the steering qualities of ships. *Int. Shipbuilding Progress* pp. 354–370.

Korte, H., J. Majohr, C. Korte, J. Ladisch, M. Wulff and B.P. Lampe (2003). Current influence on ship motions – linear approach. In: *Proc. 13. Int. Ship Control Syst. Symp.*. number 132. Orlando, Florida.

Majohr, J. (1979). *Manual of Navigation - Technical Systems.* Berlin: transpress Verlag für Verkehrswesen. [in german].

Majohr, J. (1985). Mathematical model concept for course and track controlled vessels. *Schiffbauforschung* **24**(2), 75–89. [in german].

Rosenwasser, E.N., K.Y. Polyakov and B.P. Lampe (1996). Entwurf optimaler Kursregler mit Hilfe von Parametrischen Übertragungsfunktionen. *Automatisierungstechnik* **44**(10), 487–495.

Polyakov, K. (1996). Polynomial design of sampled-data systems. I. Quadratic optimization. (14), 76–89. [in russian].

Polyakov, K. (2001*a*). Design of optimal 2-dof sampled-data systems. *Automation and Remote Control* (6), 85–94.

Polyakov, K. (2001*b*). Polynomial design of optimal sampled-data tracking systems. I. Quadratic optimization. *Automation and Remote Control* (2), 149–102.

Polyakov, K.Y., E.N. Rosenwasser and B.P. Lampe (1999). DirectSD - a toolbox for direct design of sampled-data systems. In: *Proc. IEEE Intern. Symp. CACSD'99.* Kohala Coast, Island of Hawai'i, Hawai'i, USA. pp. 357–362.

Polyakov, K.Y., E.N. Rosenwasser and B.P. Lampe (2001). Optimal design of 2-DOF sampled-data systems. In: *Proc. 13th Int. Conf. Process Control.* Strbske Pleso, SK.

Rosenwasser, E.N. and B.P. Lampe (2000). *Computer Controlled Systems: Analysis and Design with Process orientated Models.* Springer-Verlag. London.

Rosenwasser, E.N., K.Y. Polyakov and B.P. Lampe (1997). Frequency domain method for $\mathcal{H}_2$–optimization of time-delayed sampled-data systems. *Automatica* **33**(7), 1387–1392.

Youla, D.C. and J. J. Jr. Bongiorno (1976). A feedback theory of two-degree-of-freedom optimal Wiener-Hopf design. *IEEE Trans. Automat. Contr.* **AC-30**(7), 652–655.

ELSEVIER
IFAC
PUBLICATIONS
www.elsevier.com/locate/ifac

# NEURAL NETWORK AUGMENTED IDENTIFICATION OF UNDERWATER VEHICLE MODELS

**Pepijn W.J. van de Ven [1] Tor A. Johansen[2]**
**Asgeir J. Sørensen[3] Colin Flanagan[1] Daniel Toal[1]**

[1] *Department of Electronic and Computer Engineering*
*UL, Limerick, Ireland*
[2] *Department of Engineering Cybernetics*
*NTNU, Trondheim, Norway*
[3] *Department of Marine Technology*
*NTNU, Trondheim, Norway*
*pepijn.vandeven@ul.ie, tor.arne.johansen@itk.ntnu.no,*
*asgeir.sorensen@marin.ntnu.no, colin.flanagan@ul.ie,*
*daniel.toal@ul.ie*

Abstract: In this article the use of neural networks in models for underwater vehicles is discussed. Rather than using a neural network parallel to the known model to account for unmodeled phenomena in a model wide fashion, knowledge regarding the various parts of the model is used to apply neural networks for those parts of the model that are most uncertain. As an example, the damping of an underwater vehicle is identified using neural networks. The performance of the neural network based model is demonstrated for an AUV that changes its physical characteristics during a simulated intervention operation. *Copyright ©2004 IFAC*

Keywords:
Autonomous vehicles, Backpropagation, Marine systems, Neural networks, Nonlinear systems, System identification, Time-varying systems

## 1. INTRODUCTION

In recent years highly sophisticated non-linear control schemes for marine vehicles have been developed and implemented. Although modeling of marine vehicles is widely addressed, several parameters still pose uncertainties due to the absence of accurate models. Certain model parameters can be determined analytically. Other parameters, however, will need to be determined using numerical methods or identified using (scaled) model or full scale tests. Both methods can be time consuming and expensive. On top of this, many of the parameters are highly dynamic. Of prime importance in this context is the dependence of many hydrodynamic parameters and coefficients on varying velocity regimes, proximity to the sea bed, sea surface and other structures, just to mention a few. At present, models are normally only valid for a limited region of operational conditions. To overcome these problems, neural networks can be used as they offer a means of parameter identification without the necessity of detailed model knowledge. As a result they can identify the parameters of interest over the full region of operation. The application of neural networks for control and modeling (Narendra and Parthasarathy, 1990) has been given considerable attention in recent years. The reader is referred to Van de Ven *et al.* (2003) and the references therein for research endeavours applying neural networks in the field of underwater vehicle control. Rather than using a neural network to approximate unmodeled phenomena in parallel to the total vehicle model, the use of neural networks to model specific parameters in the model, is proposed. Although this approach is not new, see e.g. Psichogios and Ungar (1992) and Thompson and Kramer (1994), little attention is paid to it in the underwater vehicle literature.

## 2. ROV KINEMATICS, DYNAMICS AND HYDRODYNAMICS

In Fossen (2002) and Sørensen and Ronæss (2000) it was shown that the non-linear dynamic equations of motion of a marine vehicle in six degrees of freedom can be expressed in vector notation as:

$$\mathbf{M}\dot{\boldsymbol{\nu}} + \mathbf{C}(\boldsymbol{\nu})\boldsymbol{\nu} + \mathbf{D}(\boldsymbol{\nu})\boldsymbol{\nu} + \mathbf{g}(\boldsymbol{\eta}) = \boldsymbol{\tau} \tag{1}$$

with the kinematic equation

$$\dot{\boldsymbol{\eta}} = \mathbf{J}(\boldsymbol{\eta})\boldsymbol{\nu}, \tag{2}$$

relating the linear and angular velocity in the Earth-fixed reference frame and the body-fixed reference frame, where:

| | | |
|---|---|---|
| $\boldsymbol{\eta}$ | = | position and orientation of the vehicle in the Earth-fixed frame |
| $\boldsymbol{\nu}$ | = | linear and angular velocity of the vehicle in the body-fixed frame |
| $\mathbf{M}$ | = | inertia matrix including added mass |
| $\mathbf{C}(\boldsymbol{\nu})$ | = | matrix consisting of Coriolis and centripetal terms |
| $\mathbf{D}(\boldsymbol{\nu})$ | = | matrix consisting of damping or drag terms |
| $\mathbf{g}(\boldsymbol{\eta})$ | = | vector of restoring forces and moments due to gravity and buoyancy |
| $\boldsymbol{\tau}$ | = | vector of control inputs |

In (1) it is assumed that no water current is present. Introducing the latter with velocity $\boldsymbol{\nu}_c$ results in the added mass contribution to the Coriolis matrix and the damping matrix to be a function of the relative velocity defined as:

$$\boldsymbol{\nu}_r = \boldsymbol{\nu} - \boldsymbol{\nu}_c = [u\,v\,w\,p\,q\,r]^T - [u_c\,v_c\,0\,0\,0\,0]^T \tag{3}$$

For brevity, and without loss of generality, in this article the current velocity is assumed to be zero. A detailed derivation of the non-linear equations of motion can be found in Fossen (2002). Below a small summary of the model is given.

In the matrix $\mathbf{M}$ two inertial components are accounted for,

$$\mathbf{M} = \mathbf{M}_{RB} + \mathbf{M}_A. \tag{4}$$

The rigid body inertial matrix, $\mathbf{M}_{RB}$, represents the mass and inertia terms. Added mass is accounted for by the matrix $\mathbf{M}_A$.

For the matrix $\mathbf{C}(\boldsymbol{\nu})$, a similar discourse can be held. Both the Coriolis and the centripetal forces are functions of the rigid body mass and added mass and the velocity, $\boldsymbol{\nu}$.

$$\mathbf{C}(\boldsymbol{\nu}) = \mathbf{C}_{RB}(\boldsymbol{\nu}) + \mathbf{C}_A(\boldsymbol{\nu}). \tag{5}$$

$\mathbf{C}_{RB}(\boldsymbol{\nu})$ accounts for rigid body while $\mathbf{C}_A(\boldsymbol{\nu})$ accounts for the added mass.

In the damping matrix, $\mathbf{D}(\boldsymbol{\nu})$, four terms are combined:

$$\mathbf{D}(\boldsymbol{\nu}) = \mathbf{D}_P + \mathbf{D}_S(\boldsymbol{\nu}) + \mathbf{D}_W + \mathbf{D}_M(\boldsymbol{\nu}), \tag{6}$$

where:

| | | |
|---|---|---|
| $\mathbf{D}_P$ | = | potential damping |
| $\mathbf{D}_S(\boldsymbol{\nu})$ | = | linear and quadratic skin friction |
| $\mathbf{D}_W$ | = | wave drift damping |
| $\mathbf{D}_M(\boldsymbol{\nu})$ | = | damping due to vortex shedding |

Accurate calculation of these phenomena is difficult. Hence, often the damping is approximated by a diagonal matrix containing the linear and quadratic damping terms according to:

$$\begin{aligned}\mathbf{D}(\boldsymbol{\nu}) = &-diag\{X_u,\, Y_v,\, Z_w,\, K_p,\, M_q,\, N_r\} \\ &- diag\{X_{u|u|}|u|,\, Y_{v|v|}|v|,\, Z_{w|w|}|w|, \\ &K_{p|p|}|p|,\, M_{q|q|}|q|,\, N_{r|r|}|r|\}.\end{aligned} \tag{7}$$

Although (7) is a good approximation for decoupled motion, for manoeuvres involving movements along and about several body axes at a time, such simple models might prove to be insufficient.

## 3. IDENTIFICATION OF MODEL PARAMETERS

As mentioned, both $\mathbf{M}_{RB}$ and $\mathbf{C}_{RB}(\boldsymbol{\nu})$ can be calculated if the physical characteristics of the vehicle are known. However, proper determination of $\mathbf{M}_A$ (from which $\mathbf{C}_A(\boldsymbol{\nu})$ can be calculated) requires either the use of dedicated hydrodynamic software (Faltinsen, 1990), or using model or full scale tests to identify the added mass coefficients. In numerical calculations, the vehicle may be divided up into small sections and two dimensional added mass contributions are calculated for those sections. Consecutively, an integration over the whole body yields the three dimensional added mass parameters. In order to apply this method, which is called strip theory, the user is required to provide a detailed description of the vehicle in the form of a CAD drawing. On top of this, slender body theory should be assumed. This part of the modeling process alone can take up considerable time. For bluff bodies, however, other methods must be used. Measuring the added mass parameters using e.g. free decay tests is another option. Up to date there are, to the authors' knowledge, no methods available to perform those tests for all coupled six degrees of freedom simultaneously. On top of the above mentioned problems it should be kept in mind that no means of online updating of parameters is available from either method. This possibly even effects the analytically derived values for $\mathbf{M}_{RB}$ and $\mathbf{C}_{RB}(\boldsymbol{\nu})$ as the physical characteristics may change from mission to mission, or during one and the same mission, due to changing payloads. As an illustration to this, one might think of an ROV or an AUV sent on a salvage mission with the aim of lifting an object from the sea floor. Due to lifting the object, both mass and geometrical characteristics of the vehicle will change, thus changing $\mathbf{M}_{RB}$ and $\mathbf{C}_{RB}(\boldsymbol{\nu})$.

## 4. SYSTEM IDENTIFICATION USING NEURAL NETS

Neural networks can be applied both as control plant models and as controllers. Normally the neural network is used parallel to conventional models or controllers in a switching or output-blending fashion. In both cases, however, the neural networks make no, or only partial, use of the available *a priori* knowledge. Due to looking at the neural network as some non-linear mapping between the plant's input data and output data, knowledge regarding the dependence between parameters is lost. This might lead to an unnecessarily complicated function to be learned by the neural network. As briefly discussed in section 2 several (matrix) parameters related to the rigid-body dynamic equations of motion can be calculated accurately. It is thus intuitively appealing to use a neural network in parallel with a model of the known part of the system dynamics. To illustrate the effect of a neural network placed in parallel with the system model, the process dynamics are expressed in state-space form. To obtain a state space representation of the vehicle, (1) can be written in the following form:

$$\dot{\boldsymbol{\nu}} = \mathbf{M}^{-1}\left[\boldsymbol{\tau} - \mathbf{C}(\boldsymbol{\nu})\boldsymbol{\nu} - \mathbf{D}(\boldsymbol{\nu})\boldsymbol{\nu} - \mathbf{g}(\boldsymbol{\eta})\right]. \quad (8)$$

Taking the state vector to be $\boldsymbol{\nu}$ and the inputs to the system as: $\boldsymbol{\tau}(t)$ and $\boldsymbol{\eta}(t)$, (8) can be written in state-space form:

$$\begin{aligned} \dot{\boldsymbol{\nu}} &= \boldsymbol{\Phi}\left[\boldsymbol{\nu}(t), \boldsymbol{\tau}(t), \boldsymbol{\eta}(t)\right] \\ \mathbf{y} &= \boldsymbol{\Psi}\left[\boldsymbol{\nu}(t)\right], \end{aligned} \quad (9)$$

with $\boldsymbol{\Phi}\left[\boldsymbol{\nu}(t), \boldsymbol{\tau}(t), \boldsymbol{\eta}(t)\right]$ the right hand side of (8) and $\boldsymbol{\Psi}\left[\boldsymbol{\nu}(t)\right]$ simply $\boldsymbol{\nu}(t)$. Figure 1 shows the corresponding block diagram. Assuming that one has partial knowledge regarding the function $\boldsymbol{\Phi}$, a neural network can be used to model the unknown part of the system in parallel to the known part of the system as depicted in figure 2. In this approach one assumes: $\boldsymbol{\Phi} = \boldsymbol{\Phi}_M + \hat{\boldsymbol{\Phi}}$ where $\boldsymbol{\Phi}_M$ corresponds to the known part of $\boldsymbol{\Phi}$ and $\hat{\boldsymbol{\Phi}}$ to the unknown part approximated by the neural network. If the same assumption is made for the matrices $\mathbf{M}$, $\mathbf{C}(\boldsymbol{\nu})$, $\mathbf{D}(\boldsymbol{\nu})$ and $\mathbf{g}(\boldsymbol{\eta})$, (8) can be written as:

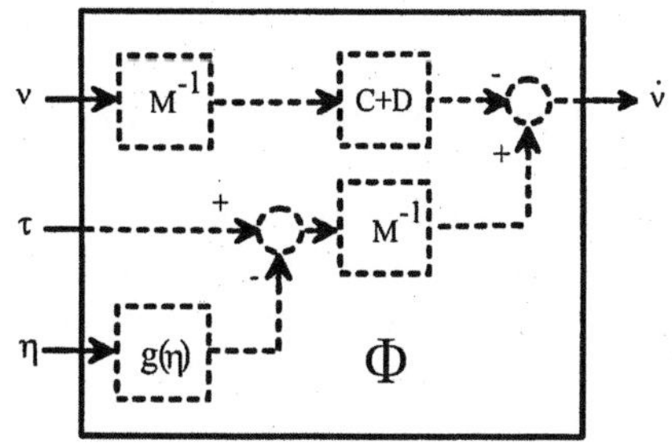


Fig. 1. State space representation of the non-linear dynamic equations of motion

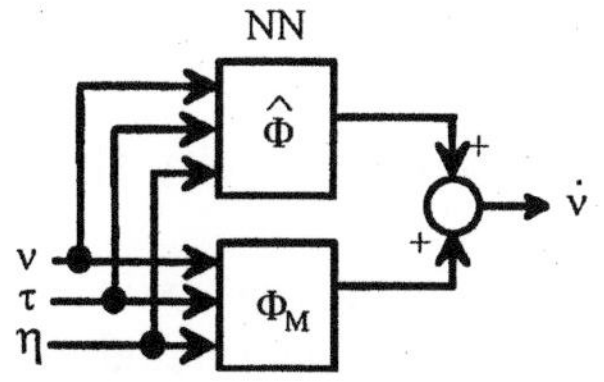


Fig. 2. State space representation of the non-linear dynamic equations with neural network in parallel to model unknown parameters

$$\begin{aligned} \dot{\boldsymbol{\nu}} = &\mathbf{M}_M^{-1}\left\{\boldsymbol{\tau} - \mathbf{C}_M(\boldsymbol{\nu})\boldsymbol{\nu} - \mathbf{D}_M(\boldsymbol{\nu})\boldsymbol{\nu} - \mathbf{g}_M(\boldsymbol{\eta})\right\} \\ &+ \mathbf{M}_M^{-1} \quad -\hat{\mathbf{C}}(\boldsymbol{\nu})\boldsymbol{\nu} - \hat{\mathbf{D}}(\boldsymbol{\nu})\boldsymbol{\nu} - \hat{\mathbf{g}}(\boldsymbol{\eta}) \\ &+ \hat{\mathbf{M}}^{-1} \quad \boldsymbol{\tau} - \mathbf{C}_M(\boldsymbol{\nu})\boldsymbol{\nu} - \hat{\mathbf{C}}(\boldsymbol{\nu})\boldsymbol{\nu} - \\ &- \mathbf{D}_M(\boldsymbol{\nu})\boldsymbol{\nu} - \hat{\mathbf{D}}(\boldsymbol{\nu})\boldsymbol{\nu} - \mathbf{g}_M(\boldsymbol{\eta}) - \hat{\mathbf{g}}(\boldsymbol{\eta}) \quad . \end{aligned} \quad (10)$$

The last three lines of (10) represent $\hat{\boldsymbol{\Phi}}$. These terms will be estimated by the neural network. Comparing this approach to the one depicted in figure 3 demonstrates that the required non-linear mapping is unnecessarily complicated. The neural networks in figure 3 namely model:

$$NN_1 = \hat{\mathbf{M}}^{-1};\ NN_2 = \hat{\mathbf{C}} + \hat{\mathbf{D}};\ NN_3 = \hat{\mathbf{g}}(\boldsymbol{\eta}), \quad (11)$$

which is a considerably easier task. Another advantage is that use can be made of known parameter features during the learning stage. Hence, the alternative configuration of the neural networks, as shown in figure 3 might prove worthwhile and is therefore used in this study.

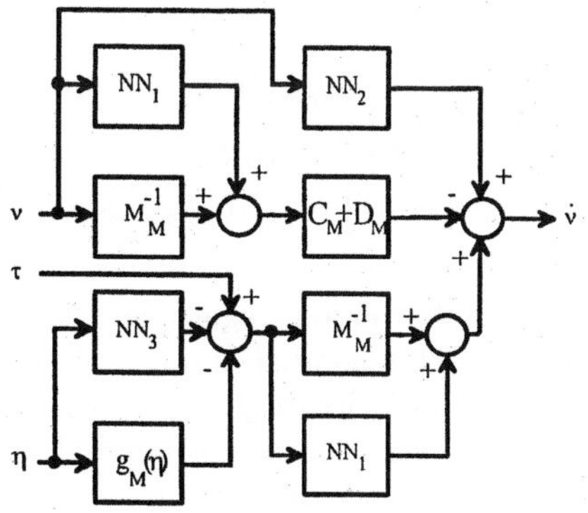


Fig. 3. State space representation of the non-linear dynamic equations with several neural networks to model unknown parameters

## 5. CASE STUDIES

In this section the viability of the above outlined method will be tested using various simulation setups. In all simulations, first offline learning was performed to initialise the neural networks. Afterwards, if improvements are expected, online training is performed. The neural networks have 12 input neurons, 5 neurons in one hidden layer and one linear output neuron. The other neurons use a hyperbolic tangent activation function. Both the hidden and the output layer use bias parameters. In the simulations the following assumptions are made:

- Acceleration and velocity of the vehicle can be measured.
- *Only* the damping matrix is unknown.
- The vehicle can be actuated in all degrees of freedom.
- Open loop simulations are performed.

The parameter that will be approximated by the neural network is the damping matrix. A first advantage of using a neural network is that strong assumptions, as discussed in Section 2 are not made. Higher order terms and coupling between various degrees of freedom can be taken into account by the neural network. To demonstrate this the damping matrix inhibits significant off-diagonal elements and non-linearities. In the first computer experiment, Section 5.1, the damping matrix will be approximated offline using velocity and acceleration data assuming zero measurement noise. Then, in Section 5.2, noise is added to the setup. Online learning will be used to decrease the state prediction error. Finally, in Section 5.3, the beneficial influence of online learning on changing vehicle dynamics will be demonstrated.

Assuming that only the damping matrix $\mathbf{D}(\boldsymbol{\nu})$ is unknown and hence, initially, not accounted for, the dynamic equation for the reference system or process plant model (ppm) and the initial approximate or control plant model (cpm) respectively can be written as:

$$\begin{aligned}\dot{\boldsymbol{\nu}}_{ppm} = &\mathbf{M}^{-1}[\boldsymbol{\tau} - \mathbf{C}(\boldsymbol{\nu}_{ppm})\boldsymbol{\nu}_{ppm} - \\ &\mathbf{D}(\boldsymbol{\nu}_{ppm})\boldsymbol{\nu}_{ppm} - \mathbf{g}(\boldsymbol{\eta}_{ppm})], \quad (12)\\ \dot{\boldsymbol{\nu}}_{cpm} = &\mathbf{M}^{-1}\ \ \boldsymbol{\tau} - \mathbf{C}(\boldsymbol{\nu}_{cpm})\boldsymbol{\nu}_{cpm} - \mathbf{g}(\boldsymbol{\eta}_{cpm})\ , \quad (13)\end{aligned}$$

To obtain information regarding the damping matrix, data is generated through simulation of the ppm model with a given input, and one-step-ahead predictions are computed from the cpm model. At every time step one assumes both systems have the same state vector: $\boldsymbol{\nu}_{ppm} = \boldsymbol{\nu}_{cpm} = \boldsymbol{\nu}$. Substituting $\boldsymbol{\nu}$ for $\boldsymbol{\nu}_{ppm}$ and $\boldsymbol{\nu}_{cpm}$ in equations 12 and 13, the difference between equations 12 and 13 becomes the damping matrix $\boldsymbol{D}(\boldsymbol{\nu})$, multiplied by the state vector $\boldsymbol{\nu}$ and the inverse of the mass matrix, $\boldsymbol{M}^{-1}$. The product $\boldsymbol{D}(\boldsymbol{\nu})\boldsymbol{\nu}$ can thus be calculated as:

$$\mathbf{D}(\boldsymbol{\nu})\,\boldsymbol{\nu} = \mathbf{M}\,[\dot{\boldsymbol{\nu}}_{cpm} - \dot{\boldsymbol{\nu}}_{ppm}]\,. \quad (14)$$

Equation (14) will be used for training of the neural networks and hence is the identification model.

### *5.1 Case Study I: Drag estimation using noise free signals*

To gather training data, a simulation of a trajectory using the process plant model is performed. The vehicle is actuated with $\boldsymbol{\tau} = [100\,100\,100\,10\,10\,10]^T$ for 15 seconds without performing control. After 15 seconds $\boldsymbol{\tau}$ is set to $[0\,0\,0\,0\,0\,0]^T$. As the vehicle used in this simulation is neutrally buoyant, its velocity gradually returns to $\boldsymbol{\nu} = [0\,0\,0\,0\,0\,0]^T$. Both velocity, i.e. the state of the system, and acceleration are recorded and consecutively used to train a neural network offline. As it is known that damping is a function of $\nu\,|\nu|$, the neural network has both $\nu$ and $|\nu|$ as inputs. The magnitudes of the velocities in the six degrees of freedom are not necessarily of the same order, which possibly leads to complications in the training algorithm. In neural networks trained with back propagation, neurons with smaller training signals tend to be dominated by neurons with larger training signals. As a result the former tend to learn slower and for those neurons the modeling error might thus not decrease properly. To prevent this, six separate networks were used for training. After identification of $\mathbf{D}(\boldsymbol{\nu})\boldsymbol{\nu}$ in (14) has been performed, the neural networks can be used in the model as shown in figure 4. The control plant model now becomes:

$$\dot{\boldsymbol{\nu}}_{cpm} = \mathbf{M}^{-1}\ \ \boldsymbol{\tau} - \mathbf{C}(\boldsymbol{\nu})\boldsymbol{\nu} - \mathbf{g}(\boldsymbol{\eta}) - \hat{\mathbf{D}}(\boldsymbol{\nu})\boldsymbol{\nu}\ . \quad (15)$$

where $\hat{\mathbf{D}}(\boldsymbol{\nu})\boldsymbol{\nu}$ indicates that this is the neural network approximation of $\mathbf{D}(\boldsymbol{\nu})\boldsymbol{\nu}$.

It should be noted that, unlike the other blocks, the block with caption $\hat{\boldsymbol{D}}(\boldsymbol{\nu})\,\boldsymbol{\nu}$ should not be interpreted as a multiplication of the inputs and the block's argument. Figure 5 shows open loop predictions made by the vehicle model using a neural network to model damping and the corresponding

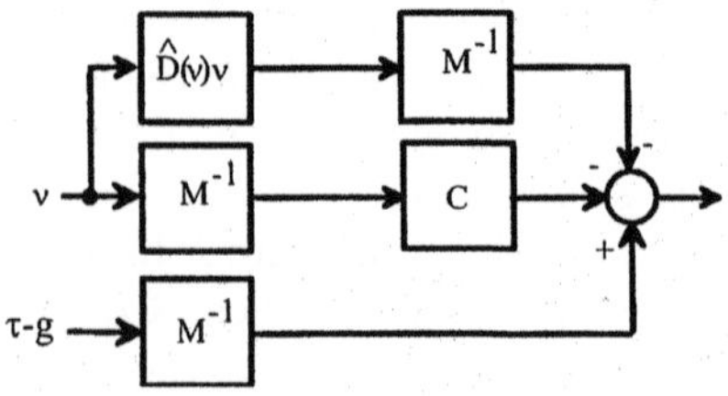


Fig. 4. Model of vehicle with the damping effects modeled by a neural network

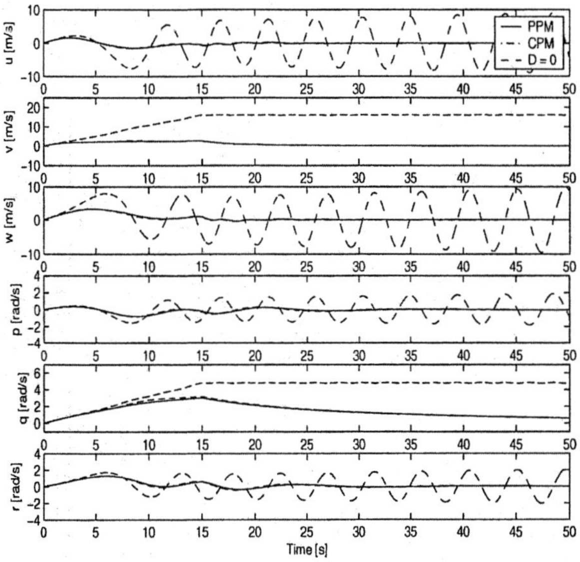


Fig. 5. Prediction of vehicle model using a NN for modeling of the damping

real trajectory. The graphs for the trajectory of the process plant model and the control plant model with neural network are lying on top of each other. It can thus be concluded that the neural networks are capable of approximating damping with sufficient accuracy for extended periods of time. To demonstrate the influence of damping on the vehicle model, figure 5 also shows the model output if the damping is simply taken equal to zero. This graph shows that, as expected, neglecting the damping leads to considerable errors in the state prediction.

### *5.2 Case Study II: Drag estimation under noisy conditions*

To investigate the ability of the neural network to perform estimation of the damping using noisy training signals, typical noise is added to the measurements. For the linear accelerations a white noise sequence with an amplitude of $0.5\,ms^{-2}$ is added. The angular accelerations are summed with a noise component of $0.05\,rads^{-2}$. The velocity measurements are assumed to be impeded with a white noise sequence, which amplitude is equal to 0.1 % of the magnitude of the velocity. It should be noted that in the simulations no noise prefiltering is assumed. This would make matters considerably easier for the neural network. Performance of the model is shown in figure 6. Clearly, the added noise affects the prediction abilities of the model. The prediction of the state can be improved by continuing the learning process online. The result of this experiment is shown in figure 7, demonstrating that online learning improves the state prediction considerably. However, in this case the neural network does not necessarily represent the damping any more. Online

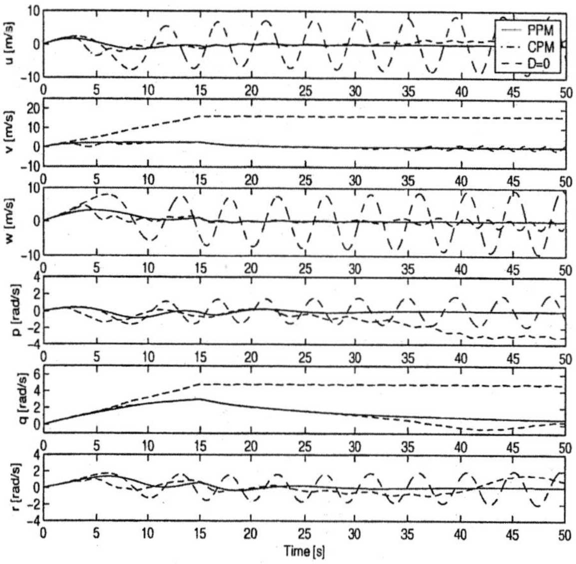


Fig. 6. Prediction of vehicle model using a NN for modeling of the damping under noisy conditions

learning results in a better local estimate of the damping as the neural network is now trained with larger weight on the latest data points. The global estimation accuracy may suffer from this.

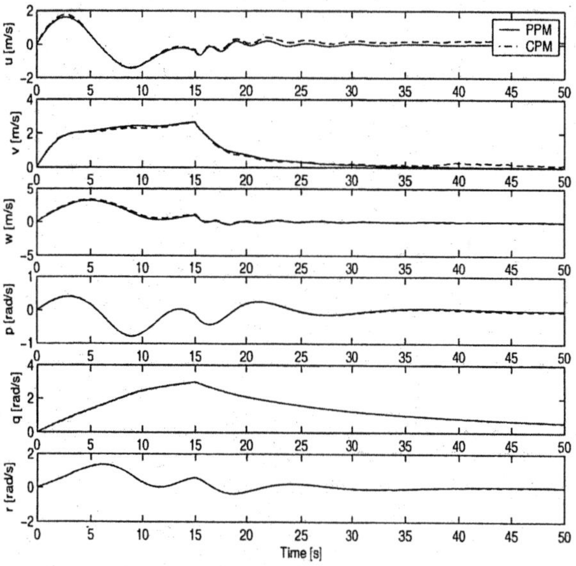


Fig. 7. Prediction of vehicle model using a NN for modeling of the damping under noisy conditions with online learning

### *5.3. Case Study III: Drag estimation for time-varying dynamics*

In this simulation an AUV with an offline trained neural network representing damping is sent on a mission. The mission is to pick up a cylinder, with a mass of 60 kg (while the vehicle mass is 95 kg) and transport it to its destination. This process starts at $t = 5\,s$ and results in several matrices to drastically be changed: changes in

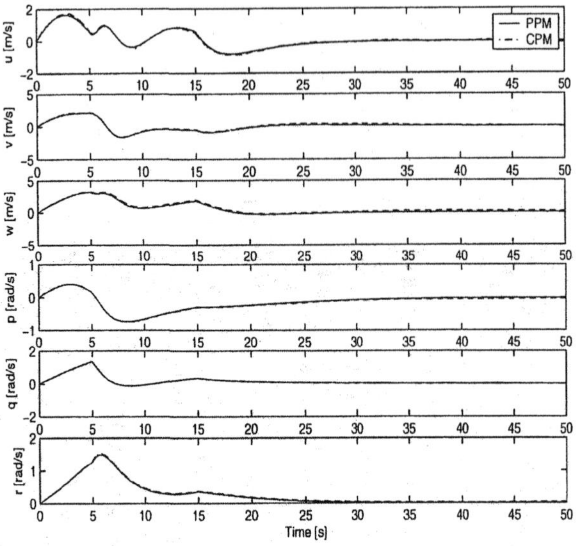


Fig. 8. Prediction of vehicle model using a NN for changing damping coefficients and using online learning

both the mass matrix, $\mathbf{M}$, and the Coriolis matrix, $\mathbf{C}(\nu)$, are assumed to be known. Changes in the damping matrix will be assumed to be unknown. Online learning is performed to adapt for the changing damping. Figure 8 shows the trajectories for the process plant model and the control plant model using the neural network to model damping. Although the change in damping is considerable with the various damping parameters changing between a factor 1.5 and 10, only small deviations from the ppm trajectory can be seen. The changes from the trajectory that would have occurred if no online updating of the neural network was performed, are shown in figure 9. Comparing these two figures demonstrates the benefit of online learning.

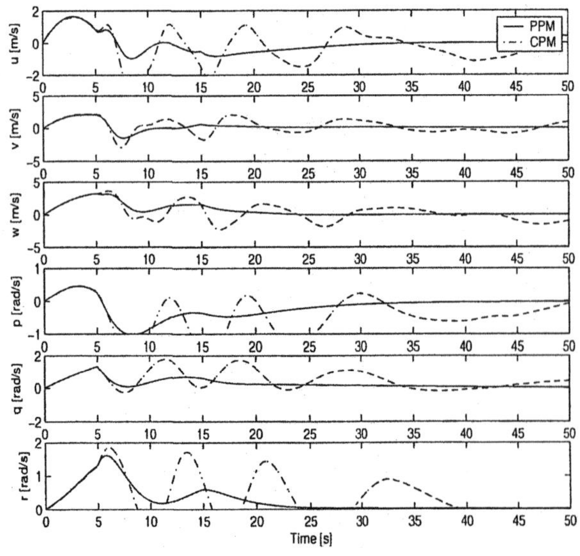


Fig. 9. Prediction of vehicle model using a NN without online learning while damping coefficients change

## 6. CONCLUSIONS

In this work, modeling of the damping of an underwater vehicle was performed using neural networks. Rather than using a neural network in parallel to the whole model (in an attempt to account for all unknown parameters in one neural network), neural networks are used to identify parts of the model that are known to be most uncertain. Under noisy conditions a neural network is trained to approximate the damping matrix. The results show that these neural networks can be used to improve the identification of poorly modeled phenomena. Online learning can be applied to alleviate the influence of noise and to adapt for changing parameters due to e.g. mission objectives. In this study, matters were simplified by assuming other parts of the model to be fully known. Future work will focuss on identification of several uncertainties at a time.

## 7. ACKNOWLEDGEMENT

This work was supported by the EC Research Directorates through a project at the Marie Curie Training Site, CyberMar at NTNU (HPMT-CT-2001-00382) and by the Irish Research Council for Science, Engineering and Technology: funded by the National Development Plan.

## REFERENCES

Faltinsen, O.M. (1990). *Sea Loads on Ships and Offshore Structures.* Cambridge University Press.

Fossen, T.I. (2002). *Marine Control Systems. Guidance, Navigation and Control of Ships, Rigs and Underwater Vehicles.* 1st ed.. Marine Cybernetics, AS.

Narendra, K.S. and K. Parthasarathy (1990). Identification and control of dynamical systems using neural networks. *IEEE Transactions on Neural Networks* **1**, 4–27.

Psichogios, D. C. and L. H. Ungar (1992). A hybrid neural network - first principles approach to process modeling. *AIChE J.* **38**, 1499–1511.

Sørensen, A. J. and M. Ronæss (2000). *The Ocean Engineering Handbook.* Chap. Mathematical modeling of dynamically positioned and thruster assisted anchored marine vessels, pp. 176–89. CRC Press.

Thompson, M. L. and M. A. Kramer (1994). Modeling chemical processes using prior knowledge and neural networks. *AIChE J.* **40**, 1328–40.

Van de Ven, P., C. Flanagan and D. Toal (2003). Artificial intelligence for the control of underwater vehicles. In: *Smart Engineering System Design, Proc. of the Artificial Neural Networks in Engineering Conf.* pp. 559–64.

ELSEVIER
IFAC
PUBLICATIONS
www.elsevier.com/locate/ifac

# OBSERVER KALMAN FILTER IDENTIFICATION OF AN AUTONOMOUS UNDERWATER VEHICLE

**A. Tiano** [*], **R. Sutton** [+], **A. Lozowicki** [§] and **W. Naeem** [+]

[*] *University of Pavia, Department of Information and Systems, Via Ferrata 1, 27100 Pavia, Italy*
[+] *University of Plymouth, Dept. of Mechanical and Marine Engineering, PL4 8AA, Plymouth, UK*
[§] *Technical University of Szczecin, ul. Piastow 17, PL-70313 Szczecin, Poland*

Abstract: This paper discusses the identification of linear multivariable models of an autonomous underwater vehicle. The OKID (Observer Kalman Identification) method is applied with the main objectives of verifying couplings between different motions and to evaluate its applicability to the design an LQG control system. The method is tested on the basis of experimental data obtained from experiments on an autonomous underwater vehicle. Some preliminary identification results are presented and commented. *Copyright © 2004 IFAC*

Keywords: system identification, parameter estimation, numerical methods, underwater vehicles.

## 1. INTRODUCTION

An increasing interest has been devoted in the recent years to the experimental determination of the dynamic behaviour of underwater vehicles by means of system identification methods. In many situations it is important that the mathematical models obtained by means of system identification methods can be directly used for control system design purposes.

This implies that linear models that take into account couplings between different motions should be determined. In fact, in such case linear control algorithms can be easily implemented and the effect of neglected dynamics can be minimized. An identification method that can cope with these requirements is OKID (Juang, 1994), that has proven to be also numerically very efficient and robust with respect to measurement noise. In this paper OKID will be applied to the depth dynamics of the autonomous vehicle Hammerhead (Naeem et Al., 2003) by using real sea trials data.

The paper is organized as follows. After the introduction, the underwater vehicle will be shortly described in section 2, where also the experiments conducted on the vehicle will be discussed. Section 3 is dedicated to the description of the identification method and section 4 is dedicated to the illustration of the preliminary identification results, while some concluding remarks are expressed in section 5.

## 2. UNDERWATER VEHICLE

The vehicle considered for identification , called *Hammerhead* (Naeem et Al., 2003), has a torpedo shaped body, is about 3.5 meters long and has a diameter of approximately 0.3 meters. The control surfaces are the two rear rudders for steering and two front hydroplanes for diving. These control surfaces are controlled by two separate onboard stepper motors and the signal to the stepper motors is sent through an umbilical cable attached to the rear end of the vehicle. The onboard sensors include an inertial navigation system (INS), a TCM2 compass, pressure sensor, global positioning system (GPS), and a shaft speed encoder. The data logged using the above mentioned sensors is summarized below:

- INS (heading, pitch, roll, linear and angular velocities)
- TCM2 compass (heading, pitch and roll angles)
- Pressure sensor (depth of the vehicle)
- GPS (cordinates of the vehicle on the sea surface, forward spewed)
- Shaft speed encoder ( vehicle speed)

The vehicle is connected through a rear umbilical cable to a control computer used to transmit various input/output signals. As a first attempt, the rudder/heading angles data were used to identify a longitudinal model, while the hydroplane angle/depth data were used to identify a lateral model (Naeem et Al. 2003).However, if cross couplings between yaw and roll and between depth and pitch result not negligible, coupled multivariable models hould be used.

During the experiments different multistep inputs were applied to the control surfaces and the output variables were recorded with a sampling frequency of 8 Hz. Some examples are shown in Fig. 1 and Fig.2.

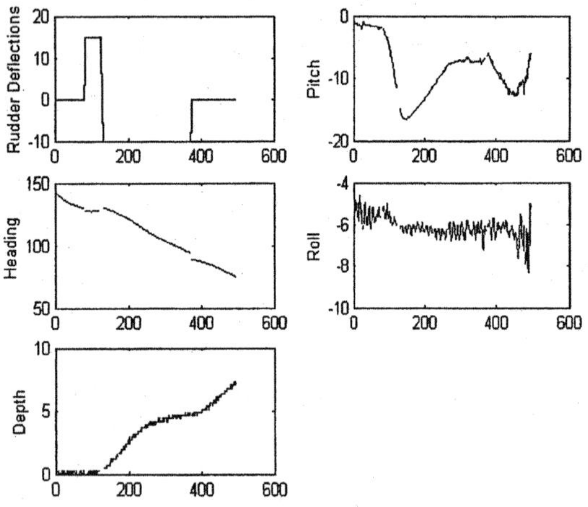


Fig 1. Example of input/output data after an identification experiment of depth dynamics.

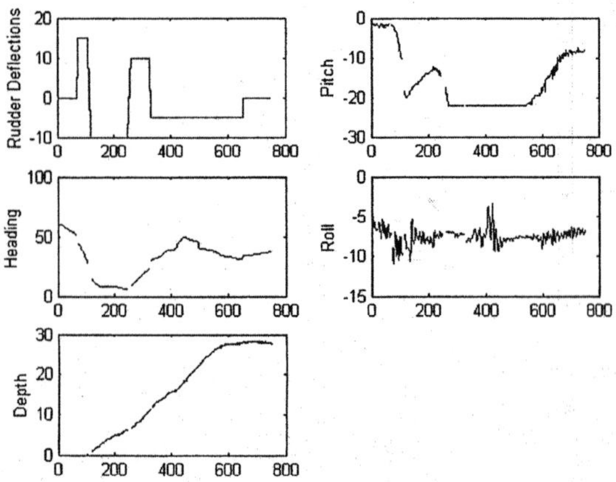


Fig 2. Example of input/output data after an identification experiment of depth dynamics.

The above figures are related to experiments aiming at the identification of the lateral dynamics, for determining the depth and pitch response to the sternplane deflections. As it can be observed , owing to the limitations of the software used for data acquisition, some missing data occurred. This inconvenience was solved by using an interpolation algorithm (Kybic et Al., 2001) deduced by a variational approach.

As it can be noticed, there exist indeed appreciable couplings between such motions. However, since this paper aims at the preliminary objective of verifying the suitability of OKID identification method for AUV dynamics, only depth response will be identified.

## 3. IDENTIFICATION METHOD

A linear time-invariant discrete time MIMO state space model describing expressing the relation between the input vector $u(k) \in R^m$, the output vector $y(k) \in R^p$ and the state vector $x(k) \in R^n$ can be expressed as :

$$\begin{aligned} x(k+1) &= Ax(k) + Bu(k) + v(k) \\ y(k) &= Cx(k) + Du(k) + e(k) \end{aligned} \tag{1}$$

Such equation includes also the effects of a disturbance vector $v(k)$ of measurement noise vector $e(k)$.

The Observer Kalman Filter Identification, (OKID) method (Juang,1994), attempts to identify, on the basis of input/output data, the smallest state space realization that is compatible with a given accuracy. For this purpose an input/output description of the above system is assumed of the form:

$$y(k+1) + \sum_{i=1}^{n} M_i^{(2)} y(k-i) = \sum_{j=1}^{n} M_j^{(1)} u(k-j) + \varepsilon(k) \tag{2}$$

where $M_i^{(2)} \in R^{pxp}, i = 1,..n$ and $M_j^{(1)} \in R^{pxm}$, $j = 1,...n$ are the Markov matrices for a model with order n and $\varepsilon(k)$ is the filter residual vector.

Equation (2) is the MIMO linear model that can be used for estimating the joint Markov parameters

$$M_k = \left[ M_k^{(1)} \quad -M_k^{(2)} \right] \in R^{px(p+m)},\ k = 1,..n \quad (3)$$

by a recursive Least Squares algorithm.

Once Markov parameters have been determined, the Hankel matrix can be constructed:

$$H(k-1) = \begin{bmatrix} M_k & M_{k+1} & .... & M_{k+s-1} \\ M_{k+1} & M_{k+2} & .... & M_{k+s} \\ . & . & .... & . \\ M_{k+s-1} & M_{k+s} & .... & M_{k+2s-2} \end{bmatrix} \quad (4)$$

By using the ERA/DC (EigenSystem Realization Algorithm with Data Correlation) (Juang et Al., 1988), it is possible to obtain a block correlation Hankel matrix composed of correlation matrices :

$$R(k) = H(k)H^T(0) \quad (5)$$

Such correlation matrices, are easily derived by Markov parameters and allow to reduce noise effects on the identification procedure.
Finally a block correlation Hankel matrix $W(k)$ associated to $R(k)$ matrices is determined, from which, by applying an SVD (Singular Value Decomposition), the unknown system order n is determined, since it is coincident with the number of positive singular values. If $W(0) = U_n S_n V_n^T$ is such SVD decomposition, it follows that a possible set of realization matrices of the identified system is:

$$\begin{aligned} A &= S_n^{-1/2} U_n^T W(1) V_n S_n^{-1/2} \\ B &= S_n^{1/2} V_n^T E_m \\ C &= E_p^T U_n S_n^{1/2} \end{aligned} \quad (6)$$

where matrices $E_m$ and $E_p$ are block selection matrices constituted by block of identity and zero matrices.

It can also be shown that it is possible to implement an observer applied to the deterministic version of linear state space equation (1), by considering the extended system :

$$\begin{aligned} x(k+1) &= \hat{A}x(k) + \hat{B}v(k) \\ y(k) &= Cx(k) + \hat{D}v(k) \end{aligned} \quad (7)$$

with the extended input $v(k) = \begin{bmatrix} u(k) \\ y(k) \end{bmatrix}$ and modified matrices :

$$\hat{A} = A + GC\ ,\ \hat{B} = B + GD - G, \hat{D} = [D \quad 0] \quad (7)$$

where all the eigenvalues of the modified system matrix $\hat{A}$ , under an observability hypothesis, can be placed in the origin. In this case, it can be recognized that the gain matrix $G$ plays a role analogous to Kalman filter in state estimation.
By using equation (6) as a state observer, it is also possible to implement an LQG controller (Juang,1994).

## 4. IDENTIFICATION RESULTS

A number of depth response experimental files have been used for identification by using the OKID method. In Fig.3 there is shown an example of sternplane input time history, while in Fig.4 the singular value magnitudes are plotted. As it can be noticed there is an appreciable drop in magnitude of the singular values between 7 and 8, indicating a system order n of 7 . It is worth noting that evidently an overestimation of the system order has been achieved. This fact derives from having neglected the coupling between depth and pitch motions. Furthermore, the depth measurement error seems to be relatively high, thus contributing to the increase of the estimated order.

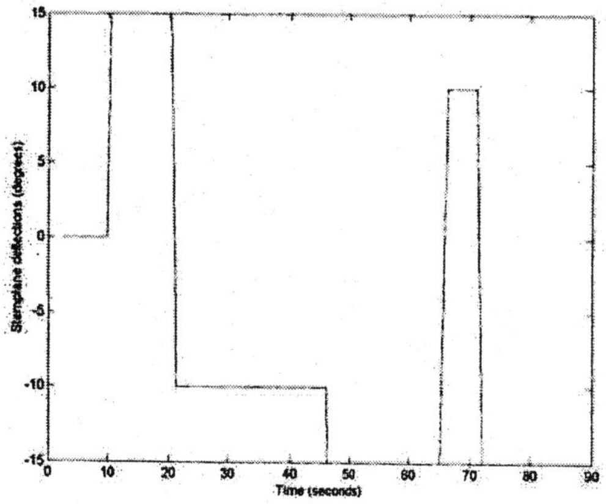


Fig. 3. Sternplane multi step input

It should be noticed, however, that as far as identification aims at determining a model for control purposes, overestimation of the model order is not a real drawback, since in this case identification is mainly used for predictive purposes. In Fig. 5 a comparison of the measured depth with the corresponding one predicted by the identified model is shown. It can be appreciated the robustness of OKID algorithm, even in the presence of a relatively high measurement noise on the depth channel.

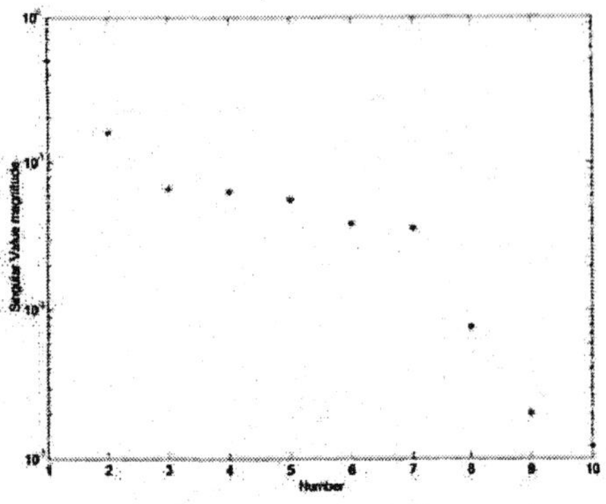

Fig. 4. SVD magnitude vs model order.

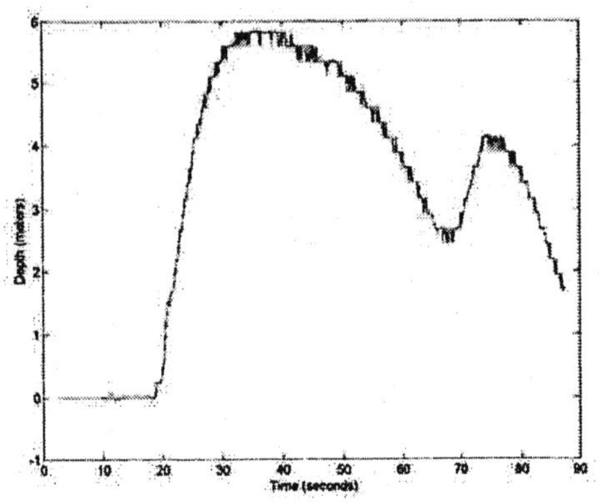

Fig. 5. Comparison between measured depth and predicted depth.

Analogous results have been obtained after the identification of different experiments.

## 5. CONCLUDING REMARKS

In this paper the OKID method has been applied to the identification of the depth dynamics of an AUV. From the preliminary results, it can be deduced that a multivariable model should be used, where the couplings between depth and pitch is directly taken into account. The results confirm also the excellent numerical capability of the algorithm in the presence of intense measurement noise.

## REFERENCES


Kybic, J., T. Blu and M. Unser (2001). Generalized sampling : a variational approach, *Technical Report Biomedical Imaging Group,* Swiss Federal Institute of Technology, Lausanne, Switzerland, 12 March 2001.

Juang , J.-N. , J.E. Cooper andJ.R. Wright (1988). An eigensystem realization algorithm using data correlations (ERA/DC) for modal parametersidentification, *Control Theory and Advanced Technology,* **Vol4** ,pp.5-14.

Juang, J.-N. (1994). *Applied System Identification*, Prentice Hall, New Jersey.

Naeem, W., R. Sutton R., and J. Chudley (2003). System identification, modelling and control of an autonomous underwater vehicle. *Proc. MCMC2003IFAC Conference,* Girona, Spain, 17-19 September 2003.

ELSEVIER
IFAC
PUBLICATIONS
www.elsevier.com/locate/ifac

# QUASI-STATIC THRUST MODEL IDENTIFICATION OF A OCEAN EXPLORER CLASS AUV

**D. Cecchi** * **A. Caiti** *,[1] **M. Innocenti** * and **E. Bovio** **

* *ISME, Interuniversity Centre of Integrated System for Marine Environment c/o DSEA University of Pisa, Via Diotisalvi 2, 56126, Pisa, Italy*
** *NATO Undersea Research Centre, Viale San Bartolomeo 400, 19138 La Spezia (SP), ITALY*

Abstract: The aim of the work is to find a simplified model for the surge equation of a slender body Autonomous Underwater Vehicle (AUV). A quasi-static thrust model is derived through a theoretical approach; estimation of the model parameters for a Ocean Explorer class AUV is pursued on real data collected during at sea tests. The obtained results are used in the design of a control scheme for the velocity autopilot of the vehicle. *Copyright © 2004 IFAC*

Keywords: Autonomous, Control, Estimation, Identification, Modelling, Vehicle

## 1. INTRODUCTION

The availability of models to simulate the behavior of vehicle like AUVs, allows for the implementation and test of autopilots and controllers without the need of heavy and time consuming at sea tests. A large amount of work can be avoided by simulating the vehicle dynamics. Often the dynamic description of the vehicle is not available or it is difficult to obtain the model parameters without specific tests (i.e. tank tests). Finding simplified model that can be estimated from general at sea maneuvers can be useful in controller and autopilot design.

Considering the velocity control of an Autonomous Underwater Vehicle (AUV), there are several methods to express the desired speed to track: a speed reference, a desired motor rotational speed or a percentual of the torque. This is the case of the Ocean Explorer (OEX) AUV available at the NATO Undersea Research Centre in La Spezia. An effective use of these modalities can be achieved after the identification, at the very least, of quasi–static models that relate vehicle forward speed and motor rotational speed. Moreover consider that the knowledge of these relations can be used by the position estimator of the vehicle under certain conditions.

The speed equations available in the literature express (in general) the surge velocity $u$ as a function of the torque $T$. During normal at sea AUV missions, measurements of the torque $T$ are not available, but it is possible to have $u$ and the motor rotational speed ($n$). The work described in this paper has the aim of deriving a simplified quasi-static model for the thrust equation of a slender body AUV. The model parameters are then identified from field data for the NATO Undersea Research Centre OEX class AUV.

The paper shows experimental results that have allowed for the identification of the OEX quasi–static thrust equation model. Consider furthermore that changing the propeller shape and/or

[1] caiti@dsea.unipi.it

size, the relation between $u$ and $n$ varies, so the identification procedure has to be repeated.

The paper is organized as follows: the next section of the paper gives some background on AUV propulsion models and on the relevant work appeared in the literature; in the third section the quasi-static model for the surge motion of a slender body AUV is proposed; the analysis of at sea test data is then illustrated, and finally a simple velocity control scheme using the identification results is showed.

## 2. PROPULSION MODELS

In this section some propulsion models available in the literature are presented. The main references are the chapter 12 of the book of Triantafyllou and Hover (Triantafyllou and Hover 2002), and the part dedicated to propulsion in (Fossen 2002). Since we are interested in the surge equation of motion, a body-fixed reference frame is assumed, and velocity and speed are always referred to the surge axis. It is also assumed, for the sake of simplicity, that shaft and motor revolution speed are equal.

In steady state the thrust is balanced by the vehicle drag:

$$R = T \tag{1}$$

where $R$ is the drag and $T$ is the thrust. The vehicle speed and the water speed seen at the propeller are related through the advance coefficient $J$:

$$J = \frac{U_p}{nD} \tag{2}$$

where $U_p$ is the water speed at the propeller, $n$ is the shaft revolution speed and $D$ is the shaft diameter. The speed $U_p$ is related to the vehicle speed:

$$U_p = U\,(1-w) \tag{3}$$

where $w$ is defined as wake fraction, $0 < w < 1$, with typical value of 0.1.

In addition, the propeller can increase the vehicle drag:

$$R_t = R\,(1-t) \tag{4}$$

where $t$ is the thrust deduction with typical values in the range $0.05 < t < 0.2$. A possible propeller description uses the following dimensionless coefficient:

$$K_T = \frac{T_0}{\rho n^2 D^4} \tag{5}$$

$$K_Q = \frac{Q_{p0}}{\rho n^2 D^5} \tag{6}$$

where the thrust $T_0$ is a function of $U_p$ and $T_0(U_p) = T$, $\rho$ is the water density, $Q_{p0} = \eta_r Q_p$ and $Q_p$ is the propeller torque ($Q_{p0}$ is the propeller torque in open water), $\eta_r$ is the rotative efficiency. $K_T$ and $K_Q$ are dependent on the speed $U_p$ and the relation is linear in a range of $U_p$.

Using the linear expression for $K_T$ and $K_Q$ the steady state of the propulsion system can be calculated (Triantafyllou and Hover 2002), and in particular the advance coefficient, the vehicle speed, the propeller torque and the shaft rotational speed.

The vehicle speed dynamic under the assumption of quasi-static propeller conditions is described by:

$$(m + m_a)\,\dot{u} = T - R \tag{7}$$

$$2\pi I_p \dot{n} = \eta_g \lambda Q_e - Q_{p_{sp}} \tag{8}$$

where $m$ is the vehicle mass, $m_a$ is the vehicle added mass, $I_p$ is the inertia seen at the propeller, $\eta_g$ is the gearbox efficiency, $\lambda$ is the gear ratio, $Q_e$ is the engine torque and $Q_{p_{sp}}$ the propeller torque when the propeller is installed in the vehicle hull. The transient behavior of the propulsion has been described by several authors.

Yoerger and coworkers in (Yoerger *et al.* 1990) proposed the following dynamic model:

$$I_e \dot{n} + k_{n|n|} n\,|n| = \tau \tag{9}$$

$$T = T\,(n, u_p) \tag{10}$$

where $I_e$ is the engine inertia, $k_{n|n|}$ is the nonlinear damping coefficient of the motor and $\tau$ is the shaft torque (control input). A more detailed model which considers the overshoot observed in experimental tests was proposed by Healey *et al.* (Healey *et al.* 1994):

$$I_e \dot{n} + k_n n = \tau - Q_p \tag{11}$$

$$m_f \dot{u}_p + d_f\,(u_p - u)\,|u_p - u| = T \tag{12}$$

The thrust and the torque are functions of $n$ and $u_p$: $T = T(n, u_p)$, $Q = Q(n, u_p)$. $k_n$ is the linear damping coefficient, $m_f$ is the mass of water in the volume covered by the propeller, $d_f$ is the quadratic damping coefficient related to $m_f$. A more general model was proposed by Blanke and Fossen (Fossen and Blanke 2000) in terms of the following equations:

$$I_e \dot{n} + k_n n = \tau - Q_p \tag{13}$$

$$m_f \dot{u}_p + d_{f0} u_p + d_f\,|u_p|\,(u_p - u_a) = T \tag{14}$$

$$(m - X_{\dot{u}})\,\dot{u} - X_u u - X_{u|u|} u\,|u| = (1-t)\,T \tag{15}$$

where $d_{f0} u_p$ is a linear damping term, $u_a$ is called advance speed and is the speed of the water entering in the propeller disc ($u_a$ is function of the vehicle speed $u$). In the surge equation $-X_u u$ is the linear laminar skin friction and $-X_{u|u|} u\,|u|$ is the nonlinear quadratic drag.

The propulsion system used in underwater vehicle is, in general, a DC motor. If motor electrical and mechanical parameters are available, it is possible to obtain $k_n$ and $\tau$:

$$\tau = \frac{K_m}{R_a} V_m \tag{16}$$

$$k_n = \frac{K_m^2}{R_a} \tag{17}$$

where $V_m$, $R_a$ and $K_m$ are the armature voltage, the armature resistance and the motor torque constant respectively.
In the following section a quasi-static analysis is performed on the surge equation (15).

## 3. SURGE EQUATION IN QUASI-STATIC CONDITION

The surge equation (15) is now considered. The motivation for this choice is that the model will be identified with at sea data from AUV cruises; the data available will allow for the identification of this equation only, and not of the complete model of the propeller (eqs. 13–15). Consider the speed surge equation (15):

$$(m - X_{\dot{u}})\,\dot{u} - X_u u - X_{u|u|} u\,|u| = \\ = (1-t)\,T \tag{18}$$

by substituting the expression (5) it has:

$$(m - X_{\dot{u}})\,\dot{u} - X_u u - X_{u|u|} u\,|u| = \\ = (1-t)\,\rho n^2 D^4 K_T\,(J\,(u_p)) \tag{19}$$

Assuming that the surge velocity derivative is negligible, it has obtained:

$$-X_u u - X_{u|u|} u\,|u| = \\ = (1-t)\,\rho D^4 K_T\,(J\,(u_p))\,n^2 \tag{20}$$

Consider now a slender body small AUV and assume that only positive forward speed is allowed. Equation (20) can be written as:

$$-X_u u - X_{u|u|} u^2 = \\ = (1-t)\,\rho D^4 K_T\,(J\,(u_p))\,n^2 \tag{21}$$

If the surge velocity $u$ varies slowly, it is possible to consider $J(u_p)$ to have a constant value. Defining:

$$\xi\,(u) = (1-t)\,\rho D^4 K_T\,(J\,(u_p)) \tag{22}$$

an expression relating the surge velocity $u$ and the shaft rotational speed $n$ can be derived:

$$X_{u|u|} u^2 + X_u u + \xi\,(u)\,n^2 = 0 \tag{23}$$

Then it has:

$$u = \frac{-X_u - \sqrt{X_u^2 - 4X_{u|u|}\xi\,(u)\,n^2}}{2X_{u|u|}} \tag{24}$$

Equation (24) is now applied to the case of the Nato Undersea Research Centre Ocean Explorer class AUV. In this particular case the linear damping coefficient $X_u$ can be neglected and the value of $X_{u|u|} < 0$ is known.
Then, from equation (24):

$$u \simeq n_e \frac{-\sqrt{-4X_{u|u|}\xi\,(u)}}{2X_{u|u|}} =$$

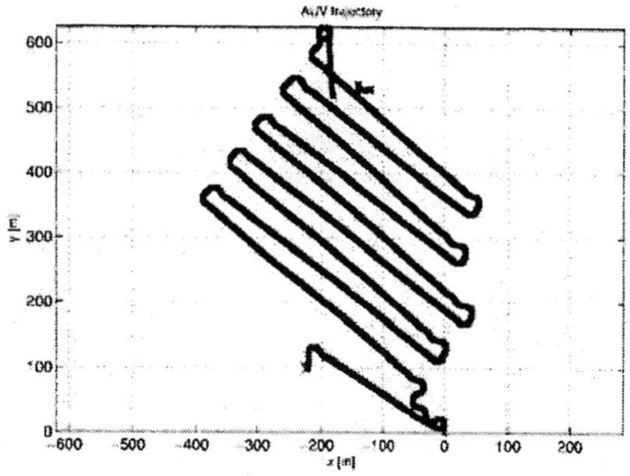


Figure 1. Typical AUV trajectory.

$$n_e \sqrt{\frac{\xi\,(u)}{\left|X_{u|u|}\right|}} \tag{25}$$

Equation (25) shows that the relation between $n$ and $u$ is linear when $X_u$ is negligible and in a range of $u$ where $\xi_u$ is approximately constant.
The knowledge of $X_{u|u|}$ can permit to estimate the thrust in steady state condition neglecting the thrust deduction $t$ without loss of generality:

$$T = -X_{u|u|} u^2 \tag{26}$$

## 4. DATA ANALYSIS

The available data are collected during sea tests of the vehicle. The measured state variables considered are the vehicle speed and the propeller rotational speed. The vehicle speed is measured by a Doppler Velocity Log (DVL) onboard the AUV. The DVL gives the three components of the vector speed in the body-fixed reference frame. In the following analysis the vehicle velocity considered is the surge speed collected during nominal straight trajectories. The objective is to find experimentally the relation between $u$ and $n$ (equation 25) that can be used in a speed control system. A typical trajectory followed by the vehicle is showed in figure (1). The data are processed as follows:

- isolation of portions of straight trajectory at the same nominal speed;
- data extraction and conversion;
- calculation of the mean values of vehicle speed and shaft rotational speed;
- estimation of the relation between $u$ and $n$.

It has been chosen to calculate the mean values of the data and then use these values to obtain the regression curve because the data distribution over the speed range is not homogeneous. The data are clustered around the values of 1, 2 2.5, 3, 3.2 knots. The mean value extraction avoid polarization phenomena in the results. To obtain the regression curve four values at speed of about 1, 2, 2.5 and 3 knots, two values at 3.2 knots have

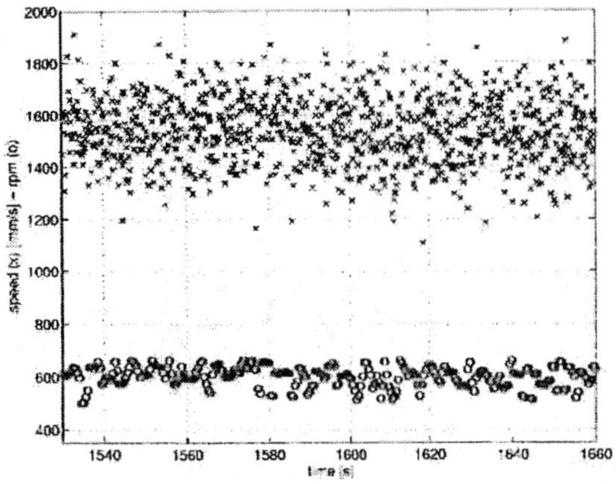


Figure 2. Surge velocity and propeller rpm.

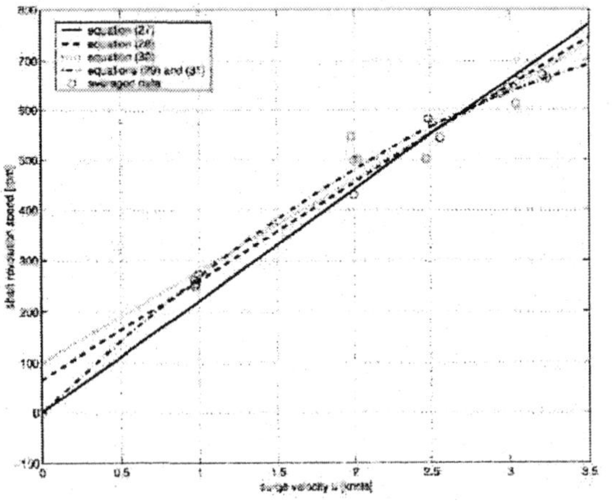

Figure 3. Surge velocity - propeller rpm relation.

been used plus the constraint of zero velocity for the vehicle in a state of rest. Typical measurements of surge velocity and propeller rotational speed in a straight trajectory is showed in figure (2). The surge speed is expressed in mm/s and the propeller speed in rpm (routes per minute). Figure (3) shows the results of the estimation of the relation between $u$ [knots] and $n$ [rpm]. Two sets of data are used in parameter estimation: one containing the rest constraint and the other without it. The regression curves are calculated by least squares procedures. The estimated equations are linear or parabolic, but it can be observed that the second order functions are quite linear in the speed range 1–3 knots.

The calculated relations including the rest constraint are:

$$n = 220.03u \tag{27}$$

$$n = 194.17u + 65.256 \tag{28}$$

$$n = -28.32u^2 + 296.85u - 1.1844 \tag{29}$$

Using the dataset without the rest constraint the following relations have been obtained:

$$n = 181.08u + 98.286 \tag{30}$$

$$n = -29.108u^2 + 300.65u - 4.8501 \tag{31}$$

Plotting the equations (29) and (31) with $u$ varying between 0 and 3.5 knots and $n$ between 0 and 800 rpm it can be observed that the two curves are practically coincident, then in figure (3) only one curve is plotted.

The second order relation is more realistic at low speed values, however operating in the range 1-3 knots the linear approximations are effective as well.

## 5. SPEED CONTROLLER

A simple PID controller has been designed using the identification results of the previous section. Figure (4) shows the implemented control scheme. The model used for the speed equation is the nonlinear surge motion. The reference input is expressed in m/s, then the estimated relation converts it in the correspondent motor rpm. The same process is made on the speed measurement. At the input of the PID controller there is a block producing a torque reference based on a modified version of equation (29) and on the knowledge of $X_{u|u|}$. Equation (29) has been modified so that the rest constraint is satisfied. Without this correction the output signal presents an offset with respect to the reference due the to zero degree term.

Vehicle physical bounds are accounted for in the model. It has to be observed that bounds may cause the presence of chattering phenomena in the surge velocity and, particularly, in the control signal. The effect is more evident when noise is inserted in the simulations. For this reason a filter has been inserted to reduce the chattering and the amount of noise in the control loop (Jalving 1994). The cutoff frequency has been chosen considering the data acquisition rate and the system time constant. The implemented filter is a second order low pass with cutoff frequency at 2 Hz.

The control scheme has been modelled as showed in figure 4 starting from the *Speed Equation* block and considering measurements available in the real vehicle and bounds on actuators. The torque value at the input of *Speed Equation* block is restricted in a definite range; inside the same block there is another limiter so that the output speed assumes really obtainable values. The block *Speed to rpm* uses the identified relations to convert linear speed into the rotational speed of the motor. The error signal is expressed in rpm, so the control scheme can be simply modified to give the reference input in rpm. The block *Conversion to torque* utilizes the knowledge of $n$ and $u$ (available measurements) to generate the estimated torque value combining equations (26) and (29). Figures (5) and (6) show the reference input and the correspondent output used in a simulation when the PID parameters are the following:

- Proportional gain $P = 25$

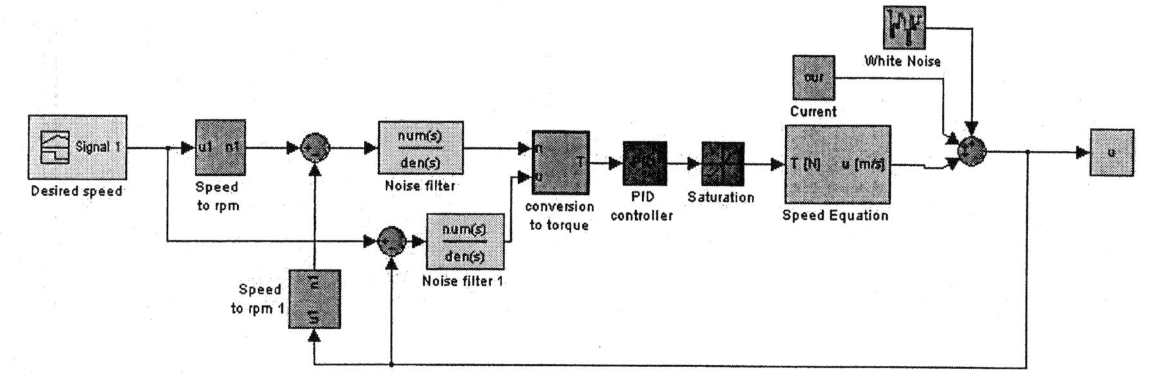


Figure 4. Control scheme.

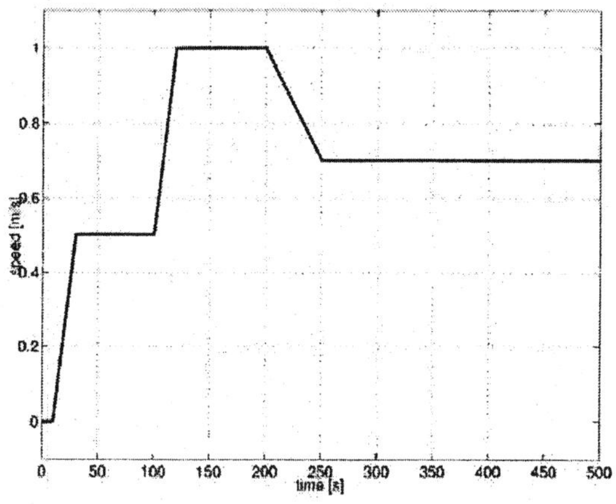


Figure 5. Desired speed.

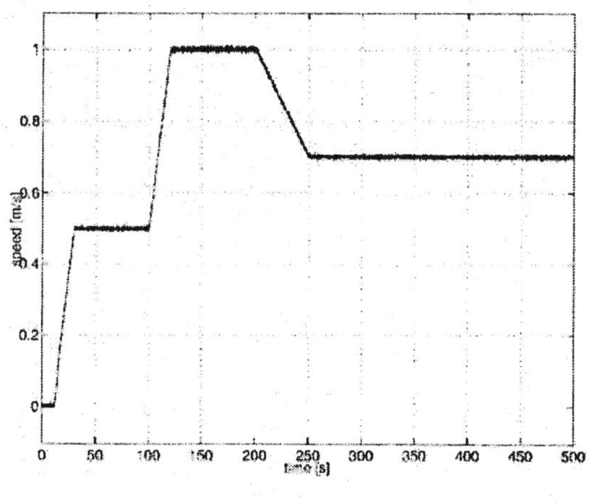


Figure 6. Vehicle speed.

- Integral gain $I = 1$
- Derivative gain $D = 1.25$
- Added pole for the causality $N = 100$

where:

$$PID(s) = P + \frac{I}{s} + \frac{Ds}{\frac{1}{Ns} + 1} \qquad (32)$$

As for the performance, it can be observed that the steady state error is null, neglecting the presence of the noise, and when the reference input is a step, then the observed overshoot is nearly 5%.

## 6. CONCLUSIONS

In this work a quasi-static surge equation is considered to derive a relation between surge velocity and propeller rotational speed. By processing data collected in sea tests, identification of the quasi-static thrust model for an Ocean Explorer class AUV has been achieved. The estimated relations have been used in the design of a simple controller scheme tested by simulation. Implementation of the controller on the real AUV is under way.

## REFERENCES

Fossen, T.I. (2002). *Marine Control Systems*. Marine Cybernetics. Trondheim, Norway.

Fossen, T.I. and M. Blanke (2000). Nonlinear output feedback control of underwater vehicle propellers using feedback form estimated axial flow velocity. *IEEE Journal of Oceanic Engineering* **25**(2), 241–255.

Healey, A.J., S.M. Rock, S. Cody, D. Miles and J.P. Brown (1994). Toward an improved understanding of thruster dynamics for underwater vehicles. In: *Proceedings of the 1994 IEEE Symposium on Autonomous Underwater Vehicle Technology*. pp. 340–352.

Jalving, B. (1994). The NDRE-AUV flight control system. *IEEE Journal of Oceanic Engineering* **19**(4), 497–501.

Triantafyllou, M.S. and F.S. Hover (2002). *Maneuvering and Control of Marine Vehicles.* Dept. of Ocean Engineering. Cambridge, MA, USA.

Yoerger, D.R., J.G. Cooke and J.E. Slotine (1990). Tthe influence of thruster dynamics on underwater vehicle behavior and their incorporation into control system design. *IEEE Journal of Oceanic Engineering* **15**(3), 167–178.

ELSEVIER
IFAC
PUBLICATIONS
www.elsevier.com/locate/ifac

# AUTOMATIC COST WEIGHT SELECTION IN MIXED SENSITIVITY $\mathcal{H}_\infty$ CONTROLLER SYNTHESIS

**Decio C. Donha** [*] **Reza M. Katebi** [**]

[*] *Mechanical Engineering Department of University of São Paulo*
*Brazil*
*e-mail: decdonha@usp.br*
[**] *ICC-Department of Electric and Eletronic Engineering University of Strathclyde, Glasgow, UK*
*e-mail: r.katebi@eee.strath.ac.uk*

Abstract:
This paper presents the tuning of a dynamic ship positioning controller. The procedure is based on a multi-criteria optimization, involving the selection of weighting functions parameters from different structure, which are then used to synthesise $\mathcal{H}_\infty$ controllers. The mixed sensitivity approach is employed in the controller synthesis. A Genetic Algorithm generates weighting function candidates and fuzzy logic is incorporated into the solution to emulate the work of a skilled designer searching for optimal solutions. Results are presented and compared to a trial-and-error approach used in a former work.


Keywords: Dynamic positioning, Genetic algorithm, Robust $\mathcal{H}_\infty$ optimal control, Multi-objective optimization.

## 1. INTRODUCTION

Advanced control theories provide the necessary framework to a straightforward synthesis of controllers for systems involving uncertainties and using all available information such as disturbances, noises and reference signals. Synthesis using the mixed sensitivity approach, based on the shaping of the well known frequency response functions of the system, is very frequent, and allows that properties like stability robustness, disturbance rejection and command response behaviour to be jointly imposed in a certain measure (Grimble,1994). However, in complex systems, where the benefits of advanced controllers are clear, the lack of a systematic and intuitive approach in selecting a large number of tuning parameters to obtain a satisfactory solution, may lead to a cumbersome and fatiguing work. The selection of appropriate weights is still very much an art demanding skill and time (Katebi et al, 1997).

The design of a control system in general involves many constraints and competing objectives. To obtain an optimal solution, a trade-off between objectives and constraints is necessary. Multi-objective fuzzy optimization techniques provide a means for the incorporation of the relative importance of competing objectives and problem restrictions and have been used in many similar contexts (Karr and Gentry, 1993), (Polkinghorne, 1996), (Sutton, 1997), and (Carlsson and Fuller, 1996).

In this work, a GA procedure is employed to tune an $\mathcal{H}_\infty$ controller of a dynamic ship positioning system, aiming at a desired performance. Results are compared to former design methods (Katebi et al, 1997). In the present case, the search procedure involves not only the determination of weighting parameters, but also the weighting function structure. Similar approaches were proposed by (Dakav et al., 1995) and (Tang et al., 1996). Genetic Algorithm (GA) provides a easy way to optimize a solution. The usual large numbers of parameters in control problems, with possible non-convex criteria, difficulties to find derivatives and the imprecise nature of some of the variables are no longer an issue (Patton and Liu, 1994).

The paper is organized as follows. Brief descriptions of the ship nonlinear model and of the $\mathcal{H}_\infty$ problem are presented in sections 2 and 3, respectively. Section 4 presents the Genetic Algorithm and the optimization procedure. Results are given and analyzed in section 5. Finally, the Conclusion is given in section 6.

## 2. SHIP MATHEMATICAL MODEL

The ship non-linear mathematical model can be described as follows (McIntyre and Katebi, 1994):

$$\begin{aligned}(m-X_{\dot{u}})\dot{u}-(m-Y_{\dot{v}})rv &= X_H+X_A+X_W+X_T\\ (m-Y_{\dot{v}})\dot{v}-(m-X_{\dot{u}})ru &= Y_H+Y_A+Y_W+Y_T\\ (I_{zz}-N_{\dot{r}})\dot{r} &= N_H+N_A+N_W+N_T \quad (1)\end{aligned}$$

where $u$, $v$ and $r$ are, respectively, the ship surge, sway and yaw velocities, in the ship reference frame; $m$ is the ship displacement and $I_{zz}$ the ship moment of inertia about $Z$; $X_{\dot{u}}$, $Y_{\dot{v}}$ and $N_{\dot{r}}$ are added masses in $X$, $Y$ and about $Z$ directions, respectively; index $H$ refers to hydrodynamic loads induced by currents; index $A$ refers to wave loads; index $W$ refers to wind loads and index $T$ refers to control forces and moments produced by thrusters.

Velocities $u_s$, $v_s$ and $\dot{\varphi}$ in an Earth-fixed reference frame are obtained by: $u_s = u\cos\psi - v\sin\psi$; $v_s = u\sin\psi + v\cos\psi$ and $\dot{\varphi} = r$, where $\psi$ is the ship heading in the ship frame.

Current loads are approximated by following expressions:

$$\begin{aligned}X_H &= -S_x(u_{sc}^2+v_{sc}^2)\cos(\psi_s-\psi_c)\\ Y_H &= -S_y(u_{sc}^2+v_{sc}^2)\sin(\psi_s-\psi_c) \quad (2)\\ N_H &= -S_n(u_{sc}^2+v_{sc}^2)\sin 2(\psi_s-\psi_c)\end{aligned}$$

where $S_x$, $S_y$ and $S_n$ are scaling factors; $u_{sc} = u_s - V_c\cos\psi_c$; $v_{sc} = v_s - V_c\sin\psi_c$ and $\psi_{sc} = \arctan v_{sc}/u_{sc}$; index $c$ refers to current and indices $sc$ refer to relative motions between the ship and current. $V_c$ is the current velocity modelled by: $\dot{V}_c = \tau(t)$, where $\tau(t)$ is an unity-variance, zero mean white-noise signal.

Wave and wind loads are introduced in the model by means of white-noises filtered by appropriate transfer functions (McIntyre and Katebi, 1994).

## 3. $\mathcal{H}_\infty$ CONTROL DESIGN

There are several ways of setting up the control problem. An usual approach to characterize the closed-loop performance objectives in the advanced control theory is the measurement of certain closed-loop transfer matrices using different matrix norms (Zhou, 1996). These norms provide a measure of how large output signals can get for certain classes of input signals, which is a measure of the gain of the system. To establish closed-loop transfer matrices, one of the most popular procedures is the mixed sensitivity loop-shaping approach, where direct bounds on system sensitivity transfer functions are considered. A mathematically convenient measure of a MIMO closed-loop matrix $T(s)$ in the frequency domain is the $\mathcal{H}_\infty$ norm defined as:

$$||T_{z\omega}||_\infty := \sup_{Re(s)>0} \overline{\sigma}(T(s)) = \max_{\omega\in\Re} \overline{\sigma}(T(jw)) \quad (3)$$

where $\overline{\sigma}(T(j\omega))$ is the largest singular value of $T(j\omega)$ over the frequency range $\omega$.

Consider the feedback control system in question represented in the standard two-port configuration shown in Figure 1, where $R$ gives the reference dynamics, $G$ the nominal plant dynamics and $K$ the controller. Exogenous inputs are the command reference $r$, the sensor noise $n$, input disturbance $d_i$ and the output disturbance $d$. Weighting functions $W_i$, $W_d$ and $W_n$ reflect the available knowledge about the input and output disturbances and measurement noise, respectively. The weighting function $W_s$ may be used to reflect requirements on the shape of the $\mathcal{H}_\infty$ controller. The weighting $W_c$ may be used to impose restrictions on the control signals, while $W_t$ may be used to modify, for example, tracking features of the system.

The first step of the $\mathcal{H}_\infty$ design procedure in this case involves the minimization of a performance index, formulated, e.g., as follows:

$$||T_{z\omega}||_\infty = \left\| \begin{matrix} W_sS \\ W_cC \end{matrix} \right\|_\infty \quad (4)$$

where $T_{zw}$ is the transfer function from

$w = [\tilde{d} \quad \tilde{n} \quad \tilde{d}_i \quad c]^T$ to $z = [z_1 \quad z_2 \quad z_3]^T$; $W_s$ and $W_c$ are weights on $S(s) = (I+GK(s))^{-1}$ and $C(s) = KS(s)$, respectively, and where $K(s)$ is the controller to be defined.

The state-space model used for control design is in the usual form as follows: $\dot{x}(t) = Ax(t)+B_1w(t)+B_2u(t)$; $z(t) = C_1x(t)+D_{11}w(t)+D_{12}u(t)$ and $y(t) = C_2x(t)+D_{21}w(t)+D_{22}u(t)$, where $x(.)$ is the state vector with a known initial state; $u(.)$ is the input vector; $w(.)$ is the dynamic disturbance; $z(.)$ is the controlled state vector; $y(.)$ is the measured state vector. System matrices are assumed to have compatible dimensions.

The following Riccati equations are associated with the $\mathcal{H}_\infty$ design problem:

$$\begin{aligned}X_\infty A+A^TX_\infty-X_\infty(\gamma^{-2}B_1B_1^T-B_2B_2^T)X_\infty+C_1^TC_1&=0\\ AY_\infty+Y_\infty A^T-Y_\infty(\gamma^{-2}C_1^TC_1-C_2^TC_2)Y_\infty+B_1B_1^T&=0\end{aligned}$$

Based on some well-known results an optimal controller $K(s)$ is defined as follows (Zhou, 1996):

$$K(s) = \left[\begin{array}{c|c} A_\infty & -Z_\infty L_\infty \\ \hline F_\infty & 0 \end{array}\right] \quad (5)$$

where: $A_\infty \equiv A + \gamma^{-2} B_1 B_1^T X_\infty + Z_\infty L_\infty C_2 + B_2 F_\infty$;

$F_\infty \equiv -B_2^T X_\infty$; $L_\infty \equiv -Y_\infty C_2^T$; $Z_\infty \equiv (I - \gamma^{-2} Y_\infty X_\infty)^{-1}$

The packed notation on the right side of equation (5) has the meaning:

$$\left[\begin{array}{c|c} A & B \\ \hline C & D \end{array}\right] := \quad C(sI - A)^{-1}B + D \quad (6)$$

The second step in an $\mathcal{H}_\infty$ design amounts to the selection of different weighting functions in order to achieve performance objectives and practical requirements such as measurement noise attenuation, disturbance rejection, tracking capability, etc.

Suitable cost weighting functions can often be found after a trial-and-error procedure (Desanj and al., 1995). However, more scientific methods of selecting weighting functions are required, when the full potential of the dynamic cost weightings are to be exploited. In this work, weighting functions are determined by a Genetic Algorithm (GA) search.

## 4. GENETIC ALGORITHM

Genetic Algorithms (GAs) are probabilistic optimization techniques that emulate the mechanisms of natural selection and evolutionary genetics. GAs operate by maintaining and modifying the characteristics of a population of individuals (possible solutions) over a number of generations. Using a Darwinian survival-of-the-fittest analogy, GAs can eliminate unfit population member characteristics and lead to a fast convergence to optimal solutions (Donha and Katebi, 2002).

Figure 2 shows a flow chart of the design procedure. A brief description of the illustrated steps follows.

*Initial Population*: An initial population of strings (individuals) must be produced. In this work binary coding was used to represent the strings.

*Decoding:* Each individual (string) of the population is decoded to produce till seven parameters for the weighting functions construction. At first, first order weightings were searched for, however, in the case of DP, which involves a MIMO problem, higher order weightings were also analyzed.. Weightings $W_s$ on $S(s)$, and $W_c$ on $C(s)$ are produced using, e.g., the following structure for each controlled motion:

$$W_s(s) = \frac{s^3 + a_2 s^2 + a_1 s + a_o}{s^3 + b_2 s^2 + b_1 s + b_o} \quad W_c(s) = \frac{s + k_3}{(k_4 s + k_5)\, h} \quad (7)$$

where $k_3$ and $k_4$ are fixed parameters; $a_i$, $b_i$ $i = 0, 1, 2$ and $h$ are the parameters to be defined by the GA algorithm.

It is then verified if the weighting functions are proper. If not the individual is discarded and replaced through the execution of the GA loop once.

*Controller Calculation, Simulation and Verifications:* If the weighting functions are proper, the algorithm tries to define an $\mathcal{H}_\infty$ controller. If no solution is available the individual is discarded and replaced by a new one performing one cycle of the GA. If an $\mathcal{H}_\infty$ controller is available, the system step response and the controller roll-off frequency are calculated. The procedure produces also the open loop frequency response, the sensitivity function, the control sensitivity function and the closed-loop frequency response. At this point informations like settling time, rise time, overshoot, actuator rate and actuator amplitude are available and may be evaluated.

*Multi-objective Fuzzy Optimization:* Fuzzy sets can be used to define vague concepts such as those found when trying to evaluate the performance of a controller. The membership function fuzzy concept is what is mainly needed here. A membership function is a curve that defines how each point in an input space (universe of discourse) is mapped to a membership value. It determines the degree to which an adjective, such as, e.g. too large in the statement: 'the overshoot is too large', truthfully describes the value of a variable and it is a measure of the designer's degree of satisfaction when faced with some result. The next stage is to fuzzify the objective functions and problem constraints and to calculate a membership value for each result produced in the previous step. The membership function for the fuzzy objective related, e.g., to the overshoot may be given by:

$$f(X,a,b,c) = max[min(\frac{X-a}{b-a}, \frac{c-X}{c-b}), 0] \quad (8)$$

where the parameters $a$ and $c$ locate the triangle basis and the parameter $b$ locates its peak. The variable $X$ is the overshoot value calculated before. If $X$ is between $a$ and $c$, a non-zero fitness value ($\mu$) is assigned to the present solution ($\mu_{fi}(X) : R^n \rightarrow [0,1]$). Otherwise, the fitness value is zero. The fitness function is a measure of the degree of satisfaction for any available solution.

*Reproduction:* The reproduction process can be subdivided into two subprocesses: Fitness Evaluation and Selection. The fitness function is what drives the evolutionary process and its purpose is to determine how well a string (individual) solves a given problem allowing for the assessment of the relative performance of each population member. In this work the fitness function was established using a convex decision-making. This technique provides a Pareto optimal solution, i.e. no unique optimal solution is found. Under the Pareto paradigm no improvement can be made in a determined objective without affecting others. The optimal decision is made by selecting the best alternative from a fuzzy decision space $A$ of alternatives.

A convex decision-making based on the concept of arithmetic mean is used here. This allows a relative weighting between stated objectives and constraints. The multi-objective fuzzy optimization and fitness function can be defuzzified and formally stated as follows:

$$max\mu_A(X) = \sum_{i=1}^{k} \alpha_i \mu_{fi}(X) + \beta_i \mu_{gi}(X)$$

$$subject\ to\ g_j(X) \leq a_j + d_j \tag{9}$$

where $X \in R^n$ is any possible solution; $\mu_\bullet$ is the fitness value of a constraint or objective; $\alpha_i$ and $\beta_i$ are weights attributed to an objective $f_i(X)$ or constraint $g_i(X)$, respectively; $a_j$ is a constrained maximum value and $d_j$ is the allowable tolerance for the fuzzy $j$-constraint. This means that an optimal solution is found in the crisp (non-fuzzy) domain, which optimizes an a-priori defined fitness function.

In this work the Tournament Selection with an Elitist Strategy (KrishnaKumar, 1994) was used to implement the selection operator (Donha and Katebi, 2002).

If the entire population has been evaluated and a suitable solution was not found a new generation is created.

*Crossover:* Reproduction cannot create new and better strings. These improvements may be achieved by Crossover and to an lesser extent by Mutation. Reproduction may proceed in tree steps as follows: a) two newly reproduced strings are random selected from the Mating Pool; b) a number of crossover positions along each string are uniformly selected at random and c) two new strings are created and copied to the next generation by swapping string characters between the crossover positions defined before.

*Mutation:* The mutation operator is a secondary mechanism of GA adaptation and is only introduced to provide a framework to ensure that a critical feature (genetic information) may be reinserted or removed from a population. Mutation generates new individuals by randomly modification of the value of a string position (gene).

*GA Search Stop and Non-linear Simulation:* The procedure will stop after a pre-defined number of generations or before that if a very good solution is found. A non-linear simulation of the process dynamics is then performed to certify that a suitable controller was found and to evaluate the features of the closed loop system.

## 5. RESULTS AND ANALYSIS

The main data of the DP ship used in the simulations are: Length overall ($L$): 68.88 $m$, Beam ($B$): 14.02 $m$, Design Draft ($D$ ): 3.96 $m$ and Design Displacement ($\nabla$): 2038.68 $m^3$.

Dynamic Positioning (DP) systems are required to minimize the energy consumption and the wear and tear of the actuators. DP systems must also avoid the high frequency thruster modulation induced by waves, wind gusts and sensor measurement noises. In addition, DP systems must tolerate sensor system transient errors and provide fast reconfiguration responses in the case of thruster failures.

The GA parameters were set as follows: number of individuals per generation $N = 30$ binary strings, each one of length $l = 54$ bits. Crossover is performed with probability $p(X) = 0.9$ with three cross-over points in each string, mutation with probability $p(x) = 0.01$ and till 500 generations are evaluated if a high degree of satisfaction is not achieved. The following relative weightings were used in the generation of the fitness function: a) for the overshoots in all motions: 0.23; b) for the rise times in all motions: 0.05; c) for the settling times in all motions: 0.05 and d) for the value of $\gamma$ : $1/\gamma$. The last weight was set to minimize the norm in expression 4, which is related to desired controller performance features. In this case only polynomial Z-curves were used to evaluate the membership grades. Running the GA, typical weighting functions found are as follows: $W_S =$

$$diag\begin{bmatrix} \frac{s^3+0s^2+0.88s+0.04}{s^3+7.88s^2+9.98s+9.22}; \\ \frac{s^3+0.1s^2+0.54s+0.02}{s^3+9.34s^2+9.70s+9.84}; \\ \frac{s^3+0.16s^2+1.32s+0.1}{s^3+9.18s^2+8.04s+2.26} \end{bmatrix}$$

and $W_C = diag\left[\frac{s+0.07}{0.02s+200}; \frac{s+0.07}{0.02s+200}; \frac{s+0.07}{(0.1s+1000)*9.92}\right]$.

In the nonlinear simulations the following weather conditions were used: Average wind velocity: $15m/s$; Absolute Wind Angle: $30^o$; Average Current Velocity: $2m/s$; Absolute Current Angle: $45^o$; Significant Wave Height: $6.5m$; Absolute Wave Angle: $60^o$.

To evaluate the dynamic positioning performance, it is verified the system capability to hold position $(X,Y) = (0,0)$ with a five degrees heading, despite the harsh weather condition.

Figures 3 and 4 illustrate typical results of inverse sensitivity weighing functions together with sensitivities and control sensitivities respectively found for low and high order weightings. The sensitivity functions were well bounded in both cases, which is an indication that robustness was achieved. However, in the case of higher order costing functions, smaller $\gamma$ values were achieved. Only the yaw costing functions are presented, since other motions showed similar patterns.

Figure 5 and 6 show, respectively, a bird's eye view of the ship position over the sea surface for a controller tuned by Katebi et al. (1997) using a trial-and-error procedure and a result using the procedure proposed here, for high order weightings. In this case, there was no significant difference of performance between the controllers. Figure 7, also from a previous work (Katebi et al.,1997), and figure 8, a result of the present work, also show similar performances of the alternative controllers. Although, the main advantage of the controller optimized by the GA is clear

| param. | *mean* | *stdv* | $m\psi$ | $dv\psi$ | *mth* | *dvth* | $\gamma$ |
|---|---|---|---|---|---|---|---|
| Katebi | 1.58 | 0.88 | 4.60 | 1.52 | 0.09 | 0.06 | **70.04** |
| Donha | 1.66 | 0.84 | 3.80 | 1.82 | 0.14 | 0.07 | **0.10** |

Table 1. Comparisons

observing table 1, which presents a more complete comparison of the controllers.

In this table, *mean* and *stdv* (standard deviation) refer to the absolute position over the sea surface; $m\psi$ is the mean value of the ship heading, with standard deviation $dv\psi$; *mth* is the mean value of the normalized thrust force and *dvth* is the normalized thrust standard deviation; $\gamma$ is the gamma final value achieved in the $\mathcal{H}_\infty$ algorithm. It is easy to conclude from table 1 that, although having very similar performances, the controller tuned by the GA outperform the controller tuned by trial-and-error; when the issue is robustness. From the $\gamma$ values achieved, it is clear that the controller tuned by trial-and-error will have an $\mathcal{H}_2$ characteristic, whereas the controller tuned by the GA is really a robust controller.

## 6. CONCLUSIONS

A simple method for the selection of suitable weighting functions for the $\mathcal{H}_\infty$ design using a multi-objective fuzzy GA for optimization has been proposed in this work. Results from a number of simulations proved that the approach can be valuable in reducing the design time necessary to tune a specific controller and that a good performance can effectively be achieved. Although GA algorithm requires no previous knowledge of the search-space, of course, the solution is accelerated if more information, such as the subset of the space of possible weighting parameters, is added to the search procedure.

The search approach developed here seems to suit very well with the $\mathcal{H}_\infty$ synthesis, mainly when multivariable problems are involved. The approach would also work in tuning of other controllers, such a PID, for example. In the case of simpler controllers, when computer burden is smaller, the procedure would be faster, and an on-line tuning seems possible.

The method also has some drawbacks. The velocity of convergence may vary from run to run due to the random generation of the initial population. The computer numerical burden is another disadvantage of the method. A parallel processing algorithm would, of course, speed-up the procedure.

A development of this work would be to establish an automatic structure selection of weighting functions.

**Acknowledgments:** The first author is grateful for the financial support from FAPESP. The second author is grateful to EPSRC Marine Directorate for their support.

## 7. REFERENCES

Carlsson, C. and Fuller R. (1996) 'Fuzzy Multiple Criteria Decision-making Recent Developments', Fuzzy Sets and Systems, Vol. 78(2), pp 139-153.

Dakav, N. V.; Whidborne, J. F. and Chipperfield A. J. (1995) '$\mathcal{H}_\infty$ Design of an EMS Control System for a Maglev Vehicle Using Evolutionary Algorithms ' Proc. GALESIA 95, Sheffield.

Desanj, D.S.; Katebi, M.R. and Grimble, M.J. (1995) '$\mathcal{H}_\infty$ Robust Autopilot Design' Proceedings of the 3rd IFAC Workshop on Control Applications in Marine Systems (CAMS'95), pp115-122, Trondheim, Norway, may.

Donha, D.C. and Katebi, M. R. (2002) 'Dynamic Positioning $\mathcal{H}_\infty$ Controller Tuning by Genetic Algorithm' Proceedings of the 15th IFAC World Congress, July, Barcelona, Spain.

Grimble, M.J. (1994) 'Robust Industrial Control', Prentice Hall, Hemal Hempstead.

Karr, C.L. and Gentry, E.J. (1993) 'Fuzzy Control of pH Using Genetic Algorithm' IEEE Trans. Fuzzy Systems, v.1, no. 1.

Katebi, M.R.; Grimble, M.J. and Zhang Y. (1997) '$\mathcal{H}_\infty$ Robust Control Design for Dynamic Ship Positioning' IEE Proc. Control Theory Appl., v. 144, no.2, pp110-120.

McIntyre, J. and Katebi, M.R. (1994) 'Nonlinear Model For DP' Industrial System and Control Ltd. report.

Patton, R. J. and Liu G. P. (1994) 'Robust Controller Design Via Eigenstructure Assignment, Genetic Algorithms and Gradient-based Optimization' IEE Proc. D, Vol. 141(3), pp202-208.

Polkinghorne, M.N.; Burns R.S. and Roberts, G.N. (1996) 'Operational Performance of an Initial Design of Self-Organizing Fuzzy Logic Autopilot' Proc. IEE Conf. Control'96, Exeter, UK.

Sutton, R. (1997) 'Optimization of Fuzzy Autopilots' 11th Ship Control Systems Symposium, v.1, pp63-76.

Tang K. S.; Man K. F. and Gu D. W. (1996) 'Structured Genetic Algorithm for Robust $\mathcal{H}_\infty$ Control System Design' IEE Trans. Industrial Electronics, Vol. 43(5) pp 575-582.

Zhou, Kemin; Glover, Keith and Doyle, J.C. (1996) 'Robust and Optimal Control', Prentice-Hall, New Jersey.

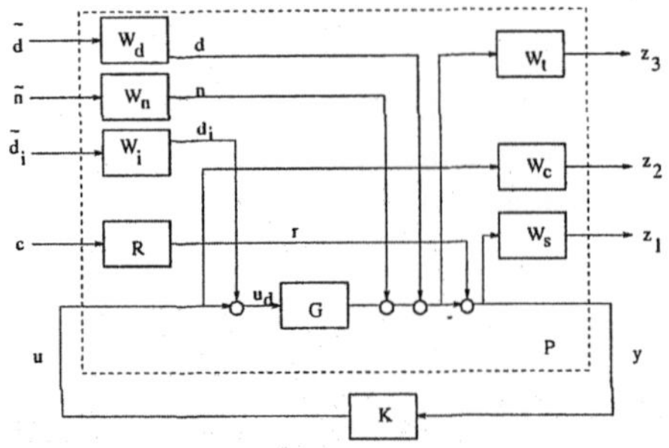


Fig. 1. Two-Port Configuration

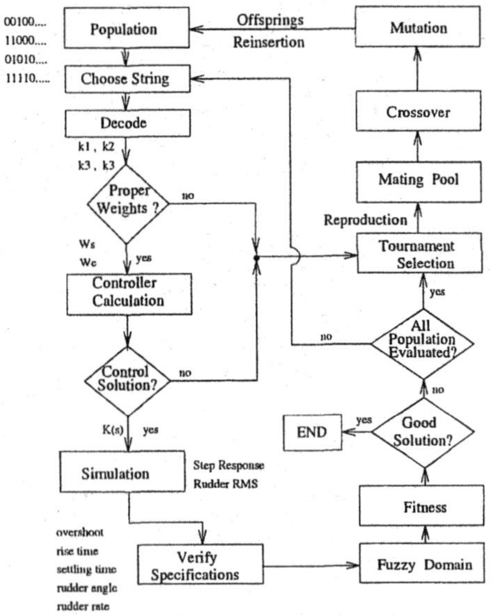


Fig. 2. Design Procedure Flow Chart Diagram

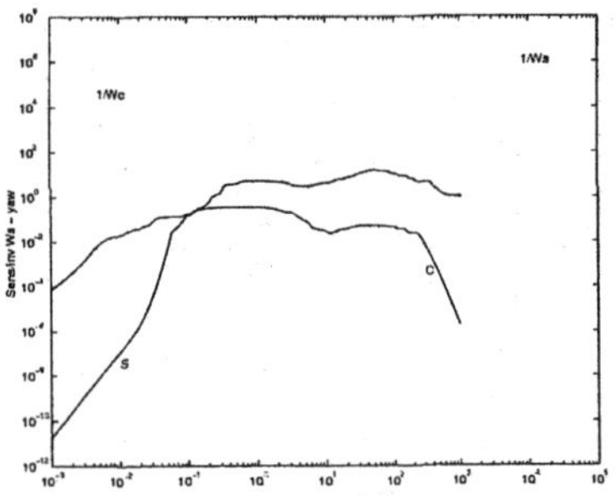

Fig. 3. Sensitivities/Inverse of Weightings - yaw -low order weighting

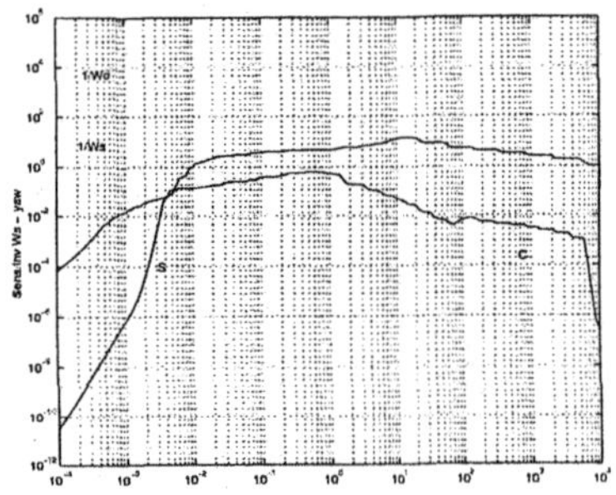

Fig. 4. Sensitivities/Inverse of Weightings - yaw -high order weighting

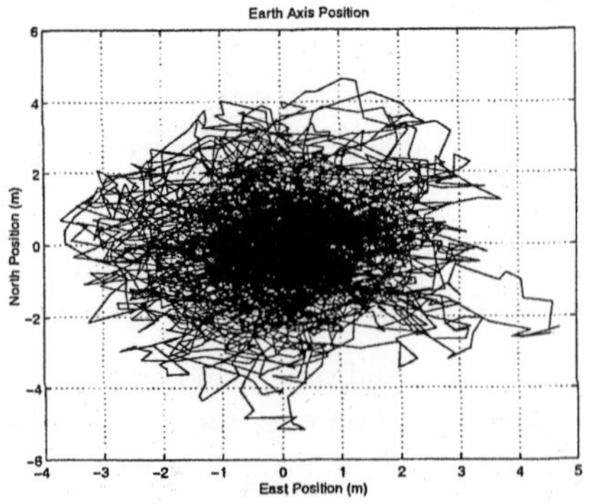


Fig. 5. Positioning at Sea Surface- by Katebi et al.

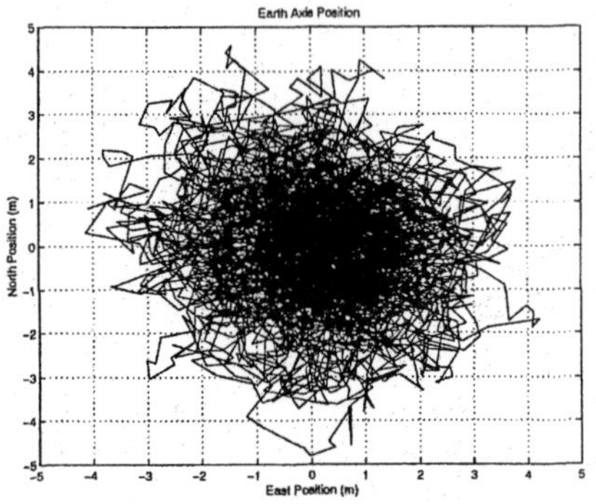


Fig. 6. Positioning at Sea Surface-high oder

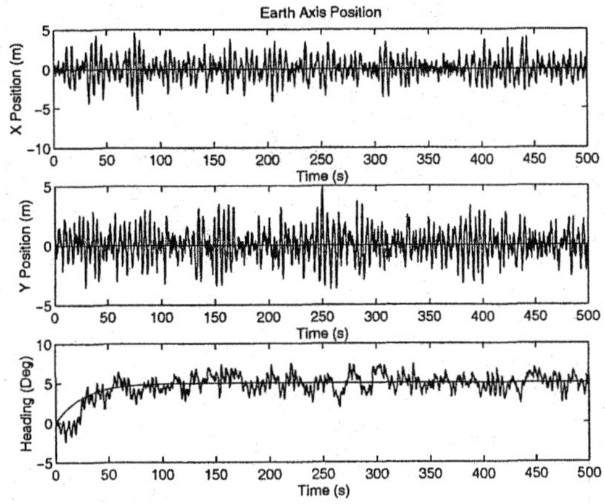


Fig. 7. Nonlinear Step Responses by Katebi et al.

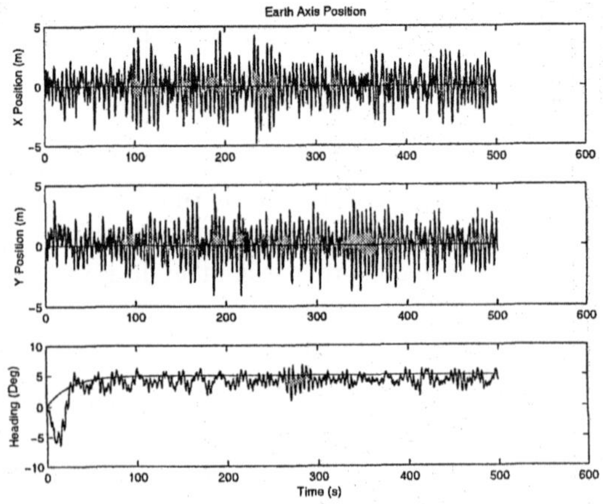


Fig. 8. Nonlinear Step Responses-high order weighting GA

ELSEVIER
IFAC
PUBLICATIONS
www.elsevier.com/locate/ifac

# FUZZY RUDDER–ROLL DAMPING SYSTEM BASED ON ANALYSIS OF THE AUTOPILOT COMMANDS

**Viorel Nicolau***, **Emil Ceanga****

*Department of Automatic Control and Electronics, "Dunarea de Jos" University of Galati, 47 Domneasca Street, Galati, 800008, Galati, ROMANIA*
*Email:* Viorel.Nicolau@ugal.ro
***Research Center in Advanced Process Control Systems, "Dunarea de Jos" University of Galati, 47 Domneasca Street, Galati, 800008, Galati, ROMANIA*
*Email:* Emil.Ceanga@ugal.ro

Abstract: For mono-variable autopilots, there are moments of time when the command of the rudder to control the yaw angle has a negative influence on roll oscillations, being a supplemental task for the roll stabilization controller. It is desirable to avoid this situation instead of trying to correct it. A fuzzy rudder-roll damping (FRRD) system is proposed, which modifies the autopilot commands, so that roll damping effects to be obtained, with small rudder angle and rudder rate values. The FRRD system can be used with any SISO autopilot with slow steering machine and with- or without other roll controller. *Copyright © 2004 IFAC*

Keywords: fuzzy systems, ship control, nonlinear systems, rudder-roll damping

## 1. INTRODUCTION

The main purpose of the steering machine (SM) is moving the rudder to control the heading of the ship in course-keeping or course-changing maneuvers. Also, the rudder can be used to reduce the roll motion induced by external disturbances.

The double problem of heading control and rudder-roll reduction has been analyzed by many authors. It is a single input-multiple output problem, since there is only one actuator to achieve two objectives, which can not be optimally at the same time.

The two objectives can be separated in the frequency domain, based on frequency characteristics of the rudder influence on yaw and roll motions. Low frequencies are used for heading control and high frequencies, for roll reduction. Thus, the problem is divided into two mono-variable control systems, as shown in Fig. 1.

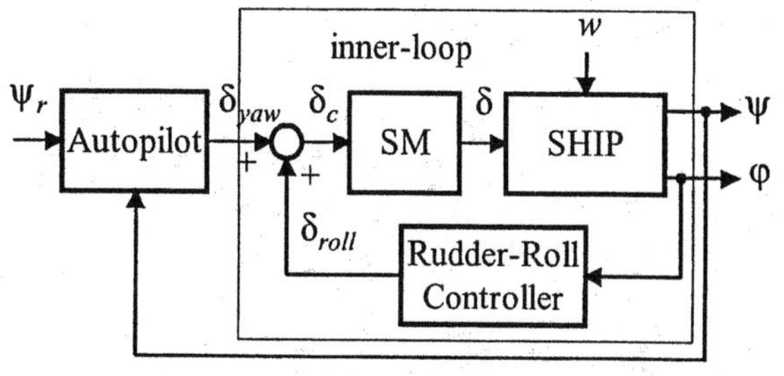


Fig. 1. Separated closed loops for yaw and roll controls

Different approaches were used to design Rudder Roll Stabilization (RRS) systems for linear or nonlinear ship models, such as: LQG method (Katebi, *et al.*, 1987), gain scheduling with automatic gain controller (AGC) (Van der Klugt, 1987) or with time-varying gain reduction (Laudval, and Fossen, 1998), adaptive control (Zhou, 1990; Van Amerongen, *et al.*, 1990), Quantitative Feedback Theory (Hearns, and Blanke, 1998), Internal Model Control (Tzeng, *et al*, 2001) etc.

The RRS systems gave good results on full-scale tests (Kallstrom, 1987; Van Amerongen, *et al.*, 1990), but they require high-speed SM and they are inefficiently at low ship speed. Hence, the RRS system can be combined with other roll stabilization systems, like fin control system (Oda, *et al.*, 2001).

The mono-variable autopilot and the rudder-roll controller take into account only one motion of the ship and ignore the other one. Inevitable there are moments of time when the command of the rudder to control the yaw angle ($\delta_{yaw}$) has a negative influence on roll oscillations.
This can be a supplemental task for the roll stabilization controller, which generates commands ($\delta_{roll}$) to compensate the negative influence of the yaw controller. It is desirable to avoid this situation instead of trying to correct it.

Therefore, it is important for autopilot to generate only rudder commands ($\delta_{yaw}$) with damping or non-increasing effects over roll movements, even with acceptable small errors of yaw angle.
One way doing this is to design MISO autopilots, using different techniques, like Sliding Mode Control (Laudval, and Fossen, 1997), Model Predictive Control (Perez, *et al.*, 2000) etc.
Another way is to modify the rudder command generated by a conventional SISO autopilot so that roll damping effects to be obtained. Such system is connected into the loop of the yaw angle and it is referred as rudder-roll damping system based on analysis of the autopilot commands.

In this paper, a fuzzy rudder-roll damping (FRRD) system is proposed, which modifies the autopilot commands when the influence on roll angle is significant, as shown in Fig. 2.

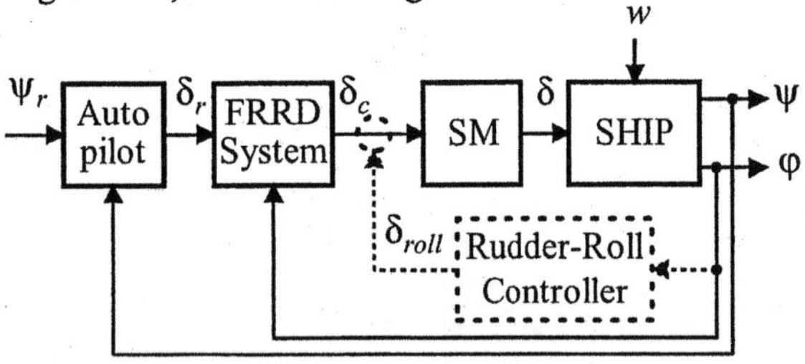


Fig. 2. The FRRD system, connected to outer loop of the yaw controller

The FRRD system can be used with any conventional autopilot and with- or without any other roll controller. Unlike the rudder-roll controller, a roll damping effect appears even for small rudder rate limits. This is important, especially for ships without roll stabilization controller and with slow steering machine.

The paper is organized as follows. In section 2, the working principle of FRRD system is described. Section 3 presents the structure of the FRRD system. In section 4, a fuzzy analyzer is designed. Section 5 provides the simulation results. Conclusions are presented in section 6.

## 2. ANALYSIS OF THE AUTOPILOT COMMANDS

In general, mono-variable autopilot has negative influence on the roll amplitude, because it takes into account only the yaw motion of the ship. Using FRRD system, roll reduction is obtained by modifying the autopilot commands when the influence on roll angle is significant.

For every discrete moment of time ($i$), the FRRD system analyzes the rudder command generated by the autopilot ($\delta_r(i)$) in correlation with roll movement characteristics (roll angle $\varphi(i-1)$ and roll rate $p(i-1)$). Then, the new rudder command ($\delta_c(i)$) is generated, as illustrated in figure below.

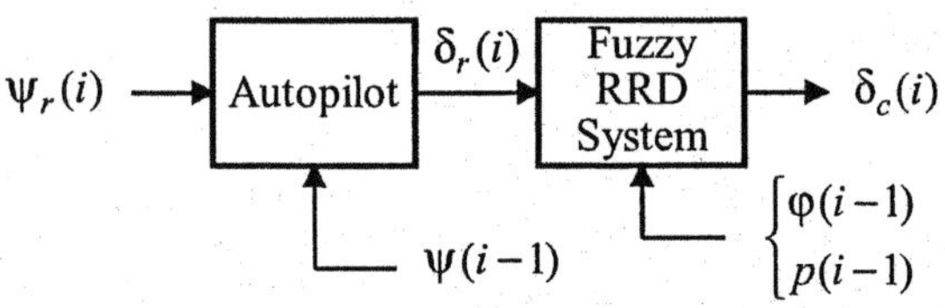


Fig. 3. Inputs and output of the FRRD system

Depending on the possible influence of the autopilot command on roll angle, the FRRD system can take one of three different actions:
a) if the effect of the rudder command on roll angle is not important, then the command from autopilot is not modified;
b) if the influence is positive, which means the new rudder command can decrease the roll angle, then the FRRD system amplifies the autopilot command;
c) if the influence is negative and the roll angle can be increased, then the FRRD system attenuates the autopilot command.

Suppose that, at the time moment $i$, the autopilot generates the rudder command $\delta_r(i)$ by modifying the last command $\delta_r(i-1)$ with an incremental value $\Delta\delta(i)$:

$$\delta_r(i) = \delta_r(i-1) + \Delta\delta(i) \qquad (1)$$

The incremental value can be positive or negative.

The FRRD system generates the new rudder command as follows:
a) if $\Delta\delta(i)$ is near zero or the effect on roll angle is not important then the autopilot command is not modified, resulting: $\delta_c(i) = \delta_r(i)$ (2)
b) if $\Delta\delta(i)$ is important and has a positive influence on roll angle, then the autopilot command is strengthened:

$$\delta_c(i) = \delta_r(i) + k_1 \cdot \Delta\delta(i) \qquad (3)$$

where $k_1 > 0$;
c) if $\Delta\delta(i)$ is important and has a negative influence on roll angle, then the autopilot command is weakened:

$$\delta_c(i) = \delta_r(i) - k_2 \cdot \Delta\delta(i) \qquad (4)$$

where $k_2 > 0$.

For example, the three situations described above are illustrated in Fig. 4, considering a positive variation of the autopilot command: $\Delta\delta(i) > 0$.

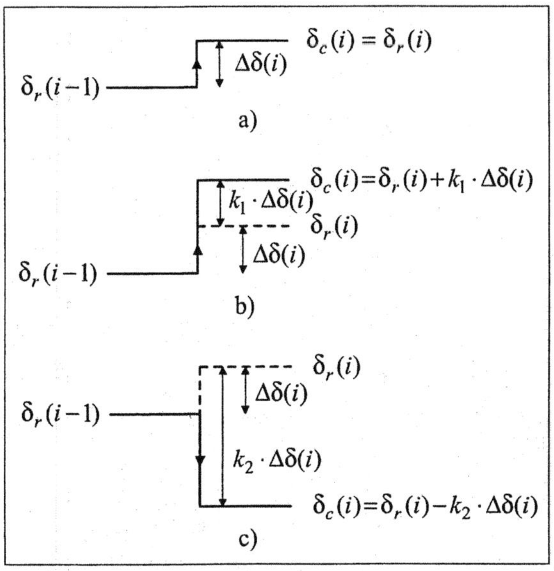


Fig. 4. Rudder command $\delta_c(i)$ generated by the FRRD system

In case a), the autopilot command is not modified. In second case, the autopilot command is strengthened and the rudder command generated by FRRD system increases. Hence, a real roll reduction is obtained without negative influence on yaw angle. In case c), the autopilot command is weakened. The rudder command can be smaller in the same direction or can be generated in opposite direction. As a result, non-increasing or damping effect on roll movement can be obtained, with small negative influence on yaw angle.
Eliminating the negative influence of the autopilot on roll angle, this system can produce important damping effects with acceptable small yaw errors.

The $k_1$ and $k_2$ parameters control both the rudder-roll damping effect and negative influence on yaw angle. Big parameter values produce important damping effects, but also they increase the rudder movements and the yaw error. The parameters are bounded by maximum values ($k_{1M}$ and $k_{2M}$), which depend on the steering machine characteristics, rudder movement policy and external disturbances:

$$k_1 \in (0, k_{1M}], \quad k_2 \in (0, k_{2M}] \qquad (5)$$

## 3. STRUCTURE OF THE FRRD SYSTEM

The three equations (2)-(4) can be condensed into one equation:

$$\delta_c(i) = \delta_r(i) + k(i)\cdot\Delta\delta(i) \qquad (6)$$

The parameter $k(i)$ has a variable form, depending on the possible influence of the autopilot command on roll angle:

$$k(i) = \begin{cases} +k_1 & \text{if the influence is positive} \\ 0 & \text{if the influence is not important} \\ -k_2 & \text{if the influence is negative} \end{cases} \qquad (7)$$

where $k_1 > 0$ and $k_2 > 0$.

In equation (6), the parameter $k(i)$ can be computed by a Fuzzy Analyzer (FA), which detects the type of influence on roll angle for every discrete moment of time ($i$). Also, the $k_1$ and $k_2$ parameters can be constants or fuzzy variables.

The structure of the FRRD system is illustrated in figure below.

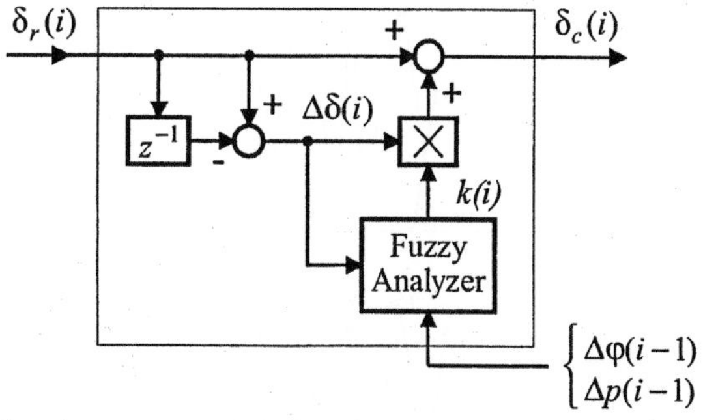


Fig. 5. Structure of the FRRD system

The Fuzzy Analyzer computes the parameter $k(i)$ based on variations of the rudder command $\Delta\delta(i)$, roll angle $\Delta\varphi(i-1)$ and roll rate $\Delta p(i-1)$. Knowing the sampling period ($T$), these variations define rudder command rate, roll rate and roll acceleration.

## 4. FUZZY ANALYZER DESIGN

The Fuzzy Analyzer has 3 inputs and 1 output, as shown in Fig. 6. The inputs are: rudder command rate (*rudder-rate*), roll acceleration (*roll-acc*) and roll rate (*roll-rate*). The output of FA is denoted *out*.

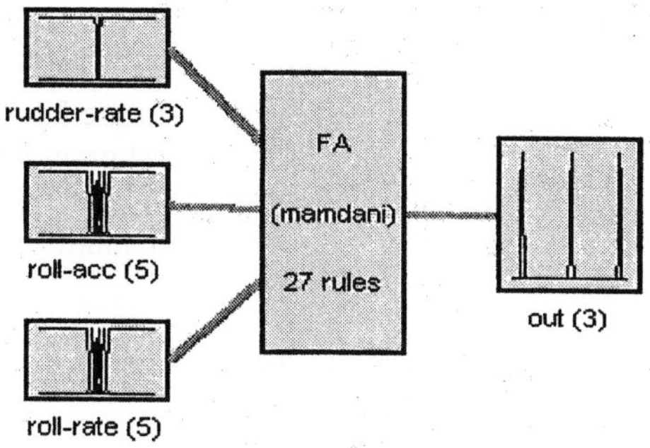


Fig. 6. Structure of the Fuzzy Analyzer

All membership functions of the FA inputs are illustrated in Fig. 7.

The *rudder-rate* fuzzy variable represents the incremental value of the autopilot command. It has only 3 fuzzy sets, showing that rudder command rate is near zero or has important positive or negative values. If the rudder command rate is near zero, then the influence on roll angle is not important and the autopilot command is not modified. The membership functions are labeled: *NEG*, *Z* and *POZ*. This fuzzy variable has a wide range of values: [-25, 25], but real values are much smaller. For simulations in section 5, they are smaller than ± 3 (deg/s).

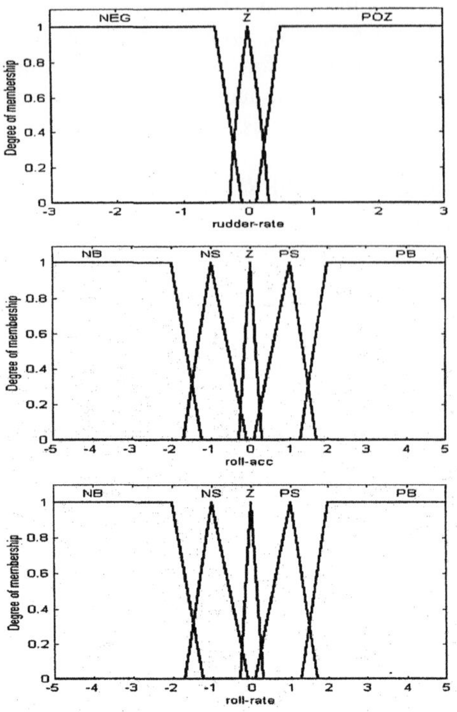


Fig. 7. Membership functions of the FA inputs

The other two inputs (*roll-acc* and *roll-rate*) have 5 fuzzy sets each, with more fine small value zones. The membership functions are the same and they are labeled: *NB*, *NS*, *Z*, *PS* and *PB*. Also, these variables have the same value range: [-10, 10], being twice larger than maximum simulation value.

Two versions of FA can be implemented, depending on the *k(i)* parameter type:
- with three constant values ($-k_2$, $0$, $+k_1$) or
- variable into the range [$-k_2$, $+k_1$].

The output of FA is denoted *out1* and *out2*, respectively. The membership functions are labeled: *-k2*, *0* and *+k1*, being singletons in first case. For a more flexible implementation, the singleton values are (*-1*, *0*, *+1*) and the gains *k2* and *k1* multiply the output *out1*.

Membership functions of the output, for both versions of FA, are illustrated below.

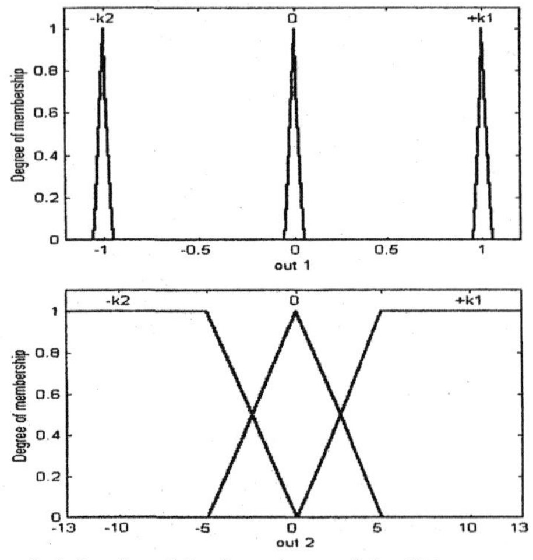


Fig. 8. Membership functions of the FA output

Table 1 Rule base of the Fuzzy Analyzer

| roll-acc \ rudder-rate | NEG | Z | POZ | roll-rate |
|---|---|---|---|---|
| | -k2 | 0 | - k2 | NB |
| | - k2 | 0 | - k2 | NS |
| NB | 0 | 0 | - k2 | Z |
| | +k1 | 0 | - k2 | PS |
| | +k1 | 0 | - k2 | PB |
| | - k2 | 0 | - k2 | NB |
| | - k2 | 0 | - k2 | NS |
| NS | 0 | 0 | - k2 | Z |
| | +k1 | 0 | - k2 | PS |
| | +k1 | 0 | - k2 | PB |
| | - k2 | 0 | 0 | NB |
| | - k2 | 0 | 0 | NS |
| Z | 0 | 0 | 0 | Z |
| | 0 | 0 | - k2 | PS |
| | 0 | 0 | - k2 | PB |
| | - k2 | 0 | +k1 | NB |
| | - k2 | 0 | +k1 | NS |
| PS | - k2 | 0 | 0 | Z |
| | - k2 | 0 | - k2 | PS |
| | - k2 | 0 | - k2 | PB |
| | - k2 | 0 | +k1 | NB |
| | - k2 | 0 | +k1 | NS |
| PB | - k2 | 0 | 0 | Z |
| | - k2 | 0 | - k2 | PS |
| | - k2 | 0 | - k2 | PB |

The complete rule base has 75 fuzzy rules and it is represented in Table 1, but the knowledge base has a set of 27 rules.

The knowledge base is generated using several expert rules, which describe the influence of the rudder command over roll angle:

*Rule no. 1:* During the time moments around the maximum values of roll angle and roll rate, the rudder command can be decreased or delayed.

*Rule no. 2*: Optimum time interval to decrease the rudder command is placed between maximum values of roll rate and roll angle.

*Rule no. 3*: During the time moments around the minimum values of roll angle and roll rate, the rudder command can be increased or delayed.

*Rule no. 4*: Optimum time interval to increase the rudder command is placed between minimum values of roll rate and roll angle.

## 5. SIMULATION RESULTS

The ship model used in simulations is a three degree-of-freedom linear system with coupled sway-yaw-roll equations. It was identified for a ship speed of 22 knots (Van der Klugt, 1987), considering the influence of the wave disturbances. The transfer functions have parameters depending on the speed of the ship and the incidence angle.

The steering machine (*SM*) is a nonlinear system, based on two-loop electro-hydraulic steering subsystem (Van Amerongen, 1982).

For roll movements, the nonlinearity of the first loop is important because it increases the phase lag and decreases the rudder force moment on roll angle. If the first loop is disregarded, simulation results are affected, being better than real values. Therefore, the nonlinear SM model is used.

To illustrate the FRRD system performance, a slow steering machine is considered, with maximum rudder deflection of ± 35 (deg) and maximum rudder rate of ± 2.5 (deg/s).

The wave disturbances are generated based on ITTC spectrum, considering a wave with significant height $h_{1/3} = 3$ (m). The wave spectrum was corrected with the ship's speed $U = 22$ (knot) and the incidence angle $\gamma = 110$ (deg).

A conventional PID autopilot is considered with- and without FRRD system.

The damping ratio is determined by the following formula (Laudval, and Fossen, 1998):

$$Reduction\ (\%) = 100 \cdot \frac{AP - FRRD}{AP}, \qquad (8)$$

where FRRD and AP are the standard deviation of roll rate with- and without the FRRD system. The positive values of *reduction* mean a damping effect, while negative values mean an increasing effect of motion amplitude. To complete the description of roll damping effect, the reduction formula can be applied for roll angle and roll acceleration.

Big values of the $k_1$ and $k_2$ parameters produce important roll reduction, but also they increase the rudder movements and the yaw error. The roll damping effect must be compared with the increasing effects of yaw and rudder movements. Therefore, the reduction formula is also computed for yaw and rudder motion characteristics (angle, rate and acceleration).

The two versions of FA (with *out1* and *out2*) are combined with two different values of the $k_1$ and $k_2$ parameters ($k_1 = k_2 = 13$ and $k_1 = k_2 = 20$), resulting 4 types of FRRD system. Simulation results are represented in Table 2.

Table 2 Simulation results of the FRRD System

| Motion | Reduction [%] with FRRD System | | | |
|---|---|---|---|---|
| | *out1* $k_1=k_2=13$ | *out2* $k_1=k_2=13$ | *out1* $k_1=k_2=20$ | *out2* $k_1=k_2=20$ |
| Roll angle | 6.1 | 6.6 | 11.7 | 11.8 |
| Roll rate | 5.9 | 6.2 | 11.2 | 11.3 |
| Roll acc. | 5.3 | 5.5 | 9.9 | 10 |
| Yaw angle | - 4.3 | - 5.2 | - 13.3 | - 10.4 |
| Yaw rate | 1.4 | 0.7 | 2.7 | 3 |
| Yaw acc. | 3.1 | 3.1 | 6 | 6.1 |
| Rudder angle | - 62.8 | - 75.5 | - 134 | - 132.7 |
| Rudder rate | - 79 | - 84.6 | - 154 | - 154.1 |
| Rudder acc. | - 42.2 | - 48 | - 72.9 | - 74.4 |

It can be observed that using the FRRD system, the yaw angle increases but the yaw rate and acceleration decrease. As a result the yaw motion is wider but smoother, which can decrease the water resistance on forward movement of the ship.

The yaw and roll angles are illustrated in Fig. 9, with- and without FRRD system (type 4: *out2*, $k_1 = k_2 = 20$). The dark lines represent the yaw and roll movements without FRRD system. The roll damping effect is visible on local extremes.

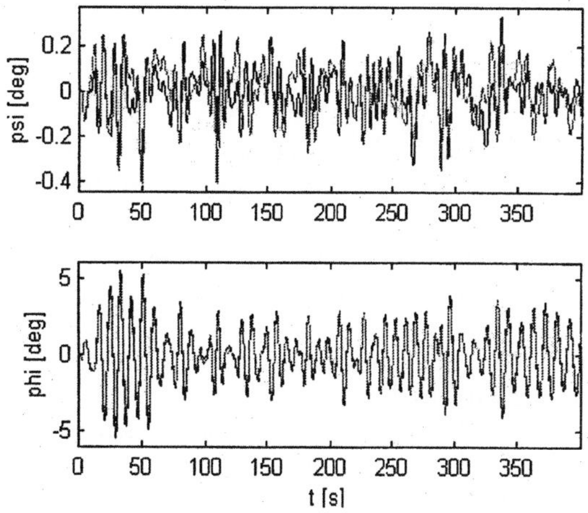


Fig. 9. Ship motions with- and without FRRD system

The roll damping effect can also be illustrated by computing the relative variation (*rel var [%]*) between roll angles without, respectively with FRRD system. The result is illustrated in Fig. 10.

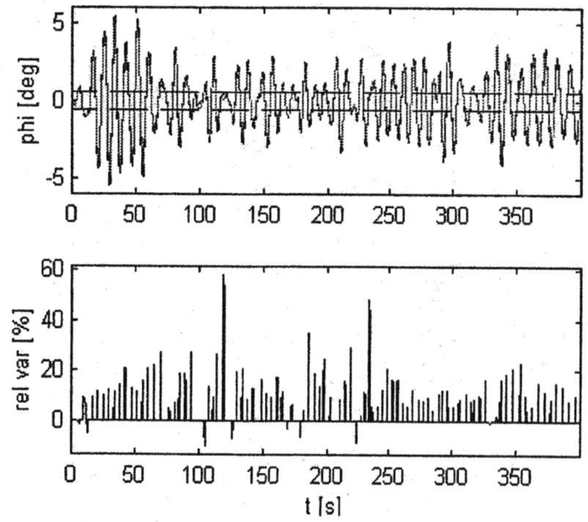


Fig. 10. Relative variation of roll angle

The relative variation was computed for all local extremes which are bigger than two limits, illustrated with horizontal dashed lines. The limits are computed as percent (10%) of the maximum and minimum values of roll angle.

Positive values of relative variation represent roll damping effect, while negative values indicate increasing roll amplitude. It can be observed that the representation has no statistical characteristics. For most local extremes, the damping effect is obviously.

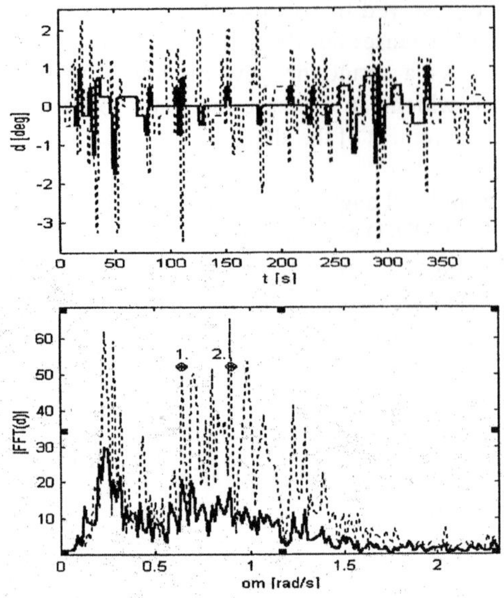


Fig. 11. Rudder commands and rudder FFT with- and without FRRD system

The rudder commands and rudder FFT with- and without FRRD system (type 4: *out2*, $k_1 = k_2 = 20$) are illustrated in Fig. 11. The dotted lines represent the rudder movement with FRRD system. Although the magnitude of rudder angle increases, it remains at small values. Thus, an important roll reduction is obtained at small rudder angle and rudder rate.
The rudder spectral components are computed with FFT. Using FRRD system, the high frequency components are amplified, especially those placed near the natural roll frequency (marked and denoted *1.*) and around the maximum component of the corrected wave spectrum (marked and denoted *2.*).

The outputs of fuzzy analyzer ($k$) are illustrated in Fig. 12, for both versions of FA, with $k_1 = k_2 = 20$. The big negative values of $k$ represent moments of time when the autopilot command has negative influence on roll angle and the command is modified. The positive values of $k$ correspond to positive influence on roll angle. There are few moments of time when the autopilot itself has a positive effect on roll angle.

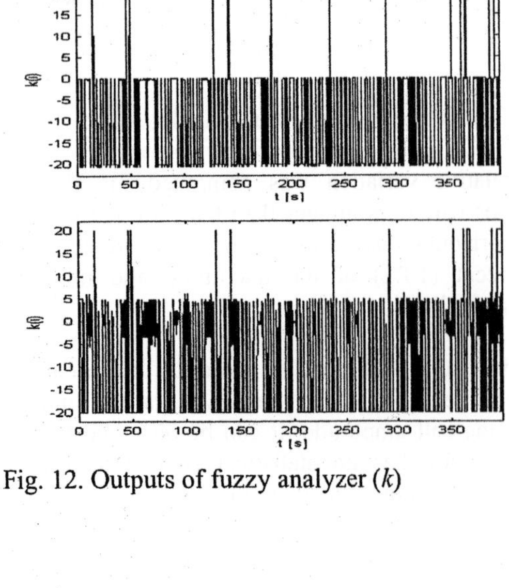


Fig. 12. Outputs of fuzzy analyzer ($k$)

## 6. CONCLUSIONS

A fuzzy rudder-roll damping (FRRD) system is proposed, which modifies the commands of SISO autopilot, so that roll damping effects to be obtained. Important roll reduction results with small rudder angle and rudder rate values. The FRRD system can be used with any SISO autopilot with slow steering machine and with- or without other roll controller.

## REFERENCES

Hearns G. and M. Blanke (1998). Quantitative Analysis and Design of a Rudder Roll Damping Controller. In: *Proceedings of IFAC Conference on Control Applications in Marine Systems (CAMS'98)*, Fukuoka, Japan.

Kallstrom, C. G. (1987). Improved Operational Effectiveness of Naval Ships by Rudder Roll Stabilization. In: *Proc. of Asian Pacific Naval Exhibition and Conf. (NAVAL'87)*, Singapore.

Katebi, M. R., D. K. Wong and M. J. Grimble (1987). LQG autopilot and rudder roll stabilization control system design. In: *Proceedings of the 8th International Ship Control Systems Symposium (SCSS'87)*, pp. 369-384, Bath, UK.

Laudval, T. and T. I. Fossen (1997). Rudder-roll damping system for a nonlinear ship model using sliding mode control. In: *Proceedings of European Control Conference (ECC'97)*.

Laudval, T. and T. I. Fossen (1998). Rudder Roll Stabilization of Ships Subject to Input Rate Saturation Using a Gain Scheduled Control Law. In:*Proc. of IFAC Conf. CAMS'98*, Fukuoka, Japan.

Oda, H., M. Kanda, T. Hyodo, K. Nakamura, H. Fukushima, S. Iwamoto, S. Takeda and K. Ohtsu (2001). The preliminary study of fin and rudder multivariate hybrid control system – Advanced rudder roll stabilization system. In: *Proceedings of IFAC Conference CAMS 2001*, Glasgow, UK.

Perez T., G.C. Goodwin and C.Y. Tzeng (2000). Model Predictive Rudder Roll Stabilization Control for Ships. In: *Proceedings of IFAC Conference on Marine Craft Maneuvering and Control (MCMC' 2000)*, Denmark.

Tzeng, C. Y., C. Y. Wu and Y. L. Chu (2001). A Sensitivity Function Approach to the Design of Rudder Roll Stabilization Controller. *J. of Marine Science and Technology*, vol.**9**, no.2, pp. 100-112.

Van Amerongen, J. (1982). *Adaptive Steering of Ships.* Ph.D.Thesis, Delft University of Technology, The Netherlands.

Van Amerongen, J., P.G.M. van der Klugt and H. Van Nauta Lempke (1990). Rudder Roll Stabilization for Ships. *Automatica*, **AUT-26**(4), pp. 679-690.

Van der Klugt, P.G.M. (1987). *Rudder Roll Stabilization.* Ph.D.Thesis, Delft University of Technology, The Netherlands.

Zhou, W. W. (1990). A New Approach for Adaptive Rudder Roll Stabilization Control. In: *Proc. of the 9th Int. SCSS'90*, pp. 1115-1125, Bethesda, MD.

ELSEVIER
IFAC
PUBLICATIONS
www.elsevier.com/locate/ifac

# SPECIALIZED LEARNING FOR SHIP INTELLIGENT TRACK-KEEPING USING NEUROFUZZY

**Yongqiang Zhuo and Grant E. Hearn** *

* *School of Engineering Sciences, University of Southampton, UK*
*zhuoyq@ship.soton.ac.uk and grant@ship.soton.ac.uk*

Abstract: An on-line trained neurofuzzy control scheme is proposed for ship track-keeping. Due to the large inertia and relatively slow responses of the ship, a single-input multi-output control strategy is developed. This specialized learning neurofuzzy controller uses the back-propagation gradient descent method to update the parameters of the network through time. With a relatively modest amount of domain knowledge of the ship behaviour, the designed scheme enables real time control of a simulated nonlinear ship manoeuvring under wind and current disturbances. *Copyright © 2004 IFAC*

Keywords: track-keeping, intelligent control, specialized learning, neurofuzzy

## 1. INTRODUCTION

Automation is becoming more and more accepted and hence more common aboard ships. By 1980 ship autopilots had mainly evolved from the classical course-keeping task (Zhang *et al.*, 1996). Using directional information taken from a gyrocompass, the autopilot is able to steer a ship on a predetermined course; a useful contribution to ship navigation in the open sea. However, in restricted or narrow waters the ship navigation task is concerned with following a specified course between two way-points. A ship equipped with a course-keeping autopilot can drift due to the influence of wind and current. Consequently the noted deviation from a desired track must be adjusted manually.

Track-keeping controllers have been designed using modern control techniques such as Linear Quadratic Gaussian (LQG) control (Holzhuter, 1990), self-tuning control (Lu *et al.*, 1990) and $H_\infty$ control (Messer and Grimble, 1993). All of these analytical control strategies exhibit dependence on a reliable model of the manoeuvring ship dynamic responses. Hydrodynamics plays a key role in these systems. Despite on going development the mathematical models are never sufficiently exhaustive to cover all possible sailing and environmental conditions. The ship dynamics has intrinsic nonlinearities that cannot be neglected during manoeuvres.

With navigational aids such as Global Positioning System (GPS) and Electronic Chart Display and Information System (ECDIS) installed on board, position measurement accuracy and reliability are increased. Hence it is possible to design a track-keeping guidance system of sufficient accuracy for ship operations in narrow waters. This paper proposes an intelligent specialized learning neurofuzzy approach to track-keeping.

### *1.1 Track-keeping Control Strategy*

A typical feedback control system block diagram consists of a plant block and a controller block. The controller block maps the plant state into a control action that should achieve a given control objective. The control objective here is to generate the rudder action necessary to stabilize ship position and attitude along a desired track.

Ship motion control is a highly nonlinear process. It involves various levels of uncertainty, primarily, due to the unpredictability of the characteristics of the environment and an insufficient knowledge of the ship dynamics. During the past few decades, fuzzy control has been successfully applied and has proved to be superior in performance to some conventional systems in many practical areas especially where the plants are poorly modelled or have nonlinear dynamics. Most of the time the controller being mimicked is an experienced human operator, whose ability to control the plant is more than satisfactory. Fuzzy logic methods provide an efficient way to cope with uncertainties and to encode and approximate numerical functions.

Although Fuzzy logic provides a feasible control method (readily capable of capturing the approximate, qualitative aspects of human knowledge and reasoning), they cannot learn. The majority of fuzzy systems developed so far are static and designed in an iterative open-loop fashion. Usually, the designer specifies a fuzzy rule base and then enters an evaluation/editing design loop. Both the performance measures and adaptation strategies are subjective. Hence if the plant dynamics and the environment change, then the performance of the 'well-designed' fuzzy systems will degrade. Therefore, an automatic learning algorithm needs to be developed for on-line adjustment of the rule bases of fuzzy systems in response to variations in operating conditions. Hence learning algorithms are used in the domain of neural networks to create fuzzy logic control from data. The learning algorithms can learn how to identify the parameters associated with the fuzzy sets and fuzzy rules.

## 2. SPECIALIZED LEARNING STRUCTURE

A neurofuzzy controller can take advantage of all the neural network controller design techniques. To obtain a desired control action there is an inverse 'general learning' control method base on off-line training. In the learning phase, a training set is obtained by generating inputs $u_k$ at random, and observing the corresponding outputs of ship state $x_k$. A major shortcoming with the inverse control scheme is minimization of the network error rather than the overall system error.

An alternative approach is to minimize the system error directly using the 'specialized learning' architecture illustrated in Figure 1. In this case the network is trained to find the plant input that drives the system output to the desired track. This is accomplished by using the error, between the desired and actual responses of the plant, to adjust the weights of the network using a steepest descent procedure; during each iteration the weights are adjusted to maximally decrease the error. This procedure requires knowledge of the plant Jacobian. This architecture can learn specifically in the region of interest, and it may be trained on-line, thus fine-tuning itself whilst actually performing useful work.

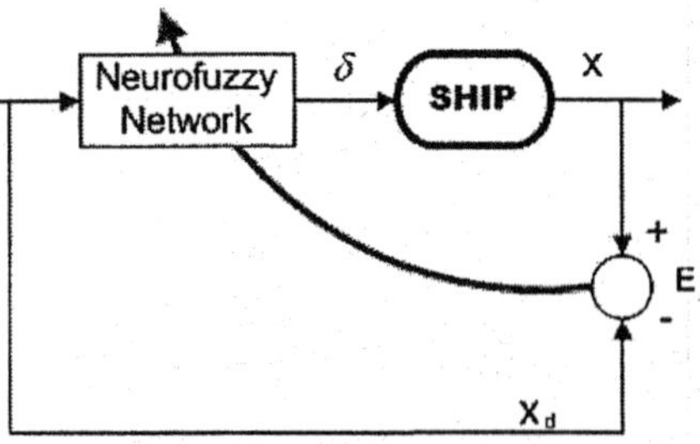


Fig. 1. The diagram for specialized learning architecture of neurofuzzy.

## 3. THE NEUROFUZZY CONTROLLER

### 3.1 Inputs and Output of the Neurofuzzy Network

The purpose of the controller is to change ship heading by manipulating the rudder to change the position of the ship. However, the inertia of a ship means it will not be able to follow rudder changes instantaneously. Thus the resulting heading changes are more gradual and provide a trajectory governed by the ship dynamics.

During course-changing, the task of the helmsman is to reduce the course error quickly with minimum over-shoot and an acceptable rate of turn. A track-keeping system can be designed by simply adding feedback from the sway position in an outer loop (Fossen, 1994). Here the track-keeping neurofuzzy network input vector consists of track error $\Delta\varepsilon$, heading error $\Delta\psi$ and yaw rate error $\Delta r$. The output of the neurofuzzy network is rudder angle $\delta$.

### 3.2 Architecture of the Proposed Neurofuzzy Approach

The proposed neurofuzzy control system is demonstrated in Figure 2.

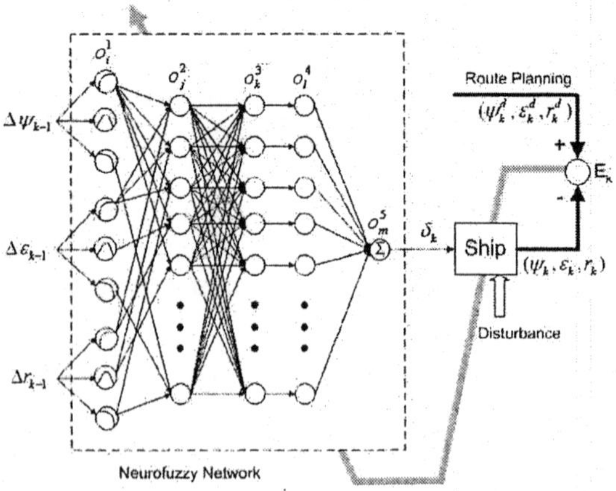


Fig. 2. The proposed neurofuzzy control system

The neurofuzzy network has 5 layers. The subscripts $i$, $j$, $k$, $l$ and $m$ define the number of neural units in each layer. The outputs of each layer are defined by:

$$O_i^1 = \mu_{A_i}(x) : i = 1, 2, ..., 9, \quad (1)$$

where $O_i^1$ is essentially the membership function of $A_i$ when $x$ assumes the input value $A_i$. The bell-shaped membership functions are defined to have the form:

$$\mu_{A_i}(x) = \frac{1}{1 + (\frac{x - c_i}{a_i})^{2b_i}}. \quad (2)$$

The parameters $a_i$, $b_i$ and $c_i$ are updated in the adaptation process.

$$O_j^2 = w_j = \mu_{A_e}(\Delta\psi) \times \mu_{A_g}(\Delta\varepsilon) \times \mu_{A_h}(\Delta r), \\ e = 1, 2, 3; g = 4, 5, 6; h = 7, 8, 9 \\ \& \ j = 1, 2, ..., 27. \quad (3)$$

$$O_k^3 = \bar{w_k} = \frac{w_k}{\sum_{k=1}^{k=27} w_k} : k = 1, 2, ..., 27. \quad (4)$$

$$O_l^4 = \bar{w_l} f_l = \bar{w_l}(u_l \Delta\psi + q_l \Delta\varepsilon + s_l \Delta r + t_l), \\ : l = 1, 2, ..., 27. \quad (5)$$

Parameters $u_l, q_l, s_l$ and $t_l$ are to be updated on the backward pass.

Finally,

$$O_m^5 = \sum_l \bar{w_l} f_l = \sum_l \frac{w_l f_l}{w_l} \\ : l = 1, 2, ..., 27 \ \& \ m = 1, \quad (6)$$

$\delta_k$ corresponding to the output of $O_m^5$.

## 4. ON-LINE SELF-LEARNING ALGORITHM

On-line learning is concerned with learning from data as the system operates (usually in real time). The data changes continuously. For on-line adaptation it exploits all available process knowledge and measurement data. In the on-line learning phase the rule premises are kept fixed and only the input membership and rule consequences are adapted.

The back-propagation Sugeno-type gradient descent method is applied for identification of the consequent parameters $(u, q, s, t)$ and the membership function parameters $(a, b, c)$ to emulate a given training data set. The parameters are updated after each data presentation via the on-line learning paradigm. Thus, there is only one epoch during each parameter update.

The error function is defined as:

$$E_k = \frac{1}{2}[\rho_1(\psi_k^d - \psi_k)^2 + \rho_2(\varepsilon_k^d - \varepsilon_k)^2 + \rho_3(r_k^d - r_k)^2] \quad (7)$$

where at time step $k$ $\psi_k$ is ship heading, $\psi_k^d$ is desired ship heading; $\varepsilon_k$ is track error, $\varepsilon_k^d$ is desired track error, normally it is zero; $r_k$ is yaw rate and $r_k^d$ is desired yaw rate. The weighting parameters $\rho_1$, $\rho_2$ and $\rho_3$ allow the assignment of the relative importance of the three parts of the error function.

The back-propagation algorithm, represented by

$$p_i(k) = p_i(k-1) - \eta \frac{\partial E_k}{\partial p_i} + \alpha \Delta p_i(k-1) \quad (8)$$

is used to train the neurofuzzy network in order to minimize the error $E_k$ defined by Equation (7). $p_i$ is the parameter to be updated, $\alpha$ is the momentum term that can increase the convergence rate. $\eta$ is the learning-rate which can be further expressed as

$$\eta = \frac{K}{\sqrt{\sum_{p_i} (\frac{\partial E}{\partial p_i})^2}}. \quad (9)$$

The parameter $K$ introduced by Jang and Sun (Jang and Sun, 1997) is the non-dimensional step size, the length of each gradient transition in the parameter space.

To update the parameters in $O_i^1$ the required gradient of $E_k$ with respect to $p_i$ in Equation (8) is determined using the chain rule, that is

$$\frac{\partial E_k}{\partial p_i} = \frac{\partial E_k}{\partial O_i^1} \frac{\partial O_i^1}{\partial p_i} \\ = \frac{\partial E_k}{\partial O_m^5} \sum_j \sum_k \sum_l \frac{\partial O_m^5}{\partial O_l^4} \frac{\partial O_l^4}{\partial O_k^3} \frac{\partial O_k^3}{\partial O_j^2} \frac{\partial O_j^2}{\partial O_i^1} \frac{\partial O_i^1}{\partial p_i} \quad (10)$$

where the required gradient of $O_i^1$ with respect to the parameters $p_i$ is determined from

$$\frac{\partial O_i^1}{\partial a_i} = \frac{2b_i(\frac{x-c_i}{a_i})^2 b_i}{a_i(1 + (\frac{x-c_i}{a_i})^2 b_i)^2}, \quad (11)$$

$$\frac{\partial O_i^1}{\partial b_i} = \frac{-log(\frac{x-c_i}{a_i})^2 (\frac{x-c_i}{a_i})^2 b_i}{(1 + (\frac{x-c_i}{a_i})^2 b_i)^2} \quad (12)$$

and

$$\frac{\partial O_i^1}{\partial c_i} = \frac{2b_i(\frac{x-c_i}{a_i})^2 b_i}{(x - c_i)(1 + (\frac{x-c_i}{a_i})^2 b_i)^2}, \quad (13)$$

with $i$=1,2,3 when $x = \varepsilon_k$; $i$=4,5,6 when $x = \psi_k$; $i$=7,8,9 when $x = r_k$;

The remaining expressions at the higher levels are:

$$\frac{\partial O_j^2}{\partial O_i^1} = \begin{cases} \mu_{A_g}(\Delta\varepsilon) \times \mu_{A_h}(\Delta r) : i = 1, 2, 3. \\ \mu_{A_e}(\Delta\varphi) \times \mu_{A_h}(\Delta r) : i = 4, 5, 6. \\ \mu_{A_e}(\Delta\varphi) \times \mu_{A_g}(\Delta\varepsilon) : i = 7, 8, 9. \end{cases} \\ : j = 1, 2, ..., 27, \quad (14)$$

$$\frac{\partial O_k^3}{\partial O_j^2} = \begin{cases} \frac{\sum w - O_k^2}{(\sum w)^2} : j = k \\ \frac{-O_k^2}{(\sum w)^2} : j \neq k \end{cases} : j, k = 1, 2, ..., 27, \quad (15)$$

$$\frac{\partial O_l^4}{\partial O_k^3} = \begin{cases} 0 : k \neq l \\ f_k : k = l \end{cases} k, l = 1, 2, ..., 27, \quad (16)$$

$$\frac{\partial O_m^5}{\partial O_l^4} = 1 \qquad : l, m = 1, 2, ..., 27 \quad (17)$$

and

$$\frac{\partial E_k}{\partial O_m^5} = \rho_1(\psi_k - \psi_k^d)\frac{\partial \psi_k}{\partial \delta_k} + \rho_2(\varepsilon_k - \varepsilon_k^d)\frac{\partial \varepsilon_k}{\partial \delta_k}$$
$$+\rho_3(r_k - r_k^d)\frac{\partial r_k}{\partial \delta_k}. \tag{18}$$

Exact calculation of the plant Jacobian $\frac{\partial \psi_k}{\partial \delta_k}$, $\frac{\partial \varepsilon_k}{\partial \delta_k}$ and $\frac{\partial r_k}{\partial \delta_k}$ will need a precise mathematical expression of the ship dynamics. Since the intelligent controller is designed for a plant that is a function of unknown form Psaltis et al. (Psaltis *et al.*, 1988) sidestepped this problem and approximated the partial derivatives as

$$\frac{\partial E_k}{\partial u_k} = \frac{E_k - E_{k-1}}{u_k - u_{k-1}}, \tag{19}$$

where $u_k$ is the input. However Ng (Ng, 1997) showed that such an approach can sometimes result in undesirable tracking when the plant output and control input change suddenly. In this work the sign of the plant Jacobian is employed, as used by Saerens and Soquet (Saerens and Soquet, 1991), Cui and Shin (Cui and Shin, 1993) and Zhang et al. (Zhang *et al.*, 1996). For the ship track-keeping problem, it is obvious that a positive increase of rudder $\delta_k$ will increase the tracking error $\varepsilon_k$, the ship heading $\psi_k$ and the ship yaw rate $r_k$. Therefore, it is natural to define

$$sign(\frac{\partial \psi_k}{\partial \delta_k}) = sign(\frac{\partial \varepsilon_k}{\partial \delta_k}) = sign(\frac{\partial r_k}{\partial \delta_k}) = +1 \tag{20}$$

so that Equation (18) assumes the form

$$\frac{\partial E_k}{\partial O_m^5} = \rho_1(\psi_k - \psi_k^d)sign(\frac{\partial \psi_k}{\partial \delta_k})$$
$$+\rho_2(\varepsilon_k - \varepsilon_k^d)sign(\frac{\partial \varepsilon_k}{\partial \delta_k}) + \rho_3(r_k - r_k^d)sign(\frac{\partial r_k}{\partial \delta_k})$$
$$= \rho_1(\psi_k - \psi_k^d) + \rho_2(\varepsilon_k - \varepsilon_k^d) + \rho_3(r_k - r_k^d). \tag{21}$$

Similarly, to update the parameters in $O_l^4$ the required gradient of $E_k$ with respect to $p_l$ in Equation (8) is determined from

$$\frac{\partial E_k}{\partial p_l} = \frac{\partial E_k}{\partial O_m^5}\frac{\partial O_m^5}{\partial O_l^4}\frac{\partial O_l^4}{\partial p_l}, \tag{22}$$

where $\frac{\partial E_k}{\partial O_m^5}$ and $\frac{\partial O_m^5}{\partial O_l^4}$ are determined from Equations (18) & (17) respectively. The required gradient of $O_l^4$ to the parameters $p_i$ is determined using

$$\frac{\partial O_l^4}{\partial u_l} = \bar{w}_l \varepsilon_k, \frac{\partial O_l^4}{\partial q_l} = \bar{w}_l \varphi_k, \frac{\partial O_l^4}{\partial s_l} = \bar{w}_l r_k. \tag{23}$$

## 5. APPLICATION OF PROPOSED SCHEME

To demonstrate the self-learning neurofuzzy control scheme, a general cargo ship of length 126.0 m with width of 20.8 m is used. The speed of the ship is 11.7 knots. In this paper, the MMG model is used as the ship maneuvering mathematical model for simulation. The model is described in some detail in the work of citeyang. Here it is sufficient to note that the surge, sway and yaw equation have the form

$$\begin{aligned}
\dot{u} &= [X_H + X_P + X_R + X_A + (m + m_y)vr]/(m + n \\
\dot{v} &= [Y_H + Y_P + Y_R + Y_A - (m + m_x)ur]/(m + m_y \\
\dot{r} &= (N_H + N_P + N_R + N_A)/(I_{zz} + J_{zz}) \\
\dot{x} &= (u + u_c)cos\psi - (v + v_c)sin\psi \\
\dot{y} &= (u + u_c)sin\psi + (v + v_c)cos\psi \\
\dot{\psi} &= r.
\end{aligned} \tag{24}$$

Here, $m, m_x$ & $m_y$, $I_{zz}$ and $J_{zz}$ are the mass of ship, the added masses of the ship, the moment of inertia of ship and the added inertia of the ship. $X, Y$ & $N$ are the fore-aft force, the lateral force and the yawing moment about z axis. $u, v$ and $r$ are forward speed, lateral speed and ship heading rate respectively. Subscripts $H, P, R$ & $A$ indicate forces and moments included relate to bare hull, propeller, rudder and wind respectively. $u_c$ and $v_c$ are current speed in the axial and transverse directions. $\psi$ is the ship heading. The organization and implementation of the NN controller and neurofuzzy controller are independent of the model used to represent the ship dynamics.

To consider the environmental disturbances, ship heading should be corrected by compensating for leeway and counteracting the current. The amount of leeway angle varies with trim and the general design of the ship's superstructure, wind direction and strength. It is complicated to calculate the leeway angle. Here only the heading correction to counteract the current is calculated in accordance with the Equation (25), that is

$$\psi_k^d = \psi_k^p + \beta. \tag{25}$$

$\psi_k^p$ is planned ship heading and this depends on way points, $\beta$ is the correcting angle to counteract the current. The value of $\beta$ can be calculated in accordance with Figure 3.

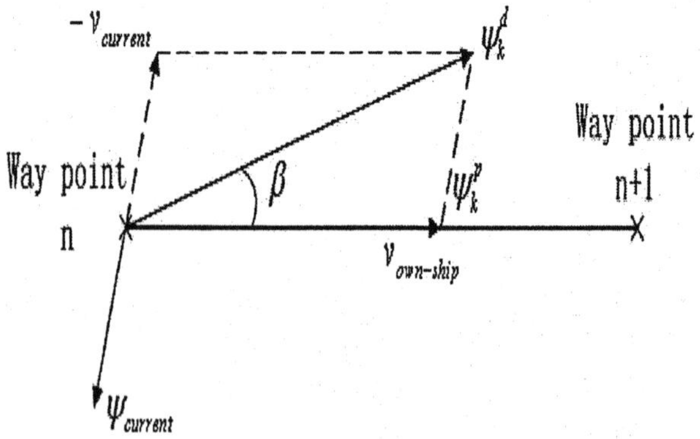


Fig. 3. Ship heading correction to counteract the current of magnitude $v_{current}$ and direction $\psi_{current}$.

A straight line route is used to test the performance of the neurofuzzy controller. In this case, the track-keeping control is simulated with and without environmental disturbances.

To judge how well the neurofuzzy controller performs, a neural controller similar in design to the work of Zhang et al. (Zhang *et al.*, 1996) is used to perform the same track-keeping tasks under the same conditions. The principal differences between the proposed method of this paper and that of Zhang et al. is:

1) The error calculation is undertaken in accordance with Equation (7). This involves track error, heading error and yaw rate error (not previously included).

2) Due to the on-line training, there is only one epoch through one time step (rather than multiple iterations within each time step).

3) Learning is facilitated using neurofuzzy approach rather than neural network.

For consistency Equation (26) is applied in both methods, whereas Zhang et al. fixed $\rho_1$, $\rho_2$ and $\rho_3$. Rather than fix $\rho_1$, $\rho_2$ and $\rho_3$ a priori they are assigned the algebraic values

$$\begin{cases} \rho_1 = 1.0 + 0.1E_\varepsilon \\ \rho_2 = 0.1 + 0.01E_\varepsilon \\ \rho_3 = 130 + 5E_\varepsilon \end{cases} \qquad (26)$$

with $E_\varepsilon = |\varepsilon_k^d - \varepsilon_k|$. This form of weighting factor is used to increase the training speed when the tracking error is large. Equation (26) effectively exaggerates the error $E_k$ of Equation (7).

Figure 4 shows the track-keeping performances without environment disturbances using these two controllers. The stepsize $K$ is 0.015. The initial condition adopted is that the track error is 50 metres, the initial heading and yaw rate are zero. The purpose of this control is to stabilize the ship positional motion, so the values of $\rho_1$, $\rho_2$ and $\rho_3$ are adjusted with the value of track error calculated by Equation (26). Whereas the neurofuzzy controller achieves the required track more quickly than the NN controller, without any overshoot, there is gentle variation in the ship heading, yaw rate and rudder activity signals in the former case.

Figure 5 shows the performances of track-keeping under environment disturbances by the use of these two controllers. The current speed is 1.5 knots and its direction is 040$^o$. The wind speed is 3 m/s and its direction is 160$^o$. Initially, the condition is adopted that the track error is 20 metres with the ship heading and yaw rate both zero. The differences between the ship heading, yaw rate and rudder activity signals are more noticeable in the case of environmental disturbance. Again the neurofuzzy controller achieves and maintains desired track much quicker. In fact if either current or wind speed are increased one found both systems can cope quite well for a wind speed of 6 m/s with current unchanged, but now there is more prolonged rudder activity with the NN controller. When the current speed is reduced from 1.5 knots to 0.5 knots the NN controller does converge to a steady track and the rudder activity is less whereas the neurofuzzy controller behaves well, quickly reaching desired track with little rudder activity.

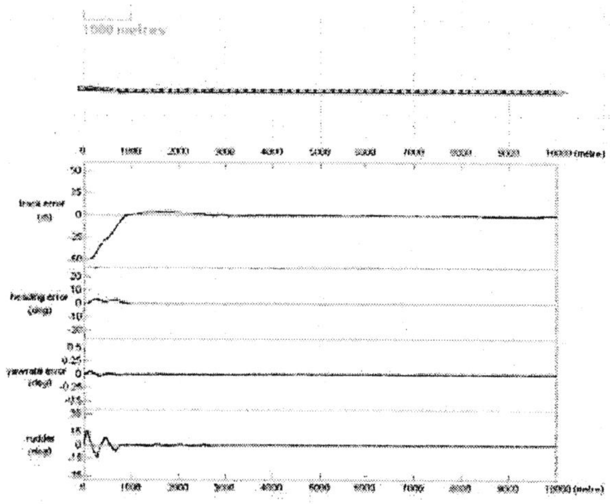

(a) ***Neurofuzzy Controller***

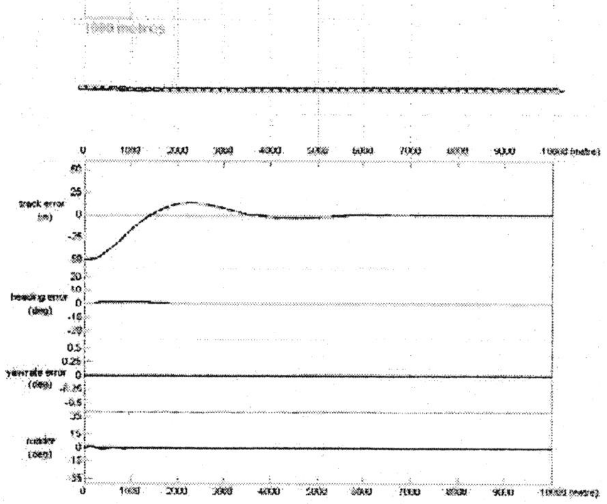

(b) ***Neural Network Controller***

Fig. 4. The track-keeping performance without environmental disturbances.

## 6. CONCLUSION

It has been demonstrated that a simulated ship can be controlled by means of a neurofuzzy controller to perform track-keeping under wind and current based environmental disturbances. The proposed specialized control strategy avoids the need for a "teacher" or an off-line training process. Together with the self-learning ability of the network, the controller copes well with different track-keeping tasks.

During the slow varying of the ship, even though there is only one epoch during each time step, this on-line training can update the network parameters effectively. The controller combines a fuzzy inference system that includes human experiences and

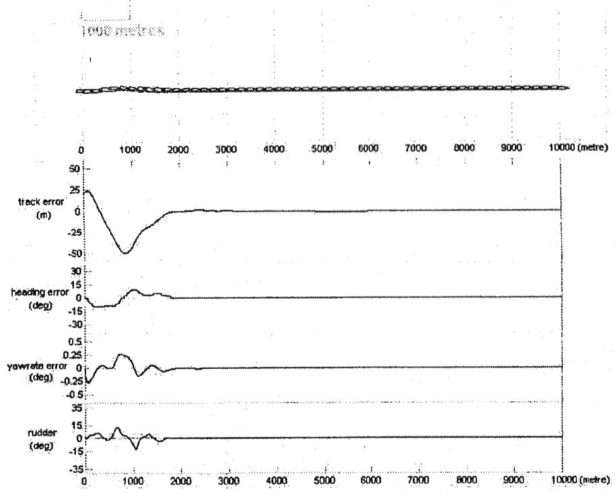


**(a) *Neurofuzzy Controller***

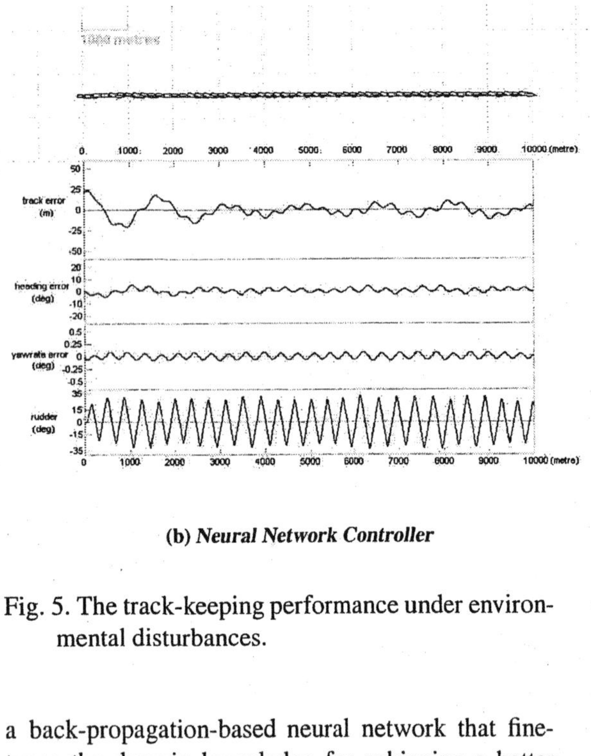


**(b) *Neural Network Controller***

Fig. 5. The track-keeping performance under environmental disturbances.

a back-propagation-based neural network that fine-tunes the domain knowledge for achieving a better performance. Comparing the performances of track-keeping by the neural network and the neurofuzzy controller, it shows that the neurofuzzy controller is superior in terms of reduced track overshoot and earlier maintenance of desired track. This is considered important for navigation in restricted waterways. When considering other tracks similar performance is advised in the work of Zhuo and Hearn (Zhuo and Hearn, n.d.).

Acknowledgments:

This work was partially supported by a School of Engineering Science Research Studentship. This is gratefully acknowledged. Referees comments to original submission were useful and have been duly taken into account in the final form of this paper.

REFERENCES

Cui, X. and K. G. Shin (1993). Direct control and coordination using neural networks. *IEEE Transactions on Systems, Man, and Cybernetics* **23(3)**, 686–697.

Fossen, T.I. (1994). *Guildance and Control of Ocean Vehicles*. John Wiley and Sons Ltd. England.

Holzhuter, T. (1990). A high precision track controller for ships. proceedings. $11^{th}$ *IFAC Triennial World Congress. Tallinn, Estonian, U.S.S.R.* pp. 425–430.

Jang, J-S.R. and C-T. Sun (1997). *Neuro-Fuzzy and Soft Computing: A Computational Approach to Learning and Machine Intelligence*. Prentice Hall.

Lu, X.R., J. H. Jiang and Y. X. Huang (1990). Design of a self-tuning adaptive track-keeping control system for ships. *Proceedings, International Conference Modelling and Control of Marine Craft. University of Exeter, UK* pp. 178–192.

Messer, A.C. and M. J. Grimble (1993). Introduction to robust ship track-keeping control design. *Transaction of the Institute of Measurement and Control* **15(3)**, 104–110.

Ng, G. W. (1997). *Application of Neural Networks to Adaptive Control of Nonlinear Systems*. Research Studies Press Ltd. England.

Psaltis, D., A. Sideris and A. A. Yamamura (1988). A multilayered neural network controller. *IEEE Control Systems Magazine* **8(2)**, 17–21.

Saerens, M. and A. Soquet (1991). Neural controller based on back-propagation algorithm. *IEE Proceedings Control Theory and Applications* **138(1)**, 55–62.

Zhang, Y., G. E. Hearn and P. Sen (1996). A neural network approach to ship track-keeping control. *IEEE Journal of Oceanic Engineering* **21(4)**, 513–527.

Zhuo, Y. and G. E. Hearn (n.d.). Ship intelligent track-keeping using self-learning neurofuzzye. *In preparation*.

ELSEVIER
IFAC
PUBLICATIONS
www.elsevier.com/locate/ifac

# MINIMUM TIME SHIP MANEUVERING USING NEURAL NETWORK AND NONLINEAR MODEL PREDICTIVE COMPENSATOR

**Naoki Mizuno*, Masaki Kuroda*, Tadatsugi Okazaki** and Kohei Ohtsu*****

**Nagoya Institute of Technology, Gokiso-cho, Shouwa-ku, Nagoya JAPAN*
***National Maritime Research Institute, 6-38-1,Shinkawa, Mitaka, Tokyo JAPAN*
****Tokyo University of Marine Science and Technology, 2-6-1, Etchujima, Koto-ku, Tokyo JAPAN*

Abstract : In this paper, a new minimum-time ship maneuvering system using neural network and nonlinear model predictive compensator is proposed. In the proposed method, the neural network is used for interpolating the pre-computed minimum-time solution for the real-time situation and the nonlinear dynamical model of the ship is used for compensating the control error caused by some modeling errors, disturbances etc. In order to investigate the effectiveness of the proposed method, computer simulations and actual sea tests are carried out using a training ship *Shioji Maru. Copyright © 2004 IFAC*

**Keyword:** ship control, minimum-time control, neural network, nonlinear model predictive control, receding horizon control

## 1. INTRODUCTION

It is very important for a ship's master to draw up a ship-handling plan before approaching a berth, leaving it, altering the heading and so on. In order to cope with these problems, they are required to derive the maximum maneuverability from the ship. The minimum-time control technique is one of the rational and effective maneuvering methods for such tedious problems if the mathematical model representing a ship's dynamics is available. However, it should be noted that the model becomes highly non-linear, especially in the case of low speed and large maneuvering motion. In this case, the solutions of minimum-time maneuvering problem can be obtained by solving non-linear two-point boundary value problems (TPBVP).

In order to take enough account of the non-linearity, Shoji and Ohtsu (1992) have formulated these problems as a non-linear, two-point boundary-value problem (TPBVP) in the calculus of variations. The problems have been solved, using the numerical method called the sequential conjugate gradient restoration (SCGR) method (Wu and Miele, 1980; Miele and Iyer, 1970) under various situations (Ohtsu and Shoji, 1994; Ohstu *et. al*,1996; Okazaki *et. al.*.2000) However, the solutions, unfortunately, do not yield on-line control laws, because, it took long computational time to obtain the optimal solution. Moreover, the geometrical and other sea conditions of maneuvering problem facing to the real ship handling scenes are various. For these problems, Okazaki *et al.*(1997) have investigated the minimum time maneuvering with neural network, but the control performance varies depend on the disturbances in real sea conditions. To overcome this problem, Mizuno *et al.*(2002) have proposed a minimum-time maneuvering system with two types of neural networks. From the experimental results, the method has some advantages compared with the previous studies. There are some papers in which the neural networks are used in order to compensate the non-linear dynamics of the ship during the tracking or berthing phase (Yao *et al.* 1996,1997; Namkyun and Hasegawa, 2001,2002). Unfortunately the on-line learning speed of the neural network is rather slow. This means that it is difficult to obtain the good transient performance for tracking or berthing in real situations. From this point of view, Mizuno *et al.* (2003) have proposed a new design method for ship's minimum time maneuvering system with neural network and non-linear model based super real-time simulator.

In this paper, we extend the above method to achieve more accurate tracking under some disturbances. The extended system is composed of a neural network based optimal solution generator and a nonlinear model predictive compensator (Mizuno *et al.*, 2003). The neural network generates the optimal solution for real situation by interpolating pre-computed minimum-time solutions for typical control conditions. The optimal solutions for the various minimum time maneuvering are numerically computed based on the sophisticated non-linear dynamical model of the ship (MMG model) and are learned off-line by the neural network for interpolation. Moreover, the same nonlinear model, which is used for the computation of the optimal solutions, is used to simulate the ship's future course. Based on the receding horizon cost function of the predicted control error caused by some disturbances, the control input for minimum time maneuvering is modified. This is a new feature of this system.

First, the solving technique of the minimum-time maneuvering problems and the mathematical model of the ship's dynamics are briefly reviewed. Next, the minimum-time parallel deviation maneuvering problem and its solutions are introduced as an example of the feasible study realized by proposed system. In the third part of this paper, a minimum-time maneuvering system with neural network and nonlinear model predictive compensator is introduced. Finally, computer simulations and on-line experiments are carried out for a training ship *Shioji Maru* (425 gross tonnage).

## 2. OUTLINE OF SOLVING TECHNIQUE OF THE MINIMUM MANEUVERING PROBLEMS

### 2.1 Formulation

The minimum time maneuvering problem is formulated to control a ship from a certain condition to another one in a minimum time. This kind of problem is considered as a two-point boundary value problem, in which an initial point as the one condition of the ship (for example, the starting point) and a terminal point as another condition of the ship (for example, the stopping point).

Such control problem might be solved using the theory of calculus of variations. However, since ship's motion has high non-linearity, it is impossible to find an analytical solution. Thus it is inevitable to adopt some numerical methods for solving the problem. In this paper, the sequential conjugate gradient-restoration (SCGR) method developed by Miele *et al.* (1992) was used.

The minimum time maneuvering problem is formulated as follows:

A performance index of this problem is defined by a functional

$$I = \int_0^1 f(x, u_c, \tau, t)dt = \int_0^1 \tau dt = \tau \qquad (1)$$

where $I$ is a scalar value, $x$ is the state vector, $u_c$ is the control vector, $t$ is the actual final time value and $\tau$ is the normalized time value.

And the solution of the problem is minimized the performance index with constrains as follows:

1) The differential constraints,

$$\dot{\mathbf{x}} - \Phi(\mathbf{x}, \mathbf{u}_c, \tau, t) = 0 \qquad 0 \le t \le 1 \qquad (2)$$

where $\Phi$ denotes a non-linear hydro dynamic model for representing ship's motions, and

2) The boundary conditions:

i) The initial ship's state,

$$[\omega(x)]_0 = 0 \qquad (3)$$

ii) The final state of the ship, specified by the function

$$[\Psi(\mathbf{x}, \tau)]_1 = 0 \qquad (4)$$

where the function $\Psi$ is a q-dimensional vector $(0 \le q \le n)$.

3) The non-differential constraints:

$$\mathbf{S}(\mathbf{x}, \mathbf{u}_c, \tau, t) = 0, 0 \le T \le 1 \qquad (5)$$

may be added, by which it is possible to set the maximum limits of rudder angle, propeller blade angle, and power of the bow and stern thrusters, to be applied.

### 2.2 Ship's motion model

*Shioji Maru* is equipped with a bow and stern thrusters, besides a rudder and a controllable pitch propeller (CPP). Her principal dimensions are shown at Table 1 and the coordinate system is shown in Figure 1.

Table 1: Principal Dimensions of *Shioji Maru*

| | |
|---|---|
| Length | 49.93 m |
| Breadth | 10.00 m |
| Tonnage | 425.0 GT |
| Propeller | CPP |
| Bow Thruster | 2.4 tons |
| Stern Thruster | 1.8 tons |

The state values are ship's position ($x$, $y$) for Xo-Yo coordinate, ship's heading $\psi$, surge speed $u$, sway speed $v$, yaw rate $r$, rudder angle $\delta$ for X-Y coordinate and CPP angle $\theta_P$ The control values are order rudder angle $\delta^*$, order CPP angle $\theta_P^*$, notch of bow thruster $bt^*$, notch of stern thruster $st^*$.

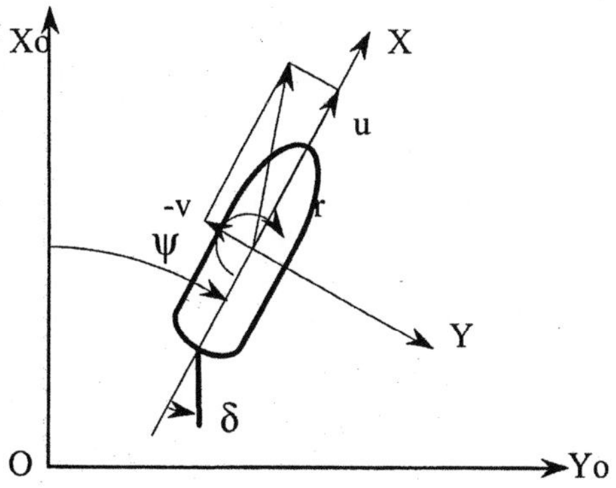


Fig. 1 Ship's coordinate in horizontal plane

Referencing to this, the sophisticated mathematical model (called MMG model) is written by

$$\dot{x} = u\cos\psi - v\sin\psi \quad (6)$$

$$m(\dot{u} - vr) = X_H + X_P + X_R \quad (7)$$

$$\dot{y} = u\sin\psi + v\cos\psi \quad (8)$$

$$m(\dot{u} - vr) = X_H + X_P + X_R \quad (7)$$

$$\dot{y} = u\sin\psi + v\cos\psi \quad (8)$$

$$m(\dot{v} + ur) = Y_H + Y_T + Y_R + Y_{Th} \quad (9)$$

$$\dot{\psi} = r \quad (10)$$

$$I\dot{r} = N_H + N_T + N_R + N_{Th} \quad (11)$$

$$\dot{\delta} = \frac{(\delta^* - \delta)}{(|\delta^* - \delta| T_{RUD} + a)} \quad (12)$$

$$\dot{\theta}_P = \frac{(\theta^*_P - \theta_P - \dot{\theta}_P T_{lp})}{T_{lp}(|\theta^*_P - \theta_P - \dot{\theta}_P T_{lp}| T_{CPP} + a)} - \frac{\dot{\theta}_P}{T_{lp}} \quad (13)$$

where $m$ and $I$ are the mass and the turning moment inertia. $T_{RUD}$, $T_{CPP}$ and $a$ are time constants. The subscripts $H$, $P$, $R$ and $T_h$ denote the highly non-linear hydrodynamic force induced by the hull, propeller, rudder and thruster. For detail of the hydrodynamic force, see (Ohtsu *et al.*, 1996).

### *2.3 Minimum-time parallel deviation maneuvering problem and the non-linearity of its solutions*

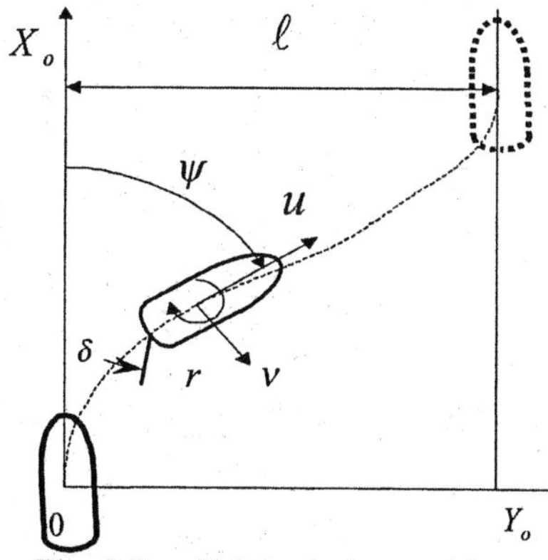


Fig. 2 Parallel deviation problem

In this paper, the minimum-time parallel deviation maneuvering problem and its solutions are considered as a simple example of the feasible study realised by proposed control system. Figure 2 shows the initial course line and terminal one of the minimum-time parallel deviation problem.

Where $\ell$ is the distance between two parallel course lines. In this case, the boundary conditions for the two-point boundary value problem and the non-differential constraint are as follows.

1) The initial ship's state

$$[x(0) \;\; y(0) \;\; u(0) \;\; v(0) \;\; r(0) \;\; \psi(0) \;\; \delta(0)]^T = given \quad (14)$$

2) The final state of the ship

$$[y(1) \;\; v(1) \;\; r(1) \;\; \psi(1) \;\; \delta(1)]^T = [y_f \;\; v_f \;\; r_f \;\; \psi_f \;\; \delta_f]^T = 0 \quad (15)$$

3) The non-differential constraints

$$\delta^*(t) - sigmoid\delta_{Dumy}(t) = 0, \; 0 \le t \le 1 \quad (16)$$

where, $\delta_{Dumy}$ is the dummy variable for the order rudder angle calculated by SCGR method without constraint. The 'sigmoid' means a kind of saturating function which guarantees the range of $\delta^*(t)$ in permissible value. Under these conditions, the minimum-time control solutions can be obtained using the SCGR method. The first examples are minimum-time deviation problems with different deviations. In this case, the ship must deviate 100m and 200m away from the initial approach line in a minimum maneuvering time, using the rudder. The ship's initial cruising speed is 12 knots, and the side ways speed must disappear and the head must be redirected on the original course after ending the deviation. Figure 3 and 4 show the optimal paths and the corresponding time histories of the rudder angles in each case.

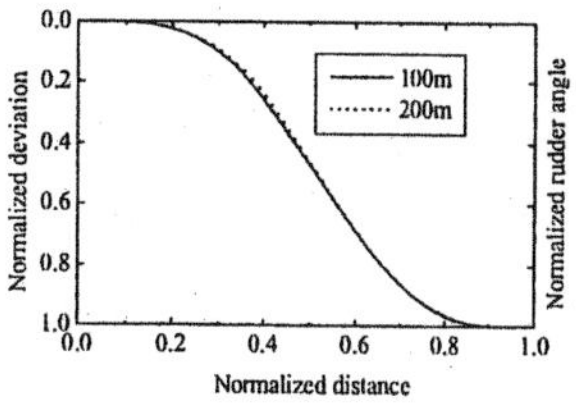


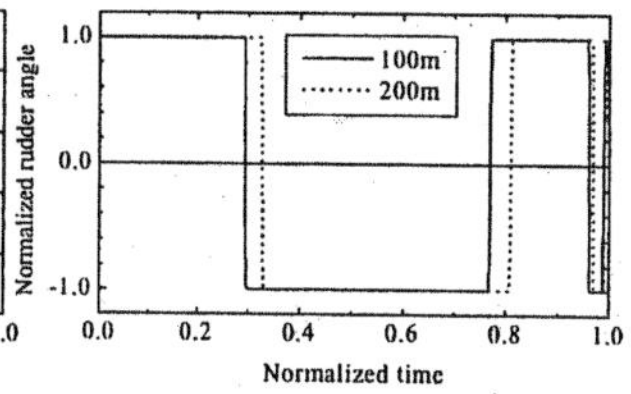


Fig. 3 The optimal path Fig. 4 The rudder angles for different course deviation

It should be noted that the time histories of the heading angles have almost the same patterns, whereas those of the rudder angles are different in each case. This means that the ship's minimum-time maneuvering system should have the real-time ability to calculate the optimal solutions.

## 3. MINIMUM-TIME MANEUVERING SYSTEM WITH NEURAL NETWORK AND NONLINEAR MODEL PREDICTIVE COMPENSATOR

### *3.1 Optimal solution generator using neural network*

In the proposed system, the real-time solution for a new maneuvering condition will be generated by interpolating the pre-computed solutions for typical conditions using neural network (NN).

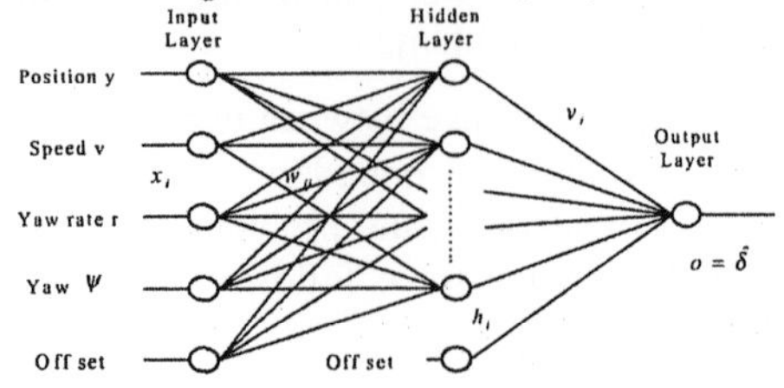


Fig. 5 Structure of 3-layered neural network

Figure 5 shows the three-layered neural network used in this research (Okazaki *et al.*,1997). The inputs $x_i$ for the neural network are the ship's state values, which include the ship's position from the terminal course line $y$, its heading $\psi$, and its speed $v$. The output of the neural network $o = \hat{\delta}$ corresponds to the rudder angle $\delta_{OPT}$ for minimum-time maneuvering solution. In Fig.5 $w_{ij}$, $v_i$ are synaptic weights between input and hidden layer, hidden and output layer respectively. The outputs of the hidden units are,

$$h_i = f(u_i) = f(\sum_i w_{ij} x_i + x_0) \tag{17}$$

And the output of the network are defined by

$$o = f(s) = f(\sum_i h_j v_j + h_0) \tag{18}$$

where, $x_0$, $h_0$ are the off sets for the inputs, and $i$ , $j$ denote the numbers of units in input and hidden layer respectively. The non-linear function $f(\cdot)$ in the unit is a sigmoid function of the form:

$$f(x) = 1/(1+e^{-x}) \tag{19}$$

As a learning algorithm for this network, the following back propagation method is adopted.

$$e = \delta_{OPT} - o \tag{20}$$

$$\Delta v_i(n) = \eta . e . f(s).(1 - f(s)).h_i + \alpha \Delta . v_i(n-1) \tag{21}$$

$$\Delta w_{ij} = \eta \cdot e . f(s).(1-f(s)).v_j).f(u_j).(1-f(u_j)).x_i + \Delta w_{ij}(n-1) \tag{22}$$

where $\Delta$ denotes the update quantity of each variables at iteration $n$ . $\eta$ and $\alpha$ are the learning rate and the momentum term. During the network training, the time series of ship state values are fed to the input layers and the network synaptic weights $w_{ij}, v_{jk}$ are updated to reduce the error between the network output signal and the minimum-time maneuvering solutions $\delta_{OPT}$ (optimal control values). After sufficient training, the neural network could make appropriate time series of control values (rudder angle $\delta$ ) for arbitrary minimum-time maneuvering within certain range of course deviation. Furthermore, the calculation time to interpolate these solutions is less than one second. In this case, the structure of the neural network (number of hidden units $j = 4$ ) is determined heuristically and verified by using AIC ( Akaike's Information Criterion ) and MDL ( Minimum Description Length Principle ).

### *3.2 Compensation of tracking error by using nonlinear model predictive compensator*

By using off-line trained neural networks, the optimal solutions for practical conditions can be computed as "closed loop configuration" in real-time and a ship's minimum-time maneuvering may be implemented by the basic structure as shown in Fig.6.

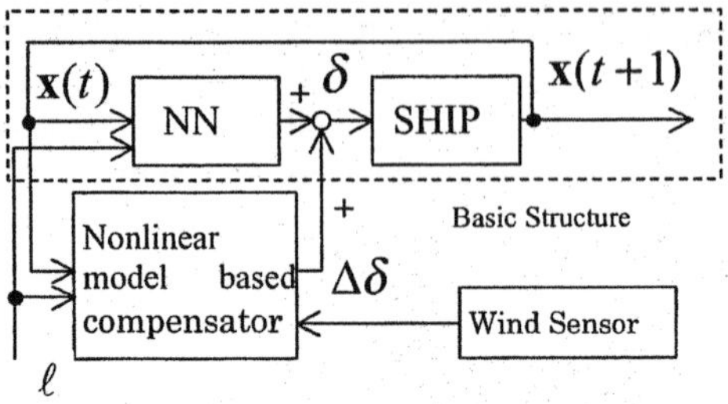


Fig. 6 Proposed minimum-time maneuvering system with nonlinear model predictive compensator

However, from practical point of view, some disturbances (for example, wind and tidal current etc.) should be taken into account. For this problem, we recommend to introduce the nonlinear model based compensator into the control system as shown in Fig. 6 (Mizuno *et al.*, 2003), because we have already construct the sophisticated non-linear dynamical model ( MMG model ) of the ship and can use it.

In the proposed system, "nonlinear model based compensator" is composed of the MMG model based ship's dynamic simulator and the search mechanism for optimal rudder compensation $\Delta\delta$ to reduce the tracking error based on the receding horizon cost function $J$ as described later. Although the MMG model is very complicated, the computational time is very short compared with the time which is required for solving the optimal solution by SCGR method with the same model. For example, the computational time for simulating the future 150 [sec] ship's behaviour is less than 1 [msec] using the computer with the Pentium III CPU at 1GHz clock. This means that we can simulate about 100 times with different conditions during the sampling period 1 [sec] in experiment.

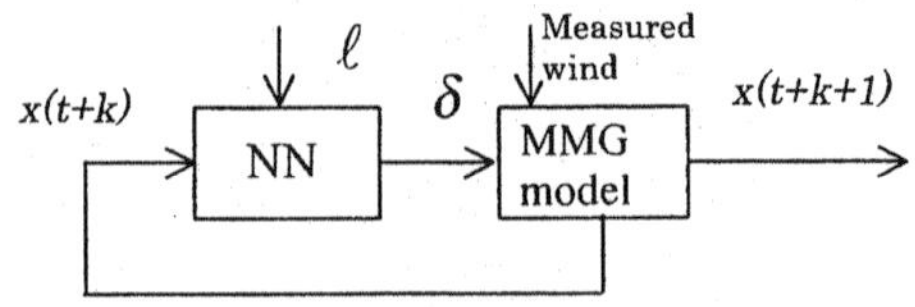


Fig. 7 Structure of MMG model based simulator

To simulate the future behaviour of the ship, we can use the neural network based optimal solution generator "NN" for generate the future control values to the ship as shown in Fig. 7, because the generation of the solution (rudder order) is performed in closed loop configuration mentioned before.

To reduce the future tracking error for the optimal course, we optimize the following receding horizon cost function $J$ (Allgower and Zheng, 2000).

$$J = \sum_{\tau=M}^{N} | Y_o(\tau) - Y_s(\tau) | \qquad (23)$$

where, $Y_o(\tau)$ denotes the optimal deviation of the minimum-time solution and $Y_s(\tau)$ is the simulated one of the controlled ship.

First, the MMG model based simulation is carried out assuming the rudder compensation $\Delta\delta$ as an appropriate setting for the future time period [t, N] and the value of $J$ is calculated as the "error area" for receding horizon [M, N] as shown in Fig. 8. Next, to optimize the cost function, the simulation and evaluation are repeated based on the linear search as shown in Fig. 9.

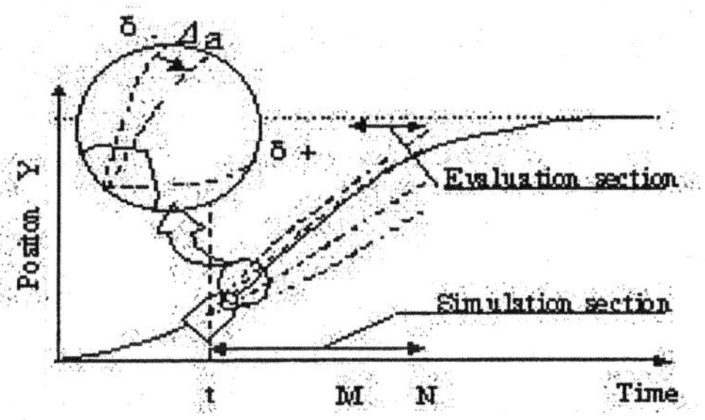


Fig. 8 Simulation and evaluation period for optimization

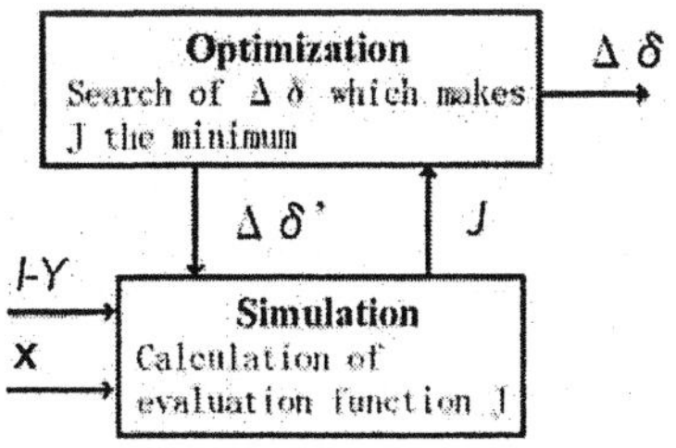


Fig. 9 Optimization mechanism by super real-time simulator

## 4. PRELIMINARY EVALUATIONS BY COMPUTER SIMULATIONS

Before implementing the proposed minimum-time maneuvering system, computer simulations of both basic and proposed scheme are performed for nonlinear dynamical model of the *Shioji Maru.* Figure 10 and 11 show the simulation results for minimum-time deviation problem with wind disturbances by using basic (NN only) and proposed (NN+model predictive compensator) control system respectively. The ship must deviate 150 m away from the initial course line. In these cases, the winds blow from the stern (180[deg]) at relative wind velocities of 12 m/sec. The neural network "NN" in both the basic and proposed maneuvering system has been off-line trained using the four minimum time solutions with $\ell = 100m$ and $200m$ for the ship speed $u = 5.0m/s$ and $6.6m/s$ under the constraint of $|\delta| \leq 5$[deg]. For the proposed method, the optimal value of the rudder compensation $\Delta\delta$ is searched in the range of $\pm 7.0$[deg] with the resolution of $0.1$[deg]. Moreover, for the safe of ship's maneuvering, the maximum rudder order is restricted as $|\delta + \Delta\delta| \leq 10.0$[deg]. Moreover, the same MMG model is used to simulate the ship's bahaviour and is used in model predictive compensator. In these figures, the dashed lines show the minimum time solution for wind free case and the solid lines are with wind disturbance. From these results, it should be noted that the error between the ship's dynamics and the actual one does not exist, whereas the path obtained using only NN is rather different from the optimal one. Thus, it is generally concluded that the control system should have a feedback compensator, in order to avoid the influence of disturbances.

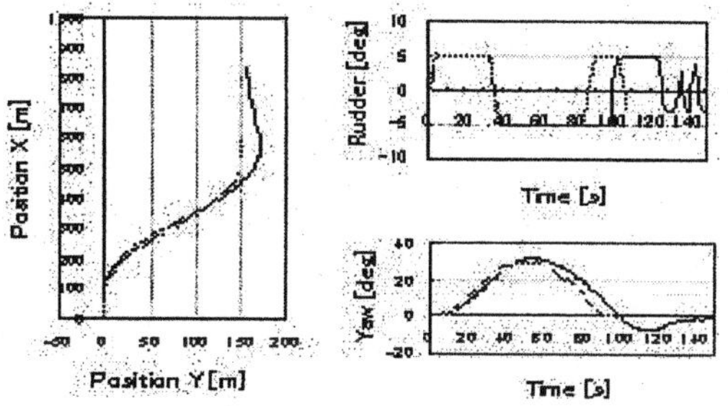


Fig. 10 Simulation results of **basic** minimum-time maneuvering system with wind

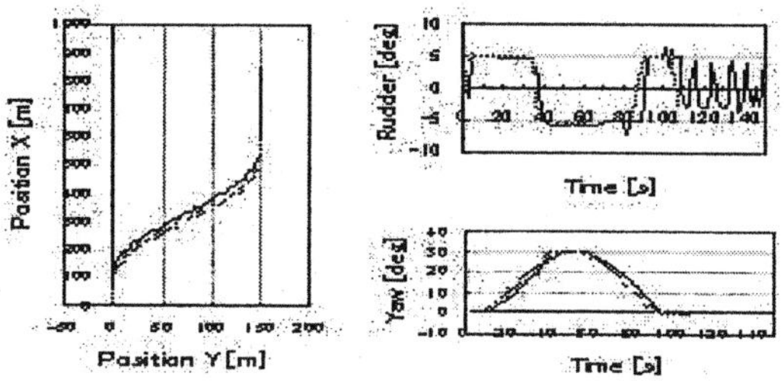


Fig. 11 Simulation results of **proposed** minimum-time maneuvering system with wind

## 5. ACTUAL SEA TEST

### *5.1 The training ship Shioji Maru*

In order to evaluate the feasibility of the proposed minimum- time maneuvering system, the actual deviation tests were carried out at sea, using the *Shioji Maru* of Tokyo University of Marine Science and Technology (Fig. 12).

Fig. 12 *Shioji Maru*

### *5.2 Real Time Control System*

Figure 13 shows the real time control system aboard the *Shioji Maru.* On board the training ship, GPS signals are available to provide accurate positions of ship.

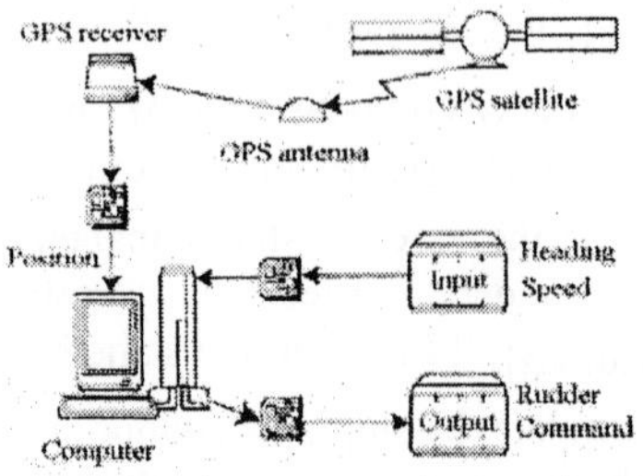


Fig. 13 Real time control systems on board

### *5.3 Experimental results*

Figure 14 shows the typical experimental result using basic scheme under wind disturbance. The left traces show ship's actual path measured through a GPS (solid line), and the optimal one (dashed line) and the right the calculated rudder angle (solid line), optimal one (dashed line), actual heading (solid line) and optimal one (dashed line).

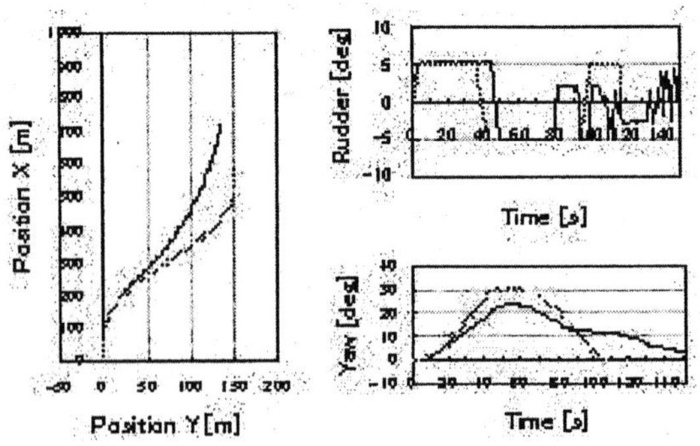


Fig. 14 The typical experimental result using **basic scheme** (150m deviation, 6m/s wind from the bow side)

Figure 15 shows the typical experimental result using proposed scheme under wind disturbance.

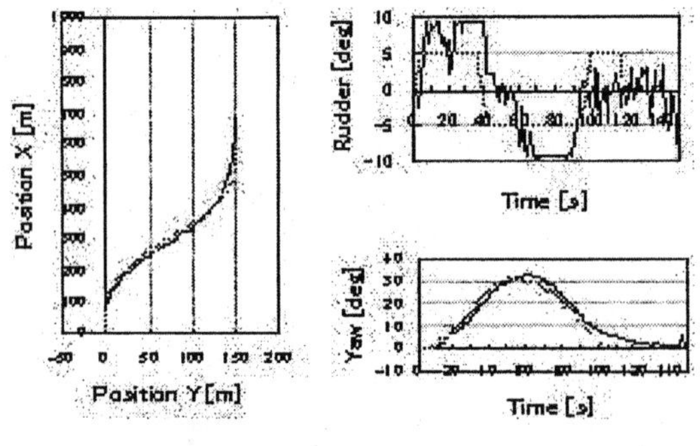


Fig. 15 The typical experimental result using **proposed scheme** (150m deviation, 8m/s wind from the bow side)

In these cases, the settings for the controller are same as the simulation. From these results, it can be concluded that the proposed minimum-time maneuvering system with NN+nonlinear model predictive compensator is feasible in actual sea conditions.

## 6. CONCLUSION

This paper presented a new practical ship's minimum-time maneuvering system with neural network and nonlinear model predictive compensator. In the minimum-time deviation problems, the system gives approximate solutions in a short computing time and good tracking performance in real situations. Moreover, the actual sea trials demonstrate the effectiveness of the proposed system.

Acknowledgments

The authors wish to thank the crewmembers of the *Shioji Maru*, for their helpful support in the actual sea tests. This work is supported in part by Grant-in-Aid for Scientific Research (c) from the Ministry of Education, Culture, Sports, Science and Technology of Japan.

## REFERENCES

Allgower T. and A. Zheng (Eds.)(2000). Nonlinear Model Predictive Control, Birkhauser.

Miele, A. and R.R. Iyer (1970)., General Technique for Solving Nonlinear two points Boundary-Value Problems via the Method of Particular Solutions, *Journal of Optimization Theory and Applications*, **Vol.5**, No.5, pp. 382-399.

Mizuno, N. *et al.*(2002). A Ship's Minimum Time Maneuvering System Using Neural Networks, *Proceedings of IECON'02.*.

Mizuno, N. *et al.*(2003). A Ship's Minimum Time Maneuvering System Using Neural Network and Non-linear Model based Super Real-Time Simulator, *Proceedings of ECC'03.*

Namkyun Im and K.Hasegawa (2001). A Study on Automatic Berthing Using Parallel Neural Controller, *Journal of the kansai society of naval architects Japan,* **No.236.**

Namkyun Im and K.Hasegawa (2002). A Study on Automatic Berthing Using Parallel Neural Controller (2nd Report), *Journal of the kansai society of naval architects Japan,* **No.237.**

Ohtsu, K. and K. Shoji (1994). Minimum time maneuvering of ships, *Proceedings of MCMC'94.*

Ohtsu, K. *et al.*(1996). Minimum Time Maneuvering of a ship, with Wind Disturbances, *Control Eng. Practice*, **Vol.4**, No.3, pp.385-392.

Okazaki, T. et al.(1997). "Study on the Minimum Time Maneuvering with Neural Network", T*he Journal of Japan Institute of Navigation*, **Vol.97**, pp.155-164.

Okazaki, T. et al.(2000). A Study of Minimum Time Berthing Solutions, *Proceedings of MCMC2000, pp. 135-139.*

Shoji, K. and K. Ohtsu (1992). Automatic berthing study by Optimal Control Theory, *Proceedings of CAMS'92*, Genova, pp.185-194.

Wu, A.K. and A. Miele (1980). Sequential Conjugate Gradient-Restoration Algorithm for optimal Control Problems with Non-Differential Constraints and General Boundary Conditions, Part 1, *Optimal Control Applications and Method.***Vol.1**, pp. 69-88.

Yao, Z. *et al.*.(1997). A multivariable Neural controller for Automatic Ship Berthing, *IEEE Control Systems.*

Yao, Z. *et al.*(1996). A Neural Network Approach to Ship Track-Keeping Control, *IEEE Journal of engineering,* **Vol.21**, No.4.

ELSEVIER
IFAC
PUBLICATIONS
www.elsevier.com/locate/ifac

# MULTI-AGENT TECHNOLOGY FOR ROBUST CONTROL OF SHIPBOARD CHILLED WATER SYSTEM

**Pavel Tichý[†], Petr Šlechta[†], Francisco P. Maturana[‡], Raymond J. Staron[‡], Kenwood H. Hall[‡], Vladimír Mařík[†], Fred M. Discenzo[‡]**

[†] *Rockwell Automation Research Center, Americka 22, 120 00 Prague, Czech Republic*
[‡] *Rockwell Automation, 1 Allen-Bradley Drive, Mayfield Hts., OH 44124-6118, USA*
*{ptichy, pslechta, fpmaturana, rjstaron, khhall, vmarik, fmdiscenzo}@ra.rockwell.com*

Abstract: This paper reports preliminary results of an ongoing research project that demonstrates distributed architecture based on agents applied in the area of industrial automation. This architecture has been built to achieve the goals of improved survivability and readiness of US Navy shipboard systems. We show benefits of multi-agent systems in the area where flexibility, survivability, and scalability are required. We present the architecture of multi-agent system, internal structure of agent, planning technique based on plan templates, fault-tolerant structure of middle-agents, and development environment for agents. *Copyright © 2004 IFAC*

Keywords: agents, artificial intelligence, automation, control systems, decentralized control, fault-tolerant systems, programmable logic controllers, robustness.

## 1. INTRODUCTION

The focus of research is to use the multi-agent system architecture to implement a survivable and reconfigurable system that will control a chilled water part of a shipboard automation for US-Navy vessels. We present requirements for the shipboard automation architecture, levels of this architecture, and a small scaled version of the chilled water system (Maturana, *et al.*, 2003). Next in the paper we focus on the design of multi-agent system and agents. We also present the fault-tolerant architecture of middle-agents that is used to ensure survivability of agent search mechanisms. Middle-agent can be defined as, for instance, 'agent that helps others to locate and connect to agent providers of services' (Klusch and Sycara, 2001) and other agents can be referred as end-agents. Next we describe the development environment that is used to build the whole control system and to download it into controllers.

## 2. SHIPBOARD AUTOMATION ARCHITECTURE

To design a highly distributed shipboard automation system, the following four fundamental requirements are considered:

- Reduced manning: This is intended to create a system with less human intervention and more intelligent components capable of making decisions on behalf of the equipment.
- Flexible distributed control: This is understood as the capability of the system to adapt its components operations to respond to dynamically changing conditions without using predefined recipes. To achieve flexible distributed control, extensions to the control and network software is requisite to enable creation of component-level intelligence that increases robustness of the automation system.
- Commercial-off-the-shelf (COTS): This aspect addresses the cost reduction and system life cycle requirements of new shipboard systems. Under the COTS scope, a ship can be maintained at any friendly location in the world.
- Reliable and survivable operation: As the system becomes more autonomous and self-determined, it is required to augment the level of diagnosability of the components in a distributed manner.

The shipboard automation architecture is divided into three levels: 1) Ship, 2) Process, and 3) Machine. The Ship-level is concerned with ship-wide resource allocation. This level has direct communication with user-level (commander) interfaces.

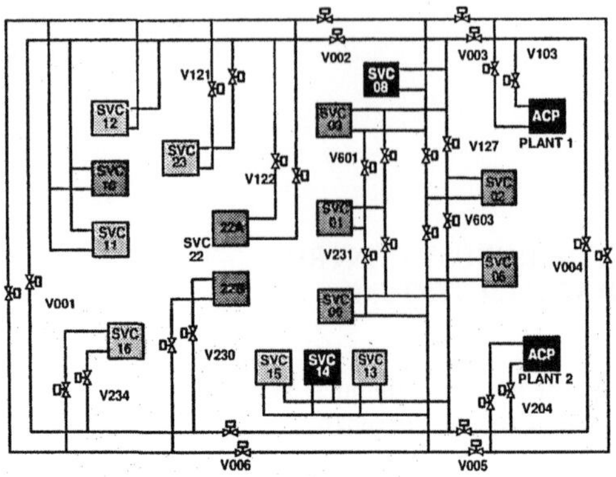


Fig. 1. RSAD Chilled Water System

The Process-level is concerned with optimizing the performance of the components and ensuring the availability of services. The Machine-level is the lowest level in the hierarchy and focuses upon control, diagnostics and reconfiguration.

The Chilled Water System (CWS) pilot system is based on the Reduced Scale Advanced Development (RSAD) model that is a reconfigurable fluid system test platform. The RSAD has an integrated control architecture, which includes Rockwell Automation technology for control and visualization. The RSAD model is presently configured as a chilled water system (see Fig. 1).

The physical layout of the chilled water system is a scaled-down form of the real ship. There is one chiller per zone, i.e., currently two plants. Essentially, there are three main components; chillers, mains, and services (consumers of cold water). There are two types of services, vital (14) and non-vital (2).

## 3. MULTI-AGENT SYSTEM ARCHITECTURE

The MAS architecture is organized according to the following characteristics:

- Autonomy: Each agent makes its own decisions and is responsible for carrying out its decisions toward successful completion
- Cooperation: Agents combine their capabilities into collaboration groups (clusters) to adapt and respond to diverse events and mission goals
- Communication: Agents share a common language for communication.
- Fault tolerance: Agents possess the capability to detect equipment failures and to isolate failures from propagating. This has special value in the detection of water leakage in CWS systems
- Pro-action: Agents periodically or asynchronously propose strategies to enhance the system performance or to prevent the system from harmful states.

Agent planning is carried out in three main phases: Creation, Commitment, and Execution. During creation, an agent initiates a collaborative decision making process (e.g., a load that will soon overheat will request cold water from the cooling service). The agents offer a solution for a specific part of the request. Then, the agents commit their resources to achieve the task in the future. Finally, the agents carry out the execution of the plans.

A primary requirement for new shipboard automation systems is the survivability of the control system. To fulfill this, we use a distributed control architecture, where the automation controllers are extended to enable creation of agents. The agents can contact any device in the network via a Job Description Language (JDL) message. The following description also applies to intra-device communication. JDL represents planning, commitment, and execution phases during the task negotiation. A second use of JDL is the encoding of plan templates. When an agent accepts a request, an instance of a plan template is created and updated with current state of agent and with data from incoming request. Information is encoded as a sequence of hierarchical actions with possible precedence constraints. If a part of the request cannot be solved locally, the agent sends a partial request to other agents as a sub-plan. The Contract Net protocol (Smith, 1980) is used to establish dynamic negotiations.

The agents can also emit messages outside of their organization. Presently, a commonly accepted language for communication among agents is the Foundation Infrastructure for Physical Agents (FIPA) Agent Communication Language (ACL). JDL is used as the content language of FIPA ACL. The FIPA standard uses a matchmaking mechanism to locate agents. Currently, our architecture includes a full implementation of Directory Facilitators (DF) that provides matchmaking and Agent Management Services (AMS) functionality.

An agent exhibits goal-oriented social behaviors to be autonomous, cooperative problem-solving entities. The Ship agent originally emits a system goal. However, there are other system goals belonging to a group of agents or cluster (Shen, *et al.*, 2001). Group goals emerge dynamically and these are agreed upon by the agents through negotiation. For instance, an agent that detects a water leakage problem in a pipe section establishes a goal to isolate the leakage, informs adjacent agents to evaluate the problem according to their data.

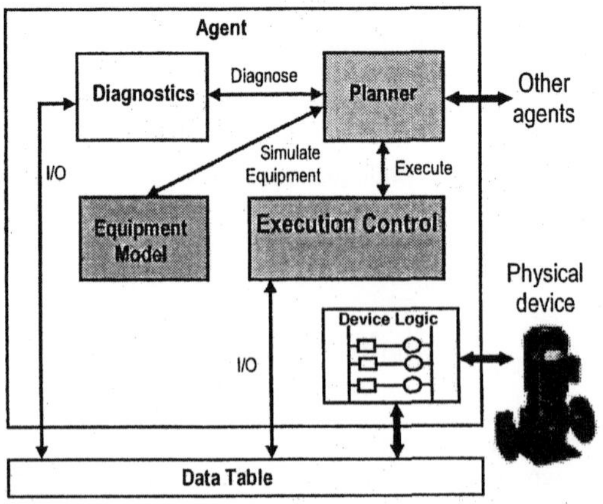


Fig. 2. Internal architecture of an agent

Fig. 2 shows the agent architecture. There are four main components:

- Planner: This component is the brain of the agent. It reasons about plans and events emerging from the physical domain;
- Equipment model: This component is a decision-making support system. The Planner evaluates various configurations using models of the physical domain;
- Execution control: This component acts as a control proxy and translates committed plans into execution control actions. It also monitors events from the control logic and translates them into response-context events to be processed by the Planner component; and
- Diagnostics: This component monitors the health of the physical device. It is programmed with a model of the physical device, where a set of input parameters is evaluated according to a model to validate the readings.

### *3.1 Planning*

Planning activities are part of the Planner module. When the Planner module receives a request for planning, it searches for a script associated with the incoming request. The system designer can for each type of agent define a set of scripts, which are associated with a specific request for planning. Each script has defined a firing condition; the script, which satisfies its firing condition in the best way, is selected. If no script has been found, the Planning Module replies with the Fail Message (this message informs the plan requester that the plan creation failed). Otherwise, the found script becomes a plan template. The plan template is a semi-plan, which describes how to solve the incoming request.

The script consists of various steps, which can be visualized as function blocks (Tichý, 2002). Each function block is processed either locally (by an agent that is processing the script), or remotely by another agent (a request is sent to this agent by the agent, which is processing the script). An example of a script shown in form of interconnected function blocks is shown in Fig. 3. When the agent has found the script, how to solve a particular request, it processes the script. Steps that can be solved locally are processed as soon as possible and for remaining steps the agent contacts middle-agent (yellow pages) and sends a request to all recommended agents.

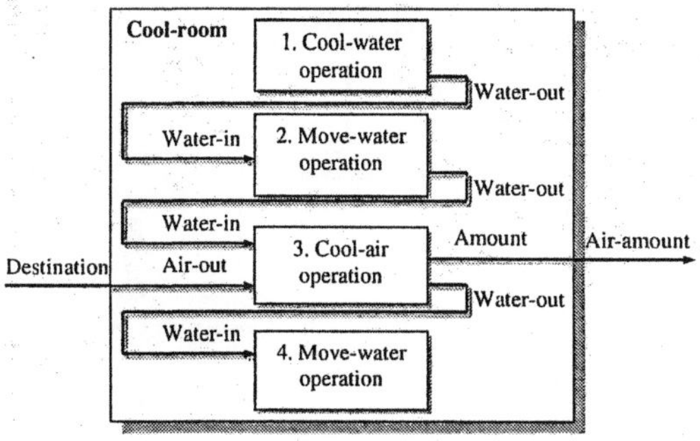


Fig. 3. Plan script example as connected function blocks

After the script is processed, the agent waits for responses to its requests. When a response comes, the agent stores the response to the corresponding step as one solution. If all responses for the step have been received (or time-out for response was reached), the agent selects the best bid from the received responses. We call this selection process 'concentration of the step'. When all steps are concentrated, the agent concentrates the whole script and sends the result back to the requester (to the agent that sent the request).

### *3.2 Distributed Diagnostics*

The diagnostic component includes a suite of data acquisition, signal processing, diagnostic, and prognostic algorithms. These algorithms describe machinery and process health such as machinery fault detection, degraded operation, and failure prediction. The algorithms and models are organized into a multi-level structure (Discenzo, *et al.*, 2002). This permits routines, such as bearing fault detection prediction algorithms, to be re-used in different agents. This architecture provides a mechanism for other agents to access information about a specific machine component.

The Diagnostic component of an agent can interrogate the diagnostic component of other agents to validate a fault hypothesis or to establish the viability of operating under extreme, possibly never anticipated, conditions. For example, a pump agent may sense a higher level of vibration and establish several fault hypotheses, such as a bad bearing or fluid cavitation. By combining of information from the diagnostic components of the motor agent and valve agent, it may be determined that cavitation is occurring.

### *3.3 Fault-Tolerant Structure of Middle-Agents*

Middle-agents are used by end-agents to locate service providers in multi-agent systems. One central middle-agent represents a single point of failure and communication bottleneck in the system. Thus a structure of middle-agents can be used to overcome these issues. We designed and implemented a structure of middle-agents called dynamic hierarchical teamworks (DHT) (Tichy, 2004) that has user-defined level of fault-tolerance and is moreover fixed scalable, i.e., the structure can be extended by fixed known cost. Several approaches have been used already to deal with these issues. Mainly the teamwork-based technique (Kumar and Cohen, 2000) has been proposed that uses a group of middle-agents where each middle-agent is connected to all other middle-agents forming a complete graph. This technique offers fault tolerance but since it uses a complete graph this structure is not fixed scalable.

Assume that a multi-agent system consists of middle-agents and end-agents. Middle-agents form a structure that can be described by the graph theory.

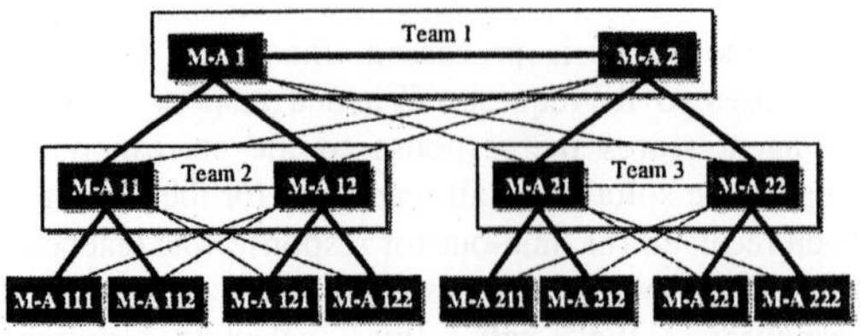


Fig. 4. Example of 3-level DHT architecture

Graph vertices represent middle-agents and graph edges represent a possibility for direct communication between two middle-agents, i.e., communication channels. The first main difference from the pure hierarchical architecture is that the DHT architecture is not restricted to have a single root of the tree that serves as a global middle-agent. The single global middle-agent easily becomes a single point of failure and possibly also a communication bottleneck. In addition, any other middle-agent that is not in a leaf position in the tree has similar disadvantages.

Therefore, to provide a more robust architecture, each middle-agent that is not a leaf in the tree should be backed up by another middle-agent. Groups of these middle-agents we call *teams* (see Fig. 4). Whenever one of the middle-agents from the team fails, other middle-agents from the team can subrogate this agent. During the normal operation of the DHT structure all middle-agents use only primary communication channels (solid lines). If a failure of a primary channel occurs then a secondary channel (dashed lines) is used instead.

To describe the DHT structure more precisely, we use graph theory (Diestel, 2000) to present the following formal definitions.

**Definition 1** *A graph G will be called a* DHT *graph if there exist non-empty sets* $V_1, \ldots, V_n \subset V(G)$ *such that they are pairwise disjoint and* $V_1 \cup \ldots \cup V_n \neq V(G)$. *In that case, the complete subgraph* $G_i$ *of the graph G induced by the set of vertices* $V_i$ *will be called a* team *of G if all of the following is satisfied:*

1) $\forall v(v \in V(G) \setminus V_1 \rightarrow \exists j \forall w(w \in V_j \rightarrow \{v, w\} \in E(G)))$[1]
2) $\forall v(v \in V(G) \wedge v \notin V_1 \cup \ldots \cup V_n) \rightarrow \exists! j \forall w(w \notin V_j \rightarrow \{v, w\} \notin E(G)))$[2]
3) $\forall j((j > 1) \wedge (j \leq n) \rightarrow \exists! k((k < j) \wedge \forall v \forall w(v \in V_j \wedge w \in V_k \rightarrow \{v, w\} \in E(G)) \wedge \forall u \forall m(u \in V_m \wedge (m < j) \wedge (m \neq k) \rightarrow \{v, u\} \notin E(G)))$[3]

**Definition 2** *The graph G is called* DHT-λ *if G is DHT and* $|V_i| = \lambda$ *for every* $i = 1, \ldots, n$, *where* $\lambda \in N$.

---

[1] For all vertices $v$ of $G$ except $V_1$ there has to be a team such that $v$ is connected to all members of this team.

[2] For all vertices $v$ that are not members of any team there are only connections to one team and there cannot be any other connection from $v$.

[3] All members of each team except $G_1$ are connected to all members of exactly one other team with lower index.

The fault tolerance of an undirected graph is measured by the vertex and edge connectivity of a graph. It can be proved (Tichy, 2004) that if the graph G is DHT-λ then the vertex connectivity $\kappa(G) = \lambda$ and the edge-connectivity $\lambda(G) = \lambda$. The DHT structure where teams consist of $\lambda$ middle-agents is therefore fault tolerant to simultaneous failure of at least $\lambda$ - 1 middle-agents and also to simultaneous failure of at least $\lambda$ - 1 communication channels.

It can be also proved (Tichý, 2004) that the graph $G$ of type DHT-λ is maximally fault tolerant, which means that there is no bottleneck in the structure of connections among nodes, i.e., middle-agents in the DHT architecture.

## 4. DEVELOPMENT ENVIRONMENT

The Development Environment (DE) is a software tool that assists the user in programming the distributed application. The development environment introduces the following dimensions into the development phase:

- allow the user to specify the physical and behavioral aspects of the application in a manner completely independent of the control system;
- enable the user to specify a multi-processor control system in a manner completely independent of the application that is to run on it;
- assist the user in combining an application with a control system;
- generate the control code and behavior descriptions for each agent in the system;
- combine the code for all agents assigned to each processor in the system;
- augment each controller automatically to handle the communications to other controllers as a result of the program distribution; and
- communicate with all the controllers involved in an application, for their programming and configuration, and for subsequent monitoring and editing.

Fig. 5 shows the general development flow through the system. One foundation for the DE is a library of components called the template library (TL).

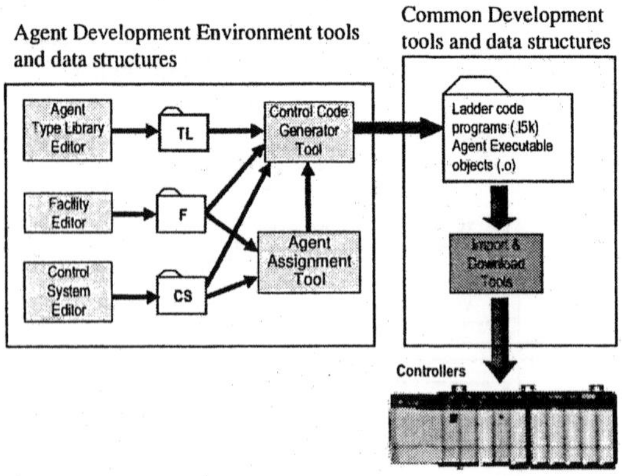


Fig. 5. Development process

The model for the control behavior supports an "object" view of the components in that, for example, inheritance is supported. The TL author can express, for example, that "A Radar is a type of CombatSystem". Each instance of Radar inherits all ladder data definitions and the CombatSystem logic.

The agent behavior, persistently stored as JDL scripts, is presented to the user as a function block diagram. The user creates a facility (F) from components of the template library and customizes their parameters. Next, the user establishes a Control System (CS) that describes all the controllers, I/O cards, and networks, plus their interconnections. After the TL, F, and CS parts are completed, the user generates and compiles the code. After the agent assignment, the user downloads the software into the controllers.

### *4.1 System Modeling and Partitioning*

To establish an agent boundary, we consider the physical proximity of the devices (active and passive) and their function and relationships. There are cases in which it makes more sense to model an agent for a standalone device. In other cases, it makes more sense to group the devices under one intelligent representative. These design decisions have a direct effect on the size and complexity of the organization. We prefer a medium size population to keep the size of the population within practicable margins. In a large organization of agents high network traffic and saturation of the controllers' memory emerge as a problem.

The water cooling plants (ACP plants) are modeled as a single agent each, and each includes pipes, valves, pumps, an expansion tank, and water-level, pressure, flow and temperature sensors. The main piping is partitioned as 'T' pipe sections. Load agents include a heat generator and a temperature sensor. Water service agents include valves and flow sensors. There are standalone valves in the main looping for the supply and return lines. This partitioning gives us a total of 52 agents.

### *4.2 Agent Capabilities*

Each agent is associated with a set of capabilities. Each capability is associated with a set of operations. The negotiation process is founded on local planning and negotiated planning (i.e., cooperation). For the local planning, the agents use their local 'world observations' to determine their next actions. The actions are translated into execution steps. The agents use the operations offered by the local capability. On the other hand, the negotiated planning is based on a capability discovery.

A chilled water system action is expressed as a chain of partial actions. For example, when a Load agent detects that it will soon overheat, it establishes an event for controlling the overheating. Since a Load agent does not have a cooling capability, it is required to ask for it as a cooling request, e.g., 'SupplyCooling'. The DF services provide the addresses of the agents that support such a capability. Water service agents will be contacted to complete the request. However, since these agents only know how to supply water, they need to propagate the request as a more refined context. Next, the request emerges as a 'CoolWater' request that is directed to the ACP Plants. Subsequently, the ACP Plant agents request for water regulators to transmit the water to the load.

## 5. RESULTS

We built a prototype in the lab. at Rockwell Automation in Cleveland, Ohio. Once the agents were completed, a set of scenarios was tested to mimic the transactions of the shipboard system. Once the agent behavior and control logic were verified successfully, we transported the software library to the RSAD facility, where we carried out the instantiation of the agents using the real equipment. The Navy provided the team with a set of operations and desirable reconfiguration scenarios to be supported by the agent organization. In our lab., the team used the agent development tool to emit the agents automatically. This code contained agents and ladder algorithms. The team tested the code on the prototype and later the same code was used on the Navy facility.

In Fig. 6, a UML sequential diagram shows the agents on the top. The vertical lines represent the time line. The horizontal lines represent inter-agent messages. CandD detects an overheating event and sends a message to its local DF (LocalDF1) to obtain possible candidates for 'CoolWater' capability. LocalDF1 sends an Inform message back to CandD with two possible candidates ACP01 and ACP02. CandD subsequently contacts them to obtain their bids for CoolWater capability.

Fig. 7 shows the structure of the decision tree required to plan for a water route from a source to a consumer and vice versa. The figure shows the water route results for moving of water from ACP01 plant to SVC22S service. The planning process is shown as a tree where successful path planning is shown as light boxes and unsuccessful as dark ones.

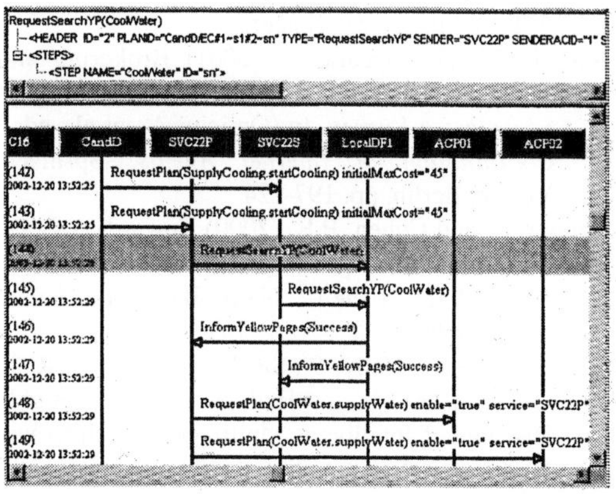


Fig. 6. Agent negotiation as UML sequential diagram

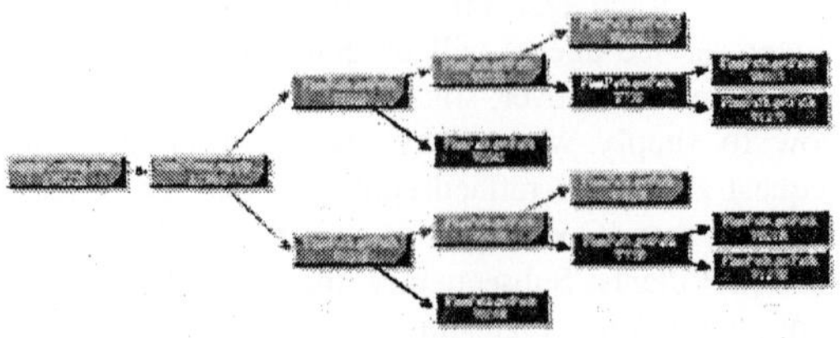

Fig. 7. Water Routing Decision Tree

## 6. CONCLUSION

The multi-agent system technology is beneficial to be used in the area where flexibility, survivability, and scalability are required. These advantages fit very well into development of shipboard systems, where the survivability is one of the main requirements. Scalability is advantageous in the case when the system is applied on a similar type of application.

We did not develop only state-of-art multi-agent system that controls the chilled water system. We also created development environment that is used to build libraries of agent templates that can be later on reused and applied in similar applications. Also a lot of issues connected with the real time control was explored and solved. The system developer is not required to fully understand all details of the multi-agent system technology but only the application specific knowledge is required to be entered to the system. The architecture and tools are presently works-in-progress and subject to continuous improvement.

## REFERENCES


Diestel, R. (2000) Graph Theory. In: *Graduate Texts in Mathematics*, **173**, Springer-Verlag, New York.

Discenzo, F.M., Rusnak, D., Hanson, L., Chung, D., and Zevchek, J. (2002) Next Generation Pump Systems Enable New Opportunities for Asset Management and Economic Optimization. In: *Fluid Handling Systems*, **5(3)**, pp. 35-42.

The Foundation for Intelligent Physical Agents (FIPA): http://www.fipa.org.

Klusch, M. and Sycara, K. (2001) Brokering and Matchmaking for Coordination of Agent Societies: A Survey. In (Omicini A., et al., eds.) *Coordination of Internet Agents*, Springer-Verlag, Berlin, pp. 197-224

Kumar, S. and Cohen, P.R. (2000) Towards a Fault-Tolerant Multi-Agent System Architecture. In: *Proceedings of the 4th ICAA*, Barcelona, Spain, pp. 459-466.

Maturana, F. P., Tichý, P., Šlechta, P., Staron, R. J., Discenzo, F. M., and Hall, K. (2003) A Highly Distributed Intelligent Multi-agent Architecture for Industrial Automation. In *Proceedings of the 3rd International/Central and Eastern European Conference on Multi-Agent Systems (CEEMAS)*, LNAI 2691, Springer-Verlag, Berlin, pp. 522-532

Shen, W., Norrie, D., and Barthès, J.P. (2001) *Multi-Agent Systems for Concurrent Intelligent Design and Manufacturing*. Taylor & Francis, London.

Smith, R.G. (1980) The Contract Net Protocol. In: *IEEE Transactions on Computers*, **C-29(12)**, pp. 1104-1113.

Tichý P. (2004) Fault Tolerant and Fixed Scalable Structure of Middle-Agents. To appear in *the 4th Computational Logic in Multi-Agent Systems Conference (CLIMA IV)*, Fort Lauderdale, FL, USA.

Tichý, P., Šlechta, P., Maturana, F., and Balasubramanian, S. (2002) Industrial MAS for Planning and Control. In (Mařík V., Štěpánková O., Krautwurmová H., Luck M., eds.) *Proceedings of Multi-Agent Systems and Applications II: 9th ECCAI-ACAI/EASSS 2001, AEMAS 2001, HoloMAS 2001*, LNAI 2322, Springer-Verlag, Berlin, pp. 280-295.

ELSEVIER
IFAC
PUBLICATIONS
www.elsevier.com/locate/ifac

# EXTENSION OF FEASIBLE REGION OF CONTROL ALLOCATION FOR OPEN-FRAME UNDERWATER VEHICLES

**E. Omerdic[1] and G.N. Roberts[2]**

[1] *University of Limerick, Ireland*
*Department of Electronic and Computer Engineering*
*email: edin.omerdic@ul.ie*

[2] *University of Wales College, Newport*
*Mechatronics Research Centre, NP20 5XR, UK*
*email: geoff.roberts@newport.ac.uk*

Abstract: Standard pseudoinverse method for solution of the control allocation problem for overactuated open-frame underwater vehicles is able to find a feasible solution only on a subset of the attainable command set. This subset is called the feasible region for pseudoinverse. Some other methods, like direct control allocation or fixed-point iteration method, are able to find the feasible solution on the entire attainable command set. A novel, hybrid approach for control allocation, proposed in this paper, is based on integration of the pseudoinverse and the fixed-point iteration method. It is implemented as a two-step process. The pseudoinverse solution is found in the first step. Then the feasibility of the solution is examined analysing its individual components. If violation of actuator constraint(s) is detected, the fixed-point iteration method is activated in the second step. In this way, the hybrid approach is able to allocate the exact solution, optimal in the $l_2$ sense, inside the entire attainable command set. This solution minimises a control energy cost function, the most suitable criteria for underwater applications.


Keywords: Control allocation, Feasible region, Pseudoinverse, Fixed-point method.

## 1. INTRODUCTION

This paper introduces a new approach associated with the control allocation problem for overactuated open-frame underwater vehicles. The work presented herein is applicable to wide class of control allocation problems, where the number of actuators is higher than the number of objectives. However, the application is focused on two remotely operated vehicles (ROVs) with different thruster configurations. The paper expands upon the work previously reported by the authors (Omerdic, *et al.*, 2003; Omerdic and Roberts, 2004).

Significant efforts have been undertaken in research community over last two decades to solve the control allocation problem for modern aircraft. Different methods were proposed such as direct control allocation (Durham, 1993), optimisation based methods using $l_2$ and $l_1$ norm (Enns, 1998), fixed-point method (Burken, *et al.*, 1999) and daisy chain control allocation (Bordignon, 1996). However, the problem of fault accommodation for underwater vehicles is closely related with the control allocation problem for aircrafts. In both cases, the control allocation problem can be defined as the determination of the actuator control values that generate a given set of desired or commanded forces and moments.

For the unconstrained control allocation problem with a control energy cost function used as optimisation criteria the optimal solution is pseudoinverse (Durham, 1993). Pseudoinverse is a special case of general inverse (GI). GI solutions have the advantage of being relatively simple to compute and allowing some control in distribution of control energy among available actuators. However, in real applications actuator constraints must be taken into account, which leads to a constrained control allocation problem. Handling of constrained controls is the most difficult problem for GI approach. In some cases, the solution obtained by the generalised inverse approach is not feasible, i.e. it violates actuator constraints. Durham (1993) demonstrated that, except in certain degenerate cases, a general inverse cannot allocate controls inside a constrained control subset $\Omega$ that will map to the entire attainable command set $\Phi$, i.e. only a subset of $\Phi$ can be covered. Two methods are suggested to handle cases where attainable control inputs cannot be allocated (Omerdic and Roberts, 2004). The first approach ($T$-approximation (truncation)) calculates a GI solution and truncates any controls (components of control vector) which exceed their limits. The second approach ($S$-approximation (scaling)) maintains the direction of the desired control input command by scaling unfeasible pseudoinverse solution to the boundary of $\Omega$ (Bordignon, 1996).

The hybrid approach will be firstly explained on the low-dimensional control allocation problem with clear geometrical interpretation. After that the problem will be expanded for the higher-dimensional case. Problem formulation is given in the second section. The third section describes the nomenclature. Pseudoinverse is described in the fourth section. The fifth section introduces hybrid approach. Application to control allocation problem for ROVs is given in the sixth section. Finally, the seventh section summarizes the concluding remarks.

## 2. PROBLEM FORMULATION

The majority of modern aircrafts and marine vessels represent overactuated systems, for which is possible to split the control design into the following steps (Härkegård, 2003):

1. REGULATION TASK: Design a control law, which specifies the total control effort to be produced (net force, moment, etc.),
2. ACTUATOR SELECTION TASK: Design a control allocator, which maps the total control effort (demand) onto individual actuator settings (thrust forces, control surface deflections, etc.).

Fig. 1. illustrates the configuration of the overall control system. The control system consists of a control law (specifying the virtual control input, $\mathbf{v} \in \Re^k$) and a control allocator (allocating the true control input, $\mathbf{u} \in \Re^m$, where $m > k$, which distributes control demand among the individual actuators). In the system, the actuators generate a total control effect, $\mathbf{v}_{sys} \in \Re^k$, which is applied as the input to system dynamics block and which determines the system behaviour. The main objective of the control allocation is to ensure that condition $\mathbf{v}_{sys} = \mathbf{v}$ is satisfied for all attainable $\mathbf{v}$.

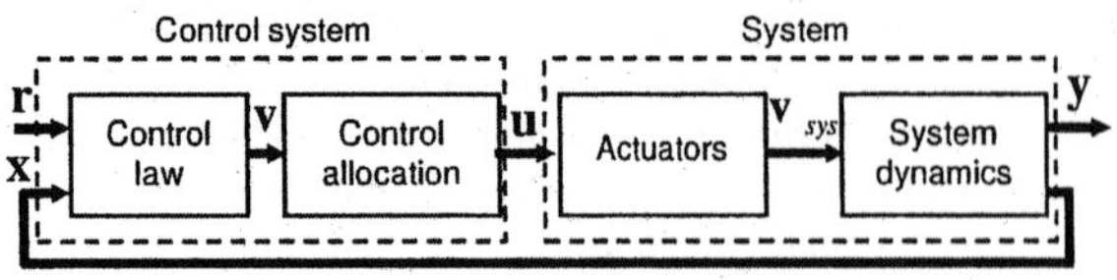


Fig. 1. The overall control system architecture.

The standard constrained linear control allocation problem can be formulated as follows: for a given $\mathbf{v}$, find $\mathbf{u}$ such that

$$\mathbf{Bu} = \mathbf{v} \tag{1}$$

$$\underline{\mathbf{u}} \le \mathbf{u} \le \overline{\mathbf{u}} \tag{2}$$

where the control effectiveness matrix $\mathbf{B}$ is a $k \times m$ matrix with rank $k$. Constraints (2) include actuator position and rate constraints, where the inequalities apply componentwise. Consider the control allocation problem $\mathbf{Bu} = \mathbf{v}$, where

$$\mathbf{u} = \begin{bmatrix} u_1 \\ u_2 \\ u_3 \end{bmatrix} \in \Re^3 \;(m = 3) \tag{3}$$

$$\mathbf{v} = \begin{bmatrix} v_1 \\ v_2 \end{bmatrix} \in \Re^2 \;(k = 2) \tag{4}$$

$$\mathbf{B} = \begin{bmatrix} \frac{1}{2} & -\frac{1}{4} & -\frac{1}{4} \\ 0 & \frac{3}{5} & -\frac{2}{5} \end{bmatrix} \tag{5}$$

$$\underline{\mathbf{u}} = \begin{bmatrix} -1 \\ -1 \\ -1 \end{bmatrix} \le \mathbf{u} = \begin{bmatrix} u_1 \\ u_2 \\ u_3 \end{bmatrix} \le \overline{\mathbf{u}} = \begin{bmatrix} 1 \\ 1 \\ 1 \end{bmatrix} \tag{6}$$

Equation $\mathbf{Bu} = \mathbf{v}$ represents system of equations

$$\begin{aligned} \frac{1}{2}u_1 - \frac{1}{4}u_2 - \frac{1}{4}u_3 &= v_1 \\ \frac{3}{5}u_2 - \frac{2}{5}u_3 &= v_2 \end{aligned} \tag{7}$$

Each equation in (7) represents a plane in the true control space $\Re^3$. The intersection of these planes is line $l$:

$$l: \quad \mathbf{p} = \begin{bmatrix} \frac{104}{77}v_1 + \frac{10}{77}v_2 + \frac{1}{4}t \\ -\frac{40}{77}v_1 + \frac{85}{77}v_2 + \frac{1}{5}t \\ -\frac{60}{77}v_1 - \frac{65}{77}v_2 + \frac{3}{10}t \end{bmatrix} \tag{8}$$

where $t$ is the parameter of the line. The constrained control subset $\Omega$, which satisfies actuator constraints (6), is a unit cube in $\Re^3$. Geometric interpretation of the control allocation problem can be obtained by reformulating the problem as follows: for a given $\mathbf{v}$, find intersection (solution set) $\Im$ of $l$ and $\Omega$. Three cases are possible:

- If $\Im$ is a segment, there is infinite number of solutions (each point that lies on the segment is solution),
- If $\Im$ is a point, there is only one solution,
- If $\Im$ is an empty set, no solution exists.

## 3. NOMENCLATURE

The following nomenclature is adopted for referring to $\Omega$ (Durham, 1993): *Boundary* of $\Omega$ is denoted by $\partial(\Omega)$. A control vector belongs to $\partial(\Omega)$ if and only if at least one of its components is at a limit. *Vertices* are the points on $\partial(\Omega)$ where each control receives a limit (min or max). In Fig. 2. vertices are denoted as $0, 1, \ldots, 7$. In the general case, the number of vertices is equal to $2^m$. Vertices are numerated using the following rule: if the vertex is represented in a binary form, then "$0$" in the $k^{th}$ position indicates that the corresponding control $u_k$ is at a minimum $\underline{u}_k$, while "$1$" indicates it is at a maximum $\overline{u}_k$. *Edges* are lines that connect vertices and that lie on $\partial(\Omega)$. In Fig. 2. edges are denoted as $01$, $02$, ..., $67$. They are generated by varying only one of the $m$ controls, while the remaining $m-1$ are at their limits, associated with the two connected vertices. In the general case, the number of edges is equal to $2^{m-1}\binom{m}{1}$. Two vertices are connected by an edge if and only if their binary representations differ in only one bit. *Facets* are plane surface on $\partial(\Omega)$ that contain two adjacent edges, i.e. two edges that have a common vertex. In Fig. 2. facets are denoted as $0132$, $0451$, ..., $7623$. In the general case, the number of facets is equal to $2^{m-2}\binom{m}{2}$. A facet is defined as the set in the control space obtained by taking all but two controls at their limits and varying the two free controls within their limits. The control effectiveness matrix $\mathbf{B}$ performs a linear transformation from the true control space $\Re^m$ to the virtual control space $\Re^k$. The image of $\Omega \subset \Re^m$ is called the attainable command set and denoted by $\Phi$. $\Phi$ is a subset of the virtual control space $\Phi_v = \{\mathbf{v} : \|\mathbf{v}\|_\infty \le 1\}$ and represents a convex polyhedron, whose boundary $\partial(\Phi)$ is the image of the facets of $\Omega$. It is important to emphasize that not all facets of $\Omega$ are mapped on the boundary $\partial(\Phi)$; most of these facets are mapped to the interior of $\Phi$.

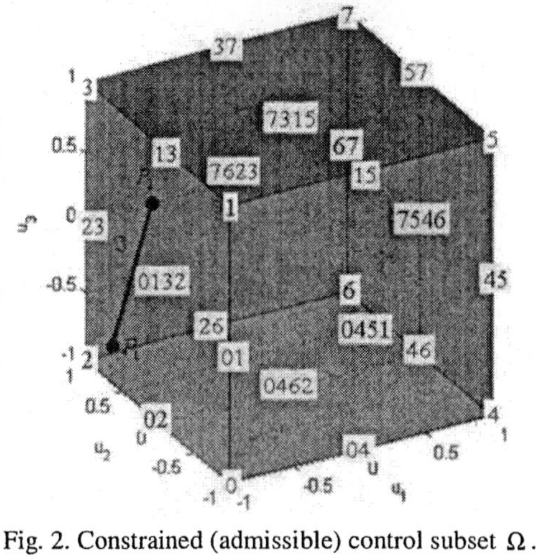


Fig. 2. Constrained (admissible) control subset $\Omega$.

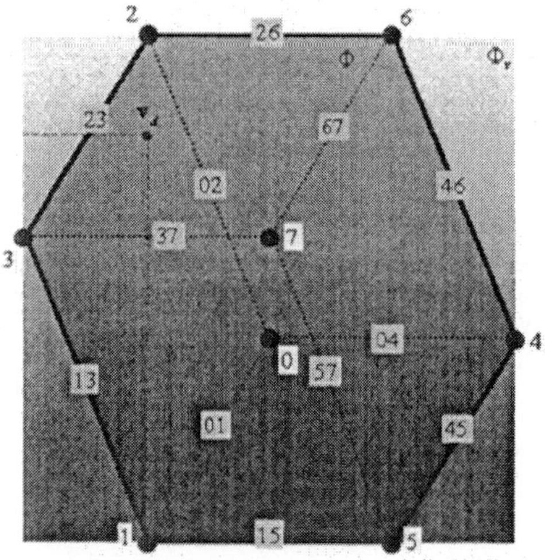


Fig. 3. Attainable command set $\Phi$.

If any $k$ columns of $\mathbf{B}$ are linearly independent (non-coplanar controls), then mapping $\mathbf{B}$ is one-to-one on $\partial(\Phi)$. The attainable command set $\Phi$ for the control allocation problem (3)-(6) is shown in Fig. 3. Images of vertices from $\partial(\Omega)$ are called *vertices* (if they lie on $\partial(\Phi)$) or *nodes* (if they lie in the interior of $\Phi$). In Fig. 3., $1$, $2$, $3$, $4$, $5$ and $6$ are vertices, while $0$ and $7$ are nodes. In a similar way, images of edges from $\partial(\Omega)$ are called edges (if they lie on $\partial(\Phi)$) or connections (if they lie in the interior of $\Phi$). In Fig. 3., 13, 23, 26, 46, 45 and 15 are edges, while 01, 02, 04, 37, 57 and 67 are connections. Images of facets that lie on $\partial(\Omega)$ are called *facets* (if they lie on $\partial(\Phi)$) or *faces* (if they lie in the interior of $\Phi$). For the problem (3)-(6), $\Phi$ is two-dimensional and there are no faces or facets. If $\Phi$ is three-dimensional, facets or faces are parallelograms. Let $\mathbf{v}_d = [-0.5 \;\; 0.6]^T$ is the desired virtual control input (Fig. 3.). Substituting $v_1 = -0.5$ and $v_2 = 0.6$ in (8) yields

$$l: \quad \mathbf{p} = \begin{bmatrix} \dfrac{-46}{77} + \dfrac{1}{4}t \\ \dfrac{71}{77} + \dfrac{1}{5}t \\ -\dfrac{9}{77} + \dfrac{3}{10}t \end{bmatrix} \tag{9}$$

The intersection $\Im$ of $l$ and $\Omega$ is a segment $\Im = P_1P_2$ (see Fig. 2.), where $P_1 = [-1 \;\; 3/5 \;\; -3/5]^T$ (for $t = t_1 = -123/77$) and $P_2 = [-1/2 \;\; 1 \;\; 0]^T$ (for $t = t_2 = 30/77$). The point $P_1$ belongs to the $0132$ facet, while $P_2$ belongs to the $7623$ facet. Each point on the segment $\Im$ is a solution. In order to extract a unique, "best" solution from $\Im$, it is necessary to introduce criteria, which is minimised by the chosen solution. By introducing criteria, the problem can be reformulated as constrained optimisation problem:

$$\min_{\mathbf{u}} \|\mathbf{u}\|_p \tag{10}$$

subject to (1) and (2). The solution depends on the choice of norm used. The most suitable criteria for underwater applications is a control energy cost function, for which case $p = 2$.

## 4. PSEUDOINVERSE

The unconstrained minimum norm allocation problem (10) subject to (1) has an explicit solution given by

$$\mathbf{u} = \mathbf{B}^{+}\mathbf{v} \tag{11}$$

where

$$\mathbf{B}^{+} = \mathbf{B}^{T}\left(\mathbf{B}\mathbf{B}^{T}\right)^{-1} \tag{12}$$

is pseudoinverse of $\mathbf{B}$ (Durham, 1993). For the constrained control allocation problem, where the constraint (2) is required to be satisfied, the solution (11) may become unfeasible, depending on the position of $\mathbf{v}$ inside $\Phi_v$. The virtual control space (square $\Phi_v = V_0V_1V_3V_2$, Fig. 4.) is mapped by the pseudoinverse to a parallelogram $\Omega_v = U_0U_1U_3U_2$ (Fig. 5a.). The intersection of the parallelogram $\Omega_v$ with the cube $\Omega$ is a convex polygon $\Omega_p = R_{13}R_{15}R_{45}R_{46}R_{26}R_{23}$, where the vertex $R_{ij}$ lies on the edge $ij$ of $\Omega$. The subset $\Phi_p \subset \Phi_v$ such that $\mathbf{B}^{+}(\Phi_p) = \Omega_p$ is a convex polygon $P_{13}P_{15}P_{45}P_{46}P_{26}P_{23}$, whose vertex $P_{ij}$ lies on the edge $ij$ of $\Phi$ (Fig. 4.). The pseudoinverse image of $\Phi$ is a polygon $\Omega_e$. Let $S = [v_1 \ \ v_2]^T$ denotes an arbitrary point from $\Phi_v$. When a point $S$ moves inside $\Phi_v$, the corresponding line $l$ moves in the true control space. For a given $S$, pseudoinverse will select the solution from $\mathfrak{I} = l \cap \Omega$ where the line $l$ intersects the parallelogram $\Omega_v$. Three characteristic cases are considered, regarding the position of $S$ relative to the partitions of $\Phi_v$ (see Fig. 4.):

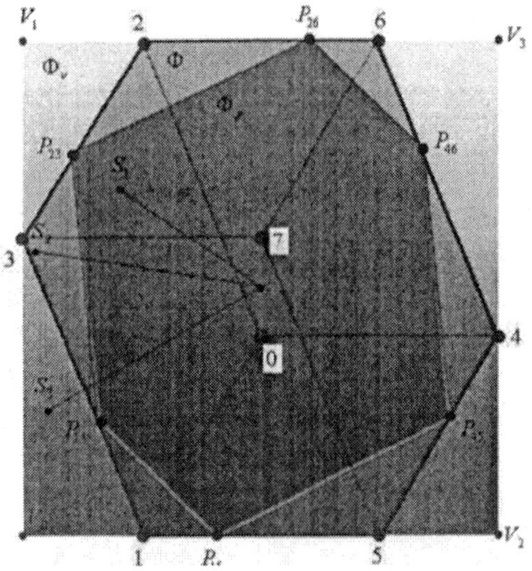

Fig. 4. Three typical cases for position of virtual control inputs relative to $\Phi_p$ and $\Phi$.

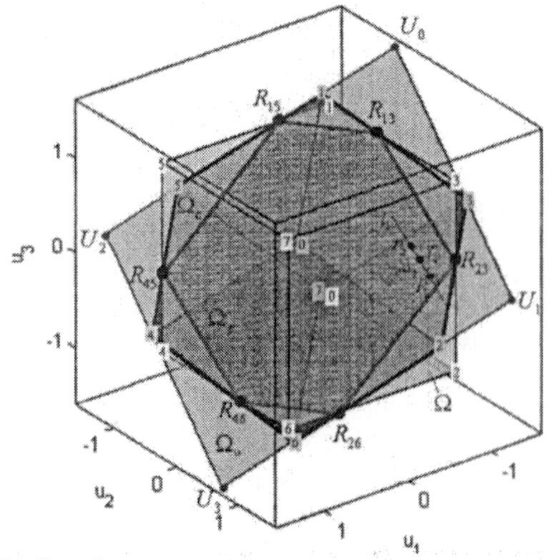

(a) Case 1: $S_1 \in \Phi_p \Rightarrow T_1 = l_1 \cap \Omega_v \in \Omega_p \subset \Omega$.

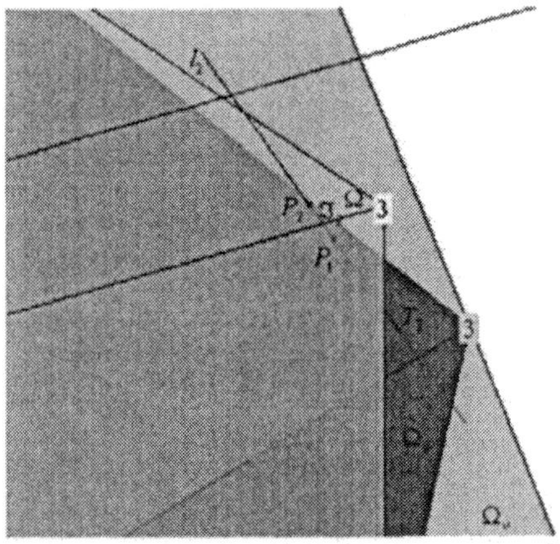

(b) Case 2: $S_2 \in \Phi \setminus \Phi_p \Rightarrow T_2 = l_2 \cap \Omega_v \notin \Omega$.

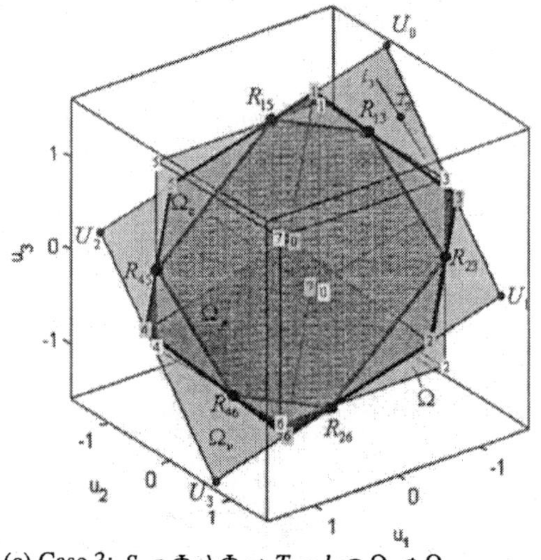

(c) Case 3: $S_3 \in \Phi_v \setminus \Phi \Rightarrow T_3 = l_3 \cap \Omega_v \notin \Omega$.

Fig. 5. Pseudoinverse solution for three typical cases.

Case 1: $S = S_1 = [-0.6 \quad 0.4]^T \in \Phi_p$,

Case 2: $S = S_2 = [-0.9375 \quad 0.1600]^T \in \Phi \setminus \Phi_p$,

Case 3: $S = S_3 = [-0.9 \quad -0.5]^T \in \Phi_v \setminus \Phi$.

In the first case, point $S_1$ lies inside $\Phi_p$ and the solution set is a segment $\mathfrak{I}_1 = P_1P_2 = l_1 \cap \Omega$, $P_1 = [-1 \ \ 14/25 \ \ -4/25]^T$, $P_2 = [-9/20 \ \ 1 \ \ 1/2]^T$

(Fig. 5a.). The pseudoinverse solution $T_1 = \mathbf{B}^+(S_1) = [-0.7584 \;\; 0.7532 \;\; 0.1299]^T$ is the point where the segment $P_1P_2$ intersects the parallelogram $\Omega_v$. This solution is feasible, since it belongs to $\Omega$. From all solutions in $\Im$, the solution $T_1$, selected by pseudoinverse, is optimal in the $l_2$ sense. In the second case, the solution set is a segment $\Im_2 = P_1P_2 = l_2 \cap \Omega$, $P_1 = [-1 \;\; 43/50 \;\; 89/100]^T$, $P_2 = [-9/20 \;\; 1 \;\; 1/2]^T$ (Fig. 5b.). The pseudoinverse solution $T_2 = \mathbf{B}^+(S_2) = [-1.2455 \;\; 0.6636 \;\; 0.5955]^T$ represents the point where the line $l_2$ intersects the parallelogram $\Omega_v$. This solution is unfeasible, since it lies on the line $l_2$ outside $\Im_2$ and does not belong to $\Omega$. Finally, in the last case the line $l_3$ does not intersect $\Omega$ and the exact solution of the control allocation problem does not exist (Fig. 5c.). However, intersection of $l_3$ and $\Omega_v$ is a point $T_3 = \mathbf{B}^+(S_3) = [-1.2805 \;\; -0.0844 \;\; 1.1234]^T$. This pseudo-inverse solution is unfeasible, since it lies outside $\Omega$.

## 5. HYBRID APPROACH

The first step in the hybrid approach algorithm is calculation of the pseudoinverse solution $\mathbf{u}$ (11). If $\mathbf{u}$ satisfies constraints (2), i.e. if $\mathbf{v} \in \Phi_p$ (Case 1), the obtained solution is feasible, optimal in the $l_2$ sense, and algorithm stops. Otherwise, the fixed-point iteration method (Burken, *et al.*, 1999) is activated, that is able to find the exact (feasible) solution $\mathbf{u}_f^*$ for $\mathbf{v} \in \Phi \setminus \Phi_p$ (Case 2) and good (feasible) approximation for $\mathbf{v} \in \Phi_v \setminus \Phi$ (Case 3). The algorithm will be explained for Case 2 in more detail. The number of iterations depends on the desired accuracy and the choice of the initial point (solution), which must be feasible. Since the (unfeasible) pseudoinverse solution $\mathbf{u}_2 = T_2$ is already found, it is natural to choose the feasible approximation of this solution as the initial point for iteration. Two choices are available: the $T$-approximation (truncation) $\mathbf{u}_{2t}^* = T_{2t}^* = [-1 \;\; 0.6636 \;\; 0.5955]^T$ is obtained from $T_2$ by truncating (clipping) components that exceed their limits. Another choice is the $S$-approximation (scaling) $\mathbf{u}_{2s}^* = T_{2s}^* = [-1 \;\; 0.5328 \;\; 0.4781]^T$ that is obtained by scaling $T_2$ by factor $f_2 = 0.8029$ such that $T_{2s}^* = \mathbf{u}_{2s}^* \in \partial(\Omega_p) \subset \partial(\Omega)$. Approximate solutions $T_{2t}^*$ and $T_{2s}^*$, and corresponding virtual control inputs $S_{2t}^*$ and $S_{2s}^*$ are shown in Fig. 6a. and 6b. Individual iterations are shown as black dots, if they start from the $T$-approximation, and as red (grey) dots, if they start from the $S$-approximation. If the desired virtual control input $\mathbf{v}_2 = S_2$ lies in $\Phi \setminus \Phi_p$, the fixed-point algorithm converges toward the exact solution $T_{2f}^* = P_1$, which lies on the solution set $\Im_2$ and has lower $l_2$ norm than any other point in $\Im_2$. The corresponding sequence in the virtual control space converges toward the desired $S_{2f}^* = S_2$.

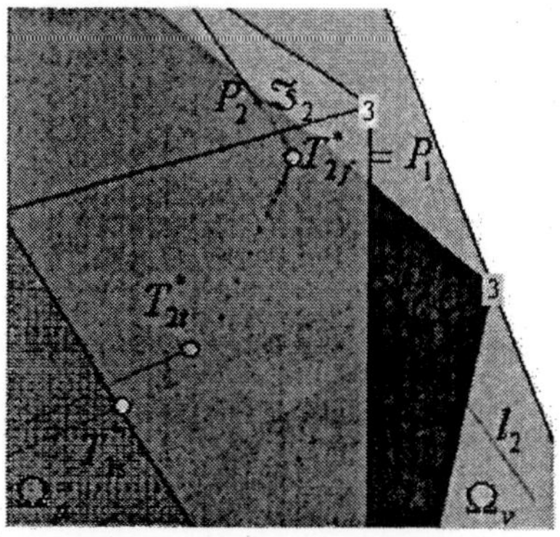


(a) If $\mathbf{v}_2 \in \Phi \setminus \Phi_p$, then $T_{2f}^* = P_1 \in \Im_2$ is the exact solution, optimal in the $l_2$ sense.

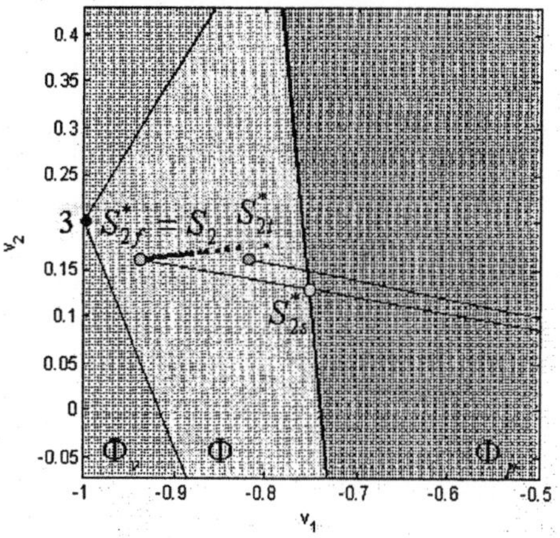


(b) If $\mathbf{v}_2 \in \Phi \setminus \Phi_p$, then $S_{2f}^* = S_2$.

Fig. 6. The fixed-point iteration method is able to find the exact solution for $\mathbf{v}_2 \in \Phi \setminus \Phi_p$.

## 6. APPLICATION TO OVERACTUATED ROVS

Two ROVs (FALCON, SeaEye Marine Ltd. and URIS, University of Girona, Fig. 7.) with different thruster configurations are used to demonstrate the performance of the proposed hybrid approach.

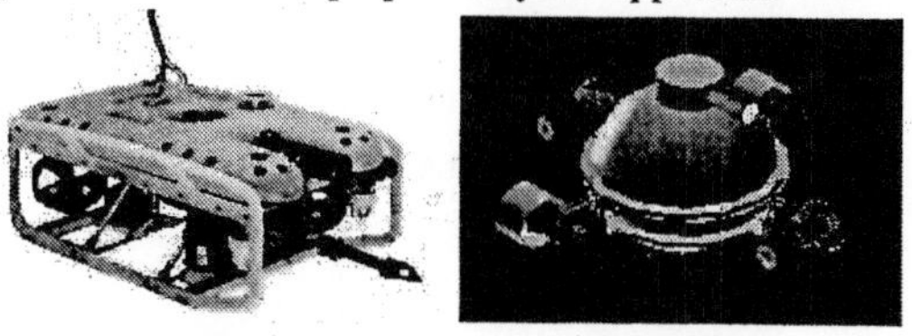

(a) FALCON (b) URIS

Fig. 7. Two ROVs with different thruster configurations (overactuated in the horizontal plane).

The normalised control allocation problem for motion in the horizontal plane is defined by (Omerdic and Roberts, 2004)

$$\text{FALCON:} \quad \underline{\mathbf{B}} = \begin{bmatrix} \frac{1}{4} & \frac{1}{4} & \frac{1}{4} & \frac{1}{4} \\ \frac{1}{4} & -\frac{1}{4} & \frac{1}{4} & -\frac{1}{4} \\ \frac{1}{4} & -\frac{1}{4} & -\frac{1}{4} & \frac{1}{4} \end{bmatrix} \tag{13}$$

URIS: $$\underline{\mathbf{B}} = \begin{bmatrix} \frac{1}{2} & \frac{1}{2} & 0 & 0 \\ 0 & 0 & \frac{1}{2} & \frac{1}{2} \\ \frac{1}{4} & -\frac{1}{4} & \frac{1}{4} & -\frac{1}{4} \end{bmatrix} \quad (14)$$

the virtual control input $$\underline{\boldsymbol{\tau}} = \begin{bmatrix} \underline{\tau}_X \\ \underline{\tau}_Y \\ \underline{\tau}_N \end{bmatrix} (k=3) \quad (15)$$

the true control input $$\underline{\mathbf{u}} = \begin{bmatrix} \underline{u}_1 \\ \underline{u}_2 \\ \underline{u}_3 \\ \underline{u}_4 \end{bmatrix} (m=4) \quad (16)$$

the actuator constraints $$\begin{bmatrix} -1 \\ -1 \\ -1 \\ -1 \end{bmatrix} \le \begin{bmatrix} \underline{u}_1 \\ \underline{u}_2 \\ \underline{u}_3 \\ \underline{u}_4 \end{bmatrix} \le \begin{bmatrix} +1 \\ +1 \\ +1 \\ +1 \end{bmatrix} \quad (17)$$

The constraint control subset $\underline{\Omega}$ is a 4D unit cube. Partitioning of the $\underline{\Phi}_v$ (3D unit cube) into regions $\underline{\Phi}_p$ and $\underline{\Phi}$ is described in (Omerdic and Roberts, 2004). In the following, the hybrid approach will be used to find the exact solution for case $\underline{\boldsymbol{\tau}}_d \in \underline{\Phi} \setminus \underline{\Phi}_p$. Let $\underline{\boldsymbol{\tau}}_d = [0.70 \;\; 0.20 \;\; 0.25]^T$ for FALCON. The pseudoinverse solution $\underline{\mathbf{u}} = [1.15 \;\; 0.25 \;\; 0.65 \;\; 0.75]^T$ (11) is unfeasible, since $\underline{u}_1 > 1$. $T$-approximation is $\underline{\mathbf{u}}_t^* = [1 \;\; 0.25 \;\; 0.65 \;\; 0.75]^T$, and $S$-approximation is $\underline{\mathbf{u}}_s^* = [1 \;\; 0.2174 \;\; 0.5652 \;\; 0.6522]^T$. They lead to approximate solutions $\underline{\boldsymbol{\tau}}_{dt}^* = [0.6625 \;\; 0.1625 \;\; 0.2125]^T$ and $\underline{\boldsymbol{\tau}}_{ds}^* = [0.6087 \;\; 0.1739 \;\; 0.2174]^T$, respectively (Fig. 8.).

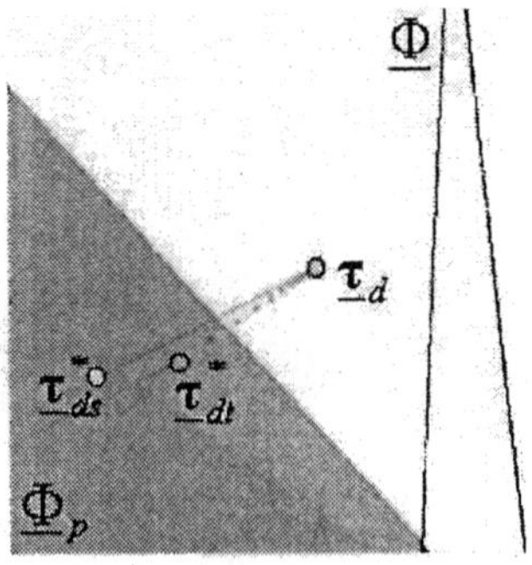


Fig. 8. Fixed-point iterations (FALCON).

Individual fixed-point iterations converge in the true control space toward the exact solution $\underline{\mathbf{u}}_f = [1.00 \;\; 0.10 \;\; 0.80 \;\; 0.90]^T$, optimal in the $l_2$ sense. Corresponding iterations in the virtual control space are shown in Fig. 8. It can be seen that they converge toward $\underline{\boldsymbol{\tau}}_d$. It is interesting to note that in this particular case iterations starting from $\underline{\boldsymbol{\tau}}_{ds}^*$ (red (grey) dots) coincide with iterations starting from $\underline{\boldsymbol{\tau}}_{dt}^*$ (black dots) after the second iteration. Similar case for URIS is shown in Fig. 9. for $\underline{\boldsymbol{\tau}}_d = [0.975 \;\; -0.025 \;\; 0.475]^T$. In most cases the number of iterations to achieve desired accuracy is smaller for initial point $\underline{\mathbf{u}}_t^*$ than $\underline{\mathbf{u}}_s^*$, i.e. $T$-approximation is better choice for the initial iteration than $S$-approximation.

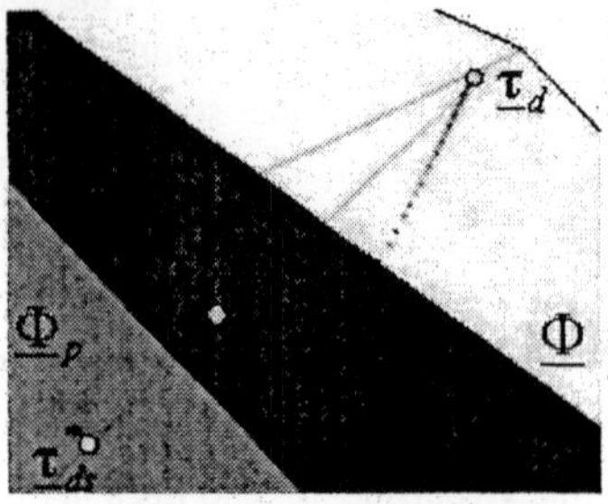


Fig. 9. Fixed-point iterations (URIS).

## 5. CONCLUDING REMARKS

A hybrid approach for control allocation is described in this paper. The approach integrates the pseudoinverse and fixed-point iteration method. Standard pseudoinverse method allocates feasible solution only on a subset of attainable command set. Introducing fixed-point iterations the feasible region is extended to the entire set and obtained solution is optimal in the $l_2$ sense. Clear geometrical interpretation of the approach is demonstrated on the low-dimensional control allocation problem. The hybrid approach is used to solve the control allocation problem for two ROVs with different thruster configuration, overactuated for motion in the horizontal plane. Results confirm that the hybrid approach is able to allocate the entire attainable command set in optimal way.

## REFERENCES


Bordignon, K.A. (1996). *Constrained Control Allocation for Systems with Redundant Control Effectors.* PhD thesis, Virginia Polytechnic Institute and State University, Blacksburg, Virginia.

Burken, J., P. Lu and Z. Wu (1999). *Reconfigurable Flight Control Designs With Applications to the X-33 Vehicle.* NASA/TM-1999-206582.

Durham, W.C. (1993). Constrained Control Allocation. *Journal of Guidance, Control, and Dynamics*, Vol. 16, No. 4, pp. 717-725.

Enns, D. (1998). Control allocation approaches. In *AIAA Guidance, Navigation, and Control Conference and Exhibit*, Boston, pp. 98-108.

Härkegård, O. (2003). *Backstepping and Control Allocation with Application to Flight Control.* PhD thesis, Department of Electrical Engineering, Linköping University, Sweden.

Omerdic, E., G.N. Roberts and P. Ridao (2003). Fault detection and accommodation for ROVs. *6$^{th}$ IFAC Conference on Manoeuvring and Control of Marine Craft (MCMC 2003)*, Girona, Spain.

Omerdic, E. and G.N. Roberts (2004). Thruster fault diagnosis and accommodation for open-frame underwater vehicles. Special feature GCUV 2003, *Control Engineering Practice*, to appear.

ELSEVIER
IFAC
PUBLICATIONS
www.elsevier.com/locate/ifac

# ROBUST CONTROL OF AN UNDERWATER ROV IN THE PRESENCE OF NONSMOOTH NONLINEARITIES IN THE ACTUATORS.

**M.L. Corradini** ** **G. Orlando** *

* *Università Politecnica delle Marche, via Brecce Bianche, 60131 Ancona, Italy.*
** *Università di Lecce, via per Monteroni, 73100 Lecce, Italy.*

Abstract. In this note a control law dealing with actuator nonlinearities is applied to the position and orientation control of a Remotely Operated Vehicle (ROV) employed by the Italian Company Snamprogetti. The proposed control law is based on sliding mode control, and guarantees robustness with respect uncertainties present both in the ROV model and in the nonlinearity considered. In order to test the effectiveness of the proposed control law, simulation results are being performed. *Copyright © 2004 IFAC*

Keywords: Robust control, Actuator nonlinearities, Underwater ROV.

## 1. INTRODUCTION

When dealing with real control problems, the designer is inevitably led to face the difficulties tied to the presence of real physical components, which often contain nonsmooth nonlinearities such as deadzones, saturation, relays, hysteresis (Kim *et al.*, 1994). In particular, actuators used in practice almost always contain static (e.g. dead zone) or dynamic (e.g. backlash) nonlinearities, whose parameters are unknown and may vary with time. Although often neglected, they are indeed present in most mechanical, hydraulic and other types of system components, and, as discussed in (Tao and Kokotovic, 1996), "Actuator and sensor nonlinearities are among the key factors limiting both static and dynamic performance of feedback control systems". As a matter of fact, these nonlinearities are particularly harmful, because they usually lead to a relevant deterioration of systems performance. Actuator nonlinearities, though expression of complex physical phenomena, can be described using piece-wise linear functions (Tao and Kokotovic, 1996) (Desoer and Shahruz, 1986). Moreover, a noticeable uncertainty has to be taken into account in models parameters, in order to match a sufficiently wide set of real situations. Therefore, a robust control design technique is claimed for.

This problem becomes particularly evident in highly uncertain plants, such as underwater vehicles. The automatic control of such vehicles, in fact, presents several difficulties, due to the non-linearity of the dynamics, to the presence of external unmeasurable disturbances, and to the high uncertainty level in the model (Fossen, 1994) (Longhi and Rossolini, 1989) (Longhi *et al.*, 1994) (Yoerger and Slotine, 1985) (Antonelli *et al.*, 2003), and references therein. Moreover, in an underwater vehicle, thrusters need often to work in adverse situations, such high values of submarine current, and it is unrealistic to assume that phenomena like dead-zone or backlash are completely absent in the actuators. Therefore to control an underwater vehicle in the presence of nonlinearities in the actuator devices seems to be a meaningful problem.

The problem of controlling systems with nonsmooth nonlinearities has been addressed in the literature using various different approaches. Neural networks (Seidl *et al.*, 1998), dithering (Desoer and Shahruz, 1986), fuzzy logic (Lewis *et al.*, 1997) (Woo *et al.*, 1998), optimal control (Ezal *et al.*, 1997) have been applied to the problem of compensating actuator nonlinearities. A major recent research thrust is within the framework of adaptive control (Tao and Kokotovic, 1996) (Grundelius and Angeli, 1996) (Tao and Kokotovic, 1995) (Tao and Kokotovic, 1994). Roughly speaking, the control scheme is based on the introduction of an inverse model of the actuator nonlinearity, updated adaptively, under the assumption of a linear plant. Variable Structure Control (VSC) has been used as well: in (Azenha and Machado, 1996), a linear plant is considered, and a describing function based model is adopted for the input nonlinearities.

An alternative approach based on sliding mode control was proposed in (Corradini and Orlando, 2002). Intrinsically nonlinear and uncertain SISO plants were addressed, with dead zone or backlash nonlinearities in the actuator.

In this note the control law proposed in (Corradini and Orlando, 2002) for the dead zone non linearity has been applied to the position and orientation control of a Remotely Operated Vehicle (ROV) employed by the Italian Company Snamprogetti for the realization of a diverless submarine structure for gas exploitation at great depths. The vehicle is equipped with four thrusts propellers, controlling its position and orientation in planes parallel to the sea surface, and is connected with the surface vessel by a supporting cable which controls the vehicle depth and provides power and communication facilities. The control system is composed of two independent parts: the first part, placed on the surface vessel, monitors the vehicle depth, and the second part controls the position and orientation of the vehicle in the dive plane.

In this paper the attention will be focussed on this second part of the control system, and the ROV dynamics will be simulated using a three-dimensional nonlinear differential equations system.

Using sliding mode control allows to consider a fully nonlinear model. The controllers performances have been tested by simulation. Results, not reported here, show the effectiveness of proposed control law.

The paper is organized as follows. Section II contains the ROV nonlinear dynamic model, while control law is described in Section 3.

## 2. PRELIMINARIES

### 2.1 ROV Nonlinear Model

The equations describing the ROV dynamics have been obtained from classical mechanics (Longhi and Rossolini, 1989) (Fossen, 1994). The ROV considered as a rigid body can be fully described with six degrees of freedom, corresponding to the position and orientation with respect to a given coordinate system. Let us consider the inertial frame $\mathbf{R}(0, x, y, z)$ and the body reference frame $\mathbf{R}_a(0_a, x_a, y_a, z_a)$ shown in Fig.1.

The ROV position with respect to $\mathbf{R}$ is expressed by the origin of the system $\mathbf{R}_a$, while its orientation by the roll, pitch and yaw angles $\psi$, $\theta$ and $\phi$ respectively.

Being the depth $z$ controlled by the surface vessel, the ROV is considered to operate on surfaces parallel to the $x - y$ plane. Accordingly the controllable variables are $x$,$y$ and the yaw angle $\phi$. It should be noticed that the roll and pitch angles $\psi$ and $\theta$ will not be considered in the dynamic model: their amplitude, in fact, has been proved to be negligible in a wide range of load conditions, and with different intensities and directions of the underwater current as well (Longhi and Rossolini, 1989).

The ROV model is described by the following system of differential equations:

$$\begin{array}{l} (M+m)\ddot{x} + H_x + R_x - T_x = 0 \\ (M+m)\ddot{y} + H_y + R_y - T_y = 0 \\ (I_z + i_z)\ddot{\phi} + M_r + M_d + M_c - M_z = 0 \end{array} \quad (1)$$

where $M$ is the vehicle mass, $m$ is the addition mass, $I_z$ is the vehicle inertia moment around the $z$ axis, $i_z$ is the addition inertia moment and $M_c$ is the resistance moment of the cable. $H_x$, $H_y$ are given by the following expressions:

$$\begin{array}{rcl} H_x &=& K(x - GV_{cx}||\mathbf{V}_c||) \\ H_y &=& K(y - GV_{cy}||\mathbf{V}_c||) \end{array} \quad (2)$$

They are the forces produced by the cable traction corresponding to a submarine current with a velocity $\mathbf{V}_c = [V_{cx}\ V_{cy}]^T$, with

$$\begin{array}{rcl} K &=& \dfrac{W}{log(1 + \frac{WL}{T_v})} \\ G &=& (L + \dfrac{T_v}{K})\rho_w C_{dc} \dfrac{D_c}{2W} \end{array} \quad (3)$$

$L$ being the cable length, $T_v$ the vehicle weight in the water, $W$ the weight for length unit of the cable, $\rho_w$ the water density, $C_{dc}$ is the drag coefficient of the cable, $D_c$ is the cable diameter.

$R_x$ and $R_y$ are the drag forces along the $x$ and $y$ axes, given by:

$$
\begin{aligned}
R_x &= \frac{1}{2}\rho_w V_x ||\mathbf{V}||[C_{d1}C_{r1}S_1|cos(\phi)|]+ \\
&+ \frac{1}{2}\rho_w V_x ||\mathbf{V}||[C_{d2}C_{r2}S_2|sin(\phi)|] \\
R_y &= \frac{1}{2}\rho_w V_y ||\mathbf{V}||[C_{d1}C_{r1}S_1|sin(\phi)|]+ \\
&+ \frac{1}{2}\rho_w V_y ||\mathbf{V}||[C_{d2}C_{r2}S_2|cos(\phi)|]
\end{aligned}
\tag{4}
$$

In formulas (4) $C_{di}$ is the drag coefficient of the $i$-th side wall ($i = 1, 2$), $C_{ri}$ the packing coefficient (depending on the geometrical characteristics of the $i$-th side wall ($i = 1, 2$)), $S_i$ is the area of the $i$-th side wall ($i = 1, 2$) and $\mathbf{V} = [V_x\ V_y]^T = [(\dot{x} - V_{cx})\ (\dot{y} - V_{cy})]^T$.

$M_d$ and $M_r$ in (1) are the components of the drag torque around the z-axis produced by the vehicle rotation and by the current, respectively, and are given by:

$$
\begin{aligned}
M_d &= \frac{1}{2}\rho_w C_d C_r S r^3 \dot{\phi}|\dot{\phi}| \\
M_r &= \frac{1}{8}\rho_w ||\mathbf{V}_c||^2 [C_{d1}C_{r1} - C_{d2}C_{r2}]\cdot \\
&\cdot\ d_1 d_2 d_3 sin(\frac{\phi - \phi_c}{2})
\end{aligned}
\tag{5}
$$

where $C_d$ is the drag coefficient of rotation, $C_r$ is the packing coefficient of rotation, $S$ is the equivalent area of rotation, $r$ is the equivalent arm of action, $d_i (i = 1 \ldots 3)$ are the vehicle dimensions along the $x_a, y_a, z_a$ axes, respectively, and $\phi_c$ is the angle between the $x$ axis and the velocity direction of the current.

This model is in agreement with models usually proposed in literature for underwater ROVs moving in the dive plane (Fossen, 1994).

Substituting (2) (3) (4) (5) in (1), the following equations are obtained:

$$
\begin{aligned}
&p_1\ddot{x} + (p_2|cos(\phi)| + p_3|sin(\phi)|)V_x|\mathbf{V}| + p_4 x+ \\
&-p_5 V_{cx}|\mathbf{V}_c| = T_x \\
&p_1\ddot{y} + (p_2|sin(\phi)| + p_3|cos(\phi)|)V_y|\mathbf{V}| + p_4 y+ \\
&-p_5 V_{cy}|\mathbf{V}_c| = T_y \\
&p_6\ddot{\phi} + p_7\dot{\phi}|\dot{\phi}| + p_8|\mathbf{V}_c|^2 sin(\frac{\phi - \phi_c}{2}) + p_9 = M_z
\end{aligned}
\tag{6}
$$

The expressions of coefficients $p_i$, $i = 1..9$, are reported in Table I below.

TABLE I
Expressions of the model parameters.

| | |
|---|---|
| $p_1$ | $M + m$ |
| $p_2$ | $\frac{1}{2}\rho_w C_{d1}C_{r1}S_1$ |
| $p_3$ | $\frac{1}{2}\rho_w C_{d2}C_{r2}S_2$ |
| $p_4$ | $W/[log(1 + \frac{WL}{T_v})]$ |
| $p_5$ | $(p_4 L + T_v)\rho_w C_{dc}\frac{D_c}{2W}$ |
| $p_6$ | $I_z + i_z$ |
| $p_7$ | $\frac{1}{2}\rho_w C_d C_r S r^3$ |
| $p_8$ | $\frac{1}{8}\rho_w [C_{d1}C_{r1} - C_{d2}C_{r2}]d_1 d_2 d_3$ |
| $p_9$ | $M_c$ |

*Assumption 2.1.* It is assumed that coefficients $p_1$ and $p_6$ do not undergo any variation, i.e. they are exactly known. Moreover, it is assumed that positive known scalars $V_{cxmax}$, $V_{cymax}$, $\rho_{pi}$, $i = 2 \ldots 5$, $i = 7 \ldots 9$ exist, such that:

$$
\begin{aligned}
&|p_i| < \rho_{pi} \\
&|V_{cx}| < V_{cxmax} \\
&|V_{cy}| < V_{cymax}
\end{aligned}
\tag{7}
$$

Moreover define $V_{cmax} = \sqrt{(V_{cxmax})^2 + (V_{cymax})^2}$.

The quantities $T_x$, $T_y$ and $M_z$ appearing in (6) are the decomposition of the thrust and the torque provided by the four vehicle propellers along the axes of $\mathbf{R}$; the corresponding decomposition with respect to the axes of $\mathbf{R}_a$ will be denoted with $T_{xa}$, $T_{ya}$ and $M_{za}$. The two triples ($T_x$, $T_y$, $M_z$) and ($T_{xa}$, $T_{ya}$, $M_{za}$) are related by the following relationships:

$$
\begin{aligned}
T_x &= cos(\phi)T_{xa} - sin(\phi)T_{ya} \\
T_y &= sin(\phi)T_{xa} + cos(\phi)T_{ya} \\
M_z &= M_{za}
\end{aligned}
\tag{8}
$$

The disposition of the 4 propellers shown in Fig.2 gives:

$$
\begin{aligned}
T_{xa} &= (T_1 + T_2 + T_3 + T_4)cos(\alpha) \\
T_{ya} &= (-T_1 - T_2 + T_3 + T_4)sin(\alpha) \\
M_{za} &= (-T_1 + T_2 - T_3 + T_4)d_a
\end{aligned}
\tag{9}
$$

with $d_a = (dxsin(\alpha) + dycos(\alpha))$ (see Fig.2).

Parameters $p_i (i = 1 \ldots 9)$ appearing in the model (6) are reported in the following table:

TABLE III
Model parameters: nominal values

| | | | |
|---|---|---|---|
| $p_1$ | 12670 $kg$ | $p_6$ | 18678 $kg\ m^2$ |
| $p_2$ | 2667 $kg\ m^{-1}$ | $p_7$ | 9200 $kg\ m^2$ |
| $p_3$ | 4934 $kgm^{-1}$ | $p_8$ | $-308.4$ $kg$ |
| $p_4$ | 417 $N\ m^{-1}$ | $p_9$ | 1492 $N\ m$ |
| $p_5$ | 46912 $kg\ m^{-1}$ | | |

The ROV model (6) can be expressed in state space form introducing the state vector $\mathbf{x} = [x_1\ x_2\ x_3\ x_4\ x_5\ x_6]^T = [x\ y\ z\ \dot{x}\ \dot{y}\ \dot{\phi}]^T$, and the input vector $\mathbf{u} = [u_1\ u_2\ u_3]^T = [T_y\ T_y\ M_z]^T$:

$$\dot{\mathbf{x}} = \mathbf{f}(\mathbf{x}) + \mathbf{B}\mathbf{u} \tag{10}$$

where:

$$\mathbf{B} = \begin{bmatrix} 0 & 0 & 0 \\ 0 & 0 & 0 \\ 0 & 0 & 0 \\ \frac{1}{p_1} & 0 & 0 \\ 0 & \frac{1}{p_1} & 0 \\ 0 & 0 & \frac{1}{p_6} \end{bmatrix} \tag{11}$$

and $\mathbf{f}(\mathbf{x}) = [x_4\ x_5\ x_6\ f_4(\mathbf{x})\ f_5(\mathbf{x})\ f_6(\mathbf{x})]^T$, with:

$$\begin{aligned} f_4(\mathbf{x}) &= -\frac{1}{p_1}\left[(p_2|\cos x_3| + p_3|\sin x_3|)\cdot\right. \\ &\quad \cdot\ \left.(x_4 - V_{cx})|\mathbf{V}| + p_4 x_1 - p_5 V_{cx}|\mathbf{V}_c|\right] \\ f_5(\mathbf{x}) &= -\frac{1}{p_1}\left[(p_2|\sin x_3| + p_3|\cos x_3|)\cdot\right. \\ &\quad \cdot\ \left.(x_5 - V_{cy})|\mathbf{V}| + p_4 x_2 - p_5 V_{cy}|\mathbf{V}_c|\right] \\ f_5(\mathbf{x}) &= -\frac{1}{p_6}\left[(p_7 x_6|x_6| + p_8|\mathbf{V}_c|^2\cdot\right. \\ &\quad \cdot\ \left.\sin\left(\frac{x_3 - \phi_c}{2}\right)\right] \end{aligned} \tag{12}$$

In view of Assumption 2.1, the uncertain but bounded parameters $p_i$, $i = 1\ldots 9$ induce the presence of uncertainties on $f_4(\mathbf{x})$, $f_5(\mathbf{x})$, $f_6(\mathbf{x})$, which are bounded as well. Denoting the induced uncertainties as $\Delta f_i(\mathbf{x})$, $i = 4\ldots 6$, one has that suitable functions $\delta_i(\mathbf{x})$, $i = 1\ldots 3$ can be found such that:

$$|\Delta f_{3+i}(\mathbf{x})| \leq \delta_i(\mathbf{x}), \quad i = 1\ldots 3 \tag{13}$$

### 2.2 Deadzone Nonlinearities

Each input of the nonlinear system (10) is supposed to be preceded by the actuating device $u_i = F_i(v_i)$, $i = 1\ldots 3$, $\mathbf{u} = [u_1\ u_2\ u_3]^T$ being the plant input not available for control. In this paper, deadzone nonlinearities have been considered to be present in the actuator. The analytical expression of the dead zone characteristic is:

$$u = F(v) = \begin{cases} m_r(v - b_r) & if \quad v \geq b_r \\ 0 & if\ b_l < v < b_r \\ m_l(v - b_l) & if \quad v \leq b_l \end{cases} \tag{14}$$

In order to lighten notation, $u_i$ and $F_i(v_i)$, $i = 1\ldots 3$, have been denoted by $u$ and $F(v)$, respectively.

*Assumption 2.2.* The coefficients of the dead zone nonlinearities are uncertain, with uncertainties bounded by known constants, i.e.:

$$\begin{aligned} m_r &= \hat{m}_r + \Delta m_r, & |\Delta m_r| &\leq \rho_{mr} \\ m_l &= \hat{m}_l + \Delta m_l, & |\Delta m_l| &\leq \rho_{ml} \\ b_r &= \hat{b}_r + \Delta b_r, & |\Delta b_r| &\leq \rho_{br} \\ b_l &= \hat{b}_l + \Delta b_l, & |\Delta b_l| &\leq \rho_{bl} \end{aligned} \tag{15}$$

## 3. ROBUST STABILIZATION OF THE ROV IN THE PRESENCE OF DEADZONE IN THE ACTUATORS

Define the following sliding surfaces:

$$\begin{aligned} &s_i = \dot{x}_i + \lambda_i x_i = x_{3+i} + \lambda_i x_i = 0, \\ &i = 1\ldots 3 \quad \lambda_i > 0 \end{aligned} \tag{16}$$

It is straightforward to verify that the achievement of a sliding motion on the three surfaces (16) ensures ROV stabilization. Moreover, define

$$w_i(\mathbf{x}) = f_{3+i}(\mathbf{x}) + \lambda_i x_{3+i} \tag{17}$$

$$r_1(\mathbf{x}) = \frac{1}{p_1} \tag{18}$$

$$r_2(\mathbf{x}) = \frac{1}{p_1} \tag{19}$$

$$r_3(\mathbf{x}) = \frac{1}{p_6} \tag{20}$$

For each component of the input vector, the following Theorem can be proved (Corradini and Orlando, 2002):

*Theorem 1.* The ROV system described by (10) is given, under Assumptions 2.1 and 2.2. The following control law:

$$v_i = v_{e,i} + v_{n,i}, \quad i = 1\ldots 3 \tag{21}$$

with

$$v_{e,i} = \begin{cases} v_{e,i}^{(1)} = \hat{b}_r - \dfrac{w_i(\mathbf{x})}{r_i(\mathbf{x})\hat{m}_r} & if\ s_i < 0 \\ v_{e,i}^{(2)} = \hat{b}_l - \dfrac{w_i(\mathbf{x})}{r_i(\mathbf{x})\hat{m}_l} & if\ s_i > 0 \end{cases} \tag{22}$$

$$v_{n,i} = \begin{cases} v_{n.i}^{(1)} = \theta_1\left[\dfrac{[\rho_{mr}|v_{e,i}^{(1)} - \hat{b}_r| + \rho_{br}(\hat{m}_r + \rho_{mr})]}{(\hat{m}_r - \rho_{mr})} + \right. \\ \quad \left. + \dfrac{\rho(\mathbf{x})}{(\hat{m}_r - \rho_{mr})r_i(\mathbf{x})}\right] \quad if\ s_i < 0 \\ v_{n,i}^{(2)} = -\theta_2\left[\dfrac{[\rho_{ml}|v_{e,i}^{(2)} - \hat{b}_l| + \rho_{bl}(\hat{m}_l + \rho_{ml})]}{(\hat{m}_l - \rho_{ml})} + \right. \\ \quad \left. + \dfrac{\rho(\mathbf{x})}{(\hat{m}_l - \rho_{ml})r_i(\mathbf{x})}\right] \quad if\ s_i > 0 \end{cases} \tag{23}$$

where:

$$\theta_1 \geq \max\{1, \alpha_1\}, \quad \theta_2 \geq \max\{1, \alpha_2\} \tag{24}$$

with:

$$\alpha_1 = \frac{(\hat{m}_r - \rho_{mr})r_i(\mathbf{x})(\hat{b}_r + \rho_{br} - v_{e,i}^{(1)})}{[\rho_{mr}|v_{e,i}^{(1)} - \hat{b}_r| + \rho_{br}(\hat{m}_r + \rho_{mr})] \cdot r_i(\mathbf{x}) + \rho(\mathbf{x})}$$

$$\alpha_2 = \frac{(\hat{m}_l - \rho_{ml})r_i(\mathbf{x})(v_{e,i}^{(2)} - \hat{b}_l + \rho_{bl})}{[\rho_{ml}|v_{e,i}^{(2)} - \hat{b}_l| + \rho_{bl}(\hat{m}_l + \rho_{ml})] \cdot r_i(\mathbf{x}) + \rho(\mathbf{x})}$$

ensures the achievement of three independent sliding motions on each (16), i.e. the asymptotic stabilization of the ROV in the presence of deadzone nonlinearities in the actuators.

## REFERENCES


Antonelli, G., F. Caccavale, S. Chiaverini and G. Fusco (2003). A novel adaptive control law for underwater vehicles. *IEEE Transactions on Control Systems Technology* **11**(2), 221–232.

Azenha, A. and J.A.T. Machado (1996). Variable structure control of robots with nonlinear friction and backlash at the joints. *Proc. 1996 IEEE Int. Conf. Robotics and Automation* **1**, 366–371.

Corradini, M.L. and G. Orlando (2002). Robust stabilization of nonlinear uncertain plants with backlash or dead zone in the actuator. *IEEE Transactions on Control Systems Technology* **10**(1), 158–166.

Desoer, C.A. and S.M. Shahruz (1986). Stability of dithered non-linear systems with backlash or hysteresis. *Int. J. Control* **43**(3), 1045–1060.

Ezal, K., P.V. Kokotovic and G. Tao (1997). Optimal control of tracking systems with backlash and flexibility. *Proc. 36th IEEE Conference on Decision and Control* **2**, 1749 –1754.

Fossen, T. I. (1994). *Guidance and control of ocean vehicles*. J.Wiley & Sons, New York.

Grundelius, M. and D. Angeli (1996). Adaptive control of systems with backlash acting on the input. *Proc. 35th IEEE Conf. Decision and Control* **4**, 4689–4694.

Kim, J.H., J.H. Park, S.W. Lee and E.K.P. Chong (1994). A two layered fuzzy logic controller for systems with deadzones. *IEEE Trans. on Industrial Electronics* **41**(2), 155 –162.

Lewis, F.L., K. Liu, R. Selmic and Li-Xin Wang (1997). Adaptive fuzzy logic compensation of actuator deadzones. *J. of Robotic Systems* **14**, 501–511.

Longhi, S. and A. Rossolini (1989). Adaptive control for an underwater vehicle: simulation studies and implementation details. *Proc. IFAC Workshop on Expert Systems and Signal Processing in Marine Automation* pp. 271– 280.

Longhi, S., G. Orlando and A. Serrani (1994). Advanced control strategies for a remotely operated underwater vehicle. *Proc. First World Automation Congress (WAC'94)* pp. 105–110.

Seidl, D.R., Sui-Lun Lam, J.A. Putman and R.D Lorenz (1998). Neural network compensation of gear backlash hysteresis in position-controlled mechanisms. *IEEE Trans. on Industry Applications* **31**(6), 1475 –14832.

Tao, G. and P.V. Kokotovic (1994). Adaptive control of system with unknown output backlash. *IEEE Transactions on Automatic Control* **40**(2), 326–320.

Tao, G. and P.V. Kokotovic (1995). Continuous-time adaptive control of systems with unknown backlash. *IEEE Transactions on Automatic Control* **40**(6), 1083–1087.

Tao, G. and P.V. Kokotovic (1996). *Adaptive control of systems with actuator and sensor nonlinearities.* Wiley.

Woo, K.T., Li-Xin Wang, F.L. Lewis and Z.X. Li (1998). A fuzzy system compensator for backlash. *Proc. 1998 IEEE Int. Conf. on Robotics and Automation* **1**, 181–186.

Yoerger, D.R. and J.J.E. Slotine (1985). Robust trajectory control of underwater vehicles. *IEEE J. Oceanic Engng* **10**(4), 462–470.


## 4. FIGURES

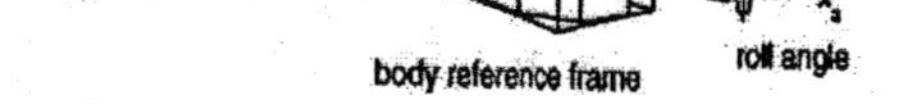

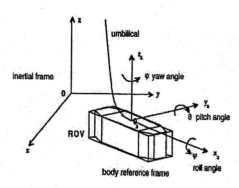

Fig.1 - Operational Configuration of the ROV

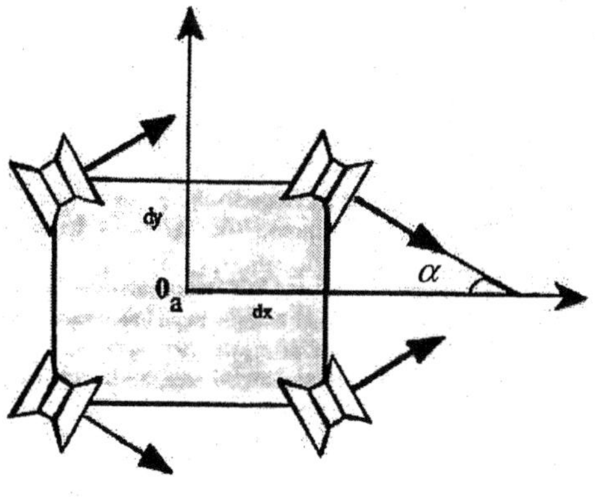


Fig.2 - ROV propellers system

ELSEVIER
IFAC
PUBLICATIONS
www.elsevier.com/locate/ifac

# MODEL PREDICTIVE CONTROL OF A REMOTELY OPERATED FLIGHT VEHICLE

**Meihong Wang and Robert Sutton**

*Marine and Industrial Dynamic Analysis Research Group*
*School of Engineering, University of Plymouth*
*Drake Circus, Plymouth, PL4 8AA, United Kingdom*
*Email: meihong.wang@plymouth.ac.uk r.sutton@plymouth.ac.uk*

Abstract: A model predictive controller design for a remotely operated flight vehicle (ROFV) is presented in this paper. The ROFV is known as Subzero III. The system characteristics of Subzero III were qualitatively examined. This provides the basis why a SISO model predictive control (MPC) scheme can be used for the control of Subzero III. The controller design was implemented with the MPC Toolbox. On-line optimisation in the use of MPC for ROFV was analysed. The influence of different tuning parameters is discussed based on simulations. Conclusions are given in the end. *Copyright ©2004 IFAC*

*Keywords:* Autonomous underwater vehicles; model predictive control; SISO; tuning; optimization

## 1. INTRODUCTION

### *1.1 Introduction of Subzero III*

Subzero III is a torpedo-shaped underwater flight vehicle that has been built by the Institute of Sound and Vibration Research (ISVR), University of Southampton (Feng *et al.*, 2003). The vehicle is 1m long with a diameter of 10cm. The payload carried is a maximum of 3kg with a target speed of 0 to 8 knots (approximately 0 to 4m/s) and a depth capability of 6m. The duration of a mission is 15minutes. The vehicle is shown in Fig. 1 (Feng *et al.*, 2003).

In Subzero III, the thruster system consists of the propeller, which is linked to a DC motor and gear box by a steel shaft with a universal coupling. The DC motor is controlled by a pulse width modulated (PWM) drive running at 800Hz. On the tail are mounted the four control surfaces: two rudders and two stern-planes. The two rudder surfaces are linked together to form a single rudder. The two stern-plane surfaces are linked together to form a single stern-plane (Feng *et al.*, 2003).

Feng and Allen (2003) have presented the development and validation of a dynamic model of Subzero III. The model is based on 6 DOF (degree of freedom) nonlinear motion equations and includes the dynamics of the actuators. The nonlinear model of Subzero III based on motion equations was implemented with Matlab/Simulink (The MathWorks; Natick, MA) (Feng and Allen, 2003).

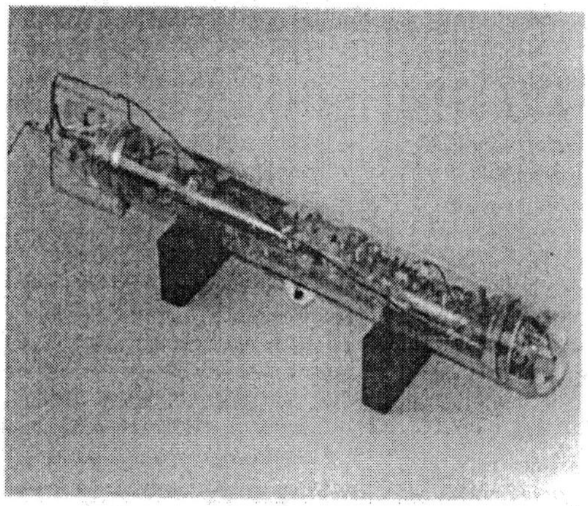

Fig. 1. *The torpedo-shaped remotely operated underwater flight vehicle - Subzero III*

### *1.2 Novel contribution and outline of the paper*

Use of MPC for the ROFV is explored in this paper. The influence of tuning parameters is discussed based on simulations. Online-optimisation, the bottleneck for the use of MPC in ROFV, is analysed. Potential solution to this problem is then proposed.

The paper is structured as follows: the MPC principle is briefly introduced in section 2. The motivation of using MPC for ROVs is also analyzed. Previous research of MPC for ROVs is also reviewed in section 2. Qualitative analysis of Subzero III is presented in section 3. Section 4 discusses the MPC implementation, the relevant on-line optimisation issue and the tuning mechanism. Simulation examples are given in section 5. Concluding remarks are then made in section 6.

## 2. MODEL PREDICTIVE CONTROL FOR ROVS

### *2.1 Introduction of MPC*

In MPC, there are four concepts very important in understanding the principles (Maciejowski, 2001):

- An internal model
- The reference trajectory
- Receding horizon principle
- On-line optimisation

The idea is that (see Fig. 2): With the internal model (as simple as step response, as complex as state-space model), the future behaviour of the process can be predicted. The deviation of the predictions away from the reference trajectory is penalised. The future optimal inputs are calculated from on-line optimisation. Only one current optimal input will be applied into the system. From this description, MPC is in fact a feedback control policy.

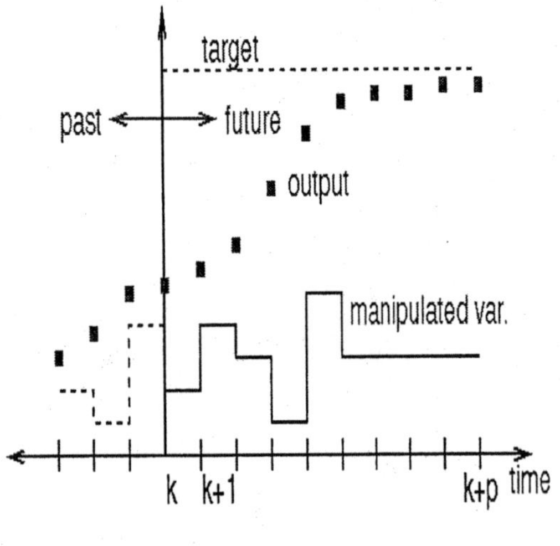


Fig. 2. Basic principle of MPC

The objective function (or the cost function) penalizes deviations of the predicted controlled outputs from the reference trajectory. The objective function also penalizes changes in the input variables.

$$V(k) = \sum_{i=H_w}^{H_p} \left\| \hat{z}(k+i \mid k) - r(k+i) \right\|_{Q(i)}^2 + \sum_{i=0}^{H_u-1} \left\| \Delta\hat{u}(k+i \mid k) \right\|_{R(i)}^2 \qquad (1)$$

where $\hat{z}(k+i \mid k)$ is predicted controlled output and $r(k+i)$ is reference trajectory; $\Delta u(k+i \mid k)$ is the changes of the input variable; $H_p$ is the prediction horizon and $H_u$ is the control horizon; $Q(i)$ and $R(i)$ are output deviation weighting matrix and input weighting matrix respectively.

### *2.2 The motivation of MPC for ROVs*

Model predictive control (MPC) has been very popular in control engineering, especially in chemical process control area, for several decades. It is a natural extension of classical optimal control. The advantages of MPC are listed as following (Maciejowski, 2001):

1. MPC can handle multivariable control problems naturally.
2. MPC can take account of actuator limitations.
3. MPC is extension of optimal control, it can handle model uncertainty effectively.

In the formulation of ROV control problem, all these problems are there:

1. ROV control is a multivariable control problem.
2. In ROV control, rudder and stern-plane have range limits (±20 degree and ±30 degree respectively).
3. In ROV control, there is model uncertainty more or less. This is more important for open-frame vehicles.

Therefore, there is strong motivation for MPC in ROV.

### *2.3 Literature review on MPC for ROVs*

Kodogiannis *et al.* (1996) proposed a scheme of using a neural network model in MPC for the depth control of an AUV. The reason for using neural network model is that neural network can model non-linear process effectively. Therefore, the MPC scheme can deal with a system containing severe non-linearity. In this case, it is for depth control since the depth channel contains severe nonlinearity. A non-linear process model was developed with Neural Network. The model was then used to design a MPC controller offline (five-step-ahead prediction and non-linear learning controller with stability assurance).

Kataebi and Grimble (1999) proposed a three-layer controller architecture for regulation, trajectory tracking, diagnosis and control reconfiguration. A model predictive control (MPC) design technique is employed at the middle layer (for guidance) to manoeuvre optimally the AUV along a desired trajectory. In the bottom layer, Hinf /H2 control is used. The advantage of using MPC is that the actuator rate constraints and a desired limit for the overshoot can be explicitly incorporated into the control design formulation (to overcome instability and overshoot).

Kwiesielewicz *et al.* (2001) presented a comparative study in which a PD controller is compared a MPC controller. The PD controller parameters are evaluated on-line with a fuzzy logic expert system. The MPC controller was designed with MPC Toolbox.

## 3. QUALITATIVE ANALYSIS OF THE SUBZERO III

To gain *a priori* knowledge for MPC design, the process is analysed qualitatively based on the first-principle model in Matlab/Simulink developed by Feng and Allen (2003).

### *3.1 Interaction between different channels*

Fig. 3 shows the step response when the rudder is used as an input under the condition that the motor command input is 522 (the motor command input is used to control the duty cycle of the DC motor, so no unit needs to be specified) and the steady-state achieved. From Fig. 3, the cross-coupling between different channels can be observed to be very weak. That is to say, when rudder is used as an input, the main (or dominant) output is heading angle, however there is a weak cross-coupling between rudder and forward speed, rudder and depth.

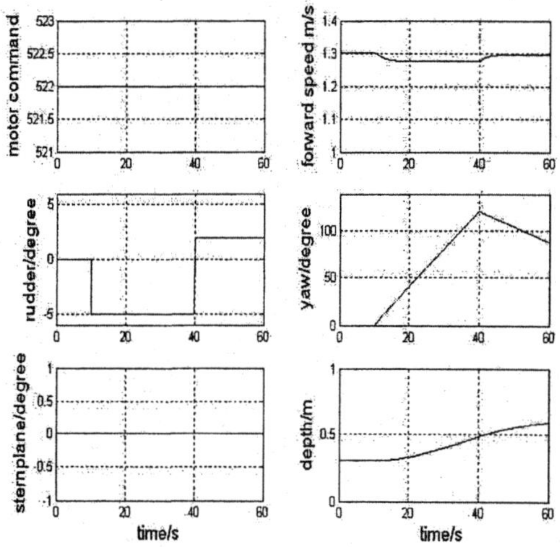


Fig. 3. *System responses when a step input is applied in the rudder and the motor command is fixed at 522.*

Fig. 4 shows the step response when the stern-plane is used as an input under the condition that the motor command input is 522 and the steady-state achieved. From Fig. 4, when stern-plane is used as an input, the main (or dominant) output is depth, but there is also a weak cross-coupling between stern-plane and forward speed.

### *3.2 Linearity*

Linear systems are those systems whose input-output relationship possesses the property of *superposition*. If a system has an input $x_1(t)$, then its output is $y_1(t)$. When its input is $x_2(t)$, then its output becomes $y_2(t)$. The *superposition* principle says that when the system has input $ax_1(t)+bx_2(t)$, its output must be $ay_1(t)+by_2(t)$ in which $a$ and $b$ could be any constants.

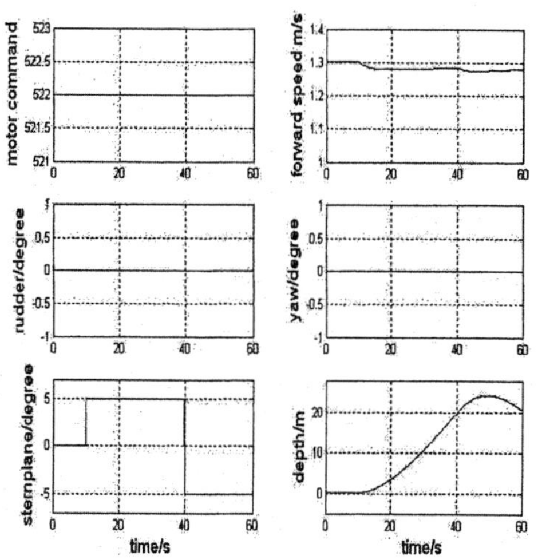


Fig. 4. *System responses when a step input is applied in the stern-plane and the motor command is fixed at 522.*

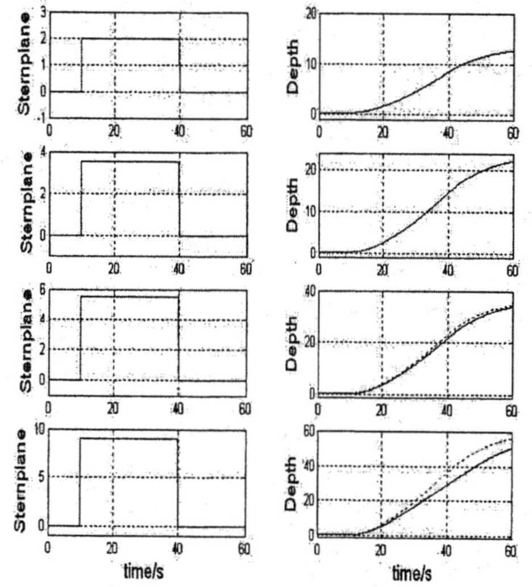


Fig. 5. *Linear or non-linear analysis of the sternplane-depth channel*

Fig. 5 was generated to further analyze whether the sternplane-depth channel is linear or non-linear. In Fig. 5, different step responses are shown when the stern-plane is used as input under the condition that the motor command input is 522 (the motor command input is used to control the duty cycle of the DC motor, so no unit needs to be specified) and that the steady-state has been achieved. In the left top panel, the stern-plane step input with a 2 degree amplitude was applied from 15seconds to 40seconds. The right top panel shows its response in depth. In the same way, the two panels in the second row show the stern-plane step input with a 3.5 degree amplitude. In the two panels of the third row, stern-plane step input with a 5.5 degree amplitude was compared with the sum of the step responses in the first and second rows. It indicates a very close fit (the dotted line *vs* the solid line). The two panels in the bottom rows show the stern-plane step input with a 9 degree amplitude. This step response is also compared with the sum of the step responses in the second and third rows. The minor difference between the two lines indicates the minor nonlinearity.

From Fig. 5, the sternplane-depth channel still can be viewed as a linear system even though there exists minor nonlinearity. The rudder – heading angle can also be analyzed in similar way and similar conclusion derived regarding linearity.

### *3.3 Conclusion on the analysis of the Subzero III process*

The analysis in this section provides the basis of why linear SISO MPC that is explored in the following sections can be used for Subzero III. It also indicates its strong integration action.

## 4. MPC IMPLEMENTATION ISSUES AND ITS TUNING MECHANISM

### *4.1 Implementation*

The MPC Toolbox (Morari and Ricker, 1995) is a convenient tool to design a MPC controller. Functions based on the step response model were used since step response models can be used for both stable and integration. As analysed before, Subzero III is an integration process. After obtaining a transfer function from system identification, *poly2tfd* and *tfd2step* were used so that the resulting step response is in MPC *step* format. *mpccon* was used to calculate the MPC controller gain with MPC *step* format model. *mpcsim* simulates closed-loop systems with saturation constraints on the manipulated variables using models in the MPC *step* format.

### *4.2 On-line optimisation*

MPC dates back to the early 1980s with applications in chemical industry and refineries. For those processes, sampling rate used is generally in minutes. On-line computation is not a big problem for such applications.

For the control of ROFV, things will be different. The sampling time for Subzero III is 0.125 second.

In the implementation with MPC Toolbox, the optimisation tool is sequential Quadrtic Programming (SQP). The SISO MPC needs online computation. However, here on-line computation is not a problem. The reasons are: (1) It is only SISO MPC; (2) No constraints were applied. All these simplified the MPC problem formulation. Therefore, the control actions obtained from the optimisation is a linear function of the predicted future errors.

### *4.3 MPC tuning mechanism*

In general, one has to choose the horizons (the prediction horizon $H_p$ and the control horizon $H_u$) and weights (output deviation weighting matrix $Q(i)$ and input weighting matrix $R(i)$) by trial-and-error using simulations to judge their effectiveness (Maciejowski, 2001).

For prediction horizon $H_p = 1$, nominal stability of the closed-loop system is guaranteed for any finite control horizon $H_u$, and time invariant output weights $Q(i)$ and input weights $R(i)$.

Decreasing $H_u$ relative to $H_p$ makes the control action less aggressive and tends to stabilize the system.

Increasing $R(i)$ always has the effect of making the control action less aggressive since increasing $R(i)$ penalizes changes in the input vector heavily.

## 5. SIMULATIONS

### *5.1 Rudder-heading angle channel*

Wang and Sutton (2004) described the linear open-loop system identification of the two channels of subzero III based on the validated Matlab/Simulink model of Subzero III. The obtained model expressed in transfer function form for the rudder-heading angle channel is

$$\frac{Y(s)}{R(s)} = \frac{-0.01873s^4 - 1.527s^3 - 63.49s^2 - 1089s - 1433}{s^5 + 24.11s^4 + 289.5s^3 + 1282s^2 + 1754s} \quad (2)$$

The sampling rate 0.125s used because it is the actual sampling time used in Subzero III. The model and the plant are assumed to be the same in the simulations. A step disturbance and no change in set-point are also assumed. Therefore, the control effect is to see how fast the MPC controller can bring the controlled output to a new steady-state.

Different cases (with different tuning parameters) were designed in the simulations to see the closed loop control effects:

Case 1: $H_p$=8; $H_u$=2; $Q(i)$=1; $R(i)$=1;

Case 2: $H_p$=8; $H_u$=2; $Q(i)$=1; $R(i)$=3;

Case 3: $H_p$=8; $H_u$=2; $Q(i)$=1; $R(i)$=0;

Case 4: $H_p$= 5; $H_u$=2; $Q(i)$=1; $R(i)$=0;

*5.2 Results and discussion*

The closed-loop control effect for the above four cases were plotted in Fig. 6 and Fig. 7.

When the left two panels in Fig. 6 (Case 1) are compared with the right panels (Case 2) in the same figure, the process output dynamics in Case 2 is much slower. This is due to a larger input weight $R(i)$ used (3 *vs* 1). This results in the control input not so aggressive.

When the left two panels in Fig. 7 (Case 3) are compared with the left two panels in the Fig. 6 (Case 1), the process output dynamics in Case 3 is much faster. This is also due to different input weight $R(i)$ used (0 *vs* 1).

When the left two panels in Fig. 7 (Case 3) are compared with the right two panels in the same figure (Case 4), the input variation in Case 4 looks more aggressive. This is due to a smaller $H_p$ used (5 *vs* 8) while $H_u$ is fixed.

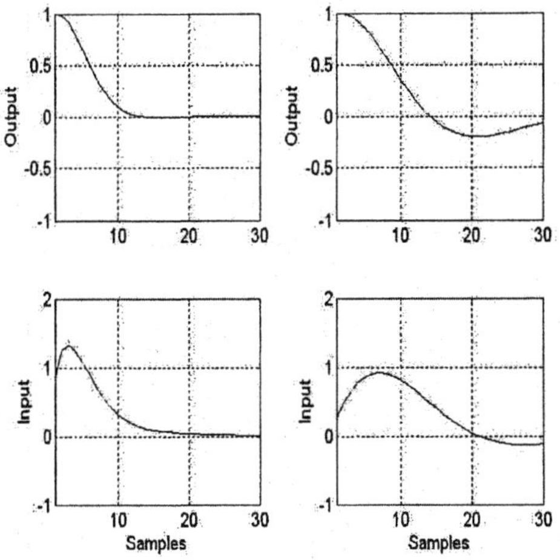


Fig. 6. *MPC closed-loop control effects; left two panels for Case 1 and right two panels for Case 2.*

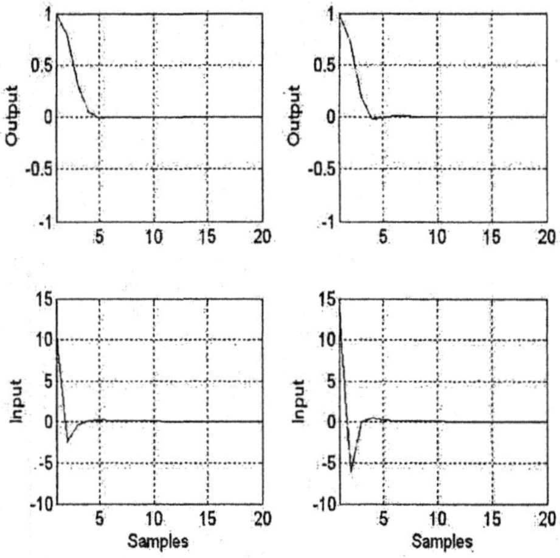


Fig. 7. *MPC closed-loop control effects; left two panels for Case 3 and right two panels for Case 4.*

*5.3 Future work*

The insight from the practice is that on-line optimisation is a key issue for use of MPC in ROVs. Therefore, future work will be concentrated on solving the on-line optimization problem.

When MPC is used for the control of ROV, one of the main obstacles is the fast dynamics. The fast dynamics impose the fast sampling requirement, which implies that there is no enough computational time to perform the online optimisation.

Bemporad *et al.* (2002) proposed the technique of multi-parametric programming to solve this problem. With this method, the states (or the estimates of the states) are viewed as a set of parameters of the quadratic programming (QP) problem. The new result obtained by this means is that the control law is continuous is states. Algorithms are also given for determining efficiently the boundaries (in the state-space) at which changes from one piecewise-linear control law to another occur. The idea is that the state-space is divided up into a manageably small number of convex pieces, one could pre-compute (off-line) the control law that should be applied in each piece, and then the MPC algorithm would consist simply of reading the appropriate gain matrix from a table lookup, depending on the current state estimate. This technique should be particularly useful in high-bandwidth applications, when high control updates are required. Pistikopoulos *et al.* (2002) proposed a similar idea of changing on-line optimization to off-line parametric optimization when MPC controller is used.

## 6. CONCLUSIONS

The paper explored the use of MPC for a ROFV based on simulations. The qualitative process analysis based on the first-principle model in Matlab/Simulink provides the basis of why linear SISO MPC can be used for the control of Subzero III. Influences from different sets of tuning parameters were observed from the simulation examples. The insight from this practice is that on-line optimisation is the bottleneck of using MPC for the control of ROV because of the fast sampling rate. Potential solution to this problem is to use the newly developed multi-parametric programming technique.

## ACKNOWLEDGEMENTS

The authors would like to thank the EPSRC for the financial support of the IMPROVES project. The authors thank Professor Robert Allen and Dr Zhengping Feng for sharing with us the first-principle model of Subzero III in Matlab/Simulink.

## REFERENCES


Bemporad, A., M. Morari, V. Dua, and E.N.Pistikopoulos (2002), The explicit linear quadratic regulator for constrained systems, *Automatica*, **38**, 3-20.

Feng, Z., Stansbridge, R., White, D., Wood, A and Allen, R. (2003), A low cost underwater flight vehicle, *1st IFAC Workshop on Guidance and Control of Underwater Vehicles*, Newport, UK, 215-219.

Feng, Z. and Allen, R. (2003), Development and validation of a dynamic model of and AUV, *1st IFAC Workshop on Guidance and Control of Underwater Vehicles*, Newport, UK, 179-184.

Katebi, MR and Grimble MJ (1999), Integrated control, guidance and diagnosis of re-configurable autonomous underwater vehicle control, *International Journal of System Science*, **30**, 1021-1032.

Kodogiannis, V.S., Lisboa, P.J.G., and Lucas, J. (1996), Neural network modelling and control for underwater vehicles, *Artificial Intelligence in Engineering*, **1**, 203-212.

Kwiesielewicz, M., Piotrowski, W., Sutton, R (2001), Predictive versus Fuzzy Control of Autonomous Underwater Vehicle, *7th IEEE International Conference on Methods and Models in Automation and Robotics*, 609-612.

Maciejowski (2001), *Predictive control with constraints,* Prentice Hall, UK.

Morari, M. and Ricker, N.L. (1995), *Model predictive control Toolbox: For use with Matlab*, The MathsWorks, Inc.

Pistikopoulos EN, Dua V, Bozinis NA, Bemporad A, Morari M(2002), On-line optimization via off-line parametric optimization tools, *Computers & Chemical Engineering*, **26**, 175-185.

Wang, M. and R. Sutton (2004), System identification of a remotely operated flight vehicle, *Journal of Marine Engineering, Science and Technology*, accepted.

ELSEVIER
IFAC
PUBLICATIONS
www.elsevier.com/locate/ifac

# ROBUST STATION-KEEPING OF UNDERWATER VEHICLES UNDER VELOCITY MEASUREMENT OFFSETS

**Andrea Serrani** *

* *Department of Electrical Engineering*
*The Ohio State University*
*Columbus, OH - USA*

Abstract: In this paper, we investigate the problem of setpoint stabilization and station keeping for Remotely Operated Underwater Vehicles. The development of automatic control schemes to achieve precise station keeping is rendered complicated by the lack of precise sensor information, and the effect of environmental disturbances. The main feature of our approach is the incorporation in the controller structure of a suitable model of the disturbances, which reduces the uncertainty in the vehicle attitude and position determination. The proposed approach is applied to the problem of robust station keeping of a specific model of a small commercial ROV, employed for pipeline inspection and maintenance of submerged offshore structure. Simulations performed on a 6 degrees-of-freedom model of a specific ROV show that the method is well suited to cope with imprecise sensor information and with disturbances due to the underwater current. *Copyright © 2004 IFAC.*

Keywords: Disturbance rejection, Lyapunov methods, autonomous vehicles.

## 1. INTRODUCTION

A crucial problem in underwater automation is given by the quality of sensor data, especially regarding the real-time position of the vehicle with respect to a reference frame. Typical information on the vehicle position comes from CCD cameras, sonar and transponders, while information on the vehicle attitude is generally retrieved from compasses and rate gyros. In an underwater environment, the performance in terms of accuracy and bandwidth may be quite poor. For example, signals available from sonar are available only at a slow sampling rate. Signals from CCD cameras have a narrow bandwidth, and should not be differentiated to retrieve the vehicle velocity to avoid delays and noise in the control loop. On the other hand, signals from rate gyros and inertial platforms must be integrated to recover vehicle velocity and angular velocity, but this induces offsets and drifts in the measurements. Moreover, it is generally unrealistic to assume that the parameters of the vehicle dynamics are accurately known. A possibly large uncertainty may affect the available information regarding the added mass coefficients, the location of the center of gravity and the center of buoyancy, as well as the coefficient of the hydrodynamic damping, as they are difficult to determine analytically if the geometry of the vehicle is complex. Even in the case in which an accurate system identification is performed, when operating in different environmental conditions and with different vehicle configurations and payloads, the gathered information may prove to be unreliable for control purposes. The goal of this paper is to employ methodologies from nonlinear regulation theory to obtain robust station-keeping for a general model of small commercial unmanned underwater vehicles.

Our approach is to employ internal models of sensor offsets to obtain rejection of the disturbance directly at the controller level, rather than at the guidance subsystem of the overall architecture. In addition, internal model-based control is shown to improve the capability of the control system to reject environmental disturbances such as the underwater current, and to deal with parametric uncertainties in the vehicle model.

### 1.1 Vehicle model

A general model for UUV dynamics is given as

$$\begin{aligned} &\dot{\boldsymbol{p}} = \boldsymbol{R}\boldsymbol{v} \\ &\dot{\boldsymbol{R}} = \boldsymbol{R}\boldsymbol{S}(\boldsymbol{\omega}) \\ &\boldsymbol{M}\dot{\boldsymbol{\nu}} + \boldsymbol{C}(\boldsymbol{\nu}_r)\boldsymbol{\nu}_r + \boldsymbol{D}(\boldsymbol{\nu}_r)\boldsymbol{\nu}_r + \boldsymbol{G}(\boldsymbol{R}) + \boldsymbol{F}(\dot{\boldsymbol{\nu}}_c) = \boldsymbol{B}\boldsymbol{\tau} \end{aligned} \tag{1}$$

(Fossen, 1994; Yuh, 1990), where $\boldsymbol{p} = (x, y, z)^{\mathrm{T}} \in \mathbb{R}^3$ are the coordinates of the origin of a body-fixed reference frane $\mathcal{F}_b$ with respect to an inertial reference frame $\mathcal{F}_i$, the matrix $\boldsymbol{R} \in SO(3)$ is the rotation matrix describing the orientation of $\mathcal{F}_b$ relative to $\mathcal{F}_i$, and $\boldsymbol{\nu} = \mathrm{col}(\boldsymbol{v}, \boldsymbol{\omega})$ is the generalized velocity of the vehicle, being $\boldsymbol{v} = (u, v, w)^{\mathrm{T}} \in \mathbb{R}^3$ and $\boldsymbol{\omega} = (p, q, r)^{\mathrm{T}} \in \mathbb{R}^3$ the translational and the rotational velocity respectively, all expressed with respect to the body-fixed coordinate frame. The relative velocity between the vehicle and the fluid is denoted by $\boldsymbol{\nu}_r = \boldsymbol{\nu} - \boldsymbol{\nu}_c$, where $\boldsymbol{\nu}_c$ is the fluid velocity expressed in the body-fixed frame. The vector $\boldsymbol{\tau} \in \mathbb{R}^6$ denotes the control input produced by the propellers, which are mapped into generalized forces and moments by the nonsingular matrix $\boldsymbol{B}$. The matrix $\boldsymbol{S}(\boldsymbol{\lambda})$ represents the cross-product, $\boldsymbol{M} = \boldsymbol{M}_{\mathrm{rb}} + \boldsymbol{M}_{\mathrm{a}}$ is the positive definite inertia matrix of the vehicle, lumping the rigid-body inertia and the added-inertia due to the displacement of the fluid, and $\boldsymbol{C}(\boldsymbol{\nu}_r)$ and $\boldsymbol{D}(\boldsymbol{\nu}_r)$ contain the terms relative to the Coriolis and centripetal forces and the hydrodynamic damping, respectively. The vector $\boldsymbol{G}(\boldsymbol{R})$ denotes the combined effect of gravity and buoyancy, while $\boldsymbol{F}(\dot{\boldsymbol{\nu}}_c)$ groups the forces induced by the acceleration of the fluid due to the underwater current. We make the simplifying assumption that the fluid is irrotational, and that the translational fluid velocity is constant in the inertial frame, that is $\boldsymbol{\nu}_c = \mathrm{col}(\boldsymbol{v}_c, \boldsymbol{0})$, and $\boldsymbol{R}\boldsymbol{v}_c =: \boldsymbol{v}_c^i = \mathrm{const}$. It is worth noting that, since $\dot{\boldsymbol{\nu}}_c = \mathrm{col}(-\boldsymbol{S}(\boldsymbol{\omega})\boldsymbol{v}_c, \boldsymbol{0})$ as a result of the previous assumption, and since the map $\boldsymbol{F}(\cdot)$ is linear, the effect of the acceleration of the fluid due to the underwater current $\boldsymbol{F}(\dot{\boldsymbol{\nu}}_c)$ vanishes when $\boldsymbol{\omega} = \boldsymbol{0}$.

### 1.2 Statement of the problem

We begin with defining a vector $\boldsymbol{\mu} \in \mathbb{R}^p$ that collects all parameters of the vehicle that are subject to uncertainty, including the unknown constant value of the underwater current velocity in the inertial frame, $\boldsymbol{v}_c^i$. The vector $\boldsymbol{\mu}$ is restricted to range over a given compact set $\mathcal{P} \subset \mathbb{R}^p$. We assume that accurate measurements of the position and the attitude parameters are available,

$$\boldsymbol{p}_{\mathrm{meas}}(t) = \boldsymbol{p}(t)\,, \qquad \boldsymbol{R}_{\mathrm{meas}}(t) = \boldsymbol{R}(t)\,, \tag{2}$$

while the measurements of the linear and angular velocities are corrupted by an unknown bias

$$\boldsymbol{\nu}_{\mathrm{meas}}(t) = \boldsymbol{\nu}(t) + \boldsymbol{\nu}_{\circ}\,, \tag{3}$$

resulting from a sensor offset or integration error from inertial measurement. Assuming that the actual value of $\boldsymbol{\nu}_{\circ} \in \mathbb{R}^6$ is unknown, but ranges over a given compact set, we can incorporate the entries of $\boldsymbol{\nu}_{\circ}$ into $\boldsymbol{\mu}$. Accordingly, we write the vehicle model given by the first two equations in (1) and equation (7) as

$$\begin{aligned} \dot{\boldsymbol{x}}(t) &= \boldsymbol{f}(\boldsymbol{x}(t), \boldsymbol{\mu}) + \boldsymbol{g}(\boldsymbol{\mu})\boldsymbol{\tau}(t) \\ \boldsymbol{x}_{\mathrm{meas}}(t) &= \boldsymbol{h}(\boldsymbol{x}(t), \boldsymbol{\mu}) \end{aligned} \tag{4}$$

with state $\boldsymbol{x} = (\boldsymbol{p}, \boldsymbol{R}, \boldsymbol{\nu}) \in \mathcal{X}$, where $\mathcal{X} = \mathbb{R}^3 \times SO(3) \times \mathbb{R}^6$, and available measurements $\boldsymbol{x}_{\mathrm{meas}} \in \mathcal{X}$ defined according to (2) and (3). For the system given in (4), we address the problem of regulating the state to a desired setpoint $\boldsymbol{x}_d = (\boldsymbol{p}_d, \boldsymbol{R}_d, \boldsymbol{0})$, using feedback from the measured variables $\boldsymbol{x}_{\mathrm{meas}}$, robustly with respect to the parametric uncertainty. Note that there is no loss in generality in assuming the desired configuration to be $(\boldsymbol{p}_d, \boldsymbol{R}_d) = (\boldsymbol{0}, \boldsymbol{I})$, as an equivalent problem of this sort can always be obtained by means of the rigid transformation $(\boldsymbol{p}, \boldsymbol{R}) \mapsto (\boldsymbol{p} - \boldsymbol{p}_d, \boldsymbol{R}\boldsymbol{R}_d^{\mathrm{T}})$. Henceforth, we will assume that this is the case. The control problem is then formulated as follows:

*Robust station-keeping problem.* For any arbitrary compact set $\mathcal{K} \subset \mathbb{R}^3 \times SO(3) \times \mathbb{R}^6$, find a smooth dynamic controller of the form

$$\begin{aligned} \dot{\boldsymbol{x}}_c(t) &= \boldsymbol{f}_c(\boldsymbol{x}_c(t), \boldsymbol{x}_{\mathrm{meas}}(t)) \\ \boldsymbol{\tau}(t) &= \boldsymbol{h}_c(\boldsymbol{x}_c(t), \boldsymbol{x}_{\mathrm{meas}}(t)) \end{aligned} \tag{5}$$

with state $\boldsymbol{x}_c \in \mathbb{R}^m$, and a compact set $\mathcal{K}_c \subset \mathbb{R}^m$ such that all trajectories of the closed loop system

$$\begin{aligned} \dot{\boldsymbol{x}}(t) &= \boldsymbol{f}(\boldsymbol{x}(t), \boldsymbol{\mu}) + \boldsymbol{g}(\boldsymbol{\mu})\boldsymbol{h}_c(\boldsymbol{x}_c(t), \boldsymbol{h}(\boldsymbol{x}(t), \boldsymbol{\mu})) \\ \dot{\boldsymbol{x}}_c(t) &= \boldsymbol{f}_c(\boldsymbol{x}_c(t), \boldsymbol{h}(\boldsymbol{x}(t), \boldsymbol{\mu})) \end{aligned}$$

originating from any initial condition $\boldsymbol{x}(0) \in \mathcal{K}$ and $\boldsymbol{x}_c(0) \in \mathcal{K}_c$ are bounded, and satisfy $\lim_{t\to\infty} \boldsymbol{x}(t) = \boldsymbol{x}_d$ for any $\boldsymbol{\mu} \in \mathcal{P}$, where $\boldsymbol{x}_d = (\boldsymbol{0}, \boldsymbol{I}, \boldsymbol{0})$.

It is easy to see that, since the point $\boldsymbol{x}_d$ is not an equilibrium for the *unforced* system (4), the controller (5) must be able to provide asymptotically the constant control input $\boldsymbol{\tau} = \boldsymbol{\tau}^{\mathrm{ss}}(\boldsymbol{\mu})$ required to maintain the desired configuration in steady-state, that is, such that $\boldsymbol{0} = \boldsymbol{f}(\boldsymbol{x}_d, \boldsymbol{\mu}) + \boldsymbol{g}(\boldsymbol{\mu})\boldsymbol{\tau}^{\mathrm{ss}}(\boldsymbol{\mu})$. It is clear that this can be achieved only if the controller dynamics embed a certain number of integrators to reproduce in steady-state the *unknown* constant input $\tau^{ss}(\mu)$. This requirement is commonly referred to as the *internal model principle*. In principle, the design of internal model-based regulators can be addressed using a large number of techniques (see, for instance, (Isidori *et al.*, 2003), and references therein). However, the results available in the literature can not be directly applied to the specific case under consideration, as the controller does not have access to a direct measurement of the regulation error $e(t) = x(t) - x_d$, since $h(x_d, \mu) \neq x_d$.

## 2. CONTROLLER DESIGN

To avoid manipulating the rotation matrix $\boldsymbol{R}$, we parameterize $SO(3)$ by means of *modified Rodrigues parameters* $\boldsymbol{\sigma} \in \mathbb{R}^3$. Letting $\boldsymbol{R}$ correspond to a rotation of an angle $\phi$ about the unit vector $\boldsymbol{\lambda}$, the modified Rodrigues parameters are defined as $\boldsymbol{\sigma} = \boldsymbol{\lambda} \tan \phi/4$. Equivalently, if $\boldsymbol{R}$ is expressed in terms of the Euler parameters $\eta = \cos\phi/2$, $\boldsymbol{\varepsilon} = \boldsymbol{\lambda} \sin\phi/2$, the modified Rodrigues parameters are given by

$$\boldsymbol{\sigma} = \frac{\boldsymbol{\varepsilon}}{1+\eta}.$$

The modified Rodrigues parameters offer a very convenient way to obtain a minimal parameterization of $SO(3)$. As a matter of fact, since it becomes singular only at $\phi = 2\pi$, the modified Rodrigues parameterization is capable of representing rotations in the range $\phi \in [0, 2\pi)$, while standard Rodrigues parameters $\boldsymbol{\rho} = \boldsymbol{\lambda} \tan\frac{\phi}{2}$ are limited to the range $[0, \pi)$, and roll, pitch and yaw angles $(\phi,\ \theta,\ \psi)$, become singular for $\theta = \pm\pi/2$. Accordingly, we replace the rotational kinematic equation in (1) with the *propagation* rule

$$\dot{\boldsymbol{\sigma}} = \tfrac{1}{2}\boldsymbol{E}(\boldsymbol{\sigma})\boldsymbol{\omega}\,, \tag{6}$$

where the matrix $\boldsymbol{E}(\boldsymbol{\sigma})$ is defined as

$$\boldsymbol{E}(\boldsymbol{\sigma}) = \left[\frac{1-\boldsymbol{\sigma}^{\mathrm{T}}\boldsymbol{\sigma}}{2}\boldsymbol{I} + \boldsymbol{S}(\boldsymbol{\sigma}) + \boldsymbol{\sigma}\boldsymbol{\sigma}^{\mathrm{T}}\right]\boldsymbol{\omega}.$$

Since $\boldsymbol{R}(\boldsymbol{\sigma}) = \boldsymbol{I}$ if and only if $\boldsymbol{\sigma} = \boldsymbol{0}$, the control objective, as far as the vehicle attitude is concerned, is translated into that of regulating $\boldsymbol{\sigma}(t)$ to the origin. The next step is the definition of an *error system* for (4). First, we write the last equation of (1) in a more compact form as

$$\boldsymbol{M}(\boldsymbol{\mu})\dot{\boldsymbol{\nu}} = \boldsymbol{\Phi}(\boldsymbol{\sigma},\boldsymbol{\nu},\boldsymbol{\mu}) + \boldsymbol{B}(\boldsymbol{\mu})\boldsymbol{\tau}\,, \tag{7}$$

where we have denoted (omitting the arguments)

$$\boldsymbol{\Phi} = -\boldsymbol{C}\boldsymbol{\nu}_r - \boldsymbol{D}\boldsymbol{\nu}_r - \boldsymbol{G} - \boldsymbol{F}.$$

Since $\boldsymbol{M}(\boldsymbol{\mu})$ is positive definite, and $\boldsymbol{B}(\boldsymbol{\mu})$ is assumed nonsingular for any $\boldsymbol{\mu} \in \mathcal{P}$, there exist positive numbers $m_1$, $m_2$, and $b_0$ satisfying

$$m_1\boldsymbol{I} \le \boldsymbol{M}(\boldsymbol{\mu}) \le m_2\boldsymbol{I}\,, \qquad b_0\boldsymbol{I} \le \boldsymbol{B}(\boldsymbol{\mu}) \tag{8}$$

for all $\boldsymbol{\mu} \in \mathcal{P}$. Let $\boldsymbol{\nu}_c^i = \mathrm{col}(\boldsymbol{v}_c^i, \boldsymbol{0})$ be the underwater current velocity at the desired vehicle attitude $\boldsymbol{R}_d = \boldsymbol{I}$, and define

$$\boldsymbol{\Delta}(\boldsymbol{\mu}) = \boldsymbol{C}(\boldsymbol{\nu}_c^i)\boldsymbol{\nu}_c^i + \boldsymbol{D}(\boldsymbol{\nu}_c^i)\boldsymbol{\nu}_c^i + \boldsymbol{G}(\boldsymbol{I}).$$

Then, equation (7) can be written as

$$\boldsymbol{M}(\boldsymbol{\mu})\dot{\boldsymbol{\nu}} = \tilde{\boldsymbol{\Phi}}(\boldsymbol{\sigma},\boldsymbol{\nu},\boldsymbol{\mu}) + \boldsymbol{B}(\boldsymbol{\mu})[\boldsymbol{\tau} - \boldsymbol{\tau}^{ss}(\boldsymbol{\mu})]\,, \tag{9}$$

where $\tilde{\boldsymbol{\Phi}}(\boldsymbol{\sigma},\boldsymbol{\nu},\boldsymbol{\mu}) := \boldsymbol{\Phi}(\boldsymbol{\sigma},\boldsymbol{\nu},\boldsymbol{\mu}) + \boldsymbol{\Delta}(\boldsymbol{\mu})$, and $\boldsymbol{\tau}^{ss}(\boldsymbol{\mu}) := \boldsymbol{B}(\boldsymbol{\mu})^{-1}\boldsymbol{\Delta}(\boldsymbol{\mu})$. Note that the entries of $\tilde{\boldsymbol{\Phi}}(\boldsymbol{\sigma},\boldsymbol{\nu},\boldsymbol{\mu})$ are continuously differentiable functions of their arguments, and that $\tilde{\boldsymbol{\Phi}}(\boldsymbol{0},\boldsymbol{0},\boldsymbol{\mu}) = \boldsymbol{0}$ for any $\boldsymbol{\mu}$. Therefore, if the "steady-state" control $\boldsymbol{\tau}^{ss}(\boldsymbol{\mu})$ was available, the problem at issue would be equivalent to that of stabilizing the equilibrium at origin for the *error system*

$$\begin{aligned} \dot{\boldsymbol{p}} &= \boldsymbol{R}(\boldsymbol{\sigma})\,\boldsymbol{v} \\ \dot{\boldsymbol{\sigma}} &= \tfrac{1}{2}\boldsymbol{E}(\boldsymbol{\sigma})\boldsymbol{\omega} \\ \boldsymbol{M}(\boldsymbol{\mu})\dot{\boldsymbol{\nu}} &= \tilde{\boldsymbol{\Phi}}(\boldsymbol{\sigma},\boldsymbol{\nu},\boldsymbol{\mu}) + \boldsymbol{B}(\boldsymbol{\mu})\boldsymbol{\tau}\,, \end{aligned} \tag{10}$$

with a guaranteed domain of attraction, robustly with respect to $\boldsymbol{\mu} \in \mathcal{P}$.

### *2.1 Design of the internal model*

We begin with defining an *internal model* of the external disturbance for the translational and rotational motion as

$$\dot{\boldsymbol{\xi}}_1 = \boldsymbol{R}^{\mathrm{T}}(\boldsymbol{\sigma})\boldsymbol{p}\,, \qquad \dot{\boldsymbol{\xi}}_2 = \boldsymbol{\sigma}\,,$$

with $\boldsymbol{\xi} = \mathrm{col}(\boldsymbol{\xi}_1, \boldsymbol{\xi}_2)$. Let $\boldsymbol{q} = \mathrm{col}(\boldsymbol{p}, \boldsymbol{\sigma})$, and write

$$\begin{aligned} \dot{\boldsymbol{\xi}} &= \boldsymbol{\Gamma}(\boldsymbol{q}) \\ \dot{\boldsymbol{q}} &= \boldsymbol{J}(\boldsymbol{q})\boldsymbol{\nu}\,, \end{aligned} \tag{11}$$

where

$$\boldsymbol{\Gamma}(\boldsymbol{q}) = \begin{pmatrix} \boldsymbol{R}^{\mathrm{T}}(\boldsymbol{\sigma})\boldsymbol{p} \\ \boldsymbol{\sigma} \end{pmatrix}, \quad \boldsymbol{J}(\boldsymbol{q}) = \begin{pmatrix} \boldsymbol{R}(\boldsymbol{\sigma}) & \boldsymbol{0} \\ \boldsymbol{0} & \frac{1}{2}\boldsymbol{E}(\boldsymbol{\sigma}) \end{pmatrix}.$$

As a preliminary step towards the design of the overall stabilizing controller, we regard the translational and angular velocities $\boldsymbol{\nu}$ as a control input for the augmented system (11) and show that there exist a state feedback control law that stabilizes the equilibrium at the origin of the resulting closed-loop system.

*Proposition 2.1.* The equilibrium at the origin for system (11) is rendered globally asymptotically, and locally exponentially stable by virtue of the control law

$$\boldsymbol{\nu} = -\boldsymbol{K}_1\boldsymbol{\xi} - \boldsymbol{K}_2(\boldsymbol{q})\boldsymbol{q}\,, \tag{12}$$

where

$$\boldsymbol{K}_1 = \begin{pmatrix} k_1\boldsymbol{I} & \boldsymbol{0} \\ \boldsymbol{0} & k_2\boldsymbol{I} \end{pmatrix}, \quad \boldsymbol{K}_2(\boldsymbol{q}) = \begin{pmatrix} k_3\boldsymbol{R}^{\mathrm{T}}(\boldsymbol{\sigma}) & \boldsymbol{0} \\ \boldsymbol{0} & k_4\boldsymbol{I} \end{pmatrix},$$

and $k_i > 0$, $i = 1, \ldots, 4$ are arbitrary design parameters.

*Proof.* Define the *Lyapunov function candidate*

$$V_0(\boldsymbol{\xi}, \boldsymbol{q}) = \frac{1}{2}\boldsymbol{\xi}^{\mathrm{T}} \boldsymbol{K}_1 \boldsymbol{\xi} + \frac{1}{2}\boldsymbol{p}^{\mathrm{T}}\boldsymbol{p} + 2\ln\left(1 + \boldsymbol{\sigma}^{\mathrm{T}}\boldsymbol{\sigma}\right)$$

which is positive definite, and radially unbounded. Easy computations show that, since

$$\boldsymbol{\sigma}^{\mathrm{T}}\boldsymbol{E}(\boldsymbol{\sigma}) = \frac{1+\boldsymbol{\sigma}^{\mathrm{T}}\boldsymbol{\sigma}}{2}\boldsymbol{\sigma}^{\mathrm{T}},$$

the derivative of $V$ along trajectories of (11) reads as

$$\frac{\partial V_0}{\partial \boldsymbol{\xi}}\dot{\boldsymbol{\xi}} + \frac{\partial V_0}{\partial \boldsymbol{q}}\dot{\boldsymbol{q}} = \boldsymbol{p}^{\mathrm{T}}\boldsymbol{R}(\boldsymbol{\sigma})[k_1\boldsymbol{\xi}_1 + v] + \boldsymbol{\sigma}^{\mathrm{T}}[k_2\boldsymbol{\xi}_2 + \omega].$$

Application of the control (12) yields

$$\frac{\partial V_0}{\partial \boldsymbol{\xi}}\dot{\boldsymbol{\xi}} + \frac{\partial V_0}{\partial \boldsymbol{q}}\dot{\boldsymbol{q}} = -k\|\boldsymbol{q}\|^2, \qquad k = \min\{k_3, k_4\}$$

which implies stability of the origin for the closed-loop system, and, according to La Salle's invariance theorem, global attractivity of the largest invariant set $\mathcal{E}$ contained in the set $\mathcal{M} = \{\boldsymbol{q} = \boldsymbol{0}\}$. Since the only trajectory of (11)-(12) entirely contained in $\mathcal{M}$ is the trivial one, global asymptotic stability of the origin follows. Local exponential stability follows from asymptotic stability of the linear approximation of the closed-loop system. ◁

### *2.2 Design of the stabilizer for the error system*

Next, we turn our attention to the design of a controller that semi-globally stabilizes the origin of the system resulting from the interconnection of the error system (10) and the internal model. In doing so, we will assume that perfect measurement of the vehicle velocity are available. Then, we will move on to the problem stated originally, in which the velocity measurements are perturbed by an unknown bias, and (9) replaces the last equation in (10). Consider the error system augmented by the internal model, and change coordinates as $\boldsymbol{z} = \boldsymbol{\nu} + \boldsymbol{K}_1\boldsymbol{\xi} + \boldsymbol{K}_2(\boldsymbol{q})\boldsymbol{q}$, to obtain a system in the form

$$\begin{aligned} \dot{\boldsymbol{\xi}} &= \boldsymbol{\Gamma}(\boldsymbol{q}) \\ \dot{\boldsymbol{q}} &= \boldsymbol{J}(\boldsymbol{q})\left[\boldsymbol{z} - \boldsymbol{K}_1\boldsymbol{\xi} - \boldsymbol{K}_2(\boldsymbol{q})\boldsymbol{q}\right] \\ \boldsymbol{M}(\boldsymbol{\mu})\dot{\boldsymbol{z}} &= \boldsymbol{\Psi}(\boldsymbol{\xi}, \boldsymbol{q}, \boldsymbol{z}, \boldsymbol{\mu}) + \boldsymbol{B}(\boldsymbol{\mu})\boldsymbol{\tau}. \end{aligned} \tag{13}$$

*Proposition 2.2.* Fix $k_i > 0$, $i = 1, \ldots, 4$, and choose $\boldsymbol{\tau}$ in (13) as

$$\boldsymbol{\tau} = -\boldsymbol{K}_3\boldsymbol{z}, \qquad \boldsymbol{K}_3 = \begin{pmatrix} K_1\boldsymbol{I} & 0 \\ 0 & K_2\boldsymbol{I} \end{pmatrix} \tag{14}$$

with $K_1 > 0$, $K_2 > 0$ scalar design parameters. Then, system (13)-(14) is semi-globally and locally exponentially stabilizable in the gains $K_1, K_2$.

*Proof.* The proof of the proposition follows from the *semi-global backstepping lemma* of Teel and Praly (Teel and Praly, 1995, Lemma 2.2), whose proof need not be repeated here. We will therefore limit ourselves to proving the necessary ingredients to apply the aforementioned result. Note that the closed-loop system (13)-(14) can be viewed as the negative feedback interconnection of the system (13) with output $\boldsymbol{y} = \boldsymbol{z}$, and the memoryless system $\boldsymbol{\tau} = \boldsymbol{K}_3\boldsymbol{y}$. System (13) has vector relative degree $r = \{1, 1, \ldots, 1\}$ with respect to $\boldsymbol{y}$, and zero dynamics given by

$$\begin{aligned} \dot{\boldsymbol{\xi}} &= \boldsymbol{\Gamma}(\boldsymbol{q}) \\ \dot{\boldsymbol{q}} &= \boldsymbol{J}(\boldsymbol{q})\left[-\boldsymbol{K}_1\boldsymbol{\xi} - \boldsymbol{K}_2(\boldsymbol{q})\boldsymbol{q}\right] \end{aligned} \tag{15}$$

which, by virtue of Proposition 2.1, has a globally asymptotically, and locally exponentially stable equilibrium at the origin. According to the converse Lyapunov theorem of Kurzweil, there exist a smooth function $V$, class-$\mathcal{K}_\infty$ functions $\alpha_1(s)$, $\alpha_2(s)$, and a class-$\mathcal{K}$ function $\alpha_3(s)$ satisfying

$$\alpha_1(\|(\boldsymbol{\xi}, \boldsymbol{q})\|) \le V(\boldsymbol{\xi}, \boldsymbol{q}) \le \alpha_2(\|(\boldsymbol{\xi}, \boldsymbol{q})\|)$$

$$\frac{\partial V}{\partial \boldsymbol{\xi}}\boldsymbol{\Gamma}(\boldsymbol{q}) + \frac{\partial V}{\partial \boldsymbol{q}}\boldsymbol{J}(\boldsymbol{q})\left[-\boldsymbol{K}_1\boldsymbol{\xi} - \boldsymbol{K}_2(\boldsymbol{q})\boldsymbol{q}\right] \le -\alpha_3(\|(\boldsymbol{\xi}, \boldsymbol{q})\|)$$

for all $(\boldsymbol{\xi}, \boldsymbol{q})$. Then, fix an arbitrary compact set $\mathcal{K} \in \mathbb{R}^6 \times \mathbb{R}^6 \times \mathbb{R}^{12}$, and consider the function

$$W(\boldsymbol{\xi}, \boldsymbol{q}, \boldsymbol{z}, \boldsymbol{\mu}) = V(\boldsymbol{\xi}, \boldsymbol{q}) + \tfrac{1}{2}\boldsymbol{z}^{\mathrm{T}}\boldsymbol{M}(\boldsymbol{\mu})\boldsymbol{z}.$$

Let $c > 0$ be defined as

$$c = \max_{\substack{(\boldsymbol{\xi}, \boldsymbol{q}, \boldsymbol{z}) \in \mathcal{K} \\ \boldsymbol{\mu} \in \mathcal{P}}} W(\boldsymbol{\xi}, \boldsymbol{q}, \boldsymbol{z}, \boldsymbol{\mu})$$

and note that, since $W(\boldsymbol{\xi}, \boldsymbol{q}, \boldsymbol{z}, \boldsymbol{\mu})$ satisfies

$$\alpha_1(\|(\boldsymbol{\xi}, \boldsymbol{q})\|) + m_1\|\boldsymbol{z}\|^2 \le W(\boldsymbol{\xi}, \boldsymbol{q}, \boldsymbol{z}, \boldsymbol{\mu}),$$

for all $(\boldsymbol{\xi}, \boldsymbol{q}, \boldsymbol{z})$ and all $\boldsymbol{\mu} \in \mathcal{P}$, the set $\Omega_c = \{W(\boldsymbol{\xi}, \boldsymbol{q}, \boldsymbol{z}, \boldsymbol{\mu}) \le c,\ \boldsymbol{\mu} \in \mathcal{P}\}$ is compact, and $\Omega_c \supset \mathcal{K}$. Moreover, denote by $\mathcal{B}_r$ the ball of radius $r$ centered at the origin, and note that for any $r > 0$ there exists $\delta > 0$ such that $\Omega_\delta = \{W(\boldsymbol{\xi}, \boldsymbol{q}, \boldsymbol{z}, \boldsymbol{\mu}) \le \delta,\ \boldsymbol{\mu} \in \mathcal{P}\} \subset \mathcal{B}_r$. Using the second of (8), it is readily seen that

$$\frac{\partial W}{\partial \boldsymbol{z}}\boldsymbol{B}(\boldsymbol{\mu})\boldsymbol{\tau} = -\boldsymbol{z}^{\mathrm{T}}\boldsymbol{B}(\boldsymbol{\mu})\boldsymbol{K}_3\boldsymbol{z} \le -b_0K\|\boldsymbol{z}\|^2$$

for all $\boldsymbol{\mu} \in \mathcal{P}$ and all $\boldsymbol{z} \in \mathbb{R}^6$, where $K = \min\{K_1, K_2\}$. Finally, continuous differentiability of $\boldsymbol{\Psi}(\boldsymbol{\xi}, \boldsymbol{q}, \boldsymbol{z}, \boldsymbol{\mu})$, and $\boldsymbol{\Psi}(\boldsymbol{0}, \boldsymbol{0}, \boldsymbol{0}, \boldsymbol{\mu}) = \boldsymbol{0}$ imply that, given any compact set $\mathcal{S}$, there exist numbers $\gamma_i > 0$ independent of $\boldsymbol{\mu}$ such that

$$\|\boldsymbol{\Psi}(\boldsymbol{\xi}, \boldsymbol{q}, \boldsymbol{z}, \boldsymbol{\mu})\| \le \gamma_1\|\boldsymbol{\xi}\| + \gamma_2\|\boldsymbol{q}\| + \gamma_3\|\boldsymbol{z}\|.$$

Therefore, application of Lemma 2.2 in (Teel and Praly, 1995) shows that, for any $\delta > 0$ and $c > 0$, there exists $K^\star > 0$ such that for all $K > K^\star$, the derivative of the Lyapunov function $W(\boldsymbol{\xi}, \boldsymbol{q}, \boldsymbol{z}, \boldsymbol{\mu})$ along trajectories of (13)-(14) satisfies

$$\dot{W}(\boldsymbol{\xi}, \boldsymbol{q}, \boldsymbol{z}, \boldsymbol{\mu}) < 0, \quad \text{for all } (\boldsymbol{\xi}, \boldsymbol{q}, \boldsymbol{z}, \boldsymbol{\mu}) \in \mathcal{S},$$

where $\mathcal{S} = \{\delta \le W(\boldsymbol{\xi}, \boldsymbol{q}, \boldsymbol{z}, \boldsymbol{\mu}) \le c,\ \boldsymbol{\mu} \in \mathcal{P}\}$. As a result, all trajectories of the closed loop system (13)-(14) originating from initial conditions in the set $\mathcal{K}$ are captured by the set $\Omega_\delta$, and ultimately confined inside $\mathcal{B}_r$. Attractivity of all trajectories in $\Omega_\delta$ to the origin follows from exponential stability of the linear approximation of (13)-(14), and the fact that a neighborhood $\mathcal{B}_r$ in which all trajectories of (13)-(14) are exponentially attracted to the origin can be determined independently of the actual value of $\mu$. ◁

*2.3 Perturbed case*

We are still left to consider the perturbed system (9), and to use the available measurements in the actual implementation of the control law. Using (2) and (3), the *implementable* control (14) becomes

$$\begin{aligned}\tau &= -K_3\nu_{\text{meas}} - K_3K_1\xi - K_3K_2(\sigma)q \\ &= -K_3\nu - K_3K_1\xi - K_3K_2(\sigma)q - K_3\nu_\circ .\end{aligned}$$

Defining the constant vector

$$w(\mu) = -K_1^{-1}\nu_\circ - K_3^{-1}K_1^{-1}\tau^{\text{ss}}(\mu)$$

and keeping in mind that the matrices $K_1$ and $K_3$ commute, we obtain

$$\tau - \tau^{\text{ss}}(\mu) = -K_3\nu - K_3K_1\tilde{\xi} - K_3K_2(\sigma)q ,$$

where we have denoted $\tilde{\xi} = \xi - w(\mu)$. Letting in turn $\tilde{z} = \nu + K_1\tilde{\xi} + K_2(q)q$, the closed-loop system in the new coordinates $(\tilde{\xi}, q, \tilde{z})$ reads as

$$\begin{aligned}\dot{\tilde{\xi}} &= \Gamma(q) \\ \dot{q} &= J(q)[\tilde{z} - K_1\tilde{\xi} - K_2(q)q] \qquad (16) \\ M(\mu)\dot{\tilde{z}} &= \Psi(\tilde{\xi}, q, \tilde{z}, \mu) - B(\mu)K_3\tilde{z} .\end{aligned}$$

It is clear that (16) has the same structure as the closed loop system (13)-(14) analyzed in the previous section. The main difference between the two systems lies in the fact that, in the original coordinates, the equilibrium has shifted from the origin to $(\xi, q, \nu) = (w(\mu), 0, 0)$. To carry over the result of Proposition 2.2 to system (16), we only need to show that semi-global stabilizability is preserved through the change of variable $\xi \mapsto \tilde{\xi}$. Note, in fact, that $\tilde{\xi}$ is a function on the gains $K_1$ and $K_2$, and this, in principle, may induce peaking and shrink the region of attraction. Since the system has relative degree one, this may happen only if the compact set of initial conditions for some of the states of the zero dynamics of the system (in our case, the state $\tilde{\xi}$) grows unbounded with $K_1$ and $K_2$. However, it is readily seen that the set of initial conditions for $\tilde{\xi}$ can be bounded independently of $K_1$ and $K_2$ if, without loss of generality, $K_3$ is chosen such that $K_3 > I$. If this is the case, a simple calculation shows that $\|w(\mu)\| \leq N$, for all $\mu \in \mathcal{P}$, where $N$ is independent of $K_3$, being

$$N = \|K_1^{-1}\| \max_{\mu\in\mathcal{P}}\{\|\nu_\circ\| + \|\tau^{\text{ss}}(\mu)\|\} .$$

Letting $R > 0$ be such that $\mathcal{K}_\xi \subset \mathcal{B}_R$, we obtain $\|\tilde{\xi}\| \leq \|\xi\| + \|w(\mu)\| \leq R+N$, $\forall \xi \in \mathcal{K}_\xi$, $\forall \mu \in \mathcal{P}$, and thus the compact set of initial conditions for the zero dynamics of the system can be bounded independently of $K_3$, once $K_1$ is fixed.

Table 1. Parameters of the vehicle

| | |
|---|---|
| Mass | 3300 $kg$ |
| Buoyancy | 34393 $N$ |
| Inertia | $I_x = 4400$, $I_y = I_z = 9900$ $kg$ $m^2$ |
| Gravity cnt | $R_g = (0\ 0\ 0.4)$ $m$ |
| Buoyancy cnt | $R_b = (0\ 0\ 0)$ $m$ |
| Add. inertia | $M_a = \text{diag}\,(695, 765, 1302, 82, 126, 87)$ |
| $D(v) = \text{diag}\,(350\|u\|, 510\|v\|, 823\|w\|, 68\|p\|, 148\|q\|, 125\|r\|)$ | |

Table 2. Controller parameters

| | | |
|---|---|---|
| $k_1 = 0.05$ | $k_2 = 0.05$ | $k_3 = 1$ |
| $k_4 = 1.2$ | $K_1 = 3000$ | $K_2 = 3000$ |

## 3. SIMULATION RESULTS

The proposed approach is applied to the problem of robust station keeping of a specific model of a small commercial ROV of the class *DOLPHIN 3K*, employed for pipeline inspection and maintenance of submerged offshore structure (Nomoto and Hattori, 1986). The model of the ROV *DOLPHIN 3K* is characterized by the set of parameters described in Table 1, for the vehicle operating at a depth of 300 m. The reader is referred to (Nomoto and Hattori, 1986) for a more comprehensive description of the vehicle model. A model of the vehicle dynamics has been derived and implemented on MATLAB-SIMULINK®. The position and attitude signals have been acquired at a sampling rate of 10 Hz, while for all the other signals a sampling rate of 100 Hz was assumed. The controller design parameters, chosen after several trials, are shown in Table 2. An initial error of about 10% has been assumed for the estimated velocity. The task is to keep the vehicle at the origin of the inertial frame, with the body frame and the inertial frame aligned, despite the effect of the underwater current velocity, and the offset in the velocity measurement. In order to evaluate the effectiveness of the internal model-based strategy, we have performed the simulations keeping the internal model disconnected at first. Figures 1 and 2 show respectively the position error and the attitude error in this case. It can be seen that the disturbance in the measurements and the effect of the underwater current determine an offset in the regulated position, which does not converge to the desired set-point. However, after turning the internal model on, we observe convergence to the desired set-point, and the fulfillment of the control objectives (figures 3 to 5). It is worth stressing that the internal model-based controller does not have access to the external disturbance (that is, the underwater current is not being measured or estimated), nor any information on the initial offset on the velocity measurement is employed.

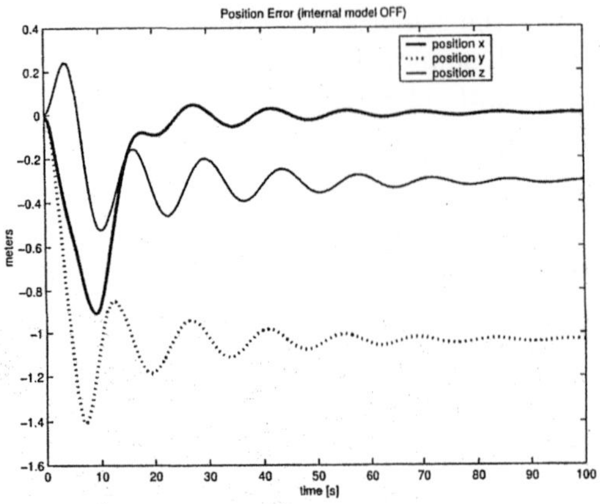


Fig. 1. Position error - internal model turned off

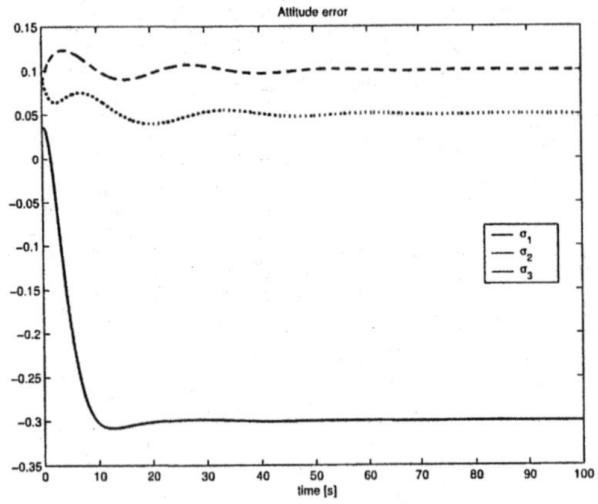


Fig. 2. Attitude error - internal model turned off

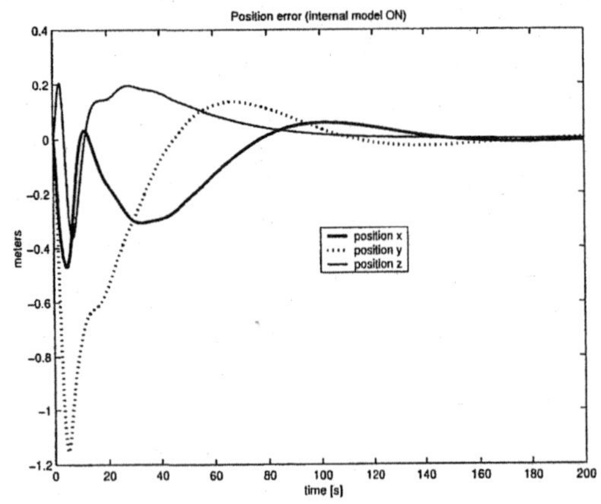


Fig. 3. Position error - internal model turned on

## 4. CONCLUSIONS

An internal model based-control approach has been proposed for the design of a set-point controller for a remotely operated underwater vehicle. Motivated by the need to cope with data affected by unknown bias, and the need to reject the disturbance exerted by the influence of the underwater current on the vehicle dynamics, we have designed a nonlinear regulator which provides robust regulation of error, with guaranteed domain of attraction. Simulation results on a model of a commercial ROV show the effectiveness of the approach. Since the low computational load of the control algorithm makes it suitable to be implemented in an on-board real-time control architecture, the proposed methodology may be of practical importance in designing reliable control schemes for station-keeping of commercial ROVs.

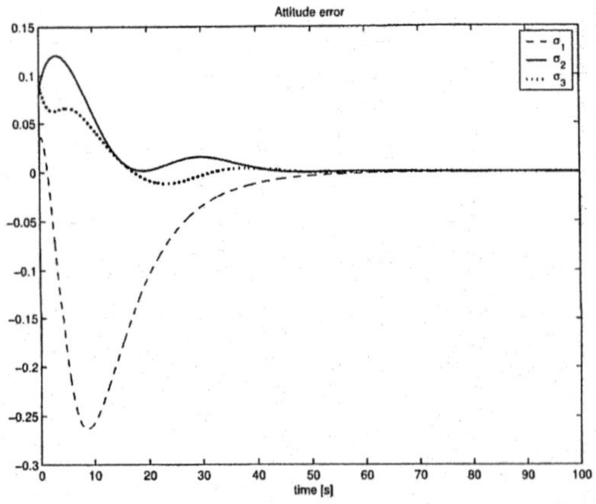


Fig. 4. Attitude error - internal model turned on

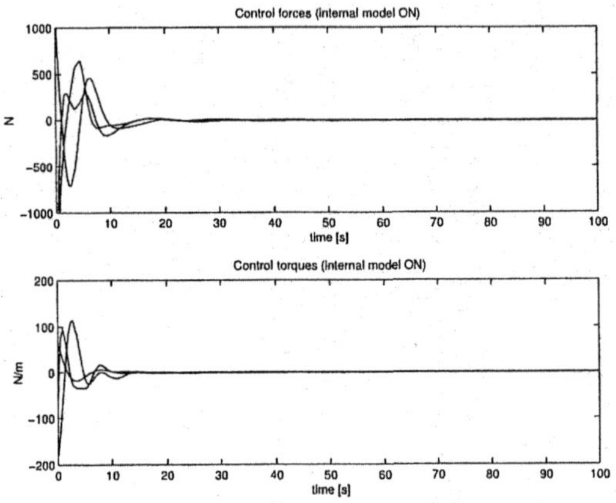


Fig. 5. Control efforts - internal model turned on

## REFERENCES


Fossen, T.I. (1994). *Guidance and Control of Ocean Vehicles.* John Wiley & Sons. New York, NY.

Isidori, A., L. Marconi and A. Serrani (2003). *Robust Autonomous Guidance: an Internal Model Approach.* Advances in Industrial Control. Springer Verlag. New York, NY.

Nomoto, M. and M. Hattori (1986). A deep ROV 'DOLPHIN 3K': Design and performance analysis. *IEEE Journal of Oceanic Engineering* **11**(3), 373–391.

Teel, A.R. and L. Praly (1995). Tools for semiglobal stabilization by partial and output feedback. *SIAM Journal of Control and Optimization* **33**(5), 1443–1488.

Yuh, J. (1990). Modeling and control of underwater robotic vehicles. *IEEE Transactions on System, Man and Cybernetics* **20**(6), 1475–1483.

ELSEVIER
IFAC
PUBLICATIONS
www.elsevier.com/locate/ifac

# THE PROBLEM OF THE COUPLING OF THE VERTICAL MOVEMENT CONTROL WITH ROLL MOVEMENT IN FAST FERRIES

**Joaquín Aranda, Rocío Muñoz-Mansilla, José Manuel Díaz**

*Dpto. Informática y Automática. UNED.*
*C/ Juan del Rosal,16. 28040 Madrid. Spain*
*e-mail: jaranda@dia.uned.es, rmunoz@dia.uned.es, josema@dia.uned.es*
*tel: 91 3987148. fax: 91 3988663*

Abstract: Previous researches have studied the attenuation of heaving and pitching motion in fast ferries. One observed problem that appears in these cases is the fact that actuators action (for instance, asymmetries T-foil), waves incidence with angle, and the control action itself cause a coupling with roll movement, and therefore an increasing in the rolling vertical component. Roll motion can be less relevant than the damped longitudinal movement, but it is significant and, in fact, it can result in unpleasant effects to passengers. In this work an analysis and design of a roll control have been carried out.


Keywords: actuators, stability, control,closed-loop, open-loop, controllers, dynamic behaviour, models, root locus, nichols chart.

## 1. INTRODUCTION

One of the objectives in the design and built of high speed crafts is passenger comfort and vehicle safety. Vertical accelerations associated with roll, pitch and heave motions are the main cause of motion sickness.

Previous researches of the work group have studied heaving and pitching motion (Cruz, *et al.*, 2000) and modeled actuators and designed different controllers, (Aranda, *et al.*, 2002a, Aranda, *et al.*, 2002b), in order to achieve heave and pitch damping and with succesfull results.

One observed problem that appears in these cases is the fact that actuators action (eg. asymmetries T-foil), and the control action itself cause a coupling with roll mode, and therefore an increasing in the roll vertical component. Roll motion may be less relevant than the damped longitudinal movement, but it is significant and, in fact, it may result in unpleasant effects to passengers together with the other motions. So, in this sense, the following phase in this investigation is In this work an analysis and design of control of roll methods have been carried out in order to reduce these additional effects. It is known that the almost lack of inherent roll dampings means that small additions to this damping can produce large reductions in the response. So, the best way of reducing it is to increase damping. A common mean of doing so is the use of special anti-rolling devices. In this case the employed devices are stabilizing fins.

Previous to designing the controller, roll motion and actuators dynamics modeling have been neccesary to do.

Two graphical methods have been probed for the control designing. Firstly, root locus has been used for the design of a PI controller. Secondly, Nichols chart has been used for the validation of the previous controller and for a new design of first order lead filters and second order filters.

In order to validate the chosen controllers, a simulink model in block diagrams of the whole system has of the closed-loop system, and gives and graphics the temporal roll response, vertical accelerations associated to roll and fins movement.

With this research, it is completed the motions analysis in diverse degrees of freedom. Further researches are focused on the analysis of motion coupling in more than one degree of freedom in order to design integral controllers that stabilise longitudinal and transversal movements and achieve course traking.

## 2. MODEL OF THE SYSTEM

Figure 1 shows a block diagram representation of a ship stabilised with active fins. Each component or block in the diagram may be considered as a 'black box' having an input and an output which are related by the block's transfer function. So, the ship block accepts an input in the form of a roll moment from the waves $M_{waves}$ and generates a roll motion output *roll.* Similarly, the stabiliser fin controller generates a demanded angle $\alpha_D$ in response to the roll motion of the ship. The fin servo mechanism responds and drives the fins to an achieved fin angle $\alpha$ and the fins convert this into a stabilising roll moment $M_{fins}$. This is substracted from the roll moment generated by the waves, thus reducing the roll motion of the ship $M_{total}$.

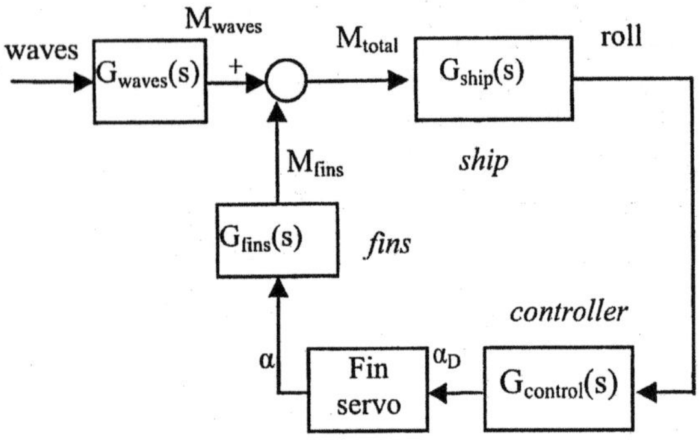


Fig 1. Block diagram for a ship with roll stabiliser fins

The transfer functions of the ship, fin servo and the fins are essentialy fixed for a given design. The fin controller transfer function is, however, adjustable and must be set up in such a way as to ensure that the fins develop roll moments which generally oppose the moments provided by the waves.

Previous the control design, a mathematical model of the ship and actuators are obtained. These are the most important tools available to the designer, since they are used to establish the best available control algorithms and to predict the behavior of the ship in the various environments in which it must operate.

### 2.1 Ship modelling

A scaled physical model (1:25) is used to experiment in the towing tank. Tests are made with various types of waves and ship speed: regular waves with frequencies between 0.393 and 1.147 rad/s, irregular waves with SSN=4, 5 and 6, 90 degrees of wave heading, and 20, 30 and 40 knots ship speed.

Model of wave to roll moment $G_{waves}(s)$ (1) and roll moment to roll $G_{ship}(s)$ (2) are identified (Aranda, *et al.*, 2003). Representation of the experimental points in Bode diagrams demonstrates that frequency response at different ship speed (20, 30 and 40 knots) are very similar. Therefore, only one model is required.

$$G_{waves}(s) = \frac{-5397s^3 - 970s^2 - 448.3s}{s^3 + 1.04s^2 + 0.43s + 2.1\cdot 10^{-3}} \quad (1)$$

$$G_{ship}(s) = \frac{-3.4\cdot 10^{-4}s^3 - 1.1\cdot 10^{-4}s^2 + 7.3\cdot 10^{-4}s^1 + 2.3\cdot 10^{-3}}{s^3 + 3.25s^2 + 1.33s + 3.65} \quad (2)$$

### 2.2 Fins Modelling

Fins consist of two active stabilizer surfaces, that generate a roll moment in order to cancel the applied moment to the ship by the seaway.

Fins are mounted on the side of a hull, and they are driven by a hydraulic system. When the fins are placed at an angle of attack, the flow over them resulting for the forward speed of the ship causes a lift. If the lifts are opposite from one side of the ship to the other, the difference results in a roll moment acting on the ship. The roll moment $M_{fins}$ (3) available from a pair of fins, each of area $A$, on a ship proceeding at speed $V$, can be derived by the Bernoulli theorem (Lewis,1989).

$$\mathrm{M}_{\mathrm{fins}} = A\frac{\partial C_L}{\partial \alpha}V^2(\alpha_p - \alpha_s)\mathrm{r} \quad (3)$$

In equation (3), $\frac{\partial C_L}{\partial \alpha}$ is the fin lift coefficient slope, $r$ is the distance along from the roll center to the center of effort on one find, $\alpha_p$ and $\alpha_s$ are the effective angles of attack of the port and starboard fins, respectively. In this case, the angle of each fin will have the same value but opposite, that is, $\alpha_s = -\alpha_p$.

Continuous linear model of the transfer function $G_{fins}(s)$ is identified. The block *fins* in figure 1 accepts the input $\alpha$ and gives the output $M_{fins.}$ As the equation (3) shows, moment generated by the fins depends on the ship speed, so three models are estimated (eq 4-6), each one for speed 20, 30 and 40 knots.

$$G_{fins20} = \frac{1459}{s + 13.5} \quad (4)$$

$$G_{fins30} = \frac{3281.9}{s + 13.5} \quad (5)$$

$$G_{fins40} = \frac{5834}{s + 13.5} \quad (6)$$

## 3. CONTROLLERS DESIGN

A control system must perform mainly three functions. The firts is to assure stability, the second is to attenuate seaway-induced motions, and the third is to assure the safety of the ship and its passengers.

Once models of the ship and actuators have been obtained, next step is the design of the control. Figure 2 shows the basic configuration of the feedback control. The block $G_{plant}(s)$ is the product of the blocks $G_{ship}(s)$ and $G_{finsV}(s)$, where $V$ refers to the ship speed.

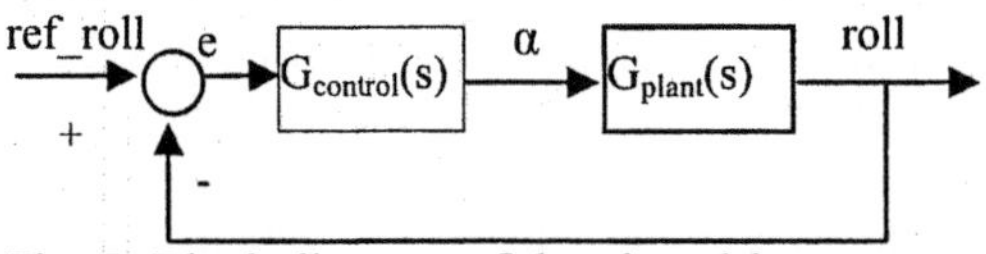


Fig. 2. Block diagram of the closed-loop system

The controller must be set up in such a way as to ensure that the fins develop roll moments which oppose the moments provided by the waves.

### *3.1 Controller Tuning*

It is necessary a fit criterium in order to find out the best controller. The main objective of the controller is to decrease vertical motions related to roll. For that, it is defined the fit function $J$ (7) as the mean value of the vertical roll acceleration measured in a simulated test.

$$J = \overline{acv} = \frac{1}{N}\sum_{i=1}^{N}|acv(t_i)| \quad (7)$$

where $acv(t_i)$ is the vertical acceleration associated with roll motion measured at time $t_i$ and 7 meters distanced to the center of gravity (8).

$$acv(t_i) = 7\cdot\frac{\Pi}{180}\frac{d^2 roll(t_i)}{dt^2} \quad (8)$$

Thus, the controller with the minimum value of $J$ will be the best controller, because this imply the maximum attenuation in vertical acceleration and, therefore, in vertical component of roll motion.

In addition, other factors must be considered in the control design :

- System stability.
- Fins saturation. $|\alpha| < \pm 15°$.
- list must no exist.
- frequency rank [0.39,1.15] rad/s

### *3.2 Methods for the design*

In this research two classical methods have been probed: Root locus and Nichols chart. Both techniques are graphical methods.

In root locus, specifications in the temporal response are translated into specifications in s-plane and therefore, controller design consist of the adjustment of the controller parameters in such a way as the poles of the system in closed loop locate in the right position. The poles of the system are the roots of the characteristic equation

$$1 + G_{plant} * G_{control} = 0 \quad (9)$$

Nichols chart analyzes the system in the frequency domain. The function plotted is the open-loop function L

$$L = G_{plant} * G_{control} \quad (10)$$

The ideas followed in this method are the same as the basic steps in QFT design (Horowitz(1992), Borghesani (1995)). The only difference is that in this case plant uncertainties are not considered. The specifications used mainly are gain and phase margins stabilily (11), output disturbance rejection (12) and control effort (13). QFT converts these closed-loop magnitude specifications into magnitude constraints on an open-loop function.

- gain and phase margins

$$\left|\frac{G_{plant}G_{control}}{1+G_{plant}G_{control}}\right| \le W_{s1} \quad (11)$$

- sensitivity reduction

$$\left|\frac{1}{1+G_{plant}G_{control}}\right| \le W_{s2} \quad (12)$$

- control effort

$$\left|\frac{G_{control}}{1+G_{plant}G_{control}}\right| \le W_{s3} \quad (13)$$

Once the controller parameters are designed by using these methods, the system in closed loop dynamic is simulated, and vertical acceleration and roll motion are measured. When the simulation is finished, it is examined if the controller accomplishes the specifications, and cost function $J$ is calculated. Finally, controller with minimum $J$ is chosen.

### *3.3 Types of controllers*

Different types of controllers have been analyzed and compared in this research. Firstly, a PI control is designed by the root locus technique. Then, Nichols chart is used to compare and examine if this control meet the fixed specifications.

In addition, a first order lead filter and a second order filter are tried by Nichols diagram with others specifications.

## 4. RESULTS

Firstly it is presented the transfer function of the controllers designed by the root locus and the cost function $J$ obtained. In addition, roll reduction percentage (14) is calculated.

$$reduction = \frac{mean(roll_{withoutCotrol}) - mean(roll_{control})}{mean(roll_{control})} \quad (14)$$

Simulations carried out in order to calculate $J$ and roll reduction use regular waves with 0.8 meters amplitude, and natural frequency between the rank [2.56, 0.39] rad/s. In addition, irregular waves with SSN= 4, 5, 6 are employed. The angle between heading and wave direction is $\beta$=90 degrees.

Results and graphics given in this paper are for the case of ship speed V = 40 knots, regular wave with frequency 1 rad/s and irregular wave SSN=5.

In order to make a comparation, $J$ is calculated for the case in which control does not work, that it, fins are located in a fixed position of 0 degrees. In this case, $J$ is independent of the speed, and its value for wave input with frequency 1 rad/s is 1.39 and for irregular wave SSN=5 is 1.25.

### 4.1 Control PI

*Root locus.* For each ship speed, a controller is determined by the locus of the root. For the particular case of V = 40 knots, figure 3 shows the root locus of the system with the controller PI $G_{PI40}(s)$. The transfer function of this controller is given by expression (15).

$$G_{PI40} = 2.2\frac{50s+1}{50s} \qquad (15)$$

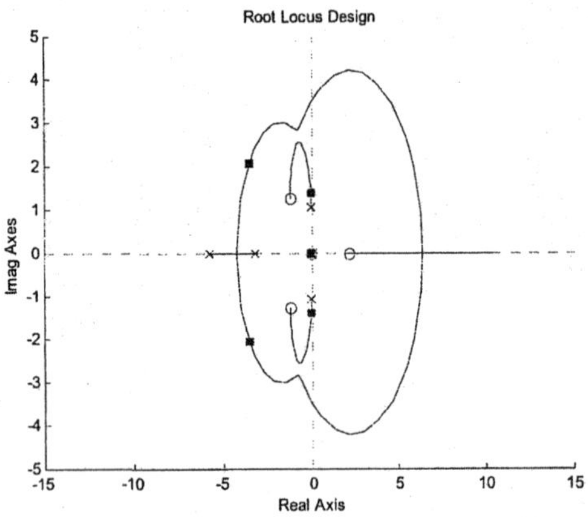


Fig. 3. Root locus of the system with $G_{PI40}$

Table 2 shows the values of $J$ and roll reduction percentage for the cases of regular wave with frequency 1 rad/s and irregular wave with SSN = 5 with this controller.

Table 2. value of J and reduction for control PI. V=40 knots

| wave type | $J$ | roll reduction (%) |
|---|---|---|
| regular $\omega$ = 1rad/s | 0.29 | 82.69 |
| irregular SSN=5 | 0.82 | 69.01 |

*Nichols chart.* The specifications fixed for Nichols chart representation for 40 knots are given by the following relation:

- gain and phase margins $W_{s1}$= 1.7
- sensivity reduction $W_{s2}$=1.9 (16)
- control effort $W_{s3}$ = 6.5

These specifications guarantee adequate gain margins, sensitivity and control effort.

Figure 4 shows the Nichols chart of the open-loop function (10) for the PI controller $G_{PI40}(s)$ with the specificacions given by (16). It is shown that this controller does not meet the specifications. Control PI is valid for conditions of phase and gain margin that are not much restrictive. This can be probed with the root locus graphic (fig. 3), in which there are two poles very close to zero. In addition, it is easy to achieve fins saturation in some frequencies.

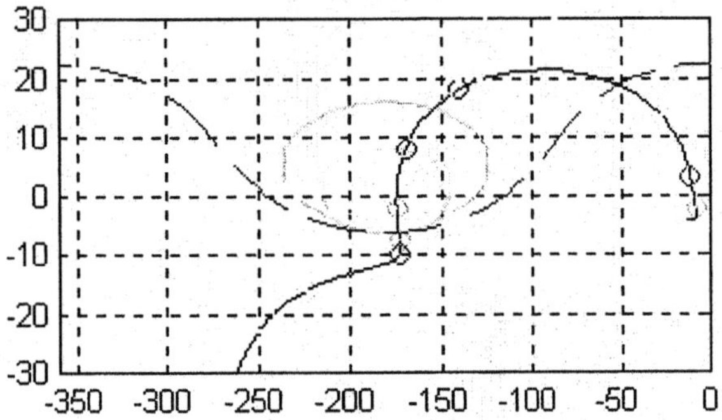

Fig. 4. Nichols chart and bounds for $G_{PI40}(s)$. The open-loop function $L$ does not meet the specifications with this controller.

Despite this fact, this controller obtains very succesful results in many cases, as it is shown in table 2. As an example, figure 5 compares the results of the simulation without controller, and with the control PI obtained for 40 knots for irregular wave SSN=5. Figure 6 shows the fins movement for this case.

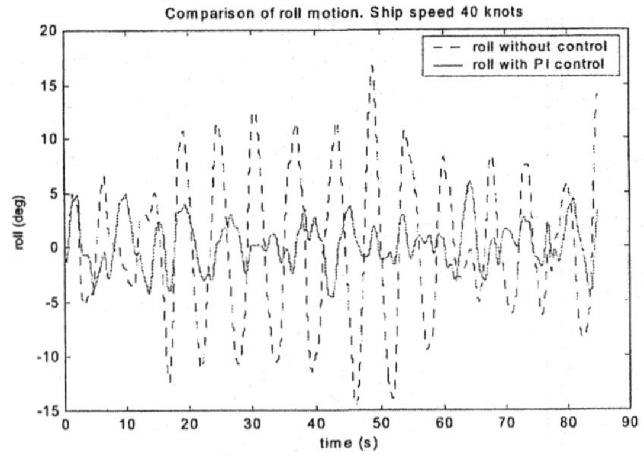


Fig. 5. Comparison of roll motion without control and with control $G_{PI40}(s)$ for irregular waves SSN=5.

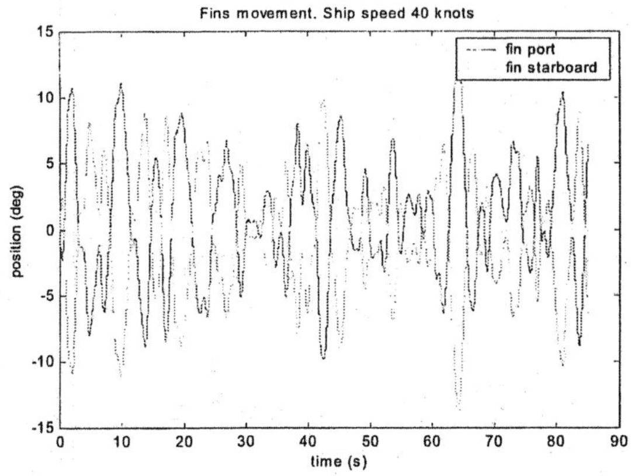


Fig. 6. Fins movement

*4.3 First order lead filter*

Now new controllers are tried by using Nichols chart in order to meet the fixed specification.

First order lead filters add a pole and a zero. Now, a controller is designed for each ship speed ($V$= 20, 30 or 40 knots) by the Nichols chart technique. For the case of 40 knots, the transfer function is given by (17)

$$G_{lead40} = \frac{1}{1.2}\frac{1.54s+1}{0.11s+1} \qquad (17)$$

Table 3 contains the values of $J$ and roll reduction percentage for the cases of waves with frequency 1 rad/s and irregular wave with SSN = 5 for this controller.

Table 3. value of J and reduction for lead filter V = 40 knots

| wave type | $J$ | roll reduction (%) |
|---|---|---|
| regular $\omega$ = 1rad/s | 0.58 | 59.53 |
| irregular SSN=5 | 1.38 | 45.41 |

Figure 7 shows the Nichols chart of the open-loop function (10) for the first order lead controller $G_{lead40}(s)$ with the specificacions given (16) for the 40 knots case. It is shown that the controller meets the specifications in the frequency rank of interest. Next, figure 8 compares the simulated roll response without controller, and with the first order lead filter. Figure 9 presents the fins movement.

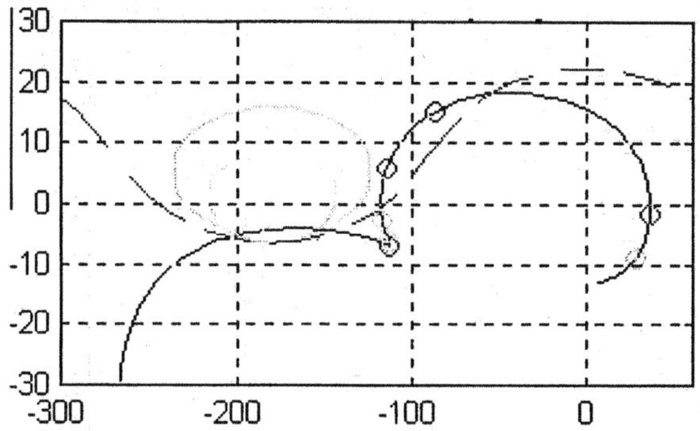


Fig. 7. Nichols chart and bounds. for $G_{lead40}(s)$. The open-loop function $L$ meets the specifications.

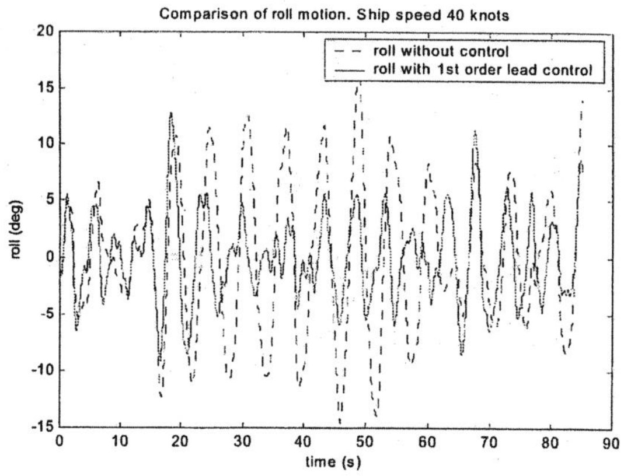


Fig. 8. Comparison of roll motion without control and with control $G_{lead40}(s)$ for irregular waves SSN=5.

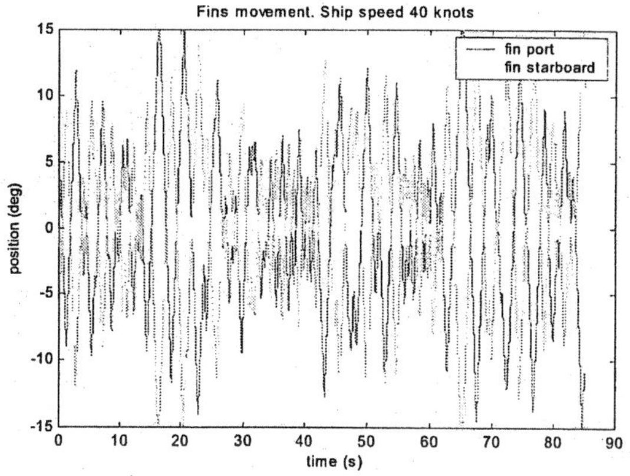


Fig. 9. Fins movement

*4.4. Second order filter*

The transfer function of the second order controller designed for the specific case of 40 knots is given by (18)

$$G_{2Or40} = \frac{2.04s^2+1.13s+2}{0.33s^2+1.27s+1} \qquad (18)$$

Table 4 shows the values of $J$ and roll reduction percentage for the cases of waves with frequency 1 rad/s and irregular wave with SSN = 5 for this controller.

Table 4. value of $J$ and reduction for second order filter V = 40 knots

| wave type | $J$ | roll reduction (%) |
|---|---|---|
| regular $\omega$ = 1rad/s | 0.49 | 65.28 |
| irregular SSN=5 | 1.49 | 40.99 |

Again, figure 10 shows the Nichols chart of the open-loop function (10) for the second order controller $G_{2Or40}(s)$ with the specificacions given (16) for the 40 knots case. It is shown that the controller meets the specifications in the frequency rank of interest. Next, figure 11 compares the simulated roll response without controller, and with the second order controller. Figure 12 presents the fins movement.

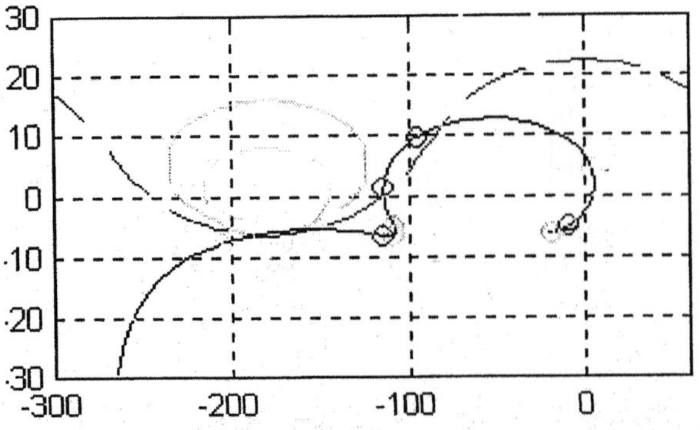


Fig. 10. Nichols chart and bounds for $G_{2Or40}(s)$. The open-loop function $L$ meets the specifications

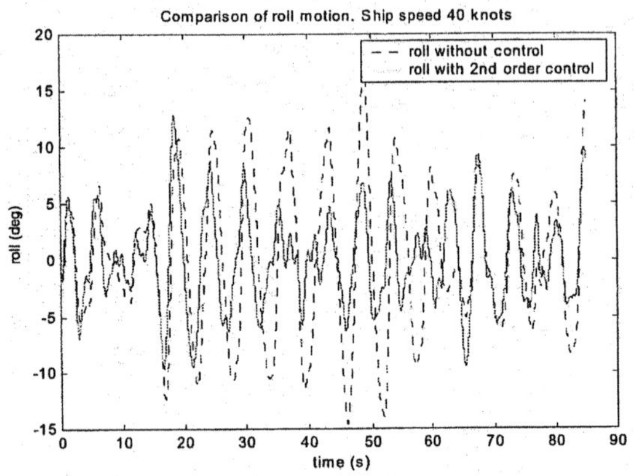


Fig. 11. Comparison of roll motion without control and with control $G_{2Or40}(s)$ for irregular waves SSN=5

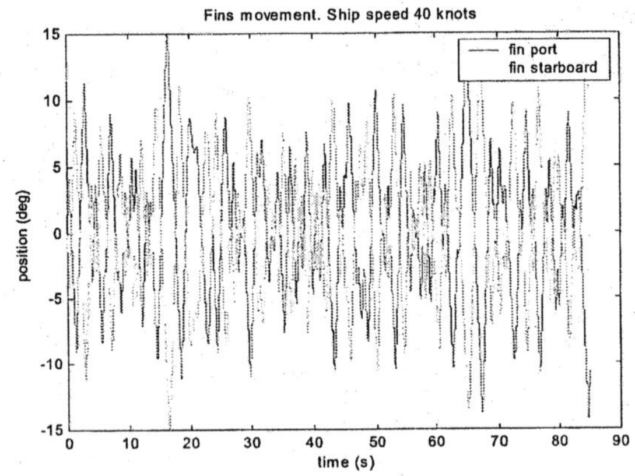


Fig. 12. Fins movement

## 5. CONCLUSIONS

In this research an analysis and design of different roll control methods have been carried out. The purpose of the controller is to reduce the roll response in a fast ferry due to lateral waves and coupling with the T-foil and flaps control. Despite the fact that roll motion is less significant than heave and pitch, its study in high crafts is justified because T-foil action for pitch control causes an increasing in vertical component of roll movement. Roll control will help pitch motion not to have adverse effects.

Previous to designing the controller, roll motion and actuators dynamics modeling have been obtained.

Several controllers have been designed by different techniques. All of these are validated by the simulation of the roll response of the feedback system. The simulations make possible a comparison and show that the control decreases the roll motion. In addition, it has been shown that simple controllers, as PI achieve significative reductions. On the contrary, these controllers can cause fins saturation for some frequencies, and besides, gain and phase margins are very little. Due to this problems, first order lead and second order filters are designed, even when the results of the PI control are rather satisfactory.

## 6. REFERENCES


Aranda, J., J.M. Díaz, P. Ruipérez, T.M. Rueda, E. López. (2002a). Decreasing of the motion sickness incidence by a multivariable classic control for a high speed ferry. *CAMS Proceeding Volume.* Pergamon Press.

Aranda, J., J.M de la Cruz, J.M. Díaz, S. Dormido Canto (2002b). QFT versus classical gain scheduling: study for a fast ferry. *15th IFAC World Congress b'02.*

Aranda, J, R. Muñoz, J.M Díaz. (2003). Roll model for control of a fast ferry. *2nd International Conference on Maritime Transport and Maritime History.*

Borguesani, C., Chait, Y., Yaniv, O.(1995). *Quantitative Feedback Yheory Toobox - for use with MATLAB.* The Mathworks Inc.,Natick, MA.

Cruz, J.M de la, J. Aranda, B. de Andrés, P. Ruipérez, J.M. Díaz, A. Marón. (2000). Identification of the vertical plane Motion Model of a High Speed craft by Model Testing in Irregular waves In:*CAMS Proceeding Volume.* (K.Kijima, TI. Fossen) 257-262. Pergamon Press.

Horowitz, I.M. (1992). *Quantitative feedback design theory (QFT),* QFT Publishers 660 South Monaco Dorkway Denver, Colorado. 80224-1229. Ph. 303-321-2839.

Lewis, E.V., (1989). *Principles of Naval Architecture.* Volume III. Society of Naval Architects and Marine Engineers.

ELSEVIER
IFAC
PUBLICATIONS
www.elsevier.com/locate/ifac

# PITCH MOVEMENT QFT CONTROL TO REDUCE THE MSI OF A TURBO FERRY

**Francisco J. Velasco*, Teresa M. Rueda*, Eloy López**, Emiliano Moyano*****

* *Dpto. Tecnología Electrónica, Ingeniería de Sistemas y Automática. Univ. de Cantabria. C/ Gamazo, 1. 39004 Santander (Spain) (velascof@unican.es, ruedat@unican.es)*
** *Dpto. Ciencias y Técnicas de la Navegación, Máquinas y Construcciones Navales. Univ. del País Vasco. C/ Mª Díaz de Haro, 68. 48920 Portugalete. Bizkaia (Spain) (cnplogae@lg.ehu.es)*

*** *Dpto. Matemática Aplicada y Ciencias de la Computación. Univ. de Cantabria. C/ Av. Los Castros s/n. 39005 Santander (Spain) (moyanoe@unican.es)*

Abstract: This article describes the tuning of a QFT controller (Quantitative Feedback Theory) designed in order to reduce the pitch movement generated in a high–speed Turbo Ferry and so reduce the MSI (Motion Sickness Incidence). The objective pursued in the design was to improve the performance of the craft for speeds of 20, 30 and 40 knots and for ssn 4 (sea state number 4), and also the minimization of the control effort. For the QFT controller tuning a linearized model of the ship was used. System specifications are robust stability, sensitivity reduction to the waves effect and minimization of the control effort. In order to control the heave movement we have used a first order controller tuned by means of genetic algorithms applied to the non-linear model of the ship. A Simulink non-linear model has been used to simulate and validate the response of the ship with the designed controllers. It is found that these controllers provide a significant reduction of the vertical acceleration and MSI for the studied cases. *Copyright© 2004 IFAC*

Keywords: Ship Control, Ship Model, High-Speed Craft, QFT Controller.

## 1. INTRODUCTION

Passengers and crew comfort and the safety of vehicles on board are two of the problems to be dealt with in the design and development of a high-speed craft. Heave and pitch motions related with the vertical acceleration of the craft are the main causes of discomfort and seasickness among passengers and crew on board.

In this paper we present the design of a QFT monovariable controller aimed to the reduction of the pitch movement, which has been found is the main component of the vertical movement in the Turbo Ferry TF-120 (figure 1). With this reduction, the vertical acceleration and the MSI is also lowered. There are not many papers dealing with this subject (Aranda *et al*, 2002).

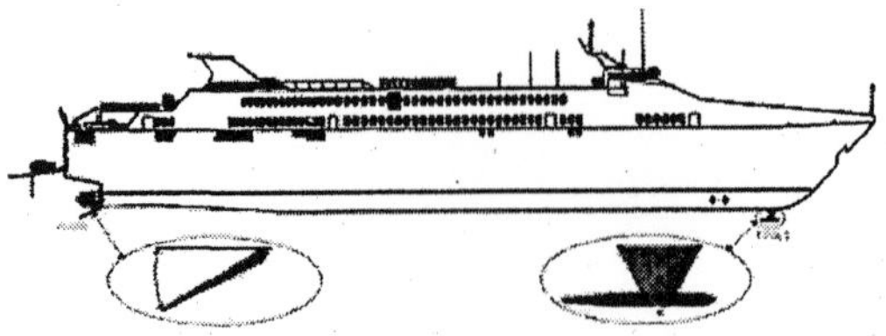

Fig. 1. High-speed craft TF-120

To tune the QFT controller a linearized ship model of reduced order has been used. System specifications are robust stability, reduction of the sensitivity of the ship to the waves force at different speeds and a minimization of the control effort.

In order to control the heave movement we have used a first order controller tuned by means of genetic algorithms.

The simulation results are shown in the table, which indicates the reduction in heave and pitch motions, the decreased in vertical acceleration and the improvement in the MSI. Here are also shown graphics in which the pitch and heave movement, vertical acceleration and the MSI reduction can be seen including the position of the actuators.

## 2. MATHEMATICAL MODEL OF THE SHIP

On the project CICYT TAP97-0607-C03 we have obtained the mathematical model referred to the ship performance for sea state of 4 and speeds of 20, 30 and 40 knots. This model implements the vertical dynamics of the ship through four linear continuous transfer SISO functions, based on data gathered by means of PRECAL software (Aranda, *et al.*, 2000; De Andrés, *et al.*, 2000). This program is a tool which integrates the hull of the ship by means of finite elements, calculating the different coefficients of the model for different speeds, waves height and incidence angles.

It was also developed a linear model to describe the behavior of the active actuators (one T-foil and two flaps) (Esteban, *et al.*, 1999) intended to reduce the vertical accelerations and movement, because the actuators counteract the effect of the waves. It is also reduced the MSI, this allows to maintain a higher speed.

The actuators model has as outlets the heave force and the pitch momentum produced by the actuators which oppose the heave force and the pitch momentum caused by the waves, in order to decrease the possible vertical acceleration which is the main cause of seasickness.
Thus, the process model is composed by the linear model of the vertical dynamics and the actuators model. To simulate and validate the response of the ship with the designed controllers has been used the non linear model of the actuators. However, for the tuning of the QFT controller has been used a linearized model. All has been integrated in a Simulink model (López, *et al.*, 2000). The modular design of this model allows to simulate the ship's movement, using different controllers, by changing the control module. The ship block diagram is composed by four transfer functions which relate: a) wave height with heave force, b) wave height with pitch momentum, c) heave force with heave movement and d) pitch momentum with pitch movement. The model of the actuators has been designed as a block which has as inputs: a) the position of the T-foil, b) the position of the flap, c) heave movement, and d) pitch movement; and as outputs: a) heave force, and b) pitch momentum. The diagrams of the corresponding blocks can be seen in figure 2.

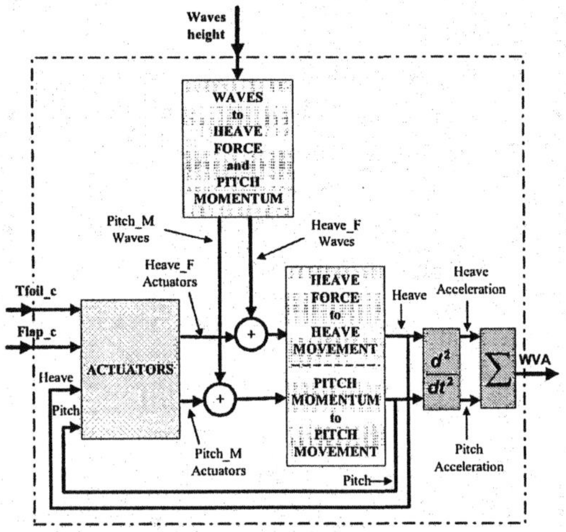


Fig. 2. Ship Model

## 3. CONTROL PROBLEM

### *3.1 Specifications*

The main aim in the design of the controller (De la Cruz, 2000) is to increase passenger comfort. To achieve that is necessary to reduce the vertical acceleration of the ship caused by the sea waves.

The ship vertical movement is a consequence of the heave and pitch motions acting together. In the Turbo Ferry TF-120 the point at which passengers are exposed to the higher vertical acceleration (worst passenger position) is situated 40 meters forward of the gravity centre of the ship. The vertical acceleration at this point can be defined as:

$$a_{V40}(t_i) = a_{VH}(t_i) + a_{VP}(t_i) = \frac{d^2 heave(t_i)}{dt^2} - 40\frac{\pi}{180}\frac{d^2 pitch(t_i)}{dt^2} \qquad (1)$$

The Worst Vertical Acceleration (WVA) is defined as the mean value of the temporal series $|a_{V40}(t_i)|$, $i$=1, ..., $N$.

The WVA is related to the MSI (percentage of passengers getting sick after two hours of motions) (Lloyd, 1989) through the following expression:

$$MSI = 100\left( 0.5 \pm erf\left(\frac{\pm \log(|\ddot{s}_3|/g) \pm \mu_{MSI}}{0.4}\right)\right) \quad (2)$$

where the function *erf(.)* is defined by the equation:

$$erf(x) = \frac{1}{\sqrt{2\pi}} \int_0^x e^{-\frac{z^2}{2}} dz \quad (3)$$

$|\ddot{s}_3|$ is the vertical acceleration at the chosen point for the half of a cycle, and $\mu_{MSI}$ is:

$$\mu_{MSI} = - \quad 0.819 \pm 2.32 \ ( \log_{10} \omega_e )^2 \quad (4)$$

with $\omega_e$ the dominant encounter frequency with waves.

The worst frequency for passengers is around 1.07 rad/sec and as the acceleration increases so does the number of sea-sicken persons (Rueda, *et al.*, 2001). The encounter frequency depends on the sea and the ship's speed. It may happen that a certain speed clearly increases the MSI, and slowing down could be necessary.

Moreover, as in any other control problem the stability of the system must be guaranteed minimizing, as much as possible, the control effort in order to avoid an excessive saturation of the actuators, which could imply a higher rate of wear and hampering of performance.

### *3.2 Controllers*

In order to design a control system, adequate to the considered multivariable model of the ship, we have taken into account some previous studies based on classical controllers (López, *et al.*, 2000). It was observed that the action of control on the T-foil produces its main effect on reducing the pitch movement, like the flap action on the heave. This agrees with the interaction analysis between the manipulated variables (T-foil and Flap positions) and the controlled variables (heave and pitch movements), being the best matching:

Flap position – Heave movement
T-foil position – Pitch movement

Having this in mind, we have designed two independent controllers of a monovariable structure.

We have also found out that in the vertical acceleration value defined in the equation (1) the influence of pitch acceleration is much higher than the heave acceleration. Thus, a QFT controller has been developed and designed in order to reduce the pitch movement. In order to reduce the heave movement a first order controller, tuned by means of genetic algorithms, has been used for the speeds of 20, 30 and 40 knots (Rueda, *et al.*, 2001).

The analysis of the values given by the relative gain matrix (RGA) indicated the existence of an interaction, not to be disregarded, between the flap position and the pitch movement and the T-foil position and the heave movement. Therefore, if we intend to work with two monovariable systems it is necessary to implement the adequate decoupling control system (figure 3).

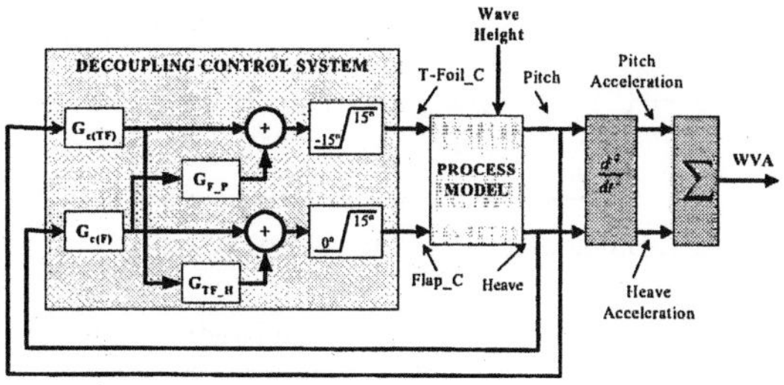


Fig. 3. Control Schema

### *3.3 QFT Controller*

For the tuning of the QTF controller a linearized model of the actuators has been used. Having in mind the transfer functions of the ship model and those of the linearized model of the actuators and using the rules of block diagram algebra it has been found out that the process model is the shown in figure 4.

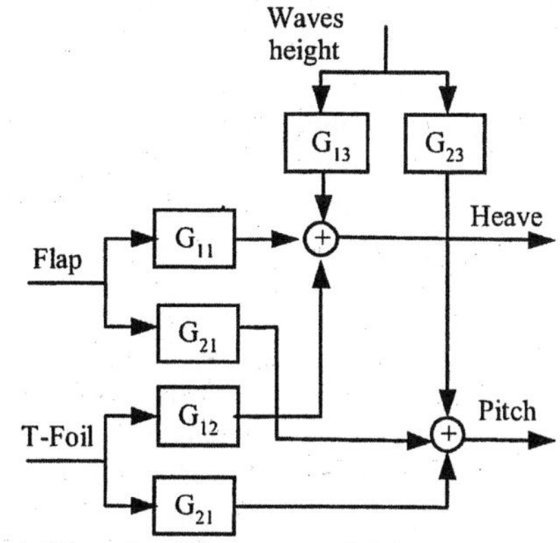


Fig. 4. Linearized process model

The transfer functions G11, G12, G21, G22 are of order 17. These values make desirable to look for some lower order model with a dynamic performance closed to that of the original model to make easier the calculations required for the controller design. On this paper we present a simplified model using the cancellation pole-zero method, beside a balanced truncation for the functions G11, G12, G21, G22. As a result functions of order 7 were obtained.

Moreover, the mathematical model of the ship TF-120 varies in according with the speed. This indicates that the process composed by the transfer functions described in the later paragraph is not a fix amount. Then an uncertainty model will be used. Thus, it becomes necessary to apply a robust technique in the design. Among the several possible ones the method QFT (Horowitz, 1992) has been chosen, with which a satisfactory balance between the degree of uncertainty in the system and the complexity of the mathematical problem involved has been achieved.

The structure of the monovariable linear system to be used for the design is (figure 5):

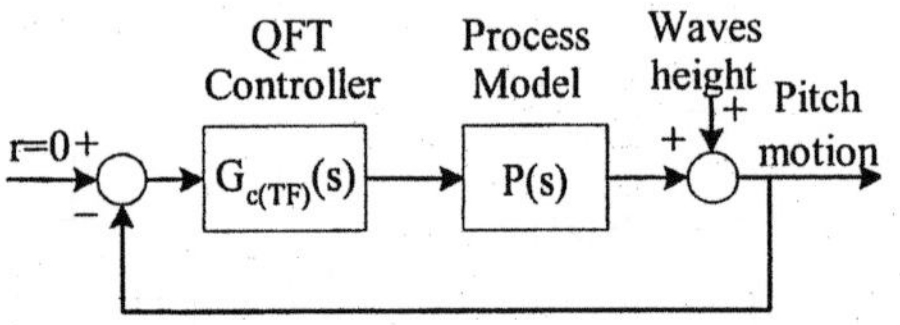


Fig. 5. Block diagram of the control system

After the development of the mathematical model the stages in the controller design can be sum up as follows:

1. The choice of the nominal plant. On this case, a ship speed of 40 Knots and a sea state of 4 have been considered, as the most common circumstances during navigations.

2. To calculate the templates or regions dependent on the frequencies represented by the uncertainty of the chosen model, using for that the Nichols Chart.
3. To define the range of the working frequencies: $\Omega = \{0.7, 0.85, 1, 1.2, 1.5, 2, 2.5, 3, 7\}$ rad/sec

4. To define the desired performance specifications. On this case, robust stability, reduction of the system sensitivity to the sea waves, and minimization of the control effort. These specifications are defined in the frequency domain (Yaniv, 1999 ) in this manner:

4a. The robust stability specification is established as a value in decibels δ which corresponds to an M-circle of the Nichols Chart which the loop function must not cross. This value is related to the values of the phase and gain margins desired in the system. In this case it is desired that the phase margin angle should be at least 45° and the gain margin 2 dB. Thus, the robust stability specification is defined by:

$$\left|\frac{P(j\omega)G_{c(TF)}(j\omega)}{1+P(j\omega)G_{c(TF)}(j\omega)}\right| \le \delta = 1.2 \qquad (5)$$

where $P(j\omega)$ is the set of plants given by the uncertainty and $G_{c(TF)}(j\omega)$ the controller.

4b. To minimize the sea waves effect on the pitch movement we take the following function:

$$\left|\frac{1}{1+P(j\omega)G_{c(TF)}(j\omega)}\right| \le Ws_2 \qquad (6)$$

with $Ws_2 = \left|\frac{1}{1+P_0(j\omega)G_1(j\omega)}\right|$. Being $P_0(j\omega)$ the nominal plant and $G_1(j\omega) = 1$.

4c. The T-.foil control is bounded by the actuator saturation limits which are in the range [-15°, 15°]. Therefore, the specification for the minimization of the control effort will be established as:

$$\left|\frac{G_{c(TF)}(j\omega)}{1+P(j\omega)G_{c(TF)}(j\omega)}\right| \le Ws_4 = 15 \qquad (7)$$

5. To obtain the bounds taking into account the desired specifications and the uncertainty of the model.

6. To adjust the loop function $L_0(j\omega)$ to fulfil the specifications. ( $L_0(j\omega) = G_{c(TF)}(j\omega)P_0(j\omega)$ ).

With the MATLAB QFT Toolbox (Borguesani, *et al.*, 1995), the results of each adjustment can be observed. It is necessary to carry out subsequently a validation of the results obtained, verifying the specifications in the frequency domain graphically.

## 4. SIMULATIONS

On the basis of the performance specifications and the plant templates and using the Nichols Chart, the nominal open-loop transfer function $L_0 = P_0G_{c(TF)}$ is adjusted, in such a way that no bounds are violated, as shown in figure 6.

The controller obtained is:

$$G_{c(TF)}(s) = \frac{3.243*10^6 s^3 + 1.008*10^7 s^2 + 2.299*10^7 s + 2.977*10^7}{s^3 + 425s^2 + 58750s + 2.625*10^6} \qquad (8)$$

With this controller, all the required specifications are fulfilled as can be seen from figures 7, 8 and 9.

The transfer function of the first order controller tuned by means of genetic algorithms in order to control the heave movement is:

$$G_{c(F)}(s) = \frac{4.5731\,s + 30.0507}{s + 4.2872} \qquad (9)$$

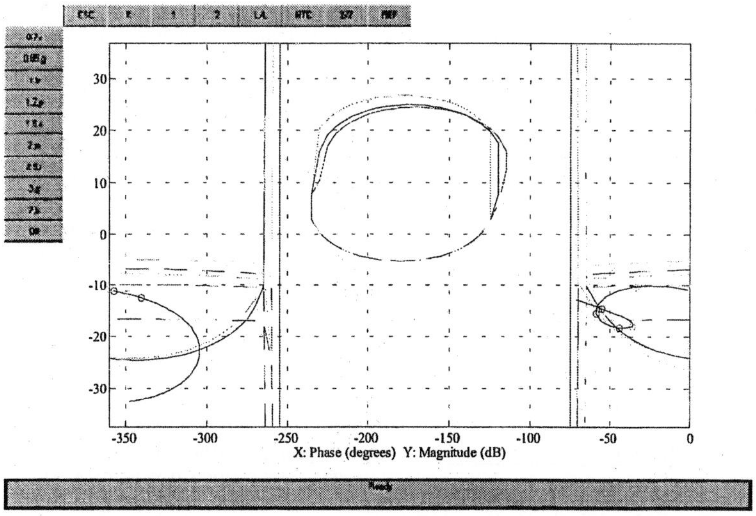


Fig. 6. Shaping of $L_0(j\omega)$ on the Nichols chart for the nominal plant.

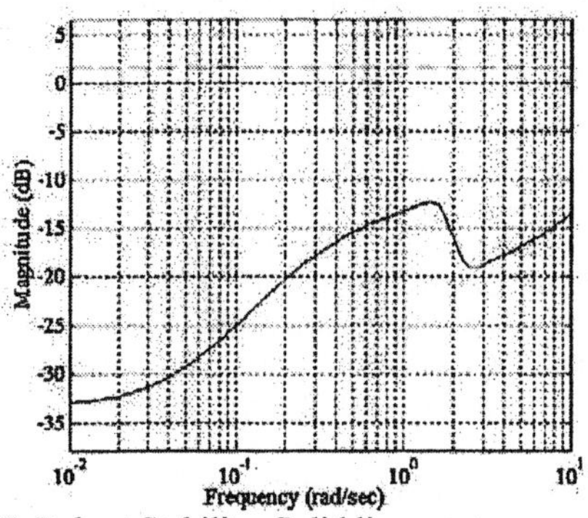


Fig. 7. Robust Stability. Solid line: system response. Dashed line: specification.

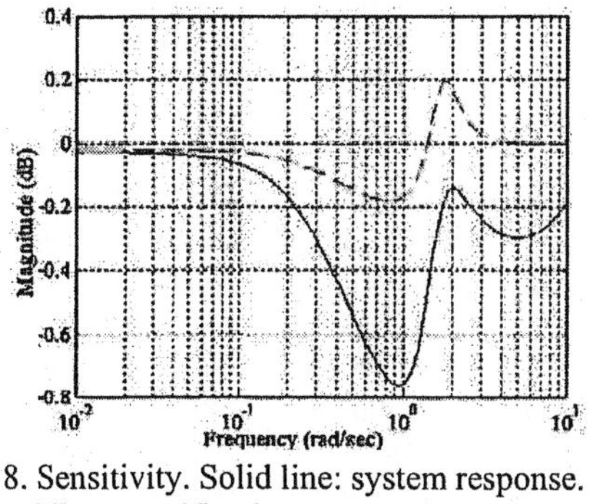


Fig. 8. Sensitivity. Solid line: system response. Dashed line: specification.

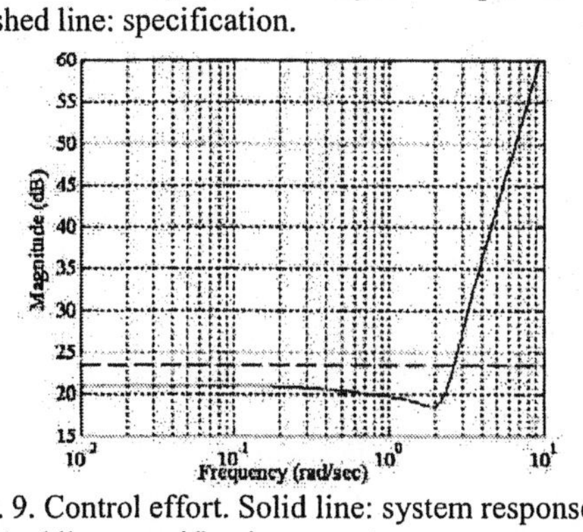


Fig. 9. Control effort. Solid line: system response. Dashed line: specification.

Table 1 presents a summary of the results obtained with the controllers studied at several speeds.

Table 1 Percentage of Improvement

| V (Knots) | *Heave Motion reduction (%)* | *Pitch Motion reduction (%)* | *WVA reduction (%)* | *MSI reduction (%)* |
|---|---|---|---|---|
| *20* | 2.8412 | 5.6262 | 6.6041 | 12.2732 |
| *30* | 6.7509 | 7.6102 | 13.1994 | 28.9029 |
| *40* | 7.1815 | 18.6393 | 19.8821 | 37.4444 |

It can be seen that a relevant reduction of the MSI is achieved for the three considered speeds, mainly for V= 40 knots which has been chosen as the nominal model in the QFT design.

To simulate the ship TF-120 vertical acceleration and MSI reduction results, the Simulink model has been used for a craft speed of 40 knots and for a sea-state of 4.

In figures 10 to 13 are shown the graphics for the heave and pitch motion, the mean vertical acceleration, actuators position (T-foil and Flap) and MSI. In the graphics can be seen that the decrease of the vertical movement was mainly produced by the reduction of the pitch movement.

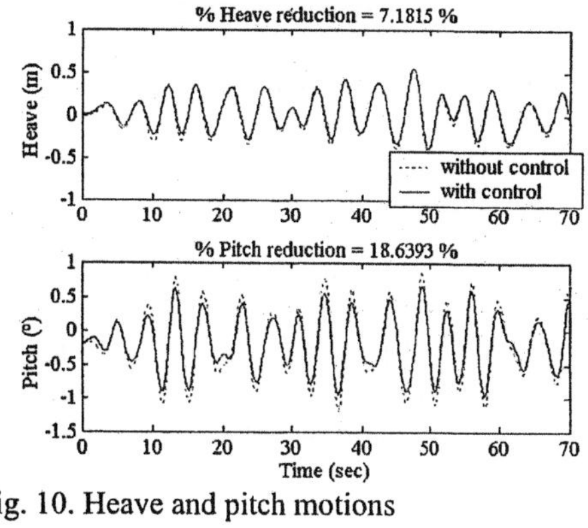


Fig. 10. Heave and pitch motions

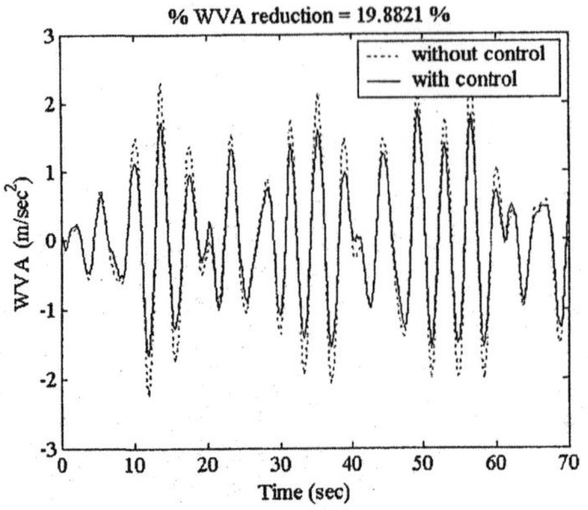


Fig. 11. Mean vertical acceleration

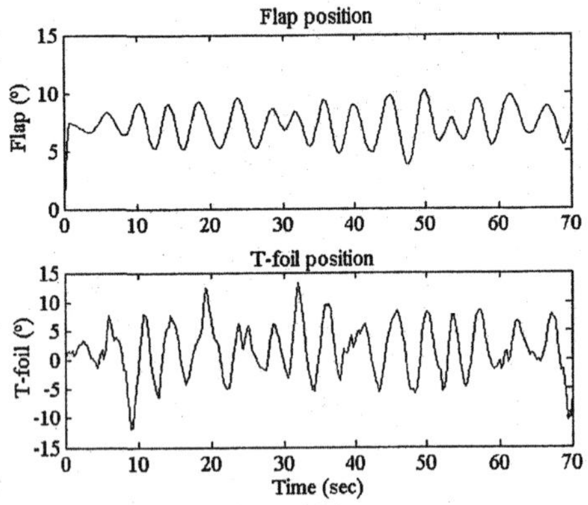


Fig. 12. T-foil and Flap position

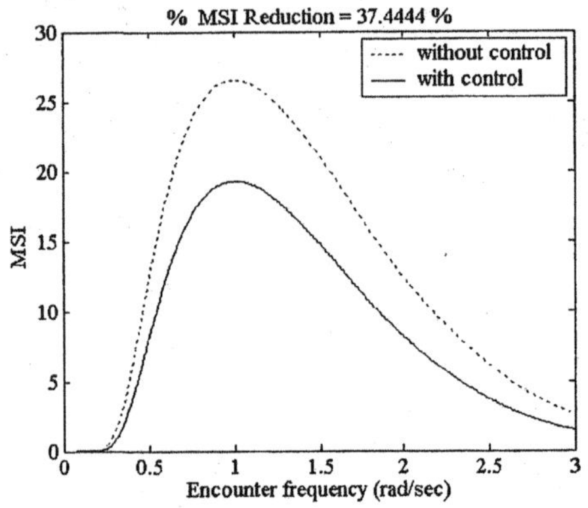


Fig. 13. MSI

## 5. CONCLUSIONS

It has been found in the studied ship that the reduction of the pitch movement plays a more significant roll in the decrease of the MSI than the heave movement.

Therefore, we have developed a QFT controller in order to reduce the pitch motion, using a linearized and reduced model of the system. To control the heave movement a first order controller tuned with genetic algorithms has been used.

The simulation of the performance of the controllers was made with a non-linear Simulink model. It was observed that a significant reduction of the MSI and of the vertical acceleration, for the three considered speeds, can be achieved without saturating the actuators. Mainly for V= 40 knots which has been chosen as the nominal model in the QFT design.

The problem of the saturation of the actuators implied that the QFT design was aimed at the minimization of the control effort.

## ACKNOWLEDGEMENTS

CICYT of Spain supported this development under contracts TAP97-0607-C03-03, DPI2000-0386-C03-03 and DPI2003-09745-C04-03.

## REFERENCES

Aranda, J., J.M. De la Cruz, J.M. Díaz, B. De Andrés, P. Ruipérez and J.M. Girón (2000). Modelling of a High Speed Craft by a Non-Linear Least Squares Method with Constraints. *5th IFAC Conference on Manoeuvring and Control of Marine Crafts MCMC2000. Aalborg.*

Aranda, J., J.M. De la Cruz, J.M. Díaz, S. Dormido, (2002). QFT Versus Classical Gain Scheduling: Study for a Fast Ferry. *15th IFAC World Congress b'02. Barcelona.*

Borguesani, C., Y. Chait and O. Yaniv (1995). *Quantitative Feedback Theory Toolbox – For use with MATLAB.* The MathWorks Inc.

De Andrés, B., S. Esteban, J.M. Girón and J.M. De la Cruz (2000). Modelling the Motions of a Fast Ferry with the Help of Genetic Algorithms. *Proc. 3rd IMACS MATHMOD. Viena.*

De la Cruz, J.M. (2000). Evaluation. *Technical report TAP97-0607. CRIBAV-01-04*

Esteban, S., J.M. Girón and J.M. De la Cruz (1999). Actuators Modelling. *Technical report TAP97-0607. CRIBAV-01-03.*

López, E., T.M. Rueda, F.J. Velasco and E. Moyano (2000). Decreasing of the Vertical Acceleration of the High-Speed Craft TF-120 by classical controllers. *Technical report TAP97-0607. CRIBAV-03-03.*

Lloyd. A.R.J.M. (1989). *Seakeeping: Ship Behaviour in Rough Weather.* RINA, London.

Rueda, T.M., E. López, E. Moyano, F.J. Velasco, J.M. Díaz, S. Esteban. (2001), Classical Controllers for a High-Speed Craft *1st International Congress on Maritime Transport. Barcelona.*

Horowitz, I.M. (1992). *Quantitative Feedback Design Theory.* QFT Publications, Boulder, Colorado

Yaniv. O. (1999). *Quantitative Feedback Design of Linear and Non-linear Control Systems.* Kluwer Academic Publishers. Norwell, Massachusetts.

ELSEVIER
IFAC
PUBLICATIONS
www.elsevier.com/locate/ifac

# SELECTION AND TUNING OF CONTROLLERS, BY EVOLUTIONARY ALGORITHMS: APPLICATION TO FAST FERRIES CONTROL

**M. Parrilla Sánchez, J. Aranda Almansa and J.M. Díaz Martínez**

*Departamento de Informática y Automática. E.T.S. de Ingeniería Informática.
Universidad Nacional de Educación a Distancia (UNED). Apdo. de Correos 60.011.
28080 Madrid. España,
{jaranda, mparrilla, josema}@dia.uned.es*

Abstract: Evolutionary algorithms have been shown very efficient tuning controllers, whose structure has been previously established. In this work, a step forward in the automation of the controllers design process will be tried, an algorithm is implemented which is able to select the appropriate controller structure and to tune it. By means of this procedure, a controller to reduce the motion sickness incidence on a high-speed ship, will be designed. The algorithm will be implemented using parallelization techniques. 

Keywords: genetic algorithms, ship control, parallel computation.

## 1. INTRODUCTION

In previous works (Aranda, et al., 2000a, b), the authors used evolutionary algorithms to tune controllers with a predetermined structure. In the current work, an evolutionary algorithm being able to select a control structure and to tune it will be built. This method is used in the problem of reducing the motion sickness incidence on a high-speed ship.

Two loops can be distinguished inside the algorithm. One is external, consisting of a genetic algorithm with chromosomes made up of binary numbers and using the mutation and crossover operators in an attempt to determine the best control structure. The other is internal, consisting of an evolutionary algorithm whose chromosomes will be made up of real numbers and using operators adapted to this kind of codification that will take charge of evaluating each one of the controller structures obtained by the external algorithm.

To select the control structure, an algorithm inspired in (Yaniv, 1999) will be used. In that reference, a technique known as 'loop shaping' is used, in which basic control blocks such as gains, simple poles and zeros, leads and lags, second order poles and zeros, and Notch blocks are used. Effects produced by each block when are included in the controller, are studied by means of Nichols diagrams.

This work expect to substitute the heuristic knowledge of the designer –needed to interpret Nichols diagrams– with an evolutionary algorithm, which gives a measure of the goodness that a certain control structure is, after find the best possible tune for that structure. That work will be made by the inner loop.

External loop will build controllers from the basic blocks previously mentioned, and will pass them to the inner loop to be tuned and evaluated.

The algorithm was implemented in C, using the library MPI, specialized in parallelization, and carrying out their execution in a Silicon Graphics Origin 2000 multiprocessor computer of the CSC of the Universidad Complutense de Madrid.

## 2. PROBLEM DEFINITION

In the CRIBAV project (Robust and Intelligent Control for High-Speed Crafts), the aim is to make robust controllers acting on some actuators (Flap and T-foil, see detail in Figure 1), absorbing pitch and heave movements of a high-speed passenger ship (de la Cruz et al. 2001).

Mathematical models for the ship and the actuators were obtained during the modeling stage (Aranda et al., 2001), and (Esteban et al., 2000).

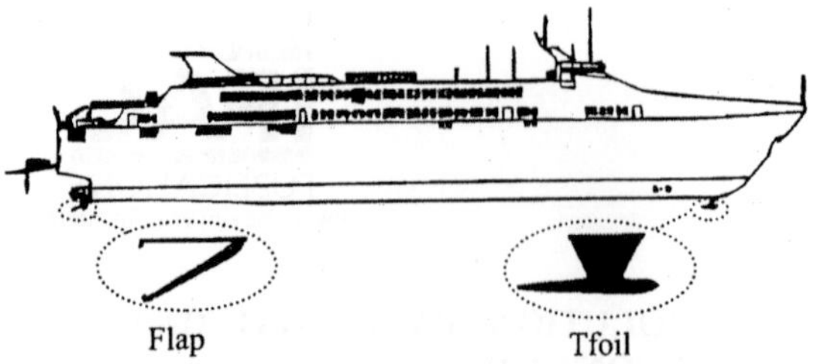


Fig. 1. Actuators detail.

### 2.1 *Process model.*

The process model is a multivariable model with 3 inputs and 2 outputs whose blocks diagram is shown in figure 2.

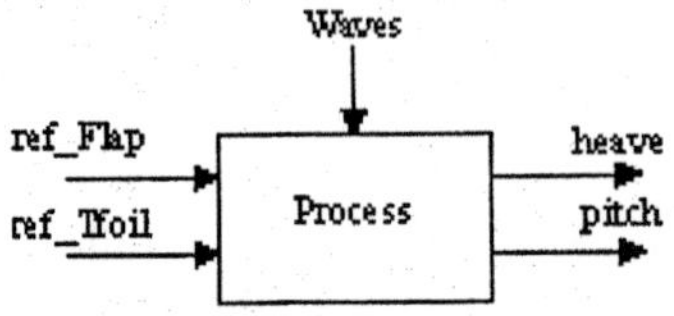


Fig. 2. Process model. ref_Flap and ref_Tfoil are the inputs to the actuators. Waves are the disturbance due to swell. A reduction of the heave and pitch output variables will be tried by the control algorithm.

Therefore, the process outputs can be obtained by the following equations:

$$heave(s) = G_{11}(s) \cdot ref\_Flap(s) + G_{12}(s) \cdot ref\_Tfoil(s) + G_{13}(s) \cdot waves(s)$$

$$pitch(s) = G_{21}(s) \cdot ref\_Flap(s) + G_{22}(s) \cdot ref\_Tfoil(s) + G_{23}(s) \cdot waves(s)$$

Expressions for transfer functions can be found in (Aranda et al., 2001), and (Esteban et al., 2000).

### 2.2 *Design specifications.*

The aim will be to reduce motion sickness incidence (MSI) (O'Hanlon and McCauley, 1974):

$$MSI = 100\left[0.5 + erf\left(\frac{\log_{10}\left(\frac{|\ddot{s}_3|}{g}\right) - \mu_{MSI}}{0.4}\right)\right]$$

where:

- $|\ddot{s}_3|$ is the averaged vertical acceleration.
- $\mu_{MSI} = -0.819 + 2.32 \cdot (\log_{10} \omega_e)^2$.
- $\omega_e$ is the encounter frequency.
- And the $erf(.)$ function is defined as

$$erf(x) = \frac{1}{\sqrt{2\pi}} \int_0^x \exp\left(-\frac{z^2}{2}\right) dz$$

To reduce the MSI, a minimization of the ship vertical accelerations, due to pitch and heave movements caused by the swell, will be tried.

The following expression for vertical acceleration, measured 40 meters away from the gravity center of the ship, will be used:

$$acv40(t_i) = a_{VH}(t_i) + a_{VP}(t_i) = \frac{d^2 heave(t_i)}{dt^2} - 40 \cdot \frac{\pi}{180} \cdot \frac{d^2 pitch(t_i)}{dt^2}$$

From the previous expression, the mean acceleration for an experiment will be calculated. Considering N sampling instants $t_i$, the expression for the mean acceleration will be:

$$J = \frac{1}{N} \sum_{i=1}^{N} |acv40(t_i)|$$

This expression will be employed by the evaluation function of the evolutionary algorithm to obtain a suitability measurement for a specific controller.

## 3. CONTROLLER DESIGN

### 3.1 *Control scheme.*

The control scheme is showned in figure 3. As can be observed, the resulting model will have 3 inputs and 4 outputs. Ref_Flap and Ref_Tfoil are also taken as outputs, to penalize those controllers giving saturations in the actuators. The waves dynamics has also been separated from the rest of the process.

Reference signals (Ref_heave and Ref_pitch) will be zero, since the ideal thing would be that heave and pitch movements didn't occur. It's a typical regulation problem, and the inputs to the controller will be the heave and pitch values with their signs changed.

The T-foil position will be in the interval [-15°, 15°], while the flap position will be in the interval [0°, 15°]. To avoid excursions toward negative values of the flap, a trim value of 7.5° will be established. That is the function of the 'Trim' block showed in the scheme of figure 3.

The controller block is detailed in figure 4. It is made up of 4 subblocks, one per each input-output pair.

The controller blocks will be obtained by the algorithm as transfer functions, and will be included in the model. After that, a simulation will be carried out to check the controller behavior as solution to the problem.

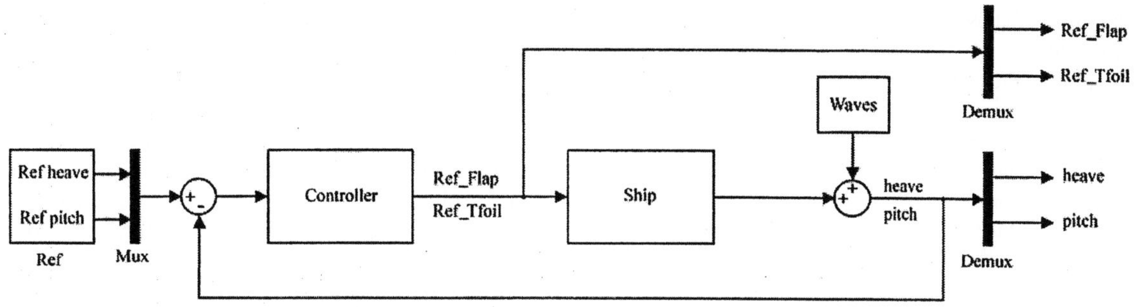


Fig. 3. Control scheme.

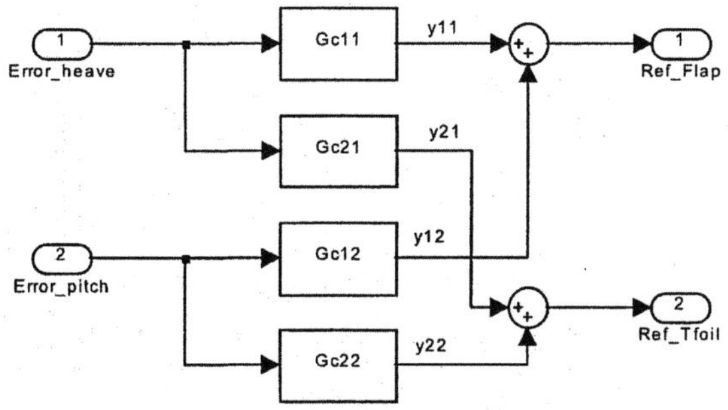


Fig. 4. Structure of the controller.

A simulation will be carried out for each controller obtained by the algorithm, and from this simulation the fitness for each chromosome or controller will be obtained.

### 3.2 *Loop shaping.*

As it was commented in the introduction, an algorithm inspired by a technique known with the name of loop shaping will be used to get the controller. By means of this technique a controller is looked for, G(s), such that certain design specifications are satisfied by the resulting open loop transfer function when placing this controller at the plant input. The controller will be formed by series connection with some of the following basic control subblocks:

1. Gain: $k$
2. Simple loop: $\dfrac{p}{s+p}$
3. Simple zero: $\dfrac{s+p}{p}$
4. Simple lead or lag: $\dfrac{s+a}{s+b}$
5. Second order pole: $\dfrac{\omega^2}{s^2+2\xi\omega s+\omega^2}$
6. Second order zero: $\dfrac{s^2+2\xi\omega s+\omega^2}{\omega^2}$
7. Notch: $\dfrac{s^2+2\xi_1\omega s+\omega^2}{s^2+2\xi_2\omega s+\omega^2}$

The procedure involve going adding these blocks to the controller transfer function, and to check its behavior by means of some technique.

In this work an optimum controller will be looked for, combining the commented basic blocks, but in this case the evolutionary algorithm will be the one that takes charge of checking the behavior by means of successive simulations.

### 3.3 *Controller synthesis algorithm.*

The algorithm used to get the controller will be made up of two loops: one of them external that, as has been commented, will take charge of obtaining the structure of the controller transfer function and another one internal that will carry out the tune of the controllers obtained by the external loop.

*Internal loop.* This loop will take charge of tuning the structures obtained by the external loop. Chromosomes will be coded as real number vectors, with as many elements as controller parameters need to be determined. The tournament selection technique and usual mutation and crossover operators for real number chromosomes will be used. The evolutionary algorithm will have the following steps:

1. The initial population is randomly chosen.
2. Chromosomes are decoded and evaluated by means of a computer-experiment simulation. The population is sorted according to fitness.
3. If the end conditions are given, the program concludes.
4. Tournament selection is used.
5. A new population is obtained from the selected chromosomes by means of mutation and crossover operators. In order to favor diversity and avoid premature convergence, a small number of immigrants are added (chromosomes randomly obtained).
6. Return to step 2.

This process will be repeated for each one of the controller structures determined by the external loop.

*External loop.* The external loop algorithm is built as a new layer or level over the internal loop just described. The objective of the external loop algorithm is to determine, by means of genetic techniques, the best control structures.

The controller will be formed by series connection with some of the basic control subblocks, previously showed in the loop shaping section.

Chromosomes will be made up of as many genes as different basic subblocks there is and each gene will be associated to one of these subblocks. Genes will take binary values, so that a 1 will indicate the presence of the corresponding subblock in the controller and a 0 will indicate its absence.

Every structure built from basic subblocks -external loop chromosomes- will be passed to the internal loop for evaluation. The internal loop will build the controller by decoding the external loop chromosome and will obtain the best values for the controller parameters. In the external loop algorithm:

1. The initial population is randomly obtained.
2. Chromosomes are evaluated by the internal loop. Population is arranged according to fitness.
3. If the end conditions are given, the program concludes.
4. Tournament selection is used.
5. A new population is obtained from the selected chromosomes by applying mutation and binary crossover operators. In order to favor diversity and avoid premature convergence, a small number of immigrants is added (chromosomes randomly obtained).
6. Return to step 2.

In short, the external loop will select the controller, while the internal loop will tune it.

### *3.3 The internal loop evaluation function.*

The internal loop evaluation function, will take charge of obtaining the fitness vector for each controller in the population. This vector will allow the population members to be sorted, so that they can be selected appropriately for reproduction.

The evaluation function will consider the following objectives:

- Stability of the collection: Plant + Controller.
- Saturations in the Flap: $0° \leq \text{value} \leq 15°$.
- Saturations in the Tfoil: $-15° \leq \text{value} \leq 15°$.
- Mean acceleration: J.

## 4. RESULTS

The program, once it had been designed, was executed for three different speed and sea state conditions. The obtained controllers were subsequently checked in an environment designed to evaluate, in an homogeneous way, different controllers obtained by different methods. The MSI reduction, regarding the ship without controller, obtained for the different conditions, are summarized in the table 1.

| Speed (Knots) | Sea State | Sickness index reduction |
|---|---|---|
| 20 | 4 | 12,7% |
| 40 | 4 | 46,4% |
| 40 | 5 | 13,6% |

Table 1: MSI reduction

Figure 5 shows the bode plot of Gc11, Gc21, Gc12 and Gc22, figure 6 shows the evolution of the total acceleration, and figure 7 shows the MSI with and without controller, for a range of encounter frequencies, also detailed for the experiment encounter frequency (2.3075 rad/s). Figure 8 shows the upper bound of the structured singular value, and it can be observed that the system has got robust stability.

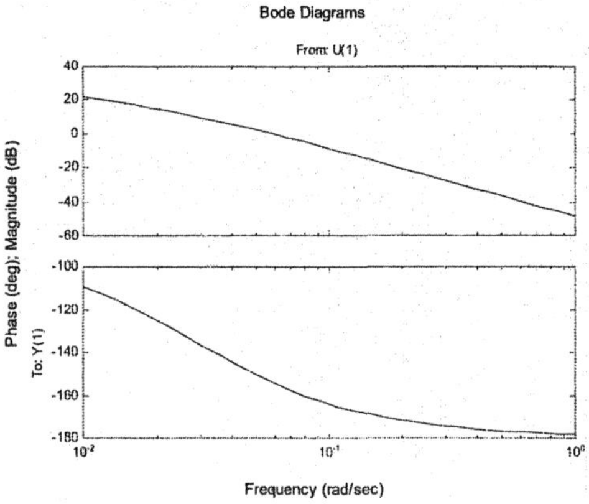


(a) Bode plot of Gc11

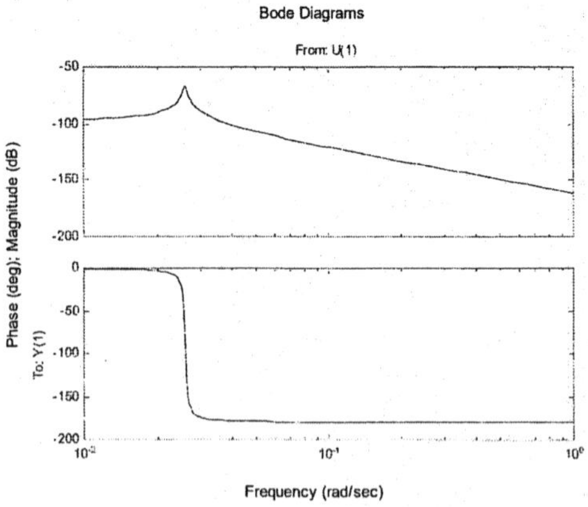


(b) Bode plot of Gc22

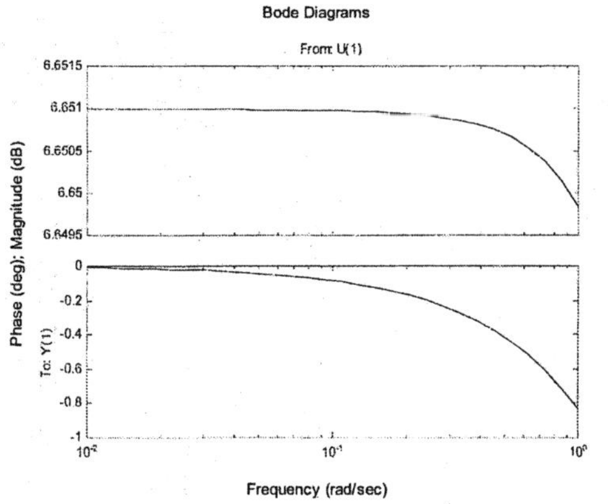


(c) Bode plot of Gc12

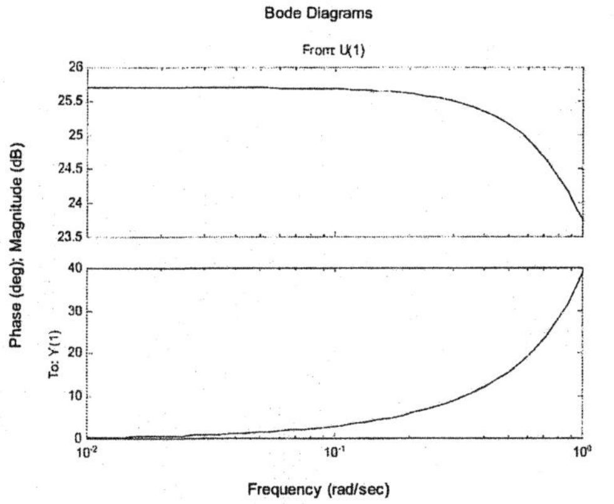


(d) Bode plot of Gc22

Fig. 5. Bode plots of controllers

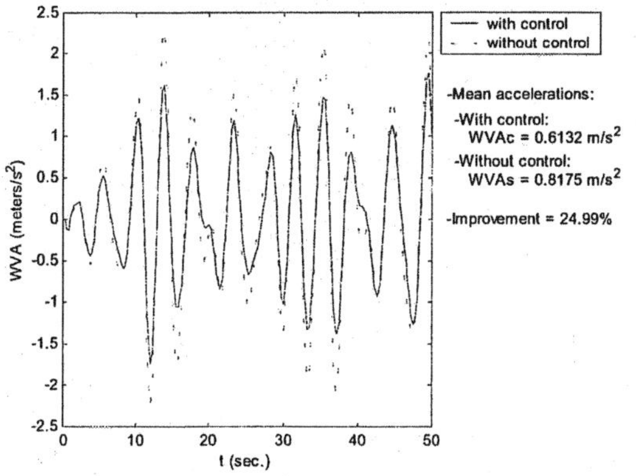


Fig. 6. Evolution of the total acceleration and mean accelerations calculation. An improvement of 24.99% is observed.

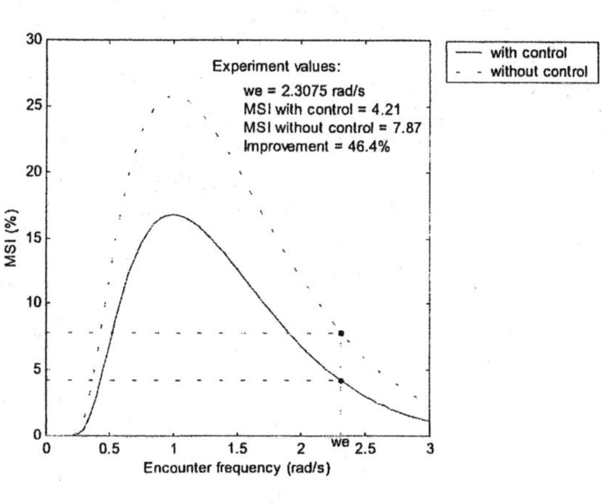


Fig. 7. Motion sickness incidence (MSI). An improvement of 46.4% in this index is observed.

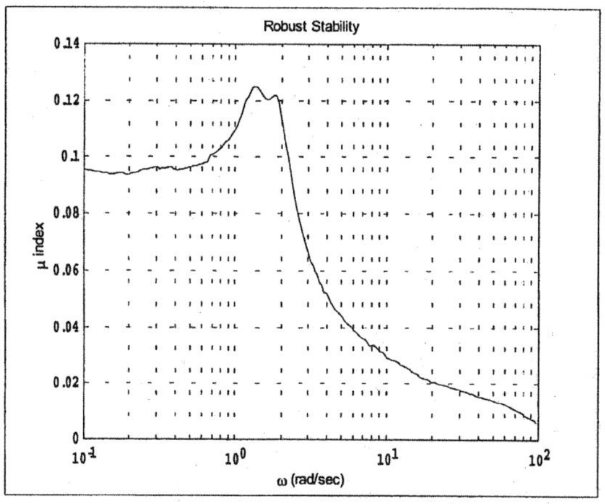


Fig. 8. μ-analysis for robust stability study

## 5. CONCLUSIONS

In this work, a step farther in the automation of the controllers design is taken. The designed algorithm has been able to select a control structure, to tune it and to satisfy the design objectives (appreciable reduction of the acceleration).

Evolutionary algorithms have been shown to be ideal for the resolution of these kinds of problems because they work simultaneously with a set of possible solutions, thereby favouring convergence towards a global optimum. In this paper we propose a way of dealing with the different objectives considered and an evolutionary algorithm that will enable some phases of the controller design to be automated.

Once the algorithm has been designed, it has been executed successively for different conditions of speed and sea state.

## 7. ACKNOWLEDGMENT

Part of this development was supported by MCyT of Spain under contract DPI2000-0386-C03-01.

The authors also would like to thank Rafael López from the Centro de Supercomputación de la Universidad Complutense de Madrid for his support and the use of the university's facilities.

## REFERENCES

Aranda, J.; de la Cruz, J. M.; Parrilla, M. y Ruipérez, P. (2000a). *Design of a Linear Quadratic Optimal Control for Aircraft Flight Control by Genetic Algorithm.* Controlo'2000: 4th Portuguese Conference on Automatic Control. Guimarães-Portugal. October-2000. Conference Proceedings.

Aranda, J.; de la Cruz, J. M.; Parrilla, M. y Ruipérez, P. (2000b). *Evolutionary Algorithms for the Design of a Multivariable Control for an Aircraft Flight Control.* AIAA Guidance, Navigation and Control Conference and Exhibit. Denver-Colorado, USA. August 2000. Conference Proceedings.

Aranda, J.; J.M. de la Cruz, J.M. Díaz, P. Ruipérez (2001). *Identification for robust control of a fast ferry* . 5th International Symposium on Quantitative Feedback Theory and Robust Frequency Domain Methods. Public University of Navarre, Pamplona, Spain. 23 - 24 August, 2001.

J.M. de la Cruz , A. Pérez de Lucas, J. Aranda , J.M. Girón-Sierra , F. Velasco, A. Marón (2001). *A research on motion smoothing of fast ferries.* IFAC Conference CAMS 2001 Control Aplications in Marine Systems. Glasgow.

S. Esteban, J.M. Girón-Sierra, J.M. de la Cruz, B. de Andres, J.M. Díaz, J. Aranda. *Fast Ferry Vertical Accelerations Reduction with Active Flaps and T-Foil.* 5th IFAC Conference on Manoeuvring and Control of Marine Crafts MCMC2000. Aalborg.

O'Hanlon, J.F. and McCauley, M.E. (1974). *Motion sickness incidence as a function of the frequency and acceleration of vertical sinusoidal motion.* Aeroespace Medicine.

Oded Yaniv. (1999) *Quantitative Feedback Design of Linear and Nonlinear Control Systems.* Kluwer Academic Publishers. 1999.

ELSEVIER
IFAC
PUBLICATIONS
www.elsevier.com/locate/ifac

# NEW RESULTS ABOUT MODEL BASED STUDY OF SEASICKNESS IN FAST FERRIES

**J.M. Giron-Sierra, B. Andres-Toro, S. Esteban,
J.M. Riola, J. Recas, J.M. De la Cruz**

*Dep.. A.C.Y.A., Fac. CC. Fisicas. Universidad Complutense de Madrid
Ciudad Universitaria, 28040 Madrid. Spain
e-mail: gironsi@dacya.ucm.es*

Abstract: As part of a research on increasing the passengers comfort in fast ferries by using moving actuators, some study was devoted to the application of control-oriented models to predict seasickness. The key idea is to apply an analogy of three filters. In a previous paper, some criteria were presented concerning the captain, on how to avoid navigation parameters inducing seasickness, and also criteria were given for the ship designer. In this paper new results are presented, linking two seasickness indexes, and offering a frequency domain method to estimate the comfort of a ship from the beginning of the ship design. *Copyright © 2004 IFAC*

Keywords: Ship control, Seakeeping, Human factors.

## 1. INTRODUCTION

Oscillating vertical motions of certain frequencies cause seasickness. This is an important aspect that should be considered in the design and operation of fast ferries and other passenger transportation systems. That is why our research on increasing the comfort of fast ferries by using moving actuators, such flaps, fins and T-foil, pays attention to the seasickness issue (Aranda, et al., 2001; Esteban, et al., 2000, 2001, 2002; Giron-Sierra, et al., 2001, 2002, 2003). In particular a recent contribution (Giron-Sierra, et al., 2003) is devoted to the filter analogy which is the basis of this paper. The research considers ship motions with head seas. A pertinent reference about ship control is (Fossen, 2002).

The motion sickness incidence index (MSI), proposed by (O'Hanlon and MacCawley, 1974), gives the percent of passengers that becomes sick as a cumulative consequence of suffering oscillating vertical accelerations. The index can be calculated from a record of vertical accelerations along time. We used this index within a ship motion simulation environment, to calculate the MSI after simulated two hour journeys. In parallel we used this index in the experimental studies with a towed model of the fast ferry.

Since we have control-oriented models of ship motions, it would be interesting to be able to derive information about seasickness impact from these models. Before a ship was built, we could know at the design stage how the ship will tend to cause seasickness, and try to improve comfort within the design constraints.

There are other ways to model seasickness impact, such the MSDV (see Lloyd, 1998). This weighted r.m.s. measure can be obtained from frequency domain plots of waves excitation, ship motions and a seasickness model in the form of a pass-band filter. This point of view suggest the approach taken in this paper. A new measure will be introduced, not based in r.m.s. values, but based in areas of frequency domain response plots. The new measure will be denoted as seasickness impact area (SIA).

Using our data, from simulations and experiments, we established a new result: a relationship between MSI and SIA. This result opens a way to calculate the MSI in the frequency domain with control-oriented models.

The paper starts with a synthesis of our previous work, considering an analogy of three chained filters. New graphical representations of this part of the study are presented. Then, the paper focus on the new result: the relationship between MSI and SIA.

## 2. WAVES, SHIP MOTION, AND SEASICKNESS

If the ship vertical motions have a frequency around 1 rad/sec., the corresponding vertical accelerations cause seasickness. We can study to which extent this circumstance occurs by considering a chain of three filters, as suggested by several authors (for instance Lloyd, 1998; Lewis, 1989). In general, the issue is a matter of frequency bands. Figure 1 shows an abstract example, with two frequency bands. There is an excitation signal, with a specific frequency band (for instance, the left frequency band in the figure). In addition, there is a system that will respond to excitations inside another frequency band (for instance, the right band in the figure). If both bands are separated (top of figure), the system will not respond to excitation. If there is some intersection, there will be some response. If both match (more or less), the system will be largely excited. The application to our case is that the excitation frequency band corresponds to encountered waves, and the response frequency band describes seasickness impact on passengers. The research of O'Hanlon and MacCawley (1974) found out that oscillating vertical accelerations with frequency around 1 rad./sec. cause seasickness.

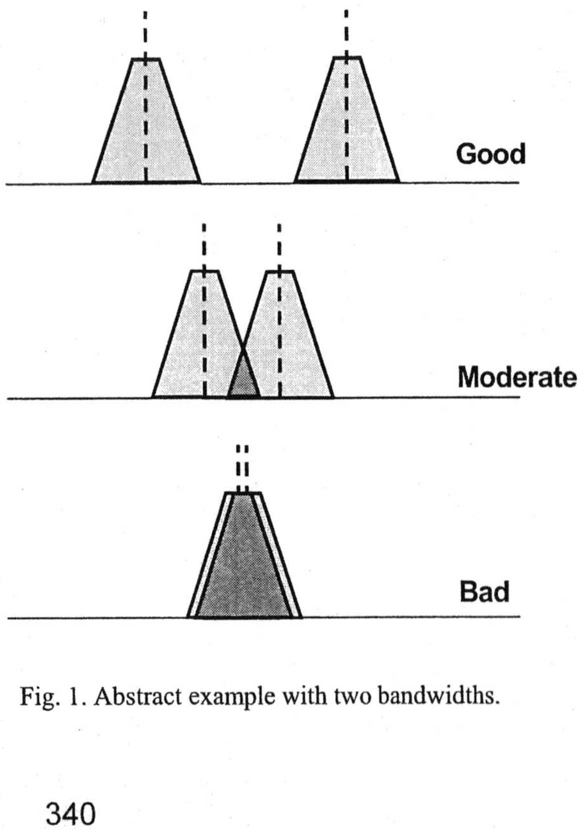


Fig. 1. Abstract example with two bandwidths.

The analogy, for this study, can be expressed as a chain of three band-pass filters (figure 2). Each filter has a bandwidth, around certain frequency values. As will be noticed when coming to specific shapes of the filter bands, these frequency values correspond to peaks of the bands.

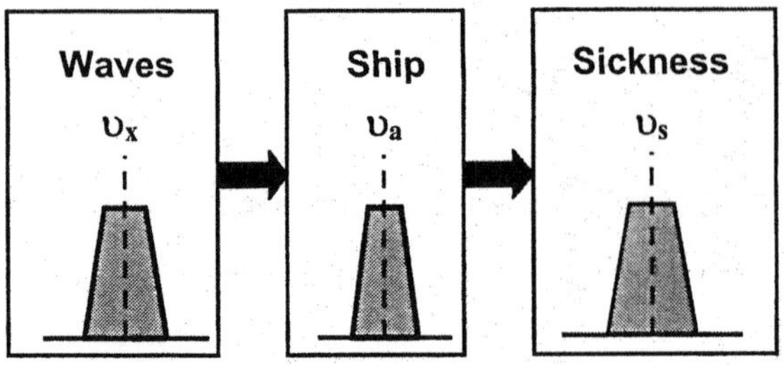


Fig. 2. The analogy: A chain of filters.

Let us consider each filter in particular.

*2.1. First filter: ocean waves, sea spectra.*

Expressions for power spectra of ocean waves can be obtained from literature (Fossen, 1994). For instance the Bretschneider spectrum.

$$S_{MPM}(\omega) = \frac{4\cdot\pi^3\cdot H_s^2}{\omega^5\cdot T^4}\cdot\exp\left(\frac{-16\cdot\pi^3}{\omega^4\cdot T^4}\right) \qquad (m^2\cdot s)$$

$$T = T_p / 1.408$$

As it is well-known, sea spectra suffer changes when seen by a moving ship (Lloyd, 1998). For instance, let us take the peak of a spectrum, let us denote $w_{op}$ the frequency of this peak. The waves with this frequency will be seen by a moving ship with another frequency $w_e$, equation (3), and the frequency of peak of the spectrum is now shifted.

$$\omega_e = \omega - \frac{\omega^2\cdot U}{g}\cdot\cos(\mu)$$

The shape of the spectra also change:

$$S(\omega_e) = \frac{S(\omega)}{1 - \dfrac{2\cdot\omega\cdot U\cdot\cos(\mu)}{g}}$$

It is a matter of frequency of encounter with waves, which depends on $U$, the speed of the ship, and μ, the heading angle (let us take 180° for heading seas). Figure 3 shows, for heading waves (μ=180°) the shifting of the Bretschneider SSN4, SSN5 and SSN6 spectra, caused by several ship's speeds.

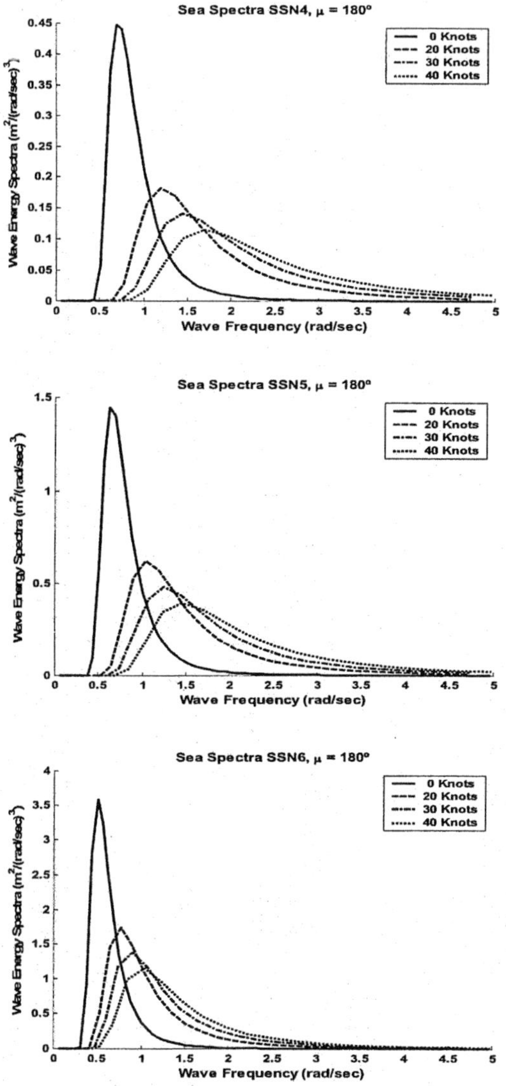


Fig. 3. Sea Spectra at several ship's speeds ($\mu$=180°)

If we take the square root of these spectra, and represent them as frequency responses, transfer functions can be obtained fitting these responses. The square root of the spectra will be used as input, since it represents to the amplitude distribution of waves.

*2.2. Second filter: ship motions, WVA.*

In a previous research, models (transfer functions) of heave and pitch motions of the fast ferry have been obtained for head seas (De Andres, et al., 2000). The models can be combined to determine the worst vertical motion (WVM) that a passenger can experiment. This WVM is located in a place near the bow. From the model of WVM we can obtain a model of the worst vertical acceleration (WVA) that a passenger can suffer. This is made according to the following equations:

$$WVM = A\cdot\sin(\omega\cdot t)\cdot dt$$

$$WVA = \int\int WVM\cdot dt = -A\cdot\omega^2\cdot\sin(\omega\cdot t)\cdot dt$$

Hence, the gain of the WVA transfer function can be obtained by multiplying the transfer function of WVM by the square of the frequency of encounter.

In (Giron-Sierra, et al., 2003) some figures of the response of ship motions and accelerations to ocean waves were presented. It is interesting to reproduce them here. Figure 4 shows the frequency response of WVM with an input of SSN5 waves, at ship's speeds of 20, 30 and 40 knots. Figure 5 do the same for WVA. These figures result from the chaining of the first and second filters.

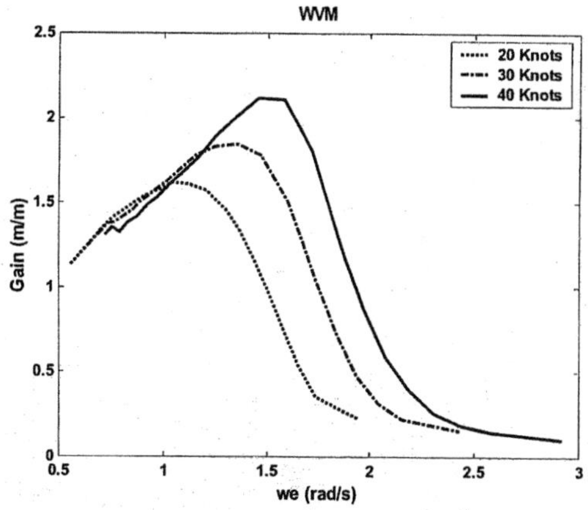


Fig. 4. WVM frequency response for SSN5&$\mu$=180°.

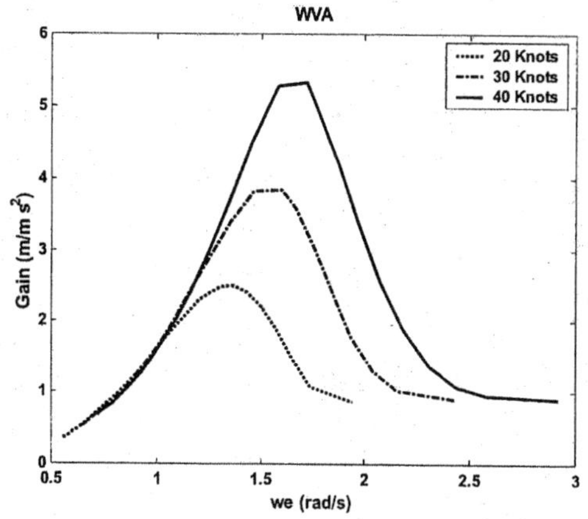


Fig. 5. WVA frequency response for SSN5&$\mu$=180°.

*2.2. Third filter: seasickness.*

There are several sources of scientific and technical information concerning seasickness. For instance the already cited study of O'Hanlon and MacCawley (1974), and the standards ISO 2631 or the British BS 6841 on whole-body vibrations. See also Lloyd (1998).

In the studies of O'Hanlon and MacCawley (1974) a mathematical model of "Motion Sickness Incidence" (MSI, an index defined as the percentage of subjects who vomited after two hours of motions) was found, as expressed in the following equation:

$$MSI = 100 \cdot \left[ 0.5 \pm erf\left( \frac{\pm \log_{10}(|a_z|/g) \mp \mu_{MSI}}{0.4} \right) \right]$$

where *erf* is the error function, $|a_z|$ is the mean vertical acceleration in a chosen place, and $\mu_{MSI}$ is:

$$\mu_{MSI} = -0.819 + 2.32(\log_{10}\omega_e)^2$$

where $\omega_e$ is the dominant frequency (rad/sec) of encounter with waves. Figure 6 shows how the MSI depends on $|a_z|$ and the frequency of encounter with waves.

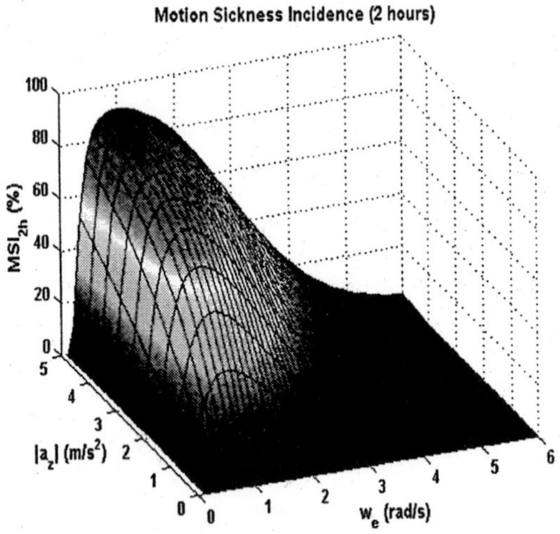


Fig. 6. The MSI in function of vertical acceleration and frequency of encounter with waves.

Another way of considering seasickness is the following. For short cruises (some hours), the equation below describes the Motion Sickness Dose Value ($MSDV_Z$ in $m/s^{3/2}$), (Anonymous, 2001) ABS guide for passenger comfort on ships, a cumulative measure of exposure to low-frequency that may be used to provide an indication of the probable incidence of motion sickness.

$$MSDV_Z = \sqrt{\int_0^T a_{zw}^2 \cdot dt}$$

where $a_{zw}^2$ is the z-axis acceleration weighted by the $W_f$ frequency weighting as defined in BS 6841:1987 and ISO 8041:1990, and $T$ is the duration of the motion.

The definition of $W_f$ is,

$$W_f(\omega) = (\omega/0.7)^{1.65} \qquad \omega < 0.7\, rad/s$$

$$W_f(\omega) = 1 \qquad 0.7 < \omega < 1.7\, rad/s$$

$$W_f(\omega) = (1.7/\omega)^{2.85} \qquad \omega > 1.7\, rad/s$$

Figure 7 shows the corresponding seasickness "filter".

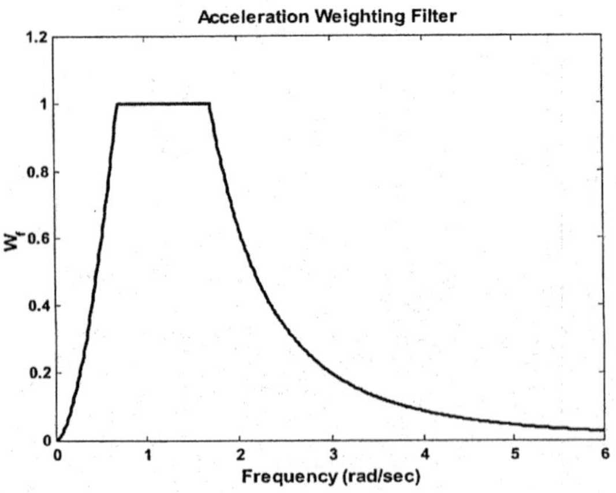


Fig. 7. Seasickness "filter" frequency response.

The curve in figure 7 can be fitted with a transfer function. The passenger can be seen as a band-pass filter: only oscillating vertical accelerations with frequency inside the band-pass cause sickness.

The MSDV can be obtained using the frequency domain expressions (Perez, et al., 2000):

$$A_{MSDV} = \sqrt{\int_0^\infty a_z^2(\omega) \cdot W_f^2(\omega) d\omega}$$

$$MSDV = p \cdot A_{MSDV} \cdot \sqrt{T}$$

These authors propose to use a coefficient $p$ that takes values:

| | |
|---|---|
| Normal activity: | 1 |
| Sat: | [0.6,1] |
| Eating: | [1.5,2.5] |
| Laid on cabin: | [0.2,0.6] |
| Outside activity: | [0.5,1] |

We define a similar frequency domain index, the seasickness impact area (SIA), which is only function of the distribution of the weighted vertical acceleration amplitude.

$$SIA = \int_0^\infty A_z(\omega) \cdot W_f(\omega) d\omega$$

It will be show that this index may be linked to the MSI by fitting an exponential function.

$$MSI(\%) = a \cdot (1 - e^{b \cdot (SIA - c)})$$

## 3. ALL THREE FILTERS

Given a sea state, and a ship's speed, the three filters of figure 2 can be used to calculate the frequency distribution of the signal that crosses the final filter, causing seasickness. This distribution can be squared, to represent its corresponding power spectrum.

Figure 8 shows the frequency response of each filter and the chain of three filters, for SSN4, with head seas, at 20, 30 and 40 knots. Figure 9 do the same for SSN5, and figure 10 the same for SSN6.

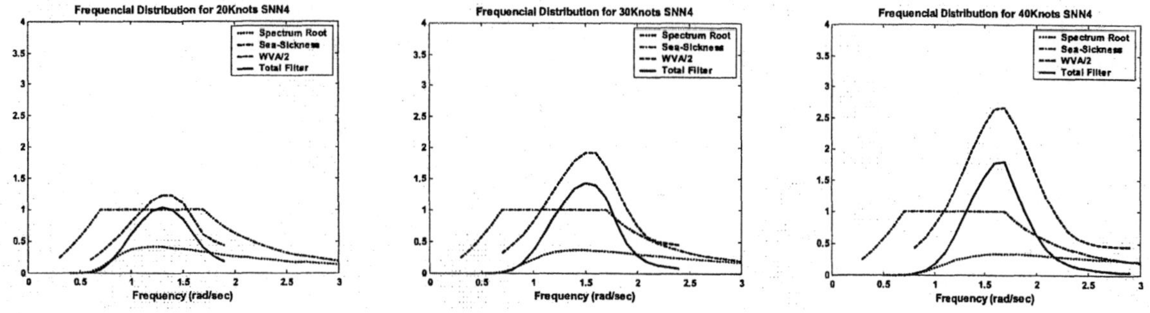


Fig. 8. The three filters at 20 , 30 and 40 knots, for SSN4.

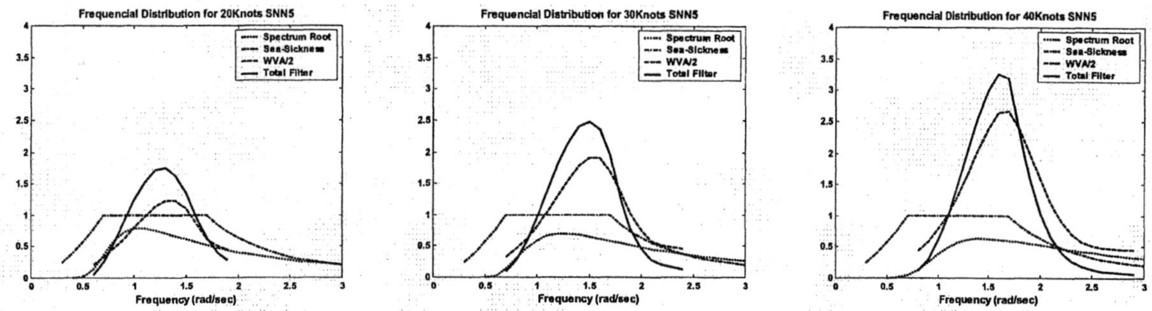


Fig. 9. The three filters at 20 , 30 and 40 knots, for SSN5.

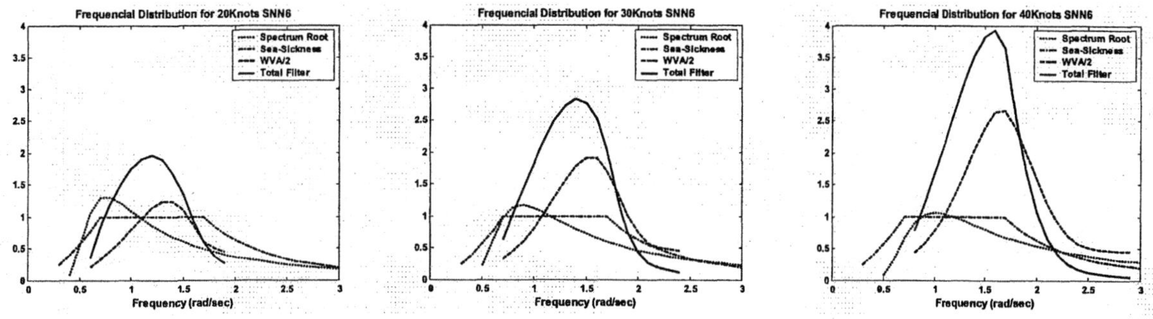


Fig. 10. The three filters at 20 , 30 and 40 knots, for SSN6.

## 4. DERIVING A NEW RESULT

Once the response of the chain of three filters has been obtained, we can compute the area under the response curves (the SIA), in the frequency domain plots. Figure 11 shows for example the curves and the areas for SSN5. The total area computation refers to nine cases, for three speeds and three sea states.

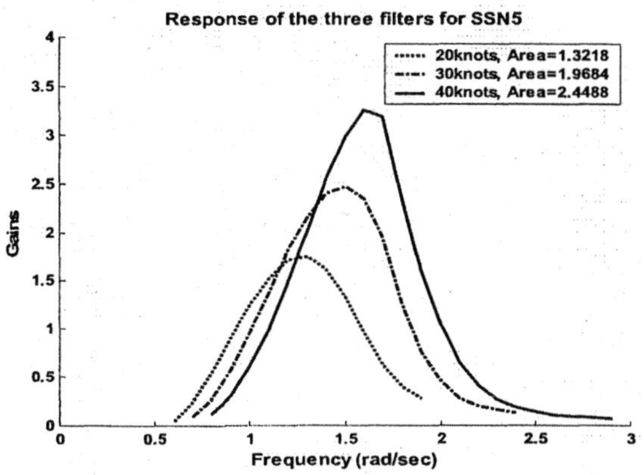


Fig. 11. Total responses and areas for SSN5 .

We have performed experimental studies with a physical model of the fast ferry with which this research is concerned. These experiments have been done in El Canal de Experiencias Hidrodinámicas de El Pardo (CEHIPAR), Madrid. As a result we keep several records of experimental data about ship motions and accelerations for SSN4, SS5 and SSN6, and ship's speeds of 20, 30 and 40 knots. That is, experimental time domain data corresponding to the same nine cases studied in the frequency domain. With these records we can calculate the MSI for the nine cases. Figure 12 shows a segment of three of the experimental records, corresponding to 20, 30 and 40 knots and SSN5.

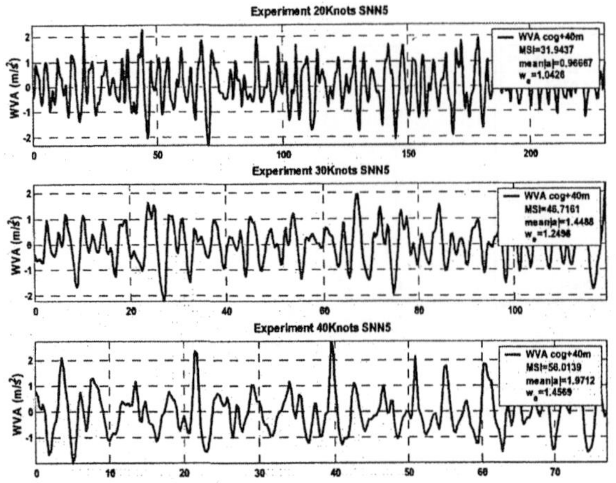


Fig. 12. Three experimental data records at SSN5.

The new result is established by plotting the MSI and the total response areas (SIA), and fitting by a simple expression. The equation of the new result is:

$$MSI = 100 \cdot e^{-0.42\,(SIA-0.5)}$$

Figure 13 shows the plot which is the basis for the expression.

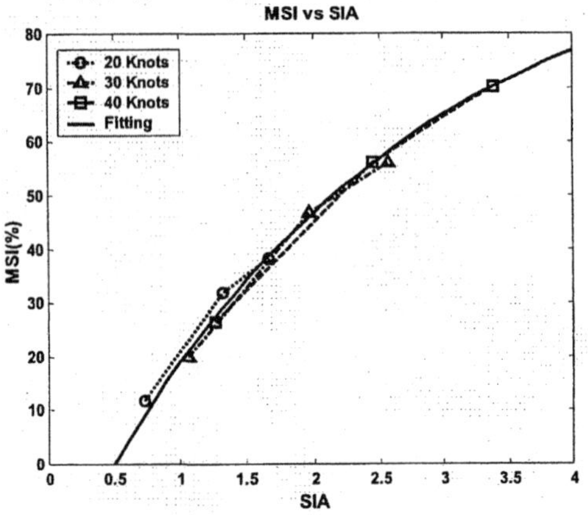


Fig. 13. MSI versus SIA plot for the nine cases.

## 5. CONCLUSIONS

In this paper a new result has been obtained, paving the way to calculate MSI from control oriented models. The part of these models concerning the ship, can be obtained from computer simulations, such PRECAL, before the ship was built.

It is hoped that this paper could yield some operational and design keys for a more comfortable transport of passengers.

## REFERENCES


Anonymus (2001). Guide for passenger comfort on Ships. *American Bureau of Shipping*.

Aranda, J., J.M. Diaz, P. Ruiperez, T.M. Rueda and E. Lopez (2001). Decreasing of the motion sickness incidence by a multivariable classic control for a high speed ferry, In *Proceedings IFAC Intl. Conf. Control Applications in Marine Systems CAMS2001*, Glasgow.

De Andres Toro, B., S. Esteban, J.M. Giron-Sierra and J.M. De la Cruz (2000). Modelling the motions of a fast ferry with the help of genetic algorithms, *In Proceedings Third IMACS MATMOD*, Vienna, 783-786.

Esteban, S., J.M. De la Cruz, J.M. Giron-Sierra, B. De Andres, J.M. Diaz and J. Aranda (2000). Fast ferry vertical acceleration reduction with active flaps and T-foil, In *Proceedings IFAC Intl. Symposium Maneuvering and Control of Marine Craft MCMC2000*, Aalborg, 233-238.

Esteban, S., B. De Andres, J.M. Giron-Sierra, O.R. Polo and E. Moyano (2001). A simulation tool for a fast ferry control design, In *Proceedings IFAC Intl. Conf. Control Applications in Marine Systems CAMS2001*, Glasgow.

Esteban, S., B. Andres-Toro, E. Besada-Portas, J.M. Giron-Sierra and J.M. De la Cruz (2002). Multiobjective control of flaps and T-foil in high speed ships, In *Proceedings IFAC 2002 World Congress*, Barcelona.

Ewing, J.A. and G.J. Goodrich (1967).The influence on ship motions of different wave spectra and the ship length, *Trans. RINA*, **109**, 47-63.

Fossen, T.I. (2002). *Marine Control Systems,* Marine Cybernetics AS, Trondheim..

Giron-Sierra, J.M., S. Esteban, B. De Andres, J.M. Diaz and J.M. Riola (2001). Experimental study of controlled flaps and T-foil for comfort improvement of a fast ferry, In *Proceedings IFAC Intl. Conf. Control Applications in Marine Systems CAMS2001*, Glasgow.

Giron-Sierra, J.M., R. Katebi, J.M. De la Cruz, S. Esteban (2002). The control of specific actuators for fast ferry vertical motions camping, In *Proceedings IEEE Intl. Conf. CCA/CACSD*, Glasgow.

Giron-Sierra, J.M., B. Andres-Toro, S. Esteban, J. Recas, E. Besada and J.M. De la Cruz (2003). Model based analysis of seasickness in a fast ferry. In *Proceedings IFAC MCMC 2003, Gerona, Spain..*

Lloyd, A.R.J.M. (1998). *Seakeeping: Ship Behaviour in Rough Weather*, A.R.M.J. Lloyd, Gosport, Hampshire, U.K.

O'Hanlon, J.F. and M.E. MacCawley (1974). Motion sickness incidence as a function of frequency and acceleration of vertical sinusoidal motion, *Aerospace Medicine.*

Perez, F., Lopez, A., and J.A. Felgueroso (2000). Movements in RO-PAX Ships. Effects on Security and Comfort, In *International Conference on Ship and Shipping Research*, Venecia.

ELSEVIER
IFAC
PUBLICATIONS
www.elsevier.com/locate/ifac

# IDENTIFICATION OF UNDERWATER VEHICLE HYDRODYNAMIC COEFFICIENTS USING FREE DECAY TESTS

**Andrew Ross * Thor I. Fossen *,** Tor Arne Johansen ****

** Centre for Ships and Ocean Structures*
*Norwegian University of Science and Technolog*
*NO-7491 Trondheim, Norway*
*** Department of Engineering Cybernetics*
*Norwegian University of Science and Technology*
*NO-7491 Trondheim, Norway*
*E-mail: andrew.ross@marin.ntnu.no, tif@itk.ntnu.no,*
*tor.arne.johansen@itk.ntnu.no*

Abstract: A new method is proposed for the experimental determination of the longitudinal and lateral hydrodynamic coefficients of a low-speed UUV. The technique presented is a development of the classical free-decay test. A body is excited with a mechanism of springs, and system identification techniques are carried out on measured data, with the intention of evaluating the added mass and linear damping of the body in decoupled longitudinal and lateral models. Simulated results are presented in order to estimate the potential accuracy of these new methods. *Copyright © 2004 IFAC.*

Keywords: Low-speed underwater vehicles, system identification, free decay test, digital differentiators

## 1. INTRODUCTION

In recent years there has been an ever increasing number of applications for unmanned underwater vehicles (UUV) in various tasks, for instance in surveying and exploration, or in missions such as the positioning of underwater laboratories (Aguiar 1997). Improvements in the evaluation of the hydrodynamic models of UUV's result in more effective control system implementation, leading to increased capability and performance in underwater operations. The efficient identification of hydrodynamic coefficients is a task which is difficult and oftentimes expensive to carry out, with many examples of how to measure or estimate them. For example, the use of towing tanks in a marine laboratory (Aage 1994) is well-established, as is the hydrodynamic modelling of underwater vehicles in computational fluid dynamics programs such as WAMIT. System identification techniques have found valuable application, for instance in Smallwood and Whitcomb (2003), Caccia *et al.* (2000), A.T Morrison III (1993), Blanke (1997).

Previous work has generally been limited by only identifying parameters in a single degree-of-freedom (DOF). For a full treatise on the classical free decay test see Faltinsen (1990). This paper advances in the area of identification by proceeding to apply techniques of system identification to a multiple-DOF model. The experiments under investigation are longitudinal free decay tests, but the results are valid for the lateral mode as well. Under investigation is whether, and to what accuracy, various hydrodynamic parameters might be estimated. Digital signal processing is applied to generate the body velocities and accelerations from only position, and parameters are estimated using linear regression.

## 2. UNDERWATER VEHICLE MODEL

### 2.1 Dynamics

This section describes the underwater vehicle dynamics, i.e. the kinematic and kinetic equations of motion.

*2.1.1. Kinematics* The kinematic model used is that of (Fossen 2002):

$$\dot{\eta} = J(\eta)\nu \tag{1}$$

Where $\eta = [x, y, z, \phi, \theta, \psi]^\top$ is the vehicle's generalised position in an inertial frame, $\nu = [u, v, w, p, q, r]^\top$ is the UUV's generalised velocity in the body frame, and $J(\eta) \in R^{6\times6}$ is the velocity transformation matrix from the body to the inertial frame.

*2.1.2. Kinetics* The dynamic equations of motion can be represented as a high-speed model for maneuvering or a low-speed model for station-keeping and low-speed maneuvering as detailed in Fossen (2002).

**High-speed model:** The nonlinear high-speed model is written:

$$(M_{RB} + M_A)\dot{\nu} + (C_{RB}(\nu) + C_A(\nu))\nu + D(\nu)\nu + g(\eta) = \tau \tag{2}$$

where $M_{RB} \in R^{6\times6}$ and $M_A \in R^{6\times6}$ are system inertia matrices for the rigid body and hydrodynamic added mass, respectively, $\mathbf{C}_{RB} \in \mathbf{R}^{6\times6}$ and $C_A \in R^{6\times6}$ are the Coriolis-centripetal terms corresponding to these, $D(\nu) \in R^{6\times6}$ is a nonlinear damping matrix, $g(\eta) \in R^{6\times1}$ is a vector of generalised gravity and buoyancy forces, and $\tau \in R^6$ is the generalised force applied

**Low-speed model:** To derive the low-speed model, it is assumed that the Coriolis-centripetal forces and non-linear damping are negligible, giving:

$$C(\nu)\nu \approx 0 \tag{3}$$

$$D(\nu)\nu \approx N\nu \tag{4}$$

where $N \in R^{6\times6}$ is a matrix of linear damping coefficients. Consequently (2) takes the form:

$$M\dot{\nu} + N\nu + g(\eta) = \tau \tag{5}$$

where

$$M = M_{RB} + M_A \tag{6}$$

**Spring forces due to the attachment device:** In the free decay experiments, it is assumed that the vehicle is attached to its surroundings using linear springs (see Figure 2) described by:

$$\tau = -K\eta \tag{7}$$

where $K \in R^{6\times6}$ is the spring stiffness matrix and $\tau \in R^6$ is the generalised force applied. The control forces (thrust) are set to zero in the experiments.

### 2.2 Decoupling into Lateral and Longitudinal Modes

The 6 DOF equations of motion can in many cases be divided into two non-interacting (or lightly interacting) subsystems. This decomposition is good for starboard-port symmetrical slender bodies, that is, bodies with large length/width ratios (Gertler and Hagen 1967, Feldman 1979, Tinker 1982).

- **Longitudinal subsystem:** states $u, w, q, x, z, \theta$
- **Lateral subsystem:** states $v, p, r, y, \phi, \psi$

The system inertia matrix can then be partitioned according to (Fossen 2002):

$$M_{\text{lon}} = \begin{bmatrix} m_{11} & m_{13} & m_{15} \\ m_{31} & m_{33} & m_{35} \\ m_{51} & m_{53} & m_{55} \end{bmatrix} \quad M_{\text{lat}} = \begin{bmatrix} m_{22} & m_{24} & m_{26} \\ m_{42} & m_{44} & m_{46} \\ m_{62} & m_{64} & m_{66} \end{bmatrix}$$

*2.2.1. Longitudinal Subsystem* Without loss of generality, we will assume that the lateral states $v, p, r, \phi$ are small, the weight $W = mg$ is equal to the buoyancy force $B$, and that the center of gravity coincides with the center of buoyancy in the $x$-direction, i.e. $x_G = x_B$, etc. Then the kinematic model in *surge, heave*, and *pitch* can be expressed according to (Fossen 2002):

$$\dot{\eta}_{\text{lon}} = J_{\text{lon}}(\eta_{\text{lon}})\nu_{\text{lon}} \tag{8}$$

with:

$$J_{\text{lon}}(\eta_{\text{lon}}) = \begin{bmatrix} \cos\theta & \sin\theta & 0 \\ -\sin\theta & \cos\theta & 0 \\ 0 & 0 & 1 \end{bmatrix} \tag{9}$$

where $\eta_{\text{lon}} = [x, z, \theta]^\top$ and $\nu_{\text{lon}} = [u, w, q]^\top$. The kinetics takes the form:

$$M_{\text{lon}}\dot{\nu}_{\text{lon}} + N_{\text{lon}}\nu_{\text{lon}} + g(\eta_{\text{lon}}) = -K_{\text{lon}}\eta_{\text{lon}} \tag{10}$$

where

$$M_{\text{lon}} = \begin{bmatrix} m - X_{\dot{u}} & -X_{\dot{w}} & mz_g - X_{\dot{q}} \\ -X_{\dot{w}} & m - Z_{\dot{w}} & -mx_g - Z_{\dot{q}} \\ mz_g - X_{\dot{q}} & -mx_g - Z_{\dot{q}} & I_y - M_{\dot{q}} \end{bmatrix}$$

$$N_{\text{lon}} = \begin{bmatrix} -X_u & -X_w & -X_q \\ -Z_u & -Z_w & -Z_q \\ -M_u & -M_w & -M_q \end{bmatrix}$$

$$g(\eta_{\text{lon}}) = \begin{bmatrix} 0 \\ 0 \\ WBG_z \sin\theta \end{bmatrix}$$

*2.2.1.1. Longitudinal Spring Stiffness Matrix* At rest $\dot{\nu}_{\text{lon}} = \nu_{\text{lon}} = 0$ the spring forces $K_{\text{lon}}\eta_{\text{lon}}$ must balance out the gravitational and buoyancy forces term, $g_{\text{lon}}(\eta_{\text{lon}})$, that is:

$$g_{\text{lon}}(\eta_{\text{lon}}) + K_{\text{lon}}\eta_{\text{lon}} = 0 \tag{11}$$

For the experimental set-up depicted in Figure (2) where $k_1$ and $k_2$ are the stiffness values of the respective springs, the $K_{\text{lon}}$ matrix becomes:

$$K_{\text{lon}} = \begin{bmatrix} k_1 + k_2 & 0 & 0 \\ 0 & k_1 + k_2 & 0 \\ k_1 r_{z1} + k_2 r_{z2} & k_1 r_{x1} + k_2 r_{x2} & 0 \end{bmatrix} \tag{12}$$

where $(r_{1x}, r_{1z})$ and $(r_{2x}, r_{2z})$ are the locations of the spring attachments on the UUV relative to the centre of gravity.

*2.2.2. Lateral Subsystem* Assume that the lateral states longitudinal states $u, w, p, r, \phi$ and $\theta$, the weight $W = mg$ is equal to the buoyancy force $B$, and that the center of gravity coincides with the center of buoyancy in the $x$-direction, i.e. $x_G = x_B$, etc. Then the kinematic model in *sway, roll,* and *yaw* can be expressed according to (Fossen 2002):

$$\dot{\eta}_{\text{lat}} = \nu_{\text{lat}} \quad (13)$$

where $\eta_{\text{lat}} = [y, \phi, \psi]^\top$ and $\nu_{\text{lat}} = [v, p, r]^\top$. The kinetics takes the form:

$$M_{\text{lat}}\dot{\nu}_{\text{lat}} + N_{\text{lat}}\nu_{\text{lat}} + g_{\text{lat}}(\eta_{\text{lat}}) = -K_{\text{lat}}\eta_{\text{lat}} \quad (14)$$

where

$$M_{\text{lat}} = \begin{bmatrix} m - Y_{\dot{v}} & -mz_g - Y_{\dot{p}} & mx_g - Y_{\dot{r}} \\ -mz_g - Y_{\dot{p}} & I_x - K_{\dot{p}} & -I_{zx} - K_{\dot{r}} \\ mx_g - Y_{\dot{r}} & -I_{zx} - K_{\dot{r}} & I_z - N_{\dot{r}} \end{bmatrix}$$

$$N_{\text{lat}} = \begin{bmatrix} -Y_v & -Y_p & -Y_r \\ -M_v & -M_p & -M_r \\ -N_v & -N_p & -N_r \end{bmatrix}$$

$$g_{\text{lat}}(\eta_{\text{lat}}) = \begin{bmatrix} 0 \\ WBG_z \sin\phi \\ 0 \end{bmatrix}$$

## 3. FREE DECAY TEST

The MIMO free decay test is performed according to flow chart in Figure 1.

### *3.1 State Measurements*

For the experiments envisaged in this paper, an underwater camera system is to be used. This system works by identifying pre-determined markings on the UUV, and measures at 20 Hz. Using cameras offers measurements of only the generalised position, $\eta$, and therefore suitable methods must be applied in order to generate the estimates of the states $\nu$ and $\dot{\nu}$.

*3.1.1. Zero Phase Differentiation* In order to reduce system complexity and cost, filtering techniques are applied to generate the unknown derivatives. In this paper a differentiator filter is applied twice in order to generate the body velocities and accelerations according to the kinematics formula (1) we can write:

$$\eta_m = \eta + w \quad (15)$$
$$\hat{\nu} = J^{-1}(\eta_m)\Lambda(\eta_m) \quad (16)$$
$$\dot{\hat{\nu}} = \Lambda(\hat{\nu}) \quad (17)$$

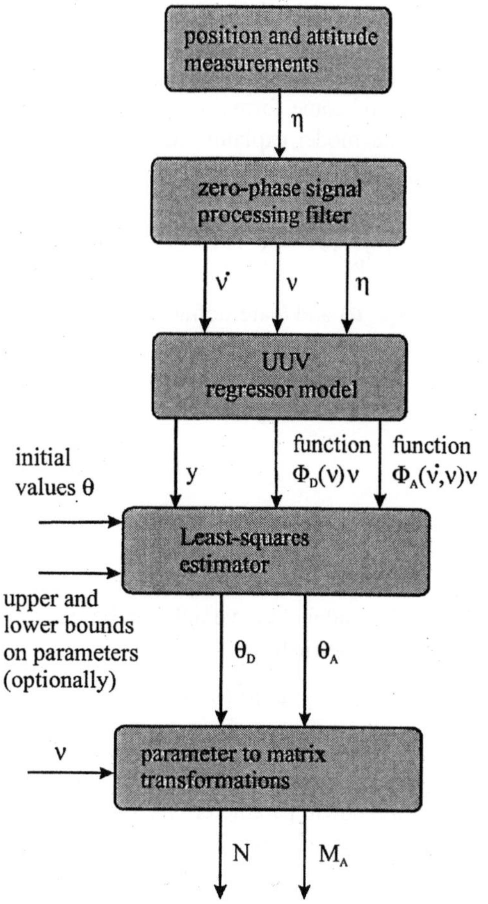


Fig. 1. Free Decay Flow Diagram

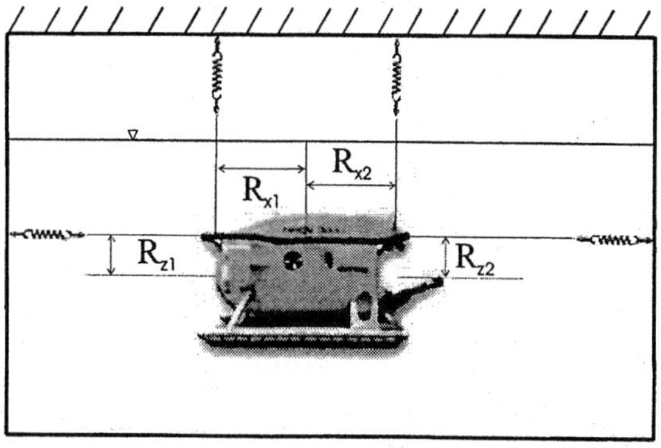


Fig. 2. Experimental set-up for longitudinal identification showing the UUV attached by 4 springs.

where $\Lambda(x)$ is a differentiation filter producing an estimate $\dot{x}$, and $w$ is an error term, assumed to be zero-mean white noise, which enters the system through the measurements $\eta_m$ of $\eta$. The differentiator is a *least squares band-limited FIR* filter of order of 301 (Oppenheim and Schafer 1989) and was generated using the Matlab program *FDATool* from the filter toolbox. Thus the filter's magnitude drops off quickly at 2Hz, and so also low-pass filters the outputs and does not differentiate high frequency noise. As the filtering is not required to be achieved in real-time, and the filter itself has linear phase properties, the phase lag is trivial to compensate for by simply time-shifting the output by $-150$ samples.

## 4. SYSTEM IDENTIFICATION

The task of system identification is essentially matching a model of some form to experimental data, in order that the model explains, in some fashion, the experimental data.

$$\min_{\theta} V = \frac{1}{2}\int_0^t (y - \Phi(\dot{\nu},\nu)\theta)^{\top} (y - \Phi(\dot{\nu},\nu)\theta)\, d\tau \quad (18)$$

where $y$, $\Phi(\dot{\nu},\nu)$ and $\theta$ are defined in their derivation at (19), and $V$ is a quadratic cost function of these variables. Since the problem is linear, the optimisation problem is solved using the standard least squares solution.

### 4.1 *Parametric Form*

The dynamic model of the vehicle can be transformed to a linear parametric form:

$$y = \Phi^T(\dot{\nu},\nu)\theta \quad (19)$$

in which $y$ is a vector consisting of measured or calculable data, $\Phi(\dot{\nu},\nu)$ is the regression matrix, and $\theta$ is the unknown parameter vector. Assume that the rigid-body system inertia matrix $M_{RB}$ and gravity/buoyancy vector $g(\eta)$ is known while added mass $M_A$ and damping $N$ are unknown. This is the usual case when modeling underwater vehicles. Consider the free decay test (low-speed) model:

$$(M_{RB} + M_A)\dot{\nu} + N\nu + g(\eta) = -K\eta \quad (20)$$

which can be written:

$$M_A\dot{\nu} + N\nu = y \quad (21)$$

where the signal $y$ is computed from the measurements and *known parameters* $K, M_{RB}$, and $g(\eta)$ according to:

$$y = -K\eta - M_{RB}\dot{\nu} - g(\eta) \quad (22)$$

The *regressor matrix* and *parameter vector* are obtained from:

$$\Phi_M^T(\dot{\nu},\nu)\theta_M := M_A\dot{\nu} \quad (23)$$

$$\Phi_N^T(\dot{\nu},\nu)\theta_N := N\nu \quad (24)$$

such that:

$$\Phi^T(\dot{\nu},\nu) = [\Phi_M^T(\dot{\nu}), \Phi_N^T(\nu)], \quad \theta = \begin{bmatrix} \theta_M \\ \theta_N \end{bmatrix} \quad (25)$$

### 4.2 *Case Study: Longitudinal Mode*

For the longitudinal mode, we get:

$$y_{\text{lon}} = -K_{\text{lon}}\eta_{\text{lon}} - M_{RB}\dot{\nu}_{\text{lon}} - g_{\text{lon}}(\eta_{\text{lon}}) \quad (26)$$

The regressor matrices take the form:

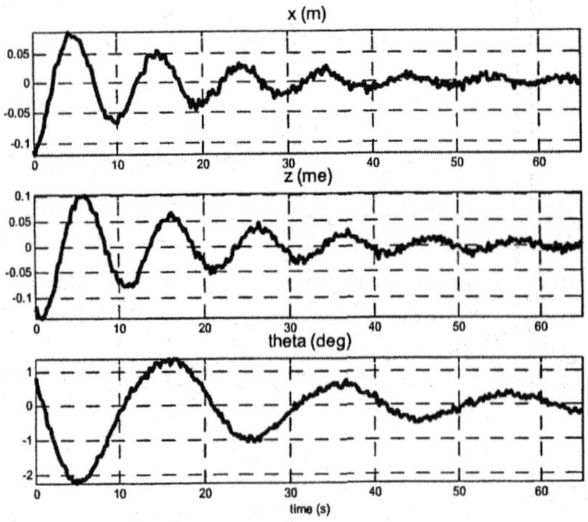


Fig. 3. Free Decay Test Showing Position $(x, z)$ and pitch angle $\theta$ versus time. The UUV is moved out of its equilibrium at $(0, 0)$ and released

$$\Phi_M^T(\dot{\nu}_{\text{lon}}) = -\begin{bmatrix} \dot{u} & \dot{w} & \dot{q} & 0 & 0 & 0 & 0 \\ 0 & \dot{u} & 0 & \dot{w} & \dot{q} & 0 & 0 \\ 0 & 0 & \dot{u} & 0 & 0 & \dot{w} & \dot{q} \end{bmatrix} \quad (27)$$

$$\Phi_N^T(\nu_{\text{lon}}) = -\begin{bmatrix} u & w & q & 0 & 0 & 0 & 0 & 0 & 0 \\ 0 & 0 & 0 & u & w & q & 0 & 0 & 0 \\ 0 & 0 & 0 & 0 & 0 & 0 & u & w & q \end{bmatrix} \quad (28)$$

and the corresponding parameter vectors are:

$$\theta_M = [X_{\dot{u}}, X_{\dot{w}}, X_{\dot{q}}, Z_{\dot{w}}, Z_{\dot{q}}, M_{\dot{q}}]^{\top} \quad (29)$$

$$\theta_N = [X_u, X_w, X_q, Z_u, Z_w, Z_q, M_u, M_w, M_q]^{\top} \quad (30)$$

where we have assumed that $M_A = M_A^{\top}$ while $N \neq N^{\top}$.

## 5. SIMULATIONS AND RESULTS

The longitudinal system was implemented in Simulink for an UUV given by the following parameters:

$$M_{RB} = \begin{bmatrix} 1000 & 0 & 200 \\ 0 & 1000 & 0 \\ 200 & 0 & 11000 \end{bmatrix} \quad (31)$$

$$M_A = \begin{bmatrix} 1000 & 0 & 100 \\ 0 & 1100 & 80 \\ 100 & 80 & 9000 \end{bmatrix} \quad (32)$$

$$N = \begin{bmatrix} 210 & 20 & 30 \\ 25 & 200 & 70 \\ 15 & 33 & 1500 \end{bmatrix} \quad (33)$$

The the spring coefficients were set to $k_1 = 300$, $k_2 = 500$.

Figure 6 shows a phase portrait of the test, with the body starting at $(0.5, 0.5)$ and spiraling inwards, and figure 3 shows the states during the same simulation. Figure 4 shows the generated states $w$ and $\dot{w}$, with close correlation, after a short transient, being evident. During identification, the transient data is discarded.

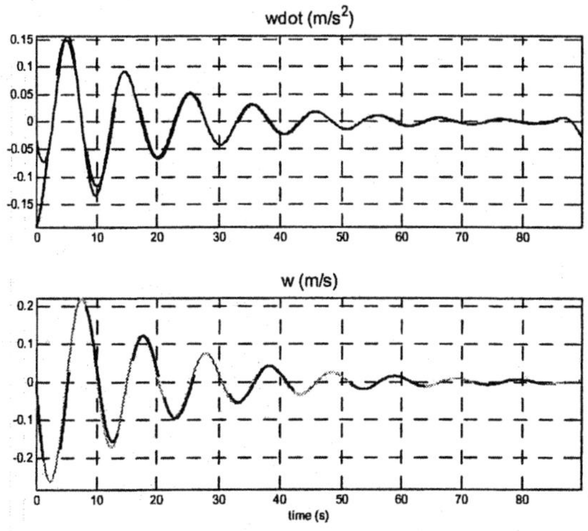


Fig. 4. Free decay test: plot showing the computed derivatives $w$ and $\dot{w}$ and their true values using the zero-phase differentiator filter.

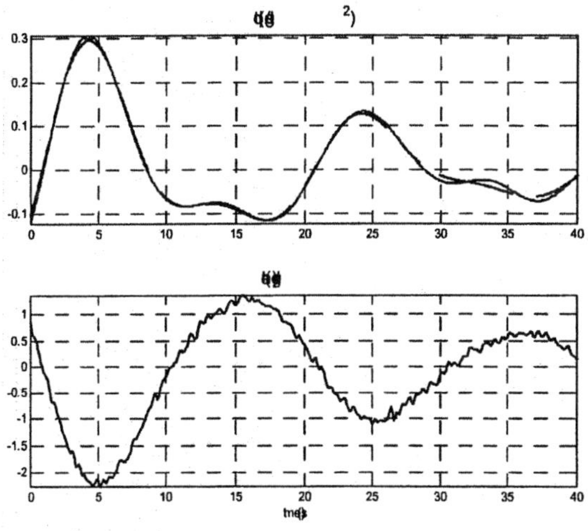

Fig. 5. Measured Pitch Angle ($\theta_m$) with $\dot{q}$ and its estimate $\hat{\dot{q}}$

By examining $\dot{q}$ in Figure 5, the closeness of $\hat{\dot{q}}$ to the actual state $\dot{q}$ is evident, and the lack of differentiated noise from $\theta$ is also clear. Firstly the system identification procedure was carried out using perfect state knowledge, that is without applying the filtering or adding noise, with the result that the parameter estimates corresponded exactly with the actual parameters. Carrying out the identification using the realistic state estimates based solely on the noisy measurements of $\eta$ led to the results shown in Table 1.

The estimates of the diagonal elements in $M_A$ and $N$ are generally very strong, especially that of $M_{\dot{q}}$ and $M_q$. Overall, the added mass matrix is very well modelled. The estimates of some off-diagonal elements in $N$ are to a lower quality, notably $Z_u$ and $Z_q$. Other off-diagonal elements are estimated extremely well, with $X_w$ and $X_q$ being particularly well evaluated. The added mass matrix is more accurately modelled than the damping matrix primarily due to prior knowledge of its structure, such as the assumption of symmetry.

| Hydrodynamic Derivative | Real | Estimate |
|---|---|---|
| $X_{\dot{u}}$ | 1000 | 928.5 |
| $X_{\dot{q}}/Z_{\dot{u}}$ | 200 | 213 |
| $Z_{\dot{w}}$ | 1100 | 1031 |
| $Z_{\dot{q}}/M_{\dot{w}}$ | 80 | 92.4 |
| $M_{\dot{q}}$ | 9000 | 8981 |
| $X_u$ | 210 | 234.9 |
| $X_w$ | 20 | 21.3 |
| $X_q$ | 30 | 32.7 |
| $Z_u$ | 25 | 49.5 |
| $Z_w$ | 200 | 208.7 |
| $Z_q$ | 70 | 41.5 |
| $M_u$ | 15 | 29.9 |
| $M_w$ | 33 | 29.5 |
| $M_q$ | 1500 | 1483.8 |

Table 1. Hydrodynamic derivatives and their estimates.

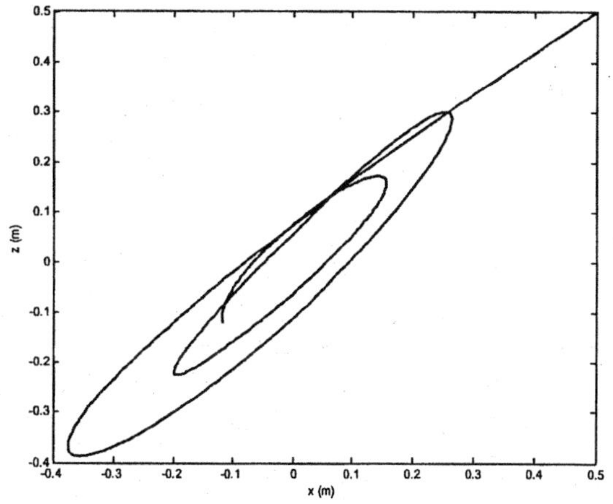


Fig. 6. Free decay test: side view showing the positions $(x, z)$. Notice that the vehicle is only moving approximately 0.50 m in each direction.

## 6. CONCLUSIONS

The new methods are presented, with simulated experimental results for the longitudinal model, demonstrating the combination of simulation with signal processing and system identification. The implications of the paper must be verified through actual experimentation to gain a clearer picture of the applicability and usefulness of the methods presented. The fact that the evaluation process converges to the correct parameters with full state knowledge implies identifiability, and so the problem is primarily one of achieving effective state measurement. The penalty of using signal processing method is that, although cheap, errors arise in state measurement, leading to the innacuracies noted in the previous section. The addition of an inertial measurement unit entirely reverses the advantages and disadvantages, in that this setup would be expensive but far more accurate. That stated, if future experiments match the simulations presented here, it can be expected that a system leading on from this will be both useful and versatile in the derivation of the hydrodynamic coefficients of low-speed UUV's.

## 7. ACKNOWLEDGEMENT

The authors are grateful to the Research Council of Norway for financial support through the Centre for Ships and Ocean Structures (CESOS) and the Strategic University Program on Computational Methods in Nonlinear Motion Control.

## REFERENCES

Aage, C., Wagner Smitt L. (1994). Hydrodynamics manoeuvrability data of a flatfish type AUV. In: *Oceans*. Vol. 3. pp. 425–430.

Aguiar, A., Pascoal A. (1997). Modelling and control of an autonomous underwater shuttle for the transport of benthic laboratories. In: *OCEANS*. Vol. 2. pp. 888–895.

A.T Morrison III, D.R. Yoerger (1993). Determination of the hydrodynamic parameters of an underwater vehicle during small scale, nonuniform, 1-dimensional translation. *OCEANS* **2**, 277–282.

Blanke, M., Tiano A (1997). Multivariable identification of ship steering and roll motions. *Transactions of the Institute of Measurement and Control* **19**(2), 62–77.

Caccia, M., G. Indiveri and G. Veruggio (2000). Modelling and identification of open-frame variable configuration unmanned underwater vehicles. *IEEE Journal of Oceanic Engineering* **25**(2), 227–240.

Faltinsen, O. M. (1990). *Sea Loads on Ships and Offshore Structures*. Cambridge University Press.

Feldman, J. (1979). DTMSRDC Revised Standard Submarine Equations of Motion. Technical Report DTNSRDC-SPD-0393-09. Naval Ship Research and Development Center. Washington D.C.

Fossen, T. I. (2002). *Marine Control Systems: Guidance, Navigation and Control of Ships, Rigs and Underwater Vehicles*. Marine Cybernetics AS. Trondheim, Norway. ISBN 82-92356-00-2.

Gertler, M. and G. R. Hagen (1967). Standard Equations of Motion for Submarine Simulation. Technical Report DTMB-2510. Naval Ship Research and Development Center. Washington D.C.

Oppenheim, A.V and R.W Schafer (1989). *Discrete-Time Signal Processing*. Prentice-Hall, Englewood Cliffs, NJ.

Smallwood, D.A and L.L Whitcomb (2003). Adaptive identification of dynamically positioned underwater robotic vehicles. *IEEE Transactions on Control Systems Technology* **11**(4), 505–515.

Tinker, S. J. (1982). Identification of Submarine Dynamics from Free-Model Test. In: *Proceedings of the DRG Seminar*. The Netherlands.

ELSEVIER
IFAC
PUBLICATIONS
www.elsevier.com/locate/ifac

# AUV DYNAMICS: MODELING AND PARAMETER ESTIMATION USING ANALYTICAL, SEMI-EMPIRICAL, AND CFD METHODS

**E. A. de Barros* [1] , A. Pascoal**, E. de Sa*****

** Department of Mechatronics Engineering and Mechanical Systems, University of São Paulo, São Paulo, SP. Brazil. Email: eabarros@usp.br*
*** Institute for Systems and Robotics (ISR) and Dept. Electrical Engineering and Computers, Instituto Superior Técnico, Lisbon, Portugal. Email: antonio@isr.ist.utl.pt*
**** National Institute of Oceanography, Dona Paula, Goa, India. Email: elgar@darya.nio.org*

Abstract: The paper addresses the problem of autonomous underwater vehicle (AUV) modeling and parameter estimation as a means to predict the expected dynamic performance of underwater vehicles and thus provide solid guidelines during their design phase. The use of analytical and semi-empirical (ASE) methods to predict the hydrodynamic derivatives of a large class of AUVs with conventional, streamlined bodies is discussed. An application is made to the estimation of the hydrodynamic derivatives of the MAYA AUV, an autonomous vehicle that is being developed under a joint Indian-Portuguese project. The estimates are used to predict the behavior of the vehicle in the vertical plane and to assess the impact of stern plane size on its expected performance. *Copyright © 2004 IFAC.*

Keywords: Autonomous Underwater Vehicles, AUV Modeling, Parameter Estimation, Computational Fluid Dynamics

## 1. INTRODUCTION

The paper addresses the problem of autonomous underwater vehicle (AUV) modeling and parameter estimation as a means to predict the expected dynamic performance of underwater vehicles and thus provide solid guidelines during their design phase, well before they can be tested at sea. The main core of the paper provides a roadmap to the use of analytical and semi-empirical methods to predict the hydrodynamic derivatives of a large class of AUVs with conventional, streamlined bodies An application is made to the estimation of a set of hydrodynamic derivatives for the MAYA AUV, an autonomous vehicle that is being developed under a joint Indian-Portuguese project, see Figure 1. These estimates are used to predict the behavior of the vehicle in the vertical plane and to assess the impact of stern plane size on its expected performance.

This work should be viewed as part of a long term research effort that aims to contribute to the development of computational methods for com-

[1] Work supported in part by the Portuguese FCT POSI Programme under framework QCA III and by project MAYA of the AdI. The work of the first author was supported in part by the Brazilian Ministry of Education through a "CAPES" scholarship.

bined plant/controller optimization (PCO), that is, for the combined design of AUVs and respective controllers to achieve increased performance at sea while meeting stringent energy requirements. See for example Silvestre *et al.* (1998), and the references therein for an introduction to combined PCO methods in the field of marine robotics. Central to the development of efficient PCO design methods is the availability of procedures to estimate the hydrodynamic parameters of an AUV before it can be actually built and tested.

Methods for parameter estimation based on the geometry and mass distribution of marine vehicles have been used for decades in the ship building industry. Important steps have also been taken in order to adapt parameter estimation methods, originated in aeronautics, to the prediction of submarine and AUV dynamics (Maeda and Tatsuta, 1989; Bohlmann, 1990). Recently, spawned by the widespread availability of powerful computers, there has also been a surge of interest in applying Computational Fluid Dynamics (CFD) methods to the prediction of stability derivatives for airplanes and marine vehicles (Humphreys, 2001). However, to the best of our knowledge, no in-depth, systematic studies have been done on the ***evaluation and validation*** of the above methods for AUV parameter estimation. As a consequence, there seems to be lacking an established approach for AUV parameter estimation allowing for the computation of the modeling errors incurred. It is therefore important to try and compare the types of estimates that are obtained with ASE and CFD methods and to later judge the precision of those estimates by resorting to towing tank experiments. This type of information will certainly play a major role during the phase of vehicle design to meet desired open loop performance requirements. At the same time, once bounds are known for the inaccuracies that are inherent to prediction methods (and thus for AUV parameter uncertainties), better control methods can be devised to explicitly deal in closed-loop with the uncertainties of the design models obtained.

Motivated by the above considerations, and as a contribution towards meeting the above goals, the paper guides the reader through the steps involved in the estimation of the parameters of slender body AUVs in the vertical plan, using information available from a number of sources, mainly the Datcom stability and control handbook (Hoak and Finck, 1978). The paper is organized as follows. Section 2 introduces the main concepts and formulas that are at the root of the ASE parameter estimation methods used to compute a full set of hydrodynamic derivatives for the MAYA AUV in the vertical plane. Section 3 discusses very briefly preliminary results of CFD analysis of the vehicle. Using the parameters ob-

Fig. 1. A diagram of the MAYA AUV (NIO design); the fin arrangement is not shown

tained in Section 2, Section 4 provides a study of the dynamics of the AUV in the vertical plane and of the impact of the position and size of the stern planes on open loop performance. Finally, Section 5 provides a critical review of the results obtained and discusses issues that warrant further research.

## 2. ESTIMATION OF HYDRODYNAMIC DERIVATIVES USING ANALYTICAL AND SEMI-EMPIRICAL(ASE) METHODS

Analytical and semi-empirical methods for the estimation of the hydrodynamic derivatives of marine vehicles are well rooted in the theory of hydrodynamics. The use of a particular method is decided by taking into consideration the physical nature of each of the parameters to be estimated, together with the underlying simplifying assumptions adopted when modeling the vehicle. In what follows, the computation of the hydrodynamic derivatives of a fully submerged body is organized by groups according to the nature of the physical phenomena involved. The presentation is restricted to those methods that were used to predict the hydrodynamic derivatives of the MAYA AUV. However, the methodology adopted for parameter estimation applies to AUVs with slender bare hulls that are solids of revolution and with fin-type control surfaces at the stern.

In what follows we restrict ourselves to motions in the vertical plane and adopt the usual notation for marine vehicles described in SNAME (1950). Methods are described for the computation of the force and moment experienced by a fully submerged *slender body* when its motion is restricted to small perturbations about steady motion in the vertical plane (that is, about the equilibrium condition that corresponds to the vehicle moving forward at a fixed speed $U$, with the angle of attack $\alpha$, pitch rate $q$, and pitch angle $\theta$ set to zero ). The end result is the set of so-called stability derivatives for the submerged body (SNAME, 1950). For the sake of clarity, some basic notation is now introduced with the help of Figure 2, where a torpedo-shaped AUV is depicted. The vehicle consists of the main hull with length $L$ and diameter $d$, together with nose and tail sections, and two stern planes that can be deflected by the same angle $\delta_e$. The vehicle´s body frame with axis $\{x, z\}$ is centered at point $O$. It is convenient to define $x^*$ as the axial coordinate of a generic point

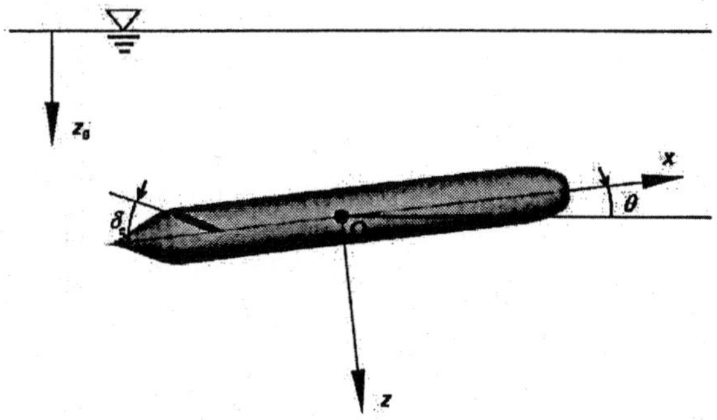


Fig. 2. Coordinate System Adopted

along the $x-$ axis of the vehicle, as measured from the nose tip, positive aft. The axial coordinate of the nose base section is denoted $x_N^*$, while that of reference point $O$ is called $x_0^*$. We warn the reader that we use the terms fins to refer to the all moving planes at the stern of the AUV. However, the hydrodynamic parameters that account for the effect of those surfaces take the subscript $W$ (from wings) because this will simplify the consultation of related, relevant literature on aerodynamics.

### *2.1 Added-Mass Coefficients*

The added-mass coefficients allow for the computation of the forces and moments exerted on a body as if it were moving in an ideal fluid (Newman, 1977). A number of reliable methods for their estimation are available in the literature. In their essence, all methods rely on the computation of a scalar velocity potential function from which the velocity of the particles (that is, the flow) around a marine vehicle can be derived. In the case of simple bodies, the flow can be obtained by combining the velocity potentials due to a distribution of sources and sinks. These results have been tabulated for a number of conventional shapes, assuming the motion takes place in an unbounded fluid. For other types of slender vehicle shapes, the three-dimensional added-mass coefficients can be approximated by a strip-theory synthesis, that is, by integrating bi-dimensional added-mass coefficients for each section along the vehicle´s length.

In the case of the MAYA AUV, the added mass coefficient in surge was computed by fitting approximately an ellipsoidal shape to the vehicle´s hull. The other added-mass coefficients for the body and fin ensemble were computed using strip-theory and the formulas for bi-dimensional added-mass coefficients of circles and finned circles (Newman, 1977). The results are tabulated in Section 4.

### *2.2 Static Coefficients*

This section describes the computation of the force and moment exerted by the fluid on a vehicle when it moves at a small angle of attack. As it is customary, the computation of the stability derivatives will be done by computing the lift, drag, and moment terms that are naturally expressed in the flow axis, and map those into the body axis, in a non-dimensional form. As a simple example, consider the expression $Z(\alpha) = -[L(\alpha)cos(\alpha) + D(\alpha)sin(\alpha)]$ that relates the body-axis force $Z$ with lift $L$ and drag $D$. Simple computations show that the hydrodynamic derivative $Z'_\alpha = \frac{1}{q}\frac{\partial Z(\alpha)}{\partial \alpha}$, where $q$ is dynamic pressure, equals $-(C_D(0) + C_{L_\alpha})$, where $C_D(0)$ is the drag coefficient for $\alpha = 0$ and the lift coefficient $C_{L_\alpha} = \frac{\partial C_L(\alpha)}{\partial \alpha}_{\alpha=0}$ denotes the derivative of the normalized lift curve $C_L(\alpha)$ at zero. Thus the importance of computing lift as a function of angle of attack and drag at zero lift angle. Identical considerations apply to the computation of the other hydrodynamic derivatives in Table 1. The organization of the section reflects the different steps involved in the computation of lift and moment, as well as drag.

#### *2.2.1. Lift on the Bare Hull*

The computation of the lift term for a slender body moving at an angle of attack with respect to the fluid requires careful consideration. In fact, if one were to make the simplifying assumption that the motion took place in an ideal fluid, then the theory of slender body hydrodynamics would show that a body with a pointy nose and tail would produce a zero lift force. However, this does not occur in practice because viscous effects induce the appearance of vortices and changes in the pressure distribution at the stern, even in the case of small angles of attack. As a result, lift occurs in a non-ideal fluid. Experimentally, it is verified that the pressure distribution at the fore part of a streamlined body agrees quite well with the ideal flow prediction(Hoerner, 1985). However, the pressure at the after body is reduced over a region that starts at a point where the vortex production or the boundary layer separation takes place and progresses all the way to the rear end of the body . The transition point at the after body is likely to occur at places where the change in the hull slope is bigger. Taking this phenomenon into account, the formula proposed by the US Air Force Datcom (Hoak and Finck, 1978) to compute lift on a slender body considers the same value of the lift coefficient as that obtained in the ideal flow case, but taking only in consideration the length of the body from the tip of the nose to the point where the ideal flow hypothesis is no longer valid.

In the case of missiles or torpedo shapes AUVs like MAYA, the region where ideal flow predictions are accurate is restricted to the nose. In this case, the bare hull lift coefficient $C_{L_{\alpha_B}}$ (normalized by the square of the vehicle length $L$), is given by

$$C_{L_{\alpha_B}} = \left(\frac{\partial C_{L_B}}{\partial \alpha}\right)_{\alpha=0} = 2(k2-k1)\frac{S_N}{L^2} \quad (1)$$

where $C_{L_B}$ denotes lift as a function of $\alpha$ and $S_N$ is the cross section area at the nose end coordinate $x^*_N$. In the above expression, $k_2 - k_1$ is the "Munk" apparent mass factor, which is a function of the fitness ratio $f = \frac{L}{d}$ of the body. For values of the fitness ratio $4 \leq f \leq 19$, the "Munk" factor can be approximated by the polynomial interpolation

$$(k_2 - k_1) = -0.0006548\, f^2 + 0.0256\, f + 0.73. \quad (2)$$

Under the same simplifying assumptions, the moment coefficient (normalized by the cubic of the vehicle length) is given by:

$$C_{m_{\alpha_B}} = \frac{2(k2-k1)}{L^3} \int_0^{x^*_N} \frac{dS_{x^*}}{dx^*}(x^* - x^*_O)\mathrm{d}x^*, \quad (3)$$

where $S_{x^*}$ denotes the body section area at $x^*$. From the above, it follows that the hydrodynamic center of the body, normalized by $L$ and expressed in body-axis coordinates, is located at $x'_B = \frac{C_{m_{\alpha_B}}}{C_{L_{\alpha_B}}}$.

### 2.2.2. Lift Produced by Small Aspect Ratio Fins

Studies on the lift force $C_L$ produced by small aspect ratio fins have been conducted by many researchers and have led to closed formula approximations that are widely available in the literature. In this study we adopt the approximation for $C_{L_{\alpha_W}} = \left(\frac{\partial C_{L_W}}{\partial \alpha}\right)_{\alpha=0}$ proposed by Whicker and Fehlner (1958) and used by Bohlmann (1990) to compute lift in submarine hydroplanes, given by

$$\frac{C_{L_{\alpha W}}}{AR} = \frac{2\pi}{2 + \sqrt{\frac{1}{\eta^2}\left(\frac{AR^2}{\cos^2 \Lambda_{c/4}} + 4\cos^2 \Lambda_{c/4}\right)}} \quad (4)$$

where $AR$ is the lift surface aspect ratio, $\Lambda_{c/4}$ is the sweep angle at one fourth of the chord length, and $\eta$ is a factor to correct for viscous effects (adopted as 0.9). The above expression yields results that are very close to the to the ones described in Hoerner (1985) for small aspect ratio surfaces. Classical results can be used to compute the hydrodynamic center of the fin (Abbot and Doenhoff, 1949) . In what follows, $x'_W$ denotes the axial coordinate of the hydrodynamic center of the AUV fins, in body-axis.

### 2.2.3. Fins and Body Combination

There is a mutual influence between the fins and the main body of an AUV due to the changes in the flow past each of these components caused by the presence of the other. To account for the impact of these effects on the total lift experienced by a fully submerged body, we follow the procedure described in Datcom (Hoak and Finck, 1978) that builds on the methodology derived in Pitts *et al.* (1957) for aircraft and missiles. The key idea is to compute the so-called interference factors between the lift surfaces and the body using slender-body theory. The first step in this procedure assumes that the fins do not deflect, the effect of the deflection being taken into account at a later stage. This step yields an approximate expression for the total (body plus fins) lift coefficient $C_{L_{\alpha(WB)}}$ (where the notation $WB$ borrows from aircraft wing-body interactions), given by

$$\begin{aligned} C_{L_{\alpha(WB)}} &= C_{L_{\alpha_B}} + C_{L_{\alpha W(B)}} + C_{L_{\alpha_{B(W)}}} \\ &= C_{L_{\alpha_B}} + (K_{W_{(B)}} + K_{B_{(W)}})(C_{L_\alpha})_e \frac{S_e}{L^2} \end{aligned} \quad (5)$$

where $S_e$ is the total exposed fin surface area, $(C_{L_\alpha})_e$ is the lift coefficient of the exposed fin surfaces, and $K_{B_{(W)}}$ and $K_{W_{(B)}}$ are the interference factors from the surfaces to the body, and from the body to the surfaces, respectively. Let $b$ be the maximum span of the fins in combination with the hull, that is, the total distance between control surface tips as if they extended inside the hull, and define

$$k = \frac{d}{b}. \quad (6)$$

Then, the interference factors can be written as

$$K_{W(B)} = \frac{2}{\pi}\frac{(1-k^4)\,\zeta_1 - k^2\zeta_2}{(1-k)^2} \quad (7)$$

and

$$K_{B(W)} = (1+k)^2 - K_{W(B)}, \quad (8)$$

where

$$\zeta_1 = \left[\frac{1}{2}\tan^{-1}\left(\frac{1}{2}(k^{-1} - k)\right) + \frac{\pi}{4}\right] \quad (9)$$

and

$$\zeta_2 = \left[(k^{-1} - k) + 2\tan^{-1} k\right]. \quad (10)$$

The estimation of the corresponding total moment coefficient $C_{m_{\alpha(WB)}}$ is given by the product of the lift coefficient computed above and the position $x'_{(WB)}$ of the new hydrodynamic center, normalized by $L$, computed by taking into consideration the interaction between the hull and the fins. To compute $x'_{(WB)}$, start by defining $x'_{W(B)}$ and $x'_{B(W)}$ as the center of the hull-lift carryover on the lift surface and the center of the fin-lift carryover on the body, respectively . The first can be taken as approximately equal to the hydrodynamic center of the fins $x'_W$. The latter is given by

$$x'_{B(W)} = \frac{1}{4} + \frac{b-d}{2c_{r_e}}\tan \Lambda_{c/4} * P, \quad (11)$$

where

$$P = -\frac{k}{1-k} + \frac{\sqrt{1-2k}\ln\left(\frac{1-k}{k} + \frac{1}{k}\sqrt{1-2k}\right) - (1-k) + \frac{\pi}{2}k}{\frac{k(1-k)}{\sqrt{1-2k}}\ln\left(\frac{1-k}{k} + \frac{1}{k}\sqrt{1-2k}\right) + \frac{(1-k)^2}{k} - \frac{\pi}{2}(1-k)}$$

where $c_{r_e}$ is the exposed tip root chord, and $\Lambda_{c/4}$ is the sweep angle at one fourth of the root chord length. This expression above assumes that the aspect ratio is greater than or equal to 4. For smaller values of the aspect ratio, the reader should consult an interpolation procedure used at Datcom. It is now possible to compute

$$x'_{(WB)} = \frac{C_{L_{\alpha W(B)}} x'_{W(B)} + C_{L_{\alpha B(W)}} x'_{B(W)} + C_{m_{\alpha B}}}{C_{L_{\alpha(WB)}}} \tag{12}$$

To capture the effects due to the deflection of the control surfaces, the methodology described in Datcom leads to the control surfaces lift and torque coefficients $C_{L_{\delta e}}$ and $C_{m_{\delta e}}$, respectively given by

$$C_{L_{\delta e}} = (k_{B(W)} + k_{W(B)})(C_{L_\alpha})_e \frac{S_e}{L^2} \tag{13}$$

and

$$C_{m_{\delta e}} = C_{L_{\delta e}} x'_W \tag{14}$$

where $\delta e$ is the deflection angle of the lift surfaces and $x'_W$ is the center location of the exposed lift surfaces. It is important to remark that $k_{B_{(W)}} + k_{W_{(B)}} = K_{W_{(B)}}$

### *2.2.4. Drag Coefficient*

This section details the computation of drag coefficient of an AUV at zero angle of attack. As discussed before, this is the only drag-related information required to compute the stability derivative $Z'_\alpha$. In this case, drag equals the pressure plus friction drag. The bare hull drag coefficient is calculated in Datcom as a function of the Reynolds number *Re*, the fitness ratio $f$, and the base diameter $d_b$. To compute the zero-lift drag coefficient $C_{D_0}$ , start by defining

$$C_D^* = C_f[1 + 60f^{-3} + 0.0025f]\frac{S_S}{L^2} \tag{15}$$

where $S_s$ is the total body wetted area and

$$C_f = \frac{0.075}{(\log \mathrm{Re} - 2)^2} + 0.00025 \tag{16}$$

is the skin friction drag coefficient, as given by the ITTC. Further define the base-drag coefficient

$$C_{D_b} = 0.029(\frac{d_b}{d})^3 (C_D^*)^{-0.5}. \tag{17}$$

where $d_b$ is the base diameter, that is, the diameter of the sternmost section of the body. Then,

$$C_{D_0} = C_D^* + C_{D_b} \tag{18}$$

Results on the fin drag coefficient are provided by Abbot and Doenhoff (1949).

### *2.3 Dynamic Coefficients*

The dynamic coefficients $C_{L_q}$ and $C_{m_q}$ of a submerged body relate the heave force and pitch moment, respectively to variations in the pitch angular velocity. Simple approximations to the coefficients can be obtained by emphasizing the contribution of the control surfaces, yielding

$$C_{L_q} = -(K_{B_{(W)}} + K_{W_{(B)}})(C_{L_\alpha})_e \frac{S_e}{L^2} x'_{W(B)} \tag{19}$$

$$C_{m_q} = -(K_{B_{(W)}} + K_{W_{(B)}})(C_{L_\alpha})_e \frac{S_e}{L^2} (x'_{W(B)})^2 \tag{20}$$

### *2.4 Stability Derivatives*

As explained before it is customary, when modeling marine vehicles, to parameterize the models in terms of so-called stability derivatives, as described in SNAME. The equivalence between non-dimensional stability derivatives and the hydrodynamic coefficients determined before is summarized in Table 1 (see also Blakelock (1991) for a lucid exposition of the subject in the aircraft area). In preparation for the study that follows, the particulars and stability derivatives of the MAYA AUV in the vertical plane are given in Tables Table 2 and Table 3, respectively. The bare hull shape is based on the geometry proposed in (Myring, 1976).

| | |
|---|---|
| $Z'_\alpha$ | $-(C_{D_0} + C_{L_{\alpha(WB)}})$ |
| $Z'_q$ | $-(C_{L_q} + X'_{\dot{u}})$ |
| $M'_\alpha$ | $C_{m_{\alpha(WB)}}$ |
| $M'_q$ | $C_{m_q}$ |
| $Z'_{\delta e}$ | $-C_{L_{\delta e}}$ |
| $M'_{\delta e}$ | $C_{m_{\delta e}}$ |

Table 1. Equivalence between stability derivatives and hydrodynamic coefficients

| | |
|---|---|
| mass ($m$) | 47.5 (Kg) |
| moment of inertia ($Iyy$) | 8.923 ($Kgm^2$) |
| reference length ($L$) | 1.64m |
| main diameter ($d$) | 0.2m |
| tail span ($b$) | 0.44m |
| tail exposed area ($S_e$) | 0.018($m^2$) |
| tail and hull combined area ($S_w$) | 0.0385($m^2$) |
| tail aspect ratio ($AR$) | 5.02 |
| tail exposed aspect ratio ($ARe$) | 3.2 |
| tail exposed taper ratio ($\lambda$) | 0.667 |
| cruising Speed ($U$) | 1.5($m/s$) |

Table 2. Particulars of Maya

| | | | |
|---|---|---|---|
| $m'$ | 21.145 | $M'_{\dot{\alpha}}$ | -0.612 |
| $I'_{yy}$ | 1.48 | $M'_{\dot{q}}$ | -1.677 |
| $X'_{\dot{u}}$ | -0.478 | $M'_\alpha$ | -2.323 |
| $Z'_{\dot{q}}$ | -0.612 | $M'_q$ | -3.221 |
| $Z'_{\dot{\alpha}}$ | -24.138 | $M'_{\delta e}$ | -8.089 |
| $Z'_\alpha$ | -68.862 | $Z'_{\delta e}$ | -30.615 |
| $Z'_q$ | -11.712 | $M'_\theta$ | -1.84 |

Table 3. The Maya Derivatives($*10^{-3}$)

## 3. CFD BASED METHODS

To estimate the hydrodynamic coefficients of MAYA, Computational Fluid Dynamics (CFD) methods are also being exploited by the National Institute of Oceanography in Goa, in India which has access to the FLOWSOLVER machine at Bangalore. At the time of writing of this paper, only results on the drag coefficient at zero-lift angle were available. Depending on the convection and turbulence scheme used, the drag coefficient(based on the maximum section area) was found to vary between 0.08 and 0.127. The same coefficient calculated using Datcom yield 0.143. A careful analysis of these results and the extension of CFD methods to compute the lift related coefficients are planned for future work. Together with real data from tank or open ocean tests, this study is expected to shed light into the accuracy of the results that can be obtained with CFD methods for AUVs of this type.

## 4. DYNAMIC ANALYSIS

The set of stability derivatives obtained above is now used to predict the dynamic behaviour of the AUV in the vertical plane and to evaluate the impact of stern plane size on its expected performance.

### 4.1 Open loop transfer functions

To study the dynamics of the AUV in open loop, it is sufficient to compute and analyze the transfer functions from stern plane deflection to pitch and heave motions. Let $\alpha(s), \theta(s)$, and $\delta_e(s)$ denote the Laplace transforms of $\alpha, \theta$, and $\delta_e$, respectively. Neglecting the surge equation, the linearized model of the AUV in the vertical plane (about its steady forward motion at trimming speed) is easily seen to be given by

$$(a_\alpha s - b_\alpha)\,\alpha(s) + (-c_\alpha s^2 - d_\alpha s)\theta(s) = Z'_{\delta e}\delta e(s) \tag{21}$$

and

$$(-c_\theta s - d_\theta)\,\alpha(s) + (a_\theta s^2 - b_\theta s - e_\theta)\theta(s) = M'_{\delta e}\delta e(s), \tag{22}$$

where the relationship between the hydrodynamic derivatives and the coefficients above are given in Table 4. Further let $\dot{z}'_0(s)$ denote the Laplace transform of depth rate in non-dimensional form, where depth is measured from the surface, positive downwards. Linearizing the depth coordinate dynamics about trimming yields

$$\dot{z}'_0(s) = \alpha(s) - \theta(s) \tag{23}$$

In order to simplify the analysis, assume that the origin $O$ of the body axis is coincident with the center of mass. From the above equations, it follows that

| | | | |
|---|---|---|---|
| $a_\alpha$ | $(m' - Z'_{\dot\alpha})L/U$ | $a_\theta$ | $(I'yy - M'_{\dot q})(L/U)^2$ |
| $b_\alpha$ | $Z'_\alpha$ | $b_\theta$ | $(M'_q - m'x'_G)L/U$ |
| $c_\alpha$ | $(Z'\dot q + m'x'_G)(L/U)^2$ | $c_\theta$ | $(M'_{\dot\alpha} + m'x'_G)L/U$ |
| $d_\alpha$ | $(Z'_q + m')L/U$ | $d_\theta$ | $M'_\alpha$ |
| | | $e_\theta$ | $M_\theta$ |

Table 4. Transfer function coefficients

$$\frac{\alpha(s)}{\delta e(s)} = \frac{N_\alpha(s)}{D(s)}, \tag{24}$$

$$\frac{\theta(s)}{\delta e(s)} = \frac{N_\theta(s)}{D(s)}, \tag{25}$$

where

$$\mathrm{N}_\alpha(s) = a_\theta Z'_{\delta e}[(1 - x'_W\frac{c_\alpha}{a_\theta})s^2 + (-\frac{b_\theta}{a_\theta} - x'_W\frac{d_\alpha}{a_\theta})s - \frac{e_\theta}{a_\theta}], \tag{26}$$

$$\mathrm{N}_\theta(s) = M'_{\delta e}a_\alpha[(1 - \frac{c_\theta}{x'_W a_\alpha})s + (-\frac{b_\alpha}{a_\alpha} - \frac{d_\theta}{a_\alpha x'_W})], \tag{27}$$

and

$$D(s) = a_\alpha a_\theta[s^3 + (\frac{-b_\alpha}{a_\alpha} - \frac{b_\theta}{a_\theta})s^2 + (-\frac{e_\theta}{a_\theta} + \frac{b_\alpha b_\theta}{a_\alpha a_\theta} - \frac{d_\alpha d_\theta}{a_\alpha a_\theta})s + \frac{b_\alpha e_\theta}{a_\alpha a_\theta}] \tag{28}$$

The denominator was simplified by neglecting the products $c_\theta c_\alpha$, $c_\theta d_\alpha$, and $c_\alpha d_\theta$ when compared to the other terms contributing to the coefficients of the same power of $s$. This is a consequence of the body symmetry characteristics, which imply small values for the added mass terms $Z'_{\dot q}$ and $M'_{\dot\alpha}$. The term $D(s)$ can be further simplified by noticing that $|d_\alpha d_\theta| \ll |-e_\theta a_\alpha + b_\alpha b_\theta|$. This follows from the following observations: for vehicles of the type considered $d_\alpha$ is generally positive, with magnitude less than half of $a_\alpha$. Furthermore, $d_\theta$ (which equals the static moment coefficient $C_{m_{\alpha(WB)}}$) has the same order of magnitude or less when compared to $e_\theta$ and $b_\theta$. Finally, $b_\alpha$ has a large magnitude when compared to the previous coefficients. With the simplifications above, $D(s)$ can be written as

$$D(s) \cong a_\alpha a_\theta(s - \frac{b_\alpha}{a_\alpha})\left[s^2 - \frac{b_\theta}{a_\theta}s + \frac{1}{a_\theta}(-e_\theta)\right]. \tag{29}$$

The first order term in $D(s)$ captures the heave dynamics, while the second order term is related to the pitch dynamics.

The small $C_{m_{\alpha(WB)}}$ hypothesis allows for a further simplification in the pitch transfer function. To see this, start by noticing that the pitch motion can be practically decoupled from the heave dynamics by neglecting the coefficient $c_\theta$ and the term $\frac{d_\theta}{a_\alpha x'_w}$ in $\mathrm{N}_\theta(s)$. The last simplification becomes clear when the last term in $\mathrm{N}_\theta(s)$ is re-written as

$$(-\frac{b_\alpha}{a_\alpha} - \frac{d_\theta}{a_\alpha x'_W}) = -\frac{b_\alpha}{a_\alpha}(1 + \gamma\frac{x'(WB)}{x'_W}), \tag{30}$$

where $0 < \gamma < 1$ due to the drag coefficient contribution in $Z'_\alpha$, see Table 1. As long as $|C_{m_{\alpha(WB)}}|$ is small enough, so is the ratio between the hydrodynamic center of the vehicle and that of the fins, and the expression above degenerates approximately to $-\frac{b_\alpha}{a_\alpha}$. As a consequence, there is an almost pole-zero cancellation (in the pitch transfer function) tied with the heave dynamics. Although the present version of Maya does not have a very small value of $|C_{m_{\alpha(WB)}}|$, the simplifications above result in reasonable approximations to the transfer functions. In fact, the poles obtained from expression (28) are

$$-1.166, -0.571 \pm 0.500i \ rads^{-1},$$

while those obtained from the simplified expression (29) are

$$-1.373, -0.467 \pm 0.637i \ rads^{-1}$$

The zero in the pitch transfer function is $-1.198$ $rads^{-1}$. From (23)-(30) and the expressions in Table 4, it is also possible to derive the transfer function from stern plane deflection to depth rate as

$$\frac{\dot{z}'_0}{\delta_e} = \frac{N_D(s)}{D(s)}, \tag{31}$$

where

$$N_D(s) \cong a_\theta Z'_{\delta e}[s^2 - \frac{L}{U}\frac{Z'_{\dot{\alpha}}x'_W}{a_\theta}s - \frac{x'_W b_\alpha + e_\theta}{a_\theta}] \tag{32}$$

By examining the signs and relative magnitude of the coefficients involved, it follows that the depth dynamics exhibit non-minimum phase characteristics.

### *4.2 Maneuvering performance*

Since the hydrodynamic derivatives depend on the overall hull geometry and fin arrangement of an AUV, they are expected to play a key role in the development of methodologies for vehicle design. In this paper, and as an illustrative example, we consider the problem of optimizing the maneuverability of an AUV in the vertical plane, when the bare-hull profile is fixed, by proper choice of the stern plane dimensions and location. Clearly, this entails the definition of a performance index to capture the maneuverability requirements in a rigorous manner, as explained below.

Suppose the objective it is to achieve a fast response in a surfacing/diving emergency maneuver (e.g. for collision avoidance), while keeping the angle of attack small so as to stay in the region of validity of the linear design model. In this case, a possible choice for the performance index is the ratio $\dot{q}'(0)/\alpha_{ss}$, where $\alpha_{ss}$ is the steady state value of the angle of attack in response to a stern plane step deflection and $\dot{q}'(0)$ is the resulting pitch acceleration at time zero. To compute this ratio, use the transfer functions defined before and apply the initial and final value theorems to the calculated step responses of pitch and heave to obtain

$$\frac{\dot{q}(0)}{\alpha_{ss}} = \frac{M'_{\delta e}b_\alpha}{a_\theta Z'_{\delta e}} = -x'_w\frac{b_\alpha}{a_\theta} \tag{33}$$

In order to simplify the analysis, suppose that the problem is limited to searching the optimal span and location of the stern planes, taken as two rectangular, non-cambered fins. Further assume that the modification of the fins does not change the center of mass of the vehicle significantly. Notice that the changing in the span of the fins affects both $a_\theta$ and $b_\alpha$. When the span is increased, the change in $b_\alpha$ is mainly due to an increase in the magnitude of the lift force, whereas the change in $a_\theta$ is mainly due to an increase in magnitude of the added-mass coefficient $M'_{\dot{q}}$. To compute the change in the latter coefficient, start by computing the sectional (heave) added mass coefficient of a finned circle, given by (Newman, 1977)

$$\rho\pi\left[\frac{d^2}{4} + \frac{((b_e/2)+d)^2(b_e/2)^2}{((b_e/2)+(d/2))^2}\right], \tag{34}$$

where $b_e = b - d$ is the total exposed fin span. It is then straightforward, using strip theory, to obtain the contribution of the fins to the new value of $M'_{\dot{q}}$. This is done by multiplying the sectional added mass coefficient above by the square of the distance from the section to the origin $O$ of the body-axis and integrating the contributions over the lift surface chord. Clearly, from (34), for a fixed position of the hydrodynamic center of the fins, the variation in $M'_{\dot{q}}$ (and therefore in $a_\theta$) is practically proportional to the square of the surface span. The change in $b_\alpha$ takes place in a more complex form, as the discussions in sections 2.2.2 and 2.2.3 show.

The performance index must also take into account the increase in energy consumption of the vehicle in steady motion (that is, with zero angle of attack) due to an increase in fin span. Clearly, this will be due to added drag. Since the fins are rectangular and have a fixed chord, the added drag force (at zero angle of attack) is proportional to the fin span. As a consequence, the energy penalty factor is inversely proportional to the fin span. In this study, the combined span of the fins was allowed to vary between $0.3m$ and $1.0m$. In the current design of MAYA, the corresponding figure is $0.44m$. With the largest span considered, the stern plane contribution to the vehicle total drag rises to about 30 percent. It is also important that the performance index should penalize the difficulties in vehicle handling that arise when the span of the surfaces assumes large values. Again, this calls for a penalty factor that is inversely proportional to the span size. Based on the above considerations, the final expression adopted for the performance index is

$$I = \frac{1}{(0.5 + 2b_e)} \frac{\dot{q}'(0)}{\alpha_{ss}}, \tag{35}$$

where the factor that multiplies the ratio $\frac{\dot{q}'(0)}{\alpha_{ss}}$ is normalized in such a way as to yield the value 1 for the current MAYA surface span, and slightly less than 0.5 for the largest span. Figure 3 is a series of plots of the performance index versus total span, for three different locations of the stern plane hydrodynamic center. The middle plot corresponds to the original location ($0.459m$ aft the center of mass), whereas the two other plots correspond to locations closer to the center of mass ($-0.204m$) and further away from it ($-0.604m$ ). The latter case yields the largest performance index. Interestingly enough, increasing the distance from the fins hydrodynamic center to the center of mass of the vehicle, allows for smaller optimal fin sizes. There is no advantage in increasing the size of the fins past the optimal value of a particular curve, because the added moment of inertia takes over and forces the performance index curve to slope down.

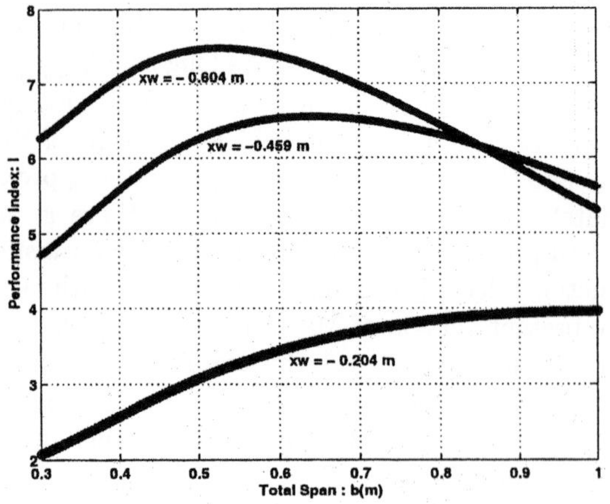


Fig. 3. Effect of stern plane span on maneuvering performance

## 5. CONCLUSIONS AND RECOMMENDATIONS FOR FUTURE RESEARCH.

The use of analytical and semi-empirical estimates (ASE) for the hydrodynamic derivatives of AUVs holds great potential to the development of powerful tools for optimal vehicle design. The estimates can be used to predict the dynamic behavior of the vehicle and to assess the impact of fin arrangement on its expected performance. It is therefore important to try and compare the types of estimates that are obtained with ASE and CFD methods and to later judge the precision of these estimates by resorting to towing tank experiments.

## ACKNOWLEDGMENT

The authors would like to thank A. Alcocer, H. Bø, and S. Fekri for their help with the text format and figures.

## REFERENCES

Abbot, I. and A. Von Doenhoff(1949). *Theory of Wing Sections.* Dover Publications Inc., New York.

Blakelock, J. (1991). *Automatic Control of Aircraft and Missiles.* Second Edition, John Wiley Sons, Inc.

Bohlmann, H. (1990). Berechnung Hydrodynamischer Koeffizienten von Ubooten zur Vohrhersage des Bewegungsverhaltens. *PhD thesis.* Institut fur Schifbau der Universitat Hamburg.

Hoak, D. and R. Finck(1978). *USAF Stability and Control Datcom.* Wright-Patterson Air Force Base, Ohio.

Hoerner, S. (1985). *Fluid Dynamic Lift.* Hoerner, Liselotte A.

Humphreys, D. (2001). Correlation and Validation of a CFD Based Hydrodynamic & Dynamic Model for a Towed Underwater Vehicle. *Proc. MTS/IEEE OCEANS 2001*, Honolulu, Hawaii.

Maeda, H. and S. Tatsuta (1989). Prediction method of hydrodynamic stability derivatives of an autonomous non-tethered submerged vehicle. *Proc. of 8th Int. Conf. on Offshore Mechanics and Artic Engineering.*

Myring, D F (1976). A theoretical study of body drag in subcritical axisymmetric flow. *Aeronautical Quarterly* **27**(3), pp. 186–194.

Newman, J N (1977). *Marine Hydrodynamics.* 9th ed., M.I.T., Cambridge, Massachusetts.

Pitts, William C, Jack N Nielsen and George E Kaattari (1957). Lift and center of pressure of wing-body-tail combinations at subsonic, transonic, and supersonic speeds. *Technical report, NACA.*

SNAME, The Society of Naval Architects and Marine Engineers (1950). Nomenclature for Treating the Motion of a Submerged Body Through a Fluid. *Technical and Research Bulletin*, No. 1-5.

Silvestre, P., A Pascoal, I Kaminer, and A Healey (1998). Combined plant/controller optimization with application to autonomous underwater vehicles. *Proc. CAMS'98, Fukuoka, Japan.*

Todd, F. (1967). Resistance and Propulsion. *Principles of Naval Architecture, Chapter VII.* John P. Comstock, Ed. New York.

Whicker, L F and L F Fehlner (1958). Free-stream characteristics of a family of low-aspect-ratio, all-movable control surfaces for application to ship design. *Technical report 933*, David Taylor Model Basin.

ELSEVIER
IFAC
PUBLICATIONS
www.elsevier.com/locate/ifac

# IDENTIFICATION OF UNDERWATER VEHICLES BY A LYAPUNOV METHOD

**Antonio Tiano**

*University of Pavia, Department of Information and Systems*
*Via Ferrata 1, 27100 Pavia, Italy*
*email:antonio@control1.unipv.it*

Abstract: This paper deals with the identification of non linear multivariable models of underwater vehicles by using a novel Lyapunov-based method. The method operates in the continuous time domain and can be applied to non linear models that are linear with respect to an unknown parameter vector. After an introduction, where the role of identification methods in the area of guidance and control of underwater vehicles is outlined, some mathematical models that are generally used for describing the dynamics of underwater vehicles are concisely presented. The main features of the identification method are then illustrated through a simulation example concerning the surge dynamics of an underwater vehicle. *Copyright © 2004 IFAC*

Keywords: system identification, parameter estimation, numerical methods, underwater vehicles.

## 1. INTRODUCTION

The application of system identification techniques to marine vehicles dynamics deals with the estimation of a number of unknown parameters that characterize their behaviour, on the basis of experimental measurements obtained either in a towing tank or during at sea trials. An increasing interest has been devoted in the last two decades to the experimental determination of the dynamical behaviour of marine vehicles by means of sea trials experiments (Abkowitz,1980). Such methods, unlike traditional naval architecture methods, are potentially capable of drastically reducing experiment time and expenses of both towing tank and at sea trials, because a multitude of parameters can be determined from a few dedicated tests. More specifically, identification plays a fundamental role in the design of guidance and control systems for underwater vehicles as well as in a number of related applications, such as, for example, simulation of vehicle's dynamics and fault detection and diagnosis. Even if underwater vehicles, since some years, are widely used in many off-shore activities, it is worth noting that the problem of adequately controlling them is still far from being solved in an optimal way. In fact, for such relatively recent and non conventional vehicles, there has not yet been developed an adequate knowledge concerning their dynamic behaviour, expressed in terms of experimentally validated mathematical models.

In the recent years identification methods based on discrete-time models (Ljung, 1987) have been intensively used in the area of engineering applications, owing to the fact that the related algorithms can be easily implemented on digital computers. It is worth noting, however, that in applications involving the identification of an intrinsically continuous-time system, such as an underwater vehicle, it is generally preferable to use a continuous-time method. In fact, given a model of a physical system described by differential equations it is more direct and much easier to use that model directly into an identification algorithm, rather than using an intermediate conversion to a discrete time form, that sometimes can be error prone. It is also important to note that a continuous-time model has a global validity, in the sense that it can be used for generating a variety of discrete-time models, by simply selecting a suitable sampling interval, to be applied, for example, to the design of control systems or of monitoring and fault detection systems. It should be remarked, furthermore, that most of the existing identification methods are generally inadequate to cope with multivariable, non linear, continuous-time models. A serious limitation is, finally, given by the high dimensionality of the parameter vector that are to be estimated.

In this paper an innovative identification method, based on a Lyapunov approach, is presented, that is potentially capable to cope with the above mentioned limitations. A concise review of the mathematical models used for describing the dynamics of underwater vehicles is presented in section 2, while section 3 is dedicated to the description of the identification method. Some simulation results concerning the identification of the surge dynamics of an UUV (Unmanned Underwater Vehicle) are presented in section 4

## 2. UNDERWATER VEHICLE MODELLING

As it can be demonstrated by a classical mechanics approach (Fossen, 1994, Fossen, 2002) , the mathematical models of a wide class of underwater vehicles can be expressed with respect to a local body reference system by a set of non linear coupled Newtonian equations of the form:

$$M(x,t,\theta)\dot{x} = f(x,t,\theta) + \tau(t) + g(x,t,\theta) \quad (1)$$

where $x(t) \in R^6$ is the vehicle's state vector, generally constituted by linear and angular velocities, i.e. $x = \begin{bmatrix} u & v & w & p & q & r \end{bmatrix}^T$ consisting of surge velocity, sway velocity, heave velocity, roll rate, pitch rate and yaw rate, $M(x,t) \in R^{6x6}$ is the body inertial matrix including hydrodynamic added masses, $f(x,t,\theta) \in R^6$ is the vector of kinematic forces and moments, $\tau(t) \in R^6$ is the vector of control forces and moments from thrusters and control surfaces, $g(x,t,\theta) \in R^6$ is a vector including all the other hydrodynamic forces and moments. The vector $\theta$ represents the set of all unknown parameters, generally constituted by hydrodynamic derivatives or other inertial coefficients. It is worth noting that the identification of the complete set of coefficients and hydrodynamic derivatives which appear in Equation (1) is a rather complex task, owing to the very high number of parameters, to nonlinearities and to space-time variant effects. The identification problem can be much more easily approached if it is assumed that the longitudinal motions, i.e. heave $z$ and pitch $q$ are decoupled from the other motions. A further simplification can be achieved if it is assumed that the vector $f(x,t)$ consists of vehicle drag forces only, the vector $g(x,t)$ consists of vehicle buoyancy forces only and if the inertia matrix is assumed to be time-space invariant and diagonal. If, in addition, we consider that for many UUVs, pitch and roll motions are negligible, it is possible to conclude that the total relevant motions are reduced to surge $u$, sway $v$, yaw $r$ and heave w. Such motions can be described, in a more compact form (Fossen, 2002) by the following set of decoupled equations:

$$\dot{x}_i = -\alpha_i x_i - \beta_i x_i |x_i| + \gamma_i \tau_i \quad i = 1,4 \quad (2)$$

where $\alpha_i$ and $\beta_i$ are the linear and quadratic drag coefficients, $\gamma_i$ are the inverse of the diagonal elements of the reduced order vehicle inertial matrix. In a quite general fashion, it is possible to express equation (2) into the following state space form :

$$\dot{x} = \varphi(x(t), \tau(t))\theta \quad (3)$$

where $t \in [0, +\infty), x \in R^n, \tau \in R^m, \theta \in R^l$ and $\varphi : R^n \times R^m \rightarrow R^{nxl}$ is a Lipschitz continuous mapping, such that the corresponding solutions are unique. The dimensions of state vector, input vector and unknown parameter vector depend on the considered motions. It is important to note that such equation is linear with respect to the parameter vector $\theta$, which makes much easier, as it will be shown, the solution of the identification problem.

## 3. IDENTIFICATION METHOD

A Lyapunov-based approach to the identification problem consists of estimating the unknown constant parameter vector $\theta$ in equation (3) by means of a model reference time-varying parameter vector $\theta_M(t)$ associated to a model reference system, that is constituted by a process subject to the same input vector and depending linearly on the parameter vector $\theta_M(t)$, according to the equation :

$$\dot{x}_M = \varphi(x_M(t), \tau(t))\theta_M(t) \quad (4)$$

where $t \in [0, +\infty), x_M \in R^n, \tau \in R^m, \theta_M \in R^l$.
By introducing the error $\Delta x(t) = x(t) - x_M(t)$ between the two state vectors and the corresponding parameter error vector $\Delta\theta(t) = \theta - \theta_M(t)$, it can be verified that the following equation for state error holds:

$$\Delta\dot{x}(t) = \varphi(x(t), \tau(t))\Delta\theta(t) + \Delta\varphi(x, x_M, \tau)\theta_M(t) \quad (5)$$

where

$$\Delta\varphi(x, x_M, \tau) = \varphi(x(t), \tau(t)) - \varphi(x_M(t), \tau(t))$$

A measure of mismatching between the given process and the model reference one, due to the parameter error, is given by:

$$e(t) = \varphi(x(t), \tau(t))\Delta\theta(t) \quad (6)$$

In order to solve the identification problem, the state error $\Delta x(t)$ and the mismatching error $e(t)$ should asymptotically tend to zero by adjusting the model reference parameter vector $\theta_M(t)$ in such a way that $\Delta\theta(t)$ tends to zero. The application of Lyapunov method to the identification problem (Lyshevski et Al., 1994) consists of defining a Lyapunov function with appropriate properties, from which convergence to zero of mismatching error $e(t)$ is implied. If there exists a Lyapunov function $V(\Delta\theta(t))$ of the form $V: R^l \to [0,+\infty)$ continuous with continuous partial derivatives with respect to $\Delta\theta$, such that :

$$\begin{aligned} &V(\Delta\theta)=0 && if\ \Delta\theta=0 \\ &V(\Delta\theta)>0 && \forall\ \Delta\theta\neq 0 \\ &V(\Delta\theta)\to\infty && as\ [\![\Delta\theta]\!]\to 0 \\ &\frac{dV(\Delta\theta(t))}{dt}<0 && \forall\ \Delta\theta\neq 0 \end{aligned} \qquad (7)$$

then $\Delta\theta(t)$ tends to zero and thus the model reference parameter vector $\theta_M(t)$ tends to the process parameter vector $\theta$. If a quadratic Lyapunov function is asssumed, it can be shown that the following differential equation can be used for determining $\theta_M(t)$:

$$\dot{\theta}_M(t) = -K(t)\varphi(x,\tau)^T(\varphi(x,\tau)\theta_M(t)-\dot{x}(t)) \qquad (8)$$

where $K(.)\in R^{l\times l}$ is a time dependent positive definite, symmetric weighting matrix, that can be chosen to ensure the convergence to zero of the parameter error $\Delta\theta(t)$. The following theorem indicates how the weighting matrix $K(t)$ can be chosen.

Theorem.1. If matrix $F(t)=\varphi(x,\tau)^T\varphi(x,\tau)$ is not singular $\forall t>0$, and if the weighting matrix $K(t)=F^{-1}(t)$ is chosen, then the solution of equation (8) globally converges to the true parameter vector $\theta$, $\forall\theta_M(0)\in R^l$. The convergence is of exponential type , as it can be deduced by the following inequality:

$$\|\Delta\theta(t)\|_2 \le e^{-t}\|\Delta\theta(0)\|_2 \qquad (9)$$

The above theorem and inequality (9) can be easily demonstrated by a direct substitution. It should be noticed that, since a perfect knowledge of the derivative of the system state vector $\dot{x}(t)$ is assumed, in realistic situations, when derivative is not available and it must be estimated by noisy measurements of state vector, it is necessary to use a suitable combination of low-pass filter and of derivative filter in order to reconstruct the derivative of state vector. A possible implementation of such filtering devices is shown in Fig.1 .

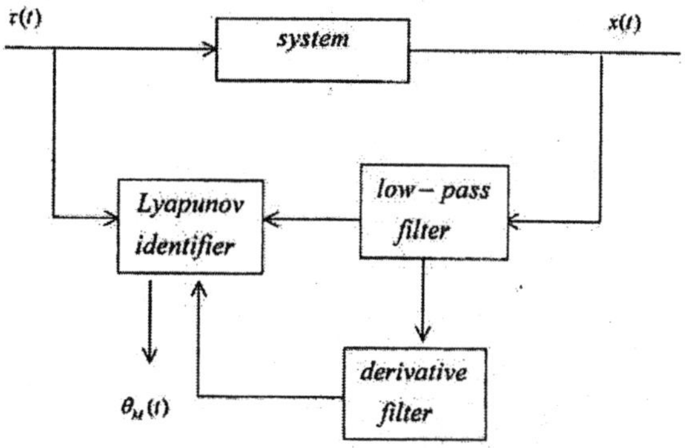


Fig. 1. Implementation of a Lyapunov identifier by use of state vector filters.

By using a Least Squares estimation algorithm, it is also possible to extend the above result in the case matrix $F(t)$ should result singular, according to the following theorem.

Theorem.2. Let matrix $F(t)=\varphi(x,\tau)^T\varphi(x,\tau)$ have $rank = p < l$, $\forall t>0$. For a set of discrete increasing time instants $\{t_k\}_{k=1}^N$ , let the corresponding set of SVDs (Singular Value Decompositions) of matrices $\{F(t_k)\}_{k=1}^N$ be determined of the form:

$$F(t_k)=U_kS_kV_k^T \qquad (10)$$

with non zero singular values $\sigma_1^{(k)}\ge\sigma_2^{(k)}\ge...\sigma_p^{(k)}>0$ of matrices $S_k$ and orthogonal matrices $U_k,V_k$, i.e. $U_kU_k^T=VV_k^T=I$. By choosing a set of weighting matrices $K(t_k)=V_kS_k^\dagger U_k^T$, where $S_k^\dagger$ are the pseudo-inverses of matrices $S_k$, the least squares estimation of $\theta_M(t_N)$ based on the discrete-time version of equation (8) converges to the unknown parameter vector $\theta$ as $N\to\infty$, if matrix $W_N=\begin{bmatrix}PV_1^T\\PV_2^T\\ \cdot\\PV_N^T\end{bmatrix}\in R^{N\cdot p\times l}$ is such that $rank(W_N^TW_N)=l$, where $P\in R^{p\times l}$ is an unitary projection matrix.

The complete proof , omitted in this paper, is based on Theorem. 1 applied to the transformed reduced dimension parameter vector $PV_k^T \theta_M(t_k)$, that is demonstrated to converge, as $N \to \infty$, to $PV_\infty^T \theta$. From this property, under the rank maximality hypothesis of matrix $W_N^T W_N$, it follows that the unknown parameter vector $\theta$ can be determined by solving a standard Least Squares regression problem as applied to a discrete number of solutions of equation (8).

## 4. SIMULATION RESULTS

The Lyapunov-based identification method has been teted by simulation of an UUV surge dynamics. It has been assumed the following mathematical model:

$$\dot{x} = -0.2x - 0.1x|x| + 0.05\tau \qquad (11)$$

where $x$ is the surge velocity and $\tau$ is the force supplied by the thruster. The considered non linear system is linear with respect to parameter vector, and has a model of the type expressed by equation (3), with :

$$\varphi(x,\tau) = [-x \quad -x|x| \quad \tau] \qquad \theta = [0.2 \quad 0.1 \quad 0.05]^T$$

State responses to different types of input signals were simulated, to which noise signals have been superimposed. Such noises have been simulated as stationary realizations of a zero mean normal process with a given standard deviation. A plot of the deterministic surge response to a step input is shown in Fig. 2

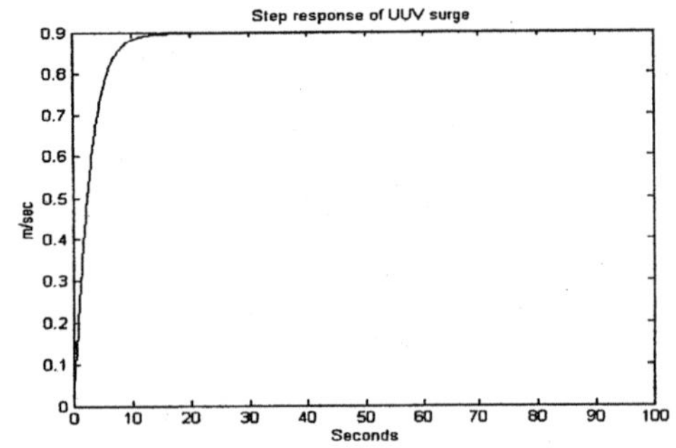


Fig. 2. Simulation of UUV surge response to a step input of 50 N.

Before using the identification algorithm, low-pass filtering and derivative filtering , as shown in Fig. 1, have been applied to the noisy state signal. In this case, as it can be easily verified, only Theorem 2 can be used, since $rank(\varphi(x,\tau)^T \varphi(x,\tau)) = 1$, while the dimension of parameter vector is 3.

Different increasing values of the state variable measurement noise $\sigma$ have been tested, starting with $\sigma = 0\ m/\sec$. The results are quite satisfactory and validate the proposed identification algorithm.

The almost immediate convergence of the normalized parameter vector $\theta_M(t)/\theta$, obtained in the deterministic case $\sigma = 0\ m/\sec$, is shown in Fig. 3, while the corresponding convergence practically to zero of Lyapunov function is shown in Fig. 4. In the simulated case of a measurement noise with a standard deviation $\sigma = 0.001\ m/\sec$, the convergence of the normalized parameter vector $\theta_M(t)/\theta$ has also been also extremely fast and accurate, as shown in Fig. 5. The corresponding convergence of Lyapunov function is shown in Fig. 6. Analogous plottings of the normalized parameter vector are shown in Fig. 6 and the corresponding Lyapunov function in Fig. 7, in the case $\sigma = 0.01\ m/\sec$.

A numerical comparison of the identified parameter vector is finally shown in Table 1.

Table 1 Comparison of identified parameter vector with measurement noise having standard deviation $\sigma$=0, $\sigma$=0.001 and $\sigma$=0.01

| | $\sigma$=0 | $\sigma$=0.001 | $\sigma$=0.01 |
|---|---|---|---|
| $\alpha$ | 0.2000 | 0.2006 | 0.1841 |
| $\beta$ | 0.1000 | 0.0990 | 0.1275 |
| $\gamma$ | 0.0050 | 0.0050 | 0.0052 |

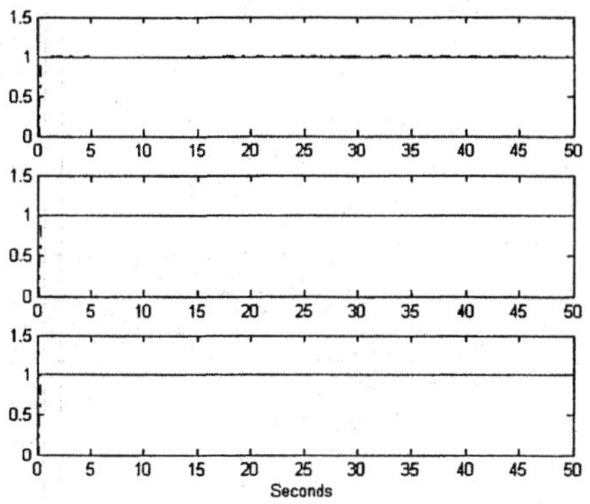


Fig. 3. Convergence of normalized identified parameter vector with a measurement noise standard deviation $\sigma = 0$.

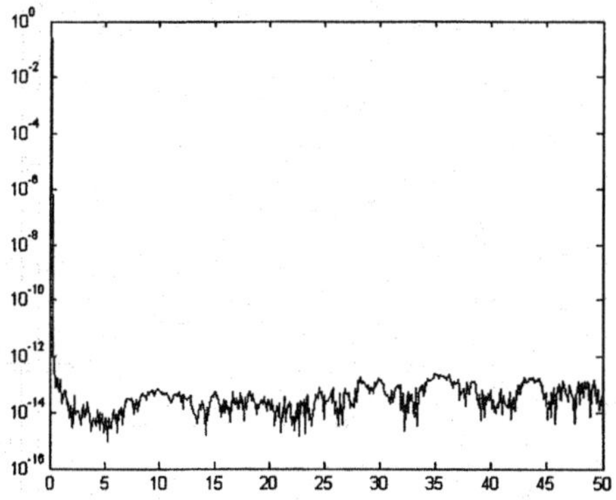


Fig. 4. Convergence of Lyapunov function with a measurement noise standard deviation $\sigma = 0$ .

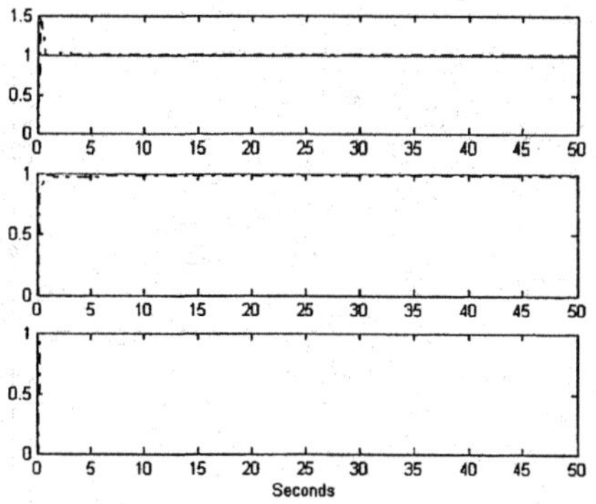


Fig. 5. Convergence of normalized identified parameter vector with a measurement noise standard deviation $\sigma = 0.001$

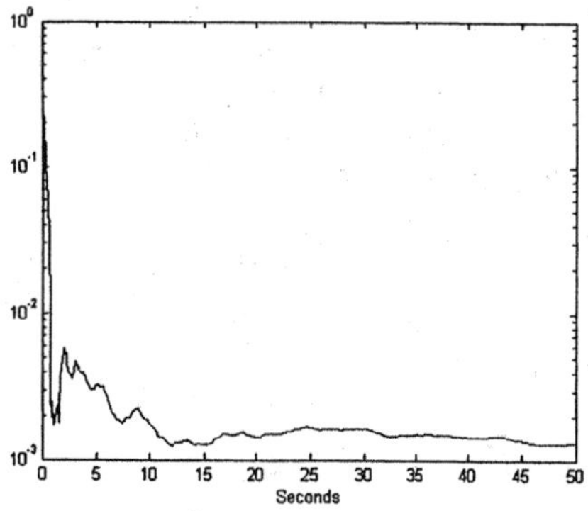


Fig. 6. Convergence of Lyapunov function with a measurement noise standard deviation $\sigma = 0.001$.

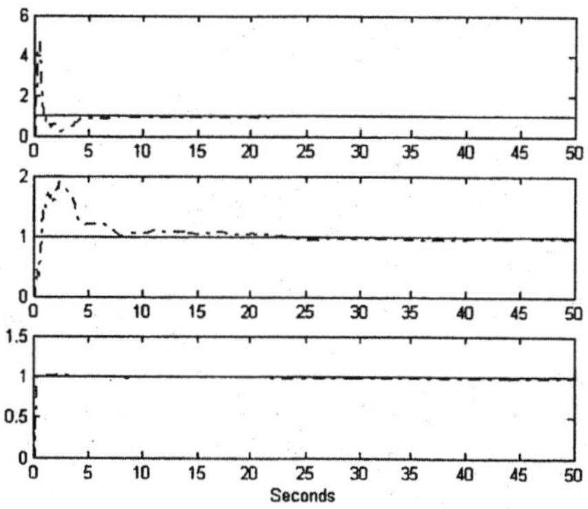


Fig. 7. Convergence of normalized identified parameter vector with a measurement noise standard deviation $\sigma = 0.01$.

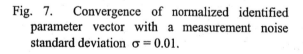

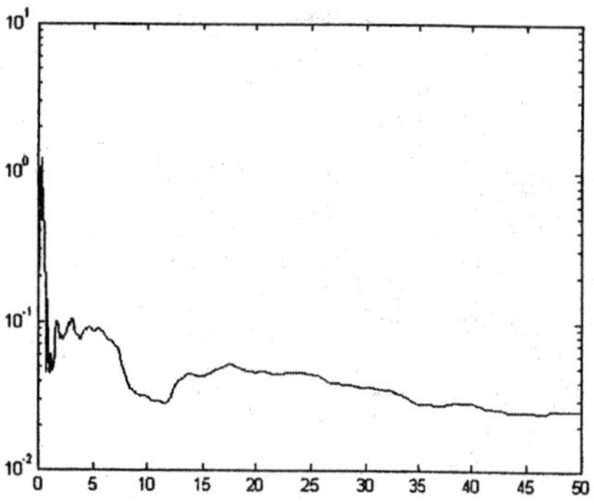


Fig. 8. Convergence of Lyapunov function with a measurement noise standard deviation $\sigma = 0.01$

## 5. CONCLUDING REMARKS

An innovative identification method for continuous time systems has been presented in this paper. The method is based on a Lyapunov approach and applies to deterministic non linear systems that are linear with respect to an unknown parameter vector. A combination of the basic identification algorithm with low-pass and derivative filters allows to cope also with systems perturbed by stochastic measurement noise. The main features of the method consist of an extremely fast and accurate convergence. The preliminary simulation tests conducted on the surge model of an UUV indicate that the Lyapunov based identification method can be used with appreciably good results. Even if the algorithm can be relatively demanding from the numerical point of view, it has the advantage of not being critically dependent on a persistent excitation condition of the input vector. A practical consequence of that is that it can be applied also during normal operation and does not require the execution of particular sequences of input vector.

## REFERENCES


Abkowitz, M.A. (1980). System identification techniques for ship manoeuvring trials. *Proc. Symposium on Control Theory and Navy Applications*, pp.337-393, Monterey, USA.

Fossen, T.I. (1994). *Guidance and Control of Ocean Vehicles*, John Wiley and Sons, New York.

Fossen, T.I. (2002). *Marine Control Systems*, Marine Cybernetics, Trondheim, Norway.

Ljung, L. (1987). *System Identification :Theory for the User*, Prentice Hall, Englewoods Clifts.

Lyshevski, S. And L.Abel, (1994). Nonlinear systems identification using the Lyapunov method, *Proc. 10th IFAC Symposium on System Identification*, Copenhagen, Denmark, 1994, Vol.1, pp. 307-312.

ELSEVIER
IFAC
PUBLICATIONS
www.elsevier.com/locate/ifac

# STUDY AND IMPLEMENTATION OF AN EKF GIB-BASED UNDERWATER POSITIONING SYSTEM

**A. Alcocer** [1] **P. Oliveira A. Pascoal**

*Institute for Systems and Robotics and*
*Department of Electrical Engineering,*
*Instituto Superior Técnico,*
*Av. Rovisco Pais, 1096 Lisboa Codex, Portugal*
*e-mail:* {alexblau,pjcro,antonio}@isr.ist.utl.pt

Abstract: The paper addresses the problem of estimating the position of an underwater target in real time. In the scenario adopted, the target carries a pinger that emits acoustic signals periodically, as determined by a very high precision clock that is synchronized with GPS, prior to system deployment. The target is tracked from the surface by using a system of four buoys equipped with hydrophones. The buoys measure the times of arrival of the acoustic signals emitted by the pinger or, equivalently, the four target-to-buoy range measurements (so-called GIB system). Due to the finite speed of propagation of sound in water, these measurements are obtained with different latencies. The paper tackles the problem of underwater target tracking in the framework of Extended Kalman Filtering by relying on a purely kinematic model of the target. The paper further shows how the differently delayed measurements can be merged using a *back and forward* fusion approach. A measurement validation procedure is introduced to deal with dropouts and outliers. Simulation as well as experimental results illustrate the performance of the filter proposed. *Copyright ©2004 IFAC.*

Keywords: Underwater Vehicles, Navigation Systems, Extended Kalman Filters

## 1. INTRODUCTION

The last decade has witnessed the emergence of Ocean Robotics as a major field of research. Remotely Operated Vehicles (ROVs) and, more recently, Autonomous Underwater Vehicles (AUVs) have shown to be extremely important instruments in the study and exploration of the oceans. Free from the constraints of an umbilical cable, AUVs are steadily becoming the tool *par excellence* to acquire marine data on an unprecedented scale and, in the future, to carry out interventions in undersea structures. Central to the operation of these vehicles is the availability of accurate vehicle navigation and/or positioning systems. The fact that electromagnetic signals do not penetrate below the sea surface makes the GPS unsuitable for underwater positioning. Hence, alternative solutions must be sought. The good propagation characteristics of sound waves in water makes acoustic positioning a viable solution.

Classical approaches to underwater vehicle positioning include Long Baseline (LBL) and Short

[1] Work supported in part by the Portuguese FCT POSI Programme under framework QCA III and by projects DREAM and MAROV of the FCT and MAYA of the AdI. The work of the first author was supported by an EC research grant in the scope of the FREESUB European Research Training Network.

Baseline (SBL) systems, to name but a few. See (Leonard *et al.*, 1998),(Larsen, 2001) and the references therein for an introduction to this challenging area. More recently, a number of methods have been proposed to "reproduce" the idea of GPS in the underwater environment. In (Youngberg, 1992) an underwater GPS concept was introduced. The system consists of surface buoys equipped with DGPS receptors that broadcast satellite information underwater, via acoustic telemetry. The underwater platform receives these messages from the buoys and computes its own position locally. Due to the technical difficulties inherent to acoustic communications, this concept has not yet materialized, as far as the authors are aware, in the form of a commercial product.

A different, yet related approach to acoustic underwater positioning has actually been implemented and is available commercially: the so-called GPS Intelligent Buoy (GIB) system (Thomas, 1998),(ACSA, 1999). This system consists of four surface buoys equipped with DGPS receivers and submerged hydrophones. Each of the hydrophones receives the acoustic impulses emitted periodically by a synchronized pinger installed on-board the underwater platform and records their times of arrival. As explained later in Section 6, the depth of the target is also available from the GIB system by coding that info in the acoustic emission pattern. The buoys communicate via radio with a central station (typically onboard a support vessel) where the position of the underwater target is computed. Due to the fact that position estimates are only available at the central station, this system is naturally suited for tracking applications.

Motivated by the latter approach to acoustic positioning, this paper addresses the general problem of estimating the position of an underwater target given a set of range measurements from the target to known buoy locations. Classically, this problem has been solved by resorting to triangulation techniques (Henry, 1978), which require that at least three range measurements be available at the end of each acoustic emission-reception cycle. This is hardly feasible in practice, due to unavoidable communication and sensor failures. It is therefore of interest to develop an estimator structure capable of dealing with the case where the number of range measurements available is time-varying. The paper shows how this problem can be tackled in the framework of Extended Kalman Filtering (EKF) whereby four vehicle-to-buoy range measurements drive a filter that relies on a simple kinematic model of the underwater target.

It is important to recall that due to the finite speed of propagation of sound in water, the range measurements are obtained at the buoys with different latencies. To overcome this problem, the paper shows how the differently delayed measurements can be merged in an EKF setting by incorporating a *back and forward* fusion approach. Simulation as well as experimental results illustrate the performance of the filter proposed.

The paper is organized as follows. Section 2 describes the problem of underwater target positioning and introduces the relevant process and measurement models. Based on the models derived, Section 3 computes the matrices that are essential to the mechanization of a solution to the positioning problem in terms of an Extended Kalman Filter (EKF). Section 4 shows how the EKF structure can be changed to accommodate latency in the measurements. Simulation and experimental results that illustrate the performance of the filter proposed are discussed in Sections 5 and 7, while Section 6 describes briefly the acoustic validation and initialization procedures that were implemented. Finally, Section 8 contains the main conclusions and describes challenging problems that warrant further research.

## 2. PROBLEM STATEMENT. FILTER DESIGN MODELS

Consider an earth fixed reference frame $\{O\}:=\{X_O, Y_O, Z_O\}$ and four (possibly drifting) buoys at the sea surface with submerged hydrophones at positions $(X_{hi}, Y_{hi}, Z_{hi})$; $i = 1, \ldots, 4$ as depicted in Figure 1. For simplicity of presentation, we restrict ourselves to the case where the target moves in a plane at a fixed known depth $Z_p$. Its position in the earth fixed frame is therefore given by vector $(x(t),\ y(t),\ Z_p)$. The problem considered in this paper can then be briefly be stated as follows: obtain estimates $(\hat{x}(t),\ \hat{y}(t))$ of the target position based on information provided by the buoys, which compute the travel time of the acoustic signals emitted periodically by a pinger installed onboard the underwater platform. The solution derived can be easily extended to the case where the target undergoes motions in three dimensional space.

### *2.1 Target (process) model*

In what follows we avoid writing explicitly the dynamical equations of the underwater target being tracked and rely on its kinematic equations of motion only. Thus, a general solution for target positioning is obtained that fits different kinds of moving bodies such as AUVs, ROVs, divers, or even marine mammals.

The following notation will be used in the sequel: $V$ is the total velocity of the vehicle in $\{O\}$, $\psi$

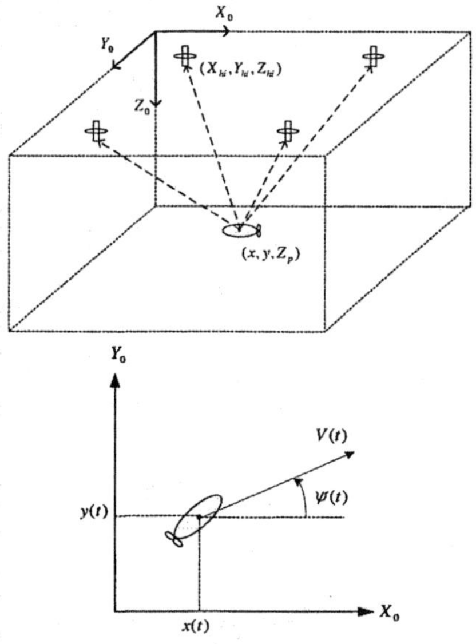


Fig. 1. Geometry of the target positioning problem.

denotes the angle between vector $V$ and $X_0$, and $r$ is the rate of variation of $\psi$. Notice that if the target moves in three dimensional space, and assuming the depth coordinated is known, tracking of its $x, y$ coordinates can be done easily by reinterpreting $V$ as the projection of the total velocity vector on its two first components. Given a continuous-time variable $w(t)$, $w(t_k)$ (abbv. $w(k)$) denotes its values taken at discrete instants of time $t_k = kh; k \in Z_+$, where $h > 0$ denotes the sampling interval. Standard arguments lead to the discrete-time kinematic model for the target

$$\begin{cases} x(k+1) & = x(k) + hV(k)\cos\psi(k) \\ y(k+1) & = y(k) + hV(k)\sin\psi(k) \\ V(k+1) & = V(k) + \xi_v(k) \\ \psi(k+1) & = \psi(k) + hr(k) + \xi_\psi(k) \\ r(k+1) & = r(k) + \xi_r(k), \end{cases} \tag{1}$$

where the inclusion of the angular rate equation for $r(k)$ captures the fact that the target undergoes motions in $\psi$ that are not measured directly and are thus assumed to be unknown. The process noises $\xi_v(k)$, $\xi_\psi(k)$, and $\xi_r(k)$ are assumed to be stationary, independent, zero-mean, and Gaussian, with constant intensities $\Xi_v$, $\Xi_\psi$, and $\Xi_r$ respectively. The above model can be written as a Linear Parametrically Varying system of the form

$$\begin{aligned} \boldsymbol{x}(k+1) &= f(\boldsymbol{x}(k), \boldsymbol{\xi}(k)) \\ &= A(\boldsymbol{x}(k))\boldsymbol{x}(k) + L\boldsymbol{\xi}(k) \end{aligned} \tag{2}$$

where

$$A(\boldsymbol{x}(k)) = \begin{bmatrix} 1 & 0 & h\cos\psi(k) & 0 & 0 \\ 0 & 1 & h\sin\psi(k) & 0 & 0 \\ 0 & 0 & 1 & 0 & 0 \\ 0 & 0 & 0 & 1 & h \\ 0 & 0 & 0 & 0 & 1 \end{bmatrix}, \quad \boldsymbol{x}(k) = \begin{bmatrix} x(k) \\ y(k) \\ V(k) \\ \psi(k) \\ r(k) \end{bmatrix} \tag{3}$$

$$L = \begin{bmatrix} 0 & 0 & 0 \\ 0 & 0 & 0 \\ 1 & 0 & 0 \\ 0 & 1 & 0 \\ 0 & 0 & 1 \end{bmatrix}, \quad \boldsymbol{\xi}(k) = \begin{bmatrix} \xi_v(k) \\ \xi_\psi(k) \\ \xi_r(k) \end{bmatrix}. \tag{4}$$

### 2.2 *Measurement model*

In the set-up adopted for vehicle positioning the underwater pinger carries a high precision clock that is synchronized with those of the buoys (and thus with GPS) prior to target deployment. The pinger emits an acoustic signal every $T$ seconds, at known instants of time. In each emission cycle, the pinger emits at discrete-time $s$, while buoys $i$; $i = 1, ..., 4$ compute their distances to the underwater unit at times $r_i \geq s$; $r_i = N_i h$, where $N_i$ is the time it takes for the acoustic signal to reach buoy $i$, modulo the sampling interval $h$. Notice that the $N_i$'s are not necessarily ordered by increasing order of magnitude, since they depend on the distance of each of the buoys to the target. Notice also that even though $z_i = z_i(s)$ refers to time $s$, its value can only be accessed at time $r_i > s$. It is therefore convenient to define $\bar{z}_i(r_i) = z_i(r_i - N_i h) = z_i(s)$, that is, $\bar{z}_i(r_i)$ is the measurement of $z_i(s)$ obtained at a later time $r_i$.

With the above notation, the noisy range measurements of each buoy are modeled as

$$z_i(s) = R_i(s) + [1 + \eta R_i(s)]\theta_i(s), \tag{5}$$

where

$$R_i^2(s) = (X_{hi} - x)^2 + (Y_{hi} - y)^2 + (Z_{hi} - Z_p)^2 \tag{6}$$

is the square of the distance from the vehicle to buoy $i$ and $x$ and $y$ denote the horizontal position of the pinger at instant $s$. In the above, $\theta_i(s)$ is a stationary, zero-mean, Gaussian white noise process with constant intensity $\Theta_i$. It is assumed that $\theta_i(s)$ and $\theta_i(s)$ are independent for $i \neq j$. The constant parameter $\eta$ captures the fact that the measurement error increases as the range grows.

The full set of available measurements available over an interrogation cycle can vary from 0 to 4, depending on the conditions of the acoustic channel. Stated mathematically, the set of $0 \leq m \leq 4$ measurements can be written as

$$\bar{z}_r^m = \bar{\mathcal{C}}[\bar{z}_1(r_1), ..., \bar{z}_4(r_4)]^T \tag{7}$$

where $\bar{\mathcal{C}} : \mathcal{R}^4 \rightarrow \mathcal{R}^m$ denotes the operator that extracts and orders the $m$ distances available according to the time-sequence at which they are computed at the buoys. For reasons that will become clear later, it is important to define

$$z^m(s) = \mathcal{C}[z_1(s), ..., z_4(s)]^T \qquad (8)$$

where $\mathcal{C} : \mathcal{R}^4 \to \mathcal{R}^m$ denotes the operator that extracts the set of $m$ distances available, as if they had been obtained at time $s$ . In an analogous manner, $z^p(s);\ p \leq m$ will denote the first $p$ components of $z^m(s)$.

## 3. EXTENDED KALMAN FILTER DESIGN

In preparation for the development of a positioning system, and taking into account the relationship between $z^m(s)$ and $\bar{z}^m_r$, consider an "ideal" situation where all or part of the $m$ measurements obtained over an interrogation cycle are available at the corresponding interrogation time $s$, as condensed in vector $z^p(s);\ p \leq m$. In this situation, given the nonlinear process and the observation models given by (1) and (5), respectively it is simple to derive an EKF structure to provide estimates of positions $x(k)$ and $y(k)$ based on measurements $z^p(s)$, where $s$ denotes an arbitrary interrogation time . The details are omitted; see for example (Anderson and Moore, 1979) or (Athans, 2003), and the references therein. Following standard practice, the derivation of an Extended Kalman Filter (EKF) for the above design model builds on the computation of the following Jacobian matrices about estimated values $\hat{x}(k)$ of the state vector $x(k)$:

$$\hat{A}(\hat{x}(k)) = \frac{\partial f(x,\xi)}{\partial x}|_{\hat{x}(k)}, \hat{L} = \frac{\partial f(x,\xi)}{\partial \xi}|_{\hat{x}(k)}, \qquad (9)$$

$$\hat{C}(\hat{x}(s)) = \frac{\partial z^p}{\partial x}|_{\hat{x}(s)}, \hat{D}(\hat{x}(s)) = \frac{\partial z^p}{\partial \theta}|_{\hat{x}(s)} \qquad (10)$$

It is straightforward to compute

$$\hat{A}(\hat{x}(k)) = \begin{bmatrix} 1 & 0 & h\cos(\hat{\psi}(k)) & -h\hat{V}(k)\sin(\hat{\psi}(k)) & 0 \\ 0 & 1 & h\sin(\hat{\psi}(k)) & h\hat{V}(k)\cos(\hat{\psi}(k)) & 0 \\ 0 & 0 & 1 & 0 & 0 \\ 0 & 0 & 0 & 1 & h \\ 0 & 0 & 0 & 0 & 1 \end{bmatrix} \qquad (11)$$

and

$$\hat{L} = \begin{bmatrix} 0 & 0 & 0 \\ 0 & 0 & 0 \\ 1 & 0 & 0 \\ 0 & 1 & 0 \\ 0 & 0 & 1 \end{bmatrix}. \qquad (12)$$

Furthermore, by defining

$$\hat{C}_i(\hat{x}(s)) = \begin{bmatrix} -\frac{X_{hi} - \hat{x}(s)}{\hat{R}_i(s)} & -\frac{Y_{hi} - \hat{y}(s)}{\hat{R}_i(s)} & 0 & 0 & 0 \end{bmatrix} \qquad (13)$$

and

$$\hat{D}_i(\hat{x}(s)) = 1 + \eta \hat{R}_i(s), \qquad (14)$$

it follows that

$$\hat{C}(\hat{x}(s)) = stack^p\{\hat{C}_j(\hat{x}(s))\} \qquad (15)$$

where $stack^p$ denotes the operation of stacking $p$ row matrices $\hat{C}_j(\hat{x}(s))$ by forcing the sequence of sub-indices $j$ to match that in $z^p(s)$. For example, if at time $s$ we have access to the distances measured by buoys 1, 3, and 2 in this order, then

$$\hat{C}(\hat{x}(s)) = \begin{bmatrix} C_1(\hat{x}(s)) \\ C_3(\hat{x}(s)) \\ C_2(\hat{x}(s)) \end{bmatrix}. \qquad (16)$$

(17)

Similarly,

$$\hat{D}(\hat{x}(s)) = diag^p\{\hat{D}_i(\hat{x}(s)\} \qquad (18)$$

where the elements of the $pxp$ diagonal matrix $\hat{D}(\hat{x}(s))$ are ordered in an analogous manner. Note that the dimensions of $\hat{C}$ and $\hat{D}$ vary according to the number of measurements that we suppose are available at time $s$. With an obvious abuse of notation, the measurement noise intensity matrix can then be written as

$$\Theta(s) = diag^p\{\Theta_j(s)\} \qquad (19)$$

where $\Theta_j(s) = \mathrm{E}\{\theta_j{}^2(s)\}$, while the process noise intensity matrix admits the representation

$$\Xi(k) = \mathrm{E}\{\xi(k)\xi^T(k)\} = \begin{bmatrix} \Xi_v(k) & 0 & 0 \\ 0 & \Xi_r(k) & 0 \\ 0 & 0 & \Xi_\psi(k) \end{bmatrix}. \qquad (20)$$

The matrices $A(\hat{x}(k))$ and $\hat{A}(\hat{x}(k))$ have an important property that will be used later:

*Property 1.* Given any nonzero positive integer $N$, define

$$\begin{cases} \alpha_1 = \alpha_1(N,k) & \triangleq \sum_{l=0}^{N} \cos(\hat{\psi}(k) + l\hat{r}(k)), \\ \alpha_2 = \alpha_2(N,k) & \triangleq \sum_{l=0}^{N} \sin(\hat{\psi}(k) + l\hat{r}(k)), \\ \beta_1 = \beta_1(N,k) & \triangleq \sum_{l=0}^{N} l\cos(\hat{\psi}(k) + l\hat{r}(k)), \\ \beta_2 = \beta_2(N,k) & \triangleq \sum_{l=0}^{N} l\sin(\hat{\psi}(k) + l\hat{r}(k)) \end{cases} \qquad (21)$$

Then it can be shown that

$$\hat{\Phi}(k + Nh, k) \triangleq \prod_{l=0}^{N} \hat{A}(\hat{x}(k + lh)) = \qquad (22)$$

$$\begin{bmatrix} 1 & 0 & h\alpha_1 & -h\hat{V}(k)\alpha_2 & -h\hat{V}(k)\beta_1 \\ 0 & 1 & h\alpha_2 & h\hat{V}(k)\alpha_1 & h\hat{V}(k)\beta_2 \\ 0 & 0 & 1 & 0 & 0 \\ 0 & 0 & 0 & 1 & hN \\ 0 & 0 & 0 & 0 & 1 \end{bmatrix} \qquad (23)$$

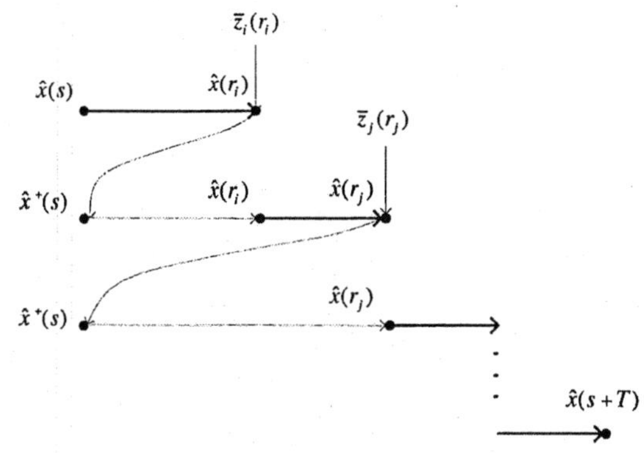


Fig. 2. *Back and forward* fusion approach. The solid line denotes the availability of real time outup filter data.

and

$$\Phi(k+Nh,k) \triangleq \prod_{l=0}^{N} A(\hat{x}(k+lh)) = \qquad (24)$$

$$\begin{bmatrix} 1 & 0 & h\alpha_1 & 0 & 0 \\ 0 & 1 & h\alpha_2 & 0 & 0 \\ 0 & 0 & 1 & 0 & 0 \\ 0 & 0 & 0 & 1 & hN \\ 0 & 0 & 0 & 0 & 1 \end{bmatrix}. \qquad (25)$$

For $N = 0$,

$$\hat{\Phi}(k,k) = \Phi(k,k) \triangleq I. \qquad (26)$$

## 4. FUSING DELAYED MEASUREMENTS WITH THE EKF

From the discussion above, the main problem to be overcome in the design on acoustic positioning system for the underwater unit is caused by the variable time-delay affecting each buoy measurement. The question then arises as to how delayed measurements can be naturally incorporated into an EKF structure. The reader will find in (Larsen *et al.*, 1998) a survey of different methods proposed in the literature to fuse delayed measurements in a linear Kalman Filter structure. In that work, a new method is also presented that relies on "extrapolating" the measurement of a variable obtained with latency to present time, using past and present estimates of the Kalman Filter. The problem tackled in this paper differs from that studied in (Larsen *et al.*, 1998) in two main aspects: the underlying estimation problem is nonlinear, and the components of the output vector at a given time $s$ are accessible with different latencies. As shown below, this problem can be tackled using a *back and forward* fusion approach which recomputes the filter estimates every time a new measurement is available. This computational complexity involved is drastically reduced by resorting to Property 1 .

In this work the estimator runs at a sampling period $h$ typically much smaller than $T$, the interrogation period of the underwater pinger. Let $s$ be an arbitrary instante of time at which the underwater pinger emits an acoustic signal and let $i \leq m$ be the buoy that first receives this signal at time $r_i = s + N_i h$. Further let $\bar{z}_i(r_i)$ be the corresponding distance. Up until time $r_i$ no new measurements are available, and a pure state and covariance prediction upatde are performed using the EKF set-up described before, leading to the predictor (see (Anderson and Moore, 1979))

$$\hat{x}(k+h) = A(\hat{x}(k))\hat{x}(k) \qquad (27)$$

$$P(k+h) = \hat{A}(\hat{x}(k))P(k)\hat{A}^T(\hat{x}(k)) + \hat{L}\Xi\hat{L}^T \qquad (28)$$

with $k = s, s+h, \ldots, r_i$. Upon reception of the first measurement $\bar{z}_i(r_i)$ available during the interrogation cycle, and assuming that the state $\hat{x}(k)$ and covariance $P(k)$ estimates at time $k = s$ have been stored, it is possible to go back to time $s$ and perform a filter state and covariance update as if measurement $\bar{z}_i(r_i)$ were available at $s$. Using the notation introduced before with $p$ set to 1, this leads to

$$\hat{x}^+(s) = \hat{x}(s) + K(s)\left[z^1(s) - \hat{z}^1(s)\right] \qquad (29)$$

$$P^+(s) = P(s) - P(s)\hat{C}^T \left[\hat{C}P(s)\hat{C}^T + \hat{D}\Theta\hat{D}^T\right]^{-1}\hat{C}P(s) \qquad (30)$$

$$K(s) = P^+(s)\hat{C}^T\left[\hat{D}\Theta\hat{D}^T\right]^{-1} \qquad (31)$$

where $\hat{z}^1(s)$ denotes the estimate of $z^1(s)$. A new prediction cycle can now be done moving forward in time until a new measurement $z_j$ is available, using (27)-(28) and starting with the updated states and covariance found in (29)-(30). Due to property 1, this prediction can be expressed in a computationally simple form. Let $r_j = s + N_j h$ be the time step at which measurement $\bar{z}_j(r_j)$ is received. Then, the prediction cycle from $s$ to $r_j$ can be computed in closed form as

$$\hat{x}(r_j) = \Phi(r_j,s)\hat{x}^+(s) \qquad (32)$$

$$P(r_j) = \hat{\Phi}(r_j,s)P^+(s)\hat{\Phi}^T(r_j,s) + \sum_{l=0}^{N_j-1} \hat{\Phi}(s+lh,s)\hat{L}\Xi\hat{L}^T\hat{\Phi}^T(s+lh,s) \qquad (33)$$

Again, upon computation of measurement $\bar{z}_j(r_j)$ it is possible to go back to time $s$ and perform a filter state and covariance update as if measurements $\bar{z}_i(r_i)$ and $\bar{z}_j(r_j)$ were available at $s$. This is done using equations (29)-(30), with the one-dimensional vector $z^1(s)$ replaced by $z^2(s)$ and matrices $\hat{C}$,$\hat{D}$, and $\Theta$ matrices recomputed accordingly. This *back and forward* structure proceeds until the $m$ measurements available over an interrogation cycle (starting at $s$ and ending at $s+T$) are dealt with. This procedure is then repeated for each interrogation cycle. The overall structure of the algorithm proposed is depicted in Figure 2.

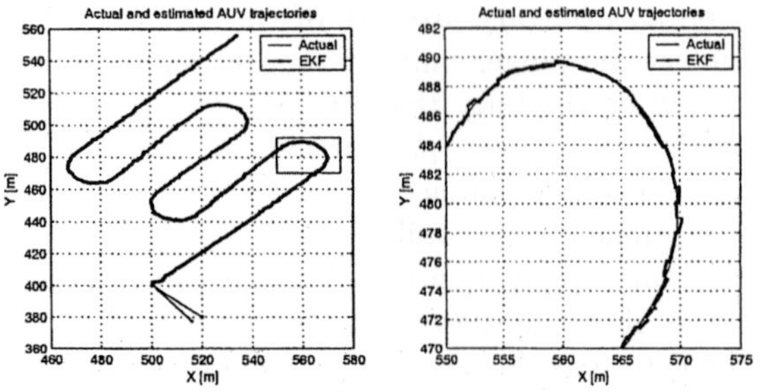


Fig. 3. Left: Actual and estimated simulated AUV trajectories. Right: idem, details of boxed area

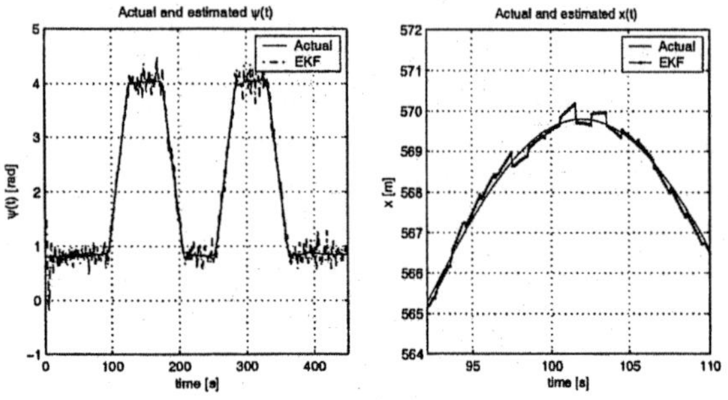


Fig. 4. Left: actual and estimated $\psi(t)$. Right: detail of actual and estimated $x(t)$.

## 5. SIMULATION SETUP AND RESULTS

In the simulations, four buoys were placed at the corners of a square with a 1Km side. The depth of the hydrophones $Z_{hi}$ was set to 5m for all the buoys. A typical target trajectory was simulated at a nominal speed of 1m/s and a turning diameter of 15m. The target maneuvered in the horizontal plane, at a constant depth $Z_p = 50$m. The range measurements were generated every $T = 1$s and corrupted according to (5) with a 0.1m standard deviation Gaussian noise. The EKF was run at a sampling period of $h = 0.1$s. The actual and estimated initial states, as well as the process and measurement noise intensities, were set to

| | |
|---|---|
| $x(0)$ | $[500\ 400\ 1\ \pi/4\ 0]^T$ |
| $\hat{x}_0$ | $[520\ 380\ 0.5\ \pi/2\ 0]^T$ |
| $P_0$ | $\text{diag}\{(20)^2\ (20)^2\ (0.5)^2\ (0.05)^2\ (0.005)^2\}$ |
| $\Xi_v$ | $(0.001)^2$ |
| $\Xi_\psi$ | $(0.005)^2$ |
| $\Xi_r$ | $(0.02)^2$ |
| $\Theta_i$ | $(0.1)^2, \quad i = 1, \ldots, 4$ |
| $\eta$ | 0.001 |

Figure 3 shows a simulation of actual and estimated $2D$ target trajectories and the details of a turning maneuver. Figure 4 shows actual and estimated $\psi(k)$ as well as the details of actual and estimated $x(k)$. Notice the 'jump' in the estimates whenever a new measurement is available. Notice also how the heading estimates change slowly in the course of a turning maneuver.

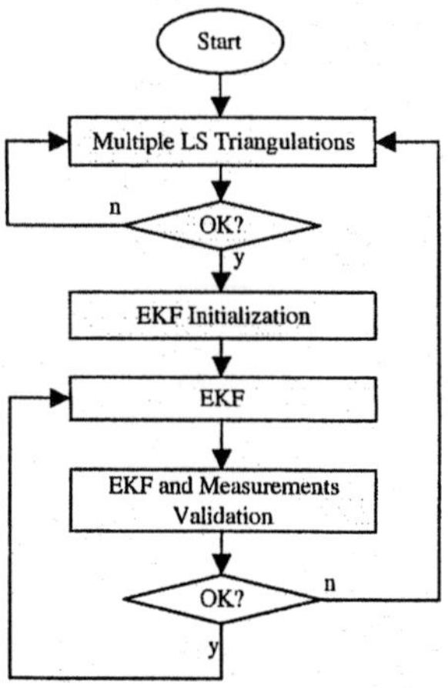


Fig. 5. Measurement validation and Initialization procedures.

## 6. MEASUREMENT VALIDATION AND EKF INITIALIZATION

In preparation for actual tests of the GIB-based system at sea, this section discusses practical issues that warrant careful consideration. As is well known, the implementation of any acoustic positioning system requires that mechanisms be developed to deal with dropouts and outliers that arise due to acoustic path screening, partial system failure, and multipath effects. See for example (Vaganay *et al.*, 1996) and the references therein for an introduction to this circle of ideas and for an interesting application to AUV positioning using a Long Baseline System. In the case of the GIB system, the problem is further complicated because of the mechanism that is used to transmit the depth of the target. In fact, the pinger onboard the vehicle emits two successive acoustic pulses during each emission cycle, the time delay between the two pulses being proportional to the pinger depth. Ideally, the data received at each buoy during each emission cycle consists of two successive pulses only. In practice, a number of pulses may be detected depending on the "quality" of the acoustic channel. For example, the data received may correspond to a number of situations that include the following: i) only the first pulse is received - a valid range measurement is acquired but the depth info is not updated, ii) only the second pulse is received - data contains erroneous information, and iii) a single pulse is received as a consequence of multipath effects - data may be discarded or taken into consideration if a model for multipath propagation is available.

In the present case, following the general strategy outlined in (Vaganay *et al.*, 1996), a two stage procedure was adopted that that includes a time-domain as well as a spatial-domain validation. Time-domain validation is done naturally in an an EKF setting by examining the residuals associated with the measurements (i.e., the difference between predicted and measured values as they arrive), and discarding the measurements with residuals that exceed a certain threshold. During system initialization, or when the tracker is not driven by valid measurements over an extended period of time, a spatial-domain validation is performed to overcome the fact that the estimate of the target position may become highly innacurate. This is done via an initilization algorithm that performs multiple Least Squares (LS) triangulations based on all possible scenarios compatible with the set of measurements received and selects the solution that produces the smallest residuals. Figure 6 shows raw and validated measurements for one of the GIB buoys. The vertical scale is presented in milliseconds to stress the fact that each of the buoys computes its distance to the pinger indirectly, by measuring the time-delay between the reception and the emission of the first acoustic pulse. Notice how the depth information is coded in the delay between two consecutive acoustic pulses. The figure on the right shows the boxed area in detail.

The diagram in Figure 5 depicts the procedure for measurement validation. In an initialization scenario, or whenever a filter reset occurs, the multiple triangulation algorithm is performed until a valid solution is obtained, that is, until the residuals of the resulting set of measurements are less than a certain threshold. Once a valid position fix is obtained, the EKF is initialized and a procedure that relies on the EKF estimates and *a piori* information about the vehicle's maximum speed and noise characteristics selects the valid measurements. The EKF will be reset if the residuals become bigger than a threshold or if the duration of a pure prediction phase (that is, the time window during which no validated measurements are available) lasts too long.

## 7. EXPERIMENTAL SETUP AND RESULTS

Experimental data were recorded during tests at sea in Sines, Portugal, during the period from 23-24th April, 2003 using a commercially available GIB system. The data were processed off-line using the positioning algorithm described. Data validation was done using the methodology described in the previous section. Four buoys were moored in an approximate square configuration with a 500 meter side. The pinger was maneuvered at an approximate depth of 5 meters. In the data subset selected for post-processing, acoustic reception at hydrophones was good and four measurements were available most of the time. However, one of the buoys had reception problems and did not commute to Differential GPS mode. This situation was easily tackled in an EKF framework by giving

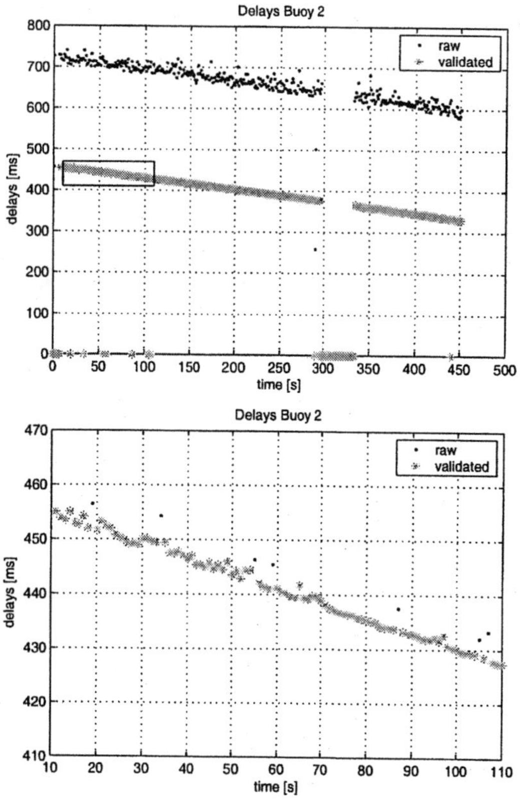


Fig. 6. Top: Raw and validated measurements from one of the buoys during tests at sea. Bottom: idem, detail of boxed area.

higher variance to the measurements provided by that buoy. The following conditions were adopted:

| | |
|---|---|
| $P_0$ | $\text{diag}\{(10)^2\ (10)^2\ (1)^2\ (0.05)^2\ (0.005)^2\}$ |
| $\Xi_v$ | $(0.01)^2$ |
| $\Xi_\psi$ | $(0.005)^2$ |
| $\Xi_r$ | $(0.025)^2$ |
| $\Theta_i$ | $(1)^2$ if DGPS, $(4)^2$ if GPS, $i = 1, \ldots, 4$ |
| $\eta$ | 0.001 |

Figure 7 shows the estimated target trajectories using EKF and a simple LS Triangulation. The EKF performs well, even with 1 or 2 measurements, whereas Triangulation is unable to compute solutions. As expected, Triangulation produces noisier estimates than the EKF solution. Figure 8 is a screenshot of the graphical interface used to report the status of the target tracking algorithm.

## 8. CONCLUSIONS AND FUTURE WORK

The paper proposed a solution to the problem of estimating the position of an underwater target in real time. The experimental set-up adopted consists of a system of four buoys that compute the times of arrival of the acoustic signals emitted periodically by a pinger installed onboard the moving platform (so-called GIB system). The positioning system fuses the vehicle-to-buoy range measurements by resorting to an

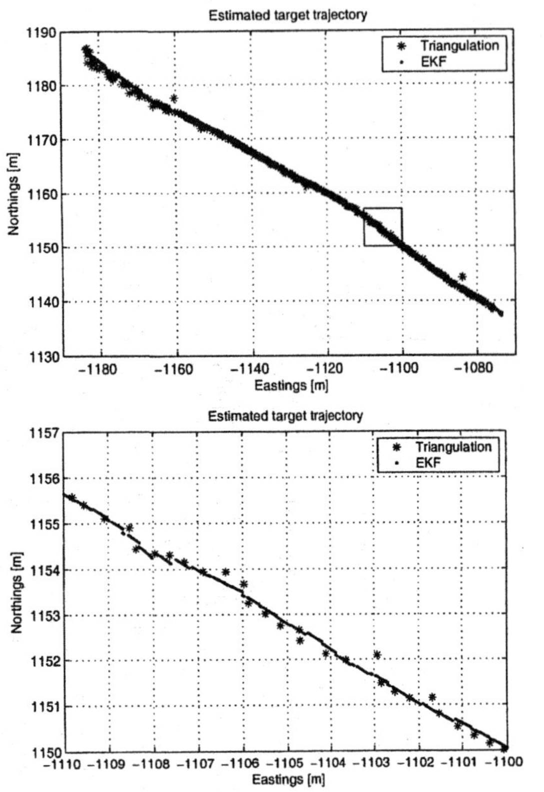


Fig. 7. Top: Estimated target trajectories with Triangulation and EKF based Estimator. Bottom: idem, zoom of boxed area.

Extended Kalman Filter (EKF)-structure that addresses explicitly the problems caused by measurement delays. By dealing directly with each buoy measurement as it becomes available, a system is obtained that exhibits far better performance than that achievable with classical triangulation schemes, where all buoy measurements are collected before an estimate of the target's position can be computed. Simulation as well as experimental results show that the proposed filter is computationally effective and yields good results, even in the presence of acoustic outliers or a reduced number of valid buoy measurements. Future work will include the study of different nonlinear filter structures for which convergence results can in principle be derived. Another interesting topic of research is how to fuse the filter estimates with other kinds of sensorial data.

REFERENCES

ACSA, ORCA (1999). *Trajectographe GIB Manuel Utilisateur.*

Anderson, B. D. O. and J. B. Moore (1979). *Optimal Filtering.* Prentice Hall.

Athans, Michael (2003). Viewgraphs and Notes on Dynamic Sthochastic Estimation, Filtering, Prediction and Smoothing. ISR/IST, Lisbon, PT.

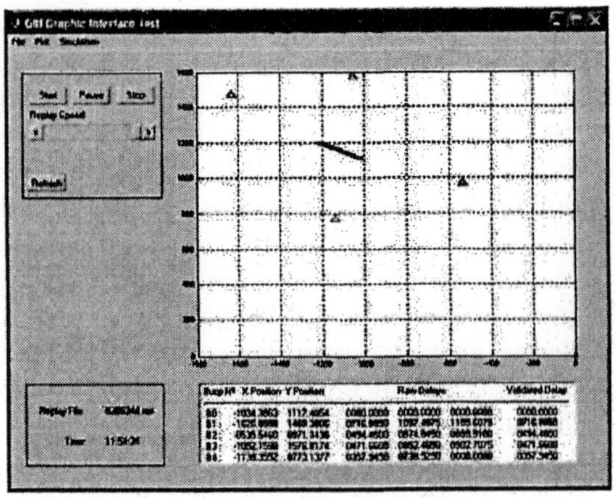

Fig. 8. Screenshot of tracking software interface.

Henry, T. D. (1978). Acoustic transponder navigation. In: *IEEE Position Location and Navigation Symp.*. pp. 237–244.

Larsen, Mikael Bliksted (2001). Autonomous Navigation of Underwater Vehicles. Ph.D. Dissertation. Department of Automation, Technical University of Denmark.

Larsen, T. D., N. K. Poulsen, N. A. Andersen and O. Ravn (1998). Incorporation of time delayed measurements in a discrete-time kalman filter. In: *Proceedings of the 37th Conference on Decision and Control, Tampa, Florida, USA*. pp. 3972–3977.

Leonard, J., A. Bennett, C. Smith and H. Feder (1998). Autonomous underwater vehicle navigation. MIT Marine Robotics Laboratory Technical Memorandum 98-1.

Thomas, H. G. (1998). Gib buoys: An interface between space and depths of the oceans. In: *Proceedings of IEEE Autonomous Underwater Vehicles, Cambridge, MA, USA*. pp. 181–184.

Vaganay, J., J. Leonard and J. Bellingham (1996). Outlier rejection for autonomous acoustic navigation.. In: *Proceedings of IEEE International Conference on Robotics and Automation, Minneapolis, MN, USA*. pp. 2174–2181.

Youngberg, James W. (1992). Method for extending GPS to Underwater Applications. US Patent 5,119,341, June 2 1992.

ELSEVIER
IFAC
PUBLICATIONS
www.elsevier.com/locate/ifac

# VISUAL-FEEDBACK POSITIONING OF A ROV

**G. Conte, S. M. Zanoli, D. Scaradozzi, S. Maiolatesi**

*Università Politecnica delle Marche*
*Department of Information, Management and Automation Engineering*
*Via Brecce Bianche – 60131 Ancona - ITALY*
*gconte@univpm.it*

Abstract: This paper consider the problem of positioning a ROV in a partially structured underwater environment using a control strategy mainly based on visual feedback. In order to reach and to maintain the desired position with respect to a known target, images taken by a CCD camera are processed and the resulting information is fused with that coming from other sensors. Low level control actions are based on PI controller, endowed with a fuzzy adaptation mechanism. The performances of the developed control scheme are evaluated and discussed on the bases of experimental results.



Keywords: ROV positioning, visual feedback control.

## 1. INTRODUCTION

In this paper we consider the problem of positioning a ROV in a partially structured underwater environment, using a simple control technique based on visual feedback and fuzzy adaptation. By means of the on-board CCD camera, the ROV has to detect a target, consisting of a spherical object of known dimension surrounded by free waters, and to position itself at an assigned distance in front of it. This task can be viewed as a basic one in survey or monitoring missions as well as in intervention ones and the possibility to perform it in an automatic way, letting the operator free to concentrate on higher level aspects of the mission, helps in making ROV operation easier, cheaper and more reliable (see Jin et al., 1996; Nguyen et al., 1988; Santos et al., 1994). In the real situation we consider, four degrees of freedom of the ROV are actuated (heave, surge, sway and yaw), while roll and pitch motions, that are not controllable, are assumed to be negligible.
In order to perform the task described above, image processing is used to detect the target and to extract the error signal employed in the control procedure. This latter essentially consists in controlling the position of the target in the image plane by moving appropriately the ROV and, as consequence, the camera. The control action is decoupled in order to regulate first the vertical position, then the horizontal one and finally the distance, using, in addition to the visual information and the knowledge of the environment (namely the shape and dimension of the target), also information coming from other on-board sensors, like compass, gyros and depth meter. This way of decoupling the control action mimics the likely behaviour of a human operator involved in the same task.
Regulation of the vertical position is chosen first because it may be assumed that the only relevant disturbance on it is the known, constant action of the restoring forces (gravity and buoyancy). The control system employs visual feedback in a first phase and, then, the depth meter signal to reach and to maintain the correct vertical position. Switching from images to the depth meter signal in the feedback loop reduces the computational burden and increases the promptness of the controller. Returning periodically to the visual feedback mode, the system is able to check performances and to correct a corrupted behaviour.
The following control action consists in orienting the ROV so that the camera points to the target. As a consequence of this motion, the target image in the image plain moves horizontally towards the central

position. As before, the control system first employs visual feedback and, then, it relies on the compass signal to reach and to maintain the correct heading. Switching from images to the compass signal in the feedback loop again reduces the computational burden and increases the promptness of the controller. After depth and heading have been stabilized around the chosen set points, horizontal displacements of the target in the image plane denote a lateral movement of the ROV due to some disturbance, like ocean current. Then, the control system counteracts appropriately in order to compensate for the effect of the disturbance.
After stabilizing the target image in the vertical and horizontal central position, the control system evaluates the distance from the target by measuring the dimension of its image and using the knowledge of the camera characteristics and of the target real dimension. Variations of the dimension in successive images denote a movement due to some disturbance, like ocean current, and the control system may counteract and, if required, move the ROV towards the target or far from it in order to reach the set point.
The control system has been developed on a modular architecture, consisting of a Supervisor and a number of lower level controllers. Implementation employs a PC-station equipped with A/D-D/A board, Windows operating system and LabView software environment. Experiments and test have been carried in a pool in order to evaluate the performances of the system. Results, as illustrated and discussed, show that the system is able to perform with acceptable reliability and accuracy in presence of disturbances which degrade image quality or influence ROV stability.

## 2. PROBLEM SETTING

In the problem we face, the ROV is navigating in an underwater environment where the main object of interest is an artificial target. This consists of a submerged, small spherical buoy, moored to the sea bottom. The diameter of the target is assumed to be known and that information is suitably made available to the control system of the ROV, in order to be used in positioning. Basically, what we are interested in is to position the ROV at a given distance from the target and in such a way that the on-board camera points to the target. Clearly, if we do not impose to the ROV a given heading - with respect to an inertial coordinate frame - there are infinitely many positions which correspond to possible solutions of the problem. However, the solution scheme we develop below can be modified to handle also situations in which heading is given. By exploiting this possibility, once the problem is solved letting the heading - with respect to an inertial coordinate frame - free and the ROV has been correctly positioned, we can force the vehicle to move around the target, while pointing at it, by imposing a different heading. In this way, we can inspect the entire surface of the target.

The ROV we consider in this work is a small class ROV DOE Phantom S2, whose thrusters (two horizontal main thrusters and two vertran ones) actuate four degrees of freedom (surge, sway, heave and yaw). The on-board sensory system consists of a CCD camera, a strain gauge depth sensor, a flux-gate compass and a low-cost, strapdown inertial measuring unit (IMU) that evaluates linear accelerations and angular velocities along and around three axes.
The hardware control architecture of the ROV integrates a standard console together with a PC station, equipped with D/A-A/C boards (see Conte et al., 2004 for a more detailed description of the control architecture). The PC performs the image processing and it implements the automatic closed loop control strategies. In addition, it records the mission data and, if required, it can process them and display the resulting information. In particular, the data collected by the IMU can be used for evaluating the orientation in space of the ROV, which can then be displayed by means of an artificial horizon. The operator can command the thrusters digitally through the PC, using the keyboard and the mouse, or analogically through the joysticks on the console. The first control mode is particularly effective for the main, horizontal thrusters, whose speed is closed loop controlled. The control variable for the vertran thrusters is voltage. Switching between the PC-based control mode and the console-based control mode is always possible, so to allow the operator to take control from the console in any situation.
The main information used in positioning the ROV comes on-line from visual images of the target, acquired by means of the on-board CCD camera. Position and orientation of the camera are fixed so that it looks forwards and its axis is aligned with that of the vehicle. Acquired, low level visual data consist of RGB signals, which are processed at different levels, using suitable numerical algorithms, by the PC in the control loop. The others sensors supply in particular information about depth (with approximately 1m resolution) and heading (with 0.1deg resolution). Information about orientation in space of the ROV, obtained from the collected by the IMU, is also used by the control system.

## 3. IMAGE PROCESSING

Typical images acquired by the on-board camera present low contrast, due to the characteristics of the underwater environment, and may be corrupted by water turbidity. However, since the ROV is navigating in essentially free waters, the target, which appears as an almost elliptical shape of contrasting colour, can be reasonably expected to be the only visible object. Therefore, after an initialization phase in which the ROV is governed by the operator until the target is in sight, the primary task of the image processing is that of detecting in the captured image an approximately elliptical region of contrasting colour (see Fisher et al., 1994 for general image processing techniques). Then, the area and the

coordinates of the centre of the elliptical region are evaluated, so to be available for generating the error signals used in the control loops. Essentially, the image processing, as well as the generation of the error signals, is performed by suitable blocks which form part of the controllers described in next section.
In details, the RGB signal coming from the camera undergoes three different processing: a Low Level Filtering, followed first by an Edge Detection and then, by a High Level Geometric Recognition. Low Level Filtering consists mainly in extracting a region of interest from the whole image and in reducing the intensity levels in order to exploit better lighting conditions and to reduce noise. Canny Edge Detection is then applied in order to select candidate contour points and to group them in connected sets by means of segmentation. Using Hough transform, High Level Geometric Recognition is performed (see Inverso et al., 2002) and the target image is recognized as an elliptical shape in the image plane (see Figure 1, 2, 3).

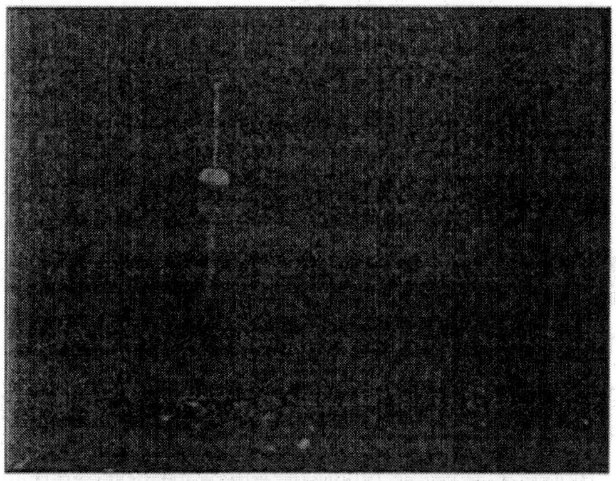

Fig. 1. Target image after the initialization phase.

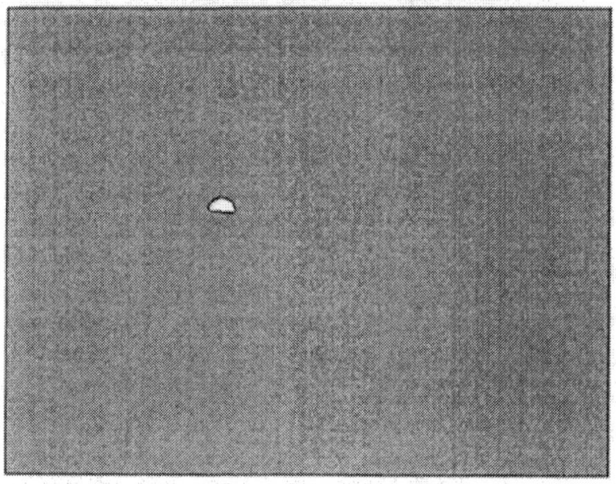

Fig. 2. Target image after filtering and edge detection (the fore ground is made grey to increase readability).

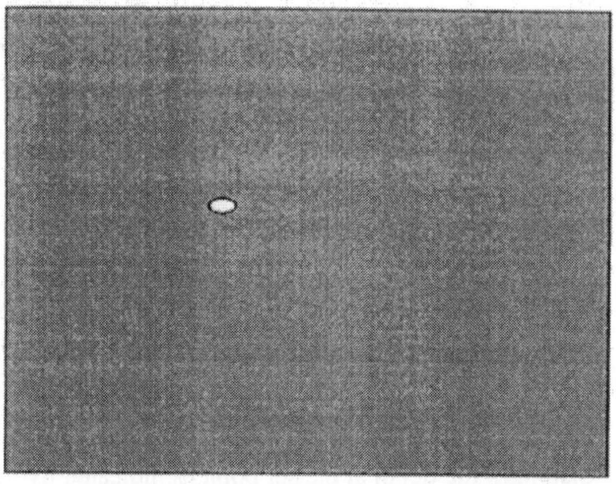

Fig. 3. Target image after geometric recognition (the fore ground is made grey to increase readability).

The coordinates (xc,yc) of the centre of the ellipse are evaluated with respect to a Cartesian coordinate frame whose origin correspond to the pixel in the upper left corner of the image, the X-axis is oriented to the right and the Y-axis is oriented downward. Coordinates and area are measured in pixels; due to the characteristics of the vision system, we have $0 \leq xc \leq 600$ and $0 \leq yc \leq 768$.

## 4. CONTROL ARCHITECTURE

The control architecture that has been developed for the present application consists of a high level Supervisor and of six lower level Controllers. The Supervisor can enable or disable the lower level Controller according to the system and mission status. Each low level Controller implement a feedback loop, where the error signal is generated by a block that processes appropriate sensory signals (coming from the camera, from the depth meter or from the compass) and the control signal is generated by a PI block. In addition, a fuzzy adaptation mechanism which adjusts the proportional gain of the PI block is implemented on one of the low level Controller.
Since the roll and pitch motions are not actuated, we will assume that the equilibrium positions in 0 are asymptotically stable and that perturbations are negligible. The first hypothesis is actually verified thank to the structural characteristics of the vehicle. In addition, by processing the IMU data, the Supervisor can evaluate the roll angle and the pitch angle and it can temporarily interrupt the mission if their values exceed a fixed threshold. After a fixed time, if the equilibrium has not been reached again, the Supervisor stops the mission and warns the operator.
During normal operation, the overall action of the control system can be described in the following way.

*Phase 1.* When the operator declares done the initialization phase and the target appears in images, the Supervisor activates the Depth Visual Controller (DVC), which evaluates an error signal $e_v(t) = yc - 384$ (recall that (xc,yc) are the coordinates of the centre of the elliptical target's image and that

$0 \leq yc \leq 768$) from the acquired images and moves the ROV vertically, in order to position the target image at mid height in the image plane. We exploit, in this phase, the stability around zero of the pitch angle of the vehicle and we assume, quite safely, that the only disturbance acting on the vertical position is due to the effects of the restoring forces (buoyancy and gravity).

*Phase 2.* When the error signal $e_v(t)$ reaches 0, the Supervisor records the value $d_0$ indicated by the depth meter and it starts a cycle of duration $t_0$, during which the DVC is disabled and the depth control is assigned to the Depth Controller (DC). This controller stabilizes the depth at the set point $d_0$ indicated by the Supervisor, namely at the depth corresponding to the condition $e_v = 0$, using, as feedback signal, that of the depth meter. At the end of the time cycle, the Supervisor enables again the DVC and it disables the DC, so to check by visual feedback the accuracy of the vertical positioning. If necessary, the DVC remains active until again $e_v(t)$ reaches 0 and the initial condition of this Phase is restored.

In addition to the above actions, at the beginning of this Phase the Supervisor activates also the Heading Visual Controller (HVC), which evaluates an error signal $e_h(t) = xc - 300$ (recall that (xc,yc) are the coordinates of the centre of the elliptical target's image and that $0 \leq xc \leq 600$) from the acquired images and modifies the yaw angle of the ROV, in order to position the target image at the centre of the image plane. It has to be remarked that the error signal employed by the HVC is generated by processing only a reduced region of interest (ROI) of the whole image, namely a central horizontal stripe of suitable height – say $2e_{vmax}$, since the previous control action guarantees that the image of the target is found there. The height of the stripe $2e_{vmax}$ is computed by the Supervisor according to the apparent dimension of the target. Use of the ROI helps in lowering the computational burden. In case the module of $e_v(t)$ turns out to be greater than evmax when the DVC is periodically enabled, the HVC is disabled until again $e_v(t)$ reaches 0 and the initial condition of this Phase is restored.

Remark that either the DVC or the DC will be active also during all the following Phases

*Phase 3.* When the error signal $e_h(t)$ reaches 0, the Supervisor records the value $a_0$ indicated by the compass and it disables the HVC, assigning the heading control to the Heading Controller (HC). This controller stabilizes the heading at the set point $a_0$ indicated by the Supervisor, namely at the angle corresponding to the condition $e_h = 0$, using, as feedback signal, that of the compass. At the same time, the Supervisor activates also the Sway Visual Controller (SVC), which also evaluates the error signal $e_h(t)$ from the acquired images and moves the ROV laterally, in order to position the target image at the centre of the image plane. In this Phase, in facts, since the heading is kept fixed by the HC, an horizontal displacement of the target image reveals the action of an ocean current on the ROV. The SVC operates by counteracting the lateral component of the force exerted by the current.

Remark that both the HC and the SVC will be active also during all the following Phases.

*Phase 4.* When, in Phase 3, all the errors signals of the active controllers equal 0, the Supervisor enables the Proximity Visual Controller (PVC), which evaluates an error signal $e_p(t)$ from the acquired images and moves the ROV forward or backward, in order to enlarge or reduce the image of the target until its area reaches a given set point, which corresponds to a desired distance between the ROV and the target. If one of the error signals of the active controllers, except that of the PVC, deviates significantly from 0 (that is for more than evmax in the case of $e_v(t)$ and for more of suitably chosen values in the case of the others), the Supervisor disables the PVC until the initial condition of this Phase are restored.

It has to be remarked that, switching between the DVC and the DC, the control system operates an elementary but effective fusion of the information coming from the camera and that coming from the depth meter. By acting on the duration $t_0$ of the cycle activated in Phase 2, the operator can practically assign more relevance to one kind of information with respect to the other.

For regulating the action of the PVC according to the situation, its proportional gain is modified on-line by a fuzzy adaptation mechanism. Decomposing the range of variation of $e_d$ into three fuzzy set, corresponding to the linguistic values Large, Medium and Small, and letting the sign of its variation be represented by the linguistic values Positive, Zero and Negative, a fuzzy mechanism of Sugeno type, with nine inferential rules has been defined. The action of the fuzzy mechanism, according to inferential rule, consists in increasing or decreasing the nominal gain, within fixed limits, or, possibly, in retaining it.

The entire control architecture is described in Figure 4.

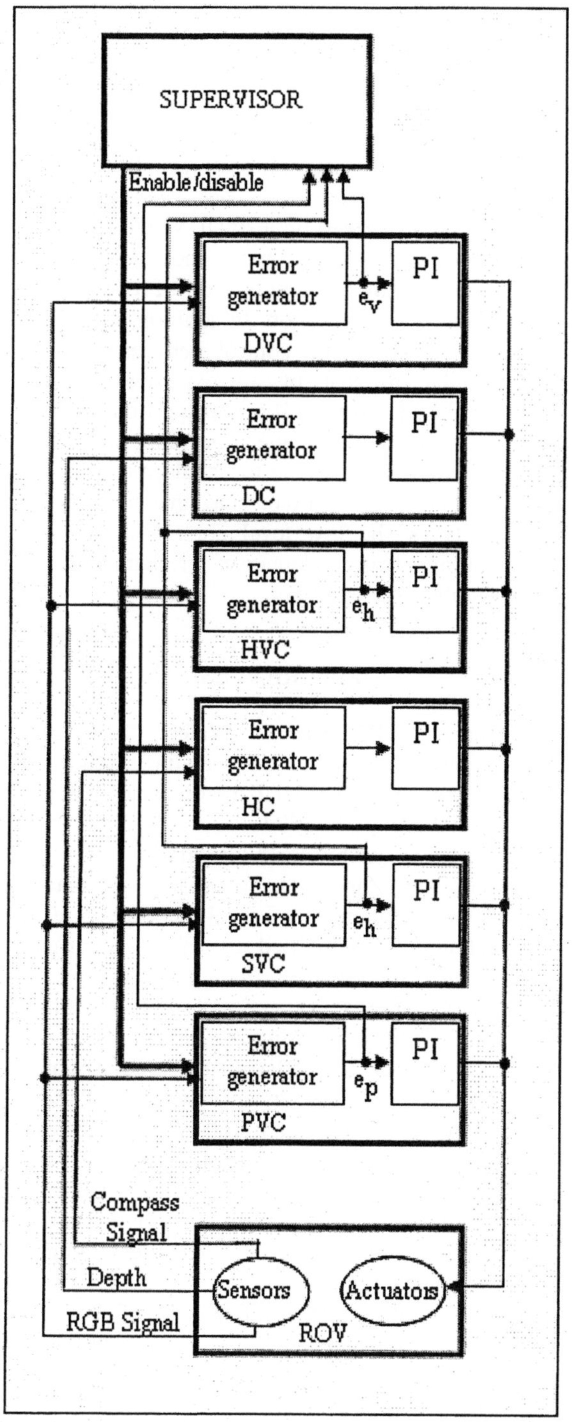


Fig. 4. Control Architecture.

## 5. EXPERIMENTAL RESULTS

Experiments for evaluating the performances of the proposed control scheme have been carried on in a pool, using as target a small with buoy with a diameter of 20cm. Figure 1 shows the buoy at a distance of about 5m. The limited depth of the pool (1.5m) did not permit to test in a satisfactory way the performances of the DVC. However, the results obtained about the performances of the HVC and of the SVC, which work in a very similar way, may reasonably be assumed to hold also for the DVC. Further experiments in a deeper pool are planned.

For the reason mentioned above, the control system practically implemented only Phase 2, 3 and 4. Figure 5 shows the behavior of the error signal $e_h(t)$ over a time interval of about 80s.

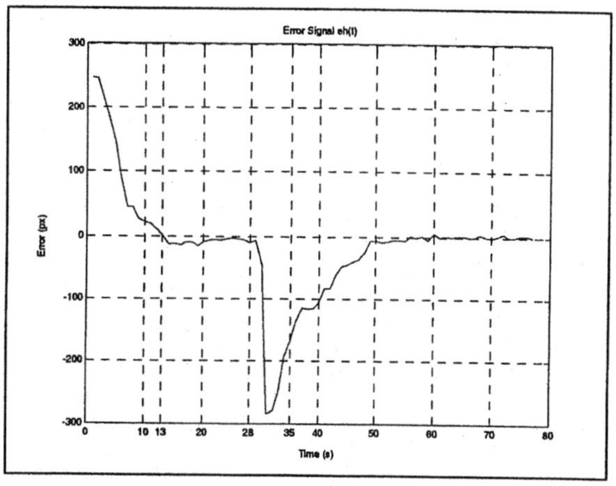


Fig. 5. Error signal in HVC and SVC ($e_h(t)$).

In the time interval [0,13] the Supervisor implements the actions characterising Phase 3; in particular the HVC is active and $e_h(t)$ is driven to 0, so that, at time 13s the vehicle is heading toward to target. Then, the Supervisor implements the actions that characterise Phase 3. In the time interval [13,80] the HVC is inactive and heading, shown in Figure 6, is controlled by the HC.

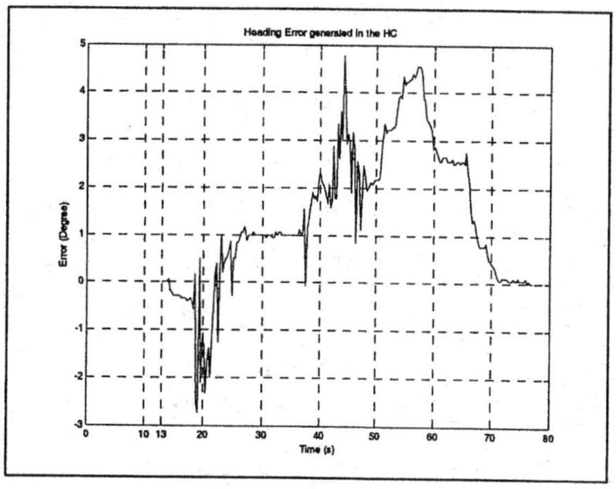


Fig. 6. Error signal in HC.

The control signal generated by the HVC and by the HC is feed to the right and left horizontal thrusters, which actuate the corresponding control action, and it is shown in Figure 7.

At time 28, a disturbance, consisting of a lateral impulsive force, has been applied to the vehicle.

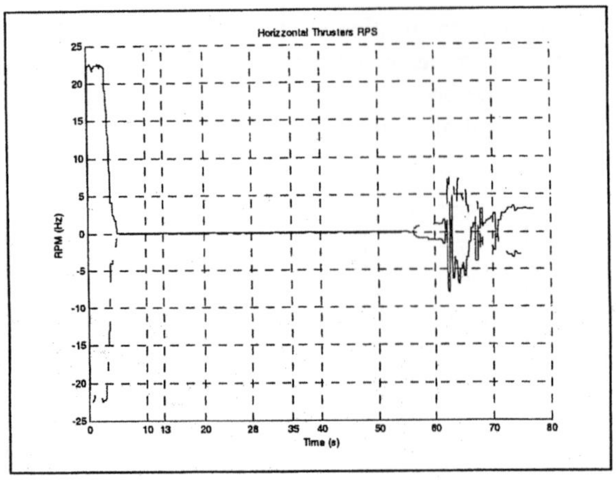


Fig. 7. Right and left horizontal thrusters command.

This causes a reaction by the SVC, whose control signal is feed to the vertran thrusters and shown in Figure 8.

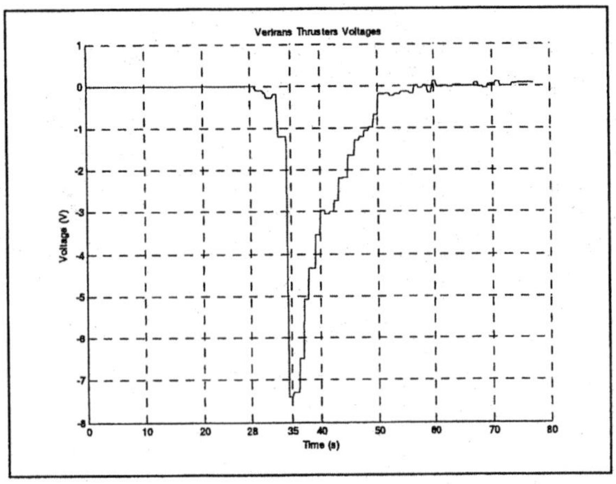


Fig. 8. Vertran thrusters command.

After a short transient, this causes again $e_h(t)$ go to 0, while heading remains constant (see Figure 5 and 6). Note that, in frames corresponding to a distance from the target of about 1m, a displacement of one pixel in the image corresponds to about 2mm of spatial displacement.
Figure 9, finally, shows the behaviour of $e_p(t)$ during Phase 4.

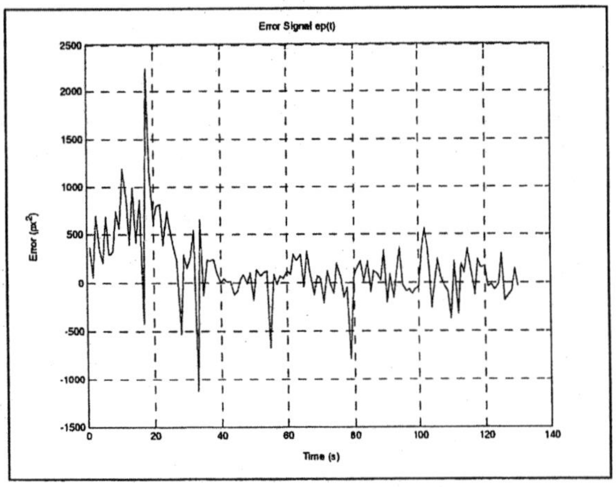


Fig. 9. Error signal in PVC ($e_p(t)$).

## 6. CONCLUSIONS

The problem of positioning an ROV in a partially structured environment has been addressed and a solution based on visual feedback techniques, sensor fusion and fuzzy adaptation has been developed, implemented and tested on a small, work class ROV. The experimental results have shown satisfactory performances from the point of view of reliability and of promptness and precision. In addition, thank to its structure, the automatic control behaves in a way quite similar to that of an experienced human operator. Further work will address, on one side, the problem of enhancing image processing capabilities, so to make the system able to handle more complex situations, and, on the other side, that of increasing adaptability of the low level control strategies by implementing fuzzy adaptation mechanisms on a larger number of controllers.

## REFERENCES


B. Fisher, S. Perkins, A. Walker, E. Wolfart, "Hypermedia Image Processing Reference", Department of Artificial Intelligence, University of Edinburgh, UK, 1994 http://www.cee.hw.ac.uk/hipr/html

Conte G., S.M. Zanoli, D. Scaradozzi, "Enhancing ROVs' Autonomy by Visual-feedback Guidance Techniques", submitted to IEEE - Journal of Oceanic Engineering

S. Inverso, "Ellipse Detection Using Randomized Hough Transform", May 20, 2002

L. Jin, X. Xu, S. Negahdaripour, C. Tsukamoto, J. Yuh, "A real-time vision-based stationkeeping system for underwater robotics applications", IEEE Proceedings of Oceans 96, 1996

H.G. Nguyen, P.K. Kaomea, P.J. Jr Heckman,, "Machine visual guidance for an autonomous undersea submersible. Underwater Imaging", SPIE, 1988

V.J. Santos, J. Sentiero, "The role of vision for underwater vehicles", Symposium on autonomous underwater vehicle technology, Cambrige, 1994

ELSEVIER
IFAC
PUBLICATIONS
www.elsevier.com/locate/ifac

# VISION-BASED ROV HORIZONTAL MOTION CONTROL: EXPERIMENTAL RESULTS

**M. Caccia** [*,1],

[*] *CNR-ISSIA Sez. di Genova, Via De Marini 6, 16149 Genova, Italy*

Abstract: The problem of high precision motion control of remotely operated vehicles (ROVs) in the proximity of the seabed through vision-based motion estimation is addressed in this paper. The proposed approach consists in the integration of a monocular vision system for the estimate of the vehicle's linear motion with a dual-loop hierarchical architecture for kinematics and dynamics control. Results obtained by operating at sea the Romeo ROV are presented, pointing out the role played by the mechanical design of the actuation and illumination systems in achieving satisfactorily performance in operating conditions. *Copyright © 2004 IFAC*

Keywords: Underwater vision, motion estimation, guidance and control, mobile robots

## 1. INTRODUCTION

The problem of accurate motion control of ROVs in the proximity of the seabed is crucial in many service, scientific and archeological applications. High accuracy and smoothness of the supplied control actions and estimated positions and speeds are basic requirements for achieving satisfying motion control performances. Accurate, reliable and high sampling rate, but quite expensive, measurement of slow horizontal motion of ROVs in the proximity of the seabed can be obtained by the combination of ring-laser gyro and acoustic devices (altimeter echo-sounders, high frequency LBL and Doppler velocimeter). On the other hand, in the case light logistic and budget are available, optical vision could supply cheap stand-alone devices for local positioning. This motivated increasing research on underwater vision for ROV motion control. For a detailed discussion of related research the reader can refer to (Caccia, 2003*c*) and the references therein. The approach presented in the following is based on the integration of the dual-loop guidance and control architecture discussed in (Caccia and Veruggio, 2000) and the vision-based motion estimation system presented in (Caccia, 2003*b*).

[1] The author is particularly grateful to Gianmarco Veruggio, Riccardo Bono, Gabriele Bruzzone, Giorgio Bruzzone and Edoardo Spirandelli for their fundamental support in fund-raising, software and hardware development and operating at sea

## 2. MODELLING AND NOMENCLATURE

### 2.1 Vehicle model

As discussed in (Fossen, 1994), the motion of marine vehicles is usually described with respect to an earth-fixed inertial reference frame $< e >$ and a moving body-fixed reference frame $< v >$, whose origin coincides with the center of gravity of the vehicle. Thus, position and orientation of the vehicle are described relative to the inertial reference frame, while linear and angular speeds are expressed relative to the body-fixed reference frame.

The vehicle kinematics nomenclature follows (see Figure 1):

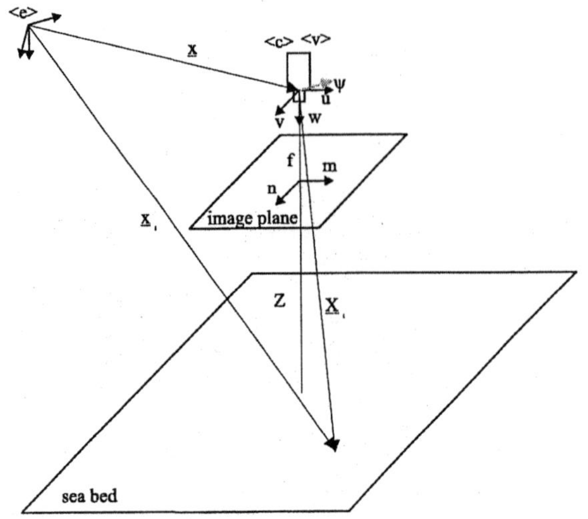


Fig. 1. Camera mounted downward-looking below the ROV: nomenclature.

$\underline{x} = [x\ y\ z]^T$: UUV position relative to the earth-fixed reference frame;
$[\varphi\ \theta\ \psi]^T$: UUV roll, pitch and yaw angles relative to the earth-fixed reference frame;
$[u\ v\ w]^T$ : UUV linear speed (surge, sway, heave) relative to the vehicle-fixed reference frame;
$[p\ q\ r]^T$ : UUV angular speed (roll, pitch and yaw rates) relative to the vehicle-fixed reference frame.
Since the vehicle is a rigid body floating in the water, it is necessary to distinguish between its velocity with respect to the water, i.e. $\underline{\xi} = [u\ v]^T$ in the horizontal plane, and its ground speed, i.e. $\underline{\xi}_G = \underline{\xi} + \underline{\xi}_C$ including the sea current, both expressed with respect to the vehicle-fixed reference frame.
The vehicle position $\underline{x}$ in the earth-fixed reference frame is related to the vehicle speed $\underline{\xi}_G = [u_G\ v_G]^T$ with respect to the ground in the body-fixed frame by the equation

$$\dot{\underline{x}} = L(\psi)\underline{\xi}_G \tag{1}$$

where $L(\psi) = \begin{bmatrix} cos\psi & -sin\psi \\ sin\psi & cos\psi \end{bmatrix}$.

### 2.2 *Camera-laser sensor model*

A camera-fixed reference frame $< c >$ is defined with the $z$-axis directed towards the scene.
The camera and image basic nomenclature follows (see Figure 1):
f: focal length;
$[m\ n]^T$ : image point in the image plane;
$[\dot{m}\ \dot{n}]$: image motion field in the image plane;
$\underline{X} = [X\ Y\ Z]$: coordinates of the generic point in the 3-D space (referred to the camera frame).
Point coordinates in the 3-D space and in the image plane are related by the camera perspective model

$$\begin{bmatrix} m \\ n \end{bmatrix} = \frac{f}{Z}\begin{bmatrix} X \\ Y \end{bmatrix} \tag{2}$$

In the case the camera is mounted downward-looking below the vehicle, the frames $< c >$ and $< v >$ are assumed to coincide. Denoting with $\underline{x}_i$ the coordinates of the generic $i^{th}$ point in the 3-D space (referred to the earth-fixed frame), then $\underline{X}_i = \underline{x}_i - \underline{x}$ and $\dot{\underline{X}}_i = -\dot{\underline{x}}$. In addition, if the vehicle pitch and roll are zero, the corresponding $Z$ axes are vertical and the laser spot $Z$ coordinates in the camera frame represent the altitude of the vehicle from the surface

### 2.3 *Rigid-body motion: motion field of a stationary scene point*

Considering a vehicle-fixed camera moving at linear and angular speed $[u\ \ v\ \ w]^T$ and $[p\ \ q\ \ r]^T$ respectively, the motion field of a generic 3D point in the camera frame is

$$\begin{bmatrix} \dot{m} \\ \dot{n} \end{bmatrix} = -\frac{f}{Z}\begin{bmatrix} u \\ v \end{bmatrix} + \frac{w}{Z}\begin{bmatrix} m \\ n \end{bmatrix} + r\begin{bmatrix} n \\ -m \end{bmatrix} + f\begin{bmatrix} -q \\ p \end{bmatrix} + \frac{pn - qm}{f}\begin{bmatrix} m \\ n \end{bmatrix} \tag{3}$$

## 3. NAVIGATION, GUIDANCE AND CONTROL ARCHITECTURE

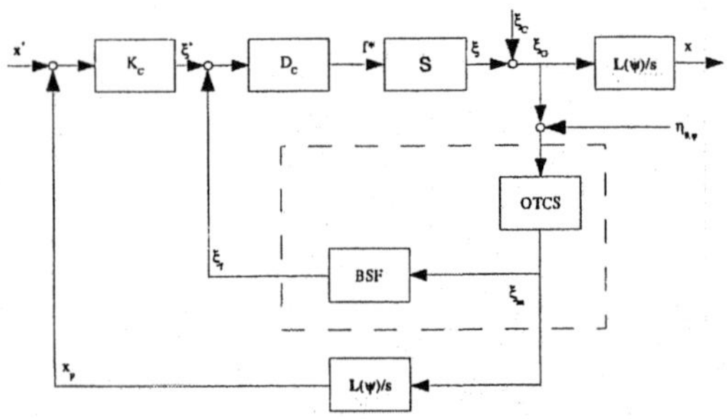


Fig. 2. NGC dual-loop architecture.

As shown in Figure 2 the systems consists of a dual-loop hierarchical guidance and control architecture of the type presented in (Caccia and Veruggio, 2000), constituted by a dynamics controller $D_c$ controlling the vehicle linear speed, i.e. surge and sway, and a kinematics controller $K_c$ handling position control. Neglecting pitch and roll, the robot position in an earth-fixed frame is obtained by integrating equation (1). As far as motion measurement and estimation is concerned, linear speed with respect to the ground in a vehicle-fixed frame is measured by an optical triangulation-correlation sensor (OTCS) of the type described in (Caccia, 2003*b*). The well-known effect of indistinguishability between small surge and sway displacements and pitch and roll rotations when a mono-camera video device for motion estimation is mounted downward-looking below a ROV (see, for instance, (Marks *et al.*, 1995)) could be modelled by adding a quasi sinusoidal disturbance $\eta_{\theta,\varphi}$ and tackled by band-stop filtering (BSF) the measured speed as discussed in (Caccia, 2003*a*). The vehicle horizontal position with respect to an earth-fixed frame is, at this stage, simply predicted by integrating the measured surge and sway multiplied by a rotation matrix $L(\psi)$ (see equation (1)).

## 4. GUIDANCE AND CONTROL

### *4.1 Dynamics control (speed control)*

Surge and sway controllers are based on the 1-DOF uncoupled model of vehicle dynamics (Caccia *et al.*, 2000*a*):

$$m_\xi \dot{\xi} = -k_\xi \xi - k_{\xi|\xi|}\xi|\xi| + f_\xi \quad (4)$$

where $\xi$, $k_\xi$, $k_{\xi|\xi|}$, $m_\xi$ and $f_\xi$ represent the linear speed with respect to the water, linear and quadratic drag coefficients, inertia included added mass, and applied force respectively.
Linearization of equation (4) about the operating point $\xi = \xi^*$ and $f = f^*(\xi^*) : \dot{\xi}(\xi^*, f^*) = 0$ results in the family of parameterized linear models

$$\dot{\xi}_\delta = -\frac{k_\xi + 2k_{\xi|\xi|}|\xi^*|}{m_\xi}\xi_\delta + \frac{1}{m_\xi}f_\delta \quad (5)$$

where $\xi_\delta = \xi - \xi^*$ and $f_\delta = f - f^*$.
Thus, according to the gain-scheduling technique presented in (Khalil, 1996), at each constant operating point $\xi^*$ the controller assumes the form

$$f = f^* + f_\delta \quad (6)$$

where

$$f^*(\xi^*) = k_\xi \xi^* + k_{\xi|\xi|}\xi^*|\xi^*| \quad (7)$$

A PI controller for the linearized system (5) determines the feedback control action

$$f_\delta = k_P e + k_I \gamma, \; \dot{\gamma} = e = \xi - \xi^* = \xi_\delta \quad (8)$$

yielding to a closed-loop linearized system with characteristic equation

$$s^2 + \left(k_\xi + 2k_{\xi|\xi|}|\xi^*| - \frac{k_P}{m_\xi}\right)s - \frac{k_I}{m_\xi} = 0 \quad (9)$$

The gains $k_P$ and $k_I$ are scheduled as functions of $\xi^*$ as it follows

$$k_P = k_\xi + 2k_{\xi|\xi|}|\xi^*| - 2m_\xi\sigma \quad (10)$$

Fig. 3. Optical triangulation-correlation sensor.

and

$$k_I = -m_\xi\left(\sigma^2 + \omega_n^2\right) \quad (11)$$

in order to obtain a desired characteristic equation for the closed-loop linearized system of the form

$$s^2 + 2\sigma s + \sigma^2 + \omega_n^2 = 0 \quad (12)$$

In operating conditions an anti-windup mechanism is implemented such that $|\gamma| \leq \eta_{MAX}$.

### *4.2 Kinematics control (position control)*

Defined a hovering task function of PI-type, $\underline{e}=(\underline{x}-\underline{x}^*)+\mu\int_0^t(\underline{x}-\underline{x}^*)\,d\tau$, the kinematics controller assumes the form

$$\xi^* = -g_P L^{-1}(\underline{x}-\underline{x}^*) - g_I L^{-1}\int_0^t(\underline{x}-\underline{x}^*)\,d\tau \quad (13)$$

with $g_P = \lambda + \mu$ and $g_I = \lambda\mu$, $\lambda > 0$ and $\mu \geq 0$.
In order to minimize wind-up effects, the integrator is enabled/disabled with an histeresis mechanism when the range from the target $r = \sqrt{(\underline{x}-\underline{x}^*)^T(\underline{x}-\underline{x}^*)}$ gets lower/higher than $I_e^{ON}/I_e^{OFF}$ respectively. In addition the proportional and integral control actions are saturated such that $|g_P L^{-1}(\underline{x}-\underline{x}^*)| \leq \xi_P^{MAX}$ and $|g_I L^{-1}\int_0^t(\underline{x}-\underline{x}^*)\,d\tau| \leq \xi_I^{MAX}$ respectively.

## 5. VISION-BASED MOTION ESTIMATION

### *5.1 The combined laser-monocular vision sensor*

The developed video system, i.e. a video camera and 4 parallel red laser diodes, for measuring range and orientation from surfaces is shown in Figure 3. The video camera is mounted inside a suitable steel canister, while the four red laser diodes are rigidly connected in the corners of a

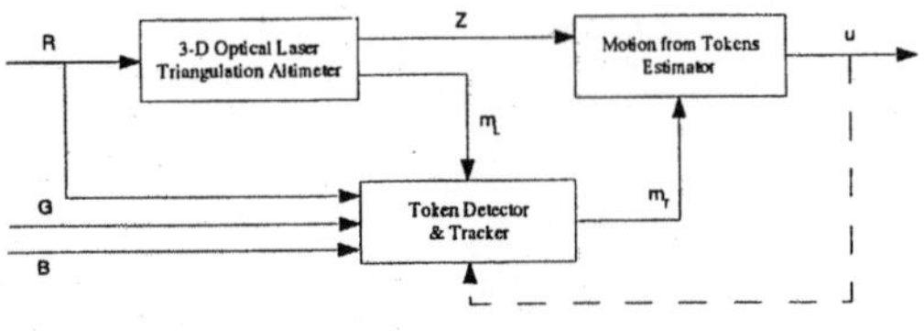


Fig. 4. OTCS architecture.

13 cm side square, with their rays perpendicular to the image plane. The selected camera is the high sensitivity (0.8 lux F1.2, 50 IRE; 0.4 lux F1.2, 30 IRE) Sony SSC-DC330 1/3" High Resolution Exwave HAD CCD Color Camera which features 480 TV lines of horizontal resolution with 768H x 594V picture elements. After calibration, a focal length $f$ of about 1063.9[pixel] has been computed. A camera-fixed reference frame $< c >$, with origin in the image center and $z$-axis directed towards the scene, is defined.

### *5.2 OTCS algorithms*

In the following a brief summary of the image processing system and algorithms used for measuring the camera (vehicle) speed is given. The system basically consists of three modules (see Figure 4):

- *3-D optical laser triangulation altimeter*: detects and tracks the laser spots in the image coordinates, estimates their spatial coordinates in the camera(vehicle)-fixed frame, and finally computes the seabed range and orientation. Since laser diodes emit a red spot only the R component of the image is processed;
- *token detector and tracker*: automatically detects and tracks areas of interest in the image: after a 2-D band-pass filtering to enhance specific spatial wavelengths, local variances are computed to evaluate contrast, and high-local variance areas are extracted as templates (see (Misu *et al.*, 1999) for details); since in the hypothesis of constant heading operations close to the seabed small rotations and inter-frame variations in scene depth occur, template tracking is performed through the computation of the highest correlation displacement in a neighborhood of the previous location, computed according to the estimated motion. Token tracking fails when the correlation gets lower than a suitable threshold.
- *motion from tokens estimator*: computes the vehicle motion in the camera-fixed reference frame from token displacements in consecutive images assuming that the image depth is supplied by the *3-D optical laser triangulation altimeter*. In the case yaw motion is considered, neglecting pitch and roll, equation (3) reduces to:

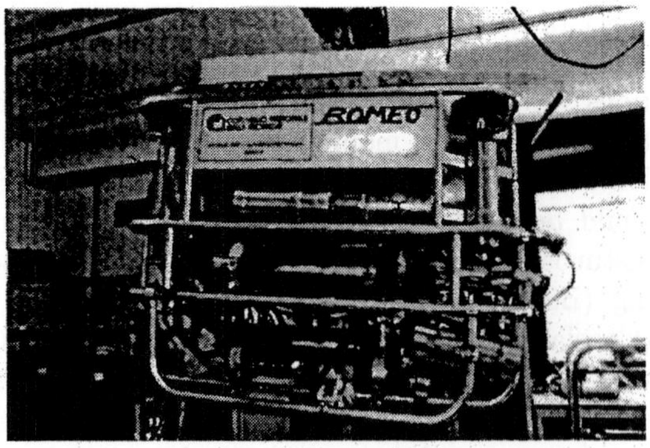

Fig. 5. Romeo benthic survey configuration.

$$\begin{bmatrix} \dot{m} \\ \dot{n} \end{bmatrix} = -\frac{f}{Z}\begin{bmatrix} u \\ v \end{bmatrix} + \frac{w}{Z}\begin{bmatrix} m \\ n \end{bmatrix} + r\begin{bmatrix} n \\ -m \end{bmatrix} \quad (14)$$

In the case the image depth is assumed to be constant (this hypothesis is reasonable given the small area covered by the image), defining the normalized speed $\tilde{u} = \frac{u}{Z}$, $\tilde{v} = \frac{v}{Z}$ and $\tilde{w} = \frac{w}{Z}$, the following overconstrained system can be obtained given $N$ tracked image templates and solved with a LS algorithm:

$$\begin{bmatrix} -f & 0 & m_1 & n_1 \\ 0 & -f & n_1 & -m_1 \\ \vdots & \vdots & \vdots & \\ -f & 0 & m_N & n_N \\ 0 & -f & n_N & -m_N \end{bmatrix} \begin{bmatrix} \tilde{u} \\ \tilde{v} \\ \tilde{w} \\ r \end{bmatrix} = \begin{bmatrix} \dot{m}_1 \\ \dot{n}_1 \\ \vdots \\ \dot{m}_N \\ \dot{n}_N \end{bmatrix} \quad (15)$$

### *5.3 Pitch and roll induced noise: band-stop filter*

As discussed in (Caccia, 2003*a*), where experimental results are presented, small oscillations in uncontrolled pitch and roll induce quasi-sinusoidal disturbance on the measured surge and sway according to the *first order* relationship

$$\begin{bmatrix} \dot{m} \\ \dot{n} \end{bmatrix} \approx -\frac{f}{Z}\begin{bmatrix} u \\ v \end{bmatrix} + \frac{w}{Z}\begin{bmatrix} m \\ n \end{bmatrix} + f\begin{bmatrix} -q \\ p \end{bmatrix} \quad (16)$$

This disturbance can be rejected by suitable band-stop filtering introducing some delay. In particular, Butterworth filters of the form $a_1 * y(n) = b_1 * x(n) + b_2 * x(n-1) + b_3 * x(n-2) + b_4 * x(n-3) + b_5 * x(n-4) - a_2 * y(n-1) - a_3 * y(n-2) - a_4 * y(n-3) - a_5 * y(n-4)$ have been designed and implemented for surge and sway.

## 6. EXPERIMENTAL SETUP

Experiments have been performed with the Romeo ROV (Caccia *et al.*, 2000*b*) equipped with the optical triangulation-correlation sensor (Caccia, 2003*b*) (see Figure 5) in the Ligurian Sea, Portofino Park area, in July 2003. As shown in Figure 6, the camera has been mounted downward-looking

Fig. 6. Optical triangulation-correlation sensor mounted below the ROV.

below the vehicle. In order to minimise the ambiguity in the conventionally defined optical flow originated by the motion of the light source together with the vehicle, i.e. with the camera (Negahdaripour, 1998), a special illumination system at diffuse light has been built [2]. Two 50 W halogen lamps covered by suitable diffusers illuminate almost uniformly the camera scene. The illumination of the camera scene, since the vehicle works in the proximity of the seabed, is not affected by the lamps mounted in front of the ROV for pilot/scientist video and photo cameras. As far as the vision-based motion estimation algorithm is concerned the reader can refer to (Caccia, 2003*b*) and (Caccia, 2003*a*).
During the experiments the ROV worked in auto-altitude using the image depth computed by the optical laser spot triangulation system as altitude measurement.
The parameters of the dynamics controller $D_C$ were $\sigma = 0.3$, $\omega_n = 0.03$ and $\eta_{MAX} = 0.2$, while the kinematics controller has been parameterized by $\lambda = 0.2$, $\mu = 0.12$, $\xi_P^{MAX} = 0.1$, $\xi_I^{MAX} = 0.05$, $I_e^{ON} = 0.12$ and $I_e^{OFF} = 0.25$. During the tests the ROV worked in auto-heading with fixed orientation: thus matrix L was the identity, i.e. $[\dot{x}\,\dot{y}]^T = [u_G\,v_G]^T$. The coefficients of the Butterworth band-stop filter are reported in (Caccia, 2003*a*).

## 7. EXPERIMENTAL RESULTS

In the experiments discussed in the following the vehicle worked at an altitude, i.e. image depth, of about 80 cm, which corresponds to a field of view of about 21x28 cm in the images shown below. The reader can refer to (Caccia, 2003*c*) for an extended presentation and discussion of the trials. Here an experiment showing the system capability in counteracting external disturbances is presented. In the examined test surge and sway force have been zeroed in order to show the effects of environmental disturbance, i.e. sea current and tether tension, on the vehicle. As shown in Figure 7 the ROV drifted, but the controller was able to drive it again over the operating point. The precision of the vision-based dead-reckoning position estimate was satisfying. It is worth noting that during trials the *integral-based* motion predictor had a drift of about 10 cm in x and y respectively after about 10 minutes.

[2] the system has been planned and built by Giorgio Bruzzone

## REFERENCES


Caccia, M. (2003*a*). Pitch and roll disturbance rejection in vision-based linear speed estimation for UUVs. In: *Proc. of MCMC 2003*. pp. 313–318.

Caccia, M. (2003*b*). Vision-based linear motion estimation for unmanned underwater vehicles. In: *Proc. of ICRA 2003*.

Caccia, M. (2003*c*). Vision-based ROV horizontal motion control: experimental results (July 2003 at-sea trials). Rob-vc. CNR-ISSIA Sez. di Genova.

Caccia, M. and G. Veruggio (2000). Guidance and control of a reconfigurable unmanned underwater vehicle. *Control Engineering Practice* **8**(1), 21–37.

Caccia, M., G. Indiveri and G. Veruggio (2000*a*). Modelling and identification of open-frame variable configuration unmanned underwater vehicles. *IEEE Journal of Oceanic Engineering* **25**(2), 227–240.

Caccia, M., R. Bono, G. Bruzzone and G. Veruggio (2000*b*). Unmanned underwater vehicles for scientific applications and robotics research: the ROMEO project. *Marine Technology Society Journal* **24**(2), 3–17.

Fossen, T.I. (1994). *Guidance and Control of Ocean Vehicles*. John Wiley and Sons. England.

Khalil, H.K. (1996). *Nonlinear systems*. Prentice Hall.

Marks, R.L., S.M. Rock and M.J. Lee (1995). Real-time video mosaicking of the ocean floor. *IEEE Journal of Oceanic Engineering* **20**(3), 229–241.

Misu, T., T. Hashimoto and K. Ninomiya (1999). Optical guidance for autonomous landing of spacecraft. *IEEE Transactions on aerospace and electronic systems* **35**(2), 459–473.

Negahdaripour, S. (1998). Revised definition of optical flow: integration of radiometric and geometric cues for dynamic scene analysis. *IEEE Transactions on Pattern Analysis and Machine Intelligence* **20**(9), 961–979.

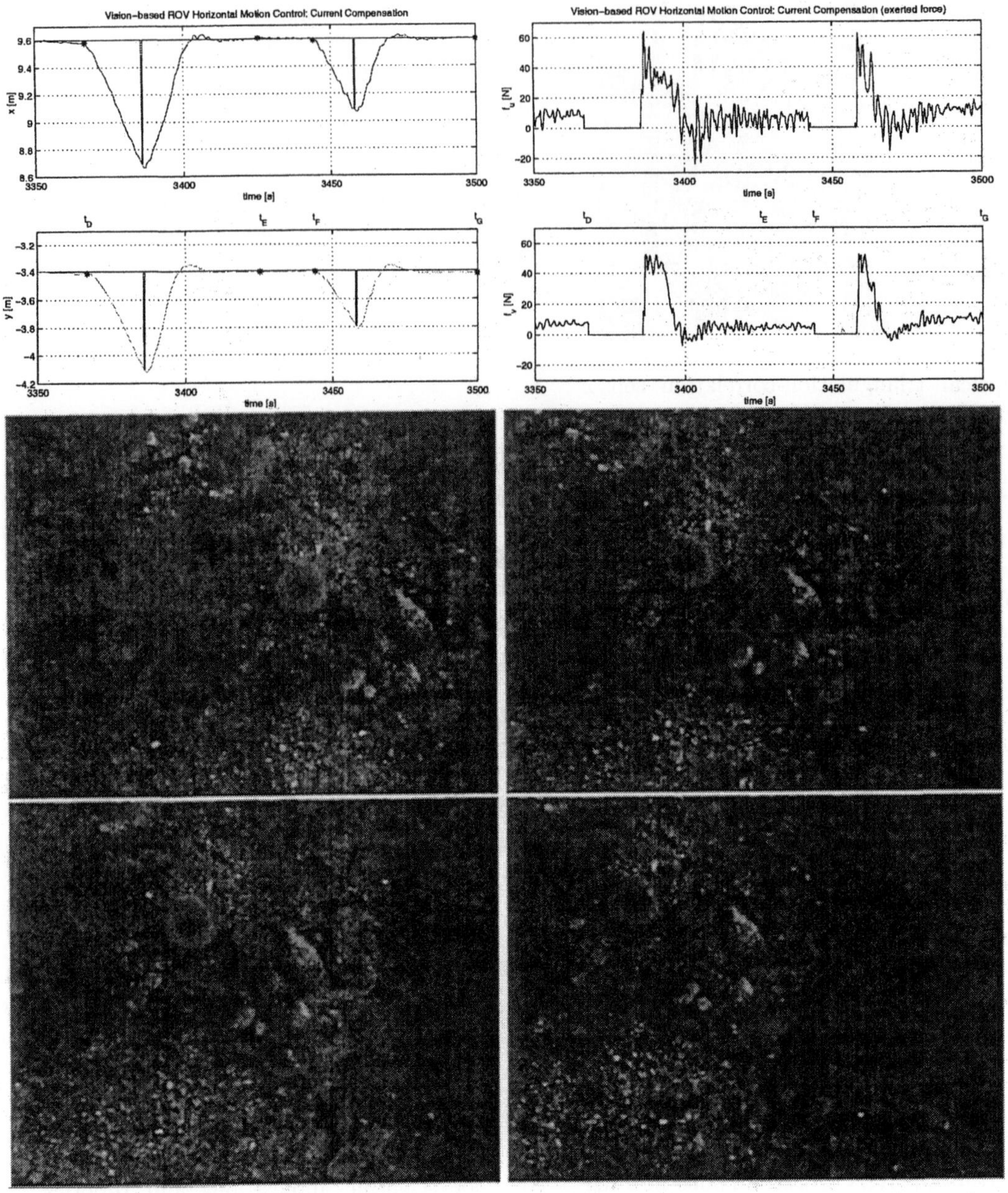


Fig. 7. ROV reference and estimated x-y coordinates and exerted control surge and sway forces: ROV camera views at time $t_D$, $t_E$, $t_F$ and $t_G$ are shown clockwise from the mid line.

ELSEVIER
IFAC
PUBLICATIONS
www.elsevier.com/locate/ifac

# DESIGN OF A HYBRID POWER/TORQUE THRUSTER CONTROLLER WITH LOSS ESTIMATION

**Øyvind N. Smogeli,* Asgeir J. Sørensen* and Thor I. Fossen****

* *Department of Marine Technology*
*Norwegian University of Science and Technology*
*NO-7491 Trondheim, Norway*
*E-mail: [oyvind.smogeli, asgeir.sorensen]@marin.ntnu.no*
** *Department of Engineering Cybernetics*
*Norwegian University of Science and Technology*
*NO-7491 Trondheim, Norway*
*E-mail: tif@itk.ntnu.no*

Abstract: A hybrid power/torque thruster control scheme is proposed, together with a concept for thrust loss estimation in moderate and extreme seas. For high-fidelity thruster control in extreme seas, estimates of the propeller load torque and losses due to waves, current, ventilation and in-and-out-of water effects are of high importance for detecting the loss incidents, optimizing the thrust production, minimizing the wear and tear of the propulsion system and limiting the power consumption. The loss estimation problem is solved by designing a propeller load torque observer and by calculating an estimate of the torque loss factor based on an expected nominal load torque. The observer is proven to be globally exponentially stable, and shows good robustness subject to modelling errors. The torque loss estimation is shown to capture the main loss effects. The hybrid controller is shown to have excellent performance in the entire operation regime, combining the best properties of torque and power control. Simulation results are presented to validate the performance of the load torque observer, the loss estimation and the hybrid controller. *Copyright © 2004 IFAC*

Keywords: Propulsion control, thrust losses, state estimation.

## 1. INTRODUCTION

High performance control systems for positioning of ships and platforms with high tolerance to fault situations are critical for performing complex marine operations during extreme environmental conditions. Presently, dynamic positioning (DP) systems have limitations in rough seas. The reasons for this are limitations in the thrust capability and reduced performance of the control system due to losses and unmodelled nonlinearities. The limitation of power in the propulsion system and loss of thrust due to ventilation, in-and-out-of water effects, transverse and in-line velocity fluctuations, cavitation and thruster-hull interactions will give poor control system performance, see for example Blanke (1981) and Minsaas *et al.* (1987).

For a marine vehicle, the purpose of low-level thruster control is to fulfill the high-level control commands from e.g. the DP system or a joystick. This is essential for the total performance of the operation, but thruster control has nevertheless received relatively little attention both in academia and in industry. For some recent results on thruster control see Fossen and Blanke (2000), Whitcomb and Yoerger (1999) and references therein. The current industrial practice in thruster control is shaft speed control for fixed pitch (FP) propellers and pitch control for controllable pitch (CP) propellers, where the DP system produces the shaft speed or pitch setpoints respectively. More sophisticated thruster control schemes based on power and torque control were introduced by Sørensen *et al.* (1997). This was extended by introducing anti-spin thruster control to handle extreme conditions caused by ventilation and in-and-out-of water effects in Smogeli *et al.* (2003). Power and torque control were shown to give increased performance in moderate seas. In addition, the power control concept was shown to have desirable properties in connection with severe thrust losses in extreme seas.

In this paper, shortcomings of the existing power and torque control concepts are discussed, and a hybrid power-torque control scheme that is valid for a larger range of operating conditions is proposed. In anti-spin thruster control, an estimate of the thrust loss due to the various loss effects is essential for optimizing the thrust production. A thruster observer providing estimates of the propeller load torque and the torque loss factor using feedback from motor torque and propeller shaft speed is presented, and simulation results are shown to validate the performance.

## 2. THRUSTER MODELING

A thruster may be modelled as an electric motor, a shaft with friction and a hydrodynamically loaded propeller:

$$\dot{Q}_m = \frac{1}{T_m}(Q_c - Q_m), \quad (1)$$

$$J\dot{\omega} = Q_m - Q_p - K_\omega \omega, \quad (2)$$

$$Q_p = f_Q(\theta, \xi). \quad (3)$$

Here, $T_m$ is the motor time constant, $Q_c$ is the commanded torque from the local thruster controller, $Q_m$ is the motor torque, $J$ is the rotational inertia of the propeller including added mass, shaft, gears and motor, $K_\omega$ is a linear friction coefficient, $\omega = 2\pi n$ is the rotational speed of the propeller, and $Q_p$ is the propeller load torque. The load torque is modelled as a nonlinear function $f_Q$ of fixed thruster parameters $\theta$ (i.e. propeller diameter, position, number of propeller blades, pitch ratio, propeller blade expanded-area ratio) and variables $\xi$ (i.e. shaft speed, relative submergence, advance speed). The actually produced propeller thrust $T_p$ is modelled in a similar manner as:

$$T_p = f_T(\theta, \xi). \quad (4)$$

The nominal thrust $T_n$ and torque $Q_n$ are typically found from open-water tests with a deeply submerged thruster, expressed by the thrust and torque coefficients $K_T$ and $K_Q$ or other equivalent mappings (Carlton, 1994):

$$T_n = K_T \rho D^4 n\,|n|\,, \quad (5)$$

$$Q_n = K_Q \rho D^5 n\,|n|\,. \quad (6)$$

Here, $D$ is the propeller diameter, and $\rho$ is the density of water. $K_T$ and $K_Q$ are in general functions of many parameters and variables, with velocity of advance $V_a$, shaft speed $n$ and thrust direction being the most important. The nominal thrust and torque coefficients $K_{T0}$ and $K_{Q0}$ are the values for zero advance speed, $V_a = 0$, which are commonly used in control systems for DP since information on $V_a$ usually is unavailable. In station-keeping, zero advance speed is a good approximation. Thrust losses may be expressed by the thrust and torque loss factors $\beta_T$ and $\beta_Q$, which express the ratio of actual to nominal thrust and torque (Minsaas *et al.*, 1987):

$$\beta_T = \frac{T_p}{T_n}, \quad \beta_Q = \frac{Q_p}{Q_n}. \quad (7)$$

This means that the actual propeller thrust and torque may be expressed as:

$$T_p = f_T(\theta, \xi) = T_n \beta_T,$$

$$Q_p = f_Q(\theta, \xi) = Q_n \beta_Q. \quad (8)$$

## 3. WEIGHT FUNCTION

A weight function satisfying certain criteria is needed in the following. The smooth weighting function $\alpha(x) : \mathbb{R} \to \mathbb{R}$, where $x$ is a variable, must satisfy the following properties:

$$\lim_{x \to 0} \alpha(x) = 1, \quad \lim_{x \to \infty} \alpha(x) = 0. \quad (9)$$

Additionally, $\alpha(x)$ should be close to zero for "large" $x$. Based on this requirement, the following weight function is proposed:

$$\alpha(x) = e^{-k|px|^r}, \quad (10)$$

where $k$, $p$ and $r$ are positive constants. Figure 1 shows $\alpha(x)$ for varying $k$, $p$ and $r$. The parameter $p$ will act as a scaling factor for the variable $x$. A small $p$ will widen the weighting function, giving a wider transition between 0 and 1. Increasing the parameter $k$ sharpens the peak about $x = 0$, whereas increasing $r$ widens it and makes the transition from 0 to 1 more steep. The weight function must show smooth behavior for all $x$. The derivative of $\alpha$ with respect to $x$ is:

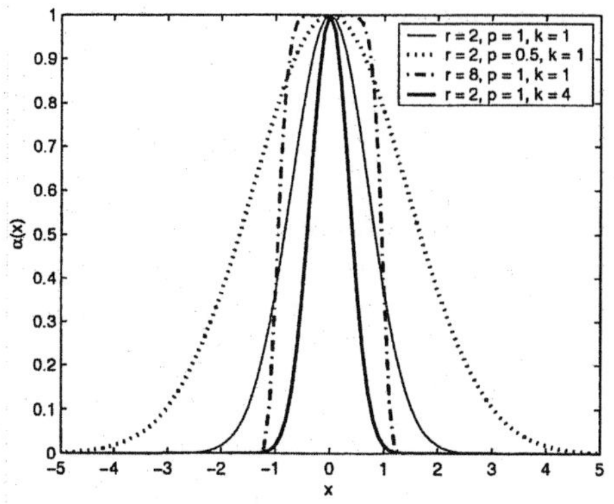


Fig. 1. Weight function $\alpha(x)$ for varying $r$, $p$, $k$.

$$\begin{aligned}\frac{d\alpha(x)}{dx} &= -k\,|px|^{r}\,e^{-k|px|^{r}}(-kpr\,|px|^{r-1}) \\ &= k^2pr\,|px|^{2r-1}\,e^{-k|px|^{r}} \\ &= k^2p^{2r}r\,|x|^{2r-1}\,\alpha(x), \qquad (11)\end{aligned}$$

which is smooth in $x$. Particularly, $d\alpha/dx = 0$ for $x = 0$ and $x \to \pm\infty$.

## 4. THRUSTER OBSERVER

### *4.1 Propeller load torque observer*

The actual propeller load torque $Q_p$ in (3) may not be measured explicitly, but may be estimated by an observer. Based on (2), the following control plant model of the thruster dynamics is proposed:

$$\begin{aligned}J\dot{\omega} &= Q_m - Q_p - K_\omega\omega, \\ \dot{Q}_p &= -\frac{1}{T_Q}Q_p + w_Q. \qquad (12)\end{aligned}$$

$Q_p$ is here treated as an exogenous disturbance, and modelled as a first order Markov process with time constant $T_Q$ driven by white noise $w_Q$ with zero mean. With the motor torque as input $u = Q_m$ and the shaft speed as measurement $y = \omega$, the state-space formulation, which is observable from $y$, is written:

$$\begin{aligned}\dot{x} &= Ax + Bu + w, \\ y &= Cx, \qquad (13)\end{aligned}$$

using the following vectors and matrices:

$$x = \begin{bmatrix}\omega \\ Q_p\end{bmatrix},\ A = \begin{bmatrix}-K_\omega/J & -1/J \\ 0 & -1/T_Q\end{bmatrix},$$
$$B = \begin{bmatrix}1/J \\ 0\end{bmatrix},\ w = \begin{bmatrix}0 \\ w_Q\end{bmatrix},\ C = \begin{bmatrix}1 & 0\end{bmatrix}. \quad (14)$$

Motivated by (13) and adding an injection term $K\tilde{y}$, the following propeller load torque observer is proposed:

$$\begin{aligned}\dot{\hat{x}} &= A\hat{x} + Bu + K\tilde{y}, \\ \hat{y} &= C\hat{x}, \\ K\tilde{y} &= KC\tilde{x} = \begin{bmatrix}k_1 & 0 \\ k_2 & 0\end{bmatrix}\tilde{x}. \qquad (15)\end{aligned}$$

Here, $\tilde{y} = y - \hat{y}$ and $\tilde{x} = x - \hat{x}$. The error dynamics found by subtracting (15) from (13) becomes:

$$\dot{\tilde{x}} = \dot{x} - \dot{\hat{x}} = (A - KC)\tilde{x} + w = F\tilde{x} + w, \quad (16)$$

where the matrix $F$ is defined as:

$$F = (A - KC) = -\begin{bmatrix}K_\omega/J + k_1 & 1/J \\ k_2 & 1/T_Q\end{bmatrix}. \quad (17)$$

The equilibrium point $\tilde{x} = 0$ of the observer estimation error is globally exponentially stable (GES) if the measurement noise $w_Q = 0$ and $F$ is Hurwitz, which is implied by positive definiteness of $(-F)$. By Sylvester's theorem $(-F)$ is positive definite if:

$$\begin{aligned}&i)\ K_\omega/J + k_1 > 0 \Rightarrow k_1 > -K_\omega/J, \\ &ii)\ (K_\omega/J + k_1)/T_Q - k_2/J > 0 \Rightarrow \\ &k_2 < (K_\omega + k_1J)/T_Q. \qquad (18)\end{aligned}$$

If $w_Q$ is a Gaussian white noise process the trajectories $\tilde{x}$ will converge to ball around the origin $\tilde{x} = 0$ and uniform ultimately boundedness (UUB) follows. The equations for implementation are:

$$\begin{aligned}\dot{\hat{\omega}} &= \frac{1}{J}(-\hat{Q}_p - K_\omega\hat{\omega} + u) + k_1(\omega - \hat{\omega}), \\ \dot{\hat{Q}}_p &= -\frac{1}{T_Q}\hat{Q}_p + k_2(\omega - \hat{\omega}). \qquad (19)\end{aligned}$$

In an alternative formulation, the propeller load torque can be modelled as a Wiener process, i.e. $\dot{Q}_p = w_Q$ in (12). The load torque observer is then implemented by replacing $(-1/T_Q)$ with 0 in (14), and choosing the gains $k_1$ and $k_2$ according to $k_1 > -K_\omega/J$ and $k_2 < 0$. Simulations indicate that this latter formulation also is adequate.

### *4.2 Torque loss calculation*

For DP operation the *expected* nominal propeller load torque $Q_{n0}$ may be calculated from (6) by feedback from the propeller shaft speed $n$ as:

$$Q_{n0} = K_{Q0}\rho D^5 n\,|n|, \quad (20)$$

where the nominal torque coefficient $K_{Q0}$ for advance speed $V_a = 0$ is used. Based on (7), the estimated torque loss factor $\hat{\beta}_Q$ is calculated from $\hat{Q}_p$ (19) and $Q_{n0}$ (20) as:

$$\hat{\beta}_Q = \frac{\hat{Q}_p}{Q_{n0}} = \frac{\hat{Q}_p}{K_{Q0}\rho D^5 n\,|n|}. \quad (21)$$

The estimate is singular for $n = 0$, which means that special precautions must be taken for the zero-crossing of $n$. Since the thrust loss is undefined for zero thrust, it makes sense to require $\lim_{n\to 0}\hat{\beta}_Q(n) = 1$. The torque loss estimate $\hat{\beta}_{Qe}$ to be implemented should therefore be defined as:

$$\hat{\beta}_{Qe} = \alpha_b(n) + (1 - \alpha_b(n))\hat{\beta}_Q, \qquad (22)$$

where $\alpha_b(n)$ is a weight function given by (10). Additionally, $\alpha_b(n)$ must be such that $\hat{\beta}_{Qe} \approx \hat{\beta}_Q$ for $|n| > \varepsilon$, where $\varepsilon$ is a small positive number. The quality of the estimate will depend on the accuracy of the thruster model. From (19) and (20), the two most important modelling parameters for $\hat{\beta}_Q$ are the rotational inertia $J$ and the nominal torque coefficient $K_{Q0}$. A modelling error in $J$ will to a large extent be compensated for by the robustness of the observer, but an error in $K_{Q0}$ will appear as a scaling error in $\hat{\beta}_Q$.

## 5. THRUSTER CONTROL

For control purposes, the nominal thrust $T_n$ and torque $Q_n$ of a propeller are commonly expressed by (5) and (6), replacing $K_T$ and $K_Q$ with the nominal coefficients $K_{T0}$ and $K_{Q0}$ (Fossen, 1994). The signed power $P_n$ is accordingly given as:

$$P_n = Q_n 2\pi\,|n|\,. \qquad (23)$$

### *5.1 Shaft speed control*

The conventional thruster control scheme is shaft speed control, which utilizes shaft speed feedback from the thruster to set the commanded motor torque $Q_{cn}$ by e.g. a PID algorithm operating on the shaft speed error. The inverse thrust characteristics, relating the desired shaft speed $n_d$ to the desired thrust $T_d$ is expressed by inverting (5):

$$n_d = sgn(T_d)\sqrt{\left|\frac{T_d}{\rho D^4 K_{T0}}\right|}. \qquad (24)$$

### *5.2 Torque control*

The object of torque control is to control the setpoint of the motor torque instead of the shaft speed. A mapping from desired thrust $T_d$ to desired torque $Q_d$ can be found from (5) and (6). The commanded motor torque $Q_{cq}$ is set equal to the desired torque, see Sørensen *et al.* (1997):

$$Q_{cq} = Q_d = \frac{K_{Q0}}{K_{T0}} D T_d. \qquad (25)$$

### *5.3 Power control*

Power control is based on controlling the power from the drive system. The desired power $P_d$ is found by inserting the desired torque from (25) and the desired shaft speed from (24) in (23). The commanded motor torque $Q_{cp}$ is calculated from $P_d$ using feedback from the measured shaft speed $n \neq 0$ according to (23), see Sørensen *et al.* (1997):

$$Q_{cp} = \frac{P_d}{2\pi|n|} = \frac{Q_d 2\pi n_d}{2\pi|n|} = \frac{K_{Q0}}{\sqrt{\rho} D K_{T0}^{3/2}} \frac{sgn(T_d)|T_d|^{3/2}}{|n|}. \qquad (26)$$

### *5.4 Hybrid power/torque control*

A significant shortcoming of the power control scheme in (26) is the singularity for zero shaft speed, $n = 0$. This means that power control should not be used close to the singular point, e.g. when commanding low thrust or changing thrust direction. For low thrust commands, torque control shows better performance in terms of constant thrust production than power control, since the mapping between thrust and torque is more direct than the mapping between thrust and power. For high thrust commands, it is essential to avoid large power transients, as these lead to higher fuel consumption and possible danger of power blackout and harmonic distortion of the power plant network. Power control is hence a natural choice for high thrust commands. This motivates the design of a hybrid power/torque control scheme, utilizing the best properties of both controllers. The commanded motor torque $Q_{ch}$ from the hybrid power/torque controller is defined as:

$$Q_{ch} = \alpha_h(n) Q_{cq} + (1 - \alpha_h(n)) Q_{cp}, \qquad (27)$$

where $\alpha_h(n)$ is a weight function given by (10), which defines the dominant regimes of the two control schemes. The shaft speed is physically limited to some max value $n_{\max}$, such that $\alpha_h(n_{\max})$ should be close to zero. The control law must show smooth behavior for all $n$. The derivative of the commanded torque with respect to $n$ is:

$$\begin{aligned}\frac{dQ_c}{dn} &= \frac{d\alpha_h(n)}{dn} Q_{cq} + \alpha_h(n)\frac{dQ_{cq}}{dn} \\ &\quad + \frac{d(1-\alpha_h(n))}{dn} Q_{cp} + (1-\alpha_h(n))\frac{dQ_{cp}}{dn} \\ &= k^2 p^{2r} r\,|n|^{2r-1}\,\alpha_h(n)\frac{K_{Q0}}{K_{T0}} D T_d + 0 \\ &\quad - k^2 p^{2r} r\,|n|^{2r-1}\,\alpha_h(n)\frac{K_{Q0} sgn(T_d)|T_d|^{3/2}}{\sqrt{\rho} D K_{T0}^{3/2}|n|} \\ &\quad + (1-\alpha_h(n))\frac{K_{Q0} sgn(T_d)|T_d|^{3/2}}{\sqrt{\rho} D K_{T0}^{3/2}(-|n|n)},\end{aligned}$$

$$\frac{dQ_c}{dn} = k^2 p^{2r} r \left|n\right|^{2r-1} \alpha_h(n) \frac{K_{Q0}}{K_{T0}} D T_d$$
$$-k^2 p^{2r} r \left|n\right|^{2r-2} \alpha_h(n) \frac{K_{Q0} sgn(T_d) |T_d|^{3/2}}{\sqrt{\rho} D K_{T0}^{3/2}}$$
$$-\frac{1-\alpha_h(n)}{|n|n} \frac{K_{Q0} sgn(T_d) |T_d|^{3/2}}{\sqrt{\rho} D K_{T0}^{3/2}}. \quad (28)$$

The first two terms contain no singularities for $r \geq 1$, however it remains to investigate the term:

$$\frac{1-\alpha_h(n)}{|n|n} = \frac{1-e^{-k|pn|^r}}{|n|n} \triangleq h(n). \quad (29)$$

The limit of $h(n)$ as $n$ tends to zero is:

$$\lim_{n\to 0} h(n) = \lim_{n\to 0} \frac{1-e^{-k|pn|^r}}{|n|n}$$
$$= \lim_{n\to 0} \frac{d/dn(1-e^{-k|pn|^r})}{d/dn(|n|n)}$$
$$= \lim_{n\to 0} \frac{-k^2 p^{2r} r \left|n\right|^{2r-1} e^{-k|pn|^r}}{2|n|}$$
$$= \lim_{n\to 0} -k^2 p^{2r} r \left|n\right|^{2r-2} e^{-k|pn|^r} = 0,$$

for $r > 1$. The hybrid controller hence shows smooth behavior with respect to $n$ as long as $r > 1$ in the function $\alpha_h(n)$ defined in (10).

## 6. RESULTS

In order to investigate the performance of the thruster observer and the hybrid controller, simulations were performed in Simulink® with MCSim (Sørensen *et al.*, 2003). A typical DP case was simulated for a supply vessel with main particulars $[L, B, T] = [80, 18, 5.6]m$ equipped with diesel-electric ducted propulsors in the aft ship. The thruster submergence relative to the free surface and the water velocity relative to the thruster was found by accounting for waves, current and vessel motion, with an assumption of undisturbed waves. Figure 2 shows the actual and estimated propeller torque found by using the observer given by (19) when the thruster is experiencing moderate losses due to in-line and cross-flow velocity fluctuations. The estimate is highly accurate, and shows good robustness subject to modelling errors. Figure 3 shows the actual and estimated propeller torque when the thruster is subject to high thrust losses due to ventilation and in-and-out-of water effects in extreme seas. The propeller torque estimate clearly follows the fast transients due to the ventilation incidents, which can be seen as sudden losses of load torque followed by fast transients when the thruster re-enters the water and ventilation terminates. Figure 4 shows the actual and estimated torque loss factors $\beta_Q$ and $\hat{\beta}_Q$ when the thruster is subject to high thrust losses. The torque loss factor estimate clearly captures the main loss events. The peaks in $\hat{\beta}_Q$ occurring at the termination of ventilation are due to the small time delay in $\hat{Q}_p$. Figure 5 shows the commanded torque from the hybrid controller, the torque controller and the power controller for a varying thrust reference, including a zero-crossing of the shaft speed. The weight function parameters were $r = 5, p = 0.12$ and $k = 4$, such that pure power control was achieved for high shaft speed and pure torque control was achieved for low shaft speed. The results illustrate that the power controller singularity in (26) is avoided by using the hybrid

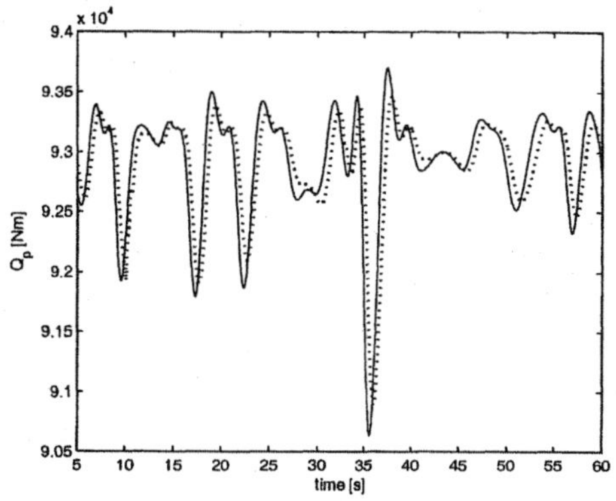


Fig. 2. Actual (solid) and estimated (dotted) propeller torque $Q_p$ in moderate seas.

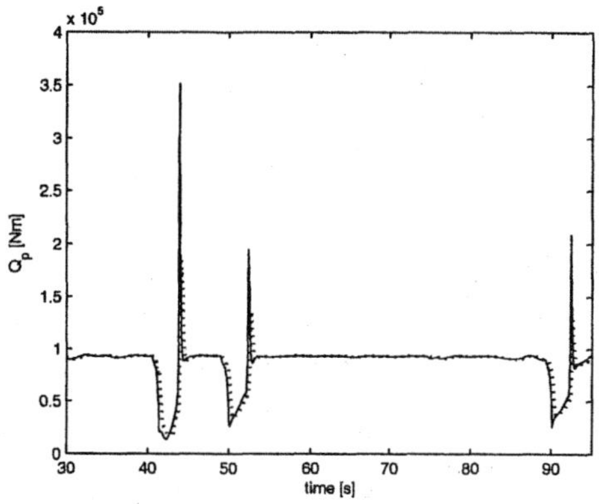


Fig. 3. Actual (solid) and estimated (dotted) propeller torque $Q_p$ when subject to high thrust losses.

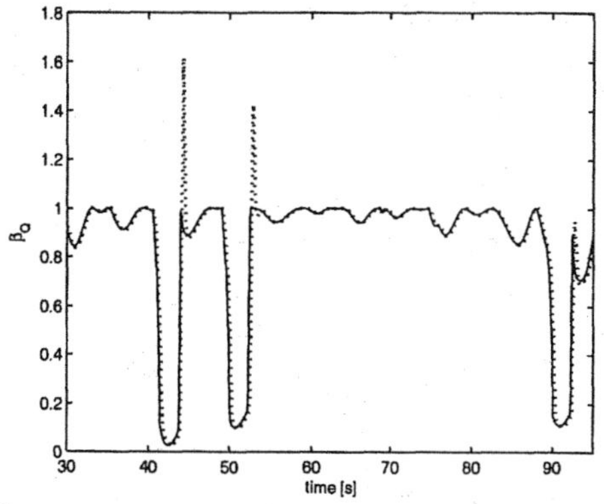


Fig. 4. Actual (solid) and estimated (dotted) torque loss factor $\beta_Q$ when subject to high thrust losses.

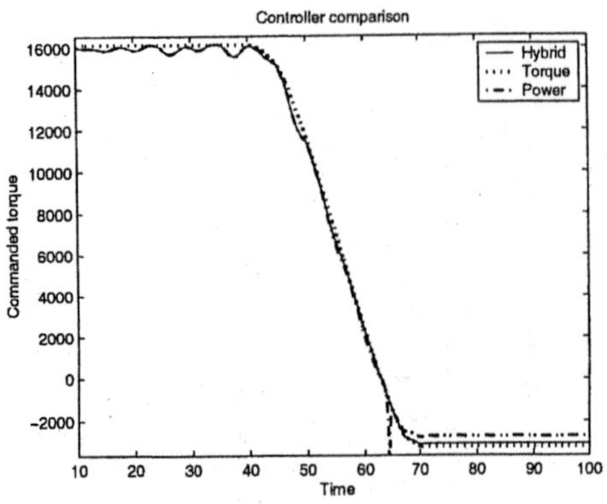


Fig. 5. Commanded torque $Q_c$ for hybrid, torque and power controller with changing thrust reference.

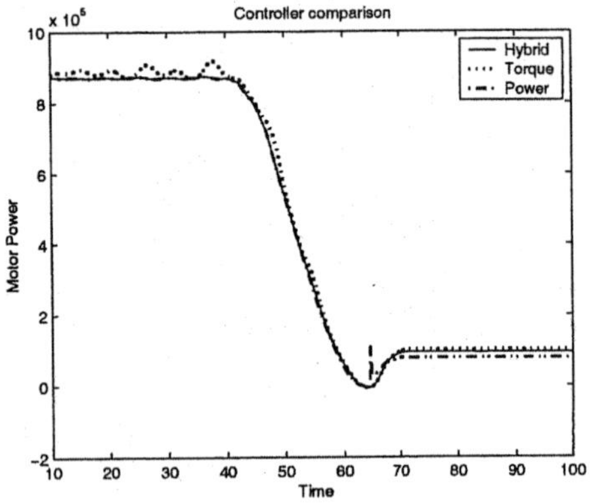


Fig. 6. Motor power for hybrid, torque and power controller with changing thrust reference.

controller (27). The corresponding motor power is plotted in Fig. 6, illustrating the advantages of the hybrid controller. For high shaft speeds, and hence high power consumption, the hybrid controller coincides with the power controller, assuring small variations in power consumption. For low shaft speeds, the controller coincides with the torque controller, assuring best possible thrust production.

## 7. CONCLUSIONS

The equilibrium point $\tilde{x} = 0$ of the observer estimation error was proven to be GES under the assumption of no measurement disturbances. For the case when plant was exposed to white noise disturbances the observer estimation error was UUB implying that the system trajectories converge to a ball around the origin. The observer showed good performance even in transient situations, and the torque loss calculation clearly captured the main effects of the thrust losses experienced in moderate and severe loss situations. The hybrid power/torque thruster controller showed good performance for the complete operating range. The best properties of both the power and the torque control schemes were utilized, and the singularity for power control at zero shaft speed was avoided. The hybrid controller and loss estimation scheme will be of high importance for further work on thruster anti-spin and fault-tolerant thruster control.

## 8. ACKNOWLEDGMENT

This work has been carried out at the *Centre for Ships and Ocean Structures* (CESOS) at NTNU in cooperation with the research project on *Energy-Efficient All-Electric Ship* (EEAES). The Norwegian Research Council is acknowledged as the main sponsor of CESOS and EEAES.

## 9. REFERENCES

Blanke, M., (1981). Ship Propulsion Losses Related to Automated Steering and Prime Mover Control. PhD dissertation, *The Technical University of Denmark,* Lyngby, Denmark.

Carlton, J.S. (1994). Marine Propellers & Propulsion. *Butterworth-Heinemann Ltd.*

Fossen, T.I. and M. Blanke (2000). Nonlinear Output Feedback Control of Underwater Vehicle Propellers Using Feedback From Estimated Axial Flow Velocity. *IEEE Journal of Oceanic Engineering,* **25** (2).

Fossen, T.I. (1994). Guidance and Control of Ocean Vehicles. *John Wiley and Sons Ltd.*

Minsaas, K.J., H.J. Thon and W. Kauczynski (1987). Estimation of Required Thruster Capacity for Operation of Offshore Vessels under Severe Weather Conditions. *PRADS 1987.*

Smogeli, Ø.N., L. Aarseth, E.S. Overå, A.J. Sørensen and K.J. Minsaas (2003). Anti-spin thruster control in extreme seas. *Proceedings of 6th IFAC Conference on Manoeuvring and Control of Marine Craft (MCMC'03),* Girona, Spain.

Sørensen, A.J., A.K. Ådnanes, T.I. Fossen and J.P. Strand (1997). A new method of thruster control in positioning of ships based on power control. *Proceedings of the 4th IFAC Conference on Manoeuvring and Control of Marine Craft (MCMC'97),* Brijuni, Croatia.

Sørensen, A.J., E. Pedersen and Ø.N. Smogeli (2003). Simulation-Based Design and Testing of Dynamically Positioned Marine Vessels. *Proceedings of International Conference on Marine Simulation and Ship Maneuverability, MARSIM'03,* August 25 - 28, Kanazawa, Japan.

Whitcomb, L.L. and D.R. Yoerger (1999). Preliminary Experiments in Model-Based Thruster Control for Underwater Vehicle Positioning. *IEEE Journal of Oceanic Engineering,* **24** (4).

ELSEVIER
IFAC
PUBLICATIONS
www.elsevier.com/locate/ifac

# ACTUATOR AND CONTROL DESIGN FOR FAST FERRY USING SEASICKNESS CRITERIA

**S. Esteban, J.M. Giron-Sierra, J. Recas,**
**J.M. Riola, B. de Andres-Toro, J.M. De la Cruz**

*Dep. A.C.Y.A., Fac. CC. Fisicas. Universidad Complutense de Madrid*
*Ciudad Universitaria, 28040 Madrid. Spain*
*e-mail: gironsi@dacya.ucm.es*

Abstract: Along our research on increasing the passengers comfort in fast ferries by using moving actuators, frequency domain models to predict seasickness caused by ship motions in different seas states, have been established. Seasickness can be alleviated by a correct control of the moving actuators, in this case flaps, fins and T-foil. The capability of seasickness prediction based on models are useful for actuator and control design focusing on minimizing seasickness. In particular, there is a frequency band where the actuation should intervene; while there are other ship motion frequencies with much less impact on seasickness, so actuators work is not needed and can be saved. The paper is devoted to actuator and control design based on seasickness models. This study considers in particular the effect of actuator area and position on seasickness alleviation. The control design turns out to be nonlinear, due to saturations. A particular actuator and control design example is studied in detail. *Copyright © 2004 IFAC*

Keywords: Ship control, Seakeeping, Nonlinear control, Human factors.

## 1. INTRODUCTION

There are some studies about seasickness which determined that it is due to oscillating vertical motions with frequencies around 1 rad./sec. In this context it is interesting to consider the results of (O'Hanlon and MacCawley, 1974), including the introduction of a motion sickness incidence index (MSI) and a mathematical model relating the MSI with oscillatory vertical accelerations.

In fast ferries there may be significant oscillatory vertical accelerations, in response to encountered waves. By using moving actuators, such flaps, fins and T-foil, it is possible to alleviate these accelerations. However the use of actuators has a cost, and the cavitation due to non-zero angles of attack of the actuators can destroy them. A correct control of the actuators motion is the main objective of our research since time ago (Aranda, et al., 2001; Esteban, et al., 2000, 2001, 2002; Giron-Sierra, et al., 2001, 2002). This is also the target of other research teams (see for instance Haywood, et al., 1995). A general view of the ride control problem is in (Ryle, 1998). A pertinent reference for ship control is (Fossen, 2002).

Part of our research has been devoted to seasickness prediction, with recent results related with a three filters approach (Giron-Sierra, et al, 2003) which allows for an MSI calculation in the frequency domain.

This paper starts by briefly considering the three filters approach. Then it establishes a mathematical framework for control study pertaining ship motions and actuators effects. It continues with a control design discussion within this framework, with the objective of focusing the control action in the frequencies of interest. Then a particular case study is detailed, with graphical results.

## 2. SEASICKNESS AND SHIP MOTIONS

With respect to seasickness, humans can be considered as "filters". Oscillating vertical accelerations crossing the filter will cause seasickness. Figure 1 shows the frequency response of this filter.

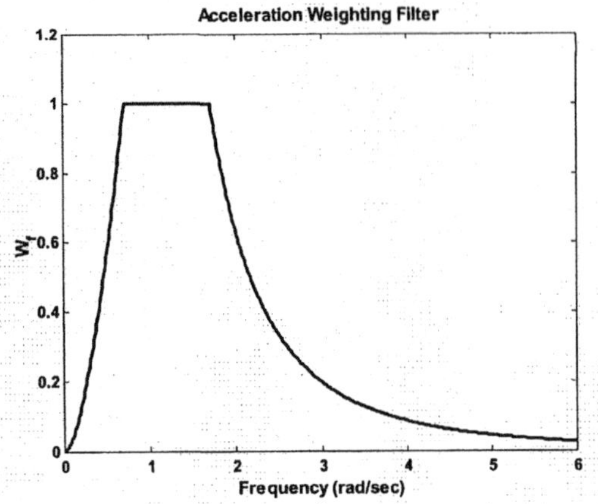


Fig.1. Seasickness "filter".

Vertical accelerations are due to the ship response to encountered waves. The ship can be considered as a low-pass filter, since short waves have almost no effect on the ship, because she lays on two or more waves, but long waves do move the ship. There are statistical descriptions of ocean waves. From these descriptions we can deduce filters, with responses to white noise having same statistical characteristics as ocean waves. Figure 2 shows a chain of three filters: waves, ship , and seasickness. If the three filters have a common band, there will exist ship motions causing seasickness. We can easily deduce the frequency response of the chain of three filters. We can also plot this response: the area under the curve can be mathematically linked to the resulting MSI (more area implies more MSI).

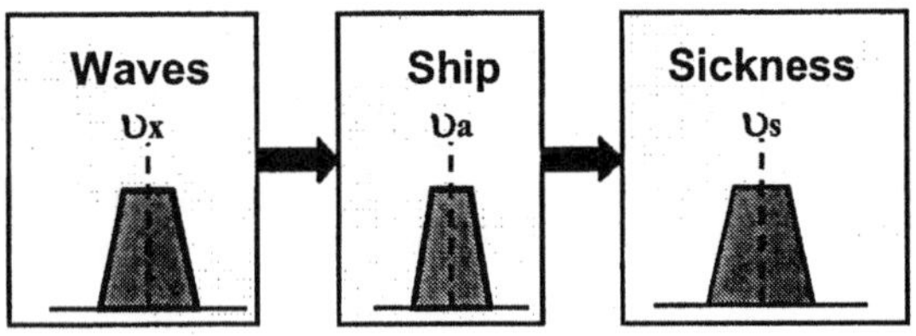


Fig.2. Chain of three filters.

The control of actuators should focus on the frequency response of the chain of three filters.

## 3. MOVING ACTUATORS AND THE SHIP

Figure 3 shows the ship with the actuators: transom flaps and a T-foil near the bow. The actuators are moved by hydraulic cylinders.

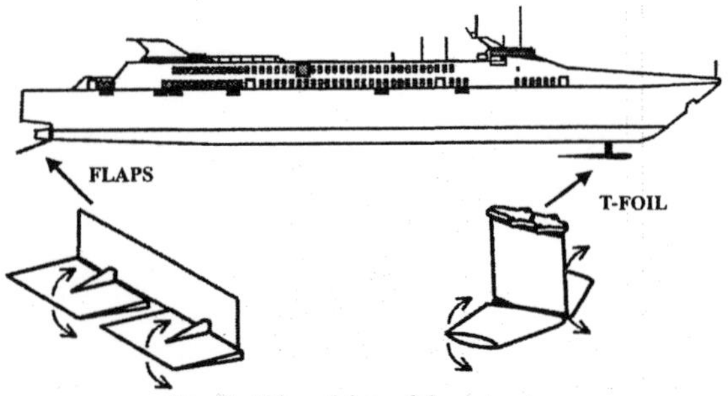


Fig.3. The ship with actuators.

Figure 4 represents the aspects of the control scenario. In our previous research we obtained, with experiments using a towed physical model in El Canal de Experiencias Hidrodinamicas de El Pardo (CEHIPAR), Madrid, control-oriented models of the pitch and heave motions of the ship with head seas. The models are decomposed into wave-to-heave force, heave force-to-heave motion, wave-to-pitch moment, and pitch moment-to-pitch motion transfer functions. Combining the models, we can obtain the worst vertical acceleration (WVA) that a passenger can experiment. The WVA is located in a place near the bow.

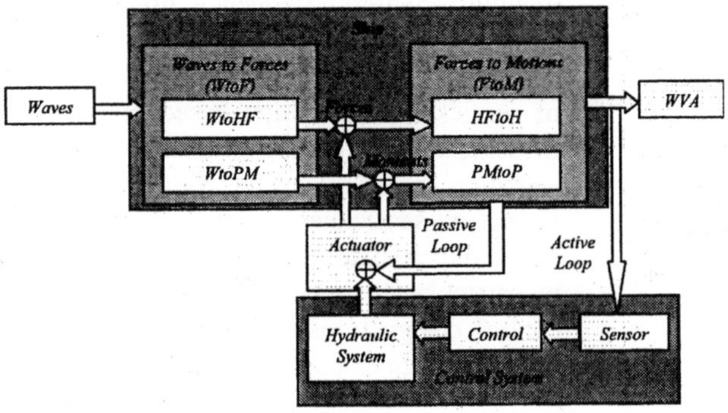


Fig.4. The control scenario.

Since the actuators are attached to the ship, ship motion cause actuator motion. And the motion of actuators, causing lift and drag changes, do have an effect on the ship. This is what is denoted as passive loop in figure 4. There are at least two parameters that we can handle to study better passive attenuation of WVA. The parameters are $d_a$, the location of the actuator, and $A$, its area. There is also an active loop in figure 4, which moves the actuators with respect to the ship, using hydraulic cylinders. Figure 5 depicts the main variables for the study of actuators action.

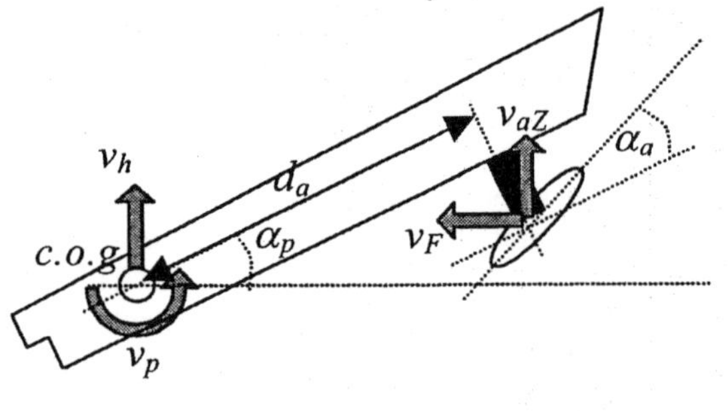


Fig. 5. Variables for actuators action study.

The equations of interest are the following:

$$v_{aZ} = v_h + v_p \cdot d_a \cdot \cos(\alpha_p) \square v_h + v_p \cdot d_a; \quad (if\ \alpha_p \downarrow)$$

$$\alpha_{ext} = -\arctan\left(\frac{v_{aZ}}{v_F}\right) \square -\frac{v_{aZ}}{v_F}; \qquad (if\ v_F \uparrow)$$

$$\alpha_{eff} = \alpha_a + \alpha_p + \alpha_{ext};$$

(1)

where $\alpha_a$ is the actuator's angle respect to the ship, $\alpha_p$ is the ship's pitch angle, $\alpha_{ext}$ is an extra-angle due to the actuator's vertical speed $v_{aZ}$, respect to the water flow speed $v_F$. To compute $v_{aZ}$, the heave speed $v_h$, and the pitch speed $v_p$ are used.

The equations for the action of the actuators are the following,

$$F_L = G(A, v_F) \cdot \alpha_{eff} = \rho \cdot A \cdot v_F^2 \cdot C_L \cdot \alpha_{eff}$$

$$F_h = F_L$$

$$M_p = d_a \cdot F_L$$

(2)

where $F_L$ is lift force, $\rho$ is the water density, $C_L$ is the lift coefficient, $F_h$ is heave force generated by the actuator, and $M_p$ is its moment with respect to the c.o.g. of the ship.

## 4. THE CONTROL PROBLEM

For the sake of clarity, let us consider only one actuator, a T-foil placed near the bow. The worst vertical motion (WVM) is measured $d_s$ meters in front of the c.o.g. The following block diagram applies (figure 6). For more details on ship model see (Esteban et. al. 2001).

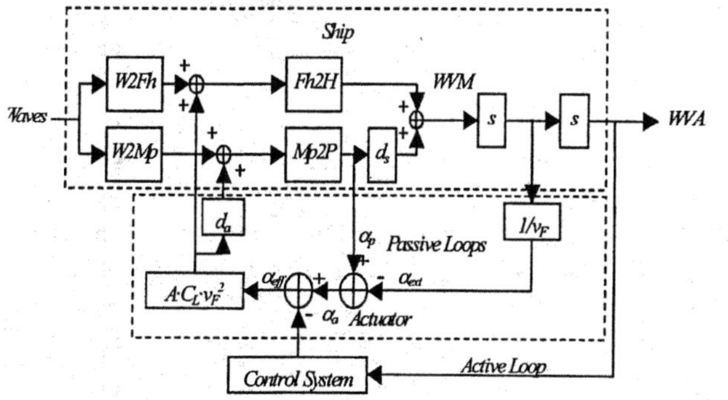

Fig.6. Block diagram of the control problem.

Using block algebra it is possible to obtain the following transfer functions:

$$\frac{WVA(s)}{Waves(s)} = \frac{(W2Fh(s)\cdot Fh2H(s) + W2Mp(s)\cdot Mp2P(s)\cdot G_1(s))\cdot s^2}{1 + \frac{G_2(s)}{V_F}\cdot s + CS(s)\cdot G_2(s)\cdot s^2};$$

$$K = A\cdot C_L \cdot V_F^2;$$

$$G_1(s) = \frac{K\cdot Fh2H(s) + d_s}{1 - K\cdot d_a \cdot Mp2P(s)};$$

$$G_2(s) = K\{Fh2H(s) + d_a \cdot Mp2P(s)\cdot G_1(s)\};$$

$$CS(s) = Control\ System\ TF;$$

(3)

The "Control System" block in figure 6 includes sensors (accelerometers), a proportional controller and the hydraulic system. The hydraulic system is an integrator loop with rate and position saturation. We shall use a method inspired in the describing function to derive a model for the adequate "Control System". There exist a limit frequency $\omega_{lim}$, so for frequencies above it, the actuators do not saturate. The value of this limit frequency is,

$$\omega_{lim} = \frac{2\cdot \pi}{\frac{4\cdot A_{sat}}{R_{sat}}}$$

(4)

Figure 7 shows the behaviour of actuators to counteract the WVA, when WVA has frequency $\omega_{lim}$.

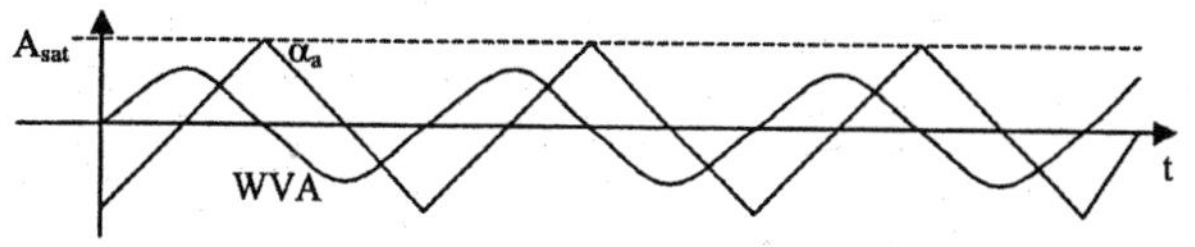

Fig.7. Control system output & WVA when $\omega = \omega_{lim}$

Figure 8 shows the behaviour of actuators when WVA has frequency less than $\omega_{lim}$.

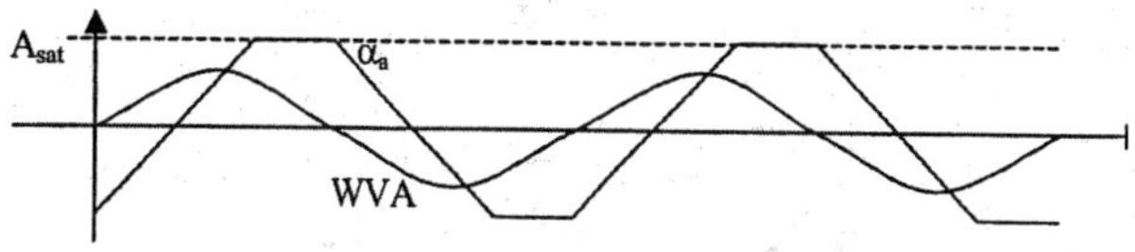

Fig.8. Control system output & WVA when $\omega < \omega_{lim}$

And figure 9 shows the behaviour of actuators when WVA has frequency larger than $\omega_{lim}$.

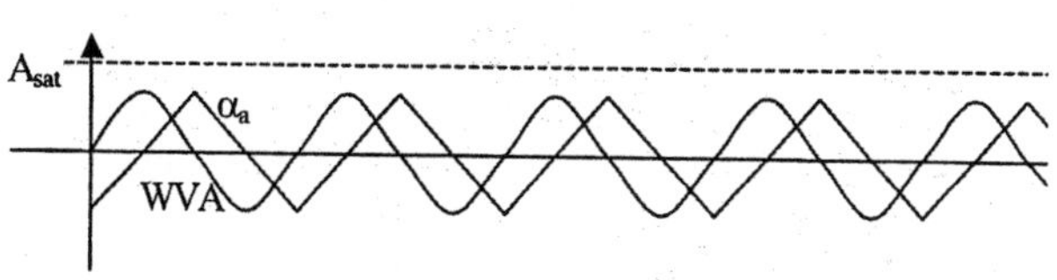

Fig.9. Control system output & WVA when $\omega > \omega_{lim}$

The control signal amplitude and phase for all cases can be described with the following representation (figure 10),

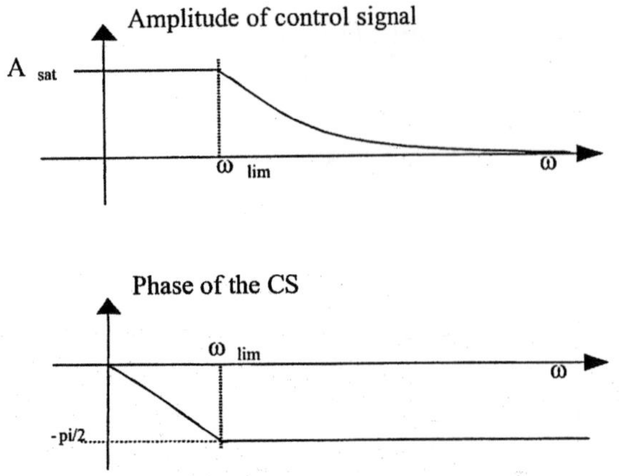


Fig. 10. Amplitude of the control signal and phase of the Control System.

In the top plot we represent the amplitude of control signal $\alpha_a$ (see figure 6). This signal is the product of the control system gain times the WVA amplitudes. The WVA amplitudes depend on amplitude of waves for each frequency, and on the corresponding gain of the ship filter.

We know the desired output of the control system, as described in figure 10. We need to obtain the gain of the proportional controller, which is inside the control system, that best approximate this desired behaviour.

Since there is a closed loop (figure 6), the control input depends on the total transfer function of the ship and the behaviour of actuators.

We shall find by iteration the gain and phase of CS(s) that best fits the curves in figure 10.

## 5. CASE STUDY

Let us take the fast ferry selected along our research. The acceleration is measured at the 15th rib, the cog is 40m aft (d=40m). The T-foil is located at d=40m., and its characteristics are: $R_{sat}$=13.5 °/s, $A_{sat}$=15 °, A=13.5 m$^2$, Cl=1.836 KN/(m$^2$·rad·(m/s)$^2$). A reasonable value for the maximun proportional gain of the control is Kcs=0.2 (input in m/s$^2$, output in rad).

Figures 11a and 11b shows that five iterations are enough to obtain a satisfactory solution. The first iteration is done with CS(s) = 0. With the amplitude of WVA and the desired control output we obtain the CS(s) gains for each frequency. We close the loop and compute the new amplitude of WVA, and so on.

Notice in figure 11a that the control system gain has been limited to 0.2 rad/(m/s$^2$) at frequencies over 2 rad./sec. This is because at frequencies over 2 rad./sec. the waves are small and do not deserve much actuator action.

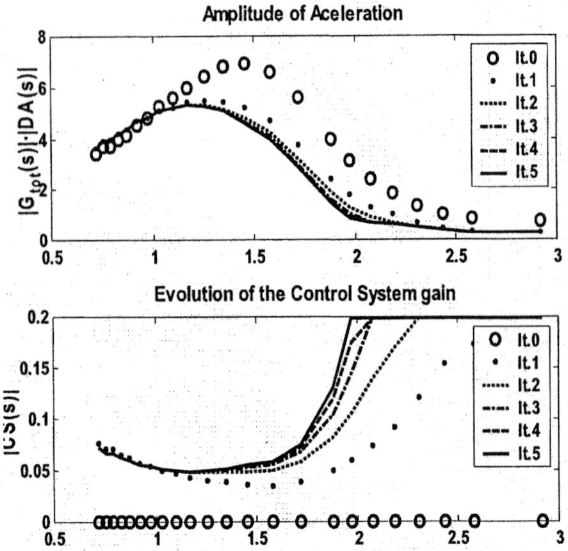


Fig. 11a. Iterations to obtain CS(s).

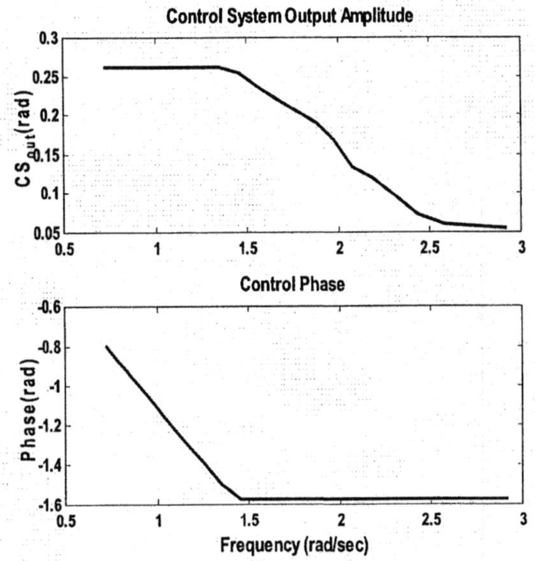


Fig. 11b. Iterations to obtain CS(s).

The research considers nine cases, combining sea states SSN4, SSN5 and SSN6, and ship's speeds of 20, 30 and 40 knots. The following figures shows, for the nine cases, the frequency response of the chain of three filters considered in section 1of this paper. The curves correspond to the ship without actuators, with fixed actuators, and with controlled moving actuators.

Figure 12 shows the results for 20 knots. Figure 13 the results for 30 knots, and figure 14 the results for 40 knots.

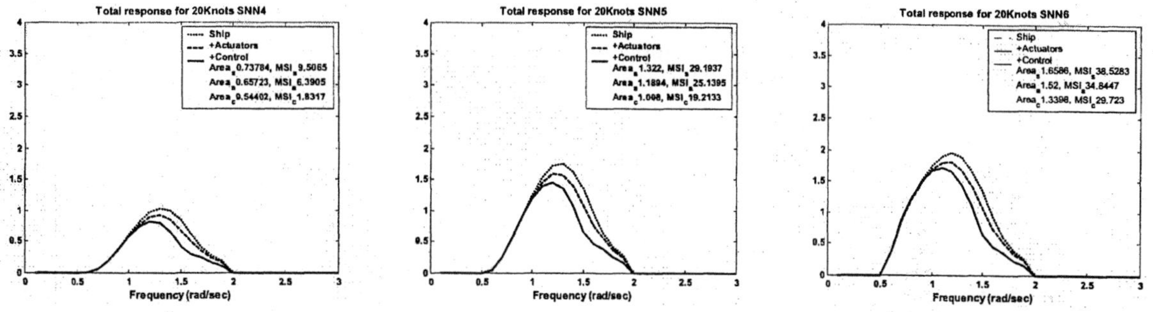


Fig. 12. Comparison of effects at 20 knots.

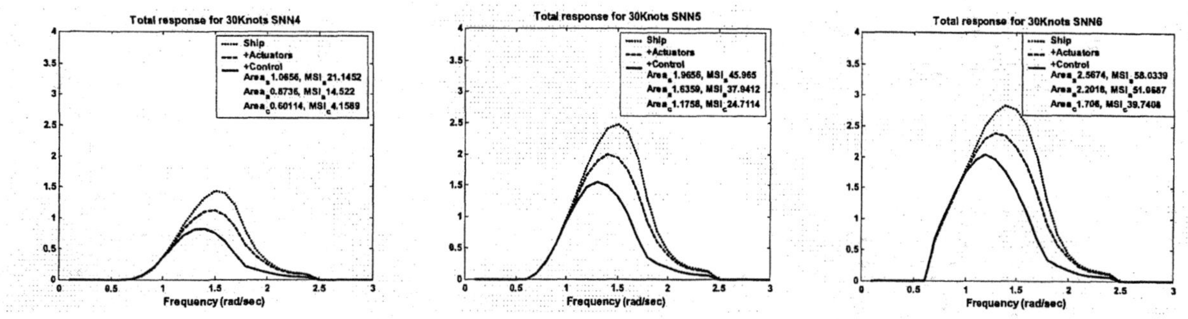


Fig. 13. Comparison of effects at 30 knots.

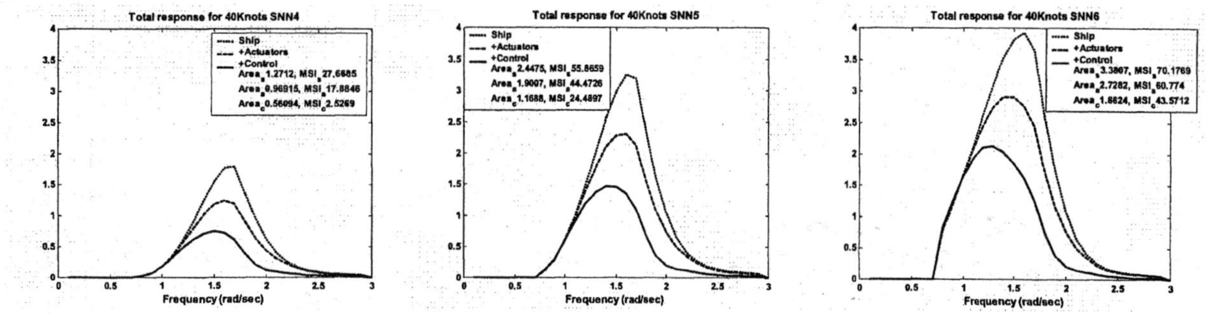


Fig. 14. Comparison of effects at 40 knots.

The higher the speed of the fast ferry, the more effective actuators are. It can be easily noticed comparing the effect of controlled actuators at 20 knots, with the curves at 40 knots.

Now it is possible to study other locations and sizes of the actuators, and compare results. For instance, figure 15 shows, at 40 knots, the effects of doubling the size of the actuators. Compare with figure 14 the MSI values of the ship with controlled actuators.

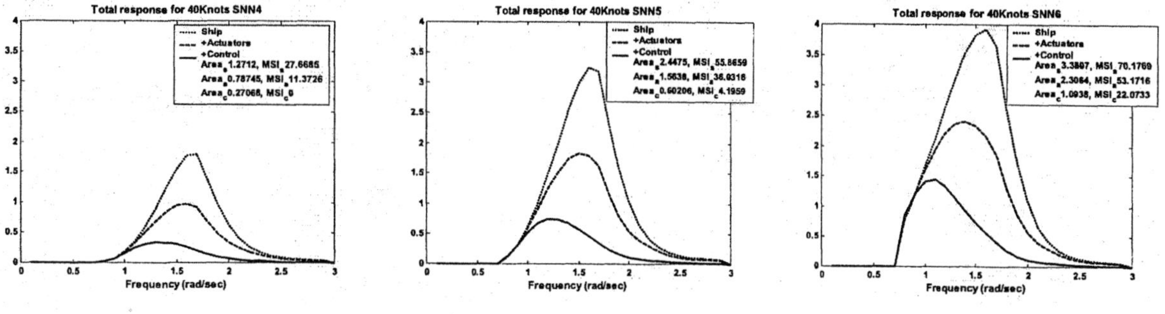


Fig. 15. Doubling the size of actuators, 40 knots.

The following figures (figure 16) display how different combinations of controlled actuator location and size make decrease or increase the MSI. This can be helpful, for instance, to choose the rib where to anchor the actuator.

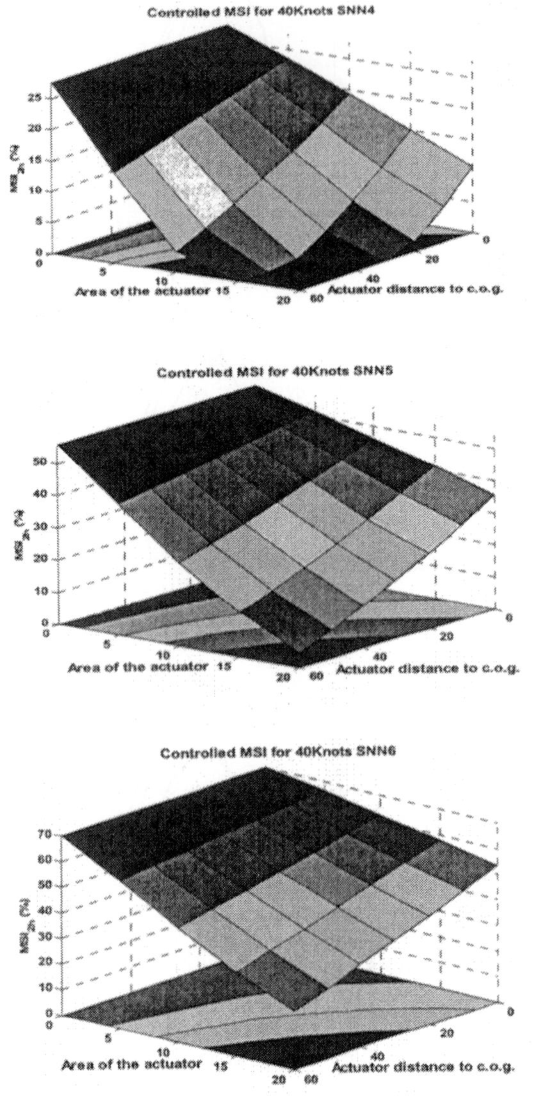


Fig. 16. MSI vs location and size of controlled actuators, 40 knots.

## 6. CONCLUSIONS

The control of moving actuators focusing on seasickness alleviation has been introduced. Also, criteria concerning actuator design has been elicited. A particular case study has been presented.

At the beginning of our research, some years ago, actuator design and location in the ship where dictated by the shipbuilder company. We were curious about other possibilities for better control results in view of the most important objective: to improve passenger comfort. Now we opened a way to make a study on this aspect. We think that the method devised for this problem may also be useful for other control cases with nonlinearities.

## REFERENCES

Aranda, J., J.M. Diaz, P. Ruiperez, T.M. Rueda and E. Lopez (2001). Decreasing of the motion sickness incidence by a multivariable classic control for a high speed ferry, In *Proceedings IFAC Intl. Conf. Control Applications in Marine Systems CAMS2001*, Glasgow.

Esteban, S., J.M. De la Cruz, J.M. Giron-Sierra, B. De Andres, J.M. Diaz and J. Aranda (2000). Fast ferry vertical acceleration reduction with active flaps and T-foil, In *Proceedings IFAC Intl. Symposium Maneuvering and Control of Marine Craft MCMC2000*, Aalborg, 233-238.

Esteban, S., B. De Andres, J.M. Giron-Sierra, O.R. Polo and E. Moyano (2001). A simulation tool for a fast ferry control design, In *Proceedings IFAC Intl. Conf. Control Applications in Marine Systems CAMS2001*, Glasgow.

Esteban, S., B. Andres-Toro, E. Besada-Portas, J.M. Giron-Sierra and J.M. De la Cruz (2002). Multiobjective control of flaps and T-foil in high speed ships, In *Proceedings IFAC 2002 World Congress*, Barcelona.

Fossen, T.I. (2002). *Marine Control Systems,* Marine Cybernetics AS, Trondheim.

Giron-Sierra, J.M., S. Esteban, B. De Andres, J.M. Diaz and J.M. Riola (2001). Experimental study of controlled flaps and T-foil for comfort improvement of a fast ferry, In *Proceedings IFAC Intl. Conf. Control Applications in Marine Systems CAMS2001*, Glasgow.

Giron-Sierra, J.M., R. Katebi, J.M. De la Cruz, and S. Esteban. (2002). The control of specific actuators for fast ferry vertical motions damping, In *Proceedings IEEE Intl. Conf. CCA/CACSD*, Glasgow.

Giron-Sierra, J.M., B. Andres-Toro, S. Esteban, J. Recas, E. Besada and J.M. De la Cruz (2003). Model based analysis of seasickness in a fast ferry. In *Proceedings IFAC MCMC 2003, Gerona, Spain.*

Haywood, A.J., A.J. Duncan, K.P. Klaka and J. Bennet (1995). The development of a ride control system for fast ferries, *Control Engineering Practice*, 695-703.

O'Hanlon, J.F. and M.E. MacCawley (1974). Motion sickness incidence as a function of frequency and acceleration of vertical sinusoidal motion, *Aerospace Medicine*.

Ryle, M. (1998). Smoothing out the ride, *The Motor Ship*, January, 23-26.

ELSEVIER
IFAC
PUBLICATIONS
www.elsevier.com/locate/ifac

# A SOFT COMPUTING METHOD FOR AN AUV NAVIGATION SYSTEM WITH PSEUDO-REAL-TIME APPLICABILITY

**D. Loebis, R. Sutton, J. Chudley**

*Marine and Industrial Dynamic Analysis Research Group,
Reynolds Building, School of Engineering, The University of Plymouth,
Drake Circus, Plymouth, PL4 8AA*

Abstract: This paper describes the implementation of a soft computing method based on fuzzy logic and multiobjective genetic algorithm techniques to adapt the parameters of an error-state complementary Kalman filter (ESCKF) to enhance the accuracy of an autonomous underwater vehicle (AUV) navigation system. In the ESCKF, inertially-derived quantities from an inertial navigation system (INS) sensor are combined with direct measurements of the same quantities by use of the global positioning system (GPS) and other aiding sensors. The backlash of the integration processes however, is that errors can grow rapidly and the values obtained therein can drift off the true value significantly. By contrast, the directly-measured data contain high frequency noise with bounded error. This instinctively suggests integrating the two sets of quantities, which is exactly what the ESCKF does. To maintain the stability and performance of the ESCKF, which is likely to deteriorate when the assumed error and noise characteristics do not reflect the true ones, a fuzzy logic based scheme is used to make these values adaptive. The choice of fuzzy membership functions for this scheme is first carried out using a heuristic approach and further refined using a multiobjective genetic algorithm method. *Copyright©2004 IFAC*

Keywords: Autonomous underwater vehicles, navigation, Kalman filters, fuzzy logic, genetic algorithm, multiobjective optimization

## 1. INTRODUCTION

The development of AUVs for scientific, military and commercial purposes in applications such as ocean surveying, unexploded ordnance hunting, and cable tracking and inspection requires the corresponding development of navigation systems. Such systems are necessary to provide knowledge of vehicle position and attitude. The need for accuracy in such systems is paramount: erroneous position and attitude data can lead to a meaningless interpretation of the collected data or even to a catastrophic failure of an AUV. A growing number of research groups around the world are developing integrated navigation systems utilising INS and GPS. However, few of these works make explicit the essential need for fusion of several INS sensors that enable the users to maintain the accuracy or even to prevent a complete failure of this part of the navigation system, before being integrated with the GPS. Several estimation methods have been used in the past for multisensor data fusion and integration purposes. To this end, the Kalman filter (KF) and its variants have been popular methods in the past and interest in developing the algorithms has continued to the present day.

However, a significant difficulty in designing a KF can often be traced to incomplete *a priori* knowledge of the process noise covariance matrix (**Q**) and measurement noise covariance matrix (**R**). In most practical applications, these matrices are initially estimated or even unknown. The problem here is that the optimality of the estimation algorithm in the KF setting is closely connected to the quality of *a priori* information about the **Q** and **R** (Mehra, 1970). It has been shown that insufficiently known *a priori* filter statistics can reduce the precision of the estimated filter states or introduce biases to their estimates. In addition, incorrect *a priori* information can lead to practical divergence of the filter. From the aforementioned it may be argued that the conventional KF with fixed **Q** and/or **R** should be replaced by an adaptive estimation formulation. In this paper, a novel fuzzy error-state complementary Kalman filter (FESCKF) is proposed. With this method, a KF with an error-state model obtained using first order Markov processes and error data analysis is used in parallel with fuzzy logic techniques to adjust **R**. A further improvement can be achieved using multiobjective genetic algorithm (MOGA) techniques, whereby the fuzzy membership functions are adjusted to produce the most optimum result.

The structure of the paper is as follows: section 2 introduces the concept of the ESCKF and the derivation of the process and measurement model and the associated noise covariance matrices. Section 3 discusses the proposed KF adaptation mechanism followed by fuzzy membership function optimization (FESCKF). Section 4 and 5 provide simulation results and concluding remarks respectively.

## 2. ESCKF MODELLING

### *2.1 The Concept of ESCKF*

Brown and Hwang (1997) discuss the advantages of the ESCKF method over the total state Kalman filter. The most important advantage is that any nonlinear relationship between the process dynamics in the inertial system and the measurement relationships can be removed in a differencing operation, and the filter becomes linear. This linearity condition is required by the Kalman filter. This condition can also lead to faster codes execution as linearisation operations are relatively slow to execute.

KF algorithms are widely available in the literature. The interested reader can refer to Brown and Hwang (1997). Works on ESCKF however, are very limited especially in the field of AUV navigation systems. An example of this work can be found in Gustaffson *et al.* (2001). Like in the KF, the ESCKF algorithm can be divided into two major parts: the measurement update and the time update. In ESCKF, the measurement update is obtained by subtracting the direct measurements from the computed version of the same quantities. By doing this, the true values cancel each other out and what remains is the difference between the measurement errors and drift errors. In the time update, the estimates are obtained by subtracting the estimated drift errors from the forward filter pass from the computed version of the same quantities.

In this paper, measurement errors from an accelerometer and a gyroscope (assembled in an inertial measurement unit (IMU)), a TCM2 electronic compass and a GPS receiver unit are estimated and modelled using first order Markov processes which are defined in the following manner:

$$\dot{x} = -\frac{1}{\tau} \cdot x + \gamma \tag{1}$$

In (1), $x$ is the error process to be modelled, $\tau$ is the time constant of the assumed Markov process and $\gamma$ is white noise. For modelling purposes, all sensor data have been collected in static conditions for a period of approximately 2.5 hours. For the same purpose, three different frames of reference are defined. The body-fixed ($b$) frame of reference is aligned to the axes of the AUV, where forward-starboard-down correspond to $x$ - $y$ - $z$. These need to be transformed to the geographical ($g$) frame of reference, where $x$ - $y$ - $z$ correspond to North-East-Down. For these particular application, the measurements in question are 3D accelerations, as well as angular rates measured by the IMU. Earth-centred Earth-fixed (ECEF) frame is where the GPS latitude and longitude are defined. The following subsections give the derivation of the process matrix (**F**) and the corresponding noise covariance matrix (**Q**)

### *2.2 Process and Noise Covariance Matrix*

The elements of the state of the ESCKF are defined as follows:

$$\mathbf{x} = \left[\, x_d^g \; y_d^g \; \psi_d^g \; x_e^g \; y_e^g \; \psi_e^g \; r_e^b \; u_e^h \; v_e^h \,\right] \tag{2}$$

In (2) the subscripts $d$ and $e$ denote drift and sensor errors respectively. Superscripts $g$, $b$ and

$h$ denote geographical frame, body-fixed-frame and horizontal frame respectively. Drift errors in position, $x_d^g$ and $y_d^g$ stem from the error in the integrated acceleration ($u_e^h$ and $v_e^h$), and compass error $\psi_e^g$. $\psi_d^g$ is heading drift error which comes from the error in the integrated yaw rate. Measurement errors in the position blend are respectively represented by the states $x_e^g$ and $y_e^g$. Finally, state $r_e^b$ represents gyroscope's yaw error.

The differential equation describing the relationship between dead reckoning position and the horizontal velocity from the integrated acceleration is given as:

$$\begin{bmatrix} \dot{x}_c^g \\ \dot{y}_c^g \end{bmatrix} = \begin{bmatrix} \cos\psi_m^g & -\sin\psi_m^g \\ \sin\psi_m^g & \cos\psi_m^g \end{bmatrix} \cdot \begin{bmatrix} u_m^h \\ v_m^h \end{bmatrix} \tag{3}$$

In (3) the subscript $c$ denotes *computed* and $m$ denotes *measured*. Expanding this into true values and errors yields:

$$\begin{bmatrix} \dot{x}^g + \dot{x}_d^g \\ \dot{y}^g + \dot{y}_d^g \end{bmatrix} = \begin{bmatrix} \cos(\psi^g + \psi_e^g) & -\sin(\psi^g + \psi_e^g) \\ \sin(\psi^g + \psi_e^g) & \cos(\psi^g + \psi_e^g) \end{bmatrix} \cdot \begin{bmatrix} u_m^h \\ v_m^h \end{bmatrix} \tag{4}$$

Expansion of (4) by applying trigonometric formulas and by assuming that the measurement error is sufficiently small whereby the relations $\cos\psi_e^g \cong 1$ and $\sin\psi_e^g \cong \psi_e^g$ holds, yields:

$$\begin{bmatrix} \dot{x}^g + \dot{x}_d^g \\ \dot{y}^g + \dot{y}_d^g \end{bmatrix} = \begin{bmatrix} \cos\psi^g - \sin\psi^g \cdot \psi_e^g & -\sin\psi^g - \cos\psi^g \cdot \psi_e^g \\ \sin\psi^g + \cos\psi^g \cdot \psi_e^g & \cos\psi^g - \sin\psi^g \cdot \psi_e^g \end{bmatrix} \cdot \begin{bmatrix} u^h \\ v^h \end{bmatrix} + \begin{bmatrix} \cos(\psi^g + \psi_e^g) & -\sin(\psi^g + \psi_e^g) \\ \sin(\psi_g + \psi_e^g) & \cos(\psi^g + \psi_e^g) \end{bmatrix} \cdot \begin{bmatrix} u_e^h \\ v_e^h \end{bmatrix} \tag{5}$$

By substituting true values into (3) and subtracting the result from (5), gives

$$\begin{bmatrix} \dot{x}_d^g \\ \dot{y}_d^g \end{bmatrix} = \begin{bmatrix} -\sin\psi^g \cdot \psi_e^g & -\cos\psi^g \cdot \psi_e^g \\ \cos\psi^g \cdot \psi_e^g & -\sin\psi^g \cdot \psi_e^g \end{bmatrix} \cdot \begin{bmatrix} u^h \\ v^h \end{bmatrix} + \begin{bmatrix} \cos(\psi^g + \psi_e^g) & -\sin(\psi^g + \psi_e^g) \\ \sin(\psi^g + \psi_e^g) & \cos(\psi^g + \psi_e^g) \end{bmatrix} \cdot \begin{bmatrix} u_e^h \\ v_e^h \end{bmatrix} \tag{6}$$

To explicitly relate drift errors in position to the accelerometer error, trivial alteration is applied to the first term of right hand side of (6). Further, by assuming that the current estimate of heading and velocities from the Kalman filter ($\hat{\psi}^g$,$\hat{u}$ and $\hat{v}$) are sufficiently close to the true heading ($\psi^g$) and velocities and also by keeping in mind that $\psi^g + \psi_e^g = \psi_m^g$, (6) becomes:

$$\begin{bmatrix} \dot{x}_d^g \\ \dot{y}_d^g \end{bmatrix} = \begin{bmatrix} -\sin\hat{\psi}^g \cdot \hat{u}^h & -\cos\hat{\psi}^g \cdot \hat{v}^h \\ \cos\hat{\psi}^g \cdot \hat{u}^h & -\sin\hat{\psi}^g \cdot \hat{v}^h \end{bmatrix} \cdot \begin{bmatrix} \psi_e^g \\ \psi_e^g \end{bmatrix} + \begin{bmatrix} \cos\psi_m^g & -\sin\psi_m^g \\ \sin\psi_m^g & \cos\psi_m^g \end{bmatrix} \cdot \begin{bmatrix} u_e^h \\ v_e^h \end{bmatrix} \tag{7}$$

As in the case with position drift, the heading drift error evolution ($\psi_d^g$) will be directly dependent on the yaw rate sensor error ($r_e^b$), because it is rotated into horizontal frame before it is integrated up to yield an alternative heading. The differential equation describing the relationship between body-fixed angular rates and horizontal heading can be written as:

$$\dot{\psi}_c^h = \frac{\sin\phi_m^h \cdot q_m^b + \cos\phi_m^h \cdot r_m^b}{\cos\theta_m^h} \tag{8}$$

In (8) $\phi$, $\theta$, $q$ and $r$ are the roll, pitch, pitch rate and yaw rate respectively. Expanding computed heading and measured yaw rate into true values plus drift and sensor errors, and using the assumptions that $q^b \cong q_m^b$, $\psi^h \cong \psi_m^h$ and $\theta \cong \theta_m^h$, true values cancel each other out. Observing that the computed heading is initialised by a TCM2 reading, the following expression gives the sought relation between drift error and yaw rate sensor error:

$$\dot{\psi}_d^g = \frac{\cos\phi_m^h}{\cos\theta_m^h} \cdot r_e^b \tag{9}$$

The rest of the diagonal elements in the process matrix describe the sensor error processes, which are modelled using a first order Markov processes. Based on the derivations and assumptions in (3) through (9), the process matrix can be written as in (10) given in the next page.

The variance of the process noise for a Markov error model can be described as in Brown and Hwang (1997),

$$variance[w_k] = (1 - e^{-\frac{2\Delta t}{\tau}}) \cdot variance[x_k] \tag{11}$$

In (11), $e^{-\frac{\Delta t}{\tau}}$ is the state transition parameter for the Markov error model. $\Delta t$ is the discrete time interval and $\tau$ is the time constant. By taking the approximation: $e^{-\beta\Delta(t)} \approx 1 - \beta \cdot \Delta t$, where $\beta = \tau^{-1}$, the following is true:

$$variance[w_k] = (2\beta\Delta t - (\beta\Delta t)^2) \cdot variance[x_k] \tag{12}$$

According to the process model, the heading drift error state represents the integrated yaw error state, in effect an integrated Markov process. The process noise covariance matrix for these two states can be defined as in Brown and Hwang (1997),

$$\mathbf{Q} = \begin{bmatrix} E[\psi_d^g \psi_d^g] & E[\psi_d^g r_e^b] \\ E[\psi_d^g r_e^b] & E[r_e^b r_e^b] \end{bmatrix} \tag{13}$$

where

$$\mathbf{F} = \begin{bmatrix} 0\,0\,0 & 0 & 0 & -\sin\hat{\psi}^g \cdot \hat{u}^h - \cos\hat{\psi}^g \cdot \hat{v}^h & 0 & \cos\psi^g_m & -\sin\psi^g_m \\ 0\,0\,0 & 0 & 0 & \cos\hat{\psi}^g \cdot \hat{u}^h - \sin\hat{\psi}^g \cdot \hat{v}^h & 0 & \sin\psi^g_m & \cos\psi^g_m \\ 0\,0\,0 & 0 & 0 & 0 & \dfrac{cos\phi^h_m}{\cos\theta^h_m} & 0 & 0 \\ 0\,0\,0 & \dfrac{-1}{\tau_x} & 0 & 0 & 0 & 0 & 0 \\ 0\,0\,0 & 0 & \dfrac{-1}{\tau_y} & 0 & 0 & 0 & 0 \\ 0\,0\,0 & 0 & 0 & \dfrac{-1}{\tau_\psi} & 0 & 0 & 0 \\ 0\,0\,0 & 0 & 0 & 0 & \dfrac{-1}{\tau_{\dot{\psi}}} & 0 & 0 \\ 0\,0\,0 & 0 & 0 & 0 & 0 & \dfrac{-1}{\tau_u} & 0 \\ 0\,0\,0 & 0 & 0 & 0 & 0 & 0 & \dfrac{-1}{\tau_v} \end{bmatrix} \quad (10)$$

$$Q_{11} = \frac{2\sigma^2}{\beta}\left[\Delta t - \frac{2}{\beta}(1-\phi) + \frac{1}{2\beta}(1-\phi^2)\right] \quad (14)$$

$$Q_{12} = Q_{21} = 2\sigma^2\left[\frac{1}{\beta}(1-\phi) + \frac{1}{2\beta}(1-\phi^2)\right] \quad (15)$$

$$Q_{22} = \sigma^2(1-\phi^2) \quad (16)$$

where $\beta$ is the inverse of the Markov time constant, $\sigma^2$ is the process noise variance of the yaw rate Markov error and $\phi = e^{-\frac{\Delta t}{\tau}}$ and defined as before.

For the first two states in the process model, $x^g_d$ and $y^g_d$, the analysis is more complicated and for this reason, the noise covariance matrix of the two states are obtained from an empirical result, and provision for the adjustment method has been made and will be reported in the future.

### *2.3 Measurement and Noise Covariance Matrix*

The measurement matrix $\mathbf{H}$ relates the available measurement updates to the element in the state vector and takes the following form:

$$\mathbf{H} = \begin{bmatrix} 1 & 0 & 0 & -1 & 0 & 0 & 0 & 0 & 0 \\ 0 & 1 & 0 & 0 & -1 & 0 & 0 & 0 & 0 \\ 0 & 0 & 1 & 0 & 0 & -1 & 0 & 0 & 0 \end{bmatrix} \quad (17)$$

The measurement noise covariance matrix $\mathbf{R}_k$ is determined empirically and given as:

$$\mathbf{R}_k = \begin{bmatrix} \sigma^2_{X-Position} & 0 & 0 \\ 0 & \sigma^2_{Y-Position} & 0 \\ 0 & 0 & \sigma^2_{Heading} \end{bmatrix} \quad (18)$$

where $\sigma^2_{X-Position}$, $\sigma^2_{Y-Position}$, $\sigma^2_{Heading}$ are the variance in X, Y direction and heading respectively. These values will be adapted using the algorithm discussed in the next section.

## 3. THE ADAPTIVE TUNING OF KALMAN FILTER ALGORITHM

Over the past few years, only a few publications in the area of adaptive Kalman filtering can be found in the literature. One of the most popular method is innovation adaptive estimation (IAE). The innovation $\mathbf{Inn}_k$ at sample time $k$ is the difference between the real measurement $\mathbf{z}_k$ received by the filter and its estimated (predicted) value $\hat{\mathbf{z}}_k$. The predicted measurement is the projection of the filter predicted states $\mathbf{x}^-_k$ onto the measurement space through the measurement matrix $\mathbf{H}_k$. Innovation represents additional information available to the filter as a result of the new measurement $\mathbf{z}_k$. The occurrence of data with statistics different from the *a priori* information will first show up in the innovation vector. For this reason the innovation sequence represents the information content in the new observation and is considered the most relevant source of information to the filter adaptation.

Herein, the IAE approach coupled with fuzzy logic techniques with membership functions designed using heuristic methods and further refined using MOGA is used to adjust the $\mathbf{R}$ matrix of the ESCKF. Initial work on this approach can be found in Loebis *et al.* (2003) .

### *3.1 Fuzzy error state complementary Kalman filter*

In this sub-section, an on-line innovation-based adaptive scheme of the ESCKF to adjust the $\mathbf{R}$ matrix employing the principles of fuzzy logic is presented. The fuzzy logic is chosen mainly because of its simplicity. This motivates the interest in the topic, as testified by related papers which have been appearing in the literature (Loebis *et al.*, 2003; Escamilla-Ambrosio and Mort, 2001).

The FESCKF proposed herein is based on the IAE approach using a technique known as covariance-matching (Mehra, 1970). The basic idea behind the technique is to make the actual value of the covariance of the innovation sequences match its theoretical value.

The actual covariance is defined as an approximation of the $\mathbf{Inn}_k$ sample covariance through averaging inside a moving estimation window of size $M$ (Mohamed and Schwarz, 1999) which takes the following form:

$$\hat{\mathbf{C}}_{Inn_k} = \frac{1}{M}\sum_{j=j_0}^{k} \mathbf{Inn}_k \cdot \mathbf{Inn}_k^T \qquad (19)$$

where $j_0 = k - M + 1$ is the first sample inside the estimation window. An empirical experiment is conducted to choose the window size $M$. From experimentation it was found that a good size for the moving window in (19) is 15. The theoretical covariance of the innovation sequence is defined as

$$\mathbf{S}_k = \mathbf{H}_k \cdot \mathbf{P}_k^- \cdot \mathbf{H}_k^T + \mathbf{R}_k \qquad (20)$$

The logic of the adaptation algorithm using covariance matching technique can be qualitatively described as follows. If the actual covariance value $\hat{\mathbf{C}}_{Inn_k}$ is is observed, whose value is within the range predicted by theory $\mathbf{S}_k$ and the difference is very near to zero, this indicates that both covariances match almost perfectly and only a small change is needed to be made on the value of $\mathbf{R}$. If the actual covariance is greater than its theoretical value, the value of $\mathbf{R}$ should be decreased. On the contrary, if $\hat{\mathbf{C}}_{Inn_k}$ is less than $\mathbf{S}_k$, the value of $\mathbf{R}$ should be increased. This adjustment mechanism lends itself very well to being dealt with using a fuzzy-logic approach based on rules of the kind:

$$IF\ \langle antecedent\rangle\ THEN\ \langle consequent\rangle \qquad (21)$$

where antecedent and consequent are of the form $\nu \epsilon O_i$, $\kappa \epsilon L_i$, $i = 1, 2, ...$ respectively. Where $\nu$ and $\kappa$ are the input and output variables, respectively, and $O_i$ and $L_i$ are the fuzzy sets.

To implement the above covariance matching technique using the fuzzy logic approach, a new variable called $\mathbf{delta}_k$, is defined to detect the discrepancy between $\hat{\mathbf{C}}_{Inn_k}$ and $\mathbf{S}_k$. The following fuzzy rules of the kind (21) are used:

$$IF\ \langle \mathbf{delta}_k \cong 0\rangle\ THEN\ \langle \mathbf{R}_k\ is\ unchanged\rangle \qquad (22)$$

$$IF\ \langle \mathbf{delta}_k > 0\rangle\ THEN\ \langle \mathbf{R}_k\ is\ decreased\rangle \qquad (23)$$

$$IF\ \langle \mathbf{delta}_k < 0\rangle\ THEN\ \langle \mathbf{R}_k\ is\ increased\rangle \qquad (24)$$

Thus $\mathbf{R}$ is adjusted according to,

$$\mathbf{R}_k = \mathbf{R}_{k-1} + \Delta\mathbf{R}_k \qquad (25)$$

where $\Delta\mathbf{R}_k$ is added or subtracted from $\mathbf{R}$ at each instant of time. Here $\mathbf{delta}_k$ is the input to the fuzzy inference system (FIS) and $\Delta\mathbf{R}_k$ is the output.

### *3.2 Fuzzy membership functions optimization*

MOGA is used here to optimize the membership functions of the FESCKF. To translate the FESCKF membership functions to a representation useful as genetic material, they are parameterised with real-valued variables. Each of these variables constitutes a gene of the chromosomes for the MOGA. Boundaries of chromosomes are required for the creation of chromosomes in the right limits so that the MOGA is not misled to some other area of search space. The technique adopted in this paper is to define the boundaries of the output membership functions according to the furthest points and the crossover points of two adjacent membership functions. In other words, the boundaries of FESCKF consist of three real-valued chromosomes (*Chs*), as in Figure 1. The

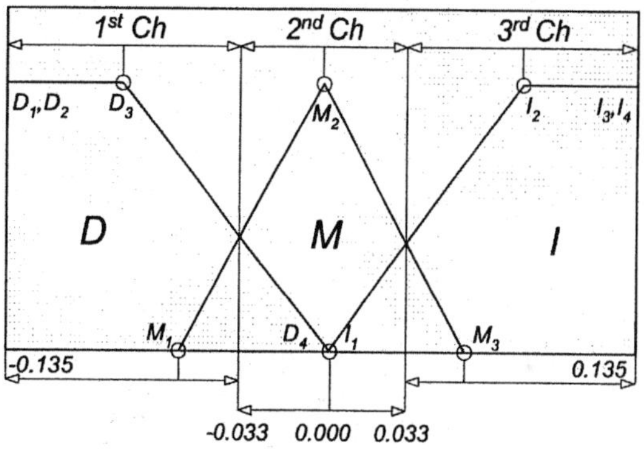


Fig. 1. Membership function and boundaries of $\mathbf{R}_k$

trapezoidal membership functions' two furthest points, -0.135 ($D_1$), -0.135 ($D_2$) and 0.135 ($I_3$), 0.135 ($I_4$) of FESCKF, remain the same in the GA's description to allow a similar representation as the fuzzy system's definition. As can be seen from Figure 1, $D_3$ and $M_1$ can change value in the $1^{st}$ *Ch* boundary, $D_4$, $M_2$ and $I_1$ in the $2^{nd}$ *Ch* boundary, and finally, $M_3$ and $I_2$ in $3^{rd}$ *Ch*.

## 4. SIMULATION RESULTS

In this section the FESCKF algorithm is applied to a set of simulated sensor data, *i.e.* latitude and longitude data from a GPS unit, 3D accelerometer and gyroscope data from four IMUs located in different parts of the vehicle, and yaw data from four TCM2s located in close proximity to the IMUs. Herein, these sensors are used to capture

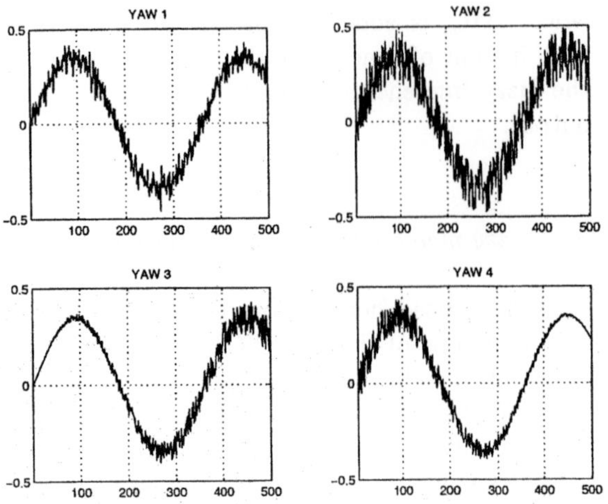


Fig. 2. Yaw output from TCM2s with (a) constant Gaussian noise1, (b) constant Gaussian noise2, (c) uniform noise increasing with time and (d) uniform noise decreasing with time

the position and attitude of the vehicle as a sinusoidal input applied to its rudder. Although there are redundancy on the 3D acceleration and attitude data, due to limited space, it is decided to focus the discussion in this paper on the fusion of the yaw data produced by the TCM2s. The yaw rates produced by the IMUs are integrated once to produce a computed version of the corresponding yaw data. As discussed in Section 2, the yaw measurement update is obtained by subtracting the yaw (TCM2s) direct measurements from the yaw (gyroscopes) computed version. Figure 2 shows the output of the TCM2s. The initial $\mathbf{R}$ was assumed to be $diag[500m^2\ 500m^2\ 0.1rad^2]$, $\hat{x}_0 = 0$, $\mathbf{P}_0 = 0.01I_9$. The value of $\mathbf{R}$ was first adapted using the FESCKF with membership functions designed heuristically and further refined using MOGA with the parameters shown in Table 1. For comparison purposes, the following performance measures were adopted:

$$J_{zv} = \sqrt{\frac{1}{n}\sum_{k=1}^{n}(za_k - z_k)^2} \qquad (26)$$

$$J_{ze} = \sqrt{\frac{1}{n}\sum_{k=1}^{n}(za_k - \hat{z}_k)^2} \qquad (27)$$

where $za_k$ is the actual value of the yaw, $z_k$ is the measured yaw, $\hat{z}_k$ is the estimated yaw at an instant of time $k$ and $n$ = number of samples. The performance comparison is presented in Table 2. It is clear that the $J_{ze}$-s of each sensor in both non-MOGA and MOGA always outperform the corresponding $J_{zv}$-s. It is also clear that the $J_{ze}$-s of MOGA case always produce a better result than the non-MOGA case. Most importantly, the $J_{ze}$-s of the fused sensor are better compared to the $J_{ze}$-s of the individual sensor.

| Parameters | Values |
|---|---|
| Number of objective functions | 5 |
| Number of generation | 200 |
| Number of individual per generation | 25 |
| Generation gap in selection operation | 0.95 |
| Rate in rate in recombination operation | 0.8 |
| rate in mutation operation | 0.09 |

Table 1. MOGA parameters

| Sensor | $J_{zv}(rad)$ | $J_{ze}(rad)$ | |
|---|---|---|---|
| | | Non-MOGA | MOGA |
| Sensor 1 | 0.0393 | 0.0379 | 0.0324 |
| Sensor 2 | 0.0799 | 0.0793 | 0.0668 |
| Sensor 3 | 0.0341 | 0.0293 | 0.0285 |
| Sensor 4 | 0.0350 | 0.0339 | 0.0299 |
| Fused | | 0.0245 | 0.0184 |

Table 2. Performance comparison

## 5. CONCLUDING REMARKS

A novel method to obtain an accurate AUV navigation system is proposed. The method is based on the ESCKF coupled with fuzzy logic to adjust the value of measurement noise covariance matrix $\mathbf{R}$. MOGA is proposed to further refine the result. The simulation results presented in this paper have shown the ability of the proposed algorithm to produce a significant improvement over the conventional method.

## REFERENCES

Brown, R. G. and P. Y. C. Hwang (1997). *Introduction to Random Signals and Applied Kalman Filtering.* 3rd Ed. John Wiley and Sons.

Escamilla-Ambrosio, P. J. and N. Mort (2001). A Hybrid Kalman Filter-Fuzzy Logic Multisensor Data Fusion Architecture with Fault Tolerant Characteristics. In: *Proc. of the 2001 International Conference on Artificial Intelligence.* Las Vegas, NV, USA. pp. 361–367.

Gustaffson, E., E. An and S. Smith (2001). A Postprocessing Kalman Smoother for Underwater Vehicle Navigation. In: *Proc. 12th International Symposium on Unmanned Untethered Submersible Technology.* New Hampshire, NH, USA. pp. 1–7.

Loebis, D., R. Sutton and J. Chudley (2003). A Fuzzy Kalman Filter for Accurate Navigation of an Autonomous Underwater Vehicle. In: *Proc. 1st IFAC Workshop on Guidance and Control of Underwater Vehicles.* Newport, South Wales, UK. pp. 161–166.

Mehra, R. K. (1970). On the Identification of Variances and Adaptive Kalman Filtering. *IEEE Transactions on Automatic Control* **AC-16(1)**, 12–21.

Mohamed, A. H. and K. P. Schwarz (1999). Adaptive Kalman Filtering for INS/GPS. *Journal of Geodesy* **73**, 193–203.

ELSEVIER
IFAC
PUBLICATIONS
www.elsevier.com/locate/ifac

# NEURAL NETWORK CONTROL SYSTEM FOR UNDERWATER ROBOTS

**Alexander A. Dyda[*], Dmitry A. Os'kin[**]**

[*]*Maritime State University,*
*Department of Automatic & Control Systems*
*50 [A] V.-Portovaya, Vladivostok, 690059 Russia*
*adyda@mail.ru, ph.+7-(4232)-284167*
[**]*Institute for Automation and Control Processes,*
*Far-Eastern Branch of Russian Academy of Science*
*5 Radio str., Vladivostok,690042 Russia*
*daoskin@mail.ru ph.+7-(4232)-220830*

Abstract. The paper is devoted to the design of the neural network based control systems for underwater robots. It is demonstrated that proposed two-stage approach makes possible to simplify control system synthesis for underwater robot and provide its dynamics close to a reference. First stage consists of the neural network correction of robot thruster dynamics. Second is devoted to neural control system synthesis of the robot as a whole. Results of mathematical simulation had confirmed the effectiveness of approach developed. *Copyright © 2004 IFAC*

Keywords: underwater robot, uncertain dynamics, multilayer neural network

## 1. INTRODUCTION

Underwater robots (UR) have wide applications in the area of ocean exploration and exploitation. As object of control, URs are characterized by the essentially nonlinear and uncertain dynamics. The control synthesis of such dynamical objects faces the difficulties. The complexity of solving of control system synthesis problem of UR space movement can be reduced by preliminary UR thruster correction, ensuring dynamics close to desirable.

Different approaches to the problem of UR thruster correction are known. In (Yorger and Slotine, 1985) ideas of variable structure system and adaptive control are used to improve the thruster dynamics. (Dyda, 1998; Dyda and Lebedev, 1996) proposed nonlinear and adaptive schemes of UR thruster correction. This methods mean the knowledge of exact structure of mathematical model of controlled object to be known. The developing such models is very complicated process. There exist some approaches which are not based on knowledge of exact models. Usage of multilayer neural networks (MNN) is one of it. (Yuh, 1990) offers the approach permitting to compensate nonlinear dynamics of UR thruster. It is noticed that MNN are capable to avoid controlled object uncertainties, and can be trained to approximate the inverse dynamics. After training, MNN can be used as a part of controller.

## 2. MULTILAYER NEURAL NETWORK AND LEARNING ALGORITHM

MNN can be presented by several layers consisting of elementary processor elements that are neurons. The neuron is characterized by a vector of weights and kind of activation function. Among layers one differs input layer, some hidden layers and target or output. The elements of each layer are connected in weights to neurons of other layers. Weights are determined in accordance with learning process.

Consider the MNN performance algorithm in more details (Behera and Gopal, 1994; Narendra and Parthasaty, 1990). The vector of input signals $V=\left(V_1^{a-1},\dots,V_k^{a-1},\dots,V_n^{a-1}\right)$ of the neuron in $a$ -th layer representing the activity vector of neurons previous ($a-1$-th) layer is weighed by a vector of weights:

$$H_b^a=\sum_{c=1}^{j} W_{b,c}^a V_b^{a-1}+s_b^a \quad (1),$$

where $W_{b,c}^a$ the weights coefficients between $b$ -th element in (a-1) -th layer and $c$ -th element a-th layer; $s_b^a$ - bias of $b$ -th element in a-th layer. Then it is exposed to nonlinear transformation by the functions of activation $g(\cdot)$:

$$V_b^a=g\left(H_b^a\right), \quad (2),$$

where $V_b^a$ the value of $b$ -th element at $a$ -th hidden layer;

for output layer:

$$O_b=H_b^a, \quad (3)$$

$g(\cdot)$ is the activation function, it is usually selected of an exponential form:

$$g(x) = 1/(1+e^{-\tau x})$$

or

$$g(x) = (1-e^{-\tau x})/(1+e^{-\tau x}) \qquad (4).$$

Thus, a problem, to be solved by MNN is to transform a space of the input vector $X$ into output $O$, i.e. realization of a functional mapping for some nonlinear vector function $O = f(X)$. This property is applied for identification of processes and dynamical systems (Narendra and Parthasaty, 1990).

It is possible to consider the learning process of MNN as a difficult optimization problem, in which an optimum solution is the set of connected weights, minimizing a MNN performance index. In a basis of learning algorithms methods of gradient decent lay. The greatest distribution has so called the back propagation method. The idea of this method consists of that the error in the output layer is rolling back for adjusting the weights connecting the previous layer and minimizes an error of the output layer. In this case the performance index is defined as a difference between desirable and real output of net.

The neural network training algorithm by the back propagation method can be represented in general case as follows (Behera and Gopal, 1994; Narendra and Parthasaty, 1990):

the value of $W^a_{b,c}$ at learning is determined according with:

$$W^a_{b,c}(t+1) = W^a_{b,c}(t) + \eta \delta^a_b V^{a-1}_c \qquad (5),$$

the value of $s^a_b$ at learning is determined as:

$$s^a_b(t+1) = s^a_b(t) + \eta \delta^a_b \qquad (6),$$

where $\eta$ the rate of learning process, (in simulation recommended $\eta$=0.001 (Behera and Gopal, 1994); $\delta^a_i$ the error of i-th element of a-th layer, is defined as:

for a output layer:

$$\delta^{out}_i = \Delta E_i \qquad (7),$$

($\Delta E_i$ the difference between desirable and current outputs of network);

for hidden layers

$$\delta^a_i = -(dg(H^a_i))/(dH^a_i)\sum_j \delta^{a+1}_j W^{a+1}_{ij} \qquad (8)$$

The algorithm of such kind should be discrete, since for recalculation of MNN's weights and process of the input/output transformation, time is required, depending on a configuration of network (numbers of layers and elements).

## 3. MNN THRUSTER DYNAMICS CORRECTION

Main purpose of UR thrusters is to generate forces and torque that are necessary to move UR along desired space trajectory. Many URs are equipped by thrusters based on direct current electric motors with propeller attached. Dynamics of UR thruster is rather complicated.

The mathematical model of UR thruster (Yorger, et. al., 1990; Yuh, 1990) can be represented as

$$\begin{cases} \dot{\Omega} = (\tau - K_c \Omega|\Omega|) J_s^{-1} \\ F = K_t \Omega|\Omega| \end{cases} \qquad (9),$$

where $\tau$ - input torque, $\Omega$ - rotation velocity of the propeller; $F$ - UR thrust; $J_s$ - the total moment of inertia (with added masses); $K_c$ - viscous friction coefficient; $K_t$ = const.

For our purpose the dynamics of corrected UR thruster should follow to process which is defined by reference model (RM). RM is proposed to choose as aperiodic block (Dyda, 1998; Dyda and Lebedev, 1996):

$$\dot{F}_d = (r(t) - F_d)/T \qquad (10),$$

where $r(t)$ - RM input signal; $F_d$ - output signal of RM – force (torque) of thruster; $T$ - time constant.

Consider the following UR thruster control law:

$$\tau = K_r r(t) + K_f F + K \qquad (11),$$

where $r(t)$ the control law parameters that are outputs of MNN.

For realization of the control, it is necessary to train MNN. To achieve this purpose, one shall use of widely known algorithm of speed gradient (Fradkov, 1990).

Determine of MNN output layer errors, from the equations (9) and (10) Taking performance index $Q = e^T pe$, where $e = F_d - F$ - error value; $p > 0$, and applying gradient method, one can derive the following MNN learning algorithm:

$$\begin{cases} \Delta K_r = -\gamma_r per(t) \\ \Delta K_f = -\gamma_f pe F \\ \Delta K = -\gamma pe \end{cases} \qquad (12),$$

where $\gamma_r, \gamma_f, \gamma$ - the coefficients in the adjusting law (selected experimentally during simulation), the symbol $\Delta$ means an increment of appropriate parameter.

Note that MNN can be learned:

1) Simultaneously with control process (on-line);
2) Before the control process (off-line).

Both kinds of learning procedures are considered in simulations.

## 4. SIMULATION RESULTS OF MNN THRUSTER CORRECTION

To check the effectiveness of MNN correction, computer simulations were carried. The parameters of UR thruster are following: $J_s = 0.024$, $K_c = 8.81 \cdot 10^{-4}$, $K_m = 0.25$, $K_w = 0.25$, $R = 0.8$, $K_t = 0.022$. Time constant of RM is equal to $T = 0.05$.

MNN was selected of the following configuration: the input layer has two inputs: $r(t)$, $F$; the hidden layer is represented by 10 neurons, the MNN has three outputs $K_r$, $K_f$, $K$. The weights are initialized by random values laying in the range (0..1).

The RM input signals were selected as following functions:

$$r(t) = \begin{cases} F_0 \cdot t, & t \le 1\,\text{sec} \\ F_0, & t > 1\,\text{sec} \end{cases} \qquad (13),$$

$$r(t) = F_0 \cdot \sin\left(\frac{\pi t}{4}\right) \qquad (14),$$

($F_0$=const)

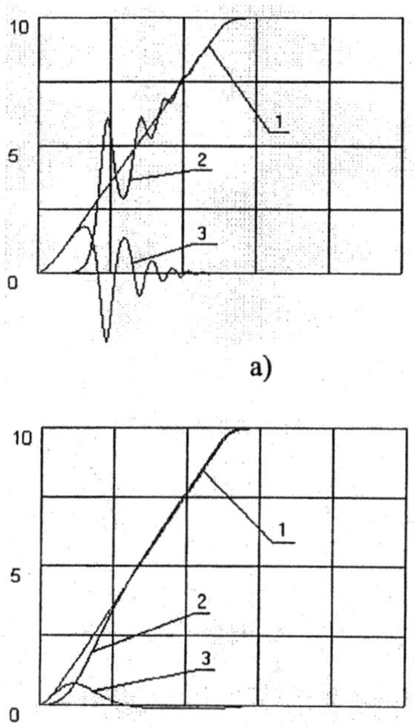


Fig.1. Correction results for (13) trajectory: a) on-line learning and control processes, b) control process after MNN has been learnt. 1 – desired thrust, 2- real thrust, 3- error.

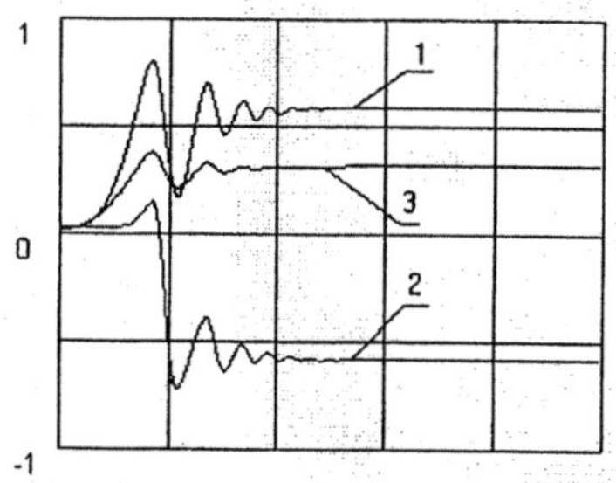


Fig.2. On-line learning process of MNN outputs with trajectory (13): 1 - $K_r$, 2 - $K_f$, 3 - $K$.

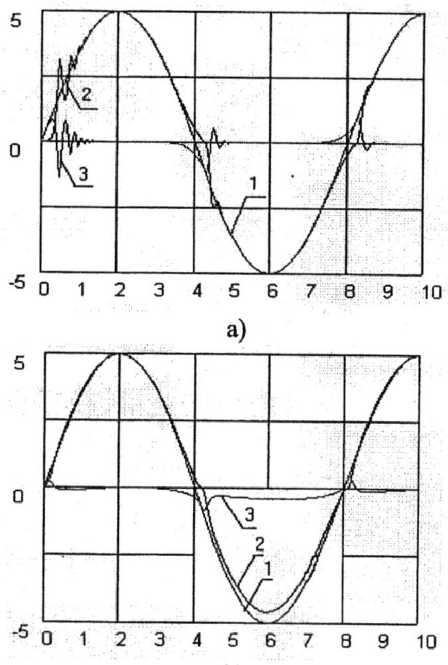


Fig.3. Simulation results with trajectory (14): a) on-line learning and control processes, b) control process after MNN has learnt. 1 – desired thrust, 2- real thrust, 3- error.

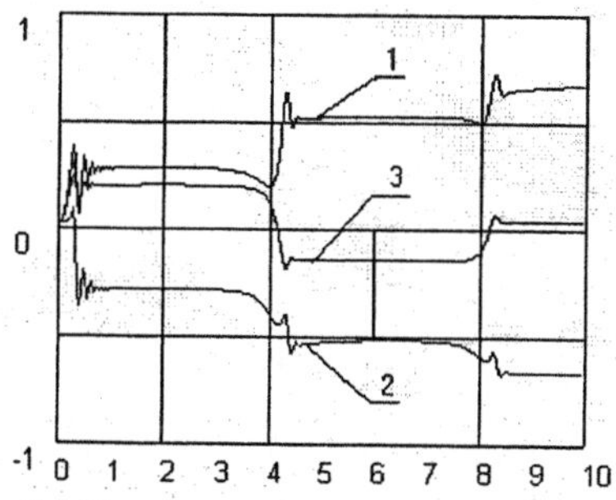


Fig.4. On-line learning process of MNN outputs with (4.2) trajectory: 1 - $K_r$, 2 - $K_f$, 3 - $K$.

On fig.1 – fig.4 the results of MNN learning for thrust correction are shown . As seen, the processes have desirable character that was determined by chosen RM. MNN corrected thruster dynamics is very close to the ideal behavior of RM

## 5. CONTROL SYSTEM SYNTHESIS FOR UNDERWATER ROBOT

Our final objective is to provide UR movement along a prescribed space trajectory. MNN correction of the UR thruster dynamics has reduced the complexity of control system synthesis on the whole. Now MNN approach will be applied to the control of UR properly. As basic mathematical UR model, the following equation of movement in horizontal plane is taken (Dyda, 1998):

$$\begin{cases} \frac{dV_X}{dt} = V_Z \omega_y r_1 - p_1 k_1 V_X |V| J_3 + \\ \qquad + p_1 J_3 F_X \\ \frac{dV_Z}{dt} = V_X \omega_y r_2 - p_2 J_1 (k_2 V_z |V| + \\ \qquad + k_3 \omega_y V_X) + F_Z p_2 J_1 \\ \frac{d\omega_y}{dt} = V_X V_Z r_3 - p_4 (k_4 V_Z V_X + \\ \qquad + k_5 \omega_y |V| + k_6 \omega_y |\omega_y|) + p_4 M_y \end{cases} \quad (15),$$

where $V_X, V_Z, \omega_y$ the linear and angular velocities, respectively; $F_x$, $F_z$, $M_y$ - control forces and torque; $M$ the UR mass; $\lambda_{11}, \lambda_{22}$ the added UR masses; $M_1 = M + \lambda_{11}$, $M_2 = M + \lambda_{22}$; $J_1, J_2, J_3$ the inertial moments of the vehicle; $YC$ the metacentric height of UR;

$k_1 = C_x \rho V_a^{2/3}/2$, $k_2 = C_z \rho V_a^{2/3}/2$,

$k_3 = C_z^{\omega y} \rho V_a^{2/3}$, $k_4 = m_y \rho V_a^{4/3}$,

$k_5 = m_y^{\omega y} \rho V_a^{4/3}$,

$k_6 = C_{my} \omega_y |\omega_y| \rho V_a^{5/3}/2$,

$p_1 = \left(J_3 M_1 - M^2 YC^2\right)^{-1}$, $p_3 = M^2 YC^2$,

$p_2 = \left(J_1 M_2 - M^2 YC^2\right)^{-1}$, $p_4 = J_2^{-1}$;

$r_1 = p_1 \left(-J_3 M_3 - p_3\right)$, $r_2 = -p_2 p_3$,

$r_3 = -p_4 \left(\lambda_{11} - \lambda_{33}\right)$, $|V| = \sqrt{V_X^2 + V_Y^2}$,

$C_x, C_z, m_y, C_{my}$ - UR parameters.

Select the following RM to control UR:

$$\dot{x}_d = \hat{A} x_d + \hat{B} r \quad (16),$$

where $x_d \in R^{n \times 1}$ - RM state coordinates vector, $x_d = (V_{xd}, V_{zd}, \omega_{yd})^T$; $r \in R^{n \times 1}$ - input trajectory; $\hat{A}, \hat{B} \in R^{n \times n}$ - diagonal matrices of coefficients in RM.

Control algorithm is chosen in the form:

$$u = k_r r + k_x x \quad (17)$$

where $u = (F_x, F_z, M_y)^T$ – UR control vector, $k_r, k_x \in R^{n \times n}$ - diagonal matrices of parameters. While neural network is used, the control law is kept without modifications, but now the outputs of MNN that must be trained are values of matrices factors $k_r, k_x$. Algorithm to learn MNN weight factors is defined as:

$$\begin{cases} \Delta k_r = -\gamma_r \, Per^T \\ \Delta k_x = -\gamma_x \, Pex^T \end{cases} \quad (18),$$

where $e = x_d - x$ the error, $\gamma_r, \gamma_x \in R^{n \times n}$ the diagonal matrices; $P = P^T > 0$ the diagonal matrices for performance index $Q = \frac{1}{2} e^T P e$.

## 6. SIMULATION RESULTS OF MNN CONTROL FOR UR

Modeling parameters of RM in (16) are accepted as follow: $\hat{A} = diag(-1/T, -1/T, -1/T)$, $\hat{B} = diag(k/T, k/T, k/T)$, where $T = 0.5\,sec$, $k = 0.5\,m/sec$.

The input trajectory is selected as

$$r_x = \begin{cases} k \cdot t, & t \le 1c \\ k, & t > 1c \end{cases} \quad (19)$$

$$r_y = r_z = 0$$

UR parameters are following: $M = 35\,kg$, $J_1 = 9.46\,kg \cdot m^2$, $J_2 = 21.12\,kg \cdot m^2$, $J_3 = 13.5\,kg \cdot m^2$, $\lambda_{11} = 8.5\,kg$, $\lambda_{22} = 12\,kg$, $YC = 0.1\,m$, $V = 0.12\,m^3$, $\rho = 1000\,kg/m^3$, $C_x = C_z = 0.12$, $m_y = 0.6$, $C_{my} = 0.15$.

Fig. 5 -10 show the simulations results, where 1 - error, 2 – real trajectory $V_x$, 3 – desired trajectory $V_{xd}$ of UR, 4 – RM input signal $r_x$.

Fig. 5 - 6 demonstrate the UR control and learning processes when thruster dynamics is ideal (it is accepted as reference model for UR thruster (10)).

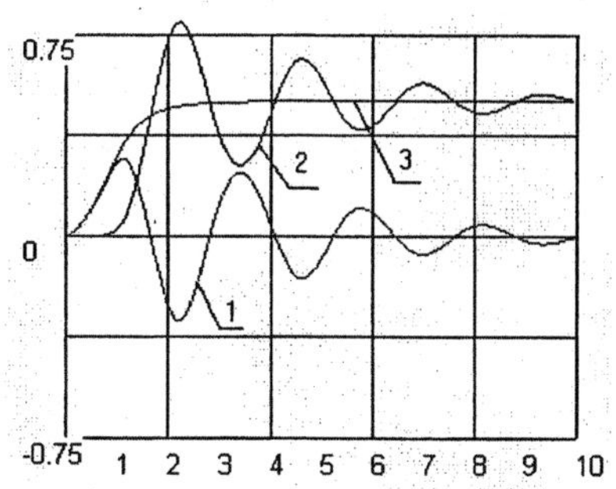

Fig.5. On-line MNN learning and control processes for UR with ideal thrusters.

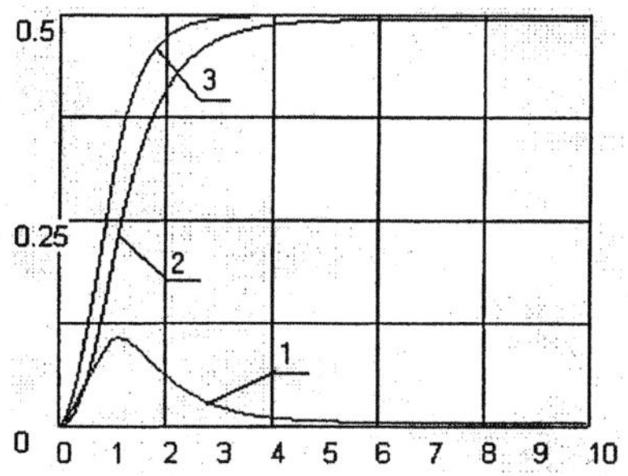

Fig.6. Control for UR with trained MNN and ideal thrusters.

Fig. 7-8 demonstrate the UR control and learning processes when thruster dynamics is not corrected.

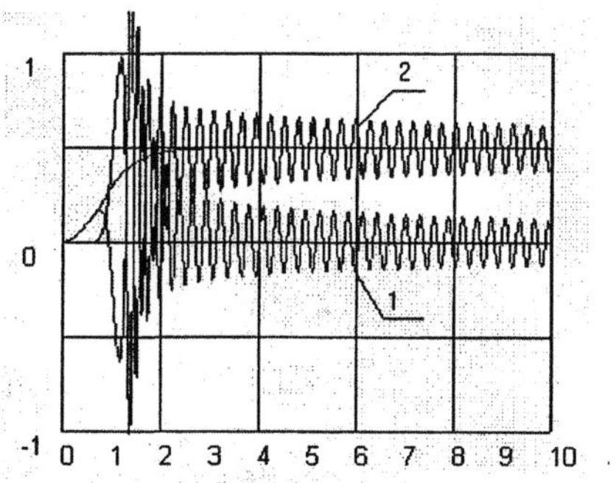

Fig.7. On-line MNN learning and control processes for UR with uncorrected thruster dynamics.

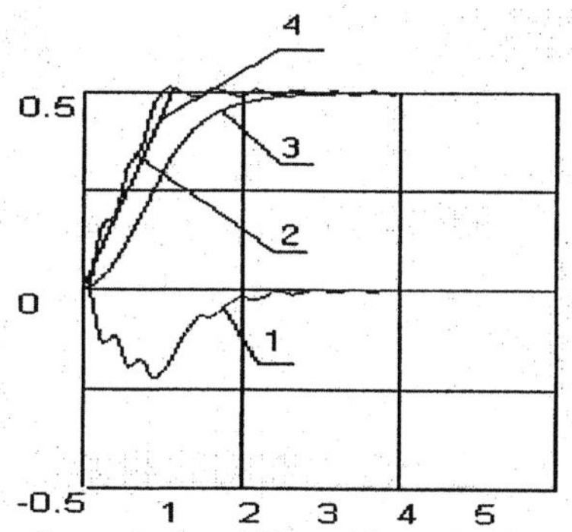

Fig.8. Control for UR with trained MNN and preliminary uncorrected thruster dynamics

Fig. 9-10 demonstrate the UR control and learning processes for MNN control system for UR, when thruster dynamics is preliminary MNN corrected before UR control process (in accordance with (11), (12)).

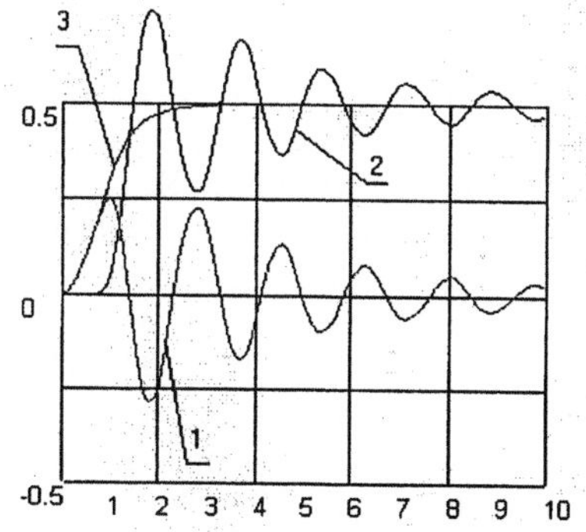

Fig.9. On-line MNN learning and control processes for UR with NN-corrected thruster dynamics.

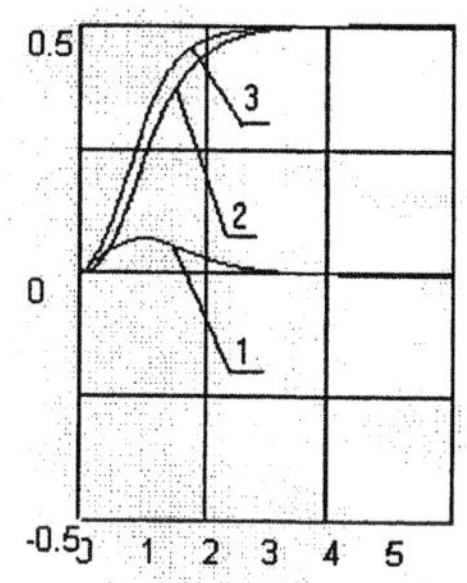

Fig.10. Control for UR with trained MNN and preliminary NN-corrected thruster dynamics

MNNs for thruster correction and UR control can be trained simultaneously or separately. Experiments had confirmed that two-stage approach is preferable when the UR control MNN learning followed this procedure for thrusters correction.

## 7. CONCLUSION

The numerical simulations have shown that quality processes can be achieved on the basis of MNN control both for UR on the whole and its subsystems (such as the thrusters). MNN control system design procedure can be more effective if it is organized in a few stages. MNNs have good perspectives in robotic applications, particularly, for underwater robot control. Neural network based approaches promise to overcome some problems of control system synthesis in conditions of high structural and parametric uncertainties. Further research is planned to carry in the direction of comparative study MNNs with other control system synthesis methods.

## REFERENCES

Behera L., Gopal M. (1994) *Adaptive manipulator trajectory control using neural networks* // Int. J. Systems SCI., Vol. 25, №. 8, pp.1249-1265.

Dyda A.A. (1998) *Synthesis of adaptive and robust control for underwater robots actuators* //

Doctoral Dissertation. Vladivostok (In Russian).
Dyda A.A., Lebedev A.V. (1996) *The nonlinear adaptive correction for the underwater robot thruster.* // Electromechanics, № 1-2, pp. 83-87 (In Russian).
Fradkov A.L. (1990) *Adaptive control in large-scale systems.*- M.: Nauka., (in Russian).
Narendra K.S., Parthasaraty K. (1990) *Identification and control of dynamical systems using neural networks* // IEEE Identification and Control of Dynamical System, Vol.1. № 1.
Yorger D.R., Slotine J.J.E. (1985) *Robust trajectory control of underwater vehicles.* IEEE J. of Oceanic Eng., vol. 10. pp.462-470.
Yorger D.R., Cook J.G., Slotine J.J.E. (1990) *The influence of thruster dynamics on underwater vehicle behavior and their incorporation into control system design*. IEEE J. of Oceanic Eng. vol. 15. pp. 167-177.
Yuh Y. (1990) *Modelling and control of underwater vehicles* IEEE J. of Trans. Syst., Man, Cybern., vol. 20, pp. 1475-1483.

ELSEVIER
IFAC
PUBLICATIONS
www.elsevier.com/locate/ifac

# MODEL PREDICTIVE CONTROL OF AN AUTONOMOUS UNDERWATER VEHICLE WITH A FUZZY OBJECTIVE FUNCTION OPTIMIZED USING A GA

**W. Naeem, R. Sutton and J. Chudley**
{wnaeem, rsutton, jchudley}@plymouth.ac.uk

*Marine and Industrial Dynamic Analysis Research Group*
*Reynolds Building, School of Engineering,*
*University of Plymouth, Drake Circus,*
*Plymouth PL4 8AA, UK*

Abstract: Recently, unmanned underwater vehicles (UUVs) have emerged as a viable tool for ocean exploration and for military purposes. This is due to the inability of human divers to reach deep sea and the hostile nature of underwater environment. UUVs are of two types, namely, remotely operated vehicles (ROVs) and autonomous underwater vehicles (AUVs). This paper is concerned with the control of an AUV. A model predictive controller is developed herein where the traditional cost function has been replaced by a fuzzy performance index which represent the goals and constraints of the problem. Since fuzzy logic is basically derived from knowledge of human expertise, it is therefore more intuitive than a conventional cost function. Moreover, the choice of aggregation operator can lead to significant reduction in tuning time which is essential for a quadratic objective function. A genetic algorithm (GA) is used as an optimization tool to evaluate the control inputs by minimization of the performance index represented by fuzzy membership values. The resulting controller is applied to an AUV simulation model obtained from system identification techniques on test trials data. Simulation results are presented that demonstrate the efficacy of the approach. *Copyright ©2004 IFAC*

Keywords: Model predictive control, genetic algorithm, fuzzy objective function, optimization and underwater vehicles

## 1. INTRODUCTION

Designing underwater robots present tremendous challenges to engineers. This is mainly due to the hostile underwater environment and the degrees of freedom of the vehicle movement. The last decade has seen a boost in underwater vehicle development for exploring the rich underwater world containing a huge number of natural resources. As the oceanographer, James Gardner says as quoted by M. Barber, (Barber, 2001)

*"We know what the surface of the moon is better than we know what the surface of the sea floor is."*

clearly giving a hint that there is still a considerable lack of research work to explore deep oceans. The main hurdle in deep sea exploration is the inability of human divers to reach these places. Underwater vehicles are thought to be a true replacement of deep sea divers for ocean surveying. In addition, they are repeatedly been used in covert missions and for mines clearing operations as the world has recently witnessed the use of the

*REMUS* underwater vehicle in the Iraqi conflict (Jordan, 2003).

AUVs are self contained craft that have onboard navigation, guidance and control systems. Thus the range of missions is only limited by the onboard power supply. The navigation system provides information related to the target and vehicle itself, using onboard sensors such as inertial navigation system (INS), compass, pressure transducer etc. This information is fed to the guidance system which by utilising some guidance law generate reference trajectories. The control system is then responsible to keep the vehicle on course as specified by the guidance system. A simple block diagram of the navigation, guidance and control system is depicted in Figure 1.

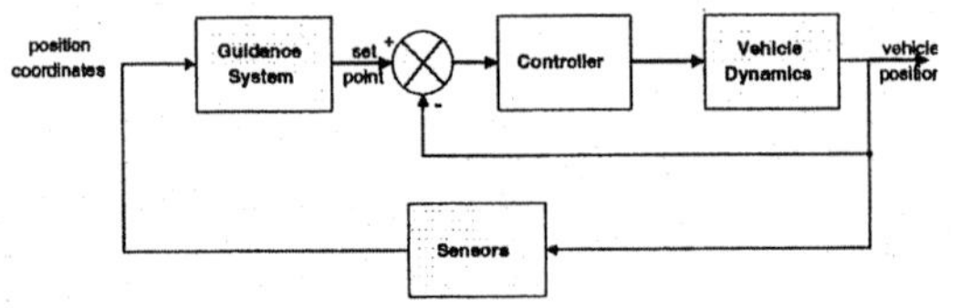


Fig. 1. Navigation, guidance and control of a vehicle

This paper is mainly concerned with the control of an AUV. The control system developed is a genetic algorithm (GA) based model predictive controller (MPC) using fuzzy decision functions. MPC was originated and has long been used in the process industry because of its strong robustness and constraint handling characteristics. As far as the authors are concerned, MPC on AUVs was first simulated by Kwiesielewicz *et al.* (2001) and was compared with a proportional derivative (PD) and an adaptive neuro-fuzzy inference system (ANFIS) tuned autopilot. The results were found to be quite promising. Then Naeem and others simulated a GA tuned MPC on an AUV for heading control (Naeem, 2002) and for subsea cable/pipeline tracking (Naeem *et al.*, 2004). The results demonstrated the robustness of the GA-MPC in the presence of sea currents. The objective function used to evaluate control actions in both cases was a simple quadratic cost function involving the output error, input and the change in input and can be seen in Equation 1.

$$
\begin{aligned}
J = &\sum_{i=1}^{H_p} e(k+i)^T Q e(k+i) \\
&+ \sum_{i=1}^{H_c} \Delta u(k+i)^T R \Delta u(k+i) \\
&+ \sum_{i=1}^{H_p} u(k+i)^T S u(k+i)
\end{aligned} \tag{1}
$$

subject to

$$
\begin{aligned}
u^l \le u(k+i) \le u^u \\
\Delta u^l \le \Delta u(k+i) \le \Delta u^u
\end{aligned}
$$

where the superscripts $l$ and $u$ represents the lower and upper bounds respectively. $Q$ is the weight on the prediction error

$$e(k) = \hat{y}(k) - w(k) \tag{2}$$

where $w(k)$ is the reference or the desired setpoint. $R$ and $S$ are weights on the change in the input $\Delta u$ and magnitude of the input $u$ respectively.

Herein, the control actions are evaluated by using fuzzy membership functions that represent the goals and constraints of the problem similar to a conventional cost function in Equation 1. Fuzzy objective functions in predictive control have been investigated (Sousa and Kaymak, 2001) and the resulting non convex optimization problem was solved using a Branch and Bound (B&B) algorithm. The work presented here is an extension of the previous study.

## 2. MODEL PREDICTIVE CONTROL

MPC refers to a class of algorithms that compute a sequence of manipulated variable adjustments in order to optimize the future behaviour of a plant. Originally developed to meet the specialised control needs of power plants and petroleum refineries, MPC technology can now be found in a wide variety of application areas including chemicals, food processing, automotive, aerospace and metallurgy (Qin and Badgewell, 2000), to name but a few.

The development of MPC can be traced back to 1978 after the publication of the paper by Richalet *et al.* (1978). They named their algorithm model predictive heuristic control (MPHC) and it was successfully applied to a fluid catalytic cracking unit main fractionator column, a power plant steam generator and a poly-vinyl chloride plant. Then Cutler and Ramaker from the Shell Oil Company in 1979 and 1980 developed their own independent MPC technology referred to as dynamic matrix control (DMC) (Cutler and Ramaker, 1980), and they showed results from a furnace temperature control application to demonstrate improved control quality. However, another form of MPC called generalised predictive control (GPC) (Clarke *et al.*, 1987*a*; Clarke *et al.*, 1987*b*) is employed in this paper. The fundamental difference between all these techniques is the type of model used and the cost function being optimized.

The process output is predicted by using a model of the process to be controlled. Any model that describes the relationship between the input and the output of the process can be used. Further if the process is subject to disturbances, a disturbance or noise model can be added to the process model. In order to define how well the predicted process output tracks the reference trajectory, a criterion function is used as defined in Equation 1. The optimal controller output sequence $u_{opt}$ over the prediction horizon is obtained by minimisation of $J$ with respect to $u$. As a result the future tracking error is minimised. If there is no model mismatch i.e. the model is identical to the process and there are no disturbances and constraints, the process will track the reference trajectory exactly on the sampling instants. The structure of an MPC is shown in Figure 2 and it consists of the following three steps.

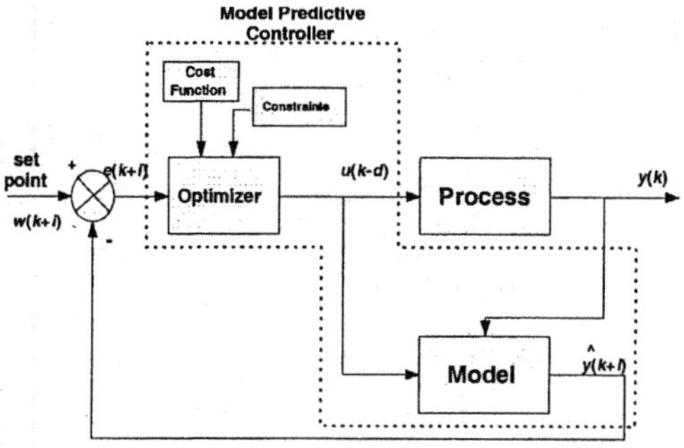


Fig. 2. Structure of a model predictive controller

(1) Explicit use of a model to predict the process output along a future time horizon (Prediction Horizon).
(2) Calculation of a control sequence along a future time horizon (Control Horizon, $H_c$), to optimize a performance index.
(3) A receding horizon strategy so that at each instant the horizon is moved towards the future, which involves the application of the first control signal of the sequence calculated at each step. The strategy is illustrated as shown in Figure 3.

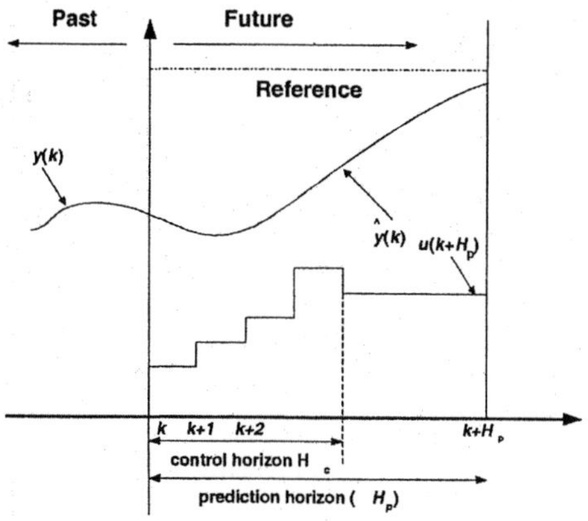


Fig. 3. Predicted output and the corresponding optimum input over a horizon $H_p$, where $u(k)$ is the optimum input, $\hat{y}(k)$ is the predicted output, and $y(k)$, process output

The selection of MPC to control an AUV is attributed to several factors. Some of them are listed below.

- The concept is equally applicable to single-input, single-output (SISO) as well as multi-input, multi-output systems (MIMO).
- MPC can be applied to linear and nonlinear systems.
- It can handle constraints in a systematic way during the controller design.
- The controller is designed at every sampling instant so disturbances can easily be dealt with.

The performance index is optimized using a GA which is described next.

### *2.1 Genetic Algorithms*

GAs inspired by Darwinian theory, are powerful non-deterministic iterative search heuristics. GAs operate on a population consisting of encoded strings, each string represents a solution. The crossover operator is used on these strings to obtain new solutions, which inherits the good and bad properties of their parent solutions. Each solution has a fitness value, solutions having higher fitness values are most likely to survive for the next generation. The mutation operator is applied to produce new characteristics, which are not present in the parent solutions. The whole procedure is repeated until no further improvement is observed or run time exceeds to some threshold, (Sait and Youssef, 1999). The flowchart of a simple GA is presented in Figure 4.

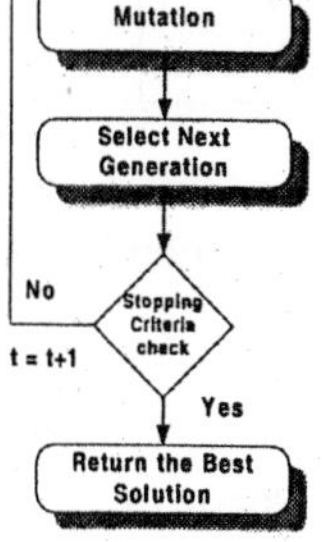

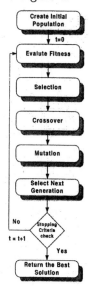

Fig. 4. Flow chart of a simple genetic algorithm

## 3. FUZZY OBJECTIVE FUNCTION

The fuzzy logic traditionally used as *if-then* rules can be translated to some design specifications using human expertise. These design specifications are represented in terms of an objective function which is more intuitive than conventional cost function. There are several other reasons to choose fuzzy membership values as an objective function to be optimized in a predictive control problem. Some of them are listed below

- Fuzzy objective functions are easy to understand
- Soft and hard constraints can be implemented using the same membership function
- No normalisation is required for the terms in the objective function as the membership function automatically maps the input space to a [0–1] interval
- Easy to tune as the weighting matrices for individual terms are not needed
- The aggregation operator normally requires only a single tuning parameter for all the terms

In this work, membership functions for the output error and input variables are considered. An exponential membership function has been elected for the output error while the input is represented by a trapezoidal membership function as shown in Figures 5 and 6 respectively. The steepness of the exponential plot can be adjusted using the $S_e$ variable and thus is an important tuning parameter. The trapezoidal membership function in Figure 6 automatically implements the soft and hard constraints, where $u_{max}$ represents the maximum allowable input and $(u_{max} - u_{constraint})$ is the input bound which is allowed but not desired.

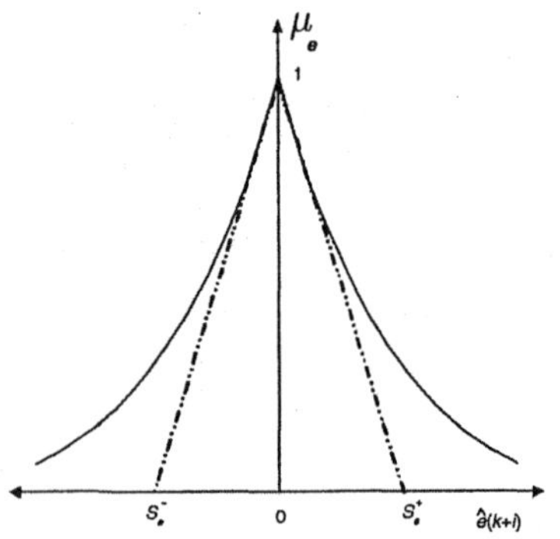


Fig. 5. Output error membership function

### 3.1 Aggregation Operator

Finally, a decision function is required which allows for interaction amongst different criteria in the objective function. A variety of aggregation operators can be chosen to be used in fuzzy predictive control such as *min* and *product t-norm*. A good account of various aggregation operators and the advantages and disadvantages of their use in

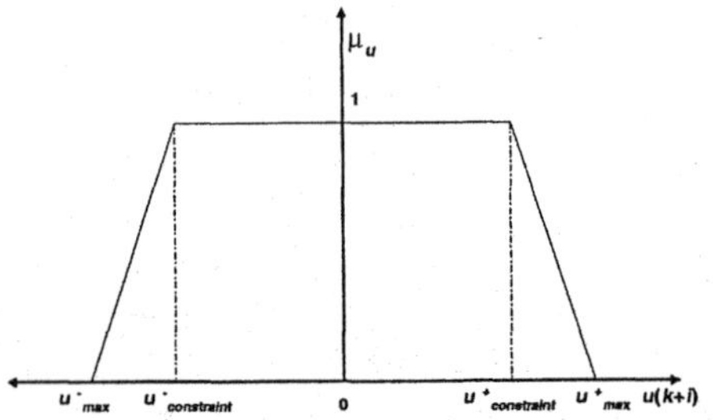


Fig. 6. Trapezoidal membership function for input variable

predictive control have been documented (Sousa and Kaymak, 2001).

Herein, the Yager *t*-norm has been chosen as the decision function since it uses only one parameter to tune the objective function and hence interact amongst different criteria. Moreover, this operator covers the entire range of *t*-norms, i.e., it goes from the drastic intersection to the minimum operator (Sousa and Kaymak, 2001). The fuzzy cost function and Yager *t*-norm are given by Equations 3 and 4 respectively.

$$\bar{\mu}_c = \sum_{i=1}^{H_p} (\bar{\mu}_e(\hat{e}(k+i)))^{w_Y} + \sum_{j=1}^{H_c} (\bar{\mu}_u(u(k+i)))^{w_Y} \tag{3}$$

$$\mu = \max(0, 1 - \bar{\mu}_c^{1/w_Y}) \tag{4}$$

where $\bar{\mu} = 1 - \mu$ and $w_Y > 0$ is the tuning parameter

## 4. SIMULATION RESULTS

The proposed control algorithm is applied to an AUV simulation model of the *Hammerhead* vehicle being developed as a combined project at the Universities of Plymouth and Cranfield, UK. The vehicle has a torpedo shaped hull approximately 3 metres long and one-third of a metre in diameter. The yaw dynamics of the vehicle has been obtained from test trials at Willen Lake, Milton Keynes, UK and system identification techniques were applied to extract a yaw-rudder channel model given by Equation 5

$$G(q) = \frac{-0.04226q^{-1} + 0.003435q^{-2}}{1 - 1.765q^{-1} + 0.7652q^{-2}} \tag{5}$$

where the data was sampled at a rate of $1Hz$ with the vehicle manoeuvring on the surface at a fixed speed of approximately 2 knots. The following steps describe the operation of a GA as an optimization tool in MPC followed by simulation results for various settings.

(1) Evaluate process outputs using the process model.

(2) Use a GA search to find the optimal control moves which optimize the cost function and satisfy process constraints. This can be accomplished as follows.
   (a) generate a set of random possible control moves.
   (b) find the corresponding process outputs for all possible control moves using the process model.
   (c) evaluate the fitness of each solution using the fuzzy cost function and the process constraints.
   (d) apply the genetic operators (selection, crossover and mutation) to produce new generation of possible solutions. Stochastic universal sampling and single point crossover is used for parents selection and mating respectively.
   (e) repeat until predefined number of generations is reached and thus the optimal control moves are determined.
(3) Apply the optimal control moves generated in step 2 to the process.
(4) Repeat steps 1 to 3 for time step $k+1$.

The hard constraints on the rudder are $\pm 22°$ therefore $u_{max}^{-}$ and $u_{max}^{+}$ in Figure 6 are taken as $-22°$ and $+22°$ respectively. However, the control is only allowed to move freely within the range $\pm 20^0$ to avoid saturation and hence any nonlinear behaviour. Therefore, $u_{constraint}^{\pm}$ in Figure 6 is set equal to $\pm 20^0$. Simulations are carried out first for a step change in heading. The vehicle is launched with an arbitrary orientation and is required to follow a specified heading. The parameters $S_e$ and $w_Y$ are chosen as 0.5 and 2 respectively whereas the GA parameters are provided in Table 1 and are selected to minimise the control effort and increase the speed of response. The step response of the closed loop system is depicted in Figure 7 showing that the vehicle is closely following the set point with little overshoot and zero steady state error. The canard demand is also shown in Figure 8 and is within the specified constraints.

Table 1. GA-MPC tuning paramters

| Parameters | Step response | Way point following |
|---|---|---|
| $H_p$ | 10 | 20 |
| $H_c$ | 1 | 1 |
| Mutation prob. | 0.008 | 0.008 |
| Crossover prob. | 0.1 | 0.1 |
| No. of generations | 1 | 3 |
| Population size | 100 | 250 |
| Insertion rate | 0.5 | 0.1 |

Next, the control law is simulated for way point following where the intent is to track all the specified way points despite the presence of disturbances. A sea current disturbance is assumed to be acting on the vehicle in the positive y-direction

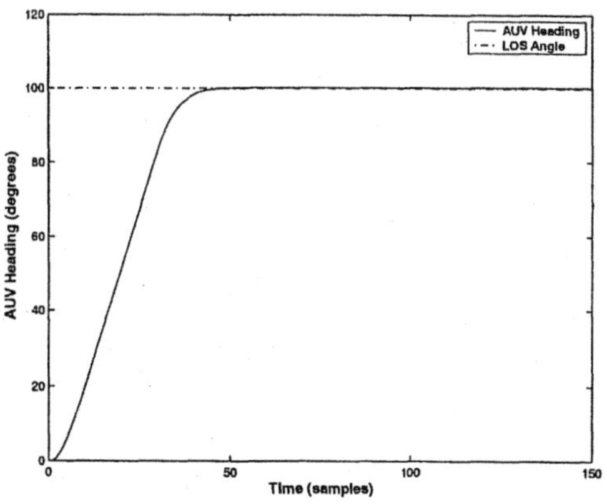


Fig. 7. Step change in heading response of the proposed controller.

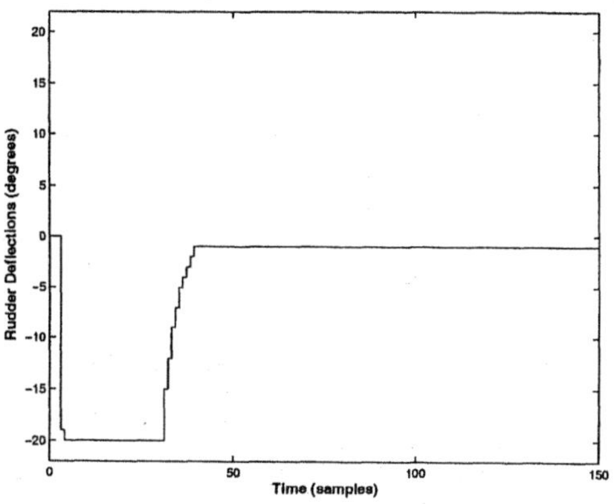


Fig. 8. Optimal rudder deflections generated by the proposed controller.

and has a magnitude of $0.5m/s$. The initial AUV coordinates in the two-dimensional frame of reference is $(0, 10)$ while the way points are chosen to be at $(100, 50)$, $(300, 50)$, $(500, 150)$ and $(500, 300)$. The next way point is selected when the vehicle enters a circle of acceptance around the way point of radius $10m$. As will be shown, the way point following is equivalent to tracking line of sight (LOS) angle between any two given way points. The GA parameters for this case are also given in Table 1 while $S_e$ and $w_Y$ are selected as 1.5 and 1.8 respectively. The resulting closed loop performance is illustrated in Figure 9 showing the affects of sea currents on vehicle's trajectory. The disturbance is striving to knock the AUV off the track, however, the controller is quite robust to follow all the way points. The AUV trajectory through the way points without any sea currents is also illustrated in Figure 9 which is closely following the ideal path. The control effort depicted in Figure 10 shows vigorous rudder movements in response to the change in vehicle's heading due to sea currents, however, it never violates the imposed constraints. The heading angle or LOS angle between the actual AUV position and way points shown in Figure 11 is varying continuously because of the addition of disturbances.

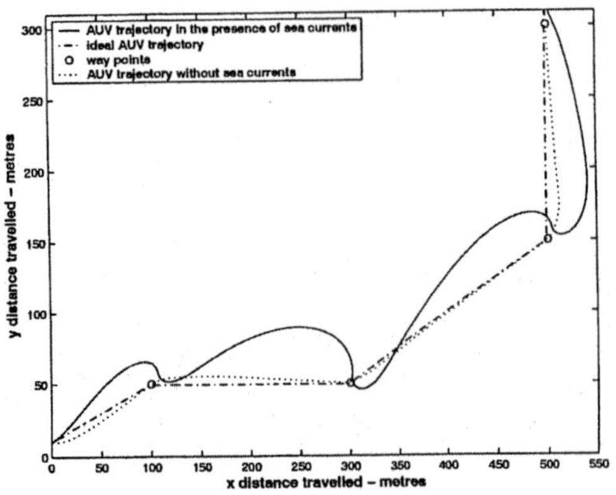


Fig. 9. AUV and target position co-ordinates with sea current disturbance in the positive y-direction.

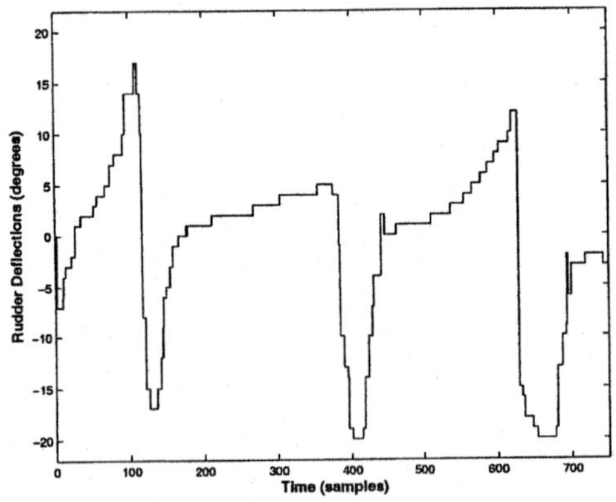


Fig. 10. Rudder deflections generated by the controller needed to track the way points with sea current disturbance in the positive y-direction.

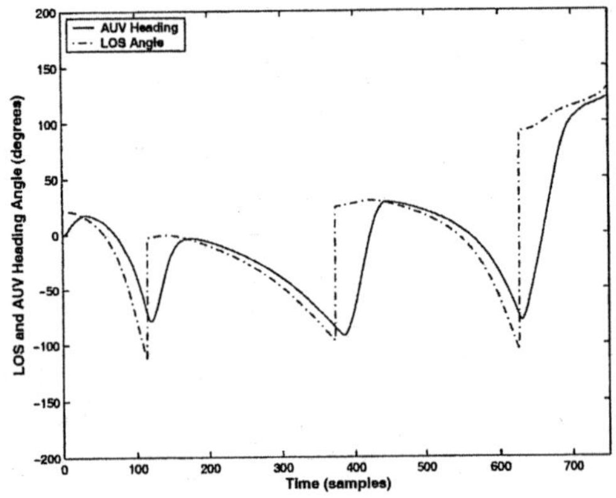


Fig. 11. Desired heading angles to the way points showing the affects of sea current disturbance

## CONCLUDING REMARKS

The traditional fuzzy logic based on if-then rules has been used for decision making in an MPC framework to control an AUV. A GA is used to optimize the resulting fuzzy objective function. The advantages of using the proposed scheme are presented. The GA optimized fuzzy MPC have been applied to an AUV simulation dynamic model based on the *Hammerhead* vehicle being developed jointly at the Universities of Plymouth and Cranfield, UK and will eventually be tested on the real vehicle.

## REFERENCES


Barber, M. (2001). *Hydrographic crew surveys underwater quake damage.* World Wide Web. `http://seattlepi.nwsource.com`.

Clarke, D. W., C. Mohtadi and P. S. Tuff (1987*a*). Generalised predictive control. Part 1: The basic algorithm. *Automatica* **23**(2), 137–148.

Clarke, D. W., C. Mohtadi and P. S. Tuff (1987*b*). Generalised predictive control. Part 2: Extensions and interpretations. *Automatica* **23**(2), 149–160.

Cutler, C. R. and B. L. Ramaker (1980). Dynamic matrix control - a computer control algorithm. In: *Proceedings of the Joint Automatic Control Conference, Paper WP5-B.* San Fransisco, CA, USA.

Jordan, K. (2003). *Remus AUV Plays Key Role in Iraq War.* World Wide Web. `http://www.diveweb.com`.

Kwiesielewicz, M., W. Piotrowski and R. Sutton (2001). Predictive versus fuzzy control of autonomous underwater vehicle. In: *Proceedings IEEE International Conference on Methods and Models in Automation and Robotics.* IEEE. Miedzyzdroje, Poland. pp. 609–612.

Naeem, W. (2002). Model predictive control of an autonomous underwater vehicle. In: *Proceedings UKACC Control 2002.* UKACC. pp. 19–23.

Naeem, W., R. Sutton and S. M. Ahmad (2004). Pure pursuit guidance and model predictive control of an autonomous underwater vehicle for cable/pipeline tracking. *IMarEST Journal of Marine Science and Environment* **C**(1), 25–35.

Qin, S. J. and T. A. Badgewell (2000). An overview of nonlinear model predictive control applications. In: *Nonlinear Model Predictive Control.* Birkhäuser. Switzerland. pp. 369–392.

Richalet, J., A. Rault, J. L. Testud and J. Papon (1978). Model predictive heuristic control: Applications to industrial processes. *Automatica* **14**, 413–428.

Sait, S. M. and H. Youssef (1999). *Iterative Computer Algorithms with Applications in Engineering, Solving Combinatorial Optimization Problems.* IEEE Computer Society.

Sousa, J. M. and U. Kaymak (2001). Model predictive control using fuzzy decision functions. *IEEE Transactions on Systems, Man, and Cybernetics-Part B: Cybernetics* **31**(1), 54–65.

ELSEVIER
IFAC
PUBLICATIONS
www.elsevier.com/locate/ifac

# HIERARCHICAL SWITCHING SCHEME FOR PID CONTROL OF UNDERWATER VEHICLES

**G. Ippoliti L. Jetto S. Longhi**

*Dipartimento di Ingegneria Informatica,
Gestionale e dell'Automazione,
Università Politecnica delle Marche,
Via Brecce Bianche, 60131 Ancona, Italy.
E-mail: g.ippoliti@diiga.univpm.it
{l.jetto, sauro.longhi}@univpm.it*

Abstract: This paper considers the position control problem of an underwater vehicle used in the exploitation of combustible gas deposits at great water depths. The vehicle is subjected to different load configurations, which introduce considerable variations of its mass and inertial parameters. The ensemble of possible vehicle configurations are known, but, it is here assumed that neither the time instants when these changes occur nor the new vehicle configuration following the change are "a priori" known. To cope with this problem, a switching scheme governed by two Decision Makers (DM) operating at two different hierarchical levels is proposed. Multiple plant models are defined to describe the different possible operative conditions of the vehicle. The task of the hierarchically higher Decision Maker ($DM_1$) is to identify the model which refers to the current vehicle configuration and to drive the control law towards that corresponding to the actual operating condition. For each fixed model of the vehicle, a family of PID regulators is defined. Each PID is designed so as to obtain a specific desired feature of the step response and the hierarchically lower Decision Maker ($DM_2$) governs the switching among them to obtain an overall time-varying control law yielding a globally satisfactory output behaviour. The performance of the doubly switched controller so obtained is evaluated by numerical simulations.


Keywords: Supervisory control, Autonomous vehicles, PID control, Switching algorithms, Control system synthesis.

## 1. INTRODUCTION

Conventional PID controllers are widely being used for their simple structure and versatility. The effectiveness of their control action is well acknowledged both through theoretical research and long-term practice in process control (Grey, 1987).

However, in many practical situations there are still some problems concerning the tuning operations which have not yet found a really satisfactory solution. The tuning of PID controllers mainly depends on well-known empirical rules, such as Ziegler-Nichols formulas and similar (Astrom and Hagglund, 1995). These methods are usually based on trial and error procedures and are influenced by the experience and ability of the operator. There is also some difficulty of tuning PID due to time delay, possible nonlinearities and time-variations due to changes in the operating conditions. On the other hand, it is well known that a poor tuning may result in an unsatisfactory step response (excessive overshot, poorly damped oscillations, too long transient response duration). Therefore, PID controllers need be regularly retuned. As accurate tuning is a time-consuming operation and as even in a simple process plant several possible PID controllers can be found, the existence of effective automatic tuning methods are of great economical importance. For this reason, since the early work of Ziegler-Nichols, several methods have been proposed for PID autotuning (see e.g. (Astrom and Hagglund, 1995) and references therein). Recently, advanced PID control schemes for non linear systems have been proposed (Adams and Rattan, 2001; Cavalcanti, 1995; Efe *et al.*, 2001; Oki *et al.*, 1997; Ruano and Azevedo, 1999; Zhi and Jingling, 1997).

An important area of application where PID regulators have been successfully employed is the control of unmanned underwater vehicles (Fossen, 2002; Jalving and Storkersen, 1994; Miyamaoto *et al.*, 2001; Park *et al.*, 2000). Although telecontrolled vehicles are generally used for relatively simple tasks, great efforts are currently being devoted at the development of underwater robots with self-governing capabilities, able to solve complex tasks in different environments and load conditions. Therefore, it appears useful to develop advanced control techniques in order to supply the underwater vehicle with the necessary "intelligence" for achieving some degree of self-governing capability.

In this regard different control strategies have been developed and efficient implementations of such controllers on real environments have been proposed. Significant solutions based on adaptive control, robust control, variable structure control, Lyapunov based control and multiple models and supervisory control have been recently investigated (see e.g. (Aguiar and Pascoal, 2001; Baldini *et al.*, 1999; Conte and Serrani, 1994; Conter *et al.*, 1989; Corradini and Orlando, 1997; Cristi *et al.*, 1990; Fossen, 1994; Ippoliti and Longhi, 2001; Kaminer *et al.*, 1991; Longhi and Rossolini, 1989; Longhi *et al.*, 1994)).

The purpose of this paper is to investigate the possibility of improving the performance of PID controlled underwater vehicles through a multiple model logic-based switching control law. The main reason for using multiple models is to ensure the existence of at least one model sufficiently close to the unknown plant (Ippoliti and Longhi, 2001; Narendra and Xiang, 2000). In this case, a multiple model approach is required because the plant parameters are assumed to be affected by sudden changes according to the different operating conditions of the vehicle. For each fixed plant model a family of PID controllers is designed. This technique is motivated by the consideration that, in general, if a single time-invariant PID is used, a non fully satisfactory closed-loop performance can be obtained, even if the PID is properly chosen. For example, a regulator producing a step response with a short rise time is like to produce a large overshoot and/or poorly damped oscillations. On the other hand, a smooth behaviour of the step response is often coupled with too long rise and settling times. To reduce the aforementioned inconveniences, it appears quite natural to design a time-varying PID controller which is suitably modified, from time to time, according to the characteristics of the produced step response (Ippoliti *et al.*, 2003).

The hybrid control scheme consists of the hierarchic connection of two supervisors $DM_1$ and $DM_2$ with a set of suitably defined PID controllers. In brief, the logic of the overall control law can be explained as follows. A set of linearized models $M_i, i = 1, \cdots N$, of the plant is defined. Each model explains the plant dynamics corresponding to a particular operating condition. For each $M_i$, a proper set $C_i$ of differently tuned PID controllers $G_i^j, j = 1, \cdots L$, is designed. For each $M_i$, supervisor $DM_2$ selects the most appropriate $G_i^j$ among the elements of $C_i$. The selection is performed minimizing the "distance" between the predicted and the desired step response. The task of supervisor $DM_1$ is to govern the switching among the different families of PID controllers $C_i, i = 1, \cdots, N$, recognizing which is the linearized model $M_i$ corresponding to the current operating condition. This last operation is accomplished minimizing some positive definite functionals of suitably defined identification errors.

For linear plants, the stability conditions for the above doubly switched control scheme can be easily deduced from the results in (Corradini *et al.*, 2004) and (Ippoliti *et al.*, 2004). Application to the considered non linear problem is justified by the possibility of performing an accurate linearization of the ROV dynamics as analyzed in (Ippoliti *et al.*, 2002) and in (Baldini *et al.*, 1999; Ippoliti and Longhi, 2001), where different approaches to multiple models control has been considered.

The switching control approach is also used in different methods and implementations such as, for example, gain scheduling (Blanchini, 2000; Rugh and Shamma, 2000) and sliding modes control (Corradini and Orlando, 1998; Utkin, 1992). All these switching control schemes can be considered as concrete examples of hybrid dynamical systems (Antsaklis and Nerode, 1998; Morse *et al.*, 1999).

The paper is organized in the following way. In Section 2, some details on the vehicle dynamics are recalled, the linearized model and its experimental identification are briefly discussed in Section 3, the proposed hierarchical switching scheme is described in Section 4. The results of the numerical simulation are reported in Section 5.

## 2. THE ROV MODEL

Reference is here made to the underwater vehicle developed by Snamprogetti (Italy) for the exploitation of combustible gas deposits at great sea depths. The Remotely Operated Vehicle (ROV) is equipped with four thrusters and connected with the surface vessel by a supporting cable. The vehicle depth is controlled by the supporting cable, while the vehicle position and orientation over planes parallel to the surface are controlled by the vehicle thrusters. For these reasons, the control system is composed of two independent parts; the first monitoring the vehicle depth and is located on the surface vessel, the second part monitoring the position and orientation of the vehicle is carried on the vehicle itself. The research activities have been mainly addressed to the development of this second part of the control system. The position error and orientation error admitted for all load configurations are, respectively, $\pm 0.2m$ and $\pm 0.035rad$. These requirements must be satisfied with a current velocity of $0.5m/sec$, a depth variable from $200m$ to $1000m$ and with different ROV load configurations.

The equations describing the ROV dynamics have been obtained from classical mechanics (Conter *et al.*, 1989; Longhi and Rossolini, 1989). The ROV, considered as a rigid body, can be fully described with six degrees of freedom, i.e. linear positions and orientations with respect to a given coordinate system. Let us consider the inertial frame $\boldsymbol{R}(0, x, y, z)$ and the body reference frame $\boldsymbol{R}_a(0_a, x_a, y_a, z_a)$ (Conter *et al.*, 1989). The ROV position with respect to $\boldsymbol{R}$ is expressed by the origin of the system $\boldsymbol{R}_a$, while its orientation by the roll, pitch and yaw angles $\psi$, $\theta$ and $\phi$ respectively. As the depth $z$ is controlled by the surface vessel, the ROV is considered to operate on surfaces parallel to the $x-y$ plane. Accordingly, the controllable variables are $x$, $y$ and the yaw angle $\phi$. It should be noticed that the roll and pitch angles $\psi$ and $\theta$ will not be considered in the dynamic model: their values have been proved to be negligible in a wide range of load conditions and with different intensities and directions of the underwater current as well (Conter *et al.*, 1989; Longhi and Rossolini, 1989) as confirmed by experimental tests (Ippoliti *et al.*, 2002).

The equations describing ROV motion are obtained by the analysis of the applied forces, these are: gravity, Archimede's thrust, cable traction, hydrodynamic forces and screw thrusts. For the mechanical structure of the ROV, the gravity force and the Archimede's thrust act only on the depth of the vehicle, therefore they are not considered in this analysis. The ROV model is described by the following system of differential equations (Conter *et al.*, 1989; Longhi and Rossolini, 1989):

$$p_1\ddot{x} + (p_2|cos(\phi)| + p_3|sin(\phi)|)V_x|\boldsymbol{V}| + p_4 x - p_5 V_{cx}|\boldsymbol{V}_c| = T_x, \tag{1}$$

$$p_1\ddot{y} + (p_2|sin(\phi)| + p_3|cos(\phi)|)V_y|\boldsymbol{V}| + p_4 y - p_5 V_{cy}|\boldsymbol{V}_c| = T_y, \tag{2}$$

$$p_6\ddot{\phi} + p_7\dot{\phi}|\dot{\phi}| + p_8|\boldsymbol{V}_c|^2 sin(\frac{\phi - \phi_c}{2}) + p_9 = M_z, \tag{3}$$

where $\boldsymbol{V}_c = [V_{cx},\ V_{cy}]^T$ is the time-invariant submarine current velocity and $\boldsymbol{V} = [V_x,\ V_y]^T = [(\dot{x} - V_{cx}),\ (\dot{y} - V_{cy})]^T$. The expressions of coefficients $p_i$, $(i = 1, \ldots, 9)$ in equations (1)–(3) and their numerical values for three different ROV configurations are reported in (Conter *et al.*, 1989; Longhi and Rossolini, 1989). These coefficients are evaluated considering the geometric an mechanical characteristics of the vehicle and the environmental and operative conditions. The quantities $T_x$, $T_y$ and $M_z$ appearing in equations (1)–(3) are the decomposition of the thrust and torque provided by the four vehicle propellers along the axes of $\boldsymbol{R}$ (Longhi and Rossolini, 1989).

## 3. THE LINEARIZED MODEL

Defining the following state, input and output variables:

$$\boldsymbol{x} = [x_1,\ x_2,\ x_3,\ x_4,\ x_5,\ x_6]^T = [x - x_0,\ y - y_0,\ \phi - \phi_0,\ \dot{x} - \dot{x}_0,\ \dot{y} - \dot{y}_0,\ \dot{\phi} - \dot{\phi}_0]^T,$$

$$\boldsymbol{u} = [T_x - T_{x0},\ T_y - T_{y0},\ M_z - M_{z0}]^T,$$

$$\boldsymbol{y} = [x - x_0,\ y - y_0,\ \phi - \phi_0]^T,$$

the linearization of (1)–(3) around an operating point $[x_0, y_0, \phi_0, \dot{x}_0, \dot{y}_0, \dot{\phi}_0, T_{x0}, T_{y0}, M_{z0}]$ results in a $3 \times 3$ transfer matrix $\boldsymbol{W}(s)$ (Conter *et al.*, 1989) and (Longhi and Rossolini, 1989). The further simplifying assumption of a diagonal $\boldsymbol{W}(s)$ can be motivated as in (Ippoliti *et al.*, 2002) and has been introduced here for testing the applicability of the proposed switching control scheme.

Each diagonal element of the linearized, decoupled model has been experimentally identified measuring the step response of the corresponding nonlinear equation of system (1)–(3) after dropping

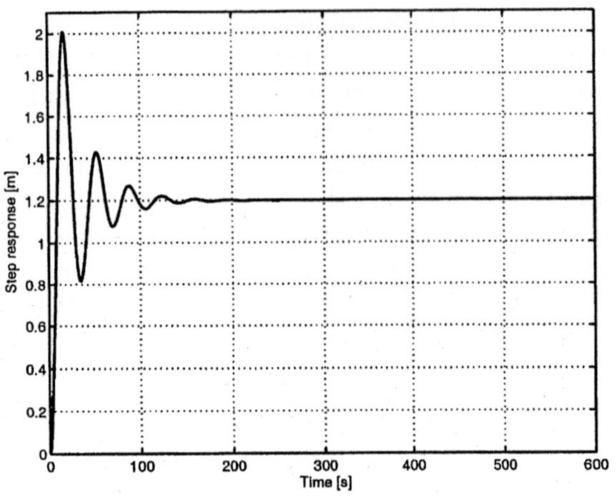


Fig. 1. Typical step response for the $x$ component of the ROV model (1)

the coupling terms. The experiment showed that each element, for the three different ROV configurations, can be well approximated by a second order system of the kind:

$$M_j(s) = \frac{K_j \omega_{n_j}^2}{s^2 + 2\zeta_j \omega_{n_j} s + \omega_{n_j}^2} \quad j = 1, 2, 3. \quad (4)$$

This is evident from the diagram of Figure 1 showing a typical step response of Equation (1) under an input thrust $T_x = 500N$.

Models (4) are characterized by three parameters: the static gain $K_j$, the natural frequency $\omega_{n_j}$ and the relative damping $\zeta_j$. These parameters can be estimated from the shape of the step response as indicated in (Astrom and Hagglund, 1995).

## 4. THE HIERARCHICAL SWITCHING CONTROLLER

The structure of the control system is defined assuming three different possible configurations of the ROV and defining a bank of three PID regulators for each of the three linearized ROV models. The structure of the proposed hierarchical controller is shown in Figure 2.

At every time instant, the selected controller $G_i^j$, $i, j = 1, \cdots, 3$, is applied to the plant $P$ and a parallel architecture of three models $M_j$ is defined. Models $M_j$ represent the different plants corresponding to the three possible different operating conditions of the ROV. These models are forced by the same control input of the plant and yield the outputs $\hat{y}_j(\cdot)$, $j = 1, \cdots, 3$, which are used to identify the model which best approximates the plant. To this purpose, the following indices are defined

$$\hat{J}_j(k) := \hat{\alpha}\hat{e}_j^2(k) + \hat{\beta} \sum_{\tau=k_0}^{k-1} \hat{\lambda}^{(k-\tau-1)} \hat{e}_j^2(\tau),$$

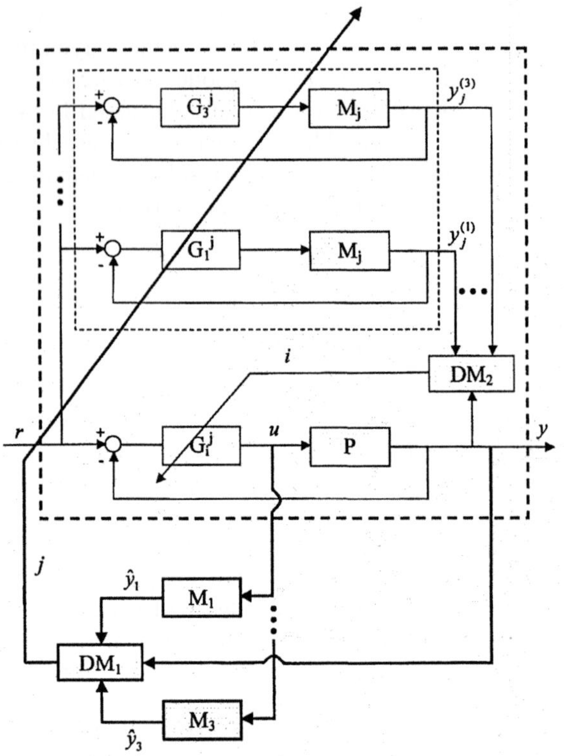


Fig. 2. Hierarchical switching controller.

$$j = 1, 2, 3, \quad k \geq 0, \; k \in \mathbb{Z}, \quad (5)$$

where: $\hat{e}_j(k) := y(k) - \hat{y}_j(k)$ are the identification errors and $\hat{\alpha} > 0$, $\hat{\beta} > 0$, $0 < \hat{\lambda} < 1$ are design parameters. Parameters $\hat{\alpha}$ and $\hat{\beta}$ determine the relative weight given by instantaneous and long-term measures respectively, while $\hat{\lambda}$ determines the memory of the index.

At every time instant, the performance indices are monitored by supervisor $DM_1$ to deduce the controller family to be applied to the plant, i.e. the controller family $G_i^{\bar{j}}$ corresponding to the model $M_{\bar{j}}$ with the minimum index $\bar{j}$.

The switching from the actual family $G_i^j$ towards $G_i^{\bar{j}}$ is performed only if the relative functionals satisfy the following hysteresis condition:

$$\hat{J}_{\bar{j}}(k) < \hat{J}_j(k)(1 - \hat{q}) \quad (6)$$

where $0 < \hat{q} < 1$ is a free design parameter and $\hat{J}_j$ is the value of the functional corresponding to the actual family $G_i^j$.

The logic driving the switching inside each fixed family of PID regulators is outlined in the following subsection.

### *4.1 The switching policy*

With reference to the linearized models (4) of Equation (1), nine different PID controllers $G_i^j$, $i, j = 1, 2, 3$, have been designed i.e. three different PID for each vehicle model $M_j$, $j = 1,2,3$ (three controller families). The corresponding third order, closed-loop systems given by the feedback connection of each $G_i^j$ with the related plant model (4) are denoted by $\Sigma_i^j, i,j = 1,2,3$. The feedback connection of each $G_i^j$ with the true plant is denoted by $S_i^j$. Following the procedure described in (Astrom and Hagglund, 1995), controllers $G_i^j$ have been designed so that the related systems $\Sigma_i^j$ have one real pole $s_i^j$ and a pair of complex conjugate poles $\alpha_i^j \pm j\beta_i^j$. The poles of the nine closed loop systems $\Sigma_i^j$ are the following:

$$s_i^j = 0.5\zeta_i^j \omega_{n_i}^j,$$
$$\alpha_i^j \pm j\beta_i^j = \zeta_i^j \omega_{n_i}^j \pm \sqrt{(1-\zeta_i^{j^2})}\omega_{n_i}^j,\ i,j = 1,2,3,$$

where $\zeta_i^j = 0.5$, $\omega_{n_i}^j = \omega_{n_j} + (i-1)0.5\omega_{n_j}$, $\omega_{n_j}$ being the natural frequencies of models (4). The distributions of closed-loop poles have been chosen to obtain three system families with different dynamic characteristics. System family $\Sigma_1^j$, $j = 1,2,3$ are expected to exhibit a smooth step response with a little overshoot but, as a counterpart, with a possible long rise time. System family $\Sigma_3^j$, $j = 1,2,3$ represents the opposite situation, it are expected to exhibit a fast dynamics, characterized by a short rise time but, as a counterpart, by a possible large overshoot. System family $\Sigma_2^j$, $j = 1,2,3$ represents an intermediate situation.

The proposed switching policy is defined in the following way. At time instant $k = 0$, a PID $G_\ell^j, \ell, j = 1,2,3$, is chosen and connected to the true plant. At the same time, and before the next output $y(k+1)$ of $S_\ell^j$ is acquired, the output prediction errors $e_j^{(i)}(k+p), p = 0, \cdots, \overline{K}$, of systems $\Sigma_i^j, i,j = 1,2,3$, related to the actual PID family, are computed. Each $e_j^{(i)}(k+p)$ is defined as $e_j^{(i)}(k+p) = r(k+p) - y_j^{(i)}(k+p)$, where $r(k)$ is the external reference, $y_j^{(i)}(k+p)$ is the predicted output of system $\Sigma_i^j$. Each $y_j^{(i)}(k+p), i,j = 1,2,3$, is computed setting the corresponding $\Sigma_i^j$ in the same initial conditions of $S_\ell^j$ at $k = 0$. The whole procedure is repeated at each time instant $k$.

The idea is to exploit the output prediction errors to foresee the future performance of $S_\ell^j$ in case the actual PID $G_\ell^j$ be changed and in case it be kept acting. To this purpose the following functionals are defined:

$$J_i^j = \beta^j e_j^{(i)^2}(k) + \gamma^j \sum_{p=1}^{\overline{K}} [e_j^{(i)^2}(k+p)(1+n^j v^j)]\lambda_j^p,$$
$$i,j = 1,2,3,$$

where: $\beta^j > 0$, $\gamma^j > 0$ and $0 \leq \lambda_j < 1$ are free design parameters which determine the relative weight of instantaneous and future long term errors, $n^j$ is a non negative integer which is increased by 1 whenever a sign change of $e_j^{(i)}(k+p)$ is observed, $v^j$ is a fixed percentage of $e_j^{(i)}(k+p)$. The term $n^j v^j$ has been introduced to penalize excessive oscillations.

At each time instant $k$ the supervisor switches the actual $G_\ell^j$ towards the controller producing the minimum index $J_i^j$. The switching is performed only if $J_i^j < J_\ell^j(1-q)$, where $0 < q < 1$, is a free design parameter and $J_\ell^j$ is the value of the functional corresponding to the actual $G_\ell^j$.

## 5. NUMERICAL RESULTS

As a consequence of the simplified diagonal form of $\boldsymbol{W}(s)$, the controller is composed of three independent SISO regulators for the $x$, $y$ and $\phi$ variables. For the sake of brevity, the numerical tests reported here only refer to the first component of the controlled variables vector, that is the $x$ variable. Namely, they refer to the decoupled Equation (1) and to the linearized models (4). The discretization of the nine different PID $G_i^j$ $i,j = 1,2,3$, has been performed according to the procedure described in (Astrom and Hagglund, 1995), assuming a sampling period of $0.5s$. For the constructive characteristics of the thruster system, the constraint on the control effort $|T_x| < 1 \cdot 10^4 N$, is imposed. The simulation tests have been carried out considering an external reference given by a piecewise constant function with step variations of its amplitude.

Several simulations have been executed considering different operative conditions for the ROV. In the sample reported in Figures 3 and 4 the ROV works at a depth of 200 $m$ with submarine current velocities $V_{cx} = V_{cy} = 0.15m/s$. The vehicle task is to carry the flow-line frame and the relative installation module to the set-point at the time instant $t = 200s$ (see Figure 3), to maintain the position after the load has been discharged at the time instant $t = 400s$ (see Figure 4) and to reach without load two different set-points at time instants $t = 600s$ and $t = 800s$ (see Figure 4).

Figures 3(a) and 4(a) show the output response produced by the hierarchical switching controller. Figures 3(b), 4(b) and 3(c), 4(c) show the switching sequences of supervisors $DM_1$ and $DM_2$ respectively.

Numerical results evidence a satisfactory performance in terms of specifications on set-point control when unexpected set-point changes occur. This is due to the hierarchically lower supervisor ($DM_2$) which generates a switched control law producing an output response with the desired features. Figure 3(a) at time instant 200$s$ and Figure 4(a) at time instants 600$s$ and 800$s$ show a step response with a short rise time and with a little overshoot. The $DM_2$ switching sequence is shown in Figures 3(c) and 4(c).

The hierarchically higher supervisor ($DM_1$) detects sudden changes in the operative conditions and identifies the vehicle configuration driving the control law towards the correct PID family. Before $DM_2$ identifies the correct configuration (see Figures 3(b) and 4(b) for the switching sequence), the output response shows a significative transient phase (Figures 3(a) and 4(a) at time instants 0$s$ and 400$s$ respectively). This transient behaviour, could be improved in a normal operating condition but is almost unavoidable in the present situation which depicts the worst possible reference scenario for a set-point tracking controller: neither the change instant nor the new operating condition are known a priori. Using a single family of PID controllers (namely dropping $DM_1$) produced very poor results when a change of the vehicle configuration occurred and the output response showed an oscillatory diverging behaviour. These results are not reported here for brevity.

This confirms the necessity of introducing a supervisor detecting the possible changes in the operating conditions.

## 6. CONCLUSIONS

This paper has presented a hierarchical switching control scheme aimed at improving the performance attainable in PID control of underwater vehicles.

This controller can operate efficiently in dynamical environments possessing a high degree of uncertainty. Multiple models are used to describe the different operative conditions and the control is performed by switching to the appropriate controller family followed by a switching inside the family for defining the appropriate time-varying controller tracking the external step reference. The main merits of the proposed technique are in terms of its simplicity and robustness with respect to different operating conditions. On the basis of numerical simulations satisfactory performance of the control system seem to be really attainable.

## REFERENCES


Adams, J.M. and K.S. Rattan (2001). Multi-stage fuzzy PID controller with a fuzzy switch. In: *Proceedings of the 2001 IEEE International Conference on Control Applications (CCA '01)*. Mexico City, Mexico. pp. 323–327.

Aguiar, Antonio Pedro and Antonio M. Pascoal (2001). Regulation of a nonholonomic autonomous underwater vehicle with parametric modeling uncertainty using lyapunov functions. In: *Proceedings of the 40th IEEE Conference on Decision and Control*. Orlando, Florida, USA. pp. 4178–4183.

Antsaklis, P. J. and A. Nerode (1998). Hybrid control systems: an introductory discussion to the special issue. *IEEE Transactions on Automatic Control* **43**(4), 457–460.

Astrom, K.J. and T. Hagglund (1995). *PID Controllers: Theory, Design, and Tuning*. ISA - The Instrumentation, Systems, and Automation Society - 2nd edition.

Baldini, M., M.L. Corradini, L. Jetto and S. Longhi (1999). A multiple-model based approach for the intelligent control of underwater remotely operated vehicles. In: *Proceedings of the 14th Triennial World Congress of IFAC*. Vol. Q. Beijing, P.R. China. pp. 19–24.

Blanchini, F. (2000). The gain scheduling and the robust state feedback stabilization problems. *IEEE Transactions on Automatic Control* **45**(11), 2061–2070.

Cavalcanti, J.H.F. (1995). Intelligent control system using PID and neural controllers. In: *Proceedings of the 38th Midwest Symposium on Circuits and Systems*. Rio de Janeiro, Brazil. pp. 425–428.

Conte, G. and A. Serrani (1994). $H_\infty$ control of a remotely operated underwater vehicle. In: *Proceedings ISOPE'94*. Osaka.

Conter, A., S. Longhi and C. Tirabassi (1989). Dynamic model and self-tuning control of an underwater vehicle. In: *Proceedings of the Eighth Int. Conf. on Offshore Mechanics and Arctic Engineering*. The Hague. pp. 139–146.

Corradini, M. L. and G. Orlando (1998). Variable structure control of discretized continuous-time systems. *IEEE Transactions on Automatic Control* **43**(9), 1329–1334.

Corradini, M.L. and G. Orlando (1997). A discrete adaptive variable structure controller for mimo systems, and its application to an underwater ROV. *IEEE Trans. Contr. Sys. Techn.* **5**(3), 349–359.

Corradini, M.L., L. Jetto and G. Orlando (2004). Robust stabilization of multivariable uncertain plants via switching control. *IEEE Transactions on Automatic Control.*

Cristi, R., A.P. Fotis and A.J. Healey (1990). Adaptive sliding mode control of autonomous underwater vehicles in the dive plane. *IEEE J. Oceanic Engng.* **15**(3), 152–160.

Efe, M.O., O. Kaynak, X. Yu and S. Iplikci (2001). Cost effective computationally intelligent control - An augmented switching manifold approach. In: *Proceedings of the 27th Annual Conference of the IEEE Industrial*

*Electronics Society. IECON '01.* Denver, CO, USA. pp. 31–36.

Fossen, T.I. (1994). *Guidance and Control of Ocean Vehicles.* J. Wiley & Sons. New York, USA.

Fossen, T.I. (2002). *Marine Control Systems. Guidance, Navigation, and Control of Ships, Rigs and Underwater Vehicles.* Marine Cybernetics. Trondheim, Norway.

Grey, J. P. (1987). A comparison of PID control algorithms. *Control Engineering* **34**(3), 102–105.

Ippoliti, G. and S. Longhi (2001). Hybrid control with switching on fixed and adaptive models for the intelligent control of underwater remotely operated vehicles. In: *9th Mediterranean Conference on Control and Automation (MED01).* Dubrovnik, Croatia.

Ippoliti, G., L. Jetto and S. Longhi (2003). Improving PID control of underwater vehicles through a hybrid control scheme. In: *Proceedings of the 1st IFAC Workshop on Guidance and Control of Underwater Vehicles (GCUV '03).* Newport, South Wales, UK.

Ippoliti, G., L. Jetto, S. Longhi and V. Orsini (2004). Adaptively switched set-point tracking PID controllers. In: *Proceedings of the 12th Mediterranean Conference on Control and Automation (MED'04).* Kusadasi, Aydin, Turkey.

Ippoliti, G., S. Longhi and A. Radicioni (2002). Modelling and identification of a remotely operated vehicle. *Journal of Marine Engineering and Technology* **A**(AI), 48–56.

Jalving, B. and N. Storkersen (1994). The control system of an autonomous underwater vehicle. In: *Proceedings of the Third IEEE Conference on Control Application.* Vol. 2. Glasgow, UK. pp. 851–856.

Kaminer, Isaac, Antonio M. Pascoal, Carlos J. Silvestre and Pramod P. Khargonekar (1991). Control of an underwater vehicle using $H_\infty$ synthesis. In: *Proceedings of the 30th IEEE Conference on Decision and Control.* Brighton, England. pp. 2350–2355.

Longhi, S. and A. Rossolini (1989). Adaptive control for an underwater vehicle: Simulation studies and implementation details. In: *Proceedings of the IFAC Workshop on Expert Systems and Signal Processing in Marine Automation.* The Technical University of Denmark, Lyngby, Copenhagen, Denmark. pp. 271–280.

Longhi, S., G. Orlando, A. Serrani and A. Rossolini (1994). Advanced control strategies for a remotely operated underwater vehicle. In: *Proceedings of the First World Automation Congress (WAC'94).* Maui, Hawaii, USA. pp. 105–110.

Miyamaoto, S., T. Aoki, T. Maeda, K. Hirokawa, T. Ichikawa, T. Saitou, H. Kobayashi, E. Kobayashi and S. Iwasaki (2001). Maneuvering control system design for autonomous underwater vehicle. In: *Proceedings of the MTS/IEEE Conference and Exhibition. OCEANS, 2001.* Honolulu, HI, USA. pp. 482–489.

Morse, A. S., C. C. Pantelides, S. S. Sastry and J. M. Schumacher (1999). Introduction to the special issue on hybrid systems. *Automatica* **35**, 347–348.

Narendra, K. S. and C. Xiang (2000). Adaptive control of discrete-time systems using multiple models. *IEEE Transactions on Automatic Control* **45**(9), 1669–1686.

Oki, T., T. Yamamoto, M. Kaneda and S. Omatsu (1997). An intelligent PID controller with a neural supervisor. In: *Proceedings of the 1997 IEEE International Conference on Systems, Man, and Cybernetics, 1997. Computational Cybernetics and Simulation.* Orlando, FL, USA. pp. 4477–4482.

Park, J., Chung Wankyun and J. Yuh (2000). Nonlinear $H_\infty$ optimal PID control of autonomous underwater vehicles. In: *Proceedings of the 2000 International Symposium on Underwater Technology. UT 00.* Tokyo, Japan. pp. 193–198.

Ruano, A. E. and A. B. Azevedo (1999). B-splines neural network assisted PID autotuning. *International Journal of Adaptive Control and Signal Processing* (13), 291–306.

Rugh, W. J. and J. S. Shamma (2000). Research on gain scheduling. *Automatica* **36**, 1401–1425.

Utkin, V. I. (1992). *Sliding Modes in Control and Optimization.* Springer–Verlag. New York, USA.

Zhi, Y. and W. Jingling (1997). Auto-tuning of PID parameters based on switch step response. In: *Proceedings of the 1997 IEEE International Conference on Intelligent Processing Systems.* Beijing, China. pp. 779–782.

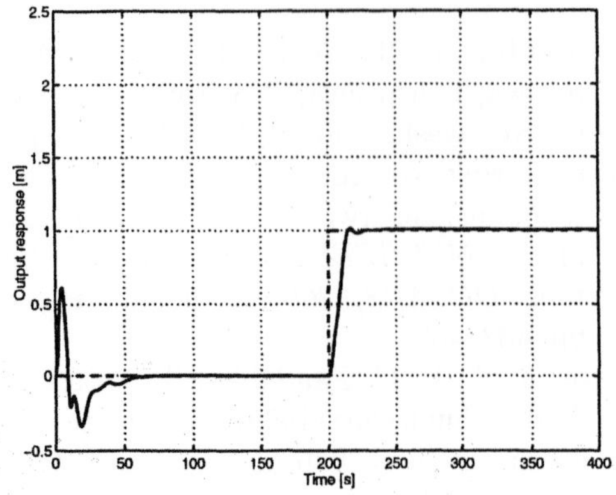


(a)

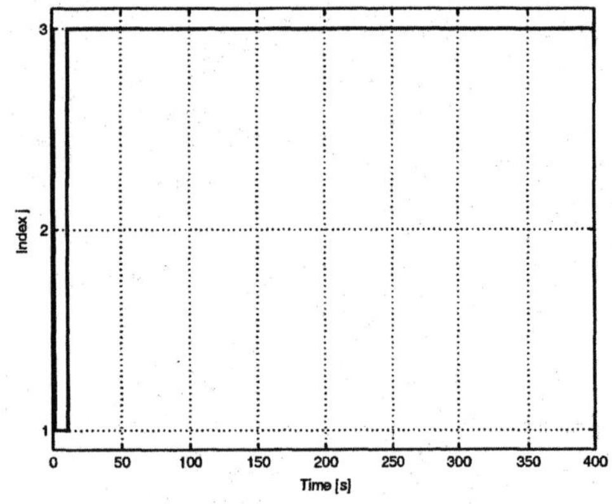


(b)

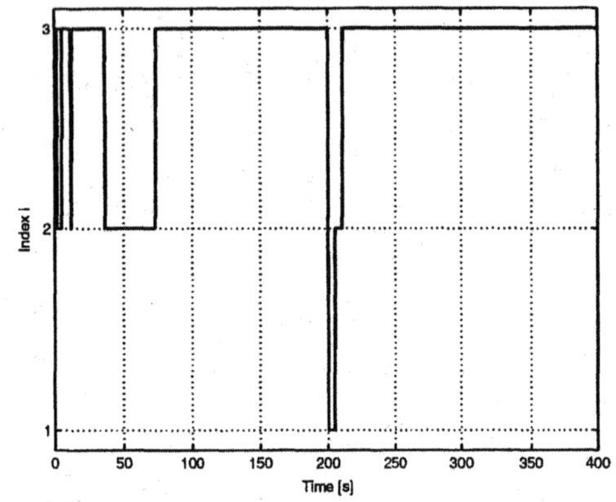


(c)

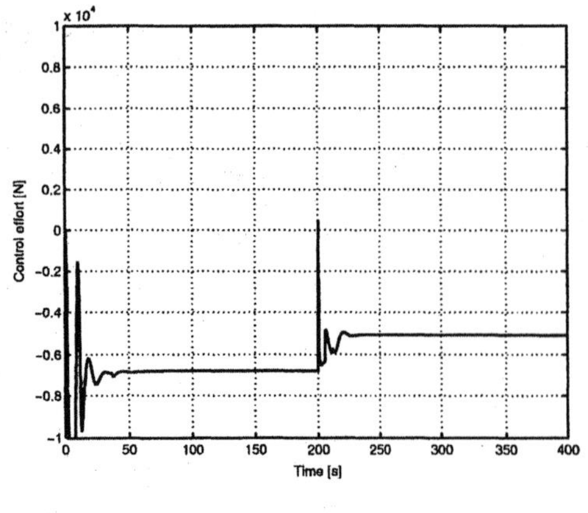


(d)

Fig. 3. Vehicle carrying the load. Part (a): Output response. Part (b): switching sequence of $DM_1$. Part (c): switching sequence of $DM_2$. Part (d): control effort.

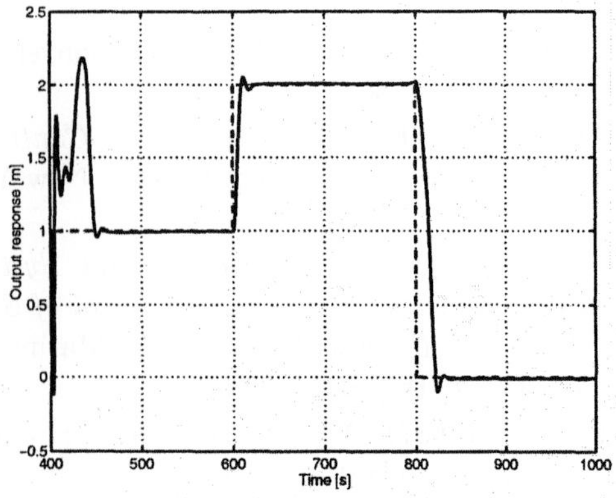


(a)

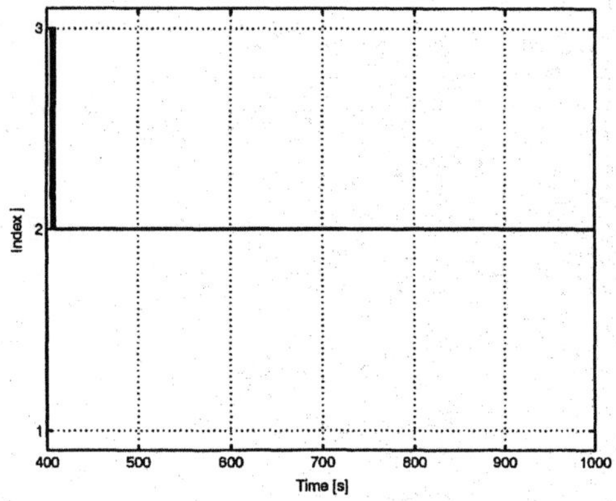


(b)

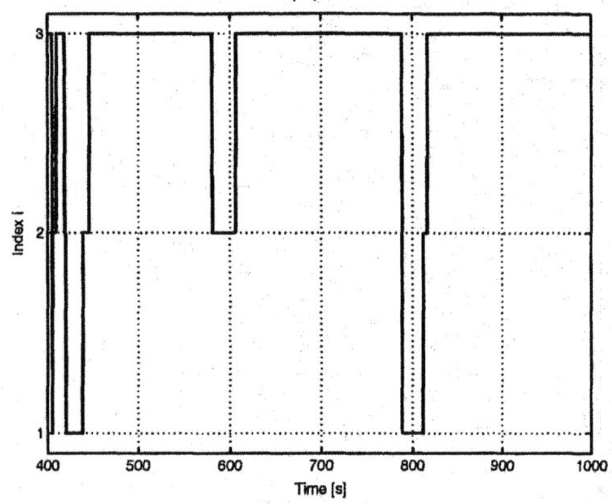


(c)

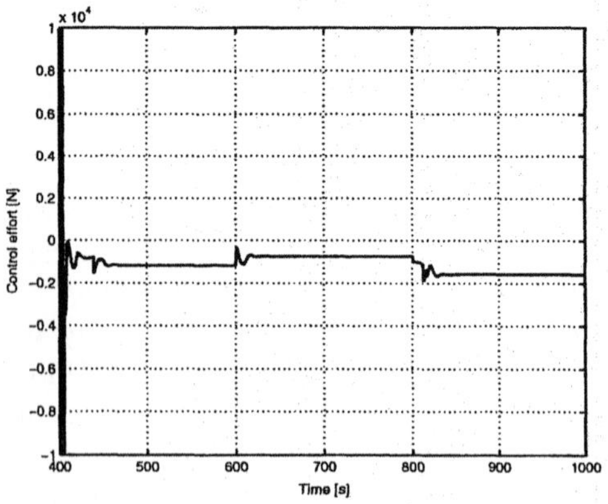


(d)

Fig. 4. Vehicle without load. Part (a): Output response. Part (b): switching sequence of $DM_1$. Part (c): switching sequence of $DM_2$. Part (d): control effort.

ELSEVIER
IFAC
PUBLICATIONS
www.elsevier.com/locate/ifac

# IMPROVED LINE-OF-SIGHT GUIDANCE FOR CRUISING UNDERWATER VEHICLES

**Vedran Bakaric[a], Zoran Vukic[b], Radovan Antonic[c]**

*[a] Brodarski institut, Av. V. Holjevca 20, 10020 Zagreb, Croatia,
e-mail: bakaric@hrbi.hr*
*[b] Department of Control and Computer Engineering in Automation, Faculty of Electrical Engineering and Computing, Unska 3, 10000 Zagreb, Croatia*
*[c] Split College of Maritime Studies, Zrinsko-Frankopanska 38, 21000 Split, Croatia*

Abstract: Guidance through waypoints is common for small autonomous marine vehicles. Guidance by the line of sight, which turns the vehicle directly towards the next waypoint without any reference path calculation, is computationally the simplest form of waypoint guidance. However, the basic algorithm gives rather poor guidance due to the missed waypoint problem, lack of sea current compensation and abrupt transitions between the consecutive waypoints. Significant path deviations and even deadlocks are possible due to these problems. Therefore, more complex algorithms are usually used in real world applications. The research reported in this paper aims to demonstrate that significant improvements of the basic line-of-sight guidance algorithm can be achieved by several intuitive, simple corrections and additions. The simplicity of the basic line-of-sight guidance algorithm is not compromised. In particular, missed waypoint detection is performed by monitoring the distance between the vehicle and the waypoint. Introduced reference heading corrections are based upon the location of the next waypoint after the one the vehicle is currently approaching, and upon the sea current direction and intensity. The results of these corrections are shown in several simulation examples. In addition, the paper includes a short discussion about the line-of-sight guidance in the diving plane. *Copyright © 2004 IFAC*

Keywords: Autonomous vehicles; Guidance systems; Cruise control; Algorithms

## 1. INTRODUCTION

The guidance system of a vehicle is the part of the motion control system which provides control references for the autopilots. At piloted vessels, this function belongs to the human operator. On the other hand, autonomous vehicles have automatic guidance. Most of such vehicles are small mobile robotic systems designed primarily for autonomous survey, inspection, reconnaissance and search over wide areas. In the marine area, their most important class are autonomous underwater vehicles.

*Waypoint guidance* scheme is common for cruising autonomous vehicles. The motion plan is here specified in the form of a list of waypoints $(x_i, y_i, z_i)$. The vehicle has to pass through them in the given order, usually with a constant forward speed reference $u_r$. The desired path between two adjacent waypoints and times of waypoint arrivals are not defined. When the vehicle arrives to a sufficiently small distance from the current waypoint, that waypoint is proclaimed achieved (reached) and the vehicle continues with the next waypoint. Besides expressing the desired vehicle motion, the waypoints can be used to delimit mission phases and to control execution of some additional commands, like forward speed changing or measurement taking (Bakaric, 2000).

Waypoint guidance methods are rather simple, computationally inexpensive and well suited for complex paths, especially when it is impossible to specify detailed motion plan in advance due to unknown or variable environment. Sudden changes in reference path specification are easily accepted when the vehicle is running. However, the guidance precision is limited because the desired vehicle motion is only vaguely defined. More complex waypoint algorithms use coordinate transformations, as in the kinematic controller designed by Aguiar and Pascoal (2002), which finds directly the reference yaw rate $r_r$ rather than the reference heading $\psi_r$. Vancek (1997) uses waypoints with specified desired crossing headings within a fuzzy guidance algorithm. This approach can achieve a rather smooth guidance, but only if the vehicle never departs the expected path, that is if the mission is fixed and the environment known and invariable. Artificial *target point*, which acts as the current waypoint but moves with the vehicle, can be constructed from the waypoint list and environment data to give robust smooth guidance (Antonelli *et al.*, 2001). To gain even more control over guidance process, the waypoints can be connected with a smooth reference path and the guidance accomplished with some *path following* guidance method (Aicardi *et al.*, 2001). Such guidance methods consider differences between the current vehicle position and orientation and the nearby reference path segment to generate heading error and lateral path error as control reference variables. If the timing precision is also important, the reference path points can be appended with desired time of arrival to form the reference trajectory (a path in space and time), but rather complex and rigid *trajectory tracking* guidance methods are still seldom needed for cruising vehicles. Instead, more research and practical interest attract

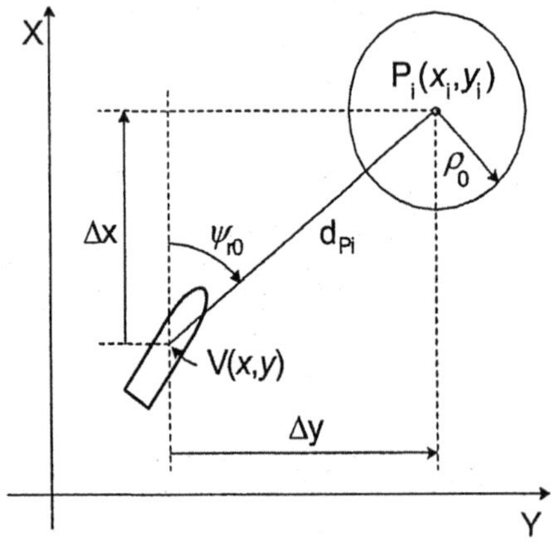


Fig. 1. Basic line of sight guidance in the steering plane. Here $\Delta x = x_i - x$ and $\Delta y = y_i - y$.

guidance methods for *obstacle-filled environments*. The guidance process here has to be reflexive, ready to quickly react on obstacle encounters, but also to include some real-time planning and optimisation, to avoid deadlocks and enable path finding between obstacles (Hyland and Taylor, 1993).

## 2. LINE OF SIGHT GUIDANCE ALGORITHM

Line of sight (LOS) guidance is the most simple of waypoint guidance methods. The guided vehicle here tries to turn itself directly towards the current waypoint $P_i(x_i, y_i, z_i)$ and to reach it with a prescribed constant forward speed $u_r = u_0$.

### *2.1 Basic line of sight guidance in the steering plane*

Basic line of sight guidance algorithm in the steering plane is described by Healey and Lienard (1993) and Fossen (1994). The algorithm has to provide reference heading $\psi_r = \psi_{r0}$ which would guide the vehicle from its current position $V(x, y)$ towards the current waypoint $P_i(x_i, y_i)$ (Fig. 1). The straightforward solution is

$$\tan(\psi_{r0}) = \frac{y_i - y}{x_i - x} \tag{1}$$

$\psi_{r0}$ can be readily computed from (1) by the four quadrant inverse tangent function, usually defined as `atan2` in mathematically oriented programming languages.

The current waypoint $P_i$ is reached, or achieved, when the vehicle enters a *circle of acceptance* around the waypoint, that is when the distance between it and the vehicle

$$d_{Pi}^2 = (x_i - x)^2 + (y_i - y)^2 \tag{2}$$

becomes smaller than the *acceptance radius* $\rho_0$, $d_{Pi} < \rho_0$. In that moment the next waypoint on the list is selected as the current waypoint. Good initial value for the acceptance radius equals two vehicle lengths.

### *2.2 Diving guidance by line of sight*

Cruising underwater vehicles often control their movement in the depth dimension by adjusting the vehicle pitch angle $\theta$ using horizontal control surfaces (elevators). Vertical manoeuvring thrusters can be added if precise depth control is needed at small forward speeds. For both cases the reference depth $z_r$ can be simply defined as the depth coordinate of the current waypoint, $z_r = z_i$. For manoeuvring using thrusters, the reference pitch angle can be set at zero, $\theta_r = 0$. For diving using control surfaces, the simplest approach is to set the reference pitch angle at some prescribed value $\theta_r = \theta_0$, which may be the maximum possible pitch angle for the considered vehicle or any smaller value, and gradually decrease its magnitude when the vehicle comes near the reference depth. When the waypoint list defines a three-dimensional reference path, the vehicle would typically firstly reach the desired depth, and then approach the current waypoint horizontally at its depth level. If the latitude and the longitude of the current waypoint were the first to be achieved, the vehicle with control fins would spiral towards it until the proper depth is also achieved.

Waypoint acceptance check must be here done simultaneously for horizontal and vertical coordinates, which can be performed in two ways. The simpler one checks besides the horizontal distance $d_{Pi}$ (2) an additional condition of depth acceptance

$$|z_i - z| < \rho_z \qquad (3)$$

where $\rho_z$ is the depth acceptance tolerance, and states that the waypoint $P_i$ is reached when both $d_{Pi} < \rho_0$ and (3) are satisfied. The more elaborated way (Healey and Lienard, 1993) is to check the weighted three-dimensional distance between the vehicle and the waypoint

$$d_{Di}^2 = (x_i - x)^2 + (y_i - y)^2 + \lambda(z_i - z)^2 \qquad (4)$$

(where $\lambda > 0$ is a weight factor) against the acceptance radius $\rho_{D0}$: the waypoint $P_i$ is proclaimed as reached when becomes $d_{Di} < \rho_{D0}$, that is, when the vehicle enters the acceptance ellipsoid. The depth dimension is not taken into account with the same weight as the horizontal dimensions, which indicates different motion dynamics and precision requirements for steering and diving. It is recommended to give less weight to the depth, that is to put $\lambda < 1$, or $\rho_z > \rho_0$.

## 3. IMPROVED LINE OF SIGHT GUIDANCE

Basic line of sight guidance algorithm, outlined in the previous chapter, has several weak points that should be amended for most practical applications. Distinct problems among them are missed waypoint handling, smooth transition between two consecutive waypoints, and rejection of sea current disturbances. Several ideas and procedures for addressing these problems for the case of steering guidance in the horizontal plane are given in this chapter.

### *3.1 Missed waypoint handling*

The missed waypoint is a waypoint that the guided vehicle passes by at a distance greater than the acceptance radius $\rho_0$ (Fig. 2). This situation may occur if two consecutive waypoints are placed too close to each other and so the vehicle has not enough space to properly turn towards the second one after

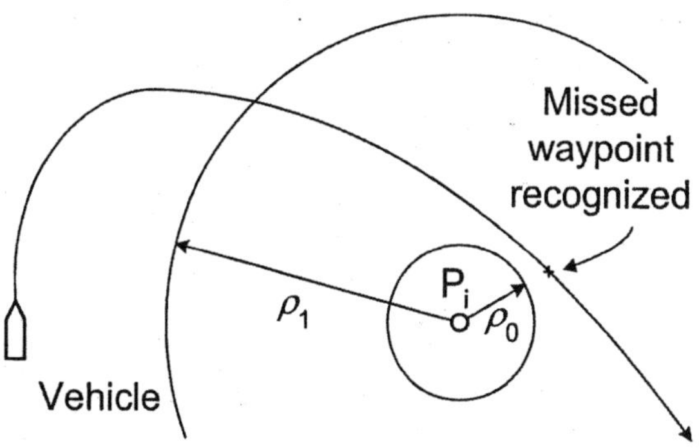


Fig. 2. Missed waypoint detection

the first was reached, but also due to a side disturbance (like a gush of current) in the critical moment of approach. Since the cruising vehicle cannot stop nor spin around in its place, it would begin to go away from the missed waypoint before it was recognized as reached, and so the guidance algorithm would initiate an abrupt turn to try a re-approach manoeuvre. If there is not enough space to turn, the vehicle can easily start to circle around the missed waypoint unable to reach it, in a temporary or permanent deadlock. Obviously, the guidance algorithm should be able to deal with such a situation. Since a re-approach manoeuvre implies delay and more vehicle manoeuvring, it should be tried only if the missed waypoint marks a place of an important action, e.g. measurement, to be performed. In all other cases the missed waypoint should be detected as such and subsequently discarded.

The most obvious method of missed waypoint detection is to recognize the moment in which the vehicle–waypoint distance $d_{Pi}$ (2) begins to increase before the waypoint is reached. However, this distance will not always monotonously decrease towards $\rho_0$ during regular approach due to measurement noise, disturbances and manoeuvres when the vehicle is still well away from the waypoint. Generally, missed waypoint detection does not make much sense if the vehicle has not yet approached the waypoint and so there is ample space to manoeuvre towards it. If the vehicle due to any reason cannot come rather near the current waypoint, then probably the particular part of the motion plan should be aborted or redefined, which is not the task of the guidance algorithm. With this remarks in mind, the following rule of missed waypoint detection can be proposed (Fig. 2). A waypoint is proclaimed as missed in the moment when the following conditions hold:

$$\begin{aligned} d_{Pi} &< \rho_1 \\ d_{Pi} &> d_{MIN} + \delta_d \end{aligned} \qquad (5)$$

where $d_{MIN}$ is memorized achieved minimum of the distance $d_{Pi}$, $\delta_d$ accuracy tolerance of distance information, and $\rho_1$ chosen maximum allowed miss distance. Initial value of $d_{MIN}$ in the moment of the current waypoint selection can be set at $\rho_1$. Vehicle's turning circle diameter (dependent on the vehicle forward speed $u$) makes a good preliminary value for $\rho_1$.

### *3.2 Reference heading correction for smooth waypoint transition*

Basic waypoint guidance scheme does not take into account the position of the next waypoint $P_{i+1}$ before the current waypoint $P_i$ is reached. As a consequence, the vehicle cannot achieve a smooth turn towards the next waypoint, but makes more manoeuvring and greater excursions from the optimal path than necessary, especially at larger path turns (Fig. 3). The next waypoint can be included into line of sight guidance in several ways. The simplest approach is probably to apply corrections on the references generated by the basic line of sight guidance. One variant of it, derived by Bakaric (2000), is described below.

Let $P_i(x_i, y_i)$ is the current waypoint, $P_{i+1}(x_{i+1}, y_{i+1})$ the next waypoint, and $V(x, y)$ the vehicle coordinates. The reference heading

$$\psi_r = \psi_{r0} + \psi_c \qquad (6)$$

is calculated as the sum of the basic line of sight heading $\psi_{r0}$ (1) and the correction term $\psi_c$, which is determined from the geometrical relation between the

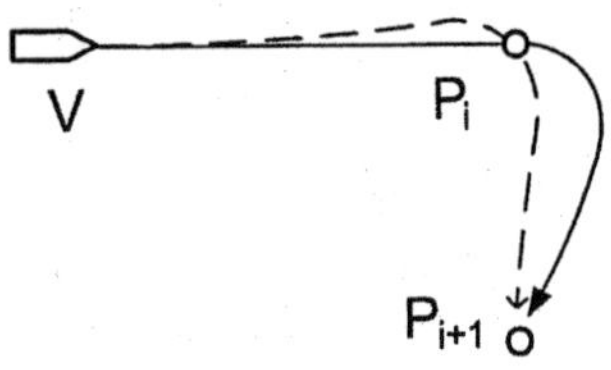


Fig. 3. Turning at waypoint with simple line of sight guidance. The dashed line represents desired, and the full one obtained vehicle motion.

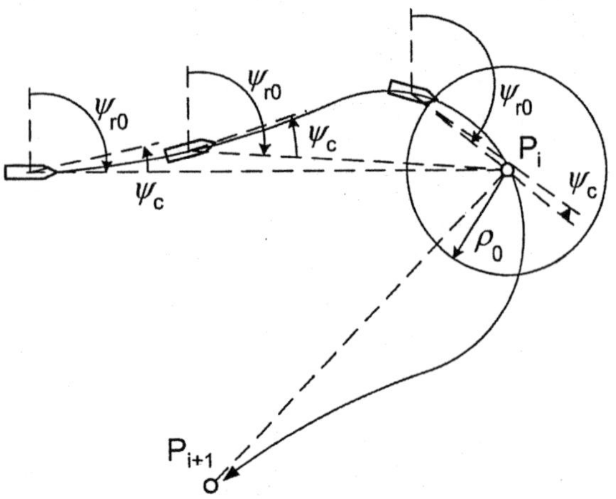


Fig. 4. Line of sight guidance with reference heading correction

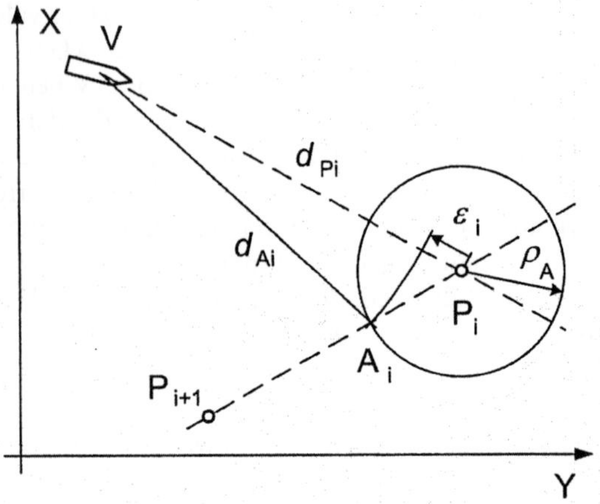


Fig. 5. Definition of the auxiliary point $A_i$ and the normalized distance difference $\varepsilon_i$. Here $\varepsilon_i < 0$.

points $P_i$, $P_{i+1}$ and V. Heading wraparound problem (e.g. 358° + 3° = 1°) should be addressed when calculating (6). As can be seen from Fig. 4, $\psi_c$ should be a function of the distance $d_{Pi}$ between V and $P_i$ (2) and of the needed heading change at waypoint $P_i$, determined by the angle $V - P_i - P_{i+1}$. Transcedent functions involved with the calculation of this variable angle can be avoided with the introduction of auxiliary point $A_i(x_{Ai}, y_{Ai})$, defined as a point at the distance $\rho_A$ from $P_i$ and in the direction of $P_{i+1}$ (Fig. 5). The coordinates of the point $A_i$ can be calculated from the geometric conditions

$$(x_{Ai}-x_i)^2 + (y_{Ai}-y_i)^2 = \rho_A^2 \qquad (7)$$

$$(y_{Ai}-y_i)(x_{i+1}-x_i) = (x_{Ai}-x_i)(y_{i+1}-y_i) \qquad (8)$$

$$\begin{aligned} \mathrm{sgn}(x_{Ai}-x_i) &= \mathrm{sgn}(x_{i+1}-x_i) \\ \mathrm{sgn}(y_{Ai}-y_i) &= \mathrm{sgn}(y_{i+1}-y_i) \end{aligned} \qquad (9)$$

as

$$\begin{aligned} y_{Ai} &= y_i + \rho_A \frac{\mathrm{sgn}(y_{i+1}-y_i)}{\sqrt{1+\left(\dfrac{x_{i+1}-x_i}{y_{i+1}-y_i}\right)^2}} \\ x_{Ai} &= x_i + \frac{x_{i+1}-x_i}{y_{i+1}-y_i}(y_{Ai}-y_i) \end{aligned} \qquad (10)$$

if $y_i \neq y_{i+1}$, or

$$\begin{aligned} x_{Ai} &= x_i + \rho_A \frac{\mathrm{sgn}(x_{i+1}-x_i)}{\sqrt{1+\left(\dfrac{y_{i+1}-y_i}{x_{i+1}-x_i}\right)^2}} \\ y_{Ai} &= y_i + \frac{y_{i+1}-y_i}{x_{i+1}-x_i}(x_{Ai}-x_i) \end{aligned} \qquad (11)$$

if $x_i \neq x_{i+1}$. Since the coordinates $(x_{Ai}, y_{Ai})$ do not depend on the vehicle position, they are calculated for each waypoint only once. The distance between the vehicle V and $A_i$ is

$$d_{Ai}^2 = (x_{Ai}-x)^2 + (y_{Ai}-y)^2 \qquad (12)$$

and the normalized difference between $d_{Ai}$ and $d_{Pi}$ is

$$\varepsilon_i = \frac{d_{Ai} - d_{Pi}}{\rho_A} \qquad (13)$$

From Fig. 5 it can be seen that $\varepsilon_i$ is a good indicator of the turn needed at the waypoint $P_i$: if the next waypoint $P_{i+1}$ lies in the direction of the current waypoint, $\varepsilon_i$ is near to 1, while $\varepsilon_i$ near −1 means that an U-turn is needed. As can be seen from the example on Fig. 4, the heading correction $\psi_c$ should be small if the current waypoint is still distant, then gradually grow to some maximum value dependent on the desired heading change as the vehicle approaches the waypoint, and finally subside to zero near the waypoint $P_i$. Consequently, the correction term $\psi_c$ can be constructed from two magnitude factors dependant on $d_{Pi}$ and $\varepsilon_i$ (Fig. 6). The sign of $\psi_c$ should be determined separately. From Fig. 4 and Fig. 5 it can be seen that the desired heading correction is to the right and hence positive if the auxiliary point $A_i$ (and also the next waypoint $P_{i+1}$) lies to the left of the orientated line from the vehicle V to the current waypoint $P_i$. The mathematical expression of this observation yields

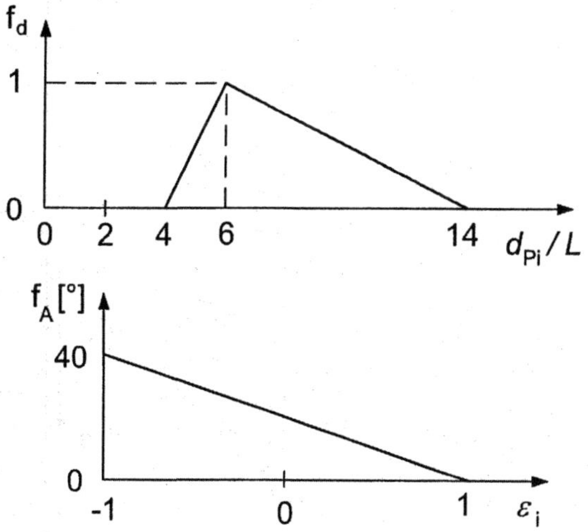


Fig. 6. Definition of two magnitude factors for the desired heading correction $\psi_c$ : (top) the distance factor $f_d(d_{Pi})$, non-dimensional ($d_{Pi}$ normalized with the vehicle length $L$; the acceptance circle radius $\rho_0 = 2L$ ); (bottom) the turn factor $f_A(\varepsilon_i)$, in degrees.

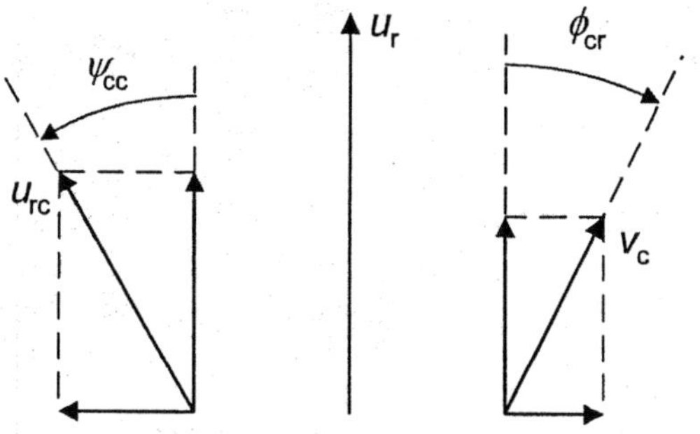


Fig. 7. Reference vector $u_r$, current vector $v_c$ and modified reference vector $u_{rc}$

$$\mathrm{sgn}(\psi_c) = \mathrm{sgn}\{(x_{Ai} - x)(y_i - y) - (x_i - x)(y_{Ai} - y)\} \qquad (14)$$

Finally, the desired heading correction is calculated as

$$\psi_c = \mathrm{sgn}(\psi_c) \cdot f_d(d_{Pi}) \cdot f_A(\varepsilon_i) \qquad (15)$$

with $f_d(d_{Pi})$ and $f_A(\varepsilon_i)$ determined by simulation experiments as in Fig. 6.

### *3.3 Compensation of sea current influence*

Encountering beam sea current, the vehicle guided by the simple line of sight guidance drifts sideways. If the current is not too strong, the line-of-sight guidance algorithm will turn the vehicle towards the waypoint and thus finally reach it, but following a winding, unoptimal path. To avoid such unwanted path deviations, the guidance algorithm has to apply additional corrections based on known or estimated current intensity $v_c$ and current direction angle $\phi_c$.

The simplest compensation method is an additional heading correction aimed to cancel the sway motion caused by the sea current. To achieve this, the reference vector $u_r$, which defines the desired vehicle motion, has to be rotated by the correction angle $\psi_{cc}$ to cancel the lateral component of the current vector $v_c \sin(\phi_{cr})$, where is $\phi_{cr} = \phi_c - \psi_r$ current direction angle in respest to the reference heading (heading correction term discussed in section 3.2 is already included in $\psi_r$). As can be seen from Fig. 7, the heading correction term $\psi_{cc}$ is

$$\sin(\psi_{cc}) = -\frac{v_c}{u_0}\sin(\phi_{cr}) \qquad (16)$$

and the corrected reference heading is calculated as $\psi_{rc} = \psi_r + \psi_{cc}$. Note that the heading correction may become impossible if the current intensity $v_c$ is greater than the forward speed reference $u_r$, which is reflected by undefined $\psi_{cc}$ in (16). Since the longitudinal component of the current vector is not compensated, the vehicle will move faster down the current, and slower up the current.

## 4. SIMULATION RESULTS

The simulations have been carried out in the MATLAB-Simulink programming language (The MathWorks, Inc.). The guided vehicle is NPS AUV II, simulation model of which is given by Healey and Lienard (1993) and Fossen (1994). The vehicle length is $L = 5.3$ m and the desired forward speed has been set at $u_r = 1.8326$ m/s. The comparison of the basic line of sight guidance with the line of sight guidance with heading correction along two selected paths is shown in Fig. 8. As can be seen, the more complex guidance algorithm (dashed line) gives well-balanced paths at various turns, while the

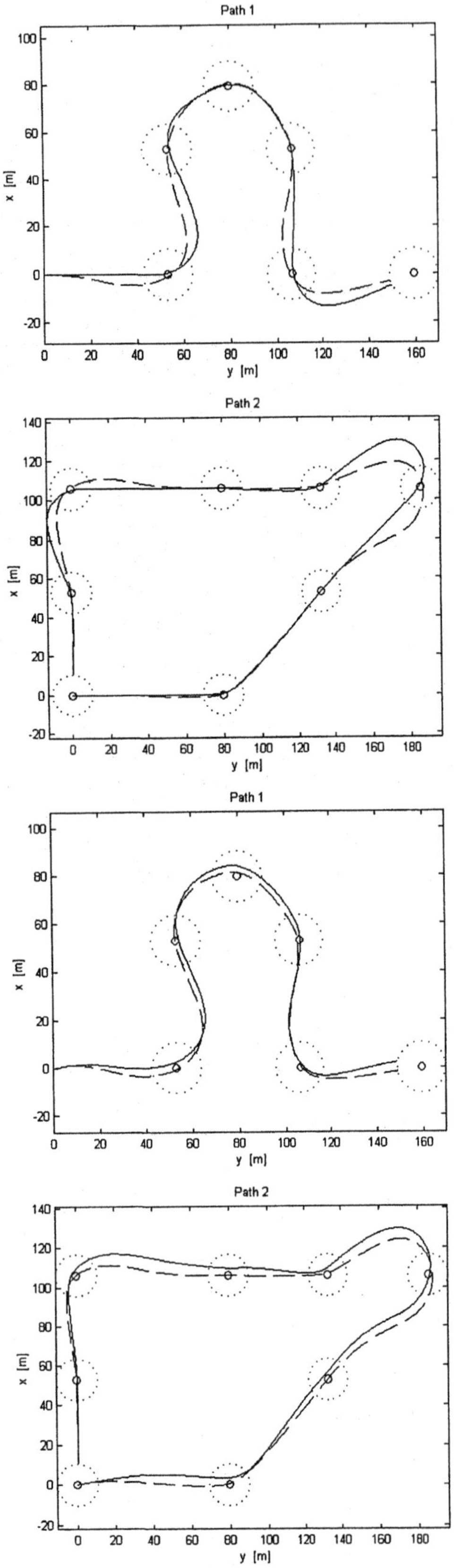


Fig. 8. Simulation results: effect of reference heading correction for smooth waypoint transition (top two panes); effect of sea current compensation (bottom two panes). In all examples the vehicle starts at the bottom left corner and goes firstly towards the right side.

simpler one (full line) guides the vehicle over unbalanced and somewhat longer trajectory at sharper turns. Fig. 8. also shows results of sea current compensation for the same paths, when the current $v_c = 0.3$ m/s , $\phi_c = 0°$ (direction along the x-axis) is introduced. Without compensation (full line) the path deviations due to current are apparent; with compensation (dashed line) they are almost completely eliminated, despite of the simplicity of the used method.

## 5. CONCLUSION

Simple enhancements presented in this paper can significantly improve the basic line-of-sight guidance algorithm through waypoints, and yet keep its original flexibility and computational simplicity. Missed waypoints are recognized and discarded by monitoring of distance changes between the vehicle and the current waypoint. Smooth transition to the next waypoint is achieved by applying a reference heading correction. Additional heading correction is used to eliminate lateral drift due to the sea current. Line of sight guidance algorithm can be also extended to include the third, diving dimension.

## REFERENCES

Aguiar, A.P. and A.M. Pascoal (2002), Way-point Tracking of Underactuated AUVs in the Presence of Ocean Currents, In: *Proc. 10th Medit. Conf. on Control and Autom. MED'02*, Lisbon, Portugal

Aicardi, M., G. Casalino, G. Indiveri, A. Aguiar, P. Encarnação and A. Pascoal (2001), A Plannar Path Following Controller for Underactuated Marine Vehicles, In: *Proc. 9th Medit. Conf. on Control and Autom. MED'01*, Dubrovnik, Croatia

Antonelli, G., S. Chiaverini, R. Finotello and R. Schiavon (2001), Real-Time Path Planning and Obstacle Avoidance for RAIS: An Autonomous Underwater Vehicle, *IEEE J. of Oceanic Eng.*, **26**, 2, pp. 216-227

Bakaric, V. (2000), *Intelligent Guidance and Control of an Unmanned Underwater Vehicle*, M. Sc. thesis, Faculty of Electrical Eng. and Computing, Zagreb (in Croatian)

Fossen, T.I. (1994), *Guidance and Control of Ocean Vehicles*, J. Wiley & Sons, Chichester

Healey, A.J. and D. Lienard (1993), Multivariable Sliding Mode Control for Autonomous Diving and Steering of Unmanned Underwater Vehicles, *IEEE J. of Oceanic Eng.*, **18**, 3, pp. 327-339

Hyland, J.C. and F.J. Taylor (1993), Mine Avoidance Techniques for Underwater Vehicles, *IEEE J. of Oceanic Eng.*, **18**, 3, pp. 340-350

Vancek, T.W. (1997), Fuzzy Guidance Controller for an Autonomous Boat, *IEEE Control Syst. Mag.*, **17**, 2, pp. 43-51

ELSEVIER
IFAC
PUBLICATIONS
www.elsevier.com/locate/ifac

# EXPERIMENTAL RESULTS ON SYNCHRONIZATION CONTROL OF SHIP RENDEZVOUS OPERATIONS

**Erik Kyrkjebø * Michiel Wondergem ****
**Kristin Y. Pettersen * Henk Nijmeijer ****

* *Dep. of Engineering Cybernetics, NTNU, Norway, E-mail: Erik.Kyrkjebo@itk.ntnu.no, Kristin.Y.Pettersen@itk.ntnu.no*
** *Dep. of Mechanical Engineering, Eindhoven Univ. of Tech., E-mail: M.Wondergem@student.tue.nl, H.Nijmeijer@tue.nl*

Abstract: The paper presents experimental results on external synchronization for underway replenishment of ships through a back-to-back comparison between experimental results and ideal simulations. The experiments illustrate exponential convergence of the closed-loop errors for position keeping, and uniform ultimate boundedness of the closed-loop errors during trajectory tracking. *Copyright ©2004 IFAC*

Keywords: Synchronization, Nonlinear control, Ship control, Observers, Tracking

## 1. INTRODUCTION

The first underway replenishment operation (UNREP) took place over a hundred years ago during an American blockade of Spanish warships outside of Cuba in 1899 (Miller and Combs, 1999), and the issue has been given much attention since in order to avoid or shorten port time for military ships. The control approaches of these rendezvous use flags and signals to communicate control commands between ships (FAS, 1999; NROTC, 2003), or some sort of tracking control of both ships to provide joint motion suitable for replenishment (Skjetne *et al.*, 2003). Kyrkjebø and Pettersen (2003) proposed to synchronize two ships based on the observed position error during operation, as opposed to using tracking control with predefined trajectories. This paper presents experimental results for this control scheme.

Synchronization is found both as a natural phenomenon in nature as in the flashing of fireflies, choruses of crickets and musical dancing, as well as the controlled synchronization of a pacemaker or a transmitter-receiver system. Systems synchronize to each other in order to coordinate their operation, and the synchronization phenomenon was early reported by Huygens (1673) who observed that a pair of pendulum clocks hanging from a lightweight beam oscillated with the same frequency. Synchronization has in the last decade attracted an increasing interest from researchers within physics, dynamical systems, circuit theory, and more lately control theory. The theory of synchronization has been applied in control problems by e.g. Blekhman (1988), Rodriguez-Angeles and Nijmeijer (2001) and Nijmeijer and Rodriguez-Angeles (2003). This paper presents experimental results on the external synchronization control scheme for underway replenishment of ships presented in Kyrkjebø and Pettersen (2003). This approach assumes that only the dynamic model of the supply ship is known. The velocities and accelerations for both ships can be estimated through nonlinear observers based on position measurements for both ships and the supply ship model. The observer-controller scheme provides semi-global exponentially stable error dynamics for position keeping, and semi-global uniform ultimate boundedness of the error dynamics during trajectory tracking.

The paper investigates the advantages, shortcomings and the practical usefulness of this synchronization control scheme using experimental data from tests with a model ship. Section 2 gives a short review of the synchronization observer-controller scheme, and in Section 3 the results are illustrated through a back-to-back comparison between ideal simulations and experimental data from a model basin. The influence of disturbances and model errors on the gain tuning is discussed in Section 4, while conclusions together with future work are presented in Section 5.

## 2. SYNCHRONIZATION CONTROL SCHEME

This section presents the synchronization control scheme, which includes an error-observer and an observer for the supply ship states. While the results presented in Kyrkjebø and Pettersen (2003) are applicable to synchronizing any two marine vehicles (ship-ship, ship-AUV, AUV-AUV), the main focus here is on the synchronization of two ships during ship rendezvous operations. The observer-controller scheme does not require information about the dynamic model of the main ship, only the dynamic model of the supply ship, and furthermore only requires position and attitude measurements for both ships. The supply ship is synchronized with the main ship through the control law, and is in fact a physical observer of the main ship states. The observer-controller scheme provides semi-global exponentially stable error dynamics for position keeping, and semi-global uniform ultimate boundedness of the error dynamics during trajectory tracking.

The general dynamic ship model in vectorial form (Fossen, 1994) can be written

$$\mathbf{M}_i(\eta_i)\ddot{\eta}_i + \mathbf{C}_i(\eta_i,\dot{\eta}_i)\dot{\eta}_i + \mathbf{D}_i(\eta_i,\dot{\eta}_i)\dot{\eta}_i + \mathbf{g}(\eta_i) = \tau_i \tag{1}$$

where $i \in s,m$ denote the supply ship and the main ship respectively. The main ship is the replenished ship receiving cargo, while the supply ship is the replenishment vessel. The matrix $\mathbf{M}$ is the matrix of inertia and added mass, and $\mathbf{D}$ the damping matrix. The damping is here assumed to be linear. $\mathbf{C}$ is the Coriolis and Centripetal matrix also including added mass effects, and represented in terms of Christoffel symbols. The vector $\mathbf{g}$ represents gravitational/buoyancy forces and moments, while $\tau$ is the vector of control torques/forces applied to the ship. The vector $\eta$ represents the Earth-fixed position and orientation of the ship. For the experiments, $\eta$ is limited to the 3 DOF manoeuvring model form (Fossen, 1994) using $\eta = [x,y,\psi] \in \mathbf{R}^2 \times SO(1)$.

### 2.1 *Synchronization objective*

The objective is to synchronize a supply ship to the actual position and velocity of the main ship, in order to transfer fuel and supplies from one ship to another at sea. This means that the supply ship is said to be synchronized to the main ship if its position/attitude and velocity coincide for all $t \geq 0$, or asymptotically for $t \to \infty$. Note that the position vector is synchronized to some offset constant reference $\eta_r$ alongside the main ship, while the velocity is synchronized to a dynamic reference $\dot{\eta}_r(t)$ that changes with the curvature of the main ship trajectory due to the difference in inner and outer curve velocity. The objective is thus to find a control law that stabilizes the error in position and velocity to zero; $(\mathbf{e},\dot{\mathbf{e}}) = (0,0)$.

### 2.2 *Synchronization rendezvous control*

*2.2.1. The synchronization controller* The synchronization controller $\tau_s$ will depend on estimated values for velocities and accelerations, and on measurements of position and attitude. The feedback control law is written as

$$\begin{aligned}\tau_s =& \mathbf{M}_s(\eta_s)\widehat{\ddot{\eta}}_m + \mathbf{C}_s\left(\eta_s,\widehat{\dot{\eta}}_s\right)\widehat{\dot{\eta}}_m \\ &+ \mathbf{D}_s\left(\eta_s,\widehat{\dot{\eta}}_s\right)\widehat{\dot{\eta}}_m + \mathbf{g}(\eta_s) - \mathbf{K}_d\widehat{\dot{\mathbf{e}}} - \mathbf{K}_p\mathbf{e}.\end{aligned} \tag{2}$$

where the synchronization errors are defined as

$$\mathbf{e} = \eta_s - \eta_m, \qquad \dot{\mathbf{e}} = \dot{\eta}_s - \dot{\eta}_m \tag{3}$$

and $\mathbf{K}_d$, $\mathbf{K}_p \in \mathbf{R}^{n\times n}$ are positive definite symmetric gain matrices.

The control law utilizes the dynamic model of the supply ship, depending on the known supply position measurements $\eta_s$ and observed velocities $\widehat{\dot{\eta}}_s$, $\widehat{\dot{\eta}}_m$ and acceleration $\widehat{\ddot{\eta}}_m$. Additional stabilizing proportional-derivative terms based upon the observed error in velocity $\widehat{\dot{\mathbf{e}}}$, and the known error in position $\mathbf{e}$, provides convergence and boundedness during replenishment. The supply ship uses the observed main ship states as reference states in the controller, and thus physically synchronizes it self to the main ship states.

Instead of using pre-calculated reference trajectories for both ships where tracking performance is vulnerable to waves, wind and currents, the reference for the supply ship is the main ship states. Hence, the performance of the operation is only dependent on the supply ship control accuracy, as opposed to the dependence of both control system when using pre-calculated trajectories.

*2.2.2. The synchronization error observer* The estimated values for the errors $\mathbf{e}$ and $\dot{\mathbf{e}}$ can be obtained through a full state nonlinear Luenberger observer

$$\begin{aligned}\tfrac{d}{dt}\widehat{\mathbf{e}} &= \widehat{\dot{\mathbf{e}}} + \mathbf{L}_{e1}\widetilde{\mathbf{e}} \\ \tfrac{d}{dt}\widehat{\dot{\mathbf{e}}} &= -\mathbf{M}_s(\eta_s)^{-1}[\mathbf{C}_s\left(\eta_s,\widehat{\dot{\eta}}_s\right)\widehat{\dot{\mathbf{e}}} \\ &\quad + \mathbf{D}_s\left(\eta_s,\widehat{\dot{\eta}}_s\right)\widehat{\dot{\mathbf{e}}} + \mathbf{K}_d\widehat{\dot{\mathbf{e}}} + \mathbf{K}_p\widehat{\mathbf{e}}] + \mathbf{L}_{e2}\widetilde{\mathbf{e}}\end{aligned} \tag{4}$$

where $\mathbf{L}_{e1}, \mathbf{L}_{e2}$ are positive definite gain matrices, and the estimated position/attitude and velocity synchronization errors are defined as

$$\widetilde{\mathbf{e}} = \mathbf{e} - \widehat{\mathbf{e}}, \qquad \widetilde{\dot{\mathbf{e}}} = \dot{\mathbf{e}} - \widehat{\dot{\mathbf{e}}}. \tag{5}$$

Note that the observers of Eqs. (4) and (5) introduces an extra correcting term in $\dot{\widetilde{\mathbf{e}}} = \widetilde{\dot{\mathbf{e}}} - \mathbf{L}_{e1}\widetilde{\mathbf{e}}$ that yields faster performance during transients, but has some negative effects on noise sensitivity.

*2.2.3. The supply ship state observer* The estimated supply ship position and velocity values $\widehat{\eta}_s$ and $\widehat{\dot{\eta}}_s$ is found using the full state nonlinear observer

$$\begin{aligned} \tfrac{d}{dt}\widehat{\eta}_s &= \widehat{\dot{\eta}}_s + \mathbf{L}_{p1}\widetilde{\eta}_s \\ \tfrac{d}{dt}\widehat{\dot{\eta}}_s &= -\mathbf{M}_s(\eta_s)^{-1}\left[\mathbf{C}_s\left(\eta_s, \widehat{\dot{\eta}}_s\right)\widehat{\dot{\mathbf{e}}}\right. \\ &\quad \left. +\mathbf{D}_s\left(\eta_s, \widehat{\dot{\eta}}_s\right)\widehat{\dot{\mathbf{e}}} + \mathbf{K}_d\widehat{\dot{\mathbf{e}}} + \mathbf{K}_p\mathbf{e}\right] + \mathbf{L}_{p2}\widetilde{\eta}_s \end{aligned} \quad (6)$$

where $\mathbf{L}_{p1}, \mathbf{L}_{p2}$ are positive definite gain matrices, and the estimated supply ship position/attitude and velocity errors are defined as

$$\widetilde{\eta}_s = \eta_s - \widehat{\eta}_s, \qquad \widetilde{\dot{\eta}}_s = \dot{\eta}_s - \widehat{\dot{\eta}}_s. \quad (7)$$

*2.2.4. The main ship state observer* The estimated main ship velocity and acceleration values $\widehat{\dot{\eta}}_m$ and $\widehat{\ddot{\eta}}_m$ are not available through direct measurement, and must be reconstructed from the position/attitude and error estimates. To compensate for the lack of a dynamic model, the velocity and acceleration values for the main ship are reconstructed based on information of the supply ship and the synchronization closed-loop system. Estimates for $\eta_m, \dot{\eta}_m$ and $\ddot{\eta}_m$ are given as

$$\begin{aligned} \widehat{\eta}_m &= \widehat{\eta}_s - \widehat{\mathbf{e}} \\ \widehat{\dot{\eta}}_m &= \widehat{\dot{\eta}}_s - \widehat{\dot{\mathbf{e}}} \\ \widehat{\ddot{\eta}}_m &= \tfrac{d}{dt}\left(\widehat{\dot{\eta}}_s - \widehat{\dot{\mathbf{e}}}\right) r \\ &= -\left(\mathbf{M}_s(\eta_s)^{-1} + \mathbf{L}_{e2}\right)\widetilde{\mathbf{e}} + \mathbf{L}_{p2}\widetilde{\eta}_s \end{aligned} \quad (8)$$

where the last relation stems from (4) and (6). The matrices $\mathbf{L}_{e1}, \mathbf{L}_{e2}, \mathbf{L}_{p1}, \mathbf{L}_{p2}$ are assumed positive definite symmetric matrices where $\mathbf{L}_{e1} = \mathbf{L}_{p1}$ and $\mathbf{L}_{e2} = \mathbf{L}_{p2}$ to simplify the stability analysis and tuning procedure.

### 2.3 Stability

The stability properties of the closed-loop error dynamics were investigated in Kyrkjebø and Pettersen (2003), and concluded with closed-loop errors being semi-global exponential convergent for position keeping, and semi-global uniformly ultimately bounded for trajectory tracking. The results are summarized in the following theorem:

*Theorem 1.* Consider the ship model (1), the controller (2) and the observers (4), (6) and (8). Under the assumption that the signals $\dot{\eta}_m(t)$ and $\ddot{\eta}_m(t)$ are bounded, i.e. that there exists bounds $\mathbf{V}_M$ and $\mathbf{A}_M$ such that

$$\begin{aligned} \sup_t \|\dot{\eta}_m(t)\| &= \mathbf{V}_M < \infty \\ \sup_t \|\ddot{\eta}_m(t)\| &= \mathbf{A}_M < \infty, \end{aligned} \quad (9)$$

and that the minimum eigenvalues of the gain matrices $\mathbf{K}_p, \mathbf{K}_d, \mathbf{L}_{p1}, \mathbf{L}_{p2}$ are chosen to satisfy a set of lower bounds, then the synchronization closed-loop error

$$\mathbf{x}^T = \left[\dot{\mathbf{e}}^T \ \mathbf{e}^T \ \widetilde{\dot{\mathbf{e}}}^T \ \widetilde{\mathbf{e}}^T \ \widetilde{\dot{\eta}}_s^T \ \widetilde{\eta}_s^T\right] \quad (10)$$

is semi-globally uniformly ultimately bounded when $(\dot{\eta}_m, \ddot{\eta}_m) \neq (\mathbf{0}, \mathbf{0})$. Furthermore, if the main ship achieve steady state $(\dot{\eta}_m, \ddot{\eta}_m) = (\mathbf{0}, \mathbf{0})$ after $t_s \geq t_0$, the synchronization closed-loop error is semi-globally exponentially convergent for $t \geq t_s$. ◇

The bounds on the velocity and acceleration of the main ship can be established based on knowledge of the desired trajectory for the main ship during replenishment, or by the limitations imposed by the maximum acceleration and velocity given by the propulsion system. The boundedness assumption of the acceleration and velocity thus has a clear physical interpretation in marine control systems.

## 3. EXPERIMENTS

Experiments were carried out to verify the theoretical results of the observer-controller scheme presented in Kyrkjebø and Pettersen (2003). In particular, the experiments aimed at obtaining an increased understanding of the proposed observer-controller scheme; its advantages and shortcomings. In order to investigate the differences between the theoretical results and practice, we present a back-to-back comparison of the experimental results with simulations under ideal conditions.

### 3.1 Experimental Setup

The experiments are carried out on the Froude scaled (1:70) model supply vessel Cybership II at the MCLab laboratory at NTNU. The length of the ship is 1.3 m and the weight is 24 kg. The ship is actuated through two rpm-controlled screws with two rudders at the stern, and an rpm-controlled tunnel bow thruster. The maximum actuated forces are 2 N in surge, 1.5 N in sway and 1.5 Nm in yaw, and the KPL thrust-allocation algorithm (Lindegaard, 2003) is used in the simulations and experiments. The position of the ship is measured with a 4 camera Proreflex motion capture system running at 15 Hz in a limited basin area of 5 m x 12 m. The basin is equipped with a DHI wave-maker system, and waves are generated using the JONSWAP distribution with a significant wave height of 0.01 m and a mean period of $T_s = 0.75$ s.

Only one model ship is available at the lab, and thus a virtual ship on a computer is playing the role of the main ship. This virtual ship is based on a theoretical ship model, and is controlled using a back-stepping controller. The limitation of only one ship implies that there is no ship interaction during the experiments, and no influence from environmental disturbances on the main ship.

The position measurement system can only measure the position of the supply vessel, and thus it is difficult to verify the velocity synchronization directly. The only states available from the experimental data are therefore $\mathbf{e}$, $\widetilde{\mathbf{e}}$, $\widetilde{\eta}_m$ and $\widetilde{\dot{\eta}}_m$.

### 3.2 Simulation Setup

The simulations serve as the ideal comparison case without modelling errors, and where no disturbances are present (no currents, wind or waves). The controller is based on a model without higher-order damping, and therefore only linear damping is included in the simulation model. The ship model is represented in the body frame as

$$\mathbf{M} = \begin{bmatrix} 25.8 & 0 & 0 \\ 0 & 33.8 & 1.0115 \\ 0 & 1.0115 & 2.76 \end{bmatrix}$$

$$\mathbf{C} = \begin{bmatrix} 0 & 0 & -33.8v-1.0115r \\ 0 & 0 & 25.8u \\ 33.8v+1.0115r & -25.8u & 0 \end{bmatrix}$$

$$\mathbf{D} = \begin{bmatrix} 0.72 & 0 & 0 \\ 0 & 0.86 & -0.11 \\ 0 & -0.11 & -0.5 \end{bmatrix}$$

### 3.3 Tuning

There is a duality in the gain tuning scheme on how to choose the $\mathbf{L}_{p2}$ gain in the observers of Equations (4) and (6); a high gain yields good velocity estimates, but also introduces measurement noise to the observed velocity, which leads to highly fluctuating control actions. A low gain results in less accurate velocity estimates, but smoother control actions.

The $\mathbf{L}_{p1}$ gain (Eq. 4 and 6) should be kept low to minimize the influence of measurement noise, but the bound on $\tilde{\eta}_s$ can be lowered by increasing the $\mathbf{K}_p$ gain (Eq. 2). The $\mathbf{K}_d$ gain is chosen to ensure that the region of attraction is large enough, and such that there is sufficient damping in the system to prevent oscillations during tracking. Thus, the semi-global validity of the scheme can be expanded by choosing the $\mathbf{K}_d$ gain larger. Note that the tuning of the controller is done in the body frame, where the gain matrices $\mathbf{K}_p$ and $\mathbf{K}_d$ are more intuitive, but are dependent on $\psi$.

### 3.4 Exponential convergence

To verify the theoretic results of exponential convergence during dynamic positioning, the main ship was held at a constant position and heading in $\eta_m = [0,-1,0]^T$, while the supply ship was synchronized to a position alongside the main ship given by $\eta_d = [0,0,0]^T$. The initial state for the supply ship was chosen as $\eta_s = \left[-1,-1.5,-\frac{\pi}{2}\right]^T$ to illustrate the convergence in all states. The same gains were used in the experiments and simulations for the observer and controller to facilitate a back-to-back comparison, and were found empirically as $\mathbf{K}_p = \text{diag}[35,15,5], \mathbf{K}_d = \text{diag}[70,40,10], \mathbf{L}_{p1} = \text{diag}[8,8,2], \mathbf{L}_{p2} = \text{diag}[10,10,10]$.

All errors are calculated and plotted in an earth-fixed North-East-Down frame. In Figure 1, the experimental results on control and observer errors are presented with plots of the transient behaviour and the steady-state. Figure 2 shows the control errors in both position and velocity during simulation. The observer

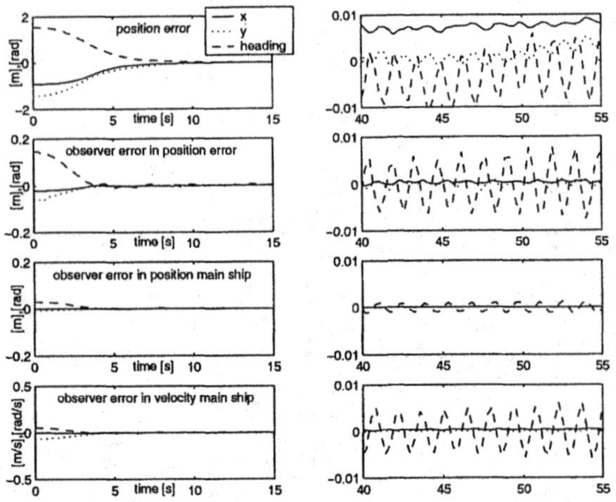


Fig. 1. Errors $\mathbf{e}$, $\tilde{\mathbf{e}}$, $\tilde{\eta}_m$ and $\tilde{\dot{\eta}}_m$ during experiment

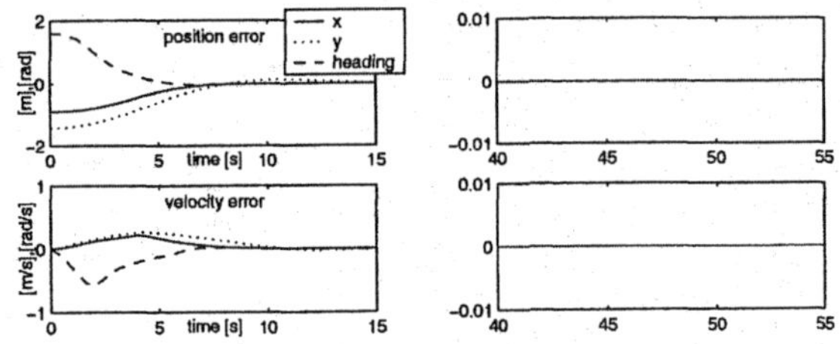


Fig. 2. Control errors $\mathbf{e}$ and $\dot{\mathbf{e}}$ during simulation

errors in the simulations converge comparable to the experiments, and are omitted from the plots due to space limitations.

The experimental results comply with the theoretical results of exponential convergence, and compare well with the simulated results. There is not much overshoot during positioning, which could otherwise have lead to dangerous situations when synchronizing to another marine structure, and this is furthermore an indication of stability margins with this set of gains for the observer-controller scheme. The settling time is sufficient for practical applications. The steady-state errors of the experiments show small persistent oscillations not found in the simulations, and this suggests that the oscillatory behaviour is caused by measurement noise influence on observer performance. The experiments were also carried out under wave disturbance, with only a slight increase in the errors as expected. In all, the experimental results for position keeping compare well with the simulations, and thus support the theoretical result of exponential convergence.

### 3.5 Ultimate boundedness

The results of ultimate boundedness of the closed-loop error during trajectory tracking were investigated using a trajectory with non-zero curvature. A trajectory with non-zero curvature is illustrative of a situation where the replenishment ships are given greater manoeuvring freedom than in a straight line experiment, and would allow a replenishment operation to be performed in close waters. The results from the non-zero curvature experiments and simulations compare well with straight-line results for the observer-controller scheme.

The forward velocity of the supply ship depends on the distance between the two ships in a curve; smaller than that of the main ship in the inner curve, and greater in the outer curve. An extreme case arises when the radius of the main ship curve is less than the distance between the two ships, where the supply ship in the inner curve would have to perform a backward movement.

The experiment shows the system behaviour during trajectory tracking. The main ship tracks a predefined curved path with a desired velocity of 0.2 m/s, corresponding to a velocity of 3.5 knots for the full scale ship. Initial states for the main ship were chosen as in Section 3.4, while the supply ship started in $\eta_s = \left[-1, 1.5, \frac{\pi}{2}\right]^T$. The same gains were used in the experiments and simulations for the observer and controller to allow for a back-to-back comparison, and were found empirically as $\mathbf{K}_p = \mathrm{diag}\,[100, 40, 10]$, $\mathbf{K}_d = \mathrm{diag}\,[30, 20, ]$, $\mathbf{L}_{p1} = \mathrm{diag}\,[8, 8, 2]$, $\mathbf{L}_{p2} = \mathrm{diag}[100, 100, 5]$.

In Figure 3 the *xy*-plot during this experiment is given. The errors during the experiment are plotted for transient behaviour and steady-state in Figure 4. In Figure 5, the control errors are shown in the body frame, while Figure 6 shows the observer errors $\tilde{e}$, $\dot{\tilde{e}}$, $\tilde{\eta}_m$ and $\dot{\tilde{\eta}}_m$ during the simulation under ideal conditions.

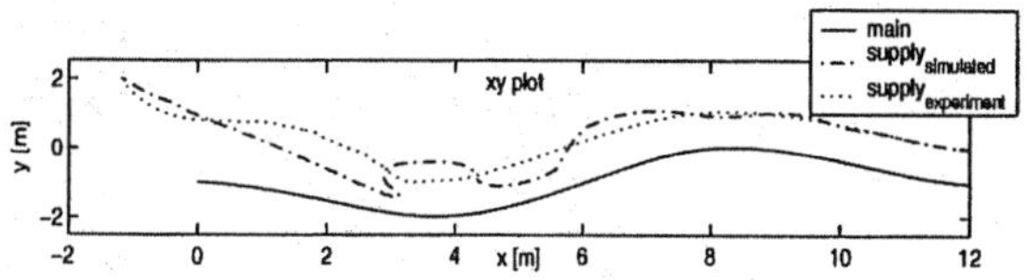


Fig. 3. *xy*-plot path

The experimental results comply with the theoretical results of ultimately boundedness of the closed-loop errors, and compare with the simulated results. Note that in the XY plot of Figure 3 and simulated results in Figs. 5 and 6, the experiments show better performance than in the simulations. This is due to the fact that the ideal simulation model of Section 3.2 is restricted to linear damping. The non-linear damping inherent in the model ship is a stabilizing effect, and thus much higher damping is needed in the controller gain $\mathbf{K}_d$ during simulations with linear damping.

There is a small bounded positive error in the *x*-position of the ship in the experiments (Fig. 4). Increasing $\mathbf{K}_p$ reduces the error, and contradicts the expected behaviour if caused by a time delay in the system. The unexpected sign of the position error still remains an unresolved issue. The observer accuracy diminishes slightly at the end of the path in Figure 4, which can be contributed to the reducing accuracy of the measurement system at the end of the basin. In all, the experimental results for trajectory tracking compare well with the simulations, and thus support

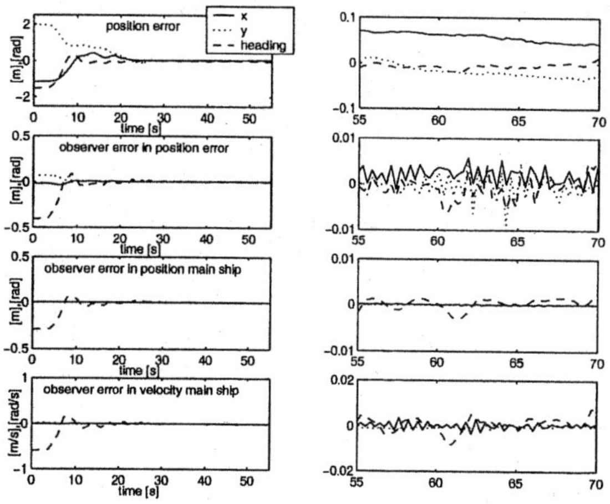


Fig. 4. Errors $\mathbf{e}$, $\tilde{\mathbf{e}}$, $\tilde{\eta}_m$ and $\dot{\tilde{\eta}}_m$ during experiment

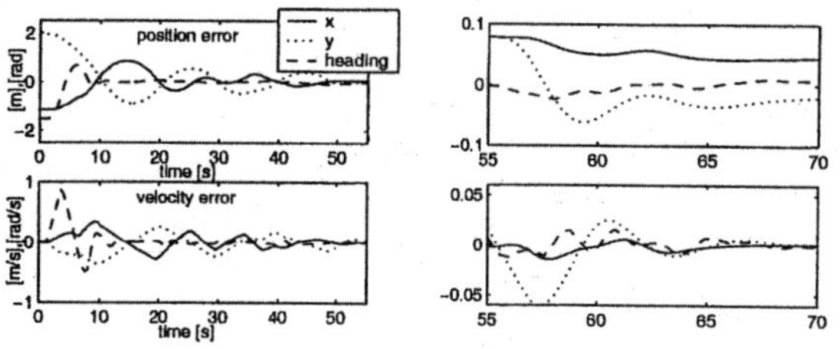


Fig. 5. Control errors $\mathbf{e}$ and $\dot{\mathbf{e}}$ during simulation

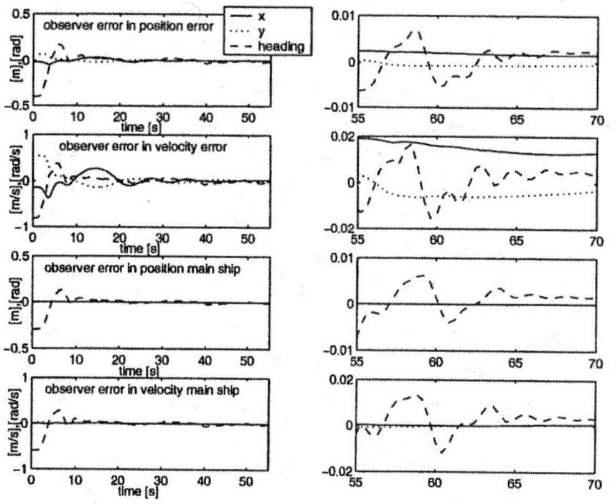


Fig. 6. Observer errors $\tilde{\mathbf{e}}$, $\dot{\tilde{\mathbf{e}}}$, $\tilde{\eta}_m$ and $\dot{\tilde{\eta}}_m$ during simulation

the theoretical result of ultimately boundedness of the closed-loop errors.

## 4. DISCUSSION

The experimental results for position keeping and trajectory tracking correspond well with the theoretical results for the observer-controller scheme as presented in Kyrkjebø and Pettersen (2003). In addition, valuable lessons are learned from the experiments in regard to the gain tuning process, the influence of measurement noise and force saturation, model errors and robustness of the scheme.

Measurement noise influence the velocity observations in the observers (with large $\mathbf{L}_{p2} = \mathbf{L}_{e2}$ in Eqs.

4 and 6), and can lead to high commanded control actions. The observer performance is affected when the commanded forces are larger than the thruster limitations, since the commanded control forces and moments are used to progress the dynamical ship model in the observer. Here, the duality of the $\mathbf{L}_{p2}$ gain is seen; a large gain may cause saturation in the forces, while a small gain may cause larger closed-loop errors.

The stabilizing nonlinear damping inherent in the model ship can be seen in the experiments of Sec.3.5, and must be compensated by increasing the $\mathbf{K}_d$ gain during simulations based on a linear damping model. The observer and controller gains $\mathbf{K}_p$, $\mathbf{K}_d$, $\mathbf{L}_{p1}$ and $\mathbf{L}_{p2}$ are optimized for either position keeping in Sec. 3.4 or trajectory tracking in Sec. 3.5, but intermediate gains that perform well for both tasks can be found. Empirical gain tuning of the observer-controller scheme is an arduous task due to the influence of observer performance on the controller performance and vice versa, and a methodic gain tuning procedure based on Nijmeijer and Rodriguez-Angeles (2003) or Section 3.3 should be adopted.

The robustness of the scheme is explored by introducing waves to the supply ship in the experiments. This does not affect the main ship, since it is a virtual ship running on a computer, and thus the results can be seen as the ability of the control scheme to suppress disturbances. The comparison between the experiments with and without disturbances during tracking is shown in Table 1 where the time-mean of the absolute error $\overline{\mathrm{E}} = \frac{1}{T}\int_0^t |\mathbf{e}|\,dt$ and the maximum of the absolute error $\mathrm{E}_{max}$ are calculated under different conditions, and the results show only small changes in performance when the supply ship is under the influence of waves.

Table 1. Mean and maximum tracking errors

| $u_d$ | $x_{body}$[m] | | $y_{body}$[m] | | heading [deg] | |
|---|---|---|---|---|---|---|
| [m/s] | mean\|e\| | max\|e\| | mean\|e\| | max\|e\| | mean\|e\| | max\|e\| |
| Without waves | | | | | | |
| 0.1 | 0.0278 | 0.0421 | 0.0029 | 0.0128 | 0.4641 | 1.2490 |
| 0.2 | 0.0548 | 0.0783 | 0.0123 | 0.0323 | 0.5214 | 2.4064 |
| 0.3 | 0.0790 | 0.1050 | 0.0367 | 0.0896 | 1.1860 | 3.7701 |
| With waves: JONSWAP $H_s$ =0.01 m, $T_s$ =0.75 s | | | | | | |
| 0.1 | 0.0293 | 0.0503 | 0.0048 | 0.0169 | 0.4412 | 1.5126 |
| 0.2 | 0.0555 | 0.0775 | 0.0146 | 0.0320 | 0.6818 | 2.2002 |
| 0.3 | 0.0790 | 0.1047 | 0.0408 | 0.0969 | 1.0600 | 4.3774 |

The robustness property is particularly useful during ship replenishment operations, where ships operating in close proximity of each other will influence each other (e.g. through Venturi-effects). Note that although the scheme is robust, it can not exceed the physical limitations of the ships. It can be seen that when the supply ship sails the outer curve with a velocity of 0.3 m/s in Table 1, the thrusters in the $y$-direction are saturated, and the errors increase.

## 5. CONCLUSIONS AND FUTURE WORK

Experimental results from tests with a model ship were back-to-back compared with ideal simulations of the synchronization based observer-controller scheme presented in Kyrkjebø and Pettersen (2003). The experimental results were found to comply well with the theoretical results of exponential convergence for position keeping and ultimate boundedness for trajectory tracking. The experiments show that the synchronization observer-controller scheme is suited for practical replenishment operations, and that the scheme is robust with respect to environmental disturbances and force saturations, and suppress the effects of model errors and measurement noise.

Future work aims at investigating the interaction effects on performance using two model ships, and to include higher order damping terms to further explore the properties of non-linear damping.

## REFERENCES


Blekhman, I.I. (1988). *Synchronization in Science and Technology*. ASME Press Translations. New York.

FAS (1999). Underway replenishment (UNREP) - navy ships. [online] http://www.fas.org/man/dod-101/sys/ship/unrep.htm. Rev: March 06 1999.

Fossen, T.I. (1994). *Guidance and Control of Ocean Vehicles*. John Wiley Ltd.

Huygens, C (1673). *Horoloquium Oscilatorium*. Paris, France.

Kyrkjebø, E. and K.Y. Pettersen (2003). Ship replenishment using synchronization control. *Proc. 6th IFAC Conference on Manoeuvring and Control of Marine Craft* pp. 286–291.

Lindegaard, Karl-Petter (2003). Acceleration Feedback in Dynamic Positioning. PhD thesis. Norwegian University of Science and Technology. Trondheim, Norway.

Miller, M.O. and J.A. Combs (1999). The next underway replenishment system. *Naval Engineers Journal* **111**(2), 45–55.

Nijmeijer, H. and A. Rodriguez-Angeles (2003). *Synchronization of Mechanical Systems*. Vol. 46. World Scientific Series on Nonlinear Science, Series A.

NROTC (2003). Underway replenishment. [online] http://www.unc.edu/depts/nrotc/classes/classinfo /NAVS52/12 Underway Replenishment.ppt. Rev: March 13, 2003.

Rodriguez-Angeles, A. and H. Nijmeijer (2001). Coordination of two robot manipulators based on position measurements only. *International Journal of Control* **74**, 1311–1323.

Skjetne, R., I.-A. F. Ihle and T. I. Fossen (2003). Formation control by synchronizing multiple maneuvering systems. In: *Proc. IFAC Conf. Manoeuvering and Control of Marine Crafts*. IFAC. Girona, Spain. pp. 280–285.

ELSEVIER
IFAC
PUBLICATIONS
www.elsevier.com/locate/ifac

# METHODOLOGY FOR DYNAMIC ANALYSIS OF OFFLOADING OPERATIONS

**Helio Mitio Morishita, Eduardo A. Tannuri, Tiago T. Bravin**

*Department of Naval Architecture and Oceanic Engineering of the University of São Paulo*

*Abstract:* The relative positioning between the Floating, Production, Storage and Offloading (FPSO) and the shuttle vessel during offloading should be analysed carefully, since the safety of the operation is of primary concern. In order to avoid collision, the shuttle vessel is kept away from the FPSO through the force of a tug-boat or by a dynamic positioning system. In both cases it is convenient to perform a preliminary study of their dynamics in order to obtain a guideline to set the best reference as well as a control approach . This paper presents a methodology for studying the dynamics and control of FPSO and shuttle vessel in tandem configuration systematically, using two software tools. The first tool allows the study of static solutions and the results have shown complex behavior, with a multiplicity of static equilibrium solutions the number and stability properties of which vary according to the combinations of the environmental conditions and vessel parameters. The second tool performs dynamic analysis of the FPSO-shuttle vessel, making it also possible to include a dynamic positioning system. Some results are compared with those obtained experimentally.


Keywords: Offloading Operation, Dynamic Positioning System, FPSO

## 1. INTRODUCTION

The FPSO systems have played an important role in the exploitation of deep-water oil in the Brazilian offshore basin. However those systems require a shuttle vessel to take the oil stored in the tanks of the FPSO to the shore from time to time. During the offloading operation the shuttle vessel and FPSO are connected to each other through a hawser in order to allow a safe oil transfer by a hose. Clearly a safe distance must be kept during this operation in order to avoid collision between the ships despite the forces due to current, wind and waves. The dynamics of the system is affected by the mooring system of the FPSO and among several alternatives, in this paper the Spread Mooring Systems (SMS) is taken into account. Regarding the tandem system both the equilibrium solutions and stability properties cannot be predicted easily since a set of twelve complex nonlinear differential equations has to be solved and their solutions depend on the relative directions and intensity of the current, wind and wave besides the parameters of the vessels (Souza Junior et al., 2000, Morishita and Souza Junior, 2001, Morishita et al., 2001a). In order to easily perform the static analysis of the system a special tool (software) was developed that solves the equations of the mathematical model numerically. This tool has allowed the study of the equilibrium points of the system systematically. Some of the fixed points obtained from the static analysis are either unstable or unacceptable (overlapping of the bodies). However convenient plot of all static solutions has been useful to understand the tendency of the stability properties

(Souza Junior and Morishita, 2002). Once the fixed points are calculated the following analysis is the dynamics of the system for which a second tool was developed.

The simulator, called Dynasim, can deal with multiple body systems, such as offloading shuttle-FPSO-monobuoy configurations, taking into account risers and mooring line effects. It can also analyze the behavior of ships equipped with Dynamic Positioning Systems (DPS). The simulator comprises several models for environmental forces (current, wind and waves), and is able to analyze 6 degrees of freedom per body. Several experimental and numerical validations have been conducted and, nowadays, Dynasim is considered an important tool for design and analysis in the Brazilian oil industry (Nishimoto et al., 2001).

Here must be emphasized that the main contribution of this methodology is to aid the selection of the shuttle tanker reference position and attitude properly. This means that keeping the ships in a safe position will require minimum force from tug-boat or less fuel consumption of the DPS.

## 2. MATHEMATICAL MODEL

Motions of the vessels in the horizontal plane are expressed in three orthogonal co-ordinate reference systems as shown in Fig. 1. The first system, OXYZ, is earth-fixed; the second and third ones, $G1x_1y_1z_1$ and $G2x_2y_2z_2$, are body-fixed in the center of gravity of the FPSO and shuttle ship, respectively. The axes of each body-fixed co-ordinate system coincide with the principal axes of inertia of the vessel. The low frequency horizontal motions of each vessel are then given by:

$$\begin{aligned}(m+m_{11})\dot{u} &= (m+m_{22})vr-(mx_g+m_{26})r^2+(m_{11}-m_{12})v_c r+X\\(m+m_{22})\dot{v} &= (m_{11}+m)ur-(mx_g+m_{26})\dot{r}+(m_{11}-m_{12})u_c r+Y\\(I_z+m_{66})\dot{r} &= -(mx_g+m_{26})(\dot{v}+ru)+N\end{aligned} \quad (1)$$

where $m$ is the mass of the vehicle; $m_{i,j}$, $i,j = 1, 2, 6$ are the added mass coefficients in surge, sway and yaw, respectively; $u$ and $v$ are the surge and sway velocities of the vehicle, respectively; $u_c$ and $v_c$ are current speeds related to GX and GY directions, respectively; $r$ is the yaw rate; $I_z$ is the moment of inertia about the GZ axis; $X$, $Y$ and $N$ represent the total external forces and moments in surge, sway and yaw directions, respectively; $x_g$ is the co-ordinate of the vessel's centre of gravity along the GX axis and the dot means time derivative of the variable. The position and heading of each vessel related to the earth-fixed co-ordinate system are obtained from the following equations:

$$\dot{x}_0=u\cos\psi-v\sin\psi\ ;\ \dot{y}_0=u\sin\psi+\cos\psi\ ;\ \dot{\psi}=r \quad (2)$$

where $\dot{x}_0$ and $\dot{y}_0$ are the components of the vessel's speed in the OX and OY axes, respectively, and $\psi$ is the vehicle heading. The components $u_c$ and $v_c$ of the current are calculated as:

$$u_c=V_c\cos(\psi_c-\psi)\ ;\ v_c=V_c\sin(\psi_c-\psi) \quad (3)$$

where $V_c$ and $\psi_c$, are the velocity and direction of the current, respectively.

The forces $X$ and $Y$, and the moment $N$ considered in this paper are due to current, wind, waves, hawser, yaw hydrodynamic damping and, in the case of the FPSO, mooring lines. Forces due to current are determined through a heuristic model based on a low aspect ratio wing theory with experimental validation (Simos et al., 2001) and the wind forces are calculated employing coefficients suggested by OCIMF (OCIMF, 1994). Forces due to waves are usually split in low and high frequency terms and the former can be considered as the sum of slow and mean drift forces. In particular, calculation of the mean drift forces considers corrections due to wave and current interaction. These forces are calculated based on sea spectra defined by their parameters, namely, significant height and period. The aero- and hydrodynamic interactions between the two vessels are not considered in this work. The forces produced by mooring lines and the hawser are calculated with catenary's equations. The high frequency components are expressed in terms of positions and attitude calculated taking into account the transfer function of the vessel and sea spectra. Therefore the total motion can be determined by adding the high frequency components to the low frequency components.

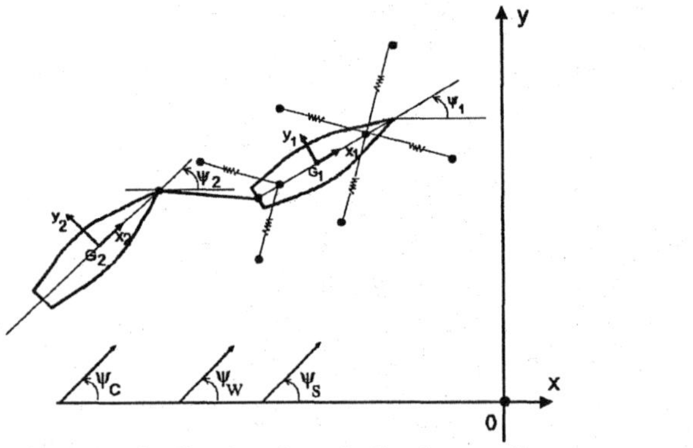


Fig. 1. Body-fixed and earth-fixed co-ordinate systems

## 3. STATIC ANALYSIS

In order to have a comprehensive knowledge of the dynamics of the system a preliminary and essential step in which the equilibria of the system are determined is required. These equilibrium solutions can be obtained by setting null all time derivative terms of the equations 1 and 2. This means the high frequency motion and slow drift forces are not considered in the static analysis. The solutions are the linear and angular positions of the vessels calculated for every set of independent parameter. These sets are defined by combining angles of incidence and speed of current and wind, angle of incidence and significant height and period of waves and the draft of the ships. Perhaps the most interesting results can be obtained by varying a specific independent parameter only within some reasonable range. For the sake of example the diagram of the equilibria of a shuttle vessel varying wind speed is shown in Fig. 2. Details of the particular of the ship are shown in Table 1 (Section 5). This picture, in which the fixed points are the heading, was assembled considering angle of incidence and speed of the current 180° and 1.0 m/s respectively and angle of incidence of the wind 20°. Furthermore, wind and wave were assumed to have the same angles of incidence, and the significant height and period of the waves were assumed to be functions of wind speed only. The offloading operation is performed by the FPSO-bow hose connection.

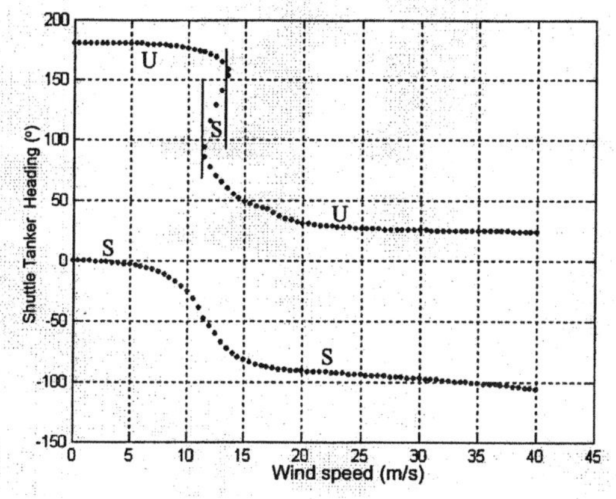


Fig. 2. Equilibrium map

Fig. 2 reveals quite interesting results since for low speed of the wind (say up to 11 m/s) there are two solutions only, being one stable and other unstable as shown in Fig. 3. For wind speed between 11 m/s and 13.5 m/s there are fold bifurcations and four solutions appear being two stable and two unstable fixed points. In Fig. 4 are shown the stable solutions for wind speed of 12.5 m/s. Increasing the speed of the wind the number of solutions is two.

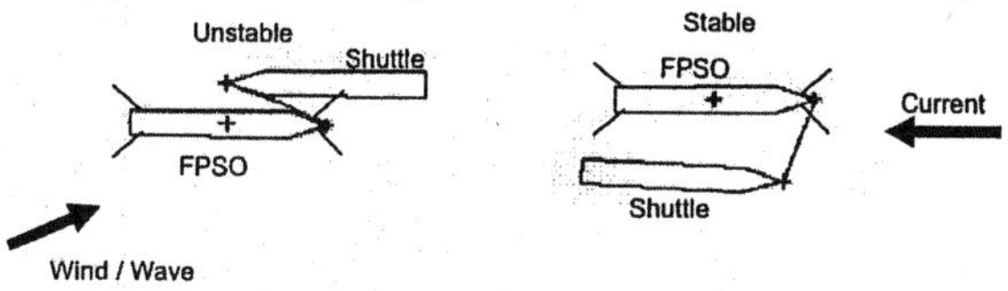


Fig. 3. Equilibrium positions for 5m/s wind speed.

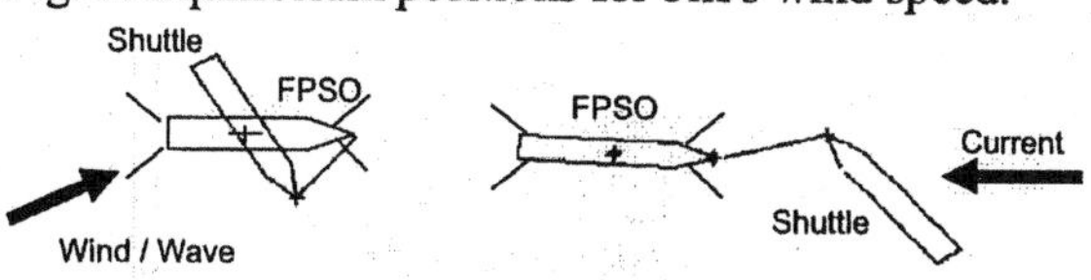


Fig. 4. Stable equilibrium positions for 12.5m/s wind speed.

Fig. 3 shows that for low wind speed the stable solution is not safe, due to the small distance between the ships. Fig. 4 reveals that for wind speed between 11,0 m/s and 13,5 m/s one of the stable solutions is unacceptable Furthermore, dynamic simulation has shown that such solution has a large base of attraction. The other stable solution, although safe, has presented a small base of attraction. Therefore in these cases, the utilization of a tug-boat or DPS is compulsory.

However, for a higher wind speed, the only stable solution is intrinsically safe. In this case additional forces can be recommended to reduce the amplitude of the oscillations or keep the positioning in case of sudden variation of the environmental condition. Here should be pointed out that the stability properties were determined through dynamic simulation.

## 4. DYNAMIC SIMULATOR

Dynasim is designed to simulate multi body systems and different positioning systems such as , mooring lines, tug boats and dynamic positioning system (DPS) can e considered according to the sort of vessel. It includes validated models for all environmental agents, including non-uniform current profile along depth, mean wind forces and gusting, unimodal or bimodal waves and interaction between current and wave (wave drift damping).

Furthermore, the simulator includes simplified models for mooring lines and riser analysis, by means of catenary's equation or using pre-defined characteristic curves of each line. The 6 degrees of freedom of each body is simulated, and the visualization can be done by means of animations, time-series plots or statistical histograms.

Concerning DPS simulator, three major blocks have been implemented, namely the wave filter, the controller and the thruster allocation, in order to represent a commercial DPS (Bray, 1998; Fossen, 1994) accurately. The wave-filter filters the high-frequency components due to first order wave forces. This filtering process is required since high frequency motions are oscillatory in nature and they should not be counteracted in order to avoid extra tear and wear of the thruster mechanism. Two types of wave filters have been considered: a conventional cascaded notch filter and a Kalman Filter. The controller determines the forces and moment required in the surge, sway and yaw directions based on the difference between reference and the filtered measurements. Dynasim allows the user to perform simulation choosing between two conventional control algorithms, namely a 3-axis uncoupled PID and a Linear Quadratic (LQ) controller. Additionally, a wind feed forward control is included. The control laws signals calculated by the controller are sent to the thrust allocation that distributes control forces among thrusters. The allocation is based on a pseudo-inverse matrix technique.

The simulator also includes models for controllable pitch propeller (cpp) and fixed pitch propellers (fpp), taking into account their characteristics curves, being able to estimate real power consumption and delivered thrust. It also evaluates time delay between command and propeller response, caused by shaft inertia (in case of fpp propellers).

More detailed description about the dynamic simulator can be found in Nishimoto et al. (2001) and Tannuri et al. (2003).

## 5. RESULTS

In this section some illustrative results concerning the utilization of the tools during the analysis of offloading operations are presented.

A real FPSO and shuttle tanker in an intermediate loading condition is considered, and their main dimensions are presented in Table 1. The FPSO is moored by a Spread Mooring System (SMS), and the simulations are performed under typical environmental conditions encountered in Campos Basin.

Table 1. FPSO and shuttle tanker properties

| Property | FPSO | Shuttle |
|---|---|---|
| Length (L) | 320.0 m | 282.0 m |
| Beam (B) | 57.3 m | 46.8 m |
| Draft (T) | 15.0 m | 9.75 m |
| Depth (D) | 29.0 m | 23.3 m |
| Mass (M) | 213000 ton | 98000 ton |

Initially, the static analysis is performed for two cases without DPS, illustrating the effect of tug-boats forces in equilibrium position of the shuttle tanker. After that, dynamical simulations are performed for the same situations, in order to predict the oscillations of the ships during the operation. For such case, it is also presented an experimental result that presented a good adherence with the simulations performed.

The second part presents the dynamical result a similar case, considering now the utilization of a DPS. The experimental comparison is also performed, and some discrepancies will be analyzed.

*Systems without DPS*

Fig.5 illustrates the first offloading condition (named condition A) analyzed. The static analysis was performed, considering wind speed varying from 0m/s to 40m/s (Figure 6). For a 19,6m/s wind, the stable equilibrium position occurs for a shuttle yaw heading equal to 120 ° approximately (related to Ox axis). At the same velocity, the unstable equilibrium occurs at -50° (Figure 7).

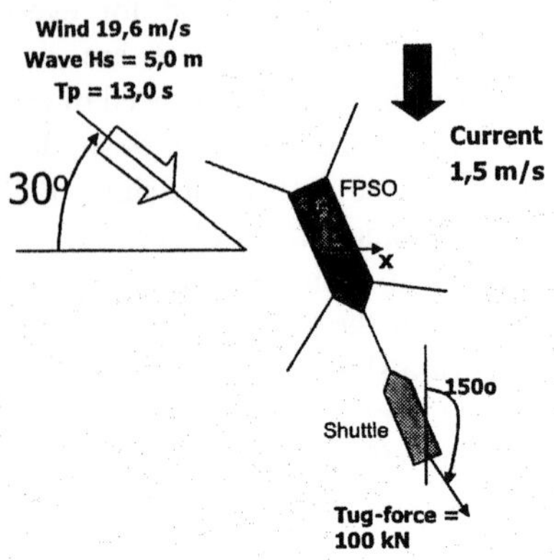


Fig. 5. Case Study A

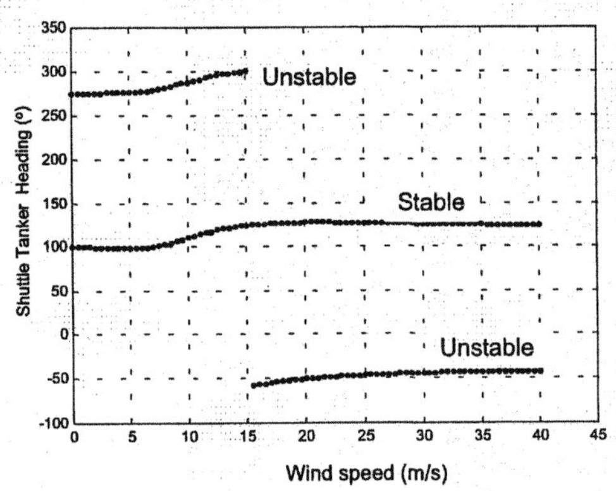


Fig. 6. Case Study A: static analysis

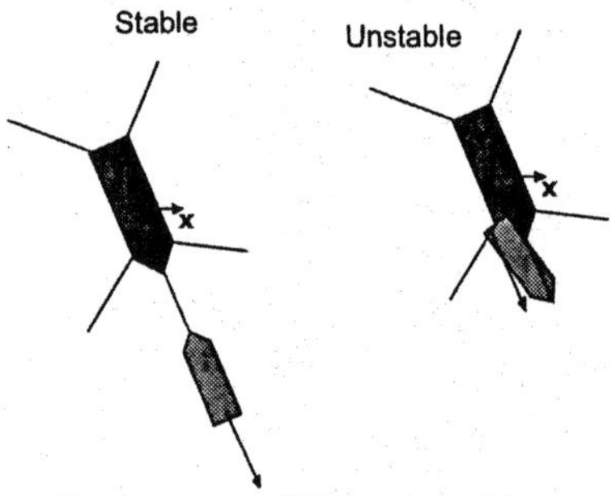


Fig. 7. Case Study A: equilibrium positions

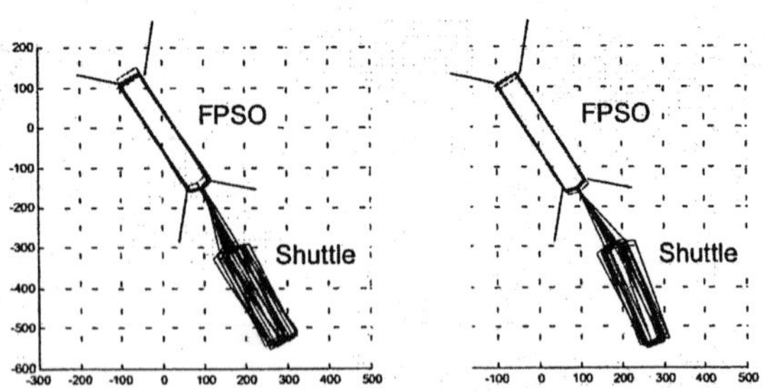


Fig. 8. Case Study A: (left) simulation result; (right) experimental result

The dynamical simulation was then performed, and the results were compared to experimental results. The towing tank tests were carried out by the supervision of Petrobras, in a certified laboratory. As can be seen in the trace plot of Figure 8 and Table 2, there is a good adherence between the simulated motion and the one measured in the experiment, with small discrepancies in three motions. The hawser tension was also compared in Figure 9, with a difference of 17% in the mean value.

Table 2. Shuttle Motion Comparison

| | X(m) | Y(m) | Yaw (°) |
|---|---|---|---|
| Experimental | 226 | -417 | 113 |
| Simulation | 220 | -419 | 118 |

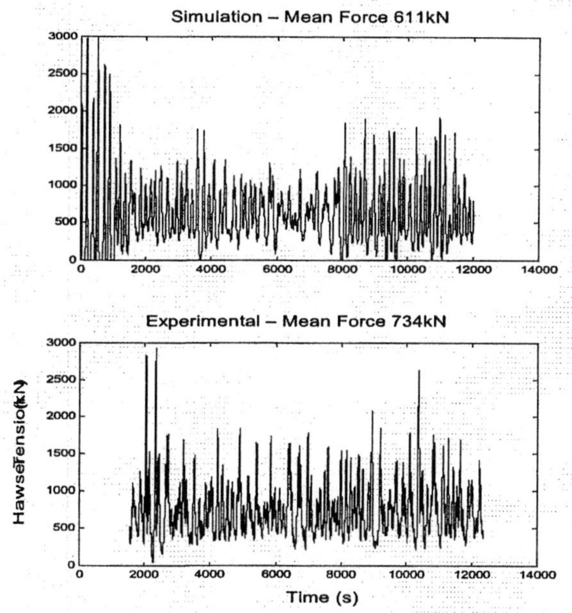


Fig. 9. Case Study A: Hawser forces

The static analysis was also performed without tug force (Figure 10). As expected, the equilibrium position does not present significant variations for wind velocities greater than 10m/s, since in those cases, the environmental forces are stronger and the influence of the 100kN tug-force decreases. Indeed, the dynamical simulation confirmed this fact, with a general behavior similar to the simulation with the tug force. Of course, even for this case the tug boat is necessary during the operation, since sudden changes in environmental conditions may induce unsafe approximations of the ships.

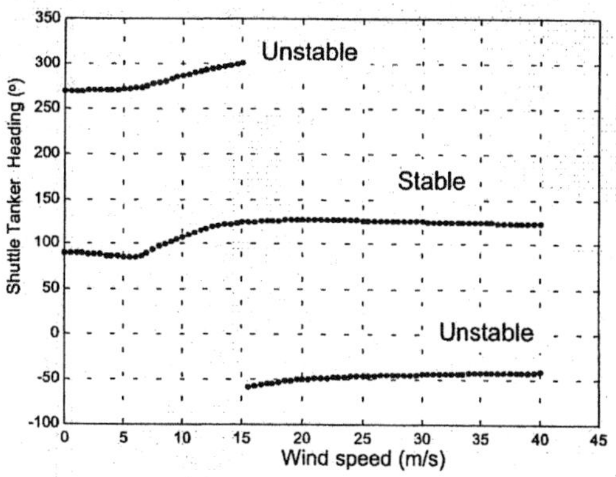


Fig. 10. Case Study A: static analysis without tug-force

The second analyzed condition (B) is presented in Figure 11. Here, it is considered the simultaneous incidence of swell and local waves, typical condition encountered in Campos Basin. The static analysis predicted the shuttle stable equilibrium angle of 104°, and it was confirmed in the dynamical simulation (Figure 12 and Table 3). Again, the experimental results presented good adherence with simulation, even in the prediction of hawser force.

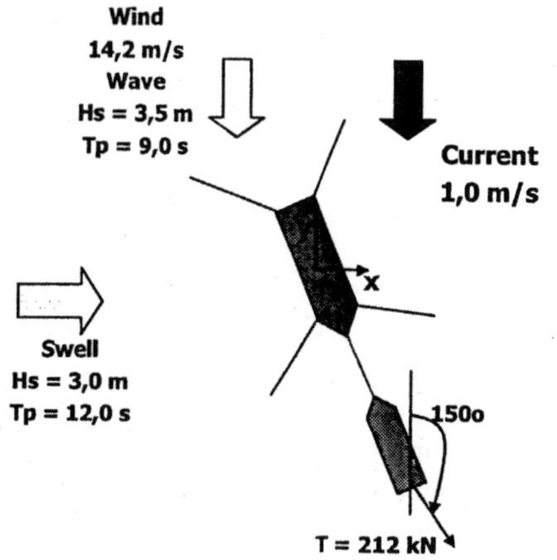


Fig. 11. Case Study B

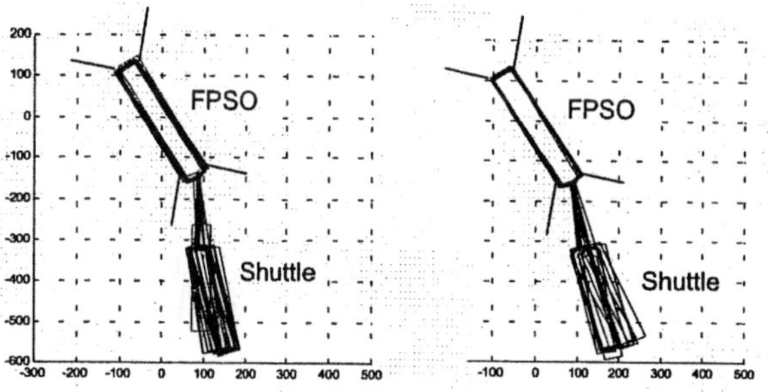


Fig. 12. Case Study B: (left) simulation result; (right) experimental result

Table3. Shuttle Motion and Hawser Force Comparison

| | X(m) | Y(m) | Yaw (°) | Hawser Force (kN) |
|---|---|---|---|---|
| Exp. | 149 | -439 | 107 | 609 |
| Simul. | 121 | -441 | 103 | 631 |

Such examples confirm that the numerical tools can be used with good accuracy to predict the behavior of passive offloading operation.

*System with DPS*

The same environmental condition of Figure 8 was tested for a DPS equipped shuttle tanker. In this particular case, the loading condition of the ships was altered, considering a ballasted shuttle tanker (M=75000ton; T=7.3m) and a full loaded FPSO (M=321000ton; T=21.2m).

The shuttle tanker is equipped with 1 main propeller, 1 tunnel stern thruster and 1 tunnel bow thruster. It is used a 3 axis uncoupled PID, associated with a cascaded notch wave filter. The set-point position is (X=125m; Y=-380m; Yaw=100°).

In Fig. 13, it can be seen that in the simulation and in the experiment, the ship was kept close to the set-point position. However, the experiment presented larger oscillations around mean position.

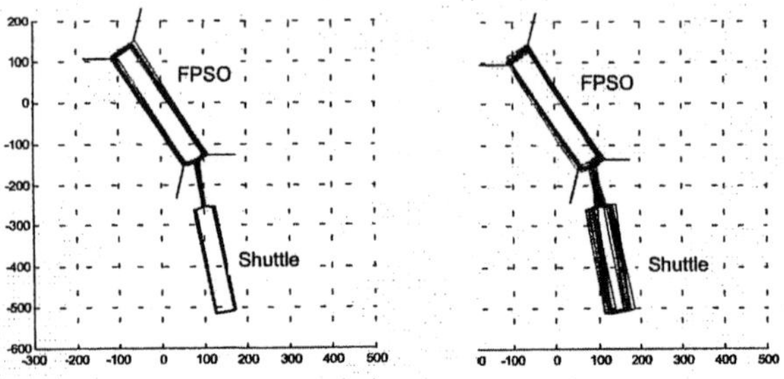


Fig. 13. Case Study C: (left) simulation result; (right) experimental result

Propellers forces are also analyzed. For example, Fig.14 contains the thrust of main propeller. It can be seen that experimental forces presents larger values than simulation results, with a discrepancy of 26% in the mean value.

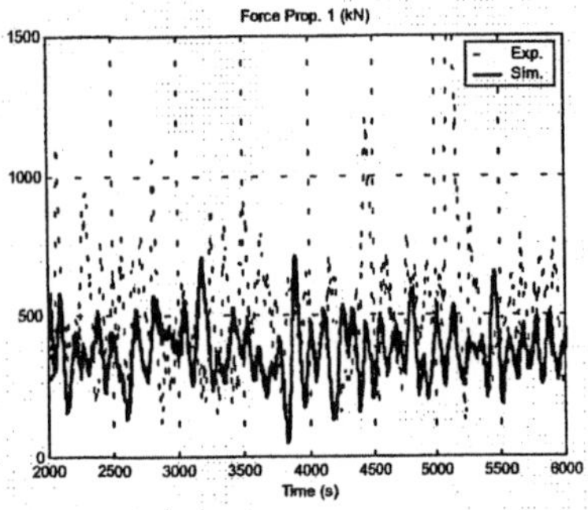


Fig. 14. Case Study C: thrust of main propeller.

Possible causes for the significant differences between experiments and simulations are still being analyzed, but it can be cited the difficulties associated with propeller forces measurement. They are estimated using the rotation of the propellers, and, in model scale, they are extremely small (approximately 1N).

## 6. CONCLUSIONS

The methodology applied in this paper to analyse the dynamics and control of offloading operation involves static and dynamical analysis. The fixed point diagrams have shown that some unexpected solution can arises depending on combinations of the environmental agents. The static solutions are then analysed in the dynamical simulator in order to predict stability properties.

In this paper illustrative examples were shown and some of then were compared with experimental results. The comparison has shown a good agreement between theoretical and experimental results confirming the accuracy of the models included in the tools except in the case of propeller forces of DPS.

## 7. ACKNOWLEDGEMENTS


The authors are grateful to the support of Petrobras, in special to Dr. Isaias Q. Masetti, due to the initial motivation and the permission to use the experimental results. The second author is grateful to FAPESP (Proc. no 02/07946-2 and 04/02402-0).


## 8. REFERENCES


Bray, D. (1998), *Dynamic Positioning,* The Oilfield Seamanship Series, Volume 9, Oilfield Publications Ltd. (OPL).

Fossen, T.I. (1994), *Guidance and Control of Ocean Vehicles*, John Wiley and Sons, Ltd

Morishita, H.M., Souza Junior, J.D.R. (2001), *Dynamical features of an autonomous two-body floating system,* Dynamical Systems and Control, FE Udwadia, and HI Weber (Eds), Gordon and Breach, London.

Morishita, H.M., Souza Junior, J.D.R., Cornet, B.J.J. (2001a), *Systematic investigation of the dynamics of a turret FPSO unit in single and Tandem Configuration*, OMAE'2001 – 20th International Conference on Offshore Mechanics and Arctic Engineering, Rio de Janeiro, Brazil.

Nishimoto, K., Fucatu, C.H., Masetti, I.Q. (2001), *Dynasim - A Time Domain Simulator of Anchored FPSO,* Proceedings of the 20th International Conference on Offshore Mechanics and Artic Engineering, OMAE, Rio de Janeiro, Brazil

OCIMF (1994), *Predictions of wind and current loads on VLCCs,* Oil Companies International Marine Forum.

Simos, A.N., Tannuri, E.A., Leite, A.J.P., Aranha, J.A.P.(2001), *A quasi-explicit hydrodynamic model for the dynamic analysis of a moored FPSO under current action,* Journal of Ship Research, Vol.45, No.4, December, pp289-301

Souza Junior, J.D.R., Morishita, H.M. (2002) *Dynamic Behavior of a Turret FPSO in Single and Tandem Configuration in Realistic Sea Environments,* OMAE'2002, 21st International Conference on Offshore Mechanics and Arctic Engineering, Norway.

Souza Junior, J.D.R., Morishita, H.M., Fernandes, C.G., and Cornet, B.J.J. (2000). *Nonlinear Dynamics and Control of a Shuttle Tanker,* Nonlinear Dynamics, Chaos, Control and Their Applications, Vol. 5, JM Balthazar, PB Gonçalves, RMFLRF Brasil, IL Caldas, and FB Rizatto (Eds), Chapter 2, pp 137-14.

Tannuri, E.A., Bravin, T.T., Pesce, C.P. (2003), *Development of a Dynamic Positioning System Simulator for Offshore Operations,* 17th International Congress of Mechanical Engineering, COBEM2003, November 10-14, São Paulo, Brazil.

ELSEVIER
IFAC
PUBLICATIONS
www.elsevier.com/locate/ifac

# COLLABORATIVE EXPLORATION FOR A TEAM OF MOBILE ROBOTS: CENTRALIZED AND DECENTRALIZED STRATEGIES

**Lucilla Giannetti, Stefano Pagnottelli, Paolo Valigi**

*Dipartimento di Ingegneria Elettronica e dell'Informazione*
*Università di Perugia*
*Via G. Duranti, 93 - 06125 Perugia – Italy*
[giannetti,pagnottelli,valigi]@diei.unipg.it

Abstract: In this paper, the problem of collaborative exploration of a wholly unknown environment by means of a team of mobile robots is studied. To allow for a quantitative comparison of different strategies, some performance indexes are introduced. Two different strategies are proposed: a decentralized strategy, based on free-market concepts, and a strategy based on the presence of a coordinating robot. The proposed strategies will be evaluated by means of simulation models. The proposed strategies are of interest also in underwater environments. *Copyright © 2004 IFAC*

Keywords: Mobile robots, multi robot exploration, robot coordination.

## 1. INTRODUCTION

The goal of underwater robotic is the design and realization of autonomous agents able to move in the depths, to recognize its position and to interact with the environment. Applications of actual interest cover, among many others, sea science (e.g., explorations concerning ocean fauna, flora or geology), *off-shore* (the industry of hydrocarbon mining), archeology and environment protection, monitoring and maintenance of deep pipelines and deep ocean telecommunication cable operations.

Multiple-vehicle operations for underwater missions using autonomous robots is expected to receive increasing attention (see, e.g, Kuroda et al. (1994); Chappell et al. (1994)): a robot team may be widely dispersed to get a high probability of detecting some interesting object (like archeological finds) or localizing the position of missing people. The coordination problem is studied also for the case of three-dimensional space motion (Smith et al. (2001)).

Working with a multi-robot system, collaborative strategies are key issues, e.g., to optimize the completion time of assigned task, and resource usage. In this case, communication among agents is a relevant aspect, which is especially crucial in underwater applications (Chappell et al. (1994)).

The collaborative exploration problem has received large attention in the literature, and several different approaches have been proposed. In Yamauchi (1999) a collaborative exploration problem is studied, for a team of mobile robots. The collaboration is based on the sharing of a single map, while goal planning for each mobile robot is carried out in a decentralized and independent manner. Each robot has its own global occupancy map, which is used to plan the local exploration. To increase the environment knowledge, robots navigate on *frontiers*, that is, regions on the boundary between unexplored and explored space.

The approach in Yamauchi (1999) is related to Yamauchi et al. (1998), where the ARIEL architecture is presented. ARIEL realizes a frontier-based exploration, and an autonomous localization, based on a continuous localization integrated with odometry. In Burgard et al. (2000, 2002) a probabilistic approach to the exploration of collaborative multiple robots is proposed. As in Yamauchi et al. (1998), the exploration is based on the concept of frontier-cell, and the selection of the goal for each robot is carried out in a coordinated manner: different frontier cells are assigned to different robots. The choice of frontier cells is based on a trade-off between the cost of reaching the target point and its utility. While the path cost is proportional to occupancy probability of crossed cells, the utility is used to coordinate robots avoiding to choose the same frontier cell. As a matter of fact, when a target point is selected for a robot, the utility of adjacent cells is reduced according to their visibility probability stated by available sensors. In Simmons et al. (2000) a central mapper is used to coordinate exploration of robot. It receives sensors information from each robot and builds a single global map based on maximum likelihood estimate: such a global map will then be transmitted to all the robots. A central coordination is adopted to assign exploration tasks to individual robots. To organize the exploration, robots sent "bid" to the central executive that tries to maximize the total expected utility of the robots by assigning them tasks. Among all "bids", the executive selects the "bid" with the highest net utility, using a simple greedy algorithm, and assigns that task to the proposer robot. Net utility is equivalent to information gain minus cost of visiting the frontier cell. While information gain is the number of unknown cells that fall within sensor range of the frontier cell, the cost is estimate computing the optimal path from the robot's current position. Whenever the best task is assigned, the bids of remaining robot are updated and the central executive repeats the procedure choosing the highest remaining net utility and assigning tasks to the relative robots until each robot have a target point to reach.

The approach proposed in Cohen (1996) is based on the use of a navigator and a set of cartographers. The cartographers move within the environment in random order. Upon detection of the target location by means of a cartographer, the navigator is notified, and it starts moving toward the target position.

The problem of task allocation and coordination among a team of mobile robots is studied in Mataric and Sukatme (2001), with extensive experiments showing the performance of the proposed schemes.

An integrated approach to the communications and navigation problem for Autonomous Underwater Vehicles (AUVs) has been studied in Singh et al. (1996) and Chappell et al. (1994), while Williams et al. (2000) a solution is proposed to the SLAM problem for underwater autonomous vehicles, which is further discussed, in a general setting, in Williams et al. (2002). The exploration problem is quite often studied together with the localization one, in the framework of multi-robot SLAM (see, among others, Liu and Thrun (2003) and references therein), and in other approaches, such as Markov localization (see, among others, Fox et al. (2000) and references therein). A relevant issue addressed in underwater application is distributed control, studied in Stilwell (2002) for team of robots.

In this paper, the exploration problem will be addressed, for a team of autonomous robots as well as the simultaneous exploration and localization problem. The major contributions of the paper are a discussion of the problem, a set of performance indexes to numerically evaluate alternative strategies, a heuristic solutions to the problem, and, finally, simulation results to compare these two strategies.

## 2. PROBLEM FORMULATION

In this papers is studied a collaborative exploration problem for a team of $N$ autonomous robots. In particular, we consider the case of a group of mobiles whose task is the exploration (and map building) of an unknown, unstructured environment. We seek for collaborative solutions to the above problem, assuming a team of identical robots. To carry out in a satisfactory manner the exploration task, the robots have also to solve a localization problem. In this paper, the focus is on the exploration problem alone, therefore assuming each robot can perfectly locate itself. The collaborative localization problem is subject of on going research, as well as the cooperative simultaneous localization and mapping one. The reference scenario for the exploration problem considered here comprises an unknown flat environment, containing fixed obstacles placed at unknown locations. The exploration task consists in scanning the whole environment, identifying all the obstacles and therefore building a complete map. To cope with this problem, the environment is cell-decomposed, according to a regular grid map. Each cell is identified by the integer $(x, y)$ coordinate of its center. Each robot is assumed to be equipped with a suitable sensing system allowing the exploration of the set $\mathcal{A}(x,y)$ containing all the cells adjacent to his current $(x,y)$ location, i.e., all the cells that can be reached by one step move along the $x$ and/or $y$ coordinate: $\mathcal{A}(x,y) = \{(\bar{x},\bar{y}) : \bar{x} = x \pm 1, \bar{y} = y \pm 1\}$. Cells along the border of the environment have reduced adjacent set. A cell $(x,y)$ is termed *visited* whenever at least one robot went through the cell, whereas a cell $(x,y)$ is termed *explored* whenever at least one robot visited a cell in its adjacent set.

To compare different exploration strategies, we introduced some performance indexes, measuring the redundancy of exploration, the number of visited cell, and the total exploration time. To formally introduce the problem, let $\mathcal{X}$ denote the environment to be explored, and let $\mathcal{E}_i$ and $\mathcal{V}_i$ denote the portion of environment explored and visited, respectively, by the $i$-th robot, $i = 1,\ldots,N$. Then, the condition under which the exploration problem is solved is given by:

$$\mathcal{X} = \cup_{i=1}^{n} \mathcal{E}_i, \tag{1}$$

which implies that the whole environment has to be explored. The performance of strategies aimed at solving the above problem can be evaluated by means of the following indices. Let $\Omega_{ij} = \mathcal{E}_i \cap \mathcal{E}_j$, $\forall i,j = 1,\ldots,N$ be the portion of environment (i.e., the set of cells) explored by both the $i$-th and the $j$-th robot. A good exploration algorithm minimizes

$$p_1 = \cup_{i,j} |\Omega_{ij}|, \tag{2}$$

where $|\Omega|$ denote the number of cell in set $\Omega$. Let $\Lambda_{ij} = \mathcal{V}_i \cap \mathcal{V}_j$, $\forall i,j = 1,\ldots,N$, be set of cells visited by both the $i$-th and the $j$-th robot. A second performance index is

$$p_2 = \cup_{i,j} |\Lambda_{ij}|. \tag{3}$$

Finally, let $\tau_i$ be the exploration time of the $i$-th robot. The time required to explore the whole environment is then given by:

$$p_3 = \max_i \tau_i. \tag{4}$$

## 3. EXPLORATION STRATEGIES

The exploration problem will be tackled by means of two heuristic strategies. The first one is based on a decentralized approach, based on free-market concepts. The second strategy, on the contrary, is based on the key role of a coordinator, taking decision affecting the behavior of the whole team.

### *3.1 Free-market collaborative exploration strategy*

In the free-market policy for collaborative exploration, the team of robots is not hierarchically structured, instead every robot has the same competence and ability. The exchange of information and the subsequent path planning process are based on the so called *free-market* concept, as in Dias and Stentz (2000). The free-market approach defines revenue and cost functions across the possible plans for executing a specified task. Free-market approach does not use centralized planning: robots are free to exchange information and services and enter into contracts among themselves, as they see fit.

Here the use of a cost function $\mathcal{C}$ is proposed, based on distance: the waiting cost is the distance between the current location of a robot and the goal location.

The revenue function $\mathcal{R}$ is based on the growth of information that the visit of the cell would determine. To represent the state of the system, for each step of exploration, an occupancy map is used. A codification is used to mark a cell like empty, occupied by obstacles, occupied by other robot or unexplored. The gain of exploration is determined by the growth of information risen by the visit of a cell, in particular we use the number of unexplored cells adjacent to the current position.

The utility of each cell is calculated as the different between cost and revenue. We did a weight mean, to give advantage to cells with a lower revenue together with limited costs. The goal of each robot is to explore a larger amount of cells with less cost.

The first question, which must be answered to design a satisfactory team, is the strategy to select the target cell. In this paper we propose a *greedy strategy*, by which the cell, which is in the center of the nearest unexplored zone, is chosen as the target point. The rationale for this choice is that this method is simpler to implement and easier to manage with respect to other methods with far away target cells. After target cell selection, the assignment of a cost is needed, to quantify revenue after visit that cell. Furthermore, it's important to take care to choose different target for each robot, to avoid collision. With a decentralized team the problem has to be faced by the local agents near the contested cell, with the choice of "local" priority.

Two rules are used in this paper:

**rule 1.** The cell at the intersection between two trajectories is reserved on a FIFO basis.

**rule 2.** The transit through the same cell is allowed one robot at a time.

### 3.2 *Coordinated collaborative exploration strategy*

The centralized strategies is based on the assumption that the team of mobile robots comprises a coordinator, which is fixed (e.g., in the origin of the global reference frame), and a number of mobile agent. The agents can alternate between two roles: the role of *explorer* and the role of *marker.* The explorers moves around the environment, according to navigation policies that will be described subsequently, while the markers allow for the explorers to localize based on measurements of relative position and orientation. The coordinator also plays the role of marker. Localization is based on triangulation, and makes use of observability results (Conticelli et al. (2000)) and a related observability index. Thus, the exploration problem is here integrated with the localization one (see Leonard and Smith. (1997); Carpenter and Medeiros (2001) for issues concerning localization in underwater environment, and Chappell et al. (1994) for issues concerning inter-robot communication).

Upon the occurrence of *role-assignment events,* (among which also the completion of an exploration period of predefined length), each robot undergoes a verification, and role re-assignment may be carried out for pair of robots. Within two such events, exploration proceeds according to a centralized policy.

The destination cell for each walker, which is always a frontier cell (see Yamauchi (1999)), is computed on the basis of cost (namely, distance from current location) and revenues (namely, number of un-explored cell encountered along the path) associated to each possible choice. The revenue of each cell takes into account the increase in information associated with a visit to the cell. The cost of each frontier cell is computed via a slightly modified version of the $A^*$ algorithm.

Simulation experience shows that the best choice for the destination to assign each walker is a frontier cell with an average value for the revenue. As a matter of facts, selecting the cell with the highest revenue will increase the probability of neglecting small sets of unexplored cells: turning back to these sets later on would increase overall completion time. Similarly, selection among destination cells with the same average revenue is carried out in order to keep high the distance among explorers. This is similar to the utility concept in Burgard et al. (2000). Additional heuristic rules are used to deal with potential collision among explorers (based on priority concepts), with specific configuration of partly identified obstacles, and other situations of interest.

## 4. PERFORMANCE EVALUATION

The exploration strategies proposed in the previous sections have been studied by means of simulation models. The simulation tool is Matlab-based, and allows for easy implementation of exploration strategies, together with the computation of the performance indexes. The tool also allow for an easy definition of the features of the environment to be explored, such as obstacle positions and dimensions. An alternative simulation tool is under consideration, based on Vaughan (2002), which is a public domain application.

The behavior of the decentralized strategy is described in the following Figures 1 through 3, where the evolution of the exploration process is illustrated. In all the figures, obstacles are in black, while in Figure 3, the lines indicate the actual paths covered by the three robots. The scenario is characterized by sparse obstacles, and the exploration is carried out by a team comprising three robots. As the exploration proceeds, the three robots distribute over the region in an autonomous manner, exploring different portions of the environment.

By following in detail the evolution of the exploration task, it is possible to appreciate the cooperation among robots, namely each robot directs himself toward non-explored regions, and remains away from other robots, thus improving the efficiency of the strategy. Simulation of other environments, such as a space with wall-like barriers, hence characterized by a number of sub-regions, shows that each robot of the team moves toward a different sub-region.

The characterization of the strategy in terms of completion time indicate that the whole is explored within 103 time unit, with a moderate value of index $p_1$ (i.e., cells explored by more than one robot) and a quite negligible value for $p_2$ (i.e., cells visited more than once).

The centralized strategy has been tested by using the same scenario considered above, i.e., an open space, with sparse obstacles. The exploration is carried out by using a team comprising four robots, two of them acting as explorers and the other as markers, according to the policy proposed in the previous section. The exploration progress is illustrated in Figure 4, where the circles indicate the robot acting as markers, while the triangles indicate robots acting as explorers.

The performance indexes for the centralized strategy are reported in the following table, for the environment considered above, and for additional environments. Notice that environments with complex obstacles yield a strong increment in the indexes $p_2$ e $p_3$.

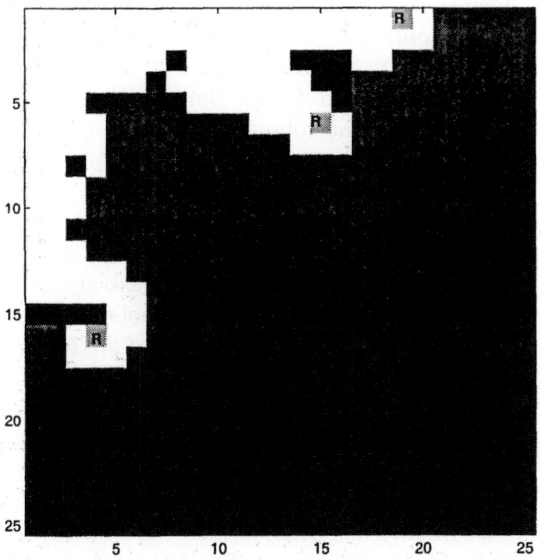


Fig. 1. Decentralized strategy: exploration in progress

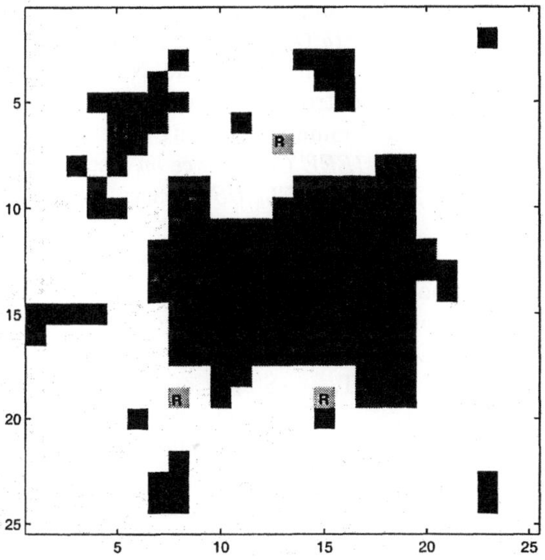


Fig. 2. Decentralized strategy: exploration in progress

| Obstacle configuration | $p_1$ | $p_2$ | $p_3$ |
|---|---|---|---|
| Sparse obstacles | 4 | 10 | 138 |
| Free space | 5 | 6 | 127 |
| Room space | 6 | 9 | 142 |
| Corridors | 1 | 35 | 160 |

Table 1. Comparing performance indexes

Notice that, in terms of completion time (index $p_3$), the centralized policy appears worst. This is due to the fact that in such a policy robots play different roles, and therefore the actual number of explorers is only two. In addition, at role interchange, a robot switching from marker to explorer state has to travel within an explored region, thus giving no contribution to exploration, and increasing both index $p_1$ and $p_2$. On the contrary, such

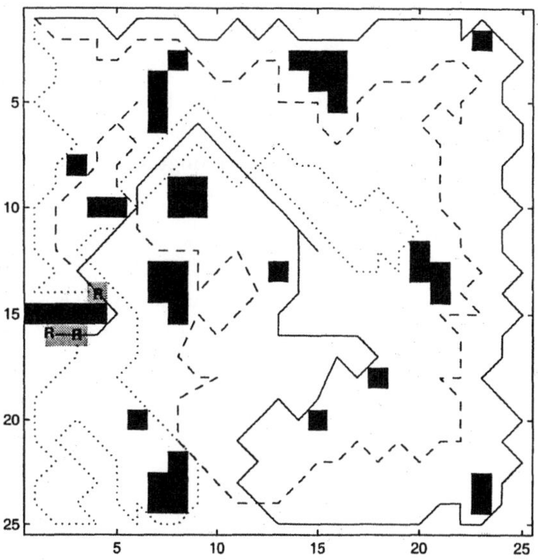


Fig. 3. Decentralized strategy: exploration completed

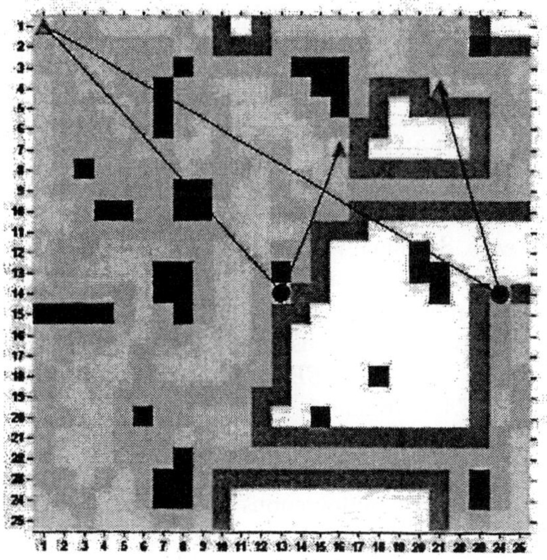

Fig. 4. Centralized strategy in progress

a centralized policy has the quite crucial advantage that localization is more efficient, due to the presence of marker robots. This is of major importance in "first-time-explored" environments, such as underwater ones.

## 5. CONCLUSIONS

Two heuristic strategies have been proposed, to tackle the problem of collaborative exploration for unknown environment, by means of a team of mobile robots. The first strategy is decentralized, and uses free-market concepts to allow robot to negotiate for exploration. The second strategy is based on the presence of a coordinator, and on a team of robots that may behave as markers or explorers, according to coordinator indications.

Simulation results illustrate the performance of the two strategies. Current research activity is considering a formal statement of the collaborative exploration problem, and other exploration policies. The design and realization of a team of mobile robots is an additional ongoing activity (of interest in indoor environments).

**Acknowledgment** This work has been supported by ASI under project T.E.M.A., contract I/R/124/02. The authors thanks Dr. Agostino Miele, who contributed to the discussion on the free-market strategy, and Dr. Massimo Apostolico, who contributed to the discussion on the coordinated strategy.

## REFERENCES


Burgard, W., M. Moors, D. Fox, R. Simmons and S. Thrun. Collaborative multi-robot exploration. In *Proceeding of the 2000 IEEE International Conference on Robotic and Automation San Francisco, CA*, April 2000.

Burgard, W., M. Moors, and F. Schneider. Collaborative exploration of unknown environments with teams of mobile robots. In M. Beetz, J. Hertzberg, M. Ghallab and M.E. Pollack, editors, *Plan-Based Control of Robotic Agents*, volume 2466 of *Lecture Notes in Computer Science*. Springer Verlag, 2002.

Carpenter, R.N. and M.R. Medeiros. Concurrent mapping and localization and map matching on autonomous underwater vehicles. In *Proc. of MTS/IEEE Conference and Exhibition OCEAN*, volume 1, pages 380–389, 2001.

Chappell, S.G., J.C. Jalbert, P. Pietryka and J. Duchesney. Acoustic communication between two autonomous underwater vehicles. In *Proceedings of the 1994 Symposium on Autonomous Underwater Vehicle Technology*, pages 462–469, 1994.

Cohen. W. Adaptive mapping and navigation by teams of simple robots. *Journal of Robotics and Autonomous Systems*, 18:411–434, 1996.

Conticelli, F., A. Bicchi and A. Balestrino. Observability and nonlinear observers for mobile robot localization. In *In IFAC Int. Symp. on Robot Control, SyRoCo 2000, Wien*, 2000.

Dias, M. B. and A. Stentz. A free market architecture for distributed control of a multirobot system. In *In Sixth Int'l Conf. on Intelligent Autonomous Systems, Venice, Italy*, pages 115–122, 2000.

Fox, D., W. Burgard, H. Kruppa and S. Thrun. A probabilistic approach to collaborative multi-robot localization. *In Special Issue of Autonomous Systems on Multi-Robot Systems*, 2000.

Kuroda, Y., T. Ura and K Aramaki. Vehicle control architecture for operating multiple vehicles. In *Proceedings of the 1994 Symposium on Autonomous Underwater Vehicle Technology*, pages 323 – 329, 1994.

Leonard, J. J. and C. M. Smith. Sensor data fusion in marine robotics. In *Proc. of the International Society of Offshore and Polar Engineering, Honolulu, HI,*, pages 100–106, 1997.

Liu Y. and Sebastian Thrun. Gaussian multi-robot SLAM. *Sumitted to NIPS 2003*, 2003.

Mataric, M.J. and G. Sukatme. Task-allocation and coordination of multiple robots for planetary exploration. In *Proceedings of the International Conference on Advanced Robotics*, 2001.

Simmons, R., D. Apfelbaum, W. Burgard, D. Fox, M. Moors, S. Thrun and H. Younes. Coordination for multi-robot exploration and mapping. In *Proceedings National Conference on Artificial Intelligence, Austin TX*, August 2000.

Singh, H., J. Catipovic, R. Eastwood, L. Freitag, H. Henriksen, F. Hover, D. Yoerger, J.G. Bellingham and B.A. Moran. An integrated approach to multiple AUV communications, navigation and docking. In *Proc. IEEE Oceanic Engineering Society OCEANS*, pages 59–94, 1996.

Smith, T.R, H. Hanßmann, and N.E. Leonard. Orientation control of multiple underwater vehicles with symmetry-breaking potentials. In *Proc.of 40th IEEE Conference on Decision and Control, Orlando, FL, USA*, pages 4598–4603, 2001.

Stilwell. D. J. Decentralized control synthesis for a platoon of autonomous vehicles. In *IEEE Int. Conf. on Robotics and Automation, Washington, DC*, pages 744–749, 2002.

Vaughan, R. T. Stage a multiple robot simulator. Technical report, Institute for Robotics and Intelligent Systems, `http//playerstage.sourceforge.net`, Technical Report IRIS-00-393, University of Southern California, 2002.

Williams, S. B. G. Dissanayake and H. Durrant-Whyte. Autonomous underwater simultaneous localisation and map building. In *IEEE Int. Conf. on Robotics and Automation*, volume 2, pages 1793–1798, 2000.

Williams, S. B. G. Dissanayake and H. Durrant-Whyte. An efficient approach to the simultaneous localisation and mapping problem. In *IEEE Int. Conf. on Robotics and Automation, Washington, DC*, volume 2, pages 2743–2748, 2002.

Yamauchi, B. Decentralized coordination for multirobot exploration. *Robotics and Autonomous Systems*, 29:111–118, 1999.

Yamauchi, B., A. Schultz, and W. Adams. Mobile robot exploration and map-building with continuous localization. In *Proceedings of the 1998 IEEE International Conference on Robotics and Automation Leuven, Belgium*, May 1998.

ELSEVIER
IFAC
PUBLICATIONS
www.elsevier.com/locate/ifac

# DESIGN AND REALIZATION OF A VERY LOW COST PROTOTYPAL AUTONOMOUS VEHICLE FOR COASTAL OCENOGRAPHIC MISSIONS

**A.Alvarez*, A. Caffaz**, A. Caiti***, G. Casalino***, E. Clerici**, F. Giorgi**, L. Gualdesi****, A. Turetta*****

** IMEDEA – Instituto Mediterraneo de Estudios Avanzados*
*C/Miguel Marques 21, 07190 Esporlas, Spain*
*** GRAALTECH s.r.l. – via Gropallo 4/10, 16122 Genova, Italy*
****ISME – Interuniversity Ctr. Integrated Systems for the Marine Environment*
*c/o DIST, University of Genova, via Opera Pia 13, 16145 Genova, Italy*
***** SACLANT Undersea Research Ctr., viale San Bartolomeo 400, 19138 La Spezia, Italy*

Abstract: The design and main features of a very low cost underwater vehicle for coastal oceanographic applications are illustrated. The key to a low cost but still effective design, in this case, is the limitation of the vehicle features and capabilities solely to those strictly required for the fullfillment of the mission. In terms of navigation and control capabilities, in particular, this allows for the use of GPS-driven algorithms when transiting from one sampling location to another, and the use of a robust closed-loop steering law automatically generating smooth transit trajectories. *Copyright © 2004 IFAC*

Keywords: Autonomous, vehicle, control.

## 1. INTRODUCTION

The last years have seen a considerable increase in field experimentations and operational applications of Autonomous Underwater Vehicles (AUVs). Despite many success stories, though, AUVs have not yet reached the technological maturity and the widespread diffusion enjoyed by their predecessors, i.e. Remotely Operated Vehicles (ROVs). Many explanations can be offered, including the obvious observation that AUVs, as a more recent development, cannot be expected to immediatley replace the market section of ROVs. It is the authors opinion, though, that two severe limitations in the diffusion of AUVs in operational scenarios are those of vehicle cost and user-friendliness. The great majority of existing AUVs have been originated directly from research prototypes, whose development was led by the general research interest of increasing the vehicle autonomy and not by specific mission needs; the absence of a well defined mission goal at the design stage has favoured prototypes able to account for several potential missions, eventually leading to conservative design choices, as for navigational capabilities, depth ranges, allowable payloads, energy consumption, safety systems. For the same reasons, AUV operation still requires in each mission a team of trained engineers, knowledgeable with the system design, in stark contrast with ROV operations. It has to be clear that the research on challenging designs and difficult tasks has favoured the considerable advancements in our understanding of underwater vehicle systems. This increase in knoweldge and in technological capabilities can now be employed to the design of special purpose vehicles, where the specific mission guides the design. It is believed that relevant reductions in cost and in operational complexity can be achieved by this approach. In the work presented in this paper, it is illustrated the design of an autonomous vehicle to be employed as a sensor platform in oceanographic cruises devoted to the investigation of ocean mesoscale motions in shallow coastal waters. Focusing on the mission has allowed to come out with a very simple and yet effective design: the resulting vehicle exhibits similarities with autonomous gliders together with some standard AUV features. In particular, the vehicle can carry a CTD package as payload; it navigates on the sea surface when in transit from one measuring station to another, and it submerges vertically when on station to perform the measurement. When on the surface, the vehicle has continuous GPS contact and land-station contact through a dedicated mobile phone link. The land station link allows for on-line modification of the mission requirements and for almost real-time data transmission. The control strategy first introduced in (Aicardi et al., 1995), and discussed in the underwater vehicle case in (Aicadi et al., 2001a, Aicardi et al., 2001b), has been implemented, accounting also for the effects of current disturbances and actuators saturation. The proposed law guarantees both robustness in the stable reaching of a target point and smoothness in the resulting trajectory from the starting point. The paper is organized as follows: in the next section, the mission objectives are discussed, in relation with other autonomous platforms described in literature. In section 3 the resulting vehicle design is reported. In the fourth section, the guidance and control laws implemented are reviewed. In section 5, some illustrative cases from the preliminary experimental trials in the bay of La Spezia, Italy, are reported. Finally conclusions are given.

## 2. BACKGROUND: OCEANOGRAPHIC MISSION IN COASTAL WATERS

Mesoscale ocean motions are characterized by length scales ranging from 100 m up to 10 km and time scales from hours up to several days. In shallow coastal environments these scales are even reduced due to the strong interactions among its physical and chemical processes and its biological population. Spatial and temporal resolution of ocean observations depends on the observing platform employed. Oceanographic observations of coastal areas have been traditionally carried out by oceanographic ships and moorings. Both systems, though, lack the required spatiotemporal resolution. For this reason new ocean observing platforms able to carry out ocean measurements at higher spatial and temporal resolutions are being experimented. These platforms include gliders (Stommel, 1989), and AUVs or AUV platoons (Schmidt et al., 1996). Gliders are autonomous vehicles designed to observe the interior of vast ocean areas over periods up to the scale of months. Structurally, gliders employ their hydrodynamic shape and small fins to induce horizontal motions while controlling their buoyancy. Besides, buoyancy control allows gliders vertical motions in the water column. In summary, changing buoyancy together with the hydrodynamic structure allow gliders to carry out zig-zag motions between the ocean surface and bottom with a net horizontal displacement. Positioning is obtained by GPS when the glider is at surface, where data transmission to the laboratory is also done. In general, the instrumentation integrated in a glider is limited by the electrical consumption and the hydrodynamic drag. Presently, gliders transport conductivity, temperature and depth (CTD) sensors with similar accuracy to those employed from oceanographic ships. The employment of gliders to sample coastal areas is quite limited due to the strong energetic processes occurring at these environments. A more adequate ocean observing platforms in these areas are the AUVs. Their hydrodynamic shape, electrical propulsion, submarine navigation and positioning allows continuous sampling of oceanographic conditions. However, acquisition of AUVs is still prohibitive for many environmental agencies: as pointed out in (Gadre et al., 2003), the complexity and cost of deployement, in particular of platoons of AUVs, is a severe limiting factor to experimental activities. Programs for development of small, easy-to-operate, AUVs, have been succesfully developed (Allen et al., 1997), while the low cost of equipment and associated infrastructure, though claimed, has still to be attained. (Gadre et al., 2003) reports on the development of miniature AUVs for experimental purposes in which the part material cost has been effectively reduced, to reach a cost of less than \$ 3000 in parts, with in-house development of most of the elctronics and mechanical components. The design of the prototype described in the present paper has also been succesfull in limiting the part costs to a comparable figure (€ 6000) using exclusively commercially off-the-shelf (COTS) components. The use of COTS components clearly allows for a much easier and reliable transition towards commercial realizations. The design presented in the following has focused on the observation that a vehicle performing mesoscale ocean observation in shallow coastal waters has to exhibit a mixture of gliders and AUVs capabilities. In particular, the vehicle can navigate as a glider, with a GPS link from surface, and it can navigate for most of the time at the sea surface; it has to dive vertically, to collect vertical CTD (Conductivity-Temperature-Depth) profiles, when on a prescribed station point; it does not need to dive more than 100-200m in depth (even less if it has just to determine the thermocline); it has to move from one station to the other faster than a glider, in order to react to the faster time scale of coastal ocean dynamics; it can communicate with a land station to transmit data immediately after a dive (almost real-time) and to receive update on the mission. All this features and requirements have been taken into account in the design described in the next section.

## 3. VEHICLE DESIGN

The mechanical structure of the vehicle is composed by two fiber-glass cylinders held together by a steel frame (Fig. 1). The propulsion and steering system of the vehicle is realized with four jet-flow pumps. A horizontal propulsion pump

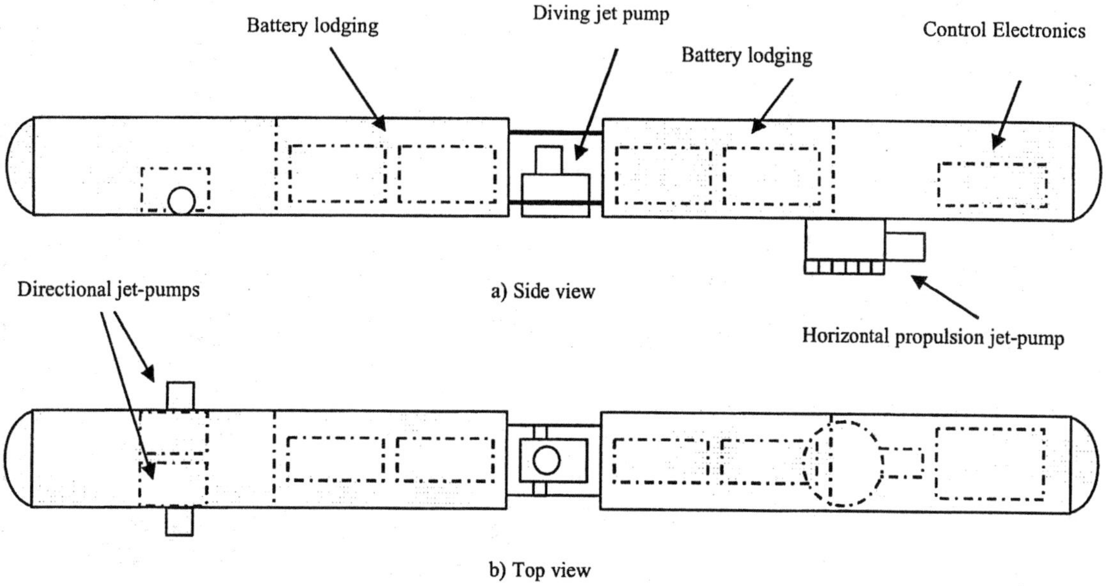


Fig. 1: Side view and top view of the vehicle

is placed in the vehicle stern, the vertical (diving) propulsion pump is placed between the two fiber-glass cyilinders, at mid-vehicle. Two steering (directional) pumps are placed at the vehicle bow, at an angle $\beta$ with respect to the vehicle axis. The total vehicle length is 3,16m, the vehicle diameter is 0.12m. This elongated and thin shape ensures robustness with respect to wave and current disturbances on the sway axis. The steel frame is geared to both cylinders opening caps. In addition to provide space for the diving pump, the steel frame allows for the transportation of the vehicle as separated sections, which are then geared together prior to the mission. Inside the cylinders are hosted the battery packs and the electronics for payload, sensors, navigation, communication and control. Connector cables between the two cylinder caps provide a bus for energy and data communication between the two sections. On the caps there are also connector openings for the payload cables. The on-board electronics has been realized with PC-104 bus boards. It comprises a 486DX2 66Mhz CPU, a flash disk, a I/O board (8 Single Ended analog input, 4 analog output, 24 digital input – A/D conversion is performed at 12 bit, with 100 kHz throughput). Additional modules include an integrated (D)-GPS receiver/GSM communication module, a digital compass, an analog pressure sensor, plus alarm signals which at the moment include battery level, depth control and diving mission time-out. The remaining channels are left for the oceanographic payload. The board output are sent to a linear power amplifier to drive the propulsion, diving and steering pumps. A block diagram of the system electronics is shown in Fig. 2. The final assembled system is shown in Fig. 3.

The vehicle weight is arranged so that the vehicle is horizontally balanced and slightly positive buoyant: in particular, the weight in air is 30.10 kg, with a 31.00 kg lift. This allows for the following procedure when on station for diving:

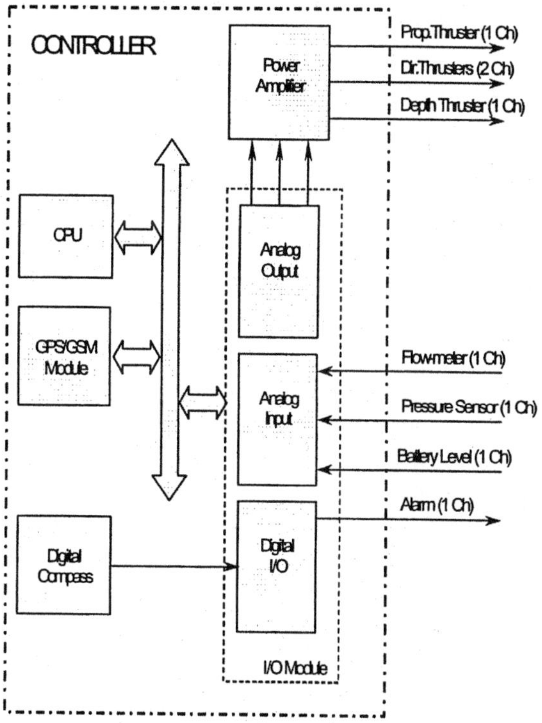


Fig. 2: Block-Diagram of the controller electronics

Fig. 3: Controller electronics, final assembly

with no horizontal propulsion, the diving jet-pump is activated and the vehicle dives down to the prescribed depth. The weight balancing of the vehicle is such that the vehicle maintains its horizontal attitude while diving vertically. When the prescribed depth is reached (as monitored from the depth sensor), the diving pump is stopped, and the vehicle resurfaces slowly just by buoyancy lift. In the resurfacing stage, the payload data can be acquired. Maximum operational diving depth of the system has been fixed at 100m. No diving speed control is required throughout the diving operation. Navigation and steering while the vehicle is on surface do require control on the pumps. The specific control laws will be described in the next section, using as input control variables the propulsion speed in the surge direction $u$ and the angular velocity $\omega$. It is clear that both these quantities are related to the actuation signals from the controller through the surge and yaw dynamic equations:

$$m\dot{u} = -\chi_u u - \chi_{u|u|} u|u| + \tau_p + \tau_s \cos\beta$$
$$J\dot{\omega} = -\chi\omega + \tau_s \sin\beta \tag{1}$$

where $u$ is the surge velocity, $\omega$ is the angular velocity, $m, J, \chi_j$ are the relevant inertia and drag parameters, including added mass coming from hydrodynamic forces, $\beta$ is the orientation angle of the steering jet-pumps with respect to the surge axis, $\tau_p, \tau_s$ are the (generalized) resulting torques exerted by the propulsion and steering jet-pumps respectively. Since the surge velocity and the yaw rate are *not* measured, use of quasi-static relations as those described in (Indiveri, 1998), or in (Cecchi et al., 2003), are employed to relate jet-pumps torques with surge velocity and yaw rate. The steering pumps, in particular, are used only for orientation purposes,and provide a much smaller torque with respect to the propulsion pump; their contribution to the surge velocity in the first of the equations in (1) can in fact be neglected for control purposes. Only the starboard or port steering pump is used at any given time.
Energy supply is given by four lead acid battery packs of 6V each, with capability of 12 A/h; the jet-flow pumps need 12V alimentation, and with 3.3 A absorption; the vehicle can accommodate 2 additional battery packs. The total autonomy of the system is estimated in 12 hours of operation. Lead acid batteries have been chosen since they are rechargeable and are a very reliable technology. Different energy sources can be employed without altering the other system characteristics. A picture of the realized vehicle prototype is shown in Fig. 4.

## 4. NAVIGATION AND CONTROL

The availability of a GPS link when on transfer between sampling stations by-passes all the traditional navigation issues of standard AUVs. More interesting is the solution to the control problem, since the vehicle manoeuvering is similar to that of a non-holonomic planar system. The control algorithm implemented is a particular instance of the general approach introduced in (Aicardi et al., 1995), and already proposed for underwater applications in (Aicardi et al., 2001a, Aicardi et al., 2001b). With respect to the above papers, the control algorithm proposed here accounts for the effect of current disturbances and for actuators saturation. The control objective defined in (Aicardi et al. 1995) is the following: given a target point to be reached with a given orientation, and the measurement of target distance and orientation error (available in our case from the GPS reading and the vehicle compass), determine the control law $(u, \omega)$ to guarantee the stable reaching of the taget position.

Fig. 4: the vehicle prototype on the pier before testing; the propulsion pump and the GSM antenna are clearly visible.

The target position is considered as the origin of a earth-fixed reference frame; a body-fixed reference frame with origin on the vehicle center of gravity is also considered. The two systems are depicted in Fig. 5. With reference to the figure, the system dynamics are described by the following set of kinematic equations:

$$\dot{e} = -u\cos\alpha \tag{2}$$

$$\dot{\alpha} = -\omega + u\frac{\sin\alpha}{e} \tag{3}$$

$$\dot{\theta} = u\frac{\sin\alpha}{e} \tag{4}$$

With respect to the model described in (Aicardi et al., 2001), here the sway equation is neglected, due to the high drag coefficient guaranteed by the elongated and thin shape of the vehicle.

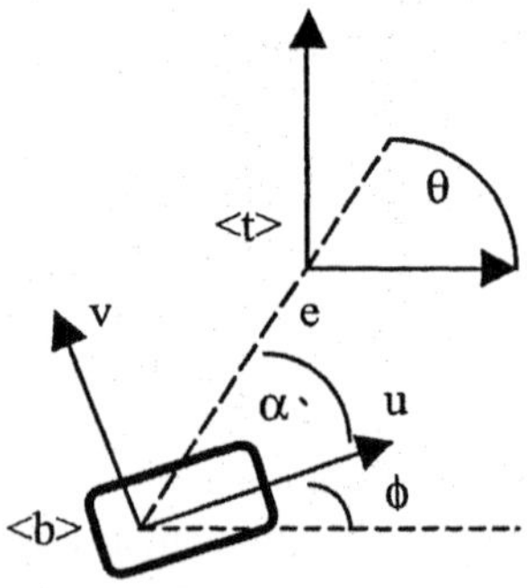


Fig. 5: Body-fixed and earth-fixed reference frames

The reaching of the target position is equivalent to the asymptotic convergence to zero of the variables $e, \alpha, \theta$. The following nominal control law has been shown to guarantee asymptotic convergence to the target position and orientation (Aicardi et al., 1995):

$$u_n = \gamma e \cos\alpha$$
$$\omega_n = \kappa\alpha + (\alpha+\theta)\frac{\gamma\sin 2\alpha}{2\alpha} \quad (5)$$

Apart its stability properties, shared by a number of other control designs, the reason for chosing this specific control law lies in the fact that the trajectories generated by this law are guaranteed to be smooth, cusp-free trajectories, very well suited to the operation of a marine vehicle. The smoothness of the generated trajectory, together with a discussion of this result in relation with the famous Brockett theorem on nonholonomic systems, is discussed for instance in the paper (Aicardi et al., 2001b). The control law of equations (5) is now modified to account for bounded current disturbances and actuator saturation. In particular, in order to cope with the current disturbances, an integral feedback action on the error variables is introduced. The implemented control law is:

$$u = \min(u_n + \tilde{\gamma}\cos\alpha \textstyle\int e, u_{max})$$
$$\omega = \begin{cases} \min(\omega_n, \omega_{max}) \text{ if } \operatorname{sgn}(\omega_n + \tilde{\kappa}\int\alpha) > 0 \\ \max(\omega_n, -\omega_{max}) \text{ if } \operatorname{sgn}(\omega_n + \tilde{\kappa}\int\alpha) < 0 \end{cases} \quad (6)$$

(where it has to be taken into account that $u \geq 0$). Actually, the integral control term on the distance between the vehicle and the target has the effect of making the vehicle almost always run at the maximum propulsion. An anti-wind up procedure has also been implemented, based on a threshold distance from the target, to avoid overshooting the target. Assuming that the modulus of the ocean current velocity is bounded by a value $c_{max}$, the control law of equation (6) can be shown to guarantee the stable reaching of the target position as long as $c_{max} < u_{max}$. Moreover, it has also to be considered that in the presence of a locally constant current disturbance, the reaching of the target point with the desired orientation is not guaranteed anymore: the vehicle will reach the target with an orientation aligned with that of the local ocean current field at the target. A thorough discussion of this effect is in (Aicardi et al., 2003a); the planning of the trajectory in terms of via point to exploit ocean current features that may favour navigation is discussed for instance in (Alvarez and Caiti, 2001). The estimated maximum propulsion speed $u_{max}$ of the vehicle in its current prototypal configuration is 3 knots, which implies that the vehicle is able to operate only with fairly limited ocean current conditions.

## 4. SEA-TRIAL RESULTS

The assembled vehicle prototype has been subject to some preliminary at-sea operational tests in the spring 2003. Objective of the tests was the verification of the functional performance of the vehicle, according with the design specification. In the testing performed, no additional equipment was available for independent measurements, so the sea trial results have to be interpreted in terms of vehicle sensor readings. Moreover, no payload was installed on the vehicle, so the capacity of the vehicle as an oceanographic platform has not been truly evaluated yet. The vehicle operation has been monitored from a lap-top PC station, through which navigation data were sent from the vehicle, and mission specification (in terms of target points) were given to the vehicle. The target points were given in geographical coordinates (lat, long) or in relative terms as distance and bearing from the current station. The lap-top station was on occasions on a support boat following the vehicle, and on other occasions on the pier close to the experimental area. The surface navigation and control system has been evaluated in terms of precision in reaching the target point by considering the fixed target point close to the harbour pier. In repeated trials, the distance error has always been evaluated in less then the vehicle length, i.e., less than 2m. This results has been very surprising, since the vehicle was running without Differential GPS, and the standard GPS error from the board specification was given as 25m. It is clear, however, that the precision in reaching the target depends only on the GPS data reception quality. Orientation error could not be reliably evaluated in these tests. The approach of the vehicle to the pier target point is shown in Fig. 6.

Fig. 6: Monitoring the vehicle navigation from the harbour pier.

The evaluation of the vehicle depth capabilites is done through a-posteriori readings of the depth vs. time data file acquired during the diving phase. It has to be remarked that the maximum water depth in the test area was 8m, so that extensive diving has not been tested. A set of five different profiles is shown in Fig. 7. The available data show good reproducibility and a constant speed in the diving phase. The resulting diving speed, however, is too slow for the oceanographic application of interest. This is possibly due to the combination of an excessive positive buoyancy together with underdimensioning of the diving pump energy. However, modifications of both aspects is possible, easy to do and cheap, and it will be pursued in the next vehicle testing. The Figs. 8, 9 show the beginning of the diving phase (with the diving pump start and with the vehicle just below the surface.

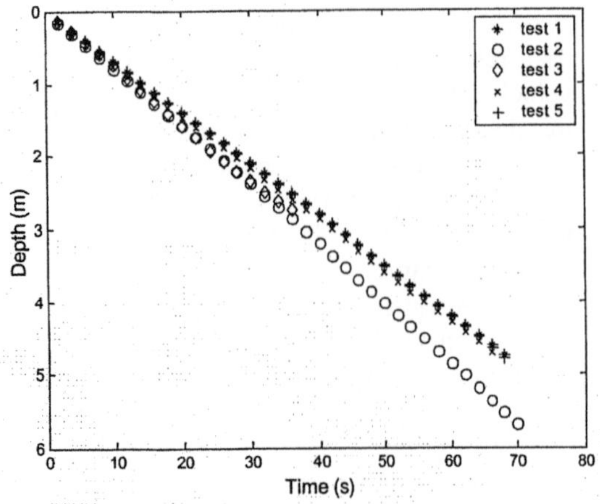


Fig. 7: Measurements of depth vs. time in 5 subsequent diving tests

Fig. 8: beginning of diving phase; the diving jet-pump at mid-vehicle gets into action

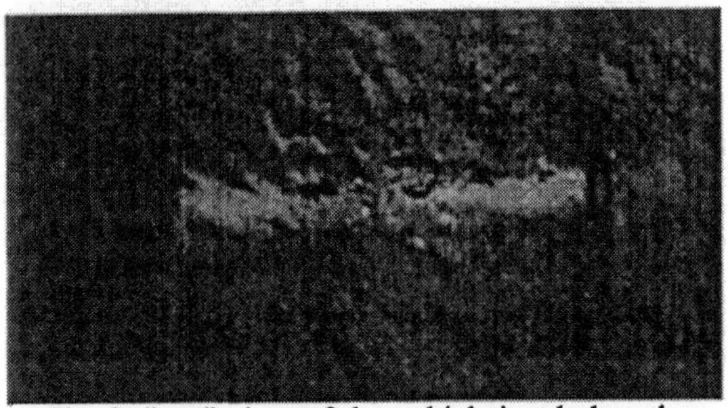

Fig. 9: "top" view of the vehicle just below the surface during the diving phase.

## 5. CONCLUSIONS

The design and realization of a very low cost autonomous vehicle for coastal oceanographic applications have been illustrated. The key to a low cost realization has been the focus on mission requirements, in contrast with the more diffuse tendency in designing multi-purpose, multi-mission vehicles. A limited number of sea trials was completed with the assembled prototype, showing a reliable functioning and qualitative behaviour as anticipated in the design stage. Some dimensioning problems, not critical for the further development of the vehicle, have been also identified Further tests are planned in the spring 2004 as part a of a complete oceanographic mission.

## ACKNOWLEDGMENTS

This work has been partially supported by University of Genova, "Young researchers" program. All the team is indebted to Pino Tonelli, that has made possible the first at sea test of the prototype. To him are warmest thanks and gratitude.

## 5. REFERENCES

Aicardi, M., G. Casalino, A. Bicchi, A. Balestrino (1995). Closed-loop steering of unicycle-like vehicles via Lyapunov techniques, *IEEE Robotics andAutomation Magazine.*, pp. 27-35, March 1995.

Aicardi, M., G. Casalino, G. Indiveri (2001a). Steering marine vehicles: a drag coefficient modulation approach, *Proc. IEEE/ASME Conf. Advanced Intelligent Mechatronics,* pp. 361-365, Como, Italy, July 2001.

Aicardi, M., G. Casalino, G. Indiveri (2001b). Closed loop time invariant control of 3D underactuated underwater vehicles, *Proc. IEEE Conf. Robotics and Automation*, pp. 903-908, Seoul, Korea, May 1995.

Allen, B., R. Stokey, N. Forrester, R. Gouldsborough, M. Purcell, C. von Alt (1997). REMUS: a small, low cost, AUV. System decription, field trials, performance results. *Proc. IEEE Conf. Oceans '97*, pp. 994-1000.

Alvarez, A., and A. Caiti (2001). Path planning of autonomous underwater vehicles in ocean environments with complex spatial variability. *Proc. GOATS'2000 – Generic Oceanographic Array Technology Conference*, Saclantcen Conference Proceedings, La Spezia, Italy

Gadre, A.S., J.J. Mach, D.J. Stilwell, C.E. Wick (2003), Design of a prototype miniature Autonomous Underwater Vehicle, *Proc. IEEE/RSJ Int. Conf. Intelligent Robots and Systems IROS 2003,* Las Vegas, USA, October 2003.

Schmidt, H., J.G. Bellingham, M. Johnson, K. Herold, D.M. Farmer, and R. Pawlowicz (1996), Real-time frontal mapping with AUVs in a coastal environment. *Proc. IEEE Conf. Oceans '96* pp. 1094-1098.

Stommel, H., (1989). The Slocum mission, *Oceanography*, vol. 2, n. 1, pp. 22-25.

ELSEVIER
IFAC
PUBLICATIONS
www.elsevier.com/locate/ifac

# SEA TRIALS OF SESAMO: AN AUTONOMOUS SURFACE VESSEL FOR THE STUDY OF THE AIR-SEA INTERFACE

**M.Caccia, R.Bono, Ga.Bruzzone, Gi.Bruzzone E.Spirandelli, G.Veruggio * , A.M. Stortini ** G. Capodaglio *****

* *CNR-ISSIA Sez. di Genova, Via De Marini 6, 16149 Genova, Italy*
** *CNR-IDPA, Calle Larga Santa Marta 2137, 30123 Venezia, Italy*
*** *Dip. Scienze Ambientali, Calle Larga Santa Marta 2137, 30123 Venezia, Italy*

Abstract: Sea trials and scientific exploitation of the SESAMO platform, an autonomous catamaran for data acquisition and sampling for biological, chemical and physical investigations on the air-sea interface are presented. The SESAMO platform is equipped with modules for the collection of surface microlayer, sampling of subsurface water, and survey in situ of water column and atmosphere parameters, along paths specified by the scientific end-users. Results of sea trials and operations during the XIX Italian Expedition to Antarctica, in January-February 2004, are presented and discussed. *Copyright © 2004 IFAC*

Keywords: Autonomous surface vessels, marine science applications.

## 1. INTRODUCTION

Research results published in the last decade pointed out the sea surface's role, or water surface film in general, as modulator for exchange of matter and energy. Thus the characterisation of the surface film, or microlayer if assumed as a portion of the sampled surface film, emerged as an important and complementary way for the evaluation of processes between surface water and low atmosphere. This motivated the development of advanced surface microlayer samplers, able to collect large quantities of water without altering the quality of the sample.

Data collection techniques are usually based on catamarans equipped with single microlayer samplers (Munster *et al.*, 1998) or with a number of devices for the evaluation of the surface and the near atmosphere, as for WHOI LADAS (Frew and Nelson, 1999). On the other hand, in the recent years progresses in the field of marine system automation and robotics have led to the development of autonomous surface craft for water monitoring as the Measuring Dolphin (Majohr *et al.*, 2000), developed by the University of Rostock, for the collection of hydrographic and bathymetric data as in the case of MIT's ACES (Manley, 1997) and AutoCat (Manley *et al.*, 2000), and for the enhancement of the acoustic communication rates with a companion AUV as in the case of DELFIM, developed by Lisbon IST-ISR (Pascoal and et al., 2000). On the basis of the above-mentioned technological background and the in-field experience gained by developing and operating MUMS [1], a radio-controlled cata-

[1] the Multi-Use Microlayer Sampler MUMS was developed in the framework of PNRA, Progetto Evoluzione e Ci-

maran equipped with a rotating drum for microlayer sampling in the Tyrrhenian (Cincinelli *et al.*, 2001) and Antarctic coastal zones and in the Venice lagoon, the SESAMO Project (Caccia *et al.*, 2003), funded in the framework of the Italian National Program of Research in Antarctica (PNRA), aimed to design, develop and test an autonomous modular platform for data collection and sampling of surface microlayer, water column and near atmosphere, in a scientific end-user predefined area. The developed platform had to satisfy basic scientific requirements in terms of quantity of collected water (up to 60 liters), manoeuvring capabilities, extreme polar environmental conditions, and pollution-avoiding characteristics. The result is the platform presented in section 2 from the mechanical, electrical and communication point of view. After a brief description of the system control architecture in section 3, basic vessel guidance and control are addressed in section 4, while results of sea trials and scientific exploitation are presented in section 5. Results are discussed in section 6 focusing on future system improvements and perspectives.

## 2. MECHANICAL AND ELECTRICAL DESIGN

### *2.1 Vessel*

**Hull**

As far as the vessel characteristics are concerned, the catamaran is, for stability with respect to roll and capability of payload transport with respect to the hydrodynamic drag, the kind of vessel usually employed in this type of applications. In addition, the redundancy in hull buoyancy makes catamarans more reliable than mono-hull vessels and, thus, to be preferred in case of autonomous craft. Thus, taking into account the above-mentioned operational needs, a catamaran characterised by a length of 2.40 m, a width of 1.80 m, a hull height of about 0.60 m and a space between the two hulls of 0.90 m, has been built. The hull, which in order to avoid contamination for organic and inorganic analities has been constructed in epoxide resina, can generate an Archimedean force of 400 Kg. The overall vehicle maximum weight in air is 300 Kg.

**Propulsion system**

The scientific requirement of operating at a very low advance speed motivated the design of a propulsion system constituted by two propellers actuated by two electrical thrusters without the use of any rudder. Steering is based on differential propeller revolution rates. Since commercial available electrical thrusters for small boats do not usually guarantee a fine velocity tuning at low speed, the choice has been to use, at least for the first catamaran prototype, ROV actuators specially designed for operating near bollard condition. Each thruster consists of a MAE M644-2530 DC motor (60 V @ 300 W) with the corresponding tachometer and rotary joint and is coupled with a three blade propeller. Each DC motor is controlled by a Mini Maestro DCD60V servo-amplifier, which performs a PID control of the thruster velocity on the basis of the error between the signal of the motor tachometer and the reference speed.

**Navigation sensors**

The vessel is equipped with a basic navigation package constituted by a GPS Ashtech GG24C at 2Hz integrated with a KVH Azimuth Gyrotrac able to compute the True North given the measured Magnetic North and the GPS-supplied geographic coordinates. A TV camera allows a robot subjective panoramic view of the operating environment. Wind direction and speed are measured by a weather station able to supply also the temperature of the atmosphere just above the sea surface.

### *2.2 Scientific payload*

The surface microlayer and sub-surface sampling systems have the structure shown in Figure 1, where all the line components are in teflon in order to avoid any contamination of the analities. The collected fluid is convoyed to the analysis

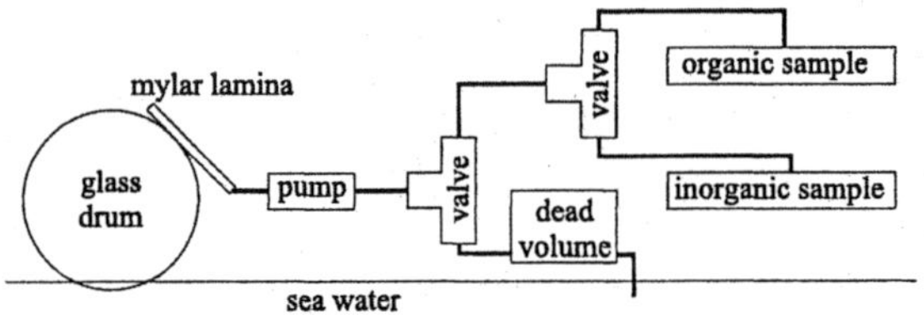


Fig. 1. Sampling line structure: surface microlayer.

or stocking bucket by a teflon-membrane pump. A three-way valve allows the deviation of the flux towards both the collecting buckets and a drain conduct, used to guarantee the washing of the lamina and the collection line. Along the drain conduct is located a kind of dead volume where probes for on-board measuring of water temperature, salinity, fluorescence and dissolved oxygen will be located.

In particular the **surface microlayer sampler** is based on the adsorption of the water film. A rotating glass drum (diameter: 33cm, length: 50cm) collects the water film on its surface from where it is removed by a mylar lamina. The drum, actioned by an electrical thruster, should rotate at very low speed, i.e. between 4 and 10 rpm.

In the case of the sub-surface layer sampling, a

cli Biogeochimici dei Contaminanti (now Contaminazione Chimica)

simple seawater intake is provided.
A view of the resulting vessel is shown in Figure 2.

Fig. 2. SESAMO catamaran during trials in Antarctica.

### 2.3 Power supply system

The power supply system is constituted by a set of four lead batteries at 12V@42Ah, which can be integrated by a set of solar panels to supply power peaks and to increase system autonomy.

### 2.4 Communications

The on-board control system of the vehicle and scientific payload communicates through a radio wireless LAN @ 1.9 Mbs, supporting robot telemetry, operator commands, and video feedback from the on-board TV camera, with an operator station located on a supporting vessel for direct at-field mission control.

## 3. CONTROL ARCHITECTURE

### 3.1 Control system

The control system is constituted by the navigation, guidance and control (NGC) system of the vessel, the data acquisition and control system (DACS) of the scientific payload, and the supervision module able to coordinate the NGC tasks with the sampling activities.
The vessel motion is controlled by the **vehicle NGC system** able to estimate the vehicle motion on the basis of GPS and compass measurements and to pilot it autonomously according to user requirements. In the first phase of the project basic automatic guidance capabilities are auto-heading and way-point navigation, e.g. line-of-sight navigation.

The **scientific payload DACS** handles the surface micro-layer and sub-surface sampling devices, controlling the pumps, valves and drum actuation system, and any additional sensor such as probes for water analysis and weather station.
The **supervision module** will coordinate the activities of the vehicle NGC system and scientific payload DACS, starting from simple actions, such as suspending water sampling during high speed maneuvers or washing the collection line at fixed time intervals, to more complex ones such as executing different maneuvers according to real-time measurements of the quality of the sample.

### 3.2 Human Computer Interface

The easiness of piloting the vehicle during operations led to the development of an operator interface, enabling the user(s) to define the required mission and monitor its execution in real-time, constituted by instances of two basic modules:

- pilot and system engineer interface: allowing a full access to the vessel and payload control system, including the capability of setting and visualizing algorithm parameters and selecting control and estimation algorithms;
- pilot and scientific end-user: providing basic navigation data, accepting and sending basic guidance commands, i.e. heading, surge, way-points, etc., according to the selected working mode, and allowing full display and access to scientific payload data and control variables.

### 3.3 Hardware and software

The on-board computing system, based on Single Board Computers and PC-104 I/O modules, is constituted by a SBC board (supporting an Intel Pentium III @ 750 MHz CPU, 4 RS-232 serial ports, 1 CompactFlash slot and Ethernet 100 Mps), and three PC104 modules supporting digital I/O, analog input and analog output respectively. Software was written in C++ using the RTKernel operating system for real-time operations and Windows 2000 for the human computer interface.

## 4. BASIC GUIDANCE AND CONTROL

### 4.1 Propulsion system modelling

The thrust exerted by each propeller can be modelled as

$$\tau = a_n n|n| = a_V V|V| \tag{1}$$

where V is the reference voltage applied to servo-amplifiers and n is the propeller revolution rate.

$V \propto n$ due to servo-amplifiers action.
Denoting with the subscripts L and R the left and right actuator respectively, the surge force $f_u$ and yaw torque $T_r$ assume the form:

$$f_u = \tau_L + \tau_R \qquad (2)$$
$$T_r = (\tau_L - \tau_R)d$$

where d is the half distance between the propellers, and the normalised control actions

$$\tilde{f}_u = \frac{f_u}{2a_V} = \frac{V_L|V_L| + V_R|V_R|}{2} \qquad (3)$$
$$\tilde{T}_r = \frac{T_r}{2a_V d} = \frac{V_L|V_L| - V_R|V_R|}{2}$$

can be defined.

### 4.2 PD-type heading control

A conventional PD controller of the vehicle heading has been designed and implemented for preliminary sea trials. The input variables are the reference and estimated heading, $\psi^*$ and $\hat{\psi}$ respectively, while the output control signal is the normalised yaw torque $\tilde{T}_r^*$.
Introducing a saturation mechanism, the resulting algorithm is:

$$\tilde{T}_{PD}^* = k_P\left(\psi^* - \hat{\psi}\right) - k_D\dot{\hat{\psi}} \qquad (4)$$
$$\tilde{T}_r^* = sat\left(\tilde{T}_{PD}^*, -T_r^{MAX}, T_r^{MAX}\right)$$

where *sat* denotes the saturation function.

### 4.3 Line-Of-Sight guidance

A line-of-sight (LOS) algorithm, generating the reference heading, has been implemented for basic way-point guidance:

$$\psi^* = atan2\left(y^* - \hat{y}, x^* - \hat{x}\right) \qquad (5)$$

where $(x^*\ y^*)$ and $(\hat{x}\ \hat{y})$ are the reference and measured coordinates of the vessel.

## 5. SEA TRIALS AND EXPLOITATION

After the launch of the SESAMO catamaran on October 22 and basic buoyancy and manoeuvrability tests, extended preliminary NGC sea trials were performed in the Genoa harbour on October 28, 2003.
Satisfactorily sea trials of the catamaran NGC and sampling systems in operating conditions and the consequent exploitation of the robotic platform for the collection of water samples were performed during the XIX Italian Expedition to Antarctica in January-February 2004.

### 5.1 Preliminary NGC trials in Genoa harbour

For safety reasons, the sampling system was not mounted on the vessel, operating in the harbour waters ploughed by service boats. Thus the catamaran was equipped with the navigation package (GPS and compass), network TV camera, basic weather station, and wireless communication system.
In conditions of strong north wind (mean of about 8 m/s with peaks up to 20 m/s), basic PD-type auto-heading and LOS guidance were tested showing satisfactory performance. The controller gains were $k_P = 4.0$ and $k_D = 6.0$ respectively with the angles expressed in degrees, and the saturation value $T_r^{MAX}$ was equal to 25.0. Here it is worth noting the classical behaviour of the LOS algorithm when the final way-point is not moved and a constant non null surge force is applied: the vehicle continues to bend in the proximity of the desired way-point (see Figure 3).

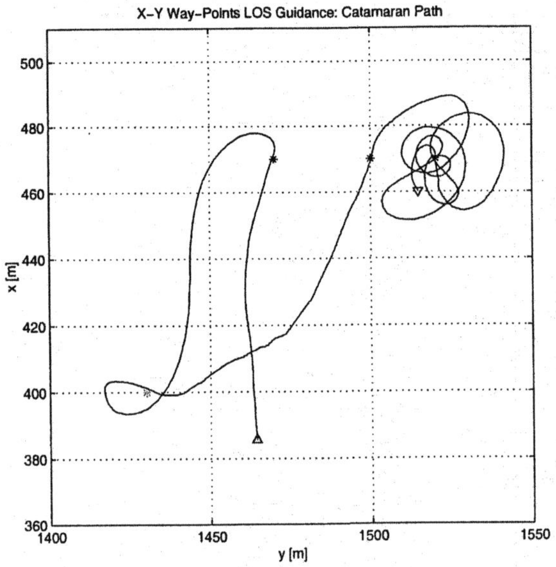


Fig. 3. Line-Of-Sight way-point guidance: catamaran path. Asterisks denote way-points; triangles the initial and final position of the vessel.

### 5.2 Trials and exploitation in Antarctica

The robotic platform was operated in the proximity of the Italian station of Terra Nova Bay ($74°41'S$, $164°03'29E$). The vehicle was operated by the Malippo, a 16m support vessel from where the catamaran was deployed from and recovered on in the proximity of the sampling area. Basic tests of wireless communications, vessel mechanics, and NGC system were performed on January 20, 2004, while the water sampling system was tested on January 26, when scientific end-user training was performed too.
The controller gains were set as in preliminary

trials in Genoa harbour. An example of auto-heading behaviour is shown in Figure 4 where the reference and measured heading, normalised control torque and surge force are plotted.

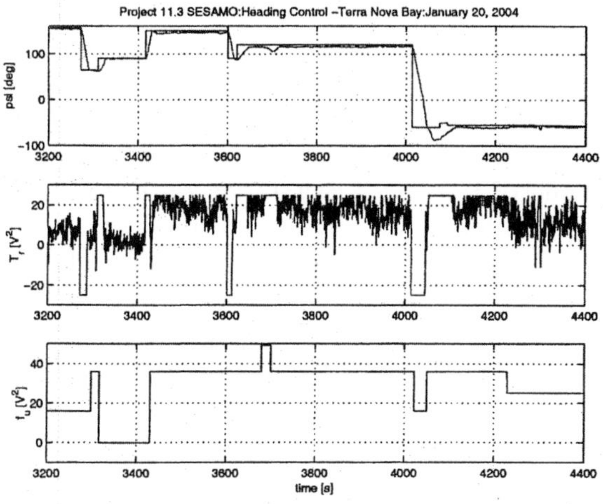


Fig. 4. PD-type auto-heading. From top to bottom: reference and measured heading, normalised control torque, normalised surge force.

As far as LOS guidance is concerned, the vessel navigated through a sequence of way-points manually set by the human operator as soon as the current one was reached by the vehicle. The applied surge force was also set by the human operator, basically according to the distance from the target point. An example of vessel path is shown in Figure 5.

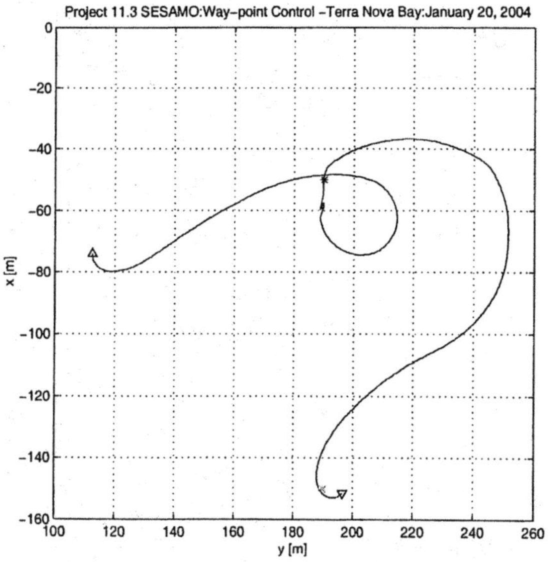


Fig. 5. Line-Of-Sight way-point guidance: catamaran path. Asterisks denote way-points; triangles the initial and final position of the vessel (Terra Nova Bay, January 20, 2004).

After satisfactorily results of sea trials the platform was exploited to collect surface microlayer and immediate sub-surface water samples in the framework of the PNRA Project 9.1 Chemical contamination. By February 10, 2004, three missions were performed collecting organic and inorganic analities with the cylinder rotating at 7 rpm as reported in Table 1. A view of the operating catamaran is shown in Figure 6. As far as the sampling strategy is concerned, the initial politics was of giving priority to the slower microlayer sampling, executing at the end of the mission the sampling of the immediate sub-surface. When the temperature got down in February, in order to minimise the risk of icing of the collecting tubes, as in the case of the sub-surface organic line on February 4, the sub-surface sampling were performed at the beginning of the mission, and the collection of organic and inorganic analities of microlayer was alternate during the time. The qualitative difference between the samples is pointed out in Figure 7, where two polycarbonate membranes used for filtering inorganic analities collected on February 9 are shown: the presence of a large amount of particulate is clearly visible on the left membrane, used for filtering the microlayer sample.

Table 1. SESAMO mission summary

| Date | Duration | $\mu$l O | $\mu$l I | ss O | ss I |
|---|---|---|---|---|---|
| Jan 20 | 1h15' | - | - | - | - |
| Jan 26 | 3h30' | 20 l | - | 20 l | 10 l |
| Jan 31 | 4h00' | 9 l | 10 l | 20 l | 10 l |
| Feb 4 | 1h15' | - | - | - | - |
| Feb 5 | 4h50' | 20 l | 6 l | - | 6 l |
| Feb 9 | 5h15' | 20 l | 11.8 l | 20 l | 10.7 l |

Fig. 6. Increasing wave motion during SESAMO sampling mission (Terra Nova Bay, February 5, 2004).

## 6. CONCLUSIONS

Basic design and results of sea trials and scientific exploitation of the SESAMO platform have been presented in this paper.
High quality sampling of surface microlayer requires, by its physical nature, a large amount of time leading to significant requests of the logistic

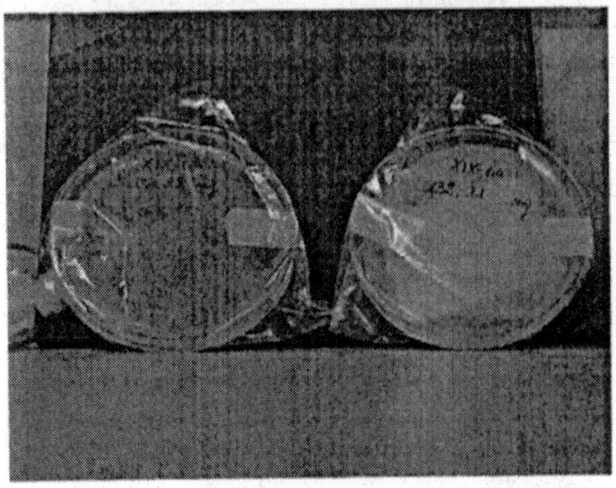

Fig. 7. Polycarbonate membranes used for filtering inorganic analities from microlayer and immediate sub-surface (Terra Nova Bay, February 9, 2004).

resources, mainly in terms of operating time of the support vessel Malippo, made available by the Italian Station of Terra Nova Bay. Since the revolution rate of the glass drum cannot be increased, because this dramatically affects the quality of the sample, the surface microlayer incoming flux can be increased by upgrading the cylinder length to 70 cm with respect to the 50 cm of the current version (the vehicle and payload mechanical frames have been designed and built to support such a device). In any case, this upgrade of the cylinder capability in collecting water should be used to compensate a reduction of the cylinder revolution rate in order to increase the quality of the sample. On the other hand, the vehicle can execute four-six hours missions without any power supply problem (if necessary, the power supply system could be immediately enhanced by mounting the solar panels on the vessel). The navigation sensors, i.e. GPS and compass, are quite reliable and precise, and the automatic guidance system guarantees accurate manoeuvrability of the vehicle. These observations suggest the possibility of increasing the vehicle autonomy, basically in terms of automatic guidance capabilities in the presence of adrift ice blocks, i.e. developing an obstacle detection and avoidance system focused on sea ice targets, in order to use the support vessel only for safe deployment and recovery of the platform, that should be operated by the end-user from a surface station located on the ground (in optical view of the operating area for communication needs). The operator ground station, constituted by the HCI notebook and the wireless communication system, can be fully portable and powered by batteries and solar panels. During the sampling the vehicle should navigate through way-points specified by the end-user, avoiding obstacles on its path. This operating scenario requires a working range of the wireless communication system of some kilometres, with respect to the about 500 m of the current one, but this requirement could be satisfied by simply substituting the antennas.

## ACKNOWLEDGMENTS

This work was partially funded by the Italian National Program of Research in Antarctica (PNRA), field of application Technology, project SESAMO. The authors wish to thank all the personnel of Terra Nova Bay station for their help and support during operations and the whole expedition. A particular mention goes to the Malippo crew, M.llo Capo G. Alessandro and M.llo Ordinario P. Leone from the Reggimento Lagunari "Serenissima", for their very kind and highly professional support during at sea trials.

## REFERENCES

Caccia, M., R. Bono, Ga. Bruzzone, Gi. Bruzzone, E. Spirandelli, G. Veruggio, G. Capodaglio and A.M. Stortini (2003). SESAMO: an autonomous surface vessel for the study and characterization of the air-sea interface. In: *Proc. of MCMC 2003*. pp. 298–303.

Cincinelli, A., A.M. Stortini, M. Perugini, L. Checchini and L. Lepri (2001). Organic pollutants in sea-surface microlayer and aerosol in the coastal environment of leghorn(tyrrhenian sea). *Marine Chemistry* **76**(1-2), 77–98.

Frew, N.M. and R.K. Nelson (1999). Spatial mapping of sea surface microlayer surfactant concentration and composition. In: *Geoscience and Remote Sensing Symposium*. Vol. 3. pp. 1472 –1474.

Majohr, J., T. Buch and C. Korte (2000). Navigation and automatic control of the Measuring Dolphin (MESSIN). In: *Proc. of MCMC 2000*. pp. 405–410.

Manley, J.E. (1997). Development of the autonomous surface craft "ACES". In: *Proc. of Oceans'97*. Vol. 2. pp. 827–832.

Manley, J.E., A. Marsh, W. Cornforth and C. Wiseman (2000). Evolution of the autonomous surface craft AutoCat. In: *Proc. of Oceans'00*. Vol. 1. pp. 403–408.

Munster, U., E. Heikkinen and J. Knulst (1998). Nutrient composition, microbial biomass and activity at the air-water interface of small boreal forest lakes. *Hydrobiologia* **363**(1/3), 261–270.

Pascoal, A. and et al. (2000). Robotic ocean vehicles for marine science applications: the european asimov project. In: *Proc. of Oceans 2000*.

ELSEVIER
IFAC
PUBLICATIONS
www.elsevier.com/locate/ifac

# AN INCREMENTAL STOCHASTIC MOTION PLANNING TECHNIQUE FOR AUTONOMOUS UNDERWATER VEHICLES

**Chiew Seon Tan, Robert Sutton, John Chudley**

*Marine and Industrial Dynamics Analysis Group*
*School of Engineering*
*The University of Plymouth,*
*PL4 8AA, UK*

Abstract: This paper presents a variant of an incremental stochastic motion planning technique for autonomous underwater vehicles (AUVs) that considers both the algebraic constraints, caused by the environment obstacles, and differential constraints, induced by the vehicle dynamics. The term kinodynamic planning is used to describe this type of motion planning. The majority of AUV path planners tend to adopt an approach where the vehicle differential constraints are neglected with the aim to simplify the path planning process. However, these techniques frequently resulted in paths that are only executable by holonomic vehicles, whereas underactuated vehicles like AUVs are unable to track the prescribed paths exactly. To circumvent this problem, an incremental stochastic planning technique based on a Rapid-exploring Random Tree (RRT) algorithm is used to take into account both types of constraints simultaneously. This paper proposes embedding the RRT algorithm with a reconnection technique to enhance the quality of the generated trajectory. Simulation results as presented below, using a 3 degree-of-freedom AUV model, show the viability of the concept. *Copyright*© 2004 *IFAC*

Keywords: Kinodynamic planning, motion planning, Rapid-exploring Random Tree, stochastic, reconnection.

## 1. INTRODUCTION

Autonomous underwater vehicles (AUVs) are not recent inventions, in fact the technology has been around since the past three decades. Nowdays, AUVs are frequently employed for sea bottom exploration, mine-hunting, seabed mapping and scientific data gathering. It is obvious that for AUVs to accomplish successfully such diverse missions, a robust and effective motion planning strategy should be forthcoming.

A majority of the proposed path planning techniques (Fogel and Fogel, 1990; Fox *et al.*, 2000) do not take into account the AUV dynamics. These paths are normally computed as the interconnection of polynomials or splines. Since these paths are independently computed without AUV dynamics, it cannot be guaranteed that they are executable in practice. Typically, very conservative constraints are imposed on the derivatives of the flight path in order to avoid violating the low-level feedback controller operating regime. To further simplify the process, the AUV body geometry is frequently neglected by shrinking it into a point via the application of the configuration space concept. This assumption is highly valid if the vehicle considered is operating in a sparse environment. Most deep sea terrain can be categorised as such. Lately, there has been a sudden paradigm shift by the scientific and naval communities from AUV deep sea exploration missions to deployment in littoral waters. The littoral zone is a subdivision of the benthic province that lies between the high and low tide marks and can be considered as an extension of the shoreline to 600 feet(183 meters) out into the water. The littoral zone is important for scientific research since it houses the bulk of ocean based organisms. Likewise, the navies have demonstrated a keen interest in exploiting the AUVs technology to complement their amphibious power projection plans. The littoral zone is an intricate area to navigate by default, with unpredictable disturbances such as internal waves, coastal currents, changing beach profile, reefs and artificial objects. Therefore, it is crucial for an AUV to exploit its dynamics to actually navigate the unknown terrain. One class of motion planning algorithm called Rapid-exploring Random Tree (RRT) is particularly well suited for this type of application.

The paper begins with a brief outline of the AUV dynamic model in Section 2. Section 3 attempts to explicate the advantages and the internal mechanisms of the RRT whilst Section 4 discusses several factors that can seriously affect the RRT performance. Section 5 introduces a tree reconnection algorithm to improve the quality of the generated trajectory whereas Section 6 elaborates upon the simulation results for both static and dynamic obstacles. The idea is also extended to a crippled AUV case in order to justify its suitability as a subcomponent to be integrated into a fault tolerant system. Lastly, Section 7 gives the concluding remarks and future extension of the study.

## 2. AUV DYNAMIC MODEL

A 3-DOF REMUS AUV dynamic model by Prestero (2001) is employed for the simulation study. In matrix form, the dynamics of the vehicle can be defined as:

$$\begin{bmatrix} m - Y_{\dot{v}_r} & 0 & 0 \\ 0 & I_{zz} - N_{\dot{r}} & 0 \\ 0 & 0 & 1 \end{bmatrix} \begin{bmatrix} \dot{v}_r \\ \dot{r} \\ \dot{\psi} \end{bmatrix} = \begin{bmatrix} Y_{v_r} & Y_r - mU_o & 0 \\ N_{v_r} & N_r & 0 \\ 0 & 1 & 0 \end{bmatrix} \begin{bmatrix} v_r \\ r \\ \psi \end{bmatrix} + \begin{bmatrix} Y_\delta \\ N_\delta \\ 0 \end{bmatrix} \delta_r(t) \quad (1)$$

**Terminology**

| | |
|---|---|
| $Y_{\dot{v}_r}$ | coefficient for added mass in sway |
| $Y_{v_r}$ | coefficient of sway force induced by side slip |
| $Y_r$ | coefficient for sway force induced by yaw |
| $N_{\dot{v}_r}$ | coefficient for added mass moment of inertia in sway |
| $N_{\dot{r}}$ | coefficient for added mass moment of inertia in yaw |
| $N_{v_r}$ | coefficient of sway moment from side slip |
| $N_r$ | coefficient of sway moment from yaw |
| $Y_\delta$ | rudder input coefficient for sway |
| $N_\delta$ | rudder input coefficient for yaw |
| $I_{zz}$ | moment of inertia at $z$ axis |
| $v_r$ | sway velocity |
| $r$ | yaw rate |
| $\delta_r$ | rudder deflection |
| $\psi$ | heading angle w.r.t inertial frame, $[0, 2\pi)$ |

This model is linearised at a constant surge velocity, $u_r$ of $1.5m/s$ (cruising speed) to avoid violating the model fidelity (Equation 1). A pole placement feedback controller is designed for the AUV to improve its dynamic response. The pitch and roll effects are neglected. Clearly, for this simple model an orthogonal transformation matrix can be used to convert the body reference velocities to the velocities in the inertial frame (Equation 2). Here, $c_\psi$ and $s_\psi$ denote $\cos(\psi)$ and $\sin(\psi)$ respectively. $[p\ q]^T$ are the surge and sway velocity while $[X\ Y]^T$ are the vehicle position (configuration) in $\mathbb{R}^2$. There is no singularity problem for this trivial case but this effect will need to be considered if the Euler angle formulation is used in $SE(3)$. By combining both Equation 1 and Equation 2, one obtains a dynamic equation of state vector, $\mathbf{x} = [u_r\ v_r\ r\ X\ Y\ \psi]^T$.

$$\begin{bmatrix} \dot{X} \\ \dot{Y} \end{bmatrix} = \begin{bmatrix} c_\psi & -s_\psi \\ s_\psi & c_\psi \end{bmatrix} \begin{bmatrix} p \\ q \end{bmatrix} \quad (2)$$

The vehicle has a maximum rudder deflection of $\pm 13.6°$ and a rudder rate limit of $18°/s$. Embedding these two components into the model resulted in a nonlinear system. The nominal dimensions of the vehicle are $1.4m$ in length and $0.3m$ in diameter. This information is required by the collision detection algorithm.

## 3. RAPID-EXPLORING RANDOM TREE

It has been known that, the approximate cell-decomposition methods such as $A^*$, dynamic programming and breath-first search are highly susceptible to the curse of dimensionality. Therefore, it is reasonable for one to concentrate on randomised algorithms (Branicky *et al.*, 2002). These algorithms do not have the completeness [1] and optimality properties of the previous algorithms. However, their robustness to curse of dimensionality tends to make them preferable in practical and real-time applications.

One version of this algorithm is the RRT (Lavalle, 1998). It is a form of incremental stochastic search technique that is devised to search efficiently non-convex high-dimensional state space, $\mathbf{X} \subset \mathbb{R}^n$ . In essence, the RRT algorithm attempts to build

[1] A property where the algorithm will return a solution if such a solution exists

a graph structure, or to be precise, a tree that describes the free state space, $\mathbf{X}_{free}$ of the system. Each node is implanted with the system state, $\mathbf{x}$, where $\mathbf{x} \in \mathbf{X}$. The tree is grown incrementally by picking the closest (Euclidean metric, $\rho$) node, $\mathbf{x}_{near}$ on the tree to the random node, $\mathbf{x}_{rand}$. Then the best constant input, $\mathbf{u}$ from a finite predetermined set $\mathbf{U}$ is chosen by propagating each input through the system differential equation, $\mathbf{f}(\mathbf{x}(t), \mathbf{u}(t), t)$ for a predetermined time increment, $\Delta t$. If no collisions are found, a new child node, $\mathbf{x}_{new}$ is added to the tree and the whole process is then repeated. Note that the time increment, $\Delta t$ is not the same as the system equation time, $\delta t$.

## 4. RRT PERFORMANCE ENHANCEMENT

### *4.1 Bias*

RRT performance can be improved significantly by the introduction of certain biasing techniques. One such technique is to employ a Gaussian distribution function such that the expected value is located at the goal state. Likewise, one can use a function to return either the goal state or a random state depending on a preset bias coefficient as implemented in this paper. Low discrepancy sequence (quasi-random) such as the Halton sequence has been argued to be more efficient. Despite the fact that its merit has been proven for the Probabilistic Road Map (PRM) method but its effect on RRT is still nonconclusive.

### *4.2 Computational Bottlenecks*

Perusing through the algorithm sequence, one will notice that the two major bottlenecks of the RRT algorithm are the nearest neighbour subroutine and collision detection subroutine. Of the two, the former, particularly the naive version which requires all the nodes to be checked, is the most computationally intensive. As such, it is prudent to substitute it with approximate nearest neighbour (ANN) algorithm by Arya *et al.* (1998) or KD-Tree which are more efficient. For the collision detection routine, this paper uses only non-convex polygons such as rectangle and circle to describe the obstacles thus avoiding the computational issues. High quality collision detection libraries will be needed if nonconvex obstacle models in three dimension are used.

### *4.3 Metric Sensitivity*

A proper metric is essential for the successful operation of RRT. Suitable metric varies from problem to problem. Nonetheless, the metric $\rho$ in the form of cost function or performance index, defined as,

$$\rho^* = min\left(\phi[\mathbf{x}(t_f), t_f] + \int_{t_0}^{t_f} \xi[\mathbf{x}(t), \mathbf{u}(t), t]dt\right) \quad (3)$$

while satisfying the differential constraint,

$$\mathbf{f}[\mathbf{x}(t), \mathbf{u}(t), t] - \dot{\mathbf{x}}(t) = 0$$

where $\rho$ is a function of $(\mathbf{x}_{near}, \mathbf{x}_{rand}, \mathbf{u})$, has been found to be a very suitable for RRT. Equation 3 assumed an obstacle free environment. It has been discovered that the RRT performance tends to degrade as $\rho$ and $\rho^*$ diverge. Typically, numerical method is used to solve the above optimal control problem. The solution can be time consuming since it entails solving a two-point boundary-value problem. In holonomic cases, the differential constraints disappeared and a simple weighted Euclidean metric can be used. On the other hand in nonholonomic cases, a weighted Euclidean metric is also frequently employed however tuning of the parameters (weights) can be nontrivial. Depending on the structure of the vehicle dynamics, incorrect weighting frequently introduces bias into RRT, diverging the search from the goal. Consequently, Cheng and Lavalle (2001) have devised several methods to render RRT less sensitive to metric effects. Two of their proposed methods are also incorporated into the following simulations. The first method is to record the used (expanded) inputs, thus avoiding any states duplications while the second method is to extract environment information concerning obstacles by recording the state collisions frequency. This information is kept in the form of constraint violation frequency (CVF). The objective here is to avoid expanding the state in the region where collision is bound to happen, hence biasing the search to the free space. Since RRT is a form of randomised algorithm, the solution obtained can be far from optimal. As such, this paper proposes a process called reconnection to mitigate this effect. A detail exposition of the process is given in Section 5.

### *4.4 Multiple Trees, Tree Pruning and Subconnection*

Other researchers advocates using multiple trees and tree pruning techniques to improve the RRT performance (Li and Shie, 2003). Indeed, multiple trees RRT version does provide fast solution but its effectiveness is limited to problems with algebraic constraints. This is caused by the difficulty of connecting the tree without obvious gap. Another interesting characteristic of the single tree RRT is its tendency to grow a few major branches at the initial states thus making connection with the terminal states very problematic. To circumvent this problem, one method is to introduce another start tree, with different time increment and metric when it is at the proximity of the goal states, thus improving the probability of connection (Kim and Ostrowski, 2003).

### *4.5 Hybrid Planner*

As mentioned in Section 4.3, the RRT tends to degrade as the $\rho$ and $\rho^*$ diverge. One promising technique initiated by Frazzoli *et al.* (2002) is to combine an optimal planner with the RRT. The optimal planner which exploits the precomputed trajectory primitives is used to plan an obstacle free path, while the RRT attempts to reroute the path if there are obstacles. Other researches prefer to merge RRT with collocation and nonlinear programming (Karatas and Bullo, 2001). The trajectory obtained shows substantial improvement compared to the individual methods.

## 5. RECONNECTION

The objective of the kinodynamic planning problem is to find a trajectory $\pi : [t_0, t_f] \rightarrow \mathbf{X}_{free}$ from an initial state $\mathbf{x}_{init}$ to a goal state $\mathbf{x}_{goal}$ within the tolerance $\tau$. However, it can also be formulated in such a way as the problem to find an input function $u : [t_0, t_f] \rightarrow \mathbf{U}$ that results in a collision free trajectory connecting both $\mathbf{x}_{init}$ and $\mathbf{x}_{goal}$. In most cases, it is also appropriate to select a path that optimises certain cost function, such as the time to reach $\mathbf{x}_{goal}$ or the control effort which corresponds to the energy consumption of the system. However, due to its randomised nature, the generated path will tend to be suboptimal.

Hence, this paper proposes a process termed as reconnection where the algorithm is initially executed to obtain a feasible trajectory which is then trimmed at a certain point and reexecuted again. This method exploits two inherent properties of RRT: (1) Its propensity to grow a few major branches from the initial point where these major branches are potential suboptimal trajectories. (2) Reconnecting the RRT entails recycling some of the residual branches thus achieving certain computational advantages compared to initiating a new tree. Certainly, two components need to be addressed: (1) The location of the trimming point and (2) the number of reruns required.

Once the first feasible trajectory is found, it is backtracked to the initial point. The trimming point is selected from 0.4 to 0.7 of the trajectory length. A value of 1.0 is equivalent to starting a new run since the whole core branch is trimmed. A too high value risk destroying important branches and a too low value will not provide substantial improvement as the RRT will attempt to just reconnect the trimmed branch. Multiple runs can be conducted, but, it has been experimentally determined that two to three runs are sufficient to obtain the best suboptimal trajectory.

## 6. SIMULATION RESULTS AND DISCUSSIONS

For these simulations, the environment is set to $200 \times 200m$ in dimension. Here, we assume an ideal case where *a priori* information of the environment is provided and there is no external disturbance from the environment. The convention here is to take the heading angle to start from the $x$ axis (inertial) and positive when turn counter clockwise. The algorithm is implemented in C on a 2.1 GHz Pentium IV machine, with 512 MB of RAM and running Windows XP. The simulations are run with 2000 maximum nodes and 4000 iterations, terminating when either criterion is reached or if a solution is found. The AUV initial states is set to [1.5 0 0 0 1] (angle in radian), while the goal state is [1.5 $\kappa$ 150 100 $\kappa$] where $\kappa$ denotes a variable (unconstrained). Goal tolerance is defined as a $5m$ radius. This accuracy can easily be achieved via a modern GPS employing a Wide Area Augmentation System (WAAS). The time increment and dynamic equation time are set to $3s$ and $0.1s$ respectively. Runge-Kutta method is used to propagate the dynamic equation. Instead of assuming a constant input for $\Delta t$, the input is linearly interpolated as it propagates through the state equations. This method allows one to employ larger time increment while easily taking into account the input rate constraint. All of the simulations use only the Euclidean distance as the metric.

Figure 1 shows the consequence of applying the reconnection algorithm. Notice the first feasible path (grey colour) has been found and trimmed. The trimming coefficient is set to 0.5 in this case. The intrinsic RRT property of selecting the nearest node resulted in the extension of the longer untrimmed trajectory (dark colour). Both the trajectories are compared to select the shortest amongst the two. Figure 2 shows a histogram plot of 100 simulation samples comparing both the enhanced RRT and a generic RRT performance. It is clear from the histogram that the enhanced algorithm returns the solution in shorter time, particularly for the first feasible trajectory. In Figure 2, it is observed that the enhanced algorithm outperforms the generic RRT in terms of time response while returning the best suboptimal trajectory, shortest distance in this case (Figure 3). The two means are compared using a t-test, it is significance at the 0.01 level alpha whilst the 99% confidence interval for the true difference in means is [36.7 60.5]. However, one must be aware that trajectory selected remains suboptimal, future extension of the algorithm is likely to take this factor into consideration. A detail descriptive statistics comparison of both the algorithms can be found in Table 1. The table indicates that the generic RRT has a higher failure rate. The failures are caused by the program reaching the predefined maximum nodes number or maximum iterations. Once again, the enhanced RRT performance is superior in comparison to the generic version.

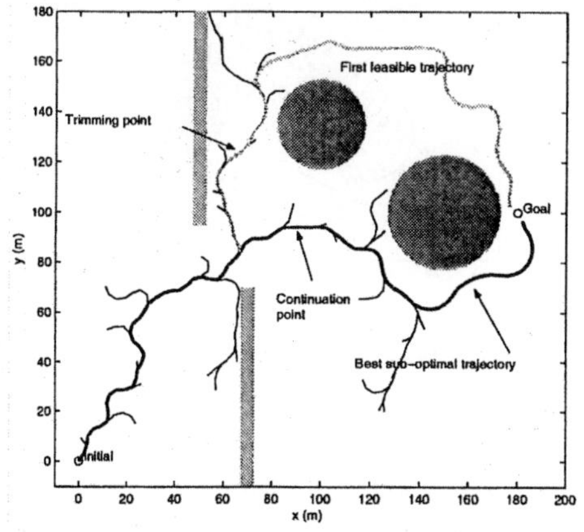


Fig. 1. Motion planning with RRT in environment populated with static obstacles. The reconnection process is also being depicted

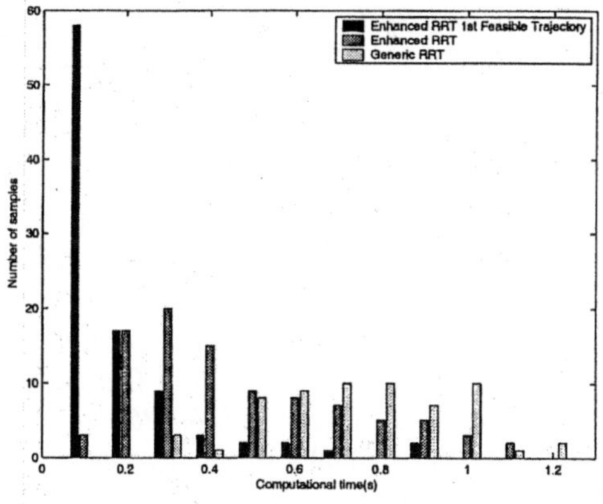


Fig. 2. Histogram plot of computational time

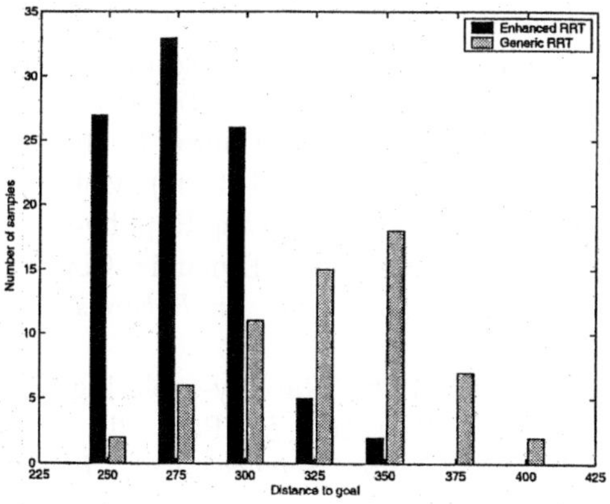


Fig. 3. Histogram plot of total distance

Figure 4 shows a more challenging environment where both the dynamic and static obstacles are presented. The dynamic obstacles are assumed to have constant velocity. The uncertainty of their position as time progresses can be considered by expanding the obstacles size through time. Figure 5 shows a feasible trajectory (dark colour) found

Table 1. Descriptive statistics collected from 100 samples run

| Parameters | Enhanced RRT | Generic RRT |
|---|---|---|
| Number of nodes used(mean) | 870.5 | 1254.5 |
| Number of nodes used(std) | 397.5 | 515.5 |
| Computational Time(mean, $s$) | 461.8 | 743.8 |
| Computational Time(std, $s$) | 245.0 | 211.2 |
| Total distance(mean, $m$) | 278.7 | 327.3 |
| Total distance(std, $m$) | 23.8 | 33.0 |
| Number of failures/100 runs | 6/100 | 39/100 |

by the RRT. To assist in visualising the dynamic effect, it is plotted with respect to time in the $z$-axis. The trimmed trajectory is not shown to avoid cluttering the view. Figure 6 illustrates a plot of the CVF magnitude. Notice an increase in CVF value of the corresponding node when it collides with an obstacle. This information allows RRT to behave in a more 'intelligent' way by avoiding extension near to the collision area.

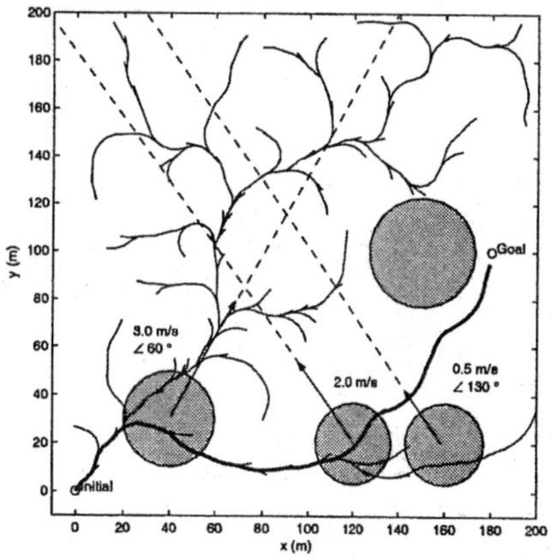


Fig. 4. Position of the static and dynamic obstacles with their corresponding velocities

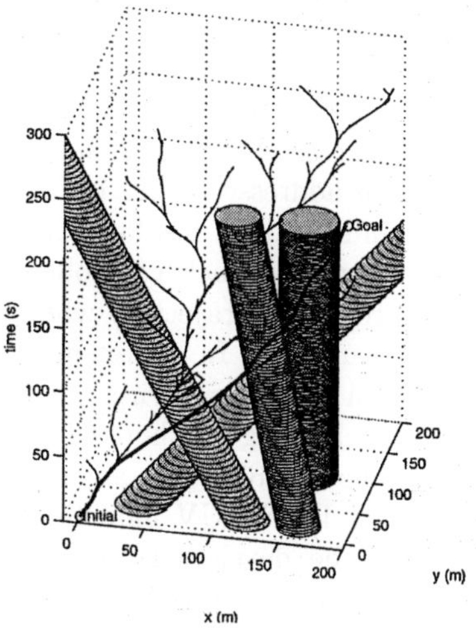


Fig. 5. Plot of configuration w.r.t time for an environment with both static and dynamic obstacles

Recently, there is a sudden increase of interest in developing systems having very high fault tolerant capacity. One of a subcomponent of these systems

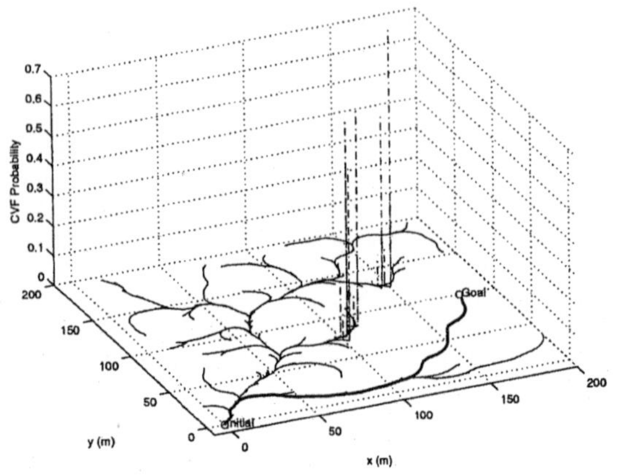


Fig. 6. Plot of the CVF superimposed with the trajectories

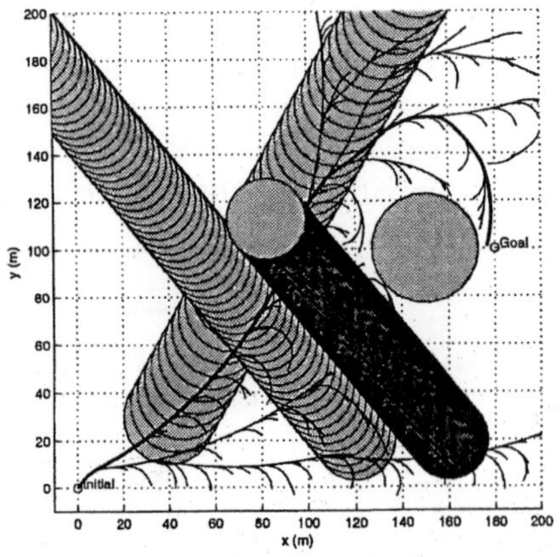


Fig. 7. Plot showing a feasible trajectory (bold) for a crippled AUV

is the fault tolerant control (FTC) system . These systems have very high reliability and is important in terms of operational cost and mission safety. Figure 7 shows a simulation run of a case where the rudder is partially jammed, due to seaweed entanglement or ice built-up. Instead of the normal rudder deflection range of $\pm 13.6°$, it is constrained to operate from $-0.5°$ to $+13.6°$. The effect can be deduced from the shape of the trajectories. The enhanced RRT algorithm found the goal in 3930 iterations and 1922 nodes in approximately $1.2s$. This clearly demonstrates RRT as a promising subunit of a FTC system.

## 7. CONCLUDING REMARKS

The objective of this paper is to introduce the enhanced RRT algorithm as a motion planner for AUVs. The algorithm is capable of generating feasible trajectories, satisfying both the algebraic and differential constraints. Its very short computational time makes it an ideal algorithm for real-time applications. The reconnection algorithm has been demonstrated to provide shorter trajectories. Nonetheless the trajectory is still suboptimal and this issue will be addressed in the future. Since the algorithm is inherently a feedforward controller, a robust low-level feedback controller is needed to track the prescribed trajectory when subjected to external disturbance. In addressing the case of navigating in an unknown environment, a sensor-based motion planning method will be merged with the enhanced RRT. The algorithm will be extended to a 6-DOF AUV model and it is anticipated that the developed algorithm will be implemented in a research AUV in the near future.

## REFERENCES


Arya, S., D. M. Mount, N. S. Netanyahu, R. Silverman and A. Y. Wu (1998). An Optimal Algorithm for Approximate Nearest Neightbour Searching Fixed Dimensions. *Journal of ACM* **45**(6), 891–923.

Branicky, M. S., S M. LaValle, K. Olson and L. Yang (2002). Deterministic vs. Probabilistic Roadmaps (summited). *IEEE Transactions on Robotics and Automation* pp. 1–13.

Cheng, P. and S. M. Lavalle (2001). Reducing Metric Sensitivity in Randomised Trajectory Design. *IEEE Int. Conference on Intelligent Robots and Systems* pp. 43–48.

Fogel, D. B. and L. J. Fogel (1990). Optimal Routing of Multiple Autonomous Underwater Vehicles Through Evolutionary Programming. *IEEE Proc. Symposium Autonomous Underwater Vehicle Technology (AUV)* pp. 44–47.

Fox, R., A. Garcia, Jr. and M. L. Nelson (2000). A Generic Path Planning Strategy for Autonomous Vehicles. *The University of Texas - Pan American, Department of Computer Science Technical Report CS-00-25.*

Frazzoli, E., M. A. Dahleh and E. Feron (2002). Real-Time Motion Planning for Agile Autonomous Vehicles. *Journal of Guidance, Control, and Dynamics* **25**(1), 116–129.

Karatas, T. and F. Bullo (2001). Randomized Searches and Nonlinear Programming in Trajectory Planning. *Pro. of The IEEE Conference on Decision and Control* **5**, 5032–5037.

Kim, J. and J. P. Ostrowski (2003). Motion Planning of Aerial Robot using Rapidly-exploring Random Trees with Dynamic Constraints. *IEEE Int. Conference on Robotics and Automation.*

Lavalle, S. M. (1998). Rapidly-exploring Random Trees: A New Tool for Path Planning. *TR 98-11 Computer Science Dept., Iowa State University.*

Li, T-Y and Y-C Shie (2003). An Incremental Learning Approach to Motion Planning with Roadmap Management. *IEEE Int. Conference on Robotics and Automation.*

Prestero, T. (2001). *Verification of a Six-Degree of Freedom Simulation Model for the REMUS Autonomous Underwater Vehicle, Master Thesis.* MIT. Massachusetts, USA.

ELSEVIER
IFAC
PUBLICATIONS
www.elsevier.com/locate/ifac

# A SWITCHING PATH FOLLOWING CONTROLLER FOR AN UNDERACTUATED MARINE VEHICLE

**Giovanni Indiveri** [*,1] **António Pascoal** [**]

[*] *DII - Department of Innovation Engineering, University of Lecce, via Monteroni, 73100 Lecce, Italy*
[**] *Instituto Superior Técnico, ISR-Torre Norte 8, Av. Rovisco Pais 1, 1049-001 Lisboa, Portugal*

Abstract: Building on previous work of Aicardi et al.(Aicardi *et al.*, 2001) a path following controller for underactuated planar vehicles is designed by adopting a polar-like kinematic model of the system. The proposed solution does not generally guarantee null asymptotic path following error, but only its boundedness below an adjustable upper threshold. However, knowledge of the spatial derivative of the paths' curvature is not necessary, thus resulting in a much easier solution to implement when compared to the alternatives available in the literature. Indeed, controllers achieving perfect path following usually require not only the use of the paths' curvature, but also of its derivative with respect to the curvilinear abscissa. Furthermore, the proposed solution may be used to solve the path following problem for unicycle models as well as underactuated marine vehicle models or to master-slave coordinated maneuvering of two vehicles. Simulation results illustrate the performance of the proposed control law. *Copyright©2004 IFAC.*

Keywords: Path following, guidance control, marine systems.

## 1. INTRODUCTION

Designing path following and trajectory tracking controllers for underactuated marine vehicles is a challenging task that has received wide attention in the past few years. Trajectory tracking deals with the case where a vehicle must track a time-parameterized reference. Path following refers to the problem of making a vehicle converge to and follow a given path, without any temporal specifications. A guidance controller to accomplish this task generates angular and linear velocity commands that are then transformed into actuator references by a lower level control system. The guidance regulator is thus designed based on a kinematic model of the vehicle only whereas the lower level actuator control system takes care of the dynamics of the plant. The necessary state estimates to implement the guidance and control levels are generated by a navigation subsystem that processes sensor outputs. Overall this architecture is known as navigation, guidance and control (NGC) in the literature. The present paper will focus on a guidance law for path following. Among the most relevant recent results in this field, a backstepping control design technique is employed in (Fossen *et al.*, 1998) for a surface vessel having an aft and a lateral propeller along the surge and sway axis, respectively. In this work, control system design is done based on a full nonlinear dynamic model of the ship. The resulting tracking control law is static (i.e. time invariant) and guarantees global convergence of

[1] Corresponding author, giovanni.indiveri@unile.it

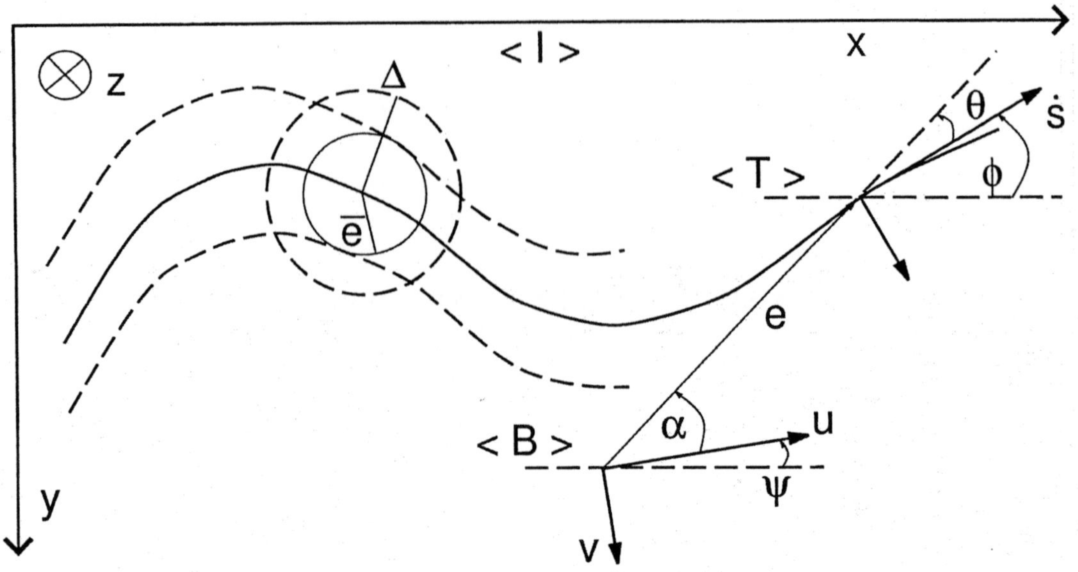


Fig. 1. The model

the position error to zero in spite of constant unknown environmental force disturbances, but the orientation of the vehicle is left in open loop. In (Pettersen and Nijmeijer, 1998*a*) the kinematics of a surface vehicle are considered and a time-varying control law for the surge and yaw inputs is designed such that the pose (i.e. position and orientation) error of the vehicle with respect to a reference trajectory of constant curvature is practically globally exponentially stabilized to zero. Remarkably it is not necessary for the reference linear speed to be nonzero, i.e. the same control law guarantees global practical stabilization if the reference trajectory should degenerate to a constant pose. In (Pettersen and Nijmeijer, 1998*b*) a similar model is considered and a static control law is designed that guarantees exponential, semi-global convergence of the surface vessel to a desired reference trajectory. The major limitation of this solution is that the reference angular velocity is required to be always nonzero, namely straight lines are not valid reference courses. In (Encarnação *et al.*, 2000*a*) a 2D path following controller is designed for the dynamic model a marine vehicle. This static solution takes explicitly into account constant but unknown currents and it guarantees global convergence of the pose error to zero. As explicitly shown in (Encarnação *et al.*, 2000*b*), reference paths are not required to have constant curvature. Remarkably the same design methodology may be extended to solve also the 3D path following problem as addressed in (Encarnação and Pascoal, 2000). In (Indiveri *et al.*, 2000) a static solution to the 2D problem of stabilizing the kinematic model of a marine vehicle on a linear course is proposed. The control law is designed exploiting a potential field -like idea which has been successfully employed to solve also the unicycle pose regulation problem (Aicardi *et al.*, 2000). Convergence to the reference linear course is global and has exponential rate, but environmental disturbances are not explicitly taken into account. In (Berge *et al.*, 1998) (Berge *et al.*, 1999) a position (not pose) tracking law is designed for the dynamic model of a surface vessel having a stern propeller and a rudder. The solution is static and explicitly takes into account a constant unknown disturbance force. Interestingly the idea behind the design of this solution is quite close to the one adopted in this paper: a *virtual reference point* (VRP) is defined at the bow or ahead of the vessel and the error variable that is globally exponentially stabilized to zero is precisely the distance between the VRP and a reference point moving along the reference path. The heading of the vessel is left in open loop, thus the vessel may rotate around if the VRP is not suitably chosen with respect to the center of mass of the vessel.

Some of the above references refer to trajectory tracking controllers and others to path following ones, but all share the ambitious control objective of perfect tracking (viz. perfect path following). This might be a quite demanding and often unrealistic goal in real applications. Indeed the resulting control laws are difficult to be experimentally implemented as they generally require knowledge not only of the reference path's curvature, but also its spatial derivative. By relaxing the control objective from perfect path following to convergence inside a "tube" of non null diameter centered on the reference path as done in (Aicardi *et al.*, 2001), the control design is considerably simplified and the synthesized law does not make use of the reference paths derivative. Indeed in the great majority of real applications perfect path following is actually not necessary and in marine application it would not be practically realizable anyhow because of the limited accuracy of the vehicles position relative to the reference path. The path following guidance law presented in (Aicardi *et al.*, 2001), although effective, forced the marine vehicle to follow the reference path with an unnatural heading, even when the reference curve was a straight line: the present paper builds on those results solving this issue. For the sake of simplicity ocean currents are not considered. The basic idea is to consider a virtual point moving on the reference path and to stabilize its distance from the marine vehicle to a target value. It turns out that the proposed controller may be also considered for master-slave coordinated guidance of two vehicles where a master vehicle replaces the virtual point on the path and the slave vehicle implements the guidance law here designed.

The paper is organized as follows. Section 2 describes the model of a marine vehicle that will serve as a focal point for the work that follows . In Section 3 the path following problem is formulated for that vehicle and in Section 4 the corresponding control design is addressed. Section 5 provides results of simulations and, finally, Section 6 contains the main conclusions.

## 2. VEHICLE MODEL

In the absence of currents, the vehicle kinematics are described by:

$$\dot{x} = u\cos\psi + v\sin\psi \tag{1}$$

$$\dot{y} = u\sin\psi - v\cos\psi \tag{2}$$

$$\dot{\psi} = r. \tag{3}$$

where, following standard notation, $u$ (surge speed) and $v$ (sway speed) are the body fixed frame $<B>$ components of the vehicle's velocity relative to the water, $x$ and $y$ are the Cartesian coordinates of its center of mass with respect to an inertial frame $< I >$, $\psi$ defines its orientation, and $r$ is the vehicle's angular speed. Notice that angles and angular speeds are positive when counter clockwise. With reference to figure (1), consider a target frame $<T>$ moving at velocity $\dot{s}$ along a desired reference path having curvature $\kappa = \omega_r/\dot{s} : |\kappa| < 1/R$ for some positive $R$, where $\omega_r = \dot{\phi}$ is the angular velocity of the target frame. The evolution of the position and orientation of the underactuated vehicle with respect to the target frame may be described by the polar like variables $e = \sqrt{x^2+y^2}, \alpha, \phi$ and $\theta$ that are defined in accordance to figure (1). Notice that by construction

$$\alpha + \psi = \phi + \theta. \tag{4}$$

The state variables are $e, \alpha, \theta$ and $v$ and their dynamics are given by:

$$\dot{e} = -u\cos\alpha + v\sin\alpha + \dot{s}\cos\theta \tag{5}$$

$$\dot{\alpha} = -r + \frac{u\sin\alpha}{e} + \frac{v\cos\alpha}{e} - \frac{\dot{s}\sin\theta}{e} \tag{6}$$

$$\dot{\theta} = -\omega_r + \frac{u\sin\alpha}{e} + \frac{v\cos\alpha}{e} - \frac{\dot{s}\sin\theta}{e} \tag{7}$$

$$\dot{v} = -a\,u\,r - k_v v - k_{v|v|}v|v| \tag{8}$$

where $a = m_{11}/m_{22}$, $k_v = |Y_v|/m_{22}$, $k_{v|v|} = |Y_{v|v|}|/m_{22}$. The symbols $m_{11}$ and $m_{22}$ capture the effect of mass and added mass terms, whereas $Y_v$ and $Y_{v|v|}$ are hydrodynamic derivatives (Fossen, 1994). Notice that not only the equations (6) and (7) for $\dot{\alpha}$ and $\dot{\theta}$ are singular in $e = 0$, but $\alpha$ and $\theta$ are themselves not defined at $e = 0$. The control inputs are surge $u$ and yaw $r$, while sway is not actuated.

## 3. PATH FOLLOWING PROBLEM FORMULATION

The standard $2D$ "perfect" path following problem formulation for the kinematic model of a vehicle may be expressed as follows:

**PF1:** Given a bounded curvature path parameterized in curvilinear coordinates and given the projection $P_p$ of the vehicle's position $P$ on the reference path (that is, the point on the path that is closest to the vehicle) , find a feedback law $u = f_u(P, P_p, \cdot)$, $r = f_r(P, P_p, \cdot)$ for the system's linear and angular velocities such that $\lim_{t\to\infty} \|P - P_p\| = 0$ and $\lim_{t\to\infty} \psi(P_p) - \phi(P) = 0$, where $\phi(P_p)$ and $\psi(P)$ are the tangent to the path and the vehicle's heading (at point $P$), respectively.

This formulation makes the problem difficult to be solved for several reasons: first of all the on line computation of the projection $P_p$ of point $P$ to the *closest* point on the reference path is generally cumbersome. Moreover with the aid of figure (2) and of equations (5 - 8) it is apparent that when $\|P - P_p\|$ and $\psi(P_p) - \phi(P)$ should approach zero, the variable $\alpha$ would converge to $\pi/2$ and the control input $u$ would completely lose authority on $e$ that would be basically driven by the ungoverned sway speed $v$. As observed in the introduction, for many practical applications the requirement of perfect path following is unrealistic and unnecessary besides the fact that controllers that solve problem PF1 usually require knowledge of the path's curvature *and* of its spatial derivative. To prevent the outlined drawbacks related to problem PF1, the following alternative formu-

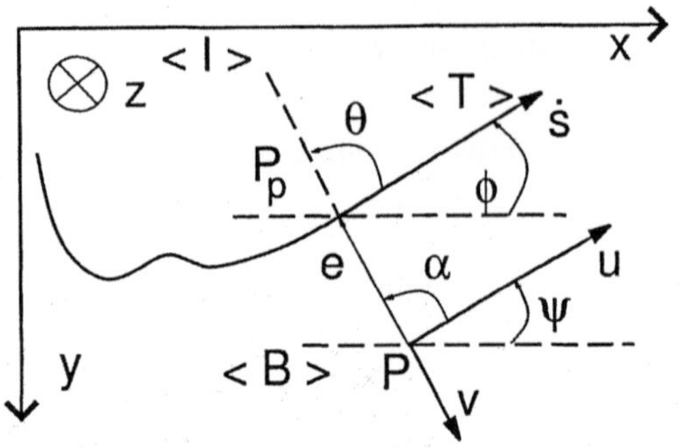


Fig. 2. Standard path following problem formulation

lation is considered:

**PF2:** With reference to figure (1), given a vehicle with body fixed frame $< B >$ and origin in point $O_B$, given a bounded curvature path and a point $O_T$ (origin of the target frame $<T>$) moving on it with assigned speed $\dot{s} : \dot{s}(t) \neq 0 \ \forall \ t$, find a feedback law for the system's linear and angular velocities $u$ and $r$ such that $\exists \ t^* \ : \ e(t) \leq \Delta \ \ \forall \ \ t \geq t^*$ for some arbitrary $\Delta > 0$ being $e = \|O_T - O_B\|$.

The above formulation does not make any use of projections of the vehicle's position on the path and it does not specify the desired behaviour of the vehicle's heading: in (Aicardi *et al.*, 2001) a solution to PF2 was found that implied $\lim_{t\to\infty} \alpha = 0$. While this guarantees to asymptotically approach maximal control authority on $e$ by $u$ and to minimize the effect of ungoverned sway speed on $e$ (see eq. (5)), it does leave $\theta$ in open loop making theoretically possible configurations with large values of $\theta$ and $\alpha \approx 0$, namely situations in which the controller tends to drive the vehicle with a velocity "opposite" to $\dot{s}$. Moreover even without such extreme situations occurring, namely even if the reference path would be such as to leave $\theta$ bounded, by example when it is a straight line, imposing $\alpha \to 0$ without guaranteeing that also $\theta \to 0$ ends up in forcing the vehicle to navigate with an unnatural heading angle (refer to figure (1) and equation (4) for a geometrical interpretation of these issues).

The most natural behaviour of the vehicle's heading once the condition $0 < e(t) \leq \Delta$ should have been reached (and not earlier) appears to be $\psi = \phi$, i.e. to navigate parallel to the target point $O_T$, exactly as commonly done when driving a car having the center lane line as a reference path. In the light of equation (4) this actually corresponds to imposing $\alpha = \theta$ when $0 < e(t) \leq \Delta$, whatever the value of $\alpha$ should be, rather than $\alpha = 0$. On the basis of these ideas, in the following sections three separate nonlinear controllers $C_1, C_2, C_3$ and a switching law $S$ between them will be designed such that:

- $C_1$ solves problem PF2 in the domain $\mathcal{D}_1 := \{(e, \alpha, \theta) : e \geq e_{th}\}$ where $e_{th} \in (0, \Delta)$ and it guarantees $\lim_{t\to\infty} \alpha(t) = 0$ if the state should continuously evolve in $\mathcal{D}_1$.
- $C_2$ solves problem PF2 in the domain $\mathcal{D}_2 := \{(e, \alpha, \theta) : e \geq e_{th} \ \cup \ |\alpha| \leq \alpha_{th}\}$ for some fixed arbitrary $\alpha_{th} \in (0, \pi/2), e_{th} \in (0, \Delta)$ and it guarantees $\lim_{t\to\infty} \alpha(t) - \theta(t) = 0$ if the state should continuously evolve in $\mathcal{D}_2$.
- $C_3$ is a bounded control law implemented in the domain $\mathcal{D}_3 := \{(e, \alpha, \theta) : e < e_{th}$ given $e_{th} \in (0, \Delta)\}$ as control laws $C_1$ and $C_2$ are not defined if $e = 0$ and could produce unacceptably large responses in its neighborhood.
- $S$ guarantees that if $e(t^*) \leq \Delta$, which will be shown to always occur by implementing law $C_1$ over a sufficiently long time, then $e(t) \leq \Delta \ \forall \ t > t^*$ in spite of any possible switching sequence to the other defined control laws.

The rationale behind this approach is to point straight at the target ($C_1$ law, $\alpha \to 0$ as in (Aicardi *et al.*, 2001)) when far from it and to "relax" the steering action ($C_2$ law, $\psi \to \phi$) trying to navigate parallel to the reference path when closer to it (and $\alpha \neq \pi/2$). In addition the switching law among the controllers must assure that $e(t)$ remains bounded by a desired threshold $\Delta$, yet strictly greater than zero. Controller $C_3$ has the only purpose of avoiding the use of laws $C_1$ and $C_2$ in a neighborhood of the polar coordinate model singularity in $e = 0$.

## 4. CONTROL DESIGN

First the design of $C_1$ is addressed:

### *4.1 $C_1$ design*

With reference to the model given by equations (5 - 8) and assuming for the moment that $e(t) > 0 \ \forall \ t$, consider the Lyapunov like function:

$$V_1 = \frac{1}{2}\left[\alpha^2 + h_1(e - \bar{e})^2\right] : h_1, \bar{e} > 0 \qquad (9)$$

and its time derivative

$$\dot{V}_1 = \alpha\left(-r + \frac{u \sin\alpha}{e} + \frac{v \cos\alpha}{e} - \frac{\dot{s}\sin\theta}{e}\right) + \\ + h_1(e - \bar{e})\left(-u\cos\alpha + v\sin\alpha + \dot{s}\cos\theta\right) \quad (10)$$

The above suggests the choice

$$C_1 : \begin{cases} r = \frac{1}{e}(u \sin\alpha + v\cos\alpha - \dot{s}\sin\theta) + \gamma_1 \alpha + \\ +h_1(e-\bar{e})\left[v\frac{\sin\alpha}{\alpha} + \dot{s}\cos\theta\,\frac{1-\cos\alpha}{\alpha}\right] \\ u = k_1(e-\bar{e})\cos\alpha + \dot{s}\cos\theta : \gamma_1, k_1 > 0 \end{cases} \quad (11)$$

that implies

$$\dot{V}_1 = -\gamma_1 \alpha^2 + k_1 h_1 (e-\bar{e})^2 \cos^2\alpha \le 0 \quad (12)$$

The $\mathcal{C}_1$ control law given by equation (11) guarantees that $\lim_{t\to\infty}\alpha(t) = 0$ and $\lim_{t\to\infty} e(t) = \bar{e}$ if the state should evolve in domain $\mathcal{D}_1 := \{(e,\alpha,\theta) : e \ge e_{th}\}$ where $e_{th} \in (0,\Delta)$.

### *4.2 $\mathcal{C}_2$ design*

Consider the following Lyapunov like function:

$$V_2 = \frac{1}{2}\left[(\alpha-\theta)^2 + h_2(e-\bar{e})^2\right] : h_2, \bar{e} > 0 \quad (13)$$

and its time derivative

$$\begin{aligned} \dot{V}_2 &= (\alpha-\theta)(-r+\omega_r) + \\ &+ h_2(e-\bar{e})(-u\cos\alpha + v\sin\alpha + \dot{s}\cos\theta) \quad (14) \end{aligned}$$

Assuming the state to be in domain $\mathcal{D}_2 := \{(e,\alpha,\theta) : e \ge e_{th} \ \cup\ |\alpha| \le \alpha_{th}\}$ where $\alpha_{th} \in (0,\pi/2)$, $e_{th} \in (0,\Delta)$, one may chose

$$\mathcal{C}_2 : \begin{cases} r = \omega_r + \gamma_2(\alpha-\theta) \ : \ \gamma_2, k_2 > 0 \\ u = k_2(e-\bar{e})\cos\alpha + \dot{s}\,\frac{\cos\theta}{\cos\alpha} + v\tan\alpha \end{cases} \quad (15)$$

that yields

$$\dot{V}_2 = -\gamma_2(\alpha-\theta)^2 + h_2 k_2 (e-\bar{e})^2\cos^2\alpha \le 0 \quad (16)$$

and guarantees that $\lim_{t\to\infty}\alpha(t) = \theta(t)$, namely that $\psi \to \phi$, and $\lim_{t\to\infty} e(t) = \bar{e}$ as long as the state stays in domain $\mathcal{D}_2$.

### *4.3 $\mathcal{C}_3$ design*

When the distance to the target $e$ should be smaller than a given threshold value $e_{th}$, i.e. in $\mathcal{D}_3 := \{(e,\alpha,\theta) : e < e_{th} \text{ given } e_{th} \in (0,\Delta)\}$, the vehicle's linear and angular velocities may be chosen to be simply equal to the target ones, namely

$$\mathcal{C}_3 : \begin{cases} r = \omega_r \\ u = \dot{s} \end{cases} \quad (17)$$

### *4.4 $\mathcal{S}$ design*

Calling $d$ a fixed positive constant to be specified later and denoting with $\mathcal{C}_j \mapsto (t_{jk}, \mathcal{C}_k)$ the transition among controllers $\mathcal{C}_j \ :\ j,k \in \{1,2,3\}$ at time $t_{jk}$, these are defined as follows:

$$\left.\begin{array}{l} |\alpha| < \frac{\alpha_{th}}{N_{ch}} \ : \ N_{ch} > 1 \\ \text{and} \\ \sqrt{\frac{2V_2(t_{\cdot 2})}{h_2}} \le d \\ \text{and} \\ e \ge e_{th} \end{array}\right\} \Rightarrow \begin{cases} \mathcal{C}_1 \mapsto (t_{12}, \mathcal{C}_2) \\ \mathcal{C}_2 \mapsto (t_{22}, \mathcal{C}_2) \\ \mathcal{C}_3 \mapsto (t_{32}, \mathcal{C}_2) \end{cases} \quad (18)$$

$$\left.\begin{array}{c} |\alpha| \ge \alpha_{th} \\ \text{and} \\ e \ge e_{th} \end{array}\right\} \Rightarrow \begin{cases} \mathcal{C}_1 \mapsto (t_{11}, \mathcal{C}_1) \\ \mathcal{C}_2 \mapsto (t_{21}, \mathcal{C}_1) \\ \mathcal{C}_3 \mapsto (t_{31}, \mathcal{C}_1) \end{cases} \quad (19)$$

$$e < e_{th} \Rightarrow \begin{cases} \mathcal{C}_1 \mapsto (t_{13}, \mathcal{C}_3) \\ \mathcal{C}_2 \mapsto (t_{23}, \mathcal{C}_3) \\ \mathcal{C}_3 \mapsto (t_{33}, \mathcal{C}_3) \end{cases} \quad (20)$$

Besides the condition on $V_2$ in (18), that as outlined below is necessary to assure boundedness of $e(t)$ over time, the switching law among $\mathcal{C}_1$ and $\mathcal{C}_2$ is driven by $\alpha$: if $|\alpha| \ge \alpha_{th} : \alpha_{th} < \pi/2$ then $\mathcal{C}_2$, being singular in $|\alpha| = \pi/2$, is inhibited and $\mathcal{C}_1$ is made active (19). The term $N_{ch}$ used in (18) to eventually hand back the control activity to $\mathcal{C}_2$ is chosen to be strictly *larger* than 1 (rather than equal to 1 which would be otherwise fine) to avoid chattering in the switching among $\mathcal{C}_1$ and $\mathcal{C}_2$.

Notice that if controller $\mathcal{C}_j$ for $j \in \{1,2\}$ is continuously active in the time interval $[\tau_{0j}, \tau_{jx}]$ for $j,x \in \{1,2\}$, then the distance to the target satisfies:

$$|e(t) - \bar{e}| \le \sqrt{\frac{2\,V_j(\tau_{0j})}{h_j}} \quad \forall\ t \in [\tau_{0j}, \tau_{jx}]. \quad (21)$$

Assuming for the moment that $e(t) \ge e_{th} \ \ \forall\ t$, namely that $\mathcal{C}_3$ is never activated, $\mathcal{C}_2$ can be active only when the state is in $\mathcal{D}_2$, i.e. where $\mathcal{C}_2$ is properly defined. Being $\mathcal{C}_2$ active, if the transition to $\mathcal{C}_1$ should never take place, then problem PF2 would be trivially solved by asymptotically converging to a distance $\bar{e}$ (eg. chose $\bar{e} < \Delta$) from the target point and (asymptotically) navigating parallel to the reference path. Similarly if $\mathcal{C}_1$ should always be active at all times, problem PF2 would be solved (as in (Aicardi *et al.*, 2001)) by asymptotically converging to a distance $\bar{e}$ (eg. chose $\bar{e} < \Delta$) from the path, but heading (asymptotically) always to the target point ($\alpha \to 0$). Notice that this situation, yet possible for specific reference paths, is highly unlikely as the convergence of $\alpha$ to zero and of $e$ to $\bar{e}$ will trigger the transition $\mathcal{C}_1 \mapsto (t_{12}, \mathcal{C}_2)$ according to (18). The likelihood of this transition can be increased by increasing the value of the $h_2$

gain. If the transition $C_1 \mapsto (t_{12}, C_2)$ should take place at time $t_{12} = t^*$ in compliance with (18), as long as $C_2$ will remain active for all $t \geq t^*$, the distance to the path will be bounded by:

$$|e(t) - \bar{e}| \leq \sqrt{\frac{2V_2(t^*)}{h_2}} \leq d \ \forall \ t \geq t^* \qquad (22)$$

At last if the control action should be switched back to $C_1$, namely if a sequence of the kind $C_1 \mapsto (t_{12}, C_2) \mapsto (t_{21}, C_1)$ should occur being $(t_{12}, C_2) \mapsto (t_{21}, C_1)$ triggered by $|\alpha|$ hitting the boundary value $\alpha_{th}$ as by (19), then the following would hold:

$$|e(t) - \bar{e}| \leq \sqrt{\frac{2V_1(t_{21})}{h_1}} \leq$$
$$\leq \sqrt{\frac{\alpha_{th}^2}{h_1} + d^2} \quad \forall \quad t \geq t_{21} \qquad (23)$$

As a consequence of this analysis, by choosing the positive design parameters $h_1, \alpha_{th}, d$ and $\bar{e}$ such that

$$\Delta \geq \bar{e} + \sqrt{\frac{\alpha_{th}^2}{h_1} + d^2} \qquad (24)$$

problem PF2 would be solved thanks to (18-19) if $e(t) \geq e_{th}$ at all times. Finally, lets consider the influence of controller $C_3$ if $e$ should be $e < e_{th}$: the transitions $C_1 \mapsto (t_{13}, C_3)$ and $C_2 \mapsto (t_{23}, C_3)$ cannot make $e > \Delta$ as by definition $e_{th} < \Delta$. To guarantee that problem PF2 is actually solved we need only to guarantee that the eventual transitions $C_3 \mapsto (t_{31}, C_1)$ and $C_3 \mapsto (t_{32}, C_2)$ leave $e$ smaller than $\Delta$. Calling

$$h_{\min} = \min\{h_1, h_2\} \qquad (25)$$

and recalling once again equation (21), it follows that:

$$|e(t) - \bar{e}| \leq \sqrt{\frac{4\pi^2}{h_{\min}} + (e_{th} - \bar{e})^2} \qquad (26)$$
$$\forall \ t \in (t_{3x}, t_{xj}) : x, j \in \{1, 2\}$$

where the term $4\pi^2$ is an upper bound of $(\alpha - \theta)^2$ and $\alpha^2$. Thus in order for the condition $e \leq \Delta$ to hold, it is sufficient that

$$\Delta \geq \bar{e} + \sqrt{\frac{4\pi^2}{h_{\min}} + (e_{th} - \bar{e})^2} \qquad (27)$$

that needs to be satisfied together with equation (24) for problem PF2 to be solved. Noticing that condition (27) would imply (24) if $d = \bar{e} - e_{th}$, the following result holds:

**Theorem:** Given the model described by equations (5-8) and

$$\alpha_{th} \in (0, \pi/2) \qquad (28)$$
$$0 < e_{th} < \bar{e} < \Delta \qquad (29)$$
$$d = \bar{e} - e_{th} \qquad (30)$$
$$h_{\min} = \min\{h_1, h_2\}, \quad h_1, h_2 > 0 \qquad (31)$$
$$\Delta \geq \bar{e} + \sqrt{\frac{4\pi^2}{h_{\min}} + (e_{th} - \bar{e})^2} \qquad (32)$$

problem PF2 is solved by the control law described in equations (11,15,17) and (18-20).

## 5. SIMULATION RESULTS

The simulation results reported refer to a reference path that was generated by a unicycle kinematic model having constant linear speed $\dot{s} = 1[m/s]$ and $\omega_r(t) = \pi/6 \exp(-4t/T) \sin(\frac{2\pi}{T} t)$ with $T = 180[s]$. In all figures concerning the simulation results, data points referring to controller $C_1$ are plotted with a circle (mostly viewed as a thick line given the resolution), and data referring to $C_2$ is plotted with a thin solid line. This allows to visually evaluate the effect of controllers $C_1$ and $C_2$ during their respective activity.

As expected, the activity of controller $C_2$ is most useful in forcing the heading of the vehicle parallel to the reference heading as opposed to the effect of $C_1$ that, while assuring maximal control authority over $e(t)$ on behalf of $u(t)$ and while preventing that $C_2$ ever reaches its singular value $|\alpha| = \pi/2$, tends to impose an unnatural sideslip to the vehicle (refer to the heading plots). The plots of $e(t)$ ($\bar{e} := 3[m]$) and of the paths (solid line for the vehicle, dashed for the reference) reveal that the control objective of converging in a $\Delta$-ball centered on the reference path is indeed achieved without danger of reaching the singular value $e = 0$ (actually $e$ is always larger than $e_{th}$ and the control law $C_3$ is never active). In the plot of $\alpha(t)$ the four boundary values $-\alpha_{th}, -\alpha_{th}/N_{ch}, \alpha_{th}/N_{ch}, \alpha_{th}$ being $\alpha_{th} = 70^o, N_{ch} = 10$ are reported (solid lines parallel to the $x$-axis) allowing to visualize the geometrical meaning of the switching law $\mathcal{S}$. Notice that if $N_{ch}$ should be continuously reduced to 1, the oscillations of $\alpha(t)$ visible from approximately $t = 15[s]$ to $t = 60[s]$, driven by the $C_1$-$C_2$ switching, would have ever increasing frequency (and decreasing amplitude) ending up in infinite frequency chattering when $N_{ch} = 1$. Hence the relevance of the $N_{ch}$ parameter. The discontinuity of signals $u$ and $r$ is due to the kinematic nature of the model adopted for the design.

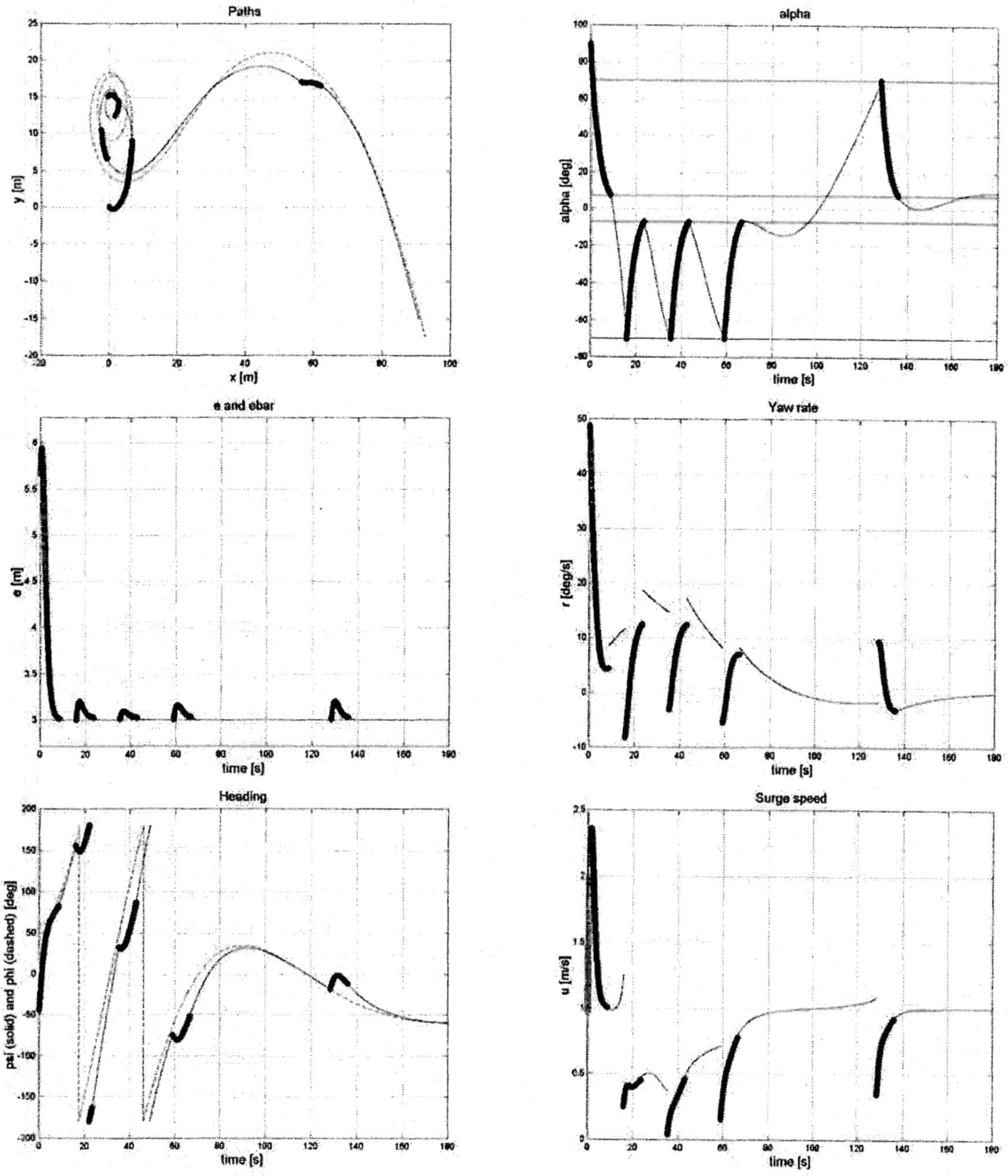


Fig. 3. Simulation results.

## 6. CONCLUSIONS

A steering controller for an underactuated marine vehicle is designed to drive and keep the vehicle within a wanted distance from a given reference path. The closed loop control solution does not require knowledge of the spatial derivative of the reference path curvature as opposed to most controllers designed for perfect path following. Once that a (virtual) target point $P_p$ moving on the desired path is defined, the steering control action is a function of the linear and angular velocities of $P_p$ and of the pose of the vehicle with respect to $P_p$ only. This makes the proposed approach suitable also for master-slave coordinated control of two vehicles, i.e. a master vehicles plays the role of the target point $P_p$ and a slave one implements the steering law here designed. Future work should focus on extending the presented approach to the case where current or wind disturbances should be present as well as consider the effect of the vehicle's dynamics on the convergence and stability properties of the overall closed loop system.

## REFERENCES

Aicardi, M., G. Cannata, G. Casalino and G. Indiveri (2000). On the stabilization of the unicycle model projecting a holonomic solution. In: *8th Int. Symposium on Robotics with Applications, ISORA 2000.* Maui, Hawaii, USA.

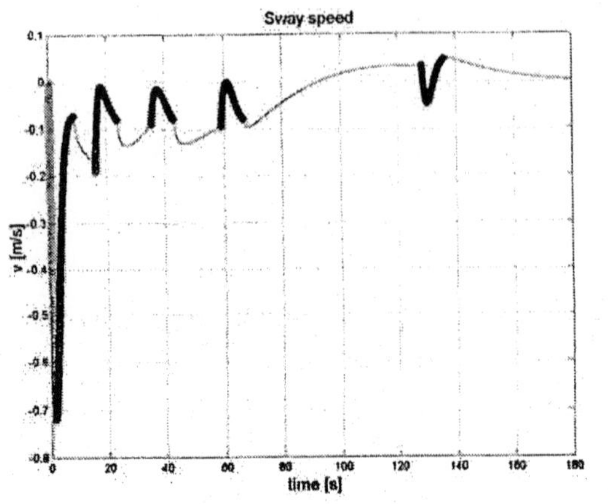


Fig. 4. Sway speed.

Aicardi, M., G. Casalino, G. Indiveri, A. Aguiar, P. Encarnação and A. Pascoal (2001). A planar path following controller for underactuated marine vehicles. In: *9th IEEE Mediterranean Conference on Control and Automation*. Dubrovnik, Croatia.

Berge, S. P., K. Ohtsu and T. I. Fossen (1998). Nonlinear control of ships minimizing the position tracking errors. In: *IFAC Conf. on Control Applications in Marine Systems, CAMS'98*. Fukuoka, Japan. pp. 83–89.

Berge, S. P., K. Ohtsu and T. I. Fossen (1999). Nonlinear control of ships minimizing the position tracking errors. *Modeling, Identification and Control* **20**(3), 177–187.

Encarnação, P., A. Pascoal and M. Arcak (2000*a*). Path following for autonomous marine craft. In: $5^{th}$ *IFAC Conference on Marine Craft Maneuvering and Control, MCMC2000*. Aalborg, Denmark. pp. 117–122.

Encarnação, P., A. Pascoal and M. Arcak (2000*b*). Path following for marine vehicles in the presence of unknown currents. In: $6^{th}$ *IFAC Symposium on Robot Control, SYROCO'00*. Vienna, Austria. pp. 469–474.

Encarnação, P. and A. Pascoal (2000). 3D path following for autonomous underwater vehicle. In: *$39^{th}$ IEEE Conference on Decision and Control CDC'2000*. Sydney, Australia. pp. 2977–2982.

Fossen, T. I. (1994). *Guidance and Control of Ocean Vehicles*. John Wiley & Sons, England.

Fossen, T. I., J. M. Godhavn, S. P. Berge and K. P. Lindegaard (1998). Nonlinear control of underactuated ships with forward speed compensation. In: *IFAC NOLCOS'98*. Enschede, The Netherlands. pp. 121–127.

Indiveri, G., M. Aicardi and G. Casalino (2000). Nonlinear time-invariant feedback control of an underactuated marine vehicle along a straight course. In: $5^{th}$ *IFAC Conference on Marine Craft Maneuvering and Control, MCMC2000*. Aalborg, Denmark. pp. 221–226.

Pettersen, K. Y. and H. Nijmeijer (1998*a*). Global practical stabilization and tracking for an underactuated ship - a combined averaging and backstepping approach. In: *IFAC Conf. on Systems Structure and Control*. Nantes, France. pp. 59–64.

Pettersen, K. Y. and H. Nijmeijer (1998*b*). Tracking control of an underactuated surface vessel. In: *$37^{th}$ Conference on Decision and Control, CDC'98*. Tampa, Florida, USA. pp. 4561–4566.

ELSEVIER
IFAC
PUBLICATIONS
www.elsevier.com/locate/ifac

# POLAR APPLICATIONS OF ROVS

**Ga.Bruzzone, M.Caccia, R.Bono, Gi.Bruzzone**
**E.Spirandelli, G.Veruggio** *

* *CNR-ISSIA Sez. di Genova, Via De Marini 6, 16149*
*Genova, Italy*

Abstract: This paper tries to perform, through a historical review of polar applications of ROVs technology carried out by CNR-IAN, a reflection upon the trends of the employment of robotics in polar environments and the influence of this application over robotic research. A particular attention will be paid on describing the polar applications of the ROVs developed by CNR-IAN in the last ten years and on giving some reports of the related tests performed in field. *Copyright © 2004 IFAC*

Keywords: Polar robotics, ROVs, marine science applications.

## 1. INTRODUCTION

In the last years the use of remotely operated vehicles (ROVs) to carry out scientific investigations in physical, chemical, geological and biological processes in the marine environment has largely increased, leading to the development of a new generation of scientific ROVs characterised by interchangeable toolsleds which can be adapted according to the specific application (Nokin, 1998)(Kirkwood, 1998)(Caccia *et al.*, 2000).

As ROV technology increased its reliability and diffusion the first experiments aimed at exploiting this technology in polar environments were performed. A pioneer work in this field was the experiment carried out by the NASA Ames Research Center in 1993 austral spring. A telepresence-controlled, underwater vehicle (TROV) was remotely operated via satellite from USA Ames labs and used to study sea floor ecology in Antarctica (Stoker *et al.*, 1995). From then on a number of ROV scientific missions in polar environments have been carried out by the prototype vehicles developed by CNR-IAN Robotlab, now Robotic group of CNR-ISSIA Genoa branch, from preliminary in-situ evaluation of scientific and technological instruments to Internet-based tele-operation by scientific end-users along the way indicated by the TROV experiment. The main idea that an underwater robot, able to carry different payloads, could be deployed and assisted at sea by a technical staff while scientists, acting in turn from their labs (located everywhere), can remotely operate it by means of Internet is, in prospect, very challenging for marine science in polar environment, where the displacement of human personnel and scientific and technological equipment is very expensive and risky. In this way it is possible on one hand to make easier technical and scientific international cooperation among partners from different countries, on the other hand to reduce costs and to minimise the number of people subjected to risks and discomforts due to long periods spent at sea or in harsh environments.

This paper tries to perform, through a historical review of polar applications of ROVs technology carried out by CNR-IAN, a reflection upon the trends of the employment of robotics in polar environments and the influence of this application over robotic research.

The paper is organised as it follows. Section 2 summarises the first campaign carried out with the Roby2 ROV in 1993-94: it focused on the scientific survey of unexplored seabed and test of technological devices in harsh environment in view of the development of SARA, the Italian Antarctic AUV. In-field experience and discussion with marine scientists generated the idea of using an ROV as a carrier for different payloads and end-users during the same Antarctic expedition in order to optimise logistics, financial, human and robotic resources. These considerations were fundamental in the design and development of Romeo, the CNR-IAN variable configuration ROV (Caccia *et al.*, 2000), and led to its exploitation in synergy with three different scientific and technological projects in the framework of the XIII Italian Expedition to Antarctica (see section 3). The first experiments of Internet-based satellite teleoperation of an ROV in polar environment, carried out in Antarctica in December 2001 and in Svalbard Islands (Arctic region) in September 2002, are discussed in section 4, pointing out the transition from the demonstration for a generic user of piloting a ROV over a remote seabed through a web browser to the execution of survey missions by scientific end-users from their labs. Current investigations focusing on even greater and more effective interaction of ROVs with the environment are introduced in section 5, where the results of the experiments carried out during the XIX Italian Expedition in Antarctica (winter 2003-2004) will be presented.

## 2. BASIC SURVEY AND TECHNOLOGY TRIALS: ROBY2

Roby2, the first ROV prototype developed by CNR-IAN (Veruggio *et al.*, 1994) was a tethered open-frame vehicle, 80 cm x 80 cm x 120 cm, weighing 250 Kg in (in air) and rated for working at a maximum depth of 200 m. It operated during the IX Italian Expedition in Antarctica (1993-94), at the Italian Terra Nova Bay (TNB) Station, with a total of 18 dives (Bono *et al.*, 1994) (see Figure 1). In co-operation with the group "Ecology and Biogeochemistry in the Southern Ocean" numerous dives were carried out in order to study the distribution of benthos along the stretch of coast from Gerlache Inlet to Adelie Cove down to a depth of 150 m, whilst other dives were performed to monitor the operating conditions of a sediment trap placed on the seabed at a depth of 40 m. The wide range of missions undertaken by Roby2 demonstrated the usefulness and versatility of underwater vehicles in the fields of oceanography, marine biology and glaciology. During each dive a complete set of data was collected and linked to high-quality video pictures. This made it possible to explore the environment in great detail. The video images recorded were also used to produce

Fig. 1. Roby2 deployment from Terra Nova Bay station wharf.

a documentary to support the Italian proposal for a Marine ASPA (Antarctic Specially Protected Area) located along the coastline between the TNB Station and Adelie Cove.

## 3. PRISMA: INTEGRATED ROBOTICS PROJECT FOR THE STUDY OF ANTARCTIC SEA

The experience gained with the experimentation of Roby2 in Antarctica led the Robotlab to the design and development of a new prototype, Romeo (Caccia *et al.*, 2000), an advanced ROV for shallow water marine science applications and research in the field of intelligent robotic vehicles. Romeo is a tethered open-frame vehicle, 90 cm x 90 cm x 130 cm, weighs 400 kg (in air), it is linked to a surface control system by a 600 m optical fibre tether and can operate to a maximum depth of 500 m. It was successfully used in the course of the XIII Italian Expedition to Antarctica, which took place from 29$^{th}$ October 1997 to 26$^{th}$ February 1998 at Terra Nova Bay (TNB) Station (Lat. 74 41' 42" S, Lon. 164 07' 23" E). During this period, in the framework of the PRISMA Project, i.e. Integrated Robotics Project for the Study of Antarctic Sea, Romeo's activity was organised in three legs. In the first leg (29$^{th}$ October - 3$^{rd}$ December 1997), the vehicle was equipped with a plankton sampler and a physical and chemical sensor package to collect data about under-ice biological processes (see Figure 2). Under-ice operations were enabled by a complex logistics consisting of an operational station positioned in a camp on the marine pack with a hole in the ice to enable the vehicle deployment and recovery. In the second leg (4$^{th}$ December1997 - 14$^{th}$ January 1998), under-ice performance of a set of acoustic devices (echo-sounders, high-frequency pencil beam profiling sonar, Doppler speed-meter) were evaluated in view of the development of SARA, the Italian Antarctic AUV (Galeazzi and Papalia, 1998). In addition, acoustic algorithms were

Fig. 2. Romeo in planckton sampling configuration (Tethis Bay ice field, Antarctica, 1997).

tested to collect data concerning the topography of the ice-canopy. In the third leg ($15^{th}$ January - $26^{th}$ February 1998), the vehicle was equipped with a payload for benthic data collection (water samplers, still camera, and CTD), and operated in the Marine Antarctic Specially Protected Area established nearby the Italian TNB Station, to continue the activity started in 1993-94 expedition by the previous prototype Roby2. A small boat named Malippo was used for the deployment and recovery of the vehicle and to host the operator station.
The need of high precision and agility in executing near bottom maneuvers, that arose from the above-mentioned applications, played a key role in motivating following investigations on modelling and identification, and navigation, guidance and control of variable configuration ROVs as discussed in (Caccia *et al.*, 2000).

## 4. INTERNET-BASED ROBOTICS

The possibility of remotely controlling an ROV operating over the seabed in a polar region through Internet was demonstrated in the framework of the projects E-Robot1 and E-Robot2 in the biennium 2001-02 in Antartica and in the Arctic region respectively. For details on system architecture and communication channels, the reader can refer to (Bruzzone *et al.*, 2003*a*). Antarctic experiments, carried out with a ground station located on the pack-ice, demonstrated the possibility for a common user, located in any corner of the world and having a standard Internet connection, of easily tele-operating an ROV in an unstructured environment. Arctic trials, carried out with the ground station mounted on-board the support vessel, verified the effectiveness of the remote control of the robot used in extreme environment for telescience applications. A preliminary Antarctic mission carried out in 2002-03 prepared all the required communication logistics to allow a full remote operability of the Romeo ROV during the XIX Italian Expedition to Antarctica in 2003-04.

### *4.1 E-Robot1*

In December 2001, in the framework of the XVII Italian Expedition to Antarctica, Romeo had the opportunity to operate in the Marine Antarctic Specially Protected Area established nearby the Italian Terra Nova Bay Station. During this period manifold experiments of Internet-based satellite teleoperation of the Romeo ROV were carried out. To allow a remote Internet user accessing to Romeo operating in Antarctica, Romeo's control system was integrated with a satellite communication system. Such a system was furnished, installed and made it to function by the Network and Telecommunication Service (SRT) department of CNR (refer to (Bruzzone *et al.*, 2002) for details). Moreover, suitable software applications were developed by the Robotlab, on one hand to effectively manage communications through the narrow and unreliable satellite channel and on the other to allow Internet users to easily access Romeo's control system. It is worth noting that in the design of the overall system a particular attention was paid to using commercial and low cost hardware and software components. In particular, standard Inmarsat channels were utilised for satellite communications and the Java programming language was used to develop the Human Computer Interface (HCI). As far as the operational logistic on field is concerned, the experiments were performed working through an artificial hole in the pack, settling a number of ice camps to support a safe and efficient activity (see Figure 3). After a number of preliminary tests carried out by tele-operating Romeo from CNR-IAN lab in Genoa, on December $17^{th}$ the first successful mission was carried out. The day after, there was a public demonstration in Rome to the presence of the press, of the CNR president and other authorities, and a large audience of undergraduate students. During the experiment, which lasted more than one hour and half, about 40 people had the possibility to use the HCI in pilot mode and remotely operate Romeo immerse in the Antarctic sea and to explore the Antarctic sea-bed rich of benthic life. Many other people using the HCI in observer mode, connected to the World Wide Web and located anywhere had the possibility to follow the experiment. Even if the video images received by the HCI had a low

Fig. 3. Operating ice camp in Tethis Bay, Antarctica, 2001. The antennas for satellite communications are clearly visible over the container.

refresh frequency (approximately 1.5 Hz) because of the narrow band of the satellite communication channel, a quite good teleoperation of Romeo was still possible. Image 4 shows a particular of the direction room of the CNR conference hall during the E-Robot1 experiment, where the HCI at work and the Romeo's live video of the Antarctic seabed transmitted by the videoconference system are visible.

Fig. 4. Direction room of the CNR conference hall during the E-Robot1 experiment.

*4.2 E-Robot2*

In September 2002, in the framework of the E-Robot2 project, Romeo operated in the Arctic Sea, more precisely in the Kongsfjord in Norway, nearby the Italian Station "Dirigibile Italia" situated in Ny-Ålesund (11° 56'E - 78° 55'N - Svalbard). The ROV was operated by means of a support vessel, named Farm, adapted with a crane to move Romeo inboard and offboard (see Figure 5). The communication system that made it possible to connect to the World Wide Web the Romeo's Control System computers located in the Arctic Region, was also in this case furnished,

Fig. 5. Romeo deployment from Farm support vessel in the Kongsfjord.

installed and made it to function by CNR-SRT (see (Bruzzone *et al.*, 2003*b*)). This allowed, for the first time, the execution of a number of missions under the control of some teams of marine scientists - from different European countries - who, without moving from their laboratories, conducted real-time experiments in the underwater environment in the Kongsfjord. In addition to demonstrate Internet-based robotics capabilities, the campaign in the Arctic Region allowed the marine scientists to conduct a set of preliminary dives in various unexplored sites in the Kongsfjord, to collect a preliminary set of data to plan a more systematic investigation in the future years. After a few problems due to bad weather condition (very strong wind) and to a fault in Romeo's gyrocompass, it was possible to operate Romeo at sea for two days ($11^{th}$ and $12^{th}$ September). A number of missions under the control of some teams of marine scientists from different European countries (the Scottish Association for Marine Science, the Norwegian College of Fishery Science of University of Tromsø, the Institute of Protein Biochemistry and the Enzymology and Institute of Biomolecular Chemistry of Naples) were executed. Using a dedicated web interface, the scientists, from their labs, were able to teleoperate Romeo and observe fauna and flora in the proximity of Kongsfjord seabed, collecting scientific data and taking photos. In the website related to the experiment (www.e-robot.it) some of the photographs taken by marine scientists during Romeo teleoperation are available.

The availability of an accurate ROV horizontal position and velocity estimation and control would facilitate the control system in managing time-varying and nondeterministic delays and the remote pilot in executing basic maneuvers such as maintaining the vehicle still in a suitable position to take photographs and/or samples. These operational requirements strongly stimulate current research on high precision vision-based horizontal motion estimation and control carried out at CNR-ISSIA Genoa branch.

## 5. ABS: ANTARCTIC BENTHIC SHUTTLE

In the course of the XIX Italian Expedition in Antarctica, that took place in winter 2003-2004, some experiments related to the Antarctic Benthic Shuttle (ABS) project were performed. The ABS is an innovative system to perform observations, data collection and sampling tasks in the harsh polar marine environment nearby TNB. A benthic module, composed of a benthic chamber for the in-situ analysis of sediment-water interactions in different conditions of ice cover, and of a time-lapse image and data acquisition system, was deployed by Romeo on the seabed (see Figure 6) below the ice-pack and recovered by the ROV after a suitable period of work (usually 24 hours). The benthic chamber, isolating a known volume of

Fig. 6. Benthic module placed on the Antarctic seabed.

sea-water and a known surface of sediments below it, allows the estimation of oxygen consumption by different benthic compartments (macro and meiofauna, bacteria), assessing their relative contribution in different environmental conditions. This innovative device was exploited in collaboration with the DIPTERIS-UNIGE (Department for the Study of the Territory and its Resources - University of Genoa) to study the structure and long-term variations in the Protected Marine Area of Baia Terra Nova, from Road Cove to Adelie Cove, as required by the Antarctic Specially Protected Area (ASPA) and the SCAR CS- EASIZ (Coastal Shelf - Ecology of the Antarctic Sea Ice Zone) programs. In order to operate as a Benthic Shuttle, Romeo was equipped with a subsystem for the docking/releasing of the benthic module, which was released on the seabed and subsequently recovered. A total of eight missions (see Table 1) were performed both through a hole in the ice-pack (see Figure 7) and from a support

Table 1. ABS missions performed in the course of the XIX Expedition.

| Position | Date | Deployment |
|---|---|---|
| 74° 41' 528 S<br>164° 01' 682 E | 21-22/12/03 | Hole in the pack |
| 74° 40' 807 S<br>164° 03' 607 E | 27-28/12/03 | Hole in the pack |
| 74° 40' 807 S<br>164° 03' 607 E | 30-31/12/03 | Hole in the pack |
| 74° 40' 807 S<br>164° 03' 607 E | 3-4/1/04 | Hole in the pack |
| 74° 41' 868 S<br>164° 01' 821 E | 9-10/1/04 | Hole in the pack |
| 74° 41' 868 S<br>164° 01' 821 E | 14-15/1/04 | Hole in the pack |
| 74° 41' 868 S<br>164° 01' 821 E | 18-19/1/04 | Hole in the pack |
| 74° 46' 375 S<br>163° 57' 537 E | 30/1/04-1/2/04 | Support vessel |

vessel, using the proven logistics used in the course of the XIII Expedition. In particular, the first seven missions were performed under-ice along the axis of Tethis Bay from the promontory in the proximity of the Italian Station to the Strand Line. Afterwards, with the progressive melting of the ice-pack, one open-sea operation from the Malippo support vessel operating in the Adelie Cove was carried out. Romeo's control system high performance in positioning, hovering and station-keeping, due to its acoustic/optical motion estimation systems, permitted a reliable and safe launch and recovery of the benthic module. However, the possibility of improving the recovery mechanism automating the hooking manoeuvre of the benthic module to the ROV by using artificial vision techniques, is being investigated.

Fig. 7. Antarctic Benthic Shuttle ready for immersion in the Antarctic sea through a hole in the pack. The benthic module installed at the bottom of Romeo is clearly visible.

## 6. CONCLUSIONS

In this paper, the polar applications of the ROVs developed by CNR-ISSIA Robotlab in the last ten years have been described. Some reports of the tests performed in field to check and evaluate the functionality, the effectiveness and the performance of such robotic systems have also been given. It is important to remark as the experimentation of these robotics systems carried out in harsh environments, as the polar ones, has continued in all these years to make arise always new needs and requirements that have highly stimulated and influenced robotic research at Robotlab. It is worth while noticing as Romeo, the latest ROV developed by Robotlab, demonstrated in manifold occasions of being a particularly reliable and flexible vehicle to use both for marine science applications and research in the field of intelligent robotic vehicles. Moreover, the experiments carried out in the framework of the E-Robot1 and E-Robot2 projects demonstrated that the current scientific and technological state-of-the-art enables common people and/or marine scientist to perform Internet teleoperation of underwater robotics systems operating at sea from the comfort of their houses/labs, by means of a simple connection to the World Wide Web and without the need of a devoted sophisticated and expensive equipment.

At the moment, the main research efforts at Robotlab are focused on one hand on improving Internet-based Romeo teleoperation by using high precision vision-based control algorithms; on the other hand on developing and testing new underwater robotic systems allowing a more strict interaction of the user with the environment (ABS project).

## ACKNOWLEDGMENTS

This work was partially funded by the Italian PNRA (Programma Nationale di Ricerche in Antartide) in the framework of the ABS (Antarctic Benthic Shuttle) project, CNR-SRT (Consiglio Nazionale delle Ricerche - Servizio Reti e Telecomunicazioni) in the framework of the project "Internet telerobotics in extreme environment with satellite support" (E-Robot1 and E-Robot2 projects) and Polarnet (CNR - Network for Polar Research). The authors are thankful to Dr. Fiorella Operto of the Science Foundation School of Robotics for the organisation of the E-Robot1 public demonstration. Moreover, they would like to thank all PNRA and Polarnet staff for their precious logistic support, all CNR-SRT staff for their technical support in the design, installing and using of the communication systems, marine scientists of the SAMS (Scottish Association for Marine Science), of the Norwegian College of Fishery Science of University of Tromsø, of the Institute of Protein Biochemistry and Enzymology and of the Institute of Biomolecular Chemistry of Naples, for their help and co-operation in experiments performing Romeo teleoperation from their laboratories, and finally Prof. Riccardo Cattaneo Vietti and Dr. Chiara Chiantore of the DIPTERIS-UNIGE (Department for the Study of the Territory and its Resources - University of Genoa) for their collaboration in the design and exploitation of the ABS benthic module.

## REFERENCES

Bono, R., G. Bruzzone, M. Caccia, F. Grassia, E. Spirandelli and G. Veruggio (1994). ROBY goes to Antarctica. In: *Proc. of OCEANS '94*. Vol. 3. pp. 621–625.

Bruzzone, Ga., R. Bono, Gi. Bruzzone, M. Caccia, M. Cini, P. Coletta, M. Maggiore, E. Spirandelli and G. Veruggio (2002). Internet-based satellite tele-operation of the romeo ROV in Antarctica. In: *Proc. of MCCA 2002*. Lisboa, Portugal.

Bruzzone, Ga., R. Bono, Gi. Bruzzone, M. Caccia, P. Coletta, E. Spirandelli and G. Veruggio (2003*a*). Internet-based tele-operation of underwater vehicles. In: *Proc. of 1st IFAC Workshop on Guidance and Control of Underwater Vehicles.*

Bruzzone, Ga., R. Bono, M. Caccia, P. Coletta and G. Veruggio (2003*b*). Internet-based satellite tele-operation of the romeo ROV in the Arctic region. In: *Proc. of MCMC 2003*. Girona, Spain. pp. 304–308.

Caccia, M., R. Bono, G. Bruzzone and G. Veruggio (2000). Unmanned underwater vehicles for scientific applications and robotics research: the ROMEO project. *Marine Technology Society Journal* **24**(2), 3–17.

Galeazzi, F. and B. Papalia (1998). SARA - an autonomous underwater vehicle for scientific surveys in Antarctica. In: *Proc. of OMAE 98*. Vol. Proc. on CD-ROM.

Kirkwood, W.J. (1998). Tiburon: science and technical results form MBARI's new ROV integrated to a SWATH platform. In: *Proc. of Oceans'98*. Vol. 3. Nice, France. pp. 1578–1583.

Nokin, M. (1998). Sea trials of the deep scientific system victor 6000. In: *Proc. of Oceans'98*. Vol. 3. Nice, France. pp. 1573–1577.

Stoker, C.R., D.R. Barch, B.P. Hine III and J. Barry (1995). Antarctic undersea exploration using a robotic submarine with a telepresence user interface. *IEEE Expert* **10**(6), 14–23.

Veruggio, G., R. Bono and M. Caccia (1994). Autonomous underwater vehicles: The Roby project. In: *Proc. of IARP 94*. pp. 141–146.

ELSEVIER
IFAC
PUBLICATIONS
www.elsevier.com/locate/i

# THE "SMART SPRING" MOUNTING SYSTEM[1]: A NEW ACTIVE CONTROL APPROACH FOR ISOLATING MARINE MACHINERY VIBRATION

**S. Daley*§ and F. A. Johnson****

** Department of Automatic Control & Systems Engineering, University of Sheffield, Sheffield, S1 3JD, UK.*
*** Consultant, Otia Tuta, Grassy Lane, Sevenoaks, Kent TN13 1PL, UK.*[2]

Abstract: A major problem, in isolating large marine machinery rafts, is how best to isolate excited resonances. These generate large forces on the hull and create a major vibration problem. The design of such mounts typically represents a compromise between providing good vibration isolation and good machinery alignment under seaway motion. This paper describes a completely new approach to the problem of isolating marine machinery by making use of digitally controlled actuators that ignore local displacements while controlling the response of the structure's rigid body modes. The paper describes how this is accomplished and gives some preliminary experimental results. *Copyright © 2004 IFAC*

## 1. INTRODUCTION

Vibration that propagates from marine propulsion and auxiliary machinery can cause significant problems associated with passenger and crew comfort and also through the generation of acoustic noise from the hull. Such acoustic noise creates a severe detection hazard in Naval vessels and is also problematic for civil vessels such as those used by fisheries research organisations.

Traditionally, the problem of vibration propagation is tackled by mounting large marine machinery on a rafted structure that is supported from the hull on a set of resilient elastomeric mounts (Crede, 1951). These passive mounts provide some isolation of machinery vibrations from the hull, however, performance markedly deteriorates in the presence of structural compliance.

A typical force transmission curve, for such a system, is shown in Figure 4 (darker curve). Here one sees that above the 5 Hz mount resonant frequency the force transmission is dominated by structural resonances. These generate large amplitude displacements at the mounts and hence transmit large forces to the hull. Further they also produce a significant and characteristic acoustic noise problem.

Passive mounting systems are designed, as far practicable, to minimise the consequences of excit structural resonances. This can done, for example, l using an array of passive mounts arranged to produ forces on the hull which approximate to a multipole order to minimise the onward acoustic transmission in the sea (Hartnell-Beavis & Swinbanks, 1976). Th exploits the fact that, with the exception of the rigid bo modes, the total momentum of each mode is zero,

$$\int \rho(x)\phi(x)dx = 0 \quad (1)$$

where $\phi(x)$ is the mode shape, $\rho(x)$ is the ma distribution and $x$ is the raft location. It follows fro equation (1), therefore, that if the distributed mou stiffness can be selected to be proportional to the ma distribution ($k(x) \propto \rho(x)$) then the total transmitt force due to a resonance is also zero, i.e.

$$\int k(x)\phi(x)dx = 0 \quad (2)$$

However, in practice this solution is not always feasib because of the large numbers of mounts required approximate the ideal continuous situation defined l equation (2) and the difficulty in accurately specifyi stiffness for a passive installation. Moreover, the meth also assumes that the hull has a high impedance wi long wavelength modes relative to the raft dimensions.

[1] "Smart Spring" is a trademark of BAE SYSTEMS
§ Corresponding author, Fax: +44-114-222-5661 Email: steve.daley@sheffield.ac.uk
[2] During the early part of the work described in this paper Dr Johnson was an employee of *BAE SYSTEMS ATC, Great Baddow, Essex, UK.*

As a result of the restricted performance of passive systems alternative designs based on active control technology are being sought. In a previous programme undertaken in the 1990's and funded by the US Office of Naval Research the authors were involved in the development of a fully active solution (Johnson & Swinbanks, 1996, Darbyshire & Kerry, 1997, Daley, 1998). This was based upon electromagnetic levitation of the machinery raft (see figure 1), however, the original concept required a large numbers of actuators and therefore represented an expensive solution.

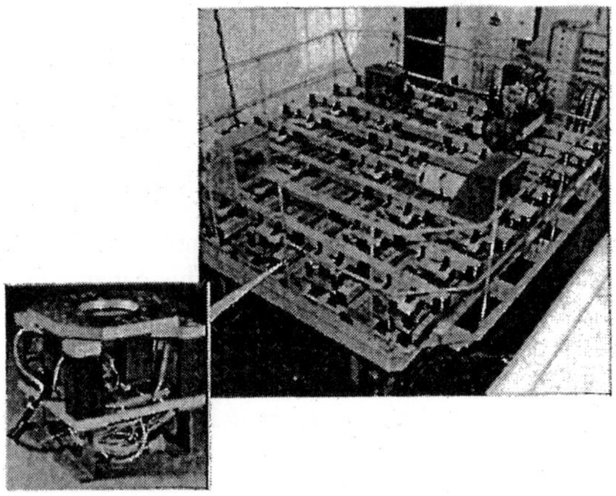

Figure 1: Prototype Active Machinery raft and tri-axial actuator (inset)

In this paper a new approach, known as the "Smart Spring" mounting system[3], which is being developed by BAE Systems is described. It is a hybrid active/passive solution that is a more practical and lower cost development of the technology illustrated in Figure 1.

## 2. "SMART SPRING" MOUNTING SYSTEM: FUNDAMENTAL PRINCIPLES

The new mounting system is based upon the use of an electromagnet combined in parallel with passive elements as shown schematically in Figure 2.

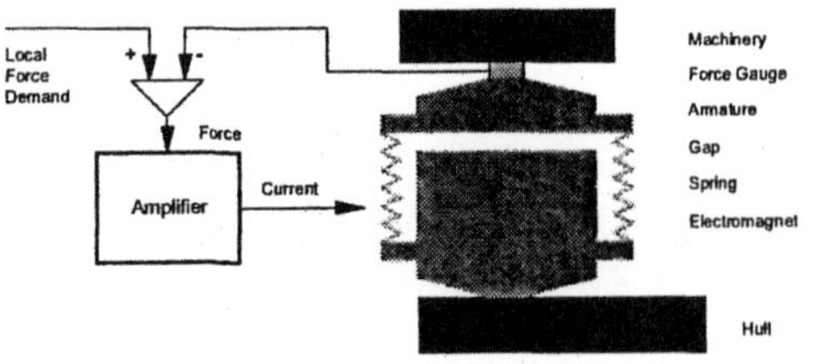


Figure 2: Local Control.

[3] This is the subject of several Patent Applications.

In order to avoid transmission of large forces at frequencies corresponding to supported structure resonances, the mounting system must fulfil a number of key requirements.

The first requirement is for an actuator that generates a reaction force on the hull that is completely independent of any local displacement of the supported structure at its attachment point. As a result *no additional force* would be generated on the hull from excited resonances. Thus the actuator for a "Smart Spring" mounting system must have effectively zero stiffness.

The second requirement is that to support the structure each actuator must also be able to generate an external demand force on the structure without affecting its zero stiffness response to local displacements of the structure.

These two objectives can be achieved by a local digital feedback control loop as illustrated in Figure 2 (the loop is described in more detail in the next section). The current in the coil of the electromagnet is manipulated by the action of the controller to ensure that force generated in the spring is cancelled and that the combined transmitted force as measured by the force gauge is identically equal to an external demand.

The third requirement is a means for generating the external demand forces, one for each supporting actuator.

Out-of-balance forces, generated by the moving machinery, result in both linear and angular displacements of the supported structure. The external demand forces are the means whereby these linear and angular displacements can be continuously opposed to return them towards their equilibrium positions in a controlled way.

However, to achieve the ideal rigid mass transmissibility curve, the external demand forces must be generated in response to displacements of the structure's six rigid body modes *only* and must not be affected by any structural resonances.

If the first two requirements of the actuator discussed above can be satisfied then this demand force can be generated using a modal control strategy for the raft (Inman, 1989). The first step in this process is to measure the displacements of the machinery raft using an array of accelerometers, or proximiters (non-contacting displacement transducers), to map the total displacements due to both rigid body modes and excited resonant modes. This data is then analysed to extract just the six rigid body displacements by exploiting the general orthogonality condition (Morse, 1948)

$$\int \rho(x)\phi_m(x)\phi_n(x)dx = \begin{cases} 0 & m \neq n \\ M_n & m = n \end{cases} \quad (3)$$

where $M_n$ is the effective mass for the nth mode and $\phi_m(x)$ and $\phi_n(x)$ are the m and nth mode shapes respectively.

In data processing terms this analysis reduces to a single matrix multiplication. For a discrete set of measurements, $\underline{x}$, the supported machinery and raft structure equations of motion can be described by

$$M\ddot{\underline{x}}+C\dot{\underline{x}}+K\underline{x}=\underline{f} \qquad (4)$$

where *M, C* and *K* represent the structure mass, damping and stiffness matrices respectively and $\underline{f}$ is a vector of applied forces. It follows that the decoupled independent modal space description

$$\ddot{\underline{p}}+\Lambda\dot{\underline{p}}+\Omega\underline{p}=\underline{f}_m \qquad (5)$$

where $\Lambda=diag(2\zeta\varpi_i)$, $\Omega=diag(\varpi_i^2)$ and $\underline{f}_m$ is a vector of modal forces, can be obtained by the transform

$$\underline{p}=V^T M^{\frac{1}{2}}\underline{x} \qquad (6)$$

where $V$ is an orthonormal matrix of eigenvectors of $M^{-\frac{1}{2}}KM^{-\frac{1}{2}}$. The matrix transform required for extracting the six rigid body modes (3 rotations and 3 translations) can therefore be obtained by utilising the rows of $V^T M^{\frac{1}{2}}$ that correspond to $\varpi_i=0$.

The final step is to use the extracted linear and angular displacements to calculate the demand forces required, at each actuator, to generate the rigid body accelerations in a way that does not introduce any cross-coupling between the linear and angular rigid body modes. These demand forces are then transmitted to the appropriate local controllers. Again this final stage, in data processing terms, reduces to a single matrix multiplication by selecting the appropriate columns of $M^{\frac{1}{2}}V$.

With this control strategy a stiffness function is applied to each of the 6 rigid body modes. If the characteristics of a damped spring are applied, then the ideal rigid body transmissibility curve is recovered. However since the stiffness function is now implemented digitally, any stabilising function can be applied. As a result it is possible to improve the isolation requirements beyond that achievable with an ideal passive system. For example Figure 3 shows the transmissibility curve for a function that falls (rather than rises) at 20dB /decade at higher frequency and also adds some phase lead around the mount resonant frequency. This produces a transmissibility that is a 40dB/decade improvement upon the ideal passive case whilst suppressing the mount resonance. When compared with the realistic case with resonance present (Figure 4) this represents a very large improvement in performance.

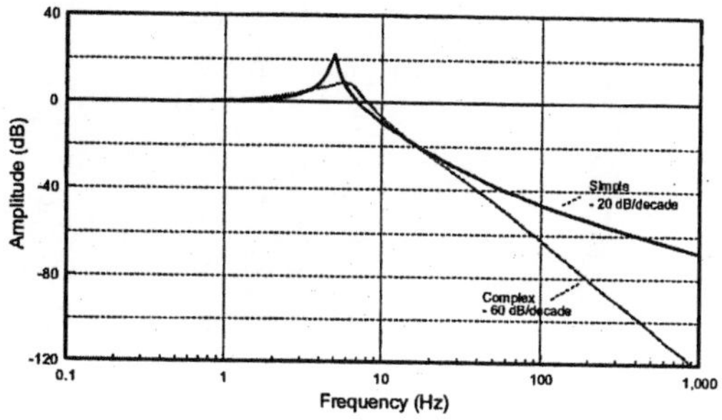


Figure 3: Comparison of a Simple with a Complex Stiffness Function for Ideal Structures.

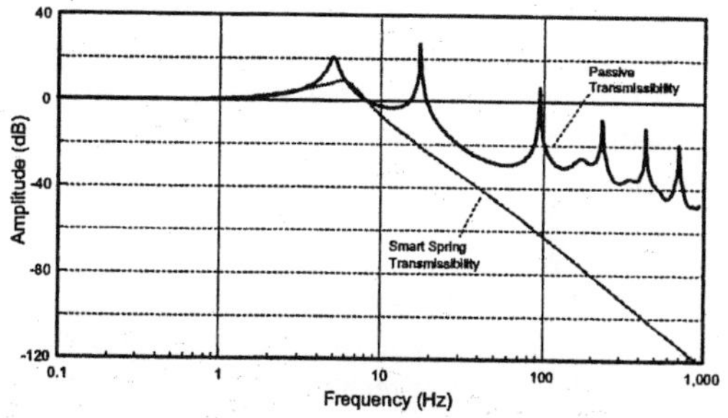


Figure 4: Improvement due to a "Smart Spring" Mounting System.

## 3. EXPERIMENTAL PROGRAMME

During 1999 an experimental programme was commenced that set out to demonstrate the fundamental principles and feasibility of the "Smart Spring" mounting system concept. Some preliminary results from this work have already appeared in the open literature (Daley, Johnson, Pearson & Dixon 2002) following a brief review of these initial results more recent developments are presented here.

A key requirement of an individual mount is that this should be able to generate a force that is independent of any local displacement but identically equal to the global demand set by the stiffness function described in the previous section.

To demonstrate the zero stiffness concept of a "Smart Spring" mounting system a prototype actuator was developed that utilised readily available components. An initial uni-axial actuator with coil steel springs is shown in figure 5. The electromagnet utilised was developed in a earlier project and was of a cylindrical design having a 4 ohm coil and a nominal lift capacity of 1500N at 5 mm and 6 Amps.

Figure 5: Prototype Actuator for a "Smart Spring" Mounting System

Force transmission across the actuator was measured using a load cell that was integrated into the design and located above the electromagnet. To enable the control system to be developed for a single actuator, the prototype was suspended from the cross-head of an electro-hydraulic test machine (Figure 6) with the armature being connected to the piston rod of an actuator.

Figure 6: Hydraulic Test Rig

A schematic diagram of the local control for an actuator for a "Smart Spring" mounting system is shown in Figure 7. This is more complex than the simple feedback control illustrated in Figure 2 and is needed to deal with the large non-linearity of an electromagnet. The controller operates on measurements of the relative displacement (or gap) between the armature and magnet face, the armature acceleration and the transmitted force to produce a demand signal for the amplifier that ensures that the transmitted force across the actuator matches the global demand force. The controller function, which is implemented on dSPACE hardware, is designed by using a mathematical model of the amplifier and combined electromagnet and spring elements. Because of the difficulties of deriving accurate parameters from physical laws, the model was developed using experimentally derived data.

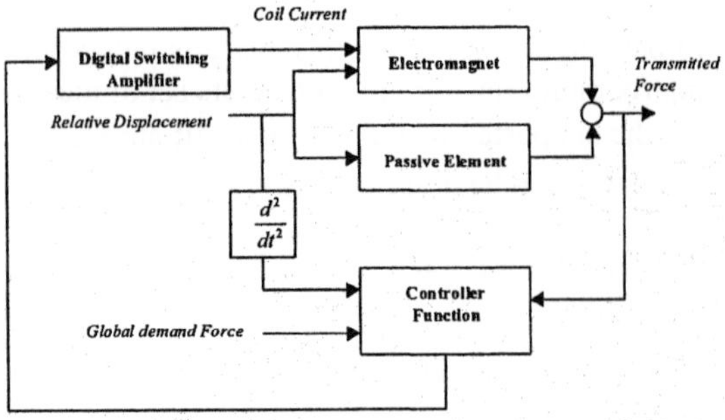


Figure 7: Schematic of a "Smart Spring" Local Control System

The initial controller design made use of a canonical system description in which the states associated with particular spring resonances (or modes) can be readily identified. This enables a state feedback law to be calculated such that the following quadratic cost function is minimised

$$J = \frac{1}{2\pi} \int_{-\infty}^{\infty} \sum_{k=1}^{n} q_k |y_k(j\omega)|^2 + q_o |e(j\omega)|^2 + r |F_d(j\omega)|^2 \, d\omega \qquad (7)$$

where, $y_k$ represent the $n$ modal components of the transmitted force $F$, $e$ is the integral of the error between the global force demand and the transmitted force, $F_d$ is the outer loop control signal (input to inverse non-linear function) and $q_k, q_o$ and $r$ are selectable weighting factors. The significance of this cost function is that the effect of specific resonances can be minimised selectively by appropriate choice of the weighting terms. This is an important feature that would allow the designer to shape the transmissibility dependent upon the specific frequencies of excitation present for any given machine.

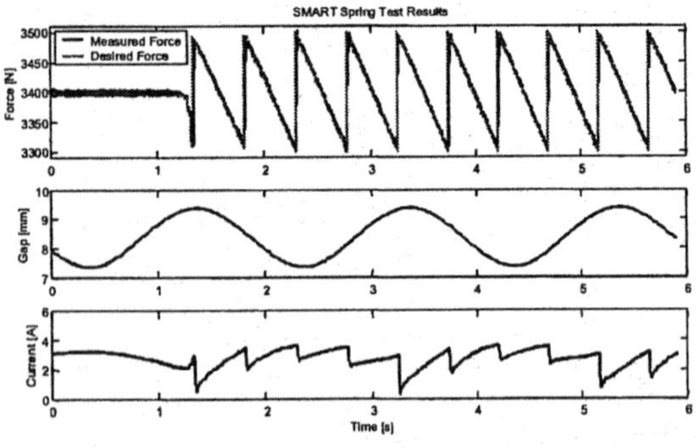


Figure 8: Actuator response for low frequency disturbance and saw-tooth demand

The ability of this controller to ensure that the force transmitted by an actuator for a "Smart Spring" mounting system is equal to a global demand whilst being independent of local displacement is show in figure 8. It can clearly be seen that despite the sinusoidal variation in displacement the force across the actuator tracks the demanded signal. The required current in the coil in order to achieve this is shown in the bottom curve of the figure.

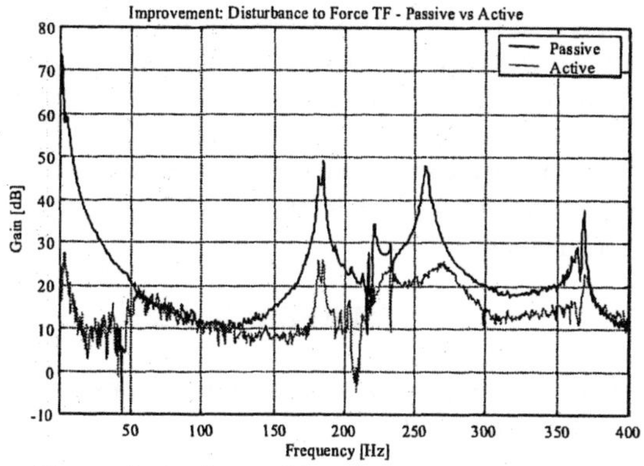


Figure 9: Active and Passive Mount Response

Figure 9 shows the frequency response function of the transmitted force for the actuator both with and without the controller enabled. It can be seen that the passive response (controller disabled) is dominated by the internal resonances of the spring. With the controller enabled the resonance is suppressed with peak force reductions of the order of 30dB being achieved. This demonstrates the feasibility of achieving the zero stiffness requirement of the mount.

Despite this success, however, the required performance necessitates a high feedback gain with the attendant problems associated with electronic noise generation. More recent work has focussed on the use of alternative materials as the passive elements with an experimental facility whereby the transmissibility can be measured directly (figure 10).

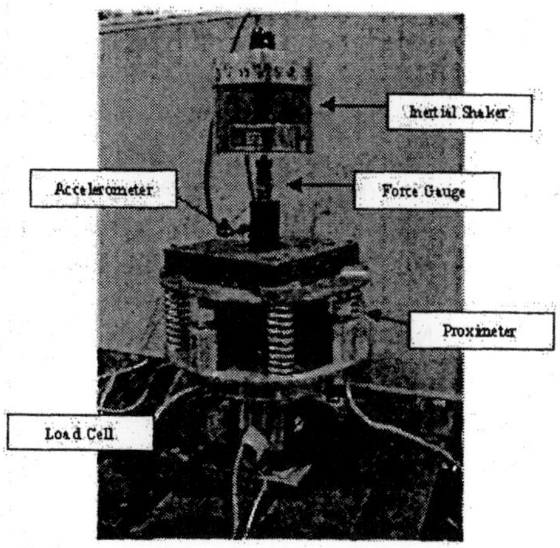


Figure 10: Single Actuator Test Facility.

The measured uncontrolled transmissibility for three different passive elements (steel coil springs, natural rubber and neoprene) using this facility are shown in figure 11.

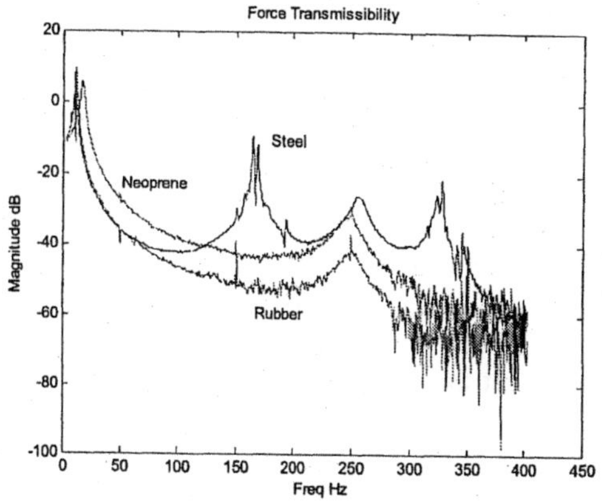


Figure 11: Force Transmissibility: Steel, Rubber and Neoprene

It will be seen from figure 11 that the steel coil springs have the lowest internal damping and the largest mount resonance. The absence of resonances in rubber and neoprene results in much lower natural force transmissibilities at frequencies above 150 Hz. It should be noted that the neoprene used has approximately twice the stiffness of the rubber and steel. In all cases, the force transmissibility rises at 250 Hz, due to the load cell resonance, but the roll-off rate, at higher frequencies, is significantly improved due to the additional high frequency isolation produced by this resonance. When the force transmissibility falls below –60 dB the measurements themselves fall within the noise floor of the load cell.

It is clear that the use of elastomers results in a better natural overall performance than steel coil springs. Natural rubber has a greater roll-off rate in the region immediately above the mount resonance (in fact it is clear that this matches the lowly damped steel) and is also often preferred in marine environments due to its superior tear strength.

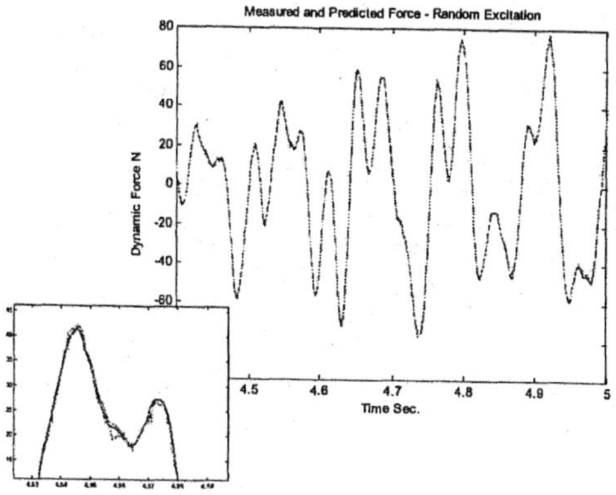


Figure 12: Measured and predicted force for rubber.

However, a critical issue is how well the performance of the elastomers can be modelled so that the composite actuator, of an elastomer in parallel with an electromagnet, can be controlled accurately by a feed forward prediction. This will then relax the dependence of the system performance on the feedback component of the controller.

The test of the modelling accuracy is shown by a comparison of the measured and predicted force outputs for a random excitation of the actuator on the test facility (Figure 10). Figure 12 shows this comparison for the rubber mount. In the figure a series of curves are shown based upon models derived in the frequency domain, time domain and a composite of the two. It has been found that the composite approach is required to achieve the necessary broad-band accuracy.

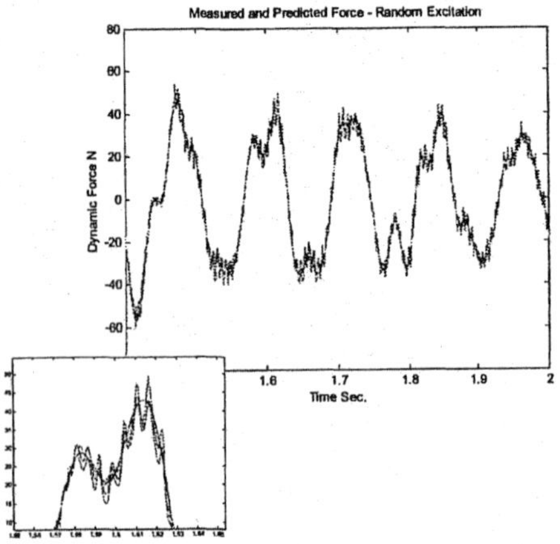


Figure 13: Measured and predicted force for steel.

The equivalent modelling results for steel are shown in figure 13. These clearly show the deleterious effects of the internal resonance on the achievable modelling accuracy.

The results achieved to date have shown the feasibility of achieving a zero stiffness actuator; a key requirement for the "Smart Spring" mounting system. The next stage planned is to support a realistic piece of machinery, of approximately 2 tonnes, on a set of digitally controlled actuators forming a complete "Smart Spring" mounting system to demonstrate the control of all six rigid body modes.

## 4. CONCLUSIONS

The work presented here shows that it is possible to design actuators for a "Smart Spring" vibration isolation mounting system that have zero stiffness in response to local displacements of their attachment points, over a bandwidth of 500 Hz, while at the same time generating an externally defined demand force. It has also been shown how such a feature enables control of a machinery raft's rigid body modes only while ignoring the consequences of excited resonances in the machinery raft's structure. The result is that the only forces then generated on the hull are the global control demand forces required to return the structure's rigid body modes to their equilibrium positions in a well defined way. This has the potential to greatly reduce the level of vibration in the hull, together with the attendant acoustic noise, over that produced with traditional mounts.

## 5. ACKNOWLEDGMENTS

We would like to acknowledge the support we have received from the U.S. Office of Naval Research for the work described in this paper. The key contributions of Phil Darbyshire and Malcolm Jay at BAE SYSTEMS ATC, Roger Dixon and Peter Knight at ALSTOM Power Technology Centre and Jonathan Sturgess of ALSTOM Research and Technology Centre are gratefully acknowledged.

## REFERENCES

Crede, C. E. (1951). *Vibration and Shock Isolation.* John Wiley. New York.

Daley, S. (1998). Algorithm architecture for sound and vibration control. *Institution of Measurement and Control Mini-symposium at UKACC Conference CONTROL'98*, Swansea. Sept. 1998.

Daley, S., Johnson, F. A., Pearson, J. P. & Dixon, R. (2004). Active vibration control for Marine Applications. *IFAC Journal Control Engineering Practice,* Volume 12, Number 4, pp465-474.

Darbyshire, E. P.& Kerry, C. J. (1997). A multi-processor computer architecture for active control. *Control Engineering Practice*, Vol. 5, No. 10, Sept 1997 pp. 1429-1434

Hartnell-Beavis, M. C. & Swinbanks, M. A. (1976). Some recent practical and theoretical developments in noise reduction in ships *Proceedings of IMarE Vibration and Noise Conference*, London, 21st October 1976, pp28-38

Inman, D. J., (1989). *Vibration with Control, Measurement and Stability.* Prentice Hall. New Jersey.

Johnson, F. A. & Swinbanks, M. A. (1996). Electromagnetic Control of Machinery Rafts. *Journal of Defence Science.* Vol1, No. 4, pp493-497

Morse, P. M..(1948). *Vibration and Sound.* McGraw-Hill. New York.

ELSEVIER
IFAC
PUBLICATIONS
www.elsevier.com/locate/ifac

# AN UNDERWATER SIMULATION ENVIRONMENT FOR TESTING AUTONOMOUS ROBOT CONTROL ARCHITECTURES *

**J. Antich, A. Ortiz**

*University of the Balearic Islands, Spain*
*Mathematics and Computer Science Department*
*{javi.antich, alberto.ortiz}@uib.es*

Abstract: In the field of robotics, a lot of valuable advantages derive from the use of simulation tools. The loss of detail, with respect to the real world, closely bound up with simulators, is compensated by a significant reduction of the effort, risks and monetary costs required to carry out a series of experiments. In this study, a 3D underwater simulation environment named $NEMO_{CAT}$ is presented as a tool to test and tune control architectures for Autonomous Underwater Vehicles (AUV). An object-oriented methodology based on the Unified Modelling Language (UML) has been applied to its development. By way of example, two different control architectures are described and validated in the simulator by using the hydrodynamic models of two real underwater vehicles called GARBI and URIS. *Copyright* © 2004 *IFAC*

Keywords: Simulator, Autonomous Underwater Vehicles, control systems.

## 1. INTRODUCTION

In the field of robotics, a lot of valuable advantages derive from the use of simulation tools. The loss of detail, with respect to the real world, closely bound up with simulators, is compensated by a significant reduction of the effort, risks and monetary costs required to carry out a series of experiments. This, however, does not mean to substitute the experimentation with prototypes as it is warned in (Arkin, 1998), but simply to complement it.

In this paper, a 3D simulator named $NEMO_{CAT}$ (Navigational Environment MOdeller, Control Architecture Tester), see fig. 1, has been developed in order to validate and tune reactive and hybrid AUV's control architectures based on schema theory (Arkin, 1989). Note that both control strategies are suitable to deal with dynamic and unstructured environments such as the submarine.

* This study has been partially supported by project CICYT-DPI2001-2311-C03-02 and FEDER fundings.

The design of this simulation tool has required the use of a software engineering methodology, the Rational Unified Process to be precise, which is usually applied together with UML. The latter is a general-purpose visual modelling language that can be used to specify, visualize, construct and document the artifacts of any complex software system such as $NEMO_{CAT}$. As for the implementation of the simulator, the C++ programming language and the OpenGL graphics library have been used.

By way of example, two control architectures which have successfully been tested in the simulator will be presented. The first one simply tries to achieve a user-defined sequence of goal points in a safe way. On the other hand, the second control system, being more specific than the previous one, is intended to locate and track autonomously an underwater cable, or pipeline, on the basis of a vision system. In this case, simulated and real results are also compared for a simple mission carried out in a water tank.

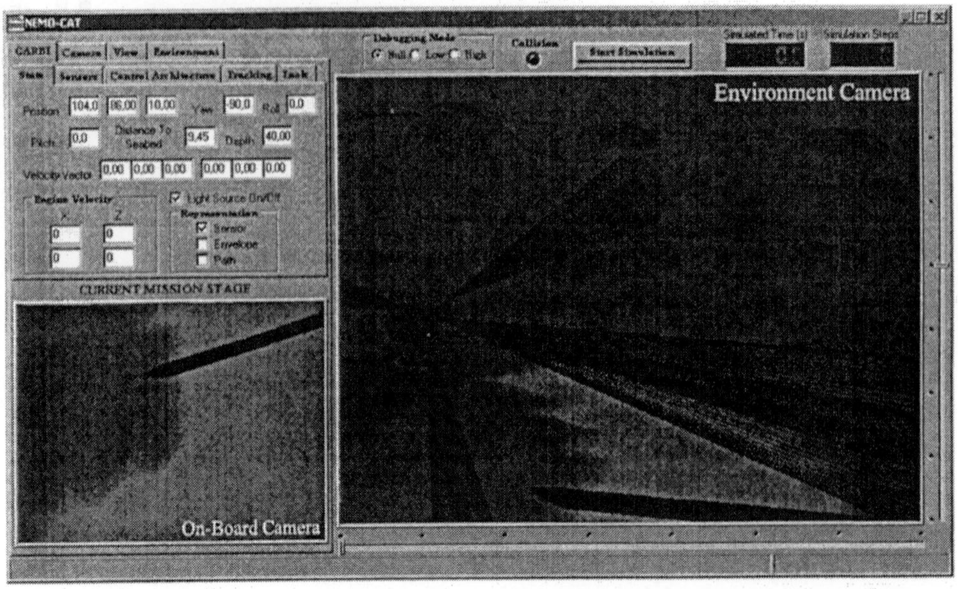


Fig. 1. A global view of the underwater simulation environment $NEMO_{CAT}$

The AUV models used in both simulations are based on the real dynamics of two vehicles designed and built by the Computer Vision and Robotics research group of the University of Girona (Spain) called GARBI and URIS. Information about the method followed to model and estimate the dynamics of such vehicles can be found in (Ridao *et al.*, 2001; Carreras *et al.*, 2003).

The rest of the paper is organized as follows: section 2 describes the virtual underwater environment through which the vehicles navigate; the dynamic model and the sensory equipment of the AUVs are discussed in section 3; section 4 summarizes the main characteristics of the control architectures supported by the simulator, while section 5 presents a specific example of each of them together with experimental results; and, finally, some conclusions are given in the last section.

## 2. THE UNDERWATER ENVIRONMENT

As it can be observed in fig. 2 (a), the seabed is modelled by means of a grid of points whose extent and resolution can be configured for the mission at hand. Initially, all the points of such grid are onto a plane which is parallel to the one defined by the $X$ and $Y$ axes. Typical elements of underwater environments such as rocks, holes and algae can be afterwards added to the seabed. To this end, the heights, or $Z$ coordinates, of some grid points are altered according to the position and shape —elliptical, in our case— of the elements incorporated. An adaptation of the well-known computer graphics algorithm called *random displacement of the midpoint* is used in order to give to those seabed deformations a natural appearance (see fig. 2 (b)).

To simulate obstacle-avoidance sensors such as sonars, the detection of rocks, holes and, even, algae is carried out in a fast and simple way through the concept of

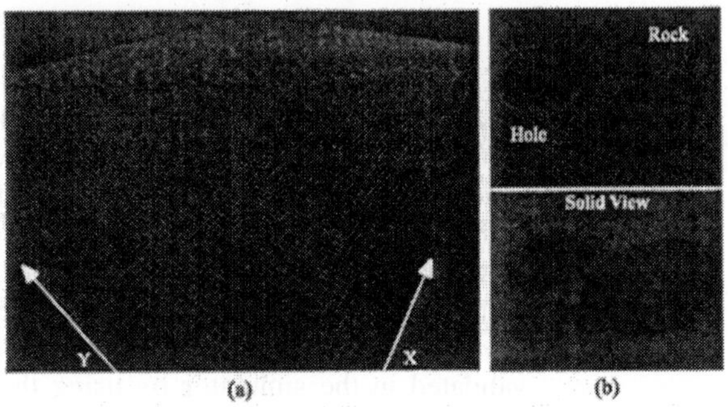


Fig. 2. (a) Model and (b) deformations of the seabed

bounding surface. It is an easy-to-compute function that approximates the shape of a certain underwater environment element.

The generality of the simulator allows modelling other elements as well. As it will be seen later, cables and pipelines can be deployed on the seabed. Interesting situations such as partial concealments, free span, etc. are obtained in a natural way.

## 3. AUTONOMOUS UNDERWATER VEHICLES

AUVs are introduced in the simulator specifying their hydrodynamic model parameters, together with their visual appearance and their sensor configuration. Since there is not a limit in relation to the maximum number of AUVs that can be simultaneously simulated, multirobot control strategies can also be studied by using $NEMO_{CAT}$.

### *3.1 Hydrodynamic Model*

The dynamics of the vehicles are assumed to obey the non-linear equations with six degrees of freedom (DOF) summarised by equation 1, according to (Fossen, 1994), where: $a$ is the acceleration vector; $M$ and $C$ are, respectively, the inertia and coriolis matrices of both the rigid body and the added mass; $\tau$ is the force and torque input vector; $g$ is the gravity and buoyancy vector; $D$ is the hydrodynamic damping matrix; and, finally, $v$ is the velocity vector. The parameters needed to solve this equation are loaded from a file at the beginning of the simulation. At the moment, the dynamic model of the underwater vehicles GARBI and URIS are available. In both cases, their parameters were estimated performing several real tests.

$$a = M^{-1} \cdot (\tau + g(\eta) - D(v) \cdot v - C(v) \cdot v) \quad (1)$$

### 3.2 *Sensory Equipment*

Three different kinds of sensors can be used by the simulator's AUVs to suitably carry out their missions: sonars, compasses and cameras. Furthermore, in applications where it is important to know the position of the vehicle, an acoustic positioning system can also be used to estimate it. A résumé of the main features of each of these sensory units is given next:

- As many *sonars* as desired can be attached to an AUV. From the location of the sonar, its orientation is computed according to the particular shape of the vehicle's hull. The beam that they generate is assumed to be conical. Its resolution and the minimum and maximum distances that can be measured are specified as parameters.
- Information about the orientation of the vehicle, the yaw angle to be precise, can be obtained by using a magnetic *compass.*
- Generic pin-hole *cameras* with six DOFs are also available to be put on board. In this way, a computer vision algorithm could be implemented in order to guide the vehicle on the basis of the captured images. Spotlights can also be attached to the vehicle to improve the brightness of the images.
- The *acoustic positioning system* incorporated into the simulator is of the so-called Long Base Line (LBL) type. Based on range measurements, it determines the position of the underwater vehicle relative to an array of transponders, at least three, deployed on the seafloor.

The possible sensory equipment of the vehicles will be extended in the near future by means of pressure, speed and water and battery charge detection sensors, as well as inertial measurement units.

## 4. SUPPORTED CONTROL SYSTEMS

Nowadays, there is a small number of fundamentally different classes of robot control methodologies (see (Coste-Manière and Simmons, 2000; Ridao *et al.*, 1999), among many others), usually embodied in particular control architectures, which can be roughly classified as: deliberative/hierarchical, behavioural/reactive, and hybrid. However, only the two last ones are suitable to deal with complex, non-structured, and changing worlds, like the majority of underwater environments. Both control approaches are briefly described in the following.

### 4.1 *Behaviour-based Control*

In this context, behaviours are the basic building blocks to carry out robotic actions, representing each of them the reaction to a stimulus. Behavioural responses are all coded as 3D vectors whose orientation denotes the direction to be followed by the vehicle while the magnitude expresses the strength of the response against other behavioral commands. The vectors generated by the architecture's active behaviours are asynchronously channelled into a coordination mechanism which can be, in the simulator, either competitive or cooperative. The first strategy selects as output the vector associated to the behaviour of highest priority. On the other hand, cooperative coordination mechanisms merge the recommendations given by all the behaviours in order to obtain a control action that represents their consensus. Both strategies can be combined making use of the concept of assemblage, which is supported by $NEMO_{CAT}$. An assemblage is a recursive structure composed by a coordination mechanism plus a set of primitive behaviours and, optionally, other assemblages. In this way, the advantages of these strategies will be able to be jointly exploited. A complete class library of simple behaviours is available in the simulator in order to accelerate to the maximum the development of these control systems.

### 4.2 *Hybrid Control*

Hybrid control systems are characterized by merging both deliberative and reactive activities. They are usually structured in three layers: planning, control execution and reactive. The former transforms the user-defined mission into a set of more simple tasks on the basis of a previous plan computed by applying some symbolic reasoning technique. The second layer supervises the accomplishment of such tasks, refining them when it is necessary. Finally, the last layer is able to give a quick response to situations that could not be suitably predicted in time by the system.

## 5. EXPERIMENTAL RESULTS

To illustrate the capabilities of the simulator, two control architectures which have successfully been implemented and tested are presented.

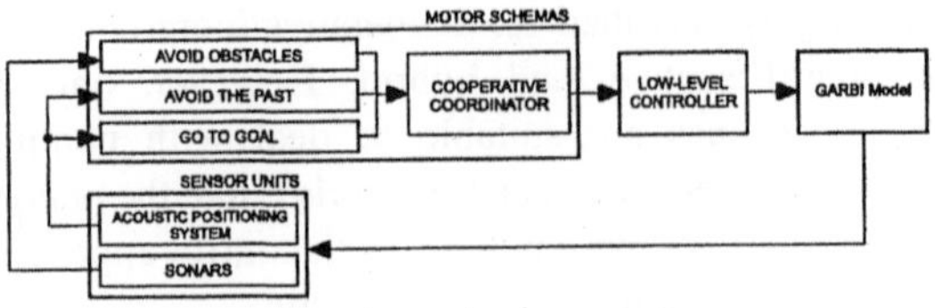


Fig. 3. Control architecture of the first example

### *5.1 A Goal-based Control Architecture*

The first proposed control system is intended to carry out a simple task: reach a user-specified sequence of goal points avoiding obstacles as well as getting stuck in any part of the environment. A purely reactive approach is, in general, enough to fulfill all these requirements. To this end, as it can be seen in fig. 3, three primitive behaviours have been used:

- *Avoid obstacles* allows the vehicle to avoid navigational barriers such as rocks, algae or, even, other possible cooperating vehicles. In this case, a vector in the opposite direction to the obstacles is generated. The magnitude of the vector is variable, according to the distance that separates the AUV from the obstacles ahead.
- *Avoid the past* is intended to avoid the well-known trapping problem characteristic of reactive strategies (Balch and Arkin, 1993). For such a purpose, a local map of the most recent AUV's path is used. When the vehicle is detected in essentially the same area for a long time, this behaviour becomes active generating a vector whose direction favours the exploration of new regions of the environment. In this case, the magnitude of the vector is proportional to the size of the area where the vehicle has been trapped into.
- *Go to Goal* drives the vehicle to a certain user-defined 3-dimensional point by generating a vector, constant in magnitude, whose direction joins the current position of the AUV with the goal point under consideration. Really, this behaviour is a bit more complex because it keeps a list of all those goal points that should be reached along the mission, following the order in which they were specified. When the vehicle gets sufficiently close to the ongoing goal point, the next one in the list is chosen.

Once both the behaviours and the mechanism required to coordinate them were selected, a mission where the vehicle had to achieve six goal points spread throughout a $200 \times 150 \times 50$-metre underwater environment was simulated by using the hydrodynamic model of the robot GARBI (see fig. 4). Three main obstacles made difficult the task of the AUV: a wall-like rock, a long and narrow tunnel and a box-shaped canyon. All of them, see fig. 5, were overcome by the vehicle which reached its high-level goal. Fig. 6 shows the activity of each behaviour along the mission.

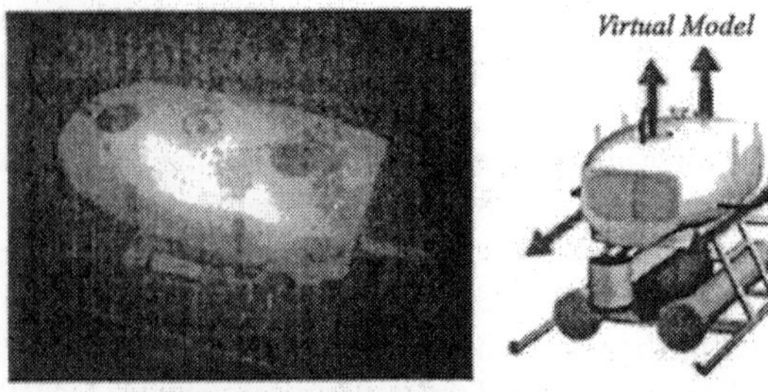


Fig. 4. The underwater vehicle GARBI

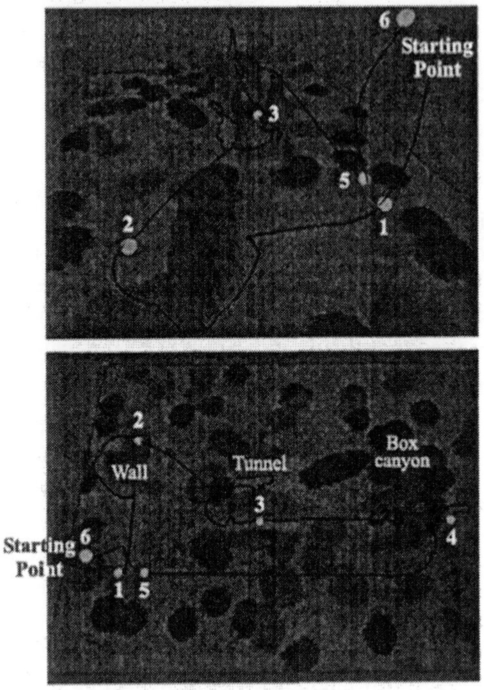


Fig. 5. Resultant AUV's trajectory. Two different views are shown

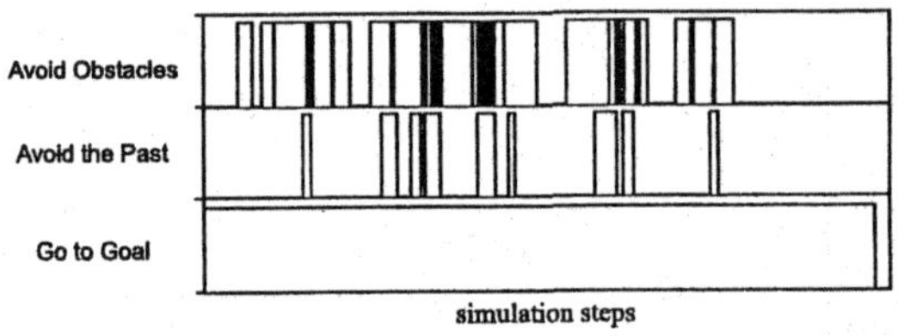


Fig. 6. Behaviour activity

### *5.2 A Control Architecture for a Vision-Guided Underwater Cable Tracker*

A control architecture for visually guiding an AUV to detect and track an underwater cable, or pipeline, laid on the seabed is presented next. For the details see (Antich and Ortiz, 2004).

In a typical mission, several different stages can be distinguished: diving, sweeping, tracking and homing. In the first one, the vehicle, after having been released from the support ship, goes down until a certain distance to the seabed is reached. The second and third stages comprise, respectively, searching for the cable in a predefined exploration area and tracking it afterwards. Finally, the AUV returns to the starting point

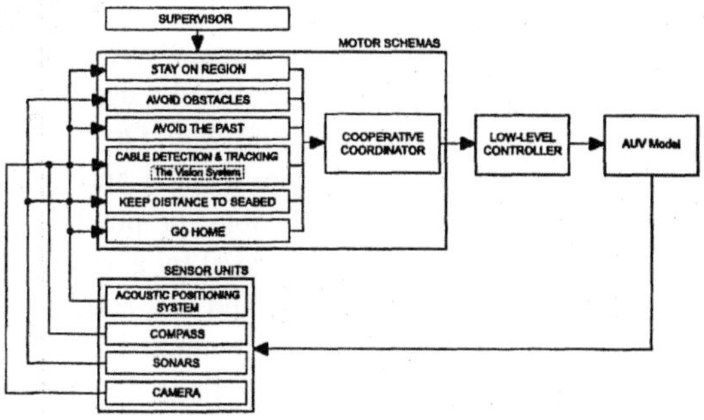


Fig. 7. Control architecture of the second example

after having achieved the limits of the exploration area while tracking the cable.

Taking into account this general way of action, the reactive control layer of the vehicle was split into six primitive behaviours (see fig. 7). Some of them have already been explained in the previous subsection so only the specific ones of this application are going to be described in the following:

- *Stay on region* prevents the AUV from straying from the area to be explored. The behaviour is exclusively activated when the vehicle is close to the limits of the exploration area. In such a case, a vector that moves the vehicle away from those limits is generated, being its magnitude directly related to the corresponding distance: the closer to the limits, the larger the magnitude.
- *Cable detection and tracking* moves the AUV strategically through the exploration area in search of a sufficient evidence of the presence of the cable. Specifically, after having acquired the working depth through a vertical path from the surface, the vehicle executes the sweeping stage performing a zigzag movement on the exploration area until the cable is found.

  Once the cable has been detected, the tracking stage starts. Two different tasks are sequentially executed: the first one tries to keep the cable oriented vertically in the images captured by the camera attached to the AUV, while the second task intends to maintain the cable in the central area of the images. In this way, improvements in both the cable visual detection and the length and smoothness of the vehicle's path are expected.

  As it can be anticipated in a real application, anomalous situations can arise. In particular, the cable can disappear from the images because the AUV's course has drifted apart from the actual cable location. In such cases, a suitable recovery mechanism is activated, consisting in making the behaviour return to its internal search state, where the vehicle acquires the aforementioned zigzag movement. However, now the area to be explored is reduced using the vehicle's trajectory during the past tracking stage. This trajectory is fitted by a straight line and a new search zone is determined computing the intersection between such line and the limits of the exploration area.

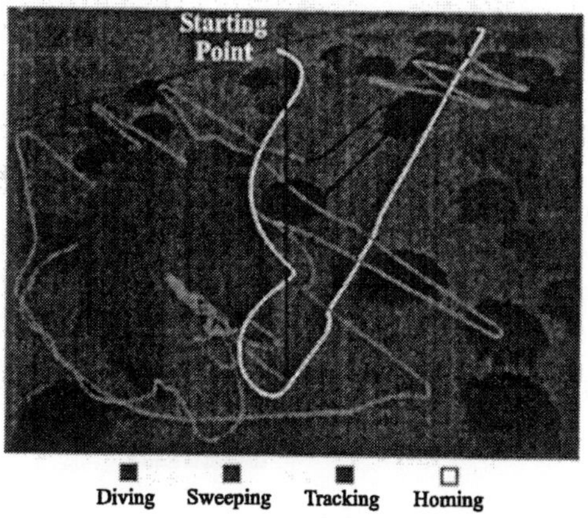


Fig. 8. Resultant AUV's trajectory

- *Keep distance to seabed* tries to keep the distance to the seabed constant in order to keep the apparent width of the cable in the images also constant. In this way, the vision subsystem can assume that the separation between both sides of the cable is nearly constant, and use this information to reduce its probability of failure. Sonars or, in case they cannot bring accurate enough measures, the acoustic positioning system, are expected to supply the required distance to the seabed.
- *Go home*, finally, makes the vehicle go to the starting point of the mission. Two different steps are carried out: first, the AUV approaches the goal point keeping a certain distance to the seabed; afterwards, it goes up until the sea surface is reached. In both cases, the magnitude of the output vector is proportional to the proximity to the intermediate/final goals considered.

In this example, a supervisor has been added to the proposed control architecture (see fig. 7). This component represents the characteristic control execution layer of hybrid control systems. From a functional point of view, it simply turns behaviours on and off depending on the mission stage where the vehicle is, in order to avoid conflicts among them.

As for the experimental results, a representative mission was simulated making use again of the hydrodynamic model of the underwater vehicle GARBI. As it can be observed in fig. 8, the AUV, after the diving stage, is trapped into a box-shaped canyon. The activation of the "avoid the past" behaviour allows the vehicle to escape from this undesirable situation. In this way, the distinctive zigzag movement of the sweeping stage can be resumed. When tracking, the cable is lost on one occasion and subsequently tracked again after a restricted search process performed on a small region of the exploration area. The mission finishes with the return of the vehicle to the starting point. The activity of the different architecture's behaviours along the mission is displayed in fig. 9.

Finally, in order to determine the accuracy of the tool under study, the same mission was carried out in both a real and the corresponding virtual underwater environ-

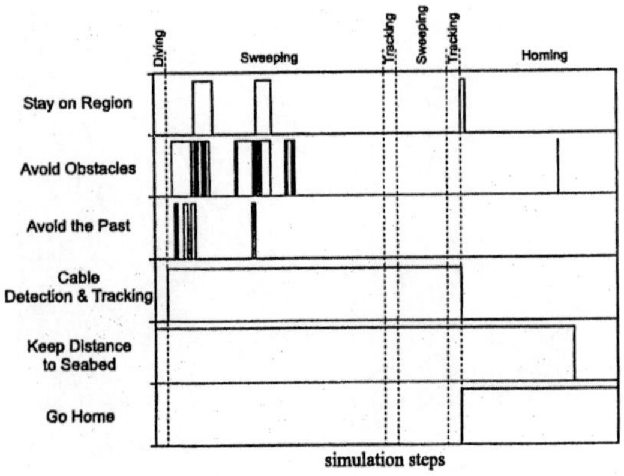


Fig. 9. Behaviour activity

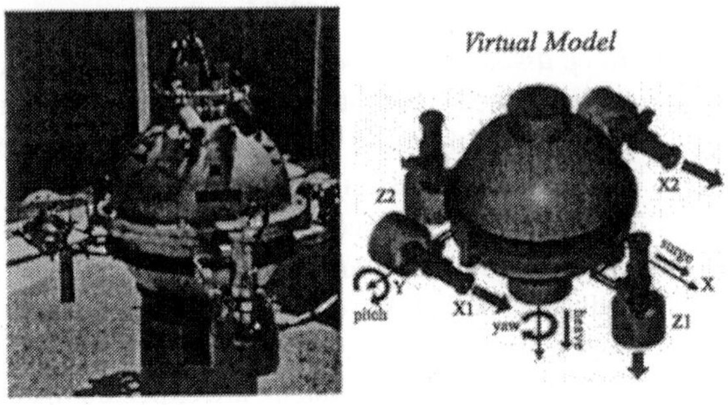


Fig. 10. The underwater vehicle URIS

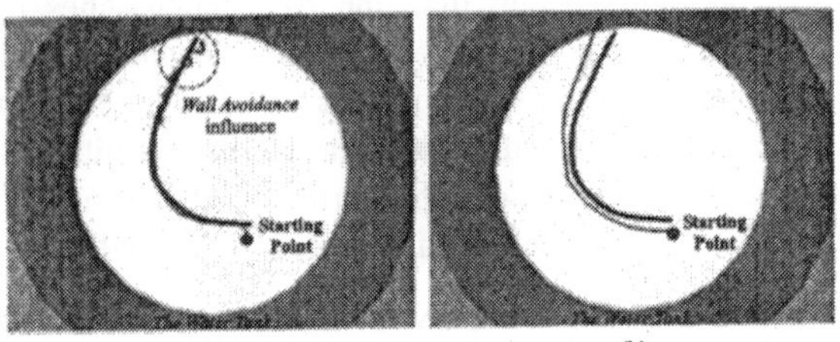


(a) (b)

Fig. 11. Comparing different kinds of results: (a) real and (b) simulated

ment. It consists in tracking a curved hosepipe laid on the floor of a water tank. The robot URIS (see fig. 10) was used in the real experiments. As for the virtual ones, the hydrodynamic model of such underwater vehicle was incorporated into the simulator. The real and simulated results obtained are respectively shown in fig. 11 (a) and (b). As it can be seen, in this case, the difference between both trajectories is not significant.

## 6. CONCLUSIONS

A 3D object-oriented simulator named $NEMO_{CAT}$ has been presented as a tool to validate and tune, before performing real experiments —or, even, in parallel—, behaviour-based and hybrid control architectures for autonomous underwater vehicles. The development of this underwater simulation environment has followed the steps defined by the Rational Unified Process using UML. In this way, future extensions of the functionality of this software are expected to be carried out spending a short time. By way of example, two particular control architectures have been successfully tested. The first one achieved a user-defined sequence of goal points in a safe way. On the other hand, the second control system located and tracked autonomously an underwater cable, or pipeline, according to the information provided by a vision system. In this last case, simulated and real results were compared for a simple mission conducted in a water tank with a hosepipe laid on the floor. Insignificant differences between the trajectories followed by the URIS robot in each of the environments were detected. In the near future, sensor error models are going to be incorporated into the simulator in order to considerably reduce those differences.

## ACKNOWLEDGEMENTS

The authors of this study would like to thank the members of the Computer Vision and Robotics research group of the University of Girona (Spain), specially to P. Ridao and M. Carreras, for providing them with the hydrodynamic model of their underwater vehicles GARBI and URIS, and for allowing them to perform real experiments with the latter. Antonio Serna, on the other hand, also helped them, taking advantage of his extensive experience in the field of software engineering, to develop an early version of the simulation environment described.

## REFERENCES

Antich, J. and A. Ortiz (2004). Development of the control architecture of a vision-guided underwater cable tracker. *Intl. Journal of Intelligent Systems (in press)*.

Arkin, R. (1989). *Neuroscience in motion: the application of schema theory to mobile robotics*. Plenum Press. New York.

Arkin, R. (1998). *Behavior-based robotics*. MIT Press.

Balch, T. and R. Arkin (1993). Avoiding the past: a simple but effective strategy for reactive navigation. In: *Proceedings of ICRA*. pp. 678–685.

Carreras, M., A. Tiano, A. El-Fakdi, A. Zirilli and P. Ridao (2003). On the identification of non linear models of unmanned underwater vehicles. In: *Proceedings of the Workshop on Guidance and Control of Underwater Vehicles*. pp. 59–64.

Coste-Manière, È. and R. Simmons (2000). Architecture, the backbone of robotic systems. In: *Proceedings of ICRA*. pp. 67–72.

Fossen, T. I. (1994). *Guidance and Control of Ocean Vehicles*. John Wiley & Sons.

Ridao, P., J. Batlle and M. Carreras (2001). Dynamics model of an underwater robotic vehicle. Technical Report IIiA 01-05-RR. UdG.

Ridao, P., J. Batlle, J. Amat and G. N. Roberts (1999). Recent trends in control architectures for autonomous underwater vehicles. *Intl. Journal of Systems Science* **30**(9), 1033–1056.

ELSEVIER
IFAC
PUBLICATIONS
www.elsevier.com/locate/ifac

# MVT: A MARINE VISUALIZATION TOOLBOX FOR MATLAB.

**Andreas Lund Danielsen, Erik Kyrkjebø, Kristin Ytterstad Pettersen**

*Department of Engineering Cybernetics, The Norwegian University of Science and Technology, Norway. Email: andreasl@stud.ntnu.no, erik.kyrkjebo@itk.ntnu.no, kristin.y.pettersen@itk.ntnu.no.*

Abstract: This paper introduces a new toolbox for use with Matlab® and the Virtual Reality Toolbox 3.0. The toolbox is intended to assist in the performance analysis of multivariable systems and to help presenting the experimental results. The toolbox displays up to 6 DOF data from simulations, experiments or measurements of marine control systems as 3D animations. The animations may be viewed on-line or saved to file.


Keywords: Marine systems, Moving objects, Performance analysis, Simulation, Software tools, User interfaces, Vehicle simulators, Virtual reality.

## 1. INTRODUCTION

Several computer tools for development and analysis of marine control systems have been introduced the last few years (Fossen, 2002a; Perez and Blanke, 2003). These support mathematical modelling, control and analysis of multivariable systems. Results from simulations, experiments and measurements of such systems are usually presented as functions of time in 2D figures.

Marine vessels in the real world imply time-varying movements in three dimensions. We therefore consider 3D animations to be a valueable way of presenting and interpreting data from such movements, as 3D animations are able to display how a vessel's position and orientation change over time.

Such software tools were developed already several years ago (Brutzman, 1995), but due to the lack of processing speed these required separate graphics workstations to display real-time 3D images. Today's modern tools for 3D animation run on ordinary computers but still require a certain level of user knowledge. We wish to provide a tool that is easily accessible and compatible with other relevant software, and thus present a toolbox that use Matlab® and VRML for 3D animation. Simulations and experiments are often developed and conducted in a Matlab® environment, and the toolbox can be used in parallel with these programs without any additional software.

The toolbox presented in this paper displays data of multivariable systems utilizing features of the Matlab® Virtual Reality Toolbox 3.0. The Virtual Reality Toolbox provides a Matlab® interface for viewing 3D models (Mathworks, 2002). The Marine Visualization Toolbox (MVT) uses this interface to animate 3D vessel models; moving and rotating the vessels according to the time-varying input data.

The input data may represent a maximum of six degrees of freedom (6 DOF), *xyz*-positions and Euler angles. Any number of vessels may be animated at the same time.

The vessels are animated in a *scene model* (e.g. a coastline or a harbor) representing the actual surroundings of the simulations, experiments or full scale measurements. Both scene and vessel models are implemented in the *Virtual Reality Modeling Language* (VRML).

This paper is organized as follows: Section 2 describes the choice of software development tools. Section 3 describes the notation and data structures in MVT, and Section 4 shows how these input data are linked to 3D vessel models and scenes. Section 5 describes how to create animations. Section 6 describes the documentation and help functions included in MVT. Finally some conclusions are drawn in Section 7.

## 2. MATLAB® AND 3D ANIMATION

Several computer tools for marine control systems are developed for use with Matlab® (Fossen, 2002a; Perez and Blanke, 2003). The motivation for developing the MVT in Matlab® is thus the seamless interface towards these tools and the software environment familiar to many potential users.

VRML is a combination of a network communication language and a 3D modeling language. VRML was developed for on-line 3D experiences across networks and interactions between multiple users (Carey and Bell, 1997). This resulted in a platform independent language, with a minimum of geometric primitives and extensive features for user interaction.

The VRML version supported by the Virtual Reality Toolbox 3.0 is VRML97, standard ISO/IEC 14772-1:1997 (MathWorks, 2002). VRML is old and already replaced by X3D in general, which provides greater flexibility and incorporates the advances in commercial graphics hardware (Web3D Consortium, 2004). The 3D rendering when using VRML is thus not optimized for most of today's computers, and the conversion from still images to movie files is a relatively slow process in MATLAB®. These limitations are weighed against the benefits of having a visualization tool closely linked with the development tool. Matlab provides the user environment familiar to most users, and is therefore the preferred choice of environment, although the toolbox would benefit from any upgrades in Matlab's 3D engine.

The Virtual Reality Toolbox 3.0 for Matlab® includes a VRML editor, a VRML viewer and m-file functions that make it possible to create, view and interact with virtual worlds from within the Matlab® workspace (MathWorks, 2002). MVT includes two libraries of VRML files; vessel models and scenes. Combinations of these are generated to resemble specific scenarios with the desired number and types of vessels located in the desired surroundings.

By following the MVT guidelines when creating VRML files, custom vessel and scene models may be used with the toolbox and this is thoroughly explained in the toolbox documentation.

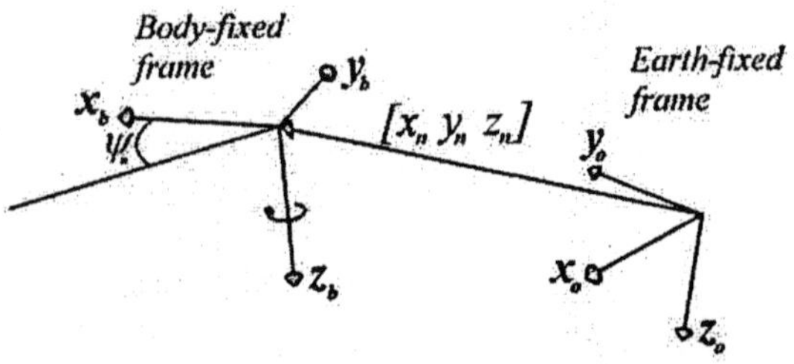


Fig. 2. Translation and rotation of a body-fixed frame with respect to the Earth-fixed frame.

## 3. COORDINATE SYSTEMS AND 6 DOF DATA

The coordinate systems used in MVT follows the convention from SNAME (1950). Figure 1 shows the positive directions of the Earth-fixed frame and the rotation axes of the body-fixed vessel frame.

In MVT the Earth fixed frame is represented by a scene model. The motion of a vessel is described by the coordinates $x$, $y$, $z$ representing the position of the vessel, and the Euler angles $\phi$, $\theta$, $\psi$ representing the orientation of the vessel, all with respect to the Earth-fixed frame, see Figure 2.

These six input variables represent the six degrees of freedom of the vessel (Fossen, 2002b). In this paper a sample of time-varying positions and rotations will be referred to as 6 DOF data. The data, in addition to a time vector that identifies the time of each sample, are the only required data inputs to create animations, see

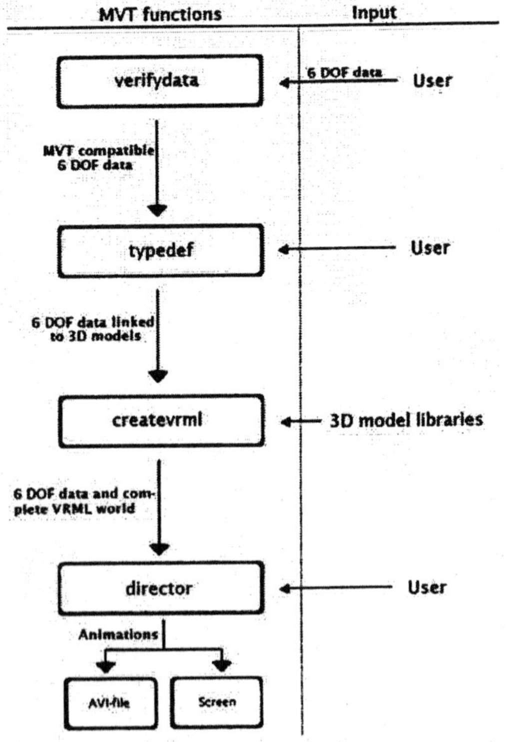


Fig. 3. Dataflow in the MVT.

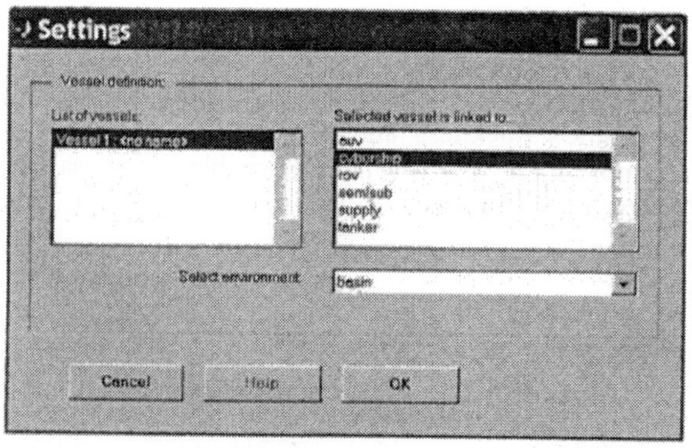


Fig. 4. The type definition GUI **typedef**.

the data flow diagram in Figure 3. Note that since animations are created frame-by-frame, velocity and acceleration data are not needed.

The input time vector, containing the time of each sample, may have variable step sizes (the difference between subsequent time samples). The only requirement is that the time is strictly increasing. When animations are displayed or saved to file the 6 DOF data are interpolated to achieve a fixed step size and the desired playback speed, using piecewise cubic hermite interpolation (the Matlab® **pchip** function).

## 4. LINKING INPUT DATA TO 3D MODELS

The toolbox includes six different VRML vessel models: an autonomous underwater vehicle (AUV), *Cybership II* (a model of a general purpose supply vessel, scale 1:70), a remotely operated vehicle (ROV), a semi submersible rig, a full-scale general purpose supply vessel and a gas tanker. Included VRML scene models are: a basin for scale experiments, a flat ocean surface and an ocean floor.

Linking the input 6 DOF data to the VRML vessel models is achieved by using the graphical user interface (GUI) **typedef** (see Figure 4 for GUI layout). In the GUI all vessels identified in the input data are listed in the left column, and the available vessel models are listed in the right column. Links are created by selecting a model for each vessel.

The established links and the 6 DOF data are passed to **createvrml**, which automatically assembles the associated VRML files into one single VRML file, see Figure 3.

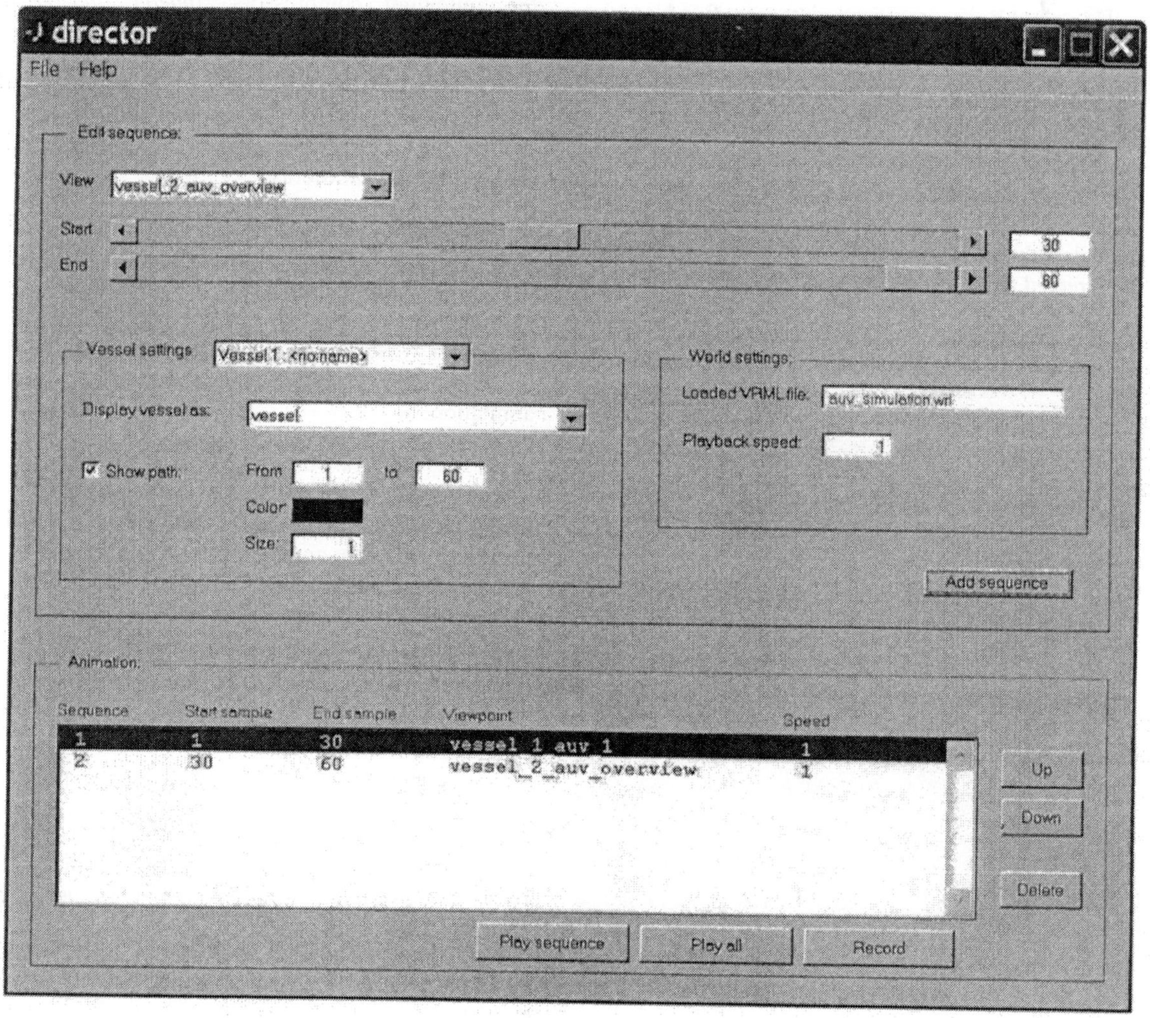


Fig. 5. The animation GUI **director**.

## 5. PLAY AND RECORD ANIMATIONS

The GUI **director** presents a way of playing and recording 3D animations (see Figure 5 for the GUI layout). The input VRML file must be a file generated by **createvrml**, see Figure 3.

An animation is composed of one or more *sequences*, and a sequence is defined by (detailed descriptions follow in the subsequent paragraphs):

- A viewpoint
- A set of properties for each vessel (e.g. visibility, path on/off)
- A time interval
- A playback speed

Each VRML model in the MVT libraries has a number of predefined viewpoints. Users are able to choose among these from the **director** GUI. When building new VRML worlds, **createvrml** assigns unique names to these viewpoints (e.g. *vessel_2_auv_overview*), making it easier to select the desired viewpoint. Only one viewpoint is defined for each sequence. Vessel viewpoints follow the vessel's position, while scene viewpoints are fixed to the Earth-fixed frame.

Vessel properties enable users to customize how vessels are displayed. Vessels from the MVT vessel library may be displayed as: the actual model, the axes of its body-fixed frame, or the vessel may not be displayed at all (see examples in Figure 6). By reducing the number of displayed vessels, focus can be directed to one particular vessel of interest.

A vessel's path is a line drawn from the vessel's first position sample through to the last position sample, see Figure 7. The size and range of vessel paths are a part of the vessel properties. The path *size* indicates the line's thickness and the range is given by two sample numbers. The path is visible for all samples between these two samples. Paths may be turned on and off.

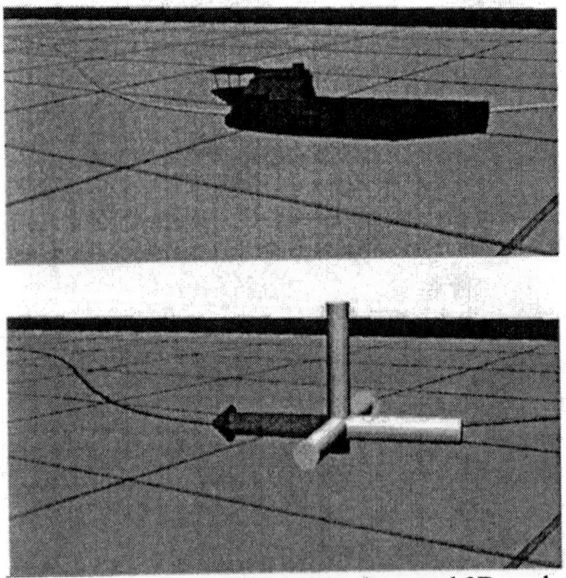

Fig. 6. *Cybership II*, displayed as the actual 3D model, and as the axes of the body-fixed frame.

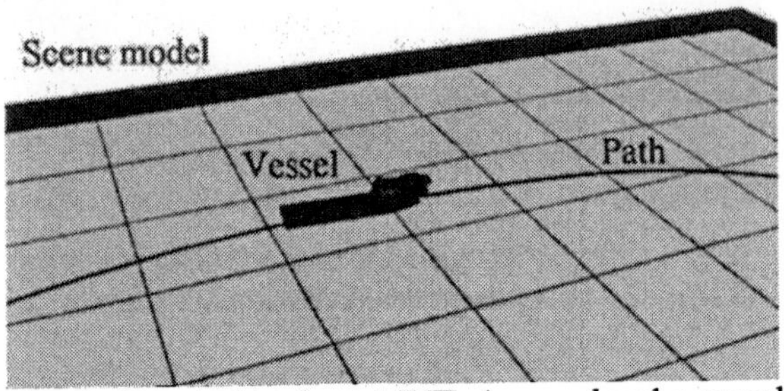


Fig. 7. Screen-shot from MVT. A *vessel* and a vessel *path* located in a *scene*.

Sequences may be viewed on-line in the VRML viewer or saved to file as AVI files. When saving, the default video compression method is *Cinepak*, but one may choose not to compress the video or select one of the following compressors: *Indeo3*, *Indeo5* and *MSVC*.

The time interval for a sequence is defined by a start and end sample that can be selected either by moving the timeline sliders or by entering the desired sample in the *start* and *end* fields.

The playback speed can be adjusted in the *Playback speed* field. The speed value is the ratio of sample time versus animation time. For instance, a speed value of 1 means that 10 seconds of 6 DOF data will produce a 10 seconds long animation. A value of 0.5 for the same set of data will produce 20 seconds of animation.

AVI files are saved with fixed frame rates. The desired playback speed is achieved by using 6 DOF data interpolation. Due to limitations in computer hardware, on-line animations may not achieve the desired frame rate, and the playback speed will in these cases slow down and become non-linear. Higher resolution and complex VRML models are software factors that might slow down the on-line playback speed.

## 6. DOCUMENTATION AND HELP

Help on MVT functions are organized as HTML pages possible to integrate with the Matlab® help browser. A printable version in PDF format is also included in the help system.

Examples show a number of different ways to use the toolbox.

Two tutorials give short lessons to help new users get started. These describe how to view and save animations based on 6 DOF data, and how to use custom VRML models with the toolbox.

## 7. CONCLUSION AND FUTURE WORK

In this paper we have presented a new toolbox for use with Matlab® and the Virtual Reality Toolbox, the Marine Visualization Toolbox (MVT). MVT presents time-varying 6 DOF data from marine operations as animations of 3D vessel models in a 3D environment. Animations can either be viewed on-line in Matlab® or saved to file for Matlab® independent viewing.

The motivation for developing the MVT is to ease the performance analysis of multivariable systems and to help presenting the experimental results. Integration with MATLAB® was desired to provide a user-interface familiar to many potential users.

Short descriptions have been given on how users may select among the included vessel models and scenes to reconstruct the specific scenario from where the 6 DOF data are retrieved (simulations, experiments, or full-scale measurements).

Help, examples and tutorials are included in the toolbox to help users get started, and to help experienced users customize the functionality, for instance how to use custom 3D models. Since the toolbox is open source, users that create custom 3D models and useful functions are encouraged to share these with others on the product web page.

The toolbox is free and can be downloaded from the *File Exchange* on the Matlab® Central, Internet address: http://www.mathworks.com/matlabcentral.

Future versions of the MVT will aim at including extensive model libraries to cover a greater area of application. Areas like the flight and automobile industries are covered simply by adding the relevant models.

## REFERENCES


Brutzman, Don (1995). Virtual World Visualization for an Autonomous Underwater Vehicle. In: *Proceedings of OCEANS '95 MTS/IEEE "Challenges of our Changing Global Environment"*, Vol. 3, pages 1592-1600.

Carey, R. and Bell, G. (1997). *The Annoted VRML 2.0 Reference Manual.* Addison-Wesley Developers Press

Fossen, T. I. (2002a). *GNC toolbox; version 1.6.*

Fossen, T. I. (2002b). *Marine Control Systems, Guidance, Navigation and Control of Ships, Rigs and Underwater Vehicles.* Marine Cybernetics.

McCarthy, M. and Descartes A. (1998). *Reality Architecture.* Prentice Hall Europe.

Mathworks Inc. (2002). *Virtual Reality Toolbox; using version 3.0.*

Perez, T. and Blanke M. (2003). DCMV a Matlab®/-Simulink® Toolbox for Dynamics and Control of Marine Vehicles. In: *Proceedings of the 6th Conference on Maneuvering and Control of Marine Crafts; Girona, Spain.*

SNAME (1950). Nomenclature for treating the motion of a submerged body through a fluid. Technical Report Bulletin 1-5, Society of Naval Architects and Marine Engineers, New York, USA.

Web3D Consortium (2004). X3D Documentation. [online] http://www.web3d.org/x3d/. Rev: March 2004.

# AUTHOR INDEX

*Continued from outside back cover*

| Title/Year of publication | Editor(s) | ISBN |
|---|---|---|
| **2002 continued** | | |
| Periodic Control Systems (W) | Bittanti & Colaneri | 0 08 043682 X |
| Modeling and Control in Environmental Issues (W) | Sano, Nishioka & Tamura | 0 08 043909 8 |
| Computer Applications in Biotechnology (C) | Dochain & Perrier | 0 08 043681 1 |
| Time Delay Systems (W) | Gu, Abdallah & Niculescu | 0 08 044004 5 |
| Control Applications in Post-Harvest and Processing Technology (W) | Seo & Oshita | 0 08 043557 2 |
| Intelligent Assembly and Disassembly (W) | Kopacek, Pereira & Noe | 0 08 043908 X |
| Adaptation and Learning in Control and Signal Processing (W) | Bittanti | 0 08 043683 8 |
| New Technologies for Computer Control (C) | Verbruggen, Chan & Vingerhoeds | 0 08 043700 1 |
| Internet Based Control Education (W) | Dormido & Morilla | 0 08 043984 5 |
| Intelligent Autonomous Vehicles (S) | Asama & Inoue | 0 08 043899 7 |
| **2003** | | |
| Proceedings of the 15th IFAC World Congress 2002 (CD + 21 vols) | Camacho, Basanez & de la Puente | 008 044184 X |
| Modeling and Control of Economic Systems (S) | Neck | 0 08 043858 X |
| Mechatronic Systems (C) | Tomizuka | 0 08 044197 1 |
| Programmable Devices and Systems (W) | Srovnal & Vlcek | 0 08 044130 0 |
| Real Time Programming (W) | Colnaric, Adamski & Wegrzyn | 0 08 044203 X |
| Lagrangian and Hamiltonian Methods in Nonlinear Control (W) | Astolfi, Gordillo & van der Schaft | 0 08 044278 1 |
| Intelligent Control Systems and Signal Processing (C) | Ruano, Ruano & Fleming | 0 08 044088 6 |
| Guidance and Control of Underwater Vehicles (W) | Roberts, Sutton & Allen | 0 08 044202 1 |
| Analysis and Design of Hybrid Systems (C) | Engell, Gueguen & Zaytoon | 0 08 044094 0 |
| Intelligent Manufacturing Systems (W) | Kadar, Monostori & Morel | 0 08 044289 7 |
| Control Applications of Optimization (W) | Gyurkovics & Bars | 0 08 044074 6 |
| Fieldbus Systems and Their Applications (C) | Dietrich, Neumann & Thomesse | 0 08 044247 1 |
| Intelligent Components and Instruments for Control Applications (S) | Almeida | 0 08 044010 X |
| Modelling and Control in Biomedical Systems (S) | Feng & Carson | 0 08 044159 9 |
| **2004** | | |
| Advances in Control Education (S) | Lindfors | 0 08 043559 9 |
| Robust Control Design (S) | Bittanti & Colaneri | 0 08 044012 6 |
| Fault Detection, Supervision and Safety of Technical Processes (S) | Staroswiecki & Wu | 0 08 044011 8 |
| Technology and International Stability (W) | Kopacek & Stapleton | 0 08 044290 0 |
| System Identification (SYSID 2003) (S) | Van den Hof, Wahlberg & Weiland | 0 08 043709 5 |
| Control Systems Design (C) | Kozak & Huba | 0 08 044175 0 |
| Robot Control (S) | Duleba & Sasiadek | 0 08 044009 6 |
| Time Delay Systems (W) | Garcia | 0 08 044238 2 |
| Control in Transportation Systems (S) | Tsugawa & Aoki | 0 08 0440592 |
| Manoeuvring and Control of Marine Craft (C) | Batlle & Blanke | 0 08 044033 9 |
| Power Plants and Power Systems Control (S) | Lee & Shin | 0 08 044210 2 |
| Automated Systems Based on Human Skill and Knowledge (S) | Stahre & Martensson | 0 08 044291 9 |
| Automatic Systems for Building the Infrastructure in Developing Countries (Knowledge and Technology Transfer) (W) | Dimirovski & Istefanopulos | 0 08 044204 8 |
| Intelligent Assembly and Disassembly (W) | Borangiu & Kopacek | 0 08 044065 7 |
| New Technologies for Automation of the Metallurgical Industry (W) | Wei Wang | 0 08 044170 X |
| Advanced Control of Chemical Processes (S) | Allgöwer & Gao | 008 044144 0 |

Printed and bound by CPI Group (UK) Ltd, Croydon, CR0 4YY
08/05/2025
01864925-0004